OF THE ELEMENTS

(1971 values for atomic weights)

Groups

VIII		Ib	IIb	IIIa	IVa	Va	VIa	VIIa	O
									Noble gases
				Representative elements (nonmetals)					2 He 4.0026
				5 B 10.81	6 C 12.011	7 N 14.0067	8 O 15.9994	9 F 18.9984	10 Ne 20.179
Transition Metals				13 Al 26.9815	14 Si 28.086	15 P 30.9738	16 S 32.06	17 Cl 35.453	18 Ar 39.948
27 Co 58.9332	28 Ni 58.71	29 Cu 63.546	30 Zn 65.38	31 Ga 69.72	32 Ge 72.59	33 As 74.9216	34 Se 78.96	35 Br 79.904	36 Kr 83.80
45 Rh 102.905	46 Pd 106.4	47 Ag 107.868	48 Cd 112.40	49 In 114.82	50 Sn 118.69	51 Sb 121.75	52 Te 127.60	53 I 126.9045	54 Xe 131.30
77 Ir 192.22	78 Pt 195.09	79 Au 196.967	80 Hg 200.59	81 Tl 204.37	82 Pb 207.2	83 Bi 208.9804	84 Po (210)	85 At (210)	86 Rn (222)
				Representative elements (metals)					

Inner-transition metals

62 Sm 150.4	63 Eu 151.96	64 Gd 157.25	65 Tb 158.925	66 Dy 162.50	67 Ho 164.930	68 Er 167.26	168		71 Lu 74.97
94 Pu (242)	95 Am (243)	96 Cm (247)	97 Bk (247)	98 Cf (249)	99 Es (254)	100 Fm (253)	10[illegible] Md (256)	102 No (254)	103 Lr (257)

HANDBOOK OF ORGANIC CHEMISTRY

John A. Dean

McGRAW-HILL BOOK COMPANY

New York St. Louis San Francisco Auckland Bogotá
Hamburg Johannesburg London Madrid Mexico
Milan Montreal New Delhi Panama
Paris São Paulo Singapore
Sydney Tokyo Toronto

Library of Congress Cataloging-in-Publication Data

Dean, John Aurie, date
Handbook of organic chemistry.

Includes index.
1. Chemistry, Organic. I. Title.
QD251.2.D43 1987 547 86-2903
ISBN 0-07-016193-3

1234567890 DOCDOC 8932109876

ISBN 0-07-016193-3

The editors for this book were Betty J. Sun and Susan Thomas; the designers were Mark E. Safran and Naomi Auerbach; and the production supervisor was Teresa F. Leaden. It was set in Times Roman by J. W. Arrowsmith Ltd.

Printed and bound by R. R. Donnelley & Sons Company.

CONTENTS

PREFACE

This handbook provides a one-volume source of factual information designed specifically for organic chemists. In it an effort has been made to select only material to meet the special needs of an organic chemist. The aim is to provide sufficient data to satisfy all general needs—the first place in which to "look it up" on the spot. Even the worker with the facilities of a comprehensive library will find this volume of value as a time saver because of the many tables of numerical data which have been especially compiled for this purpose.

The desire was to produce a compilation complete within the limits set by the economy of available space. One difficulty always faced by the editor of such a book is that he must decide which data are to be excluded in order to keep the volume from becoming unwieldy in size and too expensive for an individual to purchase. To a limited extent there has been a judicious selection of entries from LANGE'S HANDBOOK OF CHEMISTRY. These selections have been supplemented with much new material pertinent to the needs of an organic chemist, and the general coverage of organic-oriented topics has been expanded.

Descriptive properties for a basic group of 4000 organic compounds is compiled in Section 1. These follow a concise introduction to organic nomenclature, including the topic of stereochemistry. Nomenclature is consistent with the 1979 rules of the Commission on Nomenclature, International Union of Pure and Applied Chemistry. All entries are listed alphabetically according to the senior prefix of the name. The data for each organic compound include: name, structural formula, formula weight, Beilstein reference, density, refractive index, melting point, boiling point, flash point, and solubility in water and various organic solvents. Structural formulas either too complex or too ambiguous to be rendered as line formulas are grouped at the bottom of the page on which the entry appears. Alternative names, as well as trivial names in long-standing usage, are listed in their respective alphabetical order at the bottom of each page in the regular alphabetical sequence. Another feature that assists the user in locating a desired entry is the empirical formula index.

Only those inorganic compounds considered to be useful to an organic chemist are compiled in Section 2. Similarly, only the stable and radioactive nuclides most likely to interest an organic chemist, along with their properties, are listed in Section 3. Bond lengths between carbon and other elements is entirely new material. Entries for bond strengths are restricted to compounds that would interest the user. The section on physical properties has an extensive tabulation of binary and ternary azeotropes comprising approximately 850 entries.

Over 975 compounds have values listed for viscosity, dielectric constant, dipole moment, and surface tension. These are physical properties often needed by persons engaged in work with various liquid chromatographic techniques. Whenever possible, data for viscosity and dielectric constant are provided at two temperatures to permit interpolation for intermediate temperatures and also to permit limited extrapolation of the data. The dipole moments are often listed for different physical states. Values for surface tension over a range of temperatures can be calculated from two constants that can be fitted into a linear equation.

The section on thermodynamic properties contains the latest recommended values for heats of formation and Gibbs energies of formation, entropies, and heat capacities for 1500 organic compounds, many in more than one physical state. A separate tabulation contains heats of melting, vaporization, transition, and sublimation.

To aid in characterizing organic compounds, extensive tabulations of spectroscopic data cover the fields of ultraviolet-visible spectroscopy, photoluminescence, infrared spectroscopy, Raman scattering, nuclear magnetic resonance (proton, carbon-13, boron-11, nitrogen-15, fluorine-19, silicon-29, and phosphorus-31, with chemical shifts and coupling constants), electron spin resonance (hyperfine splitting constants), and ionization potentials. Also useful are the polarographic half-wave potentials, which are tabulated for over 300 compounds grouped into specific classes of compounds.

pK_a values for over 2200 organic compounds are listed. Combined with the Hammett and Taft substituent constants, these can be used to investigate many equilibrium and rate processes.

Frequently used stationary phases in gas chromatography are tabulated by relative polarity (McReynolds' constants). There is also a listing of solvents having the same density, as well as the same refractive index (useful in gradient elution in liquid-column chromatography).

There is an extensive section on polymers, rubbers, fats, oils, and waxes. A discussion of polymers and rubbers is followed by the formulas and key properties of plastic materials. For each member of the plastic families, there is a tabulation of physical, electrical, mechanical, and thermal properties and characteristics. A similar treatment is accorded the various types of rubber materials. Chemical resistance and gas permeability constants are given for rubbers and plastics. The section is completed with various constants of fats, oils, and waxes.

Every effort has been made to select the most useful and most reliable information and to record it with accuracy. Many years of occupation with the editorship of handbooks of chemical data have made the editor sensitive to the problem of inadvertent errors. It is hoped that users of this handbook will offer suggestions of material that might be included, or even excluded, in future editions and call attention to errors. These communications should be directed to the editor at his home address.

201 Mayflower Drive
Knoxville, TN 37920

JOHN A. DEAN

SYMBOLS AND ABBREVIATIONS USED IN THIS BOOK

Organic Compounds and Groups

ABS	acrylonitrile-butadiene-styrene (copolymer)
ACES	*N*-(2-acetamido)-2-aminoethanesulfonic acid
acet	acetone
ADP	adenosine-5′-phosphoric acid
alc	alcohol
ANS	anilino-8-naphthalene sulfonic acid
APDC	1-pyrrolidinecarbodithioic acid, ammonium salt
Ar	aryl group
BES	*N,N*-bis(2-hydroxyethyl)-2-aminoethanesulfonic acid
bicine	*N,N*-bis(2-hydroxyethyl)glycine
BMC	butadiene–maleic acid copolymer
BR	polybutadiene rubber
BSA	*N,O*-bis(trimethylsilyl)acetamide
BSTFA	*N,O*-bis(trimethylsilyl)trifluoroacetamide
BTMSA	bis(trimethylsilyl)acetylene
Bu	butyl
BuOH	butanol
CAB	cellulose-acetate-butyrate (polymer)
CAP	cellulose-acetate-propionate (polymer)
CAPS	3-cyclohexylamino-1-propanesulfonic acid
CHES	2-(cyclohexylamino)ethanesulfonic acid
chl	chloroform
DAP	diallyl phthalate (polymer)
DMF	dimethylformamide
DMSO	dimethyl sulfoxide
DPPH	2,2-diphenyl-1-picrylhydrazyl
EDPM	ethylene-propylene-diene rubber
EDTA	ethylenediamine-*N,N,N′,N′*-tetraacetic acid
EPA	diethyl ether, isopentane, and ethanol (5:5:2)
Et	ethyl
EtAc	ethyl acetate
eth	diethyl ether
EtOH	ethanol
FEP	fluorinated ethylene-propylene (resin)
glyc	glycerol
GR-I	government rubber I
GRN	government rubber nitrile
GRS	government rubber styrene
HDPE	high-density polyethylene
HEPES	*N*-(2-hydroxyethyl)piperazine-*N′*-ethanesulfonic acid

HOAc	acetic acid
IIR	isobutene-isoprene rubber
LDPE	low-density polyethylene
Me	methyl
MeOH	methanol
MEM	2-methoxymethyl (a radical)
MES	2-(*N*-morpholino)ethanesulfonic acid
MSTFA	*N*-methyl-*N*-(trimethylsilyl)trifluoroacetamide
NBA	*N*-bromoacetamide
NBR	nitrile-butadiene rubber
NBS	*N*-bromosuccinimide
NR	natural rubber
NTA	nitrilotriacetic acid
PB	polybutylene
PBT	poly(butylene terephthalate)
PC	polycarbonate
PCTA	poly(1,4-cyclohexanedimethylene terephthalic acid)
PCTFE	poly(chlorotrifluoroethylene)
PE	petroleum ether
PET	poly(ethylene terephthalate)
PFA	perfluoroalkoxy (resin)
Ph	phenyl
PMMA	poly(methyl methacrylate)
PO	polyolefin
POPOP	*p*-bis[2-(5-phenyloxazolyl)]benzene
PP	polypropylene
PPO	2,5-diphenyloxazole
PPS	poly(phenylene sulfide)
Pr	propyl
PS	polystyrene
PSF	polysulfone
PTFE	poly(tetrafluoroethylene)
PTMT	poly(tetramethylene terephthalate)
PVAC	poly(vinyl acetate)
PVAL	poly(vinyl alcohol)
PVB	poly(vinyl butyrate)
PVC	poly(vinyl chloride)
PVDF	poly(vinylidene fluoride)
PVF	poly(vinyl fluoride)
pyr	pyridine
R	alkyl, carbon-based radical
SAN	styrene-acrylonitrile (copolymer)
SBR	styrene-butadiene rubber
SMC	styrene–maleic acid copolymer
TAPS	3-[tris(hydroxymethyl)methylamino]-1-propanesulfonic acid
TES	2-[tris(hydroxymethyl)methylamino]-1-ethanesulfonic acid
TMS	tetramethylsilane

TMSDEA	*N*-(tetramethylsilyl)diethylamine
TMSI	*N*-(trimethylsilyl)imidazole
TNS	2-*p*-toluidinylnaphthalene-6-sulfonate
tricine	*N*-[tris(hydroxymethyl)methyl]glycine
TRIS	tris(hydroxymethyl)aminomethane
UHMWPE	ultrahigh-molecular-weight polyethylene

Units of Measure

Å	angstrom
atm	atmosphere
°C	degrees Celsius
cal	calorie
cm	centimeter
D	Debye unit
dm	decimeter
eV	electronvolt
°F	degrees Fahrenheit
G	gauss
g	gram
h	hour
Hz	hertz
J	joule
K	degrees kelvin; kelvins
kcal	kilocalorie
kJ	kilojoule
kV	kilovolt
L	liter
M	molar, molarity
m	molal, molality
MeV	million electronvolts
mho	ohm^{-1}
MHz	megahertz
min	minute
mL	milliliter
mm	millimeter
mmHg	millimeters of mercury
MN	meganewton
mN	millinewton
mol	mole
$mol \cdot L^{-1}$	moles per liter
N	newton
N	normal(ity)
n	mole fraction
nm	nanometer
S	siemens
s	second

W	watt
wt %	weight percent
yr	year
Ω·cm	ohm-centimeter

Stereochemical Locants

anti-	a manner of location with respect to a C=C or C=N bond
cis-	located on the same side of a straight chain
D-	dextrorotary
DL-	*meso* (inactive)
(*E*)-	entgegen (German: *opposite* = *trans-*)
endo-	inward-twisted
exo-	outward-twisted
gem-	geminal
L-	levorotary
m-	*meta-*
meso-	optically inactive
meta-	positioned at an oblique angle on a hexagonal ring; 1,3-
o-	*ortho-*
ortho-	positioned on adjacent carbon atoms of a hexagonal ring; 1,2-
p-	*para-*
para-	positioned at the opposite sides of a hexagonal ring; 1,4-
(*R*)-	rectus (viewing angle)
rac-	racemic (optically inactive)
(*S*)-	sinister (viewing angle)
sec-	secondary
sym-	symmetrical
syn-	a manner of location with respect to a C=C or C=N bond
tert-	tertiary
threo-	having an arrangement similar to that of threose
trans-	located on opposite sides of a straight chain
vic-	vicinal
(*Z*)-	zusammen (German: *together* = *cis-*)

Other Prefixes and Suffixes

$-d_n$	deuterium (number of atoms per molecule)
H-	substituent attached to a hydrogen atom
N-	substituent attached to a nitrogen atom
O-	substituent attached to an oxygen atom
S-	substituent attached to a sulfur atom
$-t_n$	tritium (number of atoms per molecule)
α-	alpha position
β-	beta position
γ-	gamma position
δ-	delta position
ϵ-	epsilon position
ω-	omega position (farthest from parent functional group)

Miscellaneous

A	atomic weight (mass number)
abs	absolute
alc	alcohol
alk	alkali, alkaline
amorp	amorphous
anhyd	anhydrous
aq	aqueous
BP; bp	boiling point
c	crystalline solid
ca	circa (approximately)
cgs	centimeter-gram-second (system)
conc	concentrated
C_p	specific heat (constant pressure)
D	deuterium (chemical symbol)
D	bond dissociation energy
deliq	deliquescent
dil	dilute
e^-	electron
esr	electron spin resonance
expl	explodes
ΔG_f	Gibbs free energy
g	gas, gaseous
GLC	gas-liquid chromatography
ΔH_f	enthalpy
ΔH_m	heat of melting
ΔH_t	heat of transition
ΔH_v	heat of vaporization
hyd	hydrolysis
hygr	hygroscopic
i	insoluble
ign	ignites
IP	ionization potential
IUPAC	International Union of Pure and Applied Chemistry
L	ligand; liter
liq, lq	liquid
M	metal (chemical symbol)
m	moderately strong
misc	miscible
mp	melting point
m–s	moderate to strong
n	neutron
org	organic
P	proton
pH	negative logarithm (base 10) of concentration of hydrogen ions
pK_a	negative logarithm (base 10) of the acid dissociation constant
pK_{sp}	negative logarithm (base 10) of the solubility product
ppm	parts per million

rms	root mean square
S	entropy
s	soluble
s	strong
SCE	saturated calomel electrode
sl	slightly
soln	solution
subl	sublimes
T	tritium (chemical symbol)
v	very
var	variable
vs	very strong
vw	very weak
w	weak
w–m	weak to moderate
Z	atomic number
β^+	positron
β^-	beta radiation
γ	gamma radiation
$>$	greater than
$<$	less than

SECTION 1
ORGANIC COMPOUNDS

NOMENCLATURE OF ORGANIC COMPOUNDS

The following synopisis of rules for naming organic compounds and the examples given in explanation are not intended to cover all the possible cases. For a more comprehensive and detailed description, see J. Rigaudy and S. P. Klesney, *Nomenclature of Organic Chemistry*, Sections A, B, C, D, E, F, and H, Pergamon Press, Oxford, 1979. This publication contains the recommendations of the Commission on Nomenclature of Organic Chemistry and was prepared under the auspices of the International Union of Pure and Applied Chemistry (IUPAC).

Nonfunctional Compounds

Alkanes

The saturated open-chain (acyclic) hydrocarbons (C_nH_{2n+2}) have names ending in -ane. The first four members have the trivial names *methane* (CH_4), *ethane* (CH_3CH_3 or C_2H_6), *propane* (C_3H_8), and *butane* (C_4H_{10}). For the remainder of the alkanes, the first portion of the name is derived from the Greek prefix (see Table 11-4) that cites the number of carbons in the alkane followed by -ane with elision of the terminal -a from the prefix, as shown in Table 1-1.

TABLE 1-1 Names of straight-chain alkanes

n*	Name	n*	Name	n*	Name	n*	Name
1	Methane	11	Undecane‡	21	Henicosane	60	Hexacontane
2	Ethane	12	Dodecane	22	Docosane	70	Heptacontane
3	Propane	13	Tridecane	23	Tricosane	80	Octacontane
4	Butane	14	Tetradecane			90	Nonacontane
5	Pentane	15	Pentadecane	30	Triacontane	100	Hectane
6	Hexane	16	Hexadecane	31	Hentriacontane	110	Decahectane
7	Heptane	17	Heptadecane	32	Dotriacontane	120	Icosahectane
8	Octane	18	Octadecane			121	Henicosahectane
9	Nonane†	19	Nonadecane	40	Tetracontane		
10	Decane	20	Icosane§	50	Pentacontane		

* n = total number of carbon atoms.
† Formerly called enneane.
‡ Formerly called hendecane.
§ Formerly called eicosane.

For branching compounds, the parent structure is the longest continuous chain present in the compound. Consider the compound to have been derived from this structure by replacement of hydrogen by various alkyl groups. Arabic number prefixes indicate the carbon to which the alkyl group is attached. Start numbering at whichever end of the parent structure that results in the lowest-numbered locants. The arabic prefixes are listed in numerical sequence, separated from each other by commas and from the remainder of the name by a hyphen.

If the same alkyl group occurs more than once as a side chain, this is indicated by the prefixes di-, tri-, tetra-, etc. Side chains are cited in alphabetical order (before insertion of any multiplying prefix). The name of a complex radical (side chain) is considered to begin with the first letter of its complete name. Where names of complex radicals are composed of identical words, priority for citation is given to that radical which contains the lowest-numbered locant at the first cited point of difference in the radical. If two or more side chains are in equivalent positions, the one to be assigned the lowest-numbered locant is that cited first in the name. The complete expression for the side chain may be enclosed in parentheses for clarity or the carbon atoms in side chains may be indicated by primed locants.

If hydrocarbon chains of equal length are competing for selection as the parent, the choice goes in descending order to (1) the chain that has the greatest number of side chains, (2) the chain whose side chains have the lowest-numbered locants, (3) the chain having the greatest number of carbon atoms in the smaller side chains, or (4) the chain having the least-branched side chains.

These trivial names may be used for the unsubstituted hydrocarbon only:

Isobutane $(CH_3)_2CHCH_3$	Neopentane $(CH_3)_4C$
Isopentane $(CH_3)_2CHCH_2CH_3$	Isohexane $(CH_3)_2CHCH_2CH_2CH_3$

Univalent radicals derived from saturated unbranched alkanes by removal of hydrogen from a terminal carbon atom are named by adding -yl in place of -ane to the stem name. Thus the alkane *ethane* becomes the radical *ethyl.* These exceptions are permitted for unsubstituted radicals only:

Isopropyl $(CH_3)_2CH—$	Isopentyl $(CH_3)_2CHCH_2CH_2—$
Isobutyl $(CH_3)_2CHCH_2—$	Neopentyl $(CH_3)_3CCH_2—$
sec-Butyl $CH_3CH_2CH(CH_3)—$	*tert*-Pentyl $CH_3CH_2C(CH_3)_2—$
tert-Butyl $(CH_3)_3C—$	Isohexyl $(CH_3)_2CHCH_2CH_2CH_2—$

Note the usage of the prefixes iso-, neo-, *sec*-, and *tert*-, and note when italics are employed. Italicized prefixes are never involved in alphabetization, except among themselves; thus *sec*-butyl would precede isobutyl, isohexyl would precede isopropyl, and *sec*-butyl would precede *tert*-butyl.

Examples of alkane nomenclature are

$$\begin{array}{ccccccc} \overset{4}{C}H_3 & — & \overset{3}{C}H_2 & — & \overset{2}{C}H & — & \overset{1}{C}H_3 \\ & & & & | & & \\ & & & & CH_3 & & \end{array}$$

2-Methylbutane (or the trivial name, isopentane)

$$\begin{array}{ccccccc} \overset{5}{C}H_3 & — & \overset{4}{C}H_2 & — & \overset{3}{C}H & — & CH_3 \\ & & & & | & & \\ & & & & \underset{2}{C}H_2 & — & \underset{1}{C}H_3 \end{array}$$

3-Methylpentane (not 2-ethylbutane)

$$\begin{array}{ccccccccccccccc} & & & & & & & & & & & & CH_3 & & \\ & & & & & & & & & & & & | & & \\ \overset{8}{C}H_3 & — & \overset{7}{C}H_2 & — & \overset{6}{C}H_2 & — & \overset{5}{C}H & — & \overset{4}{C}H_2 & — & \overset{3}{C}H_2 & — & \overset{2}{C} & — & \overset{1}{C}H_3 \\ & & & & & & | & & & & & & | & & \\ & & & & & & CH_2 & — & CH_3 & & & & CH_3 & & \end{array}$$

5-Ethyl-2,2-dimethyloctane (note cited order)

$$\begin{array}{ccccccccccccccc} \overset{8}{C}H_3 & — & \overset{7}{C}H_2 & — & \overset{6}{C}H & — & \overset{5}{C}H_2 & — & \overset{4}{C}H_2 & — & \overset{3}{C}H & — & \overset{2}{C}H_2 & — & \overset{1}{C}H_3 \\ & & & & | & & & & & & | & & & & \\ & & & & CH_3 & & & & & & CH_2 & — & CH_3 & & \end{array}$$

3-Ethyl-6-methyloctane (note locants reversed)

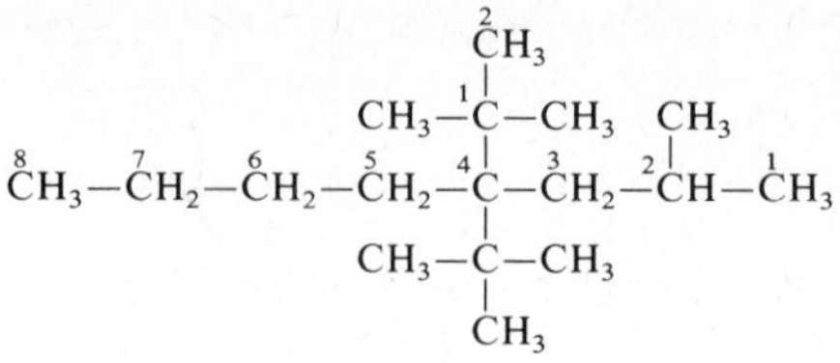

$$
\begin{array}{c}
\overset{2}{C}H_3 \\
| \\
CH_3-\overset{1}{C}-CH_3 \quad CH_3 \\
\overset{8}{C}H_3-\overset{7}{C}H_2-\overset{6}{C}H_2-\overset{5}{C}H_2-\overset{4}{C}-\overset{3}{C}H_2-\overset{2}{C}H-\overset{1}{C}H_3 \\
| \\
CH_3-C-CH_3 \\
| \\
CH_3
\end{array}
$$

4,4-Bis(1,1-dimethylethyl)-2-methyloctane
4,4-Bis-1′,1′-dimethylethyl-2-methyloctane
4,4-Bis(*tert*-butyl)-2-methyloctane

Bivalent radicals derived from saturated unbranched alkanes by removal of two hydrogen atoms are named as follows: (1) If both free bonds are on the same carbon atom, the ending -ane of the hydrocarbon is replaced with -ylidene. However, for the first member of the alkanes it is methylene rather than methylidene. Isopropylidene, *sec*-butylidene, and neopentylidene may be used for the unsubstituted group only. (2) If the two free bonds are on different carbon atoms, the straight-chain group terminating in these two carbon atoms is named by citing the number of methylene groups comprising the chain. Other carbons groups are named as substituents. Ethylene is used rather than dimethylene for the first member of the series, and propylene is retained for $CH_3-\underset{|}{C}H-CH_2-$(but trimethylene is $-CH_2-CH_2-CH_2-$).

Trivalent groups derived by the removal of three hydrogen atoms from the same carbon are named by replacing the ending -ane of the parent hydrocarbon with -ylidyne.

Alkenes and Alkynes

Each name of the corresponding saturated hydrocarbon is converted to the corresponding alkene by changing the ending -ane to -ene. For alkynes the ending is -yne. With more than one double (or triple) bond, the endings are -adiene, -atriene, etc. (or -adiyne, -atriyne, etc.). The position of the double (or triple) bond in the parent chain is indicated by a locant obtained by numbering from the end of the chain nearest the double (or triple) bond; thus $CH_3CH_2CH{=}CH_2$ is 1-butene and $CH_3C{\equiv}CCH_3$ is 2-butyne.

For multiple unsaturated bonds, the chain is so numbered as to give the lowest possible locants to the unsaturated bonds. When there is a choice in numbering, the double bonds are given the lowest locants, and the alkene is cited before the alkyne where both occur in the name. Examples:

$CH_3CH_2CH_2CH_2CH{=}CH-CH{=}CH_2$ 1,3-Octadiene

$CH_2{=}CHC{\equiv}CCH{=}CH_2$ 1,5-Hexadiene-3-yne

$CH_3CH{=}CHCH_2C{\equiv}CH$ 4-Hexen-1-yne

$CH{\equiv}CCH_2CH{=}CH_2$ 1-Penten-4-yne

Unsaturated branched acyclic hydrocarbons are named as derivatives of the chain that contains the maximum number of double and/or triple bonds. When a choice exists, priority goes in sequence to (1) the chain with the greatest number of carbon atoms and (2) the chain containing the maximum number of double bonds.

These nonsystematic names are retained:

Ethylene $CH_2{=}CH_2$

Allene $CH_2{=}C{=}CH_2$

Acetylene $HC{\equiv}CH$

An example of nomenclature for alkenes and alkynes is

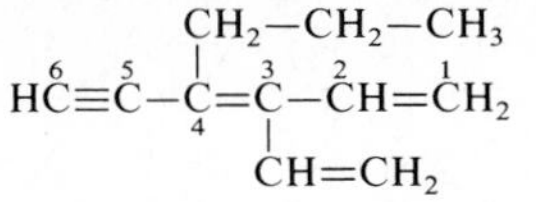

$$\overset{6}{H}C \equiv \overset{5}{C} - \underset{4}{\overset{|}{C}}(CH_2-CH_2-CH_3) = \overset{3}{C}(CH=CH_2) - \overset{2}{C}H = \overset{1}{C}H_2$$

4-Propyl-3-vinyl-1,3-hexadien-5-yne

Univalent radicals have the endings -enyl, -ynyl, -dienyl, -diynyl, etc. When necessary, the positions of the double and triple bonds are indicated by locants, with the carbon atom with the free valence numbered as 1. Examples:

$CH_2{=}CH{-}CH_2{-}$ 2-Propenyl

$CH_3{-}C{\equiv}C{-}$ 1-Propynyl

$CH_3{-}C{\equiv}C{-}CH_2CH{=}CH_2{-}$ 1-Hexen-4-ynyl

These names are retained:

Vinyl (for ethenyl) $CH_2{=}CH{-}$

Allyl (for 2-propenyl) $CH_2{=}CH{-}CH_2{-}$

Isopropenyl (for 1-methylvinyl but for unsubstituted radical only) $CH_2{=}C(CH_3){-}$

Should there be a choice for the fundamental straight chain of a radical, that chain is selected which contains (1) the maximum number of double and triple bonds, (2) the largest number of carbon atoms, and (3) the largest number of double bonds. These are in descending priority.

Bivalent radicals derived from unbranched alkenes, alkadienes, and alkynes by removing a hydrogen atom from each of the terminal carbon atoms are named by replacing the endings -ene, -diene, and -yne by -enylene, -dienylene, and -ynylene, respectively. Positions of double and triple bonds are indicated by numbers when necessary. The name *vinylene* instead of ethenylene is retained for $-CH{=}CH-$.

Monocyclic Aliphatic Hydrocarbons

Monocyclic aliphatic hydrocarbons (with no side chains) are named by prefixing cyclo- to the name of the corresponding open-chain hydrocarbon having the same number of carbon atoms as the ring. Radicals are formed as with the alkanes, alkenes, and alkynes. Examples:

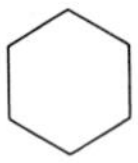

	Cyclohexane	Cyclohexyl- (for the radical)
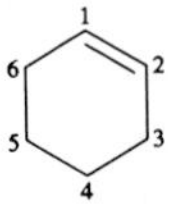 1 2 3 4 5 6	Cyclohexene	1-Cyclohexenyl- (for the radical with the free valence at carbon 1)
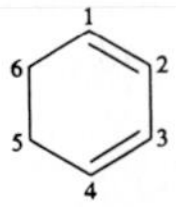 1 2 3 4 5 6	1,3-Cyclohexandiene	Cyclohexadienyl- (the unsaturated carbons are given numbers as low as possible, numbering from the carbon atom with the free valence given the number 1)

For convenience, aliphatic rings are often represented by simple geometric figures: a triangle for cyclopropane, a square for cyclobutane, a pentagon for cyclopentane, a hexagon (as illustrated) for cyclohexane, etc. It is understood that two hydrogen atoms are located at each corner of the figure unless some other group is indicated for one or both.

Monocyclic Aromatic Compounds

Except for six retained names, all monocyclic substituted aromatic hydrocarbons are named systematically as derivatives of benzene. Moreover, if the substituent introduced into a compound with a retained trivial name is identical with one already present in that compound, the compound is named as a derivative of benzene. These names are retained:

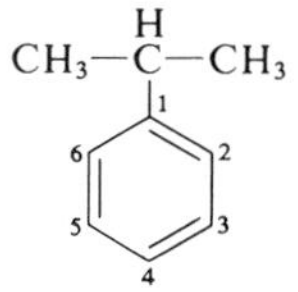

Cumene

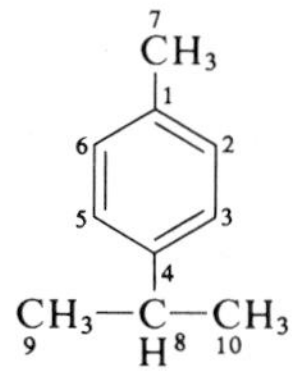

Cymene (all three forms; *para-* shown)

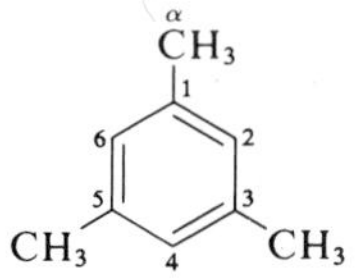

Mesitylene

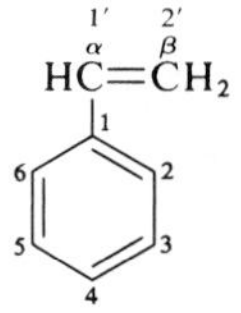

Styrene

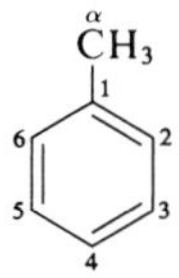

Toluene

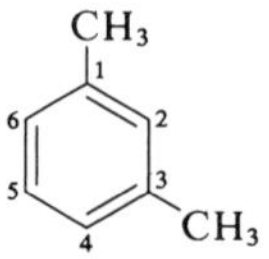

Xylene (all three forms; *meta-* shown)

The position of substituents is indicated by numbers, with the lowest locant possible given to substituents. When a name is based on a recognized trivial name, priority for lowest-numbered locants is given to substituents implied by the trivial name. When only two substituents are present on a benzene ring, their position may be indicated by *o-* (*ortho-*), *m-* (*meta-*), and *p-* (*para-*) (and alphabetized in the order given) used in place of 1,2-, 1,3-, and 1,4-, respectively.

Radicals derived from monocyclic substituted aromatic hydrocarbons and having the free valence at a ring atom (numbered 1) are named phenyl (for benzene as parent, since benzyl is used for the radical $C_6H_5CH_2—$), cumenyl, mesityl, tolyl, and xylyl. All other radicals are named as substitued phenyl radicals. For radicals having a single free valence in the side chain, these trivial names are retained:

Benzyl $C_6H_5CH_2—$
Benzhydryl (alternative to diphenylmethyl) $(C_6H_5)_2CH—$
Cinnamyl $C_6H_5CH{=}CH—CH_2—$
Phenethyl $C_6H_5CH_2CH_2—$
Styryl $C_6H_5CH{=}CH—$
Trityl $(C_6H_5)_3C—$

Otherwise, radicals having the free valence(s) in the side chain are named in accordance with the rules for alkanes, alkenes, or alkynes.

The name *phenylene* (*o-*, *m-*, or *p-*) is retained for the radical $—C_6H_4—$. Bivalent radicals formed from substituted benzene derivatives and having the free valences at ring atoms are named as substituted phenylene radicals, with the carbon atoms having the free valences being numbered 1,2-, 1,3-, or 1,4-, as appropriate.

Radicals having three or more free valences are named by adding the suffixes -triyl, -tetrayl, etc. to the systematic name of the corresponding hydrocarbon.

Fused Polycyclic Hydrocarbons

The names of polycyclic hydrocarbons containing the maximum number of conjugated double bonds end in -ene. Here the ending does not denote one double bond. Names of hydrocarbons containing five or more fixed benzene rings in a linear arrangement are formed from a numerical prefix (see Table 11-4) followed by -acene. A partial list of the names of polycyclic hydrocarbons is given in Table 1-2. Many names are trivial.

Numbering of each ring system is fixed, as shown in Table 1-2, but it follows a systematic pattern. The individual rings of each system are oriented so that the greatest number of rings are (1) in a horizontal row and (2) the maximum number of rings are above and to the right (upper-right quadrant) of the horizontal row. When two orientations meet these requirements, the one is chosen that has the fewest rings in the lower-left quadrant. Numbering proceeds in a clockwise direction, commencing with the carbon atom not engaged in ring fusion that lies in the most counterclockwise position of the uppermost ring (upper-right quadrant); omit atoms common to two or more rings. Atoms common to two or more rings are designated by adding lowercase roman letters to the number of the position immediately preceding. Interior atoms follow the highest number, taking a clockwise sequence wherever there is a choice. Anthracene and phenanthrene are two exceptions to the rule on numbering. Two examples of numbering follow:

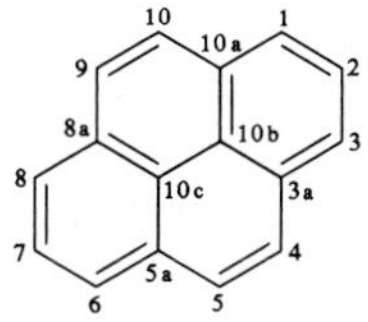

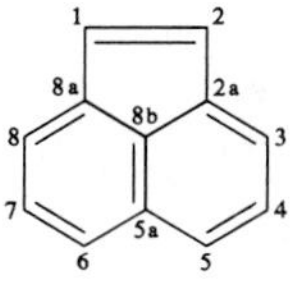

When a ring system with the maximum number of conjugated double bonds can exist in two or more forms differing only in the position of an "extra" hydrogen atom, the name can be made specific by indicating the position of the extra hydrogen(s). The compound name is modified with a locant followed by an italic capital *H* for each of these hydrogen atoms. Carbon atoms that carry an indicated hydrogen atom are numbered as low as possible. For example, 1*H*-indene is illustrated in Table 1-2; 2*H*-indene would be

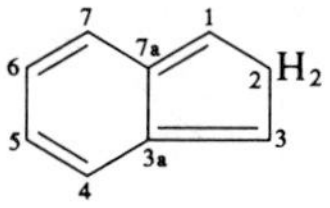

Names of polycyclic hydrocarbons with less than the maximum number of noncumulative double bonds are formed from a prefix dihydro-, tetrahydro-, etc., followed by the name of the corresponding unreduced hydrocarbon. The prefix perhydro- signifies full hydrogenation. For example, 1,2-dihydronaphthalene is

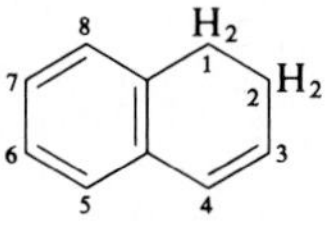

TABLE 1-2 Fused polycyclic hydrocarbons

Listed in order of increasing priority for selection as parent compound

Asterisk after a compound denotes exception to systematic numbering.

1. Pentalene
2. Indene
3. Naphthalene
4. Azulene
5. Heptalene
6. Biphenylene
7. *asym*-Indacene
8. *sym*-Indacene
9. Acenaphthylene

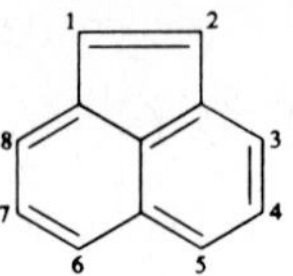

10. Fluorene
11. Phenalene
12. Phenanthrene*

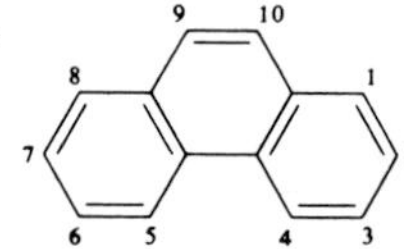

13. Anthracene*
14. Fluoranthene
15. Acephenanthrylene
16. Aceanthrylene

TABLE 1-2 Fused polycyclic hydrocarbons (*continued*)

Listed in order of increasing priority for selection as parent compound

Asterisk after a compound denotes exception to systematic numbering.

17. Triphenylene 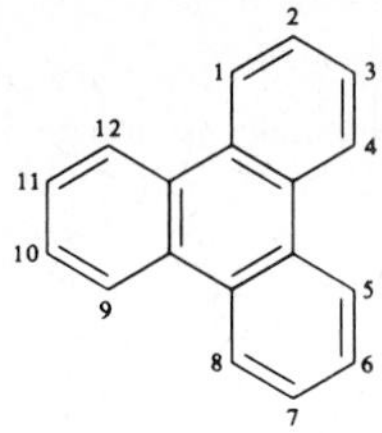	19. Chrysene 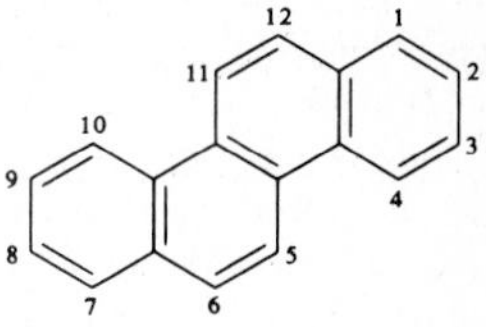
18. Pyrene 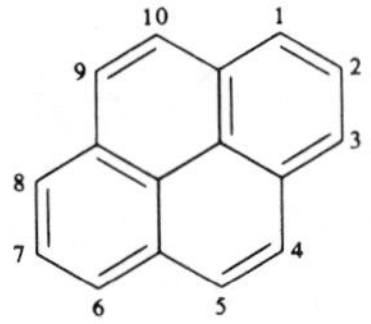	20. Naphthacene 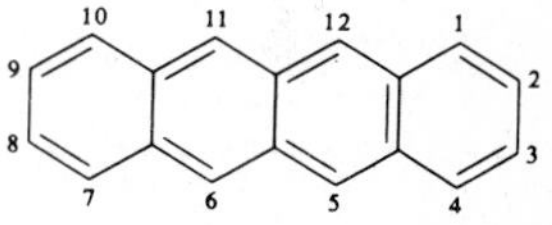

Examples of retained names and their structures are as follows:

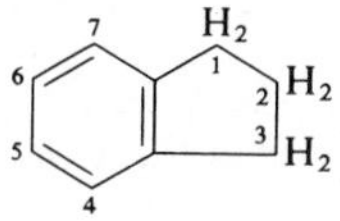

Indan

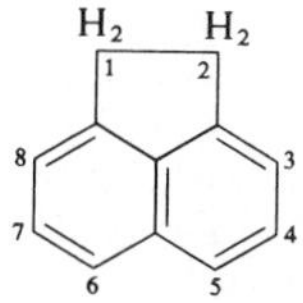

Acenaphthene

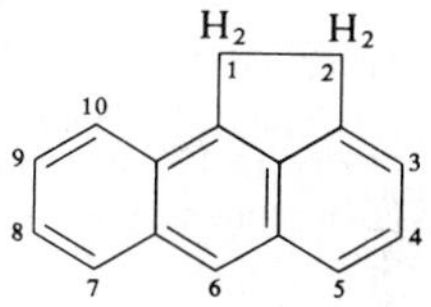

Aceanthrene

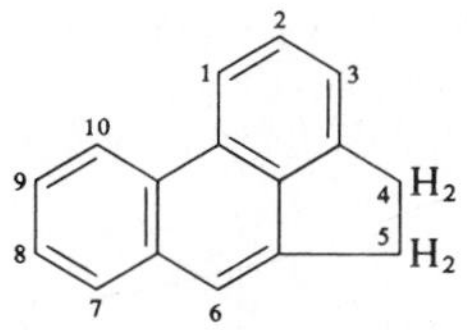

Acephenanthrene

Polycyclic compounds in which two rings have two atoms in common or in which one ring contains two atoms in common with each of two or more rings of a contiguous series of rings and which contain at least two rings of five or more members with the maximum number of noncumulative double bonds and which have no accepted trivial name (Table 1-2) are named by prefixing to the name of the parent ring or ring system designations of the other components. The parent name should contain as many rings as possible (provided it has a trivial name) and should occur as far as possible from the beginning of the list in Table 1-2. Furthermore, the attached component(s) should be as simple as possible. For example, one writes dibenzophenanthrene and not naphthophenanthrene because the attached component benzo- is simpler than naphtho-. Prefixes designating attached components are formed by changing the ending -ene into -eno-; for example, indeno- from indene. Multiple prefixes are arranged in alphabetical order. Several abbreviated prefixes are recognized; the parent is given in parentheses:

Acenaphtho- (acenaphthylene)	Naphtho- (naphthalene)
Anthra- (anthracene)	Perylo- (perylene)
Benzo- (benzene)	Phenanthro- (phenanthrene)

For monocyclic prefixes other than benzo-, the following names are recognized, each to represent the form with the maximum number of noncumulative double bonds: cyclopenta-, cyclohepta-, cycloocta-, etc.

Isomers are distinguished by lettering the peripheral sides of the parent beginning with *a* for the side 1,2, and so on, lettering every side around the periphery. If necessary for clarity, the numbers of the attached position (1,2, for example) of the substituent ring are also denoted. The prefixes are cited in alphabetical order. The numbers and letters are enclosed in square brackets and placed immediately after the designation of the attached component. Examples are

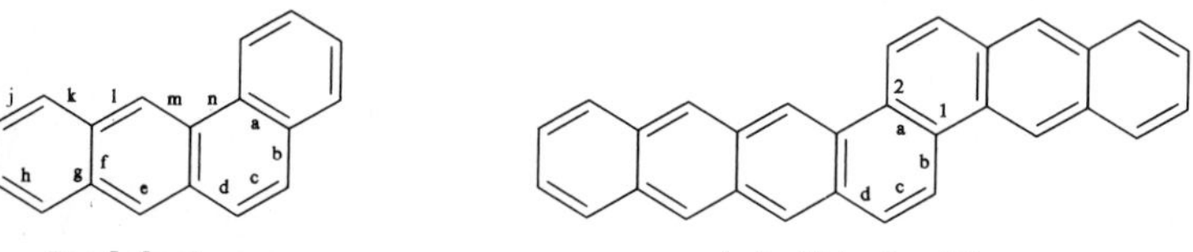

Benz[α]anthracene　　Anthra[2,1-α]naphthacene

Bridged Hydrocarbons

Saturated alicyclic hydrocarbon systems consisting of two rings that have two or more atoms in common take the name of the open-chain hydrocarbon containing the same total number of carbon atoms and are preceded by the prefix bicyclo-. The system is numbered commencing with one of the bridgeheads, numbering proceeding by the longest possible path to the second bridgehead. Numbering is then continued from this atom by the longer remaining unnumbered path back to the first bridgehead and is completed by the shortest path from the atom next to the first bridgehead. When a choice in numbering exists, unsaturation is given the lowest numbers. The number of carbon atoms in each of the bridges connecting the bridgeheads is indicated in brackets in descending order. Examples are

$$\begin{array}{lll} \overset{7}{C}H_2-\overset{1}{C}H-\overset{2}{C}H_2 \\ \quad\quad {}^{8}CH_2 \;\; CH_2{}^{3} \\ \underset{6}{C}H_2-\underset{5}{C}H-\underset{4}{C}H_2 \end{array} \qquad \begin{array}{l} \overset{9}{C}H_2-\overset{1}{C}H-\overset{2}{C}H_2-\overset{3}{C}H_2 \\ \quad\quad\quad\quad\quad\quad {}^{4}CH_2 \\ \underset{8}{C}H_2-\underset{7}{C}H-\underset{6}{C}H_2-\underset{5}{C}H_2 \end{array}$$

Bicyclo[3.2.1]octane　　Bicyclo[5.2.0]nonane

Hydrocarbon Ring Assemblies

Assemblies are two or more cyclic systems, either single rings or fused systems, that are joined directly to each other by double or single bonds. For identical systems naming may proceed (1) by placing the prefix bi- before the name of the corresponding radical or (2), for systems joined through a single bond, by placing the prefix bi- before the name of the corresponding hydrocarbon. In each case, the numbering of the assembly is that of the corresponding radical or hydrocarbon, one system being assigned unprimed numbers and the other primed numbers.

The points of attachment are indicated by placing the appropriate locants before the name; an unprimed number is considered lower than the same number primed. The name *biphenyl* is used for the assembly consisting of two benzene rings. Examples are

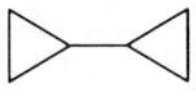

$CH_3—CH_2$

$CH_2—CH_2—CH_3$

1,1′-Bicyclopropyl or 1,1′-bicyclopropane 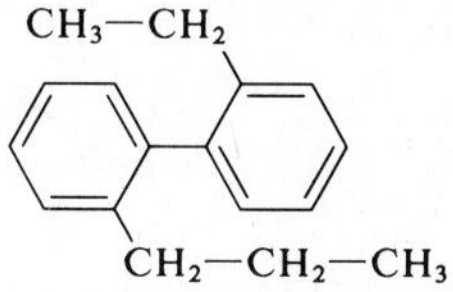   2-Ethyl-2′-propylbiphenyl

For nonidentical ring systems, one ring system is selected as the parent and the other systems are considered as substituents and are arranged in alphabetical order. The parent ring system is assigned unprimed numbers. The parent is chosen by considering the following characteristics in turn until a decision is reached: (1) the system containing the larger number of rings, (2) the system containing the larger ring, (3) the system in the lowest state of hydrogenation, and (4) the highest-order number or ring systems set forth in Table 1-2. Examples are given, with the deciding priority given in parentheses preceding the name:

(1) 2-Phenylnaphthalene
(2) and (4) 2-(2′-Naphthyl)azulene
(3) Cyclohexylbenzene

Radicals from Ring Systems

Univalent substituent groups derived from polycyclic hydrocarbons are named by changing the final *e* of the hydrocarbon name to -yl. The carbon atoms having free valences are given locants as low as possible consistent with the fixed numbering of the hydrocarbon. Exceptions are naphthyl (instead of naphthalenyl), anthryl (for anthracenyl), and phenanthryl (for phenanthrenyl). However, these abbreviated forms are used only for the simple ring systems. Substituting groups derived from fused derivatives of these ring systems are named systematically. Substituting groups having two or more free bonds are named as described in Monocyclic Aliphatic Hydrocarbons on p. 1-5.

Cyclic Hydrocarbons with Side Chains

Hydrocarbons composed of cyclic and aliphatic chains are named in a manner that is the simplest permissible or the most appropriate for the chemical intent. Hydrocarbons containing several chains attached to one cyclic nucleus are generally named as derivatives of the cyclic compound, and compounds containing several side chains and/or cyclic radicals attached to one chain are named as derivatives of the acyclic compound. Examples are

2-Ethyl-1-methylnaphthalene   Diphenylmethane
1,5-Diphenylpentane   2,3-Dimethyl-1-phenyl-1-hexene

Recognized trivial names for composite radicals are used if they lead to simplifications in naming. Examples are

1-Benzylnaphthalene   1,2,4-Tris(3-*p*-tolylpropyl)benzene

Fulvene, for methylenecyclopentadiene, and stilbene, for 1,2-diphenylethylene, are trival names that are retained.

Heterocyclic Systems

Heterocyclic compounds can be named by relating them to the corresponding carbocyclic ring systems by using replacement nomenclature. Heteroatoms are denoted by prefixes ending in *a*, as shown in Table 1-3. If two or more replacement prefixes are required in a single name, they are cited in the order of their listing in the table. The lowest possible numbers consistent with the numbering of the corresponding carbocyclic system are assigned to the heteroatoms

TABLE 1-3 Specialist nomenclature for heterocyclic systems

Heterocyclic atoms are listed in decreasing order of priority.

Element	Valence	Prefix	Element	Valence	Prefix
Oxygen	2	Oxa-	Antimony	3	Stiba-*
Sulfur	2	Thia-	Bismuth	3	Bisma-
Selenium	2	Selena-	Silicon	4	Sila-
Tellurium	2	Tellura-	Germanium	4	Germa-
Nitrogen	3	Aza-	Tin	4	Stanna-
Phosphorus	3	Phospha-*	Lead	4	Plumba-
Arsenic	3	Arsa-*	Boron	3	Bora-
			Mercury	2	Mercura-

* When immediately followed by -in or -ine, phospha- should be replaced by phosphor-, arsa- by arsen-, and stiba- by antimon-. The saturated six-membered rings corresponding to phosphorin and arsenin are named *phosphorinane* and *arsenane*. A further exception is the replacement of borin by borinane.

TABLE 1-4 Suffixes for specialist nomenclature of heterocyclic systems

Number of ring members	Rings containing nitrogen		Rings containing no nitrogen	
	Unsaturation*	Saturation	Unsaturation*	Saturation
3	-irine	-iridine	-irene	-irane
4	-ete	-etidine	-ete	-etane
5	-ole	-olidine	-ole	-olane
6	-ine†	‡	-in	-ane§
7	-epine	‡	-epin	-epane
8	-ocine	‡	-ocin	-ocane
9	-onine	‡	-onin	-onane
10	-ecine	‡	-ecin	-ecane

* Unsaturation corresponding to the maximum number of noncumulative double bonds. Heteroatoms have the normal valences given in Table 1-3.

† For phosphorus, arsenic, antimony, and boron, see the special provisions in Table 1-3.

‡ Expressed by prefixing perhydro- to the name of the corresponding unsaturated compound.

§ Not applicable to silicon, germanium, tin, and lead; perhydro- is prefixed to the name of the corresponding unsaturated compound.

TABLE 1-5 Trivial names of heterocyclic systems suitable for use in fusion names

Listed in order of increasing priority as senior ring system

Asterisk after a compound denotes exception to systematic numbering.

Structure	Parent name	Radical name	Structure	Parent name	Radical name
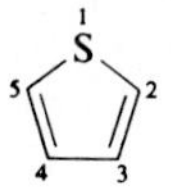	Thiophene	Thienyl	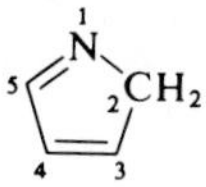	2*H*-Pyrrole	2*H*-Pyrrolyl
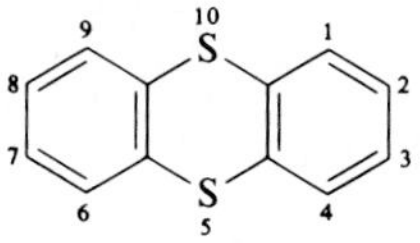	Thianthrene	Thianthrenyl	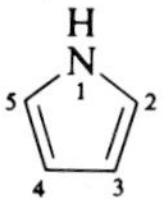	Pyrrole	Pyrrolyl
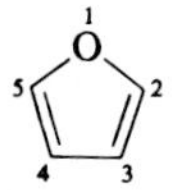	Furan	Furyl	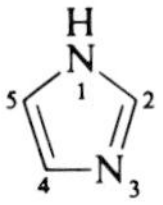	Imidazole	Imidazolyl
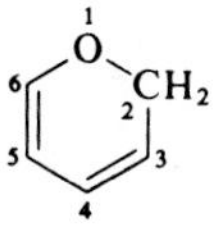	Pyran (2*H*-shown)	Pyranyl	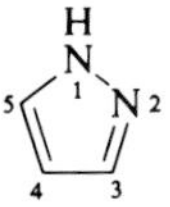	Pyrazole	Pyrazolyl
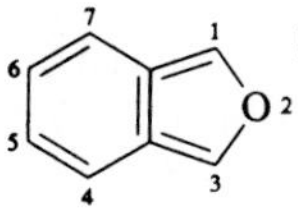	Isobenzofuran	Isobenzo-furanyl	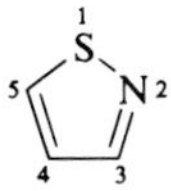	Isothiazole	Isothiazolyl
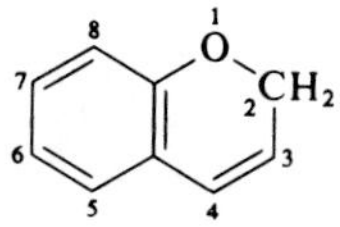	Chromene (2*H*-shown)	Chromenyl	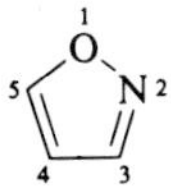	Isoxazole	Isoxazolyl
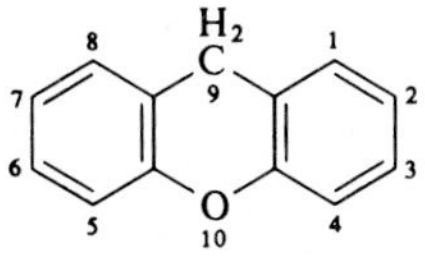	Xanthene*	Xanthenyl	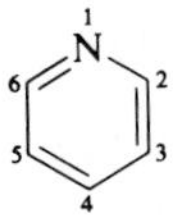	Pyridine	Pyridyl
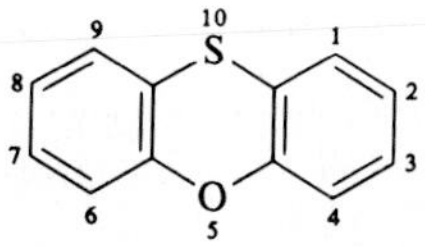	Phenoxathiin	Phenoxa-thiinyl	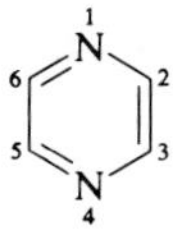	Pyrazine	Pyrazinyl
			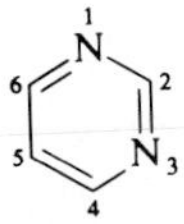	Pyrimidine	Pyrimidinyl
			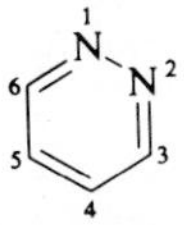	Pyridazine	Pyridazinyl

TABLE 1-5 Trivial names of heterocyclic systems suitable for use in fusion names (*continued*)

Listed in order of increasing priority as senior ring system

Asterisk after a compound denotes exception to systematic numbering.

Structure	Parent name	Radical name
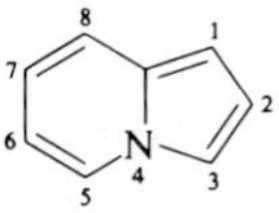	Indolizine	Indolizinyl
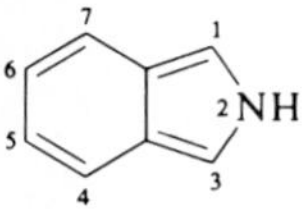	Isoindole	Isoindolyl
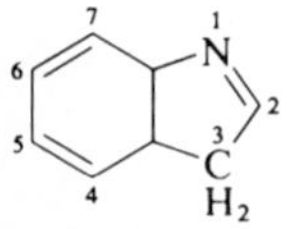	3*H*-Indole	3*H*-Indolyl
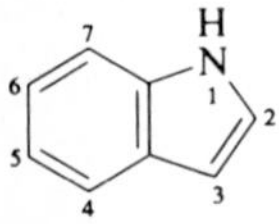	Indole	Indolyl
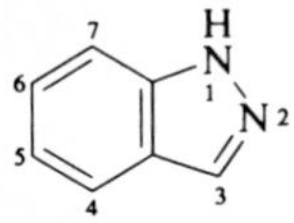	1*H*-Indazole	1*H*-Indazolyl
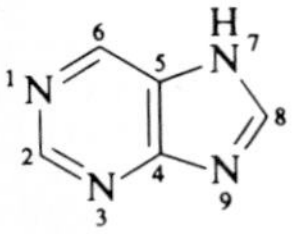	Purine*	Purinyl
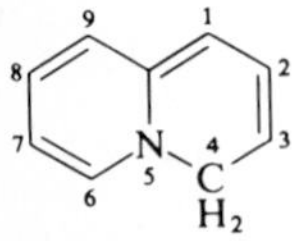	4*H*-Quinolizine	4*H*-Quinolizinyl
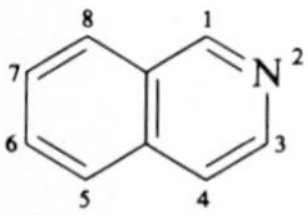	Isoquinoline	Isoquinolyl
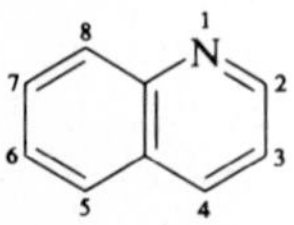	Quinoline	Quinolyl
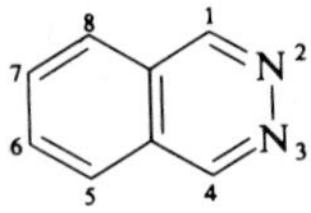	Phthalazine	Phthalazinyl
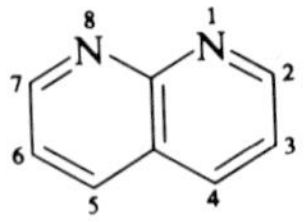	Naphthyridine (1,8-shown)	Naphthyridinyl
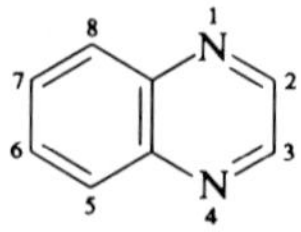	Quinoxaline	Quinoxalinyl
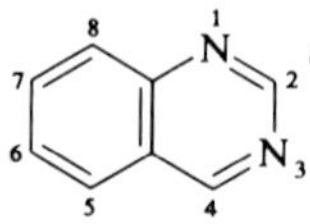	Quinazoline	Quinazolinyl
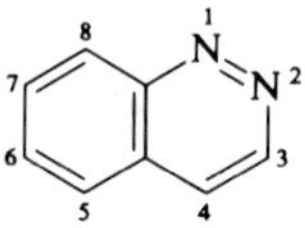	Cinnoline	Cinnolinyl
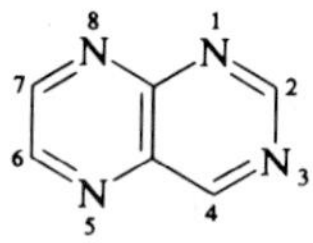	Pteridine	Pteridinyl
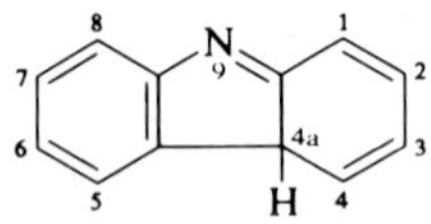	4α*H*-Carbazole*	4α*H*-Carbazolyl
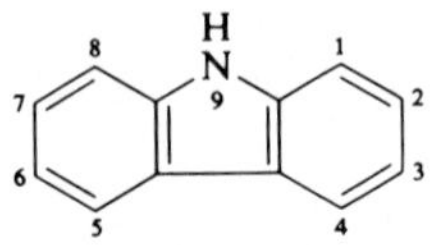	Carbazole*	Carbazolyl

TABLE 1-5 Trivial names of heterocyclic systems suitable for use in fusion names (*continued*)

Listed in order of increasing priority as senior ring system

Asterisk after a compound denotes exception to systematic numbering.

Structure	Parent name	Radical name	Structure	Parent name	Radical name
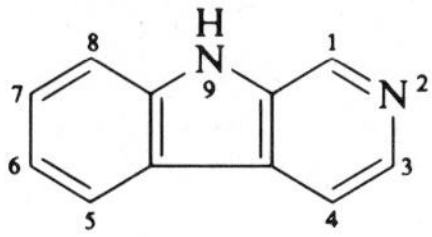	β-Carboline	β-Carbolinyl	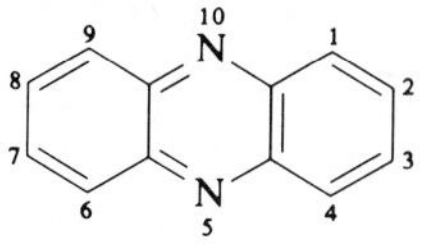	Phenazine	Phenazinyl
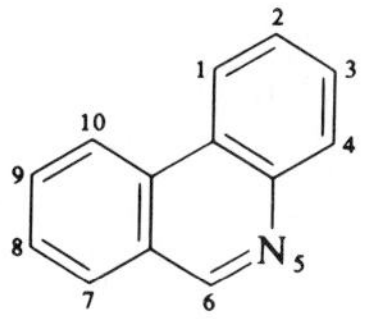	Phenanthridine	Phenanthridinyl	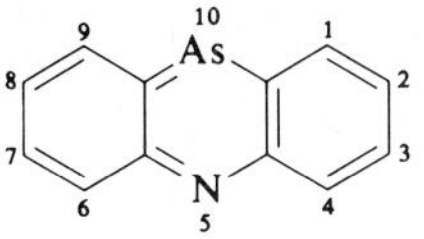	Phenarsazine	Phenarsazinyl
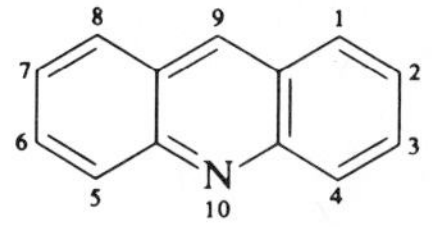	Acridine*	Acridinyl	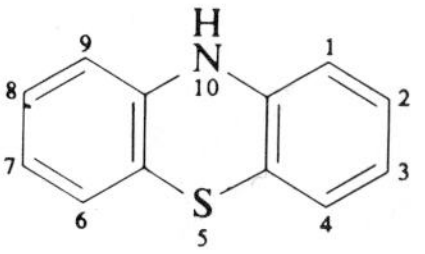	Phenothiazine	Phenothiazinyl
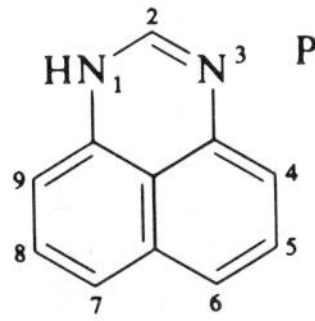	Perimidine	Perimidinyl	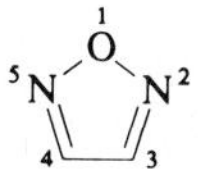	Furazan	Furazanyl
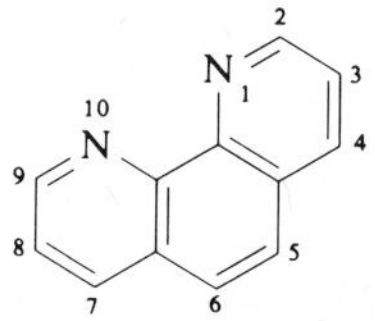	Phenanthroline (1,10-shown)	Phenanthrolinyl	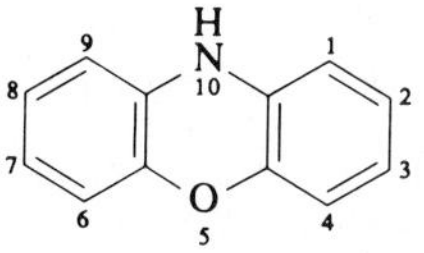	Phenoxazine	Penoxazinyl

and then to carbon atoms bearing double or triple bonds. Locants are cited immediately preceding the prefixes or suffixes to which they refer. Multiplicity of the same heteroatom is indicated by the appropriate prefix in the series: di-, tri-, tetra-, penta-, hexa-, etc.

If the corresponding carbocyclic system is partially or completely hydrogenated, the additional hydrogen is cited using the appropriate *H*- or hydro- prefixes. A trivial name from Tables 1-5 and 1-6, if available, along with the state of hydrogenation may be used. In the specialist nomenclature for heterocyclic systems, the prefix or prefixes from Table 1-3 are

TABLE 1-6 Trivial names for heterocyclic systems that are not recommended for use in fusion names

Listed in order of increasing priority

Structure	Parent name	Radical name	Structure	Parent name	Radical name
	Isochroman	Isochromanyl		Pyrazoline (3-shown*)	Pyrazolinyl
	Chroman	Chromanyl		Piperidine	Piperidyl†
	Pyrrolidine	Pyrrolinyl		Piperazine	Piperazinyl
	Pyrroline (2-shown*)	Pyrrolinyl		Indoline	Indolinyl
	Imidazolidine	Imidazolidinyl		Isoindoline	Isoindolinyl
	Imidazoline (2-shown*)	Imidazolinyl		Quinuclidine	Quinuclidinyl
	Pyrazolidine	Pyrazolidinyl		Morpholine	Morpholiny‡

* Denotes position of double bond.
† For 1-piperidyl, use piperidino.
‡ For 4-morpholinyl, use morpholino.

combined with the appropriate stem from Table 1-4, eliding an *a* where necessary. Examples of acceptable usage, including (1) replacement and (2) specialist nomenclature, are

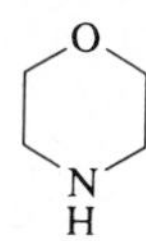

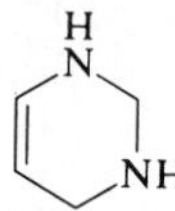

(1) 1-Oxa-4-azacyclohexane	(1) 1,3-Diazacyclohex-5-ene	(1) Thiacyclopropane
(2) 1,4-Oxazoline Morpholine	(2) 1,2,3,4-Tetrahydro-1,3-diazine	(2) Thiirane Ethylene sulfide

Radicals derived from heterocyclic compounds by removal of hydrogen from a ring are named by adding -yl to the names of the parent compounds (with elision of the final *e*, if present). These exceptions are retained:

Furyl (from furan)	Furfuryl (for 2-furylmethyl)
Pyridyl (from pyridine)	Furfurylidene (for 2-furylmethylene)
Piperidyl (from piperidine)	Thienyl (from thiophene)
Quinolyl (from quinoline)	Thenylidyne (for thienylmethylidyne)
Isoquinolyl	Furfurylidyne (for 2-furylmethylidyne)
Thenylidene (for thienylmethylene)	Thenyl (for thienylmethyl)

Also, piperidino- and morpholino- are preferred to 1-piperidyl- and 4-morpholinyl-, respectively.

If there is a choice among heterocyclic systems, the parent compound is decided in the following order of preference:

1. A nitrogen-containing component
2. A component containing a heteroatom, in the absence of nitrogen, as high as possible in Table 1-3
3. A component containing the greatest number of rings
4. A component containing the largest possible individual ring
5. A component containing the greatest number of heteroatoms of any kind
6. A component containing the greatest variety of heteroatoms
7. A component containing the greatest number of heteroatoms first listed in Table 1-3

If there is a choice between components of the same size containing the same number and kind of heteroatoms, choose as the base component that one with the lower numbers for the heteroatoms before fusion. When a fusion position is occupied by a heteroatom, the names of the component rings to be fused are selected to contain the heteroatom.

Functional Compounds

There are several types of nomenclature systems that are recognized. Which type to use is sometimes obvious from the nature of the compound. Substitutive nomenclature, in general, is preferred because of its broad applicability, but radicofunctional, additive, and replacement nomenclature systems are convenient in certain situations.

Substitutive Nomenclature

The first step is to determine the kind of characteristic (functional) group for use as the principal group of the parent compound. A characterstic group is a recognized combination of atoms that confers characteristic chemical properties on the molecule in which it occurs. Carbon-to-carbon unsaturation and heteroatoms in rings are considered nonfunctional for nomenclature purposes.

Substitution means the replacement of one or more hydrogen atoms in a given compound by some other kind of atom or group of atoms, functional or nonfunctional. In substitutive nomenclature, each substituent is cited as either a prefix or a suffix to the name of the parent (or substituting radical) to which it is attached; the latter is denoted the parent compound (or parent group if a radical).

In Table 1-7 are listed the general classes of compounds in descending order of preference for citation as suffixes, that is, as the parent or characteristic compound. When oxygen is

Table 1-7 Characteristic groups for substitutive nomenclature

Listed in order of decreasing priority for citation as principal group or parent name

Class	Formula*	Prefix	Suffix
1. Cations:		-onio-	-onium
	H_4N^+	Ammonio-	-ammonium
	H_3O^+	Oxonio-	-oxonium
	H_3S^+	Sulfonio-	-sulfonium
	H_3Se^+	Selenonio-	-selenonium
	H_2Cl^+	Chloronio-	-chloronium
	H_2Br^+	Bromonio-	-bromonium
	H_2I^+	Iodonio-	-iodonium
2. Acids:			
Carboxylic	—COOH	Carboxy-	-carboxylic acid
	—(C)OOH		-oic acid
	—C(=O)OOH		-peroxy···carboxylic acid
	—(C=O)OOH		-peroxy···oic acid
Sulfonic	$—SO_3H$	Sulfo-	-sulfonic acid
Sulfinic	$—SO_2H$	Sulfino-	-sulfinic acid
Sulfenic	—SOH	Sulfeno-	-sulfenic acid
Salts	—COOM		Metal···carboxylate
	—(C)OOM		Metal···oate
	$—SO_3M$		Metal···sulfonate
	$—SO_2M$		Metal···sulfinate
	—SOM		Metal···sulfenate
3. Derivatives of acids:			
Anhydrides	—C(=O)OC(=O)—		-carboxylic anhydride
	—(C=O)O(C=O)—		-oic anhydride
Esters	—COOR	R-oxycarbonyl-	R···carboxylate
	—C(OOR)		R···oate
Acid halides	—CO—halogen	Haloformyl	-carbonyl halide
Amides	$—CO—NH_2$	Carbamoyl-	-carboxamide
	$(C)O—NH_2$		-amide

Table 1-7 Characteristic groups for substitutive nomenclature (*continued*)

Listed in order of decreasing priority for citation as principal group or parent name

Class	Formula*	Prefix	Suffix
Hydrazides	—CO—NHNH$_2$	Carbonyl-hydrazino-	-carbohydrazide
	—(CO)—NHNH$_2$		-ohydrazide
Imides	—CO—NH—CO—	R-imido-	-carboximide
Amidines	—C(=NH)—NH$_2$	Amidino-	-carboxamidine
	—(C=NH)—NH$_2$		-amidine
4. Nitrile (cyanide)	—CN	Cyano-	-carbonitrile
	—(C)N		-nitrile
5. Aldehydes	—CHO	Formyl-	-carbaldehyde
	—(C=O)H	Oxo-	-al
	(then their analogs and derivatives)		
6. Ketones	$>$(C=O)	Oxo-	-one
	(then their analogs and derivatives)		
7. Alcohols (and phenols)	—OH	Hydroxy-	-ol
Thiols	—SH	Mercapto-	-thiol
8. Hydroperoxides	—O—OH	Hydroperoxy-	
9. Amines	—NH$_2$	Amino-	-amine
Imines	$>$NH	Imino-	-imine
Hydrazines	—NHNH$_2$	Hydrazino-	-hydrazine
10. Ethers	—OR	R-oxy-	
Sulfides	—SR	R-thio-	
11. Peroxides	—O—OR	R-dioxy-	

* Carbon atoms enclosed in parentheses are included in the name of the parent compound and not in the suffix or prefix.

replaced by sulfur, selenium, or tellurium, the priority for these elements is in the descending order listed. The higher valence states of each element are listed before considering the successive lower valence states. Derivative groups have priority for citation as principal group after the respective parents of their general class.

In Table 1-8 are listed characteristic groups that are cited only as prefixes (never as suffixes) in substitutive nomenclature. The order of listing has no significance for nomenclature purposes.

Systematic names formed by applying the principles of substitutive nomenclature are single words except for compounds named as acids. First one selects the parent compound, and thus the suffix, from the characteristic group listed earliest in Table 1-7. All remaining functional groups are handled as prefixes that precede, in alphabetical order, the parent name. Two examples may be helpful:

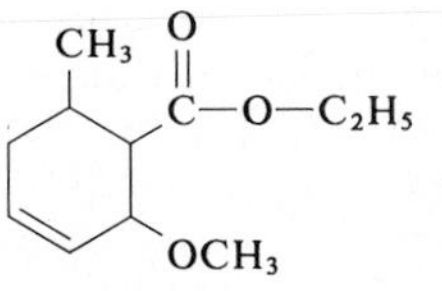

CH$_3$ C(=O)—O—C$_2$H$_5$ OCH$_3$

Structure I

$BrCH_2—CH_2—C(=O)—CH_2OH$

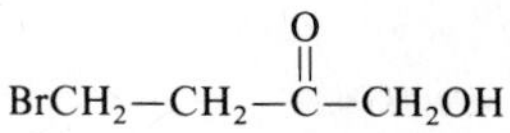

Structure II

TABLE 1-8 Characteristic groups cited only as prefixes in substitutive nomenclature

Characteristic group	Prefix	Characteristic group	Prefix
$-Br$	Bromo-	$-IX_2$	X may be halogen or a radical; dihalogenoiodo- or diacetoxyiodo-, e.g., $-ICl_2$ is dichloroido-
$-Cl$	Chloro-		
$-ClO$	Chlorosyl-		
$-ClO_2$	Chloryl-	$\geq N_2$	Diazo-
$-ClO_3$	Perchloryl-	$-N_3$	Azido-
$-F$	Fluoro-	$-NO$	Nitroso-
$-I$	Iodo-	$-NO_2$	Nitro-
$-IO$	Iodosyl-	$\geq N(=O)OH$	*aci*-Nitro-
$-IO_2$	Iodyl*	$-OR$	R-oxy-
$-I(OH)_2$	Dihydroxyiodo-	$-SR$	R-thio-
		$-SeR$ ($-TeR$)	R-seleno- (R-telluro-)

* Formerly iodoxy-.

Structure I contains an ester group and an ether group. Since the ester group has higher priority, the name is ethyl 2-methoxy-6-methyl-3-cyclohexene-1-carboxylate. Structure II contains a carbonyl group, an hydroxy group, and a bromo group. The latter is never a suffix. Between the other two, the carbonyl group has higher priority, the parent has -one as suffix, and the name is 4-bromo-1-hydroxy-2-butanone.

Selection of the principal alicyclic chain or ring system is governed by these selection rules:

1. For purely alicyclic compounds, the selection process proceeds successively until a decision is reached: (a) the maximum number of substituents corresponding to the characteristic group cited earliest in Table 1-7, (b) the maximum number of double and triple bonds considered together, (c) the maximum length of the chain, and (d) the maximum number of double bonds. Additional criteria, if needed for complicated compounds, are given in the IUPAC nomenclature rules.
2. If the characteristic group occurs only in a chain that carries a cyclic substituent, the compound is named as an aliphatic compound into which the cyclic component is substituted; a radical prefix is used to denote the cyclic component. This chain need not be the longest chain.
3. If the characteristic group occurs in more than one carbon chain and the chains are not directly attached to one another, then the chain chosen as parent should carry the largest number of the characteristic group. If necessary, the selection is continued as in rule 1.
4. If the characteristic group occurs only in one cyclic system, that system is chosen as the parent.
5. If the characteristic group occurs in more than one cyclic system, that system is chosen as parent which (a) carries the largest number of the principal group or, failing to reach a decision, (b) is the senior ring system.
6. If the characteristic group occurs both in a chain and in a cyclic system, the parent is that portion in which the principal group occurs in largest number. If the numbers are the same, that portion is chosen which is considered to be the most important or is the senior ring system.

7. When a substituent is itself substituted, all the subsidiary substituents are named as prefixes and the entire assembly is regarded as a parent radical.
8. The seniority of ring systems is ascertained by applying the following rules successively until a decision is reached: (a) all heterocycles are senior to all carbocycles, (b) for heterocycles, the preference follows the decision process described under "Heterocyclic Systems," page 1-12, (c) the largest number of rings, (d) the largest individual ring at the first point of difference, (e) the largest number of atoms in common among rings, (f) the lowest letters in the expression for ring functions, (g) the lowest numbers at the first point of difference in the expression for ring junctions, (h) the lowest state of hydrogenation, (i) the lowest-numbered locant for indicated hydrogen, (j) the lowest-numbered locant for point of attachment (if a radical), (k) the lowest-numbered locant for an attached group expressed as a suffix, (l) the maximum number of substituents cited as prefixes, (m) the lowest-numbered locant for substituents named as prefixes, hydro prefixes, -ene, and -yne, all considered together in one series in ascending numerical order independent of their nature, and (n) the lowest-numbered locant for the substituent named as prefix which is cited first in the name.

Numbering of Compounds. If the rules for aliphatic chains and ring systems leave a choice, the starting point and direction of numbering of a compound are chosen so as to give lowest-numbered locants to these structural factors, if present, considered successively in the order listed below until a decision is reached. Characteristic groups take precedence over multiple bonds.

1. Indicated hydrogen, whether cited in the name or omitted as being conventional
2. Characteristic groups named as suffix following the ranking order of Table 1-7
3. Multiple bonds in acyclic compounds; in bicycloalkanes, tricycloalkanes, and polycycloalkanes, double bonds having priority over triple bonds; and in heterocyclic systems whose names end in -etine, -oline, or -olene
4. The lowest-numbered locant for substituents named as prefixes, hydro prefixes, -ene, and -yne, all considered together in one series in ascending numerical order
5. The lowest locant for that substituent named as prefix which is cited first in the name

For cyclic radicals, indicated hydrogen and thereafter the point of attachment (free valency) have priority for the lowest available number.

Prefixes and Affixes. Prefixes are arranged alphabetically and placed before the parent name; multiplying affixes, if necessary, are inserted and *do not* alter the alphabetical order already attained. The parent name includes any syllables denoting a change of ring member or relating to the structure of a carbon chain. Nondetachable parts of parent names include

1. Forming rings: cyclo-, bicyclo-, spiro-
2. Fusing two or more rings: benzo-, naphtho-, imidazo-
3. Substituting one ring or chain member atom for another: oxa-, aza-, thia-
4. Changing positions of ring or chain members: iso-, *sec*-, *tert*-, neo-
5. Showing indicated hydrogen
6. Forming bridges: ethano-, epoxy-
7. Hydro-

Prefixes that represent complete terminal characteristic groups are preferred to those representing only a portion of a given group. For example, for the prefix $-C(=O)CH_3$, the name (formylmethyl) is preferred to (oxoethyl).

The multiplying affixes di-, tri-, tetra-, penta-, hexa-, hepta-, octa-, nona-, deca-, undeca-, and so on are used to indicate a set of *identical* unsubstituted radicals or parent compounds. The forms bis-, tris-, tetrakis-, pentakis-, and so on are used to indicate a set of identical radicals or parent compounds *each substituted in the same way.* The affixes bi-, ter-, quater-, quinque-, sexi, septi-, octi-, novi-, deci-, and so on are used to indicate the number of identical rings joined together by a single or double bond.

Although multiplying affixes may be omitted for very common compounds when no ambiguity is caused thereby, such affixes are generally included throughout this handbook in alphabetical listings. An example would be ethyl ether for diethyl ether.

Conjunctive Nomenclature

Conjunctive nomenclature may be applied when a principal group is attached to an acyclic component that is directly attached by a carbon-carbon bond to a cyclic component. The name of the cyclic component is attached directly in front of the name of the acyclic component carrying the principal group. This nomenclature is not used when an unsaturated side chain is named systematically. When necessary, the position of the side chain is indicated by a locant placed before the name of the cyclic component. For substituents on the acyclic chain, carbon atoms of the side chain are indicated by Greek letters proceeding from the principal group to the cyclic component. The terminal carbon atom of acids, aldehydes, and nitriles is omitted when allocating Greek positional letters. Conjunctive nomenclature is not used when the side chain carries more than one of the principal group, except in the case of malonic and succinic acids.

The side chain is considered to extend only from the principal group to the cyclic component. Any other chain members are named as substituents, with appropriate prefixes placed before the name of the cyclic component.

When a cyclic component carries more than one identical side chain, the name of the cyclic component is followed by di-, tri-, etc., and then by the name of the acyclic component, and it is preceded by the locants for the side chains. Examples are

$$H_3C-C_6H_{10}-\overset{\beta}{C}H_2-\overset{\alpha}{C}H_2OH$$

4-Methyl-1-cyclohexaneethanol

$$C_6H_{11}-C(CH_3)_2-CH(OH)-CH_2-CH_3$$

α-Ethyl-β,β-dimethylcyclohexaneethanol

When side chains of two or more different kinds are attached to a cyclic component, only the senior side chain is named by the conjunctive method. The remaining side chains are named as prefixes. Likewise, when there is a choice of cyclic component, the senior is chosen. Benzene derivatives may be named by the conjunctive method only when two or more identical side chains are present. Trivial names for oxo carboxylic acids may be used for the acyclic component. If the cyclic and acyclic components are joined by a double bond, the locants of this bond are placed as superscripts to a Greek capital delta that is inserted between the two names. The locant for the cyclic component precedes that for the acyclic component, e.g., indene-$\Delta^{1,\alpha}$-acetic acid.

Radicofunctional Nomenclature

The procedures of radicofunctional nomenclature are identical with those of substitutive nomenclature except that suffixes are never used. Instead, the functional class name (Table 1-9) of the compound is expressed as one word and the remainder of the molecule as another that precedes the class name. When the functional class name refers to a characteristic group that is bivalent, the two radicals attached to it are each named, and when different, they are written as separate words arranged in alphabetical order. When a compound contains more than one kind of group listed in Table 1-9, that kind is cited as the functional group or class name that occurs higher in the table, all others being expressed as prefixes.

Radicofunctional nomenclature finds some use in naming ethers, sulfides, sulfoxides, sulfones, selenium analogs of the preceding three sulfur compounds, and azides.

TABLE 1-9 Functional class names used in radicofunctional nomenclature

Groups are listed in order of decreasing priority.

Group	Functional class names
X in acid derivatives	Name of X (in priority order: fluoride, chloride, bromide, iodide; cyanide, azide; then the sulfur and selenium analogs)
—CN, —NC	Cyanide, isocyanide
>CO	Ketone; then S and Se analogs
—OH	Alcohol; then S and Se analogs
—O—OH	Hydroperoxide
>O	Ether or oxide
>S, >SO, >SO_2	Sulfide, sulfoxide, sulfone
>Se, >SeO, >SeO_2	Selenide, selenoxide, selenone
—F, —Cl, —Br, —I	Fluoride, chloride, bromide, iodide
—N_3	Azide

Replacement Nomenclature

Replacement nomenclature is intended for use only when other nomenclature systems are difficult to apply in the naming of chains containing heteroatoms. When no group is present that can be named as a principal group, the longest chain of carbon and heteroatoms terminating with carbon is chosen and named as though the entire chain were that of an acyclic hydrocarbon. The heteroatoms within this chain are identified by means of prefixes aza-, oxa-, thia-, etc., in the order of priority stated in Table 1-3. Locants indicate the positions of the heteroatoms in the chain. Lowest-numbered locants are assigned to the principal group when such is present. Otherwise, lowest-numbered locants are assigned to the heteroatoms considered together and, if there is a choice, to the heteroatoms cited earliest in Table 1-3. An example is

$$\mathrm{HO{-}\overset{13}{C}H_2{-}\overset{12}{O}{-}\overset{11}{C}H_2{-}\overset{10}{C}H_2{-}\overset{9}{O}{-}\overset{8}{C}H_2{-}\overset{7}{C}H_2{-}\underset{H}{\overset{6}{N}}{-}\overset{5}{C}H_2{-}\overset{4}{C}H_2{-}\underset{H}{\overset{3}{N}}{-}\overset{2}{C}H_2{-}\overset{1}{C}OOH}$$

13-Hydroxy-9,12-dioxa-3,6-diazatridecanoic acid

Specific Functional Groups

Characteristic groups will now be treated briefly in order to expand the terse outline of substitutive nomenclature presented in Table 1-7. Alternative nomenclature will be indicated whenever desirable.

Acetals and Acylals

Acetals, which contain the group $>C(OR)_2$, where R may be different, are named (1) as dialkoxy compounds or (2) by the name of the corresponding aldehyde or ketone followed by the name of the hydrocarbon radical(s) followed by the word *acetal.* For example, $CH_3{-}CH(OCH_3)_2$ is named either (1) 1,1-dimethoxyethane or (2) acetaldehyde dimethyl acetal.

A cyclic acetal in which the two acetal oxygen atoms form part of a ring may be named (1) as a heterocyclic compound or (2) by use of the prefix methylenedioxy for the group $-O-CH_2-O-$ as a substituent in the remainder of the molecule. For example,

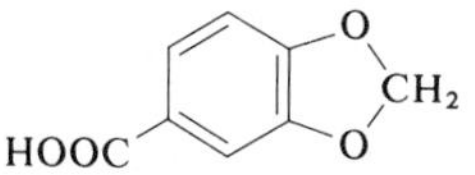

HOOC–C₆H₃(O–CH₂–O)

(1) 1,3-Benzo[*d*]dioxole-5-carboxylic acid

(2) 3,4-Methylenedioxybenzoic acid

Acylals, $R^1R^2C(OCOR^3)_2$, are named as acid esters;

$$CH_3{-}CH_2{-}CH_2{-}CH\begin{cases}O{-}CO{-}CH_3\\O{-}CO{-}CH_2{-}CH_3\end{cases}$$

Butylidene acetate propionate

α-Hydroxy ketones, formerly called acyloins, had been named by changing the ending -ic acid or -oic acid of the corresponding acid to -oin. They are preferably named by substitutive nomenclature; thus

$CH_3{-}CH(OH){-}CO{-}CH_3$ 3-Hydroxy-2-butanone (formerly acetoin)

Acid Anhydrides

Symmetrical anhydrides of monocarboxylic acids, when unsubstituted, are named by replacing the word *acid* by *anhydride.* Anhydrides of substituted monocarboxylic acids, if symmetrically substituted, are named by prefixing bis- to the name of the acid and replacing the word *acid* by *anhydride.* Mixed anhydrides are named by giving in alphabetical order the first part of the names of the two acids followed by the word *anhydride,* e.g., acetic propionic anhydride or acetic propanoic anhydride. Cyclic anhydrides of polycarboxylic acids, although possessing a heterocyclic structure, are preferably named as acid anhydrides. For example,

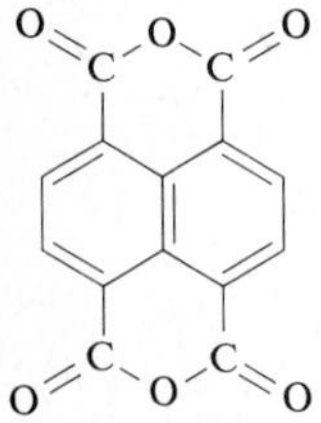

1,8;4,5-Napthalenetetracarboxylic dianhydride. (Note the use of a semicolon to distinguish the pairs of locants.)

Acyl Halides

Acyl halides, in which the hydroxyl portion of a carboxyl group is replaced by a halogen, are named by placing the name of the corresponding halide after that of the acyl radical. When another group is present that has priority for citation as principal group or when the acyl halide is attached to a side chain, the prefix haloformyl- is used as, for example, in fluoroformyl-.

Alcohols and Phenols

The hydroxyl group is indicated by a suffix -ol when it is the principal group attached to the parent compound and by the prefix hydroxy- when another group with higher priority for citation is present or when the hydroxy group is present in a side chain. When confusion may arise in employing the suffix -ol, the hydroxy group is indicated as a prefix; this terminology is also used when the hydroxyl group is attached to a heterocycle, as, for example, in the name 3-hydroxythiophene to avoid confusion with thiophenol (C_6H_5SH). Designations such as isopropanol, *sec*-butanol, and *tert*-butanol are incorrect because no hydrocarbon exists to which the suffix can be added. Many trivial names are retained. These structures are shown in Table 1-10. The radicals (RO—) are named by adding -oxy as a suffix to the name of the R radical, e.g., pentyloxy for $CH_3CH_2CH_2CH_2CH_2O$—. These contractions are exceptions: methoxy (CH_3O—), ethoxy (C_2H_5O—), propoxy (C_3H_7O—), butoxy (C_4H_9O—), and phenoxy (C_6H_5O—). For unsubstituted radicals only, one may use isopropoxy [$(CH_3)_2CH$—O—], isobutoxy [$(CH_3)_2CH_2CH$—O—], *sec*-butoxy [$CH_3CH_2CH(CH_3)$—O—], and *tert*-butoxy [$(CH_3)_3C$—O—].

TABLE 1-10 Retained trivial names of alcohols and phenols with structures

Ally alcohol	$CH_2{=}CHCH_2OH$
tert-Butyl alcohol	$(CH_3)_3COH$
Benzyl alcohol	$C_6H_5CH_2OH$
Phenethyl alcohol	$C_6H_5CH_2CH_2OH$
Ethylene glycol	$HOCH_2CH_2OH$
1,2-Propylene glycol	$CH_3CHOHCH_2OH$
Glycerol	$HOCH_2CHOHCH_2OH$
Pentaerythritol	$C(CH_2OH)_4$
Pinacol	$(CH_3)_2COHCOH(CH_3)_2$
Phenol	C_6H_5OH
Xylitol	OH (on central CH) $HOCH_2CH{-}CH{-}CH{-}CH_2OH$ OH (on 2nd C), OH (on 4th C)
Geraniol	$(CH_3)_2C{=}CHCH_2CH_2C{=}CHCH_2OH$ CH_3 (on C=)
Phytol	CH_3 (on CH) $CH_2CH_2CHCH_2CH_2CH_2CH(CH_3)_2$ | $CH_2CHCH_2CH_2CH_2C{=}CHCH_2OH$ CH_3 (on CH), CH_3 (on C=)

TABLE 1-10 Retained trivial names of alcohols and phenols with structures (*continued*)

Menthol

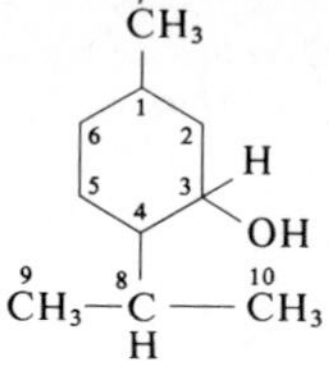

Borneol

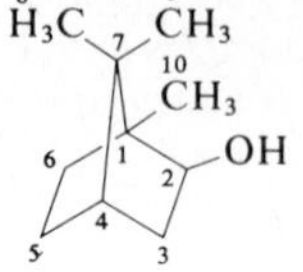

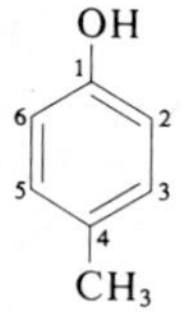

Cresol (1,4-isomer shown)

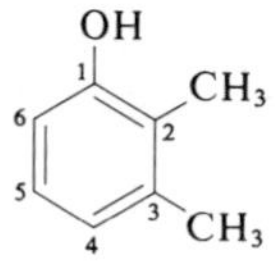

Xylenol (2,3-isomer shown)

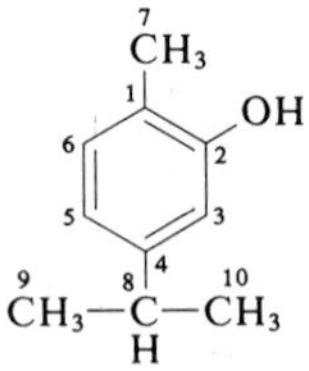

Carvacrol

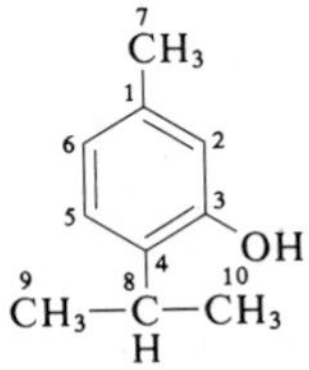

Thymol

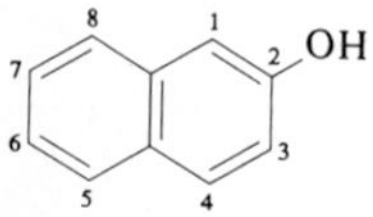

Naphthol (2-isomer shown)
2-Hydroxynaphthalene

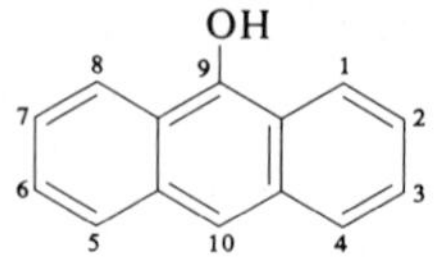

Anthrol (9-isomer shown)
9-Hydroxyanthracene

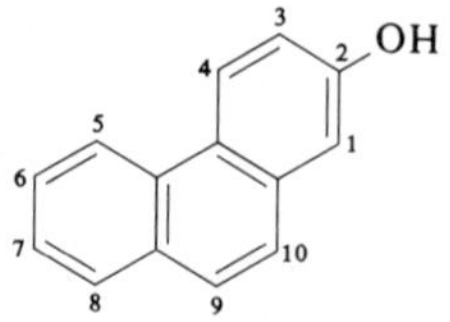

Phenanthrol (2-isomer shown)
2-Hydroxyphenanthrene

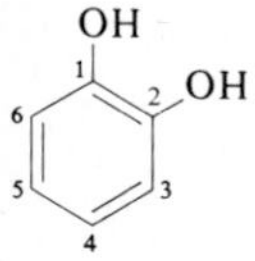

Pyrocatechol
1,2-Dihydroxybenzene

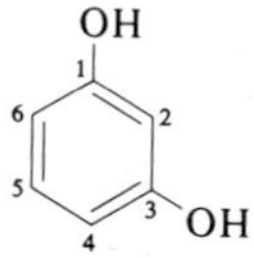

Resorcinol
1,3-Dihydroxybenzene

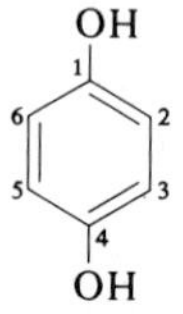

Hydroquinone
1,4-Dihydroxybenzene

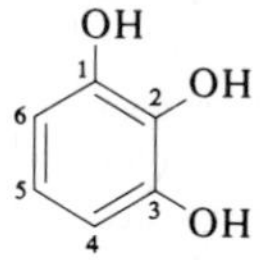

Pyrogallol
1,2,3-Trihydroxybenzene

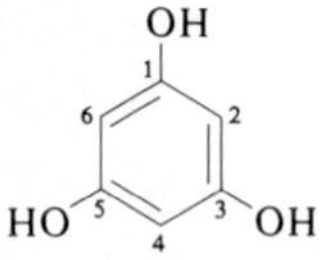

Phloroglucinol
1,3,5-Trihydroxybenzene

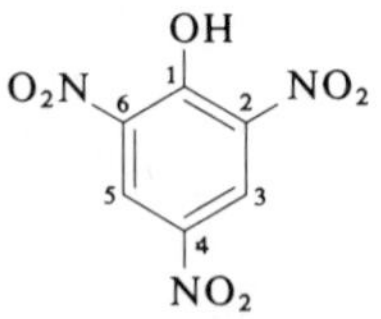

Picric acid
2,4,6-Trinitrophenol

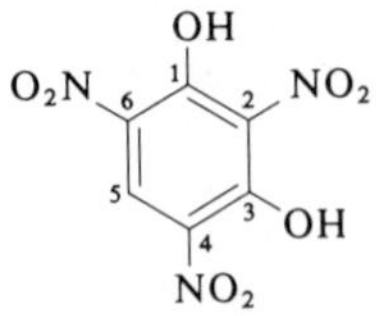

Styphnic acid
1,3-Dihydroxy-2,4,6-trinitrobenzene

Bivalent radicals of the form O—Y—O are named by adding -dioxy to the name of the bivalent radicals except when forming part of a ring system. Examples are —O—CH_2—O— (methylenedioxy), —O—CO—O— (carbonyldioxy), and —O—SO_2—O— (sulfonyldioxy). Anions derived from alcohols or phenols are named by changing the final -ol to -olate.

Salts composed of an anion, RO—, and a cation, usually a metal, can be named by citing first the cation and then the RO anion (with its ending changed to -yl oxide), e.g., sodium benzyl oxide for $C_6H_5CH_2ONa$. However, when the radical has an abbreviated name, such as methoxy, the ending -oxy is changed to -oxide. For example, CH_3ONa is named sodium methoxide (not sodium methylate).

Aldehydes

When the group —C(=O)H, usually written —CHO, is attached to carbon at one (or both) end(s) of a linear acyclic chain the name is formed by adding the suffix -al (or -dial) to the name of the hydrocarbon containing the same number of carbon atoms. Examples are butanal for $CH_3CH_2CH_2CHO$ and propanedial for $OHCCH_2CHO$.

Naming an acyclic polyaldehyde can be handled in two ways. First, when more than two aldehyde groups are attached to an unbranched chain, the proper affix is added to -carbaldehyde, which becomes the suffix to the name of the longest chain carrying the maximum number of aldehyde groups. The name and numbering of the main chain do not include the carbon atoms of the aldehyde groups. Second, the name is formed by adding the prefix formyl- to the name of the -dial that incorporates the principal chain. Any other chains carrying aldehyde groups are named by the use of formylalkyl- prefixes. Examples are

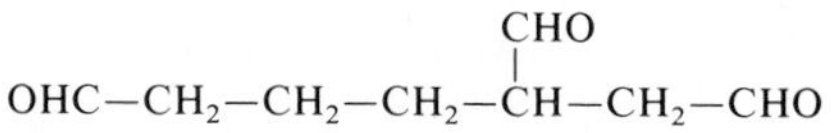

$$\begin{array}{c} \qquad\qquad\qquad\qquad\qquad\quad CHO \\ \qquad\qquad\qquad\qquad\qquad\quad | \\ OHC{-}CH_2{-}CH_2{-}CH_2{-}CH{-}CH_2{-}CHO \end{array}$$

(1) 1,2,5-Pentanetricarbaldehyde
(2) 3-Formylheptanedial

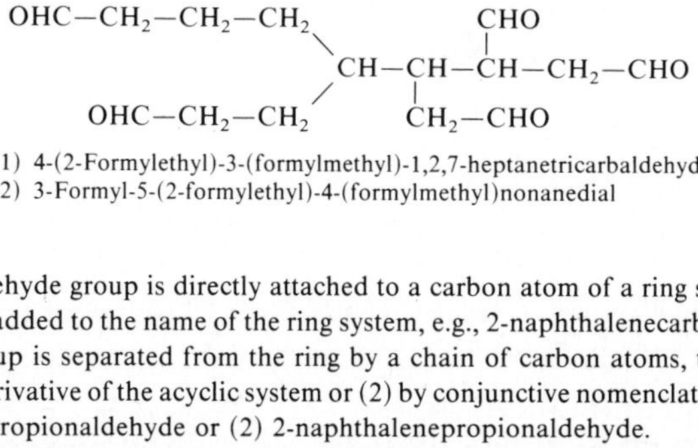

$$\begin{array}{l} OHC{-}CH_2{-}CH_2{-}CH_2\diagdown \qquad\quad CHO \\ \qquad\qquad\qquad\qquad\qquad\quad CH{-}CH{-}CH{-}CH_2{-}CHO \\ \qquad OHC{-}CH_2{-}CH_2\diagup \qquad\ \ CH_2{-}CHO \end{array}$$

(1) 4-(2-Formylethyl)-3-(formylmethyl)-1,2,7-heptanetricarbaldehyde
(2) 3-Formyl-5-(2-formylethyl)-4-(formylmethyl)nonanedial

When the aldehyde group is directly attached to a carbon atom of a ring system, the suffix -carbaldehyde is added to the name of the ring system, e.g., 2-naphthalenecarbaldehyde. When the aldehyde group is separated from the ring by a chain of carbon atoms, the compound is named (1) as a derivative of the acyclic system or (2) by conjunctive nomenclature, for example, (1) (2-naphthyl)propionaldehyde or (2) 2-naphthalenepropionaldehyde.

An aldehyde group is denoted by the prefix formyl- when it is attached to a nitrogen atom in a ring system or when a group having priority for citation as principal group is present and part of a cyclic system.

When the corresponding monobasic acid has a trivial name, the name of the aldehyde may be formed by changing the ending -ic acid or -oic acid to -aldehyde. Examples are

Formaldehyde	Acrylaldehyde (not acrolein)
Acetaldehyde	Benzaldehyde
Propionaldehyde	Cinnamaldehyde
Butyraldehyde	2-Furaldehyde (not furfural)

The same is true for polybasic acids, with the proviso that all the carboxyl groups must be changed to aldehyde; then it is not necessary to introduce affixes. Examples are

Glyceraldehyde
Glycolaldehyde
Malonaldehyde
Succinaldehyde
Phthalaldehyde (*o*-, *m*-, *p*-)

These trivial names may be retained: citral (3,7-dimethyl-2,6-octadienal), vanillin (4-hydroxy-3-methoxybenzaldehyde), and piperonal (3,4-methylenedioxybenzaldehyde).

Amides

For primary amides the suffix -amide is added to the systematic name of the parent acid. For example, $CH_3—CO—NH_2$ is acetamide. Oxamide is retained for $H_2N—CO—CO—NH_2$. The name -carboxylic acid is replaced by -carboxamide.

For amino acids having trivial names ending in -ine, the suffix -amide is added after the name of the acid (with elision of *e* for monoamides). For example, $H_2N—CH_2—CO—NH_2$ is glycinamide.

In naming the radical R—CO—NH—, either (1) the -yl ending of RCO— is changed to -amido or (2) the radicals are named as acylamino radicals. For example.

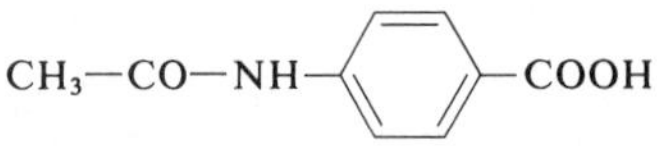

$CH_3—CO—NH—C_6H_4—COOH$ (1) 4-Acetamidobenzoic acid
(2) 4-Acetylaminobenzoic acid

The latter nomenclature is always used for amino acids with trivial names.

N-substituted primary amides are named either (1) by citing the substituents as *N* prefixes or (2) by naming the acyl group as an *N* substituent of the parent compound. For example,

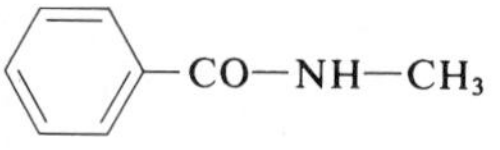

$C_6H_5—CO—NH—CH_3$ (1) *N*-Methylbenzamide
(2) Benzoylaminomethane

Amines

Amines are preferably named by adding the suffix -amine (and any multiplying affix) to the name of the parent radical. Examples are

$CH_3CH_2CH_2CH_2CH_2NH_2$	Pentylamine
$H_2NCH_2CH_2CH_2CH_2CH_2NH_2$	1,5-Pentyldiamine or pentamethylenediamine

Locants of substituents of symmetrically substituted derivatives of symmetrical amines are distinguished by primes or else the names of the complete substituted radicals are enclosed in parentheses. Unsymmetrically substituted derivatives are named similarly or as *N*-substituted products of a primary amine (after choosing the most senior of the radicals to be the parent amine). For example,

$HN(CH_2CH_2CH_2F)(CHF—CH_2CH_3)$

(1) 1,3′-Diflurodipropylamine
(2) 1-Fluoro-*N*-(3-fluoropropyl)propylamine
(3) (1-Fluoropropyl)(3-fluoropropyl)amine

Complex cyclic compounds may be named by adding the suffix -amine or the prefix amino- (or aminoalkyl-) to the name of the parent compound. Thus three names are permissible for

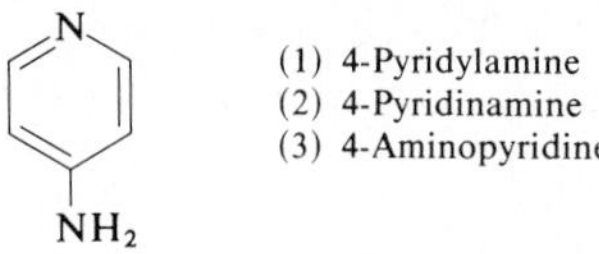

(1) 4-Pyridylamine
(2) 4-Pyridinamine
(3) 4-Aminopyridine

Complex linear polyamines are best designated by replacement nomenclature. These trivial names are retained: aniline, benzidene, phenetidine, toluidine, and xylidine.

The bivalent radical —NH— linked to two identical radicals can be denoted by the prefix imino-, as well as when it forms a bridge between two carbon ring atoms. A trivalent nitrogen atom linked to three identical radicals is denoted by the prefix nitrilo-. Thus ethylenediaminetetraacetic acid (an allowed exception) should be named ethylenedinitrilotetraacetic acid.

Ammonium Compounds

Salts and hydroxides containing quadricovalent nitrogen are named as a substituted ammonium salt or hydroxide. The names of the substituting radicals precede the word *ammonium*, and then the name of the anion is added as a separate word. For example, $(CH_3)_4N^+I^-$ is tetramethylammonium iodide.

When the compound can be considered as derived from a base whose name does not end in -amine, its quaternary nature is denoted by adding ium to the name of that base (with elision of *e*), substituent groups are cited as prefixes, and the name of the anion is added separately at the end. Examples are

$C_6H_5NH_3{}^+HSO_4^-$ Anilinium hydrogen sulfate

$[(C_6H_5NH_3)^+]_2PtCl_6^{2-}$ Dianilinium hexachloroplatinate

The names *choline* and *betaine* are retained for unsubstituted compounds.

In complex cases, the prefixes amino- and imino- may be changed to ammonio- and iminio- and are followed by the name of the molecule representing the most complex group attached to this nitrogen atom and are preceded by the names of the other radicals attached to this nitrogen. Finally the name of the anion is added separately. For example, the name might be 1-trimethylammonioacridine chloride or 1-acridinyltrimethylammonium chloride.

When the preceding rules lead to inconvenient names, then (1) the unaltered name of the base may be used followed by the name of the anion or (2) for salts of hydrohalogen acids only the unaltered name of the base is used followed by the name of the hydrohalide. An example of the latter would be 2-ethyl-*p*-phenylenediamine monohydrochloride.

Azo Compounds

When the azo group (—N=N—) connects radicals derived from identical unsubstituted molecules, the name is formed by adding the prefix azo- to the name of the parent unsubstituted molecules. Substituents are denoted by prefixes and suffixes. The azo group has priority for lowest-numbered locant. Examples are azobenzene for $C_6H_5—N=N—C_6H_5$, azobenzene-4-sulfonic acid for $C_6H_5—N=N—C_6H_5SO_3H$, and 2′,4-dichloroazobenzene-4′-sulfonic acid for $ClC_6H_4—N=N—C_6H_3ClSO_3H$.

When the parent molecules connected by the azo group are different, azo is placed between the complete names of the parent molecules, substituted or unsubstituted. Locants are placed between the affix azo and the names of the molecules to which each refers. Preference is given to the more complex parent molecule for citation as the first component, e.g., 2-aminonaphthalene-1-azo-(4′-chloro-2′-methylbenzene).

In an alternative method, the senior component is regarded as substituted by RN=N—, this group R being named as a radical. Thus 2-(7-phenylazo-2-naphthylazo)anthracene is the name by this alternative method for the compound named anthracene-2-azo-2′-naphthalene-7′-azobenzene.

Azoxy Compounds

Where the position of the azoxy oxygen atom is unknown or immaterial, the compound is named in accordance with azo rules, with the affix azo replaced by azoxy. When the position of the azoxy oxygen atom in an unsymmetrical compound is designated, a prefix *NNO*- or *ONN*- is used. When both the groups attached to the azoxy radical are cited in the name of the compound, the prefix *NNO*- specifies that the second of these two groups is attached directly to —N(O)—; the prefix *ONN*- specifies that the first of these two groups is attached directly to —N(O)—. When only one parent compound is cited in the name, the prefixed *ONN*- and *NNO*- specify that the group carrying the primed and unprimed substituents is connected, respectively, to the —N(O)— group. The prefix *NON*- signifies that the position of the oxygen atom is unknown; the azoxy group is then written as $—N_2O—$. For example,

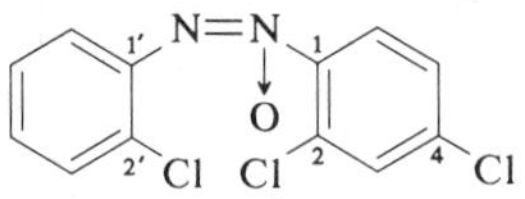

2,2′,4-Trichloro-*NNO*-azoxybenzene

Boron Compounds

Molecular hydrides of boron are called boranes. They are named by using a multiplying affix to designate the number of boron atoms and adding an Arabic numeral within parentheses as a suffix to denote the number of hydrogen atoms present. Examples are pentaborane(9) for B_5H_9 and pentaborane(11) for B_5H_{11}.

Organic ring systems are named by replacement nomenclature. Three- to ten-membered monocyclic ring systems containing uncharged boron atoms may be named by the specialist nomenclature for heterocyclic systems. Organic derivatives are named as outlined for substitutive nomenclature. The complexity of boron nomenclature precludes additional details; the text by Rigaudy and Klesney should be consulted.

Carboxylic Acids

Carboxylic acids may be named in several ways. First, —COOH groups replacing CH_3— at the end of the main chain of an acyclic hydrocarbon are denoted by adding -oic acid to the name of the hydrocarbon. Second, when the —COOH group is the principal group, the suffix -carboxylic acid can be added to the name of the parent chain whose name and chain numbering *does not include* the carbon atom of the —COOH group. The former nomenclature is preferred unless use of the ending -carboxylic acid leads to citation of a larger number of carboxyl groups as suffix. Third, carboxyl groups are designated by the prefix carboxy- when attached to a group named as a substituent or when another group is present that has higher priority for citation as principal group. In all cases, the principal chain should be linked to as many carboxyl groups as possible even though it might not be the longest chain present. Examples are

$CH_3CH_2CH_2CH_2CH_2CH_2COOH$ (1) Heptanoic acid
(2) 1-Hexanecarboxylic acid

$C_6H_{11}COOH$ (2) Cyclohexanecarboxylic acid

$$\begin{array}{c} \quad\quad\quad\quad COOH \quad\quad CH_2COOH \\ \quad\quad\quad | \quad\quad\quad\quad\quad | \\ CH_3{-}CH_2{-}CH{-}CH_2{-}CH{-}CH_2{-}COOH \end{array}$$

(3) 2-(Carboxymethyl)-1,4-hexanedicarboxylic acid

Removal of the OH from the —COOH group to form the acyl radical results in changing the ending -oic acid to -oyl or the ending -carboxylic acid to -carbonyl. Thus the radical $CH_3CH_2CH_2CH_2CO$— is named either pentanoyl or butanecarbonyl. When the hydroxyl has not been removed from all carboxyl groups present in an acid, the remaining carboxyl groups are denoted by the prefix carboxy-. For example, $HOOCCH_2CH_2CH_2CH_2CH_2CO$— is named 6-carboxyhexanoyl.

Many trivial names exist for acids; these are listed in Table 1-11. Generally, radicals are formed by replacing -ic acid by -oyl.* When a trivial name is given to an acyclic monoacid or diacid, the numeral 1 is always given as locant to the carbon atom of a carboxyl group in the acid or to the carbon atom with a free valence in the radical RCO—.

Ethers (R^1—O—R^2)

In substitutive nomenclature, one of the possible radicals, R—O—, is stated as the prefix to the parent compound that is senior from among R^1 or R^2. Examples are methoxyethane for $CH_3OCH_2CH_3$ and butoxyethanol for $C_4H_9OCH_2CH_2OH$.

When another principal group has precedence and oxygen is linking two identical parent compounds, the prefix oxy- may be used, as with 2,2′-oxydiethanol for $HOCH_2CH_2OCH_2CH_2OH$.

Compounds of the type RO—Y—OR, where the two parent compounds are identical and contain a group having priority over ethers for citation as suffix, are named as assemblies of identical units. For example, $HOOC{-}CH_2{-}O{-}CH_2CH_2{-}O{-}CH_2{-}COOH$ is named 2,2′-(ethylenedioxy)diacetic acid.

Linear polyethers derived from three or more molecules of aliphatic dihydroxy compounds, particularly when the chain length exceeds ten units, are most conveniently named by open-chain replacement nomenclature. For example, $CH_3CH_2{-}O{-}CH_2CH_2{-}O{-}CH_2CH_3$ could be 3,6-dioxaoctane or (2-ethoxy)ethoxyethane.

An oxygen atom directly attached to two carbon atoms already forming part of a ring system or to two carbon atoms of a chain may be indicated by the prefix epoxy-. For example, $CH_2{-}CH{-}CH_2Cl$ (with O bridging CH_2 and CH) is named 1-chloro-2,3-epoxypropane.

Symmetrical linear polyethers may be named (1) in terms of the central oxygen atom when there is an odd number of ether oxygen atoms or (2) in terms of the central hydrocarbon group when there is an even number of ether oxygen atoms. For example, $C_2H_5{-}O{-}C_4H_8{-}O{-}C_4H_8{-}O{-}C_2H_5$ is bis-(4-ethoxybutyl)ether, and 3,6-dioxaoctane (earlier example) could be named 1,2-bis(ethoxy)ethane.

* Exceptions: formyl, acetyl, propionyl, butyryl, isobutyryl, valeryl, isovaleryl, oxalyl, malonyl, succinyl, glutaryl, furoyl, and thenoyl.

TABLE 1-11 Names of some carboxylic acids

Systematic name	Trivial name	Systematic name	Trivial name
Methanoic	Formic	*trans*-Methylbutenedioic	Mesaconic*
Ethanoic	Acetic		
Propanoic	Propionic	1,2,2-Trimethyl-1,3-cyclopentanedicarboxylic acid	Camphoric
Butanoic	Butyric		
2-Methylpropanoic	Isobutyric*		
Pentanoic	Valeric	Benzenecarboxylic	Benzoic
3-Methylbutanoic	Isovaleric*	1,2-Benzenedicarboxylic	Phthalic
2,2-Dimethylpropanoic	Pivalic*	1,3-Benzenedicarboxylic	Isophthalic
Hexanoic	(Caproic)	1,4-Benzenedicarboxylic	Terephthalic
Heptanoic	(Enanthic)	Naphthalenecarboxylic	Naphthoic
Octanoic	(Caprylic)	Methylbenzenecarboxylic	Toluic
Decanoic	(Capric)	2-Phenylpropanoic	Hydratropic
Dodecanoic	Lauric*	2-Phenylpropenoic	Atropic
Tetradecanoic	Myristic*	*trans*-3-Phenylpropenoic	Cinnamic
Hexadecanoic	Palmitic*	Furancarboxylic	Furoic
Octadecanoic	Stearic*	Thiophenecarboxylic	Thenoic
		3-Pyridinecarboxylic	Nicotinic
Ethanedioic	Oxalic	4-Pyridinecarboxylic	Isonicotinic
Propanedioic	Malonic		
Butanedioic	Succinic	Hydroxyethanoic	Glycolic
Pentanedioic	Glutaric	2-Hydroxypropanoic	Lactic
Hexanedioic	Adipic	2,3-Dihydroxypropanoic	Glyceric
Heptanedioic	Pimelic*	Hydroxypropanedioic	Tartronic
Octanedioic	Suberic*	Hydroxybutanedioic	Malic
Nonanedioic	Azelaic*	2,3-Dihydroxybutanedioic	Tartaric
Decanedioic	Sebacic*	3-Hydroxy-2-phenylpropanoic	Tropic
Propenoic	Acrylic	2-Hydroxy-2,2-diphenylethanoic	Benzilic
Propynoic	Propiolic		
2-Methylpropenoic	Methacrylic	2-Hydroxybenzoic	Salicylic
trans-2-Butenoic	Crotonic	Methoxybenzoic	Anisic
cis-2-Butenoic	Isocrotonic	4-Hydroxy-3-methoxybenzoic	Vanillic
cis-9-Octadecenoic	Oleic		
trans-9-Octadecenoic	Elaidic	3,4-Dimethoxybenzoic	Veratric
cis-Butenedioic	Maleic	3,4-Methylenedioxybenzoic	Piperonylic
trans-Butenedioic	Fumaric	3,4-Dihydroxybenzoic	Protocatechuic
cis-Methylbutenedioic	Citraconic*	3,4,5-Trihydroxybenzoic	Gallic

Note: The names in parentheses are abandoned but are listed for reference to older literature.
* Systematic names should be used in derivatives formed by substitution on a carbon atom.

Partial ethers of polyhydroxy compounds may be named (1) by substitutive nomenclature or (2) by stating the name of the polyhydroxy compound followed by the name of the etherifying radical(s) followed by the word *ether*. For example,

$$\begin{array}{l} CH_2O-C_4H_9 \\ | \\ HCOH \\ | \\ CH_2OH \end{array}$$

(1) 3-Butoxy-1,2-propanediol

(2) Glycerol 1-butyl ether; also, 1-*O*-butylglycerol

Cyclic ethers are named either as heterocyclic compounds or by specialist rules of heterocyclic nomenclature. Radicofunctional names are formed by citing the names of the radicals R^1 and R^2 followed by the word *ether.* Thus methoxyethane becomes ethyl methyl ether and ethoxyethane becomes diethyl ether.

Halogen Derivatives

Using substitutive nomenclature, names are formed by adding prefixes listed in Table 1-8 to the name of the parent compound. The prefix perhalo- implies the replacement of all hydrogen atoms by the particular halogen atoms.

Cations of the type $R^1R^2X^+$ are given names derived from the halonium ion, H_2X^+, by substitution, e.g., diethyliodonium chloride for $(C_2H_5)_2I^+Cl^-$.

Retained are these trivial names; bromoform ($CHBr_3$), chloroform ($CHCl_3$), fluoroform (CHF_3), iodoform (CHI_3), phosgene ($COCl_2$), thiophosgene ($CSCl_2$), and dichlorocarbene radical ($\geq CCl_2$). Inorganic nomenclature leads to such names as carbonyl and thiocarbonyl halides (COX_2 and CSX_2) and carbon tetrahalides (CX_4).

Hydroxylamines and Oximes

For RNH—OH compounds, prefix the name of the radical R to hydroxylamine. If another substituent has priority as principal group, attach the prefix hydroxyamino- to the parent name. For example, C_6H_5NHOH would be named *N*-phenylhydroxylamine, but HOC_6H_4NHOH would be (hydroxyamino)phenol, with the point of attachment indicated by a locant preceding the parentheses.

Compounds of the type $R^1NH—OR^2$ are named (1) as alkoxyamino derivatives of compound R^1H, (2) as *N,O*-substituted hydroxylamines, (3) as alkoxyamines (even if R^1 is hydrogen), or (4) by the prefix aminooxy- when another substituent has priority for parent name. Examples of each type are

1. 2-(Methoxyamino)-8-naphthalenecarboxylic acid for $CH_3ONH—C_{10}H_6COOH$
2. *O*-Phenylhydroxylamine for $H_2N—O—C_6H_5$ or *N*-phenylhydroxylamine for $C_6H_5NH—OH$
3. Phenoxyamine for $H_2N—O—C_6H_5$ (not preferred to *O*-phenylhydroxylamine)
4. Ethyl (aminooxy)acetate for $H_2N—O—CH_2CO—OC_2H_5$

Acyl derivatives, RCO—NH—OH and $H_2N—O—CO—R$, are named as *N*-hydroxy derivatives of amides and as *O*-acylhydroxylamines, respectively. The former may also be named as hydroxamic acids. Examples are *N*-hydroxyacetamide for $CH_3CO—NH—OH$ and *O*-acetylhydroxylamine for $H_2N—O—CO—CH_3$. Further substituents are denoted by prefixes with *O*- and/or *N*-locants. For example, $C_6H_5NH—O—C_2H_5$ would be *O*-ethyl-*N*-phenylhydroxylamine or *N*-ethoxylaniline.

For oximes, the word *oxime* is placed after the name of the aldehyde or ketone. If the carbonyl group is not the principal group, use the prefix hydroxyimino-. Compounds with the group $\geq N—OR$ are named by a prefix alkyloxyimino- as oxime *O*-ethers or as *O*-substituted oximes. Compounds with the group $\geq C=N(O)R$ are named by adding *N*-oxide after the name of the alkylideneaminc compound. For amine oxides, add the word *oxide* after the name of the base, with locants. For example, $C_5H_5N—O$ is named pyridine *N*-oxide or pyridine 1-oxide.

Imines

The group $>C=NH$ is named either by the suffix -imine or by citing the name of the bivalent radical $R^1R^2C<$ as a prefix to amine. For example, $CH_3CH_2CH_2CH=NH$ could be named 1-butanimine or butylideneamine. When the nitrogen is substituted, as in $CH_2=N-CH_2CH_3$, the name is *N*-(methylidene)ethylamine.

Quinones are exceptions. When one or more atoms of quinonoid oxygen have been replaced by $>NH$ or $>NR$, they are named by using the name of the quinone followed by the word *imine* (and preceded by proper affixes). Substituents on the nitrogen atom are named as prefixes. Examples are

O=⟨ring⟩=NH *p*-Benzoquinone monoimine

HN=⟨ring⟩=NH *p*-Benzoquinone diimine

Ketenes

Derivatives of the compound ketene, $CH_2=C=O$, are named by substitutive nomenclature. For example, $C_4H_9CH=C=O$ is butyl ketene. An acyl derivative, such as $CH_3CH_2-CO-CH_2CH=C=O$, may be named as a polyketone, 1-hexene-1,4-dione. Bisketene is used for two to avoid ambiguity with diketene (dimeric ketene).

Ketones

Acyclic ketones are named (1) by adding the suffix -one to the name of the hydrocarbon forming the principal chain or (2) by citing the names of the radicals R^1 and R^2 followed by the word *ketone.* In addition to the preceding nomenclature, acyclic monoacyl derivatives of cyclic compounds may be named (3) by prefixing the name of the acyl group to the name of the cyclic compound. For example, the three possible names of

(2-furyl)–C(=O)–CH_2CH_3

(1) 1-(2-Furyl)-1-propanone
(2) Ethyl 2-furyl ketone
(3) 2-Propionylfuran

When the cyclic component is benzene or naphthalene, the -ic acid or -oic acid of the acid corresponding to the acyl group is changed to -ophenone or -onaphthone, respectively. For example, $C_6H_5-CO-CH_2CH_2CH_3$ can be named either butyrophenone (or butanophenone) or phenyl propyl ketone.

Radicofunctional nomenclature can be used when a carbonyl group is attached directly to carbon atoms in two ring systems and no other substituent is present having priority for citation.

When the methylene group in polycarbocyclic and heterocyclic ketones is replaced by a keto group, the change may be denoted by attaching the suffix -one to the name of the ring system. However, when $\geqslant CH$ in an unsaturated or aromatic system is replaced by a keto group, two alternative names become possible. First, the maximum number of noncumulative double bonds is added after introduction of the carbonyl group(s), and any hydrogen that remains to be added is denoted as indicated hydrogen with the carbonyl group having priority over the indicated hydrogen for lower-numbered locant. Second, the prefix oxo- is used, with the hydrogenation indicated by hydro prefixes; hydrogenation is considered to have occurred before the introduction of the carbonyl group. For example,

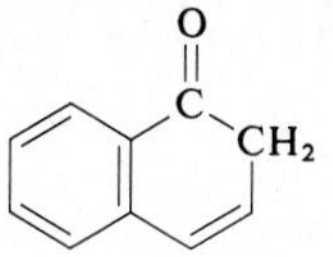

(1) 1(2*H*)-Naphthalenone
(2) 1-Oxo-1,2-dihydronaphthalene

When another group having higher priority for citation as principal group is also present, the ketonic oxygen may be expressed by the prefix oxo-, or one can use the name of the carbonyl-containing radical, as, for example, acyl radicals and oxo-substituted radicals, Examples are

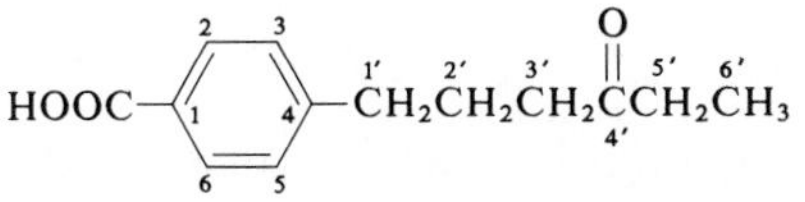

$HOOC-C_6H_4-CH_2CH_2CH_2C(=O)CH_2CH_3$ 4-(4′-Oxohexyl)-1-benzoic acid

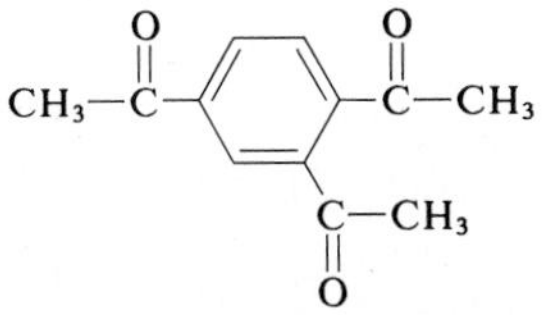

1,2,4-Triacetylbenzene

Diketones and tetraketones derived from aromatic compounds by conversion of two or four ⩾CH groups into keto groups, with any necessary rearrangement of double bonds to a quinonoid structure, are named by adding the suffix -quinone and any necessary affixes.

Polyketones in which two or more contiguous carbonyl groups have rings attached at each end may be named (1) by the radicofunctional method or (2) by substitutive nomenclature. For example,

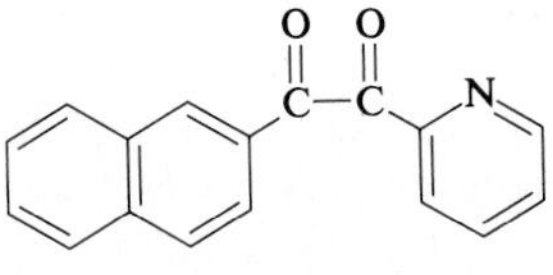

(1) 2-Naphthyl 2-pyridyl diketone
(2) 1-(2-Naphthyl)-2-(2-pyridyl)ethanedione

Some trivial names are retained: acetone (2-propanone), biacetyl (2,3-butanedione), propiophenone ($C_6H_5-CO-CH_2CH_3$), chalcone ($C_6H_5-CH=CH-CO-C_6H_5$), and deoxybenzoin ($C_6H_5-CH_2-CO-C_6H_5$).

These contracted names of heterocyclic nitrogen compounds are retained as alternatives for systematic names, sometimes with indicated hydrogen. In addition, names of oxo derivatives of fully saturated nitrogen heterocycles that systematically end in -idinone are often contracted to end in -idone when no ambiguity might result. For example,

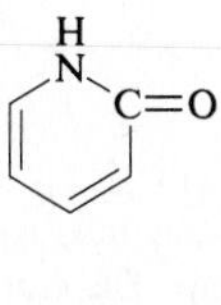

2-Pyridone	4-Pyridone	2-Quinolone	4-Quinolone
2(1*H*)-Pyridone	4(1*H*)-Pyridone	2(1*H*)-Quinolone	4(1*H*)-Quinolone

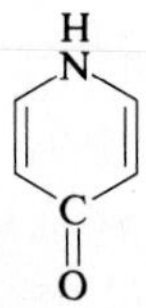

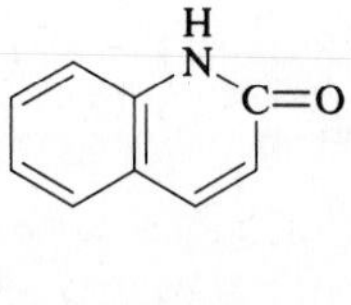

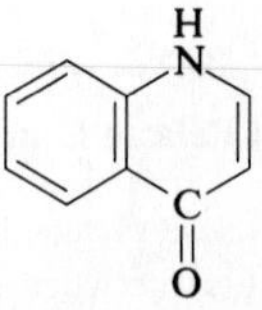

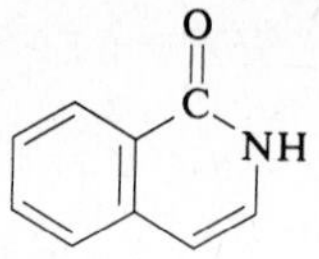

1-Isoquinolone
1(2*H*)-Isoquinolone

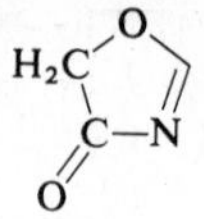

4-Oxazolone
4(5*H*)-Oxazolone

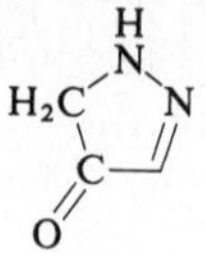

4-Pyrazolone
4(5*H*)-Pyrazolone

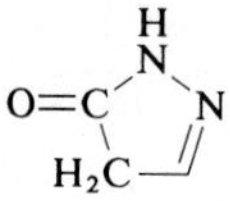

5-Pyrazolone
5(4*H*)-Pyrazolone

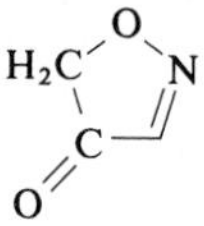

4-Isoxazoline
4(5*H*)-Isoxazolone

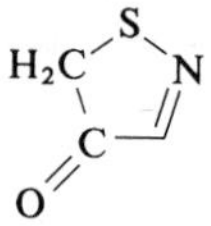

4-Thiazolone
4(5*H*)-Thiazolone

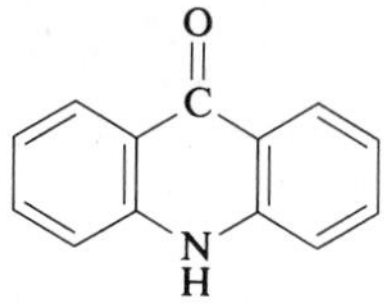

9-Acridone
9(10*H*)-Acridone

Lactones, Lactides, Lactams, and Lactims

When the hydroxy acid from which water may be considered to have been eliminated has a trivial name, the lactone is designated by substituting -olactone for -ic acid. Locants for a carbonyl group are numbered as low as possible, even before that of a hydroxyl group.

Lactones formed from aliphatic acids are named by adding -olide to the name of the nonhydroxylated hydrocarbon with the same number of carbon atoms. The suffix -olide signifies the change of $>CH\cdots CH_3$ into $>C\cdots C{=}O$ (—O—).

Structures in which one or more (but not all) rings of an aggregate are lactone rings are named by placing -carbolactone (denoting the —O—CO— bridge) after the names of the structures that remain when each bridge is replaced by two hydrogen atoms. The locant for —CO— is cited before that for the ester oxygen atom. An additional carbon atom is incorporated into this structure as compared to the -olide.

These trivial names are permitted: γ-butyrolactone, γ-valerolactone, and δ-valerolactone. Names based on heterocycles may be used for all lactones. Thus, γ-butyrolactone is also tetrahedro-2-furanone or dihydro-2(3*H*)-furanone.

Lactides, intermolecular cyclic esters, are named as heterocycles. *Lactams* and *lactims*, containing a —CO—NH— and —C(OH) = N— group, respectively, are named as heterocycles, but they may also be named with -lactam or -lactim in place of -olide. For example,

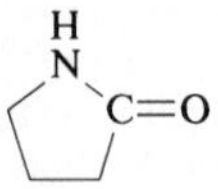

(1) 2-Pyrrolidinone
(2) 4-Butanelactam

Nitriles and Related Compounds

For acids whose systematic names end in -carboxylic acid, nitriles are named by adding the suffix -carbonitrile when the —CN group replaces the —COOH group. The carbon atom of the —CN group is excluded from the numbering of a chain to which it is attached. However, when the triple-bonded nitrogen atom is considered to replace three hydrogen atoms at the end of the main chain of an acyclic hydrocarbon, the suffix -nitrile is added to the name of the hydrocarbon. Numbering begins with the carbon attached to the nitrogen. For example, $CH_3CH_2CH_2CH_2CH_2CN$ is named (1) pentanecarbonitrile or (2) hexanenitrile.

Trivial acid names are formed by changing the endings -oic acid or -ic acid to -onitrile. For example, CH_3CN is acetonitrile. When the —CN group is not the highest priority group, the —CN group is denoted by the prefix cyano-.

In order of decreasing priority for citation of a functional class name, and the prefix for substitutive nomenclature, are the following related compounds:

Functional group	Prefix	Radicofunctional ending
—NC	Isocyano-	Isocyanide
—OCN	Cyanato-	Cyanate
—NCO	Isocyanato-	Isocyanate
—ONC	—	Fulminate
—SCN	Thiocyanato-	Thiocyanate
—NCS	Isothiocyanato-	Isothiocyanate
—SeCN	Selenocyanato-	Selenocyanate
—NCSe	Isoselenocyanato-	Isoselenocyanate

Peroxides

Compounds of the type R—O—OH are named (1) by placing the name of the radical R before the word *hydroperoxide* or (2) by use of the prefix hydroperoxy- when another parent name has higher priority. For example, C_2H_5OOH is ethyl hydroperoxide.

Compounds of the type $R^1O—OR^2$ are named (1) by placing the names of the radicals in alphabetical order before the word *peroxide* when the group —O—O— links two chains, two rings, or a ring and a chain, (2) by use of the affix dioxy to denote the bivalent group —O—O— for naming assemblies of identical units or to form part of a prefix, or (3) by use of the prefix epidioxy- when the peroxide group forms a bridge between two carbon atoms, a ring, or a ring system. Examples are methyl propyl peroxide for $CH_3—O—O—C_3H_7$ and 2,2′-dioxydiacetic acid for $HOOC—CH_2—O—O—CH_2—COOH$.

Phosphorus Compounds

Acyclic phosphorus compounds containing only one phosphorus atom, as well as compounds in which only a single phosphorus atom is in each of several functional groups, are named as derivatives of the parent structures listed in Table 1-12. Often these are purely hypothetical parent structures. When hydrogen attached to phosphorus is replaced by a hydrocarbon group, the derivative is named by substitution nomenclature. When hydrogen of an —OH group is replaced, the derivative is named by radicofunctional nomenclature. For example, $C_2H_5PH_2$ is ethylphosphine; $(C_2H_5)_2PH$, diethylphosphine; $CH_3P(OH)_2$, dihydroxy-methyl-phosphine or methylphosphonous acid; $C_2H_5—PO(Cl)(OH)$, ethylchlorophosphonic acid or ethylphosphonochoridic acid or hydrogen chlorodioxoethylphosphate(V); $CH_3CH(PH_2)COOH$, 2-phosphinopropionic acid; $HP(CH_2COOH)_2$, phosphinediyldiacetic acid; $(CH_3)HP(O)OH$, methylphosphinic acid or hydrogen hydridomethyldioxophosphate(V); $(CH_3O)_3PO$, trimethyl phosphate; and $(CH_3O)_3P$, trimethyl phosphite.

TABLE 1-12 Parent structures of phosphorus-containing compounds

Formula	Parent name	Substitutive prefix	Radicofunctional ending
H_3P	Phosphine	$H_2P—$ Phosphino-	Phosphide
H_5P	Phosphorane	$H_4P—$ Phosphoranyl-	
		$H_3P<$ Phosphoroanediyl-	
		$H_2P\equiv$ Phosphoranetriyl-	
H_3PO	Phosphine oxide		
H_3PS	Phosphine sulfide		
H_3PNH	Phosphine imide		
$P(OH)_3$	Phosphorous acid		Phosphite
$HP(OH)_2$	Phosphonous acid		Phosphonite
H_2POH	Phosphinous acid		Phosphinite
$P(O)(OH)_3$	Phosphoric acid	$P(O)\equiv$ Phosphoryl-	Phosphate(V)
$HP(O)(OH)_2$	Phosphonic acid	$HP(O)<$ Phosphonoyl-	Phosphonate
		$—P(O)OH_2$ Phosphono-	
$H_2P(O)OH$	Phosphinic acid	$H_2P(O)—$ Phosphinoyl-	Phosphinate
		$>P(O)OH$ Phosphinoco-	
		Phosphinato-	

Salts and Esters of Acids

Neutral salts of acids are named by citing the cation(s) and then the anion, whose ending is changed from -oic to -oate or from -ic to -ate. When different acidic residues are present in one structure, prefixes are formed by changing the anion ending -ate to -ato- or -ide to -ido-. The prefix carboxylato- denotes the ionic group $—COO^-$. The phrase: (metal) salt of (the acid) is permissible when the carboxyl groups are not all named as affixes.

Acid salts include the word *hydrogen* (with affixes, if appropriate) inserted between the name of the cation and the name of the anion (or word *salt*).

Esters are named similarly, with the name of the alkyl or aryl radical replacing the name of the cation. Acid esters of acids and their salts are named as neutral esters, but the components are cited in the order: cation, alkyl or aryl radical, hydrogen, and anion. Locants are added if necessary. For example,

$CH_2—CO—OC_2H_5$
$HOC—COO^-$ $\quad K^+ \quad H^+$
$CH_2—COO^-$

Potassium 1-ethyl hydrogen citrate

Ester groups in $R^1—CO—OR^2$ compounds are named (1) by the prefix alkoxycarbonyl- or aryloxycarbonyl- for $—CO—OR^2$ when the radical R^1 contains a substituent with priority for citation as principal group or (2) by the prefix acyloxy- for $R^1—CO—O—$ when the radical R^2 contains a substituent with priority for citation as principal group. Examples are

$CH_2CH_2CH_2CO—OCH_3$
$CO—OCH_3$

Methyl 3-methoxycarbonyl-2-naphthalenebutyrate

$[CH_3O—CO—CH_2CH_2\overset{+}{N}(CH_3)_3]\ Cl^-$ [(2-Methoxycarbonyl)ethyl]trimethylammonium chloride

$C_6H_5—CO—OCH_2CH_2COOH$ 3-Benzoyloxypropionic acid

The trivial name *acetoxy* is retained for the CH_3—CO—O— group. Compounds of the type $R^2C(OR^2)_3$ are named as R^2 esters of the hypothetical ortho acids. For example, $CH_3C(OCH_3)_3$ is trimethyl orthoacetate.

Silicon Compounds

SiH_4 is called silane; its acyclic homologs are called disilane, trisilane, and so on, according to the number of silicon atoms present. The chain is numbered from one end to the other so as to give the lowest-numbered locant in radicals to the free valence or to substituents on a chain. The abbreviated form silyl is used for the radical SiH_3—. Numbering and citation of side chains proceed according to the principles set forth for hydrocarbon chains. Cyclic nonaromatic structures are designated by the prefix cyclo-.

When a chain or ring system is composed entirely of alternating silicon and oxygen atoms, the parent name *siloxane* is used with a multiplying affix to denote the number of silicon atoms present. The parent name *silazane* implies alternating silicon and nitrogen atoms; multiplying affixes denote the number of silicon atoms present.

The prefix sila- designates replacement of carbon by silicon in replacement nomenclature. Prefix names for radicals are formed analogously to those for the corresponding carbon-containing compounds. Thus silyl is used for SiH_3—, silyene for —SiH_2—, silylidyne for —SiH<, as well as trily, tetrayl, and so on for free valences(s) on ring structures.

Sulfur Compounds

Bivalent Sulfur. The prefix thio-, placed before an affix that denotes the oxygen-containing group or an oxygen atom, implies the replacement of that oxygen by sulfur. Thus the suffix -thiol denotes —SH, -thione denotes —(C)=S and implies the presence of an =S at a nonterminal carbon atom, -thioic acid denotes [(C)=S]OH $\rightleftharpoons$ [(C)=O]SH (that is, the *O*-substituted acid and the *S*-substituted acid, respectively), -dithioc acid denotes [—C(S)]SH, and -thial denotes —(C)HS (or -carbothialdehyde denotes —CHS). When -carboxylic acid has been used for acids, the sulfur analog is named -carbothioic acid or -carbodithioic acid.

Prefixes for the groups HS— and RS— are mercapto- and alkylthio-, respectively; this latter name may require parentheses for distinction from the use of thio- for replacement of oxygen in a trivially named acid. Examples of this problem are 4-C_2H_5—C_6H_4—CSOH named *p*-ethyl(thio)benzoic acid and 4-C_2H_5—S—C_6H_4—COOH named *p*-(ethylthio)benzoic acid. When —SH is not the principal group, the prefix mercapto- is placed before the name of the parent compound to denote an unsubstituted —SH group.

The prefix thioxo- is used for naming =S in a thioketone. Sulfur analogs of acetals are named as alkylthio- or arylthio-. For example, $CH_3CH(SCH_3)OCH_3$ is 1-methoxy-1-(methylthio)ethane. Prefix forms for -carbothioic acids are hydroxy(thiocarbonyl)- when referring to the *O*-substituted acid and mercapto(carbonyl)- for the *S*-substituted acid.

Salts are formed as with oxygen-containing compounds. For example, C_2H_5—S—Na is named either sodium ethanethiolate or sodium ethyl sulfide. If mercapto- has been used as a prefix, the salt is named by use of the prefix sulfido- for $—S^-$.

Compounds of the type R^1—S—R^2 are named alkylthio- (or arylthio-) as a prefix to the name of R^1 or R^2, whichever is the senior.

Sulfonium Compounds. Sulfonium compounds of the type $R^1R^2R^3S^+X^-$ are named by citing in alphabetical order the radical names followed by -sulfonium and the name of the anion. For heterocyclic compounds, -ium is added to the name of the ring system. Replacement of >CH by sulfonium sulfur is denoted by the prefix thionia-, and the name of the anion is added at the end.

Organosulfur Halides. When sulfur is directly linked only to an organic radical and to a halogen atom, the radical name is attached to the word *sulfur* and the name(s) and number of the halide(s) are stated as a separate word. Alternatively, the name can be formed from R—SOH, a sulfenic acid whose radical prefix is sulfenyl-. For example, CH_3CH_2—S—Br would be named either ethylsulfur monobromide or ethanesulfenyl bromide. When another principal group is present, a composite prefix is formed from the number and substitutive name(s) of the halogen atoms in front of the syllable thio. For example, BrS—COOH is (bromothio)formic acid.

Sulfoxides. Sulfoxides, R^1—SO—R^2, are named by placing the names of the radicals in alphabetical order before the word *sulfoxide.* Alternatively, the less senior radical is named followed by sulfinyl- and concluded by the name of the senior group. For example, CH_3CH_2—SO—$CH_2CH_2CH_3$ is named either ethyl propyl sulfoxide or 1-(ethylsulfinyl)propane.

When an $>$SO group is incorporated in a ring, the compound is named an oxide.

Sulfones. Sulfones, R^1—SO_2—R^2, are named in an analogous manner to sulfoxides, using the word *sulfone* in place of *sulfoxide.* In prefixes, the less senior radical is followed by -sulfonyl-. When the $>SO_2$ group is incorporated in a ring, the compound is named as a dioxide.

Sulfur Acids. Organic oxy acids of sulfur, that is, —SO_3H, —SO_2H, and —SOH, are named sulfonic acid, sulfinic acid, and sulfenic acid, respectively. In subordinate use, the respective prefixes are sulfo-, sulfino, and sulfeno-. The grouping —SO_2—O—SO_2— or —SO—O—SO is named sulfonic or sulfinic anhydride, respectively.

Inorganic nomenclature is employed in naming sulfur acids and their derivatives in which sulfur is linked only through oxygen to the organic radical. For example, $(C_2H_5O)_2SO_2$ is diethyl sulfate and C_2H_5O—SO_2—OH is ethyl hydrogen sulfate. Prefixes *O*- and *S*- are used where necessary to denote attachment to oxygen and to sulfur, respectively, in sulfur replacement compounds. For example, CH_3—S—SO_2—ONa is sodium *S*-methyl thiosulfate.

When sulfur is linked only through nitrogen, or through nitrogen and oxygen, to the organic radical, naming is as follows: (1) *N*-substituted amides are designated as *N*-substituted derivatives of the sulfur amides and (2) compounds of the type R—NH—SO_3H may be named as *N*-substituted sulfamic acids or by the prefix sulfoamino- to denote the group HO_3S—NH—. The groups —N=SO and —N=SO_2 are named sulfinylamines and sulfonylamines, respectively.

Sultones and Sultams. Compounds containing the group —SO_2—O— as part of the ring are called -sultone. The —SO_2— group has priority over the —O— group for lowest-numbered locant.

Similarly, the —SO_2—N= group as part of a ring is named by adding -sultam to the name of the hydrocarbon with the same number of carbon atoms. The —SO_2— has priority over —N= for lowest-numbered locant.

Stereochemistry

Concepts in stereochemistry, that is, chemistry in three-dimensional space, are in the process of rapid expansion. This section will deal with only the main principles. The compounds discussed will be those that have identical molecular formulas but differ in the arrangement of their atoms in space. *Stereoisomers* is the name applied to these compounds.

Stereoisomers can be grouped into three categories: (1) Conformational isomers differ from each other only in the way their atoms are oriented in space, but can be converted into one another by rotation about sigma bonds. (2) Geometric isomers are compounds in which rotation about a double bond is restricted. (3) Configurational isomers differ from one another only in configuration about a chiral center, axis, or plane. In subsequent structural representations, a broken line denotes a bond projecting behind the plane of the paper and a wedge denotes a bond projecting in front of the plane of the paper. A line of normal thickness denotes a bond lying essentially in the plane of the paper.

Conformational Isomers

A molecule in a conformation into which its atoms return spontaneously after small displacements is termed a *conformer.* Different arrangements of atoms that can be converted into one another by rotation about single bonds are called *conformational isomers* (see Fig. 1-1). A pair of conformational isomers can be but do not have to be mirror images of each other. When they are not mirror images, they are called *diastereomers.*

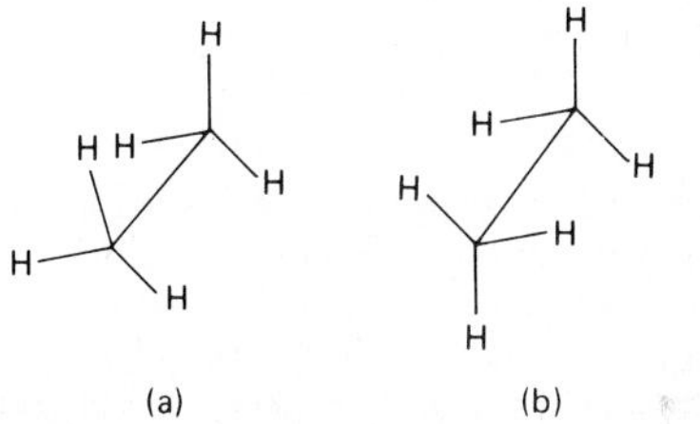

(a) (b)

FIG. 1-1 Conformations of ethane. (*a*) Eclipsed; (*b*) staggered.

Acyclic Compounds. Different conformations of acyclic compounds are best viewed by construction of ball-and-stick molecules or by use of Newman projections (see Fig. 1-2). Both types of representations are shown for ethane. Atoms or groups that are attached at opposite ends of a single bond should be viewed along the bond axis. If two atoms or groups attached at opposite ends of the bond appear one directly behind the other, these atoms or groups are described as eclipsed. That portion of the molecule is described as being in the eclipsed conformation. If not eclipsed, the atoms or groups and the conformation may be described as staggered. Newman projections show these conformations clearly.

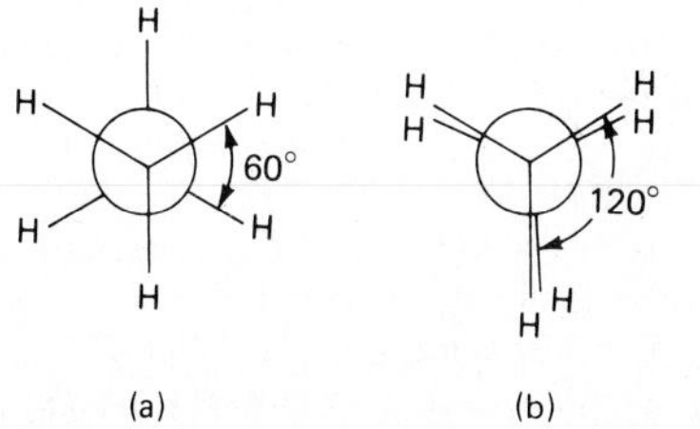

(a) (b)

FIG. 1-2 Newman projections for ethane. (*a*) Staggered; (*b*) eclipsed.

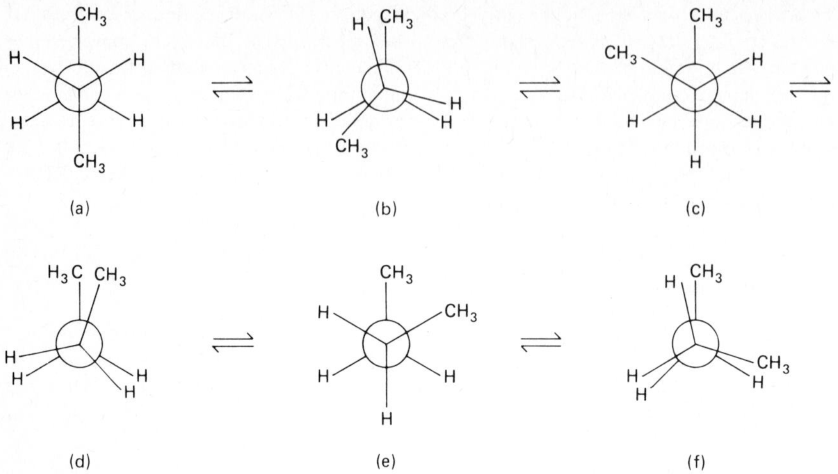

FIG. 1-3 Conformations of butane. (*a*) *Anti*-staggered; (*b*) eclipsed; (c) *gauche*-staggered; (*d*) eclipsed; (*e*) *gauche*-staggered; (*f*) eclipsed. (Eclipsed conformations are slightly staggered for convenience in drawing; actually they are superimposed.)

Certain physical properties show that rotation about the single bond is not quite free. For ethane there is an energy barrier of about 3 kcal · mol^{-1} (12 kJ · mol^{-1}). The potential energy of the molecule is at a minimum for the staggered conformation, increases with rotation, and reaches a maximum at the eclipsed conformation. The energy required to rotate the atoms or groups about the carbon-carbon bond is called *torsional energy*. Torsional strain is the cause of the relative instability of the eclipsed conformation or any intermediate skew conformations.

In butane, with a methyl group replacing one hydrogen on each carbon of ethane, there are several different staggered conformations (see Fig. 1-3). There is the *anti* conformation in which the methyl groups are as far apart as they can be (dihedral angle of 180°). There are two *gauche* conformations in which the methyl groups are only 60° apart; these are two nonsuperimposable mirror images of each other. The *anti* conformation is more stable than the *gauche* by about 0.9 kcal · mol^{-1} (4 kJ · mol^{-1}). Both are free of torsional strain. However, in a *gauche* conformation the methyl groups are closer together than the sum of their van der Waals' radii. Under these conditions van der Waals' forces are repulsive and raise the energy of conformation. This strain can affect not only the relative stabilities of various staggered conformations but also the heights of the energy barriers between them. The energy maximum (estimated at 4.8 to 6.1 kcal · mol^{-1} or 20 to 25 kJ · mol^{-1}) is reached when two methyl groups swing past each other (the eclipsed conformation) rather than past hydrogen atoms.

Cyclic Compounds. Although cyclic aliphatic compounds are often drawn as if they were planar geometric figures (a triangle for cyclopropane, a square for cyclobutane, and so on), their structures are not that simple. Cyclopropane does possess the maximum angle strain if one considers the difference between a tetrahedral angle (109.5°) and the 60° angle of the cyclopropane structure. Nevertheless the cyclopropane structure is thermally quite stable. The highest electron density of the carbon-carbon bonds does not lie along the lines connecting the carbon atoms. Bonding electrons lie principally outside the triangular internuclear lines and result in what is known as *bent bonds* (see Fig. 1-4).

Cyclobutane has less angle strain than cyclopropane (only 19.5°). It is also believed to have some bent-bond character associated with the carbon-carbon bonds. The molecule exists in a nonplanar conformation in order to minimize hydrogen-hydrogen eclipsing strain.

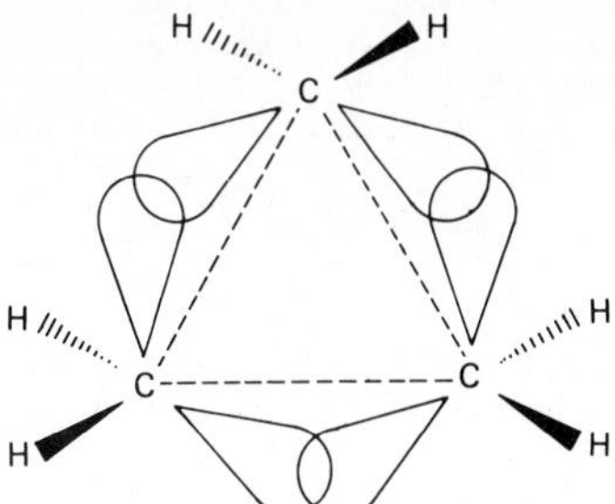

FIG. 1-4 The bent bonds ("tear drops") of cyclopropane.

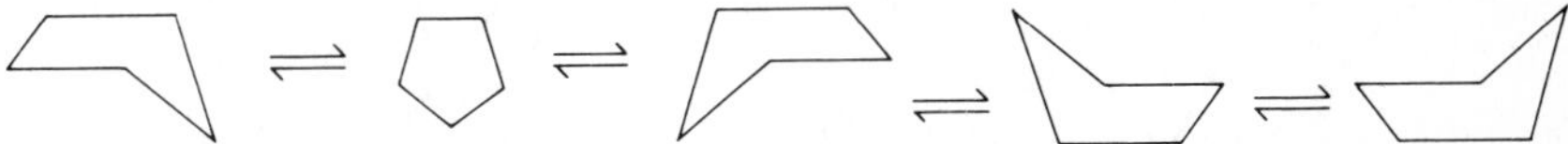

FIG. 1-5 The conformations of cyclopentane.

Cyclopentane is nonplanar, with a structure that resembles an envelope (see Fig. 1-5). Four of the carbon atoms are in one plane, and the fifth is out of that plane. The molecule is in continual motion so that the out-of-plane carbon moves rapidly around the ring.

The 12 hydrogen atoms of cyclohexane do not occupy equivalent positions. In the chair conformation six hydrogen atoms are perpendicular to the average plane of the molecule and six are directed outward from the ring, slightly above or below the molecular plane (see Fig. 1-6). Bonds which are perpendicular to the molecular plane are known as *axial bonds*, and

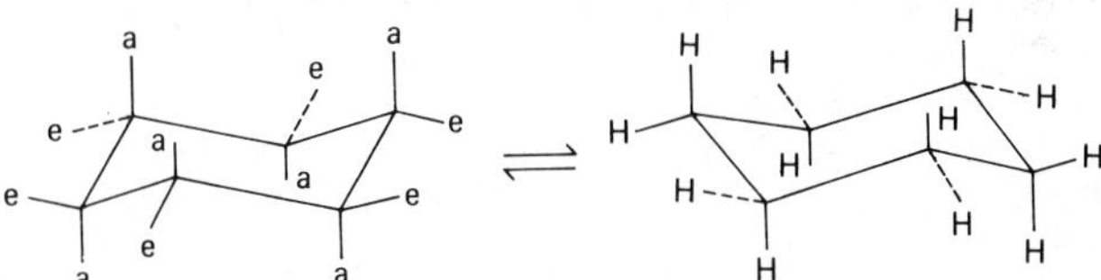

FIG. 1-6 The two chair conformations of cyclohexane: a = axial hydrogen atom and e = equatorial hydrogen atom.

those which extend outward from the ring are known as *equatorial bonds*. The three axial bonds directed upward originate from alternate carbon atoms and are parallel with each other; a similar situation exists for the three axial bonds directed downward. Each equatorial bond is drawn so as to be parallel with the ring carbon-carbon bond once removed from the point of attachment to that equatorial bond. At room temperature, cyclohexane is interconverting rapidly between two chair conformations. As one chair form converts to the other, all the equatorial hydrogen atoms become axial and all the axial hydrogens become equatorial. The interconversion is so rapid that all hydrogen atoms on cyclohexane can be considered equivalent. Interconversion is believed to take place by movement of one side of the chair structure to produce the twist boat, and then movement of the other side of the twist boat to give the other chair form. The chair conformation is the most favored structure for cyclohexane. No angle strain is encountered since all bond angles remain tetrahedral. Torsional strain is minimal because all groups are staggered.

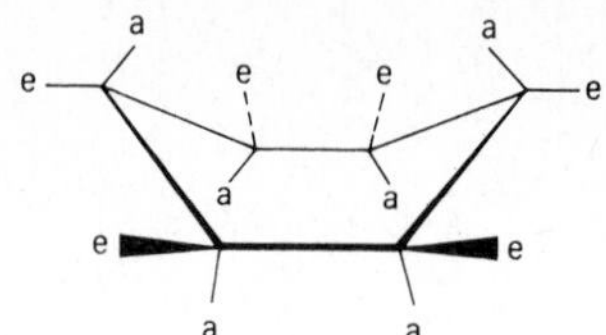

FIG. 1-7 The boat conformation of cyclohexane. $a =$ axial hydrogen atom and $e =$ equatorial hydrogen atom.

In the boat conformation of cyclohexane (Fig. 1-7) eclipsing torsional strain is significant, although no angle strain is encountered. Nonbonded interaction between the two hydrogen atoms across the ring from each other (the "flagpole" hydrogens) is unfavorable. The boat conformation is about 6.5 kcal · mol^{-1} (27 kJ · mol^{-1}) higher in energy than the chair form at 25°C.

FIG. 1-8 Twist-boat conformation of cyclohexane.

A modified boat conformation of cyclohexane, known as the twist boat (Fig. 1-8), or skew boat, has been suggested to minimize torsional and nonbounded interactions. This particular conformation is estimated to be about 1.5 kcal · mol^{-1} (6 kJ mol^{-1}) lower in energy than the boat form at room temperature.

The medium-size rings (7 to 12 ring atoms) are relatively free of angle strain and can easily take a variety of spatial arrangements. They are not large enough to avoid all nonbonded interactions between atoms.

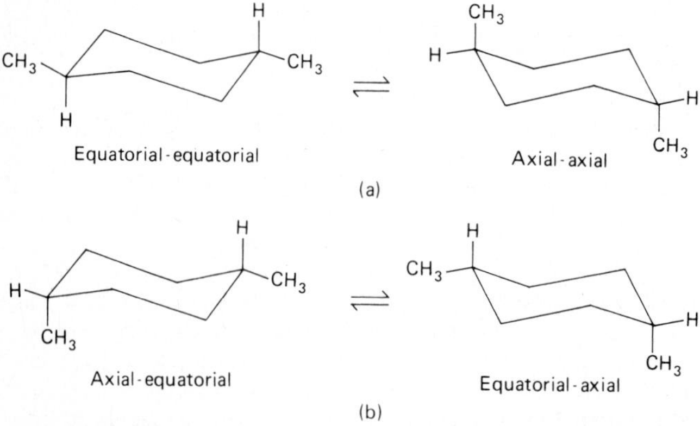

FIG. 1-9 Two isomers of 1,4-dimethylcyclohexane. (*a*) *Trans* isomer; (*b*) *cis* isomer.

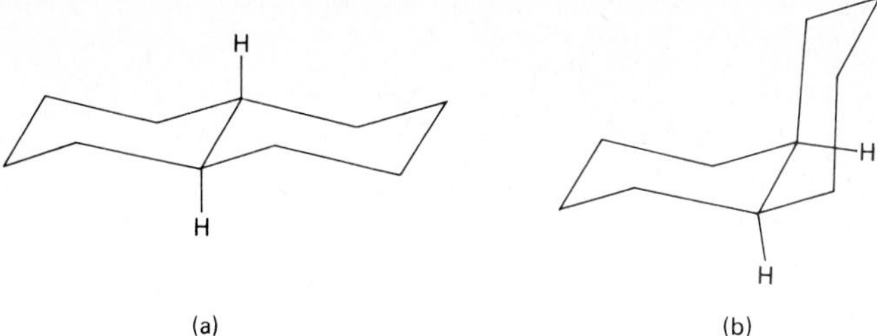

FIG. 1-10 Two isomers of decahydronaphthalene, or bicyclo[4.4.0]decane. (*a*) *Trans* isomer; (*b*) *cis* isomer.

Disubstituted cyclohexanes can exist as *cis-trans* isomers as well as axial-equatorial conformers. Two isomers are predicted for 1,4-dimethylcyclohexane (see Fig. 1-9). For the *trans* isomer the diequatorial conformer is the energetically favorable form. Only one *cis* isomer is observed, since the two conformers of the *cis* compound are identical. Interconversion takes place between the conformational (equatorial-axial) isomers but not configurational (*cis-trans*) isomers.

The bicyclic compound decahydronaphthalene, or bicyclo[4.4.0]decane, has two fused six-membered rings. It exists in *cis* and *trans* forms (see Fig. 1-10), as determined by the configurations at the bridgehead carbon atoms. Both *cis*- and *trans*-decahydronaphthalene can be constructed with two chair conformations.

Geometrical Isomerism

Rotation about a carbon-carbon double bond is restricted because of interaction between the *p* orbitals which make up the pi bond. Isomerism due to such restricted rotation about a bond is known as *geometric isomerism.* Parallel overlap of the *p* orbitals of each carbon atom of the double bond forms the molecular orbital of the pi bond. The relatively large barrier to rotation about the pi bond is estimated to be nearly 63 kcal · mol^{-1} (263 kJ · mol^{-1}).

When two different substituents are attached to each carbon atom of the double bond, *cis-trans* isomers can exist. In the case of *cis*-2-butene (Fig. 1-11*a*), both methyl groups are on the same side of the double bond. The other isomer has the methyl groups on opposite sides and is designated as *trans*-2-butene (Fig. 1-11*b*). Their physical properties are quite different. Geometric isomerism can also exist in ring systems; examples were cited in the previous discussion on conformational isomers.

For compounds containing only double-bonded atoms, the reference plane contains the double-bonded atoms and is perpendicular to the plane containing these atoms and those directly attached to them. It is customary to draw the formulas so that the reference plane is

CH_3 CH_3 CH_3 H
C=C C=C
H H H CH_3

(a) (b)

FIG. 1-11 Two isomers of 2-butene. (*a*) *Cis* isomer, bp 3.8°C, mp −138.9°C, dipole moment 0.33 D; (*b*) *trans* isomer, bp 0.88°C, mp −105.6°C, dipole moment 0 D.

perpendicular to that of the paper. For cyclic compounds the reference plane is that in which the ring skeleton lies or to which it approximates. Cyclic structures are commonly drawn with the ring atoms in the plane of the paper.

Sequence Rules for Geometric Isomers and Chiral Compounds

Although *cis* and *trans* designations have been used for many years, this approach becomes useless in complex systems. To eliminate confusion when each carbon of a double bond or a chiral center is connected to different groups, the Cahn, Ingold, and Prelog system for designating configuration about a double bond or a chiral center has been adopted by IUPAC. Groups on each carbon atom of the double bond are assigned a first (1) or second (2) priority. Priority is then compared at one carbon relative to the other. When both first priority groups are on the *same side* of the double bond, the configuration is designated as *Z* (from the German *zusammen*, "together"), which was formerly *cis*. If the first priority groups are on *opposite sides* of the double bond, the designation is *E* (from the German *entgegen*, "in opposition to"), which was formerly *trans*. (See Fig. 1-12.)

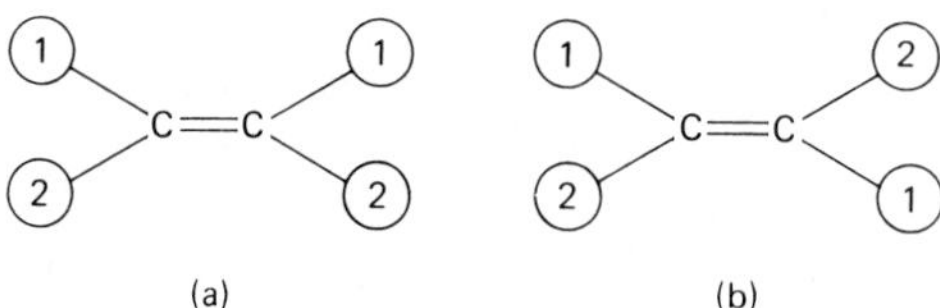

FIG. 1-12 Configurations designated by priority groups. (*a*) *Z* (*cis*); (*b*) *E* (*trans*).

When a molecule contains more than one double bond, each *E* or *Z* prefix has associated with it the lower-numbered locant of the double bond concerned. Thus (see also the rules that follow)

```
      H         H
       \       /
        C=C          H
       /    \       /
  H3C        C=C
            /    \
           H      COOH
```

(2*E*,4*Z*)-2,4-Hexadienoic acid

When the sequence rules permit alternatives, preference for lower-numbered locants and for inclusion in the principal chain is allotted as follows in the order stated: *Z* over *E* groups and *cis* over *trans* cyclic groups. If a choice is still not attained, then the lower-numbered locant for such a preferred group at the first point of difference is the determining factor. For example,

```
HOOC          H H          H
     \       /   \        /
      C=C          C=C
     /    \       /    \
   H        CH2         COOH
```

(2*Z*,5*E*)-2,5-Heptadienedioic acid

Rule 1. Priority is assigned to atoms on the basis of atomic number. Higher priority is assigned to atoms of higher atomic number. If two atoms are isotopes of the same element, the atom of higher mass number has the higher priority. For example, in 2-butene, the carbon atom of each methyl group receives first priority over the hydrogen atom connected to the same carbon atom. Around the asymmetric carbon atom in chloroiodomethanesulfonic acid, the priority sequence is I, Cl, S, H. In 1-bromo-1-deuteroethane, the priority sequence is Cl, C, D, H.

Rule 2. When atoms attached directly to a double-bonded carbon have the same priority, the second atoms are considered and so on, if necessary, working outward once again from the double bond or chiral center. For example, in 1-chloro-2-methylbutene, in CH_3 the second atoms are H, H, H and in CH_2CH_3 they are C, H, H. Since carbon has a higher atomic number than hydrogen, the ethyl group has the next highest priority after the chlorine atom.

```
Cl          CH2CH3          Cl          CH3
  \        /                  \        /
   C=C                         C=C
  /        \                  /        \
H           CH3             H           CH2CH3
```

(*Z*)-1-Chloro-2-methylbutene (*E*)-1-Chloro-2-methylbutene

Rule 3. When groups under consideration have double or triple bonds, the multiple-bonded atom is replaced conceptually by two or three single bonds to that same kind of atom. Thus, =A is considered to be equivalent to two A's, or $<^{A}_{A}$ and ≡A equals $\leftarrow^{A}_{A}A$. However, a real $<^{A}_{A}$ has priority over =A; likewise a real $\leftarrow^{A}_{A}A$ has priority over ≡A. Actually, both atoms of a multiple bond are duplicated, or triplicated, so that C=O is treated as

```
C—O
|  |
O  C
```

, that is

```
C—O
|
(O)
```

and

```
O—C
|
(C)
```

, and C≡N is treated as

```
   C————————N
  / \      / \
(N) (N)  (C) (C)
```

. A phenyl carbon becomes

```
    CH
   /
—C—C
   \
    CH
```

. Only the double-bonded atoms themselves are duplicated, not the atoms or groups attached to them. The duplicated atoms (or phantom atoms) may be considered as carrying atomic number zero. For example, among the groups OH, CHO, CH_2OH, and H, the OH group has the highest priority, and the C(O, O, H) of CHO takes priority over the C(O, H, H) of CH_2OH.

Chirality and Optical Activity

A compound is chiral (the term *dissymmetric* was formerly used) if it is not superimposable on its mirror image. A chiral compound does not have a plane of symmetry. Each chiral compound possesses one (or more) of three types of chiral element, namely, a chiral center, a chiral axis, or a chiral plane.

```
      COOH          |          COOH
       |            |           |
       C            |           C
     /   \          |         /   \
   H      OH        |       HO     H
       |            |           |
      CH3           |          CH3
               Mirror plane
```

FIG. 1-13 Asymmetric (chiral) carbon in the lactic acid molecule.

Chiral Center. The chiral center, which is the chiral element most commonly met, is exemplified by an asymmetric carbon with a tetrahedral arrangement of ligands about the carbon. The ligands comprise four different atoms or groups. One "ligand" may be a lone pair of electrons; another, a phantom atom of atomic number zero. This situation is encountered in sulfoxides or with a nitrogen atom. Lactic acid is an example of a molecule with an asymmetric (chiral) carbon. (See Fig. 1-13 on the previous page.)

A simpler representation of molecules containing asymmetric carbon atoms is the Fischer projection, which is shown here for the same lactic acid configurations. A Fischer projection

```
       COOH              COOH
        |                 |
   H ---+--- OH     HO ---+--- H
        |                 |
       CH3               CH3
```

involves drawing a cross and attaching to the four ends the four groups that are attached to the asymmetric carbon atom. The asymmetric carbon atom is understood to be located where the lines cross. The horizontal lines are understood to represent bonds coming toward the viewer out of the plane of the paper. The vertical lines represent bonds going away from the viewer behind the plane of the paper as if the vertical line were the side of a circle. The principal chain is depicted in the vertical direction; the lowest-numbered (locant) chain member is placed at the top position. These formulas may be moved sideways or rotated through 180° in the plane of the paper, but they may not be removed from the plane of the paper (i.e., rotated through 90°). In the latter orientation it is essential to use thickened lines (for bonds coming toward the viewer) and dashed lines (for bonds receding from the viewer) to avoid confusion.

Enantiomers. Two nonsuperimposable structures that are mirror images of each other are known as *enantiomers.* Enantiomers are related to each other in the same way that a right hand is related to a left hand. Except for the direction in which they rotate the plane of polarized light, enantiomers are identical in all physical properties. Enantiomers have identical chemical properties except in their reactivity toward optically active reagents.

Enantiomers rotate the plane of polarized light in opposite directions but with equal magnitude. If the light is rotated in a clockwise direction, the sample is said to be dextrorotatory and is designed as (+). When a sample rotates the plane of polarized light in a counterclockwise direction, it is said to be levorotatory and is designed as (−). Use of the designations *d* and *l* is discouraged.

Specific Rotation. Optical rotation is caused by individual molecules of the optically active compound. The amount of rotation depends upon how many molecules the light beam encounters in passing through the tube. When allowances are made for the length of the tube that contains the sample and the sample concentration, it is found that the amount of rotation, as well as its direction, is a characteristic of each individual optically active compound.

Specific rotation is the number of degrees of rotation observed if a 1-dm tube is used and the compound being examined is present to the extent of 1 g per 100 mL. The density for a pure liquid replaces the solution concentration.

$$\text{Specific rotation} = [\alpha] = \frac{\text{observed rotation (degrees)}}{\text{length (dm)} \times (\text{g}/100\text{ mL})}$$

The temperature of the measurement is indicated by a superscript and the wavelength of the light employed by a subscript written after the bracket; for example, $[\alpha]^{20}_{590}$ implies that the measurement was made at 20°C using 590 nm radiation.

Optically Inactive Chiral Compounds. Although chirality is a necessary prerequisite for optical activity, chiral compounds are not necessarily optically active. With an equal mixture of two enantiomers, no net optical rotation is observed. Such a mixture of enantiomers is said to be *racemic* and is designated as (±) and not as *dl.* Racemic mixtures usually have melting points higher than the melting point of either pure enantiomer.

A second type of optically inactive chiral compounds, *meso* compounds, will be discussed in the next section.

Multiple Chiral Centers. The number of stereoisomers increases rapidly with an increase in the number of chiral centers in a molecule. A molecule possessing two chiral atoms should have four optical isomers, that is, four structures consisting of two pairs of enantiomers. However, if a compound has two chiral centers but both centers have the same four substituents attached, the total number of isomers is three rather than four. One isomer of such a compound is not chiral because it is identical with its mirror image; it has an internal mirror plane. This is an example of a diastereomer. The achiral structure is denoted as a *meso* compound. Diastereomers have different physical and chemical properties from the optically active enantiomers. Recognition of a plane of symmetry is usually the easiest way to detect a *meso* compound. The stereoisomers of tartaric acid are examples of compounds with multiple chiral centers (see Fig. 1-14), and one of its isomers is a *meso* compound.

When the asymmetric carbon atoms in a chiral compound are part of a ring, the isomerism is more complex than in acyclic compounds. A cyclic compound which has two different asymmetric carbons with different sets of substituent groups attached has a total of $2^2 = 4$ optical isomers: an enantiometric pair of *cis* isomers and an enantiometric pair of *trans* isomers. However, when the two asymmetric centers have the same set of substituent groups attached, the *cis* isomer is a *meso* compound and only the *trans* isomer is chiral. (See Fig. 1-15.)

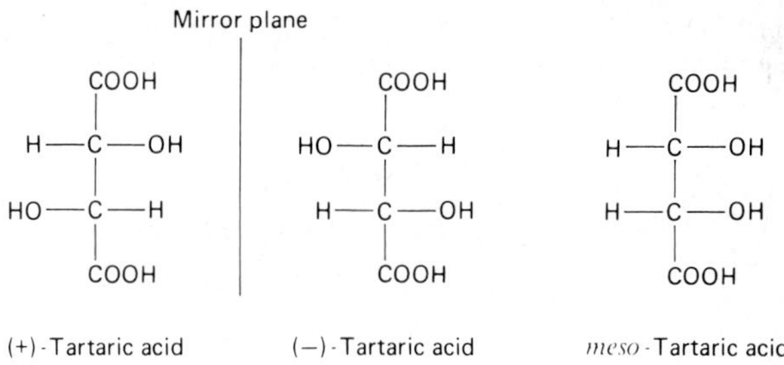

FIG. 1-14 Isomers of tartaric acid.

HOOC H H COOH HOOC COOH

H COOH HOOC H H H

Mirror plane

(a) (b)

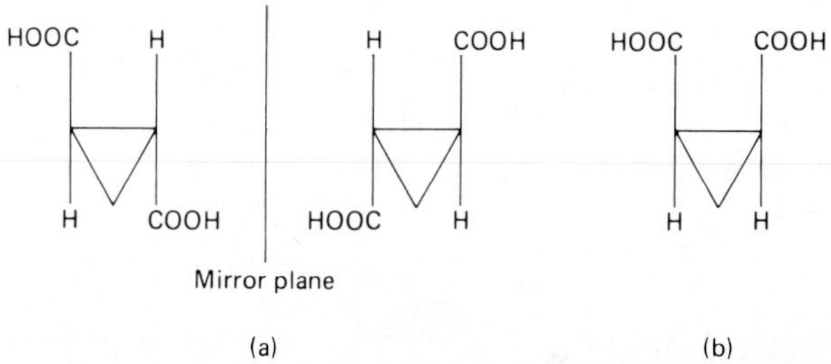

FIG. 1-15 Isomers of cyclopropane-1,2-dicarboxylic acid. (*a*) *Trans* isomer; (*b*) *meso* isomer.

Torsional Asymmetry. Rotation about single bonds of most acyclic compounds is relatively free at ordinary temperatures. There are, however, some examples of compounds in which nonbonded interactions between large substituent groups inhibit free rotation about a sigma bond. In some cases these compounds can be separated into pairs of enantiomers.

A *chiral axis* is present in chiral biaryl derivatives. When bulky groups are located at the *ortho* positions of each aromatic ring in biphenyl, free rotation about the single bond connecting the two rings is inhibited because of torsional strain associated with twisting rotation about the central single bond. Interconversion of enantiomers is prevented (see Fig. 1-16).

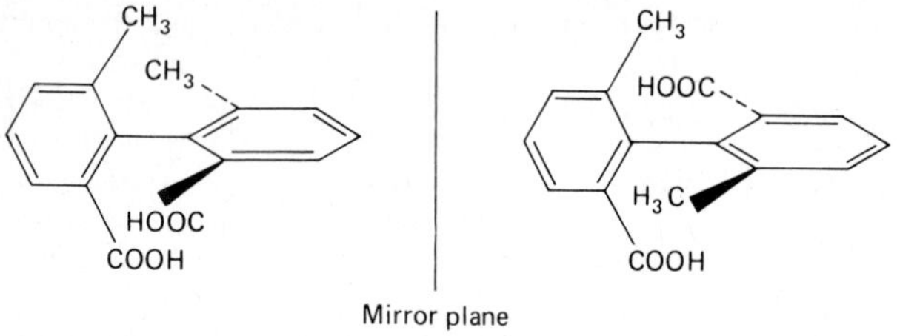

CH_3 CH_3 HOOC COOH | CH_3 HOOC H_3C COOH

Mirror plane

FIG. 1-16 Isomers of biphenyl compounds with bulky groups attached at the *ortho* positions.

For compounds possessing a chiral axis, the structure can be regarded as an elongated tetrahedron to be viewed along the axis. In deciding upon the absolute configuration it does not matter from which end it is viewed; the nearer pair of ligands receives the first two positions in the order of precedence (see Fig. 1-17). For the meaning of (*S*), see the discussion under "Absolute Configuration" on p. 1-51.

A *chiral plane* is exemplified by the plane containing the benzene ring and the bromine and oxygen atoms in the chiral compound shown in Fig. 1-18. Rotation of the benzene ring around the oxygen-to-ring single bonds is inhibited when x is small (although no critical size can be reasonably established).

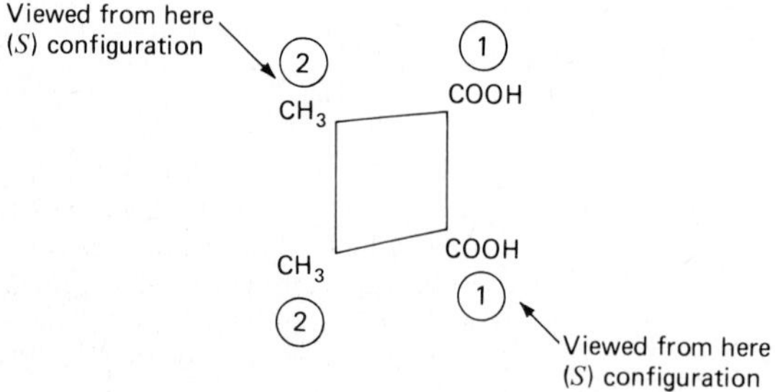

FIG. 1-17 Example of a chiral axis.

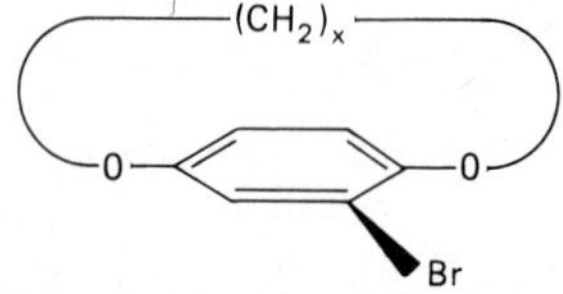

FIG. 1-18 Example of a chiral plane.

Absolute Configuration. The terms absolute stereochemistry and absolute configuration are used to describe the three-dimensional arrangement of substituents around a chiral element. A general system for designating absolute configuration is based upon the priority system and sequence rules. Each group attached to a chiral center is assigned a number, with number one the highest-priority group. For example, the groups attached to the chiral center of 2-butanol (see Fig. 1-19) are assigned these priorities: 1 for OH, 2 for CH_2CH_3, 3 for CH_3, and 4 for

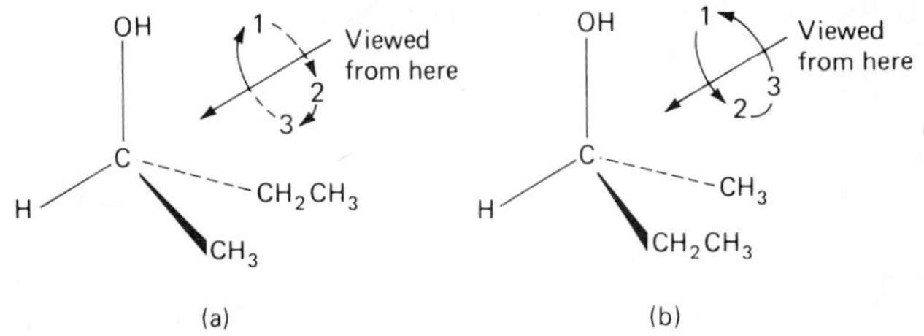

FIG. 1-19 Viewing angle as a means of designating the absolute configuration of compounds with a chiral axis. (*a*) (*R*)-2-Butanol (sequence clockwise); (*b*) (*S*)-2-butanol (sequence counterclockwise).

H. The molecule is then viewed from the side opposite the group of lowest priority (the hydrogen atom), and the arrangement of the remaining groups are noted. If, in proceeding from the group of highest priority to the group of second priority and thence to the third, the eye travels in a clockwise direction, the configuration is specified *R* (from the Latin *rectus*, "right"); if the eye travels in a counterclockwise direction, the configuration is specified *S* (from the Latin *sinister*, "left"). The complete name includes both configuration and direction of optical rotation, as for example, (*S*)-(+)-2-butanol.

The relative configurations around the chiral centers of many compounds have been established. One optically active compound is converted to another by a sequence of chemical reactions which are stereospecific; that is, each reaction is known to proceed spatially in a specific way. The configuration of one chiral compound can then be related to the configuration of the next in sequence. In order to establish absolute configuration, one must carry out sufficient stereospecific reactions to relate a new compound to another of known absolute configuration. Historically the configuration of D-(+)-2,3-dihydroxypropanal has served as the standard to which all configuration has been compared. The absolute configuration assigned to this compound has been confirmed by an X-ray crystallographic technique.

Chemical Abstracts Indexing System

When compounds of complex structure are considered, the number of name possibilities grows rapidly. To avoid having index entries for all possible names, Chemical Abstracts Service has developed what might be called the principle of inversion. The indexing system employs inverted entries to bring together related compounds in an alphabetically arranged index. The *index heading parent* from the Chemical Substance Index appears in the Formula Index in lightface before the "comma of inversion." The *substituents* follow the "comma of inversion" in alphabetical order. Any *name modification* appears on a separate line. If necessary, the chemical description is completed by citation of an associated ion, a functional derivative, a "salt with" or "compound with" terms and/or a stereochemical descriptor.

Quite naturally there is a certain amount of arbitrariness in this system, although the IUPAC nomenclature is followed. The preferred *Chemical Abstracts* index names for chemical substances have been, with very few exceptions, continued unchanged (since 1972) as set forth in the *Ninth Collective Index Guide* and in a journal article.* Any revisions appear in the updated *Index Guide*; new editions appear at 18-month intervals. Appendix VI is of particular interest to chemists. Reprints of the Appendix may be purchased from Chemical Abstracts Service, Marketing Division, P.O. Box 3012, Columbus, Ohio 43210.

PHYSICAL PROPERTIES OF PURE SUBSTANCES

TABLE 1-13 Empirical formula index for organic compounds

The alphanumeric designations are keyed to Table 1-14

Cl_2H_2Si: d226
Cl_3HSi: t251
Cl_6OSi_2: h28

C_1

$CBrClF_2$: b255
$CBrCl_3$: b361
$CBrF_3$: b363
CBr_2F_2: d75
$CClF_3$: c254
$CClNO_3S$: c241
CCl_2F_2: d170
CCl_3D: c128
CCl_3F: t236
CCl_3NO_2: t243
CCl_4O_2S: t240
CCl_4S: t239
CD_4O: m36
$CHBrCl_2$: b268
$CHBr_2Cl$: d71
$CHBr_3$: t210
$CHClF_2$: c85
$CHCl_2F$: d183
$CHCl_3$: c127
CHF_3: t300
CHF_3O_3S: t301
CHI_3: i36
CHN_3O_6: t392
CH_2BrCl: b258
CH_2Br_2: d88
CH_2Cl_2: d190
CH_2Cl_4Si: c166
CH_2I_2: d405
CH_2N_2: c286, d47
CH_2N_4: t138
CH_2O: f27
$(CH_2O)_x$: p1
CH_2O_2: f32
CH_2S_3: t441
CH_3Br: b303
CH_3Br_3Ge: m255
CH_3Cl: c138
CH_3ClHg: m298
CH_3ClO_2S: m32
CH_3Cl_3Ge: m440
CH_3Cl_3Si: t242
CH_3DO: m35
CH_3F: f18
CH_3I: i40
CH_3NO: f28
CH_3NO_2: m317, n56
CH_3NO_3: m316
CH_3N_5: a294
CH_4: m29
CH_4Cl_2Si: d199, m223
CH_4N_2O: f34, u12
$CH_4N_2O_2S$: f30
CH_4N_2S: t166
$CH_4N_4O_2$: n54
CH_4O: m34
CH_4O_2: m276
CH_4O_3S: m30
CH_4S: m33
CH_5AsO_3: m125
CH_5N: m115
CH_5NO_3S: a208
CH_5N_3: g29
CH_5N_3O: s3
CH_5N_3S: t165
CH_6N_2: m271
CH_6N_4: a180, a181
CH_6N_4O: c11
CN_4O_8: t126a

C_2

$C_2Br_2ClF_3$: d72
$C_2Br_2Cl_4$: d99
$C_2Br_2F_4$: d100
$C_2Br_2O_2$: o50
C_2ClF_3: c253
$C_2Cl_2F_3I$: d188
$C_2Cl_2F_4$: d227
$C_2Cl_2O_2$: o51
$C_2Cl_3F_3$: t255
C_2Cl_3N: t221
C_2Cl_4: t30
$C_2Cl_4F_2$: d347, d348, t27
C_2Cl_4O: t222
C_2Cl_6: h29
C_2D_3N: a30
$C_2D_4O_2$: a21
C_2D_6OS: d617
C_2F_4: t65
C_2F_6: h44
$C_2F_6O_5S_2$: t302
$C_2HBrClF_3$: b260
$C_2HBr_2F_3$: d103
C_2HBr_2N: d63
C_2HBr_3: t209
C_2HBr_3O: t205
$C_2HBr_3O_2$: t206
$C_2HClF_2O_2$: c83
$C_2HCl_2F_3$: d232
C_2HCl_3: t234
C_2HCl_3O: d141
$C_2HCl_3O_2$: t220

* *J. Chem. Doc.*, **14**(1): 3–15 (1974).

TABLE 1-13 Empirical formula index for organic compounds (*continued*)

The alphanumeric designations are keyed to Table 1-14

C_2HCl_5: p9
$C_2HF_3O_2$: t292
C_2H_2: a41
C_2H_2BrClO: b225
$C_2H_2Br_2$: d80, d81
$C_2H_2Br_2F_2$: d74
$C_2H_2Br_2O$: b224
$C_2H_2Br_2O_2$: d62
$C_2H_2Br_4$: t10
$C_2H_2ClF_3$: c252
C_2H_2ClN: c27
$C_2H_2Cl_2$: d178, d179, d180
$C_2H_2Cl_2O$: c31
$C_2H_2Cl_2O_2$: d138
$C_2H_2Cl_4$: t28, t29
$C_2H_2F_3NO$: t291
C_2H_2O: k1
$C_2H_2O_2$: g27
$C_2H_2O_3$: g28
$C_2H_2O_4$: o48, o49
C_2H_3Br: b286
C_2H_3BrO: a35
$C_2H_3BrO_2$: b221
$C_2H_3Br_2Cl_3Si$: d82
$C_2H_3Br_3O$: t208
C_2H_3Cl: c110
$C_2H_3ClF_2$: c84
C_2H_3ClO: a37
$C_2H_3ClO_2$: c24, m188
$C_2H_3Cl_3$: t230, t231
$C_2H_3Cl_3O$: t232
$C_2H_3Cl_3Si$: t256
$C_2H_3Cl_5Si$: d182
$C_2H_3DO_2$: a20
C_2H_3FO: a43
$C_2H_3FO_2$: f6
$C_2H_3F_3$: t296
$C_2H_3F_3O$: t297
C_2H_3IO: a48
$C_2H_3IO_2$: i25
C_2H_3N: a29
C_2H_3NO: m288
C_2H_3NS: m290, m429
$C_2H_3N_3$: t203
$C_2H_3N_3S_2$: a295
C_2H_4BrCl: b256
C_2H_4BrNO: b219
$C_2H_4Br_2$: d77, d78
C_2H_4ClNO: c22
$C_2H_4Cl_2$: d176, d177
$C_2H_4Cl_2O$: d197
$C_2H_4Cl_6Si_2$: b205
C_2H_4FNO: f5
$C_2H_4F_2$: d346
C_2H_4INO: i24
$C_2H_4I_2$: d404
$C_2H_4N_2$: a106
$C_2H_4N_2O_2$: o54
$C_2H_4N_2O_4$: d634
$C_2H_4N_2O_6$: e129
$C_2H_4N_2S_2$: d712
$C_2H_4N_4$: a300, d235
$C_2H_4N_4O_2$: a329
C_2H_4O: a4, e132
C_2H_4OS: t144
$C_2H_4O_2$: a19, h86, m251
$C_2H_4O_2S$: m14
$C_2H_4O_3$: h87, p59
$C_2H_4O_5S$: s24
C_2H_4S: e133
$C_2H_5AlCl_2$: e58
C_2H_5Br: b279
$C_2H_5BrNaO_2S$: b280
C_2H_5BrO: b281, b311
C_2H_5Cl: c103
C_2H_5ClHg: e168
C_2H_5ClO: c104, c156
$C_2H_5ClO_2S$: e19
C_2H_5ClS: c157
$C_2H_5Cl_2OPS$: e118
$C_2H_5Cl_2O_2P$: e117
$C_2H_5Cl_3Si$: c154, t235
C_2H_5DO: e22
C_2H_5F: f17
$C_2H_5FO_3S$: e137
C_2H_5I: i34
C_2H_5IO: i35
C_2H_5N: e134
C_2H_5NO: a5, a6, m249
$C_2H_5NO_2$: e189, g25, m182, n53
$C_2H_5NO_3$: e188
C_2H_5NS: t143
$C_2H_5N_3O_2$: b216, o53
C_2H_6: e14
C_2H_6BrN: b283
C_2H_6Cd: d503
C_2H_6ClN: c107
$C_2H_6ClNO_2S$: d611
$C_2H_6ClO_2PS$: d506
$C_2H_6Cl_2Si$: d174
C_2H_6Hg: d548
$C_2H_6N_2$: a7
$C_2H_6N_2O$: a25, m447, n79
$C_2H_6N_2O_2$: m272
$C_2H_6N_2O_4S$: a107
$C_2H_6N_2S$: m434
$C_2H_6N_4O_2$: o52
C_2H_6O: d520, e21
C_2H_6OS: d616, m18
$C_2H_6O_2$: e16, e131
$C_2H_6O_2S$: d615
$C_2H_6O_3S$: d614, m300
$C_2H_6O_4S$: d612, h114
$C_2H_6O_5S_2$: m31
C_2H_6S: d613, e20
$C_2H_6S_2$: d518, e18
C_2H_6Te: d619
C_2H_6Zn: d626
$C_2H_7AsO_2$: d486
C_2H_7ClSi: c93
C_2H_7N: d463, e59
C_2H_7NO: a163, a164
$C_2H_7NO_3S$: a161
$C_2H_7NO_4S$: a169
C_2H_7NS: a162
$C_2H_7N_5$: b134
$C_2H_7O_3P$: d543
$C_2H_8N_2$: d541, d542, e15
$C_2H_8N_2O$: h120

C_3

$C_3Br_2F_6$: d85
$C_3Cl_3NO_2$: t223
$C_3Cl_3N_3$: t254
$C_3Cl_3N_3O_3$: t238
C_3Cl_6: h31
C_3Cl_6O: h23
C_3D_6O: a27
C_3HCl_5O: p7
C_3H_2ClN: c32
$C_3H_2Cl_2O_2$: m6
$C_3H_2Cl_4$: t34

TABLE 1-13 Empirical formula index for organic compounds (*continued*)

The alphanumeric designations are keyed to Table 1-14

$C_3H_2Cl_4O$: t22
$C_3H_2Cl_4O_2$: t233
$C_3H_2F_6O$: h45
$C_3H_2N_2$: m5
$C_3H_2N_2O_3$: i6
$C_3H_2O_2$: p244
C_3H_3Br: b347
C_3H_3Cl: c233
C_3H_3ClO: a65
$C_3H_3Cl_3O$: e13
C_3H_3N: a64
$C_3H_3NOS_2$: r3
$C_3H_3NO_2$: c288
C_3H_3NS: t142
$C_3H_3N_3O_2S$: a252
$C_3H_3N_3O_3$: c300
C_3H_4: a78, p243
C_3H_4BrClO: b343, b344
C_3H_4BrN: b342
$C_3H_4Br_2$: d95
$C_3H_4Br_2O_2$: d96
C_3H_4ClN: c221
$C_3H_4Cl_2$: d221, d222
$C_3H_4Cl_2O$: c222, c223, d139
$C_3H_4Cl_2O_2$: m220
$C_3H_4Cl_3NO$: m439
$C_3H_4F_4O$: t66
$C_3H_4N_2$: i4, p248
$C_3H_4N_2O$: c287
$C_3H_4N_2OS$: t155
$C_3H_4N_2O_2$: h84
$C_3H_4N_2S$: a296
C_3H_4O: p207, p245
$C_3H_4O_2$: a63, o59, p213
$C_3H_4O_3$: e125, o60
$C_3H_4O_4$: m3
C_3H_5Br: a85, b226, b338, b339
C_3H_5BrO: b278
$C_3H_5BrO_2$: b340, b341, m143
$C_3H_5Br_3$: t212
C_3H_5Cl: c217
C_3H_5ClO: c102, c216, p219
C_3H_5ClOS: e102
$C_3H_5ClO_2$: c219, c220, e99, m183
$C_3H_5Cl_3$: t248
$C_3H_5Cl_3O$: t249
$C_3H_5Cl_3Si$: a102
C_3H_5FO: f7
$C_3H_5F_3O_3S$: m441
C_3H_5I: a92, i50
C_3H_5N: p218
C_3H_5NO: a62, c291, h169, h170
$C_3H_5NO_2$: o55
C_3H_5NS: e164, m424
$C_3H_5N_3O$: c289
$C_3H_5N_3O_9$: g21
$C_3H_5N_3S$: c293
C_3H_6: c365, p208
C_3H_6BrCl: b259
$C_3H_6BrNO_4$: b319
$C_3H_6Br_2$: d92, d93
$C_3H_6Br_2O$: d94
C_3H_6ClNO: d504
$C_3H_6Cl_2$: d218, d219
$C_3H_6Cl_2O$: d220
$C_3H_6Cl_2Si$: d200
$C_3H_6Cl_4Si$: c230
$C_3H_6I_2$: d406
$C_3H_6N_2$: a279, d507
$C_3H_6N_2O$: i7
$C_3H_6N_2O_2$: m4, m270
$C_3H_6N_2S$: a297, i5
$C_3H_6N_2OS$: a58
$C_3H_6N_6$: t202
C_3H_6O: a26, a81, e10, m449, p214, p230, t350
C_3H_6OS: m423, t164
$C_3H_6O_2$: d649, e11, e138, h89, m111, p216
$C_3H_6O_2S$: m21, m294
$C_3H_6O_3$: d398, d399, d505, L1, L2, m38, m260, t395
$C_3H_6O_3S$: p201
C_3H_6S: p209, p231, t350a
$C_3H_6S_3$: t440
C_3H_7Br: b335, b336
C_3H_7BrO: b337
C_3H_7Cl: c211, c212
C_3H_7ClO: c112, c153, c214, c215
C_3H_7ClOS: c137
$C_3H_7ClO_2$: c213
$C_3H_7ClO_2S$: p200
$C_3H_7Cl_2OP$: p239
$C_3H_7Cl_3Si$: d194, p240
C_3H_7I: i48, i49
C_3H_7N: a82, p229
C_3H_7NO: a28, d524, m110, p215
$C_3H_7NO_2$: a73, a74, a75, a76, e92, m259, n73, n74
$C_3H_7NO_2S$: c371
$C_3H_7NO_3$: i105, n75, p236, s4
$C_3H_7NO_5S$: a293
C_3H_7NS: d622
$C_3H_7NS_2$: d519
$C_3H_7O_5P$: c17
C_3H_8: p194
C_3H_8ClN: c225
$C_3H_8Cl_2Si$: c75, c151
C_3H_8IN: d551
$C_3H_8N_2O$: d625, e233
$C_3H_8N_2O_2$: e93, f29
$C_3H_8N_2S$: d623
C_3H_8O: e174, p205, p206
$C_3H_8OS_2$: d426, m308
$C_3H_8O_2$: d442, m65, p197, p198
$C_3H_8O_2S$: m20
$C_3H_8O_3$: g16
C_3H_8S: e185, p202, p203
$C_3H_8S_2$: p199
C_3H_9Al: t332
$C_3H_8BO_3$: t324
$C_3H_9B_3O_6$: t325
C_3H_9BrGe: b366
C_3H_9BrSi: b367
C_3H_9ClGe: c255
C_3H_9ClSi: c256
C_3H_9IOS: t385
C_3H_9IS: t384
C_3H_9ISi: i55
C_3H_9N: i88, m247, p223, t333
C_3H_9NO: a274, a275, a276, a277, m69, m119, t334
$C_3H_9NO_2$: a273

TABLE 1-13 Empirical formula index for organic compounds (*continued*)
The alphanumeric designations are keyed to Table 1-14

$C_3H_9N_3Si$: a324
$C_3H_9O_3P$: d553, t370
$C_3H_9O_4P$: t369
$C_3H_{10}N_2$: m248, p195, p196
$C_3H_{10}N_2O$: d43
$C_3H_{11}Br_2N_3S$: a171

C_4

$C_4Cl_2F_6$: d185
$C_4Cl_2F_8$: d206
$C_4Cl_2O_3$: d189
$C_4Cl_3F_7$: h3
C_4Cl_6: h25
$C_4D_6O_3$: a23
$C_4F_6O_3$: t293
C_4HBrO_3: b302
$C_4HCl_3N_2$: t250
$C_4HF_7O_2$: h2
C_4H_2: b379
$C_4H_2Br_2S$: d101
$C_4H_2Cl_2N_2$: d223
$C_4H_2Cl_2O_2$: f38
$C_4H_2Cl_2O_3$: d208
$C_4H_2Cl_2S$: d228
$C_4H_2F_6O_2$: t299
$C_4H_2O_3$: m2
$C_4H_2O_4$: a42
C_4H_3BrS: b356
C_4H_3ClS: c243
$C_4H_3Cl_2N_3O$: d193
C_4H_3IS: i52
C_4H_4: b410
$C_4H_4BrNO_2$: b354
$C_4H_4Br_2O_2$: d69
$C_4H_4Br_2O_4$: d98
$C_4H_4ClNO_2$: c240
$C_4H_4Cl_2$: d168
$C_4H_4Cl_2O_2$: s20
$C_4H_4Cl_2O_3$: c25
$C_4H_4N_2$: b383, p247, p251, p271, s19
$C_4H_4N_2O_2$: d401, p272
$C_4H_4N_2O_2S$: d389
$C_4H_4N_2O_3$: b1
$C_4H_4N_2O_5$: a79
$C_4H_4N_4$: d40
C_4H_4O: f40
$C_4H_4O_2$: d423
$C_4H_4O_3$: s17
$C_4H_4O_4$: f37, m1
C_4H_4S: t157
$C_4H_5BrO_4$: b353
C_4H_5Cl: c63, c70
C_4H_5ClO: c284, c367, m28
$C_4H_5ClO_2$: a87
$C_4H_5ClO_3$: e194
$C_4H_5Cl_3O_2$: e229
$C_4H_5F_3O_2$: e230
C_4H_5N: b403, c366, m27, p273
C_4H_5NO: m291
$C_4H_5NO_2$: e107, m194, s18
$C_4H_5NO_2S$: e33
$C_4H_5NO_3$: h183
C_4H_5NS: a93
$C_4H_5N_3$: a289, i11
$C_4H_5N_3O$: a200
$C_4H_5N_3OS$: a193
$C_4H_5N_3O_2$: a154, a155, c290, m325
C_4H_6: b376, b377, b493, b494
$C_4H_6Br_2O_2$: d70
C_4H_6ClN: c73
$C_4H_6Cl_2$: d165, d166, d167
$C_4H_6Cl_2O$: c74
$C_4H_6Cl_2O_2$: m222
$C_4H_6Cl_3NSi$: c295
$C_4H_6N_2$: a151, m281, m282, m283
$C_4H_6N_2O_2$: e115
$C_4H_6N_2S$: a232
$C_4H_6N_4O$: d39
$C_4H_6N_4O_3$: a77
C_4H_6O: b409, c283, d357, d545a, m24, m399
$C_4H_6O_2$: b389, b404, b405, b406, b495, b500, b501, c368, m26, m114, v2
$C_4H_6O_2S$: d369
$C_4H_6O_3$: a22, a24, m337, o56, p228
$C_4H_6O_4$: d568, s15
$C_4H_6O_4S$: m23, t151
$C_4H_6O_5$: h181, h182, o61
$C_4H_6O_6$: t1, t2
C_4H_7Br: b242, b243, b244
$C_4H_7BrO_2$: b246, b282, b310, e76, m146
C_4H_7Cl: c68, c69, c164, c165
C_4H_7ClO: b504, c67, c116, i78
$C_4H_7ClO_2$: c71, c72, e95, m190
$C_4H_7Cl_2NSi$: c292
$C_4H_7Cl_3O$: t241
$C_4H_7Cl_3O_2Si$: c13
$C_4H_7FO_2$: e136
C_4H_7N: b502, i76
C_4H_7NO: h146, i98, m25, m335, p234, p279
$C_4H_7NO_2$: m334
$C_4H_7NO_3$: a46, e195, s13
$C_4H_7NO_4$: a319, i10
C_4H_7NS: m422
$C_4H_7N_3O$: c279
C_4H_8: b398, b399, b400, c301, m386
C_4H_8BrCl: b253
$C_4H_8Br_2$: d67, d68
$C_4H_8Br_2O$: b150
$C_4H_8Cl_2$: d162, d163, d164
$C_4H_8Cl_2O$: b159, d181
$C_4H_8Cl_2Si$: a89
$C_4H_8N_2O$: a105, a150
$C_4H_8N_2O_2$: d528, s14
$C_4H_8N_2O_3$: a318, g26
$C_4H_8N_2S$: a101, t81
C_4H_8O: b396, b407, b408, b496, e3, e235, i73, m96, m377, m388, t68
C_4H_8OS: e223, t109, t167
$C_4H_8O_2$: b401, b402, b498, d648, e52, h106, i75, m392, m393, p232
$C_4H_8O_2S$: e167, m297, t108
$C_4H_8O_3$: e24, e153, h116, h127, m64, m292, m301
$C_4H_8O_3S$: m387
C_4H_8S: a95, t83
$C_4H_8S_2$: d709

TABLE 1-13 Empirical formula index for organic compounds (*continued*)
The alphanumeric designations are keyed to Table 1-14

C_4H_9Br: b240, b241, b313, b314
C_4H_9BrO: b287
C_4H_9Cl: c64, c65, c162, c163
C_4H_9ClO: c66, c111
$C_4H_9ClO_2$: c89, c105, m67
C_4H_9ClSi: c94
$C_4H_9Cl_3Si$: b486, c226
$C_4H_9Cl_3Sn$: b484
C_4H_9F: f20
C_4H_9I: i30, i31, i43, i44
C_4H_9Li: b460, b461
C_4H_9N: p274
C_4H_9NO: a326, b397, b497, d459, e53, i74, m391, m451
$C_4H_9NO_2$: a138, a139, a225, b467, b468, h115, i71, n50
$C_4H_9NO_2S$: a207
$C_4H_9NO_3$: a189, a190, a191, a192, i70, n51
C_4H_9NSi: c299
$C_4H_9N_3O_2$: c278
C_4H_{10}: b381, m378
$C_4H_{10}ClN$: d469
$C_4H_{10}ClO_2PS$: d292
$C_4H_{10}ClO_3P$: d291
$C_4H_{10}Cl_2Si$: b161, m395
$C_4H_{10}N_2$: p182
$C_4H_{10}N_2O$: a234
$C_4H_{10}N_2O_4S$: a8
$C_4H_{10}O$: b394, b395, d300, m384, m385, m396
$C_4H_{10}OS$: e156
$C_4H_{10}OS_2$: b187
$C_4H_{10}O_2$: b384, b385, b386, b387, b388, b456, d439, d440, e35, m95
$C_4H_{10}O_2S$: m433, t152
$C_4H_{10}O_2S_2$: d425, h118
$C_4H_{10}O_3$: b182, b393, t359
$C_4H_{10}O_3S$: d338
$C_4H_{10}O_4S$: d336
$C_4H_{10}S$: b391, b392, d337, i104, m381, m382, m383, m398
$C_4H_{10}S_2$: b390, d294a
$C_4H_{10}S_3$: b188
$C_4H_{10}Zn$: d344
$C_4H_{11}ClSi$: c167
$C_4H_{11}N$: b380, b420, b421, d267, d268, d522, i63
$C_4H_{11}NO$: a136, a137, a224, d315, d467, e39, e63
$C_4H_{11}NO_2$: a165, a223, d245, d441
$C_4H_{11}NO_3$: t430
$C_4H_{11}O_2PS_2$: d296
$C_4H_{11}O_3P$: d314
$C_4H_{12}BrN$: t95
$C_4H_{12}ClN$: t96
$C_4H_{12}Ge$: t111
$C_4H_{12}IN$: t97
$C_4H_{12}N_2$: b382, b455, d523, m379, m380
$C_4H_{12}N_2O$: a166
$C_4H_{12}N_2S_2$: c370
$C_4H_{12}OSi$: m108
$C_4H_{12}O_3Si$: t326a
$C_4H_{12}O_4Si$: t94
$C_4H_{12}Pb$: t114
$C_4H_{12}Si$: t122
$C_4H_{12}Sn$: t125
$C_4H_{13}N_3$: d298
$C_4H_{14}OSi_2$: t107
$C_4H_{16}O_4Si_4$: t105

C_5

C_5Cl_5N: p12
C_5Cl_6: h27
C_5D_5N: p253
$C_5H_3Br_2N$: d97
$C_5H_3ClO_2$: f48
$C_5H_3Cl_2N$: d224
C_5H_4BrN: b348, b349
C_5H_4ClN: c234
C_5H_4FN: f23
$C_5H_4F_8O$: o18
$C_5H_4N_2O_3$: n76
$C_5H_4N_4O$: h187
$C_5H_4N_4O_3$: u13
C_5H_4OS: t159
$C_5H_4O_2$: f39
$C_5H_4O_2S$: t160
$C_5H_4O_3$: c272, f42
$C_5H_5ClN_2$: a149
$C_5H_5ClN_2O_2$: c168
$C_5H_5F_3O_2$: t298
C_5H_5N: p252
C_5H_5NO: h174, h175, h176, p266
$C_5H_5NO_2$: d402, h178
$C_5H_5NO_3S$: p267
$C_5H_5N_3O_2$: a251
$C_5H_5N_3O_4$: a160
$C_5H_5N_5$: a69
C_5H_6: m166
$C_5H_6Br_2N_2O_2$: d76
$C_5H_6Cl_2N_2O_2$: d173
$C_5H_6Cl_2O_2$: d195, g15
$C_5H_6Cl_4O_2$: t229
$C_5H_6N_2$: a286, a287, a288, g14, m400, v7
$C_5H_6N_2O$: a47, a199
$C_5H_6N_2OS$: h129
$C_5H_6N_2O_2$: d392
$C_5H_6N_2O_3$: e108
C_5H_6O: m59, m253
C_5H_6OS: f44
$C_5H_6O_2$: f46
$C_5H_6O_3$: g12
$C_5H_6O_4$: c271, m246
$C_5H_6O_4S_3$: b156
C_5H_6S: m430
$C_5H_7BrO_2$: m145
$C_5H_7BrO_3$: e80
$C_5H_7ClO_3$: m184, m189
C_5H_7N: m407
C_5H_7NO: f47
$C_5H_7NO_2$: e106
C_5H_7NS: t161
$C_5H_7N_3$: a231, d44
$C_5H_7N_3O$: a194
C_5H_8: c359, m147, m148, m172, p16, p17, p18, p19, p57
$C_5H_8Br_2O_2$: e116
$C_5H_8Br_4$: p21
$C_5H_8F_4O$: m417

TABLE 1-13 Empirical formula index for organic compounds (*continued*)
The alphanumeric designations are keyed to Table 1-14

$C_5H_8N_2$: d544, d605, e162, p276
$C_5H_8N_2O$: m452
$C_5H_8N_2O_2$: d540
$C_5H_8N_4O_{12}$: p22
C_5H_8O: c357, c369, d364, e8, m173, p51
$C_5H_8O_2$: a80, e57, g13, i84, m58, m161, m162, m163, m193, m218, m299, p31, p32, p40, p40a, p50, p211
$C_5H_8O_3$: e199, m112, o58
$C_5H_8O_4$: d547, g11, m275, m415
C_5H_9Br: b265
$C_5H_9BrO_2$: e82, e83, m144
C_5H_9Cl: c79
C_5H_9ClO: c193, d602, m180, p44
C_5H_9ClOS: b440, c229
$C_5H_9ClO_2$: b439, e100, e101, i65, m187
$C_5H_9F_3O_2Si$: t383
C_5H_9N: d604, m179, p33, t80
C_5H_9NO: b458, b459, c358, e196, m409
$C_5H_9NO_2$: m118, p277
$C_5H_9NO_4$: g9
$C_5H_9N_3$: i8
C_5H_{10} c353, m158, m159, m160, p47, p48, p49
$C_5H_{10}Br_2$: d91
$C_5H_{10}ClNO$: d288
$C_5H_{10}Cl_2$: d209
$C_5H_{10}Cl_2O_2Si$: c12
$C_5H_{10}Cl_2Si$: c352
$C_5H_{10}N_2$: d293, d476
$C_5H_{10}N_2O$: d545, p183
$C_5H_{10}N_2O_3$: g10
$C_5H_{10}O$: a91, c356, d598, i108, m157, m164, m165, m174, m175, m418, p27, p41, p42, t78
$C_5H_{10}OS$: m428
$C_5H_{10}O_2$: d455, d600, e211, h140, h157, h158, i66, i87, m102, m176, m177, m178, m287, p36, p222, t70
$C_5H_{10}O_2S$: e186, m309, m420
$C_5H_{10}O_3$: d289, d454, e165, m68, m279
$C_5H_{10}O_4$: b185
$C_5H_{10}O_5$: a315, r5, x8
$C_5H_{11}Br$: b308, b325, b326
$C_5H_{11}BrO_2$: b271
$C_5H_{11}BrO_2Si$: t377
$C_5H_{11}Cl$: c92, c149, c150, c192
$C_5H_{11}ClSi$: a86
$C_5H_{11}Cl_2N$: b160
$C_5H_{11}I$: i42, i47
$C_5H_{11}N$: a90, m408, p186
$C_5H_{11}NO$: d304, d599, h168, m310, t71
$C_5H_{11}NO_2$: a256, a257, b129, e234, i81, v1
$C_5H_{11}NO_2S$: m37
$C_5H_{11}NO_3$: n58
$C_5H_{11}NS_2$: d295
$C_5H_{11}O_5P$: t371
C_5H_{12}: d594, m149, p28
$C_5H_{12}ClN$: d477
$C_5H_{12}Cl_2O_2Si$: b158
$C_5H_{12}N_2$: a272, m368, m369
$C_5H_{12}N_2O$: b490, t126
$C_5H_{12}N_2O_2$: b434, o46
$C_5H_{12}N_2S$: t124
$C_5H_{12}N_2S_2$: p275
$C_5H_{12}O$: b463, d453, d597, e212, m153, m154, m155, m156, p37, p38, p39
$C_5H_{12}OSi$: t386
$C_5H_{12}O_2$: d596, m57, p30
$C_5H_{12}O_2S$: e224
$C_5H_{12}O_3$: h142, m66, t358, t432
$C_5H_{12}O_3S$: p34
$C_5H_{12}O_4$: p20, t116
$C_5H_{12}O_5$: x7
$C_5H_{12}S$: b466, e213, m150, m151, m152, p35
$C_5H_{12}Si$: t387
$C_5H_{13}N$: a254, a255, d603, m168, m169, m170, p53
$C_5H_{13}NO$: a216, a217, a258, d474, d475, e48, i89, p224
$C_5H_{13}NOSi$: t374
$C_5H_{13}NO_2$: a176, d443, d473, d525, m224
$C_5H_{13}N_3$: t112
$C_5H_{14}N_2$: d595, p29, t115
$C_5H_{14}OSi$: e51, t379
$C_5H_{14}O_2Si$: d255
$C_5H_{15}N_3$: a175

C_6

C_6BrD_5: b231
C_6BrF_5: b324
$C_6Cl_4O_2$: t25, t26
$C_6Cl_5NO_2$: p10
C_6Cl_6: h24
C_6D_6: b10
C_6D_{12}: c314
C_6F_6: h43
C_6HBr_5O: p6
$C_6HCl_4NO_2$: t31
C_5HCl_5: p8
C_6HCl_5O: p11
$C_6H_2BrFN_2O_4$: b274
$C_6H_2Cl_2O_4$: d172
$C_6H_2Cl_3NO_2$: t242a
$C_6H_2Cl_4$: t23, t24
$C_6H_3Br_2F$: d84
$C_6H_3Br_2NO_2$: d90
$C_6H_3Br_3O$, t211
$C_6H_3ClFNO_2$: c122
$C_6H_3ClN_2O_4$: c95, c96
$C_6H_3ClN_2O_4S$: d629
$C_6H_3Cl_2NO_2$: d203, d204, d205
$C_6H_3Cl_3$: t226, t227, t228
$C_6H_3Cl_3O$: t244, t245
$C_6H_3Cl_3O_2S$: d155
$C_6H_3FN_2O_4$: d635
$C_6H_3N_3O_6$: t389, t390

TABLE 1-13 Empirical formula index for organic compounds (*continued*)
The alphanumeric designations are keyed to Table 1-14

$C_6H_3N_3O_7$: p176
C_6H_4BrCl: b249, b250, b251
$C_6H_4BrClO_2S$: b232
C_6H_4BrF: b290, b291, b292
$C_6H_4BrNO_2$: b317
$C_6H_4BrN_3O_4$: b273
$C_6H_4Br_2$: d65
$C_6H_4Br_2N_2O_2$: d89
$C_6H_4Br_3N$: t207
C_6H_4ClF: c117, c118, c119
C_6H_4ClFO: c123
C_6H_4ClI: c136
$C_6H_4ClNO_2$: c176, c177, c178, c235, c236
$C_6H_4ClNO_3$: c187
$C_6H_4ClNO_4S$: n35
$C_6H_4ClO_2P$: p112
$C_6H_4Cl_2$: d152, d153, d154
$C_6H_4Cl_2N_2O_2$: d202
$C_6H_4Cl_2O$: d210, d211, d212, d213
$C_6H_4Cl_2O_2$: d171
$C_6H_4Cl_2O_2S$: c43
$C_6H_4Cl_3N$: t224, t225
$C_6H_4Cl_4Si$: c209
$C_6H_4FNO_2$: f21
$C_6H_4F_2$: d345
$C_6H_4INO_2$: i45
$C_6H_4I_2$: d403
$C_6H_4N_2$: c296, c297, c298
$C_6H_4N_2O_2$: b43
$C_6H_4N_2O_4$: d628
$C_6H_4N_2O_5$: d637
$C_6H_4N_4$: a278
$C_6H_4N_4O_6$: t388
$C_6H_4O_2$: b59
$C_6H_5BO_2$: c21
C_6H_5Br: b230
C_6H_5BrO: b328, b329
C_6H_5BrS: b357
C_6H_5Cl: c41
C_6H_5ClHg: p128
$C_6H_5ClN_2O_2$: c173, c173a, c174, c175
C_6H_5ClO: c195, c196, c197
$C_6H_5ClO_2$: c87, c88
$C_6H_5ClO_2S$: b23
C_6H_5ClS: c244
C_6H_5ClSe: p154
$C_6H_5Cl_2N$: d142, d143, d144, d145, d146, d147
$C_6H_5Cl_2OP$: p139
$C_6H_5Cl_2O_2P$: p107
$C_6H_5Cl_2P$: d216
$C_6H_5Cl_2PS$: p140
$C_6H_5Cl_3Si$: p158
C_6H_5D: b9
C_6H_5F: f11
C_6H_5FO: f22
$C_6H_5FO_2S$: b24
$C_6H_5F_7O_2$: e140
C_6H_5I: i27
C_6H_5NO: n78, p255, p256, p257
C_6H_5NOS: t156
$C_6H_5NO_2$: n30, n83, p259, p260, p261
$C_6H_5NO_3$: h177, n60, n61
$C_6H_5NO_4$: c273
$C_6H_5N_3$: b62
$C_6H_5N_3O$: h103
$C_6H_5N_3O_4$: d627
C_6H_6: b8a
$C_6H_6AsNO_6$: h154
C_6H_6BrN: b226, b227, b228
C_6H_6ClN: c33, c34, c35
C_6H_6ClNO: a148, c142
$C_6H_6C.NO_2S$: c42
$C_6H_6Cl_2N_2$: d215
$C_6H_6Cl_6$: h26
C_6H_6FN: f9
C_6H_6HgO: p129
C_6H_6IN: i26
$C_6H_6N_2O$: e44, p254, p258
$C_6H_6N_2O_2$: n24, n25, n26
$C_6H_6N_2O_3$: a247, a248, m84
$C_6H_6N_4O_4$: d639
C_6H_6O: p64
C_6H_6OS: a57, m431
$C_6H_6O_2$: a44, d378, d379, d380, m252
$C_6H_6O_2S$: b20, t158
$C_6H_6O_3$: h147, m254, t309, t310
$C_6H_6O_3S$: b22
$C_6H_6O_4$: d461
$C_6H_6O_5S$: d383
$C_6H_6O_6$: p210
$C_6H_6O_8S_2$: d382
C_6H_6S: t162
$C_6H_7AsO_3$: b11
$C_6H_7BO_2$: b12
$C_6H_7ClN_2$: c203, c204, c205, c206
C_6H_7N: a303, a304, m401, m402, m403
C_6H_7NO: a260, a261, a262, m101, m406, p268, p269
$C_6H_7NO_2S$: b21
$C_6H_7NO_3S$: a118, a119, a120, s24
$C_6H_7NO_6S_2$: a117
C_6H_7NS: a298
$C_6H_7N_3O$: p262
$C_6H_7N_3O_2$: n67, n68, n69
$C_6H_7O_2P$: p137
$C_6H_7O_3P$: p138
$C_6H_8AsNO_3$: a115, a116
$C_6H_8Cl_2O_2$: h62, m221
$C_6H_8N_2$: a226, a227, a228, a229, a230, d238, m121, m258, p109, p110, p111, p120
$C_6H_8N_2O$: a211, o63
$C_6H_8N_2O_2S$: b25, s23
$C_6H_8N_2O_3S$: d32
$C_6H_8N_4$: p184
C_6H_8O: c332, d527, h40, m217
$C_6H_8O_2$: b378, c323, d365, h42, m215, v4
$C_6H_8O_3$: a36, d366, f43, h184
$C_6H_8O_4$: d526, d546
$C_6H_8O_6$: a317, g8, i59
$C_6H_8O_7$: c274
C_6H_9Br: b264
C_6H_9ClO: c78
$C_6H_9ClO_3$: e96, e97

TABLE 1-13 Empirical formula index for organic compounds (*continued*)
The alphanumeric designations are keyed to Table 1-14

$C_6H_9F_3O_2$: b487

C_6H_9NO: v11

C_6H_9NOS: m421

$C_6H_9NO_2$: b441

$C_6H_9NO_6$: n21

$C_6H_9N_3$: a158

$C_6H_9N_3O_2$: a159, c285, h83

C_6H_{10} c331, d490, h41, h82, m354

$C_6H_{10}N_2$: e175, p187

$C_6H_{10}N_2O_2$: c324

$C_6H_{10}N_2O_4$: d279

$C_6H_{10}N_2O_5$: a14

$C_6H_{10}N_4$: p26

$C_6H_{10}O$: c329, d26, d362, e5, e6, h78, m216, m353, m355

$C_6H_{10}O_2$: a96, c354, d360, e41, e105, e113, e169, h61, h71, h76, m352

$C_6H_{10}O_3$: d437, e54, e55, h121, p217

$C_6H_{10}O_4$: d325, d610, e17, h57, m273

$C_6H_{10}O_4S$: t154

$C_6H_{10}O_4S_2$: d711

$C_6H_{10}O_5$: d326

$C_6H_{10}O_6$: d618

$C_6H_{10}O_8$: t86

$C_6H_{10}S$: d27

$C_6H_{11}Br$: b263

$C_6H_{11}BrO_2$: b297, e77, e78, e79

$C_6H_{11}Cl$: c77

$C_6H_{11}ClO$: h73

$C_6H_{11}ClO_2$: b436, c152, e98

$C_6H_{11}Cl_3Si$: c345

$C_6H_{11}I$: i32

$C_6H_{11}N$: d25, h63, m342, m419

$C_6H_{11}NO$: c330 e220, f35, m376, o57, t360

$C_6H_{11}NO_2$: e62

C_6H_{12}: c313, d500, d501, e85, h75, m214, m350, m351

$C_6H_{12}Br_2$: d86

$C_6H_{12}ClN$: c161

$C_6H_{12}ClNO$: c113

$C_6H_{12}Cl_2$: d187

$C_6H_{12}Cl_2O$: b162

$C_6H_{12}Cl_2O_2$: b157, d169

$C_6H_{12}Cl_3O_3P$: t424

$C_6H_{12}Cl_3O_4P$: t423

$C_6H_{12}F_3NOSi$: m443

$C_6H_{12}NO_3P$: d293a

$C_6H_{12}N_2$: d45, t274

$C_6H_{12}N_2O_3$: s16

$C_6H_{12}N_2O_4S_2$: c372

$C_6H_{12}N_2S_4$: b175

$C_6H_{12}N_2Si$: t380

$C_6H_{12}N_4$: h52

$C_6H_{12}O$: a100, b491, c328, d499, d620, e88, h54, h72, h77, i72, m349, o47

$C_6H_{12}O_2$: b415, b416, b417, d502, e50, e89, e90, h66, h143, i62, m228, m305, m343, m344, m345, t79

$C_6H_{12}O_3$: d436, d458, d517, e38, e155, e157, i99, p2, p235, t69

$C_6H_{12}O_4Si$: d23

$C_6H_{12}O_6$: f36, g1, g6, i23, m11, s6

$C_6H_{12}O_7$: g4

$C_6H_{12}S$: c327

$C_6H_{13}Br$: b296

$C_6H_{13}BrO_2$: b269

$C_6H_{13}Cl$: c130

$C_6H_{13}ClO$: c131

$C_6H_{13}ClO_2$: c81

$C_6H_{13}ClO_3$: c106

$C_6H_{13}Cl_3O_3Si$: t422

$C_6H_{13}I$: i39

$C_6H_{13}N$: c335, h51, m371, m372, m373, m374

$C_6H_{13}NO$: d260, d555, e187, h145, p190

$C_6H_{13}NO_2$: a185, a186, h122, i79, L4, L5

$C_6H_{13}NO_4$: b183

$C_6H_{13}NO_4S$: m454

$C_6H_{13}NO_5$: g5, t435

C_6H_{14}: d491, d492, h55, m339, m340

$C_6H_{14}ClN$: d272

$C_6H_{14}Cl_4OSi_2$: b169

$C_6H_{14}N_2$: a182, a222, c319, c320

$C_6H_{14}N_2O$: a172, h123

$C_6H_{14}N_2O_2$: L12

$C_6H_{14}N_4O_2$: a316

$C_6H_{14}O$: b452, d418, d494, d495, d496, d497, d498, d703, e84, h68, h69, h70, m346, m347, m348

$C_6H_{14}OSi$: a97, e34, t381

$C_6H_{14}O_2$: b413, d251, d252, d493, e182, h58, h59, h60, i86, m341

$C_6H_{14}O_2S$: d706

$C_6H_{14}O_3$: b192, d253, e36, e159, h65, h173, t327

$C_6H_{14}O_4$: e130 t275

$C_6H_{14}O_4S$: d705

$C_6H_{14}O_6$: d740, m10, s5

$C_6H_{14}O_6S_2$: b189

$C_6H_{14}S$: b454, h64

$C_6H_{14}Si$: a104

$C_6H_{15}Al$: t268

$C_6H_{15}As$: t271

$C_6H_{15}B$: t273

$C_6H_{15}Bi$: t272

$C_6H_{15}ClO_2Si$: c155

$C_6H_{15}ClO_3Si$: c232

$C_6H_{15}ClSi$: b449

$C_6H_{15}Ga$: t279

$C_6H_{15}In$: t281

$C_6H_{15}N$: d412, d698, e86, e87, h80, m353a, t269

$C_6H_{15}NO$: a187, a219, a220, b422, b451, d270

$C_6H_{15}NOSi$: m442

$C_6H_{15}NO_2$: d254, e119

$C_6H_{15}NO_3$: t264

$C_6H_{15}NO_6S$: t431

$C_6H_{15}N_3$: a174

$C_6H_{15}O_3B$: t265

$C_6H_{15}O_3P$: d421, t287

$C_6H_{15}O_3PS$: t290

TABLE 1-13 Empirical formula index for organic compounds (*continued*)

The alphanumeric designations are keyed to Table 1-14

$C_6H_{15}O_4P$: t285
$C_6H_{15}P$: t286
$C_6H_{15}Sb$: t270
$C_6H_{16}Cl_2Si_2$: t106
$C_6H_{16}Cl_2Si_2$: t106
$C_6H_{16}N_2$: d302, h56, t110
$C_6H_{16}OSi$: p221
$C_6H_{16}Br_2OSi_2$: b151
$C_6H_{16}O_2Si$: d249
$C_6H_{16}O_3SSi$: m22
$C_6H_{16}O_3Si$: t266b
$C_6H_{16}Si$: t289
$C_6H_{17}NO_3Si$: a285
$C_6H_{17}NO_5S$: b181
$C_6H_{17}N_3$: i9
$C_6H_{18}LiNSi_2$: L11
$C_6H_{18}N_2Si$: b173
$C_6H_{18}N_3ClSi$: c258a
$C_6H_{18}N_3OP$: h53
$C_6H_{18}N_4$: t277
$C_6H_{18}OSi_2$: h50
$C_6H_{18}O_3Si_3$: h48
$C_6H_{19}NOSi_2$: b211
$C_6H_{19}NSi_2$: h49
C_6N_4: t38

C_7

C_7F_5N: p23
$C_7H_3BrClF_3$: b252
$C_7H_3BrF_3NO_2$: b318
$C_7H_3ClF_3NO_2$: c183, c184, c185
$Cl_7H_3ClN_2O_5$: d632
$C_7H_3ClN_2O_6$: c97
$C_7H_3Cl_3O$: d160, d161
$C_7H_4BrF_3$: b235, b236
C_7H_4ClFO: f14
$C_7H_4ClF_3$: c51, c52, c53
C_7H_4ClN: c47, c48
C_7H_4ClNO: c207
$C_7H_4ClNO_3$: n41, n42
$C_7H_4ClNO_4$: c179, c180, c181, n66
$C_7H_4Cl_2O$: c55, c56, d150
$C_7H_4Cl_2O_2$: d156, d157, d158
$C_7H_4Cl_3F$: t237
$C_7H_4Cl_4S$: t35
$C_7H_4F_3NO_2$: n88, n89
$C_7H_4F_{12}O$: d720
$C_7H_4I_2O_3$: h111
$C_7H_4N_2O_2$: n40
$C_7H_4N_2O_6$: d630, d631
$C_7H_4N_2O_7$: d640
$C_7H_4O_3S$: h104
$C_7H_4O_4S$: s25
C_7H_5BrO: b66, b229
$C_7H_5BrO_2$; b233
$C_7H_5BrO_3$: b351
$C_7H_5ClF_3N$: a144, a145, a146
$C_7H_5ClN_2$: a141
C_7H_5ClO: b67, c38,, c39
C_7H_5ClOS: p105
$C_7H_5ClO_2$: c45, c46, c46a, c238, c239, p104
$C_7H_5ClO_3$: c194
$C_7H_5Cl_2F$: c120
$C_7H_5Cl_2N$: d196
$C_7H_5Cl_2NO$: d151
$C_7H_5Cl_3$: t252, t253
C_7H_5FO: b69, f10
$C_7H_5FO_2$: f12, f13
$C_7H_5F_3$: t305
$C_7H_5F_3N_2O_2$: a244, a245
$C_7H_5F_3O$: t295
$C_7H_5F_4N$: a179
$C_7H_5IO_2$: i29
$C_7H_5IO_3$: i51
$C_7H_5I_2NO_2$: a156
C_7H_5N: b51
C_7H_5NO: b63, p123
$C_7H_5NO_3$: n27, n28
$C_7H_5NO_3S$: s1
$C_7H_5NO_4$: n37, n38, n39, p263, p264, p265
$C_7H_5NO_5$: h155
C_7H_5NS: b60, p124
$C_7H_5NS_2$: m17
$C_7H_5N_3O_2$: a241, n36, n55
$C_7H_5N_3O_2S$: a243
$C_7H_5N_3O_6$: t393
C_7H_6BrClO: b254
$C_7H_6BrNO_2$: n46
$C_7H_6BrNO_3$: h156
$C_7H_6Br_2$: b238, d102
C_7H_6ClF: c124, c125, c126, f16
C_7H_6ClNO: c40
$C_7H_6ClNO_2$: a140, c188, c189, c190, n47
$C_7H_6ClNO_3$: c141
$C_7H_6Cl_2$: c59, c60, d229, d230, d231
$C_7H_6Cl_2O$: d191, d192
$C_7H_6F_3N$: a129, a130, a131
$C_7H_6INO_2$: a205
$C_7H_6N_2$: a124, a125, a126, b38
$C_7H_6N_2O_3$: n29
$C_7H_6N_2O_4$: a240, d641, d642, d643
$C_7H_6N_2O_5$: d633, d33a
$C_7H_6N_2S$: a128, m15
C_7H_6O: b3
C_7H_6OS: t145
$C_7H_6O_2$: b44, h94, h95, h96, m241
$C_7H_6O_2S$: m16
$C_7H_6O_3$: d376, d377, f41, h99, h100, h101
$C_7H_6O_4$: d384, d385, d386
$C_7H_6O_5$: t311
$C_7H_6O_6S$: s29
C_7H_7Br: b85 b358, b359, b360
C_7H_7BrO: b237, b304, b305, b306
C_7H_7Cl: b89, c245, c246, c247
$C_7H_7ClN_4O_2$: c242
C_7H_7ClO: c57, c140, c159, c160
$C_7H_7ClO_2S$: t180
$C_7H_7ClO_3S$: m49
C_7H_7ClS: c249
$C_7H_7Cl_3Si$: b124, t196
C_7H_7F: f24, f25, f26
C_7H_7FO: f15, f19
$C_7H_7FO_2S$: t181
C_7H_7I: i53, i54

TABLE 1-13 Empirical formula index for organic compounds (*continued*)
The alphanumeric designations are keyed to Table 1-14

C_7H_7IO: i41
C_7H_7N: v9, v10
C_7H_7NO: a53, a54, a55, b4, f31
$C_7H_7NO_2$: a121, a122, a123, h97, h98, m404, m405, n85, n86, n87
$C_7H_7NO_3$: a291, a292, m81, m82, m326, m327, n44, n45
$C_7H_7NO_4S$: c16
$C_7H_7N_3$: a203, a204, m136
C_7H_8: b130, c311, t170
C_7H_8BrN: b307
C_7H_8ClN: c58, c143, c144, c145, c146, c147
C_7H_8ClNO: c138a, c139
$C_7H_8ClNO_2S$: c248
$C_7H_8Cl_2Si$: d198, m361
$C_7H_8N_2O$: a114, b72, p168
$C_7H_8N_2O_2$: d33, h166, m318, m319, m320
$C_7H_8N_2O_3$: m78, m79, m80
$C_7H_8N_2S$: p157
$C_7H_8N_4O_2$: t140
C_7H_8O: b78, c280, c281, c282, m48
C_7H_8OS: m432
$C_7H_8O_2$: d390, d391, h105, m87, m88, m89, m277
$C_7H_8O_2S$: t176
$C_7H_8O_3$: e139, f45, m307
$C_7H_8O_3S$: m127, t179
C_7H_8S: m367, p130, t150
C_7H_9ClSi: m360
C_7H_9N: b79, d606, d607, d608, d609, e214, e215, e216, m122, t184, t185, t186
C_7H_9NO: a221, b99, h126, m42, m43, m44
$C_7H_9NO_2$: d457
$C_7H_9NO_2S$: t177
$C_7H_9NO_3S$: a299
C_7H_9NS: m425, m426
$C_7H_9N_3O$: a133
C_7H_{10}: b131
$C_7H_{10}N_2$: a157, a177, a178, d478, m363, t171, t172, t173, t174
$C_7H_{10}N_2O$: m94
$C_7H_{10}N_2OS$: h130
$C_7H_{10}N_2O_2$: e176, m233
$C_7H_{10}N_2O_2S$: a213, t178
$C_7H_{10}O$: m61, m62, n108, t67
$C_7H_{10}O_2$: a40, c360
$C_7H_{10}O_3$: e12, h159, m336, t346
$C_7H_{10}O_4$: d552
$C_7H_{10}O_5$: d460
$C_7H_{10}Si$: m366
$C_7H_{11}Br$: b321
$C_7H_{11}BrO_4$: d286
$C_7H_{11}ClO$: c317
$C_7H_{11}ClO_4$: d290
$C_7H_{11}NO$: c341, h110
$C_3H_{11}NO_2$: a52
$C_7H_{11}NO_3$: m338
$C_7H_{11}NO_5$: a45
$C_7H_{11}NS$: c342
C_7H_{12}: c312, h22, m208, m209, n107
$C_7H_{12}O$: c310, c316, m205, m206, m207, m269
$C_7H_{12}O_2$: b419, c318, d356, e122
$C_7H_{12}O_3$: e171, e198
$C_7H_{12}O_4$: d317, d318, d554, d576, d577, d578, h8, m274, t128
$C_7H_{12}O_5$: g17, g18
$C_7H_{12}O_6Si$: m438
$C_7H_{12}O_7$: g3
$C_7H_{13}Br$: b262, b309
$C_7H_{13}BrO_2$: e81
$C_7H_{13}ClO$: h18
$C_7H_{13}N$: a253, d333, q5
$C_7H_{13}NO$: a322, c340
$C_7H_{13}NO_2$: a152
C_7H_{14}: c307, h19, m195
$C_7H_{14}ClN$: c114
$C_7H_{14}N_2$: d417
$C_7H_{14}N_2O$: a283
$C_7H_{14}N_2O_2$: e204
$C_7H_{14}O$: c309, c343, d571, d580, h5, h15, h16, h17, m198, m199, m200, m201, m202, m203, m204, m268
$C_7H_{14}O_2$: b482, c308, d258, e124, e173, e201, h10, i80, m75, m266, p52
$C_7H_{14}O_3$: i68
$C_7H_{14}O_6$: m257
$C_7H_{15}Br$: b293, b294
$C_7H_{15}Cl$: c129
$C_7H_{15}ClO_2$: c82
$C_7H_{15}Cl_3Si$: h21
$C_7H_{15}I$: i37
$C_7H_{15}N$: c325, d593, e205, e206, m210, m211, m212
$C_7H_{15}NO$: d471, e160, h125, m375, p188, p189
$C_7H_{15}NO_2$: p280
$C_7H_{15}NO_3$: c18, m455
$C_7H_{15}O_5P$: e120
C_7H_{16}: d572, d573, d574, d575, e200, h6, m265, t342
$C_7H_{16}BrNO_2$: a38
$C_7H_{16}ClNO_2$: a39
$C_7H_{16}N_2$: a218, m304, t372
$C_7H_{16}N_2O$: a282
$C_7H_{16}N_2O_2$: p185
$C_7H_{16}O$: d579, h12, h13, h14, m267, t343
$C_7H_{16}O_2$: d257, d331, m397
$C_7H_{16}O_2Si$: d256, e231
$C_7H_{16}O_3$: d702, t283, t326
$C_7H_{16}O_4$: t93
$C_7H_{16}S$: h9
$C_7H_{17}N$: h20, m269a
$C_7H_{17}NO$: d275
$C_7H_{17}NO_2$: b423, d274
$C_7H_{17}NO_5$: m256
$C_7H_{17}NO_6S$: t434
$C_7H_{17}NO_7S$: t433
$C_7H_{18}N_2$: d330, h7, i103, t120
$C_7H_{18}N_2O$: b174
$C_7H_{18}N_2O_2$: a281
$C_7H_{18}O_2Si$: b489

TABLE 1-13 Empirical formula index for organic compounds (*continued*)
The alphanumeric designations are keyed to Table 1-14

$C_7H_{18}O_3Si$: b488, t266a
$C_7H_{19}NOSi_2$: b210
$C_7H_{19}NSi$: d343, t378
$C_7H_{19}N_3$: d42, t426
$C_7H_{21}N_3Si$: t427

C_8

$C_8Br_4O_3$: t11
$C_8Cl_4O_3$: t32
C_8D_{10}: e68
$C_8HCl_4NO_2$: t33
$C_8H_3NO_5$: n72
$C_8H_4BrNO_2$: b298
$C_8H_4Cl_2O_2$: b14, b15, p174
$C_8H_4Cl_2O_4$: d217
$C_8H_4Cl_6$: b203
$C_8H_4F_3N$: t303
$C_8H_4F_6$: b207
$C_8H_4N_2$: d236, d237
$C_8H_4O_3$: p171
$C_8H_5Br_5$: p5
$C_8H_5ClO_4$: c210
$C_8H_5Cl_3O_3$: t246
$C_8H_5F_3O$: t294
$C_8H_5F_3O_2S$: t139
$C_8H_5F_6N$: b206
C_8H_5NO: b68
$C_8H_5NO_2$: i21, p173
$C_8H_5NO_3$: h167, i58
$C_8H_5NO_6$: n31, n32, n33, n34
C_8H_6: p82
C_8H_6BrClO: b248
C_8H_6BrN: b332
$C_8H_6Br_2O$: d64
$C_8H_6Br_4$: t12, t13
$C_8H_6ClF_3$: t304
C_8H_6ClN: c202
$C_8H_6ClNO_3$: c172
$C_8H_6Cl_2O$: d140
$C_8H_6Cl_2O_3$: d214
$C_8H_6Cl_4$: t36
$C_8H_6N_2$: q4
$C_8H_6N_2O_2$: a271, n65
$C_8H_6N_2O_6$: d638, m229
C_8H_6O: b42
$C_8H_6O_2$: b13, p172
$C_8H_6O_3$: b70, c14, f33, m240
$C_8H_6O_4$: b16, b17, m242, p170
C_8H_6S: b61
C_8H_7Br: b352
C_8H_7BrO: b222, b223
$C_8H_7BrO_2$: b330, b331
C_8H_7ClO: c28, c29, c30, p81, t191, t192, t193
C_8H_7ClOS: b91
$C_8H_7ClO_2$: b90, c201, m53, p69
$C_8H_7ClO_3$: c86, c198, m185, m186
$C_8H_7ClO_4$: c134
C_8H_7FO: f8
C_8H_7N: i18, p80, t188, t189, t190
C_8H_7NO: m9, m137, t194
$C_8H_7NO_2$: h134, n84
$C_8H_7NO_3$: n22, n23
$C_8H_7NO_3S$: t182
$C_8H_7NO_4$: a116a, m321, m322, m323, m324, n62, n63, n64
$C_8H_7NO_5$: m83
C_8H_7NS: b122, m135
$C_8H_7N_3O_2$: a153
C_8H_8: s11
C_8H_8BrNO: b220
$C_8H_8Br_2$: d79, d104, d105
C_8H_8ClNO: c23
$C_8H_8ClNO_3S$: a10
$C_8H_8Cl_2$: d233, d234
$C_8H_8Cl_2Si$: p169
$C_8H_8HgO_2$: p127
$C_8H_8N_2$: a263, m128
$C_8H_8N_2OS$: a210
C_8H_8O: a31, e9, m126, p76a
C_8H_8OS: m427, p156
$C_8H_8O_2$: b41, b97, h90, h91, h92, m45, m46, m129, m130, m131, m132, p78, p79
$C_8H_8O_2S$: t163
$C_8H_8O_3$: d371, d381, h131, h132, h138, h139, h161, m8, m50, m51, m52, m243, m278, m413, p68, t76
$C_8H_8O_4$: d21, h133
$C_8H_8O_4S$: a33
C_8H_9Br: b284, b285, b371, b372, b373, b374
C_8H_9BrO: b272, b288
$C_8H_9BrO_2$: b270
C_8H_9Cl: c108, c109, c259, c260, c261, c262
C_8H_9ClO: c91
C_8H_9N: b101, c361, i22, m450
C_8H_9NO: a18, a108, a109, a110, b96, m250
$C_8H_9NO_2$: a15, a16, a17, a214, a215, b88, d558, d559, d560, d561, e190, e217, e218, e219, m47, m116, m117, p117, t77
$C_8H_9NO_3$: a209, h164, h165, m85, n59
$C_8H_9NO_4$: d445
C_8H_{10}: e69, m245, x4, x5, x6
$C_8H_{10}N_2O$: d562
$C_8H_{10}N_4O_2$: c1, d240
$C_8H_{10}O$: b132, d581, d582, d583, d584, d585, d586, e29, e202, m105, m106, m107, m138, m139, m140, p114, p115
$C_8H_{10}O_2$: b18, d432, d433, d434, m54, p72, p113
$C_8H_{10}O_3$: c321, d447, h136, h160
$C_8H_{10}O_3S$: m437
$C_8H_{10}O_4$: d263
$C_8H_{10}S$: b106
$C_8H_{11}ClSi$: d587
$C_8H_{11}N$: b104, d479, d480, d481, d482, d483, d484, d485, e64, e65, e66, e183, e184, m141, m142, p116, t373

TABLE 1-13 Empirical formula index for organic compounds (*continued*)
The alphanumeric designations are keyed to Table 1-14

$C_8H_{11}NO$: a173, a259, a264, a265, a305, d472, e25, h117, m55, m71, m72, m73, p270
$C_8H_{11}NO_2$: d428, d429, d430
$C_8H_{11}NO_2S$: m436
$C_8H_{11}NO_3$: e135
$C_8H_{11}NO_3S$: d465
$C_8H_{11}N_5$: p94
C_8H_{12}: c346, v6
$C_8H_{12}N_2$: d239, d588, t121, x9
$C_8H_{12}N_2O_2$: d411
$C_8H_{12}N_2O_3$: d280
$C_8H_{12}N_4$: a328
$C_8H_{12}O$: e237
$C_8H_{12}O_2$: d510, e222, h186, n111
$C_8H_{12}O_3$: e197
$C_8H_{12}O_4$: d305, d316
$C_8H_{12}O_6Si$: t199
$C_8H_{12}Si$: d589
$C_8H_{13}N$: e238
C_8H_{14}: c350, d534, o17, o44, v5
$C_8H_{14}N_2$: p191
$C_8H_{14}O$: c349, d512, e7a, m263, o45
$C_8H_{14}O_2$: b462, c334, c364, d539, i69, m196
$C_8H_{14}O_3$: b418, b499, d714, e91
$C_8H_{14}O_4$: b450, d320, d335, d536a, e152, o24
$C_8H_{14}O_4S$: d621
$C_8H_{14}O_4S_2$: d710
$C_8H_{14}O_6$: d339, d340
$C_8H_{14}O_6Si$: t198
$C_8H_{15}ClO$: e148, o37
$C_8H_{15}N$: o27
$C_8H_{15}NO$: d368
$C_8H_{15}NO_2$: d470, e207, e208, e209
C_8H_{16}: c347, d508, d509, e109, o39, t365
$C_8H_{16}ClN$: c228
$C_8H_{16}O$: c348, d511, e110, e111, o34, o35, o36, o40
$C_8H_{16}O_2$: b433, c322, e145, e146, h79, i67, m262, o29, p237
$C_8H_{16}O_4$: e37, t127
$C_8H_{17}Br$: b323
$C_8H_{17}Cl$: c191
$C_8H_{17}Cl_3Si$: o43
$C_8H_{17}I$: i46
$C_8H_{17}N$: c351, d513
$C_8H_{17}NO_2$: p192
$C_8H_{17}NO_3S$: c336
$C_8H_{17}O_5P$: t288
C_8H_{18}: d535, e143, e177, e178, m261, o22, t102, t361, t362, t363
$C_8H_{18}ClNO_2$: a49
$C_8H_{18}Cl_2O_2Si_3$: d186
$C_8H_{18}Cl_2Si$: d184
$C_8H_{18}Cl_2Sn$: d136a
$C_8H_{18}F_3NOSi_2$: b213
$C_8H_{18}N_2$: c315
$C_8H_{18}N_2O$: m225, m370
$C_8H_{18}N_2O_4S$: h124
$C_8H_{18}O$: d115, d408, e147, o30, o31, o32, o33
$C_8H_{18}OSi_2$: d715
$C_8H_{18}OSn$: d137
$C_8H_{18}O_2$: d122, d537, e144, o25, o26, t364
$C_8H_{18}O_2S$: d135
$C_8H_{18}O_3$: b177, b414, d700, t282
$C_8H_{18}O_3S$: d134
$C_8H_{18}O_3Si$: t267
$C_8H_{18}O_4$: b190
$C_8H_{18}O_4S$: d131
$C_8H_{18}O_5$: t52
$C_8H_{18}S$: d132, d133, o28
$C_8H_{18}S_2$: b154, b155, d113, d114
$C_8H_{18}Si_2$: b209
$C_8H_{19}N$: d107, d407, d419, d538, e150, o41, t104
$C_8H_{19}NO$: d413
$C_8H_{19}NO_2$: b447, d247, d248
$C_8H_{19}NO_5$: b184
$C_8H_{19}O_3P$: d127
$C_8H_{20}BrN$: t49
$C_8H_{20}ClN$: t50
$C_8H_{20}Ge$: t58
$C_8H_{20}N_2$: d536, o23, t103, t278
$C_8H_{20}O_3SSi$: m19
$C_8H_{20}O_3Si$: t266
$C_8H_{20}O_4Si$: t48
$C_8H_{20}O_5P_2$: t61
$C_8H_{20}O_7P_2$: t60
$C_8H_{20}Pb$: t59
$C_8H_{20}Si$: t62
$C_8H_{20}Sn$: t64
$C_8H_{21}NO$: t51
$C_8H_{21}NOSi_2$: b208
$C_8H_{21}NO_2Si$: a280
$C_8H_{22}N_2O_3Si$: a167a, t329
$C_8H_{22}N_4$: b146
$C_8H_{22}O_2Si_2$: b212
$C_8H_{23}N_5$: t56
$C_8H_{24}Cl_2O_3Si_4$: d207
$C_8H_{24}O_2Si_3$: o21
$C_8H_{24}O_4Si_4$: o20
$C_8H_{28}N_4Si_4$: o19

C_9

$C_9F_{15}N_3$: t439
$C_9H_2Cl_6O_3$: h30
$C_9H_3Cl_3O_3$: b32
$C_9H_4O_5$: b31, c15
$C_9H_5BrClNO$: b257
$C_9H_5Br_2NO$: d87
C_9H_5ClINO: c133
$C_9H_5Cl_2N$: d225
C_9H_6BrN: b350
C_9H_6ClN: c237
C_9H_6ClNO: c135
$C_9H_6N_2O_2$: n77, t175
$C_9H_6O_2$: b56, c277
$C_9H_6O_3$: h108, h109
$C_9H_6O_4$: i16
$C_9H_6O_6$: b28, b29, b30
C_9H_7BrO: b261
C_9H_7ClO: c269
$C_9H_7ClO_2$: c76

TABLE 1-13 Empirical formula index for organic compounds (*continued*)

The alphanumeric designations are keyed to Table 1-14

$C_9H_7Cl_3O_3$: t247
C_9H_7N: i110, q3
C_9H_7NO: h179, i20
$C_9H_7NO_3$: h144, m286
$C_9H_7NO_4S$: h180
$C_9H_7N_3O_4S_2$: a250
C_9H_8: i17
$C_9H_8Cl_2O_2$: n109
$C_9H_8N_2$: m412
$C_9H_8N_2O_5$: n43
C_9H_8O: c267, i15
$C_9H_8O_2$: c268, d354
$C_9H_8O_3$: h107
$C_9H_8O_4$: a56, p126
C_9H_9BrO: b345
C_9H_9Cl: c218
C_9H_9ClO: c224
$C_9H_9ClO_3$: c200, c250
C_9H_9N: d488, m284, m285
C_9H_9NO: m93
$C_9H_9NO_2$: a9
$C_9H_9NO_2S$: t195
$C_9H_9NO_3$: a11, a12, b71
$C_9H_9N_3O$: a268
C_9H_{10}: a84, i13, m414, v3
$C_9H_{10}F_3NO_2$: m123
$C_9H_{10}N_2$: a306, p121a
$C_9H_{10}N_2O$: p150
$C_9H_{10}N_2O_2$: p83
$C_9H_{10}N_2O_3$: a132
$C_9H_{10}O$: a98, a99, c270, d361, i14, m113, p146, p147, p212, p220
$C_9H_{10}O_2$: b77, d487, e9a, e26, e27, e70, h171, h172, m39, m40, m41, m359, p74, p148
$C_9H_{10}O_2S$: b121
$C_9H_{10}O_3$: d431, e30, e31, e40, e49, e154, e221, m91, m280, m293, p75
$C_9H_{10}O_4$: d435, m289, m448
$C_9H_{10}O_8$: c355
$C_9H_{11}Br$: b300, b334, b364, b365
$C_9H_{11}BrO$: b346
$C_9H_{11}ClO_3S$: c115
$C_7H_{11}Cl_3Si$: c227, m357
$C_9H_{11}N$: a83, a201, a202, c333, t73, t82
$C_9H_{11}NO$: d464, m358, m435
$C_9H_{11}NO_2$: d466, e28, e60, e61, p84
$C_9H_{11}NO_3$: t444
C_9H_{12}: e161, i91, n91, p225, t338, t339, t340, v8
$C_9H_{12}Cl_2Si$: m356
$C_9H_{12}N_2O_4$: a249
$C_9H_{12}N_2O_6$: u14
$C_9H_{12}O$: b95, d549, d550, i106, i107, p144, p145, p238, t366, t367, t368
$C_9H_{12}O_2$: b111, e32, i85, n110, p73, p142, t354
$C_9H_{12}O_3$: m197, t320
$C_9H_{12}O_3S$: e225
$C_9H_{12}S$: p143
$C_9H_{13}N$: b483, d489, d624, e73, e172, e226, e227, i90, t335
$C_9H_{13}NO$: a267, b80, m86, n112
$C_9H_{13}NO_2$: a266
$C_9H_{13}N_3O_2$: t445
$C_9H_{14}BrN$: p162
$C_9H_{14}Br_3N$: p165
$C_9H_{14}ClN$: p163
$C_9H_{14}IN$: p164
$C_9H_{14}N_2$: n94
$C_9H_{14}O$: d529, d531, i82, t345
$C_9H_{14}OSi$: t382
$C_9H_{14}O_2Si$: d444
$C_9H_{14}O_3$: b193
$C_9H_{14}O_3Si$: p161
$C_9H_{14}O_5$: d262, d321
$C_9H_{14}O_6$: p204
$C_9H_{14}Si$: p166
$C_9H_{15}NO$: c362
$C_9H_{15}NO_2$: d570
$C_9H_{15}NO_5$: d261
$C_9H_{15}NSi$: t375
C_9H_{16}: h46
$C_9H_{16}Cl_2Si$: c333a
$C_9H_{16}N_2$: d46
$C_9H_{16}O$: d530
$C_9H_{16}O_2$: c326
$C_9H_{16}O_3$: b470
$C_9H_{16}O_4$: d303, d307, d322, d532, n95
$C_9H_{17}ClO$: n101
$C_9H_{17}N$: a88, n97
$C_9H_{17}NO$: m181
$C_9H_{17}NO_2$: e180, e181
C_9H_{18}: i94, n102, p227, t344
$C_9H_{18}NO$: t119
$C_9H_{18}N_2O_3Si$: t330
$C_9H_{18}O$: d533, n100, n103
$C_9H_{18}O_2$: e141, m332, n98
$C_9H_{18}O_3$: d111
$C_9H_{19}Br$: b320
$C_9H_{19}N$: i95, t337
$C_9H_{19}NO$: d116
$C_9H_{19}NO_2$: e121
$C_9H_{19}NO_3S$: c337
C_9H_{20}: n92, t352
$C_9H_{20}Cl_2Si$: m333
$C_9H_{20}N_2$: a301
$C_9H_{20}N_2S$: d136
$C_9H_{20}O$: n99, t353
$C_9H_{20}O_2$: b453, n96
$C_9H_{20}O_3$: d701, t284
$C_9H_{20}O_3Si$: a103
$C_9H_{20}O_4$: t415
$C_9H_{20}O_5$: t55
$C_9H_{21}BO_3$: t413
$C_9H_{21}ClO_3Si$: c231
$C_9H_{21}ClSi$: c258
$C_9H_{21}N$: n104, t414
$C_9H_{21}NO_3$: t314
$C_9H_{21}N_3$: t280
$C_9H_{21}O_3B$: t315
$C_9H_{21}O_3P$: t318
$C_9H_{22}N_2$: d327, n93
$C_9H_{22}O_3Si$: p241
$C_9H_{23}NO_3Si$: a284
$C_9H_{24}N_4$: b148

C_{10}

$C_{10}H_2O_6$: b27
$C_{10}H_4Cl_2O_2$: d201

TABLE 1-13 Empirical formula index for organic compounds (*continued*)
The alphanumeric designations are keyed to Table 1-14

$C_{10}H_6N_2$: b100
$C_{10}H_6N_2O_4$: d636
$C_{10}H_6N_2O_4S$: d48
$C_{10}H_6O_2$: n11
$C_{10}H_6O_3$: h153
$C_{10}H_6O_8$: b26
$C_{10}H_7Br$: b315
$C_{10}H_7BrO$: b316
$C_{10}H_7Cl$: c169, c170
$C_{10}H_7NO_2$: n57, n81, p125
$C_{10}H_7NO_8S_2$: n82
$C_{10}H_8$: a331, n2
$C_{10}H_8BrNO_2$: b289
$C_{10}H_8N_2$: d707
$C_{10}H_8O$: n9, n10
$C_{10}H_8O_2$: d393, d394, d395, d396, m192
$C_{10}H_8O_3$: h141
$C_{10}H_8O_3S$: n18
$C_{10}H_8O_7S_2$: h151, h152
$C_{10}H_8O_8S_2$: d397
$C_{10}H_9ClCrN_2O_3$: b144
$C_{10}H_9N$: m410, m411, n17
$C_{10}H_9NO$: a51, a239
$C_{10}H_9NO_2$: i19
$C_{10}H_9NO_3S$: a237
$C_{10}H_9NO_4S$: a195, a196, a197, a198
$C_{10}H_9NO_6$: d563
$C_{10}H_9NO_6S_2$: a235, a236
$C_{10}H_9N_3$: d708
$C_{10}H_{10}ClFO$: c121
$C_{10}H_{10}ClNO_2$: c26
$C_{10}H_{10}N_2$: a290, n4, n5
$C_{10}H_{10}N_2O$: m365
$C_{10}H_{10}O$: d363, m191, p98, p100
$C_{10}H_{10}O_2$: b64, s2
$C_{10}H_{10}O_3$: b73, m60
$C_{10}H_{10}O_4$: d590, d591, d592, h137, p155
$C_{10}H_{11}BrO$: b314a
$C_{10}H_{11}ClO_3$: c199
$C_{10}H_{11}ClO_4$: t322
$C_{10}H_{11}IO_4$: i28
$C_{10}H_{11}N$: p103
$C_{10}H_{11}NO_2$: a32, d449
$C_{10}H_{11}NO_4$: c10
$C_{10}H_{11}NO_6$: m227
$C_{10}H_{12}$: d244, t75
$C_{10}H_{12}NO$: b411
$C_{10}H_{12}N_2$: a170, b81, b103
$C_{10}H_{12}N_2O_2$: p77
$C_{10}H_{12}O$: a94, b503, e56, i77, m97, m389, m390, m394, p96, p97
$C_{10}H_{12}O_2$: e203, h112, h163, m70, m92, m98, m99, m100, p101, p102, p226
$C_{10}H_{12}O_3$: d427, e42, e166, m303, p71, p233
$C_{10}H_{12}O_4$: d448, m226, t319
$C_{10}H_{12}O_5$: d306, p242, t321
$C_{10}H_{12}O_6$: d516
$C_{10}H_{13}Br$: b301
$C_{10}H_{13}BrO$: b245
$C_{10}H_{13}Cl$: b437
$C_{10}H_{13}NO$: p131
$C_{10}H_{13}NO_2$: e46
$C_{10}H_{13}NO_2S$: b92
$C_{10}H_{13}N_5O_4$: a70
$C_{10}H_{14}$: b426, b427, b428, d282, d283, d284, i64, i100, i101, i102, t99, t100, t101
$C_{10}H_{14}NO_5PS$: p3
$C_{10}H_{14}N_2$: n20, p141
$C_{10}H_{14}N_2O$: d323, d334
$C_{10}H_{14}N_4O_4$: d400
$C_{10}H_{14}N_5O_7P$: a72
$C_{10}H_{14}O$: b472, b473, b474, b475, b476, b480, c20, c363, i92, i102a, p58, t118, t259
$C_{10}H_{14}O_2$: b435, b457, d451
$C_{10}H_{14}O_3$: c6, c9
$C_{10}H_{14}O_4$: m90, t323
$C_{10}H_{15}BrO$: b247
$C_{10}H_{15}N$: b424, d277, d278, d566, e236, i93, p99, t98
$C_{10}H_{15}NO$: d273, e1, e2
$C_{10}H_{15}NO_2$: d452, p108
$C_{10}H_{15}N_5O_{10}P_2$: a71
$C_{10}H_{16}$: a67, c2, d651, L6, L7, m456, p25, p178, p179, t5, t6, t258
$C_{10}H_{16}ClN$: b127
$C_{10}H_{16}Cl_2O_2$: d11
$C_{10}H_{16}N_2O_8$: e128
$C_{10}H_{16}O$: c3, c4, d353, d564, d565, L8, p180, p181, p246, t357
$C_{10}H_{16}OSi$: d521
$C_{10}H_{16}O_4$: c5, d266
$C_{10}H_{16}O_4S$: c7
$C_{10}H_{16}O_5$: d265, d301
$C_{10}H_{16}Si$: b128
$C_{10}H_{17}N$: a66, p278
$C_{10}H_{17}NO$: c344, m453
$C_{10}H_{18}$: d1, d2, p177
$C_{10}H_{18}N_2O_7$: h119
$C_{10}H_{18}O$: b217, b444, b445, c266, d3, g2, i60, i83, i109, L9, m13, p193, t7, t356
$C_{10}H_{18}O_2$: e112
$C_{10}H_{18}O_3$: d601, t72
$C_{10}H_{18}O_4$: b176, d9, d121, d332, d567
$C_{10}H_{18}O_4S$: d341
$C_{10}H_{18}O_6$: d422
$C_{10}H_{19}ClO$: d17
$C_{10}H_{19}N$: d12, t336
$C_{10}H_{19}NO_2$: e210
$C_{10}H_{20}$: c302, d18
$C_{10}H_{20}Br_2$: d73
$C_{10}H_{20}N_2S_4$: t63
$C_{10}H_{20}O$: b442, b443, c275, d6, d16, d358, d367, e151, m12, m306
$C_{10}H_{20}O_2$: d14, e149, e193, m63, m171
$C_{10}H_{20}O_5$: p45
$C_{10}H_{20}O_5Si$: t331
$C_{10}H_{21}Br$: b267
$C_{10}H_{21}Cl$: c80
$C_{10}H_{21}I$: i33
$C_{10}H_{21}N$: d294
$C_{10}H_{21}NO$: a233

TABLE 1-13 Empirical formula index for organic compounds (*continued*)
The alphanumeric designations are keyed to Table 1-14

$C_{10}H_{22}$: d7
$C_{10}H_{22}N_2$: d41
$C_{10}H_{22}O$: d15, d653, t74
$C_{10}H_{22}O_2$: d10, d106
$C_{10}H_{22}O_3$: d699, t420
$C_{10}H_{22}O_3S$: d13
$C_{10}H_{22}O_4$: t419
$C_{10}H_{22}O_5$: b191
$C_{10}H_{22}O_7$: d650
$C_{10}H_{23}N$: d19, d652
$C_{10}H_{23}NO$: d108
$C_{10}H_{23}NO_2$: d259
$C_{10}H_{24}N_2$: d8, t57, t113
$C_{10}H_{24}N_2O_2$: d647
$C_{10}H_{24}N_4$: b147
$C_{10}H_{24}OSi$: m109
$C_{10}H_{24}O_3Si$: m444
$C_{10}H_{24}O_6Si$: t436
$C_{10}H_{27}O_3N_3Si$: t328
$C_{10}H_{30}O_3Si_4$: d5
$C_{10}H_{30}O_5Si_5$: d4

C_{11}

$C_{11}H_4F_{20}O$: i1
$C_{11}H_7N$: c294
$C_{11}H_8O$: n1
$C_{11}H_8O_2$: h148, m313, n3
$C_{11}H_8O_3$: h149, h150
$C_{11}H_9Br$: b312
$C_{11}H_9Cl$: c158
$C_{11}H_9N$: p151
$C_{11}H_{10}$: m311, m312
$C_{11}H_{10}N_2S$: n19
$C_{11}H_{10}O$: m76, m77
$C_{11}H_{11}N$: n6
$C_{11}H_{12}N_2O$: a314
$C_{11}H_{12}N_2O_2$: t443
$C_{11}H_{12}O_2$: d359, e104, m104
$C_{11}H_{12}O_3$: e71
$C_{11}H_{13}ClO$: b431
$C_{11}H_{13}ClO_3$: c251
$C_{11}H_{13}NO$: b120
$C_{11}H_{13}NO_2$: t187
$C_{11}H_{13}NO_3$: a307, a308
$C_{11}H_{13}N_3O$: a113
$C_{11}H_{13}N_3O_3S$: d569
$C_{11}H_{14}O$: m103, p43
$C_{11}H_{14}O_2$: b429, b430, d456, e47
$C_{11}H_{14}O_3$: b412, b471, b479, e170
$C_{11}H_{14}O_4$: e158
$C_{11}H_{14}O_4Si$: d24
$C_{11}H_{15}NO$: d269
$C_{11}H_{15}NO_2$: d276, d468, e123
$C_{11}H_{16}$: b485, p24, p54
$C_{11}H_{16}N_2$: b115
$C_{11}H_{16}O$: b86, b464, b465, p56
$C_{11}H_{16}O_2$: a68
$C_{11}H_{16}O_3$: m302
$C_{11}H_{16}O_4$: d716
$C_{11}H_{17}N$: b432, e163
$C_{11}H_{17}NO$: e228
$C_{11}H_{17}NO_2$: b105
$C_{11}H_{17}O_3P$: b93
$C_{11}H_{18}O$: d308, n105, p4
$C_{11}H_{18}O_5$: d264
$C_{11}H_{19}ClO$: u11
$C_{11}H_{19}N$: a212
$C_{11}H_{20}O$: p55, u7
$C_{11}H_{20}O_2$: u9
$C_{11}H_{20}O_4$: d119, d287, d309
$C_{11}H_{21}BrO_2$: b370
$C_{11}H_{22}$: u8
$C_{11}H_{22}N_2$: d697
$C_{11}H_{22}O$: u1, u5, u6, u10
$C_{11}H_{22}O_2$: m219, u3
$C_{11}H_{22}O_4Si$: e7
$C_{11}H_{23}NO_2$: a302
$C_{11}H_{24}$: u2
$C_{11}H_{24}O$: d310, u4
$C_{11}H_{24}O_3Si$: t316
$C_{11}H_{24}O_4$: t417
$C_{11}H_{24}O_6$: p46
$C_{11}H_{24}O_6Si$: t437
$C_{11}H_{26}N_2$: d129
$C_{11}H_{26}N_2O_6$: b215

C_{12}

$C_{12}Br_{10}O$: b198
$C_{12}H_4Cl_6S_2$: b204
$C_{12}H_5ClO_3$: c171
$C_{12}H_6Br_4O_4S$: s26
$C_{12}H_6O_3$: n7
$C_{12}H_6O_{12}$: b19
$C_{12}H_7NO_2$: n8
$C_{12}H_8$: a3
$C_{12}H_8Br_2$: d66
$C_{12}H_8Cl_2OS$: b167
$C_{12}H_8Cl_2O_2S$: b166
$C_{12}H_8N_2$: p63
$C_{12}H_8N_2O_2$: a238
$C_{12}H_8N_2O_4S_2$: b195, b196
$C_{12}H_8O$: d50
$C_{12}H_8O_6$: b133
$C_{12}H_8S$: d52
$C_{12}H_9Br$: b239
$C_{12}H_9BrO$: b333
$C_{12}H_9ClO_2S$: c208
$C_{12}H_9N$: c8, d667, n16
$C_{12}H_9NO$: b74, b75, b76
$C_{12}H_9NO_2$: n48, n49
$C_{12}H_9NO_3$: n70, n71
$C_{12}H_9NS$: p66
$C_{12}H_{10}$: a2, b135
$C_{12}H_{10}ClN$: c61, c62
$C_{12}H_{10}ClO_3P$: d664
$C_{12}H_{10}ClP$: c100
$C_{12}H_{10}Cl_2Si$: d175
$C_{12}H_{10}Hg$: d677
$C_{12}H_{10}N_2$: a327
$C_{12}H_{10}N_2O$: n80, p90
$C_{12}H_{10}N_2O_2$: n52
$C_{12}H_{10}N_2O_2S$: a246
$C_{12}H_{10}N_3O_3P$: d684
$C_{12}H_{10}O$: d669, m314, m315, p133, p134
$C_{12}H_{10}OS$: d692
$C_{12}H_{10}O_2$: d388, h88, n14, n15
$C_{12}H_{10}O_2S$: d691, t153
$C_{12}H_{10}O_3$: n12
$C_{12}H_{10}O_3S$: b140
$C_{12}H_{10}O_4$: q1
$C_{12}H_{10}O_4S$: s28, t147
$C_{12}H_{10}S$: d690
$C_{12}H_{10}S_2$: d666
$C_{12}H_{10}Se_2$: d665
$C_{12}H_{11}ClNO_2P$: p136

TABLE 1-13 Empirical formula index for organic compounds (*continued*)
The alphanumeric designations are keyed to Table 1-14

$C_{12}H_{11}N$: a134, a135, b118, b119, d657
$C_{12}H_{11}NO$: n13, p70
$C_{12}H_{11}N_3$: p88
$C_{12}H_{11}O_3P$: d683
$C_{12}H_{12}$: d556, d557
$C_{12}H_{12}N_2$: b137, d675, p135
$C_{12}H_{12}N_2O$: o62
$C_{12}H_{12}N_2O_2$: b40
$C_{12}H_{12}N_2O_2S$: d36, d37
$C_{12}H_{12}N_4$: d31
$C_{12}H_{12}O$: e45
$C_{12}H_{12}O_2Si$: d689
$C_{12}H_{12}O_3$: t200
$C_{12}H_{12}O_6$: t197, t341
$C_{12}H_{13}N_3$: d34
$C_{12}H_{14}N_2O_3S$: a167
$C_{12}H_{14}N_4O_2S$: s22
$C_{12}H_{14}O_3$: e179
$C_{12}H_{14}O_4$: d329
$C_{12}H_{15}N$: d370
$C_{12}H_{15}NO$: b117
$C_{12}H_{15}N_3O_3$: t201
$C_{12}H_{16}$: c339, m213, p106
$C_{12}H_{16}O_2$: m364
$C_{12}H_{16}O_3$: d246
$C_{12}H_{17}N$: b116, c338
$C_{12}H_{17}NO$: d319, d342
$C_{12}H_{18}$: b492, c305, d415, d416, h47, p119, t442
$C_{12}H_{18}Cl_2N_4OS$: t141
$C_{12}H_{18}O$: d420, d515
$C_{12}H_{18}O_2$: b477, b478
$C_{12}H_{18}O_4$: b448
$C_{12}H_{19}N$: d414, h81
$C_{12}H_{20}O_2$: b186, b218, e103, L10
$C_{12}H_{20}O_3Si$: p160
$C_{12}H_{20}O_4$: d118
$C_{12}H_{20}O_4Si$: t9
$C_{12}H_{21}N$: t438
$C_{12}H_{21}N_3$: t429
$C_{12}H_{22}$: c306, d241
$C_{12}H_{22}O$: c304, e4
$C_{12}H_{22}O_3$: h67
$C_{12}H_{22}O_4$: d130, d324, d514, d704, d723
$C_{12}H_{22}O_{11}$: L3, m7, s21
$C_{12}H_{23}ClO$: d730
$C_{12}H_{23}N$: d242, d726
$C_{12}H_{23}NO$: a323
$C_{12}H_{24}$: d731
$C_{12}H_{24}N_2$: d696
$C_{12}H_{24}O$: c303, d733, m446, t355
$C_{12}H_{24}O_2$: d728, e114
$C_{12}H_{24}O_6$: h74
$C_{12}H_{25}Br$: b277
$C_{12}H_{25}Cl$: c101
$C_{12}H_{25}Cl_3Si$: d738
$C_{12}H_{26}$: d721
$C_{12}H_{26}O$: d351, d729, t354a
$C_{12}H_{26}O_2$: d724, d725
$C_{12}H_{26}O_3$: b152
$C_{12}H_{26}O_4$: t418
$C_{12}H_{26}O_4S$: d737
$C_{12}H_{26}S$: d727
$C_{12}H_{27}Al$: t312
$C_{12}H_{27}BO_3$: t213
$C_{12}H_{27}ClSn$: t219
$C_{12}H_{27}N$: d350, d734, t214
$C_{12}H_{27}O_3P$: t218
$C_{12}H_{27}O_4P$: t216
$C_{12}H_{27}P$: t217
$C_{12}H_{28}BrN$: t137
$C_{12}H_{28}N_2$: d722
$C_{12}H_{28}O_4Si$: t90, t136
$C_{12}H_{28}O_4Ti$: t169
$C_{12}H_{28}O_8Si$: t91
$C_{12}H_{36}O_4Si_4Ti$: t92

C_{13}

$C_{13}H_5N_3O_7$: t391
$C_{13}H_8ClNO_3$: c182
$C_{13}H_8ClNOS$: p67
$C_{13}H_8Cl_2O$: d159
$C_{13}H_8N_2O_7$: b194
$C_{13}H_8O$: f3
$C_{13}H_8OS$: t168
$C_{13}H_8O_2$: x3
$C_{13}H_9BrO$: b234
$C_{13}H_9ClO$: c49, c50
$C_{13}H_9ClO_2$: c132
$C_{13}H_9N$: a61
$C_{13}H_{10}$: f2
$C_{13}H_{10}ClNO$: a142, a143, d661
$C_{13}H_{10}Cl_2O_2$: m234
$C_{13}H_{10}N_2$: p91
$C_{13}H_{10}N_2O_3$: a242
$C_{13}H_{10}O$: b53, x1
$C_{13}H_{10}O_2$: b136, h102, p92
$C_{13}H_{10}O_3$: d387, d663, p153
$C_{13}H_{10}O_5$: t85
$C_{13}H_{11}Br$: b276
$C_{13}H_{11}Cl$: c98
$C_{13}H_{11}ClO$: c44
$C_{13}H_{11}NO$: a127, b5
$C_{13}H_{11}NO_2$: h162, p85
$C_{13}H_{11}NO_3$: p86
$C_{13}H_{12}$: d678
$C_{13}H_{12}N_2$: b54, d38, d672
$C_{13}H_{12}N_2O$: d695
$C_{13}H_{12}N_2S$: d694, t148
$C_{13}H_{12}N_4O$: p89
$C_{13}H_{12}N_4S$: d693
$C_{13}H_{12}O$: b139, d679, h113, m56, p76
$C_{13}H_{12}S$: b114
$C_{13}H_{13}ClSi$: c99
$C_{13}H_{13}N$: d680, m231, m93
$C_{13}H_{13}NO$: b107
$C_{13}H_{13}N_3$: d673
$C_{13}H_{14}N_2$: d35, m239, t349
$C_{13}H_{14}N_2O_3$: a59
$C_{13}H_{14}N_4O$: d662
$C_{13}H_{14}Si$: m232
$C_{13}H_{16}O_2$: m74
$C_{13}H_{16}O_3$: e72
$C_{13}H_{16}O_4$: d328
$C_{13}H_{17}NO_2$: e75
$C_{13}H_{20}$: p118
$C_{13}H_{20}N_2O_2$: d271
$C_{13}H_{20}O$: i56, i57
$C_{13}H_{22}ClN$: b126
$C_{13}H_{22}N_2$: d243
$C_{13}H_{22}O_2$: n106
$C_{13}H_{22}O_3Si$: b125
$C_{13}H_{26}$: t263
$C_{13}H_{26}N_2$: m237, t348

TABLE 1-13 Empirical formula index for organic compounds (*continued*)
The alphanumeric designations are keyed to Table 1-14

$C_{13}H_{26}O_2$: e232, t262
$C_{13}H_{27}Br$: b362
$C_{13}H_{28}$: t261
$C_{13}H_{28}O_4$: t416
$C_{13}H_{29}NO_4$: b170

C_{14}

$C_{14}H_6Cl_2O_2$: d148, d149
$C_{14}H_7ClO_2$: c36, c37
$C_{14}H_8ClNO_5$: c186
$C_{14}H_8O_2$: a310, p62
$C_{14}H_8O_3$: h93
$C_{14}H_8O_4$: d372, d373, d374, d375
$C_{14}H_8O_5S$: a313
$C_{14}H_8O_8$: a311, d312
$C_{14}H_9Br$: b327
$C_{14}H_9ClO_3$: c54
$C_{14}H_9Cl_5$: b168
$C_{14}H_9NO_2$: a111, a112
$C_{14}H_9NO_3$: a188
$C_{14}H_{10}$: a309, d656, p61
$C_{14}H_{10}Br_2O$: b275
$C_{14}H_{10}ClNO_3$: a147
$C_{14}H_{10}Cl_2O_4$: b163
$C_{14}H_{10}Cl_4$: b164
$C_{14}H_{10}N_2O_2$: d28, d29, d30
$C_{14}H_{10}O_2$: b34
$C_{14}H_{10}O_3$: b45, b65, x2
$C_{14}H_{10}O_4$: b138, d54, t84
$C_{14}H_{11}N$: d655, p122
$C_{14}H_{11}NOS$: a50
$C_{14}H_{12}$: d352, s9
$C_{14}H_{12}Cl_2O$: b165
$C_{14}H_{12}N_2O$: b37
$C_{14}H_{12}N_2O_2$: b35
$C_{14}H_{12}O$: a34, d22, m133, m134
$C_{14}H_{12}O_2$: b46, b83, b84, b108, b109, d654
$C_{14}H_{12}O_3$: b36, h35
$C_{14}H_{13}ClO$: c148
$C_{14}H_{13}N$: e94, i12
$C_{14}H_{13}NO$: b82
$C_{14}H_{13}NO_2$: b50
$C_{14}H_{14}$: d668
$C_{14}H_{14}N_2$: a168
$C_{14}H_{14}N_2O_3$: a330
$C_{14}H_{14}O$: d58
$C_{14}H_{14}OS$: b201
$C_{14}H_{14}O_2$: b110
$C_{14}H_{14}S_2$: b200, d57
$C_{14}H_{15}N$: d56, d670
$C_{14}H_{15}O_3P$: d61
$C_{14}H_{16}N_2$: d671
$C_{14}H_{16}O_2Si$: d438
$C_{14}H_{16}O_4$: d281
$C_{14}H_{18}O_4$: d285
$C_{14}H_{20}N_2O_6S$: m120
$C_{14}H_{20}O_5$: b39
$C_{14}H_{22}$: p132a
$C_{14}H_{22}O$: d123, d124, d125, d126
$C_{14}H_{22}O_2$: d112
$C_{14}H_{23}N$: d109, o42
$C_{14}H_{23}N_3O_{10}$: d299
$C_{14}H_{26}O_3$: h11
$C_{14}H_{26}O_4$: d409
$C_{14}H_{27}ClO$: t42
$C_{14}H_{28}$: t43, t44
$C_{14}H_{28}O_2$: t40
$C_{14}H_{29}Br$: b355
$C_{14}H_{29}Cl_3Si$: t47
$C_{14}H_{30}$: t39
$C_{14}H_{30}O$: t41
$C_{14}H_{31}N$: t45
$C_{14}H_{32}N_2O_4$: t89

C_{15}

$C_{15}H_{10}O_2$: b102, m124
$C_{15}5H_{11}NO$: d681
$C_{15}H_{12}N_2O_2$: d674
$C_{15}H_{12}O$: d355, d687
$C_{15}H_{12}O_2$: d53
$C_{15}H_{13}NO$: a13
$C_{15}H_{14}O$: d686
$C_{15}H_{14}O_2$: b49, b142, d688
$C_{15}H_{14}O_3$: b112
$C_{15}H_{16}O$: m362
$C_{15}H_{16}O_2$: i97
$C_{15}H_{17}N_3$: d713
$C_{15}H_{18}OSi$: e43
$C_{15}H_{22}O_3$: d117
$C_{15}H_{24}$: t317
$C_{15}H_{24}O$: d120
$C_{15}H_{26}O$: h185
$C_{15}H_{26}O_6$: g19
$C_{15}H_{30}N_2$: t347
$C_{15}H_{30}N_3OP$: t412
$C_{15}H_{30}O$: p14
$C_{15}H_{30}O_2$: m416
$C_{15}H_{32}$: p13
$C_{15}H_{32}O_3Si_4$: p167
$C_{15}H_{32}O_{10}$: t396

C_{16}

$C_{16}H_{10}$: b52, f1, p249
$C_{16}H_{11}NO_2$: p152
$C_{16}H_{12}N_2O_5S$: a60
$C_{16}H_{12}N_4O_9S_2$: t3
$C_{16}H_{13}N$: p132
$C_{16}H_{14}$: d658, d659, e67
$C_{16}H_{14}O$: d660
$C_{16}H_{14}O_6S$: s27
$C_{16}H_{15}NO_4$: d446
$C_{16}H_{16}O_2$: b47, b113
$C_{16}H_{16}O_3$: d450
$C_{16}H_{18}ClN_3S$: m238
$C_{16}H_{19}ClSi$: b438
$C_{16}H_{20}N_2$: d59
$C_{16}H_{20}O_2Si$: d250
$C_{16}H_{22}O_4$: d128, d410
$C_{16}H_{22}O_{11}$: g7
$C_{16}H_{26}O_3$: d732
$C_{16}H_{26}O_7$: t53
$C_{16}H_{32}$: h37
$C_{16}H_{32}O_2$: h35
$C_{16}H_{33}Br$: b295
$C_{16}H_{33}I$: i38
$C_{16}H_{33}NO$: d297
$C_{q6}H_{34}$: h4, h32
$C_{16}H_{34}O$: h36
$C_{16}H_{34}O_2$: h33
$C_{16}H_{34}S$: d646, h34
$C_{16}H_{35}N$: d645, h38
$C_{16}H_{35}O_4P$: b179
$C_{16}H_{36}BF_4N$: t20
$C_{16}H_{36}BrN$: t15
$C_{16}H_{36}ClN$: t16
$C_{16}H_{36}FN$: t17

TABLE 1-13 Empirical formula index for organic compounds (*continued*)

The alphanumeric designations are keyed to Table 1-14

$C_{16}H_{36}IN$: t19
$C_{16}H_{36}O_4Si$: t14
$C_{16}H_{36}Sn$: t21
$C_{16}H_{37}NO_4S$: t18

C_{17}

$C_{17}H_6O_7$: b55
$C_{17}H_{10}O$: b8
$C_{17}H_{12}O_3$: p121
$C_{17}H_{13}N_3O_5S_2$: p175
$C_{17}H_{16}O_4$: d60
$C_{17}H_{18}O_3$: b481
$C_{17}H_{20}N_2O$: b172
$C_{17}H_{20}N_4O_6$: r4
$C_{17}H_{21}NO_4$: c276
$C_{17}H_{22}N_2$: m235
$C_{17}H_{23}NO_3$: a320
$C_{17}H_{34}O_2$: m264
$C_{17}H_{36}$: h1
$C_{17}H_{37}N$: m230

C_{18}

$C_{18}H_9Cl_6O_4P$: t425
$C_{18}H_{10}O_6$: h85
$C_{18}H_{12}$: b6, b7, t403
$C_{18}H_{12}N_5O_6$: d685
$C_{18}H_{14}$: t4
$C_{18}H_{14}O$: d682
$C_{18}H_{14}O_8$: d55
$C_{18}H_{15}As$: t401
$C_{18}H_{15}N$: t399
$C_{18}H_{15}N_3Si$: a325
$C_{18}H_{15}O_3P$: t410
$C_{18}H_{15}O_4P$: t406
$C_{18}H_{15}P$: t407
$C_{18}H_{15}PS$: t409
$C_{18}H_{15}PSe$: t408
$C_{18}H_{15}Sb$: t400
$C_{18}H_{16}O_2$: b425
$C_{18}H_{16}Si$: t411
$C_{18}H_{18}O_3$: e74
$C_{18}H_{20}O_2$: b48
$C_{18}H_{25}NO_3$: i61
$C_{18}H_{30}O$: t215
$C_{18}H_{30}O_2$: o7
$C_{18}H_{31}N$: d735
$C_{18}H_{32}O_2$: o1
$C_{18}H_{32}O_{16}$: r1
$C_{18}H_{34}O_2$: o10, o11
$C_{18}H_{34}O_4$: d110
$C_{18}H_{36}$: d736, o8
$C_{18}H_{36}O$: o12
$C_{18}H_{36}O_2$: e142, o5
$C_{18}H_{37}Br$: b322
$C_{18}H_{37}Cl_3Si$: o15
$C_{18}H_{37}N$: o9
$C_{18}H_{37}NO$: o2
$C_{18}H_{38}$: o3
$C_{18}H_{38}O$: o6
$C_{18}H_{38}S$: o4
$C_{18}H_{39}ClSi$: t307
$C_{18}H_{39}N$: o13, t306
$C_{18}H_{39}O_7P$: t421
$C_{18}H_{40}Si$: t308

C_{19}

$C_{19}H_{15}Br$: b369
$C_{19}H_{15}Cl$: c257
$C_{19}H_{16}$: t404
$C_{19}H_{16}O$: t405
$C_{19}H_{18}BrP$: m445
$C_{19}H_{20}Br_4O_4$: i96
$C_{19}H_{20}O_4$: b87
$C_{19}H_{22}N_2O$: c265
$C_{19}H_{30}O_5$: m244
$C_{19}H_{32}$: p159
$C_{19}H_{34}ClN$: b123
$C_{19}H_{34}O_2$: m328
$C_{19}H_{36}O_2$: m330
$C_{19}H_{37}NO$: o14
$C_{19}H_{38}O_2$: m329
$C_{19}H_{40}$: n90, t117
$C_{19}H_{40}Cl_2Si$: m331

C_{20}

$C_{20}H_{10}Br_2O_5$: d83
$C_{20}H_{12}$: b57, b58, d49
$C_{20}H_{12}O_5$: f4
$C_{20}H_{14}O_4$: p65
$C_{20}H_{15}Br$: b368
$C_{20}H_{18}O_3Si$: t398
$C_{20}H_{19}N_3$: b2
$C_{20}H_{22}O_6$: t276
$C_{20}H_{24}N_2O_2$: q2
$C_{20}H_{24}O_6$: d51
$C_{20}H_{28}O_2P$: d676
$C_{20}H_{30}O_2$: a1
$C_{20}H_{31}N$: d20
$C_{20}H_{35}N$: t46
$C_{20}H_{36}O_2$: e191
$C_{20}H_{38}O_2$: e192
$C_{20}H_{40}$: i3
$C_{20}H_{40}O$: o16
$C_{20}H_{42}$: i2

C_{21}

$C_{21}H_{15}NO$: b143
$C_{21}H_{15}N_3O_3$: t397
$C_{21}H_{21}N$: t204
$C_{21}H_{22}N_2O_2$: s10
$C_{21}H_{24}O_2$: b145
$C_{21}H_{28}N_2O$: b171
$C_{21}H_{36}O$: p15
$C_{21}H_{39}N_3$: t260

C_{22}

$C_{22}H_{23}N_3O_9$: a321
$C_{22}H_{30}O_2S$: t146
$C_{22}H_{34}O_4$: b446
$C_{22}H_{39}N$: h39
$C_{22}H_{42}O_4$: d312
$C_{22}H_{44}O_2$: b469, d718
$C_{22}H_{46}$: d717
$C_{22}H_{46}O$: d719

C_{23}

$C_{23}H_{16}O_6$: m236
$C_{23}H_{26}N_2O_4$: b375

C_{24}

$C_{24}H_{16}N_2O_2$: b199
$C_{24}H_{18}$: t402
$C_{24}H_{20}BNa$: t130
$C_{24}H_{20}O_4Si$: t129
$C_{24}H_{20}Si$: t134
$C_{24}H_{20}Sn$: t135

TABLE 1-13 Empirical formula index for organic compounds (*continued*)
The alphanumeric designations are keyed to Table 1-14

$C_{24}H_{22}N_2O$: b141
$C_{24}H_{38}O_4$: b180, d313
$C_{24}H_{40}O_5$: c264
$C_{24}H_{46}O_4$: d643a
$C_{24}H_{50}$: t37
$C_{24}H_{51}N$: t394
$C_{24}H_{51}O_3P$: d424, t428
$C_{24}H_{52}O_4Si$: t87
$C_{24}H_{54}OSn_2$: b202

C_{26}

$C_{26}H_{20}$: t133
$C_{26}H_{26}N_2O_2S$: b153
$C_{26}H_{26}OSi_2$: t132
$C_{26}H_{50}O_4$: b178, d311

C_{27}

$C_{27}H_{19}NO$: b149
$C_{27}H_{42}ClNO_2$: b33
$C_{27}H_{46}O$: c263
$C_{27}H_{50}ClN$: b94

C_{28}

$C_{28}H_{22}$: t131
$C_{28}H_{31}ClN_2O_3$: r2
$C_{28}H_{32}O_2Si_3$: t123

C_{30} to C_{49}

$C_{30}H_{50}$: s8
$C_{30}H_{62}$: s7
$C_{30}H_{63}O_3P$: t313
$C_{32}H_{66}$: d739
$C_{32}H_{68}O_4Si$: t88
$C_{36}H_{75}O_3P$: d644
$C_{38}H_{30}NiO_2P_2$: b214
$C_{39}H_{74}O_6$: g20
$C_{40}H_{56}$: c19
$C_{40}H_{82}O_6P_2$: b197

C_{45} to C_{57}

$C_{45}H_{86}O_6$: g24
$C_{48}H_{40}O_4Si_4$: o38
$C_{51}H_{98}O_6$: g23
$C_{57}H_{104}O_6$: g22

TABLE 1-14 Physical constants of organic compounds

See also the special tables of fats, oils, and waxes.

Names of the compounds in the table starting on p. 1-72 are arranged alphabetically. Usually substitutive nomenclature is employed; exceptions generally involve ethers, sulfides, sulfones, and sulfoxides. Each compound is given a number within its letter classification; thus compound c196 is 3-chlorophenol. The section "Nomenclature of Organic Compounds" should be consulted to familiarize oneself with present nomenclature systems.

Synonyms or *Alternate Names* are found at the bottom of each spread in their alphabetical listing; the number following the name refers to the numerical place of this compound in the table. For example, epichlorohydrin, c102, indicates that this compound is found listed under the name 1-chloro-2,3-epoxypropane.

Formulas are presented in a semistructural form when no ambiguity is possible. Complicated systems are drawn in complete structural form and located at the bottom of each page and keyed to the number of the entry.

Beilstein Reference. In the column so headed is found the reference to the volume and page numbers of the fourth edition of Beilstein: *Handbuch der Organischen Chemie* (Springer-Verlag, New York, New York). Thus the entry 9, 202 refers to an entry in volume 9 appearing on page 202. When the volume number has a superscript attached, reference is made to the appropriate supplementary volume. For example, 12^2, 404 indicates that the compound will be found listed in the second supplement to volume 12 on page 404. The earliest Beilstein entry is listed. Supplementary information may be found in the supplements to the basic series; such coordinating references (series number, volume number, and page number of the main edition) along with the system number are found at the top of each *odd-numbered* page. Similarly, a back reference such as H 93; E II 64; E III 190 in a volume of Supplementary Series IV means that previous items on this compound are found in the same volume of the Basic Series on page 93, of Supplementary Series II on page 64, and of Supplementary Series III on page 190. The absence of a back reference implies that the compound involved is described *for the first time* in the series concerned.

Formula Weights are based on the International Atomic Weights of 1973 and are computed to the nearest hundredth.

Density values are given at room temperature unless otherwise indicated by the superscript figure; thus 0.9711^{112} indicates a density of 0.9711 for the substance at 112°C. A density of 0.899_4^{16} indicates a density of 0.899 for the substance at 16°C relative to water at 4°C.

Refractive Index, unless otherwise specified, is given for the sodium line at 589.6 nm. The temperature at which the measurement was made is indicated by the superscript figure; otherwise it is assumed to be room temperature.

Melting Point is recorded in certain cases as 250 d and in some other cases as d 250, the distinction being made in this manner to indicate that the former is a melting point with decomposition at 250°C, while the latter decomposition occurs only at 250°C and higher temperatures. Where a value such as $-2H_2O$, 120 is given, it indicates a loss of 2 mol of water per formula weight of the compound at a temperature of 120°C.

Boiling Point is given at atmospheric pressure (760 mmHg) unless otherwise indicated; thus 82^{15mm} indicates that the boiling point is 82°C when the pressure is 15 mmHg. Also, subl 550 indicates that the compound sublimes at 550°C.

Flash Point is given in degrees Celsius, usually closed cup. Because values will vary with the specific procedure employed, and sometimes the method was not stated, the values listed for the flash point should be considered only as indicative. See also Table 4-13, Properties of Combustible Mixtures in Air.

Solubility is given in parts by weight (of the formula weight) per 100 parts by weight of the solvent and at room temperature. Other temperatures are indicated by the superscript. In the case of gases, the solubility is often expressed as $5^{10°}$ mL, which indicates that at 10°C, 5 mL of the gas is soluble in 100 g of the solvent.

Abbreviations Used in the Table

abs, absolute
acet, acetone
alc, ethanol
alk, alkali (i.e., aqueous NaOH or KOH)
anhyd, anhydrous
aq, aqueous; water
as, asymmetrical
atm, atmosphere
BuOH, butanol
bz, benzene
c, cold
chl, chloroform, $CHCl_3$
conc, concentrated
d, decomposes or decomposed
D (or *d*), dextrorotatory
deliq, deliquescent
dil, dilute
diox, dioxane
DL (or *dl*), inactive (i.e., 50% D and 50% L)
DMF, dimethylformamide
EtAc, ethyl acetate
eth, diethyl ether
EtOH, ethanol, 95%
expl, explodes
glyc, glycerol
h, hot
HOAc, acetic acid
hyd, hydrolysis
hygr, hygroscopic
i, insoluble
ign, ignites
i-PrOH, isopropanol
L (or *l*), levorotatory
m, meta position
Me, methyl
MeEtKe, methyl ethyl ketone
MeOH, methanol
misc, miscible; soluble in all proportions
NaOH, aqueous sodium hydroxide
o, ortho position
org, organic
p, para position
PE, petroleum ether
pyr, pyridine
s, soluble
sec, secondary
sl, slight or slightly
soln, solution
solv, solvent
subl, sublimes
s, symmetrical
sym, symmetrical
tert, tertiary
v, very
v s, very soluble
v sl s, very slightly soluble
vac, vacuo or vacuum
vols, volumes
>, greater than
<, less than
~, approximately
α, alpha position
β, beta position
γ, gamma position
δ, delta position
ε, epsilon position
ω, omega position (farthest from parent functional group)

TABLE 1-14 Physical constants of organic compounds (*continued*)

No.	Name	Formula	Formula weight	Beilstein reference	Density	Refractive index	Melting point	Boiling point	Flash point	Solubility in 100 parts solvent
a1	(−)-Abietic acid		302.44	9^2, 424			172–175			i aq; s alc, bz, chl, eth, acet, dil alk
a2	Acenaphthene		154.21	5, 586	1.069_{95}^{95}		93.45	279		i aq; 3.2 alc; 20 bz
a3	Acenaphthlyene		152.20	5, 625	0.899_4^{16}		80–83	280		i aq; v s alc, eth
a4	Acetaldehyde	CH_3CHO	44.05	1, 594	0.8053_4^0	1.3311^{20}	−123.5	20.2	−27	misc aq, alc
a5	Acetaldoxime	$CH_3CH{=}NOH$	59.07	1, 608	0.966	1.415^{20}	46.5	114.5	38	v s aq, alc, eth
a6	Acetamide	CH_3CONH_2	59.07	2^2, 177	0.9711^{112}	1.4158^{110}	80.1	221.15		70 aq; 50 alc; s chl, hot bz
a7	Acetamidine HCl	$CH_3({=}NH)NH_2{\cdot}HCl$	94.54	2, 185			170–172			v s aq, alc; i acet, eth
a8	*N*-(2-Acetamido)-2-aminoethane-sulfonic acid	$H_2N(CO)CH_2NHCH_2$-CH_2SO_3H	182.20				>220 d			
a9	4-Acetamidobenz-aldehyde	$CH_3CONHC_6H_4CHO$	163.18	14, 38			154–156			s aq, bz; sl s alc
a10	4-Acetamidobenzene-sulfonyl chloride	$CH_3CONHC_6H_4SO_2Cl$	233.67	14, 439			149			d aq; v s alc, eth
a11	2-Acetamidobenzoic acid	$CH_3CONHC_6H_4COOH$	179.18	14, 337			185–187			sl s aq; v s alc, bz, eth, acet
a12	4-Acetamidobenzoic acid	$CH_3CONHC_6H_4COOH$	179.18	14, 432			260–262			i aq; s alc; sl s eth
a13	2-Acetamidofluorene		223.28	12, 1331			194			i aq; s alc, glycols
a14	*N*-(2-Acetamido)-iminodiacetic acid	$H_2NCOCH_2N(CH_2COOH)_2$	190.16				219 d			
a15	2-Acetamidophenol	$CH_3CONHC_6H_4OH$	151.17	13, 370			207–209			
a16	3-Acetamidophenol	$CH_3CONHC_6H_4OH$	151.17	13, 415			146–149			
a17	4-Acetamidophenol	$CH_3CONHC_6H_4OH$	151.17	13, 460	1.293_4^{21}		170			s alc, acet

a18	Acetanilide	$CH_3CONHC_6H_5$	135.17	12, 237	1.219^{15}_{4}		114.2	304	173	0.56 aq^{25}; 29 alc; 2bz; 27 chl; 25 acet; 5 eth
a19	Acetic acid	CH_3COOH	60.65	2, 96	1.0492^{20}_{4}	1.3716^{20}	16.63	117.90	40	misc aq, alc, eth, CCl_4
a20	Acetic acid-*d*	CH_3COOD	61.05		1.07	1.3715^{20}		115.5	40	misc aq, alc, eth, CCl_4
a21	Acetic-d_3, acid-*d*	CD_3COOD	64.08		1.11	1.3709^{20}		115.5	40	misc aq, alc, eth
a22	Acetic anhydride	$(CH_3CO)_2O$	102.09	2, 166	1.082^{15}_{4}	1.3904^{20}	−73.1	140.0	130	13 aq; s chl, eth
a23	Acetic anhydride-d_6	$(CD_3CO)_2O$	108.14			1.3875^{20}		65^{65mm}	54	d aq, alc
a24	Acetoacetic acid	CH_3COCH_2COOH	102.09	3, 630			36–37	d violently 100		misc aq, alc, eth
a25	Acetohydrazide	$CH_3CONHNH_2$	74.08	2, 191				129^{18mm}		
a26	Acetone	CH_3COCH_3	58.08	1, 635	0.7908^{20}_{4}	1.3588^{20}	−95.35	56.24	−20	misc aq, alc, chl,
a27	Acetone-d_6	CD_3COCD_3	64.13		0.88	1.3554^{20}		55.5	−17	
a28	Acetone oxime	$(CH_3)_2C{=}NOH$	73.10	1, 649	0.901		60–63	135		v s aq, alc, eth

ACES, a8
Acetal, d251
Acetaldehyde ammonia, a163
Acetaldehyde diethyl acetal, d251
Acetaldehyde dimethyl acetal, d439
Acetamidoacetic acid, a46
2-Acetamidopentanedioic acid, a45
Acethydrazide, a25
Acetic acid hydrazide, a25
Acetoacetanilide, a32
2-Acetoacetanisidide, a307
4-Acetoacetanisidide, a308
Acetoacetic esters, e54, e55
Acetoin, h106
Acetonaphthones, m314, m315
Acetonecarboxylic acid, a24
Acetone cyanohydrin, h146
Acetone dimethyl acetal, d453
Acetone ketal of glycerine, d517

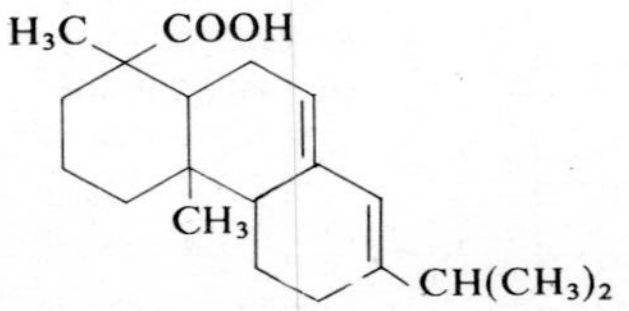

a1

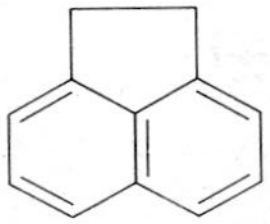

a2

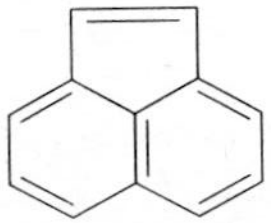

a3

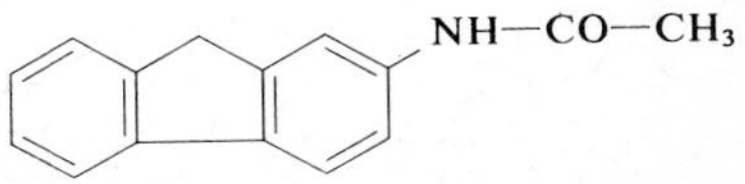

a13

TABLE 1-14 Physical constants of organic compounds (*continued*)

No.	Name	Formula	Formula weight	Beilstein reference	Density	Refractive index	Melting point	Boiling point	Flash point	Solubility in 100 parts solvent
a29	Acetonitrile	CH_3CN	41.05	2, 183	0.7857^{20}	1.3441^{20}	−43.8	81.60	5	misc aq, alc, chl
a30	Acetonitrile-d_3	CD_3CN	44.08		0.84	1.3420^{20}		80.7	5	misc aq, alc, chl
a31	Acetophenone	$C_6H_5COCH_3$	120.15	7, 271	1.0238^{25}	1.5322^{25}	19.62	202.08	82	0.55 aq; s alc, eth
a32	2-Acetylacetanilide	$C_6H_5NHCOCH_2COCH_3$	177.20	12, 518			85			sl s aq; s alc, hot bz, chl, eth, acids, alk
a33	4-Acetylbenzene-sulfonic acid, Na salt	$CH_3COC_6H_4SO_3^-Na^+$	222.02	11^2, 186			>300			
a34	4-Acetylbiphenyl	$C_6H_5C_6H_4COCH_3$	196.25	7^2, 337			116–118	325–327		i aq; v s alc, acet
a35	Acetyl bromide	CH_3COBr	122.95	2, 174	1.663_4^{16}		−96	75–77	1	d aq, alc; misc bz, chl, eth
a36	2-Acetylbutyrolactone		128.13		1.1846_4^{20}	1.4585^{20}		107^{5mm}		21 aq
a37	Acetyl chloride	CH_3COCl	78.50	2, 173	1.104_4^{20}	1.3886^{20}	−112.9	50.8	4	d aq, alc; misc bz, chl, eth
a38	Acetylcholine bromide	$(CH_3)_3NBrCH_2CH_2$-$OCOCH_3$	226.14	4^1, 428			114–116			v s aq (d hot aq); s alc; i eth
a39	Acetylcholine chloride	$(CH_3)_3NClCH_2CH_2$-$OCOCH_3$	181.66	4, 281			150–152			v s aq, alc; d hot aq; i eth
a40	2-Acetylcyclo-pentanone		126.16	7, 558	1.043	1.4905^{20}		$72\text{–}75^{80mm}$	72	
a41	Acetylene	$HC{\equiv}CH$	26.02	1, 228	0.90(*g*)		-81^{891mm}	−83.95 subl		90 aq; 14 alc; v s bz, eth; acet dissolves (25 vols (15°)
a42	Acetylenedicarboxylic acid	$HOOCC{\equiv}CCOOH$	114.06	2, 801			180 d			v s aq, alc, eth
a43	Acetyl fluoride	CH_3COF	62.04	2, 172	1.032		>−60	20		5 aq(d); misc alc, bz, eth

a44	2-Acetylfuran		110.11	17, 286	1.098	1.5065^{20}	29–30	67^{10mm}	71	
a45	N-Acetyl-L-glutamic acid	$HOOCCH_2CH_2CH(NHCOCH_3)COOH$	189.17	4^2, 908			200–1			
a46	N-Acetylglycine	$CH_3CONHCH_2COOH$	117.10	4, 354			207–209			2.7 aq^{15}; s alc; i eth
a47	N-Acetylimidazole		110.12				93–96			
a48	Acetyl iodide	CH_3COI	169.96	2, 174	2.0674_4^{20}	1.5491^{20}		108		d aq, alc; s bz, eth
a49	Acetyl-2-methyl-choline chloride	$CH_3COOCH(CH_3)CH_2$-$NCl(CH_3)_3$	195.69				171–173			v s aq, alc, chl; i eth
a50	2-Acetylphenothiazine		241.31				180–185			
a51	2-Acetylphenylaceto-nitrile	$C_6H_5CH(CN)COCH_3$	159.19	10, 699			89–92			
a52	N-Acetyl-4-piperidone		141.17		1.146	1.5026^{20}		218	>112	

Acetonylacetone, h61
Acetophenetidin, e46
Acetophenetidide, e46
p-Acetotoluide, m358
Acetoxime, a28
2-Acetoxybenzoic acid, a56
1-Acetoxy-1,3-butadiene, b378
Aceturic acid, a46
Acetylacetaldehyde dimethyl acetal, d436
Acetylacetone, p31
N-Acetylanthranilic acid, a11
Acetylbenzene, a31
Acetylcyclopropane, c369
Acetylene dichlorides, d179, d180
Acetylene tetrabromide, t10
Acetylene tetrachloride, t29
N-Acetylethanolamine, h115
3-Acetyl-6-methyl-2H-pyran-2,4-(3H)-dione, d21
2-(Acetyloxy)benzoic acid, a56
3-Acetyl-1-propanol, h158

a36 a40 a44 a47 a50 a52

TABLE 1-14 Physical constants of organic compounds (*continued*)

No.	Name	Formula	Formula weight	Beilstein reference	Density	Refractive index	Melting point	Boiling point	Flash point	Solubility in 100 parts solvent
a53	2-Acetylpyridine	$(C_5H_4N)COCH_3$	121.14	21, 279	1.080	1.5203^{20}		188–189	>112	v s alc, eth
a54	3-Acetylpyridine	$(C_5H_4N)COCH_3$	131.14	21, 279	1.102	1.5336^{20}		220	150	v s acids, alc, eth; s aq
a55	4-Acetylpyridine	$(C_5H_4N)COCH_3$	121.14	21, 279	1.095	1.5290^{20}		212	>112	
a56	Acetylsalicyclic acid	$HOOCC_6H_4OOCCH_3$	180.16	10, 67	1.35		135			0.33 aq^{25}; 20 alc; 5.9 chl; 5 eth; sl s bz
a57	2-Acetylthiophene	$(C_4H_3S)COCH_3$	126.18	17, 287	1.168_4^{22}	1.5564^{20}	10–11	214		sl s aq; misc alc, eth
a58	*N*-Acetylthiourea	$CH_3CONHC(S)NH_2$	118.16	3, 191			165–169			s hot aq, alc; sl s eth
a59	*N*-Acetyl-DL-tryptophan		246.27	22^2, 469			204–206			s aq, alc; v s eth
a60	Acid alizarin violet N		366.33	16^2, 127						
a61	Acridine		179.22	20, 459			107–110 subl 110	346		s alc, eth, CS_2, PE
a62	Acrylamide	$H_2C{=}CHCONH_2$	71.08	2, 400	1.122_4^{30}		84.5	125^{25mm}		215 aq^{30}; 86 alc^{30}; 63 acet; 2.7 chl; v s eth
a63	Acrylic acid	$H_2C{=}CHCOOH$	72.06	2, 397	1.0511^{20}	1.4224^{20}	13	140–141	54	misc aq, alc, bz, eth, chl, acet
a64	Acrylonitrile	$H_2C{=}CHCN$	53.06	2, 400	0.8060_4^{20}	1.3911^{20}	−83.7	77.4	0	7.3 aq; misc org solv
a65	Acryloyl chloride	$H_2C{=}CHCOCl$	90.51	2, 400	1.114	1.4350^{20}		72–76	16	d aq; v s chl
a66	1-Adamantanamine		151.25				206–208			sl s aq
a67	Adamantane		136.24		1.09	1.568	268 sealed tube	subl 205		

a68	1-Adamantane-carboxylic acid		180.25				174–175			
a69	Adenine		135.13	26, 420			>360 d	subl 220		0.05 aq; sl s alc; i chl, eth
a70	Adenosine		267.25	31, 27			234–236			s aq; i alc

N-Acetylsulfanilyl chloride, a10
Aconitic acid, p210
Acrolein, p207
Acrolein diethyl acetal, d258
Acrolein dimethyl acetal, d455
Acrylaldehyde, p207
1-Adamantanemethylamine, a212
Adenosine monophosphate, a72

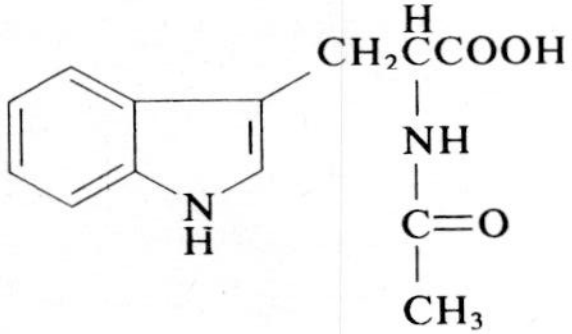

a59 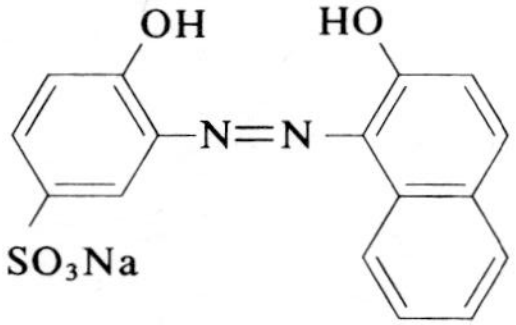 a60 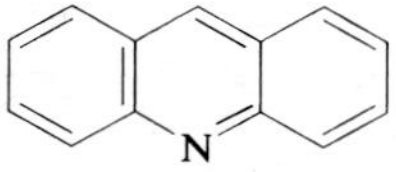 a61 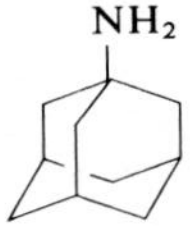 a66 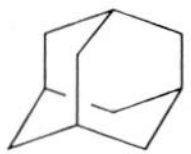 a67  a68 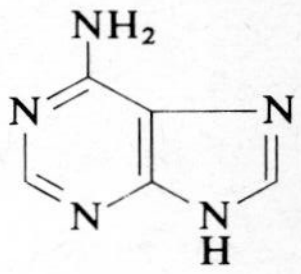 a69

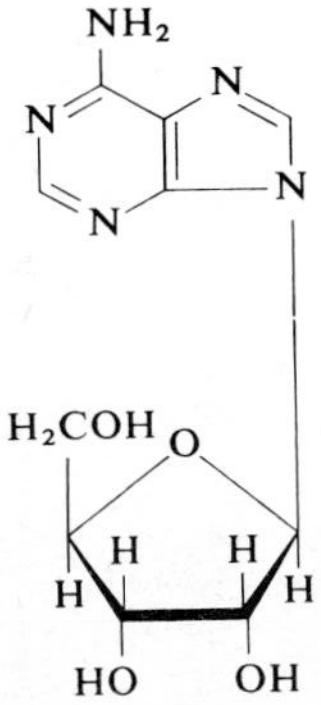

a70

TABLE 1-14 Physical constants of organic compounds (*continued*)

No.	Name	Formula	Formula weight	Beilstein reference	Density	Refractive index	Melting point	Boiling point	Flash point	Solubility in 100 parts solvent
a71	Adenosine-5′-diphosphoric acid		427.22							
a72	Adenosine-5′-phosphoric acid		347.22				200 d			v s hot aq, HCl
a73	D-α-Alanine	$CH_3CH(NH_2)COOH$	89.09	4, 385			291–293 d			
a74	DL-α-Alanine	$CH_3CH(NH_2)COOH$	89.09	4, 387	1.402		289 d	subl		16.7 aq^{25}; 8.7 alc^{25}; i eth
a75	L-α-Alanine	$CH_3CH(NH_2)COOH$	89.09	4, 381			315–316			16.7 aq^{25}; 8.7 alc^{25}; i eth
a76	β-Alanine	$H_2NCH_2CH_2COOH$	89.09	4, 401	1.437^{-5}		197–198 d			v s aq; sl s alc; i eth
a77	Allantoin		158.12	25, 474			238			0.45 aq; 0.2 alc
a78	Allene	$H_2C{=}C{=}CH_2$	40.06	1, 248	1.787	1.4168	−136.2	−34.5		
a79	Alloxan monohydrate		160.09	24, 500			253 d			s alc, acet, HOAc; sl s chl, PE, EtAc
a80	Allyl acetate	$H_2C{=}CHCH_2OCOCH_3$	100.12	2, 136	0.928^{20}_{4}	1.4040^{20}		104	6	i aq; misc alc, eth
a81	Allyl alcohol	$H_2C{=}CHCH_2OH$	58.08	2, 436	0.8540^{20}_{4}	1.4127^{20}	−50 glass	97.1	22	misc aq, alc, chl, eth
a82	Allylamine	$H_2C{=}CHCH_2NH_2$	57.10	4, 205	0.760^{20}_{20}	1.4205^{20}	−88.2	53.3	−28	misc aq, alc, chl, eth
a83	*N*-Allylaniline	$C_6H_5NHCH_2CH{=}CH_2$	133.19	12, 170	0.982^{25}	1.5630^{20}		218–220	89	i aq; s alc, eth
a84	Allylbenzene	$C_6H_5CH_2CH{=}CH_2$	118.18	5, 484	0.892^{20}_{0}	1.5122^{20}		156–157	33	i aq; s alc, eth
a85	Allyl bromide	$H_2C{=}CHCH_2Br$	120.98	1, 201	1.451^{25}_{25}	1.465^{25}	−50	70	7	i aq; misc org solv
a86	Allylchlorodimethylsilane	$H_2C{=}CHCH_2Si(CH_3)_2Cl$	134.7		0.8964^{2}	1.4195^{20}		110–112		
a87	Allyl chloroformate	$H_2C{=}CHCH_2OOCCl$	120.54		1.13	1.423		27	31	

a88	Allylcyclohexylamine	$C_6H_{11}NHCH_2CH{=}CH_2$	139.24		0.962	1.4664^{20}		66^{12mm}	53	
a89	Allyldichloromethyl-silane	$H_2C{=}CHCH_2Si(CH_3)Cl_2$	155.1		1.0758^{20}	1.4419^{20}		119–120		
a90	*N*-Allyl-*N*,*N*-dimethylamine	$H_2C{=}CHCH_2N(CH_3)_2$	85.0			1.4010^{20}		63–64		
a91	Allyl ethyl ether	$H_2C{=}CHCH_2OCH_2CH_3$	86.13	1, 438	0.7651_4^{20}	1.3881^{20}		64–66		i aq; misc alc, eth
a92	Allyl iodide	$H_2C{=}CHCH_2I$	167.98	1, 202	1.846_4^{20}		−99.3	103.1		i aq; misc alc, eth

Adenosine 5′-(trihydrogen diphosphate), a71
5′-Adenylic acid, a72
Adipic acid, h57
Adipic acid monoethyl ester, e152
Adiponitrile, d238
Adipolyl chloride, h62
ADP, a71
Alaninols, a276, a277
Alizarin, d372
Allylacetic acid, p50
Allylacetone, h78
4-Allylanisole, a94
Allyl carbamide, a105
Allyl chloride, c217
Allyl cyanide, b403
Allyldimethylchlorosilane, a86
Allyl glycidyl ether, a96
1-Allyl-2-hydroxybenzene, a98
Allyl iodide, i50

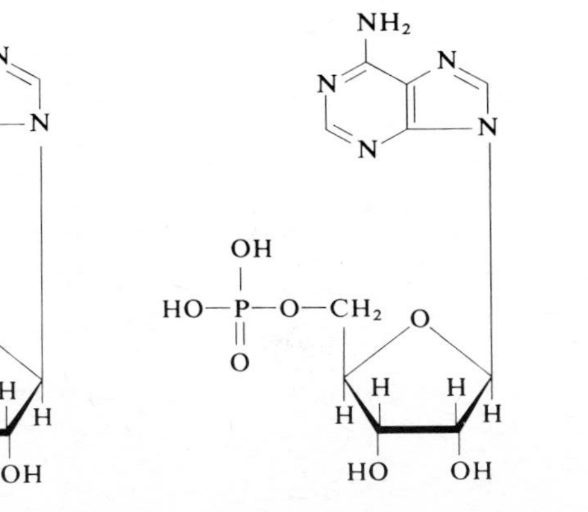

a71 a72 a77 a79

TABLE 1-14 Physical constants of organic compounds (*continued*)

No.	Name	Formula	Formula weight	Beilstein reference	Density	Refractive index	Melting point	Boiling point	Flash point	Solubility in 100 parts solvent
a93	Allyl isothiocyanate	$H_2C{=}CHCH_2NCS$	99.16	4, 214	1.013_4^{20}	1.5300^{20}	−80	150	46	0.2 aq; misc org solv
a94	1-Allyl-4-methoxybenzene	$H_2C{=}CHCH_2C_6H_4OCH_3$	148.21	6, 571	0.9645_4^{21}	1.5195^{20}		215–216		a slc, chl
a95	Allyl methyl sulfide	$H_2C{=}CHCH_2SCH_3$	88.17	1, 440	0.803	1.4714^{20}		91–93	18	
a96	1-Allyloxy-2,3-epoxypropane	$H_2C{-}CHCH_2OCH_2{-}CH{=}CH_2$ (epoxide O bridging H_2C–CH)	114.14		0.962	1.4332^{20}		154	57	
a97	Allyloxytrimethylsilane	$H_2C{=}CHCH_2\ OSi(CH_3)_3$	130.3		0.7830	1.4075^{25}		100–101		
a98	2-Allylphenol	$H_2C{=}CHCH_2C_6H_4OH$	134.18	6, 572	1.0255_{15}^{15}	1.5455^{20}	−6	220	88	s alc, eth
a99	Allyl phenyl ether	$H_2C{=}CHCH_2OC_6H_5$	134.18	6, 144	0.983_4^{15}	1.5200^{20}		192	62	i aq; s alc; misc eth
a100	Allyl propyl ether	$H_2C{=}CHCH_2OC_3H_7$	100.16	1^3, 1882	0.7670_4^{20}	1.3919^{20}		90–92	38	s alc; misc eth
a101	1-Allyl-2-thiourea	$H_2C{=}CHCH_2NHC(S)NH_2$	116.18	4, 211	1.219_{20}^{20}		78			3.3 aq; s alc; i bz; v sl s eth
a102	Allyltrichlorosilane	$H_2C{=}CHCH_2SiCl_3$	175.5		1.2011_4^{20}	1.4460^{20}		117.5		
a103	Allyltriethoxysilane	$H_2C{=}CHCH_2Si(OC_2H_5)_3$	204.3		0.9030^{20}	1.4072^{20}		176^{740mm}		
a104	Allyltrimethylsilane	$H_2C{=}CHCH_2Si(CH_3)_3$	114.27		0.7193_4^{20}	1.4074^{20}		85–86	7	
a105	Allylurea	$H_2C{=}CHCH_2NHCONH_2$	100.12	4, 209			78			v s aq, alc; v sl s eth
a106	Aminoacetonitrile	H_2NCH_2CN	56.07	4, 344				58^{15mm} d		s acids, alc
a107	Aminoacetonitrile hydrogen sulphate	$H_2NCH_2CN{\cdot}H_2SO_4$	154.14	4, 344			101	d 165		v s aq; sl s alc; i eth
a108	2′-Aminoacetophenone	$H_2NC_6H_4COCH_3$	135.17	14, 41				70^{3mm}		v sl s aq; s alc, eth
a109	3′-Aminoacetophenone	$H_2NC_6H_4COCH_3$	135.17	14, 45			98–99	289–290		
a110	4′-Aminoacetophenone	$H_2NC_6H_4COCH_3$	135.17	14, 46			106	293–295		s hot aq, alc, eth, HOAc; sl s bz

No.	Name	Formula	Formula weight	Beilstein			mp	bp		Solubility
a111	1-Aminoanthra-quinone		223.23	14, 177			253–255	subl		i aq; v s alc, bz, chl, eth, HOAc, HCl
a112	2-Aminoanthra-quinone		223.23	14, 191			295 d	subl		i aq, eth; s alc, bz
a113	4-Aminoantipyrine		203.25	24, 273			109			s aq, alc, bz; sl s eth
a114	2-Aminobenzamide	$H_2NC_6H_4CONH_2$	136.15	14, 320			110	300 sl d		v s hot aq, alc; i bz; sl s eth
a115	2-Aminobenzene-arsonic acid	$H_2NC_6H_4AsO(OH)_2$	217.06	16^1, 463			153			
a116	4-Aminobenzene-arsonic acid	$H_2NC_6H_4AsO(OH)_2$	217.06	16, 878			>300			s hot aq, alk CO_3, mineral acids
a116a	5-Aminobenzene-1,3-dicarboxylic acid	$H_2NC_6H_3(COOH)_2$	181.15	14^1, 636			>300			
a117	2-Aminobenzene-1,4-disulfonic acid	$H_2NC_6H_3(SO_3H)_2$	253.24							
a118	2-Aminobenzene-sulfonic acid	$H_2NC_6H_4SO_3H$	173.19	14, 681			d 325			1.5 aq^{15}; v sl s alc, eth

Allyl mercaptan, p209
4-Allyl-2-methoxyphenol, m99
2-Allyl-4-methylphenol, m390
2-Allyl-6-methylphenol, m389
Allyl sulfide, d27
Aluminon, a321
N-Amidinosarcosine, c278
Aminoacetaldehyde diethyl acetal, d254
Aminoacetaldehyde dimethyl acetal, d441
1-Aminoadamantane, a66
Aminoanisoles, m42, m43, m44
p-Aminoazobenzene, p88

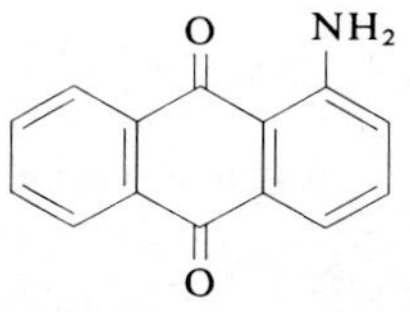

a111

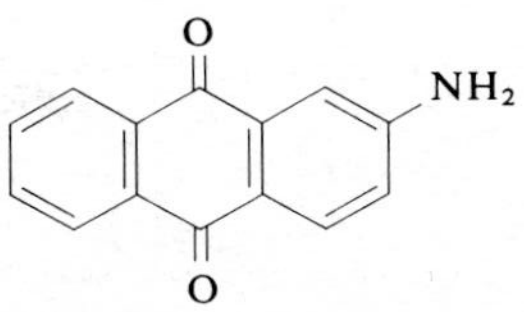

a112

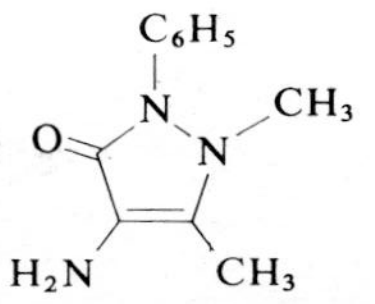

a113

TABLE 1-14 Physical constants of organic compounds (*continued*)

No.	Name	Formula	Formula weight	Beilstein reference	Density	Refractive index	Melting point	Boiling point	Flash point	Solubility in 100 parts solvent
a119	3-Aminobenzene-sulfonic acid	$H_2NC_6H_4SO_3H$	173.19		1.69					2 aq^{15}; sl s alc
a120	4-Aminobenzene-sulfonic acid	$H_2NC_6H_4SO_3H$	173.19	14, 695			d 288			1 aq^{20}; sl s hot MeOH
a121	2-Aminobenzoic acid	$H_2NC_6H_4COOH$	137.14	14, 310			144–146	subl		v s hot aq, alc, eth
a122	3-Aminobenzoic acid	$H_2NC_6H_4COOH$	137.14	14, 383	1.511^{4}		172–174			sl s aq; v s alc; s eth
a123	4-Aminobenzoic acid	$H_2NC_6H_4COOH$	137.14	14, 418	1.374		187			0.59 aq; 5.6 alc
a124	2-Aminobenzonitrile	$H_2NC_6H_4CN$	118.14	14, 322			49	268		s alc, eth
a125	3-Aminobenzonitrile	$H_2NC_6H_4CN$	118.14	14, 391			53	288–290		s hot aq; v s alc, eth
a126	4-Aminobenzonitrile	$H_2NC_6H_4CN$	118.14	14, 425			85	d		v s hot aq, alc, eth
a127	2-Aminobenzo-phenone	$H_2NC_6H_4COC_6H_5$	197.24	14, 76			108	223–226		sl s aq; s alc, eth
a128	2-Aminobenzothiazole		150.20	27, 182			132	d		v s alc, chl, eth
a129	2-Aminobenzotri-fluoride	$H_2NC_6H_4CF_3$	161.13	12^{2}, 453	1.290^{25}	1.4785^{25}	34	175	55	
a130	3-Aminobenzotri-fluoride	$H_2NC_6H_4CF_3$	161.13	12, 870	1.290	1.4800^{20}	6	187	85	
a131	4-Aminobenzotri-fluoride	$H_2NC_6H_4CF_3$	161.13	12^{3}, 2151	1.283^{27}	1.4815^{25}	38	107^{39mm}	85	
a132	*N*-(*p*-Aminobenzoyl)-glycine	$H_2NC_6H_4CONHCH_2COOH$	194.19	14^{2}, 258			198–199			i aq; s alc, bz, chl
a133	4-Aminobenzoyl hydrazide	$H_2NC_6H_4CONHNH_2$	151.17	14^{1}, 570			227			
a134	2-Aminobiphenyl	$H_2NC_6H_4C_6H_5$	169.23	12, 1317			53	299		sl s aq; s alc
a135	4-Aminobiphenyl	$H_2NC_6H_4C_6H_5$	169.23	12, 1318			54	191^{15mm}		s hot aq, alc, eth
a136	D-(+)-2-Amino-1-butanol	$CH_3CH_2CH(NH_2)CH_2OH$	89.14	4, 291	0.947^{20}	1.4521^{20}	−2	174	79	misc aq; s alc

a137	L-(−)-2-Amino-1-butanol	$CH_3CH_2CH(NH_2)CH_2OH$	89.14	4, 291	0.947^{20}	1.4525^{20}	−2	174	82	misc aq; s alc
a138	DL-2-Aminobutyric acid	$CH_3CH_2CH(NH_2)COOH$	103.12	4, 408			304	subl 300		21 aq; 0.2 hot alc
a139	4-Aminobutyric acid	$H_2NCH_2CH_2CH_2COOH$	103.12	4, 413			195.d			v s aq; i alc, eth
a140	2-Amino-4-chlorobenzoic acid	$H_2N(Cl)C_6H_3COOH$	171.58	14, 365			233			
a141	2-Amino-5-chlorobenzonitrile	$H_2N(Cl)C_6H_3CN$	152.58				99	$132^{0.5mm}$	>112	
a142	2-Amino-4′-chlorobenzophenone	$H_2NC_6H_4COC_6H_4Cl$	231.68	14^1, 389			104			
a143	2-Amino-5-chlorobenzophenone	$H_2N(Cl)C_6H_3COC_6H_5$	231.68	14, 79			100			
a144	2-Amino-5-chlorobenzotrifluoride	$H_2N(Cl)C_6H_3CF_3$	195.57	12^3, 1921	1.386	1.5069^{20}		$66–67^{3mm}$		
a145	3-Amino-4-chlorobenzotrifluoride	$H_2N(Cl)C_6H_3CF_3$	195.57		1.428	1.4975^{25}		$82–83^{9mm}$	none	
a146	5-Amino-2-chlorobenzotrifluoride	$H_2N(Cl)C_6H_3CF_3$	195.57				36			

Aminobenzenethiol, a298
Aminobenzoic acid hydrazide, a133
Aminobenzyl cyanide, a263
1-Aminobutane, b420
2-Aminobutane, b380
Aminobutanoic acids, a138, a139
3-Amino-2-butenamide, a150
4-Aminobutyraldehyde diethyl acetal, d247
α-Aminocaproic acid, a183
ε-Aminocaproic acid, a184

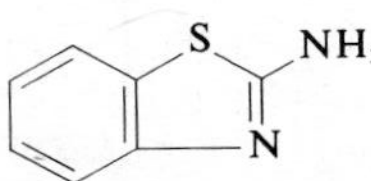

a128

TABLE 1-14 Physical constants of organic compounds (*continued*)

No.	Name	Formula	Formula weight	Beilstein reference	Density	Refractive index	Melting point	Boiling point	Flash point	Solubility in 100 parts solvent
a147	2-(3-Amino-4-chlorobenzoyl)benzoic acid	$H_2N(Cl)C_6H_3CO-C_6H_4COOH$	275.69	14, 661			171–173			
a148	2-Amino-4-chlorophenol	$H_2N(Cl)C_6H_3OH$	143.57	13, 383			138			
a149	2-Amino-5-chloropyridine	$H_2N(Cl)C_5H_3N$	128.56	22^2, 332			138	128^{11mm}		
a150	3-Aminocrotamide	$CH_3C(NH_2){=}CHCONH_2$	100.12				102			
a151	3-Aminocrotononitrile	$CH_3C(NH_2){=}CHCN$	82.11	3, 660						
a152	1-Amino-1-cyclohexanecarboxylic acid	$C_6H_{10}(NH_2)COOH$	143.19	14, 299			>300			
a153	5-Amino-2,3-dihydro-1,4-phthalazinedione		177.16	25^1, 698			319–320			
a154	2-Amino-4,6-dihydroxypyrimidine		127.10	24, 468			>300			
a155	4-Amino-2,6-dihydroxypyrimidine		127.10	24, 469			>300			
a156	4-Amino-3,5-diiodobenzoic acid	$I_2(NH_2)C_6H_2COOH$	388.93	14, 439			>300			i aq, alc
a157	2-Amino-4,6-dimethylpyridine	$(CH_3)_2(NH_2)(C_5H_2N)$	122.17	22, 435			64	235		
a158	4-Amino-2,6-dimethylpyridimide		123.16	24^2, 45			181			156 aq; 18.9 alc
a159	6-Amino-1,3-dimethyluracil		155.16	24, 471			295 d			
a160	5-Amino-2,6-dioxo-1,2,3,6-tetrahydro-4-pyrimidinecarboxylic acid		171.11	25, 264			>300			

a161	2-Aminoethane-sulfonic acid	$H_2NCH_2CH_2SO_3H$	125.15	4, 528			d > 300			6.45 aq^{12}; i abs alc
a162	2-Aminoethanethiol	$HSCH_2CH_2NH_2$	77.14	4, 286			99–100			v s aq; s alc
a163	1-Aminoethanol	$CH_3CH(OH)NH_2$	61.08				97	110 d		s aq; sl s eth
a164	2-Aminoethanol	$H_2NCH_2CH_2OH$	61.08	4, 274	1.0158^{20}	1.4539^{20}	10.52	171	93	misc aq, org solv
a165	2-(2-Aminoethoxy)-ethanol	$H_2NCH_2CH_2OCH_2CH_2OH$	105.14	4^3, 642	1.460			218–224		
a166	2-(2-Aminoethyl-amino)ethanol	$H_2NCH_2CH_2NHCH_2CH_2OH$	104.15	4, 286	1.030	1.4861^{20}		241	129	v s aq, alc; sl s eth
a167	5-(2-Aminoethyl-amino)-1-naphthalenesulfonic acid	$H_2NCH_2CH_2NHC_{10}H_6SO_3H$	266.32				> 300			
a167a	3-(2-Aminoethyl-amino)propyl-trimethoxy-silane	$H_2NCH_2CH_2NHCH_2CH_2CH_2Si(OCH_3)_3$	222.1		1.01_4^{25}	1.4418^{25}		140^{15mm}	150	

4-Amino-*m*-cresol, a221
Aminocyclohexane, c335
Aminodecane, d19
2-Amino-2-deoxyglucose, g5
2-Amino-5-diethylaminopentane, d327
2-Amino-1,5-dihydro-1-methyl-4*H*-imidazol-4-one, c279
2-Aminodiphenylamine, p135
1-Amino-1,2-diphenylethane, d670
Aminodiphenylmethane, d680
Aminoethane, e59
1-(2-Aminoethyl)amino-2-[(2-aminoethyl)-aminoethyl]aminoethane, t56

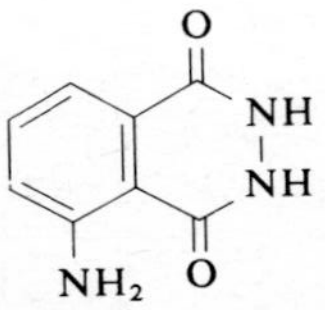

O NH NH NH₂ O

a153

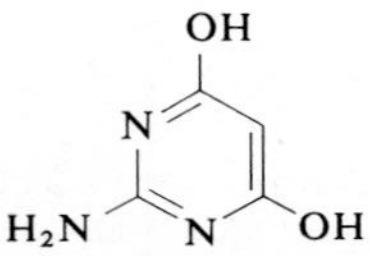

OH N H₂N N OH

a154

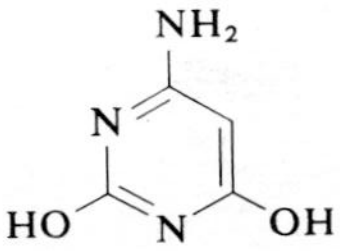

NH₂ N HO N OH

a155

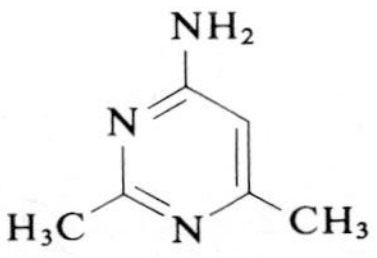

NH₂ N H₃C N CH₃

a158

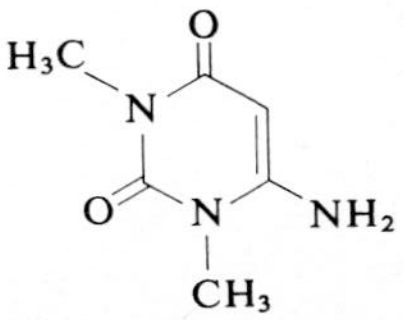

H₃C O N O N NH₂ CH₃

a159

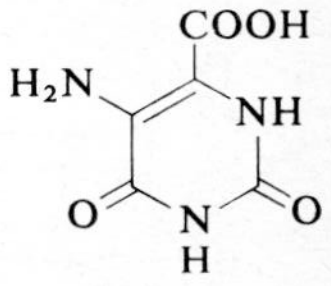

COOH H₂N NH O N O H

a160

TABLE 1-14 Physical constants of organic compounds (*continued*)

No.	Name	Formula	Formula weight	Beilstein reference	Density	Refractive index	Melting point	Boiling point	Flash point	Solubility in 100 parts solvent
a168	3-Amino-9-ethyl-carbazole		210.28	22^1, 642			98–100			
a169	2-Aminoethyl hydrogen sulfate	$H_2NCH_2CH_2OSO_3H$	141.15	4, 276			280 d			
a170	3-(2-Aminoethyl)-indole		160.22	22^1, 636			118	$137^{0.15mm}$		i aq, bz, chl, eth; s alc, acet
a171	*S*-2-Aminoethyl-isothiouronium bromide HBr		281.02				194–195			
a172	*N*-(2-Aminoethyl)-morpholine		130.19		0.992	1.4755^{20}	25.6	205	175	
a173	*p*-(2-Aminoethyl)-phenol	$HOC_6H_4CH_2CH_2NH_2$	137.18	13, 625			161–163	175^{8mm}		
a174	*N*-(2-Aminoethyl)-piperazine		129.21		0.985	1.4983^{20}	−26	222	93	
a175	*N*-(2-Aminoethyl)-1,3-propanediamine	$H_2NCH_2CH_2CH_2NHCH_2$-CH_2NH_2	117.20		0.928	1.4815^{20}			96	
a176	2-Amino-2-ethyl-1,3-propanediol	$HOCH_2C(NH_2)$-$(C_2H_5)CH_2OH$	119.16		1.099^{20}_{20}	1.490^{20}	38	152^{10mm}	74	misc aq; s alc
a177	2-(2-Aminoethyl)-pyridine	$H_2NCH_2CH_2(C_5H_4N)$	122.17	22, 434	1.021	1.5357^{20}		93^{12mm}		
a178	4-(2-Aminoethyl)-pyridine	$H_2NCH_2CH_2(C_5H_4N)$	122.17		1.012	1.5403^{20}		104^{9mm}		
a179	3-Amino-4-fluorobenzo-trifluoride	$H_2N(F)C_6H_3CF_3$	179.0			1.4608^{20}		81^{20mm}		
a180	Aminoguanidine H_2CO_3	$H_2NNHC(=NH)$-$NH_2 \cdot H_2CO_3$	136.11	3, 117			172 d			i aq; d hot aq

a181	Aminoguanidine nitrate	$H_2NNHC(=NH)$-$NH_2 \cdot HNO_3$	137.11	3, 117			137			
a182	*N*-Aminohexamethyleneimine	$C_6H_{12}N—NH_2$	114.19		0.984	1.4850^{20}		165	56	
a185	2-Aminohexanoic acid	$CH_3(CH_2)_3CH(NH_2)$-COOH	131.18	4, 433	1.172		d 327			1.15 aq^{25}; 0.42 alc
a186	6-Aminohexanoic acid	$H_2N(CH_2)_4CH_2COOH$	131.18	4, 434			204–206			v s aq; i alc
a187	6-Amino-1-hexanol	$H_2N(CH_2)_5CH_2OH$	117.19	4^2, 748			56–58	135^{30mm}		
a188	1-Amino-4-hydroxyanthraquinone		239.23	14, 268			207–209			s eth
a189	L-2-Amino-3-hydroxybutyric acid	$CH_3CH(OH)CH$-$(NH_2)COOH$	119.12	4, 514			d 255–257			v s aq; i alc, eth, chl

Aminoethylbenzenes, e65, e66
2-(2-Aminoethyl)-2-thiopseudourea, a171
1-Aminoheptane, h20
2-Aminoheptane, m269a
1-Aminohexane, h80
2-Aminohexane, m353a
p-Aminohippuric acid, a132
Aminohydroxybenzoic acids, a291, a292

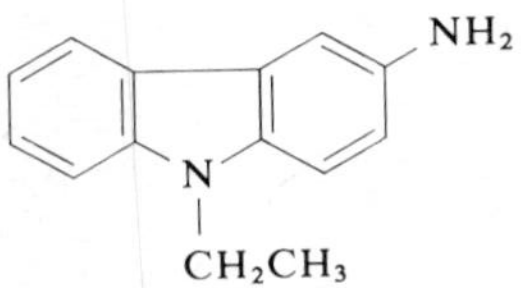

NH_2, N, CH_2CH_3

a168

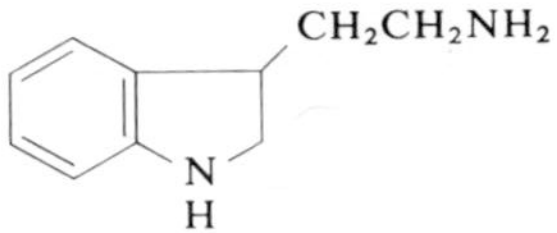

$CH_2CH_2NH_2$, N, H

a170

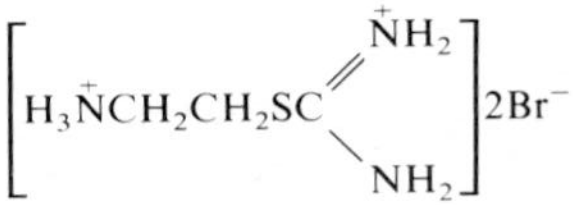

$[H_3\overset{+}{N}CH_2CH_2SC(=\overset{+}{N}H_2)NH_2]2Br^-$

a171

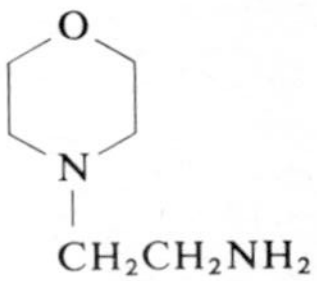

O, N, $CH_2CH_2NH_2$

a172

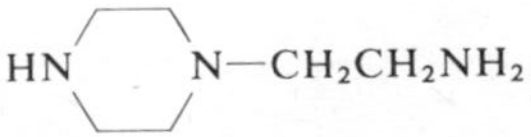

HN, $N—CH_2CH_2NH_2$

a174

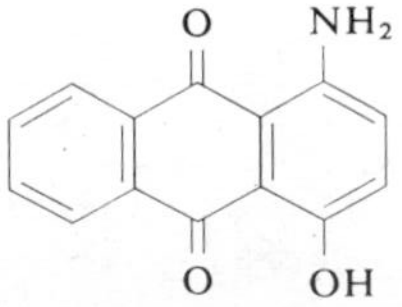

O, NH_2, O, OH

a188

TABLE 1-14 Physical constants of organic compounds (*continued*)

No.	Name	Formula	Formula weight	Beilstein reference	Density	Refractive index	Melting point	Boiling point	Flash point	Solubility in 100 parts solvent
a190	DL-2-Amino-4-hydroxy-butyric acid	$HOCH_2CH_2CH(NH_2)COOH$	119.12	4, 514			188–189			s alc
a191	L-2-Amino-4-hydroxy-butyric acid	$HOCH_2CH_2CH(NH_2)COOH$	119.12	4^3, 1636			203 d			
a192	DL-4-Amino-3-hydroxy-butyric acid	$H_2NCH_2CH(OH)CH_2COOH$	119.12	4^2, 938			202 d			s aq; sl s alc, eth
a193	4-Amino-6-hydroxy-2-mercapto-pyrimidine hydrate		161.18	24, 476			> 300			
a194	2-Amino-4-hydroxy-6-methylpyrimidine		125.13	24, 343			> 300			
a195	4-Amino-3-hydroxy-1-naphthalenesulfonic acid		239.25	14, 846			295 d			i aq, alc, bz, eth
a196	4-Amino-5-hydroxyl-1-naphthalenesulfonic acid		239.25	14, 835						sl s aq; i alc, eth
a197	5-Amino-6-hydroxy-2-naphthalenesulfonic acid		239.25							sl s hot aq; i eth
a198	6-Amino-7-hydroxy-2-naphthalenesulfonic acid		239.25	14, 849			> 300			
a199	2-Amino-3-hydroxy-pyridine	$H_2N(HO)(C_5H_3N)$	110.12	22^2, 408			172–174			

a200	4-Amino-2-hydroxy-pyrimidine		111.10	24, 314			> 300			0.77 aq; sl s alc
a201	1-Aminoindan		133.19	12, 1191	1.038_4^{15}	1.5613^{20}	1.5	97^{8mm}	94	sl s aq
a202	5-Aminoindan		133.19	12^1, 511			36	249^{745mm}		sl s aq
a203	5-Aminoindazole		133.15	25^2, 308			178			
a204	6-Aminoindazole		133.15	25, 317			206 d			
a205	2-Amino-5-iodobenzoic acid	$H_2N(I)C_6H_3COOH$	263.03	14, 373			221 d			sl s aq, PE; s alc
a207	DL-2-Amino-4-mercaptobutyric acid	$HSCH_2CH_2CH(NH_2)COOH$	135.19	4^3, 1647			232–233			
a208	Aminomethanesulfonic acid	$H_2NCH_2SO_3H$	111.12	1, 583			185 d			v s aq

2-Amino-2-(hydroxymethyl)-1,3-propanediol, t430
α-Amino-4-imidazolepropanoic acid, h83
Aminoiminomethanesulfinic acid, f30
N-(Aminoiminomethyl)-*N*-methylglycine, c278
2-Aminoisobutyric acid, a225
5-Aminoisophthalic acid, a116a
6-Amino-2,4-lutidine, a157
2-Amino-3-mercaptopropanoic acid, c371

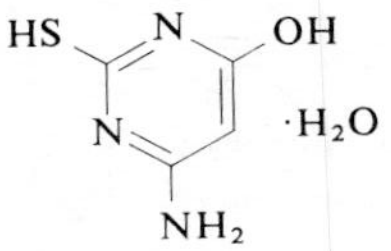

a193 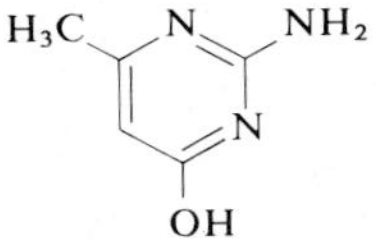 a194 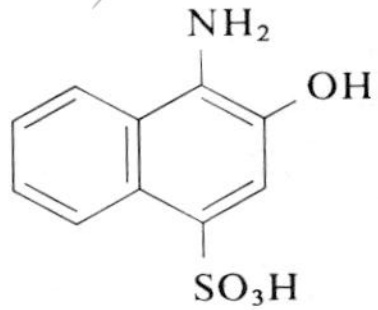 a195 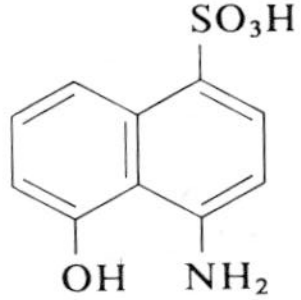 a196 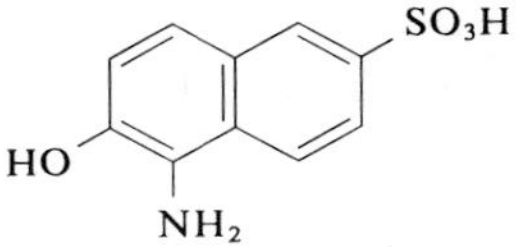 a197 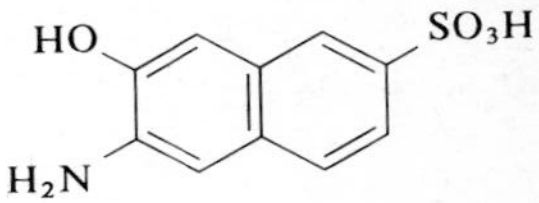 a198

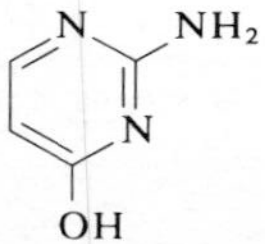

a200 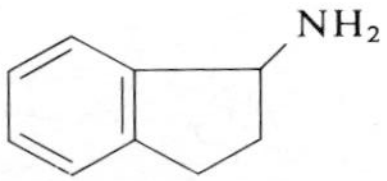 a201 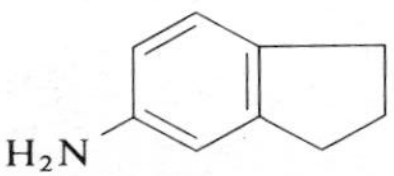 a202 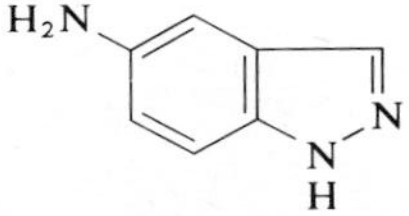 a203 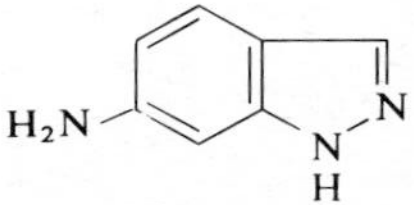 a204

TABLE 1-14 Physical constants of organic compounds (*continued*)

No.	Name	Formula	Formula weight	Beilstein reference	Density	Refractive index	Melting point	Boiling point	Flash point	Solubility in 100 parts solvent
a209	3-Amino-4-methoxy-benzoic acid	$CH_3O(NH_2)C_6H_3COOH$	167.16	14^1, 657			241			
a210	2-Amino-6-methoxy-benzothiazole		180.23	27^2, 334			165–167			
a211	5-Amino-2-methoxy-pyridine	$CH_3O(NH_2)—C_5H_3N$	124.14	22^2, 408		1.5745^{20}	31	90^{1mm}		
a212	1-(Aminomethyl)-adamantane		165.28		0.933	1.5137^{20}		$83–85^{0.3mm}$	92	
a213	4-Aminomethyl)-benzenesulfonamide	$H_2NCH_2C_6H_4SO_2NH_2$	186.25				151–152			s dil alk, dil acid
a214	2-Amino-5-methyl-benzoic acid	$H_2N(CH_3)C_6H_3COOH$	151.17	14, 481			177 d			sl s aq; s alc, eth
a215	3-Amino-4-methyl-benzoic acid	$H_2N(CH_3)C_6H_3COOH$	151.17	14, 487			166			a aq
a216	DL-2-Amino-3-methyl-1-butanol	$(CH_3)_2CHCH(NH_2)-CH_2OH$	103.17			1.4543^{20}		77^{8mm}	83	
a217	L-2-Amino-3-methyl-1-butanol	$(CH_3)_2CHCH(NH_2)-CH_2OH$	103.17		0.926	1.4548^{20}		81^{8mm}	78	
a218	2-(Aminomethyl)-1-ethylpyrrolidine		128.22		0.887	1.4665^{20}		60^{16mm}		
a219	2-Amino-3-methyl-1-pentanol	$CH_3CH_2CH(CH_3)CH-(NH_2)CH_2OH$	117.19			1.4589^{20}	30	97^{14mm}		
a220	2-Amino-4-methyl-1-pentanol	$CH_3CH(CH_3)CH_2CH-(NH_2)CH_2OH$	117.19	4, 298	0.917	1.4511^{20}		200	90	
a221	4-Amino-3-methyl-phenol	$H_2N(CH_3)C_6H_3OH$	123.16	13, 593			179			
a222	4-(Aminomethyl)-piperidine		114.19			1.4900^{20}	25	200	78	
a223	2-Amino-2-methyl-1,3-propanediol	$HOCH_2C(CH_3)-(NH_2)CH_2OH$	105.14	4, 303			110	151^{10mm}		$250\ aq^{20}$; s alc

a224	2-Amino-2-methyl-1-propanol	$(CH_3)_2C(NH_2)CH_2OH$	89.14		0.934^{20}_{20}	1.4480^{20}	30–31	165	67	misc aq; s alc, org solv
a225	2-Amino-2-methylpropionic acid	$(CH_3)_2C(NH_2)COOH$	103.12	4, 414			335 sealed tube	280 subl		v s aq
a226	(2-Aminomethyl)-pyridine	$H_2NCH_2—C_5H_4N$	108.14		1.049	1.5445^{20}		85^{12mm}		
a227	(3-Aminomethyl)-pyridine	$H_2NCH_2—C_5H_4N$	108.14		1.062	1.5510^{20}	−21	74^{1mm}	100	
a228	2-Amino-3-methyl-pyridine	$H_2N(CH_3)—C_5H_3N$	108.14	22^2, 342		1.5782^{20}	34	222		v s aq; s alc
a229	2-Amino-4-methyl-pyridine	$H_2N(CH_3)—C_5H_3N$	108.14	22^2, 342			100	230		v s aq, alc, DMF
a230	2-Amino-6-methyl-pyridine	$H_2N(CH_3)—C_5H_3N$	108.14	22^1, 633			45	209		v s aq
a231	2-Amino-4-methyl-pyrimidine		109.13	24, 84			160	subl		s hot aq; s alc
a232	2-Amino-4-methyl-thiazole		114.17	27, 159			45	232		v s aq, alc, eth
a233	2-Aminomethyl-3,5,5-trimethylcyclo-hexanol		171.29		0.969	1.4904^{20}	43–48	265	> 112	

1-Amino-2-methoxyethane, m69
α-(Aminomethyl)benzyl alcohol, a265
3-Amino-α-methylbenzyl alcohol, a264
2-Amino-3-methylpentanoic acid, i79
2-Aminomethylthiophene, t161

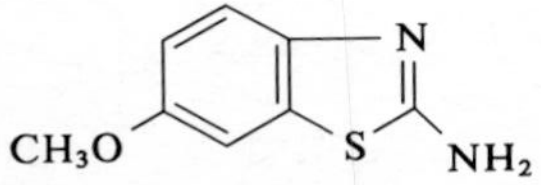

a210 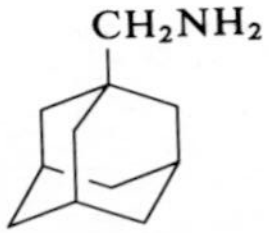 a212 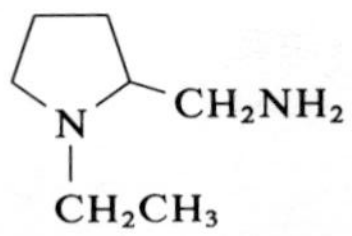 a218 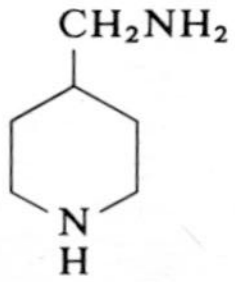 a222 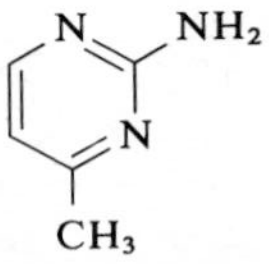 a231 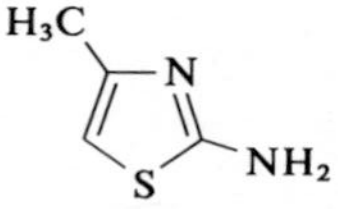 a232 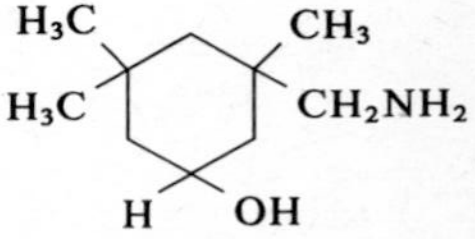 a233

TABLE 1-14 Physical constants of organic compounds (*continued*)

No.	Name	Formula	Formula weight	Beilstein reference	Density	Refractive index	Melting point	Boiling point	Flash point	Solubility in 100 parts solvent
a234	*N*-Aminomorpholine		102.14	27, 8	1.059	1.4772^{20}		168	58	
a235	2-Amino-1-5,naphthalenedisulfonic acid		303.31	14, 786			> 300			
a236	7-Amino-1,3-naphthalenedisulfonic acid		303.31	14, 784			> 300			
a237	4-Amino-1-naphthalenesulfonic acid	$H_2N—C_{10}H_6SO_3H$	223.26		1.670_4^{25}		d			0.031 aq; s dil alk
a238	4-Amino-1,8-naphthalimide		212.21	22^2, 452			360			
a239	3-Amino-2-naphthol	$H_2N—C_{10}H_6OH$	159.19	13, 685			207			
a240	2-Amino-4-nitrobenzoic acid	$H_2N(NO_2)C_6H_3COOH$	182.14	14, 374			270 d			i aq; v s alc, eth
a241	2-Amino-5-nitrobenzonitrile	$H_2N(NO_2)C_6H_3CN$	163.14	14^2, 234			200–207			
a242	2-Amino-5-nitrobenzophenone	$C_6H_5COC_6H_3$-$(NH_2)NO_2$	242.23	14, 79			166–168			
a243	2-Amino-6-nitrobenzothiazole		195.20	27^2, 232			247–249			
a244	2-Amino-5-nitrobenzotrifluoride	$H_2N(NO_2)C_6H_3CF_3$	206.12				90–92			
a245	4-Amino-3-nitrobenzotrifluoride	$H_2N(NO_2)C_6H_3CF_3$	206.12				105–106			
a246	4-Amino-4′-nitrodiphenylsulfide	$O_2NC_6H_4SC_6H_4NH_2$	246.29	13, 534			142			
a247	2-Amino-4-nitrophenol	$O_2N(NH_2)C_6H_3OH$	154.13	13^2, 192			145			

a248	4-Amino-2-nitrophenol	$O_2N(NH_2)C_6H_3OH$	154.13	13, 520			127			
a249	D-(−)-*threo*-2-Amino-1-(*p*-nitrophenyl–1,3-propanediol	$HOCH_2CH(NH_2)CH(OH)$-$C_6H_4NO_2$	212.21				163–165 163–165			
a250	2-Amino-5-(*p*-nitrophenylsulfonyl)-thiazole		285.30				222–226			
a251	2-Amino-5-nitropyridine	$H_2N—C_5H_3N—NO_2$	139.11	22^1, 631			188			sl s aq, bz, eth
a252	2-Amino-5-nitrothiazole		145.14				202 d			v sl a aq; 0.7 alc; 0.4 eth
a253	*exo*-2-Aminonorbornane		111.19		0.938	1.4807^{20}		49^{10mm}	35	

1-Aminonaphthalene, n17

1-Amino-2-naphthol-4-sulfonic acid, a195

1-Amino-2-naphthol-6-sulfonic acid, a197

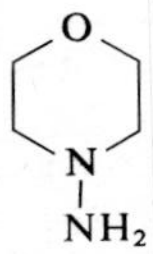

a234 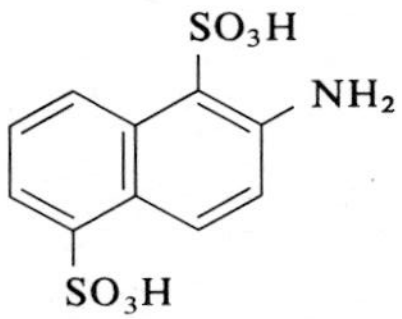 a235 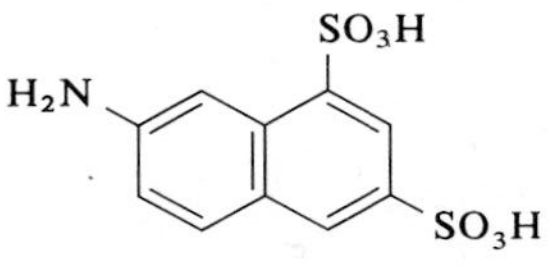 a236 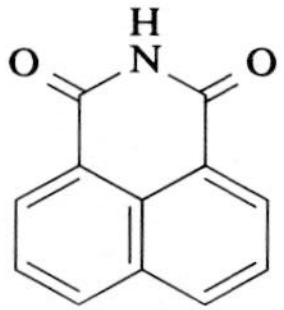 a238 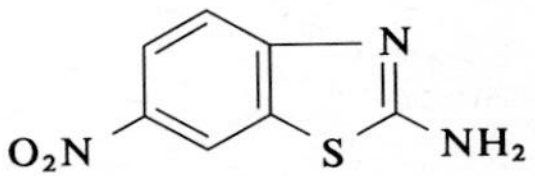 a243

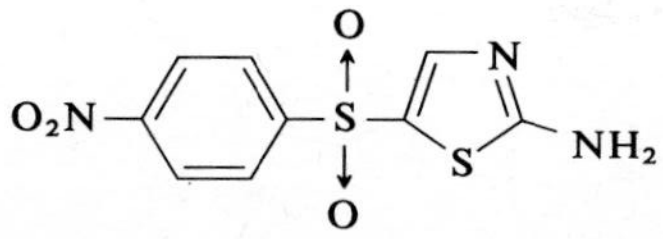

a250 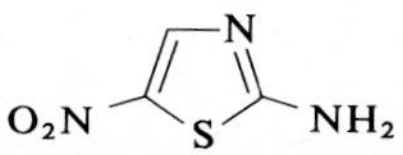 a252 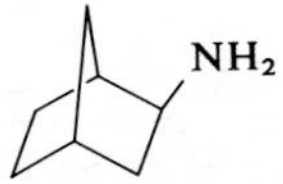 a253

TABLE 1-14 Physical constants of organic compounds (*continued*)

No.	Name	Formula	Formula weight	Beilstein reference	Density	Refractive index	Melting point	Boiling point	Flash point	Solubility in 100 parts solvent
a254	2-Aminopentane	$H(CH_2)_3CH(NH_2)CH_3$	87.17	4, 177	0.739^{20}	1.4047^{20}		91–92		s aq, alc, eth, PE
a255	3-Aminopentane	$C_2H_5CH(NH_2)C_2H_5$	87.17	4, 178	0.749^{20}_{4}	1.4055^{20}		91	1	misc aq, alc, eth
a256	DL-2-Aminopentanoic acid	$H(CH_2)_3CH(NH_2)COOH$	117.15	4, 416			303	320 subl		5.5 aq^{18}; v sl s alc, chl, eth, PE
a257	5-Aminopentanoic acid	$H_2N(CH_2)_4COOH$	117.15	4, 418			158–161			v s aq; sl s alc; i eth
a258	5-Amino-1-pentanol	$H_2N(CH_2)_5OH$	103.17	4^1, 441		1.4615^{20}	37	122^{16mm}	65	
a259	2-Aminophenethyl alcohol	$H_2NC_6H_4CH_2CH_2OH$	137.18	13^3, 1679	1.045	1.5849^{20}		148^{4mm}	> 112	
a260	2-Aminophenol	$H_2NC_6H_4OH$	109.13	13, 354			170–174			2 aq; 4.3 alc; v s eth; sl s bz
a261	3-Aminophenol	$H_2NC_6H_4OH$	109.13	13, 401			122–123	164^{11mm}		2.5 aq; v s alc, eth
a262	4-Aminophenol	$H_2NC_6H_4OH$	109.13	13, 427			190	284 d		0.65 aq; s alc, eth
a263	4′-Aminophenylacetonitrile	$H_2NC_6H_4CH_2CN$	132.17	14, 457			44	312		sl s hot aq; s alc
a264	1-(3-Aminophenyl)-ethanol	$H_2NC_6H_4CH(CH_3)OH$	137.18	13^3, 1654			68–71			
a265	2-Amino-1-phenyl-ethanol	$C_6H_5CH(CH_2NH_2)OH$	137.18	13^2, 361			56–57	160^{17mm}		v s aq; s alc
a266	1*S*,2*S*-(+)-2-Amino-1-phenyl-1,3-propanediol	$C_6H_5CH(OH)CH(NH_2)$-CH_2OH	167.21				109–113			
a267	L-2-Amino-3-phenyl-1-propanol	$C_6H_5CH_2CH(NH_2)$-CH_2OH	151.21	13^3, 1757			92–94			
a268	3-Amino-1-phenyl-2-pyrazolin-5-one		175.19				210–215			
a271	*N*-Aminophthalimide		162.15				200–202			
a272	*N*-Aminopiperidine		100.17	20, 89	0.928	1.4750^{20}		146^{730mm}	36	
a273	3-Amino-1,2-propanediol	$H_2NCH_2CH(OH)CH_2OH$	91.11	4, 301	1.175	1.4920^{20}		265^{739mm}	> 112	

a274	DL-1-Amino-2-propanol	$CH_3CH(OH)CH_2NH_2$	75.11	4, 289	0.973	1.4483^{20}	−2	160	73	s aq, alc; i eth
a275	DL-2-Amino-1-propanol	$CH_3CH(NH_2)CH_2OH$	75.11	4^1, 432	0.943	1.4495^{20}		173–176		v s aq, alc, eth
a276	L-2-Amino-1-propanol	$CH_3CH(NH_2)CH_2OH$	75.11	4^1, 432	0.965	1.4495^{20}		176	62	v s aq, alc, eth
a277	3-Amino-1-propanol	$H_2NCH_2CH_2CH_2OH$	75.11	4, 288	0.982	1.4598^{20}	12	188	79	s aq, alc
a278	2-Amino-1-propene-1,1,3-tricarbonitrile	$NCC(CN){=}C(NH_2)$-CH_2CN	132.13				171–173			s aq
a279	3-Aminopropionitrile	$H_2NCH_2CH_2CN$	70.09					185		
a280	3-Aminopropyl-(diethoxy)methyl-silane	$H_2N(CH_2)_3Si(CH_3)$-$(OCH_2CH_3)_2$	191.4		0.916_4^{20}	1.427^{20}		85– 88^{8mm}		
a281	*N*-(3-Aminopropyl)-iminodiethanol	$H_2N(CH_2)_3N$-$(CH_2CH_2OH)_2$	162.23		0.1071	1.4980^{20}		170^{2mm}	137	
a282	*N*-(3-Aminopropyl)-morpholine		144.22		0.9872_{20}^{20}	1.4761^{20}	−15	224	98	misc aq, alc, bz
a283	*N*-(3-Aminopropyl)-2-pyrrolidinone		142.20		1.014	1.5000^{20}		120–123^{1mm}	> 112	

5-Aminoorotic acid, a160
1-Aminopentane, p53
4-Aminophenyl phenyl ether, p70
Aminophenyl sulfones, d36, d37
3-Aminophthalhydrazide, a153
Aminopicolines, a228, a229, a230
1-Aminopropane, p223
2-Aminopropane, i88
3-Amino-1-propene, a82
N-(3-Aminopropyl)diethanolamine, a281

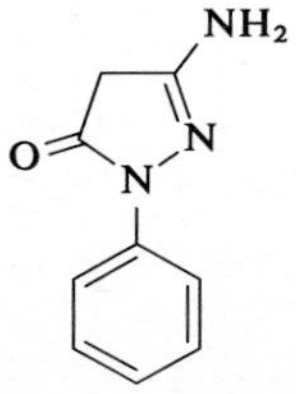

a268

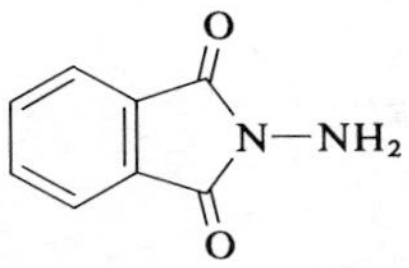

a271

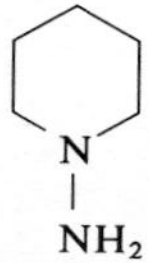

a272

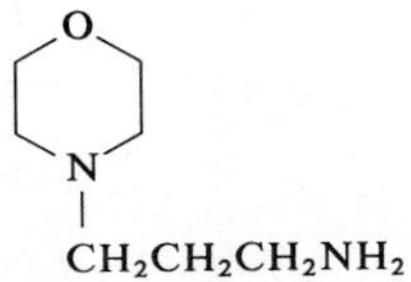

$CH_2CH_2CH_2NH_2$

a282

$CH_2CH_2CH_2NH_2$

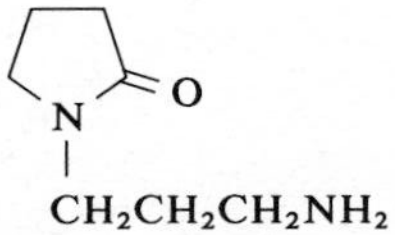

a283

TABLE 1-14 Physical constants of organic compounds (*continued*)

No.	Name	Formula	Formula weight	Beilstein reference	Density	Refractive index	Melting point	Boiling point	Flash point	Solubility in 100 parts solvent
a284	3-Aminopropyltriethoxysilane	$H_2N(CH_2)_3Si(OC_2H_5)_3$	221.37		0.9506_4^{20}	1.4225^{20}		217	96	
a285	3-Aminopropyltrimethoxysilane	$H_2N(CH_2)_3Si(OCH_3)_3$	179.2		1.01_4^{25}	1.420^{25}		80^{8mm}	104	
a286	2-Aminopyridine	$(C_5H_4N)NH_2$	94.12	22, 428			58.1	210.6	92	s aq, alc, bz, eth
a287	3-Aminopyridine	$(C_5H_4N)NH_2$	94.12	22, 431			64	248		s aq, alc, bz, eth
a288	4-Aminopyridine	$(C_5H_4N)NH_2$	94.12	22, 433			155–158	273		s aq, alc; sl s bz, eth
a289	2-Aminopyrimidine		95.11	24, 80			123–126	subl		v s aq
a290	4-Aminoquinaldine		158.20	22, 453			169	333		sl s aq; v s alc, eth, acet; s hot bz
a291	4-Aminosalicyclic acid	$H_2NC_6H_3(OH)COOH$	153.14	14, 579			147 d			0.2 aq; 4.8 alc; s dil acid, alk
a292	5-Aminosalicyclic acid	$H_2NC_6H_3(OH)COOH$	153.14	14, 579			280 d			sl s aq, alc; s acid
a293	2-Amino-3-sulfopropionic acid	$HOOCCH(NH_2)$-CH_2SO_3H	187.17	4, 533			260 d			v s aq
a294	5-Amino-1,2,3,4-tetrazole hydrate		103.08	26, 403			204 d			
a295	5-Amino-1,3,4-thiadiazole-2-thiol		133.20	27, 674			235 d			
a296	2-Aminothiazole		100.14	27, 155			93			sl s aq, alc, eth
a297	2-Amino-2-thiazole		100.14	27, 136			91–93			
a298	2-Aminothiophenol	$H_2NC_6H_4SH$	125.19	13, 397		1.6405^{20}	26	234	79	
a299	6-Amino-3-toluenesulfonic acid	$H_2NC_6H_3(CH_3)SO_3H$	187.22	14, 723			> 300			1 aq[12]; v s hot aq
a300	3-Amino-1,2,4-triazole		84.08	26, 137			159			s aq, alc, chl

a301	5-Amino-2,2,4-trimethyl-1-cyclopentanemethylamine		156.27		0.901	1.4733^{20}		221	97	
a302	11-Aminoundecanoic acid	$H_2N(CH_2)_{10}COOH$	201.31				190–192			
a303	Aniline	$C_6H_5NH_2$	93.13	12, 59	1.0217^{20}	1.5855^{20}	−5.98	184.40	70	3.5 aq^{25}; s alc, CCl_4, eth, acids
a304	Aniline hydrochloride	$C_6H_4NH_2 \cdot HCl$	129.59		1.222		198		193	100 aq; v s alc
a305	2-Anilinoethanol	$C_6H_5NHCH_2CH_2OH$	137.18	12, 182	1.085	1.5793^{20}		150–152^{10mm}	> 112	sl s aq; v s alc, chl, eth
a306	3-Anilinopropionitrile	$C_6H_5NHCH_2CH_2CN$	146.19				52–53			

6-Aminopurine, a69
2-Amino-3-pyridinol, a199
Aminopyrimidinediols, a154, a155
2-Aminosuccinamic acid, a318
Aminosuccinic acid, a319
6-Amino-2-thiouracil, a193
α-Amino-*p*-toluenesulfonamide, a213
2-Amino-1,1,3-tricyanopropene, a278
1-Aminotricyclo[$3.3.1.1^{3,7}$]decane, a66

Aminouracil, a155
2-Aminovaleric acid, a256
5-Aminovaleric acid, a257
Amyl compounds, *see* Pentyl
Amyl alcohol, p37
act-Amyl alcohol, m153
sec-Amyl alcohol, p38
tert-Amyl alcohol, m154
tert-Amylamine, d603

Amyl bromides, b325, b326
Amyl chloride, c192
Amyl iodide, i47
Amyl mercaptan, p35
Amyl methyl ketone, h15
Anethole, m97
Angelic acid, m162
Anilinesulfonic acids, a118, a119, a120
Aniline-2,5-disulfonic acid, a117

a289 a290 a294 a295 a296 a297 a300 a301

TABLE 1-14 Physical constants of organic compounds (*continued*)

No.	Name	Formula	Formula weight	Beilstein reference	Density	Refractive index	Melting point	Boiling point	Flash point	Solubility in 100 parts solvent
a307	1-(*o*-Anisidino)-1,3-butanedione	$CH_3OC_6H_4NHCOCH_2$-$COCH_3$	207.23	13^1, 117			84–85			
a308	1-(*p*-Anisidino)-1,3-butanedione	$CH_3OC_6H_4NHCOCH_2$-$COCH_3$	207.23	13^1, 177			115–117			
a309	Anthracene		178.23	5, 657	1.25_4^{27}		216.3	340		i aq; 1.5 alc; 1.6 bz; 1.2 chl; 3.1 CS_2
a310	9,10-Anthracenedione		208.22	7, 781	1.43_4^{20}		286	377	185	i aq; 0.44 alc; 0.26 bz; 0.61 chl; 0.11 eth
a311	9,10-Anthraquinone-1,5-disulfonic acid disodium salt		412.31	11, 340			> 300			s aq
a312	9,10-Anthraquinone-2,6-disulfonic acid disodium salt		412.31	11, 342			> 325			s aq
a313	9,10-Antrhaquinone-2-sulfonic acid Na salt		310.26							
a314	Antipyrine		188.23	24, 27	1.088_4^{113}		114	319^{174mm}		100 aq; 77 alc; 100 chl; 2.3 eth
a315	L-(+)-Arabinose		150.13	31, 32			160–163			100 aq
a316	L-(+)-Arginine	$H_2NC(=NH)NH(CH_2)_3$-$CH(NH_2)COOH$	174.20	4, 420			223 d			17.6 aq; sl s alc
a317	L-(+)-Ascorbic acid		176.12				190–192 d			100 aq; 3.3 alc
a318	L-(+)-Asparagine hydrate	$H_2NCOCH_2CH(NH_2)$-$COOH \cdot H_2O$	150.14	4, 484			233–235			3.6 aq^{28}; s alk acids; i alc, bz, eth

No.	Name	Formula	Mol. wt.	Beilstein			mp			Solubility
a319	L-(+)-Aspartic acid	$HOOCCH_2CH(NH_2)COOH$	133.10	4, 472			270 sealed tube			0.45 aq; i alc, eth
a320	Atropine		289.38	21, 27			114–116			0.22 aq; s bz, dil acid

Anisaldehydes, m45, m46
Anisamide, m47
Anisic acids, m50, m51, m52
Anisidines, m42, m43, m44
Anisole, m48
p-Anisoyl chloride, m53
p-Anisyl alcohol, m54
Anthraflavic acid, d375
Anthranilamide, a114
Anthranilic acid, a121
Anthranionitrile, a124
9,10-Anthraquinone, a310
APDC, p275
Araboascorbic acid, i59
Aspirin, a56
Arsanilic acids, a115, a116

a309

O O
a310

O SO_3Na NaO_3S O
a311

O SO_3Na NaO_3S O
a312

O SO_3Na O
a313

H_3C N N O H_3C
a314

HO O OH OH
a315

CH_2OH HCOH O O HO OH
a317

CH_3 N O $OCCHCH_2OH$
a320

TABLE 1-14 Physical constants of organic compounds (*continued*)

No.	Name	Formula	Formula weight	Beilstein reference	Density	Refractive index	Melting point	Boiling point	Flash point	Solubility in 100 parts solvent
a321	Aurintricarboxylic acid, triammonium salt		473.44	10^2, 775			225 d			v s aq
a322	2-Azacyclooctanone		127.19	21, 242			35–38	148^{10mm}		
a323	2-Azacyclotridecanone		197.32				150–153			
a324	Azidotrimethylsilane	$(CH_3)_3SiN_3$	115.21		0.868	1.4142^{20}	−95	95–96	23	
a325	Azidotriphenylsilane	$(C_6H_5)_3SiN_3$	301.4				83–84	$100^{0.01mm}$		
a326	1-Aziridineethanol	$(CH_2)_2{=}NCH_2CH_2OH$	87.12		1.088	1.4560^{20}		168	67	
a327	*cis*-Azobenzene	$C_6H_5N{=}NC_6H_5$	182.23	16, 8	1.20		68.3	293		i aq; s alc, eth, HOAc
a328	2,2′-Azobis(2-methylpropionitrile	$(CH_3)_2C(CN)N{=}N{-}C(CN)(CH_3)_2$	164.21	4, 563				107 d		2 EtOH; 5 MeOH; can explode in acetone
a329	Azodicarbonamide	$H_2NCON{=}NCONH_2$	116.08	3, 123			225 d			
a330	4,4′-Azoxydianisole	$CH_3OC_6H_4N{=}N(\rightarrow O){-}C_6H_4OCH_3$	258.28	16, 637			120			
a331	Azulene		128.17	5^2, 432			100.5	250		
b1	Barbituric acid		128.09	24, 467			248–252 d			s hot aq, dil acid
b2	Basic fuchsin		337.86	13, 765	1.22		d 186			0.3 aq; s alc, acids
b3	Benzaldehyde	C_6H_5CHO	106.12	7, 174	1.0447^{20}	1.5455^{20}	−26	178.9	62	0.3 aq; misc alc, eth
b4	Benzamide	$C_6H_5CONH_2$	121.14	9, 195	1.341^4		127.2	288		1.3 aq; 17 alc; 30 pyr
b5	Benzanilide	$C_6H_5CONHC_6H_5$	197.24	12, 262	1.315		163.1	117^{10mm}		i aq; 1.7 alc; sl s eth
b6	1,2-Benzanthracene		228.29	5, 718			155–157	437.6		sl s hot alc; s most other org solv

b7	2,3-Benzanthracene		228.29	5^2, 628	1.35		341	subl		sl s most org solv
b8	7*H*-Benz[*de*]-anthracen-7-one		230.27	7, 518			170			1.6 bz; 0.5 HOAc
b8a	Benzene	C_6H_6	78.11	5, 179	0.8737^{25}	1.4979^{25}	5.53	80.10	−11	0.17 aq; s most org solv
b9	Benzene-*d*	C_6H_5D	79.12			1.4980^{20}		80	−11	
b10	Benzene-d_6	C_6D_6	84.16		0.95	1.4978^{20}		79.1	−11	

Azacyclopropane, e134
Azelaic acid, n95
Azelonitrile, n94
Aziridine, e134
Azobis(isobutyronitrile), a328
4,4′-Azoxyanisole, a330
Barbitol, d280
Behenic acid, d718
Behenyl alcohol, d719
Benzalacetone, p98
Benzal bromide, d102
Benzalphthalide, b102
Benzanthrone, b8
Benzeneacetaldehyde, p76a

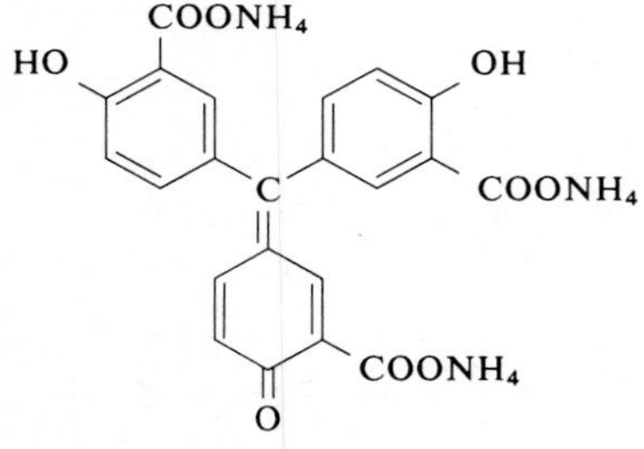

a321

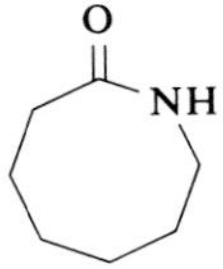

a322

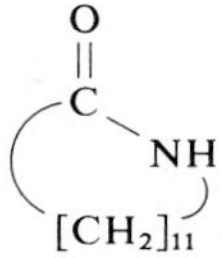

a323

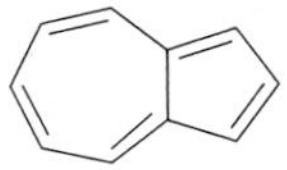

a331

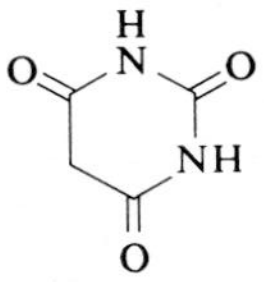

b1

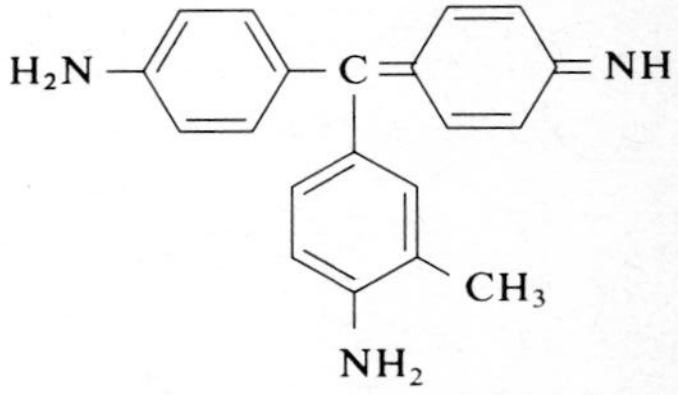

b2

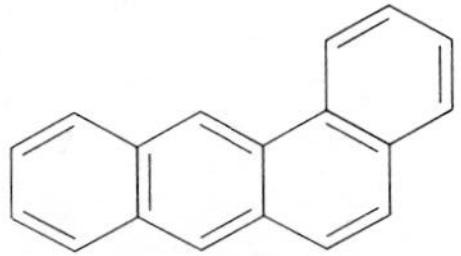

b6

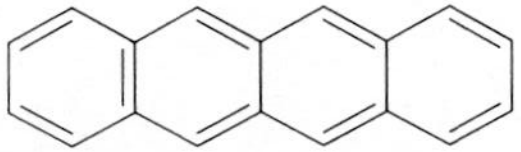

b7

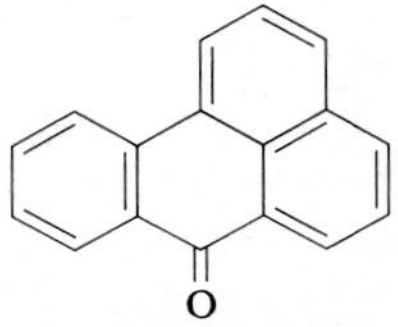

b8

TABLE 1-14 Physical constants of organic compounds (*continued*)

No.	Name	Formula	Formula weight	Beilstein reference	Density	Refractive index	Melting point	Boiling point	Flash point	Solubility in 100 parts solvent
b11	Benzenearsonic acid	$C_6H_5AsO(OH)_2$	202.04	16, 868	1.760^{25}		163 d			2.5 aq; 2 alc
b12	Benzeneboronic acid	$C_6H_5B(OH)_2$	121.93	16, 920			217 to the anhydride	$-H_2O$ on standing in air		2.6 aq; 1.8 alc; 43 eth; s bz
b13	1,4-Benzenedicarbaldehyde	$C_6H_4(CHO)_2$	134.13	7, 675			114	248		i aq; 6 bz; 17 acet; 2 eth; 14 diox; 46 MeOH
b14	1,3-Benzendicarbonyl dichloride	$C_6H_4(COCl)_2$	203.02	9, 834			43–44	276	180	73 bz; 62 CCl_4
b15	1,4-Benzenedicarbonyl dichloride	$C_6H_4(COCl)_2$	203.02	9, 844			81	266	180	37 bz; 9 CCl_4
b16	1,3-Benzenedicarboxylic acid	$C_6H_4(COOH)_2$	166.13	9, 832			345–348	subl		0.012 aq; v s alc, HOAc; i bz, PE
b17	1,4-Benzendicarboxylic acid	$C_6H_4(COOH)_2$	166.13	9, 841			subl without melting			v sl s aq, chl, eth; sl s alc; s alk
b18	1,4-Benzenedimethanol	$C_6H_4(CH_2OH)_2$	138.17	6, 919	1.100^{17}		115	143^{1mm}	188	v s aq, alc, eth
b19	Benzenehexacarboxylic acid	$C_6(COOH)_6$	342.17	9, 1008			286 d			v s aq, alc
b20	Benzenesulfinic acid	$C_6H_5S(=O)OH$	142.16	11, 2			85	100 d		sl s aq; s alc, bz, eth
b21	Benzenesulfonamide	$C_6H_5SO_2NH_2$	157.19	11, 39			152			i aq; sl s alc; s eth
b22	Benzenesulfonic acid	$C_6H_5SO_2OH$	158.18	11, 26			50–51			v s aq, alc; sl s bz

b23	Benzenesulfonyl chloride	$C_6H_5SO_2Cl$	176.62	11, 34	1.3842^{15}_{15}	1.5518	14.5	177^{100mm}	>112	i aq; s alc, eth
b24	Benzenesulfonyl fluoride	$C_6H_5SO_2F$	160.16	11^2, 23	1.3286^{20}_{4}	1.4932^{18}		203–204		s alc, eth
b25	Benzenesulfonyl hydrazide	$C_6H_5SO_2NHNH_2$	172.21	11, 52			101–103			flammable solid
b26	1,2,4,5-Benzenetetra-carboxylic acid	$C_6H_2(COOH)_4$	254.15	9, 997			276			1.5 aq; v s alc
b27	1,2,4,5-Benzenetetra-carboxylic anhydride		218.12	19, 196			283–286	397–400		
b28	1,2,3-Benzenetricar-boxylic acid dihydrate	$C_6H_3(COOH)_3 \cdot 2H_2O$	246.18	9, 976			192 d			sl s aq; v s eth
b29	1,2,4-Benzenetricar-boxylic acid	$C_6H_3(COOH)_3$	210.14	9, 977			321 d			2.1 aq; 25.3 alc; 7.9 acet; v s eth
b30	1,3,5-Benzenetricar-boxylic acid	$C_6H_3(COOH)_3$	210.14	9, 978			>330			sl s aq; v s alc; s eth
b31	1,2,4-Benzenetricar-boxylic anhydride		192.13	18, 468			161–164	245^{14mm}		50 acet; 22 EtAc

Benzeneazobenzene, a327
Benzenecarbonitrile, b51
Benzene-1,2-dicarboxylic acid, p170
Benzenemethanol, b78
Benzenethiol, t162

b27

COOH

b31

TABLE 1-14 Physical constants of organic compounds (*continued*)

No.	Name	Formula	Formula weight	Beilstein reference	Density	Refractive index	Melting point	Boiling point	Flash point	Solubility in 100 parts solvent
b32	1,3,5-Benzenetricarboxylic trichloride	$C_6H_3(COCl)_3$	265.48				35–36			
b33	Benzethonium chloride	$(CH_3)_3CCH_2C(CH_3)_2$-$C_6H_4OCH_2CH_2OCH_2$-$CH_2N^+(CH_3)_2$-$CH_2C_6H_5Cl^-$	448.10				164–166			v s aq; s alc, acet
b34	Benzil	$C_6H_5COCOC_6H_5$	210.23	7, 747	1.23_4^{15}		94.9	346		i aq; s alc, eth
b35	Benzil-α-dioxime	$C_6H_5C(=NOH)C(=NOH)$-C_6H_5								s alk
b36	Benzilic acid	$(C_6H_5)_2C(OH)COOH$	228.25	10, 342			153			sl s aq; v s alc, eth
b37	Benzil monohydrazone	$C_6H_5C(=NNH_2)COC_6H_5$	224.26	7^1, 394			150–152			
b38	Benzimidazole		118.14	23, 131			170.5	>360		sl s aq, eth; v s alc
b39	Benzo-15-crown-5		268.3				76–78			
b40	7,8-Benzo-1,3-diazaspiro-[4,5]decane-2,4-dione		216.24				268–270			
b41	1,4-Benzodioxan		136.15		1.142	1.5485^{20}		103^{6mm}	87	
b42	2,3-Benzofuran		118.14	17, 54	1.072	1.5660^{20}	<−18	175		i aq; misc bz, eth, PE
b43	Benzofurazan 1-oxide		136.11	27^1, 740			69–71			
b44	Benzoic acid	C_6H_5COOH	122.13	9, 92	1.080		122.4	132.5^{10mm}	121	0.29 aq; 43 alc; 10 bz; 22 chl; 33 eth; 33 acet
b45	Benzoic anhydride	$(C_6H_5CO)_2O$	226.23	9, 164	1.199		39–40	360		i aq; s alc, acet, chl, bz, HOAc
b46	DL-Benzoin	$C_6H_5COCHOHC_6H_5$	212.25	8, 165	1.3100_4^{20}		134–136	344		s acet; 20 pyr
b47	Benzoin ethyl ether	$C_6H_5CH(OC_2H_5)$-COC_6H_5	240.30	8, 174	1.1016_4^{17}	1.5727^{17}	61	195^{20mm}		s alc, bz, eth

b48	Benzoin isobutyl ether	$C_6H_5CH[OCH_2CH(CH_3)_2]COC_6H_5$	268.36		0.985	1.5485^{20}		$133^{0.5mm}$	85	
b49	Benzoin methyl ether	$C_6H_5CH(OCH_3)COC_6H_5$	226.28	8, 174	1.1278^{14}_{4}		48	189^{15mm}		v s alc, bz, eth
b50	α-Benzoinoxime	$C_6H_5CH(OH)C(=NOH)C_6H_5$	227.26	8, 175			151–152			sl s aq; s alc, NH_4OH
b51	Benzonitrile	C_6H_5CN	103.12	9, 275	1.0006^{25}	1.5257^{25}	−12.75	191.1	71	0.2 aq; misc alc, bz, chl, eth
b52	Benzo[*def*]phenanthrene		202.26	5, 693	1.271^{23}		156	404		i aq; s alc, eth
b53	Benzophenone	$C_6H_5COC_6H_5$	182.22	7, 411	1.1108^{15}_{4}		48.1	305		i aq; 13.3 alc; 17 eth
b54	Benzophenone hydrazone	$C_6H_5C(=NNH_2)C_6H_5$	196.25	7, 417			98	230^{55mm}		

Benzhydrazide, b72
Benzhydrol, d679
Benzhydrylamine, d680
Benzhydryl bromide, b276
Benzidine, b137
2-Benzimidazolethiol, m15
1,3-Benzodiazole, b38
1,3,2-Benzodioxaborole, c21
1,3-Benzodioxole, m241
Benzofuroxan, b43
Benzoglyoxaline, b38
Benzoic acid hydrazide, b72
o-Benzoic sulfimide, s1

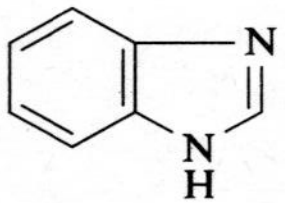
b38

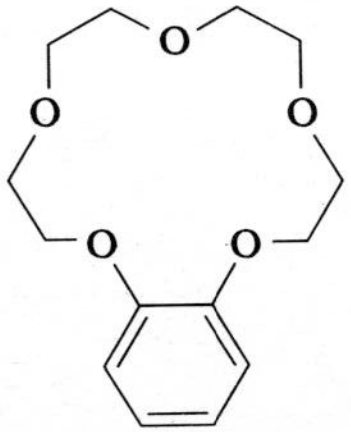
b39

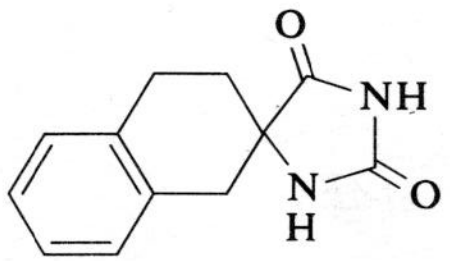
b40

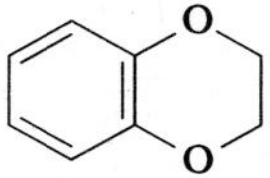
b41

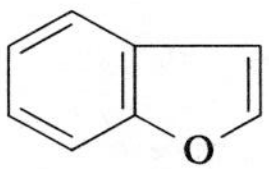
b42

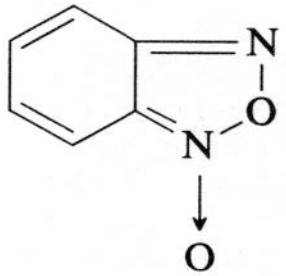
b43

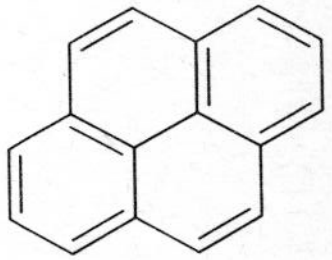
b52

TABLE 1-14 Physical constants of organic compounds (*continued*)

No.	Name	Formula	Formula weight	Beilstein reference	Density	Refractive index	Melting point	Boiling point	Flash point	Solubility in 100 parts solvent
b55	3,3′,4,4′-Benzophenonetetracarboxylic dianhydride		322.23				215–217			
b56	1-Benzopyran-4(4*H*)-one		146.15	17, 327			55–57			
b57	1,2-Benzo[*a*]pyrene		252.32				179.3	495		i aq; s bz; sl s alc
b58	4,5-Benzo[*e*]pyrene		252.32				182			i aq
b59	1,4-Benzoquinone	$O{=}C_6H_4{=}O$	108.10	7, 609	1.318_4^{20}		115.7			sl s aq; s alc, eth, hot bz, alk (with d)
b60	Benzothiazole		135.19		1.246_4^{20}	1.6379^{20}	2	231	>112	sl s aq; v s alc, CS_2
b61	Benzo[*b*]thiophene		134.20	17, 59	1.1937^{40}	1.6302^{40}	31.32	221		s alc, bz, chl, eth
b62	1,2,3-Benzotriazole		119.13	26, 38	1.238	1.6420^{20}	98.5	204^{15mm}		sl s aq; s alc, bz, chl
b63	Benzoxazole		119.12	27, 42		1.5594	30	182	58	sl s aq
b64	1-Benzoylacetone	$C_6H_5COCH_2COCH_3$	162.19	7, 680	1.090_{60}^{60}		60	260 sl d		sl s aq; v s alc, eth
b65	2-Benzoylbenzoic acid	$C_6H_5COC_6H_4COOH$	226.23	10, 747			129	265		sl s aq; v s alc, eth
b66	Benzoyl bromide	C_6H_5COBr	185.03	9, 195	1.5467^{20}			218–219	90	d aq, alc; misc eth
b67	Benzoyl chloride	C_6H_5COCl	140.57	9, 182	1.211_4^{20}	1.5525^{20}	−1.0	197.2	68	d aq, alc; misc bz, CS_2, eth
b68	Benzoyl cyanide	C_6H_5COCN	131.13	10, 659			32	206		i aq
b69	Benzoyl fluoride	C_6H_5COF	124.11	9, 181	1.140	1.4960^{20}	−28	161	48	d hot aq; v s alc, eth
b70	Benzoylformic acid	$C_6H_5COCOOH$	150.13	10, 654			69			
b71	*N*-Benzoylglycine	$C_6H_5CONHCH_2COOH$	179.18	9, 225			178–179			0.4 aq; 0.1 chl; 0.25 eth; sl s alc; i bz, PE
b72	Benzoylhydrazine	$C_6H_5CONHNH_2$	136.15	9, 319			117			

b73	3-Benzoylpropionic acid	$C_6H_5COCH_2CH_2COOH$	178.19	10, 696			116			sl s aq; s alc
b74	2-Benzoylpyridine	$C_6H_5CO-C_5H_4N$	183.21	21, 330			44	317	150	
b75	3-Benzoylpyridine	$C_6H_5CO-C_5H_4N$	183.21	21, 331			40	307	150	s alc, bz, eth
b76	4-Benzoylpyridine	$C_6H_5CO-C_5H_4N$	183.21	21, 331			71	315	150	s alc, bz, eth
b77	Benzyl acetate	$CH_3COOCH_2C_6H_5$	150.18	6, 435	1.0515^{25}	1.5232^{20}	−51.5	215.5	102	sl s aq; misc alc, eth
b78	Benzyl alcohol	$C_6H_5CH_2OH$	108.13	6, 428	1.0413^{25}	1.5371^{25}	−15.3	205.45	100	0.08 aq; misc alc, eth, eth
b79	Benzylamine	$C_6H_5CH_2NH_2$	107.16	12, 1013	0.981^{19}_{4}	1.5424^{20}	10	185	60	misc aq, alc, eth
b80	2-Benzylaminoethanol	$C_6H_5CH_2NHCH_2CH_2OH$	151.21	12, 1040	1.065	1.5435^{20}		156^{12mm}	>112	
b81	(3-Benzylamino)-propionitrile	$C_6H_5CH_2NHCH_2CH_2CN$	160.22			1.5308^{20}				

Benzoresorcinol, d387
2-Benzothiazolethiol, m17
Benzotrichloride, t252

Benzotrifluoride, t305
Benzoylamide, b4
Benzoylbenzene, b53

Benzoyl peroxide, d54
1,2-Benzphenanthrene, b6
Benzylaniline, p93

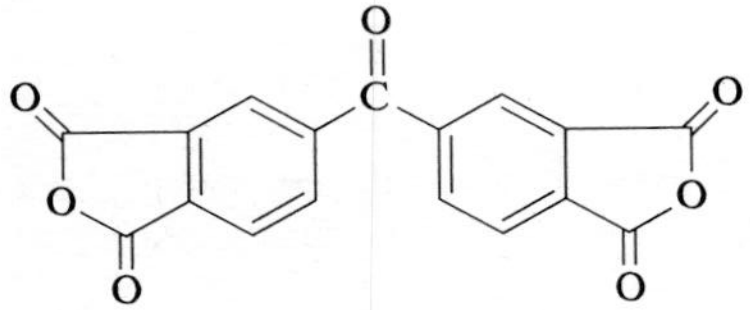

b55 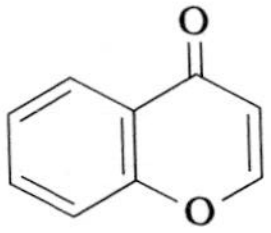 b56 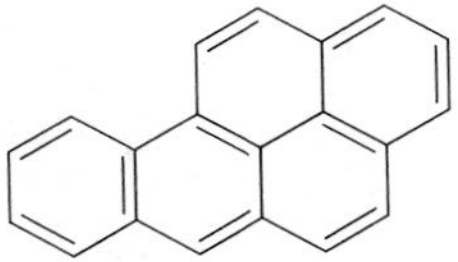 b57 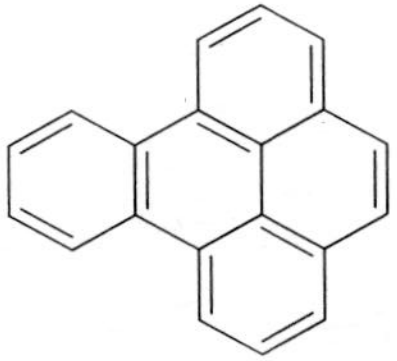 b58 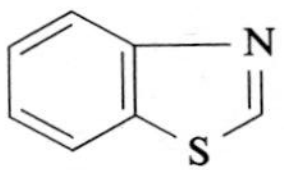 b60 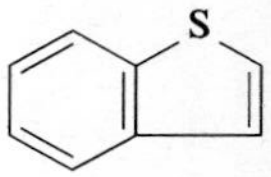 b61

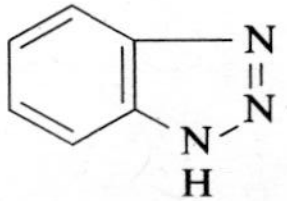

b62 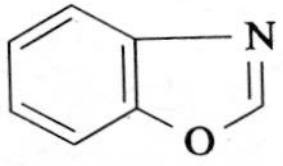 b63

TABLE 1-14 Physical constants of organic compounds (*continued*)

No.	Name	Formula	Formula weight	Beilstein reference	Density	Refractive index	Melting point	Boiling point	Flash point	Solubility in 100 parts solvent
b82	*N*-Benzylbenzamide	$C_6H_5CONHCH_2C_6H_5$	211.26				106			
b83	Benzyl benzoate	$C_6H_5COOCH_2C_6H_5$	212.25	9, 121	1.118^{25}_{4}	1.5681^{21}	19.4	323.5	147	i aq; misc alc, chl, eth
b84	2-Benzylbenzoic acid	$C_6H_5CH_2C_6H_4COOH$	212.24	9^2, 471			110–113			sl s aq; s alc, bz, chl, eth
b85	Benzyl bromide	$C_6H_5CH_2Br$	171.04	5, 306	1.438^{22}_{0}	1.5752^{20}	−3.9	198–199	86	sl d aq
b86	Benzyl-*tert*-butanol	$C_6H_5CH_2CH_2$-$C(CH_3)_2OH$	164.25	6, 548		1.5090^{20}	33	144^{85mm}	>112	
b87	Benzyl butyl 1,2-phthalate	$C_6H_5CH_2OOCC_6H_4$-$COOC_4H_9$	312.37		1.119^{25}_{25}				218	
b88	Benzyl carbamate	$C_6H_5CH_2OCONH_2$	151.17	6, 437			87–89	220 d		i aq; v s alc; sl s eth
b89	Benzyl chloride	$C_6H_5CH_2Cl$	126.59	5, 292	1.0993^{20}	1.5391^{20}	−43 to −48	179	73	i aq; misc alc, chl, eth
b90	Benzyl chloroformate	$C_6H_5CH_2OC(O)Cl$	170.60	6, 437	1.195	1.5190^{20}		103^{20mm}	91	d aq; s eth
b91	Benzyl chlorothiol formate	$C_6H_5CH_2S—COCl$	186.5		1.237^{30}_{4}	1.5711^{30}		$80^{0.13mm}$	118	
b92	*S*-Benzyl-L-cysteine	$C_6H_5CH_2SCH_2$-$CH(NH_2)COOH$	211.28	6, 465			214 d			
b93	Benzyl diethyl phosphite	$C_6H_5CH_2P(O)(OC_2H_5)_2$	228.23		1.076	1.4930^{20}		110^{2mm}	>112	
b94	Benzyldimethyl-stearylammonium chloride	$C_6H_5CH_2N[(CH_2)_{17}$-$CH_3](CH_3)_2Cl \cdot H_2O$	442.18	12^3, 2212			67–69			
b95	Benzyl ethyl ether	$C_6H_5CH_2OC_2H_5$	136.20		0.9478^{20}	1.4958^{20}		185.0		i aq; misc alc, eth
b96	*N*-Benzylformamide	$C_6H_5CH_2NHCHO$	135.17	12, 1043			60–61			
b97	Benzyl formate	$C_6H_5CH_2OOCH$	136.15		1.081^{20}_{4}			203		i aq; s alc; misc eth
b99	*O*-Benzylhydroxyl-amine	$C_6H_5CH_2ONH_2$	123.16	6, 440				119^{30mm}		

b100	Benzylidenemalononitrile	$C_6H_5CH{=}C(CN)_2$	154.17	9, 895			83–85			
b101	*N*-Benzylidenemethylamine	$C_6H_5CH{=}NCH_3$	119.17	7, 213	0.967	1.5526^{20}		80^{18mm}	>112	
b102	3-Benzylidenephthalide		222.24	17, 376			102			
b103	2-Benzyl-2-imidazoline HCl		196.68				174			v s aq, alc; s chl; v sl s eth, EtAc
b104	Benzylmethylamine	$C_6H_5CH_2NHCH_3$	138.23	12, 1019	0.939	1.5224^{20}		184.189	77	
b105	3-(*N*-Benzyl-*N*-methylamino)-1,2-propanediol	$C_6H_5CH_2N(CH_3)$-$CH_2CH(OH)CH_2OH$	195.26		1.084	1.5341^{20}		206^{30mm}	>112	
b106	Benzyl methyl sulfide	$C_6H_5CH_2SCH_3$	138.23	6, 453	1.015	1.5620^{20}		195–198	73	
b107	3-Benzyloxyaniline	$C_6H_5CH_2OC_6H_4NH_2$	199.25	13, 404			63–67			
b108	3-Benzyloxybenzaldehyde	$C_6H_5CH_2OC_6H_4CHO$	212.25	8, 73			56–58			
b109	4-Benzyloxybenzaldehyde	$C_6H_5CH_2OC_6H_4CHO$	212.25	8, 73			73–74			

Benzyl cyanide, p80
Benzyl disulfide, d57
N-Benzylethanolamine, b80
Benzyl ether, d58
Benzylideneacetone, p98
Benzylideneacetophenone, d688
Benzyl mercaptan, p130
Benzyl methyl ketone, p146
Benzyloxyamine, b99

b102

· HCl

b103

TABLE 1-14 Physical constants of organic compounds (*continued*)

No.	Name	Formula	Formula weight	Beilstein reference	Density	Refractive index	Melting point	Boiling point	Flash point	Solubility in 100 parts solvent
b110	4-Benzyloxybenzyl alcohol	$C_6H_5CH_2OC_6H_4CH_2OH$	214.26				86–87			
b111	2-Benzyloxyethanol	$C_6H_5CH_2OCH_2CH_2OH$	152.19		1.07_{20}^{20}			255.9	129	0.4 aq
b112	4-Benzyloxy-3-methoxybenzaldehyde	$C_6H_5CH_2OC_6H_3$-$(OCH_3)CHO$	242.27				63–65			
b113	4′-Benzyloxypropiophenone	$C_6H_5CH_2OC_6H_4COC_2H_5$	240.30				100–102			
b114	Benzyl phenyl sulfide	$C_6H_5CH_2SC_6H_5$	200.30	6, 454			43	197^{27mm}		i aq; sl s alc; s eth
b115	1-Benzylpiperazine		176.26		1.014	1.5467^{20}			>112	s aq, alc, eth
b116	4-Benzylpiperidine		175.28	20, 296	0.997	1.5379^{20}	7	279	>112	
b117	1-Benzyl-4-piperidone		189.26		1.021	1.5399^{20}		134^{7mm}	>112	
b118	2-Benzylpyridine	$C_6H_5CH_2—C_5H_4N$	169.23	20, 425	1.054	1.5785^{20}	10	276	125	i aq; v s alc, eth
b119	4-Benzylpyridine	$C_6H_5CH_2—C_5H_4N$	169.23	20, 426	1.061_0^{20}	1.5818^{20}		287	115	s alc; v s eth
b120	1-Benzyl-2-pyrrolidinone		175.23		1.095	1.5525^{20}			>112	
b121	(Benzylthio)acetic acid	$C_6H_5CH_2SCH_2COOH$	182.24				59–63			
b122	Benzyl thiocyanate	$C_6H_5CH_2SCN$	149.22	6, 460			43	235		i aq; s alc; v s eth
b123	Benzyltributylammonium chloride	$C_6H_5CH_2N(C_4H_9)_3{}^+Cl^-$	312.94				155 d			
b124	Benzyltrichlorosilane	$C_6H_5CH_2SiCl_3$	225.57		1.288_4^{20}	1.526^{20}		140–142^{100mm}		
b125	Benzyltriethoxysilane	$C_6H_5CH_2Si(OC_2H_5)_3$	254.40		0.986_4^{20}			170–175^{70mm}		
b126	Benzyltriethylammonium chloride	$C_6H_5CH_2N(C_2H_5)_3{}^+Cl^-$	227.78				185 d			
b127	Benzyltrimethylammonium chloride	$C_6H_5CH_2N(CH_3)_3{}^+Cl^-$	185.70	12, 1020						

b128	Benzyltrimethylsilane	$C_6H_5CH_2Si(CH_3)_3$	164.32		0.8933^{20}	1.4941^{20}		190–191		
b129	Betaine	$(CH_3)_3\overset{+}{N}CH_2COO^-$	117.15	4, 347			d >310			160 aq; 55 MeOH; 6 alc
b130	Bicyclo[2.2.1]hepta-2,5-diene		92.14		0.909^{20}	1.4707^{20}	−20	89	−21	i aq; s PE
b131	Bicyclo[2.2.1]-2-heptene		94.16				46	96	−15	s eth
b132	Bicyclo[2.2.1]-5-heptene-2-carbaldehyde		122.16		1.018	1.4883^{20}		67–70^{12mm}	51	

Benzylphenol, h113
BES, b181
Betahistine, m121

o,o-Bibenzoic acid, b138
Bibenzyl, d668
Bicine, b183

Bicyclo[4.4.0]decane, d1, d2

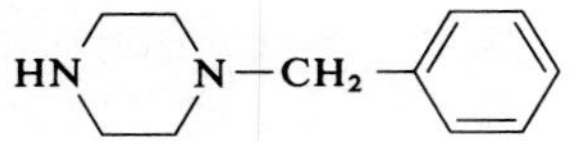

HN N—CH_2—
b115

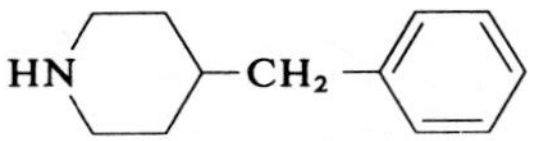

HN —CH_2—
b116

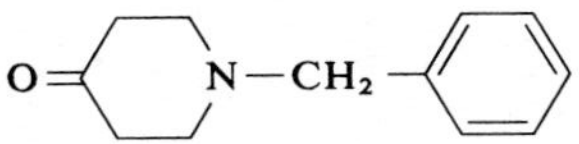

O= N—CH_2—
b117

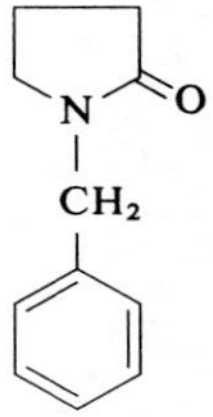

N O
CH_2
b120

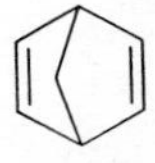

b130

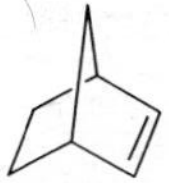

b131

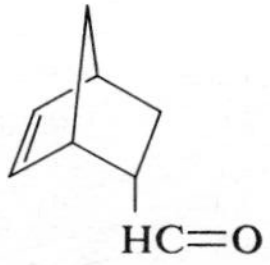

HC=O
b132

TABLE 1-14 Physical constants of organic compounds (*continued*)

No.	Name	Formula	Formula weight	Beilstein reference	Density	Refractive index	Melting point	Boiling point	Flash point	Solubility in 100 parts solvent
b133	Bicyclo[2.2.2]oct-7-ene-2,3,5,6-tetracarboxylic-2,3,5,6-dianhydride		248.19				>300			
b134	Biguanide	$H_2NC(=NH)NHC(=NH)NH_2$	101.11	3, 93			130	d 142		s aq, alc; i bz, eth
b135	Biphenyl	$C_6H_5-C_6H_5$	154.20	5, 578	0.9939^{70}	1.588^{77}	68.8	255.0		i aq; s alc, eth
b136	4-Biphenylcarboxylic acid	$C_6H_5C_6H_4COOH$	198.22	9, 671			226	subl		i aq; v s alc, eth; s bz
b137	(1,1'Biphenyl)-4,4'-diamine	$H_2NC_6H_4C_6H_4NH_2$	184.23	13, 214			128	400^{740mm}		0.04 aq; s alc; 2 eth
b138	(1,1'-Biphenyl)2,2'-dicarboxylic acid	$HOOCC_6H_4C_6H_4COOH$	242.23	9, 922			228–229			0.06 aq; s org solv
b139	4-Biphenylmethanol	$C_6H_5C_6H_4CH_2OH$	184.24	6^2, 636			101			
b140	4-Biphenylsulfonic acid	$C_6H_5C_6H_4SO_3H$	234.26				138			
b141	2-(4-Biphenylyl)-5-(4-*tert*-butylphenyl)-1,3,4-oxadiazole		354.46				138			
b142	*o*-Biphenylyl glycidyl ether		226.28				30–32	$120^{0.1mm}$		
b143	2-(4-Biphenylyl)-5-phenyloxazole		197.36				118			
b144	2,2'-Bipyridinium chlorochromate	$C_5H_4N-C_5H_4NH^+ CrClO_3^-$	292.64							
b145	2,2-Bis[*p*-(allyloxy)-phenyl]propane	$H_2C=CHCH_2OC_6H_4C(CH_3)_2C_6H_4OCH_2CH=CH_2$	308.42		1.022	1.5636^{20}			>112	

b146	*N,N'*-Bis(3-aminopropyl)-ethylenediamine	$H_2N(CH_2)_3NHCH_2CH_2$-$NH(CH_2)_3NH_2$	174.29					$118^{0.2mm}$		
b147	*N,N'*-Bis(3-aminopropyl)piperazine		200.33	23^2, 12	0.973	1.5015^{20}	15	152^{2mm}		
b148	*N,N'*-Bis(3-aminopropyl)-1,3-propanediamine	$H_2N(CH_2)_3NHCH_2CH_2$-$CH_2NH(CH_2)_3NH_2$	188.32					98–103^{1mm}		
b149	2,5-Bis(4-biphenylyl)-oxazole		373.46				240			

Bicyclo[4.3.0]nonane, h46
Biphenol, d388
Biphenylamines, a134, a135
3-(*o*-Biphenylyloxy)-1,2-epoxypropane, b143
2,2'-Bipyridine, d707
Bis(4-aminophenyl)ether, o61
1,3-Bis(aminomethyl)cyclohexane, c315
1,2-Bis(benzylamino)ethane, d59

b133

$(CH_3)_3C$

b141

$OCH_2—CH—CH_2$

b142

143

$H_2NCH_2CH_2CH_2—N$ $N—CH_2CH_2CH_2NH_2$

b147

b149

TABLE 1-14 Physical constants of organic compounds (*continued*)

No.	Name	Formula	Formula weight	Beilstein reference	Density	Refractive index	Melting point	Boiling point	Flash point	Solubility in 100 parts solvent
b150	Bis(2-bromoethyl) ether	$BrCH_2CH_2OCH_2CH_2Br$	231.92					$103\text{–}107^{20}$		
b151	1,3-Bis(bromoethyl)-tetramethyldi-siloxane	$[BrCH_2Si(CH_3)_2]_2O$	320.17		1.3918_4^{20}	1.4719^{20}		$103\text{–}104^{15mm}$		
b152	Bis(2-butoxyethyl) ether	$(C_4H_9OCH_2CH_2)_2O$	218.33		0.8853_{20}^{20}	1.4233^{20}	−60.2	254.6	47	0.3 aq; misc alc, eth, ketones, esters, CCl_4
b153	2,5-Bis(5-*tert*-butyl-2′-benzoxazolyl)-thiophene		430.57				201			
b154	Bis(*sec*-butyl) disulfide	$[CH_3CH_2CH(CH_3)]_2S_2$	178.36	1^3, 1549	0.957	1.4920^{20}		164^{739mm}	>112	
b155	Bis(*tert*-butyl) disulfide	$(CH_3)_3CSSC(CH_3)_3$	178.36	1, 379	0.909	1.4930^{20}		204	79	
b156	Bis(carboxymethyl) trithiocarbonate	$HOOCCH_2SC(=S)\text{-}SCH_2COOH$	226.29	3, 252			172–175			
b157	1,2-Bis(2-chloro-ethoxy)ethane	$(ClCH_2CH_2OCH_2-)_2$	187.07		1.197_4^{20}	1.4617		108^{8mm}		
b158	Bis(2-chloroethoxy)-methylsilane	$H(CH_3)Si\text{-}(OCH_2CH_2Cl)_2$	203.1		1.1643_4^{20}	1.4431^{20}		$95\text{–}97^{18mm}$		
b159	Bis(2-chloroethyl) ether	$ClCH_2CH_2OCH_2CH_2Cl$	143.01	1^2, 335	1.2192^{20}	1.4575^{20}	−51.7	178.8	55	i aq; s most org solv
b160	Bis(2-chloroethyl)-*N*-methylamine	$CH_3N(CH_2CH_2Cl)_2$	156.07		1.118_4^{25}		−60	75^{10mm}		v sl s aq; misc most org solv
b161	Bis(chloromethyl)-dimethylsilane	$(CH_3)_2Si(CH_2Cl)_2$	157.12	4^3, 1845	1.075_4^{20}	1.4600^{20}		160		

b162	Bis(2-chloro-1-methyl)ethyl ether	$ClCH_2CH(CH_3)OCH(CH_3)CH_2Cl$	171.07		1.1122^{20}_{20}			187.3	85	
b163	Bis(4-chlorophenoxy)-acetic acid	$(ClC_6H_4O)_2CHCOOH$	313.14				142			
b164	2,2-Bis(*p*-chlorophenyl)-1,1-dichloroethane	$(ClC_6H_4)_2CHCHCl_2$	320.05				111			
b165	1,1-Bis(4′-chlorophenyl)ethanol	$(ClC_6H_4)_2C(OH)CH_3$	267.16	6^3, 3396			69			v sl s aq; s org solv
b166	Bis(4-chlorophenyl) sulfone	$ClC_6H_4SO_2C_6H_4Cl$	287.16	6, 327				250^{10mm}		
b167	Bis(4-chlorophenyl) sulfoxide	$ClC_6H_4S(O)C_6H_4Cl$	271.17	6^1, 149			144			
b168	1,1-Bis(*p*-chlorophenyl)-2,2,2-trichloroethane	$(ClC_6H_4)_2CHCCl_3$	354.49				109			i aq; 58 acet; 78 bz; 45 CCl_4; v s pyr, diox
b169	1,3-Bis(dichloromethyl)tetramethyldisiloxane	$[Cl_2CH(CH_3)_2Si]_2O$	300.16		1.2213^{20}_{4}	1.4660^{20}		149^{40mm}		
b170	*N,N*-Bis(2,2-diethoxyethyl)-methylamine	$[(C_2H_5O)_2CHCH_2]_2NCH_3$	263.38	4, 311	0.945	1.4259^{20}		222^{244mm}	60	

Bis(3-*tert*-butyl-4-hydroxy-5-methylphenyl) sulfide, t146

Bis(2-cyanoethyl) ether, o63

b153

TABLE 1-14 Physical constants of organic compounds (*continued*)

No.	Name	Formula	Formula weight	Beilstein reference	Density	Refractive index	Melting point	Boiling point	Flash point	Solubility in 100 parts solvent
b171	4,4′-Bis(diethylamino)benzophenone	$[(C_2H_5)_2NC_6H_4]_2C{=}O$	324.47	14, 98			95			
b172	4,4′-Bis(dimethylamino)benzophenone	$[(CH_3)_2NC_6H_4]_2C({=}O)$	268.36	14, 89			172–176	d 360		i aq; s alc, warm bz
b173	Bis(dimethylamino)dimethylsilane	$[(CH_3)_2N]_2Si(CH_3)_2$	146.3		0.810^{22}	1.432^{22}	−98	128–129		
b174	1,3-Bis(dimethylamino)-2-propanol	$[(CH_3)_2NCH_2]_2CHOH$	146.23	4, 290	0.897	1.4422^{20}			>112	
b175	Bis(dimethylthiocarbamyl) disulfide	$[(CH_3)_2NC({=}S)S{-}]_2$	240.43	4, 76	1.29		155–156			s alc, eth; sl s bz, acet; i aq
b176	1,4-Bis(2,3-epoxypropoxy)butane	$[H_2C\overset{O}{\frown}CHCH_2$-$OCH_2CH_2{-}]_2$	202.25		1.049	1.4535^{20}		160^{11mm}	>112	
b177	Bis(2-ethoxyethyl) ether	$(C_2H_5OCH_2CH_2)_2O$	162.23	1^2, 519	0.907_4^{20}	1.4110^{20}	−44.3	188.4	54	v s aq, alc, org solv
b178	Bis(2-ethylhexyl) decanedioate	$CH_3(CH_2)_3CH(C_2H_5)$-$CH_2OOC(CH_2)_8COOCH_2$-$CH(C_2H_5)(CH_2)_3CH_3$	426.66		0.9119_{25}^{25}	1.4496^{25}				
b179	Bis(2-ethylhexyl) hydrogen phosphate	$[CH_3(CH_2)_3CH(C_2H_5)$-$CH_2O]_2P(O)OH$	322.43	1^4, 1786	0.965	1.4450^{20}	−60	209^{10mm}		
b180	Bis(2-ethylhexyl) *o*-phthalate	$[CH_3(CH_2)_3CH(C_2H_5)$-$CH_2OOC]_2C_6H_4$	390.57		0.9843^{20}	1.4859^{20}	−50	384	207	0.01 aq
b181	*N,N*-Bis(2-hydroxyethyl)-2-aminoethanesulfonic acid	$(HOC_2H_5)_2$-$NCH_2CH_2SO_3H$	213.25				152–154			
b182	Bis(2-hydroxyethyl) ether	$HOCH_2CH_2OCH_2CH_2OH$	106.12	1, 468	1.118_{20}^{20}	1.4460^{20}	−10.45	245	143	misc aq, alc, acet, eth

b183	*N,N*-Bis(2-hydroxy-ethyl)glycine	$(HOCH_2CH_2)_2NCH_2COOH$	163.17				192 sl d			sl s aq
b184	Bis(2-hydroxyethyl)-iminotris(hydroxy-methyl)methane	$(HOCH_2CH_2)_2$-$NC(CH_2OH)_3$	209.24				104			
b185	2,2-Bis(hydroxy-methyl)propionic acid	$(HOCH_2)_2C(CH_3)COOH$	134.13	3, 401			189–191			
b186	4,8-Bis(hydroxy-methyl)tricyclo-$[5.2.0^{2,6}]$decane		196.29				1.5280^{20}		>112	
b187	Bis(2-mercaptoethyl) ether	$(HSCH_2CH_2)_2O$	138.25		1.114		−80	217		
b188	Bis(2-mercaptoethyl) sulfide	$(HSCH_2CH_2)_2S$	154.32		1.183	1.5982^{20}		136^{10mm}	90	
b189	1,4-Bis(methane-sulfonoxy)butane	$(CH_3SO_2OCH_2CH_2—)_2$	246.30				115–117			sl hyd aq; 0.1 alc; 1.4 acet
b190	1,2-Bis(methoxy-ethoxy)ethane	$(CH_3OCH_2CH_2OCH_2—)_2$	178.23		0.990_4^{20}	1.4224^{20}	−45	216	110	misc aq
b191	Bis[2,(2-methoxyeth-oxy)ethyl] ether	$(CH_3OCH_2CH_2OCH_2$-$CH_2—)_2O$	222.28	1^3, 2107	1.0087_4^{20}	1.4330^{20}	−27	275.3	140	s aq
b192	Bis(2-methoxyethyl) ether	$(CH_3OCH_2CH_2—)_2O$	134.18		0.9440^{25}	1.4043^{25}	−68	162	70	misc aq

Bis(2-ethylhexyl) sebacate, b178
Bis(2-hydroxyethyl) sulfide, t152

2,2-Bis(hydroxymethyl)-2,2′,2″-nitrilotriethanol, b184

Bis(4-hydroxyphenyl) sulfide, t153

$HOCH_2$ CH_2OH

b186

TABLE 1-14 Physical constants of organic compounds (*continued*)

No.	Name	Formula	Formula weight	Beilstein reference	Density	Refractive index	Melting point	Boiling point	Flash point	Solubility in 100 parts solvent
b193	Bis(2-methylallyl) carbonate	$[H_2C{=}C(CH_3)CH_2O]_2$-$C({=}O)$	170.21		0.943^{20}	1.4371		202	72	
b194	Bis(4-nitrophenyl) carbonate	$(O_2NC_6H_4O)_2C({=}O)$	304.21	6^1, 120			141			
b195	Bis(3-nitrophenyl) disulfide	$O_2NC_6H_4SSC_6H_4NO_2$	308.22	6, 339			83			i aq; s alc, eth
b196	Bis(4-nitrophenyl) disulfide	$O_2NC_6H_4SSC_6H_4NO_2$	308.33	6, 340			181			sl s alc
b197	Bis(octadecyl)penta-erythritol diphosphite	$[C_{18}H_{37}OP(OCH_2)_2]_2$	721.01		0.925	1.457	40		261	
b198	Bis(pentabromo-phenyl) ether	$C_6Br_5OC_6Br_5$	969.22	6^1, 108			>300			
b199	1,4-Bis(5-phenyloxa-zol-2-yl)benzene		364.40				244			
b200	Bis(*p*-tolyl) disulfide	$CH_3C_6H_4SSC_6H_4CH_3$	246.39	6, 425			43–46			i aq; s alc; v s eth
b201	Bis(*p*-tolyl) sulfoxide	$CH_3C_6H_4S(O)C_6H_4CH_3$	230.33	6, 419			94–96			v s alc, bz, chl, eth
b202	Bis(tributyltin) oxide	$(C_4H_9)_3SnOSn(C_4H_9)_3$	596.08		1.170	1.4864^{20}		180^{2mm}	>112	
b203	1,4-Bis(trichloro-methyl)benzene	$Cl_3CC_6H_4CCl_3$	312.84	5, 385			108–110			i aq; 26 acet; 38 bz; 22 CCl_4; 33 eth; 3 MeOH
b204	Bis(2,4,5-trichloro-phenyl) disulfide	$Cl_3C_6H_4SSC_6H_4Cl_3$	425.01				140–144			
b205	1,2-Bis(trichloro-silyl)ethane	$Cl_3SiCH_2CH_2SiCl_3$	296.64		1.483^{20}_4	1.473^{20}	24.5	201–202		

b206	3,5-Bis(trifluoromethyl)aniline	$(F_3C)_2C_6H_3NH_2$	229.13		1.467	1.4335^{20}		85^{15mm}	83
b207	1,3-Bis(trifluoromethyl)benzene	$F_3CC_6H_4CF_3$	214.0		1.3790^{25}	1.3916^{25}		116	
b208	*N,O*-Bis(trimethylsilyl)acetamide	$CH_3C(=N-Si(CH_3)_3)-O-Si(CH_3)_3$	203.43		0.832_4^{20}	1.4170^{20}		73^{35mm}	11
b209	Bis(trimethylsilyl)acetylene	$(CH_3)_3SiC\equiv CSi(CH_3)_3$	170.41		0.770_4^{20}	1.413^{20}		137	2
b210	Bis(trimethylsilyl)formamide	$HC(=NSi(CH_3)_3)-O-Si(CH_3)_3$	189.41		0.885	1.4381^{20}		$54-55^{13mm}$	
b211	*N,O*-Bis(trimethylsilyl)hydroxylamine	$(CH_3)_3SiONHSi(CH_3)_3$	177.40		0.830	1.4112^{20}		$78-80^{100mm}$	28
b212	1,2-Bis(trimethylsilyloxy)ethane	$(CH_3)_3SiOCH_2CH_2OSi(CH_3)_3$	206.43		0.842	1.4034^{20}		165–166	46
b213	*N,O*-Bis(trimethylsilyl)trifluoroacetamide	$CF_3C[=NSi(CH_3)_3]OSi(CH_3)_3$	257.40		0.969	1.3939^{20}	−10	50^{14mm}	23
b214	Bis(triphenylphosphine)dicarbonylnickel	$[(C_6H_5)_3P]_2Ni(CO)_2$	639.32				209		

Bis(phenylmethyl) disulfide, d57

"Bis-tris," b184

"Bis-tris" propane, b215

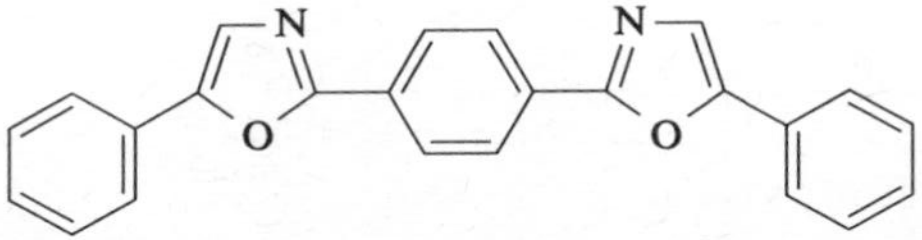

b199

TABLE 1-14 Physical constants of organic compounds (*continued*)

No.	Name	Formula	Formula weight	Beilstein reference	Density	Refractive index	Melting point	Boiling point	Flash point	Solubility in 100 parts solvent
b215	1,3-Bis[tris(hydroxymethyl)methylamino]propane	$CH_2[CH_2NHC(CH_2OH)_3]_2$	282.34	4^3, 859			170			
b216	Biuret	$H_2NCONHCONH_2$	103.08	3, 70	1.467_4^{-5}		110	d 190		v s alc; 2 aq
b217	1-Borneol		154.25	6, 72	1.011_4^{20}		204	212	65	i aq; 176 alc; s eth, bz, PE
b218	1-Bornyl acetate		196.29	6, 82	0.982	1.4626	27	224	84	sl s aq; s alc, eth
b219	*N*-Bromoacetamide	$CH_3CONBrH$	137.97	2, 181			102–105			sl s aq; v s eth
b220	*p*-Bromoacetanilide	$BrC_6H_4NHCOCH_3$	214.07	12, 642	1.717		168			i aq; s bz, chl, EtAc
b221	Bromoacetic acid	$BrCH_2COOH$	138.95	2, 213	1.934_4^{50}	1.4804^{50}	50	208		v s aq, alc, eth
b222	*α*-Bromoacetophenone	$C_6H_5COCH_2Br$	199.05	7, 283	1.647_4^{20}		50	135^{18mm}		i aq; v s alc, bz, chl, eth
b223	*p*-Bromoacetophenone	$BrC_6H_4COCH_3$	199.05	7, 283	1.647		54	255		s alc, bz, eth, HOAc
b224	Bromoacetyl bromide	$BrCH_2COBr$	201.86	2, 215	2.317_{22}^{22}	1.5480^{20}		150	none	d aq, alc
b225	Bromoacetyl chloride	$BrCH_2COCl$	157.40	2, 215	1.908	1.4960^{20}		128	none	d aq, alc
b226	2-Bromoaniline	$BrC_6H_4NH_2$	172.03	12, 631	1.578_4^{20}	1.6113^{20}	31	229		i aq; s alc, eth
b227	3-Bromoaniline	$BrC_6H_4NH_2$	172.03	12, 633	1.580_4^{20}	1.6250^{20}	16.8	251	>112	sl s aq; s alc, eth
b228	4-Bromoaniline	$BrC_6H_4NH_2$	172.03	12, 636	1.4970_4^{100}		66.3			i aq; v s alc, eth
b229	3-Bromobenzaldehyde	BrC_6H_4CHO	185.03	7, 238	1.587	1.5935^{20}		230	96	i aq; v s alc, eth
b230	Bromobenzene	C_6H_5Br	157.02	5, 206	1.4952_4^{20}	1.5580^{20}	−30.72	156.2	51	0.044 aq; 10.4 alc; misc bz, chl, PE; 71.6 eth
b231	Bromobenzene-d_5	C_6D_5Br	162.06					53^{23mm}	65	
b232	4-Bromobenzenesulfonyl chloride	$BrC_6H_4SO_2Cl$	255.52	11, 57			74.5	153^{15mm}		i aq; d alc; v s eth

b234	4-Bromobenzo-phenone	$BrC_6H_4COC_6H_5$	261.12	7, 422				82	350	i alc; sl s bz, eth
b234	4-Bromobenzo-phenone	$BrC_6H_4COC_6H_5$	261.12	7, 422				82	350	i alc; sl s bz, eth
b235	2-Bromobenzotri-fluoride	$BrC_6H_4CF_3$	225.01		1.652^{20}	1.4817^{20}		168	51	
b236	3-Bromobenzotri-fluoride	$BrC_6H_4CF_3$	225.01		1.613	1.4749^{20}		152	43	
b237	2-Bromobenzyl alcohol	$BrC_6H_4CH_2OH$	187.04	6, 445			82			s hot aq; vs alc, eth
b238	2-Bromobenzyl bromide	$BrC_6H_4CH_2Br$	249.94	5, 308		1.6193^{20}	31	129^{19mm}		d hot aq; s alc, eth
b239	4-Bromobiphenyl	$BrC_6H_4C_6H_5$	233.11	5, 580	0.9327_4^{25}		87	310		i aq; s alc, bz, eth
b240	1-Bromobutane	$CH_3CH_2CH_2CH_2Br$	137.02	1, 119	1.2686_4^{25}	1.4374^{25}	−112.4	101.6	23	i aq; s alc, bz, eth
b241	2-Bromobutane	$CH_3CH_2CHBrCH_3$	137.03	1, 119	1.2530_4^{25}	1.4360^{20}	−112.4	21		<0.1 aq; vs alc, eth
b242	1-Bromo-2-butene	$CH_3CH{=}CHCH_2Br$	135.01	1, 205	1.312	1.4765^{20}		99	11	
b243	2-Bromo-2-butene	$CH_3CH{=}C(Br)CH_3$	135.01	1, 205	1.328	1.4613^{20}		90^{740mm}	<1	
b244	4-Bromo-1-butene	$BrCH_2CH_2CH{=}CH_2$	135.01	1^1, 84	1.3230_4^{20}	1.4608^{30}		100	<1	i aq; s alc, eth

2-Bornanone, c3
Bromal, t205
Bromoacetaldehyde diethyl acetal, b269
Bromoanisoles, b304, b305, b306
p-Bromobenzenethiol, b357
4-Bromobenzyl cyanide, b332
exo-2-Bromobicyclo[2.2.1]heptane, b321
Bromobutanedioic acid, b353

b217

b218

TABLE 1-14 Physical constants of organic compounds (*continued*)

No.	Name	Formula	Formula weight	Beilstein reference	Density	Refractive index	Melting point	Boiling point	Flash point	Solubility in 100 parts solvent
b245	4-Bromobutyl phenyl ether	$C_6H_5OCH_2CH_2CH_2$-CH_2Br	229.12	6^2, 82			41–42	$153\text{–}156^{18mm}$		
b246	2-Bromobutyric acid	$CH_3CH_2CH(Br)COOH$	167.01	2, 281	1.5669^{20}_{20}	1.4720^{20}	−4	103^{10mm}	>112	6.7 aq; s alc, eth
b247	*endo*-3-Bromo-D-camphor		231.14	7^2, 101	1.449		76–78	244		i aq; 15 alc; 200 chl; 62 eth
b248	α-Bromo-*p*-chloroacetophenone	$ClC_6H_4COCH_2Br$	233.50	7, 285			96.5			
b249	2-Bromochlorobenzene	BrC_6H_4Cl	191.46	5, 209	1.6382^{25}_{4}	1.5789^{25}		204	79	i aq; v s bz
b250	3-Bromochlorobenzene	BrC_6H_4Cl	191.46	5, 209	1.6302^{20}_{4}	1.5771^{20}	−21	196	80	i aq; v s alc, eth
b251	4-Bromochlorobenzene	BrC_6H_4Cl	191.46	5, 209	1.576^{71}_{4}	1.5531^{70}	64.5	196		0.1 aq; misc MeOH, eth
b252	3-Bromo-4-chlorobenzotrifluoride	$Br(Cl)C_6H_3CF_3$	259.47		1.743^{25}	1.4973^{25}	−22	191–192		
b253	1-Bromo-4-chlorobutane	$ClCH_2CH_2CH_2CH_2Br$	171.47	5^3, 294	1.488	1.4875^{20}		82^{30mm}	60	i aq; s alc, chl, eth
b254	4-Bromo-6-chloro-*o*-cresol	$Br(Cl)C_6H_2(OH)CH_3$	221.49	6, 360			47			
b255	Bromochlorodifluoromethane	$Br(Cl)CF_2$	165.4		1.83^{21}		−160.5	−4.01		
b256	1-Bromo-2-chloroethane	$ClCH_2CH_2Br$	143.43	1, 89	1.7392^{20}_{4}	1.4917^{20}	−18.4	106.6	none	0.7 aq; misc org solv
b257	7-Bromo-5-chloro-8-hydroxyquinoline		258.51	21^1, 222			177–179			
b258	Bromochloromethane	$ClCH_2Br$	129.39	1, 67	1.923^{25}_{4}	1.480^{25}	−88	67.8	none	0.9 aq; misc MeOH, eth

b259	1-Bromo-3-chloropropane	$ClCH_2CH_2CH_2Br$	157.44	1, 109	1.472	1.486^{20}	<−50	143.5	none	0.1 aq; misc org solv
b260	2-Bromo-2-chloro-1,1,1-trifluoroethane	$HC(Br)ClCF_3$	197.4		1.8636^{25}	1.3738^{25}		50		
b261	α-Bromocinnamaldehyde	$C_6H_5CH{=}C(Br)CHO$	211.06	7, 358			66–68			
b262	Bromocycloheptane	BrC_7H_{13}	177.09	5, 29	1.2887_4^{22}	1.5052^{20}		72^{10mm}	68	i aq; v s chl, eth
b263	Bromocyclohexane	BrC_6H_{11}	163.06	5, 24	1.3264_4^{15}	1.4956^{15}		165.8	62	0.1 aq; 10 MeOH; 71 eth
b264	3-Bromocyclohexene		161.04	5^2, 40	1.3890_4^{20}	1.5292^{20}		$64–65^{15mm}$		
b265	Bromocyclopentane	BrC_5H_9	149.04	5, 19	1.3900_4^{20}	1.4881^{20}		137–139	35	
b266	Bromocyclopropane	BrC_3H_5	120.98			1.4605^{20}		69	2	
b267	1-Bromodecane	$CH_3(CH_2)_9Br$	221.19	1^2, 130	1.0658_4^{20}	1.4560^{20}	−30	238	94	i aq; v s chl, eth
b268	Bromodichloromethane	$HCBrCl_2$	163.83	1, 67	1.980^{20}	1.4964^{20}	−55	89.2	none	sl s aq; misc org solv
b269	2-Bromo-1,1-diethoxyethane	$BrCH_2CH(OC_2H_5)_2$	197.08	1, 625	1.310	1.4385^{20}		67^{18mm} 180 d	51	s hot alc

2-Bromo-*p*-cumene, b301
β-Bromocumene, b300

4-Bromodiphenyl ether, b333

Bromoethene, b286

H_3C CH_3 Br H_3C O

b247

Cl Br N OH

b257

Br

b264

TABLE 1-14 Physical constants of organic compounds (*continued*)

No.	Name	Formula	Formula weight	Beilstein reference	Density	Refractive index	Melting point	Boiling point	Flash point	Solubility in 100 parts solvent
b270	4-Bromo-1,2-dimethoxybenzene	$BrC_6H_3(OCH_3)_2$	217.07	6, 784	1.702	1.5743^{20}		256	109	
b271	1-Bromo-2,2-dimethoxypropane	$CH_3C(OCH_3)_2CH_2Br$	183.05		1.355	1.4475^{20}		87^{80mm}	40	
b272	4-Bromo-2,6-dimethylphenol	$BrC_6H_2(CH_3)_2OH$	201.07	6, 485			78			
b273	2-Bromo-4,6-dinitroaniline	$BrC_6H_2(NO_2)_2NH_2$	262.02	12, 761			154	subl		v s hot alc, hot acet
b274	3-Bromo-4,6-dinitrofluorobenzene	$BrC_6H_2(NO_2)_2F$	264.9				90–91			
b275	2-Bromo-2,2-diphenylacetyl bromide	$BrC(C_6H_5)_2COBr$	354.05	9^1, 283			63–65			
b276	α-Bromodiphenylmethane	$C_6H_5CH(Br)C_6H_5$	247.14	5, 592			40	184^{20mm}		
b277	1-Bromododecane	$CH_3(CH_2)_{11}Br$	249.24	1^2, 133	1.038	1.4580^{20}	−9	135^{6mm}	110	0.1 aq; s alc, eth
b278	1-Bromo-2,3-epoxypropane	$H_2C{-}CHCH_2Br$ (epoxy O bridging)	136.98	17, 9	1.601^{20}	1.4820^{20}	−40	134–136	56	i aq; sl s alc; s eth
b279	Bromoethane	CH_3CH_2Br	108.97	1, 88	1.4708^{15}	1.4276^{15}	−118.6	38.4	none	0.91 aq
b280	2-Bromoethanesulfonic acid sodium salt	$BrCH_2CH_2SO_3^-Na^+$	211.02	4, 7			283–285 d			
b281	2-Bromoethanol	$BrCH_2CH_2OH$	124.97	1, 338	1.7629_4^{20}	1.4920^{20}		150	40	misc aq; s org solv
b282	2-Bromoethyl acetate	$CH_3COOCH_2CH_2Br$	167.01	2^1, 57	1.514_4^{20}	1.4547^{20}	−13.8	159	71	v s aq; misc alc, eth
b283	2-Bromoethylamine HBr	$BrCH_2CH_2NH_2{\cdot}HBr$	204.90	4, 134			172–174			v s aq, alc

b284	*o*-Bromo(ethyl)-benzene	$CH_3CH_2C_6H_4Br$	185.07	5, 355	1.3566^{25}_{25}	1.5603^{20}		199		0.1 aq; misc org solv
b285	(2-Bromoethyl)-benzene	$C_6H_5CH_2CH_2Br$	185.07	5, 356	1.355	1.5563^{20}		221	89	i aq; s bz, eth
b286	Bromoethylene	$H_2C{=}CHBr$	106.96	1, 188	1.493^{20}	1.4350^{20}	−139.5	15.8		i aq; misc alc, eth
b287	2-Bromoethyl ethyl ether	$BrCH_2CH_2OCH_2CH_3$	153.02	1, 338	1.3572^{20}_{4}	1.4450^{20}		150	21	sl s aq; misc alc, eth
b288	2-Bromoethyl phenyl ether	$BrCH_2CH_2OC_6H_5$	201.07	6, 142			34	144^{40mm}	65	i aq; v s alc, eth
b289	*N*-(2-Bromoethyl)-phthalimide		254.09	21, 461			81–84			s hot aq; v s eth
b290	2-Bromofluoro-benzene	BrC_6H_4F	175.01		1.601	1.5337^{20}		156	43	
b291	3-Bromofluoro-benzene	BrC_6H_4F	175.01		1.567	1.5257^{20}		150	38	
b292	4-Bromofluoro-benzene	BrC_6H_4F	175.01	5, 209	1.593^{15}	1.5310^{15}	−17.4	151–152	60	
b293	1-Bromoheptane	$H(CH_2)_7Br$	179.11	1, 155	1.1384^{20}_{4}	1.4505^{20}	−58	180	60	i aq; v s alc, eth
b294	2-Bromoheptane	$H(CH_2)_5CH(Br)CH_3$	179.11	1, 155	1.142	1.4470^{20}		66^{21mm}	47	
b295	1-Bromohexadecane	$H(CH_2)_{16}Br$	305.35	1^2, 138	0.9991	1.4618	17.8	336	177	i aq; misc org solv
b296	1-Bromohexane	$H(CH_2)_6Br$	165.08	1, 144	1.1763^{20}_{4}	1.4475	−85	154–158	57	i aq; misc alc, eth

(Bromomethyl)benzene, b85

Bromoform, t210

b289

TABLE 1-14 Physical constants of organic compounds (*continued*)

No.	Name	Formula	Formula weight	Beilstein reference	Density	Refractive index	Melting point	Boiling point	Flash point	Solubility in 100 parts solvent
b297	DL-2-Bromohexanoic acid	$CH_3(CH_2)_3CH(Br)$-COOH	195.06	2, 325	1.370	1.4720^{20}		$136\text{–}138^{18mm}$		
b298	5-Bromoisatin		226.03	21, 453			251–253			
b300	(2-Bromoisopropyl)-benzene	$C_6H_5CH(CH_3)CH_2Br$	199.10	5^1, 191	1.316	1.5480^{20}		108^{18mm}	91	
b301	2-Bromo-4-isopropyl-1-methylbenzene	$CH_3(Br)C_6H_3CH(CH_3)_2$	213.0		1.253^{25}_{25}	1.535^{25}	−20	120		i aq; 50 MeOH; misc org solv
b302	Bromomaleic anhydride		176.96	17, 435	1.905	1.5400^{20}		215	>112	
b303	Bromomethane	CH_3Br	94.94	1, 67	1.732^0_0	1.4234^{10}	−84	3.56	none	0.1 aq; s alc, chl, eth
b304	2-Bromo-1-methoxy-benzene	$BrC_6H_4OCH_3$	187.04	6, 197	1.5018^{20}_4	1.5737^{20}	2	223	96	i aq; v s alc, eth
b305	3-Bromo-1-methoxy-benzene	$BrC_6H_4OCH_3$	187.04	6, 198	1.477	1.5635^{20}	211	93		i aq; s alc, eth
b306	4-Bromo-1-methoxy-benzene	$BrC_6H_4OCH_3$	187.04	6, 199	1.4564^{20}_4	1.5630^{20}	10	223	94	sl s aq; v s alc, eth
b307	4-Bromo-2-methyl-aniline	$CH_3(Br)C_6H_3NH_2$	186.06	12, 838			56	240		sl s aq; v s alc
b308	1-Bromo-3-methyl-butane	$(CH_3)_2CHCH_2CH_2Br$	151.05	1, 136	1.210^{15}_4	1.4409^{20}	−112	119.7	32	0.02 aq; misc alc, eth
b309	(Bromomethyl)cyclo-hexane	$C_6H_{11}CH_2Br$	177.09	5^2, 18	1.269	1.4907^{20}		$76\text{–}77^{26mm}$	57	
b310	2-Bromomethyl-1,3-dioxalane		167.01	19^2, 8	1.613	1.4817^{20}		$80\text{–}82^{27mm}$	62	
b311	Bromomethyl methyl ether	$BrCH_2OCH_3$	124.97	1, 582	1.531	1.4550^{20}		87	26	

b312	1-Bromo-2-methylnaphthalene	$BrC_{10}H_6CH_3$	221.10	5, 568	1.418	1.6484^{20}		296	>112	
b313	1-Bromo-2-methylpropane	$(CH_3)_2CHCH_2Br$	137.03	1, 126	1.2641^{20}	1.4362^{20}	−119	91.5	18	0.06 aq; misc alc, eth
b314	2-Bromo-2-methylpropane	$(CH_3)_3CBr$	137.03	1, 127	1.215^{25}_{25}	1.425^{25}	−16.2	73.1	18	i aq; misc org solv
b314a	α-Bromo-α-methylpropiophenone	$C_6H_5COC(CH_3)_2Br$	227.11	7, 316	1.350	1.5561^{20}		148^{30mm}	>112	
b315	1-Bromonaphthalene	$C_{10}H_7Br$	207.08	5, 547	1.4834^{20}_{4}	1.6580^{20}	−1	281.1	>112	misc alc, bz, chl, eth
b316	1-Bromo-1-naphthol	$BrC_{10}H_6OH$	223.07	6, 650			78	130 d		i aq; s alc, bz, eth
b317	1-Bromo-2-nitrobenzene	$BrC_6H_4NO_2$	202.01	5^1, 247	1.6245^{80}_{4}		43	261		v s alc; s bz, eth
b318	5-Bromo-2-nitrobenzotrifluoride	$O_2N(Br)C_6H_3CF_3$	270.02		1.7992^{25}	1.5180^{25}	40–44	99–100		
b319	2-Bromo-2-nitro-1,3-propanediol	$(HOCH_2)_2C(Br)NO_2$	199.99	1, 476			133			
b320	1-Bromononane	$H(CH_2)_9Br$	207.16	1^1, 63	1.084	1.4540^{20}		201	51	i aq; s chl, eth
b321	*exo*-2-Bromonorbornane		175.07		1.363	1.5148^{20}		82^{29mm}	60	
b322	1-Bromooctadecane	$H(CH_2)_{18}Br$	333.41	1^1, 69			23	216^{12mm}		i aq; s alc, eth

α-Bromoisobutyrophenone, b314a
2-Bromomesitylene, b365
α-Bromo-4-nitro-*o*-cresol, h156
α-Bromo-*p*-nitrotoluene, n46

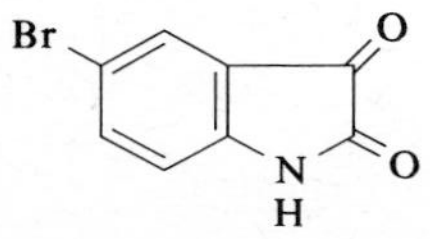

b298

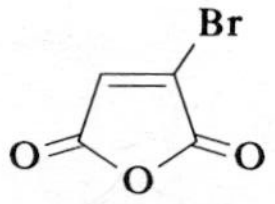

b302

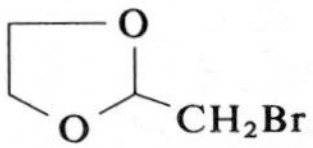

b310

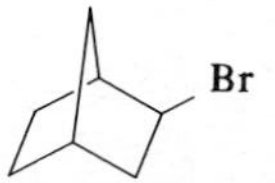

b321

TABLE 1-14 Physical constants of organic compounds (*continued*)

No.	Name	Formula	Formula weight	Beilstein reference	Density	Refractive index	Melting point	Boiling point	Flash point	Solubility in 100 parts solvent
b323	1-Bromooctane	$H(CH_2)_8Br$	193.13	1, 160	1.108_4^{25}	1.4503^{25}	−55	201	78	i aq; misc alc, eth
b324	Bromopentafluoro-benzene	BrC_6F_5	246.97		1.947^{20}	1.4490^{20}	−31	137	87	
b325	1-Bromopentane	$H(CH_2)_5Br$	151.05	1, 131	1.2237_4^{15}	1.4444^{20}	−88	129.6	31	i aq; s alc; misc eth
b326	2-Bromopentane	$CH_3CH_2CH_2CH(Br)CH_3$	151.05	1, 131	1.2039_4^{20}	1.4403^{20}		117	20	
b327	9-Bromophenanthrene		257.14	5, 671	1.409_4^{101}		54–58	190^{2mm}		i aq; s alc, eth
b328	2-Bromophenol	BrC_6H_4OH	173.01	6, 197	1.492	1.5892^{20}	6	194	42	s aq; misc chl, eth
b329	4-Bromophenol	BrC_6H_4OH	173.01	6, 198	1.5875^{80}		68	238		14 aq; v s alc, chl
b330	2-Bromo-2-phenyl-acetic acid	$C_6H_5CH(Br)COOH$	215.05	9, 451			83			
b331	*p*-Bromophenylacetic acid	$BrC_6H_4CH_2COOH$	215.05	9, 451			119			sl s aq; v s alc, eth
b332	*p*-Bromophenylaceto-nitrile	$BrC_6H_4CH_2CN$	196.05	9, 451			47–49			i aq; sl s alc; v s bz
b333	4-Bromophenyl phenyl ether	$BrC_6H_4OC_6H_5$	249.11	6^1, 105	1.423	1.6070^{20}	18	305	>112	
b334	1-Bromo-3-phenyl-propane	$C_6H_5CH_2CH_2CH_2Br$	199.10	5, 391	1.310	1.5450^{20}		220	101	
b335	1-Bromopropane	$CH_3CH_2CH_2Br$	123.00	1, 108	1.3597^{15}	1.4370^{15}	−110.1	71.0	25	0.23 aq^{30}; misc alc
b336	2-Bromopropane	$CH_3CH(Br)CH_3$	123.00	1, 108	1.3222^{15}	1.4285^{15}	−89.0	59.5	19	0.3 aq^{18}; misc alc, bz, chl, eth
b337	3-Bromo-1-propanol	$BrCH_2CH_2CH_2OH$	139.00	1, 356	1.5374_4^{20}	1.4858^{20}		62^{5mm}		s aq; misc alc, eth
b338	1-Bromo-1-propene	$CH_3CH{=}CHBr$	120.98	1, 200	1.4133_4^{20}	1.4538^{20}	−116	63	4	i aq
b339	2-Bromo-1-propene	$CH_3C(Br){=}CH_2$	120.98	1, 200	1.362_4^{20}	1.4425^{20}	−125	49	4	
b340	2-Bromopropionic acid	$CH_3CH(Br)COOH$	152.98	2, 254	1.7000^{20}	1.4750^{20}	25.7	203	100	v s aq, alc, eth
b341	3-Bromopropionic acid	$BrCH_2CH_2COOH$	152.98	2, 256	1.480		62.5		65	s aq, alc , bz, chl, eth

b342	3-Bromopropionitrile	$BrCH_2CH_2CN$	133.98	2^2, 231	1.6152_4^{20}	1.4800^{20}		78^{10mm}	98	v s alc, eth
b343	2-Bromopropionyl chloride	$CH_3CH(Br)COCl$	171.43	2, 256	1.700^{11}	1.4800^{20}		133	51	d aq; s chl, eth
b344	3-Bromopropionyl chloride	$BrCH_2CH_2COCl$	171.43	2^2, 231	1.701	1.4968^{20}		57^{17mm}	79	
b345	α-Bromopropiophenone	$C_6H_5COCHBrCH_3$	213.08	7, 302	1.430_4^{20}	1.5715^{20}		250	>112	s alc, bz, eth, acet
b346	3-Bromopropyl phenyl ether	$C_6H_5OCH_2CH_2CH_2Br$	215.10	6, 142	1.365	1.5464^{20}	10–11	130–134^{14mm}	96	
b347	3-Bromopropyne	$BrCH_2C{\equiv}CH$	118.97	1, 248	1.335	1.4905^{20}		88–90	18	
b348	2-Brompyridine	BrC_5H_4N	158.00	20, 233	1.657^{18}	1.5720^{20}		194	54	i aq; s org solv
b349	3-Bromopyridine	BrC_5H_4N	158.00	20, 233	1.645_4^{0}	1.5695^{20}	142–143	173	51	s aq; v s alc, eth
b350	3-Bromoquinoline		208.06	20, 363	1.533	1.6640^{20}	15	276	>112	s HOAc
b351	5-Bromosalicylic acid	$Br(HO)C_6H_3COOH$	217.02	10, 107			166			0.3 aq^{80}; 85 alc^{25}; 70 eth^{25}
b352	β-Bromostyrene	$C_6H_5CH{=}CHBr$	183.05	5, 477	1.422_4^{20}	1.6066^{20}	7	112^{20mm}	79	i aq; misc alc, eth
b353	Bromosuccinic acid	$HOOCCH_2CH(Br)COOH$	196.99	2, 621	2.073		172(d)			18 aq; s alc

β-Bromophenetole, b288
3-Bromopropene, a85

(3-Bromopropyl)benzene, b333

5-Bromopseudocumene, b364

b327

b350

TABLE 1-14 Physical constants of organic compounds (*continued*)

No.	Name	Formula	Formula weight	Beilstein reference	Density	Refractive index	Melting point	Boiling point	Flash point	Solubility in 100 parts solvent
b354	*N*-Bromosuccinimide		177.99	21, 380	2.098		173 sl d			1.5 aq; 14.4 acet; 3.1 HOAc; 0.02 CCl_4
b355	1-Bromotetradecane	$H(CH_2)_{14}Br$	277.30	1^2, 136	1.0124_4^{25}	1.4600^{20}	6	178^{20mm}	> 112	s alc; v s chl; misc bz, acet
b356	2-Bromothiophene	$Br—C_4H_3S$	163.04	17, 33	1.684_4^{20}	1.5860^{20}		151	60	v s acet, eth
b357	4-Bromothiophenol	BrC_6H_4SH	189.08	6, 330			76	239		
b358	2-Bromotoluene	$BrC_6H_4CH_3$	171.04	5, 304	1.422_{25}^{25}	1.552^{25}	−26	181	78	0.1 aq; misc alc, bz, chl, eth
b359	3-Bromotoluene	$BrC_6H_4CH_3$	171.04	5, 305	1.4099_4^{20}	1.5517^{20}	−39.8	183.7	60	s alc, bz, eth
b360	4-Bromotoluene	$BrC_6H_4CH_3$	171.04	5, 305	1.3959_{35}^{35}	1.5490	28.5	184.5	85	s alc, bz, eth
b361	Bromotri-chloromethane	$BrCCl_3$	198.28	1, 67	1.997_{25}^{25}	1.5063	−21	103.8	none	misc org solv
b362	1-Bromotridecane	$H(CH_2)_{13}Br$	263.27	1^2, 134	1.0262_4^{20}	1.4592^{20}	7	150^{10mm}	> 112	v s chl
b363	Bromotri-fluoromethane	$BrCF_3$	148.92	1^3, 83	1.5800_4^{20}			−57.8		v s chl
b364	5-Bromo-1,2,4-tri-methylbenzene	$BrC_6H_2(CH_3)_3$	199.10	5, 403			73	235		i aq; s alc
b365	2-Bromo-1,3,5-tri-methylbenzene	$BrC_6H_2(CH_3)_3$	199.10	5, 408	1.301	1.5511^{20}	2	225	96	i aq; s bz; v s eth
b366	Bromotrimethyl-germane	$(CH_3)_3GeBr$	197.60		1.544^{18}	1.4705^{20}	−25	113.7		
b367	Bromotrimethylsilane	$(CH_3)_3SiBr$	153.10		1.160	1.4145^{20}		79	1	
b368	Bromotriphenyl-ethylene	$(C_6H_5)_2C{=}C(Br)C_6H_5$	335.22				114–115			
b369	Bromotriphenyl-methane	$(C_6H_5)_3CBr$	323.24	5, 704			152–154	230^{15mm}		
b370	11-Bromoundecanoic acid	$Br(CH_2)_{10}COOH$	265.20	2^2, 315			51	174^{2mm}		i aq; v s alc

b371	α-Bromo-*o*-xylene	$BrCH_2C_6H_4CH_3$	185.07	5, 365	1.381^{23}	1.5742^{20}	21	223–224	82	s alc, eth
b372	α-Bromo-*m*-xylene	$BrCH_2C_6H_4CH_3$	185.07	5, 374	1.370^{23}	1.5560^{20}		185^{340mm}	82	s alc, eth
b373	2-Bromo-*p*-xylene	$BrC_6H_3(CH_3)_2$	185.07	5, 385	1.340	1.5505^{20}	9–10	199–201	79	
b374	4-Bromo-*o*-xylene	$BrC_6H_3(CH_3)_2$	185.07	5, 365	1.370^{15}_{15}	1.5560^{20}		215	80	v s alc, eth
b375	Brucine		394.45	27^2, 797			178			77 alc; 1 bz; 20 chl
b376	1,2-Butadiene	$CH_3CH{=}C{=}CH_2$	54.09	1, 249	0.676^{10}_4	1.4205^1	−136.2	10.9		misc alc, eth
b377	1,3-Butadiene	$CH_2{=}CHCH{=}CH_2$	54.09	1, 249	0.650^{-6}_4	1.4293^{-25}	−108.9	−4.4		misc alc, eth
b378	1,3-Butadienyl acetate	$CH_3C({=}O)OCH{=}CH$-$CH{=}CH_2$	112.13	2^3, 295	0.945	1.4690^{20}		60^{40mm}	33	
b379	1,3-Butadiyne	$HC{\equiv}CC{\equiv}CH$	50.06	1^3, 1056	0.7364^0_4	1.4189^5	−36	10.3		v s eth; s bz, acet
b380	2-Butanamine	$CH_3CH_2CH(NH_2)CH_3$	73.14	4, 160	0.7308^{15}_4	1.3963^{15}	−104.5	66	−19	misc aq, alc
b381	Butane	$CH_3CH_2CH_2CH_3$	58.12		0.6011^0	1.3562^{-13}	−138.3	−0.50		
b382	1,4-Butanediamine	$H_2NCH_2CH_2CH_2CH_2NH_2$	88.15	4, 264	0.877^{25}_4	1.4569^{20}	27–28	158–160	51	s aq
b383	Butanedinitrile	$NCCH_2CH_2CN$	80.09	2, 615	0.9867^{60}_4	1.4173^{60}	57.9	265–267		11.5 aq; s acet, chl, diox; sl s bz, eth
b384	1,2-Butanediol	$CH_3CH_2CH(OH)CH_2OH$	90.12	1, 477	1.006^{18}_0	1.4380^{20}		207.5	93	s aq, alc, acet
b385	1,3-Butanediol	$CH_3CH(OH)CH_2CH_2OH$	90.12	1, 477	1.0053^{20}_{20}	1.441^{20}	< -50	207.5	121	s aq, alc, acet; 9 eth

α-Bromotoluene, b85
Bromo-α,α,α-trifluorotoluenes, b235, b236
3-Bromo-1,7,7-trimethylbicyclo[2.2.1]heptane-2-one, b247
4-Bromoveratrole, b270
4-Bromo-2,6-xylenol, b272
BSA, b208
BSTFA, b213
BTMSA, b209
Busulfan, b189
α-Butadiene sulfone, d369
Butanedioic acid, s15
1,4-Butanediol diglycidyl ether, b176
1,4-Butanediol dimethanesulfonate, b189

b354

b375

TABLE 1-14 **Physical constants of organic compounds (*continued*)**

No.	Name	Formula	Formula weight	Beilstein reference	Density	Refractive index	Melting point	Boiling point	Flash point	Solubility in 100 parts solvent
b386	1,4-Butanediol	$HOCH_2CH_2CH_2CH_2OH$	90.12	1, 478	1.016_4^{25}	1.4452^{20}	120.9	230	> 112	misc aq, alc, acet; 0.3 bz; 3.1 eth; 0.9 PE
b387	*meso*-2,3-Butanediol	$CH_3CH(OH)CH(OH)CH_3$	90.12	1, 479	0.9939_4^{25}	1.4324^{35}	34.4	182	85	misc aq, alc
b388	D-(−)-2,3-Butanediol	$CH_3CH(OH)CH(OH)CH_3$	90.12	1^2, 546	0.9869_4^{25}	1.4315^{25}	19.7	180^{715mm}	85	misc aq, alc; s eth
b389	2,3-Butanedione	$CH_3C(O)C(O)CH_3$	86.09	1, 769	0.990_{15}^{15}	1.3951^{20}		88	26	25 aq; misc alc, eth
b390	1,4-Butanedithiol	$HSCH_2CH_2CH_2CH_2SH$	122.25	1, 479	1.042	1.5290^{20}		106^{30mm}	70	i aq; v s alc
b391	1-Butanethiol	$CH_3CH_2CH_2CH_2SH$	90.19	1, 370	0.8367_4^{25}	1.4403^{25}	−115.7	98.5	12	0.06 aq; v s alc, eth
b392	2-Butanethiol	$CH_3CH_2CH(SH)CH_3$	90.19	1, 373	0.8246_4^{25}	1.4338^{25}	−165	85.0	21	sl s aq; v s alc, eth
b393	1,2,4-Butanetriol	$HOCH_2CH_2CH(OH)$-CH_2OH	106.12	1, 519	1.018^{20}	1.4748^{20}		191^{18mm}	167	v s aq, alc
b394	1-Butanol	$CH_3CH_2CH_2CH_2OH$	74.12	1, 367	0.8097_4^{20}	1.3993^{20}	−88.6	117.7	35	7.4 aq; misc alc, eth
b395	2-Butanol	$CH_3CH_2CH(OH)CH_3$	74.12	1, 371	0.8069_4^{20}	1.3972^{20}	−114.7	99.5	26	12.5 aq; misc alc, eth
b396	2-Butanone	$CH_3CH_2COCH_3$	72.11	1^2, 726	0.8049_4^{20}	1.3788^{20}	−86.7	79.6	−3	24 aq; misc alc, bz, eth
b397	2-Butanone oxime	$CH_3CH_2C(=NOH)CH_3$	87.12	1^2, 730	0.9232_4^{20}	1.4428	−29.5	72^{25mm}		s aq; misc alc, eth
b398	1-Butene	$CH_3CH_2CH=CH_2$	56.10	1^3, 715	0.6255_4^{-185}	1.3962^{20}	−185.3	−6.3		i aq; v s alc, eth
b399	*cis*-2-Butene	$CH_3CH=CHCH_3$	56.10	1^3, 728	0.6213_4^{20}	1.3931^{-25}	−138.9	3.7		i aq; v s alc, eth
b400	*trans*-2-Butene	$CH_3CH=CHCH_3$	56.10	1^3, 730	0.6041_4^{20}	1.3848^{-25}	−105.6	0.88		i aq; v s alc, eth
b401	*cis*-2-Butene-1,4-diol	$HOCH_2CH=CHCH_2OH$	88.11	1^2, 567	1.0700_4^{20}	1.4793^{20}	12.5	234	128	s aq; v s alc
b402	*trans*-2-Butene-1,4-diol	$HOCH_2CH=CHCH_2OH$	88.11	1^3, 2252	1.070_4^{20}	1.4779^{20}	27.3	132		v s aq, alc
b403	3-Butenenitrile	$H_2C=CHCH_2CN$	67.09	2, 408	0.8341_4^{20}	1.4060^{20}	−87	119	21	sl s aq; misc alc, eth
b404	*cis*-2-Butenoic acid	$CH_3CH=CHCOOH$	86.09	2, 412	1.0267_4^{20}	1.4482^{14}	14	168–169		v s aq; s alc

b405	*trans*-2-Butenoic acid	$CH_3CH{=}CHCOOH$	86.09	2, 408	0.964_4^{80}	1.4228^{77}	71.4	185.0	87	54.6 aq; v s EtOH, bz, acet
b406	3-Butenoic acid	$H_2C{=}CHCH_2COOH$	86.09	2, 407	1.0091_4^{20}	1.4249^{20}	−39	163	65	s aq; misc alc, eth
b407	*cis*-2-Buten-1-ol	$CH_3CH{=}CHCH_2OH$	72.11	1, 442	0.8662_4^{20}	1.4342^{20}	−89.4	123.6	56	16.6 aq; misc alc
b408	*trans*-2-Buten-1-ol	$CH_3CH{=}CHCH_2OH$	72.11	1, 442	0.8454_4^{20}	1.4289^{20}		121.2	56	16.6 aq; misc alc
b409	3-Buten-2-one	$H_2C{=}CHCOCH_3$	70.09	1, 728	0.8636_4^{20}	1.4086^{20}		81.4	−6	v s aq, alc, acet, eth
b410	1-Buten-3-yne	$HC{\equiv}CCH{=}CH_2$	52.07	1^3, 1032	0.7095_4^1	1.4161^1		5.1		
b411	4-Butoxyaniline	$CH_3(CH_2)_3OC_6H_4NH_2$	165.24	13^2, 226	0.992	1.5343^{20}		148–149^{13mm}		
b412	4-Butoxybenzoic acid	$CH_3(CH_2)_3OC_6H_4COOH$	194.23	10^2, 93			150			
b413	2-Butoxyethanol	$CH_3(CH_2)_3OCH_2CH_2OH$	118.18	1^2, 519	0.9012_4^{20}	1.4198^{20}	−40	170.2	60	5 aq; s most org solv
b414	2-(2-Butoxyethoxy)-ethanol	$HOCH_2CH_2OCH_2$-$CH_2OC_4H_9$	162.23	1^2, 521	0.9536_{20}^{20}	1.4306^{20}	−68.1	230.4	110	misc aq, alc, bz, acet, PE, CCl_4
b415	Butyl acetate	$C_4H_9OOCCH_3$	116.16	2, 130	0.8813_4^{20}	1.3941^{20}	−73.5	126.1	37	0.43 aq; misc alc, eth; s most org solv
b416	DL-*sec*-Butyl acetate	$CH_3COOCH(CH_3)C_2H_5$	116.16	2^2, 141	0.865_4^{25}	1.3840^{25}		112.3	32	0.62 aq; s alc, eth
b417	*tert*-Butyl acetate	$(CH_3)_3COOCCH_3$	116.16	2, 131	0.8665_4^{20}	1.3853^{20}		97.8	15	i aq; misc alc, eth
b418	*tert*-Butyl aceto-acetate	$(CH_3)_3COC({=}O)CH_2$-$C({=}O)CH_3$	158.20		0.954	1.4180^{20}			60	
b419	Butyl acrylate	$H_2C{=}CHCOOC_4H_9$	128.17	2^2, 388	0.894_{16}^{25}	1.4160		148	38	i aq; s alc, eth
b420	Butylamine	$CH_3CH_2CH_2CH_2NH_2$	73.14	4, 156	0.7327_4^{25}	1.3992^{25}	−50.5	77.9	−1	misc aq, alc, eth, PE
b421	*tert*-Butylamine	$(CH_3)_3CNH_2$	73.14	4, 173	0.6951_4^{20}	1.3788^{20}	−67.5	44.4	−8	misc aq, alc
b422	2-(*tert*-Butylamino)-ethanol	$(CH_3)_3CNHCH_2CH_2OH$	117.19				42–45	90–92^{25mm}	68	

(*E*)-2-Butenal, c283
Buten-4-carboxylic acid, p50
Butopyronoxyl, b448
Butoxybenzene, b480
1-Butoxybutane, d115
Butyl alcohols, b394, b395, m385
sec-Butylamine, b380

TABLE 1-14 Physical constants of organic compounds (*continued*)

No.	Name	Formula	Formula weight	Beilstein reference	Density	Refractive index	Melting point	Boiling point	Flash point	Solubility in 100 parts solvent
b423	3-(*tert*-Butylamino)-1,2-propanediol	$(CH_3)_3CNHCH_2CH(OH)$-CH_2OH	147.22				70	92^{1mm}		
b424	4-Butylaniline	$CH_3CH_2CH_2CH_2$-$C_6H_4NH_2$	149.24	12^1, 503	0.945	1.5350^{20}		120^{15mm}	101	
b425	2-*tert*-Butylanthraquinone		264.32				100			
b426	Butylbenzene	$CH_3CH_2CH_2CH_2C_6H_5$	134.22	5, 413	0.8604^{20}_{4}	1.4898^{20}	−88	183.3	59	misc alc, bz, eth
b427	*sec*-Butylbenzene	$CH_3CH_2CH(CH_3)C_6H_5$	134.22	5, 414	0.8608^{20}_{4}	1.4902^{20}	−82.7	173.3	45	misc alc, bz, eth
b428	*tert*-Butylbenzene	$(CH_3)_3CC_6H_5$	134.22	5, 415	0.8669^{20}_{4}	1.4927^{20}	−57.9	169.1	34	misc alc, bz, eth
b429	Butyl benzoate	$C_6H_5COOC_4H_9$	178.23	9, 112	1.000^{20}	1.496	−22	250		i aq; s alc, eth
b430	4-*tert*-Butylbenzoic acid	$(CH_3)_3CC_6H_4COOH$	178.23	9, 560			167			i aq; v s alc, bz
b431	4-*tert*-Butylbenzoyl chloride	$(CH_3)_3CC_6H_4COCl$	196.68		1.007	1.5364^{20}		135^{20mm}	87	
b432	*N*-(*tert*-Butyl)benzylamine	$C_6H_5CH_2NHC(CH_3)_3$	163.27	12, 1022	0.881	1.4968^{20}		80^{5mm}	80	
b433	Butyl butyrate	$CH_3CH_2CH_2COOC_4H_9$	144.21	2, 271	0.8717^{20}_{20}	1.4035		156.9	51	i aq; misc alc, eth
b434	*tert*-Butyl carbazate	$H_2NNHCOOC(CH_3)_3$	132.16				42	$65^{0.03mm}$		
b435	4-*tert*-Butylcatechol	$(CH_3)_3CC_6H_3(OH)_2$	166.22		1.049^{60}_{25}		55	285	151	0.2 aq^{80}; 240 eth^{25}; s alc; v s acet
b436	*tert*-Butyl chloroacetate	$ClCH_2COOC(CH_3)_3$	150.61	2^3, 444	1.053	1.4230^{20}		$48–49^{11mm}$	41	
b437	4-*tert*-Butyl-1-chlorobenzene	$(CH_3)_3CC_6H_4Cl$	168.67	5, 416	1.006	1.5108^{20}	23–25	217		
b438	*tert*-Butylchlorodiphenylsilane	$(CH_3)_3CSi(C_6H_5)_2Cl$	274.87		1.057	1.5675^{20}		$90^{0.02mm}$	>112	
b439	Butyl chloroformate	$ClCOOC_4H_9$	136.58	3^2, 11	1.074^{25}_{4}	1.4114^{20}		142	25	d aq, alc; misc eth

b440	*S-tert*-Butyl chlorothioformate	$ClC(=O)SC(CH_3)_3$	152.6		1.081_4^{30}	1.4691^{30}		42.0^{10mm}	46	
b441	*tert*-Butyl cyanoacetate	$NCCOOC(CH_3)_3$	141.17			1.4200^{20}		108		
b442	2-*tert*-Butylcyclohexanol	$(CH_3)_3C—C_6H_{10}OH$	156.27		0.902		46			i aq
b443	4-*tert*-Butylcyclohexanol	$(CH_3)_3C—C_6H_{10}OH$	156.27	6^1, 18			70	115^{15mm}	105	i aq
b444	2-*tert*-Butylcyclohexanone	$(CH_3)_3C—C_6H_9(=O)$	154.25	7^3, 143	0.896	1.4565^{20}		62.5^{4mm}		
b445	4-*tert*-Butylcyclohexanone	$(CH_3)_3C—C_6H_9(=O)$	154.25	7^1, 29			50	116^{20mm}	96	i aq
b446	Butyl decyl *o*-phthalate	$C_4H_9OOCC_6H_4COO-C_{10}H_{21}$	362.51		0.994_{25}^{25}				202	
b447	*N*-Butyldiethanolamine	$C_4H_9N(CH_2CH_2OH)_2$	161.25	4, 285	0.986_{20}^{20}	1.4625^{20}	< -70	276	126	
b448	Butyl 3,4-dihydro-2,2-dimethyl-4-oxo-2*H*-pyran-6-carboxylate		226.27		1.054_{25}^{25}	1.4767^{20}		256–270	> 112	misc alc, chl, eth

Butyl bromides, b240, b241, b314
N-Butyl-1-butanamine, d107
Butyl carbitol, b414
Butyl Cellosolve, b413
Butyl chlorides, c64, c65, c163
2-*tert*-Butyl-*o*-cresol, b465
2-*tert*-Butyl-*p*-cresol, b464
tert-Butyldihydroxybenzene, b435

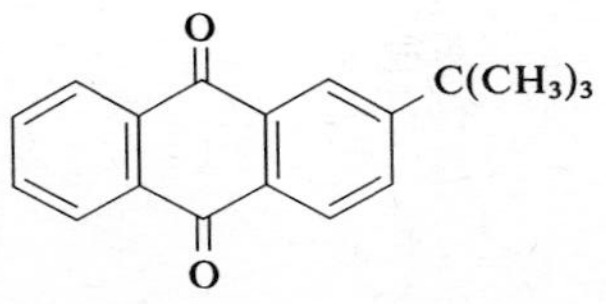

b425

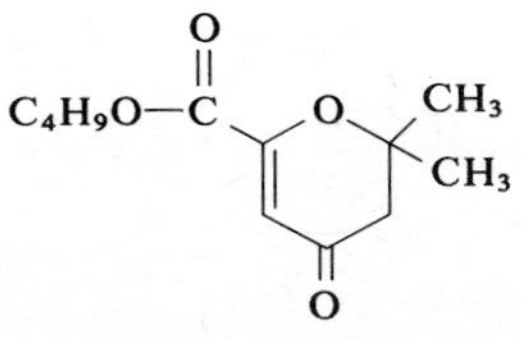

b448

TABLE 1-14 Physical constants of organic compounds (*continued*)

No.	Name	Formula	Formula weight	Beilstein reference	Density	Refractive index	Melting point	Boiling point	Flash point	Solubility in 100 parts solvent
b449	*tert*-Butyldimethyl-chlorosilane	$(CH_3)_3CSi(CH_3)_2Cl$	150.7				91.5	124–126		
b450	1,3-Butylene diacetate	$CH_3CH(OOCCH_3)CH_2$-CH_2OOCCH_3	174.20	2, 143	1.028	1.4199^{20}		99^{8mm}	85	v s aq; s alc
b451	*N*-Butylethanolamine	$HOCH_2CH_2NHC_4H_9$	117.19		0.89^{20}	1.444^{20}	−3.5	192	77	
b452	Butyl ethyl ether	$C_4H_9OC_2H_5$	102.18	1^3, 1502	0.7495^{20}_{4}	1.3818^{20}	−103	92.5		i aq; misc alc, eth
b453	2-Butyl-2-ethyl-1,3-propanediol	$HOCH_2C(C_2H_5)(C_4H_9)$-CH_2OH	160.25		0.931^{50}_{20}	1.4587^{25}	41.4	195^{100mm}		0.8 aq
b454	Butyl ethyl sulfide	$C_4H_9SC_2H_5$	118.24	1^3, 1522	0.8376^{20}_{4}	1.4491^{20}	−95.1	144.2		s chl
b455	*tert*-Butylhydrazine HCl	$(CH_3)_3CNHNH_2$ HCl	124.61	4^3, 1734			191–174			
b456	*tert*-Butylhydroper-oxide	$(CH_3)_3C—O—OH$	90.12		0.896^{20}_{4}	1.4007^{20}	4–5	$33–4^{17mm}$	62	s aq, alc, chl, eth
b457	*tert*-Butylhydro-quinone	$(CH_3)_3CC_6H_3(OH)_2$	166.22				129			
b458	Butyl isocyanate	$CH_3CH_2CH_2CH_2NCO$	99.13		0.880	1.4061^{20}		115	26	
b459	*tert*-Butyl isocyanate	$(CH_3)_3CNCO$	99.13	4, 175	0.868	1.3865^{20}		86	26	
b460	Butyllithium	$CH_3CH_2CH_2CH_2Li$	64.06					$80^{0.0001mm}$	pyro-phoric	
b461	*tert*-Butyllithium	$(CH_3)_3CLi$	64.06				subl $70^{0.1mm}$		pyro-phoric	
b462	Butyl methacrylate	$H_2C{=}C(CH_3)COOC_4H_9$	142.19		0.889^{25}_{15}	1.4220^{25}		170	49	i aq; misc alc, eth
b463	*tert*-Butyl methyl ether	$(CH_3)_3COCH_3$	88.15	1, 381	0.758	1.3685^{20}	−109	56	−10	s aq; v s alc, eth
b464	2-*tert*-Butyl-4-methylphenol	$(CH_3)_3CC_6H_3(CH_3)OH$	164.25		0.9247^{75}_{4}	1.4969^{75}	51.7	237		i aq; s org solv

b465	2-*tert*-Butyl-6-methylphenol	$(CH_3)_3CC_6H_3(CH_3)OH$	164.25			1.5195^{20}	32	230	107	
b466	Butyl methyl sulfide	$C_4H_9SCH_3$	104.21	1^3, 1521	0.8426_4^{20}	1.4477^{20}	−97.8	123.4		v s alc
b467	Butyl nitrite	C_4H_9ONO	103.12	1, 369	0.9114_4^{0}	1.3768		78	4	misc alc, eth
b468	*tert*-Butyl nitrite	$(CH_3)_3CONO$	103.12	1^2, 415	0.8671_4^{20}	1.3687^{20}		63		sl s aq; v s alc, chl, eth, CS_2
b469	Butyl octadecanoate	$CH_3(CH_2)_{16}COOC_4H_9$	340.60	2^2, 352	0.8551_4^{20}	1.4422^{25}	26.3	343	160	s alc, v s acet
b470	Butyl 4-oxopentanoate	$CH_3C(=O)CH_2$-$CH_2COOC_4H_9$	172.22		0.9735_4^{20}	1.4270^{20}		107^{6mm}	91	s alc, eth, acet
b471	*tert*-Butyl peroxybenzoate	$C_6H_5(=O)O-OC(CH_3)_3$	194.23		1.021	1.4990^{20}		$76^{0.2mm}$	93	
b472	2-*sec*-Butylphenol	$CH_3CH_2CH(CH_3)$-C_6H_4OH	150.22		0.982	1.5222^{20}	12	228	112	i aq; s alc; v s eth
b473	2-*tert*-Butylphenol	$(CH_3)_3CC_6H_4OH$	150.22	6^2, 489	0.9783_4^{20}	1.5228^{20}	−7	221–224	110	
b474	3-*tert*-Butylphenol	$(CH_3)_3CC_6H_4OH$	150.22				40–41	240		
b475	4-*sec*-Butylphenol	$CH_3CH_2CH(CH_3)$-C_6H_4OH	150.22	6, 522	0.969_4^{20}	1.5150	62	136^{25mm}	115	s hot aq, alc, eth
b476	4-*tert*-Butylphenol	$(CH_3)_3CC_6H_4OH$	150.22	6, 524	0.908_4^{114}	1.4787^{114}	100–101	237		i aq; s alc, eth
b477	2-(4-*sec*-Butylphenoxy)ethanol	$CH_3CH_2CH(CH_3)C_6H_4$-OCH_2CH_2OH	194.2		1.008^{25}		< −20	158^{10mm}	149	0.1 aq
b478	2-(4-*tert*-Butylphenoxy)ethanol	$(CH_3)_3CC_6H_4$-OCH_2CH_2OH	194.3		1.016^{25}		54	167^{10mm}	157	0.1 aq
b479	*tert*-Butyl phenyl carbonate	$C_6H_5OC(=O)OC(CH_3)_3$	194.23		1.047	1.4805^{20}		$79^{0.8mm}$		
b480	Butyl phenyl ether	$CH_3CH_2CH_2CH_2OC_6H_5$	150.22	6, 143	0.9351_4^{20}	1.4970^{20}	−19	210.3	82	

Butyl disulfides, d113, d114
1,4-Butylene bis(2,3-epoxypropyl) ether, b176
1,3-Butylene glycol methyl ether, m57
1,2-Butylene oxide, e3
Butyl ether, d115
Butyl ethyl ketone, h16
tert-Butyl fluoride, f20
Butyl glycol, b413
2,2′-(Butylimino)diethanol, b447
Butyl iodides, i30, i31, i44
Butyl levulinate, b470
Butyl mercaptans, b391, m381, m382, m383
Butyl methyl ketone, d499
tert-Butyl perbenzoate, b471

TABLE 1-14 **Physical constants of organic compounds (*continued*)**

No.	Name	Formula	Formula weight	Beilstein reference	Density	Refractive index	Melting point	Boiling point	Flash point	Solubility in 100 parts solvent
b481	4-*tert*-Butylphenyl salicylate	$HOC_6H_4COOC_6H_4$-$C(CH_3)_3$	270.31				62–64			0.i aq; 79 alc; 153 EtAc; 158 toluene
b482	Butyl propionate	$CH_3CH_2COOC_4H_9$	130.19	2, 241	0.8818^{15}	1.3982^{25}	−89.6	145.5		misc alc, eth
b483	4-*tert*-Butylpyridine	$(CH_3)_3C—C_5H_4N$	135.21	20, 252	0.915	1.4952^{20}		197	63	
b484	Butyltin chloride	$C_4H_9SnCl_3$	282.17		1.693	1.5229^{20}		93^{10mm}	81	
b485	4-*tert*-Butyltoluene	$(CH_3)_3CC_6H_4CH_3$	148.25	5, 439	0.853	1.4897^{20}		192	54	
b486	Butyltrichlorosilane	$C_4H_9SiCl_3$	191.5	4^1, 582	1.161_4^{20}	1.436^{20}		142–143		d aq, hot alc; s eth
b487	Butyl trifluoro-acetate	$CF_3COOC_4H_9$	170.1		1.0268^{22}	1.353^{22}		100.2		
b488	Butyltrimethoxysilane	$C_4H_9Si(OCH_3)_3$	178.3		0.9312_4^{20}	1.3979^{20}		164–165		
b489	*tert*-Butyl tri-methylsilyl peroxide	$(CH_3)_3C—O—O—Si$-$(CH_3)_3$	162.3		0.8219_4^{20}	1.3935^{25}	d 135	41^{41mm}		
b490	Butyl urea	$C_4H_9NHCONH_2$	116.16	4^1, 371			93–95			s aq, alc, eth
b491	Butyl vinyl ether	$C_4H_9OCH{=}CH_2$	100.16		0.7792^{20}	1.4007^{20}	−112.7	94.2	−9	0.3 aq
b492	5-*tert*-Butyl-*m*-xylene	$(CH_3)_3CC_6H_3(CH_3)_2$	162.28	5, 447	0.867	1.4946^{20}		205–206	72	
b493	1-Butyne	$CH_3CH_2C{\equiv}CH$	54.09	1, 249	0.7110_4^{-31}	1.3962^{20}	−125.7	8.1		i aq; s alc, eth
b494	2-Butyne	$CH_3C{\equiv}CCH_3$	54.09	1, 249	0.6910_4^{20}	1.3920^{20}	−32.3	17.0		i aq; s alc, eth
b495	2-Butyne-1,4-diol	$HOCH_2C{\equiv}CCH_2OH$	86.09	1^1, 261		1.450^{25}	54–58	238	152	374 aq; 83 alc; 0.04 bz; 2.6 eth; 70 acet
b496	Butyraldehyde	$CH_3CH_2CH_2CHO$	72.11	1, 662	0.8016_4^{20}	1.3791^{20}	−96.4	74.8	−6.7	7.1 aq; misc alc, eth, acet, EtAc
b497	Butyramide	$CH_3CH_2CH_2CONH_2$	87.12	2, 275			116	216		16 aq; s alc
b498	Butyric acid	$CH_3CH_2CH_2COOH$	88.11	2, 264	0.9582_4^{20}	1.3980^{20}	−5.3	163.3	77	misc aq, alc, eth
b499	Butyric anhydride	$[CH_3CH_2CH_2C(O)]_2O$	158.20	2, 274	0.9668_4^{20}	1.4130^{20}	−65.7	199.5	87	s aq, alc(d), eth

b500	3-Butyrolactone		86.09	17^1, 130	1.056	1.4109	20	73^{29mm}	60	
b501	4-Butyrolactone		86.09	17, 234	1.124_4^{25}	1.4348^{25}	−43.5	204	98	misc aq, alc, acet, bz, eth, CCl_4
b502	Butyronitrile	$CH_3CH_2CH_2CN$	69.11	2^2, 252	0.7954_4^{15}	1.3860^{15}	−111.9	117.9	16	3.3 aq; misc alc, eth
b503	Butyrophenone	$C_6H_5C(O)C_3H_7$	148.21	7, 313	1.021	1.5195^{20}	13	222	88	
b504	Butyryl chloride	$CH_3CH_2CH_2COCl$	106.55	2, 274	1.0263_4^{21}	1.4122^{20}	−89	102	21	s aq, alc(d); misc eth
c1	Caffeine		194.19	26, 461	1.23_4^{18}		238	subl 178		2.1 aq; 1.5 alc; 18 chl; 0.19 eth; 1 bz
c2	DL-Camphene		136.24	5, 156	0.8422_4^{54}	1.4551^{54}	51–52	159	36	i aq; s alc, chl, eth
c3	D-(+)-Camphor		152.23	7, 101	0.9920_4^{25}		178.8	207.4		100 alc; 100 eth; 200 chl; 250 acet
c4	DL-Camphor		152.24	7, 135			177	204	64	
c5	D-Camphoric acid		200.23	9, 745	1.186_4^{20}		186–188			4 aq; 100 alc; s chl, eth

Butyl o-phthalate, d128
Butyl propyl ketone, o36
Butyl stearate, b469
Butyl sulfate, d131
Butyl sulfides, d132, d133
Butyl sulfite, d134
Butyl sulfone, d135
Butyrolactam, p279
Cadaverine, p29
2-Camphanone, c3

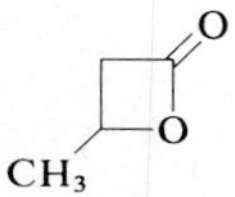

b500 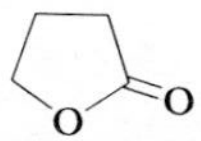 b501 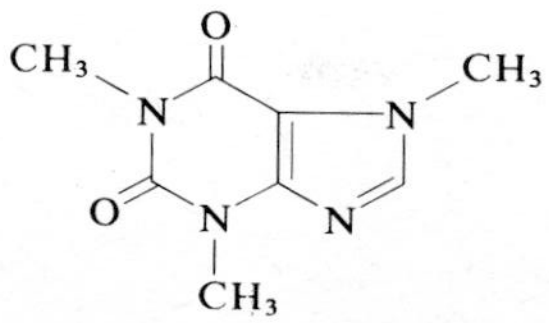 c1 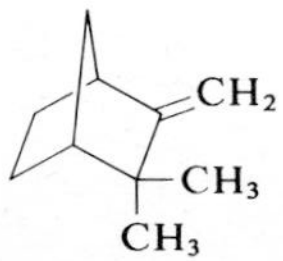 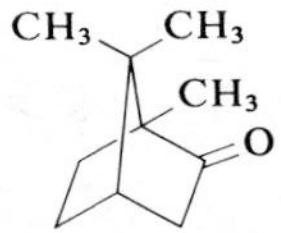 c3, c4 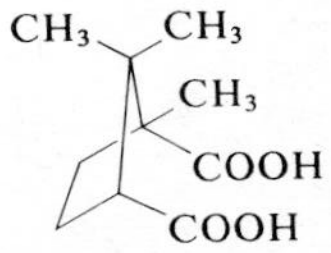 c5

TABLE 1-14 Physical constants of organic compounds (*continued*)

No.	Name	Formula	Formula weight	Beilstein reference	Density	Refractive index	Melting point	Boiling point	Flash point	Solubility in 100 parts solvent
c6	DL-Camphoric anhydride		182.22	17, 455	1.194_4^{20}		225	270		s bz; sl s aq, alc, eth
c7	D-10-Camphor-sulfonic acid hydrate		250.32	11, 316			194 d			deliq moist air; sl s HOAc, EtAc; i eth
c8	Carbazole		167.21	20, 433			245–246	355		
c9	4-Carbethoxy-3-methyl-3-cyclo-hexen-1-one		182.22	10, 631	1.078	1.4880^{20}		268–272	> 112	
c10	Carbobenzyloxy-glycine	$C_6H_5CH_2OC(=O)NH$-CH_2COOH	209.20				122			
c11	Carbohydrazide	$H_2NNHC(=O)NHNH_2$	90.09	3, 121			d 153			v s aq; i alc, bz, eth
c12	2-(Carbomethoxy)-ethylmethyl-dichlorosilane	$CH_3OC(=O)CH_2CH_2Si$-$(CH_3)Cl_2$	201.1		1.187_4^{25}	1.4439^{25}		$98–99^{25mm}$		
c13	2-Carbomethoxyethyl-trichlorosilane	$CH_3OC(=O)CH_2CH_2$-$SiCl_3$	221.6		1.325_4^{20}	1.448^{20}		$88–89^{2mm}$		
c14	2-Carboxybenz-aldehyde	$HC(=O)C_6H_4COOH$	150.13	10, 666			96–98			
			192.13	18, 468			161–164	240–245		15.5 DMF; 49.6
c15	4-Carboxy-1,2-ben-zenedicarboxylic anhydride		192.13	18, 468			161–164	240-245^{14mm}		15.5 DMF; 49.6 acet; 21.6 EtAc
c16	4-Carboxybenzene-sulfonamide	$HOOCC_6H_4SO_2NH_2$	201.20	11, 390			d 280			i aq, bz, eth; v s alc
c17	2-Carboxyethyl-phosphonic acid	$HOOCCH_2CH_2$-$P(O)(OH)_2$	154.06	4^2, 976						
c18	DL-Carnitine HCl	$(CH_3)_3NCH_2CHOH$-$CH_2COOH·HCl$	197.66				197 d			v s aq; i acet, eth

c19	*trans*-β-Carotene		536.89	30, 87	1.000^{20}_{20}		183			i aq; s bz, chl, CS_2
c20	D-(+)-Carvone		150.22	7, 153	0.965^{20}_{4}	1.4989^{20}		230	88	i aq; misc alc
c21	Catecholborane		119.92			1.5070^{20}	12	50^{50mm}		

Capric acid, d14
Caproaldehyde, h54
Caproic acid, h66
Caproic anhydride, h67
ε-Caprolactam, o57
ε-Caprolactone, h71
Capronitrile, h63
Caproyl chloride, h73
Caprylic acid, o29
Capryl alcohol, o30
Caprylaldehyde, o40

Caprylonitrile, o27
Capryloyl chloride, o37
CAPS, c337
N-(Carbamoylmethyl)iminodiacetic acid, a14
Carbamylurea, b216
Carbanilide, d695
Carbazole, d667
Carbitol, e36
Carbitol acetate, e37
Carbobenzoxy chloride, b90
4,4′-Carboxyldiphthalic anhydride, b55

N-Carbonylsulfamyl chloride, c241
Carboxybenzaldehyde, f33
(3-Carboxy-2-hydroxypropyl)
trimethylammonium hydroxide, c18
3-Carbomethoxypropionyl chloride, m189
(Carboxymethylimino)bis(ethylenenitrilo)-
tetraacetic acid, d299
(Carboxylmethyl)trimethylammonium
hydroxide, b129
3-Carboxypropyl disulfide, d710

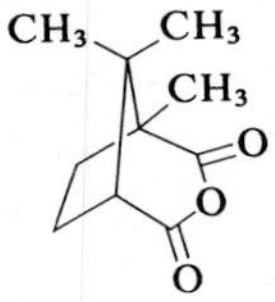

c6

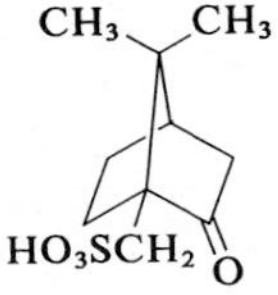

c7

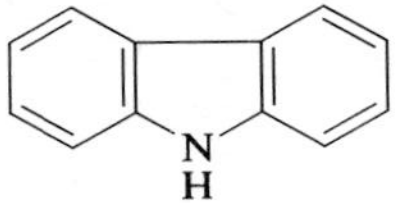

c8

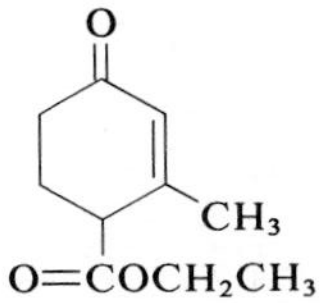

c9

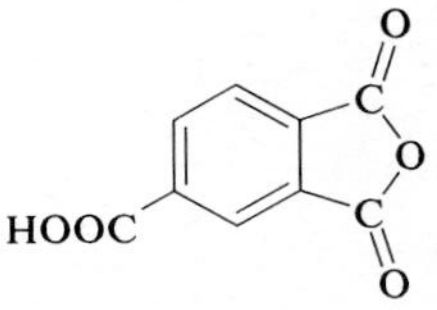

c15

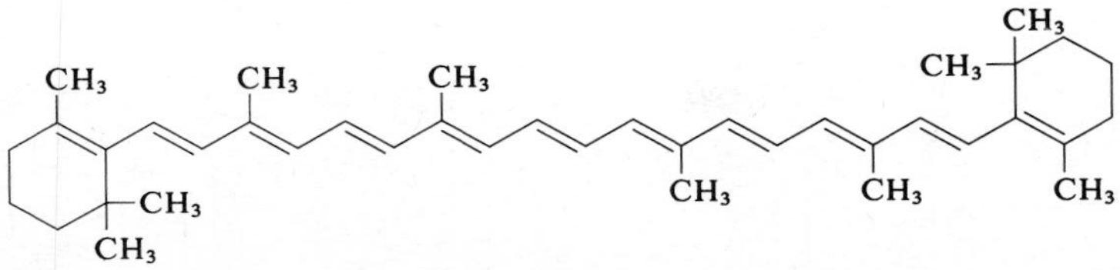

c19
[Note: 9 successive (⑊╱) units]

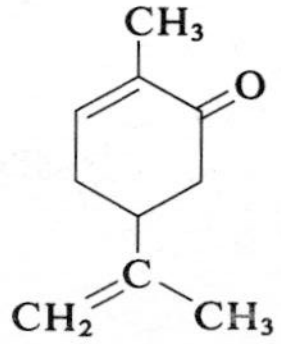

c20

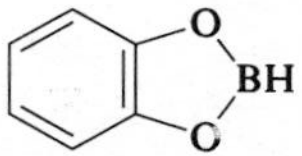

c21

TABLE 1-14 Physical constants of organic compounds (*continued*)

No.	Name	Formula	Formula weight	Beilstein reference	Density	Refractive index	Melting point	Boiling point	Flash point	Solubility in 100 parts solvent
c22	2-Chloroacetamide	$ClCH_2CONH_2$	93.51	2, 199			118	225 d		10 aq; 10 alc; sl s eth
c23	*p*-Chloroacetanilide	$ClC_6H_4NHCOCH_3$	169.61	12, 611	1.385_4^{20}		179			i aq; v s alc, eth, CS_2
c24	Chloroacetic acid	$ClCH_2COOH$	94.50	2, 194	1.580(*c*)	1.4297^{65}	63(α)	189		v s aq; s alc, bz, eth
c25	Chloroacetic anhydride	$[ClCH_2C(O)]_2O$	170.98	2, 199	1.5494_4^{20}		46	203		d aq; v s chl, eth
c26	*p*-Chloroacetoacet-anilide	$CH_3COCH_2CONHC_6H_4Cl$	211.65				134			
c27	Chloroacetonitrile	$ClCH_2CN$	75.50	2, 201	1.193	1.4225^{20}		126	47	
c28	α-Chloroaceto-phenone	$C_6H_5COCH_2Cl$	154.60	7, 282	1.324^{15}		54	245		i aq; v s alc, bz, eth
c29	*o*-Chloroaceto-phenone	$ClC_6H_4COCH_3$	154.60	7^1, 151	1.188	1.5438^{20}		228^{738mm}	88	sl s aq; s eth
c30	*p*-Chloroaceto-phenone	$ClC_6H_4COCH_3$	154.60	7, 281	1.192_4^{20}	1.5549	20–21	237	90	i aq; misc alc, eth
c31	Chloroacetyl chloride	$ClCH_2COCl$	112.94	2, 199	1.418_{25}^{25}	1.4530^{20}	−22.5	106	none	d aq, MeOH
c32	2-Chloroacrylo-nitrile	$H_2C{=}C(Cl)CN$	87.51		1.096	1.4290^{20}	−65	89	6	
c33	2-Chloroaniline	$ClC_6H_4NH_2$	127.57	12, 597	1.2125_4^{20}	1.5881^{20}	−1.94	208.8	97	0.88 aq; s alc, bz, eth
c34	3-Choroaniline	$ClC_6H_4NH_2$	127.57	12, 602	1.2150_4^{22}	1.5931^{20}	−10.4	230.5	123	i aq; s alc, bz, eth
c35	*p*-Chloroaniline	$ClC_6H_4NH_2$	127.57	12, 607	1.169_4^{77}	1.5546^{85}	72.5	232		s hot aq; v s alc, acet, eth, CS_2
c36	1-Chloroanthra-quinone		242.66	7, 787			160	subl		sl s alc; misc eth; s hot bz
c37	2-Chloroanthra-quinone		242.66	7, 787			211	subl		sl s alc, bz; i eth

c38	2-Chlorobenzaldehyde	ClC_6H_4CHO	140.57	7, 233	1.2483^{20}_{4}	1.5658	11	215	87	sl s aq; s alc, bz, eth
c39	4-Chlorobenzaldehyde	ClC_6H_4CHO	140.57	7, 235	1.196^{61}_{4}	1.552^{61}	47	214	87	s aq; v s alc, bz, eth
c40	2-Chlorobenzamide	$ClC_6H_4CONH_2$	155.58	9, 336			142–144			
c41	Chlorobenzene	C_6H_5Cl	112.56	5, 199	1.1063^{20}	1.5248^{20}	−45.3	131.7	23	0.049 aq^{30}; v s alc, bz, chl, eth
c42	4-Chlorobenzenesulfonamide	$ClC_6H_4SO_2NH_2$	191.64	11, 55			146			s hot aq, hot alc, hot eth
c43	4-Chlorobenzenesulfonyl chloride	$ClC_6H_4SO_2Cl$	211.07	11, 55			55	141^{15mm}		d aq, alc; v s bz, eth
c44	4-Chlorobenzhydrol	$ClC_6H_4CH(OH)C_6H_5$	218.68	6, 680			58–60			
c45	2-Chlorobenzoic acid	ClC_6H_4COOH	156.57	9, 334	1.544^{25}_{4}		142			0.11 aq; v s alc, eth

Cellosolve, e35
Cellosolve acetate, e38
Cetyl alcohol, h36
Cetyl bromide, b295
Chalcone, d688
CHES, c336
Chloramine T, c248
Chloranils, t25, t26
Chloranilic acid, d172
Chlorendic anhydride, h30
Chloroacetaldehyde diethyl acetal, c81
Chloroacetaldehyde dimethyl acetal, c89
Chloroacetone, c216
4-(Chloroacetyl)catechol, c86
Chloroanthranilic acid, a140
5-Chloroanthranilonitrile, a141
p-Chlorobenzenethiol, c244

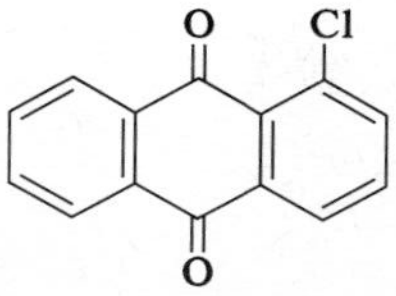

c36

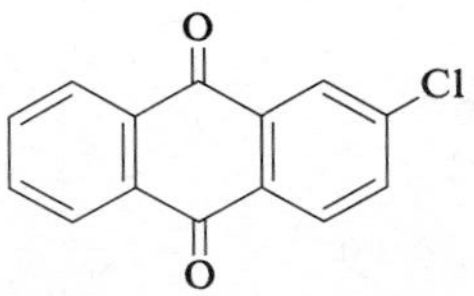

c37

TABLE 1-14 Physical constants of organic compounds (*continued*)

No.	Name	Formula	Formula weight	Beilstein reference	Density	Refractive index	Melting point	Boiling point	Flash point	Solubility in 100 parts solvent
c46	3-Chlorobenzoic acid	ClC_6H_4COOH	156.57	9, 337	1.496_4^{25}		157–158			0.04 aq; v s alc, eth
c46a	4-Chlorobenzoic acid	ClC_6H_4COOH	156.57	9, 340			241–243			0.02 aq; v s alc, eth
c47	2-Chlorobenzonitrile	ClC_6H_4CN	137.57	9, 336			46	232		s alc, eth
c48	4-Chlorobenzonitrile	ClC_6H_4CN	137.57	9, 341			93	22		s alc, bz, chl, eth
c49	2-Chlorobenzophenone	$ClC_6H_4COC_6H_5$	216.67	7, 419			44–47	300		
c50	4-Chlorobenzophenone	$ClC_6H_4COC_6H_5$	216.67	7, 419			77	196^{17mm}		s alc, acet, bz, eth
c51	2-Chlorobenzotrifluoride	$ClC_6H_4CF_3$	180.56		1.3540^{25}	1.4513^{25}	−6.4	152.3		
c52	3-Chlorobenzotrifluoride	$ClC_6H_4CF_3$	180.56		1.3311^{25}	1.4438^{25}	−56.7	137.7	36	
c53	4-Chlorobenzotrifluoride	$ClC_6H_4CF_3$	180.56		1.353^{20}	1.4463	−33.2	138.7	47	
c54	2-(4-Chlorobenzoyl)-benzoic acid	$ClC_6H_4COC_6H_4COCH$	260.68	10, 750			150			s alc, bz, eth
c55	2-Chlorobenzoyl chloride	ClC_6H_4COCl	175.01	9, 336	1.382	1.5718^{20}	−3	238	110	d aq, alc
c56	4-Chlorobenzoyl chloride	ClC_6H_4COCl	175.01	9, 341	1.377	1.5780^{20}	14	222	105	d aq, alc
c57	4-Chlorobenzyl alcohol	$ClC_6H_4CH_2OH$	142.59	6, 444			72	234		v s alc, eth
c58	4-Chlorobenzylamine	$ClC_6H_4CH_2NH_2$	141.60	12, 1074	1.164	1.5586^{20}		215	90	
c59	2-Chlorobenzyl chloride	$ClC_6H_4CH_2Cl$	161.03	5, 297	1.274	1.5591^{20}	−17	214	82	
c60	4-Chlorobenzyl chloride	$ClC_6H_4CH_2Cl$	161.03	5, 308			30	214	97	s alc; v s eth

c61	2(*p*-Chlorobenzyl)-pyridine	$ClC_6H_4CH_2—C_5H_4N$	203.67		1.390	1.5868^{20}		183^{20mm}	> 112	
c62	4-(*p*-Chlorobenzyl)-pyridine	$ClC_6H_4CH_2—C_5H_4N$	203.67		1.167	1.5900^{20}			> 112	
c63	1-Chloro-1,3-butadiene	$H_2C{=}CHCH{=}CHCl$	88.54	1^3, 949	0.9601^{20}_{4}	1.4712^{20}		68		v s chl
c64	1-Chlorobutane	$CH_3CH_2CH_2CH_2Cl$	92.57	1, 118	0.8864^{20}_{4}	1.4021^{20}	−123.1	78.44	−6	0.11 aq; misc alc, eth
c65	2-Chlorobutane	$CH_3CH_2CH(Cl)CH_3$	92.57	1, 119	0.8732^{20}_{4}	1.3971^{20}	−113.3	68.25	−15	0.1 aq; misc alc, eth
c66	4-Chloro-1-butanol	$ClCH_2CH_2CH_2CH_2OH$	108.56	1^2, 398	1.0883^{20}_{4}	1.4518^{20}		$86–89^{20mm}$	32	s alc, eth
c67	3-Chloro-2-butanone	$CH_3CH(Cl)COCH_3$	106.55	1, 669	1.055	1.4172^{20}		117	21	v s alc, eth
c68	*cis*-1-Chloro-2-butene	$CH_3CH{=}CHCH_2Cl$	90.55	1^2, 176	0.9426^{20}_{4}	1.4390^{20}		84.1	−15	s alc, acet
c69	3-Chloro-1-butene	$CH_3CH(Cl)CH{=}CH_2$	90.55	1^2, 174	0.9001^{20}_{4}	1.4155^{20}		62–65	−20	v s acet
c70	3-Chloro-1-butyne	$CH_3CH(Cl)C{\equiv}CH$	88.54	1^4, 970	0.961	1.4280^{20}		68–70	1	
c71	3-Chlorobutyric acid	$CH_3CH(Cl)CH_2COOH$	122.55	2, 277	1.186^{20}_{4}	1.4421^{20}	16.3	109^{17mm}		s alc, eth
c72	4-Chlorobutyric acid	$ClCH_2CH_2CH_2COOH$	122.55	2, 278	1.2236^{20}_{4}	1.4510^{20}	12–16	196^{22mm}	> 112	sl s aq; v s eth
c73	4-Chlorobutyronitrile	$ClCH_2CH_2CH_2CN$	103.55	2, 278	1.158	1.4413^{20}		197	85	s alc, eth
c74	4-Chlorobutyryl chloride	$ClCH_2CH_2CH_2COCl$	141.00	2, 278	1.258	1.4609^{20}		174	72	d aq, alc; s eth
c75	Chloro(chloromethyl)-dimethylsilane	$ClCH_2Si(CH_3)_2Cl$	143.09		1.086	1.4373^{20}		114^{752mm}	21	
c76	*trans-p*-Chlorocinnamic acid	$ClC_6H_4CH{=}CHCOOH$	182.61	9, 594			248–250			
c77	Chlorocyclohexane	ClC_6H_{11}	118.61	5, 21	1.000^{20}_{4}	1.4620^{20}	−44	142	28	i aq; s alc, eth
c78	2-Chlorocyclohexanone	$ClC_6H_9({=}O)$	132.59	7, 10	1.161	1.4835^{20}	23	83^{10mm}	82	s bz, eth, diox

4-Chlorobenzyl mercaptan, c249 Chlorocresols, c159, c160

TABLE 1-14 Physical constants of organic compounds (*continued*)

No.	Name	Formula	Formula weight	Beilstein reference	Density	Refractive index	Melting point	Boiling point	Flash point	Solubility in 100 parts solvent
c79	Chlorocyclopentane	ClC_5H_9	104.58	5, 19	1.0051^{20}_{4}	1.4512^{20}		114	15	i aq
c80	1-Chlorodecane	$CH_3(CH_2)_9Cl$	176.73	1, 168	0.868	1.4362^{20}	−34	223	83	i aq
c81	2-Chloro-1,1-diethoxyethane	$ClCH_2CH(OC_2H_5)_2$	152.62	1, 611	1.018	1.4157^{20}		157	29	
c82	3-Chloro-1,1-diethoxypropane	$ClCH_2CH_2CH(OC_2H_5)_2$	166.65	1, 632	0.995	1.4240^{20}		84^{25mm}	36	
c83	Chlorodifluoroacetic acid	$F_2C(Cl)COOH$	130.48	2, 201		1.3559^{20}	22.9	121.5		
c84	1-Chloro-1,1-difluoroethane	$CH_3C(Cl)F_2$	100.50		1.118^{21}		−131	−9		0.19 aq
c85	Chlorodifluoromethane	$HCClF_2$	86.47		1.209^{21}		−160	−40.8		0.30 aq
c86	α-Chloro-3′,4′-dihydroxyacetophenone	$(HO)_2C_6H_3C(=O)CH_2Cl$	186.59	8, 273			176			
c87	1-Chloro-2,4-dihydroxybenzene	$ClC_6H_3(OH)_2$	144.56	6^2, 818			107	147^{18mm}		v s aq, alc, chl, eth
c88	2-Chloro-1,4-dihydroxybenzene	$ClC_6H_3(OH)_2$	144.56	6, 849			101–102	263		v s aq; i alc; s eth
c89	2-Chloro-1,1-dimethoxyethane	$ClCH_2CH(OCH_3)_2$	124.57		1.094^{20}_{20}	1.4148^{20}		130	28	
c91	4-Chloro-3,5-dimethylphenol	$Cl(CH_3)_2C_6H_2OH$	156.61	6^2, 463			115.5	246		0.1 aq; 1 alc; s bz, eth, alk
c92	1-Chloro-2,2-dimethylpropane	$(CH_3)_3CCH_2Cl$	106.59		0.866^{20}_{4}	1.4042^{20}	−20	84.4		
c93	Chlorodimethylsilane	$(CH_3)_2Si(Cl)H$	94.62		0.852^{20}_{4}	1.3827^{20}	−111	36	−28	
c94	Chlorodimethylvinylsilane	$(CH_3)_2Si(Cl)CH{=}CH_2$	120.7		0.884^{25}_{4}	1.414^{25}		82.5		

c95	1-Chloro-2,4-dinitrobenzene	$ClC_6H_3(NO_2)_2$	202.55	5, 263	1.4982^{75}_{4}	1.5857^{60}	52–54	315	186	sl s alc; s hot alc, bz, eth
c96	1-Chloro-3,4-dinitrobenzene	$ClC_6H_3(NO_2)_2$	202.55	5, 262	1.6867^{16}	1.5870^{20}			> 112	v s eth; s alc
c97	2-Chloro-3,5-dinitrobenzoic acid	$ClC_6H_2(NO_2)_2COOH$	246.56	9, 415			198	241 explodes		0.3 aq
c98	α-Chlorodiphenylmethane	$C_6H_5CH(Cl)C_6H_5$	202.68	5^2, 600	1.140^{20}_{4}	1.5951^{20}	17	140^{3mm}	> 112	
c99	Chlorodiphenylmethylsilane	$(C_6H_5)_2Si(Cl)CH_3$	232.8		1.1277^{20}_{4}	1.5742^{20}		295		
c100	Chlorodiphenylphosphine	$(C_6H_5)_2PCl$	220.64	16, 763	1.229	1.6338^{20}		320	> 112	
c101	1-Chlorododecane	$CH_3(CH_2)_{11}Cl$	204.79		0.8673^{20}_{4}	1.4426	−9	116	93	v s alc; s bz
c102	1-Chloro-2,3-epoxypropane	$H_2C-CHCH_2Cl$ (epoxide, O bridging)	92.53	17, 6	1.1812^{20}_{4}	1.4381^{20}	−57.2	116.1	33	5.9 aq; misc alc, chl,
c103	Chloroethane	CH_3CH_2Cl	64.52	1, 82	0.9214^{0}_{4}	1.3742^{10}	−136 to −138	12.3	−43	0.45 aq^0; 48 alc; misc eth
c104	2-Chloroethanol	$ClCH_2CH_2OH$	80.52	1, 337	1.197^{20}_{4}	1.4422^{20}	−67.5	128.6	60	misc aq, alc
c105	2-(2-Chloroethoxy)-ethanol	$ClCH_2CH_2OCH_2CH_2OH$	124.57	1, 467	1.180	1.4529^{20}		81^{5mm}	90	
c106	2-[2-(2-Chloroethoxy)ethoxy]ethanol	$ClCH_2CH_2OCH_2CH_2$-OCH_2CH_2OH	168.62	1, 468	1.160	1.4580^{20}		120^{5mm}	107	
c107	2-Chloroethylamine HCl	$ClCH_2CH_2NH_2$ HCl	115.99	4, 133			146			
c108	1-Chloro-2-ethylbenzene	$ClC_6H_4C_2H_5$	140.61		1.055^{25}_{25}		−81	179.2		i aq; misc alc, eth
c109	(2-Chloroethyl)-benzene	$C_6H_5CH_2CH_2Cl$	140.61	5, 354	1.069	1.5300^{20}		84^{16mm}	66	s alc, bz, eth

Chlorodibromomethane, d71
2-Chloro-*N,N*-diethylethylamine, d272
Chlorodimethyl ether, c156
2-Chloro-*N,N*-dimethylethylamine, d469
4′-Chlorodiphenylmethanol, c44
2-Chloroethyl alcohol, c104

TABLE 1-14 Physical constants of organic compounds (*continued*)

No.	Name	Formula	Formula weight	Beilstein reference	Density	Refractive index	Melting point	Boiling point	Flash point	Solubility in 100 parts solvent
c110	Chloroethylene	$H_2C{=}CHCl$	62.50	1, 186	0.97^{-14}		−159.7	−13.9		sl s aq; s alc
c111	2-Chloroethyl ethyl ether	$ClCH_2CH_2OCH_2CH_3$	108.57	1, 337	0.989	1.4125^{20}		107	15	
c112	2-Chloroethyl methyl ether	$ClCH_2CH_2OCH_3$	94.54	1, 337	1.035	1.4111^{20}		89–90	15	
c113	*N*-(2-Chloroethyl)-morpholine HCl		186.08				186			
c114	*N*-(2-Chloroethyl)-piperidine HCl		184.11	20, 17			236			
c115	2-Chloroethyl *p*-toluenesulfonate	$CH_3C_6H_4SO_3CH_2CH_2Cl$	234.70	11^2, 45	1.294	1.5290^{20}		$153^{0.3mm}$	>112	
c116	2-Chloroethyl vinyl ether	$H_2C{=}CHOCH_2CH_2Cl$	106.55	1^2, 473	1.048	1.4370^{20}	−69.7	110	16	0.6 aq
c117	1-Chloro-2-fluoro-benzene	ClC_6H_4F	130.55	5^1, 110	1.244	1.5010^{20}	−42.5	138.5	31	s alc, eth
c118	1-Chloro-3-fluoro-benzene	ClC_6H_4F	130.55		1.219	1.4944^{20}		126	20	s alc, eth
c119	1-Chloro-4-fluoro-benzene	ClC_6H_4F	130.55	5, 201	1.226_4^{20}	1.4967^{20}	−21.5	130–131		s alc, eth
c120	2-Chloro-6-fluoro-benzyl chloride	$Cl(F)C_6H_3CH_2Cl$	179.02		1.401	1.5372^{20}				
c121	4-Chloro-4′-fluoro-butyrophenone	$FC_6H_4C({=}O)CH_2CH_2$-CH_2Cl	200.64		1.220	1.5255^{20}			110	
c122	3-Chloro-4-fluoro-nitrobenzene	$Cl(F)C_6H_3NO_2$	175.5		1.6028^{17}	1.5674^{17}	41.5	127^{17mm}		
c123	2-Chloro-4-fluoro-phenol	$Cl(F)C_6H_3OH$	146.5				23	88^{4mm}		
c124	2-Chloro-4-fluoro-toluene	$Cl(F)C_6H_3CH_3$	144.58		1.1972^{20}	1.4985^{25}		152–153		

c125	2-Chloro-6-fluoro-toluene	$Cl(F)C_6H_3CH_3$	144.58		1.191	1.5026^{20}		156	46	
c126	4-Chloro-2-fluoro-toluene	$Cl(F)C_6H_3CH_3$	144.58			1.4998^{20}		158		
c127	Chloroform	$CHCl_3$	119.39	1, 61	1.4985^{15}	1.4486^{15}	−63.59	61.7	none	0.82 aq
c128	Chloroform-*d*	$CDCl_3$	120.39		1.50	1.4445^{20}		60.9	none	
c129	1-Chloroheptane	$CH_3(CH_2)_6Cl$	134.65	1, 154	0.881^{16}_{0}	1.4250^{20}	−69	159–161	41	misc alc, eth
c130	1-Chlorohexane	$CH_3(CH_2)_5Cl$	120.62		0.8780^{20}_{4}	1.4236^{20}		134	38	i aq
c131	6-Chloro-1-hexanol	$Cl(CH_2)_6OH$	136.62		1.204	1.4557^{20}		108^{14mm}	98	sl s aq; v s alc, eth
c132	4-Chloro-4′-hydroxy-benzophenone	$ClC_6H_4C(=O)C_6H_4OH$	232.67	8^2, 187			175–178	257^{14mm}		
c133	5-Chloro-8-hydroxy-7-iodoquinoline		305.50				d 172			i alc, eth; 0.8 chl; 0.6 HOAc
c134	3-Chloro-4-hydroxy-mandelic acid	$ClC_6H_3(OH)$-$CH(OH)COOH$	202.60				145–147			
c135	5-Chloro-8-hydroxy-quinoline		179.61	21, 95			130			sl s aq HCl
c136	1-Chloro-4-iodo-benzene	ClC_6H_4I	238.46	5, 221	1.186^{57}_{4}		53–54	226–227		s alc

2-Chloroethyl ether, b159
2-Chloro-6-fluorobenzal chloride, t237
α-Chloro-4-fluorotoluene, f16
5-Chloro-2-hydroxyaniline, a148
Chlorohydroxybenzoic acids, c238, c239
1-Chloro-3-hydroxypropane, c215

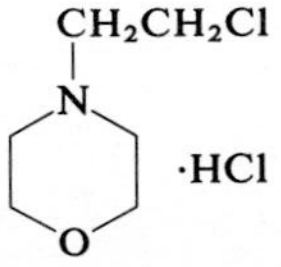

c113 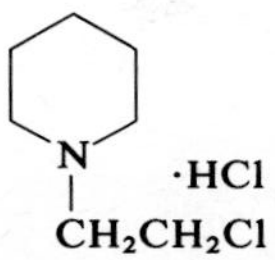 c114 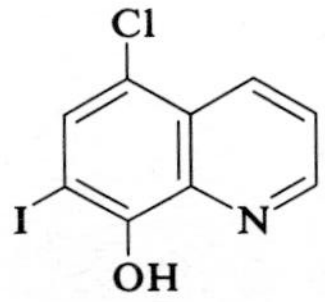 c133 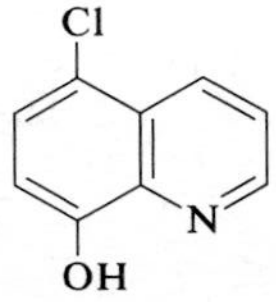 c135

TABLE 1-14 Physical constants of organic compounds (*continued*)

No.	Name	Formula	Formula weight	Beilstein reference	Density	Refractive index	Melting point	Boiling point	Flash point	Solubility in 100 parts solvent
c137	1-Chloro-3-mercapto-2-propanol	$HSCH_2CH(OH)CH_2Cl$	126.61	1^3, 2156	1.277	1.5276^{20}		$57^{1.3mm}$	97	
c138	Chloromethane	CH_3Cl	50.49	1, 59	0.92^{20}	1.3712^{-24}	−97.7	−24.22		0.48 aq^{25}; s alc; misc chl, eth, HOAc
c138a	3-Chloro-4-methoxyaniline	$ClC_6H_3(OCH_3)NH_2$	157.60	13, 511			50–55			
c139	5-Chloro-2-methoxyaniline	$ClC_6H_3(OCH_3)NH_2$	157.60	13, 383			83–85			
c140	1-Chloro-2-methoxybenzene	$ClC_6H_4OCH_3$	142.59	6, 184	1.123	1.5445^{20}		196	76	i aq; s alc, eth
c141	1-Chloro-4-methoxy-2-nitrobenzene	$CH_3O(Cl)C_6H_3NO_2$	187.58				45			s hot alc
c142	2-Chloro-6-methoxypyridine	$CH_3O(Cl)—C_5H_3N$	143.57		1.207	1.5263^{20}		186		
c143	2-Chloro-6-methylaniline	$CH_3(Cl)C_6H_3NH_2$	141.60	12^1, 388	1.152	1.5761^{20}	2	215	98	s alc
c144	3-Chloro-2-methylaniline	$CH_3(Cl)C_6H_3NH_2$	141.60	12, 836		1.5874^{20}	2	115–117^{10mm}	> 112	
c145	3-Chloro-4-methylaniline	$CH_3(Cl)C_6H_3NH_2$	141.60	12, 988		1.5830^{20}	25	238	100	
c146	4-Chloro-2-methylaniline	$CH_3(Cl)C_6H_3NH_2$	141.60	12, 835		1.5848^{20}	27	241	99	s hot alc
c147	5-Chloro-2-methylaniline	$CH_3(Cl)C_6H_3NH_2$	141.60	12, 835		1.5840^{20}	22	237	160	
c148	DL-4-Chloro-2-(α-methylbenzyl)-phenol	$C_6H_5CH(CH_3)$-$C_6H_3(Cl)OH$	232.71	6^4, 4710				155^{2mm}		
c149	1-Chloro-3-methylbutane	$(CH_3)_2CHCH_2CH_2Cl$	106.59	1, 135	0.8704^{20}_4	1.4084^{20}	−104	99	16	sl s aq; misc alc, eth

c150	2-Chloro-2-methyl-butane	$CH_3CH_2CCl(CH_3)_2$	106.59	1, 134	0.8650_4^{20}	1.4052^{20}	−73.7	85	16	i aq; s alc, eth
c151	Chloromethyldi-methylchlorosilane	$(CH_3)_2Si(Cl)CH_2Cl$	143.1		1.0865_4^{20}	1.4360^{20}		115–116		
c152	Chloromethyl 2,2-di-methylpropionate	$(CH_3)_3CCOOCH_2Cl$	150.61		1.045	1.4170^{20}			40	
c153	Chloromethyl ethyl ether	$ClCH_2OCH_2CH_3$	94.54	1^2, 645	1.04_4^{20}	1.4040^{20}		79–83		s alc; v s eth
c154	Chloromethylmethyl-dichlorosilane	$ClCH_2Si(CH_3)Cl_2$	163.5		1.2858_4^{20}	1.4500^{20}		121–122		
c155	Cloromethylmethyl-diethoxysilane	$ClCH_2Si(OC_2H_5)_2CH_3$	182.7		1.000_4^{20}	1.407^{25}		160–161		
c156	Chloromethyl methyl ether	$ClCH_2OCH_3$	80.51	1, 580	1.0703_4^{20}	1.3961^{20}	−103.5	57–59	15	d aq; s acet, CS_2
c157	Chloromethyl methyl sulfide	$ClCH_2SCH_3$	95.48		1.153	1.4963^{20}		105		
c158	1-(Chloromethyl)-naphthalene	$C_{10}H_7CH_2Cl$	176.65	5, 566		1.6380^{20}	32	169^{25mm}	> 112	
c159	4-Chloro-2-methyl-phenol	$CH_3(Cl)C_6H_3OH$	142.59	6, 359			48	225		sl s aq
c160	4-Chloro-3-methyl-phenol	$CH_3(Cl)C_6H_3OH$	142.59	6, 381			68	235		i aq; s alc, bz, chl, eth, acet
c161	4-Chloro-*N*-methyl-piperidine HCl		170.08				164			

Chloromethylbenzenes, c245, c246, c247

(Chloromethyl)oxirane, c102

Chloromethyl pivalate, c152

Cl
N
CH_3
c161

TABLE 1-14 Physical constants of organic compounds (*continued*)

No.	Name	Formula	Formula weight	Beilstein reference	Density	Refractive index	Melting point	Boiling point	Flash point	Solubility in 100 parts solvent
c162	1-Chloro-2-methyl-propane	$(CH_3)_2CHCH_2Cl$	92.57	1, 124	0.8829^{15}	1.4010^{15}	−130.3	68.9	21	0.09 aq; misc alc, eth
c163	2-Chloro-2-methyl-propane	$(CH_3)_3CCl$	92.57	1, 125	0.8474_4^{15}	1.3856^{20}	−25.4	50.8	18	sl s aq; misc alc, eth
c164	1-Chloro-2-methyl-propene	$(CH_3)_2C{=}CHCl$	90.55	1, 209	0.9186_4^{20}	1.4225^{20}		68.1	−1	misc alc, eth
c165	3-Chloro-2-methyl-propene	$ClCH_2C(CH_3){=}CH_2$	90.55	1, 209	0.9210_4^{15}	1.4272^{20}	−80	72	−10	misc alc, eth
c166	Chloromethyltri-chlorosilane	$ClCH_2SiCl_3$	183.9		1.465_4^{20}	1.4555^{20}		117–118		
c167	Chloromethyltri-methylsilane	$ClCH_2Si(CH_3)_3$	122.7	4^3, 1844	0.8861_4^{20}	1.4180^{20}		99	<1	
c168	6-(Chloromethyl)-uracil		160.56	23^1, 328			257 d			
c169	1-Chloronaphthalene	$C_{10}H_7Cl$	162.62	5, 541	1.1938_4^{20}	1.6332^{20}	−2.3	259.3	121	s alc, bz, PE
c170	2-Chloronaphthalene	$C_{10}H_7Cl$	162.62		1.1377^{71}	1.6079^{71}	59.5	256		s alc, bz, chl, eth
c171	4-Chloro-1,8-naph-thalic anhydride		232.63	17, 522			210			
c172	4′-Chloro-3′-nitro-acetophenone	$ClC_6H_3(NO_2)$-$C({=}O)CH_3$	199.60	7^3, 995			101			
c173	2-Chloro-4-nitro-aniline	$ClC_6H_3(NO_2)NH_2$	172.57	12, 733			109			sl s aq; v s alc, eth
c173a	2-Chloro-5-nitro-aniline	$ClC_6H_3(NO_2)NH_2$	172.57	12, 732			114			
c174	4-Chloro-2-nitro-aniline	$ClC_6H_3(NO_2)NH_2$	172.57	12, 729			119			v s alc, eth
c175	4-Chloro-3-nitro-aniline	$ClC_6H_3(NO_2)NH_2$	172.57	12, 731			101			v s alc; s eth

c176	1-Chloro-2-nitrobenzene	$ClC_6H_4NO_2$	157.56	5, 241	1.348		32–33	246	123	s alc, bz, eth
c177	1-Chloro-3-nitrobenzene	$ClC_6H_4NO_2$	157.56	5, 243	1.534_4^{20}		46	236	103	sl s alc; v s eth, chl
c178	1-Chloro-4-nitrobenzene	$ClC_6H_4NO_2$	157.56	5, 243	1.520		82–84	242	110	sl s alc; v s eth, CS_2
c179	2-Chloro-4-nitrobenzoic acid	$ClC_6H_3(NO_2)COOH$	201.57	9, 404			141			s hot aq, hot bz
c180	2-Chloro-5-nitrobenzoic acid	$ClC_6H_3(NO_2)COOH$	201.57	9, 403	1.608^{18}		168			sl s aq; s alc, bz, eth
c181	4-Chloro-3-nitrobenzoic acid	$ClC_6H_3(NO_2)COOH$	201.57	9, 402	1.645^{18}		183			sl s alc; s hot aq
c182	4-Chloro-3-nitrobenzophenone	$ClC_6H_3(NO_2)$-$C(=O)C_6H_5$	261.66	7^1, 230			104–105	235^{13mm}		
c183	2-Chloro-5-nitrobenzotrifluoride	$ClC_6H_3(NO_2)CF_3$	225.55		1.527	1.5083^{20}		231	98	
c184	4-Chloro-3-nitrobenzotrifluoride	$ClC_6H_3(NO_2)CF_3$	225.55		1.511	1.4893^{20}	−2.5	222	101	
c185	5-Chloro-2-nitrobenzotrifluoride	$ClC_6H_3(NO_2)CF_3$	225.55		1.526	1.4980^{20}	21–22	222–224	102	

Chloronicotinic acids, c235, c236

α-Chloronitrotoluene, n47

Chloronitro-α,α,α-trifluorotoluenes, c183, c184, c185

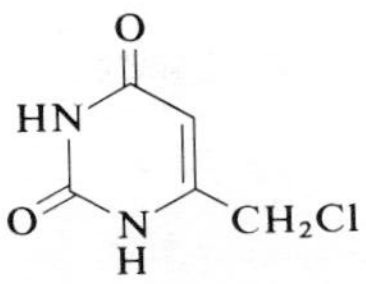

HN, O, N, H, CH_2Cl

c168

O, O, O, Cl

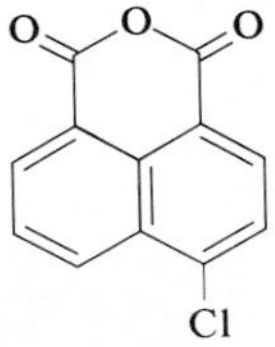

c171

TABLE 1-14 Physical constants of organic compounds (*continued*)

No.	Name	Formula	Formula weight	Beilstein reference	Density	Refractive index	Melting point	Boiling point	Flash point	Solubility in 100 parts solvent
c186	*o*-(4-Chloro-3-nitrobenzoyl)benzoic acid	$ClC_6H_3(NO_2)COC_6H_4$-COOH	305.68	10, 752			201			
c187	2-Chloro-4-nitrophenol	$ClC_6H_3(NO_2)OH$	173.56	6, 240			106			
c188	2-Chloro-4-nitrotoluene	$ClC_6H_3(NO_2)CH_3$	171.58	5, 329		1.5470^{70}	61	260		i aq; s alc, eth
c189	2-Chloro-6-nitrotoluene	$ClC_6H_3(NO_2)CH_3$	171.58	5, 327		1.5377^{70}	36	238	125	i aq
c190	4-Chloro-3-nitrotoluene	$ClC_6H_3(NO_2)CH_3$	171.58	5, 329	1.297	1.5580^{20}	7	260^{745mm}	> 112	i aq
c191	1-Chlorooctane	$CH_3(CH_2)_7Cl$	148.68	1, 159	0.875_4^{20}	1.4298^{20}	−61	183	54	i aq; v s alc, eth
c192	1-Chloropentane	$CH_3(CH_2)_4Cl$	106.60	1, 130	0.8824_4^{20}	1.4118^{20}	−99.0	98.3	12	0.02 aq; misc alc, eth
c193	5-Chloro-2-pentanone	$ClCH_2CH_2CH_2COCH_3$	120.58	1^2, 738	1.0571_4^{18}	1.4375^{20}		72^{20mm}	62	s acet, eth
c194	3-Chloroperoxybenzoic acid	$ClC_6H_4C(O)OOH$	172.57				94 d			
c195	2-Chlorophenol	ClC_6H_4OH	128.56	6, 183	1.2573_4^{25}	1.5579^{20}	9.3	175–176	63	sl s aq; v s alc, eth
c196	3-Chlorophenol	ClC_6H_4OH	128.56	6, 185	1.245_4^{45}	1.5565^{40}	33.5	214	> 112	sl s aq; s alc, eth
c197	4-Chlorophenol	ClC_6H_4OH	128.56	6, 186	1.2238_4^{78}	1.5419^{45}	43.5	220	115	sl s aq; v s alc, chl, eth
c198	4-Chlorophenoxyacetic acid	$ClC_6H_4OCH_2COOH$	186.59	6, 187			159			
c199	2-(4-Chlorophenoxy)-2-methylpropionic acid	$ClC_6H_4OC(CH_3)_2COOH$	214.65				122			
c200	DL-2-(4-Chlorophenoxy)propionic acid	$ClC_6H_4OCH(CH_3)COOH$	200.62	6^3, 695			117			

c201	4-Chlorophenylacetic acid	$ClC_6H_4CH_2COOH$	170.60	9, 448			105			v s aq, alc, eth; s bz
c202	*p*-Chlorophenylacetonitrile	$ClC_6H_4CH_2CN$	151.60	9, 448			30.5	267		
c203	2-Chloro-*p*-phenylenediamine sulfate	$H_2NC_6H_3(Cl)NH_2 \cdot H_2SO_4$	240.67	13, 117			253			
c204	4-Chloro-1,2-phenylenediamine	$ClC_6H_3(NH_2)_2$	142.59	13, 25			70			
c205	4-Chloro-1,3-phenylenediamine	$H_2N(Cl)C_6H_3NH_2$	142.59	13, 53			90			
c206	3-Chlorophenylhydrazine HCl	$ClC_6H_4NHNH_2 \cdot HCl$	179.05	15, 424			242 d			
c207	4-Chlorophenyl isocyanate	ClC_6H_4NCO	153.57	12, 616		1.5618^{20}	31	204	110	
c208	4-Chlorophenyl phenyl sulfone	$ClC_6H_4SO_2C_6H_5$	252.72	6^1, 149			94			74 acet; 44 bz; 5 CCl_4; 65 diox; 21 i-PrOH
c209	4-Chlorophenyltrichlorosilane	$ClC_6H_4SiCl_3$	246.0		1.4316_4^{20}	1.5418^{20}		115–117^{20mm}		
c210	4-Chloro-*o*-phthalic acid	$ClC_6H_3(COOH)_2$	200.58	9, 816			148			
c211	1-Chloropropane	$CH_3CH_2CH_2Cl$	78.54	1, 104	0.8985^{15}	1.3880^{20}	−122.8	46.6	18	0.27 aq; misc alc, eth
c212	2-Chloropropane	$CH_3CHClCH_3$	78.54	1, 105	0.8563^{20}	1.3777^{20}	−117.2	35	−35	0.34 aq; misc alc, eth
c213	3-Chloro-1,2-propanediol	$ClCH_2CH(OH)CH_2OH$	110.54		1.3218_4^{20}	1.4805^{20}		213	58	s aq, alc, eth
c214	1-Chloro-2-propanol	$CH_3CH(OH)CH_2Cl$	94.54	1, 363	1.115^{20}	1.4375^{20}		126–127	51	misc aq; s alc
c215	3-Chloro-1-propanol	$ClCH_2CH_2CH_2OH$	94.54	1, 356	1.1309_4^{20}	1.4460^{20}		160–162	73	

p-Chlorophenacyl bromide, b248
Chlorophenylamines, c33, c34, c35
4-Chlorophenyl sulfone, b166
4-Chlorophenyl sulfoxide, b167
Chloropicrin, t243
Chloroprene, c217

TABLE 1-14 Physical constants of organic compounds (*continued*)

No.	Name	Formula	Formula weight	Beilstein reference	Density	Refractive index	Melting point	Boiling point	Flash point	Solubility in 100 parts solvent
c216	Chloro-2-propanone	$ClCH_2COCH_3$	92.53	1, 653	1.135_4^{15}	1.4350^{20}	−44.5	119.7	7	10 aq; misc alc, chl
c217	3-Chloro-1-propene	$ClCH_2CH{=}CH_2$	76.53	1, 198	0.939_4^{20}	1.4151^{20}	−134.5	45.2	−28	0.36 aq; misc alc, chl
c218	(3-Chloropropenyl)-benzene	$C_6H_5CH{=}CHCH_2Cl$	152.62	5^2, 372		1.5845^{20}	−19	108^{12mm}	79	
c219	2-Chloropropionic acid	$CH_3CH(Cl)COOH$	108.52	2, 248	1.182	1.4345^{20}		186	101	misc aq, alc, eth
c220	3-Chloropropionic acid	$ClCH_2CH_2COOH$	108.52	2, 249			41	205	> 112	v s aq, alc, chl
c221	3-Chloropropio-nitrile	$ClCH_2CH_2CN$	89.53	2, 250	1.1443^{18}	1.4379^{20}	−50	176	75	
c222	2-Chloropropionyl chloride	$CH_3CH(Cl)COCl$	126.97	2, 248	1.308	1.4400^{20}		111	31	d aq, alc
c223	3-Chloropropionyl chloride	$ClCH_2CH_2COCl$	126.97	2, 250	1.3307^{13}	1.4570^{20}		145	61	i aq; d hot aq, hot alc; s alc; v s eth
c224	*p*-Chloropropio-phenone	$ClC_6H_4C({=}O)CH_2CH_3$	168.62	7, 301			37	97^{1mm}		
c225	3-Chloropropylamine HCl	$ClCH_2CH_2CH_2NH_2{\cdot}HCl$	130.02	4, 148			150			
c226	3-Chloropropyl-methyldichloro-silane	$Cl(CH_2)_3Si(CH_3)Cl_2$	191.6		1.2045_4^{20}	1.4580^{20}		70^{15mm}		
c227	2-Chloropropyl-(phenyl)dichloro-silane	$Cl(CH_2)_3SiCl_2(C_6H_5)$	253.6		1.241_4^{20}	1.5332^{20}		141^{10mm}		
c228	*N*-(3-Chloropropyl)-piperidine HCl		198.14	20, 18			220			

c229	3-Chloropropyl thiolacetate	$CH_3C(=O)SCH_2CH_2CH_2Cl$	152.64	2^3, 493	1.159	1.4946^{20}		84^{10mm}	77	
c230	3-Chloropropyltri-chlorosilane	$ClCH_2CH_2CH_2SiCl_3$	212.0		1.3590^{20}_{4}	1.4668^{20}		181.5	66	
c231	3-Chloropropyltri-ethoxysilane	$Cl(CH_2)_3Si(OC_2H_5)_3$	240.8		1.009^{20}_{4}	1.420^{20}		102^{10mm}		
c232	3-Chloropropyltri-methoxysilane	$Cl(CH_2)_3Si(OCH_3)_3$	198.72		1.077^{25}_{4}	1.4183^{25}		183	66	
c233	3-Chloropropyne	$ClCH_2C\equiv CH$	74.51	1, 248	1.0306^{25}_{4}	1.4349^{20}	−78	58	18	misc bz, alc, eth, EtAc
c234	2-Chloropyridine	ClC_5H_4N	113.55	20, 230	1.205^{15}	1.5320^{20}		166^{714mm}	65	sl s aq; s alc, eth
c235	2-Chloro-3-pyridine-carboxylic acid	$C_5H_3N(Cl)COOH$	157.56	22^2, 35			d 175			
c236	6-Chloro-3-pyridine-carboxylic acid	$C_5H_3N(Cl)COOH$	157.56	22, 43			200 d			
c237	2-Chloroquinoline		163.61	20, 359	1.2464^{25}_{4}	1.6259^{25}	37	267		i aq; s alc, bz, eth
c238	4-Chlorosalicyclic acid	$HO(Cl)C_6H_3COOH$	172.57	10, 101			212			
c239	5-Chlorosalicylic acid	$HO(Cl)C_6H_3COOH$	172.57	10, 102			172			
c240	*N*-Chlorosuccinimide		133.53	21, 380	1.65		150–151			1.4 aq; 0.67 alc; 2 bz; sl s chl, eth
c241	Chlorosulfonyl isocyanate	$ClSO_2NCO$	141.53		1.626	1.4467^{20}	−44	107		

β-Chloropropionaldehyde diethyl acetal, c82

3-Chloropropylene-1,2-oxide, c102

1-Chloro-2,5-pyrrolidinedione, c240

c228

c237

c240

TABLE 1-14 Physical constants of organic compounds (*continued*)

No.	Name	Formula	Formula weight	Beilstein reference	Density	Refractive index	Melting point	Boiling point	Flash point	Solubility in 100 parts solvent
c242	8-Chlorotheophylline		214.61	26, 473			d 290			s alk
c243	2-Chlorothiophene	$Cl—C_4H_3S$	118.59	17, 32	1.286	1.5483^{20}	−72	129	22	i aq; misc alc, eth
c244	4-Chlorothiophenol	ClC_6H_4SH	144.62	6, 326			51	207		
c245	2-Chlorotoluene	$ClC_6H_4CH_3$	126.59	5, 290	1.0826_4^{20}	1.5250^2	−34	159.0	47	sl s aq; v s alc, bz, chl, eth
c246	3-Chlorotoluene	$ClC_6H_4CH_3$	126.59	5, 291	1.0760_4^{19}	1.5218^{20}	−48.9	161.8	50	s alc, bz, chl; misc eth
c247	4-Chlorotoluene	$ClC_6H_4CH_3$	126.59	5, 292	1.0697_4^{20}	1.5208^{20}	7.2	162.0	49	sl s aq; s alc, bz, eth
c248	*N*-Chloro-*p*-toluenesulfonamide, Na salt	$CH_3C_6H_4SO_2NCl^-Na^+$	227.67				167 d			s aq; i bz, chl, eth
c249	4′-Chloro-1-toluenethiol	$ClC_6H_4CH_2SH$	158.65	6, 466	1.202	1.5893^{20}	20		76	
c250	4-Chloro-*o*-tolyloxyacetic acid, Na salt	$ClC_6H_3(CH_3)O$-$CH_2COO^-Na^+$	222.61	6^3, 1265			220–225			
c251	4-(4-Chloro-*o*-tolyloxy)butyric acid	$ClC_6H_3(CH_3)O$-$(CH_2)_3COOH$	228.68				99–100			
c252	Chloro-2,2,2-trifluoroethane	CF_3CH_2Cl	118.5		1.389^0	1.3090^0	−105	6.9		
c253	Chlorotrifluoroethylene	$CF_2{=}CFCl$	116.48		1.315		−158.2	−27.9		
c254	Chlorotrifluoromethane	$ClCF_3$	104.46	1^3, 42			−181	−81.5		
c255	Chlorotrimethylgermane	$(CH_3)_3GeCl$	153.16		1.2382^{22}	1.4283^{20}	−13	102		
c256	Chlorotrimethylsilane	$(CH_3)_3SiCl$	108.64		0.8580_4^{20}	1.3885^{20}	−40	57	−40	

No.	Name	Formula	Formula weight	Beilstein	Density	Refractive index	mp	bp	Flash point	Solubility
c257	Chlorotriphenylmethane	$(C_6H_5)_3CCl$	278.78	5, 700			110–112	230^{20mm}		v s bz, chl, eth
c258	Chlorotripropylsilane	$(C_3H_7)_3SiCl$	192.8		0.882_4^{20}	1.440^{20}		199–201		
c258a	Chlorotris(dimethylamino)silane	$[(CH_3)_2N]_3SiCl$	195.8		0.975_4^{20}	1.442^{20}		$62–63^{12mm}$		
c259	α-Chloro-*o*-xylene	$CH_3C_6H_4CH_2Cl$	140.61	5, 364	1.063	1.5391^{20}		199	73	i aq; misc alc, eth
c260	α-Chloro-*m*-xylene	$CH_3C_6H_4CH_2Cl$	140.61	5, 373	1.064^{20}	1.5350^{20}		195–196	75	i aq; misc alc, eth
c261	α-Chloro-*p*-xylene	$CH_3C_6H_4CH_2Cl$	140.61	5, 384		1.5330^{20}	4.5	200	75	misc alc, bz, eth, acet
c262	4-Chloro-*o*-xylene	$ClC_6H_3(CH_3)_2$	140.61	5, 363	1.047	1.5283^{20}		223	66	misc alc, bz, eth, acet
c263	Cholesterol		386.66		1.067_4^{20}		148.5	360 sl d		1.29 alc; 35 eth; 22 chl; s bz, PE
c264	Cholic acid		408.58				198			0.028 aq; 0.06 alc; 2.8 acet; 0.036 bz; 0.5 chl

α-Chlorotoluene, b89
Chlorotoluidines, c143, c144, c145, c146, c147
2-Chlorotriethylamine, d272

Chloro-α,α,α-trifluorotoluenes, c51, c52, c53
4-Chloro-α,α,α-trifluoro-*o*-toluidine, a144
α'-Chloro-α,α,α-trifluoro-*m*-xylene, t304

Chlorotrihexylsilane, t307
Chloroxylenol, c91

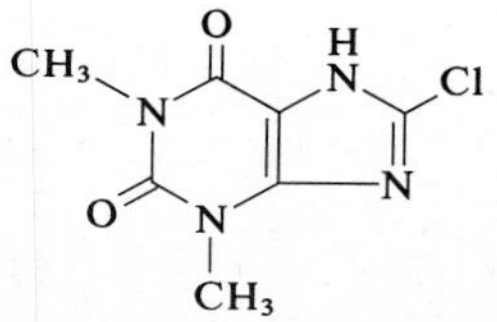

c242

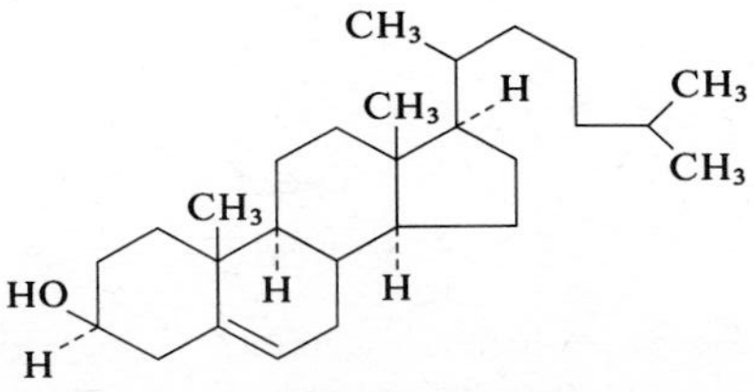

c263

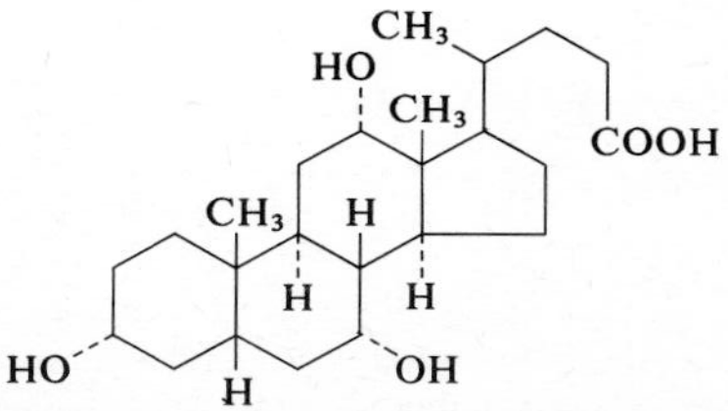

c264

TABLE 1-14 Physical constants of organic compounds (*continued*)

No.	Name	Formula	Formula weight	Beilstein reference	Density	Refractive index	Melting point	Boiling point	Flash point	Solubility in 100 parts solvent
c265	Cinchonine		294.40	23^2, 369			~260			1.4 alc; 0.9 chl; 0.2 eth
c266	1,8-Cineole		154.25	17, 23	0.921_{25}^{25}	1.4572^{20}	1.5	174.4		misc alc, chl, eth
c267	*trans*-Cinnamaldehyde	$C_6H_5CH{=}CHCHO$	132.16	7, 348	1.050_{25}^{25}	1.6219^{20}	−7.5	246	71	0.014 aq; misc alc, chl, eth
c268	*trans*-Cinnamic acid	$C_6H_5CH{=}CHCOOH$	148.16	9, 573	1.2475_4^4		134	300		0.05 aq; 16 alc; 8 chl
c269	*trans*-Cinnamoyl chloride	$C_6H_5CH{=}CHCOCl$	166.61	9^2, 390	1.1617_4^{25}	1.614^{43}	35–36	258		s hot alc, CCl_4
c270	Cinnamyl alcohol	$C_6H_5CH{=}CHCH_2OH$	134.18	6, 570	1.0397_{35}^{35}	1.5758^{33}	33	250.0		s aq; v s alc, eth
c271	Citraconic acid	$CH_3C(COOH){=}CHCOOH$	130.10	2, 768	1.62		92 d			v s aq, alc, eth; sl s chl; i bz, PE
c272	Citraconic anhydride		112.08	17, 440	1.247	1.4712^{20}	8	214	101	
c273	Citrazinic acid		155.11	22, 254			carbonizes without melting >300			i aq; s alk
c274	Citric acid	$HOOCCH_2C(OH)(COOH)$-CH_2COOH	192.12	3, 556	1.665		154			59 aq
c275	Citronellol	$(CH_3)_2C{=}CHCH_2CH_2CH$-$(CH_3)CH_2CH_2OH$	156.27	1, 451	0.8570_4^{20}	1.4556^{20}		222	79	
c276	Cocaine		303.35	22^2, 150		1.5022^{98}	98	$187^{0.1mm}$		0.17 aq; 15 alc; 140 chl; 28 eth
c277	Coumarin		146.15	17, 328	0.935_4^{20}		69	298		0.25 aq; v s alc, chl, eth

c278	Creatine	$HOOCCH_2N(CH_3)$-$C(=NH)NH_2$	131.14	4, 363			300			1.3 aq; 0.11 alc; i eth
c279	Creatinine		113.12	24, 245			255 d			8 aq; sl s alc; i eth
c280	*o*-Cresol	$CH_3C_6H_4OH$	108.14	6, 349	1.0273^{41}	1.5361^{41}	30.9	190.8	81	3.1 aq^{40}; misc alc, chl, eth; s alk
c281	*m*-Cresol	$CH_3C_6H_4OH$	108.14	6, 373	1.034_4^{20}	1.5438^{20}	12.2	202.7	86	2.5 aq^{40}; misc alc, chl, eth; s alk
c282	*p*-Cresol	$CH_3C_6H_4OH$	108.14	6, 389	1.0179^{41}	1.5312^{41}	34.8	201.9	86	2.3 aq^{40}; misc alc, chl, eth; s alk
c283	*trans*-Crotonaldehyde	$CH_3CH=CHCHO$	70.09	1, 728	0.8516^{20}	1.4373^{20}	−76.5	104.1	8	18.1 aq
c284	Crotonyl chloride	$CH_3CH=CHCOCl$	104.54	2, 411	1.091	1.4595^{20}		123	35	
c285	Cupferron	$C_6H_5N(NO)O^-NH_4^+$	155.16	16^1, 395			163–164			v s aq, alc

Chromone, b56
Chromotropic acid, d397
Chrysoldin, d31
Cinchophen, p152
Cinnamyl chloride, c218
Citral, d564, d565
Cleland's reagent, d425
2,4,6-Collidine, t373
p-Coumaric acid, h107
Cresotic acids, h138, h139
Cresylic acids, c280, c281, c282
Crotonic acid, b405
Crotononitrile, b403
Crotyl alcohols, b407, b408
Crotyl bromide, b242
Crotyl chloride, c68
12-Crown-4, t127
15-Crown-5, p45
18-Crown-6, h74
Cumene, i91
Cumic alcohol, i92
Cupron, b50

c265 c266 c272 c273 c276 c277 c279

TABLE 1-14 Physical constants of organic compounds (*continued*)

No.	Name	Formula	Formula weight	Beilstein reference	Density	Refractive index	Melting point	Boiling point	Flash point	Solubility in 100 parts solvent
c286	Cyanamide	H_2NCN	42.04	3^2, 63	1.282_4^{20}		46	83^{380mm}		78 aq; 29 BuOH; 42 EtAc; s alc, eth
c287	2-Cyanoacetamide	$NCCH_2CONH_2$	84.08	2, 589			119.5		215	25 aq; 3.1 alc
c288	Cyanoacetic acid	$NCCH_2COOH$	85.06	2, 583			65–67	108^{15mm}	107	s aq, alc, eth; sl s bz
c289	Cyanoacetohydrazide	$NCCH_2C(=O)NHNH_2$	99.09				110	d		v s aq; s alc; i eth
c290	Cyanoacetylurea	$NCCH_2C(=O)NHC(=O)$-NH_2	127.10	3, 66			214 d			
c291	2-Cyanoethanol	$NCCH_2CH_2OH$	71.08	3^2, 213	1.0588^0			106–108^{11mm}		misc aq, alc; sl s eth
c292	2-Cyanoethyldi-chloromethylsilane	$NCCH_2CH_2Si(CH_3)Cl_2$	168.1		1.202_4^{20}	1.455^{20}		63^{4mm}		
c293	1-Cyano-3-methyliso-thiourea, Na salt	$CH_3NHC(=NCN)S^-Na^+$	137.14	4, 71			290 d			
c294	1-Cyanonaphthalene	$C_{10}H_7CN$	153.18	9, 649	1.1113_{25}^{25}	1.6298^{18}	38	299		i aq; v s alc, eth
c295	3-Cyanopropyltri-chlorosilane	$NCCH_2CH_2CH_2SiCl_3$	202.6		1.280^{25}	1.465^{25}		93–94^{8mm}		
c296	2-Cyanopyridine	$NC—C_5H_4N$	104.11	22, 36		1.5288^{20}	28	215	89	s aq; v s alc, bz, eth
c297	3-Cyanopyridine	$NC—C_5H_4N$	104.11	22, 41			52	240–245		v s aq, alc,, bz, eth
c298	4-Cyanopyridine	$NC—C_5H_4N$	104.11	22, 46			80			s aq, alc, bz, eth
c299	Cyanotrimethylsilane	$(CH_3)_3SiCN$	99.21		0.783_4^{20}	1.3924^{20}	11	114–117	1	
c300	Cyanuric acid		129.08	26, 239	1.768^0		d to HO-CN			0.5 aq; s hot alc, pyr; i acet, bz, chl, eth
c301	Cyclobutane	C_4H_8	56.10	5, 17	0.7038^0	1.3752^0	−90.7	12.5		i aq; v s alc, acet
c302	Cyclodecane	$C_{10}H_{20}$	140.27			1.4707^{20}		201	1	
c303	Cyclododecanol	$C_{12}H_{23}OH$	184.32				77			

c304	Cyclododecanone	$C_{12}H_{22}(=O)$	182.31	7^2, 48	0.906		61	85^{1mm}		
c305	*trans,trans,cis*-1,5,9-cyclododecatriene		162.28		0.8925_4^{20}	1.5070^{20}	−18	231	87	
c306	*trans*-Cyclododecene		166.31		0.863	1.4822^{20}		232–245	93	
c307	Cycloheptane	C_7H_{14}	98.18	5, 29	0.811_4^{20}	1.4455^{20}	−8.0	118.8	6	v s alc, eth
c308	DL-*trans*-1,2-Cycloheptanediol	$C_7H_{12}(OH)_2$	130.19	6^3, 4086			61–63	138–139^{15mm}		
c309	Cycloheptanol	$C_7H_{13}OH$	114.19	6, 10	0.948_4^{20}	1.4760^{20}	2	185	71	sl s aq; v s alc, eth
c310	Cycloheptanone	$C_7H_{12}(=O)$	112.17	7, 13	0.9490_4^{20}	1.4611^{20}		179–181	55	i aq; v s alc; s eth
c311	1,3,5-Cycloheptatriene		92.13	5, 280	0.888	1.5211^{20}	−75.3	115.5	26	s alc, eth; v s bz, chl
c312	Cycloheptene	C_7H_{12}	96.17	5, 65	0.824_4^{20}	1.4585^{20}		114.7	−6	s alc, eth
c313	Cyclohexane	C_6H_{12}	84.16	5, 20	0.7786_4^{20}	1.4262^{20}	6.5	80.7	−18	0.01 aq; misc alc, bz, acet, eth, CCl_4
c314	Cyclohexane-d_{12}	C_6D_{12}	92.26		0.89	1.4210^{20}		78	−18	

Cyanoacetonitrile, m5
Cyanoanilines, a124, a125, a126
Cyanobenzene, b51
2-Cyanoethanol, h170
Cyanoethylene, a64
Cyanomethane, a29
2-Cyanopropene, m27
Cyanuric chloride, t254
Cyclododecane epoxide, e4
Cyclododecanone isooxime, a323
Cycloheptanone isooxime, a322
Cycloheptyl bromide, b262
Cyclohexaneacetic acid, c334
Cyclohexanecarboxylic acid chloride, c317
2,5-Cyclohexadien-1,4-dione, b59
2,5-Cyclohexadiene-1,4-dione with 1,4-benzenediol (1:1), q1
Cyclohexanemethanol, c343

OH N N HO N OH

c300

c305

c306

c311

TABLE 1-14 Physical constants of organic compounds (*continued*)

No.	Name	Formula	Formula weight	Beilstein reference	Density	Refractive index	Melting point	Boiling point	Flash point	Solubility in 100 parts solvent
c315	1,3-Cyclohexanebis-(methylamine)	$C_6H_{10}(NHCH_3)_2$	142.25						106	
c316	Cyclohexanecarb-aldehyde	$C_6H_{11}CHO$	112.17	7, 19	0.926	1.4500^{20}		163	40	
c317	Cyclohexanecarbonyl chloride	$C_6H_{11}COCl$	146.62	9, 9	1.096	1.4700^{20}		184	66	
c318	Cyclohexanecar-boxylic acid	$C_6H_{11}COOH$	128.17	7, 19	1.0480_4^{15}	1.4530^{20}	29	232.5		0.21 aq; s alc, bz, eth
c319	*cis*-1,2-Cyclohex-anediamine	$C_6H_{10}(NH_2)_2$	114.19	13, 1	0.931	1.4864^{20}		92^{18mm}		
c320	*trans*-1,2-Cyclohex-anediamine	$C_6H_{10}(NH_2)_2$	114.19	13, 1	0.931	1.4864^{20}		92^{18mm}		
c321	*cis*-1,2-Cyclohexane-dicarboxylic anhydride		154.17				34	158^{17mm}		
c322	*cis*-1,4-Cyclohexane-dimethanol	$C_6H_{10}(CH_2OH)_2$	144.21		0.978_4^{100}	1.4893^{20} super-cooled	43	288	74	misc aq, alc; 2.5 eth
c323	1,3-Cyclohexanedione	$C_6H_8(=O)_2$	112.13	7, 554	1.0861^{91}	1.4576^{102}	103–105			s aq, alc, acet, chl
c324	1,2-Cyclohexanedione dioxime	$C_6H_8(=NOH)_2$	142.16	17^2, 526			185–188			s aq
c325	Cyclohexanemthyl-amine	$C_6H_{11}CH_2NH_2$	113.20	12, 12	0.870	1.4630^{20}		145–147	43	
c326	Cyclohexanepropionic acid	$C_6H_{11}CH_2CH_2COOH$	156.23	9, 82	0.912	1.4636^{20}	14–17	275.8		
c327	Cyclohexanethiol	$C_6H_{11}SH$	116.23	6, 8	0.950	1.4921^{20}		158–160	43	
c328	Cyclohexanol	$C_6H_{11}OH$	100.16	6, 5	0.9416^{30}	1.4629^{30}	25.2	161.1	67	3.8 aq^{25}; misc alc, bz

c329	Cyclohexanone	$C_6H_{10}(=O)$	98.15	7, 8	0.9478_4^{20}	1.4510^{20}	−45 to −47	155.7	46	15 aq^{10}; s alc, eth
c330	Cyclohexanone oxime	$C_6H_{10}(=NOH)$	113.16	7, 10			89–91	206–210		s aq, eth; sl s alc
c331	Cyclohexene	C_6H_{10}	82.15	5, 63	0.8094_4^{20}	1.4464^{20}	−103.5	83.0	−12	0.02 aq; misc alc, bz, acet, eth
c332	2-Cyclohexen-1-one	$C_6H_8(=O)$	96.13	7^2, 55	0.993	1.4885^{20}	−53	168	61	v s alc
c333	2,3-Cyclohexeneo-pyridine		133.19	20^2, 176	1.025	1.5440		218	86	
c333a	[2-(3-Cyclohexenyl)-ethyl]methyldi-chlorosilane	$C_6H_9CH_2CH_2Si(CH_3)Cl_2$	223.2		1.077_4^{20}	1.481^{25}		$79–81^{2mm}$		
c334	Cyclohexylacetic acid	$C_6H_{11}CH_2COOH$	142.20	9^2, 9	1.007	1.4630^{20}	31–33	242–244	>112	sl s aq; s org solv
c335	Cyclohexylamine	$C_6H_{11}NH_2$	99.18	12, 5	0.8671^{20}	1.4593^{20}	−17.7	134.8	<32	misc aq, alc, eth, chl
c336	2-(Cyclohexylamino)-ethanesulfonic acid	$C_6H_{11}NHCH_2CH_2SO_3H$	207.29				>300			
c337	3-Cyclohexylamino-1-propanesulfonic acid	$C_6H_{11}NHCH_2CH_2$-CH_2SO_3H	221.32				>300			

Cyclohexanone cyanohydrin, h110
cis-4-Cyclohexene-1,2-dicarboximide, t77
cis-4-Cyclohexene-1,2-dicarboxylic anhydride, t76

Cyclohexene oxide, e5
N-(1-Cyclohexen-1-yl)morpholine, m453
N-(1-Cyclohexen-1-yl)pyrrolidine, p278

Cyclohexyl alcohol, c328

c321

c333

TABLE 1-14 Physical constants of organic compounds (*continued*)

No.	Name	Formula	Formula weight	Beilstein reference	Density	Refractive index	Melting point	Boiling point	Flash point	Solubility in 100 parts solvent
c338	4-Cyclohexylaniline	$C_6H_{11}C_6H_4NH_2$	175.28	12, 1209			53–56	166^{13mm}		
c339	Cyclohexylbenzene	$C_6H_{11}C_6H_5$	160.26	5, 503	0.9502_4^{20}	1.5258^{20}	5–6	239–240	98	i aq; v s alc, eth
c340	*N*-Cyclohexylformamide	$C_6H_{11}NHCHO$	127.18				38–40	137^{10mm}		
c341	Cyclohexyl isocyanate	$C_6H_{11}NCO$	125.17	12^2, 12	0.980	1.4551^{20}		168–170	48	
c342	Cyclohexyl isothiocyanate	$C_6H_{11}NCS$	141.24	12^2, 12	0.996	1.5350^{20}		219		
c343	Cyclohexylmethanol	$C_6H_{11}CH_2OH$	114.19	6, 14	0.9215_4^{25}	1.4640^{25}		181	71	s alc, eth
c344	*N*-Cyclohexyl-2-pyrrolidinone		167.25		1.026	1.495	12	284		
c345	Cyclohexyltrichlorosilane	$C_6H_{11}SiCl_3$	217.6		1.222_4^{20}	1.477^{20}		90–91^{10mm}		
c346	1,5-Cyclooctadiene		108.18	5, 116	0.8818_4^{25}	1.4905^{25}	−69	149–150	45	s CCl_4
c347	Cyclooctane	C_8H_{16}	112.22	5, 35	0.834	1.4574^{20}	14.8	151.1	30	
c348	Cyclooctanol	$C_8H_{15}OH$	128.22	6^2, 25	0.9740_4^{20}	1.4850^{20}	14–15	106–108^{22mm}	86	
c349	Cyclooctanone	$C_6H_{14}(=O)$	126.20	7, 21	0.9584_4^{20}	1.6494^{20}	41–43	195–197		
c350	Cyclooctene	C_8H_{14}	110.20	5^1, 35	0.846	1.4698^{20}	−16	145–146	25	
c351	Cyclooctylamine	$C_8H_{15}NH_2$	127.23		0.928	1.4804^{20}	−48	190	62	
c352	Cyclopentamethylenedichlorosilane		169.1		1.558_4^{20}	1.4679^{20}		169–170		
c353	Cyclopentane	C_5H_{10}	70.13	5, 19	0.7460_4^{20}	1.4065^{20}	−93.9	49.3	−37	i aq; misc alc, eth
c354	Cyclopentanecarboxylic acid	C_5H_9COOH	114.14	9, 6	1.053_4^{20}	1.4540^{20}	4	216	93	sl s aq; s MeOH
c355	*cis,cis,cis,cis*-1,2,-3,4-Cyclopentanetetracarboxylic acid	$C_5H_6(COOH)_4$	246.17	9^2, 724			192–195 d			

c356	Cyclopentanol	C_5H_9OH	86.13	6, 5	0.9488_4^{20}	1.4521^{20}	−19	140.9	51	sl s aq; s alc
c357	Cyclopentanone	$C_5H_8(=O)$	84.12	7, 5	0.9509_4^{18}	1.4366^{20}	−58	130.6	30	sl s aq; misc alc, eth
c358	Cyclopentanone oxime	$C_5H_8(=NOH)$	99.13	7, 7			53–55	196		s aq, alc, bz, chl, eth
c359	Cyclopentene	C_5H_8	68.11	5, 61	0.774	1.4228^{20}	−135.1	44.2	−28	
c360	2-Cyclopentene-1-acetic acid	$C_5H_7CH_2COOH$	126.16	9, 42	1.047	1.4675^{20}	19	93–$94^{2.5mm}$	>112	
c361	2,3-Cyclopenteneopyridine		119.17		1.018	1.5445^{20}		87–88^{11mm}	67	
c362	*N*-(1-Cyclopentene-1-yl)morpholine		153.23		0.957	1.5105^{20}		105–106^{12mm}	60	
c363	2-Cyclopentylidenecyclopentanone		150.22		1.001	1.5231^{20}		140^{20mm}	103	
c364	3-Cyclopentylpropionic acid	$C_5H_9CH_2CH_2COOH$	142.20		0.996	1.4570^{20}		130^{12mm}	46	
c365	Cyclopropane	C_3H_6	42.08	5, 15	0.720_4^{-79}		−127.4	−32.8		37 mL per 100 mL aq^{15}; v s alc, eth

Cyclohexylbenzene, p106
Cyclohexyl bromide, b263
Cyclohexyl chloride, c77
Cyclohexyl ketone, c329
Cyclohexyl mercaptan, c327
Cyclohexylmethane, m195
Cyclohexylmethyl bromide, b309
Cyclooctene oxide, e7a
Cyclopentanepropanoic acid, c364
Cyclopentene oxide, e38
Cyclopentyl bromide, b265
Cyclopentyl chloride, c79
Cyclopropyl bromide, b266
Cyclopropyl cyanide, c366

c344 c346 c252 c361 c362 c363

TABLE 1-14 Physical constants of organic compounds (*continued*)

No.	Name	Formula	Formula weight	Beilstein reference	Density	Refractive index	Melting point	Boiling point	Flash point	Solubility in 100 parts solvent
c366	Cyclopropanecarbonitrile	C_3H_5CN	67.09	9, 4	0.911^{16}	1.4207^{20}		135	32	s eth
c367	Cyclopropanecarbonyl chloride	C_3H_5COCl	104.54	9, 4	1.152	1.4522^{20}		119	23	
c368	Cyclopropanecarboxylic acid	C_3H_5COOH	86.09	9, 4	1.008	1.4380^{20}	17–19	182–184	71	sl s hot aq; s alc, eth
c369	Cyclopropyl methyl ketone	$C_3H_5COCH_3$	84.12	7, 7	0.8993^{20}_{4}	1.4241^{20}		114	21	s aq, alc, eth
c370	Cystamine dihydrochloride	$H_2NCH_2CH_2SSCH_2CH_2NH_2 \cdot 2HCl$	225.20	4, 287			217 d			
c371	L-(+)-Cysteine	$HSCH_2CH(NH_2)COOH$	121.16	4, 506			220 d			v s aq, alc; i bz, eth
c372	L-Cystine	$HOOCCH(NH_2)CH_2SSCH_2CH(NH_2)COOH$	240.30	4, 507			d 240			0.01 aq; s acid, alk; i alc
d1	*cis*-Decahydronaphthalene	$C_{10}H_{18}$	138.26	5, 92	0.8963^{20}_{4}	1.4810^{20}	−43.0	195.8	58	v s alc, chl, eth; misc most ketones, esters
d2	*trans*-Decahydronaphthalene	$C_{10}H_{18}$	138.26	5^2, 56	0.8700^{20}_{4}	1.4697^{20}	−30.4	187.3	52	see under *cis* isomer
d3	Dehydro-2-naphthol	$C_{10}H_{17}OH$	154.25	6, 67	0.996	1.4992		109^{14mm}	>112	
d4	Decamethylcyclopentasiloxane	$[-Si(CH_3)_2O-]_5$	370.8		0.9593^{20}_{4}	1.3982^{20}	−38	101^{20mm}		i aq
d5	Decamethyltetrasiloxane	$(CH_3)_3SiO[Si(CH_3)_2O]_2Si(CH_3)_3$	310.7		0.8536^{20}_{4}	1.3880^{20}	−70	194–195	86	sl s alc; s bz, PE
d6	Decanal	$H(CH_2)_9CHO$	156.27	1, 711	0.830^{15}_{4}	1.4280^{20}		207–209	85	i aq; s alc, eth
d7	Decane	$CH_3(CH_2)_8CH_3$	142.29	1, 168	0.7301^{20}_{4}	1.4119^{20}	−29.7	174.1	46	0.07 aq
d8	1,10-Decanediamine	$H_2N(CH_2)_{10}NH_2$	172.32	4, 273			62–63	140^{12mm}		
d9	Decanedioic acid	$HOOC(CH_2)_8COOH$	202.25	2, 718	1.207^{20}_{4}	1.422^{134}	134.5	295^{100mm}		0.1 aq; vs alc, esters, ketones
d10	1,10-Decanediol	$HO(CH_2)_{10}OH$	174.28	1^2, 560			72–75	170^{8mm}		sl s aq, eth; vs alc

d11	Decanedioyl dichloride	$ClC(O)(CH_2)_8COCl$	239.14	2, 719	1.1212_4^{20}	1.4678^{20}		220^{75mm}	>112	d aq, alc
d12	Decanenitrile	$CH_3(CH_2)_8CN$	153.27	2, 356	0.8295_4^{15}	1.4295^{20}	−15	235–237		misc alc, chl, eth
d13	1-Decanesulfonic acid, Na salt	$CH_3(CH_2)_9SO_3^-Na^+$	244.33	4^3, 27			300			
d14	Decanoic acid	$CH_3(CH_2)_8COOH$	172.27	2^2, 309	0.8782_4^{50}	1.4288^{40}	31.4	270		0.015 aq; s alc, chl, bz, eth, CS_2
d15	1-Decanol	$CH_3(CH_2)_9OH$	158.29	1, 425	0.8297_4^{20}	1.4371^{20}	6.9	230.2	82	i aq; s alc, eth
d16	4-Decanone	$CH_3(CH_2)_5C(=O)$-$(CH_2)_2CH_3$	156.27	1, 711	0.824_0^{20}	1.4237^{20}		207	71	i aq; misc alc, eth
d17	Decanoyl chloride	$CH_3(CH_2)_8C(=O)Cl$	190.71	2, 356	0.919	1.4410^{20}	−34.5	96^{5mm}	98	d aq, alc; s eth
d18	1-Decene	$CH_3(CH_2)_7CH=CH_2$	140.27	1^3, 858	0.7408_4^{20}	1.4215^{20}	−66.3	170.6	47	i aq; misc alc, eth
d19	Decylamine	$CH_3(CH_2)_9NH_2$	157.30	4, 199	0.787	1.4360^{20}	12–14	216–218	85	sl s aq; misc alc, bz, eth, acet
d20	Dehydroabietylamine		285.48			1.5460^{20}			>112	
d21	Dehydroacetic acid		168.15	17, 559			111–113	269.9		22 acet; 18 bz; 5 MeOH
d22	Deoxybenzoin	$C_6H_5CH_2C(=O)C_6H_5$	196.25	7, 431	1.201_4^0		55–56	320		i aq; v s alc, eth

Cymenes, i100, i101, i102
4-Cymylphenol, m362
Cysteamine, a162
Cysteic acid hydrate, a293
Cytosine, a200

2,4-D, d214
p,p′-DDT, b168
1,2-Decahydroacenaphthylene, a2
Decamethylene glycol, d10
1,10-Decadedicarboxylic acid, d723

Decyl aldehyde, d6
Decyl chloride, c80
Decyl iodide, i33
Dextrose, g6

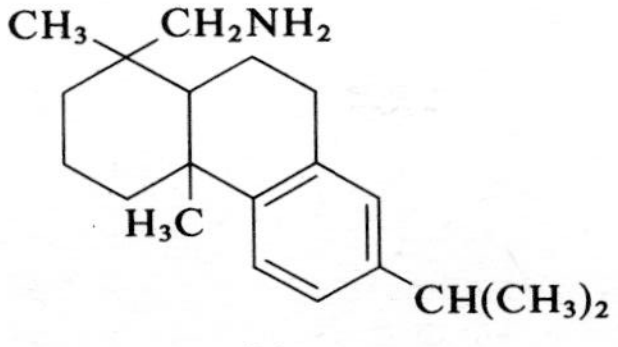

d20

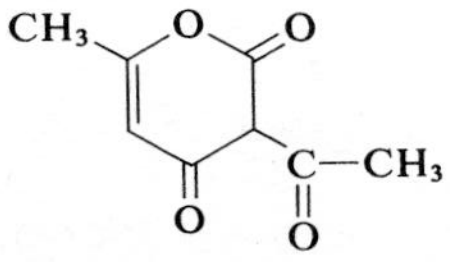

d21

TABLE 1-14 Physical constants of organic compounds (*continued*)

No.	Name	Formula	Formula weight	Beilstein reference	Density	Refractive index	Melting point	Boiling point	Flash point	Solubility in 100 parts solvent
d23	Diacetoxydimethylsilane	$(CH_3)_2Si(OOCCH_3)_2$	176.3		1.054_4^{20}	1.4030^{20}		164–166		
d24	Diacetoxymethylphenylsilane	$CH_3(C_6H_5)Si(OCOCH_3)_2$	238.3			1.487^{20}		127^{6mm}		
d25	Diallylamine	$(H_2C{=}CHCH_2)_2NH$	97.16	4, 208	0.787	1.4405^{20}	−88	111–112	15	
d26	Diallyl ether	$(H_2C{=}CHCH_2)_2O$	98.15	1^2, 477	0.805_0^{18}	1.4240^{20}		94		i aq; misc alc, eth
d27	Diallyl sulfide	$(H_2C{=}CHCH_2)_2S$	114.21	1, 440	0.8877_4^{27}	1.4889^{20}	−83	138	46	sl s aq; misc alc, eth
d28	1,2-Diaminoanthraquinone		238.25	14^1, 459			289–291			sl s alc, eth
d29	1,4-Diaminoanthraquinone		238.25	14, 197			265–268			sl s aq, alc; v s bz
d30	2,6-Diaminoanthraquinone		238.25	14, 215			>325			sl s hot aq, pyr
d31	2,4-Diaminoazobenzene HCl	$C_6H_5N{=}NC_6H_3(NH_2)_2{\cdot}HCl$	248.72	16, 383			235 d			
d32	2,5-Diaminobenzenesulfonic acid	$(H_2N)_2C_6H_3SO_3H$	188.21	14, 713			298 d			sl s aq, alc
d33	3,5-Diaminobenzoic acid	$(H_2N)_2C_6H_3COOH$	152.15	14, 453			228	$-H_2O$, 110		sl s aq; s alc, eth
d34	4,4′-Diaminodiphenylamine sulfate	$H_2NC_6H_4NHC_6H_4NH_2{\cdot}H_2SO_4$	297.33	13, 110			300			
d35	4,4′-Diaminodiphenylmethane	$H_2NC_6H_4CH_2C_6H_4NH_2$	198.27	13, 238			91–92	398	221	sl s aq; v s alc, bz, eth
d36	3,3′-Diaminodiphenyl sulfone	$H_2NC_6H_4SO_2C_6H_4NH_2$	248.30	13, 426			167–170			i aq; s alc, bz
d37	4,4′-Diaminodiphenyl sulfone	$H_2NC_6H_4SO_2C_6H_4NH_2$	248.30	13, 536			175–177			i aq; s alc, acet, HCl
d38	2,7-Diaminofluorene		196.25	13, 266			165–166			sl s aq; v s alc

d39	2,4-Diamino-6-hydroxypyrimidine		126.12	24, 469			285 d			s aq
d40	Diaminomaleonitrile	$NCC(NH_2){=}C(NH_2)CN$	108.10	4^2, 949			178–179			
d41	1,8-Diamino-*p*-menthane		170.30	13, 4	0.914	1.4805^{20}	−45	$107–125^{10mm}$	93	
d42	3,3′-Diamino-*N*-methyl-dipropyamine	$CH_3N[(CH_2)_3NH_2]_2$	145.25	4^4, 1279				$110–112^{6mm}$	102	

Diacetins, g17, g18
Diacetone acrylamide, d570
Diacetone alcohol, h143
Diacetonitrile, a151
(Diacetoxyiodo)benzene, i28
Diacetyl, b389
Diallyl, h41
2,5-Diaminoanisole, m94
1,4-Diaminobutane, b382
1,2-Diaminocyclohexanes, c319, c320
1,10-Diaminodecane, d8
p-Diaminodiphenyl, b137
3,3′-Diaminodipropylamine, i9
1,12-Diaminododecane, d722
1,2-Diaminoethane, e15
1,7-Diaminoheptane, h7
1,6-Diaminohexane, h56

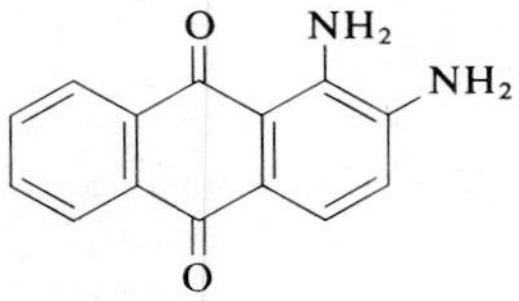

d28

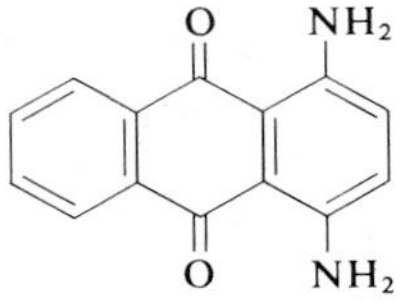

d29

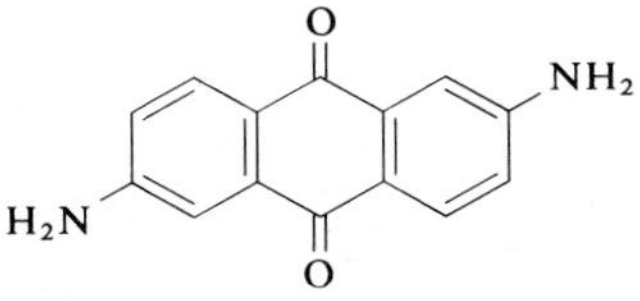

d30

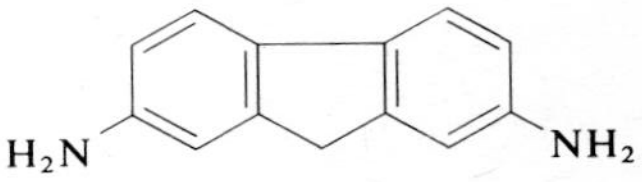

d38

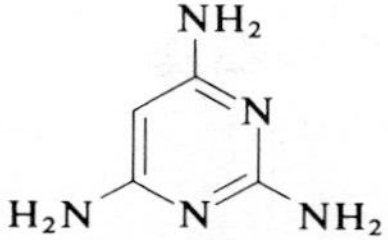

d39

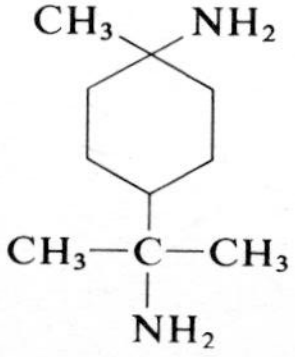

d41

TABLE 1-14 Physical constants of organic compounds (*continued*)

No.	Name	Formula	Formula weight	Beilstein reference	Density	Refractive index	Melting point	Boiling point	Flash point	Solubility in 100 parts solvent
d43	1,3-Diamino-2-propanol	$H_2NCH_2CH(OH)CH_2NH_2$	90.13	4, 290			40–45	235		
d44	2,6-Diaminopyridine	$(H_2N)_2C_5H_3N$	109.13	22^1, 647			118–120			s aq, alc
d45	1,4-Diazabicyclo-[2.2.2]octane		112.18				158	174		45 aq; 77 EtOH; 51 bz; 13 acet; 26 MeEtKe
d46	1,8-Diazabicyclo-[5.4.0]undec-7-ene		152.24		1.018	1.5219^{20}		$80^{0.6mm}$	>112	
d47	Diazomethane	$CH_2=\dot{N}=N$	42.04	23, 25			−145	−23	very explosive	s eth, diox
d48	1-Diazo-2-naphthol-4-sulfonic acid, Na salt		272.22	16, 595			166			
d49	Dibenz[*de,kl*]anthracene		252.32	5^1, 363	1.35		273–274	503		s bz; sl s alc, eth
d50	Dibenzofuran		168.20	17, 70	1.0886^{99}_4	1.6079^{99}	81–83	285		i aq; s alc, bz, eth
d51	2,3,11,12-Dibenzo-1,4,7,10,13-hexaoxacyclooctadeca-2,11-diene		360.41				162–164			
d52	Dibenzothiophene		184.26	17, 72			97.100	332–333		s aq; v s alc, bz
d53	Dibenzoylmethane	$C_6H_5C(=O)CH_2$-$C(=O)C_6H_5$	224.26	7, 769			78–79	220^{18mm}		s alc; v s eth
d54	Dibenzoly peroxide	$C_6H_5C(O)O$—$OC(O)C_6H_5$	242.23				103–106	may explode when heated		sl s aq, alc; s bz, chl, eth

d55	(−)-Dibenzoyl-L-tartaric acid hydrate	$[(C_6H_5COOCH\text{-}(COOH)-]_2 \cdot H_2O$	376.34	9, 170						
d56	Dibenylamine	$C_6H_5CH_2NHCH_2C_6H_5$	197.28	12, 1035	1.026	1.5731^{20}	−26	300	143	i aq; s alc, eth
d57	Dibenzyl disulfide	$C_6H_5CH_2SSCH_2C_6H_5$	246.39	6, 465			69	d > 270		s hot alc, bz, eth
d58	Dibenzyl ether	$C_6H_5CH_2OCH_2C_6H_5$	198.27	6, 434	1.0014_4^{20}	1.5610^{20}	3.5	298 d	135	misc alc, acet, chl, eth

1,3-Diamino-2-hydroxypropane, d43
Diaminonaphthalenes, n4, n5
1,2-Diamino-4-nitrobenzene, n68
1,4-Diamino-2-nitrobenzene, n67
1,9-Diaminononane, n93
1,8-Diaminooctane, o23
1,5-Diaminopentane, p29
2,5-Diaminopentanoic acid, o46
1,2-Diaminopropane, p195
1,3-Diaminopropane, p196
4,6-Diamino-4-pyrimidinol, d39
Diaminotoluenes, t171, t172, t173, t174
1,3-Diaminourea, c11
4,5-Diamino-*o*-xylene, d588
Diamylamine, d652
Diamyl ether, d653
Diamyl ketone, u6
1,2-Dianilinoethane, d671
Diazoacetic ester, e115
1,3-Diazole, i4
Dibenzo-18-crown-6, d51
Dibenzo[*b*,*e*]pyridine, a61
Dibenzopyrrole, d667
Dibenzoyl, b34
Dibenzyl, d668

d45

d46

N_2^+ O^- $SO_3^-\ Na^+$

d48

d49

d50

d51

d52

TABLE 1-14 Physical constants of organic compounds (*continued*)

No.	Name	Formula	Formula weight	Beilstein reference	Density	Refractive index	Melting point	Boiling point	Flash point	Solubility in 100 parts solvent
d59	*N,N'*-Dibenzylethylenediamine	$(C_6H_5CH_2NHCH_2—)_2$	240.35	12, 1067	1.024_4^{20}	1.5624^{20}	26	195^{4mm}	>112	v s alc, bz, chl, eth
d60	Dibenzyl malonate	$CH_2[COOCH_2C_6H_5]_2$	284.31	6, 436	1.137	1.5447^{20}		$188^{0.2mm}$	>112	
d61	Dibenzyl phosphonate	$(C_6H_5CH_2O)_2P(O)H$	262.25		1.187	1.5540^{20}	−5 to +5	$110^{0.01mm}$	>112	
d62	Dibromoacetic acid	$Br_2CHCOOH$	217.86	2, 218				$128–130^{16mm}$		
d63	Dibromoacetonitrile	Br_2CHCN	198.86	2, 219	2.296	1.5393^{20}		$67–69^{24mm}$	none	
d64	2,4'-Dibromoacetophenone	$BrC_6H_4C(=O)CH_2Br$	277.96	7, 285			108–110			s warm alc, eth
d65	1,4-Dibromobenzene	$C_6H_4Br_2$	235.92	5, 211	0.9641^{100}	1.5743^{100}	87.3	219	none	1.4 alc; s bz; 101 eth
d66	4,4'-Dibromobiphenyl	$BrC_6H_4C_6H_4Br$	312.00	5, 580			162–163	355–360		s bz; sl s hot alc
d67	1,3-Dibromobutane	$CH_3CH(Br)CH_2CH_2Br$	215.93	1, 120	1.800^{20}	1.5085^{20}		175	none	s chl, eth
d68	1,4-Dibromobutane	$BrCH_2CH_2CH_2CH_2Br$	215.93	1, 120	1.8080_4^{20}	1.5186^{20}	−20	198	>112	s chl
d69	1,4-Dibromo-2,3-butanedione	$BrCH_2C(=O)C(=O)-CH_2Br$	243.89	1, 774			116–117			
d70	*trans*-2,3-Dibromo-2-butene-1,4-diol	$HOCH_2C(Br)=C(Br)-CH_2OH$	245.91	1^1, 260			112–114			
d71	Dibromochloromethane	$HCClBr_2$	208.29	1, 67	2.451	1.5465^{20}	−22	120^{748mm}	none	misc alc, bz, eth
d72	1,2-Dibromo-2-chloro-1,1,2-trifluoroethane	$FCCl(Br)C(Br)F_2$	276.5		2.2478^{20}	1.4275^{20}		93–94		
d73	1,10-Dibromodecane	$Br(CH_2)_{10}Br$	300.09	1^1, 64	1.335^{30}	1.4912^{20}	27	160^{15mm}	>112	sl s alc; s eth
d74	1,2-Dibromo-1,1-difluoroethane	$CH_2BrC(Br)F_2$	223.87	1, 92	2.2238^{20}	1.4456^{20}	−61.3	93.4		i aq
d75	Dibromodifluoromethane	Br_2CF_2	209.81	1^1, 16	2.288_4^{15}	1.3999^{12}	−141.6	23–24	none	0.1 aq; misc alc, bz, chl, eth

d76	1,3-Dibromo-5,5-dimethylhydantoin		285.93				197 d			
d77	1,1-Dibromoethane	CH_3CHBr_2	187.87	1, 90	2.055_4^{20}	1.5379^{20}		113		i aq; v s alc, eth
d78	1,2-Dibromoethane	$BrCH_2CH_2Br$	187.87	1, 90	2.1802_4^{20}	1.5416^{15}	10.0	131.7	none	0.43 aq; misc alc, eth
d79	(1,2-Dibromoethyl)-benzene	$C_6H_5CH(Br)CH_2Br$	263.97	5, 356			70–74	140^{15mm}		
d80	*cis*-1,2-Dibromo-ethylene	$BrCH{=}CHBr$	185.86	1, 190	2.21_4^{17}	1.5431^{18}	−53	112.5		s alc, bz, chl, eth
d81	*trans*-1,2-Dibromo-ethylene	$BrCH{=}CHBr$	185.86	1, 190	2.246	1.5505^{18}	−6.5	108		
d82	1,2-Dibromoethyltri-chlorosilane	$BrCH_2CH(Br)SiCl_3$	321.3		2.046_4^{20}	1.537^{20}		90^{11mm}		
d83	4′,5′,-Dibromofluo-rescein		490.12	19, 228			270–273			s hot alc, HOAc
d84	2,4-Dibromo-1-fluoro-benzene	$Br_2C_6H_3F$	253.91		2.047^{20}	1.5840^{20}		105^{22mm}	92	
d85	1,2-Dibromohexa-fluoropropane	$CF_3CF(Br)C(Br)F_2$	309.83				72.8			

Dibenzyl ketone, d686

5,7-Dibromo-8-quinolinol, d87

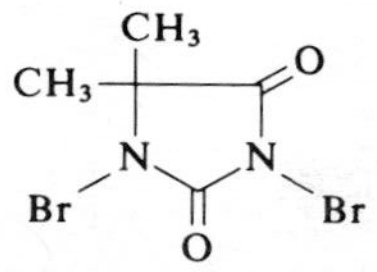

d76

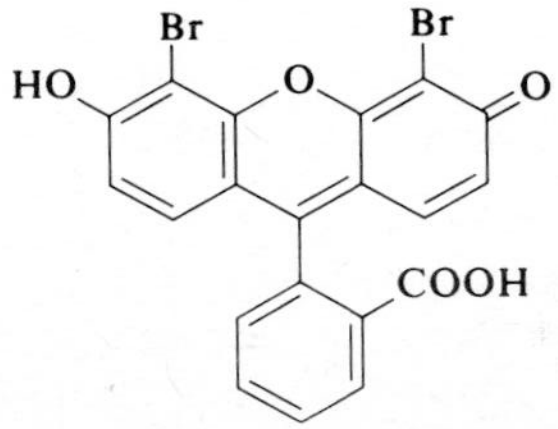

d83

TABLE 1-14 Physical constants of organic compounds (*continued*)

No.	Name	Formula	Formula weight	Beilstein reference	Density	Refractive index	Melting point	Boiling point	Flash point	Solubility in 100 parts solvent
d86	1,6-Dibromohexane	$Br(CH_2)_6Br$	243.98	1, 145	1.586_4^{18}	1.5066^{20}		243	32	misc eth
d87	5,7-Dibromo-8-hydroxyquinoline		302.96	21, 97			200–201	subl		s alc, bz; v s eth
d88	Dibromomethane	CH_2Br_2	173.85	1, 67	2.4956_4^{20}	1.5419^{20}	−52.7	96.97	none	1.15 aq; misc alc, bz, acet, chl, eth
d89	2,6-Dibromo-4-nitroaniline	$Br_2C_6H_2(NO_2)NH_2$	295.93	12, 743			206–208			sl s aq; s HOAc
d90	2,5-Dibromonitrobenzene	$Br_2C_6H_3NO_2$	280.91	5, 250	1.9581^{111}		82–84			s bz, hot alc
d91	1,5-Dibromopentane	$Br(CH_2)_5Br$	229.95	1, 131	1.6879_4^{15}	1.5092^{15}	−34	110^{15mm}	79	
d92	1,2-Dibromopropane	$CH_3CH(Br)CH_2Br$	201.90	1, 109	1.933^{20}	1.5203^{20}	−55.5	139.6	none	0.2 aq; misc alc, bz, chl, eth
d93	1,3-Dibromopropane	$BrCH_2CH_2CH_2Br$	201.90	1, 110	1.9712_4^{25}	1.5233^{20}	−34	166.8	54	0.17 aq; s alc, eth
d94	2,3-Dibromopropanol	$BrCH_2CH(Br)CH_2OH$	217.90	1, 357	2.120_4^{20}	1.5599^{20}		$95-97^{10mm}$		sl s aq; misc alc, bz, eth, acet
d95	2,3-Dibromopropene	$BrCH_2C(Br)=CH_2$	199.88	1, 201	1.9336_4^{20}	1.5470^{20}		140–143	none	
d96	2,3-Dibromopropionic acid	$BrCH_2CH(Br)COOH$	231.88	2, 258			64–66	160^{20mm}		s aq, alc, bz
d97	2,6-Dibromopyridine	$Br_2{-}C_5H_3N$	236.91	20^2, 153			118–119	255		
d98	DL-2,3-Dibromosuccinic acid	$HOOCCH(Br)CH(Br)COOH$	275.89	2, 625			167			v s aq, alc
d99	1,2-Dibromotetrachloroethane	$BrCCl_2CCl_2Br$	325.65	1, 93	2.713		220–222		none	
d100	1,2-Dibromotetrafluoroethane	$BrCF_2CF_2Br$	259.83		2.163^{25}	1.367^{25}	−110.5	47.3		
d101	2,5-Dibromothiophene	$Br_2C_4H_2S$	241.94	17, 33	2.147_{23}^{23}	1.6289^{20}	−6	221		i aq; v s alc, eth
d102	α,α-Dibromotoluene	$C_6H_5CHBr_2$	249.94	5, 308	1.510^{15}	1.6147^{20}		156^{23mm}	110	i aq; misc alc, eth
d103	1,2-Dibromo-1,1,2-trifluoroethane	$HC(Br)FC(Br)F_2$	241.8	1, 92	$2,274^{27}$	1.4191^{24}		76.5		

d104	α,α′Dibromo-*o*-xylene	$C_6H_4(CH_2Br)_2$	263.97	5, 366	1.960		92–94			sl s alc, chl, eth
d105	α,α′-Dibromo-*p*-xylene	$C_6H_4(CH_2Br)_2$	263.97	5, 385	2.012^{0}		142–143	245		v s alc, chl; s eth
d106	1,2-Dibutoxyethane	$C_4H_9OCH_2CH_2OC_4H_9$	174.28		0.8374^{20}_{20}	1.4131^{20}	−69.1	203.6		0.2 aq; misc alc, acet
d107	Dibutylamine	$(C_4H_9)_2NH$	129.25	4, 157	0.7601^{20}_{4}	1.4177^{20}	−62	159.6	33	0.47 aq; s alc, acet, eth, EtAc, PE
d108	*N,N*-Dibutylaminoethanol	$(C_4H_9)_2NCH_2CH_2OH$	173.29		0.860^{20}_{20}	1.444^{20}	<−70	227–230	93	
d109	*N,N*-Dibutylaniline	$C_6H_5N(C_4H_9)_2$	205.34	12^2, 95	0.904^{20}	1.5197^{20}		267–275	110	i aq, MeOH; s acet, bz, EtOH, EtAc, eth
d110	Dibutyl decanedioate	$C_4H_9OOC(CH_2)_8COO-C_4H_9$	314.45	2, 719	0.9366^{20}	1.4415^{20}	1.0	344–345	177	0.004 aq
d111	Di-*tert*-butyldicarbonate	$(CH_3)_3COC(O)OC-(CH_3)_3$	218.25		0.950	1.4103^{20}	23	$56^{0.5mm}$	37	
d112	2,5-Di-*tert*-butyl-1,4-dihydroxybenzene	$[(CH_3)_3C]_2C_6H_2(OH)_2$	222.33				217–219			
d113	Dibutyl disulfide	$C_4H_9SSC_4H_9$	178.36	1^2, 400	0.9383^{20}_{4}	1.4920^{20}	−71	231.2	93	i aq; misc alc, eth

Dibutyl 1,2-benzenedicarboxylate, d128

Dibutyl butanedioate, d130

Dibutyl Cellosolve, d106

d87

TABLE 1-14 Physical constants of organic compounds (*continued*)

No.	Name	Formula	Formula weight	Beilstein reference	Density	Refractive index	Melting point	Boiling point	Flash point	Solubility in 100 parts solvent
d114	Di-*tert*-butyl disulfide	$(CH_3)_3CSSC(CH_3)_3$	178.36		0.935	1.4920		229–33	93	
d115	Dibutyl ether	$C_4H_9OC_4H_9$	130.22	1, 369	0.7689^{20}_{4}	1.3992^{20}	−97.9	142.4	25	0.03 aq; misc alc, eth
d116	*N,N*-Dibutylformamide	$HC(=O)N(C_4H_9)_2$	157.26		0.864	1.4429^{20}		120^{15mm}	100	
d117	3,5-Di-*tert*-butyl-4-hydroxybenzoic acid	$[(CH_3)_3C]_2C_6H_2(OH)COOH$	250.34				206–209			
d118	Dibutyl maleate	$C_4H_9OOCCH{=}CHCOOC_4H_9$	228.28		0.9950^{20}	1.4454^{20}	<−80	d 280	135	0.05 aq
d119	Di-*tert*-butyl malonate	$CH_2(COOC(CH_3)_3)COOC(CH_3)_3$	216.27			1.4184^{20}	−6.0	93^{10mm}		
d120	2,6-Di-*tert*-butyl-4-methylphenol	$[(CH_3)_3C]_2C_6H_2(CH_3)OH$	220.36	6^3, 2073	0.894^{75}_{4}	1.4859^{75}	70	265		i aq; s alc, bz, acet
d121	Dibutyl oxalate	$C_4H_9OOC{-}COOC_4H_9$	202.25	2, 540	0.986^{20}_{20}	1.4232^{20}	−30.0	239–240	108	misc alc, ketones, PE
d122	Di-*tert*-butyl peroxide	$(CH_3)_3CO{-}OC(CH_3)_3$	146.23		0.794^{20}	1.3890^{20}	−40	110		misc acet, octane
d123	2,4-Di-*tert*-butylphenol	$[(CH_3)_3C]_2C_6H_3OH$	206.33				56.5	263.5	115	s hot alc; i alk
d124	2,6-Di-*sec*-butylphenol	$[CH_3CH_2CH(CH_3)]_2C_6H_3OH$	206.33		0.918	1.5100^{20}	−42	255–260	127	
d125	2,6-Di-*tert*-butylphenol	$[(CH_3)_3C]_2C_6H_3OH$	206.33	6^3, 2061			35–83	253	118	s hot alc; i alk
d126	3,5-Di-*tert*-butylphenol	$[(CH_3)_3C]_2C_6H_3OH$	206.33				87–89			
d127	Dibutyl phosphonate	$(C_4H_9O)_2P(O)H$	194.21	1, 187	0.995^{20}_{4}	1.4231^{20}		118^{11mm}	121	sl s (hyd) aq; misc alc, acet, eth

No.	Name	Formula	Formula weight	Beilstein reference	Density	Refractive index	Melting point	Boiling point	Flash point	Solubility
d128	Dibutyl *o*-phthalate	$C_6H_4[COOC_4H_9]_2$	278.35	9^2, 586	1.0465^{20}_{4}	1.4926^{20}	−35	340	171	0.01 aq; v s alc, bz, acet, eth
d129	*N,N*-Dibutyl-1,3-propanediamine	$C_4H_9NHCH_2CH_2CH_2$-NHC_4H_9	186.34		0.827	1.4463^{20}		205	103	
d130	Dibutyl succinate	$[C_4H_9OOCCH_2—]_2$	230.30	2^2, 551	0.9768^{20}_{4}	1.4299^{20}	−29.0	274.5		i aq; s alc, eth
d131	Dibutyl sulfate	$C_4H_9OSO_2OC_4H_9$	210.29		1.059^{25}_{4}	1.4213^{20}		130–132^{11mm}		
d132	Dibutyl sulfide	$C_4H_9SC_4H_9$	146.30	1, 370	0.839^{16}_{0}	1.4530^{20}	−75.0	188.9	76	i aq; v s alc, eth
d133	Di-*tert*-butyl sulfide	$(CH_3)_3CSC(CH_3)_3$	146.30		0.815	1.4506^{20}		151	48	
d134	Dibutyl sulfite	$(C_4H_9O)_2S(O)$	194.29	1^2, 397	0.9944^{22}_{4}	1.4310^{20}		108^{15mm}		
d135	Dibutyl sulfone	$(C_4H_9)_2SO_2$	178.29	1, 371			46	295	143	i aq; s alc, eth
d136	*N,N′*-Dibutylthiourea	$C_4H_9NHC(=S)NHC_4H_9$	188.34				63.65			i aq; s alc; sl s eth
d136a	Dibutyltin dichloride	$(C_4H_9)_2SnCl_2$	303.83				39–41	135^{10mm}	>112	
d137	Dibutyltin oxide	$(C_4H_9)_2SnO$	248.92	4^1, 588			>300			
d138	Dichloroacetic acid	$Cl_2CHCOOH$	128.94	2, 202	1.563^{20}_{4}	1.4642^{20}	9–11	193–194	>112	misc aq, alc, eth
d139	1,1-Dichloroacetone	$CH_3C(O)CHCl_2$	126.97	1, 654	1.305^{18}_{15}			150		sl s aq; s alc; misc eth
d140	2′,4′-Dichloroacetophenone	$Cl_2C_6H_3COCH_3$	189.04	7^2, 219		1.5635^{20}	33–34	145^{15mm}	>112	i aq
d141	Dichloroacetyl chloride	$Cl_2CHCOCl$	147.39	2, 204	1.5315^{16}_{4}	1.4603^{20}		107–108	66	d aq, alc; misc eth
d142	2,3-Dichloroaniline	$Cl_2C_6H_3NH_2$	162.02	12, 621		1.5969^{20}	23–24	252	>112	s alc; v s eth
d143	2,4-Dichloroaniline	$Cl_2C_6H_3NH_2$	162.02	12, 621	1.567^{20}		59.62	245		sl s aq; s alc, eth
d144	2,5-Dichloroaniline	$Cl_2C_6H_3NH_2$	162.02	12, 625			49–51	251		s alc, bz, eth

Di-*tert*-butylcresol, d120
2,5-Di-*tert*-butylhydroquinone, d112
Dibutyl ketone, n100
Dibutyl phosphite, d127
Dibutyl sebacate, d110
Dichloroacetaldehyde diethyl acetal, d169
Dichloroanisoles, d191, d192

TABLE 1-14 **Physical constants of organic compounds (*continued*)**

No.	Name	Formula	Formula weight	Beilstein reference	Density	Refractive index	Melting point	Boiling point	Flash point	Solubility in 100 parts solvent
d145	2,6-Dichloroaniline	$Cl_2C_6H_3NH_2$	162.02	12, 626			38.41			
d146	3,4-Dichloroaniline	$Cl_2C_6H_3NH_2$	162.02	12, 626			70–72	272		s alc, eth; sl s bz
d147	3,5-Dichloroaniline	$Cl_2C_6H_3NH_2$	162.02	12, 626			51–53	259^{741mm}		i aq; s alc, eth
d148	1,5-Dichloroanthra-quinone		277.11	7, 787			245–247			sl s alc, bz, acet
d149	1,8-Dichloroanthra-quinone		277.11	7, 788			202–203			sl s alc
d150	2,4-Dichlorobenz-aldehyde	$Cl_2C_6H_3CHO$	175.01	7, 236			69–73	233		i aq; s alc
d151	2,4-Dichlorobenz-amide	$Cl_2C_6H_3CONH_2$	190.03	9^3, 1376			191–194			
d152	1,2-Dichlorobenzene	$C_6H_4Cl_2$	147.01	5, 201	1.3059_4^{20}	1.5515	−17.0	180.4	65	misc alc, bz, eth
d153	1,3-Dichlorobenzene	$C_6H_4Cl_2$	147.01	5, 202	1.2884_4^{20}	1.5459	−24.8	173.1	63	0.01 aq; s alc, eth
d154	1,4-Dichlorobenzene	$C_6H_4Cl_2$	147.01	5, 203	1.2417^{60}	1.5285	53	174.1	65	s alc, bz, chl, eth
d155	2,5-Dichlorobenzene-sulfonyl chloride	$Cl_2C_6H_3SO_2Cl$	245.51	11^1, 15			36–37			d hot aq, hot alc
d156	2,4-Dichlorobenzoic acid	$Cl_2C_6H_3COOH$	191.01	9, 342			157–160			s hot aq, alc, bz, chl
d157	2,5-Dichlorobenzoic acid	$Cl_2C_6H_3COOH$	191.01	9, 342			151–154	301		sl s aq; s aq, s alc, eth
d158	3,4-Dichlorobenzoic acid	$Cl_2C_6H_3COOH$	191.01	9, 343			207–209			s hot aq, eth; v s alc
d159	4,4′-Dichlorobenzo-phenone	$(ClC_6H_4)_2CO$	251.11	7, 420			144–146	353		s hot alc; v s chl, eth
d160	2,4-Dichlorobenzoyl chloride	$Cl_2C_6H_3COCl$	209.46	9, 342	1.494	1.5297^{20}	16–18	150^{34mm}	137	d aq, alc
d161	3,4-Dichlorobenzoyl chloride	$Cl_2D_6H_3COCl$	209.46	9, 344			30–33	242	142	d aq, alc
d162	1,2-Dichlorobutane	$CH_3CH_2CH(Cl)CH_2Cl$	127.01	1^1, 38	1.118_4^{20}	1.4474^{15}		124		i aq; s chl, eth

d163	1,4-Dichlorobutane	$ClCH_2CH_2CH_2CH_2Cl$	127.01	1, 119	1.1598_4^{20}	1.4566^{20}	−38	155	40	i aq; s chl
d164	*meso*-2,3-Dichloro-butane	$CH_3CH(Cl)CH(Cl)CH_3$	127.01	1, 119	1.1025_4^{25}	1.4386^{25}	−80	115.9	18	i aq; s chl
d165	*cis*-1,4-Dichloro-2-butene	$ClCH_2CH{=}CHCH_2Cl$	125.00	1^3, 743	1.188_4^{25}	1.4887^{25}	−48	152	49	i aq; s org solv
d166	*trans*-1,4-Dichloro-2-butene	$ClCH_2CH{=}CHCH_2Cl$	125.00	1^3, 743	1.183_4^{25}	1.4861^{25}	1–3	$74–76^{40mm}$	56	i aq; s org solv
d167	3,4-Dichloro-1-butene	$ClCH_2CH(Cl)CH{=}CH_2$	125.00		1.150	1.4658^{20}	−61	123	28	
d168	1,4-Dichloro-2-butyne	$ClCH_2C{\equiv}CCH_2Cl$	122.98	1^3, 927	1.258_4^{20}	1.5048^{20}		165–168	160	
d169	1,1-Dichloro-2,2-diethoxyethane	$Cl_2CHCH(OC_2H_5)_2$	187.07	1, 614	1.138	1.4360^{20}		183–184	60	
d170	Dichlorodifluoro-methane	Cl_2CF_2	120.92	1, 61	1.486^{-30}		−158	−29.8		0.02 aq; 9 bz; 5.5 chl; 6 diox; s alc, eth
d171	4,6-Dichloro-1,3-di-hydroxypenzene	$Cl_2C_6H_2(OH)_2$	179.00	6^1, 403			104–106	254		sl s aq, bz; s eth
d172	2,5-Dichloro-3,6-di-hydroxy-*p*-benzo-quinone		208.98	8, 379			283–284			

2,6-Dichlorobenzyl chloride, t253
2,2′-Dichlorodiethyl ether, b159

5,5′-Dichloro-2,2′-dihydroxydiphenylmethane, m234

1,1-Dichlorodimethyl ether, d197
Dichlorohydrin, d220

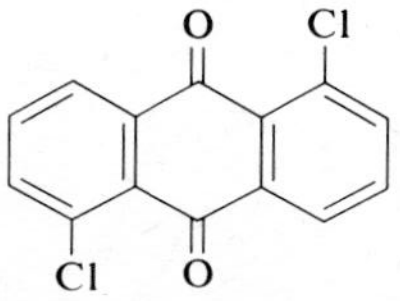

O Cl Cl O

d148

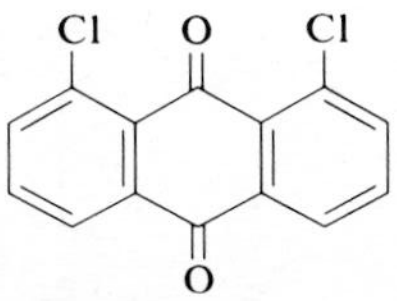

Cl O Cl O

d149

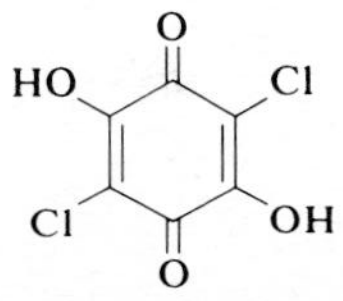

O HO Cl Cl OH O

d172

TABLE 1-14 Physical constants of organic compounds (*continued*)

No.	Name	Formula	Formula weight	Beilstein reference	Density	Refractive index	Melting point	Boiling point	Flash point	Solubility in 100 parts solvent
d173	1,3-Dichloro-3,5-dimethylhydantoin		197.02	24^2, 158			134–136			
d174	Dichlorodimethylsilane	$(CH_3)_2SiCl_2$	129.06		1.064_4^{20}	1.4038^{20}	−16	70	−16	
d175	Dichlorodiphenylsilane	$(C_6H_5)_2SiCl_2$	253.20	16, 910	1.222_4^{20}	1.582^{20}		308–309	157	d aq, alc
d176	1,1-Dichloroethane	CH_3CHCl_2	98.96	1, 83	1.1757_4^{20}	1.4164^{20}	−97.0	57.3	−5	0.51 aq; misc alc
d177	1,2-Dichloroethane	$ClCH_2CH_2Cl$	98.96	1, 84	1.2531_4^{20}	1.4448^{20}	−35.7	83.5	15	0.8 aq; misc alc, chl, eth
d178	1,1-Dichloroethylene	$H_2C{=}CCl_2$	96.94	1, 186	1.2129_4^{20}	1.4247^{20}	−122.6	31.6	−15	0.02 aq; s alc, bz, chl, eth
d179	*cis*-1,2-Dichloroethylene	$ClCH{=}CHCl$	96.94	1, 188	1.2818_4^{20}	1.4490^{20}	−80.1	60.7	6	0.7 aq; s alc, eth
d180	*trans*-1,2-Dichloroethylene	$ClCH{=}CHCl$	96.94	1, 188	1.2546_4^{20}	1.4462^{20}	−49.8	47.7	6	0.6 aq; s alc, eth
d181	2,2′-Dichloroethyl ether	$ClCH_2CH_2OCH_2CH_2Cl$	143.01	1^2, 335	1.2220_{20}^{20}	1.457^{20}		178.5	55	1.1 aq; s alc, bz, eth
d182	1,2-Dichloroethyltrichlorosilane	$ClCH_2CH(Cl)SiCl_3$	232.4		1.516_4^{25}	1.449^{25}		$82\text{–}84^{26mm}$		
d183	Dichlorofluoromethane	$FCHCl_2$	102.92	1, 61	1.345^{30}		−135	8.9		69 HOAc; 108 diox; s alc, eth; i aq
d184	Dichloroheptylmethylsilane	$C_7H_{15}Si(CH_3)Cl_2$	225.2		0.9780_4^{20}	1.4396^{25}		207–208		
d185	1,2-Dichlorohexafluorocyclobutane	$F_6C_4Cl_2$	233.0			1.3342^{25}		59–60		
d186	1,5-Dichlorohexamethyltrisiloxane	$[ClC(CH_3)_2O]_2\text{-}Si_3(CH_3)_2$	277.4		1.018_4^{20}	1.4071		184		
d187	1,6-Dichlorohexane	$Cl(CH_2)_6Cl$	155.07	1, 144	1.068	1.4568^{20}		87^{15mm}	73	s chl

d188	1,2-Dichloro-2-iodo-1,1,2-trifluoroethane	$F(I)C(Cl)C(Cl)F_2$	278.9		2.200^{20}	1.4490^{20}		100–101		
d189	Dichloromaleic anhydride		166.95	17, 434						
d190	Dichloromethane	CH_2Cl_2	84.93	1, 60	1.3255^{20}_{4}	1.4246^{20}	−96.7	40.5	none	1.3 aq; misc alc, eth
d191	2,3-Dichloro-1-methoxybenzene	$Cl_2C_6H_3OCH_3$	177.03	6^1, 102			31–33			
d192	3,5-Dichloro-1-methoxybenzene	$Cl_2C_6H_3OCH_3$	177.03	6, 190			40–42			
d193	2,4-Dichloro-6-methoxy-1,3,5-triazine		179.99				86–88	132^{49mm}		
d194	(Dichloromethyl)dimethylchlorosilane	$Cl_2CHSi(Cl)(CH_3)_2$	177.5		1.237^{20}_{4}	1.461^{20}	−49	149		
d195	2,2-Dichloro-1-methylcyclopropanecarboxylic acid	$Cl_2(C_3H_2)(CH_3)COOH$	169.01				60–65	85^{8mm}		
d196	*N*-(Dichloromethylene)aniline	$C_6H_5N{=}CCl_2$	174.03	12, 447	1.265	1.5710^{20}		106^{30mm}	79	
d197	Dichloromethyl ether	Cl_2CHOCH_3	114.96		1.271	1.4300^{20}		85	42	

Dichloroisopropyl alcohol, d220

4,4′-Dichloro-α-methylbenzhydrol, b165

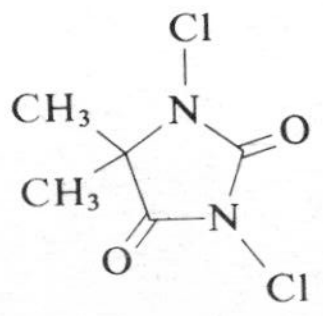

d173

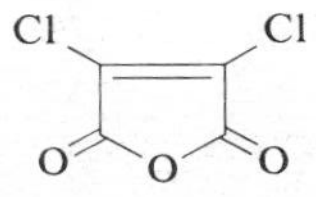

d189

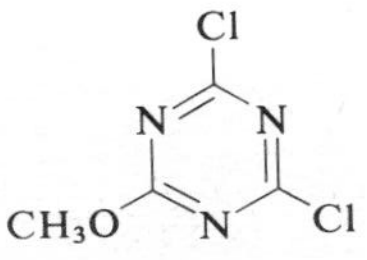

d193

TABLE 1-14 Physical constants of organic compounds (*continued*)

No.	Name	Formula	Formula weight	Beilstein reference	Density	Refractive index	Melting point	Boiling point	Flash point	Solubility in 100 parts solvent
d198	Dichloro(methyl)-phenylsilane	$C_6H_5Si(CH_3)Cl_2$	191.13		1.176	1.5190^{20}		205	82	
d199	Dichloro(methyl)-silane	$HSi(CH_3)Cl_2$	115.04	4^1, 581	1.105		−93	41	−32	
d200	Dichloro(methyl)-vinylsilane	$H_2C{=}CHSi(CH_3)Cl_2$	141.07		1.087_4^{20}	1.4300^{20}		92–93	4	
d201	2,3-Dichloro-1,4-naphthoquinone		227.05	7, 729			190–192			sl s alc, bz, eth
d202	2,6-Dichloro-4-nitroaniline	$Cl_2C_6H_2(NO_2)NH_2$	207.02	12, 735			190–192			
d203	2,3-Dichloronitro-benzene	$Cl_2C_6H_3NO_2$	192.00	5, 245	1.721^{14}		61–62	257–258		s PE
d204	2,4-Dichloronitro-benzene	$Cl_2C_6H_3NO_2$	192.00	5, 245	1.439^{80}		29–32	258		s hot alc; misc eth
d205	3,4-Dichloronitro-benzene	$Cl_2C_6H_3NO_2$	192.00	5, 246	1.456_4^{75}		41–42	255–256	123	
d206	2,3-Dichloroocta-fluorobutane	$CF_3CF(Cl)CF(Cl)CF_3$	271.0		1.6801^{20}	1.3100^{20}	−68	63		
d207	1,7-Dichloroocta-methyltetrasiloxane	$[Cl(CH_3)_2SiOSi-(CH_3)_2-]_2O$	351.6		1.011_4^{20}	1.403^{20}		222		
d208	2,3-Dichloro-4-oxo-2-butenoic acid	$ClC(CHO){=}C(Cl)COOH$	168.96	3, 727			125–128		100	sl s aq; s alc, hot bz
d209	1,5-Dichloropentane	$Cl(CH_2)_5Cl$	141.04	1, 131	1.1058_4^{15}	1.4553^{20}	−72	63^{10mm}	26	i aq; s alc, eth
d210	2,3-Dichlorophenol	$Cl_2C_6H_3OH$	163.00	6^1, 102			58.60	206		s alc, eth
d211	2,4-Dichlorophenol	$Cl_2C_6H_3OH$	163.00	6, 189			42–43	210	113	v s alc, bz, chl, eth
d212	2,5-Dichlorophenol	$Cl_2C_6H_3OH$	163.00	6, 189			56–58	211		v s alc, bz, eth
d213	2,6-Dichlorophenol	$Cl_2C_6H_3OH$	163.00	6, 190			65–68	218–220		v s alc, eth
d214	2,4-Dichlorophenoxy-acetic acid	$Cl_2C_6H_3OCH_2COOH$	221.04				138	$160^{0.4mm}$		s alc, bz, chl, eth

d215	2,5-Dichloro-*p*-phenylenediamine	$Cl_2C_6H_2(NH_2)_2$	177.03	13, 118			165 d			
d216	Dichlorophenylphosphine	$C_6H_5PCl_2$	178.99	16, 763	1.319	1.5980^{20}	−51	222	>112	
d217	4,5-Dichloro-*o*-phthalic acid	$Cl_2C_6H_2(COOH)_2$	235.02	9^1, 366			193–195			s aq; v s eth
d218	1,2-Dichloropropane	$CH_3CH(Cl)CH_2Cl$	112.99	1, 105	1.1558^{20}	1.4390^{20}	−100.4	96.4	4	0.26 aq; misc alc, bz, chl, eth
d219	1,3-Dichlopropane	$ClCH_2CH_2CH_2Cl$	112.99	1, 105	1.1878_4^{20}	1.4487^{20}	−99.5	120.5	32	v s alc, eth
d220	1,3-Dichloro-2-propanol	$ClCH_2CH(OH)CH_2Cl$	128.99	1, 364	1.3506_4^{17}	1.4835^{20}	−4	174.3	74	9.1 aq; misc alc, eth
d221	1,3-Dichloropropene	$ClCH_2CH{=}CHCl$	110.97	1, 199	1.217_4^{20}	1.470^{20}		112		i aq; s chl, eth
d222	2,3-Dichloro-1-propene	$ClCH_2C(Cl){=}CH_2$	110.97	1, 199	1.204_{25}^{25}	1.4611^{20}		94	10	misc alc; s eth
d223	3,6-Dichloropyridazine		148.98				66–69			
d224	2,6-Dichloropyridine	$Cl_2{-}C_5H_3N$	147.99	20, 231			86–88			
d225	4,7-Dichloroquinoline		198.05				84–86	148^{10mm}		
d226	Dichlorosilane	H_2SiCl_2	101.0				−122	8.3		
d227	1,2-Dichloro-1,1,2,2-tetrafluoroethane	$ClCF_2CF_2Cl$	170.93		1.470_4^{20}	1.290^{20}	−94	3.6		s alc, eth

1,1-Dichloro-2-propanone, d139

4,6-Dichlororesorcinol, d171

α,o-Dichlorotoluene, c59

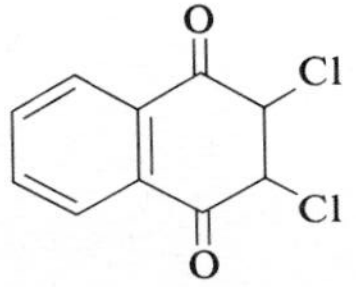

d201

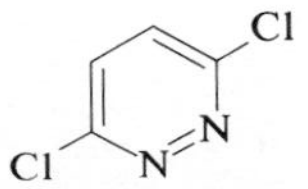

d223

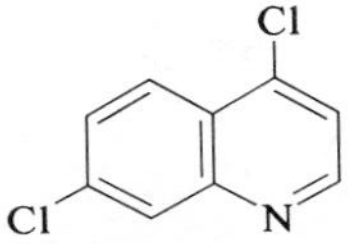

d225

TABLE 1-14 Physical constants of organic compounds (*continued*)

No.	Name	Formula	Formula weight	Beilstein reference	Density	Refractive index	Melting point	Boiling point	Flash point	Solubility in 100 parts solvent
d228	2,5-Dichlorothiophene	$Cl_2—C_4H_2S$	153.03	17, 33	1.442	1.5621^{20}	−40.5	162		i aq; misc alc, eth
d229	2,4-Dichlorotoluene	$Cl_2C_6H_3CH_3$	161.03	5, 295	1.2460^{20}_{20}	1.5454^{20}	−13	200.5	79	i aq
d230	2,6-Dichlorotoluene	$Cl_2C_6H_3CH_3$	161.03	5, 296	1.254	1.5507^{20}		196–203	82	i aq; s chl
d231	3,4-Dichlorotoluene	$Cl_2C_6H_3CH_3$	161.03	5, 296	1.251^{25}_{25}	1.5472^{20}	−14	201^{740mm}	85	i aq
d232	2,2-Dichloro-1,1,1-trifluoroethane	CF_3CHCl_2	152.9					28		
d233	α,α′-Dichloro-*p*-xylene	$C_6H_4(CH_2Cl)_2$	175.06	5, 384			100	254		22.5 acet; 20 bz; 4.5 CCl_4; 11 eth; 18 EtAc
d234	2,5-Dichloro-*p*-xylene	$Cl_2C_6H_2(CH_3)_2$	175.06	5, 384			71	222		27 acet; 44 bz; 39 eth 32 EtAc; 5 MeOH
d235	Dicyanodiamide	$H_2NC(=NH)NHCN$	84.08	3^2, 75	1.400^{25}_4		208–211			2.3 aq; 1.3 alc; i bz
d236	1,2-Dicyanobenzene	$C_6H_4(CN)_2$	128.13	9, 815			139–141			v s bz, alc; s hot eth
d237	1,3-Dicyanobenzene	$C_6H_4(CN)_2$	128.13	9, 836			158–160			s alc, bz, chl, eth
d238	1,4-Dicyanobutane	$NC(CH_2)_4CN$	108.14	2, 653	0.951	1.4380^{20}	1–3	295	>112	
d239	1,6-Dicyanohexane	$NC(CH_2)_6CN$	136.20	2, 694	0.954	1.4436^{20}	−3.5	185^{15mm}	>112	
d240	2,4-Dicyano-3-methylglutaramide	$CH_3CH[CH(CN)-CONH_2]_2$	194.19	2^2, 704			159–160			
d241	Dicyclohexyl	$C_6H_{11}C_6H_{11}$	166.31	5, 108	0.864	1.4782^{20}	3–4	227	101	7 MeOH; misc bz, acet, eth
d242	Dicyclohexylamine	$(C_6H_{11})_2NH$	181.32	12, 6	0.910	1.4842^{20}	−0.1	255.8	96	misc alc, bz, chl, eth
d243	*N,N′*-Dicyclohexylcarbodiimide	$C_6H_{11}N=C=NC_6H_{11}$	206.33				34–35	122–124^{6mm}		
d244	Dicyclopentadiene		132.21	5, 495	0.930^{25}_4	1.5050^{25}	−1	170	26	s alc, eth

d245	Diethanolamine	$HOCH_2CH_2NHCH_2CH_2OH$	105.14	4, 283	1.0881_4^{30}	1.4747^{30}	28.0	268.0	137	96 aq; 4 bz; 0.8 eth; misc MeOH, acet
d246	2,2-Diethoxyacetophenone	$C_6H_5C(=O)CH(OC_2H_5)_2$	208.26	7^1, 361	1.034	1.4995^{20}		131–134^{10mm}	110	
d247	4,4-Diethoxybutylamine	$H_2N(CH_2)_3CH(OC_2H_5)_2$	161.25	4, 319	0.933	1.4275^{20}		196	62	
d248	2,2-Diethoxy-*N*,*N*-dimethylethylamine	$(C_2H_5O)_2CHCH_2N(CH_3)_2$	161.25	4, 308	0.883	1.4129^{20}		170	45	
d249	Diethoxydimethylsilane	$(C_2H_5O)_2Si(CH_3)_2$	148.28		0.840_4^{20}	1.3811^{20}	−87	114	11	
d250	Diethoxydiphenylsilane	$(C_2H_5O)_2Si(C_6H_5)_2$	272.42		1.0329_4^{20}	1.5269^{20}		130^{2mm}		
d251	1,1-Diethoxyethane	$CH_3CH(OC_2H_5)_2$	118.18	1, 603	0.8254_4^{20}	1.3825^{20}	2.8	102.7	−21	5 aq; misc alc, eth
d252	1,2-Diethoxyethane	$C_2H_5OCH_2CH_2OC_2H_5$	118.18	1, 468	0.842	1.3922^{20}	−74	121.4	27	21 aq
d253	2,2-Diethoxyethanol	$(C_2H_5O)_2CHCH_2OH$	134.18	1, 818	0.888_4^{24}	1.4160^{20}		167	67	s alc, eth
d254	2,2-Diethoxyethylamine	$(C_2H_5O)_2CHCH_2NH_2$	133.19	4, 308	0.916	1.4170		162–163	45	
d255	Diethoxymethylsilane	$(C_2H_5O)_2SiH(CH_3)$	134.3		0.829_4^{25}	1.372^{25}		94–95		
d256	Diethoxymethylvinylsilane	$(C_2H_5O)_2Si(CH_3)CH=CH_2$	160.3		0.858_4^{20}	1.400^{20}		133–134		
d257	1,1-Diethoxypropane	$CH_3CH_2CH(OC_2H_5)_2$	132.20	1, 630	0.8232_4^{20}	1.3884^{20}		122.8	12	v s alc, eth
d258	3,3-Diethoxy-1-propene	$(C_2H_5O)_2CHCH=CH_2$	130.19	1, 727	0.854	1.4000^{20}		89–90	4	

α,*p*-Dichlorotoluene, c60

1,2-Dicyanoethane, b383

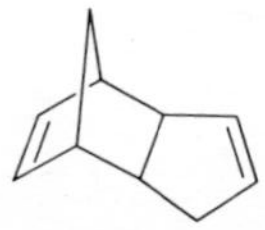

d244

TABLE 1-14 Physical constants of organic compounds (*continued*)

No.	Name	Formula	Formula weight	Beilstein reference	Density	Refractive index	Melting point	Boiling point	Flash point	Solubility in 100 parts solvent
d259	2,2-Diethoxytriethylamine	$(C_2H_5O)_2CHCH_2N(C_2H_5)_2$	189.30	4, 309	0.850	1.4189^{20}		194–195	65	
d260	*N*,*N*-Diethylacetamide	$CH_3C(=O)N(C_2H_5)_2$	115.18	4, 110	0.925	1.4401^{20}		182–186	70	
d261	Diethyl acetamidomalonate	$C_2H_5OOCCH(NHCOCH_3)COOC_2H_5$	217.22	4^2, 891			97–98	185^{20mm}		
d262	Diethyl 1,3-acetonedicarboxylate	$C_2H_5OOCCH_2COCH_2COOC_2H_5$	202.21	3, 791	1.113	1.4385^{20}		250	86	
d263	Diethyl acetylenedicarboxylate	$C_2H_5OOCC{=}CCOOC_2H_5$	170.16	2, 803	1.063	1.4426^{20}		107^{11mm}	94	
d264	Diethyl 2-acetylglutarate	$C_2H_5OOCCH_2CH_2CH(COCH_3)COOC_2H_5$	230.26		1.071	1.4386^{20}		154^{11mm}	>112	
d265	Diethyl acetylsuccinate	$C_2H_5OOCCH_2CH(COCH_3)COOC_2H_5$	216.23	3, 801	1.081	1.4346^{20}		180–183^{50mm}	>112	
d266	Diethyl allylmalonate	$C_2H_5OOCCH(CH_2CH{=}CH_2)COOC_2H_5$	200.23	2, 776	1.015	1.4304^{20}		222–223	92	
d267	Diethylamine	$(C_2H_5)_2NH$	73.14	4, 95	0.7074_4^{20}	1.3864^{20}	−50.0	55.5	−28	misc aq, alc
d268	Diethylamine HCl	$(C_2H_5)_2NH \cdot HCl$	109.60	4, 95	1.048_4^{21}			320–330		s aq, alc, chl; i eth
d269	4-(Diethylamino)benzaldehyde	$(C_2H_5)_2NC_6H_4CHO$	177.25	14^2, 25			39–41	174^{7mm}		
d270	2-Diethylaminoethanol	$(C_2H_5)_2NCH_2CH_2OH$	117.19	4, 282	0.8800^{25}	1.4389^{25}	−70	163	48	s aq, alc, bz, eth
d271	2-(Diethylamino)ethyl-4-aminobenzoate	$H_2NC_6H_4COOCH_2CH_2N(C_2H_5)_2$	236.30	14, 424			61			0.5 aq; s alc, bz, eth
d272	2-Diethylaminoethyl chloride HCl	$ClCH_2CH_2N(C_2H_5)_2 \cdot HCl$	172.10	4^2, 618			208–210			
d273	3-(Diethylamino)phenol	$(C_2H_5)_2NC_6H_4OH$	165.24	13, 408			65–69	170^{15mm}		s aq, alc, eth

d274	3-Diethylamino-1,2-propanediol	$(C_2H_5)_2NCH_2CH(OH)CH_2OH$	147.22	4, 302	0.973^{20}_{20}	1.4602^{20}	233–235		107	s aq, alc, chl, eth
d275	1-Diethylamino-2-propanol	$(C_2H_5)_2NCH_2CH(OH)CH_3$	131.22	4^2, 737	0.889	1.4255^{20}	13.5	$55–59^{13mm}$	33	s alc
d276	4-(Diethylamino)-salicylaldehyde	$(C_2H_5)_2NC_6H_3(OH)CHO$	193.25	14, 234			62–64			
d277	*N,N*-Diethylaniline	$C_6H_5N(C_2H_5)_2$	149.24	12, 164	0.9302^{25}_{4}	1.5394^{25}	−34.4	216.3	97	1 aq; sl s alc, eth
d278	2,6-Diethylaniline	$(C_2H_5)_2C_6H_3NH_2$	149.24		0.906	1.5452^{20}	3	243	123	
d279	Diethyl azodicarboxylate	$C_2H_5OOCN{=}NCOOC_2H_5$	174.16	3, 123	1.106	1.4280^{20}		106^{13mm}	26	
d280	5,5-Diethylbarbituric acid		184.19	24^2, 279	1.220		188–192			0.7 aq; 7 alc; 1.3 chl; 3.2 eth; s acet, HOAc
d281	Diethyl benzalmalonate	$C_6H_5CH{=}C(COOC_2H_5)_2$	248.28	9, 892	1.107	1.5365^{20}		215^{30mm}	> 112	
d282	1,2-Diethylbenzene	$C_6H_4(C_2H_5)_2$	134.22	5, 426	0.8800^{20}	1.5022^{20}	−31.3	183.4	49	s alc, eth
d283	1,3-Diethylbenzene	$C_6H_4(C_2H_5)_2$	134.22	5, 426	0.8640^{20}_{4}	1.4950^{20}	−83.9	181.1	50	s alc, eth
d284	1,4-Diethylbenzene	$C_6H_4(C_2H_5)_2$	134.22	5, 426	0.8620^{20}_{4}	1.4940^{20}	−42.85	183.8	56	s alc, eth
d285	Diethyl benzylmalonate	$C_6H_5CH_2CH(COOC_2H_5)_2$	250.29	9, 869	1.064	1.4868^{20}		162^{10mm}	> 112	
d286	Diethyl bromomalonate	$BrCH(COOC_2H_5)_2$	239.07	2, 594	1.4022^{25}_{4}	1.4550^{20}	−54	233–235 d		i aq; misc alc, eth

Diethyl acetal, d251
Diethylacetic acid, e90
Diethyl 2-acetylpentanedioate, d264
Diethylaminoacetaldehyde diethyl acetal, d259
3-Diethylaminopropylamine, d330
Diethyl (*Z*)-2-butenedioate, d316

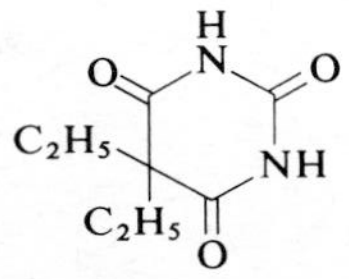

d280

TABLE 1-14 Physical constants of organic compounds (*continued*)

No.	Name	Formula	Formula weight	Beilstein reference	Density	Refractive index	Melting point	Boiling point	Flash point	Solubility in 100 parts solvent
d287	Diethyl butylmalonate	$C_4H_9CH(COOC_2H_5)_2$	216.28	2^1, 282	0.983	1.4220		235–240	93	v s alc, eth
d288	Diethylcarbamoyl chloride	$(C_2H_5)_2NCOCl$	135.59	4, 120		1.4515^{20}		187–190	75	d hot aq, hot alc
d289	Diethyl carbonate	$(C_2H_5O)_2C{=}O$	118.13	3, 5	0.9764^{20}_{4}	1.3843^{20}	−43.0	126.8	25	69 aq; misc alc, bz, eth, esters
d290	Diethyl chloromalonate	$ClCH(COOC_2H_5)_2$	194.61	2^2, 537	1.2040^{20}_{4}	1.4310^{20}		222–223		misc alc, chl, eth
d291	Diethyl chlorophosphate	$(C_2H_5O)_2P(O)Cl$	172.55	1, 332	1.194	1.4165^{20}		60^{2mm}		
d292	Diethyl chlorothiophosphate	$(C_2H_5O)_2P(S)Cl$	188.61		1.200	1.4715^{20}		45^{3mm}		
d293	Diethylcyanamide	$(C_2H_5)_2NCN$	98.15	4, 121	0.846	1.4229^{20}		186–188	69	
d293a	Diethyl cyanomethylphosphonate	$(C_2H_5O)_2P(O)CH_2CN$	177.14		1.095	1.4312^{20}		$101^{0.4mm}$	>112	
d294	*N,N*-Diethylcyclohexylamine	$C_6H_{11}N(C_2H_5)_2$	155.29	12, 6	0.850	1.4562^{20}		194–195	57	
d294a	Diethyl disulfide	$C_2H_5SSC_2H_5$	122.25	1, 347	0.998^{20}_{4}	1.5063^{20}	−101.5	154.0		sl s aq; misc alc, eth
d295	Diethyldithiocarbamic acid, Na salt	$(C_2H_5)_2NC({=}S)S^-Na^+ \cdot 3H_2O$	225.31	4^2, 613			95–99			
d296	Diethyl dithiophosphate	$(C_2H_5O)_2P(S)SH$	186.23	1, 333	1.111	1.5120^{20}		60^{1mm}		
d297	*N,N*-Diethyldodecanamide	$CH_3(CH_2)_{10}C(O)N(C_2H_5)_2$	255.45		0.847	1.4545^{20}		166^{2mm}	>112	
d298	Diethylenetriamine	$(H_2NCH_2CH_2)_2NH$	103.17	4, 255	0.9542^{20}_{20}	1.4826^{20}	−35	207.1	101	misc aq, alc, bz, eth

d299	Diethylenetriaminepentaacetic acid	$[(HOOCCH_2)_2NCH_2CH_2]_2NCH_2COOH$	393.35				220 d			
d300	Diethyl ether	$C_2H_5OC_2H_5$	74.12	1, 314	0.7134_4^{20}	1.3527^{20}	−116.3	34.6	−40	6 aq; misc alc, bz, chl
d301	Diethyl ethoxymethylenemalonate	$(C_2H_5OOC)_2C{=}CHOC_2H_5$	216.23	3, 469	1.070	1.4620^{20}		279–281	155	
d302	*N,N*-Diethylethylenediamine	$(C_2H_5)_2NCH_2CH_2NH_2$	116.21	4, 251	0.827	1.4360^{20}		145–147	30	
d303	Diethyl ethylmalonate	$C_2H_5CH(COOC_2H_5)_2$	188.22	2, 644	1.004_{20}^{20}	1.4158^{20}		$75–77^{5mm}$	88	sl s aq; v s alc, eth
d304	*N,N*-Diethylformamide	$(C_2H_5)_2NCHO$	101.15	4, 109	0.908	1.4340^{20}		176–177	60	misc aq; v s alc, eth
d305	Diethyl fumarate	$C_2H_5OOCCH{=}CHCOOC_2H_5$	172.18	2, 742	1.052_4^{20}	1.4406^{20}	1–2	218–219	91	
d306	Diethyl 3,4-furandicarboxylate	$(C_2H_5OOC)_2C_4H_2O$	212.20		1.140	1.4717^{20}		155^{13mm}	82	
d307	Diethyl glutarate	$C_2H_5OOCCH_2CH_2CH_2COOC_2H_5$	188.22	2, 633	1.022	1.4240^{20}	−23.8	237	96	0.9 aq; v s alc; s eth
d308	2,4-Diethyl-2,6-heptadienal	$H_2C{=}CHCH_2CH(C_2H_5)CH{=}C(C_2H_5)CHO$	166.27					91^{12mm}		
d309	Diethyl heptanedioate	$C_2H_5OOC(CH_2)_5COOC_2H_5$	216.28	2, 671	0.9945^{20}	1.4280^{20}	−24	192^{100mm}	> 112	i aq; s alc, eth
d310	2,4-Diethyl-1-heptanol	$CH_3CH_2CH_2CH(C_2H_5)CH_2CH(C_2H_5)CH_2OH$	172.31					109^{12mm}		

Diethyl carbitol, b177
Diethylene dioxide, d648
Diethylene glycol, b182
Diethylene glycol dibutyl ether, b152
Diethylene glycol diethyl ether, b177
Diethylene glycol dimethyl ether, b192
Diethylene glycol monobutyl ether, b414
Diethylene glycol monoethyl ether, e36
Diethylene glycol monoethyl ether acetate, e37
Diethylene glycol monomethyl ether, m66
Diethyleneimide oxide, m451
Diethyl ethoxycarbonylmethylphosphonate, t288
N,N-Diethylethanamine, t269
N,N-Diethylethanolamine, d270
Di-2-ethylhexyl adipate, d312
Di-2-ethylhexyl sebacate, d311

TABLE 1-14 Physical constants of organic compounds (*continued*)

No.	Name	Formula	Formula weight	Beilstein reference	Density	Refractive index	Melting point	Boiling point	Flash point	Solubility in 100 parts solvent
d311	Di-(2-ethylhexyl) decanedioate	$C_4H_9CH(C_2H_5)CH_2OOC(CH_2)_8COOCH_2CH(C_2H_5)C_4H_9$	426.68		0.912_4^{25}	1.451^{25}		256^{5mm}	227	i aq; s alc, bz, acet
d312	Di-(2-ethylhexyl) hexanedioate	$C_4H_9CH(C_2H_5)CH_2OOC(CH_2)_4COOCH_2CH(C_2H_5)C_4H_9$	370.57		0.925_{25}^{25}	1.4474^{20}		214^{5mm}	193	s alc, eth, acet; i aq
d313	Di-(2-ethylhexyl) *o*-phthalate	$C_6H_4[COOCH_2CH(C_2H_5)C_4H_9]$	390.56		0.981_{25}^{25}	1.4853^{20}	−50	384	207	
d314	Diethyl hydrogen phosphonate	$(C_2H_5O)_2P(O)H$	138.10	1, 330	1.079_4^{20}	1.4076^{20}		$50–51^{2mm}$	90	s aq (hyd), alc, eth
d315	*N,N*-Diethylhydroxylamine	$(C_2H_5)_2NOH$	89.14	4, 536	1.867	1.4195^{20}	−25	125–130	45	
d316	Diethyl maleate	$C_2H_5OOCCH{=}CHCOOC_2H_5$	172.18	2, 751	1.0687^{20}	1.4400^{20}	−8.8	225.3	93	1.4 aq; s alc, eth
d317	Diethyl malonate	$C_2H_5OOCCH_2COOC_2H_5$	160.17	2, 573	1.0550	1.4136^{20}	−48.9	199.3	100	2.7 aq; misc alc, eth
d318	Diethylmalonic acid	$HOOCC(C_2H_5)_2COOH$	160.17	2, 686			127	d 170–180		v s aq, alc, eth
d319	*N,N*-Diethyl-3-methylbenzamide	$CH_3C_6H_4C(=O)N(C_2H_5)_2$	191.27	9^2, 325	0.996_4^{20}	1.5212^{20}		111^{1mm}		s aq; v s alc, bz, eth
d320	Diethyl methylmalonate	$C_2H_5OOCCH(CH_3)COOC_2H_5$	174.20	2, 629	1.018_4^{20}	1.4130^{20}		198	76	sl s aq; v s alc, eth
d321	Diethyl 2-methyl-2′-oxosuccinate	$C_2H_5OOCCH(CH_3)C(=O)C(=O)OC_2H_5$	202.21	3, 794	1.073	1.4313^{20}		138^{23mm}	>112	
d322	Diethyl methylsuccinate	$C_2H_5OOCCH_2CH(CH_3)COOC_2H_5$	188.22	2, 639	1.012	1.4199^{20}		217–218		
d323	*N,N*-Diethyl-4-nitrosoaniline	$C_6H_4(NO)N(C_2H_5)_2$	178.24	12, 684			82–84			
d324	Diethyl octanedioate	$C_2H_5OOC(CH_2)_6COOC_2H_5$	230.30	2, 693	0.9822_4^{20}	1.4323^{20}	5.9	282	>112	i aq; s alc, eth

d325	Diethyl oxalate	$C_2H_5OOCCOOC_2H_5$	146.14	2, 535	1.0785_4^{20}	1.4102	−40.6	185.4	75	3.6aq (gradual d); misc alc, eth eth
d326	Diethyl oxydiformate	$[C_2H_5OC(=O)]_2O$	162.14		1.12_4^{20}	1.3980^{20}		93^{18mm}	69	s alc, esters, ketones
d327	N^1,N^1-Diethyl-1,4-pentanediamine	$CH_3CH(NH_2)(CH_2)_3$-$N(C_2H_5)_2$	158.29		0.817	1.4429^{20}		200^{753mm}	68	s aq, alc, eth
d328	Diethyl phenylmalonate	$C_6H_5CH(COOC_2H_5)_2$	236.27	9, 854	1.0950_4^{20}	1.4913^{20}	16	170^{14mm}	> 112	i aq; s alc
d329	Diethyl *o*-phthalate	$C_6H_4(COOC_2H_5)_2$	222.24	9,798	1.232_4^{14}	1.5049^{14}	−3	295	140	i aq; misc alc, eth
d330	*N,N*-Diethyl-1,3-propanediamine	$(C_2H_5)_2NCH_2CH_2$-CH_2NH_2	130.24		0.826	1.4416^{20}		159	58	
d331	2,2-Diethyl-1,3-propanediol	$(C_2H_5)_2C(CH_2OH)_2$	132.20		1.052^{20}	1.4574^{25}	61.3	125^{10mm}		25 aq; v s alc, eth
d332	Diethyl propylmalonate	$C_2H_5OOCCH(C_3H_7)$-$COOC_2H_5$	202.25	2, 657	0.987	1.4185^{20}		221–222	91	
d333	1,1-Diethyl-2-propynylamine	$HC\equiv CC(C_2H_5)_2NH_2$	111.19		0.828	1.4409^{20}		71^{90mm}	21	
d334	*N,N*-Diethyl-3-pyridine-carboxamide	$C_5H_4N{-}C(=O)N$-$(C_2H_5)_2$	178.24	22^2, 34	1.060_4^{25}	1.5240^{20}	24–26	296–300	> 112	
d335	Diethyl succinate	$C_2H_5OOCCH_2CH_2$-$COOC_2H_5$	174.20	2, 609	1.040_4^{20}	1.4200^{20}	−21	217.7	110	i aq; misc alc, eth
d336	Diethyl sulfate	$(C_2H_5O)_2SO_2$	154.18	1, 327	1.172_4^{25}	1.4004^{20}	−25	209 d	78	v aq; misc alc, eth
d337	Diethyl sulfide	$(C_2H_5)_2S$	90.19	1, 344	0.8367_4^{20}	1.4430^{20}	−103.9	92.1	−9	i aq; misc alc, eth
d338	Diethyl sulfite	$(C_2H_5O)_2S(O)$	138.19	1, 325	1.077_4^{25}			157.7		s aq(d), alc

Diethyl ketone, p42
N,N-Diethylnicotinamide, d334
O,O-Diethyl *O-p*-nitrophenyl phosphorothioate, p3
Diethyl 3-oxoglutarate, d262
Diethyl 2-pentenedioate, d307
Diethyl phosphite, d314
Diethyl phosphorochloridate, d291
Diethyl phosphorochloridothionate, d292
Diethyl pimelate, d309
Diethyl propanedioate, d317
1,1-Diethylpropargylamine, d333
Diethyl pyrocarbonate, d326
Diethyl suberate, d324

TABLE 1-14 Physical constants of organic compounds (*continued*)

No.	Name	Formula	Formula weight	Beilstein reference	Density	Refractive index	Melting point	Boiling point	Flash point	Solubility in 100 parts solvent
d339	(+)-Diethyl L-tartrate	$[-CH(OH)COOC_2H_5]_2$	206.19	3, 512	1.204_4^{20}	1.4459^{20}	17	280	93	sl s aq; misc alc, eth
d340	(−)-Diethyl D-tartrate	$[-CH(OH)COOC_2H_5]_2$	206.19	3^1, 181	1.205	1.4467^{20}		162^{19mm}	93	
d341	Diethyl 3,3′-thiopropionate	$S(CH_2CH_2COOC_2H_5)_2$	234.32		1.095	1.4655^{20}		121^{2mm}		
d342	*N,N*-Diethyl-*m*-toluamide	$CH_3C_6H_4C(=O)N(C_2H_5)_2$	191.27	9^2, 325	0.996	1.5212^{20}		111^{1mm}		
d343	*N,N*-Diethyl-1,1,1-trimethylsilylamine	$(C_2H_5)_2NSi(CH_3)_3$	145.32		0.767	1.4081^{20}		125–126	10	
d344	Diethylzinc	$(C_2H_5)_2Zn$	123.49		1.2065^{20}		−28	118		
d345	1,4-Difluorobenzene	$C_6H_4F_2$	114.09	5, 199	1.1701^{20}	1.4415^{20}	−23.7	88.9	2	
d346	1,1-Difluoroethane	CH_3CHF_2	66.05		0.909^{21}		−117	−24.7		0.32 aq
d347	1,1-Difluorotetrachloroethane	Cl_3CCClF_2	203.83	1, 86	1.649	1.413	41	91	none	sl s alc; v s eth
d348	1,2-Difluorotetrachloroethane	FCl_2CCCl_2F	203.83	1^3, 365	1.6447_4^{25}	1.413^{25}	23.8	203.8		i aq; s alc, eth
d350	Dihexylamine	$(C_6H_{13})_2NH$	185.36	4^1, 384	0.795	1.4320^{20}		192–195	95	s alc, eth
d351	Dihexyl ether	$(C_6H_{13})_2O$	186.34	1^3, 1656	0.7936_4^{20}	1.4204^{20}		226.2	77	i aq; s eth
d352	9,10-Dihydroanthracene		180.25	5, 641	0.880		108–110	312		i aq; s alc, bz, eth
d353	(+)-Dihydrocarvone		152.24	7^3, 337	0.929^{19}	1.4718^{20}		221–222	81	
d354	Dihydrocoumarin		148.16	17, 315	1.169^{18}	1.5563^{20}	25	272		sl s alc, eth; s chl
d355	10,11-Dihydro-5*H*-dibenzo-[*a,d*]cyclohepten-5-one		208.26		1.156	1.6332^{20}	32–34	$148^{0.3mm}$	>112	
d356	3,4-Dihydro-2-ethoxy-2*H*-pyran		128.17		0.957	1.4394^{20}		42^{16mm}	24	

d357	2,3-Dihydrofuran		70.09	17^3, 141	0.927	1.4239^{20}		54–55	<1	
d358	Dihydrolinalool	$(CH_3)_2C{=}CHCH_2CH_2$-$C(OH)(CH_3)CH_2CH_3$	156.27		0.925^{25}	1.433^{20}			178	
d359	3,4-Dihydro-1(2*H*)-6-methoxynaphthalenone		176.22	9^2, 889			80	171^{11mm}		
d360	3,4-Dihydro-2-methoxy-2*H*-pyran		114.14			1.4425^{20}			16	
d361	2,3-Dihydro-2-methylbenzofuran		134.18	17^1, 23	1.061	1.5308^{20}		197–198	62	
d362	5,6-Dihydro-4-methyl-2*H*-pyran		98.15	17^3, 160	0.912	1.4495^{20}		117–118	21	

Diglycine, i10
Diglycol, b182
Diglycolic acid, o61
Diglyme, b192

Dihydroanisoles, m61, m62
6,7-Dihydro-5*H*-cyclopenta[*b*]pyridine, c361
10,11-Dihydro-5*H*-dibenz[*b*,*f*]acepine, i12
2,5-Dihydro-2,5-dimethoxyfuran, d437

3,7-Dihydro-3,7-dimethyl-1*H*-pyridine-2,6-dione, t140
2,3-Dihydroindene, i13
Dihydromyrcenol, m306

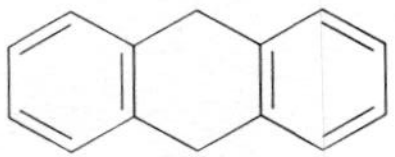

d352 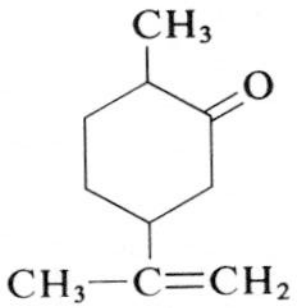 d353 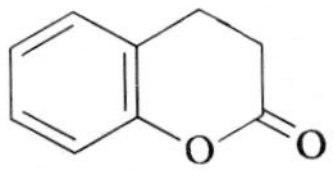 d354 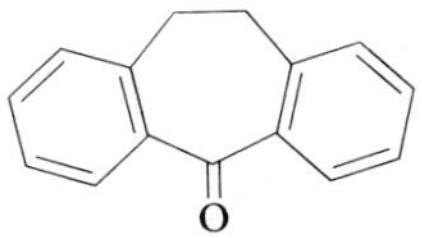 d355 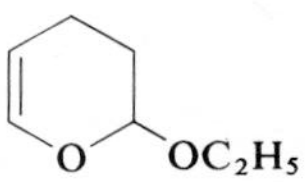 d356  d357 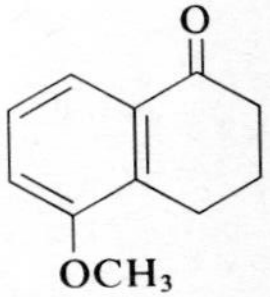 d359

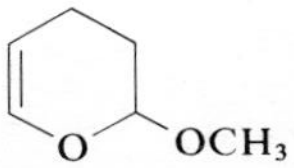

d360 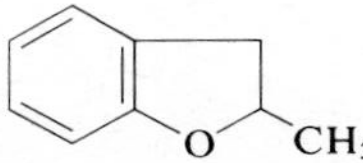 d361 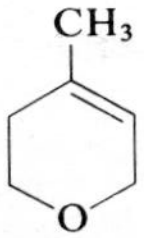 d362

TABLE 1-14 Physical constants of organic compounds (*continued*)

No.	Name	Formula	Formula weight	Beilstein reference	Density	Refractive index	Melting point	Boiling point	Flash point	Solubility in 100 parts solvent
d363	3,4-Dihydro-1(2*H*)-naphthalenone		146.19	7, 370	1.099	1.5685^{20}	5–6	116^{6mm}	>112	
d364	Dihydropyran		84.12		0.922^{19}_{15}	1.4410^{20}	−70	86	−15	s aq, alc
d365	5,6-Dihydro-2*H*-pyran-3-carbaldehyde		112.13		1.100	1.4980^{20}		78^{12mm}	77	
d366	3,4-Dihydro-2*H*-pyran-2-carboxylic acid, Na salt		150.11				242–244			
d367	Dihydroterpineol		256.27		0.907^{25}	1.4670^{20}			88	
d368	5,6-Dihydro-2,4,4,6-tetramethyl-4*H*-1,3-oxazine		141.21		0.886	1.4410^{20}		48^{17mm}		
d369	2,5-Dihydrothiophene-1,1-dioxide		118.15				64–66		>112	s aq, alc, bz, chl, eth
d370	1,2-Dihydro-2,2,4-trimethylquinoline		173.26		0.934	1.5895^{20}		$90^{0.02mm}$	101	
d371	2′,4′,-Dihydroxyacetophenone	$(HO)_2C_6H_3C(=O)CH_3$	152.15	8, 266	1.180		145–147			s warm alc, pyr, HOAc; i bz, chl, eth
d372	1,2-Dihydroxyanthraquinone		240.21	8, 439			287–289	430		s alc, bz, chl, HOAc
d373	1,4-Dihydroxyanthraquinone		240.21	8, 450			196			s alc, alk, eth
d374	1,8-Dihydroxyanthraquinone		240.21	8, 458			193–197	subl		0.005 alc; 0.2 eth; s chl
d375	2,6-Dihydroxyanthraquinone		240.21	8, 463			360 d			sl s aq, alc

No.	Name	Formula	Formula weight	Beilstein	Density	Refractive index	Melting point	Boiling point	Flash point	Solubility
d376	2,4-Dihydroxybenzaldehyde	$(HO)_2C_6H_3CHO$	138.12	8, 241			135–136	226^{22mm}		v s aq, alc, chl, eth
d377	3,4-Dihydroxybenzaldehyde	$(HO)_2C_6H_3CHO$	138.12	8, 246			153			5 aq; 79 hot alc; v s eth
d378	1,2-Dihydroxybenzene	$C_6H_4(OH)_2$	110.11	6, 759	1.344^{4}		104–106	245.5	137	43 aq; s alc, bz, chl, eth; v s pyr, alk
d379	1,3-Dihydroxybenzene	$C_6H_4(OH)_2$	110.11	6^2, 802	1.272^{15}		109–110	276	171	110 aq; 110 alc; v s eth, glyc; sl s chl
d380	1,4-Dihydroxybenzene	$C_6H_4(OH)_2$	110.11	6, 836	1.332^{15}		170–171	285–287		7 aq; v s alc, eth

4,5-Dihydro-2-(phenylmethyl)-1*H*-imidazole, b103

Dihydroresorcinol, c323

3,7-Dihydro-1,3,7-trimethyl-1*H*-purine-2,6-dione, c1

1,3-Dihydroxyacetone, d398

2,2′-Dihydroxy-2,2′-biindan-1,2′,3,3′-tetrone, h85

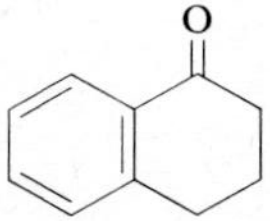

d363 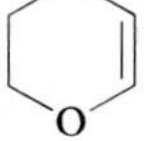 d364 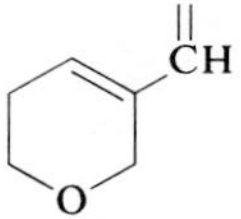 d365 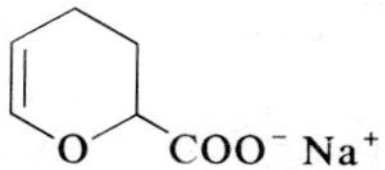 d366  d367 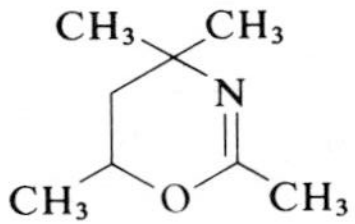 d368 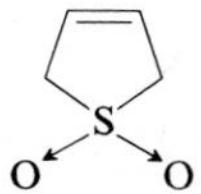 d369 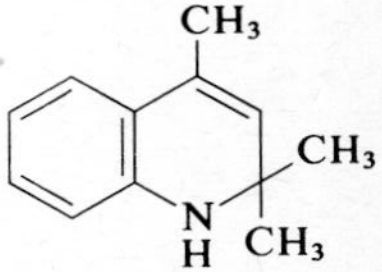 d370

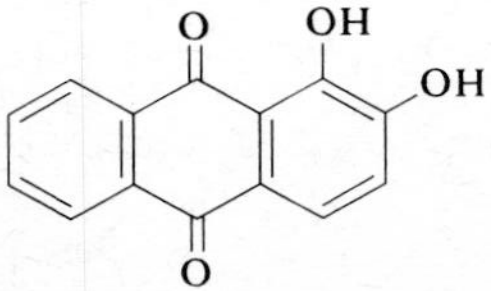

d372 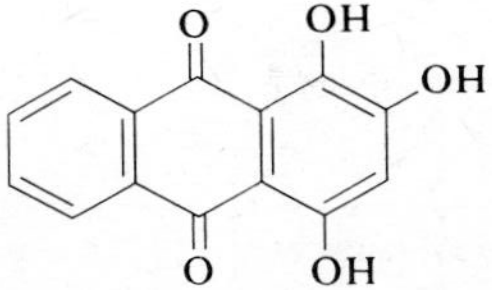 d373 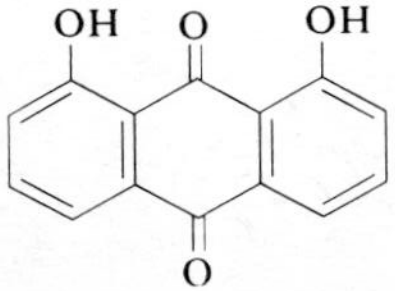 d374 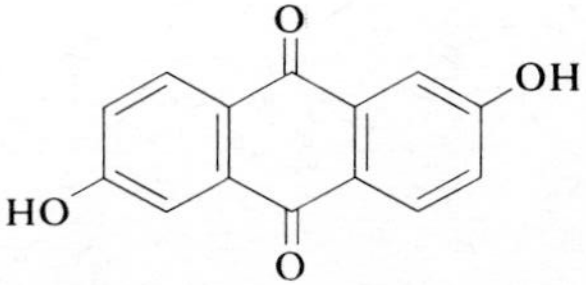 d375

TABLE 1-14 Physical constants of organic compounds (*continued*)

No.	Name	Formula	Formula weight	Beilstein reference	Density	Refractive index	Melting point	Boiling point	Flash point	Solubility in 100 parts solvent
d381	1,3-Dihydroxybenzene monoacetate	$HOC_6H_4OOCCH_3$	152.15	6, 816		1.5350^{20}		283	>112	
d382	2,5-Dihydroxy-*p*-benzenedisulfonic acid, K salt	$(HO)_2C_6H_2(SO_3^-K^+)_2$	346.43	11, 300			>300			v s aq
d383	2,5-Dihydroxy-benzenesulfonic acid, K salt	$(HO)_2C_6H_3SO_3^-K^+$	228.27	11, 300			251 d			v s aq
d384	2,4-Dihydroxybenzoic acid	$(HO)_2C_6H_3COOH$	154.12	10, 377			213			s hot aq, alc, eth
d385	2,5-Dihydroxybenzoic acid	$(HO)_2C_6H_3COOH$	154.12	10, 384			199–200			0.5 aq; s alc, eth
d386	3,5-Dihydroxybenzoic acid	$(HO)_2C_6H_3COOH$	154.12	10, 404			236 d			sl s aq; s alc, eth
d387	2,4-Dihydroxybenzo-phenone	$(HO)_2C_6H_3C(=O)C_6H_5$	214.22	8, 312			144–145			v s alc, eth, HOAc
d388	2,2′-Dihydroxybi-phenyl	$HOC_6H_4C_6H_4OH$	186.21	6, 989			110	315		s alc, bz, eth; sl s aq
d389	4,6-Dihydroxy-2-mercaptopyrimidine		144.15	24, 476			236			
d390	1,2-Dihydroxy-4-methylbenzene	$(HO)_2C_6H_3CH_3$	124.14	6, 878	1.129_4^{74}	1.5425^{74}	67–69	251		v s aq, alc, eth
d391	1,3-Dihydroxy-2-methylbenzene	$(HO)_2C_6H_3CH_3$	124.14	6, 878			115–118	264		s aq, alc, bz, eth
d392	2,4-Dihydroxy-6-methylpyrimidine		126.12	24, 342			318 d			
d393	1,5-Dihydroxynaph-thalene	$C_{10}H_6(OH)_2$	160.17	6, 980			259 d			sl s aq; s alc; v s eth

d394	1,7-Dihydroxynaphthalene	$C_{10}H_6(OH)_2$	160.17	6, 981			177–180			v s alc, eth
d395	2,3-Dihydroxynaphthalene	$C_{10}H_6(OH)_2$	160.17	6, 982			162–164			v s alc, eth
d396	2,7-Dihydroxynaphthalene	$C_{10}H_6(OH)_2$	160.17	6, 985			187 d			sl s aq; v s alc, eth
d397	4,5-Dihydroxynaphthalene-2,7-disulfonic acid	$(HO)_2C_{10}H_4(SO_3H)_2$	296.26	11, 307						v s aq; i alc, eth
d398	1,3-Dihydroxy-2-propanone	$HOCH_2C(=O)CH_2OH$	90.08	1, 846			65–71			v s aq, alc, acet, eth
d399	2,3-Dihydroxypropionaldehyde	$HOCH_2CHOCHO$	90.08	1, 845	1.455^{18}_{18}		145	$140^{0.8mm}$	> 112	3 aq; i bz, PE
d400	7-(2,3-Dihydroxypropyl)theophylline		254.25				158			33 aq; 2 alc; 1 chl
d410	3,6-Dihydroxypyridazine		112.09	24, 312			d 260			sl s hot alc; s hot aq
d402	2,3-Dihydroxypyridine	$(HO)_2C_5H_3N$	111.10	21^2, 107			245 d			

2,2′-Dihydroxydiethylamine, d245
N,N-Di(hydroxyethyl)aminoacetic acid, b183
2,2-Dihydroxy-1,3-indandione, i16
2,2-Dihydroxymethyl-1-butanol, e159
1,8-Dihydroxynaphthalene-3,6-disulfonic acid, d397
Dihydroxypropanes, p197, p198

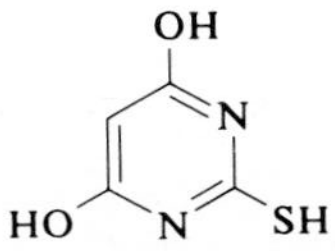

d389

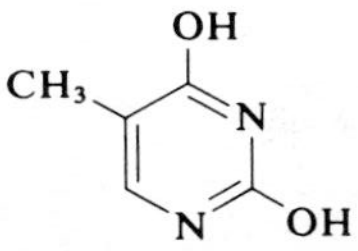

d392

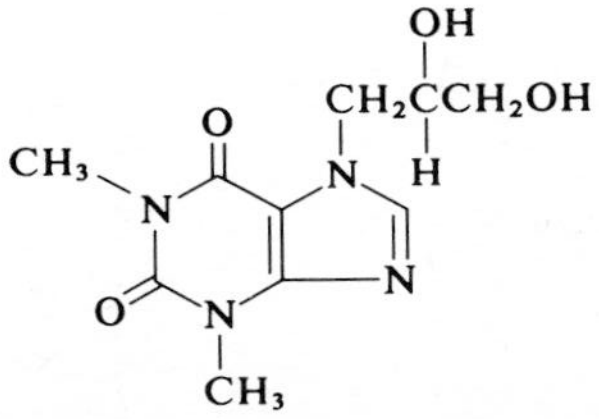

d400

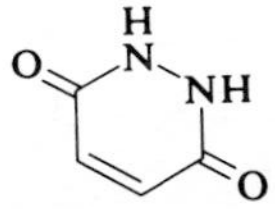

d401

TABLE 1-14 Physical constants of organic compounds (*continued*)

No.	Name	Formula	Formula weight	Beilstein reference	Density	Refractive index	Melting point	Boiling point	Flash point	Solubility in 100 parts solvent
d403	1,4-Diiodobenzene	$C_6H_4I_2$	329.91	5, 227			131–133	285		sl s alc; v s eth
d404	1,2-Diiodoethane	ICH_2CH_2I	281.86	1, 99	2.132^{10}		81	200		sl s aq; s alc, eth
d405	Diiodomethane	CH_2I_2	267.84	1, 71	3.3254^{20}_{4}	1.7411^{20}	5.6	181		0.12 aq; misc alc, bz, eth, PE
d406	1,3-Diiodopropane	$ICH_2CH_2CH_2I$	295.88	1, 115	2.5755^{20}_{4}	1.6423^{20}	−13	222		i aq; s chl, eth
d407	Diisobutylamine	$[(CH_3)_2CHCH_2]_2NH$	129.25	4, 166	0.740	1.4081^{20}	−77	137–139	29	s alc, acet, eth, EtAc
d408	Diisobutyl ether	$[(CH_3)_2CHCH_2]_2O$	130.22		0.761^{15}			122–124		i aq; misc alc, eth
d409	Diisobutyl hexanedioate	$[(CH_3)_2CHCH_2OOC\text{-}CH_2CH_2\text{—}]_2$	258.36		0.950^{25}_{25}				160	
d410	Diisobutyl *o*-phthalate	$C_6H_4[COOCH_2CH\text{-}(CH_3)_2]_2$	278.35		1.038^{25}_{25}				174	
d411	1,6-Diisocyanatohexane	$OCN(CH_2)_6NCO$	168.20	4^2, 711	1.040	1.4525^{20}		255	140	
d412	Diisopropylamine	$[(CH_3)_2CH]_2NH$	101.19	4, 154	0.7169^{20}	1.3924^{20}	−96.3	83.5	−6	11 aq
d413	2-(Diisopropylamino)-ethanol	$[(CH_3)_2CH]_2NCH_2\text{-}CH_2OH$	145.25	4^1, 430	0.826	1.4417^{20}		187–192	57	
d414	2,6-Diisopropylaniline	$[(CH_3)_2CH]_2C_6H_3NH_2$	177.29	12, 168	0.940	1.5332^{20}	−45	257	123	
d415	1,3-Diisopropylbenzene	$C_6H_4[CH(CH_3)_2]_2$	162.28	5, 447	0.856^{20}_{4}	1.4980^{20}	−63	203	76	misc alc, bz, eth, acet
d416	1,4-Diisopropylbenzene	$C_6H_4[CH(CH_3)_2]_2$	162.28	5^2, 339	0.857^{20}_{4}	1.4889^{20}		203	76	misc alc, bz, eth, acet
d417	Diisopropylcyanamide	$[(CH_3)_2CH]_2NCN$	126.20	4^3, 279	0.839	1.4270^{20}		93^{25mm}	78	
d418	Diisopropyl ether	$[(CH_3)_2CH]_2O$	102.17	1, 362	0.7258^{20}_{4}	1.3689^{20}	−86.9	68.4	−12	1.2 aq; misc alc, bz, chl, eth
d419	*N,N*-Diisopropylethylamine	$[(CH_3)_2CH]_2NC_2H_5$	129.25		0.742	1.4133^{20}		127	10	

d420	2,6-Diisopropylphenol	$[(CH_3)_2CH]_2C_6H_3OH$	178.28	6^1, 272	0.962	1.5140^{20}	18	256	> 112	
d421	Diisopropyl phosphite	$[(CH_3)_2CHO]_2P(O)H$	166.16	1, 363	0.997	1.4070^{20}		$72–75^{10mm}$	> 112	
d422	(+)-Diisopropyl L-tartrate	$[-CH(OH)COOCH(CH_3)_2]_2$	234.25	3, 517	1.114	1.4387^{20}		152^{12mm}	109	
d423	Diketene		84.07		1.073	1.4330^{20}		127	33	
d424	Dilauryl phosphite	$[CH_3(CH_2)_{11}O]_2P(O)H$	418.64		0.946	1.4520^{20}			> 112	
d425	*threo*-1,4-Dimercapto-2,3-butanediol	$HSCH_2CH(OH)CH(OH)CH_2SH$	154.25				42–43			v s aq, alc, chl, eth
d426	2,3-Dimercapto-1-propanol	$HSCH_2CH(SH)CH_2OH$	124.22		1.2385_4^{25}	1.5720^{25}		120^{15mm}	> 112	8 aq(d); s alc, eth
d427	3′,4′-Dimethoxyacetophenone	$(CH_3O)_2C_6H_3COCH_3$	180.20	8^2, 298			49–51	286–288		sl s aq, alc, eth
d428	2,4-Dimethoxyaniline	$(CH_3O)_2C_6H_3NH_2$	153.18	13, 784			34–37			s alc, bz, eth
d429	2,5-Dimethoxyaniline	$(CH_3O)_2C_6H_3NH_2$	153.18	13, 788			80–82	270 sl d		s aq, alc
d430	3,4-Dimethoxyaniline	$(CH_3O)_2C_6H_3NH_2$	153.18	13, 780			88	176^{22mm}		s hot eth
d431	3,4-Dimethoxybenzaldehyde	$(CH_3O)_2C_6H_3CHO$	166.18	8, 255			42–43	281		v s alc, eth
d432	1,2-Dimethoxybenzene	$C_6H_4(OCH_3)_2$	138.17	6, 771	1.0819^{25}	1.5232^{25}	22.5	206.3	87	sl s aq; s alc, eth
d433	1,3-Dimethoxybenzene	$C_6H_4(OCH_3)_2$	138.17	6, 813	1.055	1.5240	−55	$85–87^{7mm}$	87	s alc, bz, eth; sl s aq
d434	1,4-Dimethoxybenzene	$C_6H_4(OCH_3)_2$	138.17	6, 843	1.036_4^{68}		55–60	213		s alc; v s bz, eth

Dihydroxytoluene, d391
3,5-Diiodosalicylic acid, h111
Diisobutyl adipate, d409
Diisobutylene, t365
Diisobutyl ketone, d533
Diisopropyl ketone, d580
Diisopropylmethane, d574
Dimedone, d510
1,1-Dimethoxytrimethylamine, d525

$CH_2=C$—O
CH_2—C=O

d423

TABLE 1-14 Physical constants of organic compounds (*continued*)

No.	Name	Formula	Formula weight	Beilstein reference	Density	Refractive index	Melting point	Boiling point	Flash point	Solubility in 100 parts solvent
d435	3,4-Dimethoxybenzoic acid	$(CH_3O)_2C_6H_3COOH$	182.18	10^1, 188			180–181			0.047 aq; v s alc, eth
d436	1,1-Dimethoxy-3-butanone	$(CH_3O)_2CHCH_2COCH_3$	132.16		0.993	1.4150^{20}			49	
d437	2,5-Dimethoxy-2,5-dihydrofuran		130.14		1.073	1.4339^{20}		160–162	47	
d438	Dimethoxydiphenylsilane	$(C_6H_5)_2Si(OCH_3)_2$	244.4		1.0771_4^{20}	1.5447^{20}		161^{15mm}		
d439	1,1-Dimethoxyethane	$CH_3CH(OCH_3)_2$	90.12	1, 603	0.8502^{20}		−113	64.5		s aq, alc, chl, eth
d440	1,2-Dimethoxyethane	$CH_3OCH_2CH_2OCH_3$	90.12	1, 467	0.8629_4^{20}	1.3796^{20}	−68	85.2	1	misc aq, alc; s PE
d441	(2,2-Dimethoxy)-ethylamine	$H_2NCH_2CH(OCH_3)_2$	105.14	4^2, 758	0.965	1.4170^{20}		135^{95mm}	53	
d442	Dimethoxymethane	$CH_2(OCH_3)_2$	76.10	1, 574	0.8601_{20}^{20}	1.3534^{20}	−104.8	42.3	−17	32 aq
d443	1,1-Dimethoxy-2-methylaminoethane	$CH_3NHCH_2CH(OCH_3)_2$	119.16	4^2, 759	0.928	1.4115^{20}		140	29	
d444	Dimethoxymethylphenylsilane	$(CH_3O)_2Si(CH_3)C_6H_5$	182.3		0.993_4^{20}	1.469^{20}		199–200		
d445	1,2-Dimethoxy-4-nitrobenzene	$(CH_3O)_2C_6H_3NO_2$	183.16	6, 789	1.1888_4^{133}		95–98	230^{17mm}		v s alc, eth; s chl
d446	2,5-Dimethoxy-4′-nitrostilbene	$(CH_3O)_2C_6H_3CH{=}CH$-$C_6H_4NO_2$	285.30	6^2, 987			117–119			
d447	2,6-Dimethoxyphenol	$(CH_3O)_2C_6H_3OH$	154.17	6, 1081			53–56	261		s alc, alk; v s eth
d448	(3,4-Dimethoxy)-phenylacetic acid	$(CH_3O)_2C_6H_3CH_2COOH$	196.20	10, 409			96–98			s aq; v s alc, eth
d449	(3,4-Dimethoxy)-phenylacetonitrile	$(CH_3O)_2C_6H_3CH_2CN$	177.20	10^1, 198			62–63	171–178^{10mm}		
d450	2,2-Dimethoxy-2-phenylacetophenone	$C_6H_5C(O)C(OCH_3)_2$-C_6H_5	256.30				67–70			

d451	1,1-Dimethoxy-2-phenylethane	$C_6H_5CH_2CH(OCH_3)_2$	166.22	7, 293	1.004	1.4950^{20}		221	83	
d452	β-(3,4-Dimethoxy-phenyl-ethylamine	$(CH_3O)_2C_6H_3CH_2$-CH_2NH_2	181.24	13, 800	1.074	1.5464^{20}		188^{15mm}		
d453	2,2-Dimethoxy-propane	$(CH_3)_2C(OCH_3)_2$	104.15	1, 648	0.847	1.3780		83	4	
d454	1,1-Dimethoxy-2-propanone	$CH_3C(O)CH(OCH_3)_2$	118.13	1^1, 395	0.976	1.3978^{20}		143–147	37	
d455	3,3-Dimethoxy-1-propene	$(CH_3O)_2CHCH{=}CH_2$	102.13	1^1, 378	0.862	1.3954^{20}		89–90		
d456	1,2-Dimethoxy-4-propenylbenzene	$CH_3CH{=}CHC_6H_3$-$(OCH_3)_2$	178.23	6, 956	1.055	1.5680^{20}		262–264	> 112	
d457	2,6-Dimethoxy-pyridine	$(CH_3O)_2C_5H_3N$	139.15		1.053	1.5029^{20}		178–180	61	
d458	2,5-Dimethoxytetra-hydrofuran	$(CH_3O)_2C_4H_6O$	132.16		1.020	1.4180^{20}		145–147	35	
d459	*N,N*-Dimethylacet-amide	$CH_3C(O)N(CH_3)_2$	87.12	4, 59	0.9366^{25}	1.4356^{25}	−20	165.5	70	misc aq, alc, bz, eth
d460	Dimethyl 1,3-acetone-dicarboxylate	$[CH_3OOCCH_2]_2C{=}O$	174.15	3, 790	1.185	1.4434^{20}		150^{25mm}	> 112	
d461	Dimethyl acetylene-dicarboxylate	$CH_3OOCC{\equiv}CCOOCH_3$	142.11	2, 803	1.156	1.4470^{20}		$95–98^{19mm}$	86	
d463	Dimethylamine	$(CH_3)_2NH$	45.09	4, 39	0.680^0_4		−92.2	6.9		v s aq; s alc, eth

Dimethyl acetal, d439
Dimethylacetic acid, m393

2,3-Dimethylacrylic acids, m161, m162
3,3-Dimethylacrylic acid, m163

3,3-Dimethylallene, m148
Dimethylaminoacetaldehyde diethyl acetal, d249

d437

TABLE 1-14 Physical constants of organic compounds (*continued*)

No.	Name	Formula	Formula weight	Beilstein reference	Density	Refractive index	Melting point	Boiling point	Flash point	Solubility in 100 parts solvent
d464	4-Dimethylamino-benzaldehyde	$(CH_3)_2NC_6H_4CHO$	149.19	14, 31			74	176^{17mm}		s alc, chl, eth, HOAc
d465	*p*-(Dimethylamino)-benzenesulfonic acid, Na salt	$(CH_3)_2NC_6H_4SO_3^-Na^+$	223.23	14^3, 2023			> 300			
d466	4-Dimethylamino-benzoic acid	$(CH_3)_2NC_6H_4COOH$	165.19	14, 426			241 d			s alc; sl s eth
d467	2-(Dimethylamino)-ethanol	$(CH_3)_2NCH_2CH_2OH$	89.14	4, 276	0.8876^{20}_4	1.4294^{20}		135	40	misc aq, alc, eth
d468	2-(Dimethylamino)-ethyl benzoate	$C_6H_5COOCH_2CH_2$-$N(CH_3)_2$	193.26		1.014	1.5077^{20}		155^{20mm}		
d469	2-Dimethylamino-ethyl chloride HCl	$(CH_3)_2NCH_2CH_2Cl \cdot HCl$	144.05	4, 133			205–208			
d470	2-(Dimethylamino)-ethyl methacrylate	$H_2C{=}C(CH_3)COOCH_2$-$CH_2N(CH_3)_2$	157.22	4^3, 649	0.933	1.4391^{20}		182–192	70	
d471	4-Dimethylamino-3-methyl-2-butanone	$(CH_3)_2NCH_2CH(CH_3)$-$COCH_3$	129.20	4^1, 452	0.841	1.4250^{20}		73^{35mm}	38	
d472	3-Dimethylamino-phenol	$(CH_3)_2NC_6H_4OH$	137.18	13, 405	1.5895^{26}		82–84	265–268		v s alc, bz, eth, acet
d473	3-(Dimethylamino)-1,2-propanediol	$(CH_3)_2NCH_2CH(OH)$-CH_2OH	119.16	4, 302	1.004	1.4609^{20}		216–217	105	s aq, alc, chl, eth
d474	1-Dimethylamino-2-propanol	$CH_3CH(OH)CH_2N(CH_3)_2$	103.17		0.837	1.4193^{20}		121–127	35	
d475	3-Dimethylamino-1-propanol	$(CH_3)_2NCH_2CH_2CH_2OH$	103.17	4^1, 433	0.872	1.4360^{20}		163–164	36	
d476	3-(Dimethylamino)-propionitrile	$(CH_3)_2NCH_2CH_2CN$	98.15	4^3, 1265	0.870	1.4258^{20}	−43	171^{750mm}	62	
d477	3-Dimethyl-aminopropyl chloride HCl	$(CH_3)_2NCH_2CH_2$-$CH_2Cl \cdot HCl$	158.07	4, 148			141–144		35	

d478	4-(Dimethylamino)-pyridine	$(CH_3)_2N(C_5H_4N)$	122.17	22^2, 341			108–110			v s aq, alc, bz, chl
d479	*N,N*-Dimethyl-aniline	$C_6H_5N(CH_3)_2$	121.18	12, 141	0.9559_4^{20}	1.5584^{20}	2.5	194.2	62	v s alc, chl, eth
d480	2,3-Dimethyl-aniline	$(CH_3)_2C_6H_3NH_2$	121.18	12, 1101	0.9931^{20}	1.5685^{20}	2.5	221–222	96	sl s aq; s alc, eth
d481	2,4-Dimethyl-aniline	$(CH_3)_2C_6H_3NH_2$	121.18	12, 1111	0.980_4^{20}	1.5586^{20}		218	90	s alc, bz, eth
d482	2,5-Dimethyl-aniline	$(CH_3)_2C_6H_3NH_2$	121.18	12, 1135	0.9790_4^{21}	1.5592^{20}	11.5	218	93	al s aq; s alc, eth
d483	2,6-Dimethyl-aniline	$(CH_3)_2C_6H_3NH_2$	121.18	12, 1107	0.984^{20}	1.5601^{20}	10–12	216	91	sl s aq; s alc, eth
d484	3,4-Dimethylaniline	$(CH_3)_2C_6H_3NH_2$	121.18	12, 1103	1.076^{18}		49–51	226		sl s aq; s alc
d485	3,5-Dimethylaniline	$(CH_3)_2C_6H_3NH_2$	121.18	12, 1131	0.972_4^{20}	1.5578^{20}		104^{14mm}	93	sl s aq; s alc
d486	Dimethylarsinic acid	$(CH_3)_2As(O)OH$	137.99				195–196			v s alc; 200 aq; i eth
d487	3,4-Dimethylbenzoic acid	$(CH_3)_2C_6H_3COOH$	150.18	9^2, 353			165–167	subl		s alc, bz
d488	2,5-Dimethylbenzo-nitrile	$(CH_3)_2C_6H_3CN$	131.18	9, 535	0.957	1.5284^{20}	13–14	223^{730mm}	92	
d489	*N,N*-Dimethylbenzyl-amine	$C_6H_5CH_2N(CH_3)_2$	135.21	12, 1019	0.900	1.5011^{20}	−75	183	54	
d490	2,3-Dimethyl-1,3-butadiene	$H_2C{=}C(CH_3)C(CH_3){=}CH_2$	82.15	1^3, 991	0.7222_4^{25}	1.4362^{25}	−76.0	69.2	<1	
d491	2,2-Dimethylbutane	$CH_3CH_2C(CH_3)_3$	86.18	1, 150	0.6492^{20}	1.3688^{20}	−99.9	49.7	−28	
d492	2,3-Dimethylbutane	$(CH_3)_2CHCH(CH_3)_2$	86.18	1, 151	0.6616^{20}	1.3750^{20}	−128.5	58.0	−28	
d493	2,3-Dimethyl-2-3-butanediol	$(CH_3)_2C(OH)C(OH)(CH_3)_2$	118.18	1, 487			41.1	174.4		v s hot aq, alc, eth

3-Dimethylaminopropylamine, d595
Dimethylanisoles, d549, d550
2,4-Dimethyl-3-azapentane, d412
Dimethylbenzenes, x4, x5, x6
6,6-Dimethylbicyclo[3.1.1]hept-2-ene-2-ethanol, n105
Dimethyl (*Z*)-butenedioate, d546

TABLE 1-14 Physical constants of organic compounds (*continued*)

No.	Name	Formula	Formula weight	Beilstein reference	Density	Refractive index	Melting point	Boiling point	Flash point	Solubility in 100 parts solvent
d494	2,2-Dimethyl-1-butanol	$CH_3CH_2C(CH_3)_2CH_2OH$	102.18	1^3, 1675	0.8286_4^{20}	1.4208^{20}	<-15	136.8		sl s aq; s alc, eth
d495	2,3-Dimethyl-1-butanol	$(CH_3)_2CHCH(CH_3)$-CH_2OH	102.18	1^3, 1677	0.8300_4^{20}	1.4205^{20}		149		s alc, eth
d496	2,3-Dimethyl-2-butanol	$(CH_3)_2CHC(CH_3)_2OH$	102.18	1, 413	0.8236_4^{20}	1.4176^{20}	−10.6	118.7	29	s aq; misc alc, eth
d497	3,3-Dimethyl-1-butanol	$(CH_3)_3CCH_2CH_2OH$	102.18	1^3, 1677	0.8147^{20}	1.4120^{20}	−60	143	47	s alc, eth
d498	3,3-Dimethyl-2-butanol	$(CH_3)_3CCH(OH)CH_3$	102.18	1, 412	0.8185_4^{20}	1.4151^{20}	5.3	120.4	28	s alc; misc eth
d499	3,3-Dimethyl-2-butanone	$(CH_3)_3CCOCH_3$	100.16	1, 694	0.7250_{25}^{25}	1.3939^{25}	−52.5	106.2	23	2.5 aq; s alc, eth
d500	2,3-Dimethyl-2-butene	$(CH_3)_2C{=}C(CH_3)_2$	84.16	1, 218	0.7081_4^{20}	1.4124^{20}	−74.3	73.2	−16	s alc, eth
d501	3,3-Dimethyl-1-butene	$(CH_3)_3CCH{=}CH_2$	84.16	1, 217	0.6531_4^{20}	1.3762^{20}	−115.2	41.3	−28	
d502	3,3-Dimethylbutyric acid	$(CH_3)_3CCH_2COOH$	116.16	2, 337	0.9124_4^{20}	1.4100^{20}	6–7	190	88	s alc, eth
d503	Dimethylcadmium	$(CH_3)_2Cd$	142.48		1.9846_4^{17}	1.5488	−4.5	105.5	≥150 explodes	d aq; s PE
d504	Dimethylcarbamyl chloride	$(CH_3)_2NCOCl$	107.54	4, 73	1.168	1.4540^{20}	−33	168	68	
d505	Dimethyl carbonate	$(CH_3O)_2C{=}O$	90.08	3, 4	1.065_4^{17}	1.3682^{20}	0.5	90–91	18	i aq; misc alc, eth
d506	Dimethyl chlorothiophosphate	$(CH_3O)_2P(S)Cl$	160.56	1^1, 143	1.322	1.4819^{20}		67^{16m}		
d507	Dimethylcyanamide	$(CH_3)_2NCN$	70.09	4, 74	0.867	1.4100^{20}		161–163	58	

d508	*cis*-1,2-Dimethylcyclohexane	$(CH_3)_2C_6H_{10}$	112.22	5, 36	0.7692^{20}_{4}	1.4335^{20}	−49.9	129.7	15	i aq; s alc, bz
d509	*trans*-1,2-Dimethylcyclohexane	$(CH_3)_2C_6H_{10}$	112.22	5, 36	0.7772^{20}_{0}	1.4273^{20}	−88.2	123.4	15	i aq; s alc, eth
d510	5,5-Dimethyl-1,3-cyclohexanedione		140.18	7, 559			d 149			0.4 aq; s alc, bz
d511	2,3-Dimethylcyclohexanol	$(CH_3)_2C_6H_9OH$	128.22		0.934	1.4653^{20}			65	
d512	2,6-Dimethylcyclohexanone		126.20	7, 23	0.925	1.4460^{20}		175	51	i aq; s alc, eth
d513	2,3-Dimethylcyclohexylamine	$(CH_3)_2C_6H_9NH_2$	127.23		0.835	1.4595^{20}		160	51	
d514	Dimethyl decanedioate	$CH_3OOC(CH_2)_8COOCH_3$	230.30	2, 719	0.983^{30}_{20}	1.4335^{28}	23	144^{5mm}		i aq; s alc, eth
d515	5,7-Dimethyl-3,5,9-decatrien-2-one	$H_2C{=}CHCH_2CH(CH_3)$-$CH{=}C(CH_3)CH{=}CH$-$COCH_3$	178.28					$79^{0.05mm}$		
d516	Dimethyl 2,5-dioxo-1,4-cyclohexanedicarboxylate		228.20	10, 894			155–157			

Dimethyl 2-butynedioate, d461
Dimethyl Cellosolve, d440
Dimethylchlorosilane, c93

(*Z*)-2-Dimethylcrotonic acid, m162
Dimethyl 1,4-cyclohexanedione-2,5-dicarboxylic acid, d516

Dimethyl diphenyl sulfone 4,4′-dicarboxylate, s27

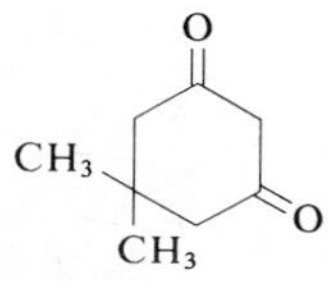

d510

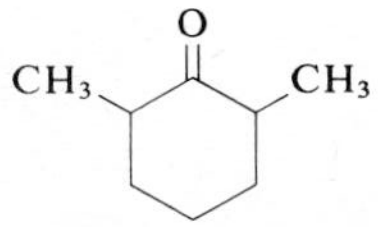

d512

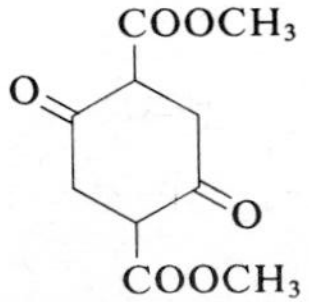

d516

TABLE 1-14 Physical constants of organic compounds (*continued*)

No.	Name	Formula	Formula weight	Beilstein reference	Density	Refractive index	Melting point	Boiling point	Flash point	Solubility in 100 parts solvent
d517	2,3-Dimethyl-1,3-dioxolane-4-methanol		132.16		1.064_4^{20}	1.4383^{20}		188–189	80	misc aq, alc, eth
d518	Dimethyl disulfide	CH_3SSCH_3	94.20	1, 291	1.046	1.5253^{20}	−84.7	109.8	24	i aq; misc alc, eth
d519	Dimethyldithiocarbamic acid dihydrate, Na salt	$(CH_3)_2NCSS^-Na^+\cdot 2H_2O$	179.24	4, 75						
d520	Dimethyl ether	$(CH_3)_2O$	46.07	1, 281	0.661^{20}		−141.5	−24.9	−41	35% aq (5 atm); 15% bz; 11.8% acet
d521	Dimethylethoxyphenylsilane	$C_2H_5O(C_6H_5)Si(CH_3)_2$	180.3		0.9263_4^{20}	1.4799^{20}		93^{35mm}		
d522	*N,N*-Dimethylethylamine	$C_2H_5N(CH_3)_2$	73.14	4, 94	0.675	1.3720^{20}	−140	36–38	−36	
d523	*N,N*-Dimethylethylenediamine	$(CH_3)_2NCH_2CH_2NH_2$	88.15	4^2, 690	0.803	1.4260^{20}			23	
d524	*N,N*-Dimethylformamide	$(CH_3)_2NCHO$	73.10	4, 58	0.9445_4^{25}	1.4282^{25}	−60.4	153.0	57	misc aq, alc, bz, eth
d525	*N,N*-Dimethylformamide dimethyl acetal	$(CH_3)_2NCH(OCH_3)_2$	119.16		0.897	1.3972^{20}		103^{720mm}	7	
d526	Dimethyl fumarate	$CH_3OOCCH{=}CHCOOCH_3$	144.13	2, 741	1.045^{106}		105	193		sl s alc, eth
d527	2,5-Dimethylfuran	$(CH_3)_2(C_4H_2O)$	96.13	17, 41	0.9000_4^{20}	1.4414^{20}	−62	93	<1	i aq; misc alc, eth
d528	Dimethylglyoxime	$CH_3C({=}NOH)C({=}NOH)CH_3$	116.12	1, 772			238–240			s alc, acet, eth, pyr
d529	2,4-Dimethyl-2,6-heptadienal	$H_2C{=}CHCH_2CH(CH_3)CH{=}C(CH_3)CHO$	138.21					47^{2mm}		
d530	2,4-Dimethyl-2,6-heptadien-1-ol	$H_2C{=}CHCH_2CH(CH_3)CH{=}C(CH_3)CH_2OH$	140.23					88^{10mm}		

d531	2,6-Dimethyl-2,5-heptadien-4-one	$(CH_3)_2C{=}CHC({=}O)$-$CH{=}C(CH_3)_2$	138.21	1, 751	0.885_4^{20}	1.4968^{21}	28	198–199	79	sl s aq; s alc, eth
d532	Dimethyl heptanedioate	$CH_3OOC(CH_2)_5COOCH_3$	188.22	2^1, 281	1.0625_4^{20}	1.4314^{20}	−21	122^{11mm}	> 112	s alc
d533	2,6-Dimethyl-4-heptanone	$[(CH_3)_2CHCH_2]_2C{=}O$	142.24	1, 710	0.806_{20}^{20}	1.4114^{20}	−41.5	168.1	48	0.06 aq; misc alc, bz, chl, eth
d534	2,5-Dimethyl-2,4-hexadiene	$(CH_3)_2C{=}CHCH{=}C$-$(CH_3)_2$	110.20	1, 259	0.7636_4^{20}	1.4741^{20}	12–14	132–134	29	i aq; s alc, eth
d535	2,5-Dimethylhexane	$(CH_3)_2CHCH_2CH_2$-$CH(CH_3)_2$	114.24	1^3, 283	0.6936_4^{20}	1.3925^{20}	−91.2	109.1	26	i aq; sl s alc; s eth
d536	2,5-Dimethyl-2,5-hexanediamine	$[(CH_3)_2C(NH_2)CH_2{-}]_2$	144.26		0.832	1.4459^{20}		64^{8mm}	62	
d536a	Dimethyl hexanedioate	$CH_3OOC(CH_2)_4COOCH_3$	174.20	1, 652	1.0600_4^{20}	1.4285^{20}	8	112^{10mm}	107	i aq; s alc, eth
d537	2,5-Dimethyl-2,5-hexanediol	$[(CH_3)_2C(OH)CH_2{-}]_2$	146.23	1, 492			86–90	214–215	126	
d538	1,5-Dimethylhexylamine	$(CH_3)_2CH(CH_2)_3$-$CH(NH_2)CH_3$	129.25		0.767	1.4209^{20}		154–156	48	
d539	2,5-Dimethyl-3-hexyne-2,5-diol	$(CH_3)_2C(OH)C{\equiv}CC(OH)(CH_3)_2$	142.20	1, 501			94–95	205–206		

Dimethyleneimine, e134
Dimethylene oxide, e132

N,N-Dimethylethanolamine, d467
Dimethyl glutarate, d576

Dimethylglutaric acids, d577, d578

d517

TABLE 1-14 **Physical constants of organic compounds (*continued*)**

No.	Name	Formula	Formula weight	Beilstein reference	Density	Refractive index	Melting point	Boiling point	Flash point	Solubility in 100 parts solvent
d540	5,5-Dimethylhydantoin		128.13	24, 289			176–178			v s aq, alc, bz, chl, eth, acet
d541	1,1-Dimethylhydrazine	$(CH_3)_2NNH_2$	60.10	4, 547	0.791_4^{22}	1.4075^{20}	−58	63.9	1	misc aq, alc, eth, PE
d542	1,2-Dimethylhydrazine	$CH_3NHNHCH_3$	60.10	4, 547	0.8274_4^{20}	1.4209^{20}		81	flammable	misc aq, alc, eth, PE
d543	Dimethyl hydrogen phosphonate	$(CH_3O)_2P(=O)H$	110.05	1, 285	1.200_4^{20}	1.4009^{20}		170–171	96	s aq(hyd); misc alc, acet, eth
d544	1,2-Dimethylimidazole		96.13	23, 66	1.084		29–30	204	92	
d545	1,3-Dimethyl-2-imidazolidinone		114.15		1.044	1.4720^{20}		108^{17mm}	80	
d545a	Dimethylketene	$(CH_3)_2C=C=O$	70.09	1, 731			−97.5	34		d aq, alc; s eth
d546	Dimethyl maleate	$CH_3OOCCH=CHCOOCH_3$	144.13	2, 751	1.1513^{20}	1.4422^{20}	−17.5	200.4		8.7 aq
d547	Dimethyl malonate	$CH_3OOCCH_2COOCH_3$	132.12	2, 572	1.154_4^{20}	1.4135^{20}	−62	180–181	90	sl s aq; misc alc, eth
d548	Dimethylmercury	$(CH_3)_2Hg$	230.66	4, 678	3.1874^{20}	1.5452^{20}		92^{740mm}		i aq; s alc, eth
d549	3,4-Dimethyl-1-methoxybenzene	$(CH_3)_2C_6H_3OCH_3$	136.19	6, 481	0.9744_4^{14}	1.5198^{14}		200		i aq; s alc, bz, eth
d550	3,5-Dimethyl-1-methoxybenzene	$(CH_3)_2C_6H_3OCH_3$	136.19	6, 493	0.9627_4^{15}	1.5107^{15}		193	65	i aq; s alc, bz, eth
d551	*N,N*-Dimethylmethyleneammonium iodide	$H_2C=N(CH_3)_2^+I^-$	185.01	4^4, 153			219 d			
d552	Dimethyl methylenesuccinate	$CH_3OOCCH_2C(=CH_2)$-$COOCH_3$	158.15	2, 762	1.1241_4^{18}	1.4442^{20}	38	208		s alc, eth
d553	Dimethyl methylphosphonate	$(CH_3O)_2P(O)CH_3$	124.08	4^1, 572	1.145	1.4130^{20}		181	43	

d554	Dimethyl methylsuccinate	$CH_3OOCCH_2CH(CH_3)COOCH_3$	160.17	2^3, 1696	1.076	1.4200^{20}		196	83	
d555	2,6-Dimethylmorpholine		115.18		0.9346^{20}	1.4470^{20}	−85	147	48	misc aq, alc, bz
d556	2,3-Dimethylnaphthalene	$(CH_3)_2C_{10}H_6$	156.23	5, 571	1.008_4^{20}		102–104	269		sl s alc; s bz, eth
d557	2,6-Dimethylnaphthalene	$(CH_3)_2C_{10}H_6$	156.23	5, 570	1.142_4^{0}		110.2	262		i aq; sl s alc
d558	1,2-Dimethyl-3-nitrobenzene	$(CH_3)_2C_6H_3NO_2$	151.17	5, 367	1.129	1.5434^{20}	7–9	245	107	i aq; s alc
d559	1,2-Dimethyl-4-nitrobenzene	$(CH_3)_2C_6H_3NO_2$	151.17	5, 368	1.139		29–31	143^{20mm}		i aq; s alc
d560	1,3-Dimethyl-2-nitrobenzene	$(CH_3)_2C_6H_3NO_2$	151.17	5, 378	1.112	1.5220^{20}	14–16	225^{744mm}	87	i aq; s alc
d561	1,3-Dimethyl-4-nitrobenzene	$(CH_3)_2C_6H_3NO_2$	151.17	5, 378	1.117	1.5497^{20}	2	237–239	107	s alc, bz, chl, eth
d562	*N,N*-Dimethyl-4-nitrosoaniline	$(CH_3)_2NC_6H_4NO$	150.18	12, 677			86	flammable solid		i aq; s alc, eth
d563	Dimethyl 2-nitro-1,4-phthalate	$O_2NC_6H_3(COOCH_3)_2$	239.18	9, 826			72–75			

Dimethyl isophthalate, d591
1,4*a*-Dimethyl-7-isopropyl-1,2,3,4,4*a*,9,10,10*a*-octahydro-1-phenanthrenemethylamine, d20
Dimethyl itaconate, d552
2,2-Dimethyl-3-methylenenorbornane, c2
6,6-Dimethyl-2-methylenenorpinene, p179

d540

d544

d545

d555

TABLE 1-14 Physical constants of organic compounds (*continued*)

No.	Name	Formula	Formula weight	Beilstein reference	Density	Refractive index	Melting point	Boiling point	Flash point	Solubility in 100 parts solvent
d564	*cis*-3,7-Dimethyl-2,6-octadienal		152.24		0.8888_4^{20}	1.4898^{20}		229	101	misc alc, eth, glyc
d565	*trans*-3,7-Dimethyl-2,6-octadienal		152.24		0.8869_4^{20}	1.4869^{20}		229	101	misc alc, eth, glyc
d566	3,7-Dimethyl-2,6-octadienenitrile		149.24		0.853	1.4753^{20}			> 112	
d567	Dimethyl octanedioate	$CH_3OOC(CH_2)_6COOCH_3$	202.25	2, 693	1.0210_4^{20}	1.4325^{20}	−4.8	268		i aq; s alc
d568	Dimethyl oxalate	$CH_3OOCCOOCH_3$	118.08	2, 534	1.148^{54}	1.379^{80}	50–54	163.5	75	6 aq; s alc, eth
d569	N^1-(4,5-Dimethyl-oxazol-2-yl)-sulfanilamide		267.31				193–194			s aq, acids, alk
d570	*N*-(1,1-Dimethyl-3-oxobutyl)acryl-amide	$CH_2{=}CHC({=}O)NHC(CH_3)_2CH_2COCH_3$	169.23				57–58	120^{8mm}		
d571	2,3-Dimethylpentanal	$CH_3CH_2CH(CH_3)CH(CH_3)CHO$	114.19		0.832	1.4132^{20}			58	
d572	2,2-Dimethylpentane	$CH_3CH_2CH_2C(CH_3)_3$	100.21	1, 157	0.674_4^{20}	1.3824^{20}	−123.8	79.2	15	i aq; s alc, eth
d573	2,3-Dimethylpentane	$CH_3CH_2CH(CH_3)CH(CH_3)_2$	100.21	1^2, 120	0.6951_4^{20}	1.3920^{20}	glass	89.8	−6	i aq; s alc, eth
d574	2,4-Dimethylpentane	$(CH_3)_2CHCH_2CH(CH_3)_2$	100.21		0.6727_4^{20}	1.3815^{20}	−119.2	80.5		s alc, eth
d575	3,3-Dimethylpentane	$CH_3CH_2C(CH_3)_2CH_2CH_3$	100.21	1, 158	0.6933_4^{20}	1.3905^{20}	−134.4	86.1	−6	i aq; s alc, eth
d576	Dimethyl pentanedioate	$CH_3OOC(CH_2)_3COOCH_3$	160.17	2, 633	1.0934_4^{15}	1.4234^{20}		$94–95^{13mm}$	102	v s alc, eth
d577	2,2-Dimethyl-pentanedioic acid	$HOOCC(CH_3)_2CH_2CH_2COOH$	160.17	2, 676			83–85			v s aq, alc, chl
d578	3,3-Dimethyl-pentanedioic acid	$(CH_3)_2C(CH_2COOH)_2$	160.17	2, 684			100–103			v s aq, alc, eth

d579	2,4-Dimethyl-3-pentanol	$(CH_3)_2CHCH(OH)$-$CH(CH_3)_2$	116.20	1, 417	0.829_4^{20}	1.4254^{20}	< 70	140	37	sl s aq; s alc, eth
580	2,4-Dimethyl-3-pentanone	$(CH_3)_2CHC(=O)$-$CH(CH_3)_2$	114.19	1, 703	0.8062_4^{20}	1.3986^{20}	−80	124	15	misc alc, eth; s bz
d581	2,3-Dimethylphenol	$(CH_3)_2C_6H_3OH$	122.17	6, 480		1.5420^{20}	75	218		v s alc, bz, chl, eth
d582	2,4-Dimethylphenol	$(CH_3)_2C_6H_3OH$	122.17	6, 486	1.0276_4^{14}	1.5390^{20}	27	210–212	> 112	v s alc, bz, chl, eth
d583	2,5-Dimethylphenol	$(CH_3)_2C_6H_3OH$	122.17	6, 494	0.965^{80}		74.5	211.5		v s alc, bz, chl, eth
d584	2,6-Dimethylphenol	$(CH_3)_2C_6H_3OH$	122.17	6, 485			49.0	203	73	v s alc, bz, chl, eth
d585	3,4-Dimethylphenol	$(CH_3)_2C_6H_3OH$	122.17	6, 480	1.064_4^{28}		62.5	225		v s alc, bz, chl, eth
d586	3,5-Dimethylphenol	$(CH_3)_2C_6H_3OH$	122.17	6, 492	1.008_4^{28}		64–68	219.5		v s alc, bz, chl, eth
d587	Dimethylphenylchlorosilane	$(CH_3)_2Si(Cl)C_6H_5$	170.7		1.032_4^{20}	1.508^{20}		192–193		
d588	4,5-Dimethyl-*o*-phenylenediamine	$(CH_3)_2C_6H_2(NH_2)_2$	136.20	13, 179			127–129			
d589	Dimethylphenylsilane	$(CH_3)_2Si(H)C_6H_5$	136.3		0.8891_4^{20}	1.4995^{20}		156–157		

3,7-Dimethyl-6-octen-1-ol, c275
Dimethylolpropionic acid, b185
Dimethyl 3-oxoglutarate, d460
1,5-Dimethyl-2-phenyl-4-aminopyrazolone, a113
2,3-Dimethyl-1-phenyl-3-pyrazolin-5-one, a314

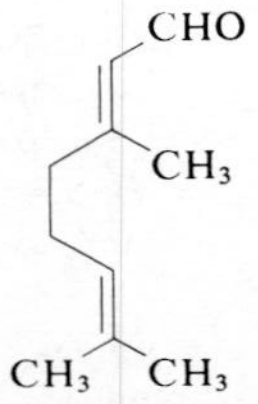

d564

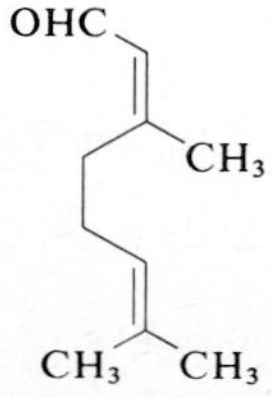

d565

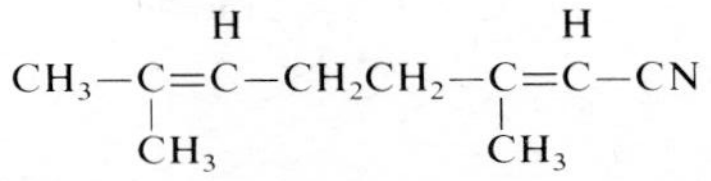

$CH_3-C(CH_3)=CH-CH_2CH_2-C(CH_3)=CH-CN$

d566

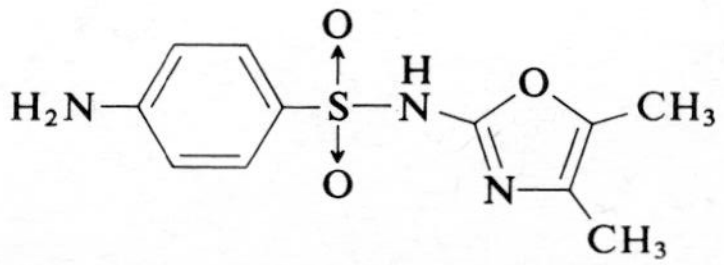

d569

TABLE 1-14 Physical constants of organic compounds (*continued*)

No.	Name	Formula	Formula weight	Beilstein reference	Density	Refractive index	Melting point	Boiling point	Flash point	Solubility in 100 parts solvent
d590	Dimethyl *o*-phthalate	$C_6H_4(COOCH_3)_2$	194.19	9, 797	1.1940^{20}_{20}	1.515^{21}	5.5	283.7	146	0.4 aq; misc alc, chl, eth; i PE
d591	Dimethyl *m*-phthalate	$C_6H_4(COOCH_3)_2$	194.19	9, 834	1.194^{20}_{4}	1.5168^{20}	67–68	282		i aq
d592	Dimethyl *p*-phthalate									alc; s eth
d593	2,6-Dimethyl-piperidine		113.20	20, 108	0.840	1.4394^{20}		127	11	
d594	2,2-Dimethylpropane	$(CH_3)_4C$	72.15		0.613^{0}	1.3476^{6}	−16.6	9.5		
d595	*N,N*-Dimethyl-1,3-propanediamine	$(CH_3)_2N(CH_2)_3NH_2$	102.18		0.812	1.4350^{20}		123	35	
d596	2,2-Dimethyl-1,3-propanediol	$(CH_3)_2C(CH_2OH)_2$	104.15	1, 483	1.11^{25}		127–128	208–210		180 aq; 12 bz; 60 acet; v s alc, eth
d597	2,2-Dimethyl-1-propanol	$(CH_3)_3CCH_2OH$	88.15	1, 406	0.812^{20}_{4}		52–54	113.1	36	3.6 aq; misc alc, eth
d598	2,2-Dimethylpropion-aldehyde	$(CH_3)_3CCHO$	186.25		0.793	1.3794^{20}	6	74^{730mm}	<1	
d599	2,2-Dimethyl propionamide	$(CH_3)_3CC(O)NH_2$	101.15	2, 320			154–157	212		
d600	2,2-Dimethyl-propionic acid	$(CH_3)_3CCOOH$	102.13	2, 319	0.905^{50}	1.3931^{37}	35.5	163.8	63	2.5 aq; v s alc, eth
d601	2,2-Dimethylpro-pionic anhydride	$[(CH_3)_3CC(O)]_2O$	186.25	2, 320	0.918	1.4092^{20}		193	57	
d602	2,2-Dimethyl-propionyl chloride	$(CH_3)_3CCOCl$	120.58	2, 320	0.979	1.4120^{20}		105–106	<1	d aq, alc; v s eth
d603	1,1-Dimethyl-propylamine	$CH_3CH_2C(CH_3)_2NH_2$	87.17	4, 179	0.731^{25}_{4}	1.3996^{20}	−105	77	65	misc aq, alc, eth

d604	1,1-Dimethyl-2-propynylamine	$HC\equiv CC(CH_3)_2NH_2$	83.13		0.790	1.4235^{20}		79–80	< 1	
d605	3,5-Dimethylpyrazole		96.13	23, 74			108	218		s aq; v s bz, eth
d606	2,4-Dimethylpyridine	$(CH_3)_2(C_5H_3N)$	107.16	20, 244	0.927_4^{25}	1.4991^{20}	< −60	158.3	37	17 aq; v s alc, bz, eth
d607	2,6-Dimethylpyridine	$(CH_3)_2(C_5H_3N)$	107.16	20, 244	0.9200_4^{25}	1.4956^{25}	−6.0	143–144	33	43 aq^{45}; s alc, eth
d608	3,4-Dimethylpyridine	$(CH_3)_2(C_5H_3N)$	107.16	20, 246	0.954_4^{25}	1.5100^{25}	−12	164	53	sl s aq; s alc, eth
d609	3,5-Dimethylpyridine	$(CH_3)_2(C_5H_3N)$	107.16	20, 246	0.939_4^{25}	1.5033^{25}	−9	170	53	s aq, alc, eth
d610	Dimethyl succinate	$CH_3OOCCH_2CH_2COOCH_3$	146.14	2, 609	1.202_4^{18}	1.4190^{20}	19.5	195–200	85	0.83 aq; 2.9 alc
d611	Dimethylsulfamoyl chloride	$(CH_3)_2NSO_2Cl$	143.59	4, 84	1.337	1.4518^{20}		114^{75mm}		
d612	Dimethyl sulfate	$(CH_3O)_2SO_2$	126.13	1, 283	1.3322_4^{20}	1.3874^{20}	−31.8	188 d	83	2.8 aq(hyd); s acet, bz, diox, eth
d613	Dimethyl sulfide	$(CH_3)_2S$	62.13	1, 288	0.846_4^{21}	1.4354^{20}	−98.3	37.3	−36	2 aq; s alc, eth
d614	Dimethyl sulfite	$(CH_3O)_2SO$	110.13	1, 282	1.294	1.4083^{20}		126–127	30	
d615	Dimethyl sulfone	$(CH_3)_2SO_2$	94.13	1, 289			109	238	143	v s aq, alc, acet
d616	Dimethyl sulfoxide	$(CH_3)_2SO$	78.13	1, 289	1.100_4^{20}	1.4783^{20}	18.5	189.0	95	s alc, acet, bz, chl
d617	Dimethyl-d_6 sulfoxide	$(CD_3)_2SO$	84.18		1.18	1.4758^{20}		55^{5mm}	95	

Dimethyl phosphite, d543
Dimethyl pimelate, d532
Dimethyl propanedioate, d547
1,1-Dimethylpropargylamine, d604
N′-(4,6-Dimethyl-2-pyrimidinyl)sulfanilamide, s22
Dimethyl sebacate, d514
Dimethyl suberate, d567

d593

d605

TABLE 1-14 Physical constants of organic compounds (*continued*)

No.	Name	Formula	Formula weight	Beilstein reference	Density	Refractive index	Melting point	Boiling point	Flash point	Solubility in 100 parts solvent
d618	(+)-Dimethyl L-tartrate	$CH_3OOCCH(OH)CH(OH)COOCH_3$	178.14	3, 510	1.328_4^{20}		48–50	163^{23mm}		s aq; 200 alc[15]; v s bz
d619	Dimethyltelluride	$(CH_3)_2Te$	157.68	1, 291			−10	91–92		d aq; v s alc; i eth
d620	2,5-Dimethyltetrahydrofuran	$(CH_3)_2(C_4H_6O)$	100.16	17, 14	0.833	1.4041		90–92	26	
d621	Dimethyl 3,3′-thiodipropionate	$(CH_3OOCCH_2CH_2)_2S$	206.26		1.198	1.4740^{20}		148^{18mm}	> 112	
d622	*N,N*-Dimethylthioformamide	$(CH_3)_2NC(S)H$	89.16	4, 70	1.047	1.5757^{20}		58^{1mm}	99	
d623	*N,N′*-Dimethylthiourea	$(CH_3NH)_2C{=}S$	104.18	4, 70			60–62			v s aq, alc, acet
d624	*N,N*-Dimethyl-*p*-toluidine	$CH_3C_6H_4N(CH_3)_2$	135.21	12, 902	0.937	1.5458^{20}		211	83	
d625	1,3-Dimethylurea	$(CH_3NH)_2C{=}O$	88.11	4, 65			101–104	268–270		v s aq, alc; i eth
d626	Dimethylzinc	$(CH_3)_2Zn$	95.45		1.386_4^{11}		−40	46	ignites in air	misc bz, PE; s eth
d627	2,4-Dinitroaniline	$(O_2N)_2C_6H_3NH_2$	183.12	12, 747	1.615^{14}		188			i aq; 0.75 alc
d628	1,3-Dinitrobenzene	$C_6H_4(NO_2)_2$	168.11	5, 258	1.575_4^{18}		89–90	300–303		0.05 aq; 2.7 alc; v s bz, chl, EtAc
d629	2,4-Dinitrobenzenesulfenyl chloride	$(O_2N)_2C_6H_3SCl$	234.62	6^2, 316			96			s bz, HOAc; d alc
d630	3,4-Dinitrobenzoic acid	$(O_2N)_2C_6H_3COOH$	212.12	9, 413			166	subl		0.7 aq; v s alc, eth
d631	3,5-Dinitrobenzoic acid	$(O_2N)_2C_6H_3COOH$	212.12	9, 413			207			1.9 hot aq; v s alc; sl s bz, eth
d632	3,5-Dinitrobenzoyl chloride	$(O_2N)_2C_6H_3COCl$	230.56	9, 414			69.5	196^{11mm}		d aq, alc; s eth
d633	2,6-Dinitro-*p*-cresol	$(O_2N)_2C_6H_2(OH)CH_3$	198.13	6, 414			77–79 (anhyd)			

d633a	4,6-Dinitro-*o*-cresol	$(O_2N)_2C_6H_2(OH)CH_3$	198.13	6, 368			87.5			v s alc, acet, eth, alk
d634	1,1-Dinitroethane	$CH_3CH(NO_2)_2$	120.07	1, 102	1.3503_{24}^{24}			185–186		s alc, eth
d635	2,4-Dinitro-1-fluorobenzene	$FC_6H_3(NO_2)_2$	186.10	5, 262		1.5690^{20}	26	178^{25mm}	> 112	s bz, eth, glyc
d636	1,5-Dinitro-naphthalene	$C_{10}H_6(NO_2)_2$	218.17	5, 558			216–217	subl		s bz; v s eth; sl s alc
d637	2,4-Dinitrophenol	$(O_2N)_2C_6H_3OH$	184.11	6, 251	1.683		112–114			s alc, bz; 16 EtAc; 36 acet; 5 chl; 20 pyr
d638	2,4-Dinitrophenyl-acetic acid	$(O_2N)_2C_6H_3CH_2COOH$	226.15	9, 459			169–175			s alc, eth
d639	2,4-Dinitrophenyl-hydrazine	$(O_2N)_2C_6H_3NHNH_2$	198.14	15, 489			~200	flammable solid		sl s aq, alc; s acid
d640	3,5-Dinitrosalicyclic acid	$(O_2N)_2C_6H_2(OH)COOH$	228.12	10, 122			169–172			s aq; v s alc, eth
d641	2,4-Dinitrotoluene	$CH_3C_6H_3(NO_2)_2$	182.14	5, 339	1.321^{71}	1.442	64.66	300 sl d		1.2 alc; 9 eth
d642	2,6-Dinitrotoluene	$CH_3C_6H_3(NO_2)_2$	182.14	5, 341	1.2833^{111}	1.479	64–66			s alc
d643	3,4-Dinitrotoluene	$CH_3C_6H_3(NO_2)_2$	182.14	5, 341	1.2594^{111}		54–57			i aq; s alc
d643a	Dinonyl hexanedioate	$C_9H_{19}OOC(CH_2)_4$-$COOC_9H_{19}$	398.63		0.917_{25}^{25}				218	
d644	Dioctadecyl phosphite	$(C_{18}H_{37}O)P(O)H$	586.97				57–59			
d645	Dioctylamine	$(C_8H_{17})_2NH$	241.46	4, 196			14–16	298	>112	i aq; v s alc, eth
d646	Dioctyl sulfide	$(C_8H_{17})_2S$	258.51	1, 419	0.842	1.4610^{20}		180^{10mm}	>112	
d647	4,9-Dioxa-1,12-dodecanediamine	$H_2N(CH_2)_3O(CH_2)_4$-$O(CH_2)_3NH_2$	204.32		0.962	1.4609^{20}		134–136^{4mm}	>112	

Dimethyl terephthalate, d592
2,3-Dimethylvaleraldehyde, d571
2,4-Dinitrochlorobenzene, c95
3,4-Dinitrochlorobenzene, c96
3,5-Dinitro-1-toluic acid, m229
Dioctyl phthalates, b180, d313
6,8-Dioxabicyclo[3.2.1]octan-7-one, h184

TABLE 1-14 **Physical constants of organic compounds (*continued*)**

No.	Name	Formula	Formula weight	Beilstein reference	Density	Refractive index	Melting point	Boiling point	Flash point	Solubility in 100 parts solvent
d648	1,4-Dioxane		88.10	19, 3	1.0329^{20}_{4}	1.4224^{20}	11.7	101.2	12	misc aq, alc, bz, chl, eth, PE
d649	1,3-Dioxolane		74.08	19^2, 3	1.060^{20}_{4}	1.4000^{20}	−95	74–75	<1	misc aq; s alc, eth
d650	Dipentaerythritol	$(HOCH_2)_3CCH_2OCH_2$-$C(CH_2OH)_3$	254.28				215–218			
d651	Dipentene		136.24	5, 137	0.8402^{21}_{4}	1.4739^{20}		176	42	i aq; misc alc
d652	Dipentylamine	$(C_5H_{11})_2NH$	157.29	4^1, 378	0.777	1.4272		195–202	39	v s alc, eth
d653	Dipentyl ether	$(C_5H_{11})_2O$	158.29	1^1, 193	0.7833^{20}_{4}	1.4120^{20}	−69.4	186.8	63	misc alc, eth; s acet
d654	Diphenylacetic acid	$(C_6H_5)_2CHCOOH$	212.25	9, 673	1.258^{15}_{15}		148	195^{5mm}		s hot aq, alc, chl, eth
d655	Diphenylacetonitrile	$(C_6H_5)_2CHCN$	193.25	9, 674			76	181^{12mm}		s alc, eth
d656	Diphenylacetylene	$C_6H_5C{\equiv}CC_6H_5$	178.23	5, 656	0.990		60–61	300		v s hot alc, eth
d657	Diphenylamine	$(C_6H_5)_2NH$	169.23	12, 174	1.160		53–54	302	152	45 alc; v s bz, eth
d658	*cis,cis*-1,4-Diphenyl-1,3-butadiene	$C_6H_5CH{=}CH$-$CH{=}CHC_6H_5$	206.29	5, 676	0.9697^{101}_{4}	1.6347^{101} (He line)	70.5			s bz, chl, eth, PE
d659	*cis,trans*-1,4-Diphenyl-1,3-diene	$C_6H_5CH{=}CH$-$CH{=}CHC_6H_5$	206.29	5, 676	0.9974^{22}_{4}	1.6053^{22}	88	$133^{0.1mm}$		s alc, bz, eth, chl
d660	1,3-Diphenyl-2-buten-1-one	$C_6H_5C(O)CH{=}C(C_6H_5)CH_3$	222.27	7^2, 433	1.1080^{15}_{4}	1.6343^{20}	−30 glass	246^{50mm}		i aq; s alc, eth
d661	Diphenylcarbamoyl chloride	$(C_6H_5)_2NCOCl$	231.68				82–84			
d662	1,5-Diphenylcarbohydrazide	$(C_6H_5NHNH)_2C{=}O$	242.28	15, 292			168–171			s hot alc, acet, HOAc
d663	Diphenyl carbonate	$(C_6H_5O)_2C{=}O$	214.22	6, 158			80–81	302–306		s hot alc, bz, eth
d664	Diphenyl chlorophosphate	$(C_6H_5O)_2P(O)Cl$	268.64	6, 179	1.296	1.5500^{20}		314^{272mm}	>112	

d665	Diphenyl diselenide	$C_6H_5SeSeC_6H_5$	312.13	6, 346	1.557_4^{80}		61–64			s hot alc
d666	Diphenyl disulfide	$C_6H_5SSC_6H_5$	218.34	6, 323	1.353_4^{20}		58–60	310		s alc, bz, eth; i aq
d667	Diphenylenimine		167.21	20, 433	1.10_4^{18}		246	355		0.8 bz; 3 eth; 16 pyr; 11 acet; i aq
d668	1,2-Diphenylethane	$C_6H_5CH_2CH_2C_6H_5$	182.27	5, 598	0.995_4^{20}	1.5338	52.5	284		s alc; v s chl, eth
d669	Diphenyl ether	$C_6H_5OC_6H_5$	170.21	6, 146	1.0661_4^{30}	1.5763^{30}	26.9	258.3	115	s alc, bz, eth, HOAc
d670	1,2-Diphenylethyl-amine	$C_6H_5CH_2CH(C_6H_5)NH_2$	197.28	12, 1326	1.020	1.5802^{20}		311	>112	
d671	*N,N'*-Diphenylethyl-enediamine	$C_6H_5NHCH_2CH_2NHC_6H_5$	212.30	12, 543			67.5	228–330		v s alc, eth
d672	*N,N'*-Diphenyl-formamidine	$C_6H_5N{=}CHNHC_6H_5$	196.25	12, 236			138–141			s eth; v s chl
d673	1,3-Diphenyl-guanidine	$C_6H_5NHC({=}NH)NHC_6H_5$	211.27	12, 369	1.13		150	d 170		s alc, hot bz, chl

3,6-Dioxa-1,8-octanediol, t275
4,8-Dioxa-1,11-undecanediol, t415
(2,5-Dioxo-4-imidazolidinyl)urea, a77
1,3-Dioxolane-2-one, e125
3,4-Dioxypentane, d440
Dipentyl ketone, u6

Diphenic acid, b138
Diphenylacetone, d686
Diphenylcarbamyl chloride, d661
sym-Diphenylcarbazide, d662
Diphenyldiazone, a327
Diphenylethanedione, b34

Diphenylethanedione dioxime, b35
1,2-Diphenylethene, s9
Diphenylethyne, d656
Diphenylglycolic acid, b36
Diphenylglyoxime, b35

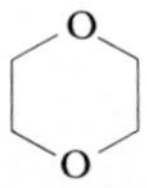

d648

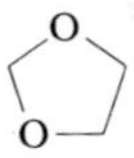

d649

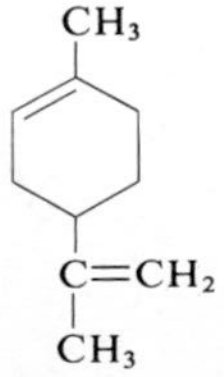

CH_3
$C{=}CH_2$
CH_3
d651

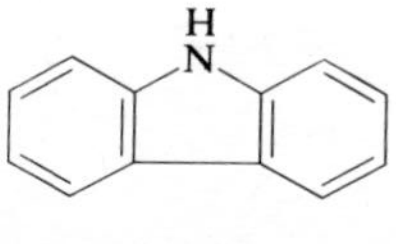

H
N
d667

TABLE 1-14 Physical constants of organic compounds (*continued*)

No.	Name	Formula	Formula weight	Beilstein reference	Density	Refractive index	Melting point	Boiling point	Flash point	Solubility in 100 parts solvent
d674	5,5-Diphenyl-hydantoin		252.27	24, 410			295–298			i aq; 1.7 alc; 3.3 acet
d675	1,2-Diphenyl-hydrazine	$C_6H_5NHNHC_6H_5$	184.24	15, 123	1.158_4^{16}		123–126			v s alc; sl s bz
d676	Diphenyl isooctyl-phosphite	$(CH_3)_2CH(CH_2)_5PH(OC_6H_5)_2$	346.40		1.044	1.522				
d677	Diphenylmercury	$(C_6H_5)_2Hg$	354.81	16, 946	2.318^4		124–125	d > 306		s chl; sl s hot alc
d678	Diphenylmethane	$C_6H_5CH_2C_6H_5$	168.24	5^2, 498	1.3421_4^{10}	1.5768	25.9	264.5	>112	v s alc, bz, chl, eth
d679	Diphenylmethanol	$(C_6H_5)_2CHOH$	184.24	6, 678			66.7	298		0.05 aq; v s alc, chl, eth
d680	1,1-Diphenylmethyl-amine	$C_6H_5CH(NH_2)C_6H_5$	183.25	12, 1323	1.0635_4^{22} supercooled	1.5956^{99}	34	295	>112	sl s aq
d681	2,5-Diphenyloxazole		221.26	27, 78			72–73	360		
d682	2,6-Diphenylphenol	$(C_6H_5)_2C_6H_3OH$	246.31	6^3, 3631			100–102			
d683	Diphenyl phosphite	$(C_6H_5O)_2P(=O)H$	234.19	6^1, 94	1.223	1.5575^{20}	12	219^{26mm}	176	
d684	Diphenylphosphoryl azide	$(C_6H_5O)_2P(=O)N_3$	275.20		1.277	1.5518^{20}		$157^{0.17mm}$	>112	
d685	2,2-Diphenyl-1-picrylhydrazyl		394.32	16^2, 363			127 d			
d686	1,3-Diphenyl-2-propanone	$C_6H_5CH_2C(=O)CH_2C_6H_5$	210.28	7, 445	1.2		32–34	330		i aq; v s alc, eth
d687	1,3-Diphenyl-2-propen-1-one	$C_6H_5CH=CHC(=O)C_6H_5$	208.26	7, 478	1.0712_4^{62}	1.6458^{62}	57–58	208^{25mm}		v s bz, chl, eth
d688	2.2-Diphenyl-propionic acid	$CH_3C(C_6H_5)_2COOH$	226.28	9^2, 474			175–177	300		s alc; v s bz, eth
d689	Diphenylsilanediol	$(C_6H_5)_2Si(OH)_2$	216.31	16, 909			140 d		53	
d690	Diphenyl sulfide	$(C_6H_5)_2S$	186.28	6, 299	1.118_{15}^{15}	1.6327^{20}	−40	296	>112	misc bz, eth, CS_2
d691	Diphenyl sulfone	$(C_6H_5)_2SO_2$	218.27	6, 300			128–129	379		i aq; s hot alc, bz

d692	Diphenyl sulfoxide	$(C_6H_5)_2S{=}O$	202.28	6, 300			69–71	207^{13mm}		
d693	Diphenylthio-carbazone	$C_6H_5N{=}NC(S)NH$-NHC_6H_5	256.33	16, 26			168 d			i aq; v s chl, CCl_4
d694	1,3-Diphenylthio-urea	$C_6H_5NHC(S)NHC_6H_5$	228.32	12, 394	1.32		154			i aq; v s alc, eth
d695	1,3-Diphenylurea	$C_6H_5NHC(O)NHC_6H_5$	212.25	12, 352	1.239		238	260 d		0.015 aq; s eth, HOAc
d696	1.2-Dipiperidino-ethane		196.34	20[1], 19	0.916	1.4876^{20}	−0.5	265	110	
d697	Dipiperidinomethane		182.31		0.915	1.4820^{20}		123^{15mm}		
d698	Dipropylamine	$(C_3H_7)_2NH$	101.19	4, 138	0.7375_4^{20}	1.4043^{20}	−39.6	109.2	17	4 aq; v s alc, eth, PE
d699	Dipropylene glycol butyl ether	$CH_3CH(OH)CH_2OCH_2$-$CH(OC_4H_9)CH_3$	190.3		0.917_{25}^{25}	1.425^{25}		229	113	

5,5-Diphenyl-2,4-imidazolidinedione, d674
Diphenyl ketone, b53
Diphenyl oxide, d669
Diphenylphosphorochloridate, d664
1,3-Diphenyl-1,3-propanedione, d53
sym-Diphenylthiourea, t148
Dipicolinic acid, p265
Di-2-propenylamine, d25
Dipropyl adipate, d704
Dipropylene glycol, h173

d674

d681

d685

d696

d697

TABLE 1-14 **Physical constants of organic compounds (*continued*)**

No.	Name	Formula	Formula weight	Beilstein reference	Density	Refractive index	Melting point	Boiling point	Flash point	Solubility in 100 parts solvent
d700	Dipropylene glycol ethyl ether	$CH_3CH(OH)CH_2OCH_2$-$CH(OC_2H_5)CH_3$	162.2		0.930_{25}^{25}	1.419^{25}		388	90	
d701	Dipropylene glycol isopropyl ether	$CH_3CH(OH)CH_2OCH_2$-$CH[OCH(CH_3)_2]CH_3$	176.2		0.878_{25}^{25}	1.421^{25}		80.1	90	
d702	Dipropylene glycol methyl ether	$CH_3CH(OH)CH_2OCH_2$-$CH(OCH_3)CH_3$	148.2		0.951_{20}^{20}	1.419^{20}	−117	188.3	85	
d703	Dipropyl ether	$(C_3H_7)_2O$	102.18	1, 354	0.7466^{20}	1.3803^{20}	−123.2	89.6	4	0.4 aq
d704	Dipropyl hexane-dioate	$C_3H_7OOC(CH_2)_4$-$COOC_3H_7$	230.30	2^2, 574	0.9790_4^{20}	1.4314^{20}	−20	144^{10mm}		i aq; s alc, eth
d705	Dipropyl sulfate	$(C_3H_7O)_2SO_2$	182.24	1, 354	1.106_4^{20}		d 140	120^{20mm}		v s PE
d706	Dipropyl sulfone	$(C_3H_7)_2SO_2$	150.24	1, 359	1.028_4^{50}		28–30	270	126	
d707	2,2′-Dipyridyl		156.19	23, 199			69.7	273		0.5 aq; v s alc, bz, chl, eth, PE
d708	2,2′-Dipyridylamine		171.20	22^1, 630			89–90	222^{50mm}		
d709	1,3-Dithiane		120.24				53–55		90	
d710	4,4′-Dithiobutyric acid	$HOOC(CH_2)_3SS(CH_2)_3$-$COOH$	238.32	3, 312			110			
d711	3,3′-Dithiopropionic acid	$HOOCCH_2CH_2SSCH_2$-CH_2COOH	210.27				157–159			
d712	Dithiooxamide	$H_2NC(=S)C(=S)NH_2$	120.20	2, 565			170 d	subl		sl s aq; s alc; i eth
d713	1,3-Di-*o*-tolyl-guanidine	$(CH_3C_6H_4NH)_2C=NH$	239.32	12, 803	1.10_4^{20}		176–178			s hot alc, eth
d714	1,5-Di(vinyloxy)-3-oxapentane	$(CH_2=CHOCH_2CH_2)_2O$	158.20		0.975^{29}	1.445		81^{10mm}		
d715	1,3-Divinyltetra-methyldisiloxane	$[CH_2=CHSi(CH_3)_2]_2O$	186.39		0.811_4^{20}	1.412^{20}	−99.7	139		
d716	3,9-Divinyl-2,4,8,10-tetraoxaspiro[5.5]-undecane		212.25		1.251		40–54	120^{2mm}	110	

d717	Docosane	$CH_3(CH_2)_{20}CH_3$	310.61	1, 174	0.7782^{45}	1.4358^{45}	44.4	369		i aq; sl s alc; v s eth
d718	Docosanoic acid	$CH_3(CH_2)_{20}COOH$	340.60	2, 391	0.8221_4^{100}	1.4270^{100}	80–82	206^{60mm}		0.2 alc; 0.19 eth
d719	1-Docosanol	$CH_3(CH_2)_{21}OH$	326.61	1, 431			65–72	$180^{0.22mm}$		sl s eth; s alc, chl
d720	1*H*,1*H*,7*H*-Dodecafluoro-1-heptanol	$HCF_2(CF_2)_5CH_2OH$	332.0		1.7616^{20}	1.3180^{20}		169–170		
d721	Dodecane	$CH_3(CH_2)_{10}CH_3$	170.41	1, 171	0.7490_4^{20}	1.4216^{20}	−9.6	216.28	71	
d722	1,12-Dodecanediamine	$H_2N(CH_2)_{12}NH_2$	200.37	4, 273			62–65		155	
d723	Dodecanedioic acid	$HOOC(CH_2)_{10}COOH$	230.30	2, 729			128–130	245^{10mm}		
d724	1,2-Dodecanediol	$CH_3(CH_2)_9CH(OH)$-CH_2OH	202.34	1^3, 2237			58–60			
d725	1,12-Dodecanediol	$HOCH_2(CH_2)_{10}CH_2OH$	202.34	1^2, 562			81–84	189^{12mm}		
d726	Dodecanenitrile	$CH_3(CH_2)_{10}CN$	181.32	2, 363	0.827	1.4360^{20}		198^{100mm}	>112	misc alc, bz, chl, eth
d727	1-Dodecanethiol	$CH_3(CH_2)_{11}SH$	202.40		0.845_{20}^{20}	1.4587^{20}		266–283	87	i aq; s alc, eth
d728	Dodecanoic acid	$CH_3(CH_2)_{10}COOH$	200.32	2, 359	0.869_4^5	1.4183^{82}	44	225^{100mm}		i aq; 100 alc; v s bz,, eth
d729	1-Dodecanol	$CH_3(CH_2)_{11}OH$	186.34	1, 428	0.8308_4^{25}	1.4413^{25}	23.8	259	>112	i aq; s alc, eth
d730	Dodecanoyl chloride	$CH_3(CH_2)_{10}COCl$	218.77	2, 363	0.946	1.4459^{20}		134^{11mm}	>112	
d731	1-Dodecene	$CH_3(CH_2)_9CH=CH_2$	168.32	1, 225	0.7584_4^{20}	1.4294^{20}	−35.2	213.4	77	s alc, eth, PE

Dipropyl ketone, h17
Distearylpentaerythritoldiphosphite, b197
Disulfiram, t63
2,3-Dithiabutane, d518
5,6-Dithiadecane, d113
3,4-Dithiahexane, d294
2,2′-Dithiodiethanol, h118
1,4-Dithiothreitol, d425
Dithizone, d693
Divinylene oxide, f40
DMSO, d616
Dodecyl alcohol, d729

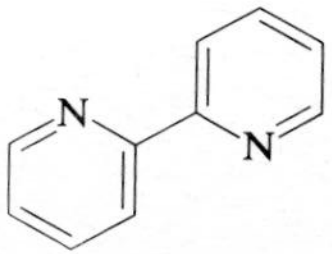

d707

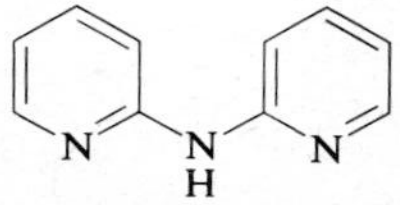

d708

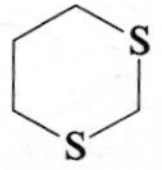

d709

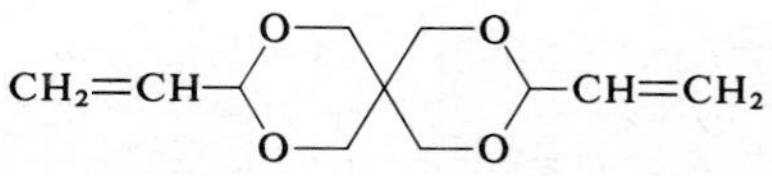

$CH_2=CH$– –$CH=CH_2$

d716

TABLE 1-14 Physical constants of organic compounds (*continued*)

No.	Name	Formula	Formula weight	Beilstein reference	Density	Refractive index	Melting point	Boiling point	Flash point	Solubility in 100 parts solvent
d732	2-Dodecen-1-yl-succinic anhydride		266.38					180^{5mm}	177	
d733	Dodecanal	$CH_3(CH_2)_{10}CHO$	184.32	1, 714	0.835	1.4344^{20}		185^{100mm}	101	
d734	Dodecylamine	$CH_3(CH_2)_{11}NH_2$	185.36	4, 200			28–30	247–249	>112	misc alc, bz, chl, eth
d735	4-Dodecylaniline	$CH_3(CH_2)_{11}C_6H_4NH_2$	261.46	12^3, 2776			40–41	220–221^{15mm}		
d736	Dodecylcyclohexane	$CH_3(CH_2)_{11}C_6H_{11}$	252.50		0.8250	1.4580^{20}	12	$131^{0.8mm}$		
d737	Dodecyl sulfate, Na salt	$CH_3(CH_2)_{11}OSO_3^-Na^+$	288.38				204–207			10 aq
d738	Dodecyltrichloro-silane	$CH_3(CH_2)_{11}SiCl_3$	303.8			1.458^{20}		155^{10mm}		
d739	Dotriacontane	$CH_3(CH_2)_{30}CH_3$	450.88	1, 177	0.8124^{20}_{4}	1.4364^{70}	68–70	467		sl s alc, bz, eth
d740	Dulcitol		182.17	1, 544	1.47^{20}		188–189	275^{1mm}		3.3 aq; sl s alc
e1	D-Ephedrine	$CH_3NHCH(CH_3)CH(OH)$-C_6H_5	165.24	13, 637			119	225		v s alc, eth
e2	L-Ephedrine	$CH_3NHCH(CH_3)CH(OH)$-C_6H_5	165.24	13, 636			34	255		5 aq; v s alc; s chl
e3	1,2-Epoxybutane	$CH_3CH_2CH{-}CH_2$ (—O— bridge)	72.11	17^2, 17	0.8297^{20}	1.3840^{20}	−150	63.2	−17	6 aq; misc alc, bz, chl, eth
e4	1,2-Epoxycyclo-dodecane		182.31		0.939	1.4773^{20}				
e5	1,2-Epoxycyclo-hexane		98.15	17, 21	0.970	1.4520^{20}		129–130	27	v s alc, bz, eth
e6	1,4-Epoxycyclohexane		98.15		0.969	1.4480^{20}		119^{713mm}	12	
e7	2-(3,4-Epoxycyclo-hexyl)ethyltri-methoxysilane		246.37		1.070^{25}_{4}	1.449^{25}		310	146	

e7a	1,2-Epoxycyclooctane		126.20				53–56	55^{5mm}	56	
e8	1,2-Epoxycyclopentane		84.12	17, 21	0.964	1.4336^{20}		102	10	
e9	1,2-Epoxyethylbenzene	$C_6H_5CH{-}CH_2$ (–O–)	120.15	17, 49	1.0523^{16}_{4}	1.5338^{20}	−37	194	79	i aq; s alc, eth
e9a	1,2-Epoxy-3-phenoxypropane	$C_6H_5OCH_2CH{-}CH_2$ (–O–)	150.18				2			
e10	1,2-Epoxypropane	$CH_3CH{-}CH_2$ (–O–)	58.08	17, 6	0.859^{0}_{4}	1.3660^{20}	−112.1	34.2	−37	41 aq; misc alc, eth
e11	2,3-Epoxy-1-propanol	$H_2C{-}CHCH_2OH$ (–O–)	74.08	17, 104	1.1143^{25}_{4}	1.4315^{20}		$66^{2.5mm}$	81	misc aq

Dodecyl aldehyde
DPPH, d685
Durene, t101
Durenol, t118
EDTA, e128

Eicosane, i2
1-Eicosene, i3
Elaidic acid, o11
Embonic acid, m236
Enanthic acid, h10

Epibromohydrin, b278
Epichlorohydrin, c102
1,4-Epoxybutane, t68
Epoxyethane, e132
1,3-Epoxypropane, t350

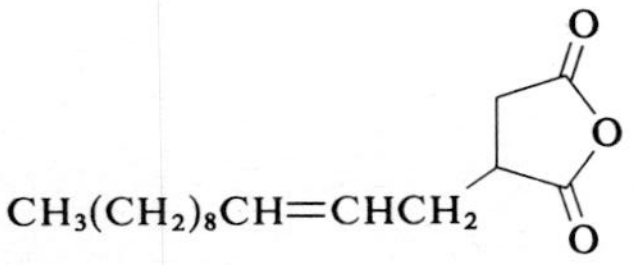

$CH_3(CH_2)_8CH{=}CHCH_2$ O O O

d732

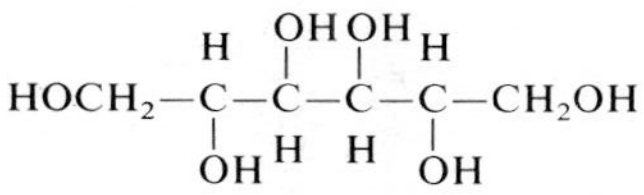

H OH OH H
$HOCH_2{-}C{-}C{-}C{-}C{-}CH_2OH$
OH H H OH

d740

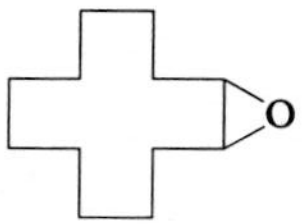

O

e4

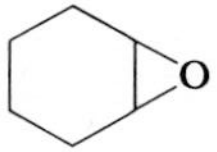

O

e5

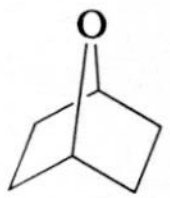

O

e6

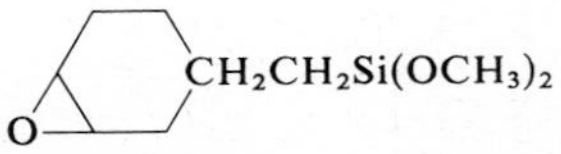

$CH_2CH_2Si(OCH_3)_2$ O

e7

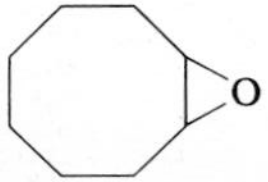

O

e7a

O

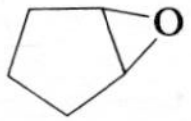

e8

TABLE 1-14 Physical constants of organic compounds (*continued*)

No.	Name	Formula	Formula weight	Beilstein reference	Density	Refractive index	Melting point	Boiling point	Flash point	Solubility in 100 parts solvent
e12	2,3-Epoxypropyl methacrylate	$H_2C{=}C(CH_3)COO\text{-}CH_2CH{-}CH_2$ (epoxide, –O– bridge)	142.15		1.042	1.4494^{20}		189	83	
e13	1,2-Epoxy-3,3,3-trichloropropane	$Cl_3CCH{-}CH_2$ (epoxide, –O– bridge)	161.42	17^2, 14	1.495	1.4778^{20}		151^{745mm}	66	
e14	Ethane	CH_3CH_3	30.07	1, 80	0.5462^{-88} 1.0493^{0}_{4}, $g \cdot L^{-1}$		−183.3	−88.6		4.7 mL aq; 46 mL alc[4]
e15	1,2-Ethanediamine	$H_2NCH_2CH_2NH_2$	60.10	4, 230	0.8977^{20}_{4}	1.4568^{20}	8.5	117.3	33	misc aq, alc; i bz
e16	1,2-Ethanediol	$HOCH_2CH_2OH$	62.07	1, 465	1.1135^{20}_{4}	1.4318^{20}	−12.6	197.3	110	misc aq, alc, glyc, pyr
e17	1,2-Ethanediol diacetate	$CH_3COOCH_2CH_2OOCCH_3$	146.14	2, 142	1.1043^{20}	1.4150^{20}	−31	190.2	82	misc alc, eth
e18	1,2-Ethanedithiol	$HSCH_2CH_2SH$	94.20	1, 471	1.123^{24}	1.5580^{20}		146	50	v s alc, alk
e19	Ethanesulfonyl chloride	$CH_3CH_2SO_2Cl$	128.57	4^2, 526	1.357^{22}	1.4339^{20}		171		d aq, alc; v s eth
e20	Ethanethiol	CH_3CH_2SH	62.13	1, 340	0.8315^{25}	1.420^{25}	−147.9	35.0	−17	0.7 aq; s alc, eth
e21	Ethanol	CH_3CH_2OH	46.07	1, 292	0.7894^{20}_{4}	1.3614^{20}	−114.5	78.3	8	misc aq, alc, eth, chl
e22	Ethanol-*d*	CH_3CH_2OD	47.08	1^3, 1287	0.801	1.3595^{20}		78.8	12	misc aq, alc, eth
e24	Ethoxyacetic acid	$CH_3CH_2OCH_2COOH$	104.11	3, 233	1.1021^{20}_{4}	1.4190^{20}		97^{11mm}	97	s aq, alc, eth
e25	4-Ethoxyaniline	$CH_3CH_2OC_6H_4NH_2$	137.18	13, 436	1.0652^{16}_{4}	1.5609^{20}	4	250	115	i aq; s alc
e26	2-Ethoxybenzaldehyde	$CH_3CH_2OC_6H_4CHO$	150.18	8, 43	1.074	1.5422^{20}	20	136^{24mm}	107	misc alc, eth
e27	4-Ethoxybenzaldehyde	$CH_3CH_2OC_6H_4CHO$	150.18	8, 73	1.080^{25}_{25}	1.5584^{20}	13–14	255	>112	v s alc, bz, eth
e28	2-Ethoxybenzamide	$CH_3CH_2OC_6H_4CONH_2$	165.19	10, 93			132–133			sl s aq; s alc, eth
e29	Ethoxybenzene	$CH_3CH_2OC_6H_5$	122.17	6, 140	0.967^{20}_{4}	1.5074^{20}	−29.5	170.0		0.12 aq; misc alc, eth
e30	2-Ethoxybenzoic acid	$CH_3CH_2OC_6H_4COOH$	166.18	10, 64	1.105	1.5400^{20}	19.4	174^{15mm}	>112	sl s aq

e31	4-Ethoxybenzoic acid	$CH_3CH_2OC_6H_4COOH$	166.18	10, 156			197–199			sl s hot aq
e32	2-Ethoxybenzyl alcohol	$CH_3CH_2OC_6H_4CH_2OH$	152.19	6, 893		1.5321^{20}		265		
e33	Ethoxycarbonyl isothiocyanate	$CH_3CH_2OC(=O)NCS$	131.15	3^3, 279	1.112	1.5000^{20}		56^{18mm}	50	
e34	Ethoxydimethylvinyl-silane	$(CH_3)_2Si(OC_2H_5)$-$CH=CH_2$	130.3		0.790_4^{20}	1.398^{20}		99^{710mm}		
e35	2-Ethoxyethanol	$CH_3CH_2OCH_2CH_2OH$	90.12	1, 467	0.9295^{20}	1.4075^{20}	−59	134.8	48	misc aq, alc, eth, acet
e36	2-(2-Ethoxyethoxy)-ethanol	$C_2H_5OCH_2CH_2OCH_2$-CH_2OH	134.18	1^2, 520	0.9841_4^{25}	1.4254^{25}	−55	201.9	96	misc aq, alc, bz, chl, acet, pyr
e37	2-(2-Ethoxyethoxy)-ethyl acetate	$C_2H_5OCH_2CH_2OCH_2$-CH_2OOCCH_3	176.21		1.0096^{20}	1.4213^{20}	−25	218.5	110	misc aq, alc, eth, oils
e38	2-Ethoxyethyl acetate	$CH_3COOCH_2CH_2$-OCH_2CH_3	132.16	2^2, 155	0.9749_4^{20}	1.4023^{20}	−61.7	156.3	57	29 aq; misc alc, eth
e39	2-Ethoxyethylamine	$CH_3CH_2OCH_2CH_2NH_2$	89.14	4^2, 718	0.8512_4^{20}	1.4101^{20}		107	21	misc aq, alc, eth
e40	3-Ethoxy-4-hydroxy-benzaldehyde	$C_2H_5OC_6H_3(OH)CHO$	166.18	8, 256			76–78			s eth, glycols; 50 alc
e41	3-Ethoxymethacrolein	$C_2H_5OCH=C(CH_3)CHO$	114.15	1^4, 4082	0.960	1.4792^{20}		78–81^{14mm}	35	
e42	4-Ethoxy-3-methoxy-benzaldehyde	$C_2H_5OC_6H_3(OCH_3)CHO$	180.20	8, 256			59–60			s alc, bz, chl, eth
e43	Ethoxymethyldi-phenylsilane	$(C_6H_5)_2Si(CH_3)OC_2H_5$	242.4		1.018_4^{20}	1.544^{20}		$122^{0.3mm}$		
e44	Ethoxymethylene-malononitrile	$CH_3CH_2OCH=C(CN)_2$	122.13	3^1, 162			64–66	160^{12mm}		

Eschenmoser's salt, d551
Estragole, a94
Ethanal, a4
Ethanediamide, o54
Ethanenitrile, a29

Ethanoic acid, a19
Ethanolamine, a164
Ethenyl acetate, v2
Ethenylbenzene, s20
4-Ethoxy-*m*-anisaldehyde, e42

1-Ethoxybutane, b452
Ethoxyethane, d300
2-Ethoxyethyl ether, b177
Ethoxyformic anhydride, d326

TABLE 1-14 Physical constants of organic compounds (*continued*)

No.	Name	Formula	Formula weight	Beilstein reference	Density	Refractive index	Melting point	Boiling point	Flash point	Solubility in 100 parts solvent
e45	1-Ethoxynaphthalene	$C_{10}H_7OCH_2CH_3$	172.23	6, 606	1.060_4^{20}	1.6040^{20}	5.5	280	>112	i aq; v s alc, eth
e46	*N*-(4-Ethoxyphenyl)-acetamide	$CH_3CH_2OC_6H_4NHCOCH_3$	179.21	13^2, 244			134–135			0.076 aq; 6.7 alc; 7.1 chl; 1.1 eth; s glyc
e47	*trans*-2-Ethoxy-5-(1-propenyl)phenol	$C_2H_5OC_6H_3(CH{=}CH\text{-}CH_3)OH$	178.23	6^2, 918						
e48	3-Ethoxypropylamine	$C_2H_5OCH_2CH_2CH_2NH_2$	103.17	4^3, 739	0.861	1.4178^{20}		136–138	32	
e49	3-Ethoxysalicyl-aldehyde	$C_2H_5OC_6H_3(OH)CHO$	166.18	8^2, 267			86–88	263–264		
e50	2-Ethoxytetrahydro-furan	$C_2H_5O(C_4H_7O)$	116.16	17^4, 1020	0.908	1.4140^{20}		170–172	16	
e51	Ethoxytrimethyl-silane	$(CH_3)_3SiOC_2H_5$	118.3		0.7573_4^{20}	1.3742^{20}				
e52	Ethyl acetate	$CH_3COOC_2H_5$	88.11	2, 125	0.9006_4^{20}	1.3724^{20}	−84	77.1	−3	9.7 aq; misc alc, acet, chl, eth
e53	Ethyl acetimidate HCl	$CH_3C({=}NH)OC_2H_5{\cdot}HCl$	123.58	2, 182			112–114			
e54	Ethyl acetoacetate (enol)	$CH_3COCH{=}C(OH)OC_2H_5$	130.15	3, 632	1.0119^{10}	1.4480^{10}	−44	180.8	84	1.9 aq; misc alc, chl
e55	Ethyl acetoacetate (keto)	$CH_3COCH_2COOC_2H_5$	130.15	3, 632	1.0368^{10}	1.4224^{10}	−39	180.8	84	12 aq; misc alc, chl
e56	*p*-Ethylacetophenone	$C_2H_5C_6H_4COCH_3$	148.21	7^4, 1101	0.993	1.5293^{20}	−20.6	114^{11mm}	90	
e57	Ethyl acrylate	$CH_2{=}CHCOOCH_2CH_3$	100.12	2, 399	0.9405_4^{20}	1.4068^{20}	−71.2	99.5	15	1.5 aq; s alc, eth
e58	Ethylaluminum dichloride	$CH_3CH_2AlCl_2$	126.95		1.207^{50}		32	113^{50mm}		
e59	Ethylamine	$CH_3CH_2NH_2$	45.09	4, 87	0.689_{15}^{15}		−81.0	16.6	−17	misc aq, alc, eth
e60	Ethyl 2-amino-benzoate	$H_2NC_6H_4COOCH_2CH_3$	165.19	14, 319	1.088^{15}		14	266–268		i aq; s alc, eth

e61	Ethyl 4-aminobenzoate	$H_2NC_6H_4COOCH_2CH_3$	165.19	14, 422			88–90	310		0.04 aq; 20 alc; 50 chl; 25 eth; s dil acid
e62	Ethyl 3-aminocrotonate	$CH_3C(NH_2){=}CH\text{-}COOCH_2CH_3$	129.16	3, 654	1.021_4^{20}		33–35	210–215		i aq; s alc, bz, eth
e63	2-(Ethylamino)ethanol	$CH_3CH_2NHCH_2CH_2OH$	89.14	4, 282	0.914_4^{20}	1.4402^{20}	−90	170	71	v s aq, alc, eth
e64	*N*-Ethylaniline	$C_6H_5NHCH_2CH_3$	121.18	12, 159	0.958_{25}^{25}	1.5559^{20}	−63	204.5	85	i aq; misc alc, eth
e65	2-Ethylaniline	$CH_3CH_2C_6H_4NH_2$	121.18	12^2, 584	0.983_4^{22}	1.5590^{20}	−44	210	97	sl s aq; v s alc, eth
e66	4-Ethylaniline	$CH_3CH_2C_6H_4NH_2$	121.18	12, 1090	0.975_4^{22}	1.5542^{20}	−5	216	85	sl s aq; v s alc, eth
e67	2-Ethylanthraquinone		236.27	7^1, 425			108–111			
e68	Ethylbenzene-d_{10}	$C_6D_5CD_2CD_3$	116.25			1.4920^{20}		134.6	31	
e69	Ethylbenzene	$C_6H_5CH_2CH_3$	106.17	5^2, 274	0.8670_4^{20}	1.4959^{20}	−95.0	136.2	20	0.01 aq; misc alc, bz, chl, eth
e70	Ethyl benzoate	$C_6H_5COOCH_2CH_3$	150.18	9, 110	1.050_4^{25}	1.5052^{20}	−34.7	212.4	84	0.05 aq; misc alc, chl, bz, eth, PE
e71	Ethyl benzoylacetate	$C_6H_5C({=}O)CH_2\text{-}COOCH_2CH_3$	192.21	10, 674	1.110	1.5338^{20}		265 d	140	i aq; misc alc, eth
e72	Ethyl 2-benzylacetoacetate	$CH_3COCH(CH_2C_6H_5)\text{-}COOC_2H_5$	220.27	10, 674	1.036	1.4996^{20}		276	>112	
e73	*N*-Ethylbenzylamine	$C_6H_5CH_2NHC_2H_5$	135.21	12, 1020	0.909	1.5117^{20}		194	66	
e74	Ethyl (2-benzyl)-benzoylacetate	$C_6H_5COCH(CH_2C_6H_5)\text{-}COOC_2H_5$	282.34	10, 764	1.110	1.5567^{20}		270^{80mm}	>112	

α-Ethoxy-α-phenylacetophenone, b47
3-Ethoxy-1-propene, a90
Ethyl alcohols, e21, e22
Ethylaldehyde, a4
Ethyl anthranilate, e60
Ethyl benzenecarboxylate, e70
Ethyl 3-benzenepropenoate, e104
α-Ethylbenzyl alcohol, p144

C_2H_5

e67

TABLE 1-14 Physical constants of organic compounds (*continued*)

No.	Name	Formula	Formula weight	Beilstein reference	Density	Refractive index	Melting point	Boiling point	Flash point	Solubility in 100 parts solvent
e75	Ethyl *N*-benzyl-*N*-cyclopropylcarbamate	$C_6H_5CH_2N(C_3H_5)$-$COOCH_2CH_3$	219.28		0.997	1.5104^{20}			>112	
e76	Ethyl bromoacetate	$BrCH_2COOCH_2CH_3$	167.01	2, 214	1.506^{20}_{20}	1.4510^{20}		159	47	i aq; misc alc, eth
e77	Ethyl 2-bromobutyrate	$CH_3CH_2CH(Br)$-$COOCH_2CH_3$	195.06	2^2, 255	1.329^{20}_{20}	1.4470^{20}		177 d	58	i aq; misc alc, eth
e78	Ethyl 4-bromobutyrate	$BrCH_2CH_2CH_2$-$COOCH_2CH_3$	195.06	2, 283	1.363	1.4559^{20}		82^{10mm}	90	
e79	Ethyl 2-bromoisobutyrate	$(CH_3)_2C(Br)$-$COOCH_2CH_3$	195.06	2, 296	1.329^{20}_{4}	1.4446^{20}		67^{11mm}	60	i aq; misc alc, eth
e80	Ethyl 3-bromo-2-oxopropionate	$BrCH_2C(=O)$-$COOCH_2CH_3$	195.02	3^2, 409	1.554	1.4695^{20}		100^{10mm}	98	
e81	Ethyl 2-bromopentanoate	$CH_3(CH_2)_2CH(Br)$-$COOCH_2CH_3$	209.09	2, 302	1.226	1.4486^{20}		190–192	77	i aq; misc alc, eth
e82	Ethyl 2-bromopropionate	$CH_3CH(Br)COOCH_2CH_3$	181.03	2, 255	1.447^{20}_{20}	1.4470^{20}	159–160	51		i aq; misc alc, eth
e83	Ethyl 3-bromopropionate	$BrCH_2CH_2COOCH_2CH_3$	181.03	2, 256	1.4123^{18}_{4}	1.4569^{18}		136^{50mm}	79	i aq; misc alc, eth
e84	2-Ethyl-1-butanol	$(C_2H_5)_2CHCH_2OH$	102.18	1, 412	0.8330^{20}	1.4224^{20}	−114.4	146.5	58	0.63 aq
e85	2-Ethyl-1-butene	$(C_2H_5)_2C=CH_2$	84.16	1^3, 814	0.6696^{20}_{4}	1.3967^{20}	−131.5	64.7		
e86	*N*-Ethylbutylamine	$CH_3(CH_2)_3NHCH_2CH_3$	101.19	4, 157	0.740^{20}_{4}	1.4050^{20}		108	18	
e87	2-Ethylbutylamine	$(C_2H_5)_2CHCH_2NH_2$	101.19	4, 192	0.776^{20}_{20}			121–125	21	s aq, alc, acet, eth
e88	2-Ethylbutyraldehyde	$(C_2H_5)_2CHCHO$	100.16	1, 693	0.8162^{20}_{20}	1.4018^{20}	−89	116.7	21	0.31 aq
e89	Ethyl butyrate	$CH_3CH_2CH_2COOCH_2CH_3$	116.16	2, 270	0.879^{20}_{4}	1.3928^{20}	−98.0	121.6	29	0.49 aq; misc alc, eth
e90	2-Ethylbutyric acid	$(C_2H_5)_2CHCOOH$	116.16	2, 333	0.9225^{20}_{20}	1.4133^{20}	−15	194.2	99	0.22 aq
e91	Ethyl butyrylacetate	$CH_3(CH_2)_2C(O)CH_2$-$COOC_2H_5$	158.20	3, 684	1.001	1.4295^{20}		104^{22mm}	78	

e92	Ethyl carbamate	$H_2NCOOCH_2CH_3$	89.09	3, 22	1.056		49–50	182–184		200 aq; 125 alc; 111 chl; 67 eth
e93	Ethyl carbazate	$H_2NNHCOOCH_2CH_3$	104.11	3, 98			44–47	110^{22mm}		
e94	*N*-Ethylcarbazole		195.27	20, 436			66–68			
e95	Ethyl chloroacetate	$ClCH_2COOCH_2CH_3$	122.55	2, 197	1.1498_4^{20}	1.4227^{20}	−26	144–146	65	i aq; misc alc, eth
e96	Ethyl 2-chloroaceto-acetate	$CH_3C(=O)CH(Cl)-COOC_2H_5$	164.59	3, 662	1.190	1.4430^{20}		107^{14mm}	50	i aq; s alc, eth
e97	Ethyl 4-chloroaceto-acetate	$ClCH_2C(=O)CH_2-COOC_2H_5$	164.59	3, 663	1.218_4^{17}	1.4520^{20}		115^{14mm}	96	i aq; misc alc, eth
e98	Ethyl 4-chloro-butyrate	$ClCH_2CH_2CH_2COOC_2H_5$	150.61	2, 278	1.0754_4^{20}	1.4306^{20}		186	51	s alc, acet, eth
e99	Ethyl chloroformate	$ClCOOC_2H_5$	108.52	3, 10	1.1403_4^{20}	1.3941^{20}	−81	95	2	misc alc,, bz, chl, eth
e100	Ethyl 2-chloropro-pionate	$CH_3CH(Cl)COOC_2H_5$	136.58	2, 248	1.087_4^{20}	1.4185^{20}		147–148	38	i aq; misc alc, eth
e101	Ethyl 3-chloropro-pionate	$ClCH_2CH_2COOC_2H_5$	136.58	2, 250	1.1086_4^{20}	1.4249^{20}		163	54	misc alc, eth
e102	Ethyl chorothio-formate	$ClC(=O)SCH_2CH_3$	124.59	3, 134	1.195	1.4820^{20}		132	30	
e103	Ethyl chrysanth-emumate		196.29	9^2, 45	0.906	1.4600^{20}		112^{10mm}		

Ethyl benzyl ether, b95
Ethyl bromide, b279
Ethyl 2-bromo-2-methylpropanoate, e79
Ethyl bromopyruvate, e80
Ethyl bromovalerate, e81
Ethyl butyl ether, b452
Ethyl butyl ketone, h16
Ethyl caprate, e114
Ethyl caproate, e145
Ethyl caprylate, e193
Ethyl chloride, c103
Ethyl chloroglyoxylate, e194

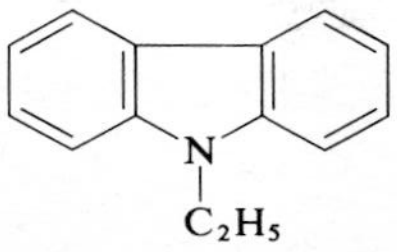
e94

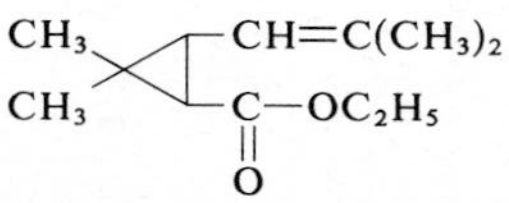
e103

TABLE 1-14 Physical constants of organic compounds (*continued*)

No.	Name	Formula	Formula weight	Beilstein reference	Density	Refractive index	Melting point	Boiling point	Flash point	Solubility in 100 parts solvent
e104	Ethyl *trans*-cinnamate	$C_6H_5CH{=}CHCOOCH_2CH_3$	176.22	9^2, 385	1.0495^{20}_4	1.5598^{20}	12	271.0		misc alc, eth; i aq
e105	Ethyl crotonate	$CH_3CH{=}CHCOOCH_2CH_3$	114.14	2, 411	0.9175^{20}_4	1.4248^{20}		138	2	i aq; s alc, eth
e106	Ethyl cyanoacetate	$NCCH_2COOCH_2CH_3$	113.12	2, 585	1.0564^{25}_4	1.4156^{20}	−22.5	206.0	110	i aq; misc alc, eth
e107	Ethyl cyanoformate	$NCCOOCH_2CH_3$	99.09	2, 547	1.003^{20}_4	1.3820^{20}		116	24	
e108	Ethyl cyano(hydroxy-imino)acetate	$NCC({=}NOH)COOCH_2CH_3$	142.12	3, 775			130–132			
e109	Ethylcyclohexane	$C_6H_{11}CH_2CH_3$	112.22	5, 35	0.7879^{20}	1.4330^{20}	−111.3	131.8	18	
e110	*cis*-2-Ethylcyclo-hexanol	$CH_3CH_2C_6H_{10}OH$	128.22	6^2, 26	0.927^{21}_4	1.4646^{20}		$74–79^{12mm}$	68	i aq
e111	4-Ethylcyclohexanol	$CH_3CH_2C_6H_{10}OH$	128.22	6^2, 26	0.889	1.4625^{20}		84^{10mm}	77	
e112	Ethyl cyclohexyl-acetate	$C_6H_{11}CH_2COOCH_2CH_3$	170.25	9, 14	0.948	1.4439^{20}		212	80	
e113	Ethyl cyclopropane-carboxylate	$C_3H_5COOCH_2CH_3$	114.14	9, 4	0.960	1.4197^{20}		129–133	18	
e114	Ethyl decanoate	$CH_3(CH_2)_8COOCH_2CH_3$	200.32	2, 356	0.862^{20}	1.4248^{20}		245	102	misc alc, chl, eth
e115	Ethyl diazoacetate	$N_2CH_2COOCH_2CH_3$	114.10	3^1, 211	1.0852^{18}_4	1.4588^{18}	−22	141^{720mm} explodes when heated	26	misc alc, bz, eth
e116	Ethyl 2,3-dibromo-propionate	$BrCH_2CH(Br)COO-CH_2CH_3$	259.94	2, 259	1.788^{16}_4	1.4986^{20}		214	91	s alc, eth
e117	Ethyl dichloro-phosphate	$CH_3CH_2OP(O)Cl_2$	162.94	1, 332	1.373	1.4338^{20}		65^{10mm}		
e118	Ethyl dichloro-thiophosphate	$CH_3CH_2OP(S)Cl_2$	179.01	1, 333	1.353	1.5040^{20}		$55–68^{10mm}$		
e119	*N*-Ethyldiethanol-amine	$CH_3CH_2N(CH_2CH_2OH)_2$	133.19	4, 284	1.014	1.4665^{20}	−50	246–252	123	
e120	Ethyl diethoxy-phosphinylformate	$(C_2H_5O)_2P(O)COOC_2H_5$	210.17	3^2, 103	1.110	1.4230^{20}		135^{13mm}		

e121	Ethyl 3-(diethylamino)propionate	$(C_2H_5)_2NCH_2CH_2COOC_2H_5$	173.26	4, 404	0.881	1.4253^{20}		84^{12mm}	7	
e122	Ethyl 3,3-dimethylacrylate	$(CH_3)_2C{=}CHCOOC_2H_5$	128.17	2, 433	0.9247_4^{20}	1.4350^{20}		155	33	
e123	Ethyl 2-dimethylaminobenzoate	$(CH_3)_2NC_6H_4COOC_2H_5$	193.25		1.061	1.5425^{20}			98	
e124	Ethyl 2,2-dimethylpropionate	$(CH_3)_3CCOOCH_2CH_3$	130.19	2^2, 280	0.8584_4^{18}	1.3922^{18}		118.2		s alc, eth
e125	Ethylene carbonate		88.06	19, 100	1.3208^{25}	1.4199^{40}	36.4	238	160	misc aq
e128	Ethylenediamine-*N,N,N′,N′*-tetraacetic acid	$(HOOCCH_2)_2NCH_2CH_2N(CH_2COOH)_2$	292.24				245 d			0.05 aq
e129	Ethylene dinitrate	$O_2NOCH_2CH_2ONO_2$	152.07		1.496_{15}^{15}	1.499^{15}	−22	106^{19mm}		
e130	2,2′-(Ethylenedioxy)bisethanol	$HOCH_2CH_2OCH_2CH_2OCH_2CH_2OH$	150.17		1.1274_4^{15}	1.4578^{15}	−72	285	166	misc aq, alc, bz
e131	Ethylene glycol	$HOCH_2CH_2OH$	62.07	1, 465	1.1135_4^{20}	1.4319^{20}	−13	197.6	110	misc aq, alc, acet, glc, HOAc, pyr; sl s eth; i bz, chl
e132	Ethylene oxide	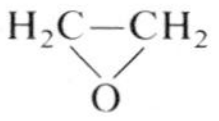	44.05		0.891_4^0	1.3597^7	−112.44	10.6	−18	misc aq; s alc, eth

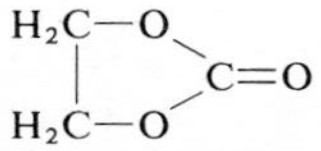

e125

TABLE 1-14 Physical constants of organic compounds (*continued*)

No.	Name	Formula	Formula weight	Beilstein reference	Density	Refractive index	Melting point	Boiling point	Flash point	Solubility in 100 parts solvent
e133	Ethylene sulfide	$H_2C—CH_2$ (ring with S)	60.12	17^2, 12	1.010	1.4935^{20}		55–56	10	sl s alc, eth
e134	Ethylenimine	$H_2C—CH_2$ (ring with NH)	43.07		0.8321^{25}_{4}	1.4123^{25}	−78.0	56	−24	misc aq; sl s alc
e135	Ethyl (ethoxymethyl-ene)cyanoacetate	$C_2H_5OCH{=}C(CN)$-$COOC_2H_5$	169.18	3, 470			51–53	190^{30mm}		
e136	Ethyl fluoroacetate	$FCH_2COOC_2H_5$	106.10	2, 193	1.0926^{21}	1.3755^{20}		119^{753mm}	30	s aq
e137	Ethyl fluoro-sulfonate	$FSO_2OC_2H_5$	128.12					23–25^{12mm}	32	
e138	Ethyl formate	$HCOOC_2H_5$	74.08	2, 19	0.917^{20}_{4}	1.3599^{20}	−79.4	54.2	−28	12 aq; misc alc, eth
e139	Ethyl 2-furoate		140.14	18, 275	1.117^{20}_{4}		33–36	196	70	i aq; s alc, eth
e140	Ethyl heptafluoro-butyrate	$CF_3CF_2CF_2COOC_2H_5$	242.09		1.394^{20}	1.3030^{20}		94–96		
e141	Ethyl heptanoate	$CH_3(CH_2)_5COOC_2H_5$	158.24	2^2, 295	0.8685^{20}_{4}	1.4144^{15}	−66	187		s alc, eth
e142	Ethyl hexadecanoate	$CH_3(CH_2)_{14}COOC_2H_5$	284.48	2^2, 336	0.8577^{25}_{4}	1.4347^{34}	22	191^{10mm}		s alc, eth
e143	3-Ethylhexane	$CH_3CH_2CH_2CH(C_2H_5)_2$	114.24	1^3, 478	0.7136^{20}_{4}	1.4016^{20}		118.5		sl s alc; s eth
e144	2-Ethyl-1,3-hexanediol	$C_3H_7CH(OH)CH$-$(C_2H_5)CH_2OH$	146.23		0.9325^{22}_{4}	1.4530^{22}	−40	244.2	129	0.6 aq; s alc
e145	Ethyl hexanoate	$CH_3(CH_2)_4COOC_2H_5$	144.21	2, 323	0.871^{20}_{4}	1.4075^{20}	−67	166–168	49	i aq; misc alc, eth
e146	2-Ethylhexanoic acid	$CH_3(CH_2)_3CH(C_2H_5)$-$COOH$	144.21	2, 349	0.9077^{20}_{20}	1.4241^{20}	−118.4	227.6	127	0.25 aq
e147	2-Ethyl-1-hexanol	$CH_3(CH_2)_3CH(C_2H_5)$-CH_2OH	130.23		0.9344^{20}_{20}	1.4231^{20}	−76	184.3	77	0.07 aq; s alc, bz, chl
e148	2-Ethylhexanoyl chloride	$CH_3(CH_2)_3CH(C_2H_5)$-$COCl$	162.66	2^2, 304	0.939	1.4335^{20}		68^{11mm}	69	
e149	2-Ethylhexyl acetate	$CH_3(CH_2)_3CH(C_2H_5)$-CH_2OOCCH_3	172.27		0.8718^{20}_{20}	1.4204^{20}	−93	198.6	82	0.03 aq; misc alc

e150	2-Ethylhexylamine	$CH_3(CH_2)_3CH(C_2H_5)$-CH_2NH_2	129.31		0.792^{20}_{20}			165–169	57	i aq; s alc, acet, eth
e151	2-Ethylhexyl vinyl ether	$CH_3(CH_2)_3CH(C_2H_5)$-$CH_2OCH{=}CH_2$	156.26		0.8102^{20}_{20}	1.4273^{20}	<100 glass	177.7		0.01 aq
e152	Ethyl hydrogen hexanedioate	$HOOC(CH_2)_4COOC_2H_5$	174.20	2^1, 277		1.4387^{20}	28–29	180^{18mm}	>112	
e153	Ethyl hydroxyacetate	$HOCH_2COOC_2H_5$	104.11	3, 236	1.087^{15}_{4}			160		v s alc, eth
e154	Ethyl 4-hydroxybenzoate	$HOC_6H_4COOC_2H_5$	166.18	10, 159			116	297 d		0.07 aq; v s alc, eth
e155	Ethyl 3-hydroxybutyrate	$CH_3CH(OH)CH_2COOC_2H_5$	132.16	3, 309	1.017^{20}_{4}	1.4205^{20}		170	64	s aq, alc
e156	Ethyl 2-hydroxyethyl sulfide	$HOCH_2CH_2SCH_2CH_3$	106.19	1^2, 525	1.020^{20}_{20}	1.4869^{20}		184.5		s eth
e157	Ethyl 2-hydroxyisobutyrate	$(CH_3)_2C(OH)COOC_2H_5$	132.16	3, 315	0.965	1.4078^{20}		150	44	d hot aq
e158	Ethyl 4-hydroxy-3-methoxyphenylacetate	$HOC_6H_3(OCH_3)CH_2$-$COOC_2H_5$	210.23	10^1, 198			44–47	180–185^{14mm}		

Ethylene glycol monoacetate, h116
Ethylene glycol monobutyl ether, b413
Ethylene glycol monoethyl ether, e35
Ethylene glycol monomethyl ether, m65
Ethylene glycol monomethyl ether acetate, m68
Ethylene iodide, d404
Ethylene iodohydrin, i35

1,8-Ethylenenaphthalene, a2
Ethylenethiourea, i5
Ethylene trichloride, t231
Ethyleneurea, i7
N-Ethylethanamine, d267
Ethyl *N*-ethylcarbamate, e234
Ethylethylene oxide, e3

Ethyl fluoride, f17
2-Ethylhexyl alcohol, e147
Ethyl homovanillate, e158
N-Ethyl-*N*-(2-hydroxyethyl)-3-toluidine, e228
Ethyl 2-hydroxy-2-methylpropanoate, e157

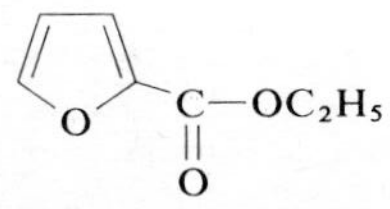

e139

TABLE 1-14 Physical constants of organic compounds (*continued*)

No.	Name	Formula	Formula weight	Beilstein reference	Density	Refractive index	Melting point	Boiling point	Flash point	Solubility in 100 parts solvent
e159	2-Ethyl-2-(hydroxymethyl)-1,3-propanediol	$CH_3CH_2C(CH_2OH)_3$	134.18	1^3, 2349			60–62	159–161^{2mm}		
e160	*N*-Ethyl-3-hydroxypiperidine		129.20		0.970	1.4754^{20}		93–95^{15mm}	47	
e161	5-Ethylidene-2-norbornene		120.20		0.893	1.4895			38	
e162	2-Ethylimidazole		96.13				79–81			
e163	2-Ethyl-6-isopropylaniline	$(CH_3)_2CHC_6H_3$-$(C_2H_5)NH_2$	163.26		0.949			249		
e164	Ethyl isothiocyanate	CH_3CH_2NCS	87.14	4, 123	1.003_4^{18}	1.5142^{18}	−6	130–132	32	i aq; misc alc, eth
e165	Ethyl L-(+)-lactate	$CH_3CH(OH)COOC_2H_5$	118.13	3, 264	1.0328^{20}	1.4124^{20}	−26	154.5	70	misc aq, alc, eth, esters, PE
e166	Ethyl DL-mandelate	$C_6H_5CH(OH)COOC_2H_5$	180.21	10, 202			37	253–255		
e167	Ethyl 2-mercaptoacetate	$HSCH_2COOC_2H_5$	120.17	3, 255	1.0964^{15}	1.4571^{20}		54^{12mm}	47	s alc, eth
e168	Ethylmercury chloride	CH_3CH_2HgCl	165.13		3.5		192	subl		0.78 eth; 2.6 chl
e169	Ethyl methacrylate	$H_2C{=}C(CH_3)COOC_2H_5$	114.14	2, 423	0.909_{15}^{25}	1.4116^{25}		118	49	i aq; s alc, eth
e170	Ethyl 4-methoxyphenylacetate	$CH_3OC_6H_4CH_2COOC_2H_5$	194.23	10^1, 83	1.097	1.5075^{20}		138^{7mm}	46	
e171	Ethyl 2-methylacetoacetate	$CH_3C({=}O)CH(CH_3)$-$COOC_2H_5$	144.17	3, 679	1.019_4^{20}	1.4182^{20}		187	62	i aq; s alc, eth
e172	*N*-Ethyl-*N*-methylaniline	$C_6H_5N(CH_3)C_2H_5$	135.21	12, 162	0.9193_4^{55}	1.5474^{20}		203–205		i aq; misc alc, eth
e173	Ethyl 3-methylbutyrate	$(CH_3)_2CHCH_2COOC_2H_5$	130.19	2^2, 275	0.868_{20}^{20}	1.3962^{20}	−99.3	134.7	26	0.2 aq; misc alc, bz
e174	Ethyl methyl ether	$CH_3CH_2OCH_3$	60.09	1, 314	0.725_0^0			10.8		s aq; misc alc, eth

No.	Name	Formula	Mol wt	Beilstein	Density	Refractive index	mp	bp	Flash point	Solubility
e175	2-Ethyl-4-methylimidazole		110.16	23^2, 72	0.975	1.4995^{20}		292–295	137	
e176	Ethyl 4-methyl-5-imidazolecarboxylate		154.17	25^1, 534			204–206			
e177	3-Ethyl-2-methylpentane	$(C_2H_5)_2CHCH(CH_3)_2$	114.24	1^3, 489	0.7193_4^{20}	1.4040^{20}	−115.0	115.7		i aq; sl s alc; s eth
e178	3-Ethyl-3-methylpentane	$(C_2H_5)_3CCH_3$	114.24		0.7274_4^{20}	1.4078^{20}	−90.9	118.3		i aq; s eth
e179	Ethyl 3-methyl-3-phenylglycidate		206.24		1.09_4^{15}	1.508^{20}				
e180	Ethyl 1-methyl-2-piperidinecarboxylate		171.24	22^1, 485	0.975	1.4519^{20}		92–96^{11mm}	73	

Ethyl 2-hydroxypropionate, e165
Ethylidene bromide, d77
Ethylidene chloride, d176
Ethylidene dimethyl ether, d439
Ethylidene fluoride, d346
2,2′-Ethyliminodiethanol, e119
Ethyl iodide, i34

Ethyl isonicotinate, e219
Ethyl isonipecotate, e209
Ethyl isopropylacetate, e173
Ethyl isothiocyanatoformate, e33
Ethyl isovalerate, e173
Ethyl levulinate, e198
Ethyl linoleate, e191

Ethyl mercaptan, e20
Ethyl 3-methylcrotonate, e122
Ethyl methyl ketone, b396
Ethyl 1-methylnipecotate, e180
Ethyl 2-methyl-4-oxo-2-cyclohexene-1-carboxylate, c9
Ethyl 1-methylpipecolinate, e181

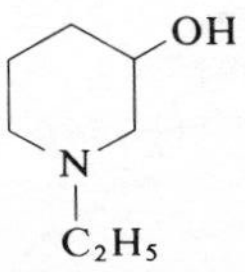

e160 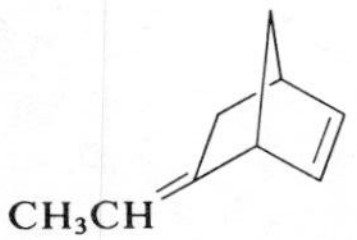 e161 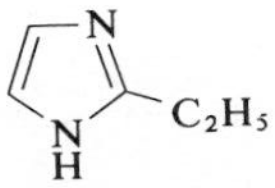 e162 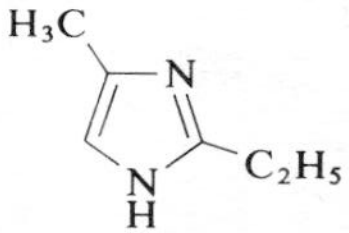 e175 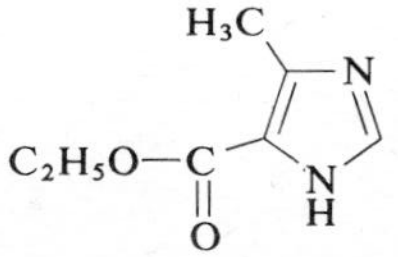 e176 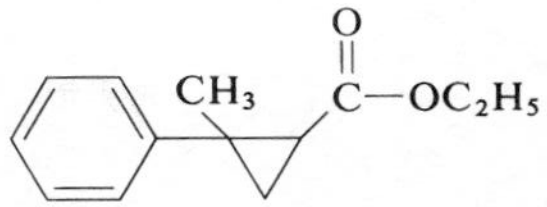 e179 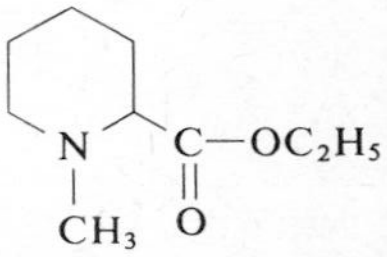 e180

TABLE 1-14 Physical constants of organic compounds (*continued*)

No.	Name	Formula	Formula weight	Beilstein reference	Density	Refractive index	Melting point	Boiling point	Flash point	Solubility in 100 parts solvent
e181	Ethyl 1-methyl-3-piperidinecarboxylate		171.24		0.954	1.4510^{20}		89^{11mm}		
e182	2-Ethyl-2-methyl-1,3-propanediol	$HOCH_2C(C_2H_5)(CH_3)$-CH_2OH	118.18	1, 487			41–44	226		
e183	3-Ethyl-4-methylpyridine	$C_2H_5(CH_3)C_5H_3N$	121.18	20^2, 163	0.9286_4^{17}			198		s alc, eth; sl s aq
e184	5-Ethyl-2-methylpyridine	$C_2H_5(CH_3)C_5H_3N$	121.18	20, 248	0.9184_4^{23}	1.4974^{20}		178	66	s alc, bz, eth, acid
e185	Ethyl methyl sulfide	$CH_3CH_2SCH_3$	76.15	1, 343	0.8422^{20}	1.4403^{20}	−105.9	66.7	49	i aq; misc alc, eth
e186	Ethyl (methylthio)-acetate	$CH_3SCH_2COOC_2H_5$	134.20		1.043	1.4587^{20}			59	
e187	*N*-Ethylmorpholine		115.18	27^1, 203	0.916_{20}^{20}	1.4410^{20}	−63	139	27	misc aq, alc, eth
e188	Ethyl nitrate	$CH_3CH_2ONO_2$	91.13	1, 329	1.100_4^{25}	1.3849^{22}	−94.6	87.7	flammable	1 aq; misc alc, eth
e189	Ethyl nitrite	CH_3CH_2ONO	75.07	1, 329	0.90_{15}^{15}			17		misc alc, eth
e190	4-Ethylnitrobenzene	$C_2H_5C_6H_4NO_2$	151.17	5, 358	1.118	1.5445^{20}	−32	245–246	>112	v s alc, eth
e191	Ethyl (*Z*,*Z*)-9,12-octadecadienoic acid	$H(CH_2)_5CH{=}CHCH_2$-$CH{=}CH(CH_2)_7COOC_2H_5$	308.51	2^2, 461	0.8846_4^{16}	1.4675^{20}		193^{6mm}	>112	misc DMF, oils
e192	Ethyl *cis*-9-octadecenoate	$CH_3(CH_2)_7CH{=}CH$-$(CH_2)_7COOC_2H_5$	310.52	2, 467	0.869_4^{20}	1.445^{25}	<−15	216^{15mm}		i aq; misc alc, eth
e193	Ethyl octanoate	$CH_3(CH_2)_6COOC_2H_5$	172.27	2, 348	0.878^{17}	1.4166^{20}	−47	206–208	75	i aq; misc alc, eth
e194	Ethyl oxalyl chloride	$CH_3CH_2OC({=}O)COCl$	136.53	2, 541	1.2223_4^{20}	1.4164^{20}		135	41	d aq, alc; s bz, eth
e195	Ethyl oxamate	$CH_3CH_2OC({=}O)CONH_2$	117.10	2, 544			114–116			s aq, eth; i bz
e196	2-Ethyl-2-oxazoline		99.13		0.982	1.4370^{20}	−62	128.4	29	

e197	Ethyl 2-oxocyclopentanecarboxylate	$(O{=})C_5H_7COOC_2H_5$	156.18		1.054	1.4485^{20}		102^{11mm}	>112	
e198	Ethyl 4-oxopentanoate	$CH_3C({=}O)CH_2CH_2$-$COOC_2H_5$	144.17	3, 675	1.012_{20}^{20}	1.4222^{20}		205–206		v s aq; misc alc
e199	Ethyl 2-oxopropionate	$CH_3C({=}O)COOC_2H_5$	116.12	3, 616	1.060_4^{16}	1.408^{16}		144	45	sl s aq; misc alc, eth
e200	3-Ethylpentane	$(C_2H_5)_3CH$	100.20	1^3, 441	0.6982_4^{20}	1.3934^{20}	−118.6	93.5		i aq; s alc, eth
e201	Ethyl pentanoate	$CH_3(CH_2)_3COOC_2H_5$	130.19	2, 301	0.877_4^{20}	1.3732^{20}	−91.3	145.5		0.2 aq; misc alc, eth
e202	4-Ethylphenol	$CH_3CH_2C_6H_4OH$	122.17	6, 472	1.011_4^{25}	1.5239	47.0	218–219		i aq; misc alc, eth
e203	Ethyl phenylacetate	$C_6H_5CH_2COOC_2H_5$	164.20	9, 434	1.0333_4^{20}	1.4980^{20}		226	77	i aq; misc alc, eth
e204	Ethyl *N*-piperazinocarboxylate		158.20	23^2, 9	1.080	1.4765^{20}		273	>112	
e205	1-Ethylpiperidine		113.20	20, 17	0.8237_4^{20}	1.4440^{20}		131	18	
e206	2-Ethylpiperidine		113.20	20, 104	0.850	1.4510^{20}		143	31	s aq
e207	Ethyl 2-piperidinecarboxylate		157.21	22, 7	1.006	1.4562^{20}		216–217	46	
e208	Ethyl 3-piperidinecarboxylate		157.21		1.012	1.4601^{20}		104^{7mm}	90	

Ethyl nicotinate, e218
Ethyl nipecotate, e208
Ethyl oleate, e192

Ethyl pentyl ketone, o35
Ethyl phenyl ether, e29
Ethyl picolinate, e217

Ethyl pipecolinate, e207
Ethylpiperidinol, e160

e181 e187 e196 e204 e205 e206 e207 e208

TABLE 1-14 Physical constants of organic compounds (*continued*)

No.	Name	Formula	Formula weight	Beilstein reference	Density	Refractive index	Melting point	Boiling point	Flash point	Solubility in 100 parts solvent
e209	Ethyl 4-piperidinecarboxylate		157.21		1.010	1.4591^{20}		204	80	s aq, alc, bz, eth
e210	Ethyl *N*-piperidinepropionate		185.27	20, 62	0.927	1.4545^{20}		217–219	87	
e211	Ethyl propionate	$CH_3CH_2COOC_2H_5$	102.13	2, 240	0.891_4^{20}	1.3839^{20}	−73.9	99.1	12	2 aq; misc alc, eth
e212	Ethyl propyl ether	$CH_3CH_2OCH_2CH_2CH_3$	88.15	1, 354	0.739_4^{20}	1.3695^{20}	−79	62–63	32	sl s aq; misc alc, eth
e213	Ethyl propyl sulfide	$CH_3CH_2SCH_2CH_2CH_3$	104.21	1^3, 1432	0.8270_4^{20}	1.4462^{20}	−117.0	118.5		s alc
e214	2-Ethylpyridine	$CH_3CH_2C_5H_4N$	107.16	20, 241	0.937	1.4964^{20}		149	29	sl s aq; s alc, eth
e215	3-Ethylpyridine	$CH_3CH_2C_5H_4N$	107.16	20, 242	0.954	1.5015^{20}		162–165	48	v s alc, eth; sl s aq
e216	4-Ethylpyridine	$CH_3CH_2C_5H_4N$	107.16	20, 243	0.9404_4^{22}	1.5009^{20}		168	47	sl s aq; s alc, eth
e217	Ethyl 2-pyridinecarboxylate		151.17	22, 35	1.1194^{20}	1.5088^{20}	2	240–241	107	misc aq, alc, eth
e218	Ethyl 3-pyridinecarboxylate		151.17	22, 39	1.1070^{20}	1.5040^{20}	8–9	23–224	93	v s aq, alc, eth; s bz
e219	Ethyl 4-pyridinecarboxylate		151.17	22^2, 37	1.009_4^{15}	1.5009^{20}	23	220	87	i aq; s alc, bz, chl
e220	1-Ethyl-2-pyrrolidinone		113.16		0.992	1.4652^{20}		97^{20mm}	76	
e221	Ethyl salicylate	$C_6H_4(OH)COOC_2H_5$	166.18	10, 73	1.131_4^{20}	1.5219^{20}	2–3	231–234	107	misc alc, eth; sl s aq
e222	Ethyl sorbate	$CH_3CH{=}CHCH{=}CH\text{-}COOC_2H_5$	140.18	2, 484	0.959	1.4942^{20}		195.5	69	
e223	*S*-Ethyl thioacetate	$CH_3C(=O)SCH_2CH_3$	104.16	2, 232	0.976_4^{28}	1.4503^{28}		116–117		i aq; v s alc, eth
e224	3-Ethylthio-1,2-propanediol	$C_2H_5SCH_2CH(OH)CH_2OH$	136.21		1.095	1.5065^{20}			>112	
e225	Ethyl 4-toluenesulfonate	$CH_3C_6H_4SO_2OC_2H_5$	200.26	11, 99	1.166_4^{45}	1.5110^{20}	33	173^{15mm}	157	i aq; s alc, eth
e226	*N*-Ethyl-*m*-toluidine	$CH_3C_6H_4NHC_2H_5$	135.21	12, 857	0.957	1.5451^{20}		221	89	

No.	Name	Formula	Mol. wt.	Beilstein	Density	Refr. index	mp	bp	Flash pt	Solubility
e227	6-Ethyl-*o*-toluidine	$CH_3CH_2C_6H_3(CH_3)NH_2$	135.21		0.968	1.5525^{20}	−33	231	89	
e228	2-(*N*-Ethyl-*m*-toluidino)ethanol	$CH_3C_6H_4N(C_2H_5)CH_2CH_2OH$	179.26		1.019	1.5540^{20}		115^{1mm}		
e229	Ethyl trichloroacetate	$Cl_3CCOOC_2H_5$	191.44	2, 209	1.383_4^{20}	1.4447^{20}		168	65	i aq; s alc, eth
e230	Ethyl trifluoroacetate	$F_3CCOOC_2H_5$	142.08	2^2, 186	1.194	1.3068^{20}		60–62	−1	
e231	Ethyl (trimethylsilyl)acetate	$(CH_3)_3SiCH_2COOC_2H_5$	160.29		0.876	1.4153^{20}		156–159	35	
e232	Ethyl undecanoate	$CH_3(CH_2)_9COOC_2H_5$	214.35	2, 358	0.859	1.4280^{20}		105^{4mm}	>112	i aq; s org solv
e233	Ethylurea	$CH_3CH_2NHC(=O)NH_2$	88.11	4, 115	1.213^{18}		93–96			v s aq; 80 alc; i eth
e234	*N*-Ethylurethane	$CH_3CH_2NHCOOC_2H_5$	117.15	4, 114	0.981_4^{20}	1.4211^{20}		85^{20mm}	75	63 aq
e235	Ethyl vinyl ether	$CH_3CH_2OCH=CH_2$	72.11	1, 433	0.7531^{20}	1.3754^{20}	−115.8	35.7	−17	0.9 aq
e236	*N*-Ethyl-2,3-xylidine	$(CH_3)_2C_6H_3NHC_2H_5$	149.24	12, 1101	0.917	1.5468^{20}		227–228	71	
e237	1-Ethynyl-1-cyclohexanol	$C_6H_{10}(C\equiv CH)OH$	124.18	6^2, 100	0.967_{20}^{20}		30–31	180	62	2.4 aq; misc alc, bz, acet, ketones, PE

Ethyl pivalate, e124
1-Ethyl-1-propanol, p39
Ethyl propenoate, e57
1-Ethylpropylamine, a255

Ethyl pyruvate, e199
2-(Ethylthio)ethanol, e156
Ethyl thioglycolate, e167
Ethyltriethoxysilane, t266

Ethyl trimethylacetate, e124
Ethyl vanillin, e40
Ethyne, a41
Ethynylbenzene, p82

e209

e210

e217

e218

e219

e220

TABLE 1-14 Physical constants of organic compounds (*continued*)

No.	Name	Formula	Formula weight	Beilstein reference	Density	Refractive index	Melting point	Boiling point	Flash point	Solubility in 100 parts solvent
e238	1-Ethynylcyclohexylamine	$C_6H_{10}(C{\equiv}CH)NH_2$	123.30		0.913	1.4817^{20}		66^{20mm}	42	
f1	Fluoranthene		202.26	5, 685	1.252^{0}_{4}		107–110	384		sl s alc; s bz, eth
f2	Fluorene		166.22	5, 625	1.203^{0}_{4}		114.8	295		v s HOAc; s bz, eth
f3	9-Fluorenone		180.21	7, 465	1.1300^{99}_{4}	1.6369^{99}	82–85	342		s alc, bz; v s eth
f4	Fluorescein		332.31	19, 222			314 d			s hot alc, hot HOAc, alk; i bz, chl, eth
f5	Fluoroacetamide	$FCH_2C(O)NH_2$	77.06	2, 193			107 subl			v s aq; s acet
f6	Fluoroacetic acid	FCH_2COOH	78.04	2, 193			33	165		sl s aq, alc
f7	Fluoroacetone	$CH_3C(O)CH_2F$	76.07		1.054	1.3700		75	7	
f8	*p*-Fluoroacetophenone	$FC_6H_4COCH_3$	138.14		1.138	1.5110^{20}		196	71	
f9	*p*-Fluoroaniline	$FC_6H_4NH_2$	111.12	12, 597	1.1725^{20}_{4}	1.5395^{20}	−1.9	187	73	sl s aq; s alc, eth
f10	*o*-Fluorobenzaldehyde	FC_6H_4CHO	124.11	7^1, 132	1.178	1.5220^{20}	−44.5	91^{46mm}	55	
f11	Fluorobenzene	C_6H_5F	96.11	5, 198	1.0240^{20}_{4}	1.4657^{20}	−42.2	84.7	−12	0.15 aq; misc alc
f12	*o*-Fluorobenzoic acid	FC_6H_4COOH	140.11	9, 333	1.460^{25}_{4}		123–125			sl s aq; s alc, eth
f13	*p*-Fluorobenzoic acid	FC_6H_4COOH	140.11	9, 333	1.479^{25}_{4}		182.6			0.1 aq; s alc, eth
f14	*p*-Fluorobenzoyl chloride	FC_6H_4COCl	158.56	9^1, 137	1.342	1.5296^{20}	9	82^{20mm}	82	
f15	*o*-Fluorobenzyl alcohol	$FC_6H_4CH_2OH$	126.13	6^1, 222	1.173	1.5136^{20}			90	
f16	*p*-Fluorobenzyl chloride	$FC_6H_4CH_2Cl$	144.58		1.207	1.5130^{20}		82^{26mm}	60	
f17	Fluoroethane	CH_3CH_2F	48.06	1, 82	0.00220^{0}		−143.2	−37.7		198 mL aq; v s alc, eth

f18	Fluoromethane	CH_3F	34.04	1, 59	1.1951 g/L		−141.8	−78.4		166 mL aq; v s alc, eth
f19	4-Fluoro-1-methoxybenzene	$FC_6H_4OCH_3$	126.13	6^1, 98	1.114	1.4877^{20}	−45	157	43	s eth
f20	2-Fluoro-2-methylpropane	$(CH_3)_3CF$	76.11	1^4, 286			−77	12.1	−12	
f21	1-Fluoro-4-nitrobenzene	$FC_6H_4NO_2$	141.10	5, 241	1.3300_4^{20}	1.5312^{20}	21	205	83	i aq; s alc, eth
f22	4-Fluorophenol	FC_6H_4OH	112.10	6, 183			46–48	185	68	
f23	2-Fluoropyridine	FC_5H_4N	97.09	20^1, 80	1.128	1.4680^{20}		126	28	
f24	*o*-Fluorotoluene	$FC_6H_4CH_3$	110.13	5, 290	1.0014^{17}	1.4716^{17}	−62.0	114.4	12	v s alc, eth
f25	*m*-Fluorotoluene	$FC_6H_4CH_3$	110.13	5, 290	0.9974^{20}	1.4691^{20}	−87.7	116.5	9	s alc, eth
f26	*p*-Fluorotoluene	$FC_6H_4CH_3$	110.13	5, 290	0.9975^{20}	1.4688^{20}	−56.7	116.6	40	s alc, eth
f27	Formaldehyde	$H_2C{=}O$	30.03	1, 558	0.815_4^{-20}	0.8153^{-20}	−92	−19.5		122 aq; s alc, eth
f28	Formamide	$HC({=}O)NH_2$	45.04	2, 26	1.1334_4^{20}	1.4475^{20}	2.6	111^{20mm}	154	misc aq, alc, acet
f29	Formamidine acetate	$HC({=}NH)NH_2{\cdot}HOOCCH_3$	104.11				158 d			
f30	Formamidinesulfinic acid	$H_2NC({=}NH)S(O)OH$	108.12	3^1, 36			126 d			

Eugenol, m99
Fenchone, t357
Fenchyl alcohol, t356
Ferulic acid, h137
2,7-Fluorenediamine, d38
N-9H-(2-Fluorenyl)acetamide, a13
Fluorotrichloromethane, t236
Fluothane, b260
Formic acid hydrazide, f34

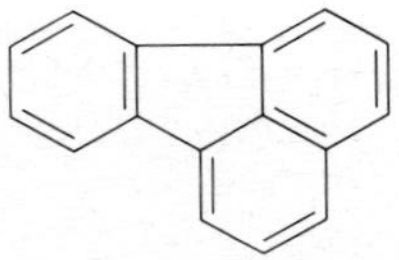

f1

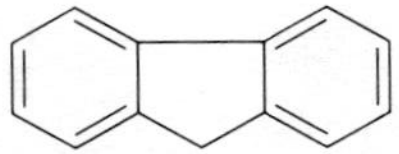

f2

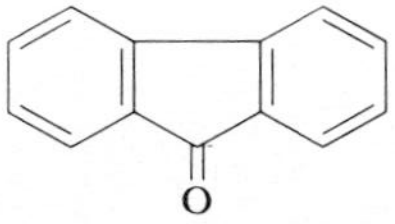

f3

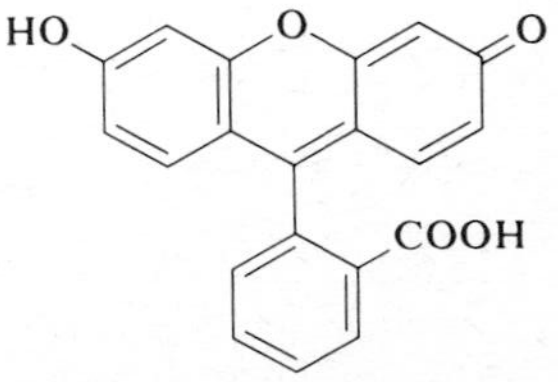

f4

TABLE 1-14 Physical constants of organic compounds (*continued*)

No.	Name	Formula	Formula weight	Beilstein reference	Density	Refractive index	Melting point	Boiling point	Flash point	Solubility in 100 parts solvent
f31	Formanilide	C_6H_5NHCHO	121.14	12, 230	1.144		47	271		2.5 aq
f32	Formic acid	$HCOOH$	46.03	2, 8	1.220_4^{20}	1.3714^{20}	8.5	100.8	68	misc aq, alc, eth
f33	2-Formylbenzoic acid	$C_6H_4(HCO)COOH$	150.13	10, 666	1.404		98			s aq; v s alc, eth
f34	Formylhydrazine	$HC(=O)NHNH_2$	60.06	2, 93			54–56			v s alc, chl, eth; s bz
f35	*N*-Formylpiperidine		113.16	20, 45	1.019	1.4780^{20}		222	91	
f36	D-(−)-Fructose		180.16	31, 321						v s aq; 6.7 alc; s pyr
f37	Fumaric acid	$HOOCCH=CHCOOH$	116.07	2, 737	1.635_4^{20}		287	subl 200		0.6 aq; 9 alc; 0.7 eth
f38	Fumaroyl dichloride	$ClC(=O)CH=CH-C(=O)Cl$	152.96	2, 743	1.408^{20}	1.4988^{20}		161–164	73	d aq, alc
f39	2-Furaldehyde		96.09	17^2, 305	1.1598_4^{20}	1.5262^{20}	−36.5	161.8	68	8 aq; misc alc, eth
f40	Furan		68.07	17, 27	0.9371_4^{20}	1.4214^{20}	−85.6	31.4	−35	1 aq; misc alc, eth
f41	2-Furanacrylic acid		138.12	18, 300			141	286		0.2 aq; 1.1 bz; s alc, eth, HOAc
f42	2-Furancarboxylic acid		112.08	18, 272			133–134	230–232		4 aq; s alc; v s eth
f43	2,5-Furandimethanol		128.13	17^1, 90			74–76			
f44	2-Furanmethanethiol		114.17	17^2, 116	1.132	1.5304^{20}		155	45	
f45	Furfuryl acetate		140.14	17^2, 115	1.1175_4^{20}	1.4618^{20}		175–177	65	i aq; s alc, eth
f46	Furfuryl alcohol		98.10	17, 112	1.1285_4^{20}	1.4868^{20}	−14.6	170.0	65	misc aq(d); v s alc
f47	Furfurylamine		97.12	18, 584	1.0995_4^{20}	1.4900^{20}	−70	145–146	45	misc aq; s alc, eth
f48	2-Furoyl chloride		130.53	18, 276	1.324	1.5310^{20}	−2	170	85	d aq, alc; s eth
g1	D-(+)-Galactose		180.16	31, 295			167			200 aq; s pyr; sl s alc
g2	Geraniol	$(CH_3)_2C=CHCH_2CH_2-C(CH_3)=CHCH_2OH$	154.25	1, 457	0.8894_4^{20}	1.4760^{20}		230	76	i aq; misc alc, eth
g3	α-D-Glucoheptonic acid γ-lactone		208.17				145–148			s aq

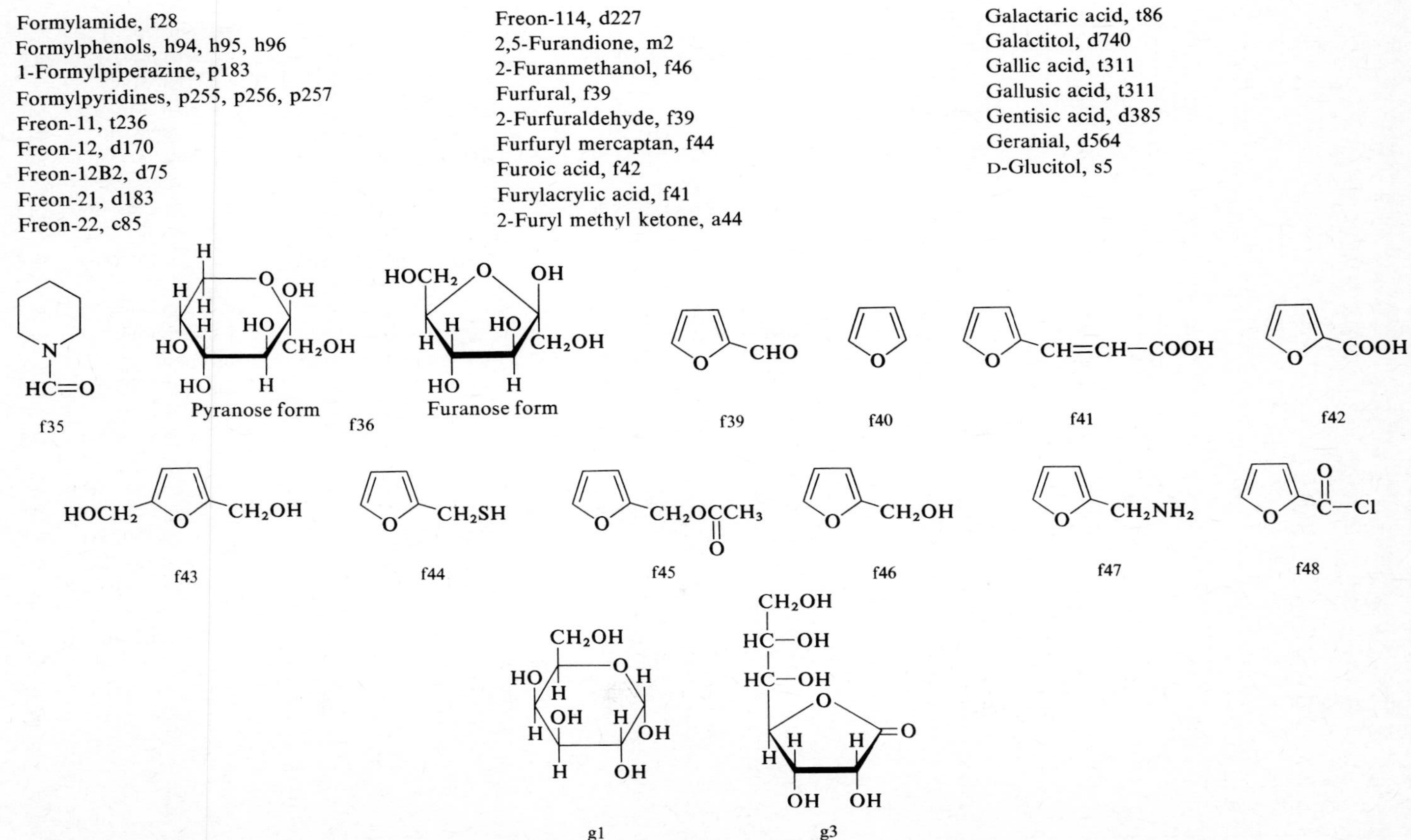
HC=O
N
f35
Pyranose form
Furanose form
f36
CHO
f39
f40
CH=CH—COOH
f41
COOH
f42
HOCH₂
CH₂OH
f43
CH₂SH
f44
CH₂OCCH₃
f45
CH₂OH
f46
CH₂NH₂
f47
C—Cl
f48
g1
g3

TABLE 1-14 Physical constants of organic compounds (*continued*)

No.	Name	Formula	Formula weight	Beilstein reference	Density	Refractive index	Melting point	Boiling point	Flash point	Solubility in 100 parts solvent
g4	D-Gluconic acid		196.16	3, 542			131			v s aq; sl s alc; i eth
g5	D-Glucosamine		179.17	1, 902			88(α)			v s aq; i chl, eth
g6	α-D-(+)-Glucose		180.16	31, 83	1.5620_4^{18}		146			91 aq; 0.83 MeOH; s pyr
g7	α-D-Glucose penta-acetate		390.34	31, 119			109–111			0.15 aq; 1.3 alc; 3 eth
g8	D-Glucurono-3,6-lactone		176.12				176–178			27 aq; 2.8 MeOH
g9	L-Glutamic acid		147.13	4, 488	1.538_4^{20}		d 247	subl 200		0.8 aq; i alc, eth
g10	L-Glutamine		146.15	4, 491			d 185			4 aq; 0.0035 MeOH; i bz, chl, eth, acet
g11	Glutaric acid	$HOOCCH_2CH_2CH_2COOH$	132.12	2, 631	1.429_4^{20}	1.4188^{106}	97.5	200^{20mm}		64 aq; v s alc, eth; s bz, chl; sl s PE
g12	Glutaric anhydride		114.10	17, 411			52–55	150^{10mm}		
g13	Glutaric dialdehyde	$OCHCH_2CH_2CH_2CHO$	100.12	1, 776		1.3730^{20}	−6	187–189 d		misc aq, alc
g14	Glutaronitrile	$NCCH_2CH_2CH_2CN$	94.12	2, 635	0.9888^{23}	1.4345^{20}	−29	286	112	s aq, alc, chl; i eth
g15	Glutaryl dichloride	$ClC(=O)CH_2CH_2CH_2$-$C(=O)Cl$	169.01	2, 634	1.324	1.4720^{20}		216–218	106	d aq, alc; s eth
g16	Glycerol	$HOCH_2CH(OH)CH_2OH$	92.09	1, 502	1.2613^{20}	1.4746^{20}	18.18	182^{20mm}		misc aq, alc; 0.2 eth
g17	Glyceryl 1,2-diacetate	$HOCH_2CH(OOCCH_3)$-CH_2OOCCH_3	176.17	2, 147	1.184_4^{16}	1.1173^{15}	40	172^{40mm}		s aq, alc, bz, eth
g18	Glyceryl 1,3-diacetate	$CH_3COOCH_2CH(OH)$-CH_2OOCCH_3	176.17	2, 290	1.179^{15}	1.4395^{20}	42	172^{40mm}		s aq, alc, bz, chl
g19	Glyceryl tris-(butyrate)		302.37	2, 273	1.032_4^{20}	1.4359^{20}	−75	305–310	173	i aq; v s alc, eth
g20	Glyceryl tris-(dodecanoate)		639.02	2, 363	0.894_4^{60}	1.4404^{60}	46			v s bz, eth; sl s alc

g21	Glyceryl tris-(nitrate)	$O_2NOCH_2CH(ONO_2)$-CH_2ONO_2	227.09	1, 516	1.594_4^{20}	1.4786^{12}	13.3	160^{5mm}	expl 270	0.18 aq; 54 alc; misc eth
g22	Glyceryl tris-(oleate)		885.46	4, 468	0.915_4^{15}	1.4621^{40}	−4 to −5	235^{15mm}		s chl, eth, CCl_4
g23	Glyceryl tris-(palmitate)		807.35	2, 373	0.8663_4^{80}	1.4381^{80}	65–66	310–320		v s bz, chl, eth
g24	Glyceryl tris-(tetradecanoate)		723.18	2, 367	0.885_4^{60}	1.4428^{60}	57			v s alc, bz, chl

Glutaraldehyde, g13
Glyceraldehyde, d399
Glycerol dichlorohydrin, d220
Glycerol α-monochlorohydrin, c213
Glyceryl triacetate, p204
Glyceryl tris(laurate), g20
Glyceryl tris(myristate), g24

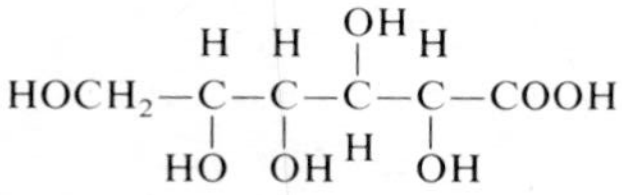

$HOCH_2-CH(OH)-CH(OH)-C(OH)H-CH(OH)-COOH$

g4

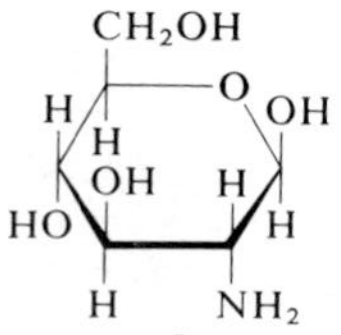

g5

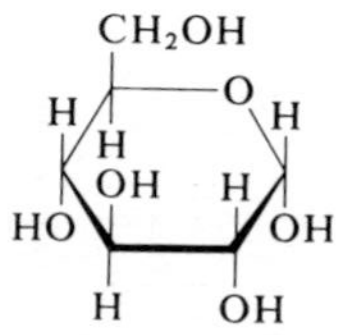

g6

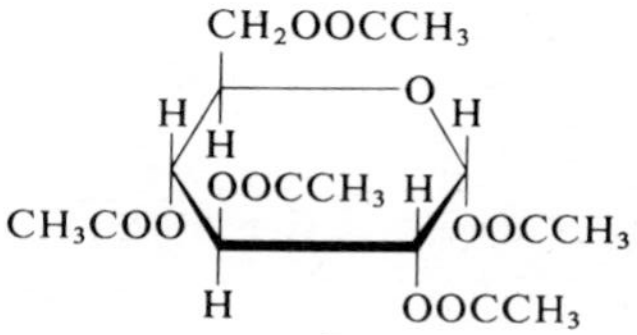

g7

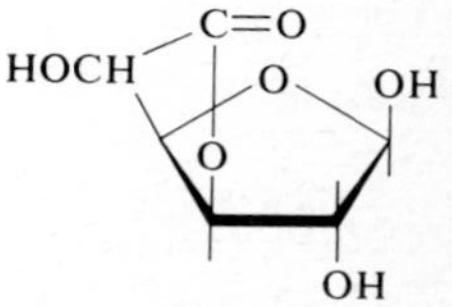

g8

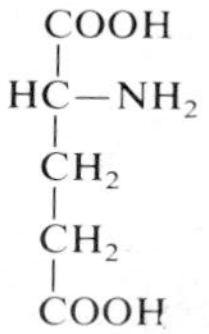

$HOOC-CH(NH_2)-CH_2-CH_2-COOH$

g9

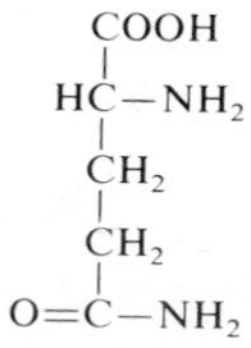

$HOOC-CH(NH_2)-CH_2-CH_2-C(=O)-NH_2$

g10

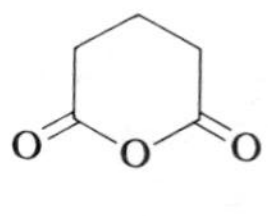

g12

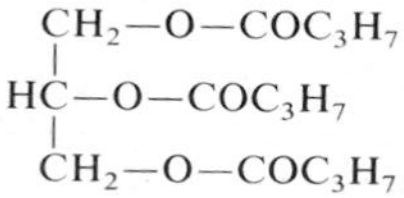

$CH_2-O-COC_3H_7$
$HC-O-COC_3H_7$
$CH_2-O-COC_3H_7$

g19

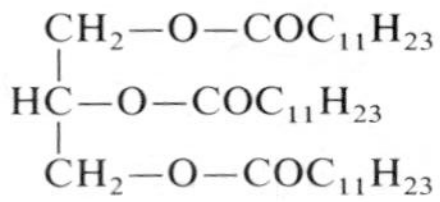

$CH_2-O-COC_{11}H_{23}$
$HC-O-COC_{11}H_{23}$
$CH_2-O-COC_{11}H_{23}$

g20

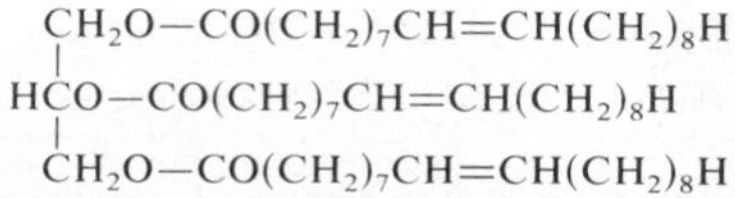

$CH_2O-CO(CH_2)_7CH=CH(CH_2)_8H$
$HCO-CO(CH_2)_7CH=CH(CH_2)_8H$
$CH_2O-CO(CH_2)_7CH=CH(CH_2)_8H$

g22

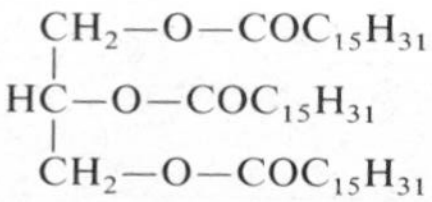

$CH_2-O-COC_{15}H_{31}$
$HC-O-COC_{15}H_{31}$
$CH_2-O-COC_{15}H_{31}$

g23

$CH_2-O-COC_{13}H_{27}$
$HC-O-COC_{13}H_{27}$
$CH_2-O-COC_{13}H_{27}$

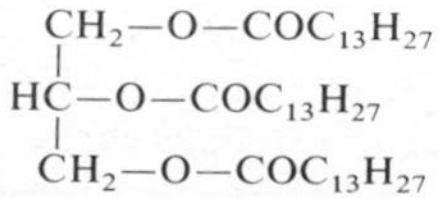

g24

TABLE 1-14 Physical constants of organic compounds (*continued*)

No.	Name	Formula	Formula weight	Beilstein reference	Density	Refractive index	Melting point	Boiling point	Flash point	Solubility in 100 parts solvent
g25	Glycine	H_2NCH_2COOH	75.07	4, 333	1.1607		d 233			25 aq; 0.6 pyr; i eth
g26	*N*-Glycylglycine	$H_2NCH_2C(=O)NHCH_2$-COOH	132.12	4, 371			d 262			s hot aq; sl s alc
g27	Glyoxal	$HC(=O)CHO$	58.04	1, 759	1.29_4^{20}	1.3826^{20}	15	51		violent reaction aq; s anhyd solv; mixtures with air may explode
g28	Glyoxylic acid	$HC(=O)COOH$	74.04	3, 594			98			v s aq; sl s alc, eth
g29	Guanidine	$H_2NC(=NH)NH_2$	59.07	3, 82			~60	d 160		v s aq, alc
h1	Heptadecane	$CH_3(CH_2)_{15}CH_3$	140.41	1, 173	0.7767^{22}	1.4360^{25}	22.0	302.2	148	s eth; sl s alc
h2	Heptafluorobutyric acid	$CF_3CF_2CF_2COOH$	214.04		1.645			120		
h3	Heptafluoro-2,3,3-trichlorobutane	$CF_3CCl_2CF(Cl)CF_3$	287.5		1.7484^{20}	1.3530^{20}	4	98		
h4	2,2,4,4,6,8,8-Heptamethylnonane	$(CH_3)_3CCH_2C(CH_3)_2CH_2$-$CH(CH_3)CH_2C(CH_3)_3$	226.45		0.793	1.4391^{20}		240		
h5	Heptanal	$CH_3(CH_2)_5CHO$	114.19	1^2, 750	0.8216_4^{15}	1.4285^{20}	−43	153	35	misc alc, eth; sl s aq
h6	Heptane	$CH_3(CH_2)_5CH_3$	100.21	1, 154	0.6838_4^{20}	1.3877^{20}	−90.6	98.4	−1	s alc, chl, eth
h7	1,7-Heptanediamine	$H_2N(CH_2)_7NH_2$	130.24	4, 271			27–29	147–149	87	
h8	Heptanedioic acid	$HOOC(CH_2)_5COOH$	160.17	2, 670	1.329^{15}		105.8	212^{10mm}		5 aq; v s alc, eth
h9	1-Heptanethiol	$CH_3(CH_2)_6SH$	132.27	1, 415			−43.2	176.9	46	i aq
h10	Heptanoic acid	$CH_3(CH_2)_5COOH$	130.19	2, 338	0.9181_4^{20}	1.4221^{20}	−7.5	223.0	>112	0.25 aq; s alc, eth
h11	Heptanoic anhydride	$[CH_3(CH_2)_5CO]_2O$	242.36	2, 340	0.932_4^{20}	1.4332^{20}	−12.4	268	>112	i aq; s alc, eth
h12	1-Heptanol	$CH_3(CH_2)_6OH$	116.20	1, 414	0.8219_4^{20}	1.4242^{20}	−34.6	175.8	73	misc alc, eth
h13	2-Heptanol	$CH_3(CH_2)_4CH(OH)CH_3$	116.20	1, 415	0.8193_4^{20}	1.4210^{20}		160	41	0.35 aq; s alc, bz, eth

h14	3-Heptanol	$CH_3(CH_2)_3CH(OH)$-CH_2CH_3	116.20)-	1^1, 205	0.818	1.4214^{20}		66^{20mm}	54	sl s aq
h15	2-Heptanone	$CH_3(CH_2)_4COCH_3$	114.19	1, 699	0.8197^{15}_{4}	1.4116^{15}	−35	151	47	s alc, eth
h16	3-Heptanone	$CH_3(CH_2)_3C(=O)$-CH_2CH_3	114.19	1, 699	0.8197^{20}_{20}	1.4085^{20}	−36.7	147.8	41	0.43 aq; s alc, eth
h17	4-Heptanone	$CH_3CH_2CH_2(O)CH_2$-CH_2CH_3	114.19	1, 699	0.821^{15}_{4}	1.4068^{20}	−32.1	143.7	48	0.53 aq; misc alc, eth
h18	Heptanoyl chloride	$CH_3(CH_2)_5COCl$	148.63	2, 340	0.960^{20}	1.4300^{20}		173	57	d aq, alc; s eth
h19	1-Heptene	$CH_3(CH_2)_4CH=CH_2$	98.90	1, 219	0.6970^{20}	1.3999^{20}	−118.9	93.6	−1	0.1 aq; s alc, eth
h20	1-Heptylamine	$CH_3(CH_2)_6NH_2$	115.22	4, 193	0.777	1.4243^{20}	−23	154–156	35	s alc, acet, eth, PE
h21	Heptyltrichloro-silane	$CH_3(CH_2)_6SiCl_3$	233.7		1.087^{20}_{4}	1.4439^{25}		211–212		
h22	1-Heptyne	$CH_3(CH_2)_4C\equiv CH$	96.17	1, 256	0.733	1.4075^{20}	−81	99–100	22	
h23	Hexachloroacetone	$Cl_3CC(=O)CCl_3$	264.75	1, 657	1.743	1.5112^{20}	−30	66^{6mm}	none	sl s aq; s acet
h24	Hexachlorobenzene	C_6Cl_6	284.78	5, 205	2.044^{24}		231	323–326		s bz, chl, eth
h25	Hexachloro-1,3-butadiene	$Cl_2C=CClCCl=CCl_2$	260.76	1, 250	1.655	1.5550^{20}	−19	210–220	none	s alc, eth
h26	1,2,3,4,5,6-Hexa-chlorocyclohexane	$C_6H_6Cl_6$	290.83	5^2, 11	1.87^{20}		113			s bz, chl
h27	Hexachlorocyclo-1,3-pentadiene		272.77		1.701^{25}_{4}	1.5644^{20}	−10	239	none	

Glycidol, e11
Glycidyl methacrylate, e12
Glycinonitrile, a106
Glycolaldehyde, h86
Glycolaldehyde diethyl acetal, d253
Glycolic acid, h87
Glycol methacrylate, h121
Glyoxaline, i4
Guaiacol, m87
Heliotropin, m240
Heliotropyl alcohol, m243
Hemimellitene, t338
Hemimellitic acid, b28
Heptaldehyde, h5
sec-Heptyl alcohol, h13
Heptyl bromide, b293
Heptyl chloride, c129
Heptyl iodide, i37
Heptyl mercaptan, h9
Hexachloro-2-propanone, h23

Cl Cl Cl Cl Cl Cl

h27

TABLE 1-14 Physical constants of organic compounds (*continued*)

No.	Name	Formula	Formula weight	Beilstein reference	Density	Refractive index	Melting point	Boiling point	Flash point	Solubility in 100 parts solvent
h28	Hexachlorodisiloxane	$Cl_3SiOSiCl_3$	284.9				−35	137		
h29	Hexachloroethane	Cl_3CCCl_3	236.74	1, 87	2.091_4^{20}		187–188			s alc, bz, chl, eth
h30	1,4,5,6,7,7-Hexachloro-5-norbornene-2,3-dicarboxylic anhydride		370.83				235–239			
h31	Hexachloropropene	$Cl_3CC(Cl)=CCl_2$	248.75	1, 200	1.765	1.5480^{20}		210	none	
h32	Hexadecane	$CH_3(CH_2)_{14}CH_3$	226.45	1, 172	0.7733_4^{20}	1.4345^{20}	18.2	286.8	135	misc eth
h33	1,2-Hexadecanediol	$CH_3(CH_2)_{13}CH(OH)$-CH_2OH	258.45	1^3, 2244			72–74			
h34	1-Hexadecanethiol	$CH_3(CH_2)_{15}SH$	258.51	1, 430	0.840	1.4720^{20}	18–20	184^{7mm}	101	sl s alc; s eth
h35	Hexadecanoic acid	$CH_3(CH_2)_{14}COOH$	256.43	2, 370	0.852_4^{62}	1.4273^{80}	63–64	215^{15mm}		s hot alc, chl, eth
h36	1-Hexadecanol	$CH_3(CH_2)_{15}OH$	242.45	1, 429	0.8116^{60}	1.4355^{60}	49.3	344	135	s alc, chl, eth
h37	1-Hexadecene	$CH_3(CH_2)_{13}CH=CH_2$	224.43	1, 226	0.783_4^{20}	1.4401	4.1	274	132	s alc, eth, PE
h38	1-Hexadecylamine	$CH_3(CH_2)_{15}NH_2$	241.46	4, 202			40–42	330	140	v s alc, eth; s bz, chl
h39	4-Hexadecylaniline	$CH_3(CH_2)_{15}C_6H_4NH_2$	317.56	12, 1186			51–52	254–255^{15mm}		
h40	2,4-Hexadienal	$CH_3CH=CHCH=CHCHO$	96.13	1^2, 809	0.898^{20}	1.5386^{20}		76^{30mm}	67	
h41	1,5-Hexadiene	$H_2C=CHCH_2CH_2CH=CH_2$	82.15	1, 253	0.6923_4^{20}	1.4042^{20}	−140.7	59.5	<1	s alc, eth
h42	2,4-Hexadienoic acid	$CH_3CH=CHCH=$ $CHCOOH$	112.13	2, 483			134.5	119^{10mm}	127	0.2 aq; 13 alc; 9 acet; 2.3 bz; 11 diox; 1 CCl_4
h43	Hexafluorobenzene	C_6F_6	186.05		1.6182^{20}	1.3781^{20}	5.1	80.3	10	
h44	Hexafluoroethane	F_3CCF_3	138.01	1^3, 132	1.590^{-78}		−100.1	−78.3		sl s alc, eth
h45	1,1,1,3,3,3-Hexafluoro-2-propanol	$(CF_3)_2CHOH$	168.04		1.596^{25}	1.2750^{20}	−3	58.2	4	s aq, bz, CCl_4
h46	*cis*-Hexahydroindane		124.23	5, 82	0.876	1.4702	−53	167	23	s eth
h47	Hexamethylbenzene	$C_6(CH_3)_6$	162.28	5, 450			165.6	264		v s bz; s acet, eth

h48	Hexamethylcyclotri-siloxane	$[-Si(CH_3)_2O-]_3$	222.48	4^3, 1884			64	133–135	35	
h49	1,1,1,3,3,3-Hexa-methyldisilazane	$(CH_3)_3SiNHSi(CH_3)_3$	161.40		0.774_4^{20}	1.4071^{20}		126	22	
h50	Hexamethyldi-siloxane	$(CH_3)_3SiOSi(CH_3)_3$	162.38		0.764_4^{20}	1.3775^{20}	−67	101	−1	
h51	Hexamethyleneimine		99.18	20, 94	0.880	1.4631^{20}		138^{749mm}	18	
h52	Hexamethylene-tetramine		140.19	1, 583	1.331^{-5}		subl 263		250	67 aq; 8 alc; 10 chl
h53	Hexamethylphosphor-amide	$[(CH_3)_2N]_3P(O)$	179.20		1.027^{20}	1.4588^{20}	7.2	233	105	misc aq
h54	Hexanal	$CH_3(CH_2)_4CHO$	100.16	1^2, 745	0.8335_4^{20}	1.4035^{20}		131	32	v s alc, eth; sl s aq
h55	Hexane	$CH_3(CH_2)_4CH_3$	86.18	1, 142	0.6594_4^{20}	1.3749^{20}	−95.4	68.7	−23	misc alc, chl, eth
h56	1,6-Hexanediamine	$H_2N(CH_2)_6NH_2$	116.21	4, 269			42	205	81	v s aq; sl s alc, bz
h57	1,6-Hexanedioic acid	$HOOC(CH_2)_4COOH$	146.14	2, 649	1.360_4^{25}		152	337.5	196	1.4 aq; v s alc; s acet

$\alpha,\alpha,\alpha,\alpha',\alpha',\alpha'$-Hexachloro-*p*-xylene, b203
Hexadecyl mercaptan, h34
$\alpha,\alpha,\alpha,\alpha',\alpha',\alpha'$-Hexafluoro-3,5-xylidine, b206
Hexahydroaniline, c335
Hexahydro-2*H*-azepin-2-one, o57
Hexahydrobenzaldehyde, c316
Hexahydrobenzoic acid, c318
Hexahydrobenzylamine, c325
Hexahydrophthalic acid, c321
Hexahydropyridine, p186
Hexamethylenediamine, h56
Hexamethylene diisocyanate, d411
Hexamethylene glycol, h59
Hexamethylene oxide, o47
Hexamethylethane, t102
2,6,10,15,19,23-Hexamethyl-2,6,10,14,18,22-tetracosahexene, s8
2,6,10,15,19,23-Hexamethyltetracosane, s7

h30

h46

h51

h52

TABLE 1-14 Physical constants of organic compounds (*continued*)

No.	Name	Formula	Formula weight	Beilstein reference	Density	Refractive index	Melting point	Boiling point	Flash point	Solubility in 100 parts solvent
h58	DL-1,2-Hexanediol	$CH_3(CH_2)_3CH(OH)CH_2OH$	118.18	1^1, 251	0.951	1.4425^{20}		223–234	> 112	
h59	1,6-Hexanediol	$HO(CH_2)_6OH$	118.18	1, 484	0.958	1.4579^{25}	42.8	243–250	101	v s aq, alc
h60	2,5-Hexanediol	$CH_3CH(OH)CH_2CH_2$-$CH(OH)CH_3$	118.18	1, 485	0.9617^{45}_{16}	1.4465^{20}	−50 glass	220.8	101	s aq, alc, eth
h61	2,5-Hexanedione	$CH_3COCH_2CH_2COCH_3$	114.14	1, 788	0.973	1.4260^{20}	−6	191.4	70	misc aq, alc, eth
h62	Hexanedioyl dichloride	$ClC(=O)(CH_2)_4COCl$	183.03	2, 653	1.259	1.4706^{20}		105^{2mm}	> 112	
h63	Hexanenitrile	$CH_3(CH_2)_4CN$	97.16	2, 324	0.8052^{20}	1.4069^{20}	−80.3	163.6	43	i aq; s alc, eth
h64	1-Hexanethiol	$CH_3(CH_2)_5SH$	118.24	1^3, 1659	0.8424^{20}_4	1.4496^{20}	−80.5	152.7	20	i aq; v s alc, eth
h65	1,2,6-Hexanetriol	$HOCH_2CH(OH)(CH_2)_3$-CH_2OH	134.17		1.1063^{20}_{20}	1.4771	−32.8	178^{5mm}	79	misc alc, acet; i bz
h66	Hexanoic acid	$CH_3(CH_2)_4COOH$	116.16	2, 321	0.9265^{20}_4	1.4168^{20}	−4.0	205.7	104	1.1 aq; v s alc, eth
h67	Hexanoic anhydride	$[CH_3(CH_2)_4C(=O)]_2O$	214.31	2, 324	0.926	1.4280^{20}	−41	246–248	> 112	s alc
h68	1-Hexanol	$CH_3(CH_2)_5OH$	102.18	1, 407	0.8186^{20}_4	1.4182^{20}	−51.6	157.5	60	8 aq; misc bz, eth; s alc
h69	2-Hexanol	$CH_3(CH_2)_3CH(OH)CH_3$	102.18	1, 408	0.8108^{25}_4	1.4128^{25}	−47	139.9	41	sl s aq; s alc, eth
h70	3-Hexanol	$CH_3CH_2CH_2CH(OH)$-CH_2CH_3	102.18	1, 408	0.8193^{20}_4	1.4160^{20}		135	41	
h71	6-Hexanolactone		114.14	17^2, 290	1.030	1.4630^{20}		97^{15mm}	109	
h72	2-Hexanone	$CH_3(CH_2)_3COCH_3$	100.16	1, 689	0.8209^{20}_4	1.4024^{20}	−56.9	127.2	35	v s alc, eth
h73	Hexanoyl chloride	$CH_3(CH_2)_4COCl$	134.61	2, 324	0.9754^{20}_4	1.4263^{20}	−87	153	79	d aq, alc; s eth
h74	1,4,7,10,13,16-Hexa-oxacyclooctadecane		264.32				40			
h75	1-Hexene	$CH_3(CH_2)_3CH=CH_2$	84.16	1, 215	0.6732^{20}	1.3879^{20}	−139.8	63.5	−26	0.005 aq
h76	*trans*-3-Hexenoic acid	$CH_3CH_2CH=CHCH_2$-$COOH$	114.14	2, 435	0.963	1.4398^{20}	11–12	119^{22mm}	> 112	
h77	*trans*-2-Hexen-1-ol	$CH_3CH_2CH_2CH=CH$-CH_2OH	100.16	1^2, 486	0.849	1.4343^{20}		158–160	54	

h78	5-Hexen-2-one	$H_2C{=}CHCH_2CH_2COCH_3$	98.15	1, 734	0.847	1.4197^{20}		128–129	23	
h79	Hexyl acetate	$CH_3(CH_2)_5OOCCH_3$	144.21	2, 132	0.860_{20}^{20}	1.4090^{20}	−80	168–170	37	0.13 aq; v s alc, eth
h80	Hexylamine	$CH_3(CH_2)_5NH_2$	101.19	4, 188	0.763_4^{25}	1.4180^{20}	−23	131–132	8	sl s aq; misc alc, eth
h81	4-Hexylaniline	$CH_3(CH_2)_5C_6H_4NH_2$	177.29	12^3, 2759				146–148^{17mm}		
h82	1-Hexyne	$H(CH_2)_4C{\equiv}CH$	82.14	1^3, 977	0.7152_4^{20}	1.3989^{20}	−131.9	71.3		i aq; s alc, eth
h83	L-Histidine		155.16	25, 513			d 285			41 aq; v sl s alc
h84	Hydantoin		100.08	24, 242			220			s alc, alk; sl s eth
h85	Hydrindantin		322.27	8^1, 631			100	d 252		v sl s aq
h86	Hydroxyacetaldehyde	$HOCH_2CHO$	60.05	1, 817	1.366^{100}		93–94	110^{12mm}		v s aq, alc; sl s eth
h87	Hydroxyacetic acid	$HOCH_2COOH$	76.05	3, 228			80			

D-*erythro*-Hex-2-enoic acid γ-lactone, i59
Hexyl alcohol, h68
sec-Hexyl alcohol, e84
sec-Hexylamine, m353a
Hexylbenzene, p119
Hexyl bromide, b296
Hexyl chloride, c130
Hexylene glycol, m341
Hexyl iodide, i39
Hexyl methyl ketone, o34
Hexyl propyl ketone, d16
Hippuric acid, b71
Histamine, i8
Homocysteine, a207
Homopiperidine, h51
Homoserines, a190, a191
Homoveratric acid, d448
Homoveratrylamine, d452
Hydracrylonitrile, h170
2-Hydrazinoethanol, h120
Hydrazobenzene, d675
Hydrindene, i13
Hydrocinnamic acid, p148
Hydroquinone, d380
Hydroquinone dimethyl ether, d434
Hydroquinonesulfonic acid, d383
Hydroxyacetanilides, a15, a16, a17

h71

h74

h83

h84

h85

TABLE 1-14 Physical constants of organic compounds (*continued*)

No.	Name	Formula	Formula weight	Beilstein reference	Density	Refractive index	Melting point	Boiling point	Flash point	Solubility in 100 parts solvent
h88	1′-Hyydroxy-2′-acetonaphthone	$C_{10}H_6(OH)COCH_3$	186.21	8, 149			98–100	325 sl d		i aq; v s bz; s HOAc
h89	Hydroxyacetone	$HOCH_2COCH_3$	74.08	1^1, 84	1.082	1.4315^{20}	−17	145–146	56	misc aq, alc, eth
h90	*o*-Hydroxyacetophenone	$HOC_6H_4COCH_3$	136.15	8, 85	1.131_4^{21}	1.5584^{20}	4–6	213^{717mm}	> 112	misc alc, eth; sl s aq
h91	*m*-Hydroxyacetophenone	$HOC_6H_4COCH_3$	136.15	8, 86	1.100^{100}	1.535^{100}	87–88	296		s aq; v s alc, bz, eth
h92	*p*-Hydroxyacetophenone	$HOC_6H_4COCH_3$	136.15	8, 87	1.109^{100}		106–107	147^{3mm}		v s alc, eth; sl s aq
h93	1-Hydroxyanthraquinone		224.22	8, 338			196–198			
h94	2-Hydroxybenzaldehyde	$C_6H_4(OH)CHO$	122.12	8, 31	1.167_4^{20}	1.5718^{20}	−7	196.7	76	1.7 aq^{86}; s alc, eth
h95	3-Hydroxybenzaldehyde	$C_6H_4(OH)CHO$	122.12	8, 58			100–102	191^{50mm}		s alc, bz, eth; sl s aq
h96	4-Hydroxybenzaldehyde	HOC_6H_4CHO	122.12	8, 64	1.129_4^{130}		117–119	subl		1 aq; 70 acet; 4 bz; v s alc, eth
h97	2-Hydroxybenzaldehyde oxime	$C_6H_4(OH)CH{=}NOH$	137.14	8, 49			57	d		v s alc, bz, eth, acid
h98	2-Hydroxybenzamide	$C_6H_4(OH)CONH_2$	137.14	10, 87			140	d 270		0.2 aq; s alc, chl, eth
h99	2-Hydroxybenzoic acid	$C_6H_4(OH)COOH$	138.12	10, 43	1.443_4^{20}		157–159	211^{20mm}		0.2 aq; 37 alc; 33 eth; 33 acet; 2 chl; 0.7 bz
h100	3-Hydroxybenzoic acid	$C_6H_4(OH)COOH$	138.12	10, 134	1.473		201–203			0.8 aq; 10 eth
h101	4-Hydroxybenzoic acid	HOC_6H_4COOH	138.12	10, 149	1.468^4		214–215			0.2 aq; v s alc; 23 eth

h102	*p*-Hydroxybenzophenone	$HOC_6H_4COC_6H_5$	198.22	8^2, 184			132–135			v s alc, eth; sl s aq
h103	1-Hydroxybenzotriazole		135.13	26, 41			155–158			
h104	6-Hydroxy-1,3-benzoxathiol-2-one		168.17	19^4, 2508			158–160			
h105	2-Hydroxybenzyl alcohol	$HOC_6H_4CH_2OH$	124.13	6, 891	1.161^{25}		86–87	subl 100		6.6 aq; v s alc, chl, eth; s bz
h106	3-Hydroxy-2-butanone	$CH_3COCH(OH)CH_3$	88.10	1, 827	0.997	1.4171^{20}	15	148	50	misc aq, alc; sl s eth
h107	*p*-Hydroxycinnamic acid	$HOC_6H_4CH{=}CHCOOH$	164.16	10, 297			210–213			s alc, eth; sl s aq
h108	4-Hydroxycoumarin		162.14	17, 488			213 d			s aq, alc, eth
h109	7-Hydroxycoumarin		162.14	18, 27			226–228	subl		v s alc, chl, alk, HOAc
h110	1-Hydroxy-1-cyclohexanecarbonitrile	$C_6H_{10}(OH)CN$	125.17	10, 5	1.031	1.4576^{20}	29		60	
h111	2-Hydroxy-3,5-diiodobenzoic acid	$I_2C_6H_2(OH)COOH$	389.91	10, 113			235 d			v s alc, eth; i bz, chl
h112	2′-Hydroxy-4′,6′-dimethylacetophenone	$(CH_3)_2C_6H_2(OH)COCH_3$	164.20				53–57			

2-Hydroxybenzenemethanol, h105
m-Hydroxybenzotrifluoride, t295
2-Hydroxybiphenyl, p133
4-Hydroxybiphenyl, p134
Hydroxybutanedioic acids, h181, h182

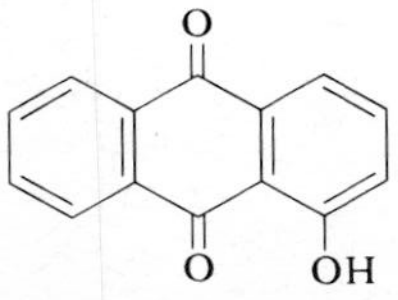

h93

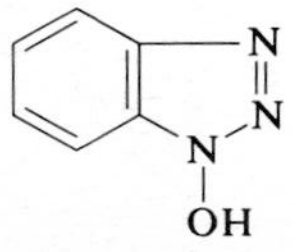

h103

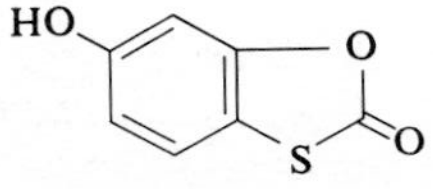

h104

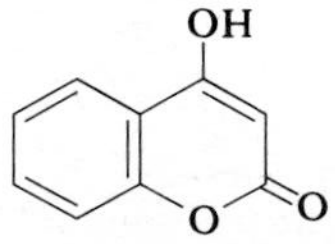

h108

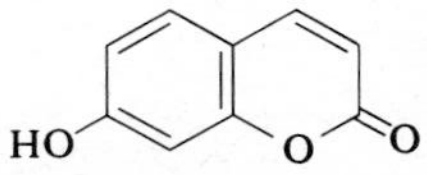

h109

TABLE 1-14 Physical constants of organic compounds (*continued*)

No.	Name	Formula	Formula weight	Beilstein reference	Density	Refractive index	Melting point	Boiling point	Flash point	Solubility in 100 parts solvent
h113	2-Hydroxydiphenylmethane	$C_6H_5CH_2C_6H_4OH$	184.24	6, 675			20.6	312		s alc, chl, eth, alk
h114	2-Hydroxyethanesulfonic acid, Na salt	$HOCH_2CH_2SO_3^-Na^+$	148.11	4^3, 42			191–194			v s aq
h115	*N*-(2-Hydroxyethyl)-acetamide	$HOCH_2CH_2NHCOCH_3$	103.12	4^1, 430	1.1233^{20}_{20}	1.4575^{20}	63–65	d	176	misc aq: sl s bz
h116	2-Hydroxyethyl acetate	$CH_3COOCH_2CH_2OH$	104.11	2, 141	1.108^{15}			181–186	102	misc aq, alc, chl, eth
h117	3-(α-Hydroxyethyl)-aniline	$CH_3CH(OH)C_6H_4NH_2$	137.18	13^3, 1654			68–71			
h118	2-Hydroxyethyl disulfide	$HOCH_2CH_2SSCH_2CH_2OH$	154.25	1, 471	1.261	1.5655^{20}	25–27	$158^{3.5mm}$		
h119	*N*-(2-Hydroxyethyl)-ethylenediamine-*N,N,N'*-triacetic acid	$HOOCCH_2N(CH_2CH_2OH)$-$CH_2CH_2N(CH_2COOH)_2$	278.26				212 d			
h120	2-Hydroxyethyl-hydrazine	$HOCH_2CH_2NHNH_2$	76.10	4^1, 562	1.119		−70	220	73	misc aq; s alc
h121	2-Hydroxyethyl methacrylate	$HOCH_2CH_2OOC$-$C(CH_3){=}CH_2$	130.14		1.034	1.4515^{20}		$67^{3.5mm}$	97	
h122	*N*-(β-Hydroxyethyl)-morpholine		131.18	27, 7	1.083	1.4760^{20}		227	99	misc aq
h123	*N*-(β-Hydroxyethyl)-piperazine		130.19	23^2, 6	1.061	1.5065^{20}		246	> 112	
h124	*N*-(2-Hydroxyethyl)-piperazine-*N'*-ethanesulfonic acid		238.31				234 d			

h125	4′-(2-Hydroxyethyl)-piperidine		129.20	21^2, 10	1.0059^{15}_{4}			199–202		
h126	2′-(2-Hydroxyethyl)-pyridine	$HOCH_2CH_2C_5H_4N$	123.16	21, 50	1.093	1.5368^{20}		116^{9mm}	92	v s aq, alc, chl
h127	2-Hydroxyisobutyric acid	$(CH_3)_2C(OH)COOH$	104.11	3, 313			77–80	$84^{1.5mm}$		v s aq, alc, eth
h129	4-Hydroxy-2-mercapto-6-methyl-pyrimidine		142.18	24^3, 1289			330 d			
h130	4-Hydroxy-2-mercapto-6-propyl-pyrimidine		170.23				219–221			0.1 aq; 1.7 alc; 1.7 acet; v s alk; i bz
h131	2-Hydroxy-3-methoxybenzaldehyde	$CH_3OC_6H_3(OH)CHO$	152.15	8, 240			40–42	265–266		v s alc, eth; sl s aq
h132	4-Hydroxy-3-methoxybenzaldehyde	$CH_3OC_6H_3(OH)CHO$	152.15	8, 247	1.056		80–81	285		1 aq; s alc, chl, pyr
h133	4-Hydroxy-3-methoxybenzoic acid	$CH_3OC_6H_3(OH)COOH$	168.15	10, 392			210			0.12 aq; v s alc

Hydroxyethanal, h86
3-(α-Hydroxyethyl)aniline, a264
N-(2-Hydroxyethyl)-3-aza-1,5-pentanediol, t264
N-(2-Hydroxyethyl)ethyleneimine, a326
N-(2-Hydroxyethyl)piperidine, p188
2-(2-Hydroxyethyl)piperidine, p189
O-Hydroxyethylresorcinol, h160
2-Hydroxyisobutyronitrile, h146
2-Hydroxy-3-methyl-2-cyclopenten-1-one, m215

O, N, CH_2CH_2OH
h122

HN, N—CH_2CH_2OH
h123

CH_2CH_2OH, N, N, $CH_2CH_2SO_3H$
h124

CH_2CH_2OH, N, H
h125

OH, N, H_3C, N, SH
h129

OH, N, C_3H_7, N, SH
h130

TABLE 1-14 Physical constants of organic compounds (*continued*)

No.	Name	Formula	Formula weight	Beilstein reference	Density	Refractive index	Melting point	Boiling point	Flash point	Solubility in 100 parts solvent
h134	4-Hydroxy-3-methoxybenzonitrile	$CH_3OC_6H_3(OH)CN$	149.15	10, 398			85–87			
h135	2-Hydroxy-4-methoxybenzophenone	$CH_3OC_6H_3(OH)COC_6H_5$	228.25	8, 312			66	155^{5mm}		v s alc, chl, eth
h136	4-Hydroxy-3-methoxybenzyl alcohol	$CH_3OC_6H_3(OH)CH_2OH$	154.17	6, 1113			113–115			
h137	4-Hydroxy-3-methoxycinnamic acid	$CH_3OC_6H_3(OH)CH{=}CH$-COOH	194.19	10, 436			174			s hot aq, alc, eth, EtAc; sl s bz, PE
h138	2-Hydroxy-3-methylbenzoic acid	$CH_3C_6H_3(OH)COOH$	152.15	10, 220			165–166			s alc, chl, eth, alk
h139	2-Hydroxy-4-methylbenzoic acid	$CH_3C_6H_3(OH)COOH$	152.15	10, 233			177			s alc, chl, eth, alk
h140	4-Hydroxy-3-methyl-2-butanone	$HOCH_2CH(CH_3)COCH_3$	102.13	1^1, 422	0.993	1.4340^{20}		92^{15mm}	78	
h141	7-Hydroxy-4-methylcoumarin		176.17	18, 31			194–195			s alc, HOAc; sl s eth
h142	2-Hydroxymethyl-2-methyl-1,3-propanediol	$HOCH_2C(CH_3)(CH_2OH)_2$	120.09	1, 520			199–203			
h143	4-Hydroxy-4-methyl-2-pentanone	$(CH_3)_2C(OH)CH_2COCH_3$	116.16		0.9385^{20}	1.4235^{20}	−42.8	169	12	misc aq
h144	*N*-(Hydroxymethyl)-phthalimide		177.16	21, 475			142–145			sl s aq, alc, bz
h145	4-Hydroxy-*N*-methylpiperidine		115.18	21^1, 188		1.4775^{20}	29–31	200	> 112	
h146	2-Hydroxy-2-methylpropanenitrile	$(CH_3)_2C(OH)CN$	85.10	3, 316	0.9267^{25}_4	1.3992^{20}	−19	95	63	s aq, alc, chl, eth

h147	3-Hydroxy-2-methyl-4-pyrone		126.11				161–162			1.2 aq; v s hot aq; s alc, alk; sl s bz, eth
h148	2-Hydroxy-1-naphthaldehyde	$C_{10}H_6(OH)CHO$	172.18	8, 143			82–85	192^{27mm}		
h149	1-Hydroxy-2-naphthalenecarboxylic acid	$C_{10}H_6(OH)COOH$	188.18	10, 331			191–192			v s alc, bz, eth, alk
h150	3-Hydroxy-2-naphthalenecarboxylic acid	$C_{10}H_6(OH)COOH$	188.18	10, 333			222–223			v s alc, eth; s bz, chl
h151	2-Hydroxy-3,6-naphthalenedisulfonic acid, disodium salt	$C_{10}H_5(OH)(SO_3^-Na^+)_2$	348.25	11, 288						v s aq, alc; i eth
h152	4-Hydroxy-2,7-naphthalenedisulfonic acid, disodium salt	$C_{10}H_5(OH)(SO_3^-Na^+)_2$	348.25	11, 227			> 300			
h153	2-Hydroxy-1,4-naphthoquinone		174.16	8, 300			d 185			s HOAc

3-Hydroxymethylpiperidine, p190

1-Hydroxy-2-napthoic acid, h149

3-Hydroxy-2-naphthoic acid, h150

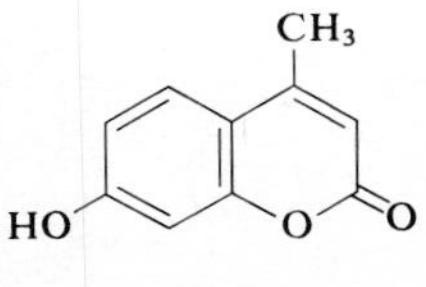

h141

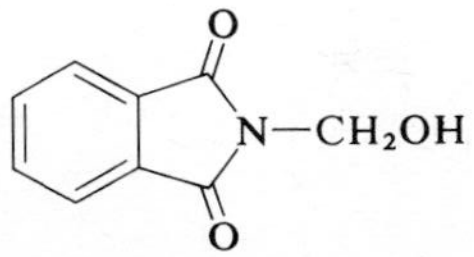

h144

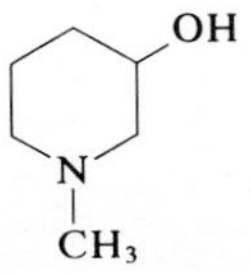

h145

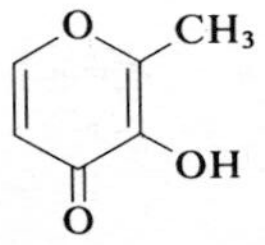

h147

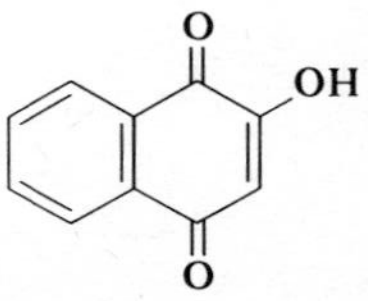

h153

TABLE 1-14 Physical constants of organic compounds (*continued*)

No.	Name	Formula	Formula weight	Beilstein reference	Density	Refractive index	Melting point	Boiling point	Flash point	Solubility in 100 parts solvent
h154	4-Hydroxy-3-nitro-benzenearsonic acid	$HOC_6H_3(NO_2)$-$AsO(OH)_2$	263.04	16^1, 456			> 300			v s alc, acet, HOAc, alk; sl s aq; i eth
h155	3-Hydroxy-4-nitro-benzoic acid	$HOC_6H_3(NO_2)COOH$	183.12	10, 146			229–231			
h156	2-Hydroxy-5-nitro-benzyl bromide	$HOC_6H_3(NO_2)CH_2Br$	232.04	6, 367			147–149			
h157	5-Hydroxy-1-pentanal	$HO(CH_2)_4CHO$	102.13		1.055	1.4530^{20}		115^{15mm}	> 112	s aq
h158	5-Hydroxy-2-pentanone	$CH_3COCH_2CH_2CH_2OH$	102.13	1, 831	1.007^{20}_4	1.4372^{20}		144^{100mm}	93	misc aq; s alc, eth
h159	4-Hydroxy-3-penten-2-one acetate	$CH_3COOC(CH_3)=CH$-$COCH_3$	142.15			1.4525^{20}		195	75	
h160	2-(*m*-Hydroxy-phenoxy)ethanol	$HOC_6H_4OCH_2CH_2OH$	154.17				83–86			
h161	4-Hydroxyphenyl-acetic acid	$HOC_6H_4CH_2COOH$	152.15	10, 190			149–151	subl		v s alc, eth; sl s aq
h162	2-Hydroxy-*N*-phenyl-benzamide	$HOC_6H_4CONHC_6H_5$	213.14	12, 500			136			v s alc, bz, chl, eth
h163	4-(*p*-Hydroxphenyl)-2-butanone	$HOC_6H_4CH_2CH_2COCH_3$	164.20				82–83			
h164	D-(−)-*p*-Hydroxy-phenylglycine	$HOC_6H_4CH(NH_2)COOH$	167.16	14^1, 659			240 d			sl s aq, alc, bz, acet
h165	*N*-(*p*-Hydroxy-phenyl)glycine	$HOC_6H_4NHCH_2COOH$	167.16	13, 488			220–248 d			s alk, acid; v sl s aq, alc, acet, bz, chl, eth
h166	1-(3-Hydroxyphenyl)-urea	$HOC_6H_4NHCONH_2$	152.15	13, 417			182–184			

h167	*N*-Hydroxyphthalimide		163.13	21, 500			233 d			
h168	*N*-Hydroxypiperidine		101.15	20, 80			37–40	111^{55mm}		
h169	2-Hydroxypropionitrile	$CH_3CH(OH)CN$	71.08	3^2, 209	0.9834^{25}	1.4027^{25}	−34	103^{50mm}	77	misc aq, alc; s eth
h170	3-Hydroxypropionitrile	$HOCH_2CH_2CN$	71.08	3, 298	1.0404^{25}_{4}	1.4256^{20}	−46	228	> 112	misc aq, alc, acet; 2.3 eth; i bz, PE
h171	*o*-Hydroxypropiophenone	$HOC_6H_4COCH_2CH_3$	150.18	8, 102	1.094	1.5480^{20}		115^{15mm}	> 112	v s alc, eth; sl s aq
h172	*p*-Hydroxypropiophenone	$HOC_6H_4COCH_2CH_3$	150.18	8, 102			148			v s alc, eth; sl s aq
h173	1-(2-Hydroxy-1-propoxy)-2-propanol	$CH_3CH(OH)CH_2OCH_2CH(OH)CH_3$	134.18		1.0252^{20}_{20}	1.4440^{20}		231.8	138	misc aq, alc
h174	2-Hydroxypyridine	HOC_5H_4N	95.10	21, 43			105–107	280–281		s aq, alc, bz; sl s eth
h175	3-Hydroxypyridine	HOC_5H_4N	95.10	21, 46			126–129	151^{3mm}		v s aq, alc; sl s eth
h176	4-Hydroxypyridine	HOC_5H_4N	95.10	21, 48				230^{12mm}		v s aq; i alc, bz, eth
h177	2-Hydroxypyridine-5-carboxylic acid	$HO(C_5H_3N)COOH$	139.11	22, 215			> 300			sl s aq, alc, eth

6-Hydroxynicotinic acid, h177
α-Hydroxy-α-phenylacetophenone, b46
3-(*p*-Hydroxyphenyl)alanine, t444
2-Hydroxy-2-phenylbenzeneacetic acid, b36
3-Hydroxy-1-propanesulfonic acid γ-sultone, p201
2-Hydroxypropanoic acids, L1, L2
1-Hydroxy-2-propanone, h89
3-Hydroxypropionitrile, c291

N—OH

h167

N—OH

h168

TABLE 1-14 Physical constants of organic compounds (*continued*)

No.	Name	Formula	Formula weight	Beilstein reference	Density	Refractive index	Melting point	Boiling point	Flash point	Solubility in 100 parts solvent
h178	3-Hydroxypyridine-*N*-oxide	$(HO)C_5H_4N{=}O$	111.10				190–192			
h179	8-Hydroxyquinoline		145.16	21, 91			76	267		v s alc, acet, bz, chl
h180	8-Hydroxyquinoline-5-sulfonic acid		225.22	22, 407			213 d			v s aq; sl s alc, eth
h181	DL-Hydroxysuccinic acid	$HOOCCH(OH)CH_2COOH$	134.09	3, 435			131–133			56 aq; 45 EtOH: 18 acet; 0.8 eth; 23 diox; i bz
h182	L-Hydroxysuccinic acid	$HOOCCH(OH)CH_2COOH$	134.09	3, 419			100			36 aq; 87 EtOH; 2.7 eth; 61 acet; 75 diox
h183	*N*-Hydroxysuccinimide		115.09	21, 380			93–95			v s aq
h184	6-Hydroxytetrahydropyran-2-carboxylic acid lactone		128.13		1.226	1.4593^{20}				
h185	3-Hydroxy-3,7,11-trimethyl-1,6,10-dodecatriene	$H_2C{=}CHC(OH)(CH_3)$-$CH_2CH_2CH{=}C(CH_3)$-$CH_2CH_2CH{=}C(CH_3)_2$	222.37		0.8760_4^{25}	1.4769^{25}		114^{1mm}	96	s abs alc
h186	3-Hydroxy-2,2,4-trimethyl-3-pentenoic acid β-lactone		140.18		0.947	1.4380^{20}	−18	170	62	
h187	Hypoxanthine		136.11	26, 416			d 150			0.25 aq; s alk, acid
i1	1*H*,1*H*,11*H*-Icosafluoro-1-undecanol	$HCF_2(CF_2)_9CH_2OH$	531.1				95–97	181^{200mm}		
i2	Icosane	$CH_3(CH_2)_{18}CH_3$	282.56	1, 174	0.7777^{37}	1.4346^{40}	36.4	343.8	> 112	
i3	1-Icosene	$CH_3(CH_2)_{17}CH{=}CH_2$	280.54	1^3, 881			28.7	342.4		

i4	Imidazole		68.08	23, 45			90–91	257	145	v s aq, alc, chl, eth
i5	2-Imidazolidinethi-one		102.16	24, 4			203–204			2 aq; s alc, pyr; i bz, acet, chl, eth
i6	Imidazolidinetrione		114.06				230	subl 100		5 aq; s alc
i7	2-Imidazolidone		86.09	24, 16			131			v s aq, hot alc
i8	2-(4-Imidazolyl) ethylamine		111.15	25, 315			83–84	209^{18mm}		v s aq, alc, hot chl
i9	3,3′-Iminobispropyl-amine	$H_2NCH_2CH_2CH_2NHCH_2CH_2CH_2NH_2$	131.22		0.938	1.4810^{20}	−14	151^{50mm}	118	
i10	Iminodiacetic acid	$HOOCCH_2NHCH_2COOH$	133.10	4, 365			243 d			2 aq; v sl s bz, eth
i11	Iminodiacetonitrile	$NCCH_2NHCH_2CN$	95.11	4, 367			77			s aq, alc: sl s eth
i12	Iminodibenzyl		195.27				105–108			
i13	Indan		118.18		0.9639^{20}_{4}	1.5360^{20}	−51.4	176.5	50	s alc, chl, eth; i aq
i14	5-Indanol		134.18	6, 575			51–53	255		v s alc, eth; sl s aq
i15	1-Indanone		132.16	7, 360	1.1090^{45}_{4}	1.561^{45}	40–42	243–245		s alc, eth; sl s aq

5-Hydroxyvaleraldehyde, h157
Imidodicarbonic diamide, b216

Indalone, b448

Indanamines, a201, a202

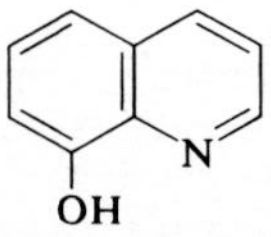

h179 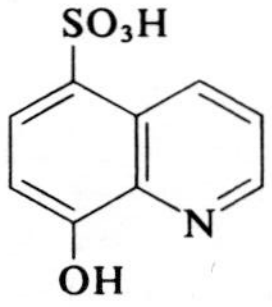 h180 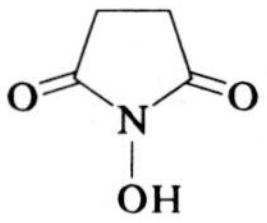 h183 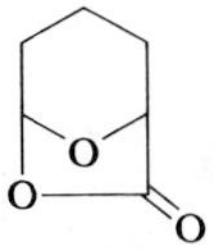 h184 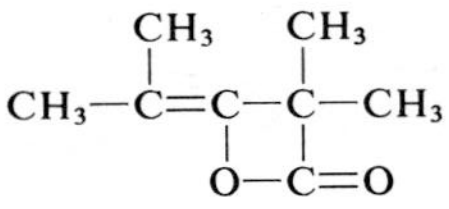 h186 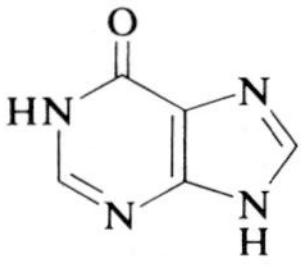 h187  i4 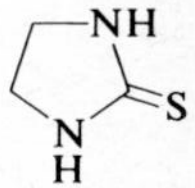 i5

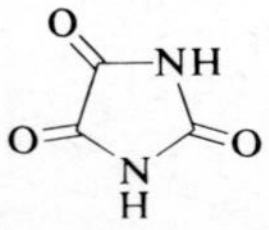

i6 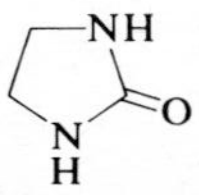 i7 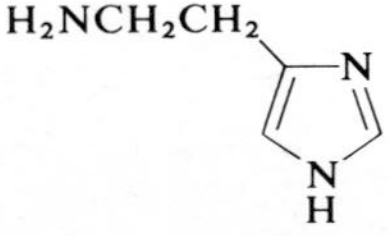 i8 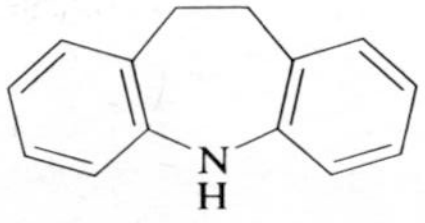 i12 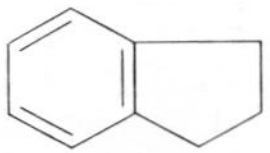 i13 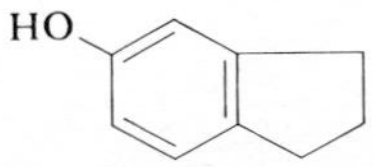 i14 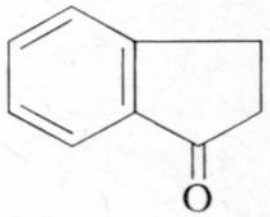 i15

TABLE 1-14 **Physical constants of organic compounds (*continued*)**

No.	Name	Formula	Formula weight	Beilstein reference	Density	Refractive index	Melting point	Boiling point	Flash point	Solubility in 100 parts solvent
i16	1,2,3-Indantrione hydrate		178.14				d 241			
i17	Indene		116.16	5, 515	0.9968^{20}_{4}	1.5762^{20}	−1.8	181.6	78	misc alc, bz, chl, eth
i18	Indole		117.15	20, 304	1.0643	1.609^{60}	52	253		s hot aq, bz, eth
i19	Indole-3-acetic acid		175.19	22, 66			168–170			v s alc; s acet, eth
i20	Indole-3-carbaldehyde		145.16	21, 313			195–198			
i21	Indole-2,3-dione		147.13	21, 432			203.5 d			s hot aq, hot alc, alk
i22	Indoline		119.17	20, 257	1.063	1.5906^{20}		221	92	sl s aq
i23	Inositol		180.16	6^2, 1157	1.752		225–227			14 aq; sl s aq; i eth
i24	Iodoacetamide	ICH_2CONH_2	184.96	2, 223			91–93			s hot aq
i25	Iodoacetic acid	ICH_2COOH	185.95	2, 222			82–83			s aq, alc; v sl s eth
i26	3-Iodoaniline	$IC_6H_4NH_2$	219.03	12, 670	1.821	1.6820^{20}	25	146^{15mm}	> 112	i aq; s alc, eth
i27	Iodobenzene	C_6H_5I	204.01	5, 215	1.8383^{25}_{4}	1.621^{18}	−30	188.3	74	misc alc, chl, eth
i28	Iodobenzene diacetate	$C_6H_5I(OOCCH_3)_2$	322.10				163–165			
i29	2-Iodobenzoic acid	IC_6H_4COOH	248.02	9, 363	2.249^{25}_{4}		162			s alc, eth; sl s aq
i30	1-Iodobutane	$CH_3CH_2CH_2CH_2I$	184.02	1, 123	1.616^{20}_{4}	1.4999^{20}	−103.5	129–130	33	i aq; s alc, eth
i31	2-Iodobutane	$CH_3CH_2CH(I)CH_3$	184.02		1.592^{20}_{4}	1.4991^{20}	−104.0	118–120	28	i aq; s alc, eth
i32	Iodocyclohexane	$C_6H_{11}I$	210.06	5^2, 13	1.626^{15}_{15}	1.5472^{20}		180		i aq; s eth
i33	1-Iododecane	$CH_3(CH_2)_9I$	268.18	1, 168	1.257^{20}_{4}	1.4827^{20}		132^{15mm}		i aq; s alc, eth
i34	Iodoethane	CH_3CH_2I	155.97	1, 96	1.9358^{20}	1.5137	−110.9	72.4	none	0.4 aq; misc alc, bz, chl, eth
i35	2-Iodoethanol	ICH_2CH_2OH	171.97	1, 339	2.2197^{20}_{4}	1.5694^{20}		75^{5mm}	65	s aq; v s alc, eth
i36	Iodoform	CHI_3	393.73	1, 73	4.008		120–123		none	1.4 alc; 10 chl; 13 eth; v s bz, acet
i37	1-Iodoheptane	$CH_3(CH_2)_6I$	226.10	1, 155	1.373^{20}_{4}	1.4900^{20}	−48.2	204	78	i aq; s alc, eth
i38	1-Iodohexadecane	$CH_3(CH_2)_{15}I$	352.35	1, 172	1.121	1.4806^{20}		206–207^{10mm}		

i39	1-Iodohexane	$CH_3(CH_2)_5I$	212.08	1, 146	1.437^{20}_{4}	1.4926^{20}		179.5	61	i aq
i40	Iodomethane	CH_3I	141.94	1, 69	2.2789^{20}_{4}	1.5308^{20}	−66.5	42.4	none	1.4 aq; misc alc, eth
i41	4-Iodomethoxybenzene	$IC_6H_4OCH_3$	234.04	6, 208			48–50	237^{726mm}		s hot alc, eth
i42	1-Iodo-3-methylbutane	$(CH_3)_2CHCH_2CH_2I$	198.06	1^3, 367	1.509^{20}_{4}	1.4939^{20}		147.5		misc alc, eth; sl s aq
i43	1-Iodo-2-methylpropane	$(CH_3)_2CHCH_2I$	184.02	1, 128	1.603^{20}_{4}		−93.5	119		i aq; misc alc, eth
i44	2-Iodo-2-methylpropane	$(CH_3)_3CI$	184.02	1^3, 326	1.571^{0}_{0}	1.4918^{20}	−38.2			d aq; misc alc, eth
i45	1-Iodo-3-nitrobenzene	$IC_6H_4NO_2$	249.01	5, 253	1.9477^{50}_{4}		36–38	280		i aq; s alc, eth
i46	1-Iodooctane	$CH_3(CH_2)_7I$	240.13	1, 160	1.330^{20}_{4}	1.4889^{20}	−45.9	221		s alc, eth
i47	1-Iodopentane	$CH_3(CH_2)_4I$	198.06	1, 133	1.512^{20}_{4}	1.4954^{20}	−85.6	154.5	79	sl s aq; s alc, eth
i48	1-Iodopropane	$CH_3CH_2CH_2I$	169.99	1, 113	1.7489^{20}	1.5058^{20}	−101	102.5	none	0.1 aq; misc alc, eth

Indonaphthene, i17

4-Iodoanisole, i41

5-Iodoanthranilic acid, a205

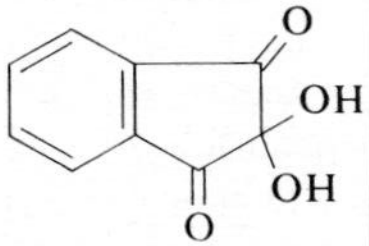

i16

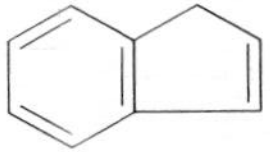

i17

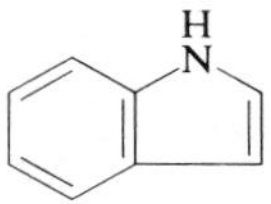

i18

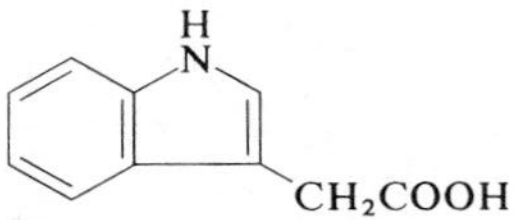

CH_2COOH

i19

CHO

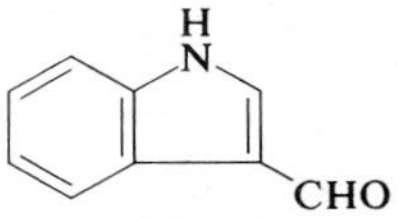

i20

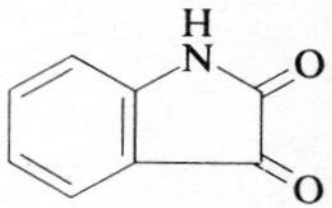

i21

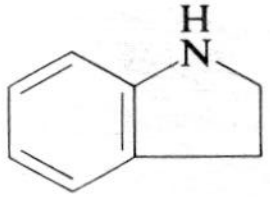

i22

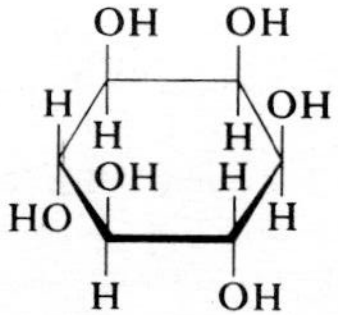

i23

TABLE 1-14 Physical constants of organic compounds (*continued*)

No.	Name	Formula	Formula weight	Beilstein reference	Density	Refractive index	Melting point	Boiling point	Flash point	Solubility in 100 parts solvent
i49	2-Iodopropane	$(CH_3)_2CHI$	169.99	1, 114	1.7025_4^{20}	1.4992^{20}	−90.0	89.5	none	0.14 aq; misc alc, bz, chl, eth
i50	3-Iodo-1-propene	$ICH_2CH{=}CH_2$	167.97	1^3, 114	1.845_4^{22}	1.5540^{21}	−99	1–3		misc alc, chl, eth
i51	5-Iodosalicylic acid	$IC_6H_3(OH)COOH$	264.02	10, 112			189–191			v s alc; i bz, chl
i52	2-Iodothiophene		210.04	17, 34	1.902	1.6520^{20}	−40	73^{15mm}	71	v s eth
i53	2-Iodotoluene	$IC_6H_4CH_3$	218.04	5, 310	1.713	1.6079^{20}		211	90	i aq; s alc, eth
i54	3-Iodotoluene	$IC_6H_4CH_3$	218.04	5, 311	1.698	1.6040^{20}		80^{10mm}	82	i aq; misc alc, eth
i55	Iodotrimethylsilane	$(CH_3)_3SiI$	200.10		1.406_4^{20}	1.4710^{20}		106	<1	
i56	α-Ionone		192.30	7, 168	0.932^{20}	1.4980^{20}		124^{11mm}	104	s alc, bz, chl, eth
i57	β-Ionone		192.30	7, 167	0.946^{17}	1.521^{17}		140^{18mm}	>112	s alc, bz, chl, eth
i58	Isatoic anhydride		163.13	27, 264			233 d			sl s aq, hot alc, acet
i59	D-(−)-Isoascorbic acid		176.12				169 d			s aq, alc, acet, pyr
i60	DL-Isoborneol		154.25	6^2, 80			212	subl		v s alc, chl, eth
i61	2-Isobutoxy-1-isobutoxycarbonyl-1,2-dihydroquinoline		303.40		1.022	1.5230^{20}		$140^{0.2mm}$	>112	
i62	Isobutyl acetate	$(CH_3)_2CHCH_2OOCCH_3$	116.16	2, 131	0.8745^{20}	1.3902^{20}	−98.9	118.0	25	0.7 aq; v s alc
i63	Isobutylamine	$(CH_3)_2CHCH_2NH_2$	73.14	4, 163	0.724_4^{20}	1.3972^{20}	−84.6	67.7	−26	misc aq, alc, acet, eth
i64	Isobutylbenzene	$C_6H_5CH_2CH(CH_3)_2$	134.22	5, 414	0.8673_4^{20}	1.4855^{20}	−51.5	172.8	55	misc alc, eth
i65	Isobutyl chloroformate	$ClCOOCH_2CH(CH_3)_2$	136.58	3, 12	1.053	1.4070^{20}		128.8	26	misc bz, chl, eth
i66	Isobutyl formate	$HCOOCH_2CH(CH_3)_2$	102.13	2, 21	0.8854_4^{20}	1.3855^{20}	−94.5	98.4	10	1 aq; misc alc, eth
i67	Isobutyl isobutyrate	$(CH_3)_2CHCH_2OOCCH(CH_3)_2$	144.22	2, 291	0.8542^{20}	1.3999^{20}	−80.7	147.5		0.5 aq; misc alc
i68	Isobutyl lactate	$CH_3CH(OH)COOCH_2CH(CH_3)_2$	146.19	3^2, 188	0.971_{20}^{20}	1.4181^{25}		96^{40mm}		

i69	Isobutyl methacrylate	$H_2C{=}C(CH_3)COOCH_2$-$CH(CH_3)_2$	142.19		0.882^{25}_{15}	1.4170^{25}		155	45	misc alc, eth
i70	Isobutyl nitrate	$(CH_3)_2CHCH_2ONO_2$	119.12		1.015^{20}_{4}	1.4028^{20}		123–125		i aq; misc alc, eth
i71	Isobutyl nitrite	$(CH_3)_2CHCH_2ONO$	103.12	1, 377	0.870^{22}_{4}	1.3715^{22}		67	4	misc alc; sl s aq(d)
i72	Isobutyl vinyl ether	$(CH_3)_2CHCH_2OCH{=}CH_2$	100.16		0.7702^{20}_{20}	1.3961^{20}	−132.3	83.4		0.2 aq
i73	Isobutyraldehyde	$(CH_3)_2CHCHO$	72.11	1, 671	0.7988^{20}_{4}	1.3723^{20}	−65.9	63–64	−40	11 aq; misc alc, bz, acet, chl, eth
i74	Isobutyramide	$(CH_3)_2CHCONH_2$	87.12	2, 293	1.013	127–129	216–220			
i75	Isobutyric acid	$(CH_3)_2CHCOOH$	88.11	2, 288	0.950^{20}_{4}	1.3925^{20}	−46	154	55	17 aq; misc alc, chl, eth

Isatin, i21
Isethionic acid, h114
Isoamyl acetate, i80
Isoamyl alcohol, m155
sec-Isoamyl alcohol, m156
Isoamyl bromide, b308
Isoamyl idodide, i42
Isoamyl nitrite, i81
Isobutane, m378
Isobutene, m386
α-Isobutoxy-α-phenylacetophenone, b48
Isobutylacetylene, m355
Isobutyl alcohol, m384
Isobutyl bromide, b313
Isobutyl chloride, c162
Isobutyl chlorocarbonate, i65
Isobutyl 1,2-dihydro-2-isobutoxy-1-quinolinecarboxylate, i61
Isobutyl ether, d408
Isobutyl heptyl ketone, t355
Isobutyl mercaptan, m382
Isobutyraldehyde, m377
Isobutyramide, m391
Isobutyric acid, m393

i52

i56

i57

i58

i59

i60

i61

TABLE 1-14 Physical constants of organic compounds (*continued*)

No.	Name	Formula	Formula weight	Beilstein reference	Density	Refractive index	Melting point	Boiling point	Flash point	Solubility in 100 parts solvent
i76	Isobutyronitrile	$(CH_3)_2CHCN$	69.11	2, 294	0.7704^{20}	1.3734^{20}	−71.5	103.8	3	v s alc, eth; sl s aq
i77	Isobutyrophenone	$C_6H_5COCH(CH_3)_2$	148.21	7, 316	0.988^{20}	1.5172		217	84	
i78	Isobutyryl chloride	$(CH_3)_2CHCOCl$	106.55	2, 293	1.017	1.4073^{20}	−90	91–93	1	d aq, d alc; s eth
i79	L-Isoleucine	$CH_3CH_2CH(CH_3)CH(NH_2)COOH$	131.18	4, 454			d 284	subl 168		4 aq; sl s hot alc
i80	Isopentyl acetate	$CH_3COOCH_2CH_2CH(CH_3)_2$	130.19	2, 132	0.876_4^{15}	1.4007^{20}	−78.5	142.0	25	0.25 aq; misc alc, eth
i81	Isopentyl nitrite	$(CH_3)_2CHCH_2CH_2ONO$	117.15	1, 402	0.872	1.3860^{20}		99	10	misc alc, eth; sl s aq
i82	Isophorone		138.21	7, 65	0.923	1.4759^{20}	−8.1	215.2	84	1.2 aq
i83	DL-Isopinocampheol		154.25	6, 67			35–36	217		
i84	Isopropenyl acetate	$CH_3COOC(CH_3)=CH_2$	100.12	2^2, 278	0.909	1.4005^{20}		94	18	
i85	2-Isopropoxyphenol	$(CH_3)_2CHOC_6H_4OH$	152.19	6^3, 4209	1.030	1.5157^{20}		$100–102^{11mm}$		
i86	1-Isopropoxy-2-propanol	$CH_3CH(OH)CH_2OCH(CH_3)_2$	118.1		0.879_{25}^{25}	1.407^{25}		47.9	49	
i87	Isopropyl acetate	$(CH_3)_2CHOOCCH_3$	102.13	2, 130	0.870_4^{20}	1.3773^{20}	−73.4	88.2	16	3 aq; misc alc, eth
i88	Isopropylamine	$(CH_3)_2CHNH_2$	59.11	4, 152	0.686_4^{25}	1.3711^{25}	−101	32.4	−17	misc aq, alc, eth
i89	2-Isopropylamino-ethanol	$(CH_3)_2CHNHCH_2CH_2OH$	103.17	4, 282	0.8970_4^{20}	1.4395^{20}		75^{11mm}		misc aq, alc, eth
i90	2-Isopropylaniline	$(CH_3)_2CHC_6H_4NH_2$	135.2		0.966			222		
i91	Isopropylbenzene	$C_6H_5CH(CH_3)_2$	120.20	5, 393	0.864_4^{20}	1.4915^{20}	−96.0	152.4	46	s alc, bz, eth
i92	4-Isopropylbenzyl alcohol	$(CH_3)_2CHC_6H_4CH_2OH$	150.22	6^3, 1911	0.982^{15}	1.5206^{20}	28	248.4	> 112	misc alc, eth; i aq
i93	*N*-Isopropylbenzyl-amine	$C_6H_5CH_2NHCH(CH_3)_2$	149.24		0.892	1.5025^{20}		200	87	
i94	Isopropylcyclohexane	$C_6H_{11}CH(CH_3)_2$	126.24	5, 41	0.8023_4^{20}	1.4399^{20}	−90	155	35	v s alc, eth
i95	*N*-Isopropylcyclo-hexylamine	$C_6H_{11}NHCH(CH_3)_2$	141.26		0.859	1.4480^{20}		60^{12mm}	33	

i96	4,4′-Isopropylidene-bis[2-(2,6-dibromophenoxy)-ethanol]	$(CH_3)_2C[C_6H_2(Br)_2$-$OCH_2CH_2OH]_2$	632.01				107			
i97	4,4′-Isopropylidene-diphenol	$(CH_3)_2C[C_6H_4OH]_2$	228.29	6, 1011			153–156	220^{4mm}		
i98	Isopropyl isocyanate	$(CH_3)_2CHCNO$	85.11	4, 155	0.866	1.3825^{20}		74–75	−2	
i99	Isopropyl *S*-(−)-lactate	$(CH_3)_2CHOOC$-$CH(OH)CH_3$	132.16	3, 282	0.998^{20}_{20}	1.4082^{25}		166–168		s aq, alc, eth

Isocapronitrile, m342
Isocinchomeronic acid, p264
Isocrotonic acid, b404
Isodurene, t100
Isoeugenol, m98
Isohexane, m339
Isoleucinol, a219
Isoniazid, p262
Isonicotinaldehyde, p257
Isonicotinic acid, p261
Isonicotinic acid hydrazide, p262
Isonicotinonitrile, c298
Isooctane, t362
Isopentane, m149
Isopentyl alcohol, m155
Isopentyl isovalerate, m171
Isophorone, t345
Isophthalic acid, b16
Isophthalonitrile, d237
Isophthaloyl dichloride, b14
Isoprene, m147
Isopropanolamine, a274
Isopropenyl acetate, p211
Isopropenylacetylene, m166
4-Isopropenyl-1-cyclohexene-1-carbaldehyde, p58
Isopropenyl methyl ether, m96
Isopropylacetylene, m172
Isopropylacrylic acid, m352
Isopropyl alcohol, p206
Isopropyl chloride, c212
Isopropyl cyanide, i76
Isopropyl ether, d418
Isopropylethylene, m160
Isopropylidone acetone, m353
Isopropyl iodide, i49

H_3C CH_3 CH_3 O

i82

CH_3 OH CH_3 CH_3

i83

TABLE 1-14 **Physical constants of organic compounds (*continued*)**

No.	Name	Formula	Formula weight	Beilstein reference	Density	Refractive index	Melting point	Boiling point	Flash point	Solubility in 100 parts solvent
i100	2-Isopropyl-1-methylbenzene	$CH_3C_6H_4CH(CH_3)_2$	134.21	5, 419	0.8766^{20}_{4}	1.5006^{20}	−71.5	178.2		misc alc, eth
i101	3-Isopropyl-1-methylbenzene	$CH_3C_6H_4CH(CH_3)_2$	134.21	5, 419	0.8610^{20}_{4}	1.4930^{20}	−63.75	175.1		misc alc, eth
i102	4-Isopropyl-1-methylbenzene	$CH_3C_6H_4CH(CH_3)_2$	134.21	5, 420	0.8573^{20}_{4}	1.4909^{20}	−67.9	177.1	47	misc alc, eth
i102a	2-Isopropyl-5-methylphenol	$CH_3C_6H_3(OH)$-$CH(CH_3)_2$	150.22	6, 532	0.925^{80}_{4}		49–51	232		i aq; v s alc, chl, eth
i103	N^1-Isopropyl-2-methyl-1,2-propanediamine	$(CH_3)_2C(NH_2)CH_2NH$-$CH(CH_3)_2$	130.24		0.822	1.4269^{20}		147–149	90	
i104	Isopropyl methyl sulfide	$(CH_3)_2CHSCH_3$	90.18	1, 367			−101.5	84.7		
i105	Isopropyl nitrate	$(CH_3)_2CHONO_2$	105.09	1^3, 1465	1.036^{19}_{19}	1.3912^{16}		102.1		
i106	2-Isopropylphenol	$(CH_3)_2CHC_6H_4OH$	136.19	6, 504	1.012^{20}	1.5259^{20}	15–16	212–213	107	misc alc, eth
i107	4-Isopropylphenol	$(CH_3)_2CHC_6H_4OH$	136.19	6, 505	0.990^{20}		59–61	212		316 alc; 350 eth
i108	Isopropyl vinyl ether	$(CH_3)_2CHOCH{=}CH_2$	86.13		0.753^{20}_{4}	1.3849^{20}	−140	5–6		
i109	Isopulegol		154.25	6, 65	0.911	1.4725^{20}		91^{12mm}	78	v sl s aq
i110	Isoquinoline		129.16	20, 380	1.0910^{30}_{4}	1.6208^{30}	26.5	243.2	107	sl s aq; s acid
k1	Ketene	$H_2C{=}C{=}O$	42.04	1, 724			−151	−41		s acet, eth; d aq
L1	DL-Lactic acid	$CH_3CH(OH)COOH$	90.08	3, 268	1.249^{15}_{4}		16.8	122^{14mm}		s aq, alc; i chl
L2	L-(+)-Lactic acid	$CH_3CH(OH)COOH$	90.08	3, 261	1.2060^{25}_{4}	1.4392^{20}	53	119^{12mm}	> 112	v s aq, alc, eth
L3	α-Lactose		342.30	31, 408	1.525^{20}		219 d			17 aq; i alc, eth
L4	DL-Leucine	$(CH_3)_2CHCH_2$-$CH(NH_2)COOH$	131.18	4, 447			d 332	subl 293		1 aq; 0.13 alc; i eth
L5	L-Leucine	$(CH_3)_2CHCH_2$-$CH(NH_2)COOH$	131.18	4, 437	1.293^{18}		d 293	subl 145		2.4 aq; 0.07 alc; 1 HOAc; i eth

L6	(+)-Limonene		136.24	5, 133	0.8411_4^{20}	1.4715	−96.5	175–176	53	misc alc, eth
L7	(−)-Limonene		136.24	5, 136	0.844	1.4706^{20}	−96.5	175–176	48	misc alc, eth
L8	(+)-Limonene oxide		152.24	17, 44	0.929	1.4661^{20}		114^{50mm}	65	
L9	Linalool		154.25	1, 462	0.865^{15}	1.4615^{20}		199	76	misc alc, eth
L10	Linalyl acetate		196.29	2, 141	0.895^{20}	1.451		220 d	84	misc alc, eth
L11	*N*-Lithiohexamethyl-disilazane	$(CH_3)_3SiN(Li)Si(CH_3)_3$	167.3				70–72	115		

i109

i110

L3

L6, L7

L8

L9

L10

TABLE 1-14 Physical constants of organic compounds (*continued*)

No.	Name	Formula	Formula weight	Beilstein reference	Density	Refractive index	Melting point	Boiling point	Flash point	Solubility in 100 parts solvent
L12	L-(+)-Lysine	$H_2N(CH_2)_4$-$CH(NH_2)COOH$	146.19	4, 435			d 224			v s aq; sl s alc; i eth
m1	Maleic acid	HOOCCH=CHCOOH	116.07	2, 748	1.590		138–139			79 aq; 70 alc; 8 eth
m2	Maleic anhydride		98.06	17, 432	1.48		52.8	202.0	103	a sq (to acid), alc (to ester); 227 acet; 53 chl; 50 bz; 112 EtAc
m3	Malonic acid	$HOOCCH_2COOH$	104.06	2, 566	1.63		135 d			154 aq; 42 alc; 8 eth
m4	Malonodiamide	$H_2NCOCH_2CONH_2$	102.09	2, 582			168–170			9 aq; i alc, eth
m5	Malononitrile	$NCCH_2CN$	66.06	2, 589	1.049		32–34	220	112	13 aq; 40 alc; 20 eth
m6	Malonyl dichloride	$ClCOCH_2COCl$	140.95	2^1, 252	1.4486_4^{19}	1.4620^{20}		53^{19mm}	47	d hot aq; s eth
m7	D-(+)-Maltose hydrate		342.30	31, 386	1.540^{17}		102–103	d 130		v s aq; sl s alc; i eth
m8	D-Mandelic acid	$C_6H_5CH(OH)COOH$	152.15	10, 197	1.300_4^{20}		119	d		16 aq; 100 alc; s eth
m9	Mandelonitrile	$C_6H_5CH(OH)CN$	133.15	10, 193	1.117	1.5315^{20}	−10	d 170		v s alc, chl, eth; i aq
m10	Mannitol		182.17	1, 534	1.52^{20}		166–168	$290^{3.5mm}$		18 aq; 1.2 alc; i eth
m11	D-(+)-Mannose		180.16	31, 284	1.54^{20}		128–130			250 aq; 28 pyr; 0.8 alc
m12	L-Menthol		156.27	6, 28	0.890_{15}^{15}	1.458^{25}	43–45	212	93	v s alc, chl, eth, PE
m13	L-Menthone		154.25	7, 38	0.895_4^{20}	1.4510^{20}	−6	207	69	misc alc, eth; sl s aq
m14	Mercaptoacetic acid	$HSCH_2COOH$	92.12	3, 245	1.325	1.5030^{20}	−16.5	96^{5mm}	> 112	misc aq, alc, bz, eth
m15	2-Mercaptobenzimidazole		150.20	24, 119			303–304			sl s aq; s alc

m16	*o*-Mercaptobenzoic acid	HSC_6H_4COOH	154.19	10, 125			164–165			v s alc, HOAc
m17	2-Mercaptobenzothiazole		167.25	27, 185	1.42_4^{20}		180–181	d		2 alc; 1 eth; 10 acet; 1 bz; s alk; i aq
m18	2-Mercaptoethanol	$HSCH_2CH_2OH$	78.13	1, 470	1.1143_4^{20}	1.5006^{20}		156.9	73	misc aq, alc, bz, eth
m19	2-Mercaptoethyltriethoxysilane	$HSCH_2CH_2Si(OC_2H_5)_3$	224.38		0.988_4^{20}	1.432^{20}		210	104	

Luminol, a153
2,6-Lupetidine, d593
β-Lutidine, e215
Lutidines, d606, d607, d608, d609
Maleic hydrazide, d401
Malic acids, h181, h182
Malonaldehyde bis(dimethyl acetal), t93
Malonamide nitrile, c287
Malonic acid diamide, m4
Malonylurea, b1
Melamine, t202
Mellitic acid, b19
MEM chloride, m67
Menadione, m313
1,8-Mentanediamine, d41
p-Mentha-1,8-diene, d651
p-Mentha-6,8-dien-2-one, c20
Mercaptobenzene, t162

m2

m7

m10

m11

m12

m13

m15

m17

TABLE 1-14 Physical constants of organic compounds (*continued*)

No.	Name	Formula	Formula weight	Beilstein reference	Density	Refractive index	Melting point	Boiling point	Flash point	Solubility in 100 parts solvent
m20	3-Mercapto-1,2-propanediol	$HSCH_2CH(OH)CH_2OH$	108.16	1, 519	1.295_{14}^{14}	1.5243^{20}		118^{5mm}	> 112	misc alc; v s acet
m21	2-Mercaptopropionic acid	$CH_3CH(SH)COOH$	106.14	3, 289	1.220_4^{15}	1.4809^{20}	10	117^{16mm}	87	misc aq, alc, eth, acet
m22	(3-Mercaptopropyl)-trimethoxysilane	$HS(CH_2)_3Si(OCH_3)_3$	196.34		1.039_4^{20}	1.4416^{20}		93^{40mm}	48	
m23	Mercaptosuccinic acid	$HOOCCH_2CH(SH)COOH$	150.15	3, 439			152–154			50 aq; 50 alc; s eth
m24	Methacrylaldehyde	$H_2C{=}C(CH_3)CHO$	70.09	1, 731	0.8304_4^{20}	1.4160^{20}	−81	69	−15	6 aq; misc alc, eth
m25	Methacrylamide	$H_2C{=}C(CH_3)CONH_2$	85.11	2^2, 399			109–111			s alc; sl s eth
m26	Methacrylic acid	$H_2C{=}C(CH_3)COOH$	86.09	2, 421	1.0153_4^{20}	1.4314^{20}	16	163	76	9 aq; misc alc, eth
m27	Methacrylonitrile	$H_2C{=}C(CH_3)CN$	67.91	2, 423	0.8001_4^{20}	1.4007^{20}	−35.8	90.3	12	2.6 aq; misc acet, bz
m28	Methacryloyl chloride	$H_2C{=}C(CH_3)COCl$	104.54	2^2, 394	1.070	1.4447^{20}		95–96	2	
m29	Methane	CH_4	16.04	1, 56	0.4240^{bp} 0.7168 g/L (gas)		−182.5	−161.5		3.3 mL aq; 47 mL alc
m30	Methanesulfonic acid	CH_3SO_3H	96.10	4, 4	1.4812_4^{18}	1.4303^{20}	20	167^{10mm}	> 112	1.5 bz; misc aq
m31	Methanesulfonic anhydride	$(CH_3SO_2)_2O$	174.19				71	138^{10mm}		v s aq(d)
m32	Methanesulfonyl chloride	CH_3SO_2Cl	114.55	4, 5	1.4805_4^{18}	1.4518^{20}	−32	161	110	s alc, eth
m33	Methanethiol	CH_3SH	48.11	1, 288	0.8665_4^{20}		−123.0	6.0		2.3 aq; v s alc, eth
m34	Methanol	CH_3OH	32.04	1, 273	0.7913_4^{20}	1.3284^{20}	−97.7	64.7	11	misc aq, alc, bz, chl, eth

m35	Methanol-*d*	CH_3OD	33.05	1^3, 1186	0.8127^{20}_4	1.3270^{20}	−110	65.5	11	misc aq, alc, eth
m36	Methanol-d_4	CD_3OD	36.07	1^3, 1187	0.888	1.3256^{20}		65.4	11	misc aq, alc, eth
m37	DL-Methionine	$CH_3SCH_2CH_2$-$CH(NH_2)COOH$	149.21	4^2, 938	1.340		281 d			3 aq; i eth; v sl s alc
m38	Methoxyacetic acid	CH_3OCH_2COOH	90.08	3, 232	1.174	1.4158^{20}		202–204	> 112	misc aq, alc, eth
m39	*o*-Methoxyacetophenone	$CH_3OC_6H_4COCH_3$	150.18	8, 85	1.090^{20}_4	1.5393^{20}		131^{18mm}	108	
m40	*m*-Methoxyacetophenone	$CH_3OC_6H_4COCH_3$	150.18	8, 86	1.094	1.5410^{20}		239–241	110	s aq
m41	*p*-Methoxyacetophenone	$CH_3OC_6H_4COCH_3$	150.18	8, 87	1.082^{41}_4	1.5335^{20}	36–38	154^{26mm}		v s alc, eth
m42	2-Methoxyaniline	$CH_3OC_6H_4NH_2$	123.16	13, 358	1.098^{15}_{15}	1.5730^{20}	5	225	98	i aq; misc alc, eth
m43	3-Methoxyaniline	$CH_3OC_6H_4NH_2$	123.16	13, 404	1.096	1.5794^{20}	1	251	> 112	s alc, acid; sl s aq
m44	4-Methoxyaniline	$CH_3OC_6H_4NH_2$	123.16	13, 435	1.087		60	243		v s alc; sl s aq
m45	2-Methoxybenzaldehyde	$CH_3OC_6H_4CHO$	136.15	8, 43	1.127	1.560^{20}	35–36	236	117	sl s alc, bz; i eth
m46	4-Methoxybenzaldehyde	$CH_3OC_6H_4CHO$	136.15	8, 67	1.119	1.5713^{20}	−1	248	108	misc alc
m47	4-Methoxybenzamide	$CH_3OC_6H_4CONH_2$	151.17	10^2, 100			164–167	295		s aq; v s alc; sl s eth
m48	Methoxybenzene	$C_6H_5OCH_3$	108.14	6, 138	0.9942^{20}	1.5170^{20}	−37.5	153.8	51	1 aq; misc alc, eth
m49	4-Methoxybenzenesulfonyl chloride	$CH_3OC_6H_4SO_2Cl$	206.65	11, 243			40–43			d aq; s alc, eth
m50	2-Methoxybenzoic acid	$CH_3OC_6H_4COOH$	152.15	10, 64	1.180		100	200		0.5 aq; v s alc, eth

MES, m454
Mesidine, t335
Mesitol, t368
Mesitylene, t340
Mesityl oxide, m353
Mesoxylurea, a79
Mesyl chloride, m32
Metanilic acid, a119
Methacholine chloride, a49
Methacrolein, m24
Methallyl alcohol, m388
Methallyl chloride, c165
Methanal, f27
Methanoic acid, f32
Methenamine, h52
Methone, d510
4-Methoxy-2-butanone dimethyl acetal, t326

TABLE 1-14 Physical constants of organic compounds (*continued*)

No.	Name	Formula	Formula weight	Beilstein reference	Density	Refractive index	Melting point	Boiling point	Flash point	Solubility in 100 parts solvent
m51	3-Methoxybenzoic acid	$CH_3OC_6H_4COOH$	152.15	10, 137			104	172^{10mm}		s hot aq, alc, eth
m52	4-Methoxybenzoic acid	$CH_3OC_6H_4COOH$	152.15	10, 154	1.385^{4}		185	275–280		0.04 aq; v s alc, chl
m53	4-Methoxybenzoyl chloride	$CH_3OC_6H_4COCl$	170.60	10, 163		1.5810^{20}	22	145^{14mm}	87	i aq(d); s alc(d); s bz, acet
m54	4-Methoxybenzyl alcohol	$CH_3OC_6H_4CH_2OH$	138.17	6, 897	1.109^{25}_{4}	1.5442^{20}	23–25	259	> 112	i aq; s alc, eth
m55	4-Methoxybenzyl-amine	$CH_3OC_6H_4CH_2NH_2$	137.18	13, 606	1.050^{15}	1.5462^{20}		236–237	> 112	v s aq, alc, eth
m56	2-Methoxybiphenyl	$CH_3OC_6H_4C_6H_5$	184.24	6, 672	1.023	1.6105^{20}		274	> 112	
m57	3-Methoxy-1-butanol	$CH_3OCH(CH_3)CH_2$-CH_2OH	104.15		0.9229^{20}_{20}	1.4145^{20}	−85	161.1	46	misc aq
m58	4-Methoxy-3-buten-2-one	$CH_3OCH{=}CHCOCH_3$	100.12		0.982	1.4660^{20}		200	63	
m59	1-Methoxy-1-buten-3-yne	$CH_3OCH{=}CHC{\equiv}CH$	82.10		0.906^{20}_{4}	1.4818^{20}		122–125	8	v s org solv
m60	4-Methoxycinnamic acid	$CH_3OC_6H_4CH{=}CHCOOH$	178.19	10, 298			172–187			s CCl_4
m61	1-Methoxy-1,3-cyclo-hexadiene		110.16	6^{3}, 367	0.929	1.4885^{20}		40^{15mm}	26	
m62	1-Methoxy-1,4-cyclo-hexadiene		110.16	6^{3}, 367	0.940	1.4819^{20}		148–150	36	
m63	7-Methoxy-3,7-di-methyloctanal	$(CH_3)_2C(OCH_3)CH_2$-$CH_2CH(CH_3)CH_2CHO$	186.30		0.877	1.4374^{20}		$60^{0.45mm}$	98	
m64	2-Methoxy-1,3-dioxolane		104.11	19^{4}, 617	1.092	1.4091^{20}		129–130	31	
m65	2-Methoxyethanol	$CH_3OCH_2CH_2OH$	76.10	1, 467	0.9646^{20}	1.4021^{20}	−85.1	124.6	46	misc aq

m66	2-(2-Methoxyethoxy)-ethanol	$CH_3OCH_2CH_2OCH_2$-CH_2OH	120.15		1.035_4^{20}	1.4264^{20}	−50	194.1	83	misc aq, alc, bz, eth, ketones
m67	2-Methoxyethoxy-methyl chloride	$CH_3OCH_2CH_2OCH_2Cl$	124.57		1.091	1.4270^{20}		50^{13mm}	> 112	
m68	2-Methoxyethyl acetate	$CH_3COOCH_2CH_2OCH_3$	118.13	2, 141	1.0049^{20}	1.4022^{20}	−65.1	144.5	43	misc aq
m69	2-Methoxyethylamine	$CH_3OCH_2CH_2NH_2$	75.11	4^2, 718	0.864	1.4054^{20}		95	9	v s aq, alc
m70	1-Methoxy-2-indanol		164.20	6, 970		1.5482^{20}		146^{11mm}	> 112	
m71	2-Methoxy-5-methyl-aniline	$CH_3OC_6H_3(CH_3)NH_2$	137.18	13^2, 388			52–54	235		s aq; v s alc, bz, eth
m72	3-Methoxy-4-methyl-aniline	$CH_3OC_6H_3(CH_3)NH_2$	137.18	13, 574			51–54	250–252		
m73	4-Methoxy-2-methyl-aniline	$CH_3OC_6H_3(CH_3)NH_2$	137.18	13^2, 330	1.065	1.5647^{20}	13–14	248–249	> 112	s alc
m74	(4*S*,5*S*)-(−)-4-Methoxymethyl-2-methyl-5-phenyl-2-oxazoline		205.26			1.5155^{20}		$79^{0.05mm}$		
m75	4-Methoxy-4-methyl-2-pentanone	$(CH_3)_2C(OCH_3)CH_2$-$COCH_3$	130.18		0.906	1.4181^{25}			61	misc aq
m76	1-Methoxynaph-thalene	$C_{10}H_7OCH_3$	158.20	6, 606	1.090	1.6220^{20}		135^{12mm}	> 112	

Methoxyethane, e174

2-Methoxyethoxychloromethane, m67

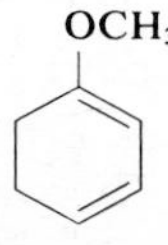

OCH_3
m61

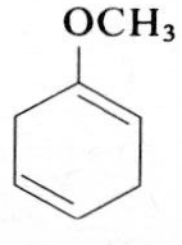

OCH_3
m62

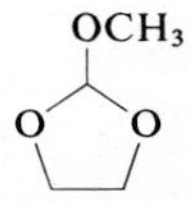

OCH_3 O O
m64

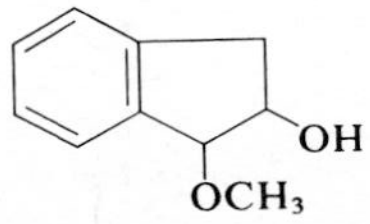

OH OCH_3
m70

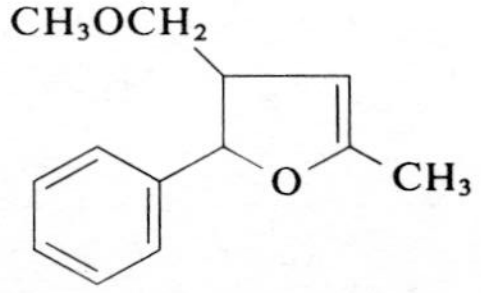

CH_3OCH_2 O CH_3
m74

TABLE 1-14 Physical constants of organic compounds (*continued*)

No.	Name	Formula	Formula weight	Beilstein reference	Density	Refractive index	Melting point	Boiling point	Flash point	Solubility in 100 parts solvent
m77	2-Methoxynaphthalene	$C_{10}H_7OCH_3$	158.20	6, 640			72	272		s bz, eth, CS_2
m78	2-Methoxy-4-nitroaniline	$CH_3OC_6H_3(NO_2)NH_2$	168.15	13, 390			138–140			
m79	2-Methoxy-5-nitroaniline	$CH_3OC_6H_3(NO_2)NH_2$	168.15	13, 389	1.207^{156}		117–119			s alc, hot bz, HOAc
m80	4-Methoxy-2-nitroaniline	$CH_3OC_6H_3(NO_2)NH_2$	168.15	13, 521			123–126			sl s aq; s alc, eth
m81	2-Methoxynitrobenzene	$CH_3OC_6H_4NO_2$	153.14	6, 217	1.2527_4^{20}	1.5619^{20}	9.4	277	> 112	0.17 aq; s alc, eth
m82	4-Methoxynitrobenzene	$CH_3OC_6H_4NO_2$	153.14	6, 230	1.233		54	260		i aq; v s alc, eth
m83	4-Methoxy-3-nitrobenzoic acid	$CH_3OC_6H_3(NO_2)COOH$	197.15	10, 181			186–189			
m84	2-Methoxy-5-nitropyridine	$CH_3OC_5H_3N(NO_2)$	154.13	21^2, 33			108–109			
m85	4-Methoxy-2-nitrotoluene	$CH_3OC_6H_3(NO_2)CH_3$	167.16	6, 411	1.207	1.5525^{20}	17	267	> 112	
m86	*p*-Methoxyphenethylamine	$CH_3OC_6H_4CH_2CH_2NH_2$	151.21	13, 626		1.5379^{20}		138^{20mm}		
m87	2-Methoxyphenol	$CH_3OC_6H_4OH$	124.14	6, 768	1.112 (liquid)	1.5429	28	205	82	1.5 aq; misc alc, eth
m88	3-Methoxyphenol	$CH_3OC_6H_4OH$	124.14	6, 813	1.131	1.5510^{20}	< −17.5	115^{5mm}	> 112	misc alc, eth; sl s aq
m89	4-Methoxyphenol	$CH_3OC_6H_4OH$	124.14	6, 843			55–57	243		v s bz; s alk
m90	3-(4-Methoxyphenoxy)-1,2-propanediol	$CH_3OC_6H_4OCH_2CH(OH)CH_2OH$	198.22	6^3, 4411			76–80			

m91	4-Methoxyphenylacetic acid	$CH_3OC_6H_4CH_2COOH$	166.18	10, 190			86–88	140^{3mm}		i aq; v s alc; s eth
m92	*o*-Methoxyphenylacetone	$CH_3OC_6H_4CH_2COCH_3$	164.20	8^3, 397	1.054	1.5250^{20}		130^{10mm}	> 112	s alc, eth
m93	(*o*-Methoxyphenyl)-acetonitrile	$CH_3OC_6H_4CH_2CN$	147.18	10, 188			65–68	143^{15mm}		s hot bz
m94	2-Methoxy-*p*-phenylenediamine sulfate	$CH_3OC_6H_3(NH_2)_2 \cdot H_2SO_4$	236.26	13^3, 1349			283 d			
m95	1-Methoxy-2-propanol	$CH_3OCH_2CH(OH)CH_3$	90.1		0.919^{20}_{20}	1.4021^{20}	−97	120.1	38	misc aq, acet, bz, eth
m96	2-Methoxypropene	$CH_3C(OCH_3){=}CH_2$	72.11	1, 435	0.753	1.3820^{20}		34–36	−18	
m97	*trans*-1-Methoxy-4-(1-propenyl)benzene	$CH_3OC_6H_4CH{=}CHCH_3$	148.21	6, 566	0.9883^{20}_4	1.5615^{20}	21.4	237	90	misc chl, eth; 50 alc; s bz, EtAc
m98	2-Methoxy-4-propenylphenol	$CH_3OC_6H_3(OH)$-$CH{=}CHCH_3$	164.20	6, 955	1.087^{20}_4	1.5748^{20}	−10	266	> 112	misc alc, eth; sl s aq
m99	2-Methoxy-4-(2-propenyl)phenol	$CH_3OC_6H_3(OH)$-$CH_2CH{=}CH_2$	164.20	6, 961	1.0664^{20}_4	1.5408^{20}	−9.2	255	> 112	misc alc, chl, eth; s HOAc, alk; i aq
m100	*p*-Methoxypropiophenone	$CH_3OC_6H_4COCH_2CH_3$	164.20	8, 103	1.071	1.5465^{20}	27–29	273–275	> 112	
m101	2-Methoxypyridine	$CH_3OC_5H_4N$	109.13	21, 44	1.038	1.5029^{29}		142	32	misc aq
m102	2-Methoxytetrahydrofuran		102.13	17^4, 1019	0.972	1.4119^{20}		105–107	7	

α-Methoxy-α-phenylacetophenone, b49

6-Methoxytetralin, m103

Methoxy-1-tetralone, d359

O OCH_3

m102

TABLE 1-14 Physical constants of organic compounds (*continued*)

No.	Name	Formula	Formula weight	Beilstein reference	Density	Refractive index	Melting point	Boiling point	Flash point	Solubility in 100 parts solvent
m103	6-Methoxy-1,2,3,4-tetrahydronaphthalene		162.23	6^2, 537		1.5402^{20}		90^{1mm}	> 112	
m104	6-Methoxy-1-tetralone		176.22	9^2, 889			77–79	171^{11mm}		
m105	2-Methoxytoluene	$CH_3OC_6H_4CH_3$	122.17	6, 352	0.9851_{15}^{15}	1.5161^{20}		170–172	51	i aq; v s alc, eth
m106	3-Methoxytoluene	$CH_3OC_6H_4CH_3$	122.17	6, 376	0.9697_{25}^{25}	1.5131^{20}		175–176	54	s alc, bz, eth; i aq
m107	4-Methoxytoluene	$CH_3OC_6H_4CH_3$	122.17	6, 392	0.969_{25}^{25}	1.5112^{20}		174	53	s alc, eth; i aq
m108	Methoxytrimethylsilane	$CH_3OSi(CH_3)_3$	104.2		0.7560_4^{20}	1.3678^{20}		57–58		
m109	Methoxytripropylsilane	$CH_3OSi(C_3H_7)_3$	188.4		0.822_4^{20}	1.428^{20}		83^{12mm}		
m110	*N*-Methylacetamide	$CH_3CONHCH_3$	73.10	4, 58	0.9460^{35}	1.4253^{35}	30.6	206		s aq
m111	Methyl acetate	CH_3COOCH_3	74.08	2, 224	0.9342_4^{20}	1.3619^{20}	−98.1	56.3	−16	24 aq; misc alc, eth
m112	Methyl acetoacetate	$CH_3COCH_2COOCH_3$	116.12	3, 632	1.0747^{20}	1.4186^{20}	−80	171.7	70	50 aq; misc alc
m113	*p*-Methylacetophenone	$CH_3C_6H_4COCH_3$	134.18	7, 307	1.0051	1.5328^{20}	22–24	226	92	i aq; v s alc, eth
m114	Methyl acrylate	$H_2C{=}CHCOOCH_3$	86.09	2, 399	0.9561_4^{20}	1.4117^{18}	−76.5	80.2	6	6 aq; s alc, eth
m115	Methylamine	CH_3NH_2	31.06	4, 32	0.699_4^{-11}		−93.5	−6.3	0	959 mL aq; 10.5 bz; s alc; misc eth
m116	Methyl 2-aminobenzoate	$H_2NC_6H_4COOCH_3$	151.17	14, 317	1.68_4^{19}	1.5820^{20}	24	256	104	sl s aq; v s alc, eth
m117	2-(*N*-Methylamino)-benzoic acid	$CH_3NHC_6H_4COOH$	151.17	14, 323			170–172 d			0.2 aq; s alc, eth
m118	Methyl 3-aminocrotonate	$CH_3C(NH_2){=}CHCOOCH_3$	115.13	3, 632			81–83			

m119	2-(Methylamino)-ethanol	$CH_3NHCH_2CH_2OH$	75.11	4, 276	0.937^{20}	1.4387^{20}		155–156	72	misc aq, alc, eth
m120	4-Methylamino-phenol sulfate	$(CH_3NHC_6H_4OH)_2 \cdot H_2SO_4$	344.39	13, 442			260 d			4 aq; sl s alc; i eth
m121	2-(Methylamino)-pyridine	$CH_3NHC_5H_4N$	108.14	22^1, 629	1.052^{29}_{29}	1.5785^{20}	15	201	87	s aq; v s alc, eth
m122	*N*-Methylaniline	$C_6H_5NHCH_3$	107.16	12, 135	0.989^{20}_4	1.5704^{20}	−57	196	73	sl s aq; s alc, eth
m123	*N*-Methylanilinium trifluoroacetate	$C_6H_5NHCH_3 \cdot HOOCCF_3$	221.18				65–66			
m124	2-Methylanthra-quinone		222.24	7, 809			177	subl		v s bz; s alc, eth
m125	Methylarsonic acid	$CH_3AsO(OH)_2$	139.96	4, 613			161			v s aq; s alc
m126	4-Methylbenz-aldehyde	$CH_3C_6H_4CHO$	120.15	7, 297	1.0194^{17}_4	1.5447^{20}		205	80	misc alc, eth; sl s aq
m127	Methyl benzene-sulfonate	$C_6H_5SO_2OCH_3$	172.20	11^2, 20	1.2889^0_4	1.5151^{20}	−4	154^{20mm}		v s alc, chl, eth
m128	2-Methylbenz-imidazole		132.17	23, 145			176–177			s alk, hot aq; sl s alc
m129	Methyl benzoate	$C_6H_5COOCH_3$	136.15	9, 109	1.0933^{15}_4	1.5205^{15}	−12.1	199.5	82	0.2 aq; misc alc, eth

Methylal, d442
Methyl alcohol, m34
Methylaminoacetaldehyde dimethyl acetal, d443
α-(1-Methylaminoethyl)benzyl alcohols, e1, e2
2-Methyl-*p*-anisidine, m71
4-Methyl-*m*-anisidine, m72
5-Methyl-*o*-anisidine, m73
Methylanisoles, m105, m106, m107
Methyl anthranilate, m116
Methylanthranilic acids, a214, a215
N-Methylanthranilic acid, m117
Methylbenzene, t170
4-Methylbenzenesulfonic acid, t179

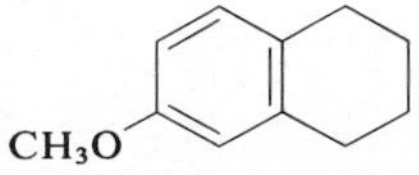

CH_3O

m103

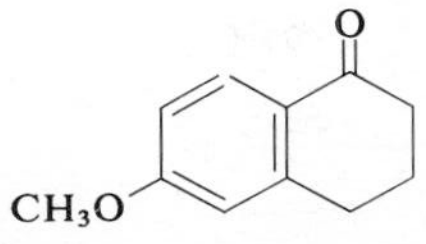

CH_3O O

m104

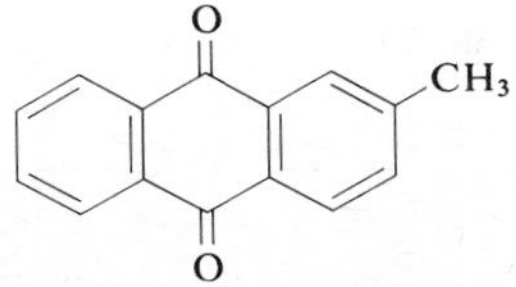

O O CH_3

m124

N N H CH_3

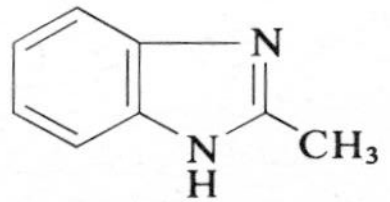

m128

TABLE 1-14 Physical constants of organic compounds (*continued*)

No.	Name	Formula	Formula weight	Beilstein reference	Density	Refractive index	Melting point	Boiling point	Flash point	Solubility in 100 parts solvent
m130	2-Methylbenzoic acid	$CH_3C_6H_4COOH$	136.15	9, 462	1.062		107–108	258–259		sl s aq; v s alc
m131	3-Methylbenzoic acid	$CH_3C_6H_4COOH$	136.15	9, 475	1.054		111–113	263		0.09 aq; v s alc
m132	4-Methylbenzoic acid	$CH_3C_6H_4COOH$	136.15	9, 483			180–182	274–275		v s alc, eth
m133	2-Methylbenzophenone	$CH_3C_6H_4COC_6H_5$	196.25	7, 439	1.083	1.5958^{20}	<−18	309–311	>112	v s alc, org solv
m134	4-Methylbenzophenone	$CH_3C_6H_4COC_6H_5$	196.25	7, 440			59–60	326		v s bz, eth
m135	2-Methylbenzothiazole		149.22	27, 46	1.173	1.6170^{20}	12–14	238	102	s alc, HOAc; i aq
m136	5-Methyl-1*H*-benzotriazole		133.15	26, 58			80–82	210–212^{12mm}		
m137	2-Methylbenzoxazole		133.15	27, 46	1.121	1.5497^{20}	8.5–10	178	75	
m138	α-Methylbenzyl alcohol	$C_6H_5CH(CH_3)OH$	122.17	6, 475	1.0191_4^{13}	1.5211^{20}	21	204	85	v s alc; s bz, chl
m139	3-Methylbenzyl alcohol	$CH_3C_6H_4CH_2OH$	122.17	6, 494	0.916^{17}	1.5334^{20}	<−20	217		5 aq; s alc, eth
m140	4-Methylbenzyl alcohol	$CH_3C_6H_4CH_2OH$	122.17	6, 498			59–61	217		s alc, eth; sl s aq
m141	DL-α-Methylbenzylamine	$C_6H_5CH(CH_3)NH_2$	121.18	12, 1094	0.940	1.5254^{20}		185	79	
m142	4-Methylbenzylamine	$CH_3C_6H_4CH_2NH_2$	121.18	12, 1141	0.952	1.5340^{20}	12–13	195	75	
m143	Methyl bromoacetate	$BrCH_2COOCH_3$	152.98	2, 213	1.616	1.4586^{20}		52^{15mm}	62	s alc
m144	DL-Methyl 2-bromobutyrate	$CH_3CH_2CH(Br)COOCH_3$	181.04	2, 282	1.573			137–138^{50mm}		

m145	Methyl 4-bromo-crotonate	$BrCH_2CH{=}CHCOOCH_3$	179.02		1.522	1.4980^{20}		85^{13mm}	91	
m146	Methyl 2-bromo-propionate	$CH_3CH(Br)COOCH_3$	167.01	2, 253	1.497	1.5420^{20}		51^{19mm}	51	s alc
m147	2-Methyl-1,3-butadiene	$H_2C{=}C(CH_3)CH{=}CH_2$	68.12	1, 252	0.681_4^{20}	1.4216^{20}	−145.9	34. 1	−53	misc alc, eth
m148	3-Methyl-1,2-butadiene	$CH_3C(CH_3){=}C{=}CH_2$	68.12	1, 252	0.694_4^{20}	1.4179^{20}	−113.6	40.9	−12	
m149	2-Methylbutane	$CH_3CH_2CH(CH_3)_2$	72.15	1, 134	0.6197^{20}	1.3537^{20}	−159.9	27.9	−56	0.005 aq; misc alc
m150	2-Methyl-1-butanethiol	$CH_3CH_2CH(CH_3)CH_2SH$	104.22	1^2, 421	0.848	1.4465^{20}		119.0	19	s alc, eth; i aq
m151	2-Methyl-2-butanethiol	$CH_3CH_2C(CH_3)_2SH$	104.22	1^1, 196	0.842	1.4385^{20}	−103.9	99.1	−1	s alc, eth; i aq
m152	3-Methyl-1-butanethiol	$(CH_3)_2CHCH_2CH_2SH$	104.22	1, 405	0.835_4^{20}	1.4432^{20}	−133.5	118.4	18	misc alc, chl, eth
m153	2-Methyl-1-butanol	$CH_3CH_2CH(CH_3)CH_2OH$	88.15	1, 388	0.816_4^{20}	1.4100^{20}	<−70	128	50	3 aq; misc alc, eth
m154	2-Methyl-2-butanol	$CH_3CH_2C(CH_3)_2OH$	88.15	1, 388	0.8090^{20}	1.4050^{20}	−9.0	102.0	21	11 aq; misc alc, bz, chl, eth
m155	3-Methyl-1-butanol	$(CH_3)_2CHCH_2CH_2OH$	88.15	1, 392	0.8129_4^{15}	1.4085^{15}	−117.2	132.0	45	2 aq; misc alc, bz, chl, eth, PE, HOAc

α-Methylbenzyl alcohol, p114
N-Methylbenzylamine, b104
Methylbenzyl bromides, b371, b372
Methylbenzyl chlorides, c259, c260, c261
Methylbis(2-chloroethoxy)silane, b158
N-Methylbis(2-chloroethyl)amine, b160
Methyl bromide, b303
3-Methyl-1-buten-1-carboxylic acid, m352

m135

m136

m137

TABLE 1-14 Physical constants of organic compounds (*continued*)

No.	Name	Formula	Formula weight	Beilstein reference	Density	Refractive index	Melting point	Boiling point	Flash point	Solubility in 100 parts solvent
m156	3-Methyl-2-butanol	$(CH_3)_2CHCH(OH)CH_3$	88.15	1, 391	0.8179^{20}	1.4096^{20}		111.5	26	2.8 aq; misc alc, eth
m157	3-Methyl-2-butanone	$(CH_3)_2CHCOCH_3$	86.13	1, 682	0.802^{20}_{4}	1.3890	92	94–95		misc alc, eth
m158	2-Methyl-1-butene	$CH_3CH_2C(CH_3){=}CH_2$	70.14	1, 211	0.6504^{20}_{4}	1.3777^{20}	−137.6	31.2		misc alc, eth
m159	2-Methyl-2-butene	$CH_3CH{=}C(CH_3)_2$	70.14	1, 211	0.6620^{20}_{4}	1.3878^{20}	−133.8	38.6	−45	misc alc, eth; i aq
m160	3-Methyl-1-butene	$(CH_3)_2CHCH{=}CH_2$	70.14	1^3, 797	0.6272^{20}_{4}	1.3638^{20}	−168.5	20.1		misc alc, eth
m161	(*E*)-2-Methyl-2-butenoic acid	$CH_3CH{=}C(CH_3)COOH$	100.12	2, 430	0.969	1.4342^{81}	64	198.5		s alc, eth; v s hot aq
m162	(*Z*)-2-Methyl-2-butenoic acid	$CH_3CH{=}C(CH_3)COOH$	100.12	2, 428	0.983^{47}_{4}	1.4437^{47}	45	185		s alc, eth; v s hot aq
m163	3-Methyl-2-butenoic acid	$(CH_3)_2C{=}CHCOOH$	100.12	2, 432	1.006^{24}		69	194–195		s aq, alc, eth
m164	2-Methyl-3-buten-2-ol	$(CH_3)_2C(OH)CH{=}CH_2$	86.13	1, 444	0.8672^{20}_{20}	1.4160^{20}	2.6	98–99	13	
m165	3-Methyl-3-buten-1-ol	$H_2C{=}C(CH_3)CH_2CH_2OH$	86.13		0.853	1.4337^{20}			36	
m166	2-Methyl-1-buten-3-yne	$H_2C{=}C(CH_3)C{\equiv}CH$	66.10	1^1, 126		1.4140^{20}	−113	32	−6	
m168	*N*-Methylbutylamine	$CH_3CH_2CH_2CH_2NHCH_3$	87.17	4, 157	0.736	1.3995^{20}	−75	91	<1	
m169	1-Methylbutylamine	$CH_3CH_2CH_2CH(CH_3)NH_2$	87.17	4, 177	0.7384^{20}_{4}	1.4029^{20}		91	35	misc aq, alc, eth
m170	2-Methylbutylamine	$CH_3CH_2CH(CH_3)CH_2NH_2$	87.17	4^3, 342	0.738	1.4116^{20}		94–97	3	
m171	3-Methylbutyl 3-methylbutyrate	$(CH_3)_2CHCH_2CH_2OOCCH_2CH(CH_3)_2$	172.27	2, 312	0.8541^{25}	1.4100^{25}		194.0		misc alc, eth
m172	3-Methyl-1-butyne	$(CH_3)_2CHC{\equiv}CH$	68.12	1, 251	0.666^{20}_{4}	1.3740^{20}	−89.8	26.4		misc alc, eth
m173	2-Methyl-3-butyn-2-ol	$(CH_3)_2C(OH)C{\equiv}CH$	84.12	1^1, 235	0.8672^{20}_{20}	1.4209^{20}	2.6	104–105	25	misc aq, acet, bz

m174	2-Methylbutyraldehyde	$CH_3CH_2CH(CH_3)CHO$	86.13	1^1, 352	0.804	1.3919^{20}		90–92	4	
m175	3-Methylbutyraldehyde	$(CH_3)_2CHCH_2CHO$	86.13	1, 684	0.785^{20}_{20}	1.3882^{20}	−51	92–93	19	misc alc, eth; sl s aq
m176	Methyl butyrate	$CH_3CH_2CH_2COOCH_3$	102.13	2^4, 786	0.898^{20}_{4}	1.3879^{20}	−85	102	14	1.4 aq; misc alc, eth
m177	2-Methylbutyric acid	$CH_3CH_2CH(CH_3)COOH$	102.13	2^4, 888	0.936	1.4055^{20}		176.5	>112	
m178	3-Methylbutyric acid	$(CH_3)_2CHCH_2COOH$	102.13	2, 309	0.9308^{20}_{4}	1.4033^{20}	−30.0	176.5	70	4 aq; s alc, chl, eth
m179	3-Methylbutyronitrile	$(CH_3)_2CHCH_2CN$	83.13	2^2, 278	0.7925^{19}_{4}	1.3927^{20}	−101	129		misc alc, eth
m180	3-Methylbutyryl chloride	$(CH_3)_2CHCH_2COCl$	120.58	2, 315	0.985^{20}_{4}	1.4161^{20}		115–117	18	d aq, alc; s eth
m181	1-(3-Methylbutyryl)-pyrrolidine		155.24		0.938	1.4710^{20}			104	
m182	Methyl carbamate	$H_2NCOOCH_3$	75.07	3, 21	1.136^{56}_{4}		52–54	177		220 aq; 73 alc; s eth
m183	Methyl chloroacetate	$ClCH_2COOCH_3$	108.52	2, 197	1.238^{20}_{20}	1.4220^{20}	−33	130–132	57	i aq; misc alc, eth

(*Z*)-2-Methyl-2-butenedioic acid, c271
Methyl 2-buten-1-oate, m193
3-Methylbutyl acetate, i80
2-Methylbutylamine, a254
Methyl *tert*-butyl ether, b463
Methyl *tert*-butyl ketone, h72
2-Methylbutyl isovalerate, m171
Methyl caprate, m219
Methyl caproate, m266
Methyl carprylate, m332
Methyl carbazate, m272
Methyl carbitol, m66
4-Methylcatechol, d390
Methyl Cellosolve, m65
Methyl Cellosolve acetate, m68
β-Methylchalcone, d660
Methyl chlorocarbonate, m188

(pyrrolidine ring) N
$O{=}C{-}CH_2CH(CH_3)_2$
m181

TABLE 1-14 Physical constants of organic compounds (*continued*)

No.	Name	Formula	Formula weight	Beilstein reference	Density	Refractive index	Melting point	Boiling point	Flash point	Solubility in 100 parts solvent
m184	Methyl 2-chloroacetoacetate	$CH_3COCH(Cl)COOCH_3$	150.56		1.236	1.4465^{20}	−32.7	137	71	
m185	Methyl *m*-chlorobenzoate	$ClC_6H_4COOCH_3$	170.60	9, 338		1.4923^{20}	21	101^{12mm}		
m186	Methyl *p*-chlorobenzoate	$ClC_6H_4COOCH_3$	170.60	9, 340	1.382^{20}		44			s alc
m187	Methyl 4-chlorobutyrate	$ClCH_2CH_2CH_2COOCH_3$	136.58	2, 278	1.1268^{14}	1.4321^{20}		176	59	v s eth; s alc, acet
m188	Methyl chloroformate	$ClCOOCH_3$	94.50	3, 9	1.223^{20}_{4}	1.3865^{20}		71	<1	misc alc, bz, chl, eth
m189	Methyl 3-(chloroformyl)propionate	$CH_3OOCCH_2CH_2COCl$	150.56	2^2, 553	1.223	1.4402^{20}		65^{3mm}	73	
m190	Methyl 2-chloropropionate	$CH_3CH(Cl)COOCH_3$	122.55	2, 248	1.075	1.4193^{20}		132–133	36	s alc
m191	2-Methylcinnamaldehyde	$C_6H_5CH{=}C(CH_3)CHO$	146.19	7, 369	1.0407^{17}_{4}	1.6045^{20}		149^{27mm}	79	
m192	6-Methylcoumarin		160.17	17, 337			75–76	303^{725mm}		
m193	Methyl crotonate	$CH_3CH{=}CHCOOCH_3$	100.12	2, 410	0.9444^{20}_{4}	1.4242^{20}		121	4	v s alc, eth; i aq
m194	Methyl cyanoacetate	$NCCH_2COOCH_3$	99.09	2, 584	1.1225^{25}	1.4166^{25}	−13.1	205.1	110	misc alc, eth
m195	Methylcyclohexane	$C_6H_{11}CH_3$	98.19	5, 29	0.7694^{20}	1.4231^{20}	−126.6	100.9	−3	
m196	Methyl cyclohexanecarboxylate	$C_6H_{11}COOCH_3$	142.20	9^1, 5	0.9954^{16}_{4}	1.4445^{20}		183	60	i aq; s alc, eth
m197	4-Methyl-1,2-cyclohexanedicarboxylic anhydride		168.19		1.162	1.4774^{20}				
m198	1-Methylcyclohexanol	$C_6H_{10}(CH_3)OH$	114.19	6, 11	0.9251^{25}	1.4587^{25}	26	168	67	i aq; s bz, chl
m199	(*Z*)-2-Methylcyclohexanol	$C_6H_{10}(CH_3)OH$	114.19	6^2, 17	0.9340^{20}_{4}	1.4654^{20}	7	165	58	misc alc, eth

m200	(*E*)-2-Methylcyclohexanol	$C_6H_{10}(CH_3)OH$	114.19	6, 11	0.9247_4^{20}	1.4616^{20}	−4	165.5	58	misc alc; s eth
m201	(*Z*)-3-Methylcyclohexanol	$C_6H_{10}(CH_3)OH$	114.19	6, 12	0.9155^{20}	1.4572^{20}	−6	94	62	misc alc, eth
m202	(*E*)-3-Methylcyclohexanol	$C_6H_{10}(CH_3)OH$	114.19	6, 12	0.9214^{20}	1.4580^{20}	−1	84	62	
m203	(*Z*)-4-Methylcyclohexanol	$CH_3C_6H_{10}OH$	114.19	6, 14	0.9122_4^{20}	1.4614^{20}		171	70	misc alc, eth
m204	(*E*)-4-Methylcyclohexanol	$CH_3C_6H_{10}OH$	114.19	6, 14	0.9118_4^{21}	1.4559^{20}		173–175	70	misc alc; s eth
m205	2-Methylcyclohexanone	$CH_3C_6H_9(=O)$	112.17	7, 14	0.925_4^{20}	1.4478^{20}		162–163	46	i aq; s alc, eth
m206	3-Methylcyclohexanone	$CH_3C_6H_9(=O)$	112.17	7, 15	0.9155_4^{20}	1.4460^{20}		168–169	51	i aq; s alc, eth
m207	4-Methylcyclohexanone	$CH_3C_6H_9(=O)$	112.17	7, 18	0.916_4^{20}	1.4455^{20}		169–171	40	i aq; s alc, eth
m208	1-Methyl-1-cyclohexene		96.17	5, 66	0.809_4^{20}	1.4502^{20}	−121	111	−3	i aq; s alc, eth
m209	4-Methyl-1-cyclohexene		96.17	5, 67	0.799	1.4412^{20}	−115.5	102	−1	i aq; s alc, eth
m210	*N*-Methylcyclohexylamine	$C_6H_{11}NHCH_3$	113.20	12, 6	0.868	1.4560^{20}		149	29	

Methyl chloroform, t230

(*E*)-2-Methylcrotonic acid, m161

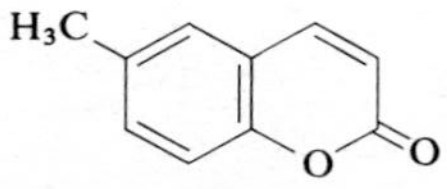

m192

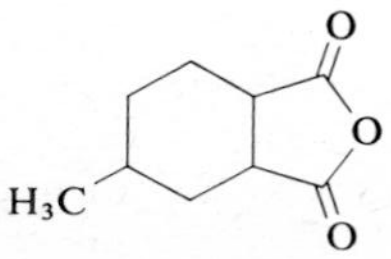

m197

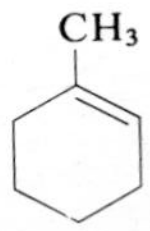

m208

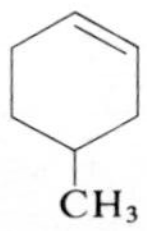

m209

TABLE 1-14 Physical constants of organic compounds (*continued*)

No.	Name	Formula	Formula weight	Beilstein reference	Density	Refractive index	Melting point	Boiling point	Flash point	Solubility in 100 parts solvent
m211	3-Methylcyclohexylamine	$C_6H_{10}(CH_3)NH_2$	113.20	12, 10	0.855	1.4525^{20}		150^{730mm}	22	
m212	4-Methylcyclohexylamine	$C_6H_{10}(CH_3)NH_2$	113.20	12, 12	0.855	1.4531^{20}		151–154	26	
m213	Methylcyclopentadiene dimer		160.26		0.941	1.4976^{20}	−51	200	26	
m214	Methylcyclopentane	$C_5H_9CH_3$	84.16	5, 27	0.7487^{20}	1.4097^{20}	−142.4	71.8	−27	0.013 aq
m215	3-Methyl-1,2-cyclopentanedione		112.13	7^1, 310			105–107			
m216	2-Methylcyclopentanone		98.15	7^2, 13	0.9200^{20}_{4}	1.4347^{20}	−76	139–140		s aq; v s alc, eth
m217	3-Methyl-2-cyclopenten-1-one		96.13	7^1, 46	0.971	1.4780^{20}		74^{15mm}	65	
m218	2-Methylcyclopropanecarboxylic acid		100.12	9, 6	1.027	1.4395^{20}		191^{745mm}	87	
m219	Methyl decanoate	$CH_3(CH_2)_8COOCH_3$	186.30	2, 356			−18	223–224		i aq; misc alc, eth
m220	Methyl dichloroacetate	$Cl_2CHCOOCH_3$	142.97	2, 203	1.3808^{19}	1.4421^{20}	−52	143	80	i aq; s alc
m221	Methyl 2,2-dichloro-1-methylcyclopropanecarboxylate		183.03		1.245	1.4639^{20}		74^{8mm}	74	
m222	Methyl 2,3-dichloropropionate	$ClCH_2CH(Cl)COOCH_3$	157.00	2^1, 111	1.3282^{20}_{4}	1.4447^{20}		92^{50mm}	42	s alc
m223	Methyldichlorosilane	CH_3SiHCl_2	115.0		1.1047^{20}_{4}	1.4222^{20}	−93	41	−25	
m224	*N*-Methyldiethanolamine	$CH_3N(CH_2CH_2OH)_2$	119.16	4, 284	1.0377^{20}	1.4685^{20}		246–248	126	misc aq, alc

m225	*O*-Methyl-*N*,*N'*-diisopropylurea	$(CH_3)_2CHNHC(OCH_3){=}NCH(CH_3)_2$	158.25		0.871	1.4358^{20}		$50–52^{0.1\,mm}$	35	
m226	Methyl 3,4-dimethoxybenzoate	$(CH_3O)_2C_6H_3COOCH_3$	196.20	10, 396			57–60	283		
m227	Methyl 4,5-dimethoxy-2-nitrobenzoate	$(CH_3O)_2C_6H_2(NO_2)COOCH_3$	241.20	10, 403			141–144			
m228	Methyl 2,2-dimethylpropionate	$(CH_3)_3CCOOCH_3$	116.16	2^1, 139	0.873	1.3880^{20}		101–103	−1	misc alc, eth; sl s aq
m229	2-Methyl-3,5-dinitrobenzoic acid	$CH_3C_6H_2(NO_2)_2COOH$	226.15	9, 474			205–207			
m230	*N*-Methyldioctylamine	$(C_8H_{17})_2NCH_3$	255.49	4^3, 381		1.4424^{20}	−30.1	165^{15mm}		
m231	*N*-Methyldiphenylamine	$(C_6H_5)_2NCH_3$	183.26	12, 180	1.048_4^{20}	1.6193^{20}	−7.6	135^{6mm}		i aq; s alc, eth
m232	Methyldiphenylsilane	$(C_6H_5)_2Si(H)CH_3$	198.3		0.997_4^{20}	1.569^{20}				
m233	*N*,*N'*-Methylenebisacrylamide	$H_2C{=}CHCONHCH_2NHCOCH{=}CH_2$	154.17				>300			
m234	2,2'-Methylenebis(4-chlorophenol)	$CH_2[C_6H_3(Cl)OH]_2$	269.13				177–178			100 EtOH; 100 eth; s PE

Methyl 4,6-dimethyl-2-oxo-2*H*-pyran-5-carboxylate, m289

Methyldinitrophenols, d633, d633a

Methyl enanthate, m262

Methylene bromide, d88

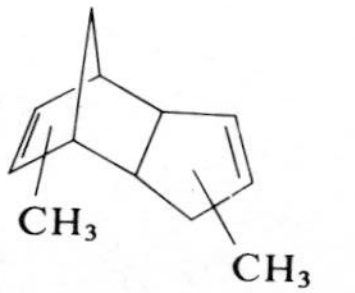

m213 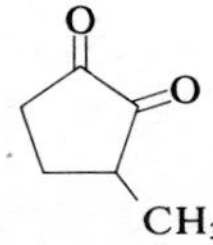 m215 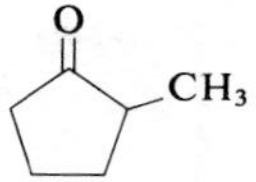 m216 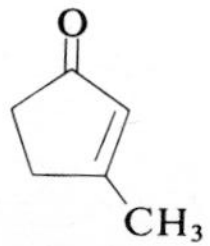 m217 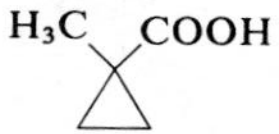 m218 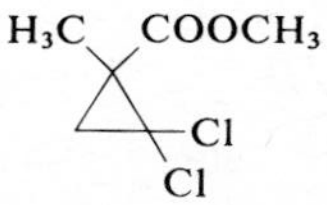 m221

TABLE 1-14 Physical constants of organic compounds (*continued*)

No.	Name	Formula	Formula weight	Beilstein reference	Density	Refractive index	Melting point	Boiling point	Flash point	Solubility in 100 parts solvent
m235	4,4′-Methylenebis-(*N,N*-dimethyl-aniline	$CH_2[C_6H_4N(CH_3)_2]_2$	254.38	13, 239			90			
m236	4,4′-Methylenebis-(3-hydroxy-2-naphthoic acid	$CH_2[C_{10}H_5(OH)COOH]_2$	388.38	10, 575			d > 280			i aq, alc, eth, bz; sl s chl; s pyr
m237	1,1′-Methylenebis(3-methylpiperidine)	$CH_2(CH_3C_5H_9N)_2$	210.37		0.887	1.4734^{20}		160^{50mm}	110	
m238	Methylene blue		373.90	27, 393			190 d			4 aq; 1.3 alc; s chl
m239	4,4′-Methylenedi-aniline	$CH_2(C_6H_4NH_2)_2$	198.26	13, 238			92		221	
m240	3,4-Methylenedi-oxybenzaldehyde		150.13	19, 115			37	264		0.2 aq; v s alc, eth
m241	1,2-Methylendi-oxybenzene		122.12	19, 20	1.064	1.5398		173	55	
m242	3,4-Methylenedi-oxybenzoic acid		166.13	19, 269			229	subl 210		sl s aq, chl, alc, eth
m243	3,4-Methylenedi-oxybenzyl alcohol		152.14	19, 67			53–55			
m244	3,4-Methylenedioxy-6-propylbenzyldi-ethyleneglycol butyl ether		338.45		1.05	1.50^{20}		180^{1mm}	171	misc alc, bz, freons
m245	5-Methylene-2-nor-bornene		106.17		0.981	1.4819^{20}			4	
m246	Methylenesuccinic acid	$H_2C{=}C(COOH)CH_2$-COOH	130.10	2, 760	1.573		162 d			8.2 aq; 20 alc; v s bz, chl, eth, PE
m247	*N*-Methylethylamine	$CH_3CH_2NHCH_3$	59.11	4^2, 589	0.690	1.3760		35	−12	v s aq, alc

m248	*N*-Methylethylenediamine	$CH_3NHCH_2CH_2NH_2$	74.13	4[1], 415	0.841	1.4395^{20}		114–116	41	
m249	*N*-Methylformamide	$HCONHCH_3$	59.07	4, 58	0.9988^{25}	1.4300^{25}	−3.8	180–185	98	misc aq
m250	*N*-Methylformanilide	$C_6H_5N(CH_3)CHO$	135.17	12, 234	1.095	1.5593^{20}	8–13	244	126	
m251	Methyl formate	$HCOOCH_3$	60.05	2, 18	0.9815^{15}	1.3465^{15}	−99.0	31.5	−32	23 aq; misc alc
m252	5-Methylfuraldehyde		110.11	17, 289	1.1072_4^{18}	1.5263^{20}		187	72	s aq; v s alc; misc eth
m253	2-Methylfuran		82.10	17, 36	0.915_4^{20}	1.4332^{20}	−88	63–66	−26	0.3 aq
m254	Methyl furoate		126.11	18, 274	1.179^{20}	1.4862^{20}		181	73	s alc, eth; sl s aq
m255	Methylgermanium tribromide	CH_3GeBr_3	327.35		2.6337_4^{20}	1.5770^{20}		168		

Methylene bromochloride, b258
Methylene chloride, d190
4,4′-Methylenedianiline, d35
Methylene dimethyl ether, d442
Methylene iodide, d405
1,1′-Methylenedipiperidine, d697
β-Methylene-*β*-propiolactone, d423
(*E*)-3,6-*endo*-Methylene-1,2,3,6-tetrahydronaphthaloyl dichloride, n109
Methyl ethyl ketone, b396
Methyl fluoroform, t296
Methyl 2-furancarboxylate, m254
5-Methylfurfural, m252
α-Methyl-D-glucopyranoside, m257

$[(CH_3)_2N \ldots N(CH_3)_2]$ $Cl^- \cdot 3H_2O$

m238

CHO

m240

m241

COOH

m242

CH_2OH

m243

C_3H_7 $CH_2OCH_2CH_2OCH_2CH_2OC_4H_9$

m244

CH_2

m245

CH_3 CHO

m252

CH_3

m253

$COOCH_3$

m254

TABLE 1-14 Physical constants of organic compounds (*continued*)

No.	Name	Formula	Formula weight	Beilstein reference	Density	Refractive index	Melting point	Boiling point	Flash point	Solubility in 100 parts solvent
m256	*N*-Methyl-D-glucamine		195.22				128.129			100 aq; 1.2 alc
m257	α-Methylglucoside		194.19	31, 179	1.46_4^{30}		168	$200^{0.2mm}$		63 aq; i alc, eth
m258	DL-2-Methylglutaronitrile	$NCCH_2CH_2CH(CH_3)CN$	108.14	2, 656	0.950			125^{10mm}	126	
m259	*N*-Methylglycine	CH_3NHCH_2COOH	89.09	4, 345			d 212			42 aq; sl s alc
m260	Methyl glycolate	$HOCH_2COOCH_3$	90.08	3, 236	1.168_4^{18}		74	151		s aq; misc alc, eth
m261	2-Methylheptane	$CH_3(CH_2)_4CH(CH_3)_2$	114.23	1, 161	0.6978_4^{20}	1.3974^{20}	−109.0	117.7	4	s eth; sl s alc
m262	Methyl heptanoate	$CH_3(CH_2)_5COOCH_3$	144.22	2, 339	0.8815_4^{20}	1.4115^{20}	−55.8	173.8	52	s alc, eth; sl s aq
m263	6-Methyl-5-hepten-2-one	$(CH_3)_2C{=}CHCH_2CH_2$-$COCH_3$	126.20	1^2, 797	0.855_4^{16}	1.4392^{20}	−67	73^{18mm}	50	misc alc, eth
m264	Methyl hexadecanoate	$CH_3(CH_2)_{14}COOCH_3$	270.46	2, 372			28	196^{15mm}		s alc, chl, eth
m265	2-Methylhexane	$CH_3(CH_2)_3CH(CH_3)_2$	100.21	1, 156	0.6786^{20}	1.3849^{20}	−118.3	90.1	−3	s alc; misc eth
m266	Methyl hexanoate	$CH_3(CH_2)_4COOCH_3$	130.19	2, 323	0.9038_4^{0}	1.4038^{23}	−71	151	54	v s alc, eth
m267	5-Methyl-2-hexanol	$(CH_3)_2CHCH_2CH_2$-$CH(OH)CH_3$	116.20	1, 416	0.814_4^{20}	1.4176^{20}		150	46	s alc, eth; i aq
m268	5-Methyl-2-hexanone	$(CH_3)_2CHCH_2CH_2$-$COCH_3$	114.19	1^2, 756	0.888_4^{20}	1.4062^{20}		141	41	0.5 aq; misc alc, eth
m269	5-Methyl-3-hexen-2-one	$(CH_3)_2CHCH{=}CHCOCH_3$	112.17			1.4400^{20}				
m269a	1-Methylhexylamine	$H(CH_2)_5CH(NH_2)CH_3$	115.22	4, 194	0.7665^{18}	1.4175^{20}		144	54	sl s aq; s alc, eth
m270	1-Methylhydantoin		114.10	24, 244			157	subl		s aq, alc; 3 eth
m271	Methylhydrazine	CH_3NHNH_2	46.07	4^2, 957	0.866	1.4235^{20}	−52.4	87.5	21	misc aq, alc; s PE
m272	Methyl hydrazinocarboxylate	$H_2NNHCOOCH_3$	90.08	3^1, 46			70–73	108^{12mm}		
m273	Methyl hydrogen glutarate	$HOOCCH_2CH_2CH_2$ $COOCH_3$	146.14	2^2, 565	1.169	1.4381^{20}		151^{10mm}	>112	

m274	Methyl hydrogen hexanedioate	$HOOC(CH_2)_4COOCH_3$	160.17	2, 652	1.081	1.4401^{20}	8–9	162^{10mm}	>112	s alc
m275	Methyl hydrogen succinate	$HOOCCH_2CH_2COOCH_3$	132.12	2, 608			56–59	151^{20mm}		v s aq, alc, eth
m276	Methyl hydroperoxide	CH_3OOH	48.04	1^2, 270	1.997^{15}_{4}	1.3642^{15}		38^{65mm}		misc aq, alc, eth; s bz
m277	Methylhydroquinone		124.14	6, 874			125–128			
m278	Methyl 4-hydroxybenzoate	$HOC_6H_4COOCH_3$	152.15	10, 158			126–128	270 d		v s alc, eth, acet
m279	Methyl 2-hydroxyisobutyrate	$(CH_3)_2C(OH)COOCH_3$	118.13	3^2, 223	1.023	1.4112^{20}		137	42	v s aq, alc
m280	Methyl 4-hydroxyphenylacetate	$HOC_6H_4CH_2COOCH_3$	166.18	10, 191			57–60	$162–163^{5mm}$		
m281	1-Methylimidazole		82.11	23, 46	1.030	1.4970^{20}	−60	198	92	misc aq
m282	2-Methylimidazole		82.11	23, 65			143	268		
m283	4-Methylimidazole		82.11	23, 69			46–48	263	>112	

N-Methylguanidine acetic acid, c278
4-Methylhexahydrophthalic anhydride, m197
Methyl hydroxyacetate, m260
Methyl 4-hydroxy-3-methoxybenzoate, m448
Methyl 2-hydroxypropionate, m292
2,2′-Methyliminodiethanol, m224
2,2′-Methyliminobis(acetaldehyde diethyl acetal), b170

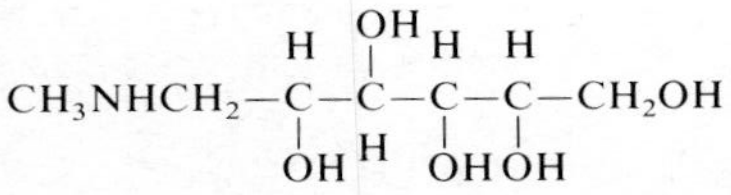

m256 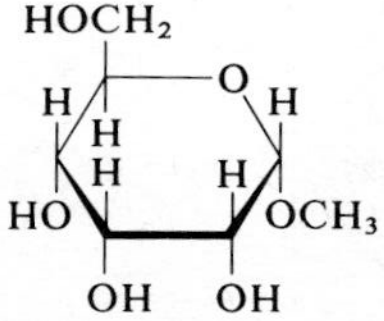 m257 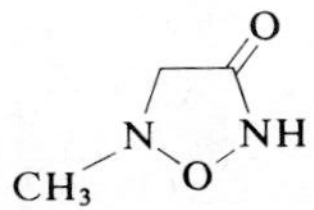 m270 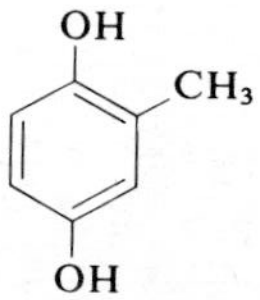 m277 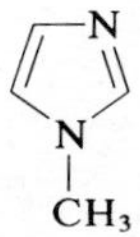 m281 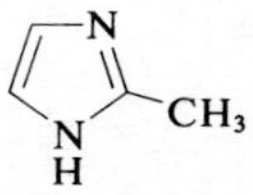 m282 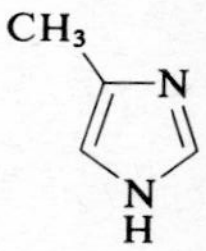 m283

TABLE 1-14 Physical constants of organic compounds (*continued*)

No.	Name	Formula	Formula weight	Beilstein reference	Density	Refractive index	Melting point	Boiling point	Flash point	Solubility in 100 parts solvent
m284	2-Methyl-1*H*-indole		131.18	20, 311	1.07_4^{20}		58–60	273		v s alc, eth; s hot aq
m285	3-Methyl-1*H*-indole		131.18	20, 315			95	266		s hot aq, alc, bz
m286	*N*-Methylisatoic anhydride		177.16	27, 265			165 d			
m287	Methyl isobutyrate	$(CH_3)_2CHCOOCH_3$	102.13	2, 290	0.891^{20}	1.3840^{20}	−84	93	<1	misc alc, eth; sl s aq
m288	Methyl isocyanate	CH_3NCO	57.05	4, 77	0.967	1.3695^{20}	−17	37–39	−18	s aq
m289	Methyl isodehyd-acetate		182.18	18, 410			60–63	167^{14mm}		
m290	Methyl isothio-cyanate	CH_3NCS	73.12	4, 77	1.069	1.5258^{37}	35–36	119	32	v s alc, eth; sl s aq
m291	5-Methylisoxazole		83.09	27, 16	1.018	1.4386^{20}		122	30	
m292	Methyl lactate	$CH_3CH(OH)COOCH_3$	104.10	3, 280	1.088_4^{20}	1.4131^{20}	~66	144.8	52	misc aq(d), alc, eth
m293	Methyl mandelate	$C_6H_5CH(OH)COOCH_3$	166.18	10, 202	1.1756^{20}		51–54	135^{12mm}		s aq, alc, bz, chl
m294	Methyl mercapto-acetate	$HSCH_2COOCH_3$	106.14		1.187	1.4657^{20}		43^{10mm}	30	s alc, eth
m297	Methyl 3-mercapto-propionate	$HSCH_2CH_2COOCH_3$	120.17	3^2, 214	1.085	1.4640^{20}		55^{14mm}	60	
m298	Methylmercury chloride	CH_3HgCl	251.10		4.06^{25}		170			
m299	Methyl methacrylate	$H_2C{=}C(CH_3)COOCH_3$	100.12	2^2, 398	0.9433^{20}	1.4146^{20}	−48.2	100.3	10	1.6 aq; s ketones, esters, CCl_4
m300	Methyl methane-sulfonate	$CH_3SO_2OCH_3$	110.13	4, 4	1.2943_4^{20}	1.4138^{20}		202–203	104	20 aq; 100 DMF
m301	Methyl methoxy-acetate	$CH_3OCH_2COOCH_3$	104.11	3, 236	1.0511_4^{20}	1.3964^{20}		130	35	v s alc, eth; sl s aq

m302	Methyl 1-methoxybicyclo[2.2.2]oct-5-ene-2-carboxylate		196.25		1.086	1.4886^{20}		105^{17mm}	103	
m303	Methyl 4-methoxyphenylacetate	$CH_3OC_6H_4CH_2COOCH_3$	180.20	10, 191	1.135	1.5165^{20}		158^{19mm}	36	
m304	1-Methyl-4-(methylamino)piperidine		128.22		0.882	1.4672^{20}			55	
m305	Methyl 3-methylbutyrate	$(CH_3)_2CHCH_2COOCH_3$	116.16	2^2, 274	0.881_4^{20}	1.3800^{25}		116–117		sl s aq; misc alc, eth
m306	2-Methyl-6-methylene-2-octanol	$C_2H_5C(=CH_2)(CH_2)_3C(CH_3)_2OH$	156.27		0.784	1.4431^{20}		84^{10mm}	76	

Methyl iodide, i40
Methyl isoamyl ketone, m268
Methyl isobutenyl ketone, m353
Methyl isobutyl ketone, m349
Methyl isonicotinate, m405
Methyl isopentyl ketone, m268
Methyl isovalerate, m305
2-Methyllactic acid, h127
Methyl linoleate, m328
Methyl mercaptan, m33
Methylmercaptoanilines, m425, m426
4-Methylmercaptobenzaldehyde, m427
Methylmercaptophenols, m431a, m432
7-Methyl-3-methylene-1,6-octadiene, m456
1-Methyl-4-(1-methylethenyl)cyclohexane, d651
5-Methyl-2-(1-methylethyl)cyclohexanol, m12
5-Methyl-2-(1-methylethyl)cyclohexanone, m13

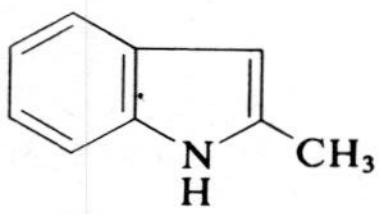

m284

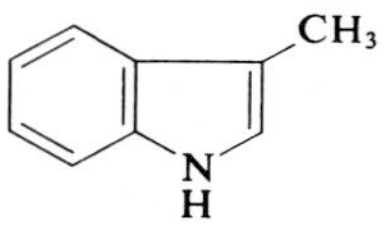

m285

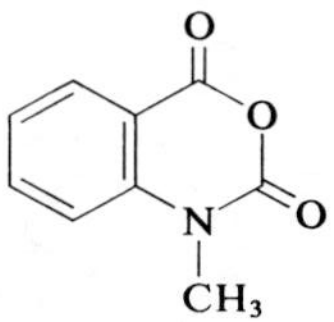

m286

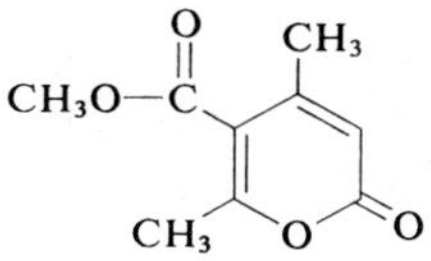

m289

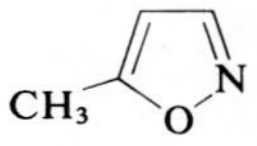

m291

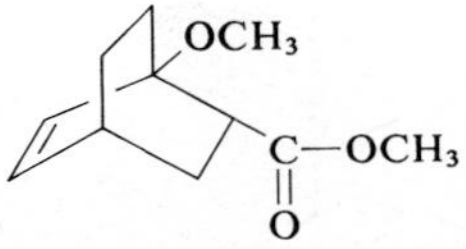

m302

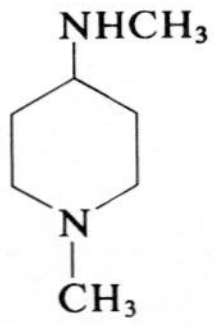

m304

TABLE 1-14 Physical constants of organic compounds (*continued*)

No.	Name	Formula	Formula weight	Beilstein reference	Density	Refractive index	Melting point	Boiling point	Flash point	Solubility in 100 parts solvent
m307	Methyl 2-methyl-3-furancarboxylate		140.14		1.116	1.4730^{20}		75^{20mm}	63	
m308	Methyl *S*-methylthiomethyl sulfoxide	$CH_3S(=O)CH_2SCH_3$	124.22		1.191	1.5487^{20}		$95^{2.5mm}$	>112	
m309	Methyl 3-(methylthio)propionate	$CH_3SCH_2CH_2COOCH_3$	134.20		1.077	1.4650^{20}		75^{13mm}	72	
m310	*N*-Methylmorpholine		101.15	27, 6	0.920	1.4349^{20}	−66	116	23	s aq, alc, eth
m311	1-Methylnaphthalene	$C_{10}H_7CH_3$	142.20	5, 566	1.025_4^{14}	1.6159^{20}	−30.5	244.7	82	v s alc, eth
m312	2-Methylnaphthalene	$C_{10}H_7CH_3$	142.20	5, 567	1.029_4^{20}	1.6026^{40}	34.6	241.4		v s alc, eth
m313	2-Methyl-1,4-naphthoquinone		172.18	7^2, 656			105–107			1.4 alc; 10 bz; s chl
m314	Methyl 1-naphthyl ketone	$C_{10}H_7COCH_3$	170.21	7, 401	1.1336_4^0	1.6284^{20}	12	296–298		s alc, eth; i aq
m315	Methyl 2-naphthyl ketone	$C_{10}H_7COCH_3$	170.21	7, 402			53–55	300–301		sl s alc; s CS_2
m316	Methyl nitrate	CH_3ONO_2	77.04	1, 284	1.2075_4^{20}	1.3748^{20}	−83.0	64 explodes		sl s aq; s alc, eth
m317	Methyl nitrite	CH_3ONO	61.04	1, 284	0.991 (liquid)			−17.35		s alc, eth
m318	2-Methyl-4-nitroaniline	$CH_3C_6H_3(NO_2)NH_2$	152.15	12, 846	1.586_4^{140}		131–133			v s alc; s bz
m319	2-Methyl-5-nitroaniline	$CH_3C_6H_3(NO_2)NH_2$	152.15	12, 844			104–107			s alc, acet, eth
m320	4-Methyl-2-nitroaniline	$CH_3C_6H_3(NO_2)NH_2$	152.15	12, 100			115–116			v s alc; s eth
m321	Methyl 2-nitrobenzoate	$O_2NC_6H_4COOCH_3$	181.15	9, 372	1.280	1.5350^{20}	−13	$106^{0.1mm}$	>112	s alc, eth

m322	2-Methyl-3-nitrobenzoic acid	$CH_3C_6H_3(NO_2)COOH$	181.15	9, 471			182–184			
m323	4-Methyl-3-nitrobenzoic acid	$CH_3C_6H_3(NO_2)COOH$	181.15	9, 502			187–190			
m324	5-Methyl-2-nitrobenzoic acid	$CH_3C_6H_3(NO_2)$–COOH	181.15	9, 482			134–136			
m325	2-Methyl-5-nitroimidazole		127.10	23^1, 23			252–254			
m326	3-Methyl-2-nitrophenol	$CH_3C_6H_3(NO_2)OH$	153.14	6, 385			35–39			
m327	4-Methyl-2-nitrophenol	$CH_3C_6H_3(NO_2)OH$	153.14	6, 412	1.240_4^{40}	1.574^{40}	32–35	125^{22mm}		v s alc, eth
m328	Methyl 9,12-octadecadienoate	$CH_3(CH_2)_4CH{=}CHCH_2$-$CH{=}CH(CH_2)_7COOCH_3$	294.46		0.8886_4^{18}	1.4593^{25}	−35	212^{16mm}		misc DMF
m329	Methyl octadecanoate	$CH_3(CH_2)_{16}COOCH_3$	298.51	2, 379			38–39	215^{15mm}	>112	s alc, eth
m330	Methyl *cis*-9-octadecenoate	$CH_3(CH_2)_7CH{=}CH$-$(CH_2)_7COOCH_3$	296.50	2, 467	0.879_4^{18}	1.4521^{20}	19.9	168^{2mm}		misc abs alc, eth
m331	Methyloctadecyldichlorosilane	$C_{18}H_{37}Si(CH_3)Cl_2$	367.5		0.930_4^{20}			$185^{2.5mm}$		

Methyl 2-methyllactate, m279
Methyl methyl-2-propenoate, m299
Methyl methylsulfinylmethyl sulfide, m308
Methyl myristate, m416
Methyl nicotinate, m404
4-Methyl-3-nitroanisole, m85
Methyl 6-nitrovertrate, m227
Methyl nonyl ketone, u5
Methyl oleate, m330

m307

m310

m313

m325

TABLE 1-14 **Physical constants of organic compounds (*continued*)**

No.	Name	Formula	Formula weight	Beilstein reference	Density	Refractive index	Melting point	Boiling point	Flash point	Solubility in 100 parts solvent
m332	Methyl octanoate	$CH_3(CH_2)_6COOCH_3$	158.24	2, 348	0.8775_4^{20}	1.4160^{25}	−40	192.9		v s alc, eth; i aq
m333	Methyloctyldichloro-silane	$C_8H_{17}Si(CH_3)Cl_2$	227.3		0.976_4^{20}	1.444^{20}		94^{6mm}		
m334	3-Methyl-2-oxazo-lidinone		101.11		1.170	1.4541^{20}	15	87–90^{1mm}	>112	
m335	2-Methyl-2-oxazo-line		85.11	27, 11	1.005	1.4340^{20}		110	20	
m336	Methyl 2-oxocyclo-pentanecarboxylate		142.15	10, 597	1.145	1.4560^{20}		105^{19mm}	>112	
m337	Methyl 2-oxo-propionate	$CH_3C(=O)COOCH_3$	102.09	3, 616	1.130	1.4065^{20}		134–137	39	misc alc, eth; sl s aq
m338	Methyl 2-oxo-1-pyrrolidineacetate		157.17		1.131	1.4719^{20}			110	
m339	2-Methylpentane	$CH_3CH_2CH_2CH(CH_3)_2$	86.18	1, 148	0.6532^{20}	1.3725^{20}	−153.7	60.3	−23	
m340	3-Methylpentane	$(CH_3CH_2)_2CHCH_3$	86.18	1, 149	0.6643^{20}	1.3765^{20}	<−50 glass	63.3	−6	
m341	2-Methyl-2,4-pentanediol	$(CH_3)_2C(OH)CH_2$-$CH(OH)CH_3$	118.18	1, 486	0.9216_4^{20}	1.4270^{20}	<−50 glass	198.3	101	misc aq
m342	4-Methylpentane-nitrile	$(CH_3)_2CHCH_2CH_2CN$	97.16	2^2, 290	0.8035_4^{20}	1.4061^{20}	−51.1	153.5		s alc; misc eth
m343	Methyl pentanoate	$CH_3(CH_2)_3COOCH_3$	116.16	2, 301	0.875	1.3962^{20}		128	22	sl s aq; misc alc, eth
m344	2-Methylpentanoic acid	$CH_3CH_2CH_2CH(CH_3)$-$COOH$	116.16	2^2, 288	0.9242_{20}^{20}	1.4135^{20}	−85 glass	196.4	107	1.3 aq
m345	3-Methylpentanoic acid	$CH_3CH_2CH(CH_3)$-CH_2COOH	116.16	2, 331	0.9262^{20}	1.4159^{20}	−42	196–198	85	s alc, eth
m346	2-Methyl-1-pentanol	$CH_3CH_2CH_2CH(CH_3)$-CH_2OH	102.18	1, 409	0.8242^{20}	1.4190^{20}		148.0	50	s alc, eth

m347	3-Methyl-3-pentanol	$(CH_3CH_2)_2C(CH_3)OH$	102.18	1, 411	0.8281^{20}	1.4186^{20}	<−38	122.4	46	misc alc, eth; sl s aq
m348	4-Methyl-2-pentanol	$(CH_3)_2CHCH_2$-$CH(OH)CH_3$	102.18	1.410	0.8080^{20}	1.4112^{20}	−90	131.7	41	1.6 aq
m349	4-Methyl-2-pentanone	$(CH_3)_2CHCH_2COCH_3$	100.16	1, 691	0.8006_4^{20}	1.3958^{20}	−83.5	115.7	13	1.7 aq; misc alc, bz, eth
m350	2-Methyl-1-pentene	$CH_3CH_2CH_2$-$C(CH_3){=}CH_2$	84.16	1^1, 90	0.6799_4^{20}	1.3920^{20}	−135.7	62.1	−26	s alc
m351	2-Methyl-2-pentene	$CH_3CH_2CH{=}C(CH_3)_2$	84.16	1, 217	0.6865_4^{20}	1.4003^{20}	−135.1	67.3	−23	s alc
m352	4-Methyl-2-pentenoic acid	$(CH_3)_2CHCH{=}$$CHCHCOOH$	114.14	2^2, 406	0.9529	1.4489	35	115^{20mm}	46	i aq; v s alc
m353	4-Methyl-3-penten-2-one	$(CH_3)_2C{=}CHCOCH_3$	98.15	1, 736	0.8548_4^{20}	1.4458^{20}	−42	129.5	30	3.1 aq
m353a	1-Methylpentylamine	$CH_3(CH_2)_3$-$CH(NH_2)CH_3$	101.19	4, 190	0.767_4^{20}		−19	116–118	13	s aq, alc, PE
m354	4-Methyl-1-pentyne	$(CH_3)_2CHCH_2C{\equiv}CH$	82.15		0.7041_4^{20}	1.3930^{20}	−104.8	61.2		
m355	3-Methyl-1-pentyn-3-ol	$CH_3CH_2C(CH_3)(OH)$-$C{\equiv}CH$	98.15	1^2, 506	0.8688_4^{20}	1.4318^{20}	−30.6	121–122	38	13 aq; misc bz, acet, PE, EtAc; s eth
m356	Methyl-(2-phenethyl)dichlorosilane	$C_6H_5CH_2CH_2Si(CH_3)Cl_2$	219.2		1.111_4^{20}	1.510^{20}		99^{6mm}		

o-Methylolphenol, h105
2-Methyloxacyclopropane, p230
Methyl oxirane, p230
Methyl palmitate, m264
Methyl pentyl ketone, h15

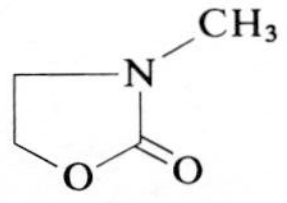

m334

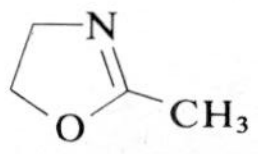

m335

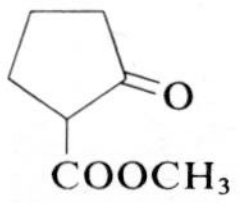

m336

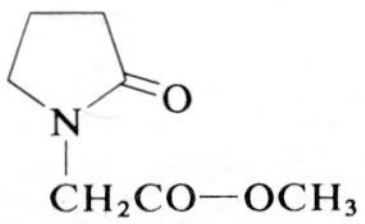

m338

TABLE 1-14 Physical constants of organic compounds (*continued*)

No.	Name	Formula	Formula weight	Beilstein reference	Density	Refractive index	Melting point	Boiling point	Flash point	Solubility in 100 parts solvent
m357	(1-Methylphenethyl)-trichlorosilane	$C_6H_5CH(CH_3)CH_2SiCl_3$	253.6		1.226_4^{20}	1.515^{20}		116^{10mm}		
m358	*N*-(4-Methylphenyl)-acetamide	$CH_3C_6H_4NHCOCH_3$	149.19	12^2, 501	1.212^{15}		153	307		s alc, EtAc, HOAc
m359	Methyl phenylacetate	$C_6H_5CH_2COOCH_3$	150.18	9, 434	1.044	1.5075^{20}		215	90	i aq; misc alc, eth
m360	Methylphenylchloro-silane	$C_6H_5(CH_3)Si(H)Cl$	156.7		1.054_4^{20}	1.571^{20}		113^{100mm}		
m361	Methylphenyldi-chlorosilane	$C_6H_5Si(CH_3)Cl_2$	191.1		1.187_4^{20}			205–206		
m362	*p*-(1-Methyl-2-phenylethyl)phenol	$C_6H_5CH_2CH(CH_3)$-C_6H_4OH	212.29				73	335		
m363	1-Methyl-1-phenyl-hydrazine	$C_6H_5N(CH_3)NH_2$	122.17	15, 117	1.038_4^{22}	1.5834^{20}		118^{21mm}	96	misc alc, bz, chl, eth
m364	1-Methyl-3-phenyl-propyl acetate	$C_6H_5CH_2CH_2CH(CH_3)$-$OOCCH_3$	192.26	6^1, 258	0.991			$74^{0.05mm}$	>112	
m365	3-Methyl-1-phenyl-2-pyrazolin-5-one		174.20	24, 20			130	287^{265mm}		
m366	Methylphenylsilane	$C_6H_5Si(CH_3)H_2$	122.1		0.889_4^{20}	1.506^{20}		139–240		
m367	Methyl phenyl sulfide	$C_6H_5SCH_3$	124.21	6, 297	1.058	1.5852^{20}	−15	188		i aq; s alc
m368	*N*-Methylpiperazine		100.17		0.903	1.4655^{20}		138	42	v s aq, alc, eth
m369	2-Methylpiperazine		100.17	23, 17			65–67	155.6	22	78 aq; 37 acet; 32 bz
m370	4-Methyl-1-piper-azinepropanol		158.25	23^3, 123		1.4835^{20}	28–30	120–121^{9mm}		
m371	*N*-Methylpiperidine	$C_5H_{10}N—CH_3$	99.19	20, 19	0.816	1.4378^{20}		106–107	<1	v s aq; misc alc, eth
m372	2-Methylpiperidine	$CH_3C_5H_9NH$	99.19	20, 95	0.844	1.4459^{20}	−5	119	8	v s aq; misc alc, eth

m373	3-Methylpiperidine	$CH_3C_5H_9NH$	99.19	20, 100	0.845	1.4470^{20}		126	<1	v s aq
m374	4-Methylpiperidine	$CH_3C_5H_9NH$	99.19	20, 101	0.838	1.4458^{20}		124	7	v s aq
m375	1-Methyl-3-piperidinemethanol		129.20	21^2, 8	1.013	1.4772^{20}		140–145	94	
m376	1-Methyl-4-piperidone		113.16	21^2, 215	0.920	1.4614^{20}			60	
m377	2-Methylpropanal	$(CH_3)_2CHCHO$	72.11	1, 671	0.7891^{20}	1.3727^{20}	−65	64.1		9 aq; misc alc, bz, chl, eth
m378	2-Methylpropane	$(CH_3)_3CH$	58.12	1, 124	0.557^{20}		−159.6	−11.7		13 mL aq; 1320 mL alc; 2890 mL eth
m379	*N*-Methyl-1,3-propanediamine	$H_2NCH_2CH_2CH_2NHCH_3$	88.15	4^1, 419	0.844	1.4468^{20}		139–141	35	
m380	2-Methyl-1,2-propanediamine	$(CH_3)_2C(NH_2)CH_2NH_2$	88.15	4, 266	0.841	1.4410^{20}			23	
m381	1-Methyl-1-propanethiol	$CH_3CH_2CH(SH)CH_3$	90.19	1, 373	0.8246^{25}_{4}	1.4338^{25}	−165	84–85	21	sl s aq; v s alc, eth

Methylphenols, c280, c281, c282
Methyl-*m*-phenylene diisocyanate, t175
Methyl phenyl ether, m48
Methyl phenyl ketone, a31
2-Methyl-2-phenylpropane, b428
Methyl γ-picolinate, m405
Methylpiperidinol, h145
Methyl pivalate, m228
1-Methyl-1-propanethiol, b392
Methyl propargyl ether, m399

m365

m368

m369

$CH_3—N\langle\rangle N—CH_2CH_2CH_2OH$

m370

m375

m376

TABLE 1-14 Physical constants of organic compounds (*continued*)

No.	Name	Formula	Formula weight	Beilstein reference	Density	Refractive index	Melting point	Boiling point	Flash point	Solubility in 100 parts solvent
m382	2-Methyl-1-propanethiol	$(CH_3)_2CHCH_2SH$	90.19	1, 378	0.8357_4^{20}	1.4396^{20}	−79	88.5	−9	v s alc, eth
m383	2-Methyl-2-propanethiol	$(CH_3)_3CSH$	90.19	1, 383	0.7943_4^{25}	1.4198^{25}	1.1	64.2	−26	i aq
m384	2-Methyl-1-propanol	$(CH_3)_2CHCH_2OH$	74.12	1, 373	0.8016^{20}	1.3958^{20}	−108	107.9	27	10 aq; misc alc, eth
m385	2-Methyl-2-propanol	$(CH_3)_3COH$	74.12	1, 379	0.7858_4^{20}	1.3877^{20}	25.8	82.4	15	misc aq, alc, eth
m386	2-Methylpropene	$(CH_3)_2C{=}CH_2$	56.10	1, 207	0.6266_4^{-140}		−140.4	−6.9		v s alc, eth
m387	2-Methyl-2-propene-1-sulfonic acid, Na salt	$H_2C{=}C(CH_3)CH_2SO_3^-Na^+$	158.15				>300			
m388	2-Methyl-2-propen-1-ol	$H_2C{=}C(CH_3)CH_2OH$	72.11	1, 443	0.857	1.4250^{20}		113–115	33	
m389	4-Methyl-2-(2-propenyl)phenol	$CH_3C_6H_3$-$(CH_2CH{=}CH_2)OH$	148.21	6^1, 287	0.980	1.5385^{20}		238	101	
m390	6-Methyl-2-(2-propenyl)phenol	$CH_3C_6H_3$-$(CH_2CH{=}CH_2)OH$	148.21	6^1, 287	0.992	1.5381^{20}		231–233	94	
m391	*N*-Methylpropionamide	$CH_3CH_2CONHCH_3$	87.12		0.9305^{25}	1.4345^{25}	−30.9	148		
m392	Methyl propionate	$CH_3CH_2COOCH_3$	88.11	2, 239	0.915_4^{20}	1.3770^{20}	−88	79.7	−2	6 aq; misc alc, eth
m393	2-Methylpropionic acid	$(CH_3)_2CHCOOH$	88.11	2, 288	0.950_4^{20}	1.3930^{20}	−46.1	154.7	55	23 aq; misc alc, chl, eth
m394	4′-Methylpropiophenone	$CH_3C_6H_4COCH_2CH_3$	148.21	7, 317	0.993	1.5280^{20}	7.2	239	96	
m395	Methylpropyldichlorosilane	$CH_3CH_2CH_2Si(CH_3)Cl_2$	157.1		1.04_4^{25}	1.425^{25}		125		
m396	Methyl propyl ether	$CH_3CH_2CH_2OCH_3$	74.12	1, 354	0.738^{20}			39.1		sl s aq; misc alc, eth

m397	2-Methyl-2-propyl-1,3-propanediol	$C_3H_7C(CH_3)(CH_2OH)_2$	132.20	1^1, 254			53–55	230		
m398	Methyl propyl sulfide	$CH_3SCH_2CH_2CH_3$	90.18	1^3, 1432	0.8424^{20}	1.4442^{20}	−113.0	95.5		s aq
m399	Methyl 2-propynyl ether	$CH_3OCH_2C\equiv CH$	70.09	1, 454	0.830	1.3961^{20}		61–62	<1	
m400	2-Methylpyrazine		94.12	23, 94	1.030	1.5042^{20}	−29	135	50	v s aq, alc, eth
m401	2-Methylpyridine	$CH_3C_5H_4N$	93.13	20, 234	0.950_4^{15}	1.5010^{20}	−67	128–129	26	v s aq; s alc, eth
m402	3-Methylpyridine	$CH_3C_5H_4N$	93.13	20, 239	0.961_4^{15}	1.5068^{20}	−18.3	143.5	36	s aq, alc, eth
m403	4-Methylpyridine	$CH_3C_5H_4N$	93.13	20, 240	0.957_4^{15}	1.5058	3.8	143–145	56	s aq, alc, eth
m404	Methyl 3-pyridinecarboxylate	$(C_5H_4N)COOCH_3$	137.14	22, 39			39	209		s aq, alc, bz
m405	Methyl 4-pyridinecarboxylate	$(C_5H_4N)COOCH_3$	137.14	22, 46	1.001	1.5122^{20}	8.5	207–209	82	
m406	1-Methyl-2-pyridone		109.13	21, 268	1.112	1.5690^{20}	7	250^{740mm}		
m407	N-Methylpyrrole		81.2	20, 163	0.914	1.4875^{20}	−57	113	15	i aq; misc alc, eth

2-Methylpropenenitrile, m27
2-Methylpropenoic acid, m26
2-Methylpropionaldehyde, i73
2-Methylpropionamide, i74
2-Methylpropionic acid, i75
2-Methylpropionitrile, i76
1-Methylpropyl acetate, b416
2-Methylpropyl acetate, i62
2-Methyl-2-propylamine, b421
2-Methylpropylamine, i63
(1-Methylpropyl)benzene, b427
(2-Methylpropyl)benzene, i64
2-Methylpropyl formate, i66
2-Methylpropyl lactate, i68
Methyl propyl ketone, p41
2-Methylpropyl 2-methylpropanoate, i67
Methyl pyruvate, m337
Methyl pyridyl ketones, a53, a54, a55

m400 m406 m407

TABLE 1-14 Physical constants of organic compounds (*continued*)

No.	Name	Formula	Formula weight	Beilstein reference	Density	Refractive index	Melting point	Boiling point	Flash point	Solubility in 100 parts solvent
m408	*N*-Methylpyrrolidine		85.15	20, 4	0.819_4^{20}	1.4247^{20}		80–81	−21	misc aq, eth
m409	*N*-Methyl-2-pyrrol-idinone		99.13	21^2, 213	1.0279^{25}	1.4680^{25}	−24.4	202	95	misc aq, alc, bz, eth
m410	2-Methylquinoline		143.19	20, 387	1.058	1.6108^{20}	−2	248	79	i aq; s chl, eth
m411	4-Methylquinoline		143.19	20, 395	1.0826_4^{20}	1.6200^{20}	9–10	261–263	>112	misc alc, bz, eth
m412	2-Methylquinoxaline		144.18	23^1, 44	1.118	1.6156^{20}	180–181	245–247	107	misc aq
m413	Methyl salicylate	$HOC_6H_4COOCH_3$	152.15	10, 70	1.1831^{20}	1.5240^{20}	−8.6	223.0	110	0.7 aq; s chl, eth; misc alc, HOAc
m414	α-Methylstyrene	$C_6H_5C(CH_3){=}CH_2$	118.18	5, 484	0.909	1.5375^{20}	−23.2	165.5	45	
m415	Methylsuccinic acid	$HOOCCH_2CH(CH_3)COOH$	132.12	2, 636	1.411	1.4303	110–112	d		66 aq; v s alc, eth
m416	Methyl tetra-decanoate	$CH_3(CH_2)_{12}COOCH_3$	242.40	2^2, 326	0.855	1.4362^{20}	18.4	323	>112	misc alc, bz, eth
m427	2-Methyl-3,3,4,4-tetrafluoro-2-butanol	$HCF_2CF_2C(CH_3)_2OH$	160.11		1.282	1.3524^{20}		117	73	
m418	2-Methyltetra-hydrofuran		86.13	17, 12	0.860	1.4056^{20}		78–80	−11	
m419	1-Methyl-1,2,3,6-tetrahydropyridine		97.16		0.837	1.4570^{20}		113–114	8	
m420	3-Methyltetrahydro-thiophene-1,1-dioxide		134.20		1.191	1.4772^{20}		276	>112	
m421	4-Methyl-5-thiazole-ethanol		143.21		1.196	1.5508^{20}		135^{7mm}		
m422	2-Methyl-2-thiazoline		101.17	27, 13	1.067	1.5200^{20}	−101	145	37	
m423	Methyl thioacetate	CH_3COSCH_3	90.14			1.4628		98	10	s alc, eth
m424	(Methylthio)aceto-nitrile	CH_3SCH_2CN	87.14		1.039	1.4826^{20}		63^{15mm}	67	

m425	2-(Methylthio)-aniline	$CH_3SC_6H_4NH_2$	139.22	13, 399	1.111	1.6239^{20}		234	>112	
m426	3-(Methylthio)-aniline	$CH_3SC_6H_4NH_2$	139.22	13^1, 141	1.130	1.6423^{20}		165^{16mm}	>112	
m427	4-(Methylthio)-benzaldehyde	$CH_3SC_6H_4CHO$	152.22	8^1, 533	1.144	1.6452^{20}		90^{1mm}		
m428	3-(Methylthio)-2-butanone	$CH_3CH(SCH_3)COCH_3$	118.20	1^4, 3993	0.975	1.4710^{20}		50–54^{20mm}	44	
m429	Methyl thiocyanate	CH_3SCN	73.12	3, 175	1.068^{20}	1.4697^{20}	−51	130–133	38	i aq; misc alc, eth
m430	3-Methylthiophene		98.17	17, 38	1.016	1.5180^{20}	−69.0	115.4	11	i aq; misc alc, eth
m431	5-Methyl-2-thiophenecarbaldehyde		126.18	17^1, 151	1.170	1.5825^{20}		114^{25mm}	87	
m432	4-(*S*-Methylthio)-phenol	$CH_3SC_6H_4OH$	140.20	6^1, 419			83–85	153–156^{20mm}		

1-Methyl-2-(3-pyridyl)pyrrolidine, n20
Methylresorcinol, d391
Methylsalicyclic acids, h138, h139
Methyl stearate, m329
Methylsuccinyl chlcride, m189
Methylsulfonic acid, m30
Methyl theobromine, c1
3-Methyl-2-thiabutane, i104
Methyl thienyl ketone, a57
Methyl thioglycolate, m294

m408 m409 m410 m411 m412 m418 m419 m420

m421 m422 m430 m431

TABLE 1-14 Physical constants of organic compounds (*continued*)

No.	Name	Formula	Formula weight	Beilstein reference	Density	Refractive index	Melting point	Boiling point	Flash point	Solubility in 100 parts solvent
m433	3-Methylthio-1,2-propanediol	$CH_3SCH_2CH(OH)CH_2OH$	122.19		1.164	1.5160^{20}			>112	
m434	*N*-Methylthiourea	$CH_3NHC(=S)NH_2$	90.15	4, 70			119–121			v s aq, alc
m435	*N*-Methyl-*o*-toluamide	$CH_3C_6H_4CONHCH_3$	149.19	9, 465	1.168^{15}		69–71			
m436	*N*-Methyl-*p*-toluenesulfonamide	$CH_3C_6H_4SO_2NHCH_3$	185.25	11, 105			76–79			
m437	Methyl *p*-toluenesulfonate	$CH_3C_6H_4SO_2OCH_3$	186.23	11, 99			27.5			
m438	Methyltriacetoxysilane	$CH_3Si(OOCCH_3)_3$	220.3	4^3, 1896	1.175_4^{20}	1.408^{20}		88^{3mm}		
m439	Methyl 2,2,2-trichloroacetimidate	$Cl_3CC(=NH)OCH_3$	176.43	2, 212	1.425	1.4780^{20}		149	none	
m440	Methyltrichlorogermane	CH_3GeCl_3	193.98		1.730			111		
m441	Methyl trifluoromethanesulfonate	$CF_3SO_2OCH_3$	164.10		1.450	1.3244^{20}		94–99	38	
m442	*N*-Methyl-*N*-trimethylsilylacetamide	$CH_3CON(CH_3)Si(CH_3)_3$	145.3	4^4, 4011	1.439_4^{20}	0.901^{20}		154		
m443	*N*-Methyl-*N*-(trimethylsilyl)trifluoroacetamide	$CF_3CON(CH_3)Si(CH_3)_3$	199.25		1.075	1.3802^{20}		132	25	
m444	Methyltripropoxysilane	$CH_3Si(OC_3H_7)_3$	220.4		0.88_4^{20}	1.4085^{20}		83^{13mm}		
m445	(Methyl)triphenylphosphonium bromide	$[CH_3P(C_6H_5)_3]^+Br^-$	357.24				230–233			

m446	2-Methylundecanal	$CH_3(CH_2)_8CH(CH_3)$-CHO	184.32		0.830_4^{15}	1.4321^{20}		271	93	s alc, eth
m447	Methyl urea	$CH_3NHCONH_2$	74.08	4, 64	1.204		101–102	d		v s aq, alc; i eth
m448	Methyl vanillate	$CH_3OC_6H_3(OH)COOCH_3$	182.18	10, 396			64–65	285–287		s hot alc, hot PE
m449	Methyl vinyl ether	$CH_3OCH{=}CH_2$	58.08	1^3, 1857	0.7511_4^{20}	1.3947	−112	5.5	−56	0.8 aq; v s alc
m450	2-Methyl-5-vinyl-pyridine		119.17		0.898	1.5437^{20}		100^{50mm}	65	
m451	Morpholine		87.12	27, 5	1.007_4^{20}	1.4542^{20}	−4.9	128.9	35	misc aq, alc, bz, eth
m452	4-Morpholine-carbonitrile		112.12		1.109	1.4730^{20}		$73^{0.6mm}$	104	
m453	*N*-Morpholino-1-cyclohexene		167.25		0.995	1.5128^{20}		117–122	68	
m454	2-(*N*-Morpholino)-ethanesulfonic acid		195.24				>300			

4-Methyl-2-thiouracil, h120
Methyltrichlorosilane, t242
Methyltriethoxysilane, t266a
Methyl trimethylacetate, m228
Methyltris(2-methoxyethoxy)silane, t436
β-Methylumbelliferone, h141
Methyluracil, d392
Methyl urethane, m182
Methyl valerate, m343
3-Methylvaleric acid, m345
4-Methylvaleronitrile, m342
Methyl veratrate, m226
Michler's ketone, b172
Monoethyl adipate, e152
Monoglyme, d440
Monomethyl adipate, m274
Monomethyl glutarate, m273
Monomethyl succinate, m275
Mordent violet 5, a60
4-Morpholinoethanol, h122
1-Morpholinocyclopentene, c362

$CH_2{=}CH$, N, CH_3 — m450

O, N, H — m451

O, N, CN — m452

O, N — m453

O, $N{-}CH_2CH_2SO_3H$ — m454

TABLE 1-14 Physical constants of organic compounds (*continued*)

No.	Name	Formula	Formula weight	Beilstein reference	Density	Refractive index	Melting point	Boiling point	Flash point	Solubility in 100 parts solvent
m455	3-(*N*-Morpholino)-1,2-propanediol		161.20		1.157		37–38	191^{30mm}	>112	
m456	*β*-Myrcene	$(CH_3)_2C{=}CHCH_2CH_2$-$C({=}CH_2)CH{=}CH_2$	136.24	1, 264	0.794_4^{20}	1.4709^{20}		166–168	39	s alc, chl, eth, HOAc
n1	1-Naphthaldehyde	$C_{10}H_7CHO$	156.18	7, 400	1.150_4^{20}	1.6520^{20}	1–2	161^{15mm}	>112	s alc, eth
n2	Naphthalene	$C_{10}H_8$	128.17	5, 531	1.162_4^{20}	1.5821^{100}	80.2 subl above mp	217.7	78	0.3 aq; 7 alc; 33 bz; 50 chl
n3	1-Naphthalenecarboxylic acid	$C_{10}H_7COOH$	172.18	9, 647			160–162	300		v s hot alc, eth
n4	1,5-Naphthalenediamine	$C_{10}H_6(NH_2)_2$	158.20	13, 203			185–187			s hot aq, hot alc
n5	1,8-Naphthalenediamine	$C_{10}H_6(NH_2)_2$	158.20	13, 204	1.1265_4^{99}	1.6828^{99}	66.5	205^{12mm}		sl s aq; s alc, eth
n6	1-Naphthalenemethylamine	$C_{10}H_7CH_2NH_2$	157.22	12, 1316	1.073	1.6429^{20}		290–293	>112	
n7	1,8-Naphthalic anhydride		198.18	17, 521			267–269			sl s HOAc
n8	1,8-Naphthalimide		197.19	21, 527			300			sl s alc; i bz, eth, aq
n9	1-Naphthol	$C_{10}H_7OH$	144.17	6, 596	1.0954_4^{99}	1.6206^{99}	96	288		v s alc, bz, chl, eth
n10	2-Naphthol	$C_{10}H_7OH$	144.17	6, 627	1.217^4		121–123	285–286	161	0.1 aq; 125 alc; 6 chl; 77 eth; s alk
n11	1,4-Naphthoquinone		158.16	7, 724	1.422		128	subl <100		s bz, chl, eth, alk
n12	(2-Naphthoxy)acetic acid	$C_{10}H_7OCH_2COOH$	202.21	6, 645			155–157			
n13	2-(1-Naphthyl)-acetamide	$C_{10}H_7CH_2CONH_2$	185.23	9, 666			181–183			i aq; s bz, CS_2

n14	1-Naphthyl acetate	$C_{10}H_7OOCCH_3$	186.21	6, 608			43–46			s alc, eth
n15	1-Naphthylacetic acid	$C_{10}H_7CH_2COOH$	186.21	9, 666			135	d		3.3 alc; v s chl, eth
n16	1-Naphthylaceto-nitrile	$C_{10}H_7CH_2CN$	167.21	9, 667		1.6192^{20}	33–35	194^{18mm}	>112	s alc
n17	1-Naphthylamine	$C_{10}H_7NH_2$	143.18	12, 1212	1.123^{25}_{25}	1.6703	50	301	157	0.2 aq; v s alc, eth
n18	2-Naphthylsulfonic acid	$C_{10}H_7SO_3H$	208.23	11, 171	1.441^{25}		91	d		77 aq; s alc, eth
n19	1-(1-Naphthyl)-2-thiourea	$C_{10}H_7NHC(=S)NH_2$	202.28	12, 1241			198			0.6 aq; 2.4 acet; s alc
n20	Nicotine		162.24	23, 117	1.0097^{20}_{4}	1.5282^{20}	−79	123^{17mm}		misc aq; v s alc, eth, eth, PE

MSTFA, m443
Mucic acid, t86
Mucochloric acid, d208
Myristolyl chloride, t42
Myristic acid, t40
Myristyl alcohol, t41
Myristyl bromide b355
Naphthacene, b7
1-Naphthaleneacetamide, n13
1-Naphthaleneacetonitrile, n16
Naphthalenediols, d393, d394, d395, d396
1-Naphthol-2-carboxylic acid, h149
3-Naphthol-2-carboxylic acid, h150
1-Naphthol-3,6-disulfonic acid, h152
2-Naphthol-3,6-disulfonic acid, h151
1-Naphthonitrile, c294
(2-Naphthoxy)acetic acid, n12
N-1-Naphthylaniline, p132
Natural orange 6, h153
NBA, b219
NBS, b354
Neohexane, d491
Neohexene, d501
Neopentane, d594
Neopentyl alcohol, d597
Neopentyl glycol, d596
Neral, d565
Nerolidol, h185
Nicotinaldehyde, p256
Nicotinic acid, p260
Nicotinonitrile, c297

m455

n7

n8

n11

n20

TABLE 1-14 Physical constants of organic compounds (*continued*)

No.	Name	Formula	Formula weight	Beilstein reference	Density	Refractive index	Melting point	Boiling point	Flash point	Solubility in 100 parts solvent
n21	Nitrilotriacetic acid	$N(CH_2COOH)_3$	191.14	4, 369			246 d			0.1 aq; s hot alc
n22	*m*-Nitroacetophenone	$O_2NC_6H_4COCH_3$	165.15	7, 288			76–78	202		s alc, eth
n23	*p*-Nitroacetophenone	$O_2NC_6H_4COCH_3$	165.15	7, 288			78–80	202		s alc
n24	*o*-Nitroaniline	$O_2NC_6H_4NH_2$	138.13	12, 687	1.442^{15}		69–70	284		s hot aq, alc, chl
n25	*m*-Nitroaniline	$O_2NC_6H_4NH_2$	138.13	12, 698	1.43		114	306		0.1 aq; 5 alc; 6 eth
n26	*p*-Nitroaniline	$O_2NC_6H_4NH_2$	138.13	12, 711	1.437^{14}		146	260^{100mm}	165	4 alc; 3.3 eth; s bz
n27	3-Nitrobenzaldehyde	$O_2NC_6H_4CHO$	151.12	7, 250	1.2792^{20}_{4}		58	164^{23mm}		s alc, chl, eth
n28	4-Nitrobenzaldehyde	$O_2NC_6H_4CHO$	151.12	7, 256	1.496		106–107			s alc, bz, HOAc
n29	2-Nitrobenzamide	$O_2NC_6H_4CONH_2$	166.12	9, 373	1.462^{32}_{4}		174–178	317		s hot aq, hot alc, eth
n30	Nitrobenzene	$C_6H_5NO_2$	123.11	5, 233	1.205^{15}_{4}	1.5546^{15}	5.8	210.8	87	v s alc, bz, eth
n31	2-Nitrobenzene-1,4-dicarboxylic acid	$O_2NC_6H_3(COOH)_2$	211.13	9, 851			270–272			
n32	3-Nitrobenzene-1,2-dicarboxylic acid	$O_2NC_6H_3(COOH)_2$	211.13	9, 823			216 d			2 aq; v s hot alc
n33	4-Nitrobenzene-1,2-dicarboxylic acid	$O_2NC_6H_3(COOH)_2$	211.13	9, 828			163–166			v s aq, alc; s eth
n34	5-Nitrobenzene-1,3-dicarboxylic acid	$O_2NC_6H_3(COOH)_2$	211.13	9, 840			260–261			0.15 aq; v s alc, eth
n35	2-Nitrobenzene-sulfonyl chloride	$O_2NC_6H_4SO_2Cl$	221.62	11, 67			65–67			s eth; d hot aq, alc
n36	6-Nitrobenzimidazole		163.14	23, 135			207–209			s alc, acid
n37	2-Nitrobenzoic acid	$O_2NC_6H_4COOH$	167.12	9, 370	1.58		146–148			0.7 aq; 33 alc; 22 eth
n38	3-Nitrobenzoic acid	$O_2NC_6H_4COOH$	167.12	9, 376	1.494		142			0.3 aq; 33 alc; 40 acet
n39	4-Nitrobenzoic acid	$O_2NC_6H_4COOH$	167.12	9, 389	1.58		242.8			9 alc; 2 eth; 5 acet

n40	4-Nitrobenzonitrile	$O_2NC_6H_4CN$	148.12	9, 397			146–149			s HOAc; sl s aq, alc
n41	3-Nitrobenzoyl chloride	$O_2NC_6H_4COCl$	185.57	9, 381			32–35	275–278		d aq, alc; v s eth
n42	4-Nitrobenzoyl chloride	$O_2NC_6H_4COCl$	185.57	9, 394			75	205^{105mm}		d aq, alc; s eth
n43	*N*-(*p*-Nitrobenzoyl)-glycine	$O_2NC_6H_4CONHCH_2COOH$	224.17	9, 395			131–133			
n44	3-Nitrobenzyl alcohol	$O_2NC_6H_4CH_2OH$	153.14	6, 449			30–32	180^{3mm}		s aq, alc, eth
n45	4-Nitrobenzyl alcohol	$O_2NC_6H_4CH_2OH$	153.14	6, 450			92–94	185^{12mm}		v s alc, eth; sl s aq
n46	4-Nitrobenzyl bromide	$O_2NC_6H_4CH_2Br$	216.04	5, 334			98–100			2 alc; v s eth
n47	4-Nitrobenzyl chloride	$O_2NC_6H_4CH_2Cl$	171.58	5, 329			70–73			8 alc; s eth
n48	2-Nitrobiphenyl	$O_2NC_6H_4C_6H_5$	199.21	5, 582	1.44_4^{25}	1.613^{25}	36.7	325	179	s alc, acet, CCl_4
n49	4-Nitrobiphenyl	$O_2NC_6H_4C_6H_5$	199.21	5, 583			112–114	340		sl s alc; v s eth
n50	1-Nitrobutane	$CH_3CH_2CH_2CH_2NO_2$	103.18	1, 123	0.975_{20}^{20}	1.4112	−81.3	152.8		sl s aq; misc alc, eth
n51	3-Nitro-2-butanol	$CH_3CH(NO_2)CH(OH)CH_3$	119.12	1, 373	1.1296_4^{25}	1.4414^{20}		92^{10mm}	91	
n52	2-Nitrodiphenylamine	$O_2NC_6H_4NHC_6H_5$	214.22	12, 690			76–78			i aq; s alc

Ninhydrin, i16
Nioxime, c324
2,2′,2″-Nitrilotriethanol, t264
1,1′,1″-Nitrilotris(2-propanol), t314
2-Nitro-*p*-anisidine, m79
5-Nitro-*o*-anisidine, m78
Nitroanisoles, m81, m82
4-Nitrobenzyl cyanide, n65
Nitrocresols, m326, m327

O_2N N N H

n36

TABLE 1-14 Physical constants of organic compounds (*continued*)

No.	Name	Formula	Formula weight	Beilstein reference	Density	Refractive index	Melting point	Boiling point	Flash point	Solubility in 100 parts solvent
n53	Nitroethane	$CH_3CH_2NO_2$	75.07	1, 99	1.0528^{20}_{20}	1.3920^{20}	−90	114.1	30	4.5 aq; misc alc, eth; s alk, chl
n54	1-Nitroguanidine	$O_2NNHC(=NH)NH_2$	104.07	3, 126			d 225			0.4 aq; sl s MeOH
n55	5-Nitro-1*H*-indazole		163.14	23, 129			207–209			s alc, bz, eth, acet
n56	Nitromethane	CH_3NO_2	61.04	1, 74	1.1322^{25}_{4}	1.3795^{25}	−28.4	101.2	35	11 aq; s alc, eth
n57	1-Nitronaphthalene	$C_{10}H_7NO_2$	173.17	5, 553	1.223		59–60	304		s alc; v s chl, eth
n58	3-Nitro-2-pentanol	$CH_3CH_2CH(NO_2)$-$CH(OH)CH_3$	133.15	1, 385	1.0818^{25}_{4}	1.4430^{20}		100^{10mm}	90	
n59	2-Nitrophenethyl alcohol	$O_2NC_6H_4CH_2CH_2OH$	167.16	6, 218	1.190	1.5637^{20}	2	267	>112	
n60	2-Nitrophenol	$O_2NC_6H_4OH$	139.11	6, 213	1.495		44–45	214–216		s alc, bz, eth, alk
n61	4-Nitrophenol	$O_2NC_6H_4OH$	139.11	6, 226	1.495		112–114	279		s aq; v s alc, chl, eth
n62	4-Nitrophenyl acetate	$O_2NC_6H_4OOCCH_3$	181.15	6, 233			77–79			s aq; v s alc, bz, eth
n63	2-Nitrophenylacetic acid	$O_2NC_6H_4CH_2COOH$	181.15	9, 454			139–142			s hot aq, alc
n64	4-Nitrophenylacetic acid	$O_2NC_6H_4CH_2COOH$	181.15	9, 455			153			s alc, bz, eth
n65	4-Nitrophenylaceto-nitrile	$O_2NC_6H_4CH_2CN$	162.15	9, 456			117			s alc, eth
n66	4-Nitrophenyl chloroformate	$O_2NC_6H_4OOCCl$	201.57	6^1, 120			77–79	162^{19mm}		
n67	2-Nitro-*p*-phenylene-diamine	$O_2NC_6H_3(NH_2)_2$	153.14	13, 120			137–140			
n68	4-Nitro-*o*-phenylene-diamine	$O_2NC_6H_3(NH_2)_2$	153.14	13, 29			199–201			s acid
n69	4-Nitrophenyl-hydrazine	$O_2NC_6H_4NHNH_2$	153.14	15, 468			156 d			s alc, chl, eth, hot bz

n70	2-Nitrophenyl phenyl ether	$O_2NC_6H_4OC_6H_5$	215.21	6^2, 222	1.2539^{22}	1.575^{20}	<−20	184^{8mm}		s alc, eth
n71	4-Nitrophenyl phenyl ether	$O_2NC_6H_4OC_6H_5$	215.21	6, 232			53–56	320		s bz, eth
n72	3-Nitrophthalic anhydride		193.11	17, 486			163–165			sl s aq, bz
n73	1-Nitropropane	$CH_3CH_2CH_2NO_2$	89.09	1, 115	1.0009^{20}	1.4016^{20}	−104.0	131.2	33	1.4 aq; misc alc, eth
n74	2-Nitropropane	$(CH_3)_2CHNO_2$	89.09	1, 116	0.9876^{20}	1.3949^{20}	−91.3	120.3	37	1.7 aq; misc alc, eth
n75	2-Nitro-1-propanol	$CH_3CH(NO_2)CH_2OH$	105.09	1, 358	1.1841_4^{25}	1.4379^{20}		99^{10mm}	100	s aq, alc, eth
n76	4-Nitropyridine-*N*-oxide	$O_2NC_5H_4N(O)$	140.10				159–162			
n77	8-Nitroquinoline		174.16	20, 373			89–91			s alc, bz, eth; i aq
n78	Nitrosobenzene	C_6H_5NO	107.11	5, 230			67–69	59^{18mm}		i aq; s alc
n79	*N*-Nitrosodimethylamine	$(CH_3)_2NNO$	74.08	8, 84	1.0048_4^{20}	1.4368^{20}		151–153	61	v s aq, alc, eth
n80	*p*-Nitrosodiphenylamine	$C_6H_5NHC_6H_4NO$	198.22				144–145			v s alc, bz, chl, eth
n81	1-Nitroso-2-naphthol	$C_{10}H_6(NO)OH$	173.16	7, 712			109–110			3 alc; s bz, eth, alk

Nitroglycerin, g21
5-Nitroisophthalic acid, n34
3-Nitrophenyl disulfide, b195
4-Nitrophenyl disulfide, b196
4-(*p*-Nitrophenylthio)aniline, a246
3-Nitro-*o*-phthalic acid, n32
4-Nitro-*o*-phthalic acid, n33
N-Nitrosophenylhydroxylamine, c285

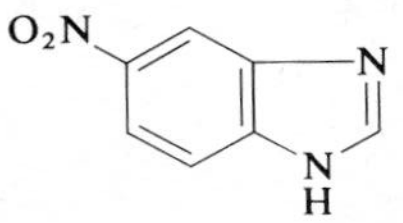

n55

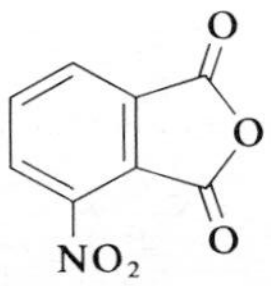

n72

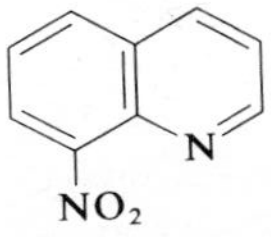

n77

TABLE 1-14 Physical constants of organic compounds (*continued*)

No.	Name	Formula	Formula weight	Beilstein reference	Density	Refractive index	Melting point	Boiling point	Flash point	Solubility in 100 parts solvent
n82	1-Nitroso-2-naphthol-3,6-disulfonic acid, di-Na salt hydrate		377.26	11^2, 190			>300			2.5 aq; sl s alc
n83	4-Nitrosophenol	ONC_6H_4OH	123.11	7, 622			d 126			s aq; v s alc, eth; explodes on contact with conc acid, alk, or fire
n84	β-Nitrostyrene	$C_6H_5CH{=}CHNO_2$	149.15	5, 478			58	250		s alc; v s eth
n85	2-Nitrotoluene	$CH_3C_6H_4NO_2$	137.14	5, 318	1.1622^{19}_{15}	1.5472^{20}	−10	222	106	s alc, bz
n86	3-Nitrotoluene	$CH_3C_6H_4NO_2$	137.14	5, 321	1.1581^{20}_{4}	1.5459^{20}	15.5	231.9	101	misc alc, eth; s bz
n87	4-Nitrotoluene	$CH_3C_6H_4NO_2$	137.14	5, 323	1.392		53–54	238	106	s alc, bz, chl, eth
n88	2-Nitro-α,α,α-trifluorotoluene	$CF_3C_6H_4NO_2$	191.11	5^2, 251			31–32	105^{20mm}		v s alc, bz
n89	3-Nitro-α,α,α-trifluorotoluene	$CF_3C_6H_4NO_2$	191.11	5, 327	1.436^{16}_{4}	1.4715^{20}	−2.4	200–205	87	s alc, eth
n90	Nonadecane	$CH_3(CH_2)_{17}CH_3$	268.51	1, 174	0.7776^{32}_{4}	1.4335^{38}	31.9	330.6	168	s eth; sl s alc
n91	1,8-Nonadiyne	$HC{\equiv}C(CH_2)_5C{\equiv}CH$	120.20	1^2, 248	0.8159^{21}_{4}	1.4492^{20}	−21	55^{13mm}	41	
n92	Nonane	$CH_3(CH_2)_7CH_3$	128.26	1, 165	0.7176^{20}_{4}	1.4054^{20}	−53.5	150.8	31	s abs alc, eth
n93	1,9-Nonanediamine	$H_2N(CH_2)_9NH_2$	158.29	4, 272			37–38	258		
n94	Nonanedinitrile	$NC(CH_2)_7CN$	150.23	2, 709	0.929	1.4460^{20}		176^{11mm}	>112	v s alc, bz, eth
n95	1,9-Nonanedioic acid	$HOOC(CH_2)_7COOH$	188.22	2, 707	1.029^{20}_{4}		106.5	286^{100mm}		0.24 aq; v s alc; 3 eth
n96	1,9-Nonanediol	$HO(CH_2)_9OH$	160.26	1, 493			45–47	177^{15mm}		
n97	Nonanenitrile	$CH_3(CH_2)_7CN$	139.24	2, 354	0.821^{15}_{4}	1.4260^{20}	−34.2	224.0	81	s alc, eth
n98	Nonanoic acid	$CH_3(CH_2)_7COOH$	158.24	2, 352	0.906^{20}_{4}	1.4330^{20}	12.5	254	100	s alc, chl, eth
n99	1-Nonanol	$CH_3(CH_2)_8OH$	144.26	1, 423	0.8274^{20}_{4}	1.4338^{20}	−5.5	213.1	75	0.6 aq; misc alc, eth
n100	5-Nonanone	$(C_4H_9)_2CO$	142.24	1, 710	0.806^{20}_{4}	1.4190^{20}	−50	187	60	misc alc, eth

n101	Nonanoyl chloride	$CH_3(CH_2)_7COCl$	176.69	2, 353	0.946_4^{15}	1.4377^{20}	−60.5	215.4	81	d aq, alc: s eth
n102	1-Nonene	$H(CH_2)_7CH{=}CH_2$	126.24	1^2, 202	0.7292^{20}	1.4157^{20}	−81.4	146.9	46	
n103	Nonyl aldehyde	$CH_3(CH_2)_7CHO$	142.24	1, 708	0.827_{19}^{19}	1.4240^{20}		185	63	
n104	Nonylamine	$CH_3(CH_2)_8NH_2$	143.27	4, 198	0.782	1.4330^{20}		201	62	sl s aq; s alc, eth
n105	Nopol		166.26		0.973	1.4930^{20}		230–240	98	
n106	Nopyl acetate		210.3		0.9805^{25}	1.4721^{20}				
n107	Norbornane		96.17	5^2, 45			82–84			s alc
n108	2-Norbornanone		110.16	7, 57			88–91	168–172	33	
n109	*trans*-5-Norbornene-2 2,3-dicarbonyl dichloride		219.07		1.349	1.5165^{20}		118^{11mm}	110	

Nitroso-R-salt, n82
Nitroterephthalic acid, n31
2-Nitro-*p*-toluidine, m320
4-Nitro-*o*-toluidine, m318
5-Nitro-*o*-toluidine, m319
4-Nitroveratrole, d445
Nitroxylenes, d558, d559, d560, d561
Nonyl alcohol, n99
2,5-Norbornadiene, b130
exo-2-Norbornanamine, a253
5-Norbornen-2-carbaldehyde, b132
Norbornene, b131

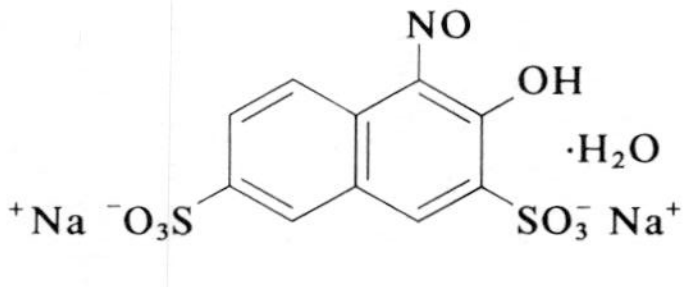

n82

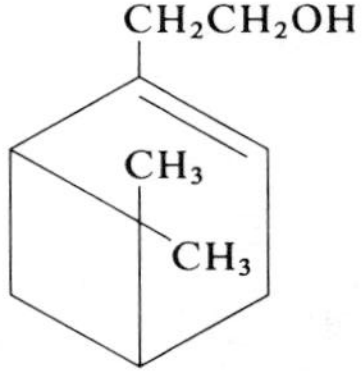

n105

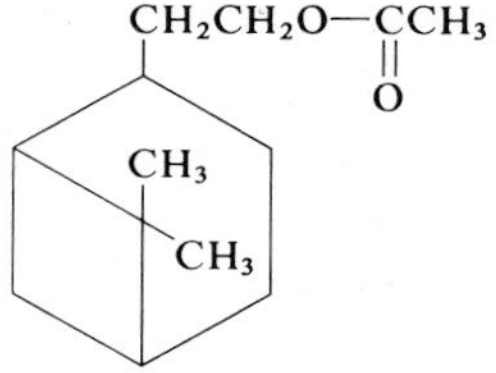

n106

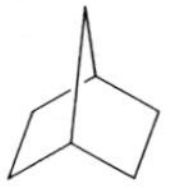

n107

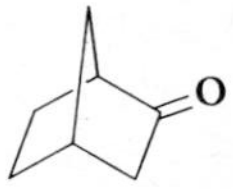

n108

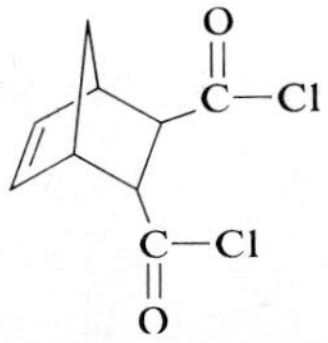

n109

TABLE 1-14 Physical constants of organic compounds (*continued*)

No.	Name	Formula	Formula weight	Beilstein reference	Density	Refractive index	Melting point	Boiling point	Flash point	Solubility in 100 parts solvent
n110	5-Norbornen-2-yl acetate		152.19		1.044	1.4700^{20}		76^{14mm}	62	
n111	*exo*-2-Norbornyl formate		140.18		1.048	1.4622^{20}		67^{16mm}	53	
n112	(+)-Norephedrine HCl		187.67	13^2, 371			174–176			
o1	(*Z*,*Z*)-9, 12-Octadecadienoic acid	$CH_3(CH_2)_4CH{=}CHCH_2$-$CH{=}CH(CH_2)_7COOH$	280.44	2, 496	0.9025^{20}_{4}	1.4699^{20}	−5	230^{16mm}		v s eth; misc PE; s abs alc
o2	Octadecanamide	$CH_3(CH_2)_{16}CONH_2$	283.50	2, 384			108–109	251^{12mm}		s hot alc, hot eth
o3	Octadecane	$CH_3(CH_2)_{16}CH_3$	254.50	1, 173	0.7767^{28}_{4}	1.4367^{28}	28.2	316.7	165	s acet, eth; sl s alc
o4	1-Octadecanethiol	$CH_3(CH_2)_{17}SH$	286.57			1.4648	29–31	360	185	s eth; sl s alc
o5	Octadecanoic acid	$CH_3(CH_2)_{16}COOH$	284.48	2, 377	0.847^{70}	1.4299^{80}	70	383		4.9 alc; 20 bz; 50 chl; 3.9 acet
o6	1-Octadecanol	$CH_3(CH_2)_{17}OH$	270.50	1, 431	0.8123^{58}_{4}	1.4388^{20}	57.9	203^{10mm}		s alc, eth
o7	9,12,15-Octadecatrienoic acid	$CH_3(CH_2CH{=}CH)_3CH_2$-$(CH_2)_6COOH$	278.44	2, 499	0.914^{18}_{4}	1.4800^{20}		230^{17mm}	>112	s alc, bz, eth
o8	1-Octadecene	$CH_3(CH_2)_{15}CH{=}CH_2$	252.49	1, 226	0.791^{18}_{4}	1.4439^{20}	17.7	314.9	148	s hot acet
o9	9-Octadecen-1-amine	$CH_3(CH_2)_7CH{=}CH$-$(CH_2)_8NH_2$	267.50		0.813	1.4578^{20}			154	
o10	(*Z*)-9-Octadecenoic acid	$CH_3(CH_2)_7CH{=}CH$-$(CH_2)_7COOH$	282.47	2, 463	0.8906^{20}_{4}	1.4571^{20}	4	286^{100mm}		misc alc, eth; s bz, chl
o11	(*E*)-9-Octadecenoic acid	$CH_3(CH_2)_7CH{=}CH$-$(CH_2)_7COOH$	282.47	2^2, 441	0.851^{79}	1.4308^{99}	44–45	288^{100mm}		s bz, chl, eth
o12	(*Z*)-9-Octadecen-1-ol	$CH_3(CH_2)_7CH{=}CH$-$(CH_2)_8OH$	268.49	1, 453	0.849^{20}_{4}	1.4610^{20}	13–19	195^{8mm}	>112	s alc, eth
o13	Octadecylamine	$CH_3(CH_2)_{17}NH_2$	269.52	4, 196	0.777^{27}		50–52	232^{32mm}	148	s alc, bz, eth
o14	Octadecyl isocyanate	$CH_3(CH_2)_{17}NCO$	295.51		0.847	1.4501^{20}		170^{2mm}	185	

o15	Octadecyltri-chlorosilane	$CH_3(CH_2)_{17}SiCl_3$	387.94		0.984	1.4602^{20}		223^{10mm}	89	
o16	Octadecyl vinyl ether	$CH_3(CH_2)_{17}OCH{=}CH_2$	296.54		0.821_4^{30}	1.4440^{30}	28	187^{5mm}	177	
o17	1,7-Octadiene	$H_2C{=}CH(CH_2)_4CH{=}CH_2$	110.20		0.746	1.4221^{20}		114–121	9	
o18	1*H*,1*H*,5*H*-Octa-fluoro-1-pentanol	$HCF_2CF_2CF_2CF_2CH_2OH$	232.08		1.6647^{20}	1.3190^{20}		140–141	74	
o19	Octamethylcyclo-tetrasilazane	$[-(CH_3)_2SiNH-]_4$	292.7		0.95^{22}	1.458^{25}		224–225		
o20	Octamethylcyclo-tetrasiloxane	$[-(CH_3)_2SiO-]_4$	296.62		0.9558^{20}	1.3968^{20}	17.5	175	90	
o21	Octamethyltri-siloxane	$[(CH_3)_3SiO]_2$-$Si(CH_3)_2$	236.0		0.8200^{20}	1.3848^{20}	~ −80	152–153	38	s bz, PE; sl s alc
o22	Octane	$CH_3(CH_2)_6CH_3$	114.23	1, 159	0.7025_4^{20}	1.3974^{20}	−56.8	125.7	15	s eth; sl s alc
o23	1,8-Octanediamine	$H_2N(CH_2)_8NH_2$	144.26	4, 271			50–52	225	165	
o24	1,8-Octanedioic acid	$HOOC(CH_2)_6COOH$	174.20	2, 691			140–144	230^{15mm}		0.16 aq; 0.6 eth; s alc
o25	1,2-Octanediol	$CH_3(CH_2)_5CH(OH)$-CH_2OH	146.23	1^3, 2217			36–38	132^{10mm}	> 112	

Norbornylene, b131
Norcamphor, n107
Norleucine, a185
Norvaline, a256
NTA, n21
Octadecyl bromide, b322
Oxtadecyl mercaptan, o4
2,3,4,6,7,8,9,10-Octahydropyrimido[1,2-*a*]azepine, d46
Octaldehyde, o40
Octamethylene glycol, o26
Octanal, o40
1,8-Octanedicarboxylic acid, d9

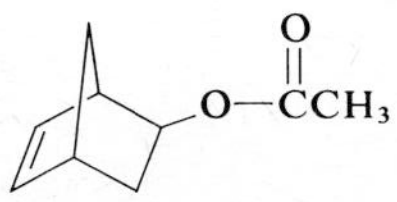

O
O—CCH₃

n110

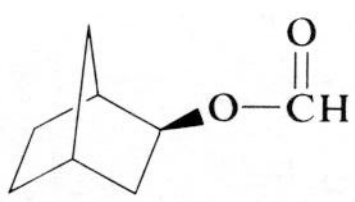

O
O—CH

n111

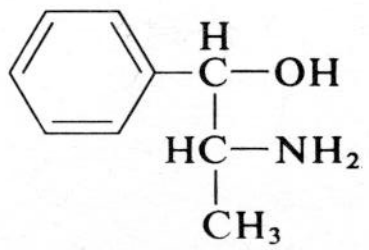

H
C—OH
HC—NH₂
CH₃

n112

TABLE 1-14 Physical constants of organic compounds (*continued*)

No.	Name	Formula	Formula weight	Beilstein reference	Density	Refractive index	Melting point	Boiling point	Flash point	Solubility in 100 parts solvent
o26	1,8-Octanediol	$HO(CH_2)_8OH$	146.23	1, 490			59–61	172^{20mm}		v s alc; sl s aq, eth
o27	Octanenitrile	$CH_3(CH_2)_6CN$	125.22	2, 349	0.8135^{20}	1.4202^{20}	−45.6	205.2	73	s eth; sl s alc
o28	1-Octanethiol	$CH_3(CH_2)_7SH$	146.30	1^3, 1710	0.843	1.4525^{20}	−49.2	199.0	68	s alc
o29	Octanoic acid	$CH_3(CH_2)_6COOH$	144.21	2, 347	0.9088_4^{20}	1.4279^{20}	16.6	239.3	110	0.07 aq; v s alc, chl, eth, PE
o30	1-Octanol	$CH_3(CH_2)_7OH$	130.23	1, 418	0.8258_4^{20}	1.4296^{20}	−15.0	195.2	81	0.06 aq; misc alc, chl, eth
o31	DL-2-Octanol	$CH_3(CH_2)_5CH(OH)CH_3$	130.23	1, 419	0.8207_4^{20}	1.4202^{20}	−38.6	179–180	71	0.08 aq; misc alc, eth
o32	DL-3-Octanol	$CH_3(CH_2)_4CH(OH)$-CH_2CH_3	130.23	1^1, 208	0.8216^{20}	1.4262^{20}		174–176	65	
o33	4-Octanol	$CH_3(CH_2)_3CH(OH)$-$CH_2CH_2CH_3$	130.23		0.8192^{20}	1.425^{20}		176.6	71	
o34	2-Octanone	$CH_3(CH_2)_5COCH_3$	128.22	1, 704	0.819_4^{20}	1.4150^{20}	−16	173	62	i aq; misc alc, eth
o35	3-Octanone	$CH_3(CH_2)_4COCH_2CH_3$	128.22	1, 706	0.8220_4^{20}	1.4150^{20}		167–168	46	i aq; misc alc, eth
o36	4-Octanone	$CH_3(CH_2)_3COCH_2$-CH_2CH_3	128.22	1, 706	0.809	1.4139^{20}		164	45	
o37	Octanoyl chloride	$CH_3(CH_2)_6COCl$	162.66	2, 348	0.955_{15}^{15}	1.4350^{20}	< −70	195	75	d aq, alc; s eth
o38	Octaphenylcyclo-tetrasiloxane	$[-(C_6H_5)_2SiO-]_4$	793.2		1.185			340^{1mm}		s alc, bz, HOAc
o39	1-Octene	$CH_3(CH_2)_5CH{=}CH_2$	112.22	1, 221	0.7149_4^{20}	1.4087^{20}	−101.7	121.3	21	i aq; misc alc, eth
o40	Octyl aldehyde	$CH_3(CH_2)_6CHO$	128.22	1, 704	0.821_4^{20}	1.4183^{20}	12–15	163.4	51	sl s aq; misc alc
o41	Octylamine	$CH_3(CH_2)_7NH_2$	129.25	4, 196	0.782	1.4290^{20}	−5 to −1	175–177	62	i aq; s alc, eth
o42	4-Octylaniline	$CH_3(CH_2)_7C_6H_4NH_2$	205.35	12, 1185				175^{13mm}		
o43	Octyltrichlorosilane	$CH_3(CH_2)_7SiCl_3$	247.7		1.07^{20}	1.447^{20}		226^{730mm}		
o44	1-Octyne	$CH_3(CH_2)_5C{\equiv}CH$	110.19	1, 258	0.7457^{20}	1.4159^{20}	−79.3	126.2		i aq; s alc, eth
o45	1-Octyn-3-ol	$CH_3(CH_2)_4CH(OH)$-$C{\equiv}CH$	126.20		0.864	1.4410^{20}			63	

o46	L-(+)-Ornithine	$H_2N(CH_2)_3CH(NH_2)$-COOH	132.16	4, 420			142			v s aq, alc; sl s eth
o47	Oxacycloheptane		100.16		0.890	1.440^{20}		122	10	
o48	Oxalic acid	HOOCCOOH	90.04	2, 502	1.90_4^{17}		189 d			9.5 aq; 24 alc; 1.3 eth
o49	Oxalic acid dihydrate	$HOOCCOOH \cdot 2H_2O$	126.07	2, 502	1.653_4^{19}		$-2H_2O$, 102			14 aq; 40 alc; 1 eth
o50	Oxalyl bromide	BrCO—COBr	215.84			1.5220		103^{720mm}	none	
o51	Oxalyl chloride	ClCO—COCl	126.93	2, 542	1.488_4^{13}	1.4340^{13}	−12	64	none	s eth; violent d aq, alc
o52	Oxalyl dihydrazide	H_2NNHCO— $CONHNH_2$	118.10	2, 559			240 d			s hot aq; sl s alc, eth
o53	Oxamic hydrazide	H_2NCO — $CONHNH_2$	103.08	2, 559			218 d			s alk; sl s aq; i eth
o54	Oxamide	H_2NCO—$CONH_2$	88.07	2, 545	1.667_4^{20}		d 350			sl s hot aq, alc
o55	2-Oxazolidone		87.08	27, 135			86–89	220^{48mm}		
o56	2-Oxobutyric acid	$CH_3CH_2C(=O)COOH$	102.09	3, 629	1.200_4^{17}	1.3972^{20}	32–34	82^{16mm}		v s aq, alc; v sl s eth

tert-Octylamine, t104
Octyl bromide, b323
Octyl cyanide, n97
Octyl chloride, c191
Octyl iodide, i46
Oleic acid, o11
Oleyl alcohol, o12
Oleylamine, o9

Orthanolic acid, a118
7-Oxabicyclo[2.2.1]heptane, e6
7-Oxabicyclo[4.1.0]heptane, e5
6-Oxabicyclo[3.1.0]hexane, e9
2-Oxabicyclo[6.1.0]nonane, e7a
Oxacyclobutane, t350
Oxacyclopentane, t68
Oxalylurea, i6

1,4-Oxathiane, t167
Oxepane, o47
Oxetane, t350
Oxirane, e132
2-Oxo-10-bornanesulfonic acid, c7
3-Oxobutanoic acid, a24
3-Oxobutyraldehyde dimethyl acetal, d436

O NH O

o47 o55

TABLE 1-14 Physical constants of organic compounds (*continued*)

No.	Name	Formula	Formula weight	Beilstein reference	Density	Refractive index	Melting point	Boiling point	Flash point	Solubility in 100 parts solvent
o57	2-Oxohexamethyleneimine		113.16	21^2, 216	1.02_4^{75}	1.4935	69.2	180^{50mm}		84 aq
o58	4-Oxopentanoic acid	$CH_3COCH_2CH_2COOH$	116.12	3, 671	1.1447_4^{25}	1.4396^{20}	33–35	245.8	137	v s aq, alc, bz, eth
o59	2-Oxopropionaldehyde	CH_3COCHO	72.06	1, 762	1.0455^{24}	1.4209^{20}		72	none	s aq, alc
o60	2-Oxopropionic acid	$CH_3COCOOH$	88.06	3, 608	1.267_4^{15}	1.4315^{20}	11.8	165 d	82	misc aq, alc, eth
o61	2,2′-Oxydiacetic acid	$HOOCCH_2OCH_2COOH$	134.09	3, 234			142–145	d		v s aq, alc; sl s eth
o62	4,4′-Oxydianiline	$H_2NC_6H_4OC_6H_4NH_2$	200.24	13, 441			190 d			
o63	3,3′-Oxydipropionitrile	$NCCH_2CH_2OCH_2CH_2CN$	124.14		1.043	1.4405^{20}		$112^{0.5mm}$	>112	
p1	Paraformaldehyde	$(CH_2O)_x$		1, 566			156 d		71	slowly s aq; s alk; i alc, eth
p2	Paraldehyde	$[-CH(CH_3)O-]_3$	132.16	19, 385	0.9984^{15}	1.4049^{20}	12.5	124		11 aq; misc alc, chl
p3	Parathion	$(C_2H_5O)_2P(=S)(O)$-$C_6H_4NO_2$	291.27		1.26_4^{25}	1.5370^{25}	6	375		v s alc, bz, eth
p4	DL-Patchenol		166.26	6^2, 64	0.987	1.5045^{20}		234–238	107	
p5	Pentabromoethylbenzene	$CH_3CH_2C_6Br_5$	500.67	5, 357			137–139			
p6	Pentabromophenol	C_6Br_5OH	488.62	6, 206			223–226	subl		sl s alc, eth
p7	Pentachloroacetone	$Cl_2CHCOCCl_3$	230.31	1, 656	1.690	1.4967^{20}	21 anhyd	192	none	i aq; v s acet
p8	Pentachlorobenzene	C_6HCl_5	250.34	5, 205	1.8342^{16}		82–85	275–277		v s bz, chl, eth
p9	Pentachloroethane	Cl_2CHCCl_3	202.30	1, 87	1.6712_4^{25}	1.5030^{20}	−29.0	160.5	none	0.05 aq; misc alc, eth
p10	Pentachloronitrobenzene	$C_6Cl_5NO_2$	295.34	5, 247	1.718_4^{25}		140–143			s bz, chl
p11	Pentachlorophenol	C_6Cl_5OH	266.34	6, 194	1.978_4^{22}		190–191	310 d		v s alc; s bz; 148 eth
p12	Pentachloropyridine	C_5Cl_5N	251.33	20, 232			124–126			

p13	Pentadecane	$CH_3(CH_2)_{13}CH_3$	212.42	1, 172	0.7684_4^{20}	1.4319^{20}	9.9	270.6	132	v s alc, eth
p14	8-Pentadecanone	$[CH_3(CH_2)_6]_2CO$	226.40	1, 717			41–43	178		s alc
p15	3-Pentadecylphenol	$C_{15}H_{31}C_6H_4OH$	304.52				45–48	195^{1mm}		
p16	1,2-Pentadiene	$CH_3CH_2CH{=}C{=}CH_2$	68.12	1, 251	0.6926_4^{20}	1.4209^{20}	−137.3	44.9		
p17	(*E*)-1,3-Pentadiene	$CH_3CH{=}CHCH{=}CH_2$	68.12	1, 251	0.6760^{20}	1.4301^{20}	−87.5	42.0	−28	
p18	(*Z*)-1,3-Pentadiene	$CH_3CH{=}CHCH{=}CH_2$	68.12	1, 251	0.6910^{20}	1.4363^{20}	−140.8	44.1	−28	
p19	1,4-Pentadiene	$H_2C{=}CHCH_2CH{=}CH_2$	68.12	1, 251	0.6608_4^{22}	1.3888^{20}	−148.3	26.0	4	
p20	Pentaerythritol	$C(CH_2OH)_4$	136.15	1, 528	1.38_4^{25}	1.548	260	subl		6 aq; v sl s alc; i eth
p21	Pentaerythrityl tetrabromide	$C(CH_2Br)_4$	387.76	1, 142			158–160	305–306		
p22	Pentaerythrityl tetranitrate	$C(CH_2ONO_2)_4$	316.15	1^2, 602	1.773_4^{20}		140		sensitive to shock; explodes on percussion	acet; sl s eth, alc

3-Oxo-*N*-phenylbutanamide, a32
2,2′-Oxybis(chloroethane), b159
1,1′-Oxybis(2-methylpropane), d408
1,1′-Oxybis(pentane), d653
3,3′-Oxybis(1-propene), d26
2,2′-Oxydiethanethiol, b187
2,2′-Oxydiethanol, b182
Palmitic acid, h35
Pamoic acid, m236
Parabanic acid, i6
Pelargonaldehyde, n103
Pelargonic acid, n98
Pelargonoyl chloride, n101
Pelargononitrile, n97
Pentabromophenyl ether, b198
Pentaerythritol diformal, t128

N
H
O
o57

CH_3
CH_3
$CHCH_2OH$
p4

TABLE 1-14 Physical constants of organic compounds (*continued*)

No.	Name	Formula	Formula weight	Beilstein reference	Density	Refractive index	Melting point	Boiling point	Flash point	Solubility in 100 parts solvent
p23	Pentafluorobenzonitrile	C_6F_5CN	193.07		1.532	1.4425^{20}		185–190	29	
p24	Pentamethylbenzene	$C_6H(CH_3)_5$	148.25	5, 443	0.917_4^{20}	1.527^{20}	54.4	231		v s alc, bz
p25	1,2,3,4,5-Pentamethylcyclopentadiene		136.24		0.870	1.4733^{20}		58^{13mm}	44	
p26	1,5-Pentamethylenetetrazole		138.17	26^2, 213			59–61	194^{12mm}		
p27	Pentanal	$CH_3CH_2CH_2CH_2CHO$	86.13	1, 676	0.8095_4^{20}	1.3942^{20}	−92	102–103	12	1.4 aq; misc alc, eth
p28	Pentane	$CH_3CH_2CH_2CH_2CH_3$	72.15	1, 130	0.6262_4^{20}	1.3575^{20}	−129.7	36.1	−49	misc alc, eth
p29	1,5-Pentanediamine	$H_2N(CH_2)_5NH_2$	102.18	4, 266	0.873_4^{25}	1.4591^{20}	−129.7	178–180	62	s aq, alc; sl s eth
p30	1,5-Pentanediol	$HO(CH_2)_5OH$	104.15	1, 481	0.9941^{20}	1.4494^{20}	−15.6	242.5	125	s aq, alc; sl s eth
p31	2,3-Pentanedione	$CH_3CH_2COCOCH_3$	100.11	1, 776	0.957	1.4068^{20}	−52	110–112	19	
p32	2,4-Pentanedione	$CH_3COCH_2COCH_3$	100.11	1, 777	0.9721^{25}	1.4510^{20}	−23.1	140.6	40	17 aq; misc alc, eth
p33	Pentanenitrile	$CH_3CH_2CH_2CH_2CN$	83.13	2, 301	0.8035_4^{15}	1.3991^{15}	−96.8	141.3	40	i aq; s alc, eth
p34	1-Pentanesulfonic acid, Na salt	$CH_3(CH_2)_4SO_3^-Na^+$	174.19	4^3, 23			> 300			4 aq
p35	1-Pentanethiol	$CH_3(CH_2)_4SH$	104.22	1, 384	0.840	1.4460^{20}	−75.7	126.6	18	i aq; misc alc, eth
p36	Pentanoic acid	$CH_3(CH_2)_3COOH$	102.13	2, 299	0.9390_4^{20}	1.4080^{20}	−33.7	185.5	88	2.4 aq; v s alc, eth
p37	1-Pentanol	$CH_3(CH_2)_4OH$	88.15	1, 383	0.8148_4^{20}	1.4100^{20}	−78.9	137.8	32	2.7 aq; misc alc, eth
p38	2-Pentanol	$CH_3CH_2CH_2CH(OH)CH_3$	88.15	1, 384	0.8393_4^{20}	1.4064^{20}	glass	119.0	40	16.6 aq; misc alc, eth
p39	3-Pentanol	$CH_3CH_2CH(OH)CH_2CH_3$	88.15	1.385	0.8150_4^{25}	1.4079^{25}	−69	115.6	40	5.2 aq; s alc, eth
p40	γ-Pentanolactone		100.12	17, 235	1.057	1.4330	−31	207–208	81	
p40a	δ-Pentanolactone		100.12	17, 235	1.079	1.4575^{20}		$60^{0.5mm}$	100	

p41	2-Pentanone	$CH_3CH_2CH_2COCH_3$	86.13	1, 676	0.8095^{20}	1.3903	−77.8	101.7	7	misc acet, bz, eth, PE
p42	3-Pentanone	$CH_3CH_2COCH_2CH_3$	86.13	1, 679	0.8143^{20}	1.3923^{20}	−39.0	102.0	12	3.4 aq
p43	Pentanophenone	$C_6H_5CO(CH_2)_3CH_3$	162.23	7, 327	0.988	1.5143^{20}		107^{5mm}	102	s alc, eth
p44	Pentanoyl chloride	$CH_3CH_2CH_2CH_2COCl$	120.58	2, 301	1.016	1.4216^{20}		125–127	23	
p45	1,4,7,10,13-Pentaoxacyclopentadecane	$[-CH_2CH_2O-]_5$	220.27			1.4615^{20}		$135^{0.2mm}$		
p46	3,6,9,12,15-Pentaoxahexadecanol	$CH_3O(CH_2CH_2O)_4CH_2CH_2OH$	252.31		0.933	1.4500^{20}		$133^{0.005mm}$	> 112	
p47	1-Pentene	$CH_3CH_2CH_2CH{=}CH_2$	70.14	1, 210	0.6410_4^{20}	1.3714^{20}	−165.2	30.0		misc alc, bz, eth
p48	(*E*)-2-Pentene	$CH_3CH_2CH{=}CHCH_3$	70.14	1.210	0.6482_4^{20}	1.3793^{20}	−140.2	36.3	−45	misc alc, eth
p49	(*Z*)-2-Pentene	$CH_3CH_2CH{=}CHCH_3$	70.14	1, 210	0.6503_4^{20}	1.3830^{20}	−151.4	36.9		misc alc, eth
p50	4-Pentenoic acid	$H_2C{=}CHCH_2CH_2COOH$	100.11	2, 425	0.9843_4^{18}	1.4341^{18}	< −18	187–189		sl s aq; s alc, eth
p51	3-Penten-2-one	$CH_3CH{=}CHCOCH_3$	84.12	1, 732	0.8624_4^{20}	1.4405^{20}		121–124	21	s aq
p52	Pentyl acetate	$CH_3(CH_2)_4OOCCH_3$	130.19	2, 131	0.8753^{20}	1.4028^{20}	< −100	149.2	23	0.17 aq
p53	Pentylamine	$CH_3(CH_2)_4NH_2$	87.17	4, 175	0.752	1.4110^{20}	−55	104	4	v s aq; misc alc, eth
p54	Pentylbenzene	$CH_3(CH_2)_4C_6H_5$	148.25	5, 434	0.8594_4^{20}	1.4885^{20}	−78.3	202.2	65	s alc; misc bz, eth

Pentalin, p9
Pentamethylene glycol, p30
Pentamethylene oxide, t78
Pentamethylethyl alcohol, t343
1,5-Pentanedicarboxylic acid, h8
Pentanedinitrile, g14
Pentanedioic acid, g11
2,5,8,11,14-Pentaoxapentadecane, b191
Penetetic acid, d299
sec-Pentylamine, a254
tert-Pentylamine, d603

p25

p26

p40

p40a

TABLE 1-14 Physical constants of organic compounds (*continued*)

No.	Name	Formula	Formula weight	Beilstein reference	Density	Refractive index	Melting point	Boiling point	Flash point	Solubility in 100 parts solvent
p55	4-*tert*-Pentylcyclohexanone		168.28	7^3, 173	0.920	1.4677^{20}		125^{16mm}	104	
p56	4-*tert*-Pentylphenol	$CH_3CH_2C(CH_3)_2$-C_6H_4OH	164.25	6, 548	0.962_4^{20}		93	262.2		s alc, eth
p57	1-Pentyne	$CH_3CH_2CH_2C{\equiv}CH$	68.11	1, 250	0.6901_4^{20}	1.3852^{20}	−105.7	40.2		v s alc; misc eth
p58	L-Perillaldehyde		150.22	7, 158	0.9645_4^{20}	1.5072^{20}		105^{10mm}	95	
p59	Peroxyacetic acid	$CH_3C(=O)OOH$	76.05		1.226_4^{15}		0.1	105 explodes 110		v s aq, alc, eth
p60	Petroleum ether	principally pentanes and hexanes			0.640			35–80	−40	misc bz, chl, eth, CCl_4
p61	Phenanthrene		178.23	5, 667	1.179^{25}		100	340		1.6 alc; 50 bz; 30 eth
p62	9,10-Phenanthrenedione		208.22	7, 796	1.405^4		209–211			s bz, eth, hot alc
p63	1,10-Phenanthroline		180.21	23, 227			117			1.4 bz; s alc, acet
p64	Phenol	C_6H_5OH	94.11	6, 110	1.0576_4^{41}	1.5418^{41}	40.9	181.8	79	6.7 aq; 8.2 bz; v s alc, chl, eth, alk
p65	Phenolphthalein		318.33	18, 143	1.299_4^{25}		258–262			8.2 alc; 1 eth
p66	Phenothiazine		199.28	27, 63			185.1	371		v s bz; s eth; sl s alc
p67	Phenothiazine-10-carbonyl chloride		261.73	27, 66			168–171			
p68	Phenoxyacetic acid	$C_6H_5OCH_2COOH$	152.15	6, 161			98	285 sl d		1.3 aq; v s alc, bz, HOAc, CS_2, eth
p69	Phenoxyacetyl chloride	$C_6H_5OCH_2COCl$	170.60	6, 162	1.235	1.5340^{20}		225–256		d aq, alc; s eth
p70	*p*-Phenoxyaniline	$C_6H_5OC_6H_4NH_2$	185.23	13, 438			82–84	189^{14mm}		s hot aq; v s alc, eth

p71	2-Phenoxybutyric acid	$CH_3CH_2CH(OC_6H_5)$-COOH	180.20	6, 163			79–83	258		sl s aq
p72	2-Phenoxyethanol	$C_6H_5OCH_2CH_2OH$	138.17	6, 146	1.102^{22}_{4}	1.5370^{20}	14	245.2	110	s aq; v s alc, eth
p73	1-Phenoxy-2-propanol	$C_6H_5OCH_2CH(OH)CH_3$	152.19	6^1, 85	1.063^{25}_{4}	1.523^{20}	13–18	240	135	
p74	Phenoxy-2-propanone	$C_6H_5OCH_2COCH_3$	150.18	6, 151	1.097	1.5210^{20}		230	85	
p75	DL-2-Phenoxy-propionic	$CH_3CH(OC_6H_5)COOH$	166.18	6, 163			116–119	265		s alc; sl s aq

Peracetic acid, p59
Perdeuterocyclohexane, c314
Perylene, d49
Phenacetin, e46
Phenacyl bromide, b222
Phenacyl chloride, c28
9,10-Phenanthraquinone, p62
Phenazone, a314
1,2,4-Phenenyl triacetate, t197
Phenethyl alcohol, p115
sec-Phenethyl alcohol, m138
Phenethylamine, p116
Phenethyl bromide, b285
Phenethyl chloride, c109
p-Phenetidine, e25
Phenetole, e29
Phenoxyacetone, p74
4-Phenoxybutyl bromide, b245

CH_3—C(CH_3)—CH_2CH_3 ; O

p55

CHO ; CH_3—C=CH_2

p58

p61

O ; O

p62

N ; N

p63

O ; O ; OH ; OH

p65

H ; N ; S

p66

COCl ; N ; S

p67

TABLE 1-14 Physical constants of organic compounds (*continued*)

No.	Name	Formula	Formula weight	Beilstein reference	Density	Refractive index	Melting point	Boiling point	Flash point	Solubility in 100 parts solvent
p76	3-Phenoxytoluene	$C_6H_5OC_6H_4CH_3$	184.24	6, 377	1.051	1.5727^{20}		271–273	> 112	
p76a	Phenylacetaldehyde	$C_6H_5CH_2CHO$	120.15	7, 292	1.027_{25}^{25}	1.5273^{20}	33–34	195	86	sl s aq; s alc, eth
p77	2-(2-Phenyl-acetamido)-acetaldoxime	$C_6H_5CH_2CONHCH_2$-$CH{=}NOH$	192.22				147–151			
p78	Phenyl acetate	$C_6H_5OOCCH_3$	136.15	6, 152	1.073_4^{20}	1.5030^{20}		196	76	misc alc, eth, chl
p79	Phenylacetic acid	$C_6H_5CH_2COOH$	136.15	9, 431	1.091_4^{77}		76.5	265.5		s hot aq, alc, eth
p80	Phenylacetonitrile	$C_6H_5CH_2CN$	117.15	9, 441	1.0214_{15}^{15}	1.5233^{20}	−23.8	233.5	101	i aq; misc alc, eth
p81	Phenylacetyl chloride	$C_6H_5CH_2COCl$	154.60	9, 436	1.169	1.5325^{20}		95^{12mm}		d aq, alc
p82	Phenylacetylene	$C_6H_5C{\equiv}CH$	102.14	5, 511	0.9300_4^{20}	1.5470^{20}	−44.9	142.4	31	misc alc, eth
p83	Phenylacetylurea	$C_6H_5CH_2CONHCONH_2$	178.19				212–216			sl s alc, bz, chl, eth
p84	L-3-Phenyl-α-alanine	$C_6H_5CH_2CH(NH_2)COOH$	165.19	14, 495			d 283			3 aq; s hot alc; i eth
p85	2-(Phenylamino)-benzoic acid	$C_6H_5NHC_6H_4COOH$	213.24	14, 327			185 d			s hot alc
p86	Phenyl 4-amino-salicylate	$H_2NC_6H_3(OH)COOC_6H_5$	229.24				153			0.7 mg aq
p88	*p*-Phenylazoaniline	$C_6H_5N{=}NC_6H_4NH_2$	197.24	16^1, 310			128	> 360		v s alc, bz, chl, eth
p89	Phenylazoformic acid 2-phenylhydrazide	$C_6H_5N{=}NCONHNHC_6H_5$	240.27	16, 24			156–159 d			
p90	*p*-Phenylazophenol	$C_6H_5N{=}NC_6H_4OH$	198.23	16, 96			155–157	230^{20mm}		v s alc, eth
p91	2-Phenylbenzimi-dazole		194.24	23, 230			291			s abs alc; sl s bz, chl
p92	Phenyl benzoate	$C_6H_5COOC_6H_5$	198.22	9, 116	1.235		70	314		v s hot alc; sl s eth
p93	*N*-Phenylbenzylamine	$C_6H_5CH_2NHC_6H_5$	183.25	12, 1023	1.061		27–38	306–307		s alc, chl, eth
p94	1-Phenylbiguanide	$C_6H_5NHC({=}NH)NH$-$C({=}NH)NH_2$	177.21				144–146			v s aq, alc

p96	1-Phenyl-2-butanone	$CH_3CH_2COCH_2C_6H_5$	148.21	7, 314	0.998	1.5122^{20}		112^{15mm}	90	s alc; misc eth; i aq
p97	4-Phenyl-2-butanone	$C_6H_5CH_2CH_2COCH_3$	148.21	7, 314	0.989	1.5122^{20}		235	98	s alc, eth
p98	(*E*)-4-Phenyl-3-buten-2-one	$C_6H_5CH{=}CHCOCH_3$	146.19	7, 364	1.0097^{45}_{4}	1.5836^{45}	41.5	261	65	v s alc, bz, chl, eth
p99	4-Phenylbutylamine	$C_6H_5CH_2CH_2CH_2$-CH_2NH_2	149.24	12, 1165	0.944	1.5196^{20}		124^{17mm}	101	
p100	2-Phenyl-3-butyn-2-ol	$CH_3C(OH)(C_6H_5)C{\equiv}CH$	146.19	6^2, 559			51–52	217–218		0.8 aq; s alc, bz, acet
p101	2-Phenylbutyric acid	$CH_3CH_2CH(C_6H_5)COOH$	164.20	9^2, 356			42–44	270–272		s bz, eth
p102	4-Phenylbutyric acid	$C_6H_5CH_2CH_2CH_2COOH$	164.20	9, 539			50–52	165^{10mm}		s alc, eth
p103	DL-2-Phenylbutyronitrile	$CH_3CH_2CH(C_6H_5)CN$	145.21	9, 541	0.974	1.5086^{20}		114–115^{15mm}	>112	
p104	Phenyl chloroformate	C_6H_5OOCCl	156.57					71^{9mm}		
p105	*S*-Phenyl chlorothioformate	C_6H_5SCOCl	172.6		1.269^{30}_{4}	1.5786^{30}	−14	101^{10mm}	116	

Phenylacetaldehyde dimethyl acetal, d451
N-Phenylacetamide, a18
Phenylacetone, p146
2-Phenylacetoacetonitrile, a51
α-*P*henylacetophenone, d22
β-Phenylacrylic acid, c268
γ-Phenylallyl alcohol, c270
o-Phenylanisole, m56
N-Phenylanthranilic acid, p85
Phenylarsonic acid, b11
Phenylazoformic acid 2-phenylhydrazide, d662
N-Phenylbenzamide, b5
Phenylbenzene, b135
Phenylbenzoic acid, b136
p-Phenylbenzyl alcohol, b139
Phenylboric acid, b12
1-Phenylbutane, b426
2-Phenylbutane, b427
1-Phenyl-1,3-butanedione, b64
4-Phenyl-*sec*-butyl acetate, m364
Phenyl Cellosolve, p72
Phenyl chloride, c41

N
N
H

p91

TABLE 1-14 Physical constants of organic compounds (*continued*)

No.	Name	Formula	Formula weight	Beilstein reference	Density	Refractive index	Melting point	Boiling point	Flash point	Solubility in 100 parts solvent
p106	Phenylcyclohexane	$C_6H_5C_6H_{11}$	160.26	5, 503	0.9427^{20}	1.5263^{20}	7.0	240.1	98	v s alc, eth
p107	Phenyl dichlorophosphate	$C_6H_5OP(O)Cl_2$	210.98	6, 179	1.412	1.5230^{20}		241–243	> 112	
p108	*N*-Phenyldiethanolamine	$C_6H_5N(CH_2CH_2OH)_2$	181.24	12, 183	1.120^{60}_{20}		56–58	350 sl d		5 aq; v s alc; 29 eth; 25 bz
p109	*o*-Phenylenediamine	$C_6H_4(NH_2)_2$	108.14	13, 6			103–104	256–258		v s alc, chl, eth
p110	*m*-Phenylenediamine	$C_6H_4(NH_2)_2$	108.14	13^1, 10	1.139^{15}_{15}		62–63	234–237		s aq, alc, acet, chl
p111	*p*-Phenylenediamine	$H_2NC_6H_4NH_2$	108.14	13, 61			145–147	267	68	1 aq; s alc, chl, eth
p112	*o*-Phenylene phosphorochloridite		174.52	27, 809	1.466	1.5712^{20}		80^{20mm}	> 112	
p113	1-Phenyl-1,2-ethanediol	$C_6H_5CH(OH)CH_2OH$	138.17	6, 907			66–68	272–274		v s aq, alc, bz, eth, chl, HOAc
p114	1-Phenylethanol	$CH_3CH(C_6H_5)OH$	122.17	6, 475	1.0150^{20}_{20}	1.5211^{20}	21.4	203.9		2.3 aq
p115	2-Phenylethanol	$C_6H_5CH_2CH_2OH$	122.17	6, 478	1.018^{25}_{25}	1.5317^{20}	−27	221	102	2 aq; misc alc, eth
p116	2-Phenylethylamine	$C_6H_5CH_2CH_2NH_2$	212.28	12, 1096	0.9640^{25}_{4}	1.5332^{20}		195	90	s aq; v s alc, eth
p117	D-(−)-α-Phenylglycine	$C_6H_5CH(NH_2)COOH$	151.17	14, 460			305–310			
p118	1-Phenylheptane	$C_6H_5(CH_2)_6CH_3$	176.30	5, 451	0.860	1.4842^{20}		233	95	
p119	1-Phenylhexane	$C_6H_5(CH_2)_5CH_3$	162.28	5^2, 337	0.861	1.4865^{20}	−61	226	83	misc eth
p120	Phenylhydrazine	$C_6H_5NHNH_2$	108.14	15^2, 44	1.0978^{20}_{4}	1.6070^{20}	19.5	243.5 d	88	misc alc, bz, chl, eth
p121	Phenyl 3-hydroxy-2-naphthoate	$C_{10}H_6(OH)COOC_6H_5$	264.28	10, 335			129–132	261^{160mm}		
p121a	2-Phenyl-2-imidazoline		146.19	23, 154			94–99			
p122	2-Phenylindole		193.25	20, 467			17	250^{10mm}		
p123	Phenyl isocyanate	C_6H_5NCO	119.12	12, 437	1.0956^{20}_{4}	1.5350^{20}	−30	162–163	55	d aq, alc; s eth
p124	Phenyl isothiocyanate	C_6H_5NCS	135.19	12, 453	1.1288^{25}_{4}	1.6497^{20}	−21	221	87	i aq; s alc, eth

p125	N-Phenylmaleimide		173.17	21, 400			89–90	163^{12mm}		s alc, chl, eth
p126	Phenylmalonic acid	$C_6H_5CH(COOH)_2$	180.16				155 d			
p127	Phenylmercury(II) acetate	$C_6H_5HgOOCCH_3$	336.74				149			0.17 aq; s alc, bz, acet
p128	Phenylmercury(II) chloride	C_6H_5HgCl	313.15				250–252			s bz, eth, pyr
p129	Phenylmercury(II) hydroxide	C_6H_5HgOH	294.70	16, 952			190 d			
p130	Phenylmethanethiol	$C_6H_5CH_2SH$	124.21	6, 453	1.058^{20}			194–195	70	
p131	N-Phenylmorpholine		163.22	27, 6			57	268		1.0 aq; v s hot alc
p132	N-Phenyl-1-naphthylamine	$C_{10}H_7NHC_6H_5$	219.29	12, 1224			60–62	226^{15mm}		s alc, bz, chl, eth
p132a	1-Phenyloctane	$C_6H_5(CH_2)_7CH_3$	190.33	5, 453	0.8572_4^{20}	1.4840^{20}	−36	261–263	107	misc eth
p133	2-Phenylphenol	$C_6H_5C_6H_4OH$	170.21	6^2, 623	1.213		57	282	123	s alc, chl, eth, alk
p134	4-Phenylphenol	$C_6H_5C_6H_4OH$	170.21	6, 674			164–165	305	165	s alc, chl, eth, alk

2-Phenylcinchoninic acid, p152
α-Phenyl-o-cresol, h113
Phenylethane, e69
Phenylethanenitrile, p80
1-Phenylethanol, m138
Phenylethanolamine, a265
N-Phenylethanolamine, a305
Phenylethene, s11
N-Phenylformamide, f31
Phenylglyoxylic acid, b70
Phenylglyoxylonitrile, b68
1-Phenyl-1-hydroxy-2-methylaminopropanes, e1, e2
2,2′-(Phenylimino)diethanol, p108
p-(2-Phenylisopropyl)phenol, m362
Phenyl-α-methylstyryl ketone, d660
1-Phenylpentane, p54

p112 p121 p122 p125 p131

TABLE 1-14 Physical constants of organic compounds (*continued*)

No.	Name	Formula	Formula weight	Beilstein reference	Density	Refractive index	Melting point	Boiling point	Flash point	Solubility in 100 parts solvent
p135	*N*-phenyl-*p*-phenyl enediamine	$C_6H_5NHC_6H_4NH_2$	184.24	13, 76			73–75			
p136	Phenyl *N*-phenyl-phosphoramido-chloridate	$C_6H_5NHP(=O)(Cl)OC_6H_5$	267.66	12, 588			132–134			
p137	Phenylphosphinic acid	$C_6H_5PH(O)OH$	142.09	16, 791			83–85			
p138	Phenylphosphonic acid	$C_6H_5P(O)(OH)_2$	158.09	16, 803			163–166			
p139	Phenylphosphonic dichloride	$C_6H_5P(O)Cl_2$	194.99	16, 804	1.375	1.5600^{20}	3	258	>112	
p140	Phenylphosphono-thioic dichloride	$C_6H_5P(S)Cl_2$	211.05	16, 807	1.360	1.6244^{20}		205^{130mm}		
p141	*N*-Phenylpiperazine		162.24		1.0621_4^{20}	1.5875^{20}		286	>112	i aq; misc alc
p142	2-Phenyl-1,2-propanediol	$CH_3C(C_6H_5)(OH)CH_2OH$	152.19	6, 930			44–45	160–162^{26mm}	>112	
p143	3-Phenyl-1-propanethiol	$C_6H_5CH_2CH_2CH_2SH$	152.26	6^1, 253	1.010	1.5494^{20}		109^{10mm}	90	
p144	1-Phenyl-1-propanol	$C_6H_5CH(OH)CH_2CH_3$	136.19	6, 502	0.9915_4^{25}	1.5169^{23}		219		misc alc, bz
p145	3-Phenyl-1-propanol	$C_6H_5CH_2CH_2CH_2OH$	136.19	6, 503	1.008	1.5257^{20}	−18	235	109	s aq; misc alc, eth
p146	1-Phenyl-2-propanone	$C_6H_5CH_2COCH_3$	134.18	7^2, 233	1.0157_4^{20}	1.5160^{20}	27	100^{13mm}	84	v s alc, eth; misc bz
p147	2-Phenylpropion-aldehyde	$CH_3CH(C_6H_5)CHO$	134.18	7^2, 237	1.009_4^{20}	1.5175^{20}		202–205	69	i aq; s alc
p148	3-Phenylpropionic acid	$C_6H_5CH_2CH_2COOH$	150.18	9, 508	1.047_4^{100}		47–48	280		0.6 aq; s bz, alc, chl, eth, HOAc, PE
p150	1-Phenyl-3-pyrazol-idinone		162.19	24, 2			121			10 hot aq; hot alc; s alk, acid

p151	2-Phenylpyridine	$C_6H_5C_5H_4N$	155.20	20, 424		1.6242^{20}		268–270	> 112	s alc, eth
p152	2-Phenyl-4-quinolinecarboxylic acid		249.27	22, 103			214–215			0.8 alc; 1 eth; 0.3 chl
p153	Phenyl salicylate	$C_6H_4(OH)COOC_6H_5$	214.22	10, 76	1.25		41–43	173^{12mm}		17 alc; 66 bz; s acet, chl, eth; 0.015 aq
p154	Phenylselenenyl chloride	C_6H_5SeCl	191.52	6^3, 1110			63–65	120^{20mm}		
p155	Phenylsuccinic acid	$HOOCCH_2$-$CH(C_6H_5)COOH$	194.19	9, 865			167–169	$-H_2O$, > 168		s hot aq, alc, eth
p156	*S*-Phenyl thioacetate	$C_6H_5SCOCH_3$	152.22			1.5720^{20}		100^{6mm}	79	
p157	1-Phenyl-2-thiourea	$C_6H_5NHC(S)NH_2$	152.22	12, 388	1.3		154			0.25 aq; s alc, alk
p158	Phenyltrichlorosilane	$C_6H_5SiCl_3$	211.56	16, 911	1.329^{20}	1.5230^{20}		201	91	

2-Phenylpropane, i91
3-Phenyl-2-propenoic acid, c268
3-Phenyl-2-propen-1-o1, c270
3-Phenyl-2-propenoyl chloride, c269
3-Phenylpropyl alcohol, p145
Phenyl propyl ketone, b503
3-Phenylpropyl mercaptan, p143
Phenyl sulfide, d690
Phenyl sulfone, d691
Phenylsulfonic acid, b22
Phenyl sulfoxide, d692
(Phenylthio)acetic acid, t163
Phenyl thiocarbamide, p157
α-Phenyl-*o*-toluic acid, b84
Phenyl *m*-tolyl ether, p76
Phenyl trifluoromethyl ketone, t294

p141

p150

p152

TABLE 1-14 Physical constants of organic compounds (*continued*)

No.	Name	Formula	Formula weight	Beilstein reference	Density	Refractive index	Melting point	Boiling point	Flash point	Solubility in 100 parts solvent
p159	1-Phenyltridecane	$C_6H_5(CH_2)_{12}CH_3$	260.47		0.8555^{20}_4	1.4814^{20}	10	346	>112	
p160	Phenyltriethoxy-silane	$C_6H_5Si(OC_2H_5)_3$	240.38	16, 911	0.996	1.4604^{20}		$113^{1.0mm}$	42	
p161	Phenyltrimethoxy-silane	$C_6H_5Si(OCH_3)_3$	198.3		1.064^{20}_4	1.4734^{20}		211		
p162	Phenyltrimethyl-ammonium bromide	$[C_6H_5N(CH_3)_3]^+Br^-$	216.13	12^2, 88			210 d			v s aq; s hot alc
p163	Phenyltrimethyl-ammonium chloride	$[C_6H_5\ N(CH_3)_3]^+Cl^-$	171.67	12, 158			237 subl			s aq; v s alc; sl s chl
p164	Phenyltrimethyl-ammonium iodide	$[C_6H_5N(CH_3)_3]^+I^-$	263.12	12^2, 88			175			s aq, alc; sl s acet
p165	Phenyltrimethyl-ammonium tribromide	$[C_6H_5N(CH_3)_3]^+Br_3^-$	375.95				114–116			
p166	Phenyltrimethylsilane	$C_6H_5Si(CH_3)_3$	150.30	16^1, 525	0.873	1.4907^{20}		168–170	44	
p167	Phenyltris(tri-methylsiloxy)silane	$[(CH_3)_3SiO]_3SiC_6H_5$	372.8		0.970^{25}_4	1.459^{25}		264–266	121	
p168	Phenylurea	$C_6H_5NHCONH_2$	136.15	12, 346	1.302		145–147	238		s hot aq, hot alc, eth
p169	Phenylvinyldi-chlorosilane	$H_2C{=}CH(C_6H_5)SiCl_2$	203.2		1.196^{25}_4	1.534^{25}		$87^{1.5mm}$		
p170	*o*-Phthalic acid	$C_6H_4(COOH)_2$	166.13	9, 791	1.593^{20}_4		206–208			0.6 aq; 10 alc; 0.5 eth; v sl s chl
p171	Phthalic anhydride		148.12	17, 469	1.53		130.8	285 subl		0.6 aq(d); s alc
p172	Phthalide		134.13	17, 310	1.164^{99}_4		72–74	290		s alc
p173	Phthalimide		147.13	21, 458			238	subl		v s alk; v sl s bz, PE
p174	*o*-Phthaloyl dichloride	$C_6H_4(COCl)_2$	203.02	9, 805	1.409^{20}	1.5684^{20}	15–16	280–282	>112	d aq, alc; s eth

p175	Phthalylsulfa-thioazole		403.44				272 d			s alk; sl s alc; i chl
p176	Picric acid	$(O_2N)_3C_6H_2OH$	229.11	6, 265	1.763_4^{20}		122–123	explodes >300		1.3 aq; 8.2 alc; 10 bz; 2.9 chl; 1.6 eth
p177	(*Z*)-Pinane		138.3	5, 93	0.839_4^{20}	1.4616^{20}	−50	167–168		
p178	(+)-α-Pinene		136.24	5, 146	0.8591_4^{20}	1.4660^{20}	−55	155–156	32	misc alc, eth
p179	(−)-β-Pinene		136.24	5, 154	0.8590^{20}	1.4666^{20}	−61.5	166	32	
p180	α-Pinene oxide		152.24	5, 152	0.964	1.4690^{20}		103^{50mm}	65	
p181	β-Pinene oxide		152.24	17^2, 44	0.976	1.4765^{20}		100^{27mm}	66	

Phloroglucinol, t310
Phorone, d531
Phthalaldehydic acid, f33
m-Phthalic acid, b16
p-Phthalic acid, b17
Phthalonitrile, d236
Picolinaldehyde, p255
Picolines, m401, m402, m403
Picolinic acids, p259, p261
Picolinonitrile, c296
Picolylamines, a226, a227
Picramide, t388
Pimelic acid, h8
Pinacol, d493
Pinacolone, d499
Pinacolyl alcohol, d498
3-Pinanol, i83

p171

p172

p173

HOOC C(=O)–N(H)–C6H4–SO2N(H)–thiazole

p175

p178
p179

p180

p181

TABLE 1-14 Physical constants of organic compounds (*continued*)

No.	Name	Formula	Formula weight	Beilstein reference	Density	Refractive index	Melting point	Boiling point	Flash point	Solubility in 100 parts solvent
p182	Piperazine		86.14	23, 4		1.446^{113}	108–110	145–146	109	v s aq; 50 alc; i eth
p183	1-Piperazinecarbaldehyde		114.15		1.107	1.5094^{20}		$97^{0.5mm}$	101	
p184	1,4-Piperazinedicarbonitrile		136.16	23^{1}, 5			167–170			
p185	3-(1-Piperazinyl)-1,2-propanediol		160.22				73–77	$133^{0.1mm}$		
p186	Piperidine		85.15	20, 6	0.8659^{15}	1.4525^{20}	−10.5	106.4	4	misc aq; s alc, bz, chl
p187	1-Piperidinecarbonitrile		110.16	20, 56	0.951	1.4705^{20}		102^{10mm}	97	
p188	*N*-Piperidineethanol		129.20	20, 25	0.9732^{25}_{25}	1.4804^{20}		200–202	68	misc aq; s alc
p189	2-Piperidineethanol		129.20	21, 2	1.010^{17}		38–40	234	102	v s aq, alc, eth
p190	3-Piperidinemethanol		115.18	21^{2}, 8	1.026			$107^{3.5mm}$	> 112	
p191	1-Piperidinepropionitrile		138.21		0.933	1.4695^{20}		111^{16mm}		
p192	3-Piperidino-1,2-propanediol		159.23	20, 34			77–80			
p193	*trans*-Piperitol		154.3		0.9178^{25}	1.4729^{20}				
p194	Propane	$CH_3CH_2CH_3$	44.10	1, 104	0.5842^{-42}	1.3397^{-42}	−187.7	−42.1		6.5 mL aq; 790 mL alc; 926 mL eth; 1300 mL chl; 1450 mL bz
p195	1,2-Propanediamine	$CH_3CH(NH_2)CH_2NH_2$	74.13	4, 257	0.878^{15}	1.4460^{20}		119.7	33	misc aq, bz; s alc, eth
p196	1,3-Propanediamine	$H_2NCH_2CH_2CH_2NH_2$	74.13	4, 261	0.884^{25}_{4}	1.4575^{20}	−12	140	48	misc alc, eth; s aq
p197	1,2-Propanediol	$CH_3CH(OH)CH_2OH$	76.10	1, 472	1.0364^{20}_{4}	1.4331^{20}	−60	188	107	misc aq, acet, chl; s alc, eth
p198	1,3-Propanediol	$HOCH_2CH_2CH_2OH$	76.10	1, 475	1.0597^{20}_{4}	1.4396^{20}	−26.7	214.4	79	misc aq, alc

No.	Name	Formula	Formula weight	Beilstein reference	Density	Refractive index	Melting point	Boiling point	Flash point	Solubility
p199	1,3-Propanedithiol	$HSCH_2CH_2\ CH_2SH$	108.23	1, 476	1.0772_4^{20}	1.5405^{20}	−79	169	40	misc alc, bz, eth, chl
p200	1-Propanesulfonyl chloride	$CH_2CH_2CH_2SO_2Cl$	142.60	4, 8	1.2864_4^{15}			66^{8mm}		d hot aq, hot alc
p201	1,3-Propane sultone		122.14		1.392		30–33	180^{30mm}		

Pipecolines, m372, m373, m374
1-Piperazineethanol, h123
1-Piperidinecarboxyaldehyde, f35
Piperonal, m240
Piperonyl alcohol, m243
Piperonyl butoxide, m244
Piperonylic acid, m242
Pivalaldehyde, d598
Pivalamide, d599
Pivalic acid, d600
Pivalic anhydride, d601
Pivaloyl chloride, d602
Pivaloyloxymethyl chloride, c152
POPOP, b199
PPO, d681
Prehnitene, t99
Procaine, d271
Proline, p277
Propadiene, a78
1-Propanal, p214
1,3-Propanedicarboxylic acid, g11
Propanedioic acid, m3
1,2-Propanediol cyclic carbonate, p228

p182 p183 p184 p185 p186 p187 p188 p189 p190

p191 p192 p193 p201

TABLE 1-14 **Physical constants of organic compounds (*continued*)**

No.	Name	Formula	Formula weight	Beilstein reference	Density	Refractive index	Melting point	Boiling point	Flash point	Solubility in 100 parts solvent
p202	1-Propanethiol	$CH_3CH_2CH_2SH$	76.16	1, 359	0.836_4^{25}	1.4380^{20}	−113.1	67.7	−20	s alc, eth
p203	2-Propanethiol	$CH_3CH(SH)CH_3$	76.16	1, 367	0.809_4^{25}	1.4255^{20}	−130.5	52.6	−34	misc alc, eth; sl s aq
p204	1,2,3-Propanetriol triacetate	H_3CCOO-$CH(CH_2OOCCH_3)_2$	218.21	2, 147	1.596_4^{20}	1.4302^{20}	−78	258–260	148	7.2 aq; misc alc, bz, chl, eth
p205	1-Propanol	$CH_3CH_2CH_2OH$	60.10	1, 350	0.8037_4^{20}	1.3856^{20}	−126.2	97.2	15	misc aq, alc, eth
p206	2-Propanol	$(CH_3)_2CHOH$	60.10	1, 360	0.7855_4^{20}	1.3772^{20}	−89.5	82.4	22	misc aq, alc, chl, eth
p207	2-Propenal	$H_2C{=}CHCHO$	56.07	1, 725	0.8389^{20}	1.4017^{20}	−87.0	52.7	−18	21 aq; s alc, eth
p208	Propene	$H_2C{=}CHCH_3$	42.08	1, 196	0.6104_4^{-48}	1.3567^{-40}	−185.2	−47.7		45 mL aq; 1200 mL alc; 500 mL acet
p209	2-Propene-1-thiol	$H_2C{=}CHCH_2SH$	74.15	1, 440	0.925_4^{23}			67–68	21	misc alc, eth
p210	(*Z*)-1,2,3-Propene-tricarboxylic acid		174.11	2, 849			d 200			50 aq; s alc; sl s eth
p211	1-Propen-2-yl acetate	$H_2C{=}C(OOCCH_3)CH_3$	100.12		0.909	1.4000^{20}		97	18	
p212	*o*-Propenylphenol	$CH_3CH{=}CHC_6H_4OH$	134.18	6^1, 279	1.044	1.5754^{20}		230–231	90	
p213	*β*-Propiolactone		72.06		1.1460_4^{20}	1.4131^{20}	−33.4	162.3	70	37 aq(hyd); misc alc (reacts), bz, eth, acet
p214	Propionaldehyde	CH_3CH_2CHO	58.08	1, 629	0.8071_4^{20}	1.3646^{19}	−81	48–49	−9	30 aq; misc alc, eth
p215	Propionamide	$CH_3CH_2CONH_2$	73.10	2, 243	0.9597_4^{80}	1.4160^{110}	79	222.2		v s aq, alc, chl, eth
p216	Propionic acid	CH_3CH_2COOH	74.09	2, 234	0.9934_4^{20}	1.3865^{20}	−21	140.8	51	misc aq; s alc, chl, eth
p217	Propionic anhydride	$[CH_3CH_2C({=}O)]_2O$	130.14	2, 242	1.0125_4^{20}	1.4047^{20}	−45	167	73	d aq; s alc, chl, eth
p218	Propionitrile	CH_3CH_2CN	55.08	2, 245	0.7818_4^{20}	1.3658^{20}	−92.8	97.2	6	10 aq; misc alc, eth

p219	Propionyl chloride	CH_3CH_2COCl	92.53	2, 243	1.065_4^{20}	1.4051^{20}	−94	80	11	d aq, alc
p220	Propiophenone	$C_6H_5COCH_2CH_3$	134.18	7^2, 231	1.0105_4^{20}	1.5258^{20}	18.6	218.0	87	misc bz, eth, abs alc
p221	Propoxytrimethyl-silane	$CH_3CH_2CH_2OSi(CH_3)_3$	132.3		0.768_4^{20}	1.384^{20}		100^{735mm}		
p222	Propyl acetate	$CH_3CH_2CH_2OOCCH_3$	102.13	2, 129	0.836_4^{20}	1.3844^{20}	−92	101.6	12	2.3 aq; misc alc, eth
p223	Propylamine	$CH_3CH_2CH_2NH_2$	59.11	4, 136	0.7173^{20}	1.3882^{20}	−83.0	47.9	−37	misc aq, alc, eth
p224	2-(Propylamino)-ethanol	$C_3H_7NHCH_2CH_2OH$	103.17	4, 282	0.900	1.4415^{20}		182^{746mm}	78	
p225	Propylbenzene	$CH_3CH_2CH_2C_6H_5$	120.20	5, 390	0.8621_4^{20}	1.4912^{20}	−99.6	159.2	47	s alc, eth
p226	Propyl benzoate	$C_6H_5COOCH_2CH_2CH_3$	164.20	9, 112	1.0232^{20}	1.5003^{20}	−51.6	231.2		i aq; s alc, eth
p227	Propylcyclohexane	$CH_3CH_2CH_2C_6H_{11}$	126.24	5^2, 23	0.7929_4^{20}	1.4370^{20}	−94.9	156.7		s bz, eth
p228	Propylene carbonate		102.09		1.2041_4^{20}	1.4210^{20}	−55	240	132	v s aq, alc, bz, eth
p229	Propyleneimine	$CH_3CH{-}CH_2$ (ring with NH)	57.09		0.8017^{25}	1.4084^{25}		66.0		misc aq, alc, PE

H, COOH, C=C, $HOOCCH_2$, COOH

p210

p213

CH_3

p228

$CH_3{-}CH{-}CH_2$, N, H

p229

TABLE 1-14 Physical constants of organic compounds (*continued*)

No.	Name	Formula	Formula weight	Beilstein reference	Density	Refractive index	Melting point	Boiling point	Flash point	Solubility in 100 parts solvent
p230	Propylene oxide	$CH_3CH{-}CH_2$ (ring via O)	58.08	17, 6	0.8287^{20}	1.3660^{20}	−112.1	37–38	−37	41 aq; misc alc, eth
p231	Propylene sulfide	$CH_3CH{-}CH_2$ (ring via S)	102.18	1, 354	0.736	1.3800^{20}	−123	88–90	4	
p232	Propyl formate	$CH_3CH_2CH_2OOCH$	88.10	2, 21	0.9006_4^{20}	1.3769^{20}	−92.9	80.9	−3	2 aq; misc alc, eth
p233	Propyl 4-hydroxybenzoate	$HOC_6H_4COOCH_2CH_2CH_3$	180.20	10, 160			86–87			0.05 aq; v s alc, eth
p234	Propyl isocyanate	$CH_3CH_2CH_2NCO$	85.11	4^1, 366	0.908	1.3970^{20}		83–84	26	
p235	Propyl lactate	$CH_3CH(OH)COOC_3H_7$	132.16	3, 265	0.996_{20}^{20}	1.4167^{25}		86^{40mm}		s aq, alc, eth
p236	Propyl nitrate	$CH_3CH_2CH_2ONO_2$	105.09	1, 355	1.0538_4^{20}	1.3976^{20}	−100	110.1	23	s alc, eth
							may explode on heating			
p237	2-Propylpentanoic	$(CH_3CH_2CH_2)_2CHCOOH$	144.21	2, 350	0.921	1.4250^{20}		220		
p238	*o*-Propylphenol	$CH_3CH_2CH_2C_6H_4OH$	136.19	6, 499	1.015^{20}	1.5279^{20}		224–226	93	s alc, eth
p239	Propylphosphonic dichloride	$CH_3CH_2CH_2P(O)Cl_2$	160.97	4, 596	1.290	1.4643^{20}		$88{-}90^{50mm}$	> 112	
p240	Propyltrichlorosilane	$CH_3CH_2CH_2SiCl_3$	177.53	4, 630	1.1851_4^{20}	1.429^{20}		123–124	2	
p241	Propyltriethoxysilane	$C_3H_7Si(OC_2H_5)_3$	206.4		0.892_4^{20}	1.396^{20}		179–180		
p242	Propyl 3,4,5-trihydroxybenzoate	$(HO)_3C_6H_2COOC_3H_7$	212.20				150			0.35 aq; 1 alc; 83 eth
p243	Propyne	$CH_3C{\equiv}CH$	40.06	1, 246	0.691_4^{-20}	1.3725^{-20}	−102.8	−23.2		v s alc; 3000 mL eth
p244	2-Propynoic acid	$HC{\equiv}CCOOH$	70.05	2, 477	1.138_4^{20}	1.4320^{20}	9	102^{200mm}	58	s aq, alc, eth
p245	2-Propyn-1-ol	$HC{\equiv}CCH_2OH$	56.06	1, 454	0.9715_4^{20}	1.4320^{20}	−51.8	113.6	33	misc aq, alc, bz, chl
p246	(+)-Pulegone		152.24	7, 81	0.9346_4^{15}	1.4850^{20}		224	82	misc alc, chl, eth

p247	Pyrazine		80.09	23, 91	1.031_4^{61}	1.4953^{61}	53	115–116		v s aq, alc, eth
p248	Pyrazole		68.08	23, 39		1.4203	70	186–188		s aq, alc, bz, eth
p249	Pyrene		202.26	5, 693			150–151			
p251	Pyridazine		80.09	23, 89	1.1035_4^{25}	1.5230^{23}	−8	208	85	misc aq, bz; v s alc, eth
p252	Pyridine	C_5H_5N	79.10	20, 181	0.9782_4^{25}	1.5067^{25}	−41.6	115.2	20	misc aq, alc, eth
p253	Pyridine-d_5	C_5D_5N	84.14		1.05	1.5079^{20}		114.4	20	
p254	2-Pyridinealdoxime	$(C_5H_4N)CH{=}NOH$	122.13	21^1, 288			110–112			
p255	2-Pyridinecarbaldehyde	$(C_5H_4N)CHO$	107.11	21^1, 287	1.126	1.5370^{20}		181	54	
p256	3-Pyridinecarbaldehyde	$(C_5H_4N)CHO$	107.11	21^1, 288	1.135	1.5493^{20}		97^{15mm}	60	
p257	4-Pyridinecarbaldehyde	$(C_5H_4N)CHO$	107.11	21, 287	1.172	1.5440^{20}		78^{12mm}	54	s aq, eth
p258	3-Pyridinecarbamide	$(C_5H_4N)CONH_2$	122.13	22, 40	1.400	1.466	130–133			100 aq; 66 alc
p259	Pyridine-2-carboxylic acid	$(C_5H_4N)COOH$	123.11	22, 33			134–136	subl		s aq, alc, bz
p260	Pyridine-3-carboxylic acid	$(C_5H_4N)COOH$	123.11	22, 38	1.473		236.6	subl		1.4 aq; s alk

Propylene oxide, e10
Propyl gallate, p242
Propyl iodide, i48
Propyl mercaptan, p202
6-Propyl-2-thiouracil, h130
Protocatechualdehyde, d377
Pseudocumene, t339
Pyrene, b52
3,6-Pyridazinediol, d401
2.3-Pyridinediol, d402

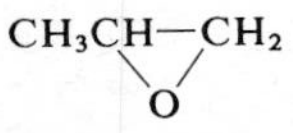
p230

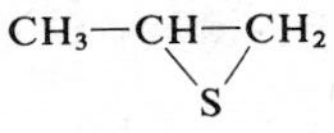
p231

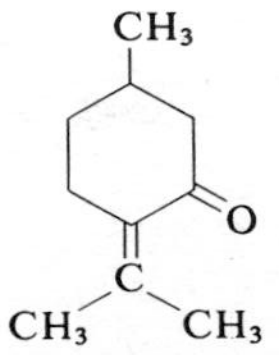
p246

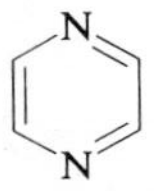
p247

p248

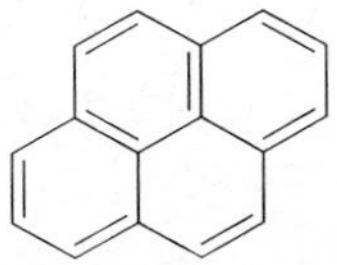
p249

p251

TABLE 1-14 Physical constants of organic compounds (*continued*)

No.	Name	Formula	Formula weight	Beilstein reference	Density	Refractive index	Melting point	Boiling point	Flash point	Solubility in 100 parts solvent
p261	Pyridine-4-carboxylic acid	$(C_5H_4N)COOH$	123.11	22, 45			319	260^{15mm}		0.52 aq; i alc, bz, eth
p262	4-Pyridinecarboxylic hydrazide	$(C_5H_4N)CONHNH_2$	137.14	22^1, 504			171.4			14 aq; 2 alc; 0.1 chl
p263	2,3-Pyridinedicarboxylic acid	$(C_5H_3N)(COOH)_2$	167.12	22, 150			190 d			0.56 aq; s alk
p264	2,5-Pyridinedicarboxylic acid	$(C_5H_3N)(COOH)_2$	167.12	22, 153			236–237	subl d		s hot acid
p265	2,6-Pyridinedicarboxylic acid	$(C_5H_3N)(COOH)_2$	167.12	22, 154			250 d			sl s aq; v sl s alc
p266	Pyridine-*N*-oxide	$C_5H_5N(O)$	95.10	20^2, 131			66	270		
p267	3-Pyridinesulfonic acid	$(C_5H_4N)SO_3H$	159.16	22, 387			> 300			v s aq
p268	2-Pyridylmethanol	$(C_5H_4N)CH_2OH$	109.13	21^1, 203	1.131	1.5420^{20}		113^{16mm}		v s aq, alc, eth
p269	3-Pyridylmethanol	$(C_5H_4N)CH_2OH$	109.13	21, 50	1.124	1.5445^{20}		154^{28mm}		v s aq, eth
p270	3-(3-Pyridyl)-1-propanol	$(C_5H_4N)CH_2CH_2CH_2OH$	137.18		1.045	1.5295^{20}				
p271	Pyrimidine		80.09	23, 89	1.016	1.5035^{20}	20–22	123–124	31	misc aq; s alc, eth
p272	2,4(1*H*,3*H*)-Pyrimidinedione		112.09	24, 312			335			0.3 aq; s alk
p273	Pyrrole		67.09	20, 159	0.9691^{20}_{4}	1.5102^{20}	−23.4	129.8	38	4.5 aq; v s alc, eth
p274	Pyrrolidine		71.12	20, 4	0.8520^{22}_{4}	1.4431^{20}	−57.8	88–89	2	misc aq; s alc, chl, eth
p275	1-Pyrrolidinecarbodithioic acid, ammonium salt		164.29				153–155			
p276	1-Pyrrolidinecarbonitrile		96.13		0.954	1.4690^{20}		$77^{1.8mm}$	107	

p277	L-(−)-2-Pyrrolidinecarboxylic acid		115.13	22, 2			d 220			162 aq; 66 abs alc
p278	1-Pyrrolidino-1-cyclohexene		151.25		0.940	1.5225^{20}		115^{15mm}	39	
p279	2-Pyrrolidinone		85.11	21, 236	1.116_4^{25}	1.486^{25}	25	245	145	misc aq, alc, bz, chl, eth, EtAc
p280	3-(*N*-Pyrrolidino)-1,2-propanediol		145.20	20^1, 4			46–48	158^{30mm}		
q1	Quinhydrone		218.20	7, 617	1.401_4^{20}		171			s hot aq, alc, eth

Pyridinols, h174, h175, h176
3-Pyridinol *N*-oxide, h178
2(1*H*)-Pyridone, h174
2-(2-Pyridyl)pyridine, d707
Pyrocatechol, d378
Pyrogallol, t309
Pyromellitic acid, b26
Pyromellitic dianhydride, b27
Pyromucic acid, f42
Pyromucic aldehyde, f39
Pyrrolidinediethiocarbamate, p275
Pyruvic acid, o60
Pyruvic aldehyde, o59
Pyruvic aldehyde dimethyl acetal, d454
Quinaldine, m410

p271 p272 p273 p274 p275 p276 p277 p278 p279 p280

q1

TABLE 1-14 Physical constants of organic compounds (*continued*)

No.	Name	Formula	Formula weight	Beilstein reference	Density	Refractive index	Melting point	Boiling point	Flash point	Solubility in 100 parts solvent
q2	Quinine		324.44			1.625	177 d			125 alc; 1.2 bz; 83 chl
q3	Quinoline		129.16	20, 339	1.095_4^{20}	1.6273^{20}	−14.9	237	101	0.6 aq; misc alc, eth
q4	Quinoxaline		130.15	23, 176	1.1334_4^{48}	1.6231^{48}	29–30	229.5		v s aq, alc, bz, eth
q5	Quinuclidine		111.19	20, 144			156 sealed tube			v s aq, alc, eth
r1	D-Raffinose pentahydrate		594.52	31, 462			80	d 118		14 aq; 10 MeOH
r2	Rhodamine B		479.02	19, 346			165			v s aq, alc
r3	Rhodanine		133.19	27, 242	0.868		170			v s hot aq, alc, eth
							may explode on rapid heating			
r4	Riboflavin		376.37				d 278			v s alk(d); i eth
r5	D-(−)-Ribose		150.13	1^1, 434			87			s aq; sl s alc
s1	Saccharin		183.19	27, 168			229–230			0.34 aq; 3 alc; 8 acet
s2	Safrole		162.19	19, 39	1.095^{20}	1.5370^{20}	11.2	232–234	97	v s alc; misc chl eth
s3	Semicarbazide	$H_2NNHCONH_2$	75.07	3, 98			96			v s aq, alc; i eth

Quinizarin, d373
Quinolinic acid, p263
8-Quinolinol, h179
p-Quinone, b59
Resacetophenone, d371
Resorcinol, d379
Resorcinol dimethyl ether, d433
Resorcinol monoacetate, d381
Resorcinol sulfide, t147
α-Resorcylaldehyde, d376
α-Resorcycylic acid, d386
β-Resorcylic acid, d384
Ribofuranose, r5
Rosaniline, b2
Rubeanic acid, d712
Salicyl alcohol, h105
Salicylaldehyde, h94
Salicylaldoxime, h97
Salicylamide, h98
Salicylanilide, h162
Salicylic acid, h99
Salol, p153
Sarcosine, m259
Sebacic acid, d9
Sebacoyl chloride, d11
Semioxamazide, o53

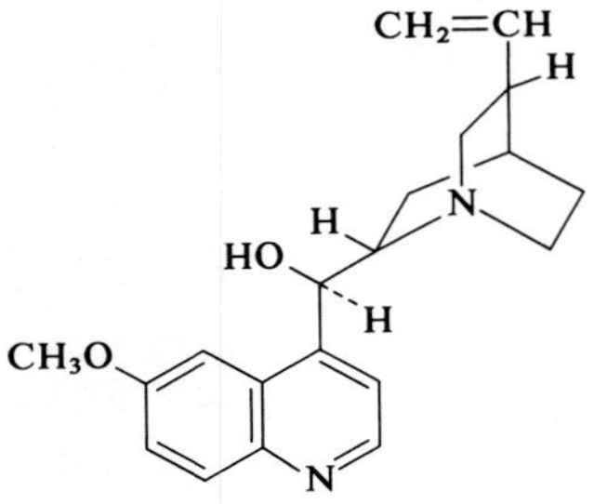

q2

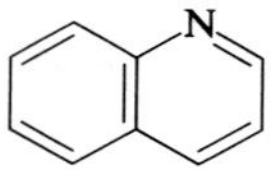

q3

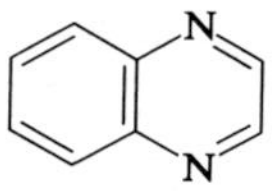

q4

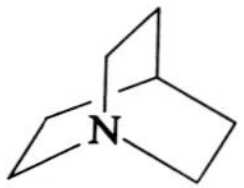

q5

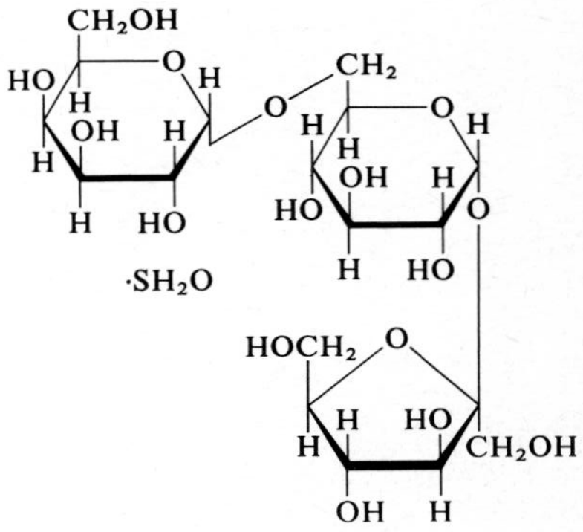

r1

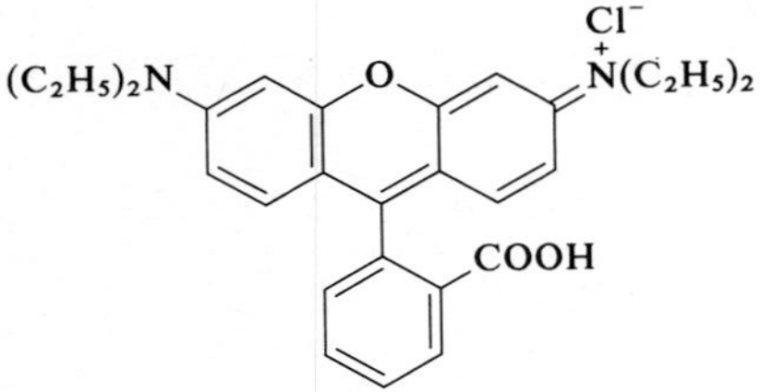

r2

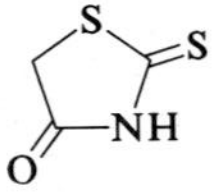

r3

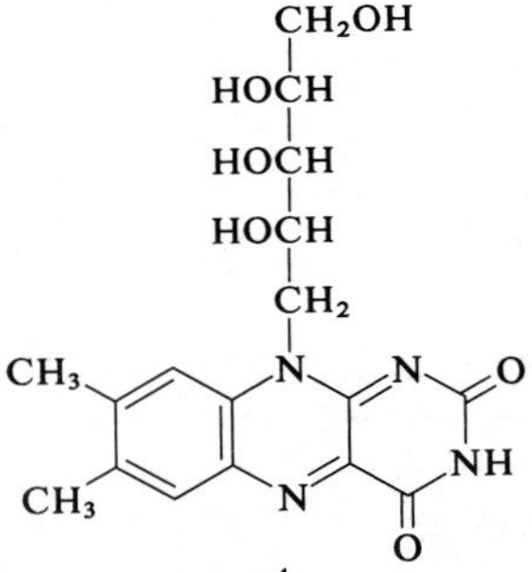

r4

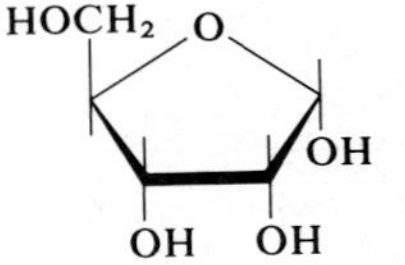

r5

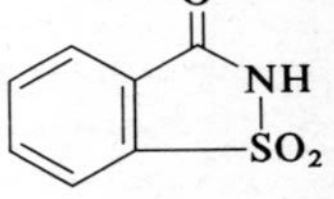

s1

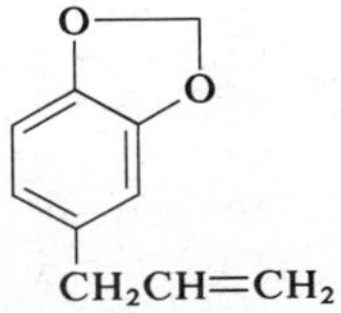

s2

TABLE 1-14 Physical constants of organic compounds (*continued*)

No.	Name	Formula	Formula weight	Beilstein reference	Density	Refractive index	Melting point	Boiling point	Flash point	Solubility in 100 parts solvent
s4	L-Serine	$HOCH_2CH(NH_2)COOH$	105.09	4, 505			222 d			s aq; v sl s alc, eth
s5	D-Sorbitol		182.17	1, 533	1.472^{-5}		110–112			83 aq; s hot alc, acet
s6	L-(−)-Sorbose		180.16	1, 927	1.65^{15}		165			55 aq; v sl s alc
s7	Squalane	$[(CH_3)_2CH(CH_2)_3$-$CH(CH_3)(CH_2)_3$-$CH(CH_3)CH_2CH_2—]_2$	422.80	1^1, 72	0.810	1.4530^{15}	−38	350	218	s bz, chl, eth, PE
s8	Squalene	$\{CH_3[C(CH_3)=CHCH_2$-$CH_2]_2C(CH_3)=CHCH_2—\}_2$	410.73	1^1, 130	0.8584^{20}_{4}	1.4965^{20}	−75	285^{25mm}	200	v s eth, acet, PE
s9	*trans*-Stilbene	$C_6H_5CH=CHC_6H_5$	180.25	5, 630	0.970		124	206–207		v s bz, eth
s10	L-Strychnine		334.42	27^2, 723	1.36^{20}_{4}		284–286	270^{5mm}		6.2 alc; 20 chl; 0.55 bz; 15 mg aq
s11	Styrene	$C_6H_5CH=CH_2$	104.15	5, 474	0.9060^{20}	1.5468^{20}	−30.6	145.1	31	s alc, acet, eth
s13	Succinamic acid	$H_2NCOCH_2CH_2COOH$	117.10	2, 614			153–156			s aq; sl s alc; i eth
s14	Succinamide	$H_2NCOCH_2CH_2CONH_2$	116.12	2, 614			260 d	125 subl		0.45 aq; i alc, eth
s15	Succinic acid	$HOOCCH_2CH_2COOH$	118.09	2, 601	1.552		187–190	235 d		7.7 aq; 5.4 alc; 2.8 acet; 0.88 eth; i bz
s16	Succinic acid 2,2-dimethylhydrazide	$HOOCCH_2CH_2CONH$-$N(CH_3)_2$	160.17				154–155			11 aq; 2.5 acet; 5 MeOH
s17	Succinic anhydride		100.07	17, 407			119.6	261		s alc, chl; v sl s eth
s18	Succinimide		99.09	21, 369	1.41		125–127	287		33 aq; 4 alc; i eth
s19	Succinonitrile	$NCCH_2CH_2CN$	80.09	2, 615	0.985		46–48	265–267	> 112	
s20	Succinyl chloride	$ClCOCH_2CH_2COCl$	154.98	2, 613	1.395^{15}_{4}	1.473^{15}	17	192–193	76	d aq, alc; s bz
s21	Sucrose		342.30	31, 424	1.587^{25}_{4}		192 d			200 aq; 0.59 alc
s22	Sulfamethazine		278.34				198–201			0.15 aq; s alk
s23	Sulfanilamide	$H_2NC_6H_4SO_2NH_2$	172.21	14, 698			164–166			0.76 aq; 2.7 alc; 20 acet; s acid, alk

s24	Sulfoacetic acid	HO_3SCH_2COOH	140.11	4, 21			84–86	245 d		s aq, alc; i eth, chl
s25	*o*-Sulfobenzoic acid cyclic anhydride		184.17	19, 110				186^{18mm}		s bz, chl, eth; i aq
s26	4,4′-Sulfonylbis-(2,6-dibromophenol)	$[HO(Br)_2C_6H_2]_2SO_2$	565.88	6, 865			289–292			

Senecioic acid, m163
Skatole, m285
Sodium tetraphenylborate, t130
Solketal, d517
Sorbic acid, h42
Sorbic aldehyde, h40
Stearamide, o2
Stearic acid, o5
Stearyl bromide, b322
Styrene dibromide, d79
Styrene glycol, p113
Styrene oxide, e9
Suberic acid, o24
Suberonitrile, d239
Succinic acid monoamide, s13
Succinonitrile, b383
Succinyl dihydrazide, s16
Sulfanilic acid, a120
N-Sulfinylaniline, t155
3-Sulfoalanine, a293
Sulfolane, t108
3-Solfolene, d369
Sulfonyldianilines, d36, d37

CH_2OH – HCOH – HOCH – HCOH – HCOH – CH_2OH

s5

CH_2OH – C=O – HOCH – HCOH – HCOH – CH_2OH

s6

s10

s17

s18

s21

s22

s26

TABLE 1-14 Physical constants of organic compounds (*continued*)

No.	Name	Formula	Formula weight	Beilstein reference	Density	Refractive index	Melting point	Boiling point	Flash point	Solubility in 100 parts solvent
s27	4,4′-Sulfonylbis-(methyl benzoate)	$(CH_3OOCC_6H_4)_2SO_2$	334.35	10^2, 109			195–196			
s28	4,4′-Sulfonyl-diphenol	$(HOC_6H_4)_2SO_2$	250.27	6, 861	1.3663^{15}		245–247			s alc, eth, acet; i aq
s29	5-Sulfosalicyclic acid	$HO_3SC_6H_3(OH)COOH$	254.21	11, 411			120			v s aq, alc; s eth
t1	D-(−)-Tartaric acid		150.09	3, 520	1.7598^{20}_{4}		168–170			139 aq; 33 alc; 0.4 eth
t2	*meso*-Tartaric acid hydrate	$HOOCCH(OH)$-$CH(OH)COOH\ xH_2O$	150.09	3, 528	1.666^{20}_{4}		140			125 aq
t3	Tartrazine		534.37	25, 252						v s aq
t4	*p*-Terphenyl	$C_6H_5C_6H_4C_6H_5$	230.31	5, 695			212–213	383		
t5	α-Terpinene		136.24	5, 126	0.8375^{20}_{4}	1.4775^{20}		174	46	misc alc, eth
t6	γ-Terpinene		136.24	5, 128	0.853^{15}_{4}	1.4754^{16}		183	51	
t7	Terpinen-4-ol		154.25	6, 55	0.9338^{20}_{4}	1.4820^{20}	36.4	219	79	v s alc, eth
t9	Tetraallyloxysilane	$(H_2C{=}CHCH_2O)_4Si$	256.4		0.9824^{20}_{4}	1.4336^{20}		114^{12mm}		
t10	1,1,2,2,-Tetrabromo-ethane	$Br_2CHCHBr_2$	345.67	1, 94	2.9529^{25}	1.6323^{25}	0.0	243.5	none	misc alc, eth; 0.07 aq
t11	Tetrabromophthalic anhydride		463.72	17, 485			274–276			sl s bz; i aq, alc
t12	α,α,α′,α′-Tetra-bromo-*o*-xylene	$C_6H_4(CHBr_2)_2$	421.77	5, 367			114–116			v s chl
t13	α,α,α′,α′-Tetra-bromo-*m*-xylene	$C_6H_4(CHBr_2)_2$	421.77	5, 375			105–108			v s bz, chl
t14	Tetrabutoxysilane	$(C_4H_9O)_4Si$	320.5		0.899^{20}_{4}	1.413^{20}		115^{3mm}		
t15	Tetrabutylammonium bromide	$(C_4H_9)_4N^+Br^-$	322.38				103–104			
t16	Tetrabutylammonium chloride	$(C_4H_9)_4N^+Cl^-$	277.92	4^3, 292			83–86			

t17	Tetrabutylammonium fluoride trihydrate	$(C_4H_9)_4N^+F^- \cdot 3H_2O$	315.52	4^3, 292			62–63			
t18	Tetrabutylammonium hydrogen sulfate	$(C_4H_9)_4N^+HSO_4^-$	339.54				169–171			
t19	Tetrabutylammonium iodide	$(C_4H_9)_4N^+I^-$	369.38	4, 157			145–148			sl s aq; s alc, eth
t20	Tetrabutylammonium tetrafluoroborate	$(C_4H_9)_4N^+BF_4^-$	329.28	4^3, 293			160–162			
t21	Tetrabutyltin	$(C_4H_9)_4Sn$	347.15		1.057	1.4742^{20}	−97	145^{10mm}	107	
t22	1,1,3,3-Tetrachloro-acetone	$Cl_2CHCOCHCl_2$	195.86	1, 656	1.624_4^{15}	1.497^{18}		182^{745mm}	none	v s acet, chl
t23	1,2,3,4-Tetrachloro-benzene	$C_6H_2Cl_4$	215.89	5, 204			46–47	254	> 112	v s eth; sl s alc
t24	1,2,4,5-Tetrachloro-benzene	$C_6H_2Cl_4$	215.89	5, 205	1.858^{22}		138–140	240–246	> 112	s bz, chl, eth

Sylvan, m253
Sylvic acid, a1
2,4,5-T, t246
TAPS, t434

Taurine, a161
Terephthaldehyde, b13
Terephthaldicarboxaldehyde, b13
Terephthalic acid, b17

Terephthaloyl chloride, b15
TES, t431
Tetracene, b7

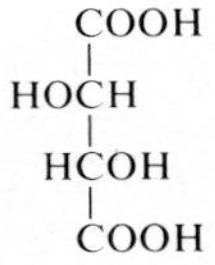

t1

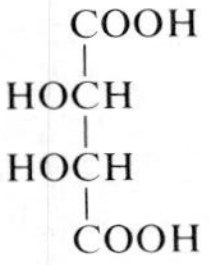

t2

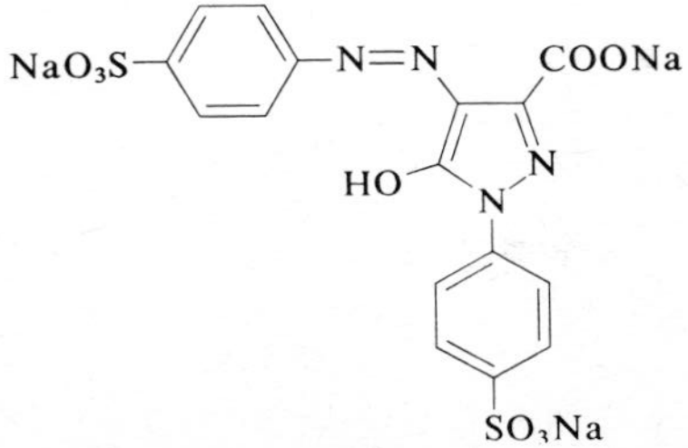

t3

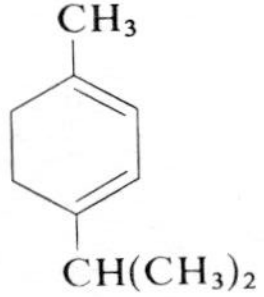

t5

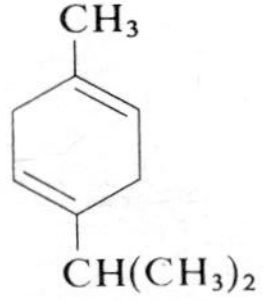

t6

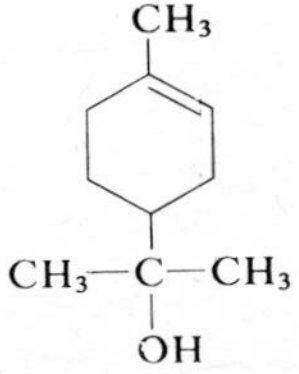

t7

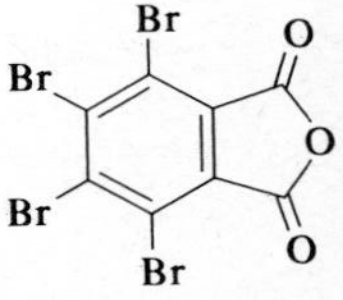

t11

TABLE 1-14 Physical constants of organic compounds (*continued*)

No.	Name	Formula	Formula weight	Beilstein reference	Density	Refractive index	Melting point	Boiling point	Flash point	Solubility in 100 parts solvent
t25	Tetrachloro-*o*-benzoquinone	$C_6Cl_4(=O)_2$	245.88	7, 602			127–129			
t26	Tetrachloro-*p*-benzoquinone	$C_6Cl_4(=O)_2$	245.88	7, 602			290	subl		s eth; sl s chl; i aq
t27	Tetrachloro-1,2-difluoroethane	$Cl_2CFCFCl_2$	203.83		1.6447^{25}	1.4130^{25}	26.0	92.8		0.012 aq
t28	1,1,1,2-Tetrachloroethane	$ClCH_2CCl_3$	167.85	1, 86	1.598	1.4819^{20}		130	none	0.02 aq; misc alc
t29	1,1,2,2-Tetrachloroethane	$Cl_2CHCHCl_2$	167.85	1, 86	1.5866_4^{25}	1.4910^{25}	−43.8	146.3	none	0.3 aq; misc alc, chl, eth, PE
t30	Tetrachloroethylene	$Cl_2C{=}CCl_2$	165.83	1, 187	1.6230_4^{20}	1.5057^{20}	−22.4	121.1	none	misc alc, chl, eth
t31	2,3,5-Tetrachloronitrobenzene	$HC_6Cl_4NO_2$	260.89	5, 247	1.744_4^{25}		98–101	304		s alc, bz, chl
t32	Tetrachlorophthalic anhydride		285.90	17, 484			254–258	371		d hot aq; sl s eth
t33	3,4,5,6-Tetrachlorophthalimide		284.91	21, 505			>300			
t34	1,1,2,3-Tetrachloro-2-propene	$ClCH{=}C(Cl)CHCl_2$	179.86	1^1, 83	1.530	1.5163^{20}	165		none	
t35	2,3,5,6-Tetrachlorothioanisole	$HC_6Cl_4SCH_3$	262.0				59–61			
t36	2,4,5,6-Tetrachloro-*m*-xylene	$C_6Cl_4(CH_3)_2$	243.95	5, 373			220–222			
t37	Tetracosane	$CH_3(CH_2)_{22}CH_3$	338.66	1, 175	0.7786^{51}	1.4283^{70}	51.1	391		9.4 chl; s eth
t38	Tetracyanoethylene	$(NC)_2C{=}C(CN)_2$	128.09				200	subl 120		
t39	Tetradecane	$CH_3(CH_2)_{12}CH_3$	198.40	1, 171	0.7627_4^{20}	1.4290^{20}	5.9	253.5		v s alc, eth
t40	Tetradecanoic acid	$CH_3(CH_2)_{12}COOH$	228.38	2, 365	0.8528_4^{70}	1.4273^{70}	58.5	250^{100mm}		v s bz, chl, eth; s alc

t41	1-Tetradecanol	$CH_3(CH_2)_{13}OH$	214.39	1, 428	0.8151^{50}	1.4358^{50}	37.8	264		s eth; sl s alc
t42	Tetradecanoyl chloride	$CH_3(CH_2)_{12}COCl$	246.82	2, 368			−1	168^{15mm}		d aq, alc; s eth
t43	1-Tetradecene	$CH_3(CH_2)_{11}CH{=}CH_2$	196.38	1, 226	0.775_4^{15}	1.4351^{20}	−12.9	251.2	115	v s alc, eth
t44	7-Tetradecene	$CH_3(CH_2)_5CH{=}CH\text{-}(CH_2)_5CH_3$	196.38		0.764	1.4351^{20}		250	99	
t45	1-Tetradecylamine	$CH_3(CH_2)_{13}NH_2$	213.41	4, 201			40–42	162^{15mm}		
t46	4-Tetradecylaniline	$CH_3(CH_2)_{13}C_6H_4NH_2$	213.41	12^3, 2780			46–49			
t47	Tetradecyltrichloro-silane	$CH_3(CH_2)_{13}SiCl_3$	331.8			1.382		156^{3mm}		
t48	Tetraethoxysilane	$(CH_3CH_2O)_4Si$	208.33		0.934_4^{20}	1.383^{20}	−77	165.8	46	d aq; s alc
t49	Tetraethylammonium bromide	$(CH_3CH_2)_4N^+Br^-$	210.16	4, 104	1.397_4^{20}		287 d			v s aq, alc, acet, chl
t50	Tetraethylammonium chloride	$(CH_3CH_2)_4N^+Cl^-$	165.71	4, 104	1.0801_4^{21}		37.5			141 aq; s alc; 8.2 chl
t51	Tetraethylammonium hydroxide	$(CH_3CH_2)_4N^+OH^-$	147.26	4, 103						misc aq
t52	Tetraethylene glycol	$(HOCH_2CH_2OCH_2\text{-}CH_2)_2O$	194.23	1, 468	1.125_{20}^{20}	1.4590^{20}	−6	307.8	176	misc aq, alc, bz, eth
t53	Tetraethylene glycol dimethacrylate	$[H_2C{=}C(CH_3)COOCH_2CH_2OCH_2CH_2]_2O$	330.37		1.08			220^{1mm}	62	
t55	Tetraethylene glycol monomethyl ether	$CH_3O(CH_2CH_2O)_3CH_2CH_2OH$	208.26		0.987	1.4453^{20}		166^{11mm}	>112	

Tetraethyl orthosilicate, t48

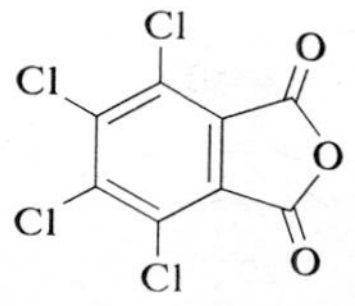

t32

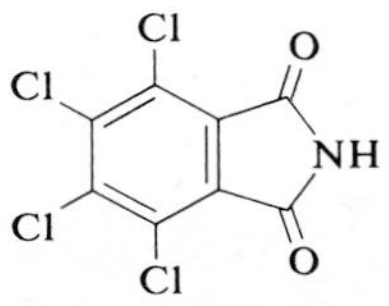

t33

TABLE 1-14 Physical constants of organic compounds (*continued*)

No.	Name	Formula	Formula weight	Beilstein reference	Density	Refractive index	Melting point	Boiling point	Flash point	Solubility in 100 parts solvent
t56	Tetraethylenepentamine	$(H_2NCH_2CH_2NHCH_2CH_2)_2NH$	189.31		0.999^{20}_{20}	1.5055^{20}	−40	340	185	misc aq, alc, eth
t57	*N,N,N′,N′*-Tetraethylethylenediamine	$(C_2H_5)_2NCH_2CH_2N(C_2H_5)_2$	172.32	4, 251	0.808	1.4343^{20}		189–192	58	
t58	Tetraethylgermanium	$(C_2H_5)_4Ge$	188.84	4, 631	1.1989		−90	165.5		s alc, eth; i aq
t59	Tetraethyllead	$(C_2H_5)_4Pb$	323.45	4, 639	1.653^{20}_4	1.5198^{20}	−136	152^{291mm}		s bz; misc eth
t60	Tetraethyl pyrophosphate	$[(C_2H_5O)_2P(O)]_2O$	290.20		1.185^{20}_4	1.4196^{20}	d 170			d aq; misc alc, bz, chl
t61	Tetraethyl pyrophosphite	$[(C_2H_5O)_2P]_2O$	258.19		1.057	1.4341^{20}		81^{1mm}	>112	
t62	Tetraethylsilane	$(C_2H_5)_4Si$	144.34	4^2, 1007	0.762^{20}_4	1.4246^{20}		153–155		i aq
t63	Tetraethylthiuram disulfide	$[(C_2H_5)_2NC(=S)S-]_2$	296.54	4, 122	1.30		70			3.8 alc; 7.1 eth; s bz, acet, chl; 0.02 aq
t64	Tetraethyltin	$(C_2H_5)_4Sn$	234.94	4, 632	1.199^{20}_4		−112	181		i aq; s eth
t65	Tetrafluoroethylene	$F_2C=CF_2$	100.02	1^3, 638	1.1507^{-40}		−131.2	−75.6		i aq
t66	2,2,3,3-Tetrafluoro-1-propanol	$HCF_2CF_2CH_2OH$	132.06		1.4853^{20}_4	1.3197^{20}	−15	109–110	49	
t67	1,2,3,6-Tetrahydrobenzaldehyde	C_6H_9CHO	110.16	7^1, 48	0.940	1.4745^{20}		163–164	57	
t68	Tetrahydrofuran		72.11	17, 10	0.8892^{20}_4	1.4072^{20}	−108.5	66	−17	misc aq, alc eth, PE
t69	2,5-Tetrahydrofurandimethanol		132.16		1.1542^{25}_4	1.4766^{25}	<−50	265		misc aq, alc, bz, chl; s eth
t70	Tetrahydro-2-furanmethanol		102.13	17^2, 106	1.0524^{20}	1.4520^{20}	<−80	178	83	misc aq, alc, bz, chl, eth, acet
t71	Tetrahydro-2-furanmethylamine		101.15	18^2, 415	0.980	1.4560^{20}		154^{744mm}	45	

t72	2-(Tetrahydrofuryl-oxy)tetrahydro-pyran		186.25		1.030	1.4606^{20}			97	
t73	1,2,3,4-Tetrahydro-isoquinoline		133.19	20, 275	1.064	1.5668^{20}	−30	232–233	98	
t74	Tetrahydrolinalool	$(CH_3)_2CHCH_2CH_2CH_2$-$C(CH_3)(OH)CH_2CH_3$	158.28		0.925^{25}	1.433^{20}				
t75	1,2,3,4-Tetrahydro-naphthalene	$C_{10}H_{12}$	132.21	5, 491	0.9702_4^{20}	1.5414^{20}	−35.8	207.6	77	misc alc, bz, chl, eth, acet, PE
t76	*cis*-1,2,3,6-Tetra-hydrophthalic anhydride		152.15	17, 462			101–102			
t77	*cis*-1,2,3,6-Tetra-hydrophthalimide		151.17				134–138			
t78	Tetrahydropyran		86.14		0.8814_4^{20}	1.4211^{20}	−45	88	−20	misc aq, alc, eth
t79	Tetrahydropyran-2-methanol		116.16		1.0254^{20}	1.4580^{20}	−70 glass	187	93	misc aq, alc, bz, eth

Tetraglyme, b191
1,2,3,4-Tetrahydrobenzene, c331
Tetrahydrodicyclopentadiene, t258
Tetrahydro-2,5-dimethoxyfuran, d458
Tetrahydrofurfurylamine, t71
Tetrahydrofurfuryl alcohol, t70

t68 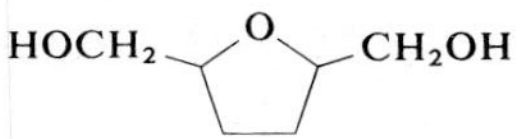 t69 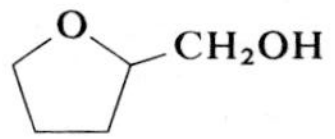 t70 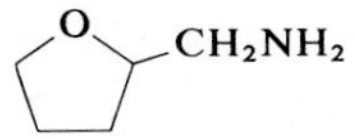 t71 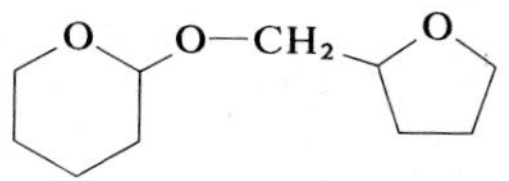 t72 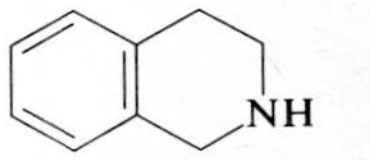 t73

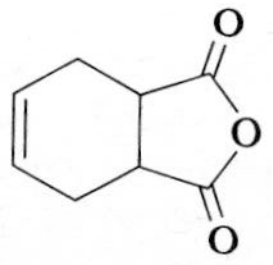

t76 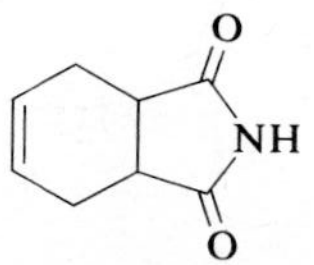 t77  t78 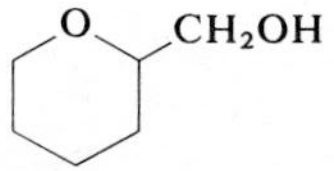 t79

TABLE 1-14 Physical constants of organic compounds (*continued*)

No.	Name	Formula	Formula weight	Beilstein reference	Density	Refractive index	Melting point	Boiling point	Flash point	Solubility in 100 parts solvent
t80	1,2,3,6-Tetrahydropyridine		83.13	20^3, 1912	0.911	1.4800^{20}	−48	108	16	
t81	3,4,5,6-Tetrahydropyrimidinethiol		116.19	24, 5			210–212			
t82	1,2,3,4-Tetrahydroquinoline		133.19	20, 262	1.061	1.5924	15–16	249	100	s aq; misc alc, eth
t83	Tetrahydrothiophene		88.17	17^1, 5	0.9987^{20}	1.5048^{20}	−96.2	120.9	12	misc alc, eth; i aq
t84	1,4,9,10-Tetrahydroxyanthracene		242.23	8, 431			147–149			
t85	2,2′,4,4′-Tetrahydroxybenzophenone	$[(HO)_2C_6H_3]_2C{=}O$	246.22	8, 496			200–203			
t86	Tetrahydroxyhexanedioic acid	$HOOC[CH(OH)]_4COOH$	210.14	3, 581			230 d			0.003 aq; s alk
t87	Tetrakis(2-ethylbutoxy)silane	$[CH_3CH_2CH(C_2H_5)CH_2O]_4Si$	432.8		0.892^{20}_4	1.430^{20}		171^{2mm}		
t88	Tetrakis(2-ethylhexoxy)silane	$[CH_3(CH_2)_3CH(C_2H_5)CH_2O]_4Si$	549.95		0.880^{20}_4	1.4388^{20}		194^{1mm}	190	
t89	*N,N,N′,N′*-Tetrakis(*p*-hydroxypropyl)ethylene diamine	$\{[CH_3CH(OH)CH_2]_2NCH_2—\}_2$	292.42	4^4, 1685	1.013	1.4812^{20}		175–$181^{0.8mm}$		
t90	Tetrakis(isopropoxy)silane	$[(CH_3)_2CHO]_4Si$	264.4		0.877^{20}_4	1.385^{20}		64^{5mm}		
t91	Tetrakis(2-methoxyethoxy)silane	$(CH_3OCH_2CH_2O)_4Si$	328.4		1.079^{20}_4	1.422^{20}		182^{10mm}		
t92	Tetrakis(trimethylsiloxy)titanium	$[(CH_3)_3SiO]_4Ti$	404.7		0.900^{20}_4	1.427^{20}		110^{10mm}		

t93	1,1,3,3-Tetramethoxy-propane	$[(CH_3O)_2CH]_2CH_2$	164.20		0.997	1.4081^{20}		183	54	
t94	Tetramethoxysilane	$(CH_3O)_4Si$	152.2	1, 287	1.052_4^{20}	1.368^{20}		121–122	20	
t95	Tetramethyl-ammonium bromide	$(CH_3)_4N^+Br^-$	154.06	4, 51	1.56		d > 230	subl > 360		55 aq
t96	Tetramethyl-ammonium chloride	$(CH_3)_4N^+Cl^-$	109.60	4, 51	1.169_4^{20}		d > 230	subl > 300		s aq, hot alc
t97	Tetramethyl-ammonium iodide	$(CH_3)_4N^+I^-$	201.06		1.829		d 230			sl s aq; v s abs alc
t98	*N,N*,3,5-Tetra-methylaniline	$(CH_3)_2C_6H_3N(CH_3)_2$	149.24	12, 1131	0.913	1.5443^{20}		226–228	90	
t99	1,2,3,4-Tetramethyl-benzene	$C_6H_2(CH_3)_4$	134.22	5, 430	0.905_4^{20}	1.5187^{20}	−6.2	205.0	68	misc alc, eth
t100	1,2,3,5-Tetramethyl-benzene	$C_6H_2(CH_3)_4$	134.22	5, 430	0.8906_4^{20}	1.5134^{20}	−23.7	198.0	63	s alc; v s eth
t101	1,2,4,5-Tetramethyl-benzene	$C_6H_2(CH_3)_4$	134.22	5, 431	0.838_4^{81}		79.2	196.8	73	v s alc, bz, eth

6,7,8,9-Tetrahydro-5*H*-tetrazoloazepine, p26
Tetrahydrothiophene 1,1-dioxide, t108
Tetrahydrothiophene oxide, t109
Tetrahydroxyadipic acid, t86
Tetralin, t75
β-Tetralonehydantoin, b40
N,N,N′,N′-Tetramethyldiaminomethane, t115
N,N,N′,N′-Tetramethyl-1,3-diamino-2-propanol, b174

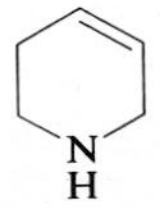

t80 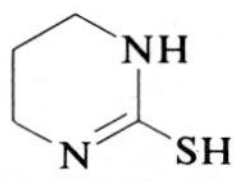 t81 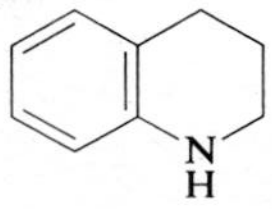 t82  t83 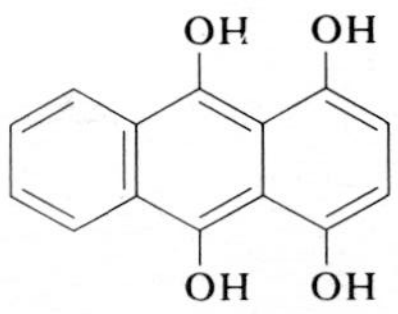 t84

TABLE 1-14 Physical constants of organic compounds (*continued*)

No.	Name	Formula	Formula weight	Beilstein reference	Density	Refractive index	Melting point	Boiling point	Flash point	Solubility in 100 parts solvent
t102	2,2,3,3-Tetramethylbutane	$(CH_3)_3CC(CH_3)_3$	114.23	1, 165	0.656^{-120}		−120.7	106.5	<1	
t103	*N,N,N',N'*-Tetramethyl-1,4-butanediamine	$(CH_3)_2N(CH_2)_4N(CH_3)_2$	144.26	4, 265	0.786^{20}	1.4280^{20}		169	46	s aq, alc, eth
t104	1,1,3,3-Tetramethylbutylamine	$(CH_3)_3CCH_2C(CH_3)_2NH_2$	129.25	4, 198	0.805	1.4240^{20}		137–143	32	s alc, eth, PE; i aq
t105	1,3,5,7-Tetramethylcyclotetrasiloxane	$[-SiH(CH_3)O-]_4$	240.5		0.9912_4^{20}	1.3870^{20}	−69	134–135		
t106	1,1,4,4-Tetramethyl-1,4-dichlorodisilylethylene	$[(CH_3)_2Si(Cl)CH_2-]_2$	215.3				37	198^{734mm}	68	
t107	Tetramethyldisiloxane	$[(CH_3)_2SiH]_2O$	134.3		0.757_4^{20}	1.370^{20}		71^{731mm}		
t108	Tetramethylene sulfone		120.71	17^1, 5	1.2614_4^{30}	1.4820^{30}	27.6	285	165	misc aq, acet, bz
t109	Tetramethylene sulfoxide		104.17		1.158	1.5200^{20}			>112	
t110	*N,N,N',N'*-Tetramethylethylenediamine	$(CH_3)_2NCH_2CH_2N(CH_3)_2$	116.21	4, 250	0.770	1.4179^{20}	−55	120–122	10	
t111	Tetramethylgermanium	$(CH_3)_4Ge$	132.73		1.006^0	1.3871^{20}	−88	43.4		
t112	1,1,3,3-Tetramethylguanidine	$[(CH_3)_2N]_2C{=}NH$	115.18					163		
t113	*N,N,N',N'*-Tetramethyl-1,6-hexanediamine	$[(CH_3)_2N(CH_2)_3-]_2$	172.32	4^1, 423	0.806	1.4359^{20}		209–210	73	
t114	Tetramethyllead	$(CH_3)_4Pb$	267.33	4, 639	1.995_4^{20}		−27.5	110		misc alc, eth

t115	*N,N,N′,N′*-Tetramethylmethanediamine	$(CH_3)_2NCH_2N(CH_3)_2$	102.18	4, 54	0.749^{20}	1.4005		85	<1	
t116	Tetramethyl orthocarbonate	$C(OCH_3)_4$	136.15	3^2, 4	1.023	1.3845^{20}	−5	114	6	
t117	2,6,10,14-Tetramethylpentadecane	$[(CH_3)_2CH(CH_2)_3$-$CH(CH_3)CH_2]_2CH_2$	268.53		0.7827_4^{20}	1.4379^{20}	−100	167^{11mm}		s bz, chl, eth, PE
t118	2,3,5,6-Tetramethylphenol	$(CH_3)_4C_6HOH$	150.22	6, 547			108–110	250		
t119	2.2,6,6-Tetramethylpiperidino-*N*-oxy-(free radical)		156.25				36–38		67	
t120	*N,N,N′,N′*-Tetramethyl-1,3-propanediamine	$(CH_3)_2N(CH_2)_3N(CH_3)_2$	130.24	4, 262		1.4234^{20}		145–146	31	
t121	Tetramethylpyrazine		136.20	23, 99			84–86	190		
t122	Tetramethylsilane	$(CH_3)_4Si$	88.23	4, 625	0.6411_4^{20}	1.3585^{20}	−99.5	26.5	−27	v s alc, eth
t123	1,2,2,3-Tetramethyl-1,1,3,3-tetraphenyltrisiloxane	$[(C_6H_5)_2Si(CH_3)O]_2$-$Si(CH_3)_2$	484.8		1.07_4^{20}	1.551^{25}		$235^{0.5mm}$	221	

2,2,5,5-Tetramethyl-3,4-dithiahexane, d114
Tetramethylene chlorobromide, b253
Tetramethylethylene glycol, d493
Tetramethylene oxide, t68
Tetramethylene sulfide, t83
Tetramethylolmethane, p20
2,2,4,4-Tetramethyl-3-thiapentane, d133
Tetramethylthiuram disulfide, b175

t108 t109 t119 t121

TABLE 1-14 Physical constants of organic compounds (*continued*)

No.	Name	Formula	Formula weight	Beilstein reference	Density	Refractive index	Melting point	Boiling point	Flash point	Solubility in 100 parts solvent
t124	1,1,3,3-Tetramethyl-2-thiourea	$(CH_3)_2NC(=S)N(CH_3)_2$	132.23	4^1, 336			75–77	245		
t125	Tetramethyltin	$(CH_3)_4Sn$	178.83	4, 631	1.3149^{25}	1.5201	−54.8	78		
t126	1,1,3,3-Tetramethylurea	$(CH_3)_2NC(=O)N(CH_3)_2$	116.16	4, 74	0.9687_4^{20}	1.4493^{25}	−1.2	176	65	misc aq, alc, chl, eth
t126a	Tetranitromethane	$C(NO_2)_4$	196.03	1, 80	1.6229_4^{25}	1.4358^{25}	13.5	126	>112	v s alc, eth, alk
t127	1,4,7,10-Tetraoxacyclododecane		176.21		1.089	1.4621^{20}	16		>112	
t128	2,4,8,10-Tetraoxaspiro[5.5]undecane		160.17	19, 436			52–55	$83^{1.5mm}$		
t129	Tetraphenoxysilane	$(C_6H_5O)_4Si$	400.5		1.141_4^{60}	1.554^{60}	48–49	237^{1mm}		
t130	Tetraphenylboron sodium	$(C_6H_5)_4B^-Na^+$	342.23				>300			v s aq, acet; s chl
t131	1,1,4,4-Tetraphenyl-1,3-butadiene	$(C_6H_5)_2C=CHCH=C(C_6H_5)_2$	358.49	5, 750			207–209			
t132	1,1,3,3-Tetraphenyl-1,3-dimethyldisiloxane	$[(C_6H_5)_2Si(CH_3)]_2O$	410.7		1.076_4^{25}	1.5866^{26}	50	$215^{0.5mm}$	193	
t133	Tetraphenylethylene	$(C_6H_5)_2C=C(C_6H_5)_2$	332.45	5, 743			222–224	420		
t134	Tetraphenylsilane	$(C_6H_5)_4Si$	336.5		1.078_4^{20}		236–237	228^{3mm}		
t135	Tetraphenyltin	$(C_6H_5)_4Sn$	427.11		1.490^0		226	>420	110	
t136	Tetrapropoxysilane	$(C_3H_7O)_4Si$	264.4		0.916_4^{20}	1.401^{20}		94^{5mm}		
t137	Tetrapropylammonium bromide	$(CH_3CH_2CH_2)_4N^+Br^-$	266.27	4^1, 364			270 d			s aq
t138	1*H*-Tetrazole		70.06	26, 346			156–158	subl		s aq, alc, acet
t139	2-Thenoyltrifluoroacetone		222.18				40–44	98^{8mm}		

t140	Theobromine		180.17	26, 457			357	subl 290		0.05 aq; 0.045 alc; s alk; i bz, chl, eth
t141	Thiamine HCl		337.27				d 248			100 aq; 1 alc
t142	Thiazole		85.13	27, 15	1.200^{17}	1.5375^{20}		117–118	22	s alc, eth; sl s aq
t143	Thioacetamide	$CH_3C(=S)NH_2$	75.13	2, 232			112–114			16 aq; sl s alc, eth
t144	Thioacetic acid	$CH_3O—SH$	76.12	2, 230	1.065	1.4630^{20}	<−17	88–91	<1	s aq; misc alc, eth
t145	Thiobenzoic acid	$C_6H_5CO—SH$	138.19	9, 419	1.174	1.6020^{20}	15–18	d	>112	misc eth; v s alc; i aq

Tetrantoin, b40
2,5,8,13-Tetraoxadodecane, b190
3,6,9,12-Tetraoxatridecanol, t55
Tetraphene, b6
2-Thenoic acid, t160
2-Thiabutane, e185
Thiacyclobutane, t350a
1-Thia-3-cyclopentene, 1,1-dioxide, d369
3-Thiaheptane, b454
2-Thiahexane, b466
3-Thiahexane, e213
Thianaphthene, b61
5-Thianonane, d132
2-Thiapentane, m398
3-Thiapentane, d337
Thioanisole, m367
2-Thiobarbituric acid, d389

t127

t128

t138

t139

t140

t141

t142

TABLE 1-14 Physical constants of organic compounds (*continued*)

No.	Name	Formula	Formula weight	Beilstein reference	Density	Refractive index	Melting point	Boiling point	Flash point	Solubility in 100 parts solvent
t146	4,4′-Thiobis(2-*tert*-butyl-6-methylphenol)		358.54				127	316^{40mm}	240	
t147	4,4′-Thiobis(1,3-dihydroxybenzene)	$[(HO)_2C_6H_3]_2S$	250.27	6^3, 6291			175–177			
t148	Thiocarbanilide	$C_6H_5NHCSNHC_6H_5$	228.32	12, 394	1.32^{24}		154			v s alc, eth
t150	*p*-Thiocresol	$HSC_6H_4CH_3$	124.21	6, 416			43–44	195	68	s alc, eth; i aq
t151	2,2′-Thiodiacetic acid	$(HOOCCH_2)_2S$	150.15	3, 253			129			s aq, alc
t152	2,2′-Thiodiethanol	$(HOCH_2CH_2)_2S$	122.19	1, 470	1.1824_4^{20}	1.5203^{20}	−16	282	110	misc aq, alc; sl s eth
t153	4,4′-Thiodiphenol	$(HOC_6H_4)_2S$	218.27	6, 860			150–155			
t154	3,3′-Thiodipropionic acid	$(HOOCCH_2CH_2)_2S$	178.21				134			3.4 aq; v s alc
t155	2-Thiohydantoin		116.14	24, 260			231 d			sl s aq; i alc, eth
t156	*N*-Thionylaniline	$C_6H_5N{=}SO$	139.18	12, 578	1.236	1.6270^{20}		200		
t157	Thiophene	C_4H_4S	84.14	17, 29	1.0573_4^{25}	1.5257^{25}	−38.2	84.2	−1	misc alc, eth; i aq
t158	2-Thiopheneacetic acid	$(C_4H_3S)CH_2COOH$	142.18	18, 293			63–67	160^{22mm}		
t159	2-Thiophenecarbaldehyde	$(C_4H_3S)CHO$	112.15	17, 285	1.200	1.5900^{20}		198	77	s eth
t160	2-Thiophenecarboxylic acid	$(C_4H_3S)COOH$	128.15	18, 289			128.5	260		s aq, chl; v s alc, eth
t161	2-Thiophenemethylamine	$(C_4H_3S)CH_2NH_2$	113.19	18^4, 7096	1.103	1.5569^{20}		99^{28mm}	73	
t162	Thiophenol	C_6H_5SH	110.18	6, 294	1.0766^{20}	1.5897^{20}	−14.9	169.1	50	v s alc; misc bz, eth
t163	Thiophenoxyacetic acid	$C_6H_5SCH_2COOH$	168.21	6, 313			64–66			

t164	Thiopropionic acid	$CH_3CH_2CO{-}SH$	90.14	2, 264	1.014	1.4640^{20}		108–110	11	
t165	3-Thiosemicarbazide	$H_2NC({=}S)NHNH_2$	91.14	3, 195			182–184			s aq, alc
t166	Thiourea	$H_2NC({=}S)NH_2$	76.12	3, 180	1.045		176–178			9 aq; s alc; sl s eth
t167	1,4-Thioxane		104.17	19, 3	1.114	1.5095^{20}		147	42	
t168	Thioxanthen-9-one		212.27	17, 357			211	273^{715mm}		v s bz, chl, hot HOAc
t169	Titanium(IV) isopropoxide	$Ti[OCH(CH_3)_2]_4$	284.26	1^2, 382	0.955	1.4654^{20}	18–20	218^{10mm}	22	
t170	Toluene	$C_6H_5CH_3$	92.14	5, 280	0.8660^{20}_4	1.4969^{20}	−95.0	110.6	7	misc alc, chl, eth, acet, HOAc
t171	2,4-Toluenediamine	$CH_3C_6H_3(NH_2)_2$	122.17	13, 124			97–99	283.5		s hot aq, alc, eth
t172	2,5-Toluenediamine	$CH_3C_6H_3(NH_2)_2$	122.17	13, 144			64	273–274		v s aq, alc, eth
t173	2,6-Toluenediamine	$CH_3C_6H_3(NH_2)_2$	122.17	13, 148			104–106			s aq, alc
t174	3,4-Toluenediamine	$CH_3C_6H_3(NH_2)_2$	122.17	13, 148			88–90	156^{18mm}		v s aq
t175	Toluene-2,4-diisocyanate	$CH_3C_6H_3(NCO)_2$	174.16	13, 138	1.2244^{20}_4	1.5689^{20}	20–21	251	121	d aq, alc; misc bz, acet, eth

Thiocarbanilide, d694
2,2′-Thiodiethanethiol, b188
Thiodiethylene glycol, t152
Thiodiglycol, t152
Thiodiglycolic acid, t151
Thioethanolamine, a162
1-Thioglycerol, m20
Thioglycolic acid, m14
Thiolactic acid, m21
Thiomalic acid, m23
4,4′-Thioresorcinol, t147
Thiosalicyclic acid, m16
Thiosinamine, a101
2-Thioxo-4-thiazolidinone, r3
Threonine, a189
Tiglic acid, m161
Tioxolone, h104
TMS, t122
TMSDEA, t378
TMSI, t380
Tolazoline, b103
p-Tolualdehyde, m126
Toluenethiols, p130, t150

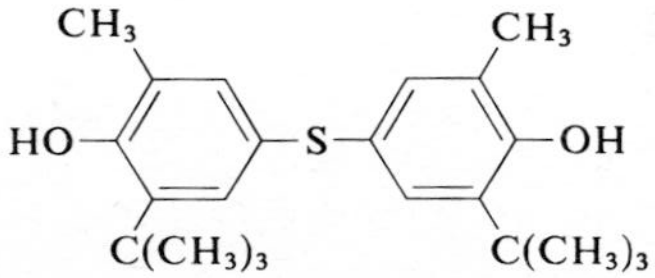

t146

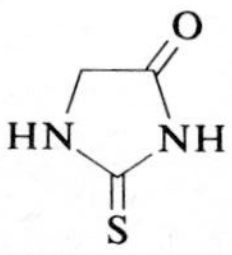

t155

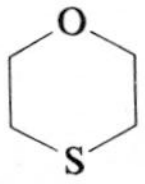

t167

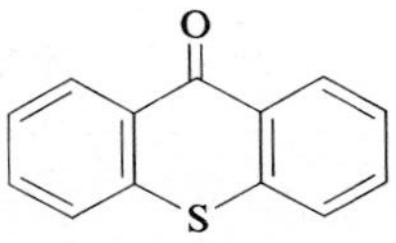

t168

TABLE 1-14 Physical constants of organic compounds (*continued*)

No.	Name	Formula	Formula weight	Beilstein reference	Density	Refractive index	Melting point	Boiling point	Flash point	Solubility in 100 parts solvent
t176	*p*-Toluenesulfinic acid	$CH_3C_6H_4SO_2H$	172.20	11, 9			85			v s alc, eth
t177	*p*-Toluenesulfonamide	$CH_3C_6H_4SO_2NH_2$	171.22	11, 104			137–140			0.2 aq; 3.6 alc
t178	*p*-Toluenesulfonyl-hydrazide	$CH_3C_6H_4SO_2NHNH_2$	186.23	11^2, 66			110 d			
t179	*p*-Toluenesulfonic acid	$CH_3C_6H_4SO_3H$	172.20	11, 97				140^{20mm}		67 aq; s alc, eth
t180	*p*-Toluenesulfonyl chloride	$CH_3C_6H_4SO_2Cl$	190.65	11, 103			69–71	134^{10mm}		v s alc, bz, eth; i aq
t181	*p*-Toluenesulfonyl fluoride	$CH_3C_6H_4SO_2F$	174.19	11^2, 54			41–42	112^{16mm}		
t182	*p*-Toluenesulfonyl isocyanate	$CH_3C_6H_4SO_2NCO$	197.21			1.4355^{20}		144^{10mm}		
t184	*o*-Toluidine	$CH_3C_6H_4NH_2$	107.16	12, 772	0.9984^{20}	1.5725^{20}	−16.1	200.4	85	1.7 aq; s alc, eth
t185	*m*-Toluidine	$CH_3C_6H_4NH_2$	107.16	12, 853	0.989^{20}_{4}	1.5681^{20}	−30.4	203.4	85	misc alc, eth
t186	*p*-Toluidine	$CH_3C_6H_4NH_2$	107.16	12, 880	1.046^{20}_{4}	1.5532^{59}	43.8	200.6	88	7.4 aq; v s alc, eth
t187	1-(*o*-Toluidino)-1,3-butanedione	$CH_3C_6H_4NHCOCH_2$-$COCH_3$	191.23	12, 823			104–106	143		
t188	*o*-Tolunitrile	$CH_3C_6H_4CN$	117.15	9, 466	0.9955^{20}_{4}	1.5279^{20}	−13	205.2	84	i aq; misc alc, eth
t189	*m*-Tolunitrile	$CH_3C_6H_4CN$	117.15	9, 477	0.976^{15}	1.5256^{20}	−23	210	86	0.09 aq; v s alc, eth
t190	*p*-Tolunitrile	$CH_3C_6H_4CN$	117.15	9, 489	0.9785^{30}_{4}		29.5	217.6		i aq; v s alc, eth
t191	*o*-Toluoyl chloride	$CH_3C_6H_4COCl$	154.60	9, 464	1.185	1.5549^{20}		90^{12mm}	76	
t192	*m*-Toluoyl chloride	$CH_3C_6H_4COCl$	154.60	9, 477	1.173	1.5485^{20}		86^{5mm}	76	
t193	*p*-Toluoyl chloride	$CH_3C_6H_4COCl$	154.60	9, 484	1.169	1.5535^{20}	−2	225–257	82	
t194	*m*-Tolyl isocyanate	$CH_3C_6H_4NCO$	133.15	12, 864	1.033	1.5305^{20}		76^{12mm}	65	s alc, eth; i aq
t195	(*p*-Tolylsulfonyl)-methyl isocyanide	$CH_3C_6H_4SO_2CH_2NC$	195.24				114–115			
t196	*p*-Tolyltrichloro-silane	$CH_3C_6H_4SiCl_3$	225.6		1.3^{20}_{4}			218–220		

No.	Name	Formula	Formula weight	Beilstein reference	Density	Refractive index	Melting point	Boiling point	Flash point	Solubility
t197	1,2,4-Triacetoxybenzene	$C_6H_3(OOCCH_3)_3$	252.22	6, 1089			98–100			
t198	Triacetoxyethylsilane	$C_2H_5Si(OOCCH_3)_3$	234.3		1.1428_4^{20}	1.4123^{20}		$107–108^{8mm}$		
t199	Triacetoxyvinylsilane	$(CH_3COO)_3SiCH{=}CH_2$	232.3		1.167_4^{20}	1.423^{20}		113^{1mm}	104	
t200	1,3,5-Triacetylbenzene	$C_6H_3(COCH_3)_3$	204.23	7, 866			160–162		17	
t201	Triallyl-*s*-triazine-2,4,6(1*H*,3*H*,5*H*)-trione		249.27			1.5129^{20}		152^{4mm}	>112	
t202	2,4,6-Triamino-1,3,5-triazine		126.12	26, 245	1.573^{250}		>250	subl		sl s aq; i alc, eth
t203	1*H*-1,2,4-Triazole		69.07	26, 13			119–121	260 d		s aq, alc
t204	Tribenzylamine	$(C_6H_5CH_2)_3N$	287.41	12, 1038	0.991_4^{95}		91–94		65	s hot alc, eth
t205	Tribromoacetaldehyde	Br_3CCHO	280.76	1, 626	2.665	1.5850^{20}		174	65	s aq, alc, chl, eth
t206	Tribromoacetic acid	Br_3CCOOH	296.76	2, 220			130–133	245 d		s aq, alc, eth
t207	2,4,6-Tribromoaniline	$Br_3C_6H_2NH_2$	329.83	12, 663	2.35		120–122	300		s hot alc, chl, eth
t208	2,2,2-Tribromoethanol	Br_3CCH_2OH	282.77	1^2, 338			80–81	93^{10mm}		2 aq; s alc, bz, eth
t209	1,1,2-Tribromoethylene	$BrCH{=}CBr_2$	264.74	1, 191	1.708^{21}	1.6247^{25}		162.5		

Toluic acids, m130, m131, m132
α-Tolunitrile, p80
p-Tolylacetamide, m358
Triacetin, p204
1,3,5-Triazine-2,4,6-triol, c300
Tributyl borate, t213

$CH_2{=}CHCH_2$–N, N–$CH_2CH{=}CH_2$, N–$CH_2CH{=}CH_2$ (triazinetrione ring, O, O, O)

t201

H_2N, N, NH_2, N, N, NH_2

t202

H, N, N, N

t203

TABLE 1-14 Physical constants of organic compounds (*continued*)

No.	Name	Formula	Formula weight	Beilstein reference	Density	Refractive index	Melting point	Boiling point	Flash point	Solubility in 100 parts solvent
t210	Tribromomethane	$CHBr_3$	252.77	1, 68	2.9031^{15}	1.6005^{15}	8.1	149.6	none	0.3 aq; misc eth, MeOH
t211	2,4,6-Tribromophenol	$Br_3C_6H_2OH$	330.82	6, 203	2.55		94–96	244		s alc, chl, eth; i aq
t212	1,2,3-Tribromopropane	$BrCH_2CH(Br)CH_2Br$	280.78	1, 112	2.4114^{15}		16–17	219–221		s alc, eth
t213	Tributoxyborane	$(C_4H_9O)_3B$	230.16	1^2, 398	0.8580^{20}	1.4092^{20}	−70	233.5	93	hyd aq
t214	Tributylamine	$(C_4H_9)_3N$	185.36	4, 157	0.7784^{20}_{4}	1.4283^{20}	−70	216–217	63	v s alc, eth; s acet
t215	2,4,6-Tri-*tert*-butylphenol	$[(CH_3)_3C]_3C_6H_2OH$	262.44		0.864^{27}_{4}		131	278		
t216	Tributyl phosphate	$(C_4H_9O)_3P(O)$	266.32	1^2, 397	0.972^{25}	1.4226^{25}	<-80	289 d	146	0.04 aq; misc org solv
t217	Tributylphosphine	$(C_4H_9)_3P$	202.32	4^2, 971	0.812	1.4619^{20}		150^{50mm}	40	
t218	Tributyl phosphite	$(C_4H_9O)_3P$	250.32	1^1, 187	0.925^{20}_{4}	1.4326^{20}		125^{7mm}	121	misc alc, bz, eth, PE
t219	Tributyltin chloride	$(C_4H_9)_3SnCl$	325.49		1.200	1.4905^{20}		173^{25mm}	>112	
t220	Trichloroacetic acid	Cl_3CCOOH	163.39	2, 206	1.629^{61}_{4}		57–58	196–197		120 aq; v s alc, eth
t221	Trichloroacetonitrile	Cl_3CCN	144.39	2, 212	1.4403^{25}_{4}	1.4409^{20}		85.7	none	
t222	Trichloroacetyl chloride	Cl_3CCOCl	181.83	2, 210	1.629	1.4689^{20}		114–116	none	
t223	Trichloroacetyl isocyanate	$Cl_3CC(=O)NCO$	188.40			1.4809^{20}		85^{20mm}	65	
t224	2,4,5-Trichloroaniline	$Cl_3C_6H_2NH_2$	196.46	12, 627			93–95	270		s alc
t225	2,4,6-Trichloroaniline	$Cl_3C_6H_2NH_2$	196.46	12, 627			73–75	262		s alc, eth
t226	1,2,3-Trichlorobenzene	$C_6H_3Cl_3$	181.45	5, 203	1.69^{25}_{25}		52.6	221	113	v s bz, CS_2
t227	1,2,4-Trichlorobenzene	$C_6H_3Cl_3$	181.45	5, 204	1.446^{25}	1.5707^{20}	17	214	110	misc bz, eth, PE

No.	Name	Formula	Formula weight	Beilstein reference	Density	Refractive index	Melting point	Boiling point	Flash point	Solubility
t228	1,3,5-Trichloro-benzene	$C_6H_3Cl_3$	181.45	5, 204		1.5662^{19}	63.4	208.5	107	v s bz, eth, PE
t229	2,2,2-Trichloro-1,1-dimethylethyl chloroformate	$ClCOOC(CH_3)_2CCl_3$	239.92				30–32	83–84^{14mm}	none	
t230	1,1,1-Trichloro-ethane	CH_3CCl_3	133.41	1, 85	1.3376_4^{20}	1.4379^{20}	−30.4	74.0	none	0.13 aq; s bz, eth
t231	1,1,2-Trichloro-ethane	$ClCH_2CHCl_2$	133.41	1, 85	1.4416_4^{20}	1.4711^{20}	−36.6	113.5	none	0.4 aq; misc alc, eth
t232	2,2,2-Trichloro-ethanol	Cl_3CCH_2OH	149.40	1, 338	1.557_{20}^{20}	1.4885^{20}	17.8	151		8 aq; misc alc, eth
t233	2,2,2-Trichloro-ethyl chloroformate	$ClCOOCH_2CCl_3$	211.86		1.539	1.4703^{20}		171–172	none	
t234	1,1,2-Trichloro-ethylene	$ClCH{=}CCl_2$	131.39	1, 187	1.4649_4^{20}	1.4775^{20}	−84.8	86.7	none	0.1 aq; misc alc, chl, eth
t235	Trichloroethylsilane	$C_2H_5SiCl_3$	163.5		1.2373_4^{20}	1.4256^{20}	−106	100.5	27	
t236	Trichlorofluoro-methane	Cl_3CF	137.4		1.485^{21}	1.384^{20}	−111	23.8		0.14 aq; s alc, eth
t237	α,α,2-Trichloro-6-fluorotoluene	$ClC_6H_3(F)CHCl_2$	213.47	5^3, 701	1.446	1.5506^{20}		228–230	>112	
t238	Trichloroisocyanuric acid		232.41	25, 256			249–251			

Tributyrin, g19
β,β,β-Trichloroethoxycarbonyl chloride, t233
Trichloromethane, c127
Trichlorophenylsilane, p158
3,3,3-Trichloropropylene oxide, e13

t238

TABLE 1-14 Physical constants of organic compounds (*continued*)

No.	Name	Formula	Formula weight	Beilstein reference	Density	Refractive index	Melting point	Boiling point	Flash point	Solubility in 100 parts solvent
t239	Trichloromethane-sulfenyl chloride	Cl_3CSCl	185.89	3, 135	1.700_4^{20}	1.5436^{20}		146–148	none	
t240	Trichloromethane sulfonyl chloride	Cl_3CSO_2Cl	217.88	3^2, 16			139			s alc, eth
t241	1,1,1-Trichloro-2-methyl-2-propanol	$(CH_3)_2C(OH)CCl_3$	177.46	1, 382			99	167		s alc, bz, chl, eth
t242	Trichloromethyl-silane	CH_3SiCl_3	149.48		1.275_4^{20}	1.4108^{20}	−90	66	5	
t242a	1,2,4-Trichloro-5-nitrobenzene	$Cl_3C_6H_2NO_2$	226.45	5, 246	1.790^{20}		49–55	288		v s bz, eth
t243	Trichloronitro-methane	Cl_3CNO_2	164.38	1, 76	1.6558_4^{20}	1.4611^{20}	−64	112		misc alc, bz; s eth
t244	2,4,5-Trichloro-phenol	$Cl_3C_6H_2OH$	197.45	6^2, 180			67	253		615 acet; 163 bz; 525 eth; s alc; i aq
t245	2,4,6-Trichloro-phenol	$Cl_3C_6H_2OH$	197.45	6, 190	1.4901_4^{75}		69	246	none	525 acet; 113 bz; 354 eth; v s alc; i aq
t246	(2,4,5-Trichloro-phenoxy)acetic acid	$Cl_3C_6H_2OCH_2COOH$	255.49	6^3, 702			153			s alc; v sl s aq
t247	2-(2,4,5-Trichloro-phenoxypropionic acid	$Cl_3C_6H_2O$-$CH(CH_3)COOH$	269.51				181.6			0.14 aq; 16 acet; 0.16 bz; 7.1 eth
t248	1,2,3-Trichloro-propane	$ClCH_2CH(Cl)CH_2Cl$	147.43	1, 106	1.3880^{20}	1.4834^{20}	−14.7	156.9	82	misc alc, eth; i aq
t249	1,1,1-Trichloro-2-propanol	$CH_3CH(OH)CCl_3$	163.43	1, 365			50	162	82	2.9 aq; v s alc, eth

t250	2,4,6-Trichloro-pyrimidine		183.43	23, 90		1.5700^{20}	23–25	210–215	>112	
t251	Trichlorosilane	$HSiCl_3$	135.45		1.3417_4^{20}	1.400^{20}	−1.28	31–32	−20	d aq; s bz, chl
t252	α,α,α-Trichlorotoluene	$C_6H_5CCl_3$	195.48	5, 300	1.3756_4^{20}	1.5570^{20}	−5.0	220.8	97	s alc, bz, eth
t253	α,2,6-Trichloro-toluene	$Cl_2C_6H_3CH_2Cl$	195.48				36–39	119^{14mm}		v s alc, eth
t254	2,4,6-Trichloro-1,3,5-triazine		184.41	26, 35			148	190^{720mm}		i aq; s alc
t255	1,1,2-Trichloro-trifluoroethane	$Cl_2CFCClF_2$	187.38	1^3, 157	1.5635^{25}	1.3557^{25}	−36.4	47.6	none	0.017 aq
t256	Trichlorovinyl-silane	$H_2C{=}CHSiCl_3$	161.49		1.243_4^{20}	1.4300^{20}	−95	90–93	−9	
t258	Tricyclo[5.2.1.$0^{2,6}$]-decane		136.24	5, 164			77–79	193	40	
t259	Tricyclo[5.2.1.$0^{2,6}$]-decan-8-one		150.22	7^2, 133	1.063	1.5025^{20}		132^{30mm}		
t260	1,3,5-Tricyclohexyl-hexahydro-*s*-triazine		333.57				74–75	97^{6mm}		

Tricine, t435

Tricyclo[3.3.1.$1^{3,7}$]decane, a67

Tricyclo[5.2.1.$0^{2,6}$]decane-4,8-dimethanol, b186

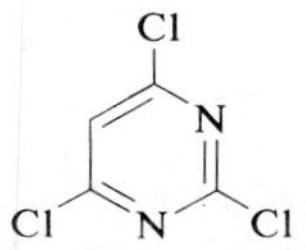

t250

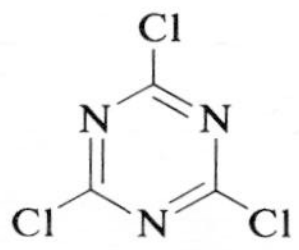

t254

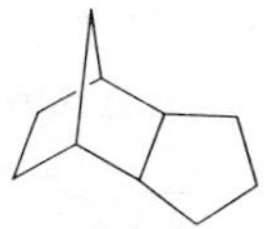

t258

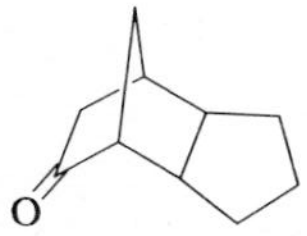

t259

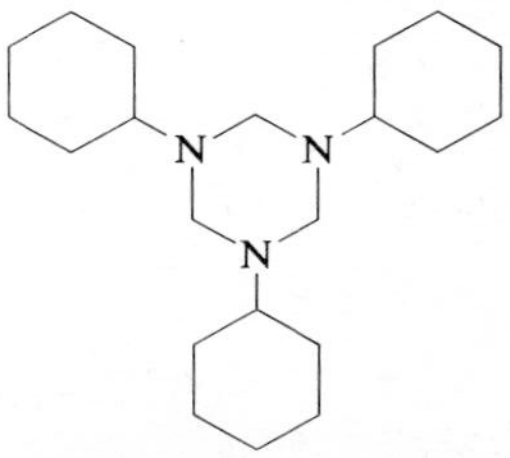

t260

TABLE 1-14 Physical constants of organic compounds (*continued*)

No.	Name	Formula	Formula weight	Beilstein reference	Density	Refractive index	Melting point	Boiling point	Flash point	Solubility in 100 parts solvent
t261	Tridecane	$CH_3(CH_2)_{11}CH_3$	184.37	1, 171	0.7563_4^{20}	1.4256^{20}	−5.4	235.4	79	v s alc, eth
t262	Tridecanoic acid	$CH_3(CH_2)_{11}COOH$	214.35	2, 364			41–42	236^{100mm}		v s alc, eth; i aq
t263	1-Tridecene	$CH_3(CH_2)_{10}CH{=}CH_2$	182.35	1, 225	0.7653^{20}	1.4334^{20}	−23.1	232.8	79	s alc; v s eth
t264	Triethanolamine	$(HOCH_2CH_2)_3N$	149.19	4, 285	1.1242_4^{20}	1.4835^{25}	21.6	335.4	185	misc aq, alc, acet; 4.5 bz; 1.6 eth; s chl
t265	Triethoxyborane	$(CH_3CH_2O)_3B$	145.99	1, 335	0.864_{20}^{20}	1.3740^{20}		117–118	11	d aq
t266	Triethoxyethyl-silane	$(C_2H_5O)_3SiC_2H_5$	192.3		0.8963_4^{20}	1.3955^{20}		158–159		
t266a	Triethoxymethyl-silane	$CH_3Si(OC_2H_5)_3$	178.30	4, 629	0.895_4^{20}	1.3845^{20}		141–143	23	s alc
t266b	Triethoxysilane	$(C_2H_5O)_3SiH$	164.28	1, 334	0.875_4^{20}	1.3762		131.5	26	
t267	Triethoxyvinyl-silane	$(C_2H_5O)_3SiCH{=}CH_2$	190.32		0.903_4^{20}	1.3978^{20}		160–161	34	
t268	Triethylaluminum	$(C_2H_5)_3Al$	114.17	4, 643	0.832^{25}		−58	194		d aq, air
t269	Triethylamine	$(C_2H_5)_3N$	101.19	4, 99	0.7326_4^{25}	1.3980^{25}	−114.7	89.6	−6	5.5 aq; misc alc, eth; s acet, EtAc
t270	Triethylantimony	$(C_2H_5)_3Sb$	208.94	4, 618	1.324^{16}	1.42	−29	159.5		
t271	Triethylarsine	$(C_2H_5)_3As$	162.11	4, 602	1.150_4^{20}			140^{736mm}		i aq; misc alc, eth
t272	Triethylbismuthine	$(C_2H_5)_3Bi$	296.17	4, 622	1.82			107^{79mm}		i aq; v s alc, eth
							explodes when heated in air			
t273	Triethylborane	$(C_2H_5)_3B$	98.00	4, 641	0.6961^{23}		−92.9	95		i aq; d air
t274	Triethylenediamine		112.18				158	174		45 aq; 13 acet; 77 alc; 51 bz
t275	Triethylene glycol	$(HOCH_2CH_2OCH_2{-})_2$	150.17	1, 468	1.1274_4^{15}	1.4578^{15}	−4.3	285	165	misc aq, alc, bz
t276	Triethylene glycol dibenzoate	$(C_6H_5COOCH_2$-$CH_2OCH_2{-})_2$	358.39		1.2715^{30}	1.5252^{50}	47			
t277	Triethylenetetramine	$(H_2NCH_2CH_2NHCH_2{-})_2$	146.24	4, 255	0.982	1.4971^{20}	12	266–267	143	

t278	*N,N,N'*-Triethyl-ethylenediamine	$(C_2H_5)_2NCH_2CH_2NHC_2H_5$	144.26	4^2, 691	0.804	1.4311^{20}		55^{13mm}	32	
t279	Triethylgallium	$(C_2H_5)_3Ga$	156.91		1.0576^{30}		−82.3	142.6		
t280	1,3,5-Triethylhexa-hydro-*s*-triazine		171.29	26, 2	0.894	1.4595^{20}		207–208		
t281	Triethylindium	$(C_2H_5)_3In$	202.01		1.260^{20}	1.538^{20}	−32	144		
t282	Triethyl ortho-acetate	$CH_3C(OC_2H_5)_3$	162.23	2, 129	0.8847_4^{25}	1.3950^{25}		142	55	misc alc, chl, eth
t283	Triethyl ortho-formate	$HC(OC_2H_5)_3$	148.20	2, 20	0.891_4^{20}	1.3919^{20}	−76	146	30	d aq; s alc, eth
t284	Triethyl ortho-propionate	$CH_3CH_2C(OC_2H_5)_3$	176.26	2, 240	0.876	1.3995^{20}		155–160	60	v s alc, eth
t285	Triethyl phosphate	$(C_2H_5O)_3P(O)$	182.16	1, 332	1.0725^{19}	1.4045^{20}		215–216		s aq(d), alc, eth
t286	Triethylphosphine	$(C_2H_5)_3P$	118.16	4, 582	0.800_4^{15}		−88	129	pyro-phoric	i aq; misc alc, eth
t287	Triethyl phosphite	$(C_2H_5O)_3P$	166.2	1, 330	0.969_4^{20}	1.4131^{20}		65^{24mm}	55	i aq(hyd); misc alc, acet, bz, eth, PE
t288	Triethyl phosphono-acetate	$(CH_3CH_2O)_2P(O)-CH_2COOC_2H_5$	224.19	4^1, 573	1.130	1.4310^{20}		145^{9mm}	>112	

Tridecylbenzene, p159
3-Triethoxysilylpropylamine, a284
Triethyl borate, t265
Triethylenediamine, d45
Triethylene glycol, e130
Triethylene glycol dimethyl ether, b190
O,O,O-Triethyl phosphorothioate, t290

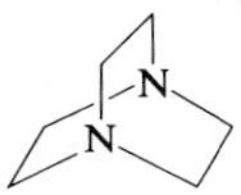
t274

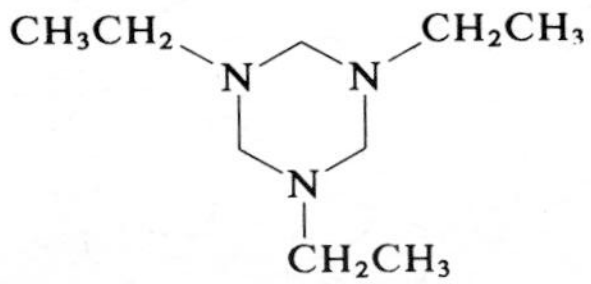

t280

TABLE 1-14 Physical constants of organic compounds (*continued*)

No.	Name	Formula	Formula weight	Beilstein reference	Density	Refractive index	Melting point	Boiling point	Flash point	Solubility in 100 parts solvent
t289	Triethylsilane	$(C_2H_5)_3SiH$	116.28	4, 625	0.731_4^{20}	1.412^{20}		107–108		i aq; misc alc, eth
t290	Triethyl thiophosphate	$(C_2H_5O)_3P(S)$	198.22	1, 333	1.082	1.4480^{20}		100^{16mm}	107	
t291	2,2,2-Trifluoroacetamide	CF_3CONH_2	113.04	2^2, 186			75	162.5		
t292	Trifluoroacetic acid	CF_3COOH	114.02	2^2, 186	1.4890^{20}	1.2850^{20}	−15.3	71.8		misc aq
t293	Trifluoroacetic anhydride	$[CF_3C(O)]_2O$	210.03	2^2, 186	1.487	>1.30	−65	39		
t294	α,α,α-Trifluoroacetophenone	$C_6H_5COCF_3$	174.12		1.240	1.4595^{20}		165–166	41	
t295	α,α,α-Trifluoro-*m*-cresol	$CF_3C_6H_4OH$	162.11	6^1, 187	1.333	1.4588^{20}	−1.8	178–179	73	
t296	1,1,1-Trifluoroethane	CH_3CF_3	84.04				−111.3	−47.3		
t297	2,2,2-Trifluoroethanol	CF_3CH_2OH	100.04		1.3842_4^{20}	1.2907^{22}	−43.5	74.1	29	
t298	2,2,2-Trifluoroethyl acrylate	$CF_3CH_2OOCCH{=}CH_2$	154.0		2.142_4^{25}	1.3981^{25}		46^{125mm}		
t299	2,2,2-Trifluoroethyl trifluoroacetate	$CF_3CH_2OOCCF_3$	196.0		1.4725_4^{18}	1.2812^{18}	−65.5	55		
t300	Trifluoromethane	HCF_3	70.01	1, 59	1.52^{-100}		−155.2	−82.2		75 mL aq; 500 mL alc
t301	Trifluoromethanesulfonic acid	CF_3SO_3H	150.07	3^3, 34	1.695^{25}	1.3250^{25}	34	162	none	v s aq; misc eth
t302	Trifluoromethanesulfonic anhydride	$(CF_3SO_2)_2O$	282.13	3^4, 35	1.677	1.3212^{20}		84		d aq, alc
t303	3-(Trifluoromethyl)benzonitrile	$CF_3C_6H_4CN$	171.12	9, 478	1.2813^{20}	1.4505^{20}	14.5	189	72	

t304	3-(Trifluoromethyl)-benzyl chloride	$CF_3C_6H_4CH_2Cl$	194.59		1.254	1.4605		70^{12mm}		
t305	α,α,α-Trifluoro-toluene	$C_6H_5CF_3$	146.11	5, 290	1.1886^{20}	1.4145^{20}	−29	102	12	
t306	Trihexylamine	$[CH_3(CH_2)_5]_3N$	269.52	4, 188				263–265	>112	v s alc, eth; i aq
t307	Trihexylchloro-silane	$[CH_3(CH_2)_5]_3SiCl$	319.1		0.871_4^{20}	1.456^{20}		155^{5mm}		
t308	Trihexylsilane	$[CH_3(CH_2)_5]_3SiH$	284.60			1.448^{20}		160^{5mm}		
t309	1,2,3-Trihydroxy-benzene	$C_6H_3(OH)_3$	126.11	6, 1071	1.45		131–133			59 aq; 77 alc; 62 eth
t310	1,3,5-Trihydroxy-benzene	$C_6H_3(OH)_3$	126.11	6, 1092			218–220	subl d		1 aq; 10 alc; s eth
t311	3,4,5-Trihydroxy-benzoic acid	$(HO)_3C_6H_2COOH$	170.12	10, 470			d 235			1.1 aq; 17 alc; 1 eth; 20 acet; i bz, chl, PE
t312	Triisobutylaluminum	$[(CH_3)_2CHCH_2]_3Al$	198.33		0.781^{25}		6	86^{10mm}	pyro-phoric	
t313	Triisodecyl phosphite	$[(CH_3)_2CH(CH_2)_7O]_3P$	502.80		0.886_{15}^{25}	1.454^{25}	<0	$180^{0.1mm}$	235	
t314	Triisopropanolamine	$[CH_3CH(OH)CH_2]_3N$	191.27		0.9996_{20}^{50}		46	305.4	152	v s aq
t315	Trisisopropoxyborane	$[(CH_3)_2CHO]_3B$	188.08	1, 363	0.815	1.3764^{20}		139–141	17	
t316	Triisopropoxyvinyl-silane	$[(CH_3)_2CHO]_3Si\text{-}CH{=}CH_3$	232.4		0.863_4^{25}	1.396^{25}		179–181		
t137	1,3,5-Triisopropyl-benzene	$C_6H_3[CH(CH_3)_2]_3$	204.36	5, 458	0.845	1.4884^{20}		232–236	86	
t318	Triisopropyl phosphite	$[(CH_3)_2CHO]_3P$	208.24	1, 363	0.914_4^{20}	1.4101^{20}		64^{11mm}	73	i aq(sl hyd)

2,2,2-Trifluoroethyl mesylate, m441
2-(Trifluoromethyl)aniline, a129
3-(Trifluoromethyl)aniline, a130
m-Trifluoromethylphenol, t295
4,4,4-Trifluoro-1-(2-thienyl)-1,3-butanedione, t139
α,α,α-Trifluorotoludidines, a129, a130, a131
α,α,α-Trifluorotolunitrile, t303
Triglyme, b190
Tri-(2-hydroxyethyl)amine, t264
1,2,6-Trihydroxyhexane, h65
Triiodomethane, i36

TABLE 1-14 **Physical constants of organic compounds (*continued*)**

No.	Name	Formula	Formula weight	Beilstein reference	Density	Refractive index	Melting point	Boiling point	Flash point	Solubility in 100 parts solvent
t319	3,4,5-Trimethoxy-benzaldehyde	$(CH_3O)_3C_6H_2CHO$	196.20	8, 391			73–75	165^{10mm}		
t320	1,2,3-Trimethoxy-benzene	$C_6H_3(OCH_3)_3$	168.19	6, 1081	1.112		43–45	241		
t321	3,4,5-Trimethoxy-benzoic acid	$(CH_3O)_3C_6H_2COOH$	212.20	10, 481			168–171	227^{10mm}		v s alc, eth; s chl
t322	3,4,5-Trimethoxy-benzoyl chloride	$(CH_3O)_3C_6H_2COCl$	230.65	10, 487			79–81	185^{18mm}		
t323	3,4,5-Trimethoxy-benzyl alcohol	$(CH_3O)_3C_6H_2CH_2OH$	198.22	6, 1159	1.233	1.5459^{20}		228^{25mm}	>112	
t324	Trimethoxyborane	$(CH_3O)_3B$	103.91	1, 287	0.920_4^{23}	1.3568^{20}	−34	67–68	−1	hyd aq; misc alc, eth
t325	Trimethoxyboroxine	$[-OB(OCH_3)-]_3$	173.53		1.195	1.3996^{20}	10	130	10	
t326	1,3,3-Trimethoxy-butane	$(CH_3O)_2C(CH_3)CH_2$-CH_2OCH_3	148.20	1^3, 3214	0.940	1.4096^{20}		63^{20mm}	45	
326a	Trimethoxy(methyl)-silane	$CH_3Si(OCH_3)_3$	136.23		0.9548_4^{20}	1.3696^{20}		102–103	21	
t327	1,3,3-Trimethoxy-propane	$CH_3OCH_2CH_2$-$CH(OCH_3)_2$	134.18	1, 820	0.942	1.4004^{20}		45–46^{17mm}		
t328	(Trimethoxysilyl)-propyldiethylene-triamine	$(CH_3O)_3Si(CH_2)_3NH$-$CH_2CH_2NHCH_2CH_2NH_2$	265.4		1.03_4^{20}	1.463^{20}				
t329	*N*-[3-Trimethoxy-silyl)propyl]-ethylenediamine	$(CH_3O)_3Si(CH_2)_3NH$-$CH_2CH_2NH_2$	222.4		1.010	1.4450^{20}		146^{15mm}	>112	
t330	*N*-(Trimethoxysilyl-propyl)imidazole		230.3		1.00_4^{20}	1.45^{25}				
t331	3-(Trimethoxysilyl)-propyl methacrylate	$(CH_3O)_3SiCH_2CH_2$-$CH_2OOCC(CH_3){=}CH_2$	249.3		1.045_4^{20}	1.429^{25}		190	92	

t332	Trimethylaluminum	$(CH_3)_3Al$	72.09	4, 643	0.752^{20}	1.432^{12}	15.4	20^{8mm}	pyro-phoric	s alk; v sl s alc
t333	Trimethylamine	$(CH_3)_3N$	59.11	4, 43	0.636		−117.1	2.9	−6	41 aq; misc alc; s bz, chl, eth
t334	Trimethylamine-*N*-oxide	$(CH_3)_3N(O)$	75.11				257			s aq, MeOH
t335	2,4,6-Trimethyl-aniline	$(CH_3)_3C_6H_2NH_2$	135.21	12, 1160	0.963	1.5510^{20}		233	96	
t336	1,3,3-Trimethyl-6-azabicyclo[3.2.1]-octane		153.27		0.902	1.4716^{20}		194	75	
t337	3,3,5-Trimethyl-1-azacycloheptane		141.26		0.852	1.4563^{20}		180	67	
t338	1,2,3-Trimethyl-benzene	$C_6H_3(CH_3)_3$	120.20	5, 399	0.894_4^{20}	1.5139^{20}	−25.4	176.1	48	i aq; s alc, eth
t339	1,2,4-Trimethyl-benzene	$C_6H_3(CH_3)_3$	120.20	5, 400	0.8756_4^{20}	1.5048^{20}	−43.9	169.4	48	s alc, bz, eth
t340	1,3,5-Trimethyl-benzene	$C_6H_3(CH_3)_3$	120.20	5, 406	0.8637_4^{20}	1.4994^{20}	−44.7	164.7	44	misc alc, bz, eth

Trimellitic acid, b29
Trimesic acid, b30
Trimesoyl chloride, b32
Trimethylacetaldehyde, d598
Trimethylacetamide, d599
Trimethylacetic acid, d600
Trimethylacetic anhydride, d601
Trimethylacetyl chloride, d602
endo-1,7,7-Trimethylbicyclo[2.2.1]heptan-2-o1, b217

$N—CH_2CH_2CH_2Si(OCH_3)_2$

t330 t336 t337

TABLE 1-14 Physical constants of organic compounds (*continued*)

No.	Name	Formula	Formula weight	Beilstein reference	Density	Refractive index	Melting point	Boiling point	Flash point	Solubility in 100 parts solvent
t341	Trimethyl 1,2,4-benzenetricarboxylate	$C_6H_3(COOCH_3)_3$	252.22	9^1, 429	1.261	1.5214^{20}	38–40	194^{12mm}	>112	
t342	2,2,3-Trimethylbutane	$(CH_3)_2CHC(CH_3)_3$	100.20	1^2, 121	0.6901^{20}_{4}	1.3894^{20}	−24.9	80.9		s alc, eth
t343	2,3,3-Trimethyl-2-butanol	$(CH_3)_3CC(CH_3)_2OH$	116.20	1^2, 447	0.8380^{25}_{4}	1.4233^{22}	15–17	130.5		misc alc, eth
t344	1,1,3-Trimethylcyclohexane	$C_6H_9(CH_3)_3$	126.24			1.4296^{20}		136.6		
t345	3,5,5-Trimethylcyclohex-2-ene-1-one		138.2	7, 65	0.925^{20}_{20}	1.478^{20}	−8.1	215.2	96	1.2 aq
t346	2,2,6-Trimethyl-1,3-dioxen-4-one		142.15	19^3, 1604	1.088	1.4622^{20}	12–13	65–67^{2mm}		
t347	4,4′-Trimethylenebis-(1-methylpiperidine)		238.42		0.896	1.4820^{20}	13	215^{50mm}	110	
t348	4,4′-Trimethylenedipiperidine		210.37				65–68			
t349	4,4′-Trimethylenedipyridine		198.27				57–60			
t350	Trimethylene oxide		58.08	17, 6	0.8930^{25}_{4}	1.3895^{25}		50	<1	misc aq
t350a	Trimethylene sulfide		74.15	17^1, 3	1.025^{20}	1.5102^{20}	−73.3	95.0	<1	
t352	2,2,5-Trimethylhexane	$(CH_3)_2CHCH_2CH_2$-$C(CH_3)_3$	128.26	1^3, 516	0.7072^{20}	1.3997^{20}	−105.8	124.1		v s org solv
t353	3,5,5-Trimethyl-1-hexanol	$(CH_3)_3CCH_2CH(CH_3)$-CH_2CH_2OH	144.25		0.8236^{20}_{4}	1.4300^{25}	<−70	194		s alc, eth
t354	Trimethylhydroquinone	$(CH_3)_3C_6H(OH)_2$	152.19	6, 931			172–174			s aq; v s alc, bz, eth

t354a	2,6,8-Trimethyl-4-nonanol	$(CH_3)_2CHCH_2CH(CH_3)$-$CH_2CH(OH)CH_2$-$CH(CH_3)_2$	186.33		0.8193			225	93	
t355	2,6,8-Trimethyl-4-nonanone	$(CH_3)_2CHCH_2CH(CH_3)$-$CH_2C(O)CH_2CH(CH_3)_2$	184.31		0.818_{20}^{20}		−75	218.4		
t356	α-(−)-1,3,3-Trimethyl-2-norbornanol		154.25	6, 70	0.9641_{4}^{20}		48	201	73	s alc, eth

Trimethyl borate, t324
Trimethylchlorosilane, c256
α,α,4-Trimethyl-3-cyclohexene-1-methanol, t7
3,5,5-Trimethylcyclohex-2-en-1-one, i82
1,2,2-Trimethyl-1,3-cyclopentanedicarboxylic acid, c5
Trimethylene chlorobromide, b259
Trimethylene chlorohydrin, c215
Trimethylenediamine, p196
Trimethylene dibromide, d93
Trimethylene glycol, p198
Trimethylethylene, m159
Trimethylgermanium bromide, b366
3,3,5-Trimethylhexahydroazepine, t337

t345

t346

t347

t348

t349

t350

t350a

α form
t356

TABLE 1-14 Physical constants of organic compounds (*continued*)

No.	Name	Formula	Formula weight	Beilstein reference	Density	Refractive index	Melting point	Boiling point	Flash point	Solubility in 100 parts solvent
t357	(+)-1,3,3-Trimethyl-2-norbornanone		152.24	7, 96	0.948^{18}	1.4635^{18}	5–6	192–193	52	v s alc, eth
t358	Trimethyl orthoacetate	$CH_3C(OCH_3)_3$	120.15	1^2, 128	0.9428^{25}_{4}	1.3859^{25}		105		v s alc, eth
t359	Trimethyl orthoformate	$HC(OCH_3)_3$	106.12	2, 19	0.9676^{20}_{4}	1.3790^{20}		100.6	15	
t360	2,4,4-Trimethyl-1-oxazoline		113.16		0.887	1.4213^{20}		112–113	12	
t361	2,2,3-Trimethylpentane	$(CH_3)_3CCH(CH_3)$-CH_2CH_3	114.23	1^1, 62	0.7160^{20}_{4}	1.4030^{20}	−112.3	109.8		s eth; sl s alc
t362	2,2,4-Trimethylpentane	$(CH_3)_2CHCH_2C(CH_3)_3$	114.23	1^2, 127	0.6919^{20}_{4}	1.3915^{20}	−107.4	99.2	−7	s bz, chl, eth
t363	2,3,4-Trimethylpentane	$(CH_3)_2CH[CH(CH_3)]_2CH_3$	114.24	1^3, 500	0.7190^{20}_{4}	1.4042^{20}	−109.2	113.5		s alc, org solv
t364	2,2,4-Trimethyl-1,3-pentanediol	$(CH_3)_2CHCH(OH)$-$C(CH_3)_2CH_2OH$	146.22	1^3, 2225	0.928^{55}_{15}	1.4513^{15}	46	229	113	1.8 aq; 75 alc; 22 bz; 25 acet
t365	2,4,4-Trimethyl-1-pentene	$(CH_3)_3CCH_2$-$C(CH_3)=CH_2$	112.22	1^3, 849	0.7150^{20}_{4}	1.4112^{20}	−93	101.4	<1	
t366	2,3,5-Trimethylphenol	$(CH_3)_3C_6H_2OH$	136.19	6, 518			92–95	230–231		
t367	2,3,6-Trimethylphenol	$(CH_3)_3C_6H_2OH$	136.19				62–64			
t368	2,4,6-Trimethylphenol	$(CH_3)_3C_6H_2OH$	136.19	6, 158			68–71	220		
t369	Trimethyl phosphate	$(CH_3O)_3P(O)$	140.08	1, 286	1.197	1.3958^{20}	−46	197	none	100 aq; s alc
t370	Trimethyl phosphite	$(CH_3O)_3P$	124.08	1, 285	1.046^{20}_{4}	1.4080^{20}	<−78	111–112	40	d aq; misc alc, acet, bz, PE

No.	Name	Formula	Formula weight		Density	Refractive index	Melting point	Boiling point	Flash point	Solubility
t371	Trimethyl phosphonoacetate	$(CH_3O)_2P(O)CH_2COCH_3$	182.11		1.125	1.4370^{20}		$118^{0.85mm}$	>112	
t372	1,2,4-Trimethylpiperazine		128.22		0.851_{25}^{25}	1.4480^{25}	<−50	151^{746mm}		s aq, alc, acet, bz
t373	2,4,6-Trimethylpyridine	$(C_5H_2N)(CH_3)_3$	121.18	20, 250	0.9166_4^{22}	1.4979^{20}	−43	170.5	57	3.5 aq; misc eth; s alc, bz, chl
t374	*N*-(Trimethylsilyl)acetamide	$CH_3CONHSi(CH_3)_3$	131.25				52–54	185–186	57	
t375	*N*-(Trimethylsilyl)aniline	$(CH_3)_3SiNHC_6H_5$	165.3		0.940_4^{20}	1.522^{20}		207–208		
t377	Trimethylsilyl bromoacetate	$BrCH_2COOSi(CH_3)_3$	211.14		1.284	1.4421^{20}		$57–58^{9mm}$	28	
t378	*N*-(Trimethylsilyl)diethylamine	$(CH_3)_3SiN(C_2H_5)_2$	145.33					127^{738mm}		
t379	2-(Trimethylsilyl)ethanol	$(CH_3)_3SiCH_2CH_2OH$	118.25		0.825	1.4246^{20}		$71–73^{35mm}$	50	
t380	*N*-(Trimethylsilyl)imidazole		140.26		0.956	1.4751^{20}		99^{14mm}	80	
t381	3-(Trimethylsilyloxy)allene	$(CH_3)_3SiOCH_2CH{=}CH_2$	130.3		0.7830_4^{30}	1.4075^{25}		100–102		

Trimethylolpropane, e159
Trimethylsilyl cyanide, c299

Trimethylsilyldiethylamine, d343
Trimethylsilyl iodide, i55

Trimethylsilylnitrile, c299

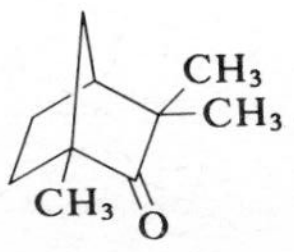

t357

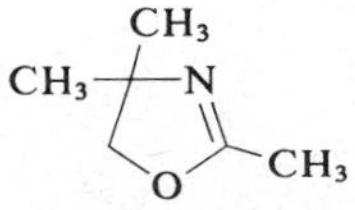

t360

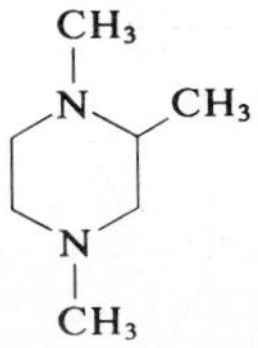

t372

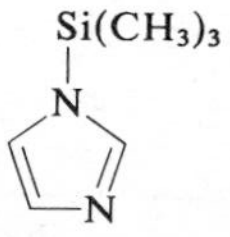

t380

TABLE 1-14 Physical constants of organic compounds (*continued*)

No.	Name	Formula	Formula weight	Beilstein reference	Density	Refractive index	Melting point	Boiling point	Flash point	Solubility in 100 parts solvent
t382	Trimethylsilyl-phenoxide	$(CH_3)_3SiOC_6H_5$	166.3		0.9256_4^{20}	1.4782^{20}	−55	81^{23mm}		
t383	Trimethylsilyl trifluoroacetate	$(CH_3)_3SiOOCCF_3$	186.2		1.077_4^{20}	1.3880^{20}		89–90		
t384	Trimethylsulfonium iodide	$[(CH_3)_3S]I$	204.07				215–220	subl		
t385	Trimethylsulf-oxonium iodide	$[(CH_3)_3S(O)]I$	220.07				175 d			
t386	Trimethylvinyl-oxysilane	$(CH_3)_3SiOCH{=}CH_2$	116.2		0.772_4^{20}	1.389^{20}		74–75		
t387	Trimethylvinyl-silane	$(CH_3)_3SiCH{=}CH_2$	100.2		0.690_4^{20}	1.3920^{20}		55	<1	
t388	2,4,6-Trinitro-aniline	$(O_2N)_3C_6H_2NH_2$	228.12	12, 763	1.762^{14}		188–190	explodes		s hot acet; sl s alc
t389	1,2,4-Trinitro-benzene	$C_6H_3(NO_2)_3$	213.11	5, 271	1.73^{16}		61–62	explodes		5.5 alc; 7.1 eth; i aq
t390	1,3,5-Trinitro-benzene	$C_6H_3(NO_2)_3$	213.11	5, 271	1.688_4^{20}		122.5	explodes		0.035 aq; 1.9 alc; 1.5 eth; 6.2 bz
t391	2,4,7-Trinitro-9-fluorenone		315.20	7^2, 410			175–176			v s bz, acet; sl s aq
t392	Trinitromethane	$HC(NO_2)_3$	151.04	1, 79	1.597_4^{24}		15	47^{22mm}		s aq, alk
t393	2,4,6-Trinitro-toluene	$(O_2N)_3C_6H_2CH_3$	227.13	5, 347	1.654_4^{20}		80.1	explodes		1.5 alc; 4 eth; s bz, acet; 0.01 aq
t394	Trioctylamine	$[CH_3(CH_2)_7]_3N$	353.68	4, 196	0.809	1.4485^{20}		365–367	>112	
t395	*s*-Trioxane		90.08	19, 381	1.170^{65}		64	115	45	17.2 aq; v s alc, bz, eth, EtAc
t396	Tripentaerythritol		372.41				245 d			
t397	2,4,6-Triphenoxy-*s*-triazine		357.37				232–234			

t398	Triphenoxyvinyl-silane	$(C_6H_5O)_3SiCH{=}CH_2$	334.5		1.130^{25}_{4}	1.562^{25}		210^{7mm}		
t399	Triphenylamine	$(C_6H_5)_3N$	245.33	12, 181	0.774^{0}_{0}		125–127	347–348		s acet, eth; sl s alc
t400	Triphenylantimony	$(C_6H_5)_3Sb$	353.07	16, 891	1.4343^{25}		52–54	377		v s bz, eth; sl s alc
t401	Triphenylarsine	$(C_6H_5)_3As$	306.24	16, 828	1.2225^{48}	1.6139^{48}	60–62	233^{14mm}		v s bz, eth; s alc
t402	1,3,5-Triphenyl-benzene	$(C_6H_5)_3C_6H_3$	306.41	5, 737	1.205		172–174	460		v s bz; s abs alc, eth
t403	Triphenylene		228.29	5, 720	1.302		199	425		s alc; v s bz, eth
t404	Triphenylmethane	$(C_6H_5)_3CH$	244.34	5, 698	1.0134^{99}_{4}		93.4	360		v s hot alc, eth; 49 chl; 7 bz; s PE; i aq
t405	Triphenylmethanol	$(C_6H_5)_3COH$	260.34	6, 713	1.199^{0}_{4}		164.2	360		v s alc, bz, eth; i aq
t406	Triphenyl phosphate	$(C_6H_5O)_3P(O)$	326.29	6, 179			49–51	244^{10mm}	223	misc alc; s bz, acet, chl, eth; i aq

2,4,6-Trinitrophenol, p176
Triolein, g22

Trioxymethylene, t395
Tripalmitin, g23

Triphenylmethyl bromide, b369

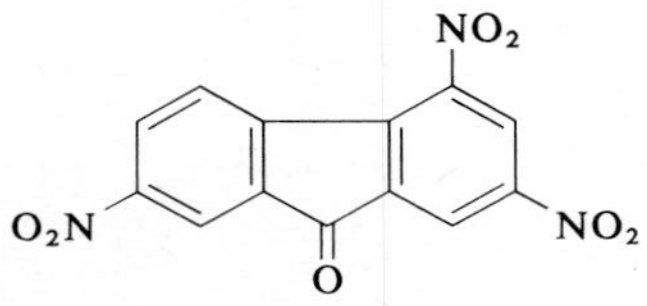

O_2N NO_2 NO_2 O

t391

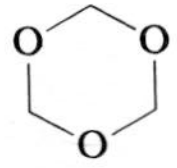

t395

$(HOCH_2)_3CCH_2OCH_2C(CH_2OH)_2CH_2OCH_2C(CH_2OH)_3$

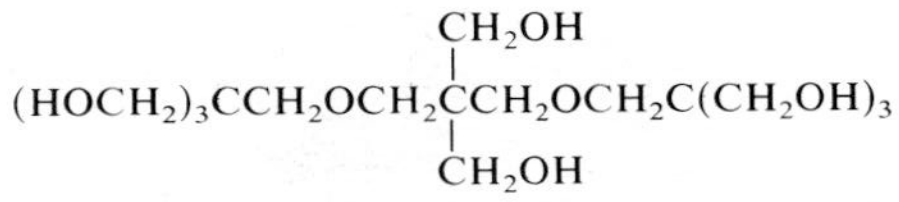

t396

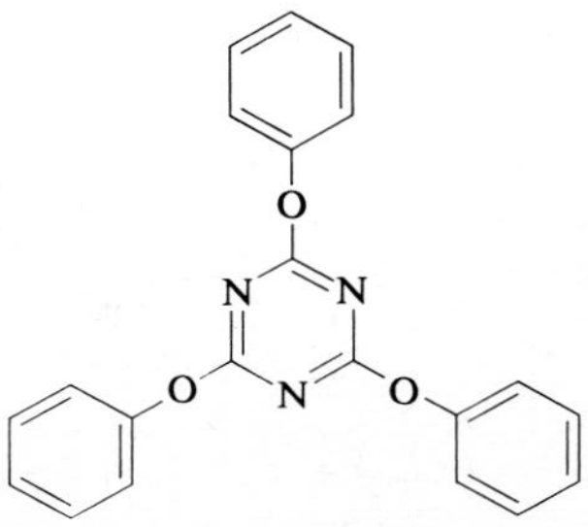

t397

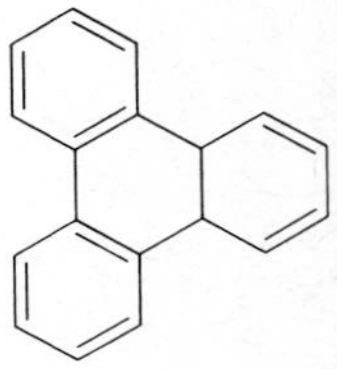

t403

TABLE 1-14 Physical constants of organic compounds (*continued*)

No.	Name	Formula	Formula weight	Beilstein reference	Density	Refractive index	Melting point	Boiling point	Flash point	Solubility in 100 parts solvent
t407	Triphenylphosphine	$(C_6H_5)_3P$	262.29	16, 759	1.075_4^{81}		80.5	377	181	v s eth; s bz; chl, HOAc; sl s alc; i aq
t408	Triphenylphosphine selenide	$(C_6H_5)_3P(Se)$	341.25				187–189			
t409	Triphenylphosphine sulfide	$(C_6H_5)_3P(S)$	294.36	16, 784			162–164			
t410	Triphenyl phosphite	$(C_6H_5O)_3P$	310.29	6, 177	1.184_{15}^{25}	1.5903^{20}	22–24	360	218	s alc, bz, chl, eth
t411	Triphenylsilane	$(C_6H_5)_3SiH$	260.41	16^2, 605			42–44	152^{2mm}		
t412	Tripiperidino-phosphine oxide		299.40	20, 88			40–42	273^{50mm}		
t413	Tripropoxyborane	$(CH_3CH_2CH_2O)_3B$	188.08	1^2, 369	0.8576_4^{20}	1.3948^{20}		175		v s alc; misc eth
t414	Tripropylamine	$(CH_3CH_2CH_2)_3N$	143.27	4, 139	0.753	1.4160^{20}	−93	155–158	36	s aq, alc, eth
t415	Tripropylene glycol	$H(OCH_2CH_2CH_2)_3OH$	192.3		1.018	1.442^{25}		267.2	141	s aq
t416	Tripropylene glycol butyl ether	$HO(CH_2CH_2CH_2O)_3$-$(CH_2)_3CH_3$	248.4		0.934_{25}^{25}	1.430^{25}		276	135	
t417	Tripropylene glycol ethyl ether	$HO(CH_2CH_2CH_2O)_3$-CH_2CH_3	220.3		0.948_{25}^{25}	1.427^{25}		486	132	
t418	Tripropylene glycol isopropyl ether	$HO(CH_2CH_2CH_2O)_3$-$CH(CH_3)_2$	234.8		0.942_{25}^{25}	1.428^{25}		112.7	124	
t419	Tripropylene glycol-methyl ether	$HO(CH_2CH_2CH_2O)_3CH_3$	206.3		0.967_{25}^{25}	1.428^{25}	−42	242.4	127	misc aq, alc, eth
t420	Tripropyl orthoformate	$HC(OCH_2CH_2CH_3)_3$	190.28		0.8805_4^{20}	1.4072^{20}		108^{5mm}		
t421	Tris(butoxyethyl) phosphate	$(C_4H_9OCH_2CH_2O)_3P(O)$	398.48		1.006	1.4359^{20}		228^{4mm}	>112	
t422	Tris(2-chloro-ethoxy)silane	$(ClCH_2CH_2O)_3SiH$	267.6		1.2886_4^{20}	1.4577^{20}		118^{2mm}		

t423	Tris(2-chloroethyl) phosphate	$(ClCH_2CH_2O)_3P(O)$	285.49	1², 337	1.390	1.4721^{20}		330	232	
t424	Tris(2-chloroethyl) phosphite	$(ClCH_2CH_2O)_3P$	269.49		1.353_4^{20}	1.4863^{20}		115^{2mm}	190	misc alc, bz, eth
t425	Tris(2,6-dichlorophenyl) phosphate	$(Cl_2C_6H_3O)_3P(O)$	533.09				208–210			
t426	Tris(dimethylamino)methane	$CH[N(CH_3)_2]_3$	145.25		1.4360^{20}		42–43^{12mm}			
t427	Tris(dimethylamino)methylsilane	$[(CH_3)_2N]_3SiCH_3$	175.4		0.850_4^{22}	1.432^{22}	−11	56^{17mm}		
t428	Tris(2-ethylhexyl) phosphite	$[CH_3(CH_2)_3CH(CH_2CH_3)CH_2O]_3P$	418.6		0.902_4^{20}	1.4494^{20}		$164^{0.3mm}$	185	i aq
t429	Tris(heptafluoropropyl)-*s*-triazine		585.1		1.7158^{25}	1.7158^{25}		165		
t430	Tris(hydroxymethyl)aminomethane	$(HOCH_2)_3CNH_2$	121.14	4, 303			172	220^{10mm}		

Triphenylsilyl azide, a325
Tripropyl borate, t413
Tripropylsilyl chloride, c258
TRIS, t430
Tris(dimethylamino)silyl chloride, c258a
1,1,1-Tris(hydroxymethyl)ethane, h142

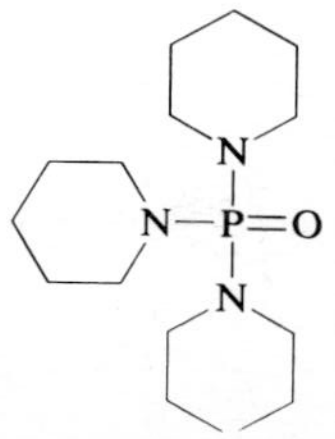
t412

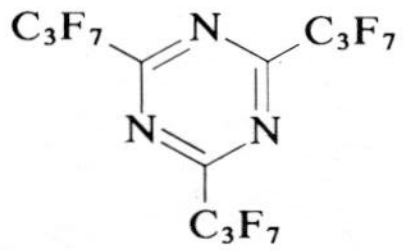
t429

TABLE 1-14 Physical constants of organic compounds (*continued*)

No.	Name	Formula	Formula weight	Beilstein reference	Density	Refractive index	Melting point	Boiling point	Flash point	Solubility in 100 parts solvent
t431	2-[Tris(hydroxy-methyl)methyl-amino]-1-ethane-sulfonic acid	$(HOCH_2)_3CNHCH_2$-CH_2SO_3H	229.25				223–225			
t432	1,1,1-Tris(hydroxy-methyl)ethane	$CH_3C(CH_2OH)_3$	120.15	1, 520						
t433	3-[*N*-Tris(hydroxy-methyl)methyl-amino]-2-hydroxypropane-sulfonic acid	$(HOCH_2)_3CNHCH_2$-$CH(OH)CH_2SO_3H$	259.3				226			
t434	3-[Tris(hydroxy-methyl)-methyl-amino]-1-propanesulfonic acid	$(HOCH_2)_3CNHCH_2CH_2$-CH_2SO_3H	243.28				240 d			
t435	*N*-[Tris(hydroxy-methyl)methyl]-glycine	$(HOCH_2)_3CNHCH_2COOH$	179.17				184 d			
t436	Tris(2-methoxy-ethoxy)methyl-silane	$CH_3Si(OCH_2CH_2OCH_3)_3$	268.4		1.045_4^{20}	1.420^{20}		145^{16mm}		
t437	Tris(2-methoxy-ethoxy)vinylsilane	$H_2C{=}CHSi(OCH_2$-$CH_2OCH_3)_3$	280.38		1.034_4^{25}	1.427^{25}		284–286	65	
t438	Tris(2-methylallyl)-amine	$[H_2C{=}C(CH_3)CH_2]_3N$	173.91	4^3, 462	0.794	1.4575^{20}		83–85^{15mm}	53	
t439	Tris(pentafluoro-ethyl)-*s*-triazine		435.1		1.6506^{25}	1.3131^{25}		121–122		

t440	1,3,5-Trithiane		138.27	19, 382			216–218			s bz; sl s alc, eth
r441	Trithiocarbonic acid	$(HS)_2C(S)$	110.21	3, 221	1.483_4^{20}	1.8225^{20}	−26.9	57.8		d aq, alc; sl s eth
t442	1,2,4-Trivinylcyclo- hexane	$(H_2C{=}CH)_3C_6H_9$	162.28		0.836	1.4780^{20}		88^{20mm}	68	
t443	L-(−)-Tryptophan		204.23	22, 546			280–285 d			1 aq; s hot alc, alk; i eth, chl
t444	L-Tyrosine	$(HO)C_6H_4CH_2$- $CH(NH_2)COOH$	181.19	14, 605			>300 d			0.03 aq; 0.01 alc; s alk; i eth
t445	L-Tyrosine hydrazide	$HOC_6H_4CH_2CH(NH_2)$- $CONHNH_2$	195.22	14^1, 665			196–198			
u1	Undecanal	$CH_3(CH_2)_9CHO$	170.30	1, 712	0.825	1.4322^{20}	−4	115^{5mm}	96	i aq; s alc, eth
u2	Undecane	$CH_3(CH_2)_9CH_3$	156.31	1, 170	0.7402_4^{20}	1.4173^{20}	−25.6	195.9	60	i aq; misc alc, eth
u3	Undecanoic acid	$CH_3(CH_2)_9COOH$	186.30	2, 358	0.8907	1.4294^{45}	28.5	228^{160mm}		s alc, chl, eth; i aq
u4	1-Undecanol	$CH_3(CH_2)_{10}OH$	172.31	1, 427	0.8324^{20}	1.4402^{20}	15.9	242.8	>112	0.02 aq; s alc
u5	2-Undecanone	$CH_3(CH_2)_9COCH$	170.30	1, 173	0.829	1.4280^{20}	11–12	231–232	88	s alc, bz, chl, eth, acet; i aq
u6	6-Undecanone	$CH_3(CH_2)_4$- $CO(CH_2)_4CH_3$	170.30	1, 174	0.831	1.4280^{20}	14.6	228	88	i aq; v s alc, eth
u7	10-Undecenal	$H_2C{=}CH(CH_2)_8CHO$	168.28		0.810	1.4427^{20}			92	
u8	1-Undecene	$CH_3(CH_2)_8CH{=}CH_2$	154.29	1, 225	0.763_4^{20}	1.4261^{20}	−49.2	192.7		i aq; misc alc, eth

Tris(7-methylnonyl) phosphite, t313
Trityl alcohol, t405
Tryptamine, a170
Tyramine, a173
Umbelliferone, h109
Undecyl alcohol, u4
Undecyl-10-en-1-oic acid, u9
Undecylenic aldehyde, u7
Undecylic aldehyde, u1

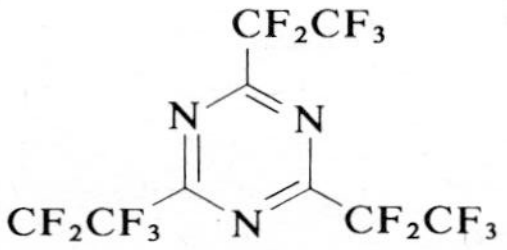

t439

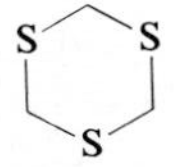

t440

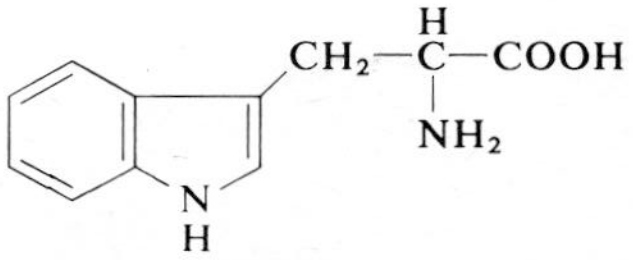

t443

TABLE 1-14 Physical constants of organic compounds (*continued*)

No.	Name	Formula	Formula weight	Beilstein reference	Density	Refractive index	Melting point	Boiling point	Flash point	Solubility in 100 parts solvent
u9	10-Undecenoic acid	$H_2C{=}CH(CH_2)_8COOH$	184.28	2, 458	0.907_4^{24}	1.4493^{20}	24.5	137^{2mm}	148	s alc, chl, eth; i aq
u10	10-Undecen-1-ol	$H_2C{=}CH(CH_2)_9OH$	170.30	1.452	0.850^{15}	1.4500^{20}	−2	245	93	
u11	10-Undecenoyl chloride	$H_2C{=}CH(CH_2)_8COCl$	202.73	2, 459	0.944	1.4532^{20}		122^{10mm}	93	
u12	Urea	$(H_2N)_2CO$	60.06	3, 42	1.32_4^{18}		132.7	d > mp		100 aq; 20 alc
u13	Uric acid		168.11	26, 513	1.893^{20}		>300	d		s alk; i aq, alc, eth
u14	Uridine		244.20	31, 23			165			s aq; hot alc, pyr
v1	L-Valine	$(CH_3)_2CH$-$CH(NH_2)COOH$	117.15	4, 427	1.230		315	subl		8.8 aq; v sl s alc, eth
v2	Vinyl acetate	$H_2C{=}CHOOCCH_3$	86.09	2^1, 63	0.9318_4^{20}	1.3959^{20}	−92.8	72.5	−6	2 aq; misc alc, eth
v3	5-Vinylbicyclo-[2.2.1]-2-heptene		120.19		0.84	1.4802	−80	141		
v4	Vinyl crotonate	$CH_3CH{=}CHCOOCH{=}CH_2$	112.13	2^3, 1263	0.940	1.4488^{20}		50^{10mm}	27	
v5	Vinylcyclohexane	$C_6H_{11}CH{=}CH_2$	110.20	5^1, 35		1.4463^{20}		128	21	
v6	4-Vinyl-1-cyclohexene		108.18	5^1, 63	0.830_4^{20}	1.4640^{20}	−101	126–127	20	
v7	1-Vinylimidazole		94.12	23^4, 569	1.039	1.5308^{20}		78–79^{13mm}	81	
v8	5-Vinyl-2-norbornene		120.20		0.841	1.4802^{20}	−80	141	27	
v9	2-Vinylpyridine	$(C_5H_4N)CH{=}CH_2$	105.14	20, 256	0.975	1.5490^{20}		158–159	43	v s alc, chl, eth
v10	4-Vinylpyridine	$(C_5H_4N)CH{=}CH_2$	105.14	20^2, 170	0.975	1.5500^{20}		65^{15mm}	51	sl s hot aq, hot alc
v11	*N*-Vinyl-2-pyrrolidinone		111.14		0.980	1.5120^{20}		93^{13mm}	93	
x1	Xanthene		182.22	17, 73			101	310–312		s bz, eth; sl s alc, aq
x2	Xanthen-9-carboxylic acid		226.23	12^2, 279			217 d			s hot alc, eth
x3	9-Xanthenone		196.21	17, 354			174	350^{730mm}		0.5 alc; v s chl

Uracil, p272
5-Ureidohydantoin, a77
Urethane, e92
Valeraldehyde, p27
Valeric acid, p36
γ-Valerolactone, p40
Valerone, d533
Valeronitrile, p33
Valeryl chloride, p44
Valinols, a216, a217
Vanillic acid, h133
Vanillin, h132
o-Vanillin, h131
Vanillyl alcohol, h136
Veratraldehyde, d431
Veratric acid, d435
Veratrole, d432
Veronal, d280
Vinylacetic acid, b406
Vinyl bromide, b286
Vinyl 2-butenoate, v4
Vinyl chloride, c110
Vinylidene chloride, d178
Vinyltrimethylsilane, t387
Vinyltris(2-methoxyethoxy)silane, t437
Vitamin B_1, t141
Vitamin B_2, r4
Vitamin C, a317
Xanthone, x3

u13

u14

v3

v6

v7

v8

v11

x1

x2

x3

TABLE 1-14 Physical constants of organic compounds (*continued*)

No.	Name	Formula	Formula weight	Beilstein reference	Density	Refractive index	Melting point	Boiling point	Flash point	Solubility in 100 parts solvent
x4	*o*-Xylene	$C_6H_4(CH_3)_2$	106.17	5, 362	0.8802^{20}_{4}	1.5054^{20}	−25.2	144.4	32	misc alc, eth; 0.017 aq
x5	*m*-Xylene	$C_6H_4(CH_3)_2$	106.17	5, 370	0.8684^{15}_{4}	1.4972^{20}	−47.9	139.1	25	misc alc, eth; 0.02 aq
x6	*p*-Xylene	$C_6H_4(CH_3)_2$	106.17	5, 382	0.8611^{20}_{4}	1.4958^{20}	13.3	138.4	30	v s eth; s alc; 0.02 aq
x7	Xylitol	$HOCH_2(CHOH)_3CH_2OH$	152.15	1, 531			95–97			s aq
x8	D-(+)-Xylose		150.13	31, 47	1.535^{0}		144–145			117 aq; s hot alc, pyr
x9	*m*-Xylylenediamine	$C_6H_4(CH_2NH_2)_2$	136.20	13, 186	1.032	1.5709^{20}		265^{745mm}	>112	

Xylene-α,α'-diol, b18
Xylenols, d581, d582, d583, d584, d585, d586
o-Xylyl bromide, b371
Xylyl chlorides, c259, c260, c261
p-Xylylene glycol, b18

x8

SECTION 2
INORGANIC COMPOUNDS

TABLE 2-1 Physical constants of inorganic compounds

Names, while following the IUPAC nomenclature, are often alphabetized by the central atom to facilitate their location. For example, an entry such as **Aluminum** (tetra-) carbide, tri- would be for the compound tetraaluminum tricarbide. Solvates are listed under the entry for the anhydrous salt. Acid salts are entered as hydrogen.

Formula Weights are based on the International Atomic Weights of 1973 and are computed to the nearest hundredth.

Refractive Index, unless otherwise specified, is given for the sodium line at 589.6 nm.

Density values are given at room temperature unless otherwise indicated by the superscript figure; thus, 2.487^{15} indicates a density of 2.487 for the substance at 15°C. For gases the values are given as grams per liter ($g \cdot L^{-1}$).

Melting Point is recorded in certain cases as 250 d and in some other cases d 250, the distinction being made in this manner to indicate that the former is a melting point with decomposition at 250°C, while in the latter decomposition only occurs at 250°C and higher temperatures. Where a value such as $-6H_2O$, 150 is given, it indicates a loss of 6 mol of water per formula weight of the compound at a temperature of 150°C.

Boiling Point is given at atmospheric pressure (760 mm of mercury) unless otherwise indicated; thus 82^{15mm} indicates that the boiling point is 82°C when the pressure is 15 mmHg. Also, subl 550 indicates that the compound sublines at 550°C.

Solubility is given in parts by weight (of the formula weight) per 100 parts by weight of the solvent (water unless otherwise specified) and at room temperature. Other temperatures are indicated by superscript. The symbols of the common mineral acids represent aqueous solutions of these acids.

Abbreviations Used in the Table

a, acid
abs, absolute
acet, acetone
alc, alcohol
alk, alkali (aq NaOH)
anhyd, anhydrous
aq, aqueous
aq reg, aqua regia
atm, atmosphere
bz, benzene
c, solid state
ca, approximately
chl, chloroform
conc, concentrated
cub, cubic
d, decompose(s)
dil, dilute
DMF, dimethylformamide
eth, diethyl ether
EtOH, ethanol
expl, explodes, explosive
fcc, face-centered cubic
g, gas
glyc, glycerol
h, hot
hex, hexagonal
hyd, hydrolysis
i, insoluble
ign, ignites
lq, liquid
MeOH, methanol
min, mineral
misc, miscible
org, organic
PE, petroleum ether
pyr, pyridine
s, soluble
satd, saturated
sl, slightly
soln, solution
solv, solvent(s)
subl, sublimes
tetr, tetragonal
THF, tetrahydrofuran
tr, transition
v, very
vac, vacuo or vacuum
viol, violently
>, greater than
α, alpha position

TABLE 2-1 Physical constants of inorganic compounds

Name	Formula	Formula weight	Density	Melting point, °C	Boiling point, °C	Solubility in 100 parts solvent
Aluminum	Al	26.98	2.70	660.1	2450	s HCl, H_2SO_4, alk
acetylacetonate	$Al(C_5H_7O_2)_3$	324.31	1.27	subl 193 (vac)	314	i aq; v s alc; s bz, eth
ammonium bis(sulfate) 12-water	$AlNH_4(SO_4)_2 \cdot 12H_2O$	453.33	1.64	−12 H_2O, 250	d >280	15 aq; i alc
bis(acetylsalicylate)	$Al(OOCC_6H_4O\text{-}COCH_3)_2OH$	402.30				v sl s aq, alc, eth
bromide	$AlBr_3$	266.71	2.64^{10}	97.5	253.3	d viol aq; s alc, acet, bz, CS_2
butoxide, *sec-*	$Al(C_4H_9O)_3$	246.33	0.967		$200–206^{30mm}$	v s org solv (flash point 27°C)
butoxide, *tert-*	$Al(C_4H_9O)_3$	246.33	1.025_0^{20}	subl 180		v s org solv
(tetra-) carbide, tri-	Al_4C_3	143.96	2.36	2100	d >2200	d to CH_4 in aq (fire hazard)
chlorate	$Al(ClO_3)_3$	277.35				v s aq; s alc
chloride	$AlCl_3$	133.34	2.44	$194^{2.5 atm}$	subl 181	70 aq (viol); 100^{12} abs alc; s CCl_4, eth; sl s bz
chloride 6-water	$AlCl_3 \cdot 6H_2O$	241.43	2.40	d 100		83^{20} aq; 25 abs alc; s eth
ethoxide	$Al(C_2H_5O)_3$	162.14	1.142_0^{20}	134	205^{14mm}	s hot aq (d); v sl a alc, eth
fluoride	AlF_3	83.98	2.882_4^{25}	1040	subl 1276	0.56^{25} aq; i a, alk, alc, acet
hydroxide	$Al(OH)_3$	78.00	2.42	−H_2O, 300		i aq; s a, alk
iodide	AlI_3	407.71	3.98^{25}	191	360	s aq(d); s alc, CS_2, eth
isopropoxide	$Al(C_3H_7O)_3$	204.25	1.0346_0^{20}	118.5	135^{10mm}	d aq; s alc, bz, chl, PE
nitrate 9-water	$Al(NO_3)_3 \cdot 9H_2O$	375.13		73	d 135	64^{25} aq; 100 alc; s acet
oxide	Al_2O_3	101.96	3.965	2054	2980	i aq; v sl s a, alk
phenoxide	$Al(C_6H_5O)_3$	306.27	1.23	d 265		d aq; s alc, chl, eth
potassium bis(sulfate 12-water	$AlK(SO_4)_2 \cdot 12H_2O$	474.39	1.757^{20}	−9H_2O, 92	−12H_2O, 200	11.4^{20} aq
propoxide	$Al(C_3H_7O)_3$	204.25	1.0578_0^{20}	106	248^{14mm}	d aq; s alc
sodium bis(sulfate) 12-water	$AlNa(SO_4)_2 \cdot 12H_2O$	458.28	1.675^{20}	61		110^{15} aq

TABLE 2-1 Physical constants of inorganic compounds (*continued*)

Name	Formula	Formula weight	Density	Melting point, °C	Boiling point, °C	Solubility in 100 parts solvent
Aluminum						
stearate	$Al(C_{18}H_{35}O_2)_3$	877.42	1.010	103		i aq; s alc, bz, alk
sulfate	$Al_2(SO_4)_3$	342.15	2.710	d 770		36.4^{20} aq; sl s alc
sulfate 18-water	$Al_2(SO_4)_3 \cdot 18H_2O$	666.45	1.69^{17}	d 86.5		87^{0} aq; i alc
tetrahydroborate	$Al(BH_4)_3$	71.53		−64.5	44.5	d aq
Amidosulfuric acid	H_2NSO_3H	97.09	2.126	205	d	14.7 aq
Ammonia	NH_3	17.03	0.7188^{20} $g \cdot L^{-1}$	−77.75	−33.42	89.9 aq; 13.2^{20} alc; s eth, org solv
$-d_3$ or [^{2}H]	ND_3 or N^2H_3	20.05	0.8437^{20} $g \cdot L^{-1}$	−74.33	−31.05	
Ammonium						
acetate	$NH_4C_2H_3O_2$	77.08	1.17^{20}	114	d	148^{4} aq; 7.9^{15} MeOH; s alc
benzoate	$NH_4C_7H_5O_2$	139.16	1.260	d 198	subl 160	20^{15} aq; 2.8 alc; s glyc; i eth
boranate, tetrafluoro-	NH_4BF_4	104.84	1.87^{15}	subl		25^{16} aq
bromide	NH_4Br	97.95	2.429	452 (under pressure)	d 397 (vac)	76^{20} aq; s acet, alc, eth
carbamate	NH_4COONH_2	78.07		subl 60		v s aq; sl s alc; i eth
carbonate 1-water	$(NH_4)_2CO_3 \cdot H_2O$	114.10		d 20		100^{15} aq; i alc
cerate(IV), hexanitrato-	$(NH_4)_2[Ce(NO_3)_6]$	548.23				135^{20} aq; s alc, HNO_3
chloride	NH_4Cl	53.49	1.527	subl 340		26^{15} aq; 0.6^{19} abs alc; i acet, eth
chromate	$(NH_4)_2CrO_4$	152.08	1.91^{12}	d 180		34^{20} aq; sl s MeOH, acet; i alc
chromium(III) bis(sulfate) 12-water	$NH_4Cr(SO_4)_2 \cdot 12H_2O$	478.34	1.72	94		7.2^{0} aq
citrate	$(NH_4)_3C_6H_5O_7$	243.22	1.48	d		100 aq; sl s alc
copper(II) tetrachloride 2-hydrate	$Cu(NH_4)_2Cl_4 \cdot 2H_2O$	277.46	1.993	$-2H_2O$, 110	d >120	40^{20} aq; s alc
dichromate(VI)	$(NH_4)_2Cr_2O_7$	252.06	2.155_4^{25}	d 170		36^{20} aq; s alc (flammable)
dithiocarbamate	$NH_4S—CS—NH_2$	110.19	1.451_4^{20}	d 99		v s aq; s alc; sl s eth

diuranate(VI)	$(NH_4)_2U_2O_7$	624.22				v sl s aq, alk; s acids
fluoride	NH_4F	37.04	1.009^{25}	subl		100^{0} aq; s alc
formate	NH_4OOCH	63.06	1.280	116	d 180	143^{20} aq; s alc, eth
hexadecanoate	$NH_4OOC(CH_2)_{14}CH_3$	273.45		21–22		s aq; sl s bz; i alc, acet
hexafluoroaluminate	$(NH_4)_3AlF_6$	195.10	1.78	d >100		v s aq
hydrogen carbonate	NH_4HCO_3	79.06	1.58	d 35	subl	22^{20} aq; i alc, acet
hydrogen citrate	$(NH_4)_2HC_6H_5O_7$	226.19	1.48			100 aq; sl s alc
hydrogen difluoride	NH_4HF_2	57.04	1.50	125.6		v s aq; sl s alc
hydrogen oxalate 1-water	$NH_4HC_2O_4 \cdot H_2O$	125.08	1.556	$-H_2O$, 170		s aq; i bz, eth
hydrogen phosphate	$(NH_4)_2HPO_4$	132.05	1.619	d 155		69^{20} aq; i alc, acet
hydrogen phosphate, di-	$NH_4H_2PO_4$	115.03	1.803^{19}	d 190		37^{20} aq; sl s alc; i acet
hydrogen sulfate	NH_4HSO_4	115.11	1.78	146.9	d 350	100 aq; i alc, acet
hydrogen sulfide	NH_4HS	51.11	1.17	d 25		128^{0} aq; s alc; sl s acet; i bz
hydrogen sulfite	NH_4HSO_3	99.10	2.03	subl 150 (in N_2)		72^{0} aq
hydroxide	NH_4OH	35.05		−77		misc aq
iodide	NH_4I	144.95	2.514^{25}	subl 551	220 (vac)	172^{20} aq; v s alc, acet
iron(II) bis(sulfate) 6-water	$Fe(NH_4)_2(SO_4)_2 \cdot 6H_2O$	392.14	1.864_4^{20}	d 100		36^{20} aq; i alc
molybdate(VI)(6-) 4-water, hepta-	$(NH_4)_6Mo_7O_{24} \cdot 4H_2O$	1235.86	2.498	$-H_2O$, 90	d 190	43 aq; s a; i alc
nitrate	NH_4NO_3	80.04	1.725^{25}	169.6	210^{11mm}	192^{20} aq; 3.8^{20} alc; 17^{20} MeOH
octadecanoate	$NH_4OOC(CH_2)_{16}CH_3$	301.50		21–22		sl a aq; s alc; i acet
octanoate	$NH_4OOCC_7H_{15}$	161.24		d on standing		v s aq, alc, acet; sl s eth
oxalate 1-water	$(NH_4)_2C_2O_4 \cdot H_2O$	142.11	1.50	d 70		5.1^{20} aq
palladate(II) tetrachloro-	$(NH_4)_2PdCl_4$	284.29	2.170	d		v s aq; i abs alc
perchlorate	NH_4ClO_4	117.50	1.95	d 240		22^{20} aq; s MeOH; sl s alc, acet
peroxodisulfate	$(NH_4)_2S_2O_8$	228.18	1.982	d 120	expl 180	58^{0} aq
phosphate, hexafluoro-	NH_4PF_6	163.00	2.180_4^{12}	d		75^{20} aq; s alc, acet
phosphinate	$NH_4PH_2O_2$	83.03	1.634	200	d 240	100 aq; 5 alc; i acet
picrate	$NH_4C_6H_2N_3O_7$	246.14	1.719	d	expl 423	1.1^{20} aq; sl s alc
platinate(IV), hexachloro-	$(NH_4)_2PtCl_6$	443.89	3.065	d		0.5^{20} aq
silicate, hexafluoro-	$(NH_4)_2SiF_6$	178.14	2.011	d		18.6^{20} aq; i alc, acet

TABLE 2-1 Physical constants of inorganic compounds (*continued*)

Name	Formula	Formula weight	Density	Melting point, °C	Boiling point, °C	Solubility in 100 parts solvent
Ammonium						
sulfamate	$NH_4SO_3NH_2$	114.13		131	d 160	v s aq; sl s alc
sulfate	$(NH_4)_2SO_4$	132.14	1.769^{20}	d >280		43.5^{25} aq; i alc, acet
sulfide	$(NH_4)_2S$	68.14		d		v s aq; s alc
DL-tartrate	$(NH_4)_2C_4H_4O_6$	184.15	1.601	d		58^{15} aq; sl s alc
tetraborate 4-water	$(NH_4)_2B_4O_7{\cdot}4H_2O$	263.44				s aq; i alc
thiocyanate	NH_4SCN	76.12	1.305	149.6	d 170	128^0 aq; v s alc; s acet
thiosulfate	$(NH_4)_2S_2O_3$	148.20	1.679	d 150		v s aq
vanadate(V)(1-)	NH_4VO_3	116.98	2.326	d 200		0.48^{20} aq
Antimony						
(III) chloride	$SbCl_3$	228.11	3.14_4^{20}	73.4	223.5	10^{20} aq; s alc, bz, chl
(V) chloride	$SbCl_5$	299.02	2.336_4^{20}	3.5	140	d aq; s HCl, chl, CCl_4
(III) fluoride	SbF_3	178.75	4.379_{20}^{20}	292	376	444^{20} aq
(V) fluoride	SbF_5	216.74	2.99^{23}	8.3	141	d viol aq; s HOAc; forms solids with alc, bz, CS_2, eth
hydride	SbH_3	124.77	4.36^{15}	−91.5	−18.4	20^0 mL aq; s CS_2
(III) oxide	Sb_2O_3	291.50	5.2	655	1425	v sl a aq; s HCl, KOH
(V) oxide	Sb_2O_5	323.50	2.78	$-O_2$, >300		v sl s aq; sl s warm KOH, eth
potassium oxide tartrate 0.5-water	$K(SbO)C_4H_4O_6{\cdot}0.5H_2O$	333.93	2.607	d 100		8.3^{20} aq; 6.7 glyc; i alc
(III) sulfide	Sb_2S_3	339.69	4.64	546		0.002^{20} aq d; s H_2SO_4
(V) sulfide	Sb_2S_5	403.82				i aq; s HCl d, NaOH
Argon	Ar	39.95	1.7824 $g \cdot L^{-1}$	−189.38	−185.87	3.36^{20} mL aq
Arsenic	As	74.92	5.72	$817^{28\,atm}$	subl 612	i aq; s HNO_3
(III) chloride	$AsCl_3$	181.28	2.1497_4^{25}	−16	130.2	d aq; misc chl, CCl_4, eth; s alc
(III) oxide dimer	As_4O_6	395.68	4.15	313	465	1.8^{20} aq; s alc

(V) oxide	As_2O_5	229.84	4.32	d 800		66^{20} aq; s alc
(III) sulfide	As_2S_3	246.04	3.46	300–325	707	i aq; s alk; slowly s hot HCl
Barium						
acetate 1-water	$Ba(C_2H_3O_2)_2 \cdot H_2O$	273.46	2.19	d 150		76^{20} aq; 0.14 alc
benzenesulfonate	$Ba(O_3SC_6H_5)_2$	451.70				s aq; sl s alc
carbonate	$BaCO_3$	197.35	4.43	d 1360		0.002 aq; s a
chlorate 1-water	$Ba(ClO_3)_2 \cdot H_2O$	322.26	3.18	$-H_2O$, 120	$-O_2$, 250	34^{20} aq
chloride	$BaCl_2$	208.25	3.856	962	2029	36^{20} aq
fluoride	BaF_2	175.34	4.89	1368	2272	0.16^{20} aq
hydrogen phosphate	$BaHPO_4$	233.32	4.165^{15}	d 410		0.01 aq; s a
hydroxide 8-water	$Ba(OH)_2 \cdot 8H_2O$	315.48	2.18^{16}	78		3.9^{20} aq
manganate(VI)(2-)	$BaMnO_4$	256.28	4.85			v sl s aq
nitrate	$Ba(NO_3)_2$	261.35	3.24	575	d	9^{20} aq
nitrite 1-water	$Ba(NO_2)_2 \cdot H_2O$	247.37	3.173^{20}	d 115		73^{20} aq; i alc
oxide	BaO	153.34	5.72	2013	3088	3.5^{20} aq
perchlorate 3-water	$Ba(ClO_4)_2 \cdot 3H_2O$	390.29	2.74	d 400		198^{25} aq; s MeOH; sl s alc, acet
permanganate	$Ba(MnO_4)_2$	375.21	3.77	d 200		62^{11} aq
peroxide	BaO_2	169.34	4.96	450	$-O_2$, 800	1.5^{0} aq
sulfate	$BaSO_4$	233.40	4.50^{15}	1580		0.0002 aq
sulfide	BaS	169.40	4.25^{15}	2227		7.9^{22} aq d
sulfite	$BaSO_3$	217.40		d		0.02^{20} aq
thiocyanate 2-water	$Ba(SCN)_2 \cdot 2H_2O$	289.53	2.286^{18}	d 160		170^{25} aq
thiosulfate 1-water	$BaS_2O_3 \cdot H_2O$	267.48	3.5^{18}	d 220		0.21^{20} aq
Beryllium	Be	9.01	1.86	1277	2484	i aq; s a, alk
bromide	$BeBr_2$	168.83	3.465^{25}	506–509	521	v s aq; s alc; 19 pyr
chloride	$BeCl_2$	79.92	1.899^{25}	399	482	42 aq; s alc, eth, CS_2, pyr; i bz
fluoride	BeF_2	47.01	1.986_4^{25}	552	1175	v s aq but slow
hydride	BeH_2	11.03		$-H_2$, 220		d slowly aq; d rapidly a
hydroxide	$Be(OH)_2$	43.03	1.92	134 d		s hot conc a, alk
iodide	BeI_2	262.82	4.2	480	482	hyd aq; s alc, eth, CS_2
oxide	BeO	25.01	3.01	2408(α)	3787	s conc H_2SO_4
sulfate 4-water	$BeSO_4 \cdot 4H_2O$	177.14	1.713^{11}	$-4H_2O$, 270	d 580	39^{20} aq; i alc

TABLE 2-1 Physical constants of inorganic compounds (*continued*)

Name	Formula	Formula weight	Density	Melting point, °C	Boiling point, °C	Solubility in 100 parts solvent
Bismuth						
chloride, tri-	$BiCl_3$	315.34	4.75	ca 232	447	d aq; s HCl, alc, eth, acet
fluoride, penta-	BiF_5	303.98	5.4^{25}	151	230	d viol aq giving O_3
hydroxide	$Bi(OH)_3$	260.00	4.36	$-H_2O$, 100		d aq; s a
(III) nitrate 5-water	$Bi(NO_3)_3 \cdot 5H_2O$	485.07	2.83	d 30		d aq; s, acet
(III) oxide	Bi_2O_3	495.96	8.76	817	1890	i aq; s a
Boron						
bromide, tri-	BBr_3	250.57	2.695^0	−46.0	91.3	d aq
choride, tri-	BCl_3	117.19	1.35_4^{12}	−107	18	d aq, alc
fluoride, tri-	BF_3	67.81	$2.99\ g \cdot L^{-1}$	−127.1	−100.4	105^0 mL aq; s bz, chl, CCl_4
fluoride-1-diethyl ether	$BF_3 \cdot O(C_2H_5)_2$	141.94	1.125_4^{25}	−60.4	125.7	d aq
fluoride-1-methanol	$BF_3 \cdot CH_3OH$	131.89	1.203		59^{4mm}	
oxide	B_2O_3	69.62	2.46	450	2065	sl a aq
Bromine	Br_2	159.81	3.1028	−7.3	58.75	3.6^{20} aq; v s alc, chl, eth, CS_2
fluoride, tri-	BrF_3	136.90	2.803^{25}	8.77	125.74	d viol aq; d alk
Cadmium						
acetate	$Cd(C_2H_3O_2)_2$	230.50	2.341	256	d	v s aq
chloride	$CdCl_2$	183.32	4.047	568	961	120^{25} aq
iodide	CdI_2	366.21	5.670^{30}	387	796	85^{20} aq; s alc, acet, eth
oxide	CdO	128.40	8.15	subl 1497		i aq; s a
sulfate-water (3/8)	$3CdSO_4 \cdot 8H_2O$	769.56	3.09	$-H_2O$, 40	forms monohydrate 80	94.4^{25}; i alc
sulfide	CdS	144.46	4.82 hex		sub 1380 (in N_2)	3.13^{18} aq; s a
Calcium						
acetate	$Ca(C_2H_3O_2)_2$	158.17	d >160			37^0 aq; i alc, acet, bz
arsenate(V)	$Ca_3(AsO_4)_2$	398.08	3.620			0.013^{25} aq

bromide	$CaBr_2$	199.90	3.353	765	806–812	143^{20} aq; v s alc, acet
carbide, di-	CaC_2	64.10	2.22		2300	d aq giving C_2H_2
carbonate	$CaCO_3$	100.09	2.930	d 900		0.0013^{20}; s a
chlorate	$Ca(ClO_3)_2$	206.99		340		178 aq; s alc, acet
chloride	$CaCl_2$	110.99	2.15	772	1940	75^{20}; s alc, acet
chloride 6-water	$CaCl_2{\cdot}6H_2O$	219.08	1.71	$-6H_2O$, 200		536^{20} aq; s alc
citrate 4-water	$Ca(C_6H_6O_7){\cdot}4H_2O$	570.51		$-4H_2O$, 120		0.85^{18} aq; 0.0065^{18} alc
cyanamide	$CaCN_2$	80.11	2.29^{20}_{4}	1340	subl 1150	i aq; no known solv
cyanide	$Ca(CN)_2$	92.12		d 350		d aq
diphosphate	$Ca_2P_2O_7$	254.10	3.09	1230		i aq; s a
fluoride	CaF_2	78.08	3.180	1418	2510	0.002^{20} aq; sl s a
formate	$Ca(OOCH)_2$	130.12	2.015	d		16.6^{20} aq; i alc
glycerophosphate	$Ca[C_3H_5(OH)_2]PO_4$	210.16		d >170		1.7^{20} aq; i alc
hydrogen phosphate, di- 1-water	$Ca(H_2PO_4)_2{\cdot}H_2O$	252.07	2.220^{18}_{4}	$-H_2O$, 109	d 203	1.8^{30} aq
hydroxide	$Ca(OH)_2$	74.09	2.24	$-H_2O$, 522		0.17^{10} aq; s a
hypochlorite	$Ca(OCl)_2$	142.99	2.35	100 d		d aq evolving Cl_2; i alc
iodate 6-water	$Ca(IO_3)_2{\cdot}6H_2O$	497.98		d 35		0.24^{20} aq; i alc
lactate 5-water	$Ca(C_3H_5O_3)_2{\cdot}5H_2O$	308.30		$-3H_2O$, 100	$-5H_2O$, 120	5.4^{15} aq; v sl s alc
nitrate	$Ca(NO_3)_2$	164.09	2.504^{18}	561		152^{30} aq
nitrite 4-water	$Ca(NO_2)_2{\cdot}4H_2O$	204.15	1.674^{0}_{0}	$-2H_2O$, 44		84.5^{18} aq; sl s alc
oleate	$Ca(C_{18}H_{33}O_2)_2$	603.01		83–84	d 140	0.04 aq; s bz, chl; v sl s alc
oxide	CaO	56.08	3.25	2927	3500	0.13^{25} aq; s a
palmitate	$Ca(C_{16}H_{31}O_2)_2$	550.93		d 155		0.003 aq; sl s bz, chl; i alc, eth
pantothenate (vitamin B_3)	$Ca[O_2CH_2CH_2BHO$-$CH(OH)C(CH_3)_2$-$CH_2OH]_2$	476.55		d 195–196		35 aq; sl s alc, acet
peroxide	CaO_2	72.08	2.92^{25}_{4}	d 275		sl s aq; s a
phenoxide	$Ca(OC_6H_5)_2$	226.28				sl s aq, alc
phosphate	$Ca_3(PO_4)_2$	310.18	3.14	1730		0.03^{25}; s a; i alc
salicylate 2-water	$Ca(C_7H_5O_3)_2{\cdot}2H_2O$	350.34		$-2H_2O$, 120	d 240	2.8^{15} aq; 0.015^{16} EtOH
selenate 2-water	$CaSeO_4{\cdot}2H_2O$	219.07	2.68^{20}_{4}	$-2H_2O$, 200	d 698	9.2^{25} aq
stearate	$Ca(C_{18}H_{35}O_2)_2$	607.04		179–180		0.004^{15} aq; s hot pyr; i chl, eth

TABLE 2-1 Physical constants of inorganic compounds (*continued*)

Name	Formula	Formula weight	Density	Melting point, °C	Boiling point, °C	Solubility in 100 parts solvent
Calcium						
succinate 3-water	$CaC_4H_6O_4 \cdot 3H_2O$	212.22				1.28^{20} aq; s a; i alc
sulfate	$CaSO_4$	136.14	2.960	1400		0.20 aq; s a
sulfate hemihydrate	$CaSO_4 \cdot 0.5H_2O$	145.15		$-H_2O$, 163		0.3^{20} aq; s a, glyc
sulfate 2-water	$CaSO_4 \cdot 2H_2O$	172.17	2.32	$-2H_2O$, 163		0.26^{20} aq; s a, glyc
sulfite 2-water	$CaSO_3 \cdot 2H_2O$	156.17		$-2H_2O$, 100		0.004 aq; s a; sl s alc
DL-tartrate 4-water	$CaC_4H_4O_6 \cdot 4H_2O$	260.21		$-4H_2O$, 200		0.0045^{25} aq; sl s alc
tetrahydridoaluminate	$Ca(AlH_4)_2$	102.10				ign moist air; d viol aq, alc
thiocyanate 3-water	$Ca(SCN)_2 \cdot 3H_2O$	210.29		d 160		150 aq; v s alc
Carbon						
(graphite)	C	12.01	2.25^{20}	$4000^{63.5\,atm}$	3930	i aq, alc
bromide, tetra-	CBr_4	331.65	3.42	90.1	190	i aq; s alc, chl, eth
chloride, tetra-	CCl_4	153.82	1.5867^{20}_{20}	−22.9	76.7	i aq; s alc, chl, eth
hydride, tetra-	CH_4	16.04	0.415^{-164}	−182.48	−161.49	i aq; s bz
iodide, tetra-	CI_4	519.63	4.34	d 171		sl hyd aq; s alc, bz, eth
oxide, mon-	CO	28.01	0.793 (lq) 1.250 $g \cdot L^{-1}$ (gas)	−205.05	−191.49	2.1 mL aq; a alc, bz
oxide di-	CO_2	44.01	1.56^{-79} (c) 1.975 $g \cdot L^{-1}$	−56.2 solid subl	−78.44	31^{15} mL aq
(tri-) oxide, di-	C_3O_2	68.03	1.114^{0}_{4}	−112.19	6.4	d aq to malonic acid
selenide, di-	$CaSe_2$	169.93	2.663^{25}_{4}	−43	125.1	i aq; d alc, pyr; misc CCl_4; s acet, eth
sulfide, di-	CS_2	76.14	1.261^{22}	−111.6	46.26	0.29^{20} aq; s alc, eth

Carbonic acid	$H_2CO_3(CO_2+H_2O)$	62.03				known in soln only
Carbonyl						
chloride	$COCl_2$	98.92	1.392	−127.8	7.6	hyd aq; s bz
fluoride	COF_2	66.01	1.139^{-114}	−114.0	−83.3	hyd aq
sulfide	COS	60.07	1.073^{0} $g \cdot L^{-1}$	−138.81	−50.23	54^{20} mL aq; s alc, CS_2
Cerium						
(III) chloride	$CeCl_3$	246.48	3.92	8.7	1730	100^{20} aq; 30 alc; s acet
(IV) fluoride	CeF_4	216.12	4.80	>650	d >550	i aq; s a
(IV) oxide	CeO_2	172.13				i aq; s a
(IV) sulfate	$Ce(SO_4)_2$	332.24	3.91	d 195		hyd aq; s H_2SO_4
Cesium						
bromide	$CsBr$	212.81	4.44	635	1300	107^{18} aq
carbonate	Cs_2CO_3	325.82		d 610		260^{15} aq; 11^{20} alc; s eth
chloride	$CsCl$	168.36	3.988	645	1324	187^{20} aq; 34^{25} MeOH; v s alc
fluoride	CsF	151.90	4.115	703	1231	322^{18} aq
hydroxide	$CsOH$	149.91	3.675	272	990	386^{15} aq; s alc
iodide	CsI	259.81	4.510	621	ca 1280	77^{20} aq; s EtOH; i acet
nitrate	$CsNO_3$	194.91	3.685_4^{20}	414	d 849	23^{20} aq; s acet; v sl s alc
oxalate	$Cs_2C_2O_4$	353.82	3.230^{15}			313 aq
selenate	Cs_2SeO_4	408.77	4.4528_4^{20}			244^{12} aq
sulfate	Cs_2SO_4	361.87	4.243	1019		179^{20} aq; i alc, acet, pyr
Chlorine						
fluoride, tri-	ClF_3	92.45	1.825^{11}	−76.28	11.74	hyd viol aq; glass wool and org matter ign
(di-) oxide	Cl_2O	86.91	3.02^{2}	−120.6	2.1	3.5^{20} aq (hyd to HClO); s CCl_4
oxide, di-	ClO_2	67.46	1.642^{0}	−59.6	10.9	11.2^{10} aq
(di-) oxide, hepta-	Cl_2O_7	182.90	1.805^{25}	−91.5	83.6	d aq; expl on concussion or contact with flame or I_2
Chlorosulfonic acid	HSO_3Cl	116.52	1.753_4^{20}	−80	158	d viol aq to $HCl+H_2SO_4$
Chromium						
(II) acetate	$Cr(C_2H_3O_2)_2$	170.10				sl s aq, alc; i eth
carbonyl, hexa-	$Cr(CO)_6$	220.06	1.77^{18}	d 130	expl 210	i aq, alc, eth

TABLE 2-1 Physical constants of inorganic compounds (*continued*)

Name	Formula	Formula weight	Density	Melting point, °C	Boiling point, °C	Solubility in 100 parts solvent
Chromium						
(II) chloride	$CrCl_2$	122.90	2.878	815	1300	v s aq
(III) chloride	$CrCl_3$	158.35	2.76^{15}	877	subl 947	i aq, alc, acet, eth
(III) fluoride	CrF_3	108.99	3.8	1100	subl	i aq, alc; s HF
(III) nitrate 9-water	$Cr(NO_3)_3 \cdot 9H_2O$	400.15		60	d 100	208^{15} aq; s alc
(III) oxide	Cr_2O_3	152.02	5.21	2330	3000	i aq, alc
(IV) oxide	CrO_2	83.99	4.89	$-O_2$, 300		i aq; s HNO_3
(VI) oxide	CrO_3	99.99	2.70	198	d 250	167^{20} aq; may ign org materials
(III) phosphate 6-water	$CrPO_4 \cdot 6H_2O$	255.06	2.121^{14}	100		i aq; v s a, alk; sl s HOAc
(II) sulfate 7-water	$CrSO_4 \cdot 7H_2O$	274.17				23^{0} aq
(III) sulfate 18-water	$Cr_2(SO_4)_3 \cdot 18H_2O$	716.45	1.7	d 100		220^{20} aq
Chromyl						
chloride	CrO_2Cl_2	154.90	1.92	−96.5	117	d aq; s eth
Cobalt						
(II) acetate 4-water	$Co(C_2H_3O_2)_2 \cdot 4H_2O$	249.08	1.705^{19}	$-4H_2O$, 140		s aq; 2.1^{15} MeOH
(III) acetate	$Co(C_2H_3O_2)_3$	236.07		d 100		s aq, alc, HOAc
(II) bromide	$CoBr_2$	218.75	4.909_4^{25}	678 (in N_2)		112^{20} aq
(II) carbonate	$CoCO_3$	118.94	4.13	d		0.18^{15} aq; s a
(II) chloride	$CoCl_2$	129.84	3.356	740	1087	53^{20} aq
(II) fluoride	CoF_2	96.93	4.46	1127	1739	1.36^{20} aq; s a
(III) fluoride	CoF_3	115.93	3.88			d aq; i alc, bz, eth
(II) hydroxide	$Co(OH)_2$	92.95	3.597_4^{15}	d		0.0018 aq; s a
(II) iodide	CoI_2	312.74	5.68	505 d	570 (vac)	203 aq
(II nitrate 6-water	$Co(NO_3)_2 \cdot 6H_2O$	291.04	1.87	55	d 74	155^{30} aq; v s alc
(II) oxalate	CoC_2O_4	146.95	3.021_4^{25}	d 250		0.002^{18} aq; s a
(II) oxide	CoO	74.93	6.45	1805		i aq; s a

(II,III) oxide	Co_3O_4	240.80	6.07	d 900		i aq; v sl s a
(II) sulfate 7-water	$CoSO_4 \cdot 7H_2O$	281.10	2.03_4^{25}	96.8	$-7H_2O$, 420	65^{20} aq; sl s alc
Copper						
(II) acetate hydrate	$Cu(C_2H_3O_2)_2 \cdot H_2O$	199.65	1.882	115	d 240	8 aq; 0.48 MeOH; sl s eth, glyc
(II) acetate-metaarsenite	$Cu(C_2H_3O_2)_2 \cdot 3Cu(AsO_2)_2$	1013.77				i aq; s a, NH_4OH
(I) bromide	CuBr	143.45	4.98	488	1318	v sl s aq; s a
(II) bromide	$CuBr_2$	223.31	4.710_4^{20}	498		126 aq; s alc, acet, pyr; i bz, eth
(II) chlorate 6-water	$Cu(ClO_3)_2 \cdot 6H_2O$	338.53		65	d 100	242^{18} aq; v s alc; s acet
(I) chloride	CuCl	98.99	4.14	430	1212	0.024 aq; s HCl
(II) chloride	$CuCl_2$	134.44	3.386_4^{25}	d 300		73^{20} aq; s alc, acet
(II) chloride 2-water	$CuCl_2 \cdot 2H_2O$	170.47	2.54	$-2H_2O$, 100	d	76^{25} aq; v s alc; s acet
(I) cyanide	CuCN	89.56	2.92	473 (in N_2)	d	0.00026 aq; s HCl, KCN
(II) fluoride	CuF_2	101.54	4.23	770	1449	0.075 aq; s a
formate	$Cu(OOCH)_2$	153.55	1.831			12.5 aq
hydroxide	$Cu(OH)_2$	97.55	3.368	160		i aq; s a
(I) iodide	CuI	190.44	5.62	588	1207	i aq; s HCl, KI
(II) nitrate 3-water	$Cu(NO_3)_2 \cdot 3H_2O$	241.60	2.05	114.5	d 170	138^{0} aq; v s alc
(II) oleate	$Cu(OOCC_{17}H_{33})_2$	626.43				i aq; sl s alc; s eth
(I) oxide	Cu_2O	143.08	6.0	1236	$-O_2$, 1800	i aq; s HCl
(II) oxide	CuO	79.54	6.315_4^{14}	d 1122		i aq, alc; s a
(II) perchlorate	$Cu(ClO_4)_2$	262.43	2.225^{23}	82.3	d 130	146^{30} aq; s eth; i bz, CCl_4
(II) stearate	$Cu(OOCC_{17}H_{35})_2$	630.46		ca 250		i aq, alc, eth; s pyr, hot bz
(II) sulfate	$CuSO_4$	159.61	3.603	805 d		14.3^{0} aq
(II) sulfate 5-water	$CuSO_4 \cdot 5H_2O$	249.68	2.284_4^{16}	$-5H_2O$, 150		32^{20}aq; s MeOH, glyc; sl s EtOH
Cyanogen	NC—CN	52.04	$2.335\ g \cdot L^{-1}$	−27.84	−21.15	420^{20} mL aq; 230 mL alc
azide	$NC—N_3$	68.04		detonates		s acetonitrile; can be handled safely only in solv
bromide	CNBr	105.93	2.015_4^{20}	51.4	61.35	v s aq, alc, eth
chloride	CNCl	61.48	1.186	−6.90	13.0	s aq, alc, eth
iodide	CNI	152.92		146–147	subl 140	s aq, alc, eth
Deuterium	D_2 or $^2H_2$	4.03	0.169^{mp} (lq)	−252.89	−248.24	sl s aq
oxide	D_2O or ^2H_2O	20.03	1.1056^{20}	3.82	101.43	misc aq

TABLE 2-1 Physical constants of inorganic compounds (*continued*)

Name	Formula	Formula weight	Density	Melting point, °C	Boiling point, °C	Solubility in 100 parts solvent
Disulfuryl dichloride	$S_2O_5Cl_2$	215.03	1.818^{11}	−37.5	152.5	d aq, a
Diphosphoric(V) acid	$H_4P_2O_7$	117.98		61		s aq
Fluorine	F_2	38.00	1.554^{25} g · L^{-1}	−219.70	−188.20	d aq viol
Fluoroboric acid	HBF_4	87.81		d 130		v s aq
Fluorosulfonic acid	HSO_3F	100.07	1.743^{15}	−87.3	165.5	s aq
-*d* or [^{2}H]	DSO_3F or $^2HSO_3F$	101.08		−89	163	s aq
Germane	GeH_4	76.62	1.523^{-142}	−165.9	−88.5	sl s hot HCl
Gold						
(III) chloride	$AuCl_3$	303.33	3.9	254 d	subl 265	68^{20} aq
Helium	He	4.00	0.1784^{0} g · L^{-1} 0.1249 (lq)	$-272.2^{25\,atm}$	−268.935	0.861^{20} mL aq
Hydrazine	H_2NNH_2	32.05	1.0083^{20}	1.54	113.8	misc aq, alc
hydrate	$H_2NNH_2{\cdot}H_2O$	50.16	1.038^{21}	−51.7	119.4	misc aq, alc
Hydrazinium						
(1+) chloride	H_2NNH_3Cl	68.51		92.6	d 240	v s aq
(2+) chloride	ClH_3NH_3Cl	104.97	1.4226^{20}	198	d 200	v s aq; sl s alc
(2+) sulfate	$(H_3NNH_3)SO_4$	130.13	1.378	254	d	3.4^{20} aq; i alc
Hydrogen	H_2	2.02	0.0899 g · L^{-1}	−259.76	−252.76	1.9 mL aq
azide	HN_3	43.03	1.126	−80	37	v s aq (v expl)
borate(1−)	HBO_2	43.83	2.486	236		v sl s aq
borate(3−), ortho-	H_3BO_3	61.83	1.435^{15}	171.0	d 300	6.4^{30} aq
bromide	HBr	80.92	3.388^{20} g · L^{-1} 2.160^{-66} (lq)	−86.81	−66.71	193^{25} aq; s alc
bromide	48% HBr + H_2O		1.49 g · L^{-1}	−11	126	v s aq (constant boiling)
bromide-*d*	DBr or ^{2}HBr	81.92	3.39^{20} g · L^{-1}	−87.46	−66.5	v s aq

chloride	HCl	36.46	1.526^{20} g · L^{-1} 1.187^{-85} (lq)	−114.18	−85.00	72^{20} aq
chloride	20.24% $HCl + H_2O$		1.097		110	v s aq (constant boiling)
cyanide	HCN	27.06	0.901 g · L^{-1} 1.2675^{10} (lq)	−13.24	25.70	v s aq
fluoride	HF	20.01	0.922^{0} g · L^{-1} 0.957^{19} (lq)	−83.57	19.52	v s aq
fluoride	35.35% $HF + H_2O$				120	v s aq (constant boiling)
iodide	HI	127.92	5.37^{20} g · L^{-1} 2.799^{-35} (lq)	−50.79	−35.35	70^{0} aq
iodide	57% $HI + H_2O$		1.70^{15}		127	v s aq (constant boiling)
nitrate	HNO_3	63.02	1.5027	−41.59	83	v s aq
nitrate	69% $HNO_3 + H_2O$		1.41^{20}		120.5	misc aq (constant boiling)
oxide	H_2O	18.02	1.000^{4}	0.00	100.00	
oxide-d_2	D_2O or ^2H_2O	20.03	1.1045	3.82	101.43	misc aq
perchlorate 2-water	$HClO_4 \cdot 2H_2O$	136.49	1.67^{20}	−17.8	203	v s aq (commercial 72% a)
periodate(1−)	HIO_4	191.91		subl 110	d 138	440^{25} aq
periodate(5−)	H_5IO_6	227.94		130	d 140	113 aq
peroxide	H_2O_2	34.02	1.4649^{0}	−0.40	151.2	misc aq; s alc, eth
phosphate(V)(1−)	HPO_3	79.98	2.2–2.5	subl		s aq
phsophate(V)(3−)	H_3PO_4	98.00	1.88	42.3	d 213	v s aq (commercial 85% a)
phosphide	PH_3	34.00	1.529 g · L^{-1}	−133.81	−87.78	26^{17} mL aq; s alc, eth
selenide	H_2Se	80.98	2.12_4^{-42}	−65.73	−42	9.5^{20} mL aq
sulfide	H_2S	34.08	1.1906 g · L^{-1}	−85.52	−60.33	0.334^{25} mL aq
telluride	H_2Te	129.63	6.234 g · L^{-1}	−49	−2	d aq
tungstate(VI)(2−)	H_2WO_4	249.86	5.5	$-H_2O$, 100		i aq; s alk, HF
Hydroxylamine	$HONH_2$	33.03	1.332	33.1	58^{22mm}	s aq, alc
Hydroxylammonium						
chloride	$HONH_3Cl$	69.49	1.680^{20}	150.5	d	83^{17} aq; 4.4^{20} alc
sulfate	$(HONH_3)_2SO_4$	164.14		d 170		69^{20} aq

TABLE 2-1 Physical constants of inorganic compounds (*continued*)

Name	Formula	Formula weight	Density	Melting point, °C	Boiling point, °C	Solubility in 100 parts solvent
Iodic acid	HIO_3	175.91	4.629^{0}	d 110 to H_5IO_6	d 195 to I_2O_5	310^{16} aq
Iodine	I_2	53.82	4.660^{20}	113.60	184.24	0.029^{20} aq; s alc, bz, chl, CS_2, CCl_4, eth
bromide	IBr	206.81	4.4157^{0}	42	116.d	s aq, alc, eth
chloride	ICl	162.36	3.20	27.38	97.8	d aq; s alc, eth
chloride, tri-	ICl_3	233.26	3.202	101 d		d aq; s alc, bz, eth
fluoride, penta-	IF_5	221.90	3.252	8.5	102	d aq
fluoride, hepta-	IF_7	259.89	2.8^{6} g · L^{-1}	4.5	5.5	
(di-) oxide, penta-	I_2O_5	333.81	4.799^{25}	d 275		187^{13} aq
Iron	Fe	55.85	7.86	1537	2872	i aq; s a
(II) bromide	$FeBr_2$	215.67	4.636	691	934	117^{20} aq
(III) bromide	$FeBr_3$	295.57		subl		s aq
carbonyl, penta-	$Fe(CO)_5$	195.00	1.49	−21	103	i aq; s alc, bz, eth
(II) chloride	$FeCl_2$	126.75	3.16^{25}	677	1024	63^{20} aq; v s alc, acet; i eth
(III) chloride	$FeCl_3$	162.21	2.898	304	332	74^{0} aq
(III) ferrate(II), hexacyano-	$Fe_4[Fe(CN)_6]_3$	859.25	1.80	d		i aq; s HCl
(II) fluoride	FeF_2	93.84	4.09	1100	1837	sl s aq; a a
(III) fluoride	FeF_3	112.84	3.87	subl 927		0.091^{25} aq; s a; i alc, bz
(II) iodide	FeI_2	309.66	5.315	587	1093	s aq
(III) nitrate 9-water	$Fe(NO_3)_3 \cdot 9H_2O$	404.02	1.684^{21}	47	d 100	138^{20}
(II) oxalate 2-water	$FeC_2O_4 \cdot 2H_2O$	179.90	2.28	d 150–160		0.044^{18} aq; s a
(II) oxide	FeO	71.85	5.7	1377	d 3414	i aq; s a
(III) oxide	Fe_2O_3	159.69	5.24	1462 d		i aq; s HCl
(II, III) oxide	Fe_3O_4	231.54	5.1	1597		i aq; s a
(II) sulfate 7-water	$FeSO_4 \cdot 7H_2O$	278.04	1.89			48^{20} aq

(III) sulfate	$Fe_2(SO_4)_2$	399.88	3.097^{18}	d 1178		sl s aq (hyd); sl s alc
(III) sulfate 9-water	$Fe_2(SO_4)_3 \cdot 9H_2O$	562.01	2.1	d 175		440 aq
Krypton	Kr	83.80	3.736 $g \cdot L^{-1}$	−157.2	−153.4	5.94^{20} mL aq
Lead	Pb	207.21	11.34 (fcc)	327.50	1753	i aq; s HNO_3
(II) acetate 3-water	$Pb(C_2H_3O_2)_2 \cdot 3H_2O$	379.33	2.55	d 200		46^{15} aq
(IV) acetate	$Pb(C_2H_3O_2)_4$	443.37	2.228^{17}	175		d aq; s chl
(II) azide	$Pb(N_3)_2$	291.23		expl 350		0.023^{18} aq; s HOAc
(II) carbonate	$PbCO_3$	267.20	6.6	d 340		i aq; s a, alk
(II) chromate(VI)(2−)	$PbCrO_4$	323.18	6.12^{15}	844	d	i aq; s a
(IV) fluoride	PbF_4	283.21	6.7	600		hyd aq
(II) nitrate	$Pb(NO_3)_2$	331.23	4.53^{20}	d 200		56^{20} aq; 1.3 MeOH
(II) oleate	$Pb(C_{18}H_{33}O_2)_2$	770.12				i aq; s alc, bz, eth
(II) oxide	PbO	223.21	9.53	886	1516	0.0017^{20}; s HNO_3
(IV) oxide	PbO_2	239.21	9.375	d 752		i aq; s HCl
(II) phosphate	$Pb_3(PO_4)_2$	811.59	6.9	1014		i aq; s HNO_3, alk
(II) stearate	$Pb(C_{18}H_{35}O_2)_2$	774.15		ca 125		0.05^{35} aq; s hot alc
(II) sulfate	$PbSO_4$	303.28	6.2	1090		0.004 aq
Lithium	Li	6.94	0.535^{20}	180.6	1340	d aq to LiOH
aluminate, tetrahydrido-	$LiAlH_4$	37.95	0.917	d 125		d aq, alc; 30 eth (flammable)
amide	$LiNH_2$	22.96	1.178^{18}	374	d 430	d aq; i bz, eth
benzoate	$LiC_7H_5O_2$	128.05		>300		33 aq; 7.7 alc
boronate	$LiBH_4$	21.79	0.666	268	d 380	d aq; s eth, THF
bromate	$LiBrO_3$	134.85	3.62			179^{20} aq
bromide	LiBr	86.84	3.464	550	1289	164 aq; s alc, eth
carbonate	Li_2CO_3	73.89	2.11^{0}	720	d	1.3^{20} aq; i alc; s a
chloride	LiCl	42.40	2.068	610	1383	77^{20} aq; s alc, acet
fluoride	LiF	25.94	2.640^{20}	846	1717	0.13^{25} aq; s a
hydride	LiH	7.95	0.780	688.7	d 950	d aq; no known solv (flammable)
hydroxide	LiOH	23.95	2.54	471.2	1626	12.4^{20} aq
iodide	LiI	133.84	4.061	467	1178	165^{20} aq; v s alc
iodide 3-water	$LiI \cdot 3H_2O$	187.89	3.5	73	$-3H_2O$, 300	200 aq; 200 alc
nitrate	$LiNO_3$	68.94	2.38	261		70^{20} aq; s alc

TABLE 2-1 Physical constants of inorganic compounds (*continued*)

Name	Formula	Formula weight	Density	Melting point, °C	Boiling point, °C	Solubility in 100 parts solvent
Lithium						
perchlorate	$LiClO_4$	106.40	2.43^{25}	236	d 400	56^{20} aq
sulfate	Li_2SO_4	109.88	2.22	860		34.5^{20} aq
Magnesium	Mg	24.31	1.74^{20}	650	1105	i aq; s a
amide	$Mg(NH_2)_2$	56.37	1.39^{25}_{4}	ign in air		d viol aq giving NH_3
bromide	$MgBr_2$	184.13	3.72	711	1158	101^{20} aq
bromide 6-water	$MgBr_2 \cdot 6H_2O$	292.22	2.00	165 d		160^{20} aq; s alc
carbonate	$MgCO_3$	84.32	2.958	d 402		0.01 aq; s a
chloride	$MgCl_2$	95.23	2.41	714	1437	54.6^{20} aq
hydride	MgH_2	26.34	1.45	d 287 (vac)	ign air	d viol aq, alc
hydroxide	$Mg(OH)_2$	58.33	2.36	268 d		i aq; s a
oleate	$Mg(C_{18}H_{33}O_2)_2$	293.61				i aq; s alc, eth, PE
oxide	MgO	40.52	3.58	2825	3260	i aq; s a
perchlorate	$Mg(ClO_4)_2$	223.23	2.21^{20}	d 251		49.6 aq
sulfate 7-water	$MgSO_4 \cdot 7H_2O$	246.49	1.67	$-6H_2O$, 120	$-7H_2O$, 250	27.2 aq; s alc
sulfite 6-water	$MgSO_3 \cdot 6H_2O$	212.47	1.725	$-6H_2O$, 200	d	66^{25} aq
Manganese						
acetate 4-water	$Mn(C_2H_3O_2)_2 \cdot 4H_2O$	245.08	1.589			38^{50} aq; s alc
bromide 4-water	$MnBr_2 \cdot 4H_2O$	286.82		54 d		200 aq; s alc
carbonate	$MnCO_3$	114.94	3.125	d		0.0065^{25} aq; s a
(di-) carbonyl, deca-	$Mn_2(CO)_{10}$	389.99	1.75^{25}	155 (CO atm)	d 110	i aq; s org solv
chloride 4-water	$MnCl_2 \cdot 4H_2O$	197.91	2.01	$-4H_2O$, 198		143 aq; s alc; i eth
(III) fluoride	MnF_3	111.93	3.54	d 600		hyd aq; s a
nitrate 6-water	$Mn(NO_3)_2 \cdot 6H_2O$	287.05	1.8	25.8		v s aq, alc
(IV) oxide	MnO_2	86.94	5.026	d 530		i aq; s HCl
sulfate hydrate	$MnSO_4 \cdot H_2O$	169.01	2.95	$-H_2O$, 400		70^{20} aq

Name	Formula	Formula weight	Density	Melting point	Boiling point	Solubility
Mercury	Hg	200.59	13.594^{20}	−38.86	356.60	i aq; s HNO_3
(II) acetate	$Hg(C_2H_3O_2)_2$	318.70	3.28	178		25^{10} aq; 7.5^{15} MeOH
(II) bromide	$HgBr_2$	360.44	6.05	241	subl > 241	0.56^{20} aq; 20^{25} alc
(I) chloride	Hg_2Cl_2	472.09	7.150	subl 382	d	0.00027 aq; s aqua regia
(II) chloride	$HgCl_2$	271.52	5.44	277	304	6.6^{20} aq; 33 alc; 4 eth
(II) cyanide	$Hg(CN)_2$	252.65	3.996	d 320		9.3^{20} aq; 8 alc; 25 MeOH
(II) fluoride	HgF_2	238.61	8.95^{15}	645	647	hyd aq; s HF
(II) iodide	HgI_2	454.45	6.28	259	350	0.006^{25} aq; 1 alc; 1.7 acet
(II) nitrate	$Hg(NO_3)_2$	324.63	4.3	79	d	v s aq; s acet
(II) oxide	HgO	216.61	11.14	d 476		0.005^{25} aq; s a
(I) sulfate	Hg_2SO_4	497.29	7.56	d		0.06^{25} aq; s HNO_3
(II) sulfate	$HgSO_4$	296.68	6.47	d		d aq; s a
(II) sulfide, red	HgS	232.68	8.10	subl 583		i aq; s aqua regia
Molybdenum						
carbonyl, hexa-	$Mo(CO)_6$	264.02	1.96	subl 102	156.4	s bz
(V) chloride	$MoCl_5$	273.21	2.928	194	264	hyd aq; s conc a
(VI) oxide	MoO_3	143.95	4.696^{26}	801	1155	0.22^{28} aq; s alk, NH_3
sulfide, di-	MoS_2	160.08	5.06^{15}	2375	subl 450	i aq; s aqua regia
Molybdic acid hydrate	$H_2MoO_4 \cdot H_2O$	179.97	3.124^{15}	$-H_2O$, 70		0.133^{18}; s alk
Molybdic phosphoric acid	$H_7[P(Mo_2O_7)_6] \cdot 28H_2O$	2365.71	2.53	78		hyd aq
Neon	Ne	20.18	0.8899^{0} $g \cdot L^{-1}$	−248.6	−246.1	1.05^{20} mL aq
Nickel	Ni	58.71	8.90	1455	2920	i aq; s HNO_3
acetylacetonate	$Ni(C_5H_7O_2)_2$	256.93	1.455^{17}	229	235	s aq, alc, bz, chl
bromide	$NiBr_2$	218.53	5.098	963	subl	131^{20} aq
chloride 6-water	$NiCl_2 \cdot 6H_2O$	237.70				111^{20} aq
dimethylglyoxime	$Ni(HC_2H_6N_2O_2)_2$	288.91		subl 250		i aq; s abs alc, a
formate 2-water	$Ni(OOCH)_2 \cdot 2H_2O$	184.78	2.154^{20}	$-2H_2O$, 130	d 180	s aq; i alc
nitrate 6-water	$Ni(NO_3)_2 \cdot 6H_2O$	290.81	2.05	56.7	136.7	150^{20} aq
sulfate 6-water	$NiSO_4 \cdot 6H_2O$	262.86	2.07	53.3		40^{20} aq

TABLE 2-1 Physical constants of inorganic compounds (*continued*)

Name	Formula	Formula weight	Density	Melting point, °C	Boiling point, °C	Solubility in 100 parts solvent
Niobium						
(V) chloride	$NbCl_5$	270.20	2.75	204	250	s HCl, CCl_4
(V) fluoride	NbF_5	187.91	2.70_4^{80}	80	235	hyd aq, alc
(V) oxide	Nb_2O_5	265.82	4.6	1512		i aq; s HF, hot H_2SO_4
Nitrogen	N_2	28.01	1.165^{20} $g \cdot L^{-1}$	−210.00	−195.81	1.52^{20} mL aq
[^{15}N]	$^{15}N_2$	30.01	1.25^{20} $g \cdot L^{-1}$	−209.95	−195.73	
chloride, tri-	NCl_3	120.37	1.653^{20}	−27	71	i aq; s bz, CS_2, CCl_4
(di-) oxide	N_2O	44.02	1.8433^{20} $g \cdot L^{-1}$	−90.85	−88.47	130^0 mL aq; s alc
oxide	NO	30.01	1.2488^{20} $g \cdot L^{-1}$	−163.64	−151.76	7^0 mL aq
(di-) oxide, tetra-	N_2O_4	92.02	1.447_4^{20}	−9.3	21.10 d	d aq; s HNO_3, H_2SO_4, chl
(di-) oxide, penta-	N_2O_5	108.01	2.05^{15}	30	47.0	s aq, chl
Nitrosyl						
chloride	NOCl	65.47	1.592^{-5}	−61.5	−5.5	hyd aq
fluoride	NOF	49.01	2.788^{20} $g \cdot L^{-1}$	−132.5	−59.9	hyd aq
Nitryl						
chloride	NO_2Cl	81.46	2.81^{100} $g \cdot L^{-1}$	−145	−13.5	d aq
fluoride	NO_2F	65.00	2.7^{20} $g \cdot L^{-1}$	−166.0	−72.4	d aq
Osmium oxide, tetra-	OsO_4	254.20	4.91	40.6	130.0	7.24^{25} aq; 375^{25} CCl_4
Oxygen	O_2	32.00	1.331^{20} $g \cdot L^{-1}$	−218.75	−182.96	36^{25} mL aq

Ozone	O_3	48.00	1.998^{20} $g \cdot L^{-1}$	−192.5	−110.50	49.4^{0} mL aq
Palladium	Pd	106.4	12.023	1550	2940	s hot HNO_3, H_2SO_4
acetate	$Pd(C_2H_3O_2)_2$	224.49		205 d		i aq, alc; s acet, chl
chloride	$PdCl_2$	177.30	4.0^{18}	680	d 680	s aq
nitrate	$Pd(NO_3)_2$	230.42		d		hyd aq; s HNO_3
oxide	PdO	122.40	8.70^{20}	870 d		i aq, a
Perchloryl fluoride	ClO_3F	102.46	0.637	−147.74	−46.67	
Phosphine	PH_3	34.00	1.529 $g \cdot L^{-1}$	−133.81	−87.78	26^{17} mL aq; s alc, eth
Phosphinic acid	HPH_2O_2	66.00	1.493^{19}	26.5	d 50	s aq
Phosphonic acid	H_2PHO_3	82.00	1.651^{21}	ca 73	d 180	v s aq, alc
Phosphoric acid						
meta-	HPO_3	79.98	2.2–2.5			slowly hyd aq; s alc
ortho-	H_3PO_4	98.00	1.88	42.35		
commercial 85% acid			1.685	anhyd 150	to $H_4P_2O_7$ ca 200; to $HPO_3 > 300$	v s aq
fluoro-	H_2PO_3F	99.99	1.818	−80		
Phosphorus						
(white)	P (P_4 molecules)	30.97	1.828	44.2	280.3	i aq; 0.025 alc; 1 eth; 2.5 chl, bz; 1.25 CS_2
(red)	P	30.97	2.34	597	subl 416	i aq (ign in air 260°)
bromide, tri-	PBr_3	270.73	2.85^{15}	−40.5	173.2	d aq, alc; s acet, CS_2
bromide, penta-	PBr_5	430.56	3.46^{20}	d 100		d aq; s CCl_4, CS_2
chloride, tri-	PCl_3	137.35	1.575^{20}	−91	75	d aq, alc; s bz, chl
chloride, penta-	PCl_5	208.27	2.119^{20}	subl 100	166 d	hyd aq; s CCl_4, CS_2
fluoride, penta-	PF_5	125.98	5.805 $g \cdot L^{-1}$	−93.8	−84.6	hyd aq
(tetra-) oxide, hexa-	P_4O_4	219.90	2.136^{20}_{4}	24	175 (in N_2)	hyd aq; s bz, CS_2
(tetra-) oxide, deca-	P_4O_{10}	283.88	2.30	340	subl 360	d aq; s H_2SO_4
(tetra-) selenide, tri-	P_4Se_3	360.80	1.31	245	360–400	hyd aq; s bz, chl, acet
(tetra-) sulfide, deca-	P_4S_{10}	444.54	2.09	288	514	hyd aq; s alk, CS_2

TABLE 2-1 Physical constants of inorganic compounds (*continued*)

Name	Formula	Formula weight	Density	Melting point, °C	Boiling point, °C	Solubility in 100 parts solvent
Phosphoryl chloride, tri-	$POCl_3$	153.35	1.645^{25}	2	105	d aq, alc
Platinic(IV) acid 6-water, hexachloro-	$H_2PtCl_6 \cdot 6H_2O$	517.92	2.431	60		v s aq, alc
Platinum	Pt	195.09	21.45^{20}	1770	3824	i aq; s aqua regia, fused alk
Platinum						
(II) chloride	$PtCl_2$	266.00	6.05	d 581		i aq; s HCl, NH_4OH
(IV) oxide	PtO_2	227.09	10.2	450		i aq, aqua regia
Potassium	K	39.10	0.856^{20}	63.7	765.5	d to KOH aq; s a
acetate	$KC_2H_3O_2$	98.14	1.57^{25}	292		256^{20} aq; 34 alc
bismuthate(4−), heptaiodo-	K_4BiI_7	1253.82				d aq; s alk iodide soln
borate, tetrahydrido-	KBH_4	53.95	1.11	d 497		21^{25} aq; 3.5^{20} MeOH
bromate	$KBrO_3$	167.01	3.27^{17}	350	d 370	6.9^{20} aq
bromide	KBr	119.01	2.75	734	1398	65^{20} aq; 0.4 alc
carbonate	K_2CO_3	138.20	2.29	901	d	111^{20} aq; i alc
chlorate	$KClO_3$	122.55	2.238^{20}	368	d 368	7.3^{20} aq; 2 glyc
chloride	KCl	74.56	1.988	771	1437	34^{20} aq; 7 glyc
chromate(VI)	K_2CrO_4	194.20	2.732^{18}	975		64^{20} aq; i alc
citrate hydrate	$K_3C_6H_5O_7 \cdot H_2O$	324.42	1.98	d 230		167^{15} aq
cobaltate(III) 1.5-water, hexanitrito-	$K_3[Co(NO_2)_6] \cdot 1.5H_2O$	479.30		d 200		0.089^{17} aq; v sl s alc
cyanate	KOCN	81.11	2.048	d 700–900		s aq; sl s alc
cyanide	KCN	65.12	1.52^{16}	622	1625	50 aq
dichromate(VI)	$K_2Cr_2O_7$	294.19	2.676^{25}	398	d 500	12.3^{20} aq
disulfate(IV)	$K_2S_2O_5$	222.32				s aq (flammable if ground)

Potassium						
ethyldithiocarbonate	KC_2H_5OCSS	160.30	1.558^{22}	d 200		v s aq
ferrate(III), hexacyano-	$K_3[Fe(CN)_6]$	329.26	1.89	d		84^{20} aq (slow)
fluoride	KF	58.10	2.481	858	1517	95^{20} aq
formate	KOOCH	84.10	1.91	167.5	d 168	337^{20} aq
gluconate	$KC_6H_{11}O_7$	234.24		d 180		v s aq; i alc, bz, chl
hydride	KH	40.11	1.43	417 d		d aq
hydrogen arsenate, di-	KH_2AsO_4	180.02	2.867	288		19^{6} aq; 63 gly; i alc
hydrogen carbonate	$KHCO_3$	100.11	2.17	d 100–200		34^{20} aq
hydrogen difluoride	KHF_2	78.11	2.37	238.7	d 478	30^{20} aq; s alc
hydrogen bisiodate	$KH(IO_3)_2$	389.92				1.3^{15} aq
hydrogen oxalate	KHC_2O_4	128.11	2.044	d		2.5 aq
hydrogen bisoxalate dihydrate, tri-	$KH_3(C_2O_4)_2 \cdot 2H_2O$	254.20	1.836	d		1.8^{13} aq
hydrogen phosphate	K_2HPO_4	174.18		d		150 aq
hydrogen phosphate, di-	KH_2PO_4	136.09	2.338	400	d	22.6^{20} aq
hydrogen phthalate	$KHC_8H_4O_4$	204.22	1.636^{25}	d		10.2 aq; sl s alc
hydrogen sulfate	$KHSO_4$	136.17	2.24	197	d	48^{20} aq
hydrogen tartrate	$KHC_4H_4O_6$	188.18	1.956			0.5^{20} aq
hydroxide	KOH	56.11	2.044	406	1320	112^{20} aq; 33 alc
iodate	KIO_3	214.02	3.89^{25}	560 d		8.1^{20} aq; i alc
iodide	KI	166.02	3.12	681	1345	144^{20} aq; 4.5 alc; 1.2 acet
manganate(VI)	K_2MnO_4	197.12		d 190		s aq (stable in KOH)
nitrate	KNO_3	101.10	2.109^{16}	334.3	d 400	32^{20} aq; 0.16 alc; s glyc
nitrite	KNO_2	85.10	1.915	441	d 250	306^{20} aq
oxalate hydrate	$K_2C_2O_4 \cdot H_2O$	184.24	2.127^{4}	$-H_2O$, 160	d	36^{20} aq
periodate	KIO_4	230.01	3.618^{15}_{4}	582 d		0.42^{20} aq
permanganate	$KMnO_4$	158.03	2.703	d 240		6.34^{20} aq
peroxide	K_2O_2	110.20		490		d
peroxodisulfate	$K_2S_2O_8$	270.32	2.477	d 100		5.3^{20} aq
phenolsulfonate hydrate	$KC_6H_4(OH)SO_3 \cdot H_2O$	240.28	1.87			s aq, alc
phosphate	K_3PO_4	212.28	2.564^{17}	1340		92^{20} aq
selenocyanate	KSeCN	144.08		d 100		s aq

TABLE 2-1 Physical constants of inorganic compounds (*continued*)

Name	Formula	Formula weight	Density	Melting point, °C	Boiling point, °C	Solubility in 100 parts solvent
Potassium						
silicate(2−)	K_2SiO_3	154.29		976		s aq
silicate, hexafluoro-	K_2SiF_6	220.25	2.27	d		sl s aq
sodium tartrate 4-water	$KNaC_4H_4O_6 \cdot 4H_2O$	282.23	1.790	70–80	d 220	54^{15} aq
sorbate	$KC_6H_7O_2$	150.22	1.363	d 270		110^{20} aq
stannate(IV) 3-water	$K_2SnO_3 \cdot 3H_2O$	298.94	3.197	$-3H_2O$, 140		100^{20} aq
sulfate	K_2SO_4	174.27	2.662	1067	1670	11^{20} aq; 1.3 glyc; i alc
sulfite dihydrate	$K_2SO_3 \cdot 2H_2O$	194.30		d		106^{20} aq
thiocarbonate	K_2CS_3	186.41		d		v s aq
thiocyanate	KSCN	97.18	1.886^{14}	173	d 500	217^{20} aq; 200 acet; 8 alc
						acet; 8 alc
thiosulfate	$K_2S_2O_3$	190.33		d 400		155^{20} aq
titanate(IV), oxobis-(oxalato)diaqua-	$K_2[TiO(C_2O_4)_2(H_2O)_2]$	354.18				v s aq
Rhenium(VII) sulfide	Re_2S_7	596.85	4.866	d 460		i aq; s HNO_3
Rhodium(III) chloride	$RhCl_3$	209.28		d 450	subl 850	i aq; s KOH, KCN
Rubidium						
chloride	RbCl	120.94	2.76	715	1381	91^{20} aq; 1.1 MeOH
iodide	RbI	212.37	3.55	640	1304	144^{18} aq
nitrate	$RbNO_3$	147.47	3.11	310		53^{20} aq
sulfate	Rb_2SO_4	267.03	3.613^{20}	1060		48^{20} aq
Ruthenium						
(III) chloride	$RuCl_3$	207.47	3.11	d 500		i aq; s HCl, alc
(IV) oxide	RuO_2	133.07	6.97	d		i aq; s fused alk
Selenic acid	H_2SeO_4	144.98	2.9508_4^{15}	58	260	567^{20} aq (viol)
Selenium	Se	78.96	4.81^{20}	221	685	s CS_2, KOH, KCN
(IV) oxide	SeO_2	110.96	3.954_{15}^{15}	340	subl 315	38^{14} aq; 10^{12} MeOH

(di-) sulfide, hexa-	Se_2S_6	350.28	2.44	121.5		i aq; 1.2 bz; s CS_2
(tetra-) sulfide, tetra-	Se_4S_4	444.08	3.20	113		i aq; 0.04 bz; s CS_2
Silane	SiH_4	32.09	0.68^{-185}	−184.7	−111.9	d aq slowly
Silicon	Si	28.09	2.33^{25}	1415	2680	s HF+HNO_3, fused alk oxides
carbide	SiC	40.07	3.217	subl 2700	d 2972	s fused alk
chloride	$SiCl_4$	169.89	1.48^{20}	−70	57.6	hyd aq; s bz, CCl_4, eth
isothiocyanate, tetra-	$Si(NCS)_4$	260.40		143.8	314.2	d aq
oxide, di- (quartz)	SiO_2	60.08	2.64–2.66	1423	2230	i aq; s HF
oxide–tungsten trioxide-water (1/12/26) (silico-tungstic acid)	$SiO_2 \cdot 12WO_3 \cdot 26H_2O$	3310.66				v s aq, alc
telluride, tri-	Si_2Te_3	438.97		892		
Silver	Ag	107.87	10.49^{15}	960.15	2164	i aq; s HNO_3
acetate	$AgC_2H_3O_2$	166.92	3.259^{15}	d		1.04^{20} aq
azide	AgN_3	149.89		252	297	i aq; s KCN, HNO_3 (expl)
carbonate	Ag_2CO_3	275.77	6.077	d 220		0.003^{30} aq
chlorate	$AgClO_3$	191.34	4.430^{20}	231	d 270	15.3^{20} aq
chloride	AgCl	143.34	5.56	455	1564	0.00019 aq; s NH_4OH
chromate(VI)	Ag_2CrO_4	331.77	5.625^{25}			0.002^{20} aq; s HNO_3, NH_4OH
cyanide	AgCN	133.90	3.95	d 320		i aq; s KCN
fluoride	AgF	126.88	5.852^{16}	435	1150	172^{20} aq
(II) fluoride	AgF_2	145.87	4.57	690	d 700	hyd viol aq
iodate	$AgIO_3$	282.80	5.525^{20}	>200	d	0.004^{20} aq
iodide	AgI	234.80	5.683^{30}	558	1505	i aq; s KCN
nitrate	$AgNO_3$	169.89	4.352^{19}	210	d 440	216^{20} aq
nitrite	$AgNO_2$	153.89	4.453	d 140		0.41^{25} aq
oxide	Ag_2O	231.76	7.22^{25}	d 200		0.002^{25} aq
(II) oxide	AgO	123.88	7.483^{25}	d 100		i aq; s alk
permanganate	$AgMnO_4$	226.81	4.49	d		0.9 aq; d alc
phosphate, ortho-	Ag_3PO_4	418.62	6.370	849		0.006 aq
sulfate	Ag_2SO_4	311.83	5.45^{30}	660	d 1085	0.80^{20} aq

TABLE 2-1 Physical constants of inorganic compounds (*continued*)

Name	Formula	Formula weight	Density	Melting point, °C	Boiling point, °C	Solubility in 100 parts solvent
Sodium	Na	22.99	0.968^{20}	97.82	881.4	d aq to NaOH
acetate	$NaC_2H_3O_2$	82.04	1.528	324		46.5^{20} aq
aluminate, tetrachloro-	$NaAlCl_4$	191.80		151		s aq
amide	$NaNH_2$	39.02		210	subl 400	d viol aq
aurate(III) dihydrate, tetrachloro-	$NaAuCl_4 \cdot 2H_2O$	397.80	1.6	d 100		166^{20} aq
azide	NaN_3	65.01	1.846^{20}	d		41^{20} aq; 0.3 alc
benzoate	$NaC_6H_5O_2$	144.11				63^{25} aq; 1.3 alc
bismuthate(V)(1−)	$NaBiO_3$	280.00		d		i aq; d a
boranate	$NaBH_4$	37.84	1.074	497 d		55^{25} aq; 4 alc; 1.4 pyr; 5 DMF
borate, tetra-	$Na_2B_4O_7$	201.27	2.367	742.5		2.6^{20} aq
borate, tetrafluoro-	$NaBF_4$	109.82	2.47^{20}	384	d	108^{27} aq
bromate	$NaBrO_3$	150.91	3.339^{17}	380 d		36^{20} aq
bromide	NaBr	102.91	3.205^{18}	747	1447	90^{20} aq; 6 alc; 16 MeOH
carbonate	Na_2CO_3	106.00	2.533	850.0	d	21.5^{20}; s glyc
carbonate 10-water	$Na_2CO_3 \cdot 10H_2O$	286.14	1.46	34		50 aq; s glyc
chlorate	$NaClO_3$	106.45	2.489	248	d 350	96^{20} aq; 0.77 alc; 25 glyc
chloride	NaCl	58.45	2.164^{20}	801	1465	36^{20} aq; 10 glyc
chlorite	$NaClO_2$	90.45		d 180–200		34^{17} aq
chromate(VI)	Na_2CrO_4	161.97	2.723	792		84^{20}
citrate 2-water	$Na_3C_6H_5O_7 \cdot 2H_2O$	294.10		$-2H_2O$, 150		77^{25} aq
cobaltate(III), hexanitrito-	$Na_3[Co(NO_2)_6]$	403.98				v s aq
cyanate	NaOCN	65.01	1.893^{20}	550		s aq d; 0.22^{0} alc
cyanide	NaCN	49.02		562	1530	58.7^{20} aq
cyanoborohydride	$NaBH_3CN$	62.84		d 242		(flammable solid)
dichromate(VI) 2-water	$Na_2Cr_2O_7 \cdot 2H_2O$	298.00	2.348^{25}_{4}	356 anhyd	d 400	208^{20} aq
diethyldithiocarbamate	$NaS_2CN(C_2H_5)_2$	225.31		94 anhyd		s aq, alc

dimethylarsonate 3-water	$NaO_2As(CH_3)_2 \cdot 3H_2O$	214.03		60	$-3H_2O$, 120	200 aq; 40 alc
diphosphate(V)	$Na_4P_2O_7$	265.90	2.45	988		2.26^{0} aq
dithionate 2-water	$Na_2S_2O_6 \cdot 2H_2O$	242.13	2.189	$-2H_2O$, 110	d 267	6.05^{20} aq
dithionate(III) (hydrosulfite)	$Na_2S_2O_4$	174.13		d		22^{20} aq
dodecylsulfate (laurate)	$NaO_3SOC_{12}H_{25}$	288.38				10 aq
ethoxide	$NaOC_2H_5$	68.06		>300		d aq; s abs alc
ethylenebis(aminodiacetate) (EDTA)	$Na_4C_2H_4N_2(C_2H_3O_2)_4$	380.20				103 aq
ethylsulfate	$NaO_3SOC_2H_5$	148.11				
ferrate(II) 10-water, hexacyano-	$Na_4[Fe(CN)_6] \cdot 10H_2O$	484.07	1.458	$-10H_2O$, 82	d 435	18.8^{20} aq
ferrate(III) 2-water, pentacyanonitrosyl- (nitroprusside)	$Na_2[Fe(CN)_5NO] \cdot 2H_2O$	297.65	1.72			40^{16} aq
fluoride	NaF	41.99	2.78	996	1787	4^{20} aq; i alc
formate	NaOOCH	68.02	1.919	253 d		81^{20} aq; s glyc; sl s alc
gluconate	$NaC_6H_{11}O_7$	218.13				59^{25} aq; sl s alc; i eth
glycerophosphate	$Na_2C_3H_5(OH)_2PO_4$	216.03		d 130		60 aq; i alc
hydride	NaH	24.00	1.396	d 425		d viol aq, alc
hydrogen carbonate	$NaHCO_3$	84.01	2.20	$-CO_2$, 270		9.6^{20} aq; i alc
hydrogen phosphate hydrate, di-	$NaH_2PO_4 \cdot H_2O$	137.99	2.040	$-H_2O$, 100	d 200	71^{0} aq
hydrogen phosphate 7-water	$Na_2HPO_4 \cdot H_2O$	268.07	1.679	d		185^{40} aq
hydrogen sulfate	$NaHSO_4$	120.07	2.435	315	d	28.5^{25} aq; d alc
hydrogen sulfite	$NaHSO_3$	104.06	1.48	d		29 aq; 1.4 alc
hydrogen sulfide 2-water	$NaHS \cdot 2H_2O$	92.09		55	d	s aq, alc, eth
hydroxide	NaOH	40.01	2.130^{25}	322	1557	108^{20} aq; 14 abs alc; 24 MeOH; s glyc
hydroxymethanesulfinate dihydrate	$NaO_2SCH_2OH \cdot 2H_2O$	154.12		63–64		v s aq; i abs alc, bz, eth
hypochlorite	NaClO	74.44				53^{20} aq (anhyd v expl)

TABLE 2-1 Physical constants of inorganic compounds (*continued*)

Name	Formula	Formula weight	Density	Melting point, °C	Boiling point, °C	Solubility in 100 parts solvent
Sodium						
iodate	$NaIO_3$	197.90	4.227^{20}	d		8.1^{20} aq
iodide	NaI	149.92	3.667^{0}	660	1304	178^{20}
lactate	$NaOOCCHOHCH_3$	112.07		d		misc aq, alc
methoxide	$NaOCH_3$	54.03		>300		d aq; s alc
molybdate dihydrate	$Na_2MoO_4 \cdot 2H_2O$	241.95	3.28	687	$-2H_2O$, 100	65^{20} aq
nitrate	$NaNO_3$	85.01	2.257	308	d 380	88^{20} aq
nitrite	$NaNO_2$	69.00	2.168^{0}	271	d 320	81^{20} aq
oxalate	$Na_2C_2O_4$	134.01	2.27			3.4^{20} aq
oxide	Na_2O	61.98	2.27	1132	d 1950	d aq to NaOH
perchlorate	$NaClO_4$	122.44	2.499	468		201^{20}
periodate	$NaIO_4$	213.91	3.865^{16}_{4}	d 300		10.3^{20} aq
peroxide	Na_2O_2	77.99	2.805	675	d	v s aq (d)
peroxoborate 4-water	$NaBO_3 \cdot 4H_2O$	153.88		d 60		2.5 aq
peroxodisulfate(VI)	$Na_2S_2O_8$	238.13		d		55 aq
phosphate 12-water	$Na_3PO_4 \cdot 12H_2O$	380.12	1.62	73.4	$-11H_2O$, 100	28.3^{15} aq
platinate(IV) 6-water, hexachloro-	$Na_2PtCl_6 \cdot 6H_2O$	561.88	2.50	$-6H_2O$, 110		v s aq; s alc
propionate	$NaOOCCH_2CH_3$	96.07				100^{25} aq; 4.1^{25} alc
salicylate	$NaC_7H_5O_3$	160.11				95^{20} aq; 11 alc; 25 glyc
selenate(VI)	Na_2SeO_4	188.94	3.098			27^{20} aq
silicate, hexafluoro-	Na_2SiF_6	188.05	2.679	red heat		0.44^{0} aq; i alc
stannate(IV) 3-water	$Na_2SnO_3 \cdot 3H_2O$	266.71		d 140		50^{0} aq
stearate	$NaOOCC_{17}H_{35}$	306.47		d		sl s aq
sulfate	Na_2SO_4	142.06	2.664	884		19.5^{20}
sulfate 10-water	$Na_2SO_4 \cdot 10H_2O$	322.19	1.464	32.4	$-10H_2O$, 100	36^{15} aq

sulfide	Na_2S	78.05	1.856_4^{14}	950		15.7^{20} aq
sulfite	Na_2SO_3	126.06	2.633^{15}	d		26^{20} aq
tartrate dihydrate	$Na_2C_4H_4O_6 \cdot 2H_2O$	230.08	1.818	$-2H_2O$, 120		29^{6} aq
tetraphenylborate	$NaB(C_6H_5)_4$	342.24				s aq, acet
thiocyanate	NaSCN	81.07		287		134^{20} aq
thiosulfate	$Na_2S_2O_3$	158.11	2.345			s aq; i alc
thiosulfate 5-water	$Na_2S_2O_3 \cdot 5H_2O$	248.18	1.685	$-5H_2O$, 100		70^{20} aq (d slowly)
tungstate(VI) dihydrate	$Na_2WO_4 \cdot 2H_2O$	329.86	3.245	$-2H_2O$, 100		88^{0}
Strontium						
carbonate	$SrCO_3$	147.64	3.70	$-CO_2$, 1172		0.001^{25} aq; s a
chloride	$SrCl_2$	158.52	3.052	874	2058	52.9^{20} aq
chromate(VI)	$SrCrO_4$	203.64	3.895^{15}			0.09^{20} aq; s HCl
hydroxide	$Sr(OH)_2$	121.64	3.625	375 (in H_2)	$-H_2O$, 710	1.77^{20} aq
Sulfamic acid	H_2NSO_3H	97.09	2.126	d 200		14.7 aq
Sulfinyl						
bromide	$SOBr_2$	207.88	2.67	−49.5	139.7	d aq
chloride	$SOCl_2$	118.98	1.656^{15}	−104.5	75.8	hyd aq
fluoride	SOF_2	86.06	3.0^{-44}	−110	−43.8	d aq; s bz, chl, eth
Sulfonyl						
chloride	SO_2Cl_2	134.98	1.6674^{20}	−46	69.3	d aq; s bz
fluoride	SO_2F_2	102.07	3.72 $g \cdot L^{-1}$	−135.8	−55.38	4 mL aq; 24 mL alc; 136 mL CCl_4; 210 mL toluene
Sulfur	S	32.07	1.92	106.8	444.60	i aq; 23^{0} CS_2; s alc, bz
	S_8	256.51	1.96^{20}	115.21	444.60	i aq; 23^{0} CS_2; s alc, bz
(di-) chloride, di-	S_2Cl_2	135.03	1.688_{15}^{15}	−80	138.1	hyd aq
fluoride, tetra-	SF_4	108.07	1.919^{-73}	−121	−38	d viol aq; s bz
fluoride, hexa-	SF_6	146.07	1.88^{-50} $g \cdot L^{-1}$	−50.8	subl 63.8	sl a aq; s alc, KOH
oxide, di	SO_2	64.07	2.716^{20} $g \cdot L^{-1}$; 1.46^{-10} (lq)	−75.47	−10.01	3937^{20} mL aq; 25 mL alc
oxide, tri (III)	SO_3	80.07	1.9225^{20}	16.86	43.4	slowly v s aq

TABLE 2-1 **Physical constants of inorganic compounds (*continued*)**

Name	Formula	Formula weight	Density	Melting point, °C	Boiling point, °C	Solubility in 100 parts solvent
Sulfuric acid	H_2SO_4	98.08	1.8318^{20}	10.38	335.5	v s aq
chloro-	$HOSO_2Cl$	116.52	1.753^{20}	−80	152	d viol aq
fluoro-	FSO_2OH	100.07	1.726^{25}	−88.98	162.6	d viol aq
Tantalum	Ta	180.95	16.69	2985	5513	i aq; s HF, fused alk
(V) fluoride	TaF_5	275.95	4.74^{20}	95–97	229	s aq
Tellurium	Te	127.60	6.24^{20}	450	1009	i aq; s HNO_3, KOH
Thallium	Tl	204.37	11.85	303.5	1487	i aq; s HNO_3
(III) acetate sesquihydrate	$Tl(C_2H_3O_2)_3 \cdot 1.5H_2O$	408.53		182 d		
(I) bromide	TlBr	284.31	7.54	460	825	0.05^{20} aq; s alc
(I) chloride	TlCl	239.85	7.004^{30}	429	816	0.33^{20} aq
(I) ethoxide	$TlOC_2H_5$	249.43	3.493^{20}	−3	d 130	sl s alc; s eth
(I) fluoride	TlF	223.39	8.23^{4}	322	700	78^{15} aq
(I) nitrate	$TlNO_3$	266.40	5.556	206	430	9.6^{20} aq
(III) nitrate 3-water	$Tl(NO_3)_3 \cdot 3H_2O$	444.43		102–103		s aq
(I) oxide	Tl_2O	424.78	9.52^{16}	300	1080	v s aq (d); s a
(III) oxide	Tl_2O_3	456.78	10.19^{22} (hex)	717	$-O_2$, 875	i aq; s a
(I) sulfate	Tl_2SO_4	504.85	6.77	632	d	4.9^{20} aq
Thiocarbonyl chloride	$CSCl_2$	114.98	1.509^{15}		73.5	d aq; s eth
Thiocyanogen	$(SCN)_2$	116.16		ca −2		d aq; s alc, CS_2, eth
Thionyl, *see* **Sulfinyl**						
Tin (silver-white, tetr)	Sn	118.69	7.28	231.89	2623	i aq; s HCl, H_2SO_4
(gray, cub)			5.75	stable −161 to 13.2		
(IV) bromide	$SnBr_4$	438.36	3.35^{33}	30	207	hyd aq; s acet
(II) chloride	$SnCl_2$	189.61	3.95	247	652	84^{0} aq; s alc, eth

(IV) chloride	$SnCl_4$	260.53	2.226	−34	115	s aq, eth
(II) diphosphate(V)	$Sn_2P_2O_7$	411.32	4.009^{16}			i aq; s conc a
(II) fluoride	SnF_2	156.70	4.57^{25}	213		30 aq
(IV) fluoride	SnF_4	194.70	4.780^{19}	subl 705		hyd aq
(IV) oxide	SnO_2	150.70	6.95	1630	subl 1900	i aq
(II) sulfide	SnS	150.77	5.08	881	1210	i aq; s conc HCl
(IV) sulfide	SnS_2	182.83	4.5	765		i aq; d by aqua regia
(II) zirconate(IV), hexafluoro-	$SnZrF_6$	323.92	4.21			s aq
Titanium	Ti	47.90	4.507	1660	3318	s hot a, HF
(III) chloride	$TiCl_3$	154.27	2.71	subl 831 (vac)	d 5000	s aq, alc
(IV) chloride	$TiCl_4$	189.73	1.726	−24.10	136.4	s cold aq, alc
hydride, di	TiH_2	49.92	3.752	d 400		
(IV) isopropoxide	$Ti[OCH(CH_3)_2]_4$	284.26	0.9711^{20}	ca 20	220^{10mm}	
(IV) oxide (rutile)	TiO_2	79.90	4.23	1857		s HF
(III) sulfate	$Ti_2(SO_4)_3$	384.00				s HCl
Trisulfuryl dichloride	$ClSO_2OSO_2OSO_2Cl$	295.09	1.90^{20}	18.7	61^{3mm}	
Tungsten						
(VI) chloride	WCl_6	396.57	2.721^{282}	281.5	340.5	hyd aq; s CS_2, CCl_4
(VI) oxide	WO_3	231.86	7.16	1472	1837	i aq; s hot alk
sulfide, di-	WS_2	247.98	7.5^{10}	d 1250		s HNO_3+HF
Uranyl						
(VI) acetate 2-water	$UO_2(C_2H_3O_2)_2 \cdot 2H_2O$	422.13	2.893^{15}	$-2H_2O$, 110	d 275	7.7^{15} aq
nitrate 6-water	$UO_2(NO_3)_2 \cdot 6H_2O$	502.13	2.807^{13}	60.2	d 100	155^{20} aq
Vanadium						
(III) oxide	V_2O_3	149.00	4.87	2067		i aq; s HNO_3+HF
(V) oxide	V_2O_5	181.90	3.35	670	1690	0.80 aq; s a, alk
(IV) oxide sulfate	$VOSO_4$	163.00				v s aq
Xenon	Xe	131.30	5.8971^{0} $g \cdot L^{-1}$	−111.8	−108.10	10.8^{20} mL aq
fluoride, di-	XeF_2	169.30	3.13^{25}	129.0	subl 114	2.5^{0} aq

TABLE 2-1 Physical constants of inorganic compounds (*continued*)

Name	Formula	Formula weight	Density	Melting point, °C	Boiling point, °C	Solubility in 100 parts solvent
Xenon						
fluoride, tetra-	XeF_4	207.30	3.03^{25}	117.1	subl 116	hyd aq; s F_3CCOOH
fluoride, hexa-	XeF_6	245.30	3.411^{25}	49.5	75.6	hyd aq
Zinc	Zn	65.37	7.14^{25}	419.6	911	i aq; s a, alk
acetate dihydrate	$Zn(C_2H_3O_2)_2 \cdot 2H_2O$	219.49	1.735	237		41.6^{20} aq; 3.3 alc
bromide	$ZnBr_2$	225.21	4.22	402	650	446^{20} aq; 200 alc; s eth
carbonate	$ZnCO_3$	125.38	4.398	$-CO_2$, 300		0.02^{25} aq; s a, alk
chloride	$ZnCl_2$	136.29	2.907^{25}	318	732	395^{20} aq; 77 alc; 50 glyc
chromate(VI)	$ZnCrO_4$	181.36	3.40			i aq; s a
cyanide	$Zn(CN)_2$	117.42	1.852	d 800		0.058^{18} aq; s KCN, alk
fluoride	ZnF_2	103.38	5.00^{25}	872	1500	1.6^{20} aq
iodide	ZnI_2	319.22	4.736^{25}	446	730	432^{20} aq; 50 glyc
nitrate 6-water	$Zn(NO_3)_2 \cdot 6H_2O$	297.47	2.065^{14}	36.4	$-6H_2O$, 131	146^{0} aq
oxide	ZnO	81.37	5.67	1970		i aq; s a, alk
peroxide	ZnO_2	97.38	3.00	d 150		i aq; d slowly
p-phenolsulfonate 8-water	$Zn[C_6H_4(OH)SO_3]_2 \cdot 8H_2O$	555.83		$-8H_2O$, 120		63 aq; 56 alc
phosphate(V)	$Zn_2(PO_4)_2$	386.05	3.998^{15}	900		i aq; s a, NH_4OH
phosphide	Zn_3P_2	258.09	4.55	>420	subl 1100 (in H_2)	d aq; s bz, CS_2; d viol HCl
propionate	$Zn(OOCCH_2CH_3)_2$	211.52				32 aq; 2.8 alc
silicate 6-water, hexafluoro-	$ZnSiF_6 \cdot 6H_2O$	315.54	2.104	d 100		v s aq
stearate	$Zn(OOCC_{17}H_{35})_2$	632.33		ca 120		i aq, alc, eth; s bz
sulfate	$ZnSO_4$	161.44	3.54	1200		53.8^{20} aq
sulfate 7-water	$ZnSO_4 \cdot 7H_2O$	287.54	1.957	$-7H_2O$, 280	d 500	96^{20} aq; 40 glyc; i alc

sulfide	ZnS	97.43	4.087	1722		i aq; s a
thiocyanate	$Zn(SCN)_2$	181.53				0.14^{18} aq; s alc
Zirconium	Zr	91.22	6.52^{30}	1852	4504	s aqua regia
(IV) chloride	$ZrCl_4$	233.05	2.803^{15}	437	subl 334	hyd aq; a alc, eth
chloride oxide 8-water	$ZrCl_2O \cdot 8H_2O$	322.25	1.91	$-8H_2O$, 210		s aq
hydroxide	$Zr(OH)_4$	159.25	3.25	$-2H_2O$, 500		s a
(IV) oxide	ZrO_2	123.22	5.85	2677	4275	s hot H_2SO_4, HF slowly
silicate(4−)	$ZrSiO_4$	183.31	4.56	d 1538		very inert
sulfate 4-water	$Zr(SO_4)_2 \cdot 4H_2O$	355.41	3.22^{16}	anhyd 380		52.5^{18} aq

SECTION 3

PROPERTIES OF ATOMS, RADICALS, AND BONDS

NUCLIDES

TABLE 3-1 Table of nuclides

Explanation of column headings

Nuclide. Each nuclide is identified by its atomic number Z, equal to the number of protons in the nucleus; the corresponding symbol for that element; and the mass number A, equal to the sum of the numbers of protons Z and neutrons N in the nucleus. Thus, $A = Z + N$, or $N = A - Z$. The m following the mass number (e.g., ^{69m}Zn) indicates an isomer of that nuclide.
Half-Life. For the radioactive nuclides this time period corresponds to that during which loss by disintegration of 50% of the nuclide occurs. The units of time are designated by year (yr), day (d), hour (h), minute (min), and second (s).
Natural Abundance. The isotopic abundances listed are on an "atom percent" basis for the stable nuclides present in naturally occurring elements in the earth's crust.
Thermal Neutron Absorption Cross Section. The ease with which a given nuclide can absorb a thermal neutron (energy $\leq \frac{1}{40}$ eV) and become of a different nuclide is indicated by the cross section, given here in units of barns (1 barn $= 10^{-24}$ cm^2). If the mode of reaction is other than (n, γ), it is so indicated, for example, (n, p) or (n, α), where n = neutron, p = proton, γ = gamma ray, and α = alpha particle (^4_2He).
Major Radiations. In this column are listed the principal mode(s) of decay and the energies of the emanating radiations in million electronvolts (MeV). The gamma-ray (γ) intensities, where given, are given to the nearest whole percentage in parentheses following the numerical energy value for that particular γ. In most cases the radiations listed should be sufficient for identification of the particular nuclide. The following designations are used: negatron (β^-), positron (β^+), conversion electron (e^-), gamma ray (γ), and alpha particle (α).

Nuclide				Thermal neutron absorption	
Symbol	Mass	Half-life	Natural abundance, %	cross section, barns	Major radiations
^1H	1.007 825		99.985	0.332	
^2H	2.014 102		0.015	0.000 5	
^3H	3.016 050	12.26 yr			β^-, 0.018 6; no γ
^6Li	6.015 125		7.42	953(n,α)	
^7Li	7.016 004		92.58	0.037	
^7Be	7.016 929	53.6 d		54,000(n,p)	γ, 0.477(10)
^9Be	9.012 186		100	0.009	
^{10}Be	10.013 534	2.5×10^6 yr			β^-, 0.555; no γ
^{10}B	10.012 939		19.7	3837(n,α)	
^{11}B	11.009 305		80.3	0.005	
^{11}C	11.011 432	20.34 min			β^+, 0.97; γ, 0.511
^{13}C	13.003 354		1.108	0.000 9	
^{14}C	14.003 242	5730 yr			β^-, 0.156; no γ
^{13}N	13.005 738	9.96 min			β^+, 1.20; γ, 0.511
^{14}N	14.003 074		99.635	1.81(n,p)	
^{19}O	19.003 578	29.1 s			β^-, 4.60; γ, 0.197(97), 1.37(59)
^{18}F	18.000 937	109.7 min			β^+, 1.74; γ, 0.511
^{22}Na	21.994 437	2.62 yr			β^+, 1.820, 0.545; γ, 0.511, 1.275(100)

TABLE 3-1 Table of nuclides (*continued*)

Nuclide				Thermal neutron absorption cross section, barns	
Symbol	Mass	Half-life	Natural abundance, %		Major radiations
^{23}Na	22.934 473		100	0.53	
^{24}Na	23.990 962	14.96 h			β^-, 4.17, 1.389; γ, 0.511, 1.275(100)
^{25}Mg	24.985 839		10.11	0.3	
^{28}Mg	27.983 875	21.2 h			β^-, 0.46; e^-, 0.03; γ, 0.031(96), 0.40(30), 0.95(30), 1.35(70)
^{26}Al	25.986 891	7.4×10^5 yr			β^-, 8.5; γ, 0.511, 1.12(4), 1.81(100)
^{27}Al	26.981 539		100	0.235	
^{28}Al	27.981 905	2.31 min			β^-, 2.85; γ, 1.780 (100)
^{30}Si	29.973 763		3.12	0.11	
^{31}Si	30.975 349	2.62 h			β^-, 1.48; γ, 1.26
^{31}P	30.973 765		100	0.19	
^{32}P	31.973 909	14.28 d			β^-, 1.710
^{33}P	32.971 728	24.4 d			β^-, 0.248; no γ
^{34}S	33.967 865		4.22	0.27	
^{35}S	34.969 031	87.9 d			β^-, 0.167; no γ
^{38}S	37.971 230	2.87 h			β^-, 3.0, 1.1; γ, 1.88(95)
^{35}Cl	34.968 851		75.53	44	
^{36}Cl	35.968 309	3.08×10^5 yr		100	β^-, 0.714; γ, 0.511
^{37}Cl	36.965 898		24.47	0.4	
^{38}Cl	37.968 005	37.29 min			β^-, 4.91; γ, 1.60(38)
^{39}Cl	38.968 008	55.5 min			β^-, 3.45, 2.18, 1.91; γ, 0.246(44)
^{37}Ar	32.966 772	35.1 d			Cl X rays
^{40}K	39.964 000	1.26×10^9 yr	0.118	70	β^-, 1.314; β^+, 0.483; γ, 1.460(11)
^{41}K	40.961 832		6.77	1.2	
^{42}K	41.962 406	12.36 h			β^-, 3.52; γ, 0.31, 1.524(18)
^{44}Ca	43.955 490		2.06	0.7	
^{45}Ca	44.956 189	165 d			β^-, 0.252
^{47}Ca	46.954 538	4.535 d			β^-, 1.98, 0.67; γ, 0.49(5), 0.815(5), 1.308(74)
^{46}Sc	45.955 919	83.9 d			β^-, 1.48, 0.357, γ, 0.889(100), 1.120(100)

TABLE 3-1 Table of nuclides (*continued*)

Nuclide		Half-life	Natural abundance, %	Thermal neutron absorption cross section, barns	Major radiations
Symbol	Mass				
^{44}Ti	43.959 572	48 yr			γ, 0.068(90), 0.078(98); e^-, 0.065, 0.073
^{48}V	47.952 259	16.0 d			β^+, 0.696; γ, 0.511, 0.945(10), 0.983(100), 1.312(97), 2.241(3)
^{49}V	48.949 522	330 d			Ti X rays
^{50}Cr	49.946 054		4.31	17	
^{51}Cr	50.944 768	27.8 d			γ, 0.320(9); e^-, 0.315
^{54}Mn	53.940 362	303 d			γ, 0.835(100); e^-, 0.829
^{55}Mn	54.938 050		100	13.3	
^{56}Mn	55.938 910	2.576 h			β^-, 2.85; γ, 0.847(99), 1.811(29), 2.110(15)
^{54}Fe	53.939 617		5.84	2.9	
^{55}Fe	54.938 299	2.60 yr			Mn X rays
^{58}Fe	57.933 282		0.31	1.1	
^{59}Fe	58.934 878	45.6 d			β^-, 1.57, 0.475; γ, 0.143(1), 0.192 (3), 1.095(56), 1.292 (44)
^{57}Co	56.936 296	270 d			γ, 0.014(9), 0.122(87), 0.136(11), 0.692; e^-, 0.115, 0.129
^{58}Co	57.935 761	71.3 d			β^+, 0.474; γ, 0.511, 0.810(99), 0.865(1), 1.67(1)
^{59}Co	58.933 189		100	19	
^{60}Co	59.933 813	5.263 yr		6	β^-, 1.48, 0.314; γ, 1.173(100), 1.332(100)
^{62}Ni	61.928 342		3.66	15	
^{63}Ni	62.929 664	92 yr			β^-, 0.067; no γ
^{64}Ni	63.927 958		1.16	1.5	
^{65}Ni	64.930 072	2.564 h			β^-, 2.13; γ, 0.368(5), 1.115(16), 1.481(25)
^{63}Cu	62.929 592		69.1	4.5	
^{64}Cu	63.929 759	12.80 h			β^-, 0.573; β^+, 0.656; e^-, 1.33; γ, 0.511, 1.34(1)

TABLE 3-1 Table of nuclides (*continued*)

Nuclide		Half-life	Natural abundance, %	Thermal neutron absorption cross section, barns	Major radiations
Symbol	Mass				
^{64}Zn	63.929 145	$>8 \times 10^{15}$ yr	48.89	0.46	
^{65}Zn	64.929 234	245 d			β^+, 0.327; e^-, 1.106; γ, 0.511, 1.115(49)
^{68}Zn	67.924 857		18.56	1.0	
^{69m}Zn		13.8 h			γ, 0.439(95); e^-, 0.429
^{71}Ge	70.924 956	11.4 d			Ga X rays
^{75}As	74.921 595		100	4.5	
^{76}As	75.922 397	26.4 h			β^-, 2.97; γ, 0.559(43), 0.657(6), 1.22(5) 1.44(1), 1.789, 2.10(1)
^{77}As	76.920 645	38.7 h			β^-, 0.68; γ, 0.086, 0.239(3), 0.522(1)
^{75}Se	74.922 525	120.4 d			γ, 0.066(1), 0.097(1), 0.121(17), 0.136(57), 0.265(60), 0.280(25), 0.401(12); e^- 0.085, 0.095, 0.109, 0.124, 0.253
^{79}Br	78.918 329		50.52	8.5	
^{80}Br	79.918 536	17.6 min			β^-, 2.00; β^+, 0.87; γ, 0.511, 0.618(7), 0.666(1)
^{81}Br	80.916 292		49.48	3	
^{82}Br	81.916 802	35.34 h			β^-, 0.444; γ, 0.554(66), 0.619(41), 0.698(27), 0.777(83), 0.828(25), 1.044(29), 1.317(26), 1.475(17)
^{85}Kr	84.912 523	10.76 yr		<15	β^-, 0.67; γ, 0.514
^{86}Rb	85.911 193	18.66 d			β^-, 1.78; γ, 1.078(9)
^{85}Sr	84.912 989	64.0 d			γ, 0.514(100); e^-, 0.499
^{90}Y	89.907 163	64.0 h			β^-, 2.27; no γ
^{95}Nb	94.906 832	35.0 d		~7	β^-, 0.160; γ, 0.765(100)

TABLE 3-1 Table of nuclides (*continued*)

Nuclide		Half-life	Natural abundance, %	Thermal neutron absorption cross section, barns	Major radiations
Symbol	Mass				
^{99}Mo	98.907 720	66.7 h			β^-, 1.23; γ, 0.041(12), 0.181(7), 0.372(1), 0.740(12), 0.780(4)
^{99m}Tc		6.049 h			γ, 0.140(90); e^-, 0.110
^{103}Ru	102.906 306	39.5 d			β^-, 0.70, 0.21; γ, 0.497(88), 0.610(6)
^{108}Pd	107.903 891		26.7	12	
^{109}Pd	108.905 954	13.47 h			β^-, 1.028; γ, 0.088(5), 0.129, 0.31, 0.41, 0.60, 0.64
^{109}Ag	108.904 756		48.65	89	
^{110m}Ag		39.2 s			γ, 0.088(5)
^{111}Ag	110.905 316	7.5 d			β^-, 1.05; γ, 0.247(1), 0.342(6)
^{109}Cd	108.904 928	453 d			γ, 0.088; e^-, 0.062
^{115}Cd	114.905 431	53.5 h			β^-, 1.11; γ, 0.230(1), 0.262(2), 0.49(10), 0.53(26)
^{113m}In		99.8 min			γ, 0.393(64); e^-, 0.365, 0.389
^{114}In	113.904 905	72 s			β^-, 1.988; β^+, 0.42; γ, 1.299
^{113}Sn	112.905 187	115 d			γ, 0.255(2)
^{121}Sb	120.903 816		57.25	6	
^{122}Sb	121.905 183	2.80 d			β^-, 1.97; β^+, 0.56; γ, 0.584(66), 0.686(3), 1.14(1) 1.26(1)
^{123}Sb	122.904 213	1.3×10^{16} yr	42.75	3.3	
^{124}Sb	123.905 973	60.4 d		2000	β^-, 2.31; γ, 0.603(97), 0.644(7), 0.72(14), 0.967(2), 1.048(2), 1.31(3), 1.37(5), 1.45(2), 1.692(50), 2.088(7)
^{125}Sb	124.905 232	2.71 yr		<20	β^-, 0.61; e^-, 0.114, 0.395; γ, 0.176(6), 0.427(31), 0.463(10), 0.599(24),

TABLE 3-1 Table of nuclides (*continued*)

Nuclide				Thermal neutron absorption	
Symbol	Mass	Half-life	Natural abundance, %	cross section, barns	Major radiations
					0.634(11), 0.66(3)
^{132}Te	131.908 523	77.7 h			β^-, 0.22; e^-, 0.197; γ, 0.053(17), 0.230(90)
^{125}I	124.904 578	60.2 d		9	γ, 0.035(7); e^-, 0.030
^{127}I	126.904 470		100	6.4	
^{128}I	127.905 838	24.99 min			β^-, 2.12; γ, 0.441(14), 0.528(1), 0.743, 0.969
^{131}I	130.906 127	8.05 d		~0.7	β^-, 0.806, 0.606; e^-, 0.330; γ, 0.080(3), 0.284(5), 0.364(82), 0.637(7), 0.723(2)
^{132}I	131.907 981	2.26 h			β^-. 2.12; γ, 0.24(1), 0.52(20), 0.67(44), 0.773(89), 0.955(22) 1.14(6), 1.28(7), 1.40(14), 1.45(1), 1.91(1), 1.99(1)
^{133}Xe	132.905 815	5.270 d		190	β^-, 0.346; e^-, 0.045, 0.075; γ, 0.081(37)
^{131}Cs	130.905 466	9.70 d			Xe X rays
^{134}Cs	133.906 823	2.046 yr		136	β^-, 0.662; γ, 0.57(23), 0.605(98), 0.796(99), 1.038(1), 1.168(2), 1.365(3)
^{137}Cs	136.906 770	30.0 yr		0.11	β^-, 1.176, 0.514; e^-, 0.624, 0.656; γ, 0.662(85)
^{131}Ba	130.906 716	12.0 d			γ, 0.124(28), 0.216(19), 0.25(5), 0.373(13), 0.496(48), 0.60(3); e^-, 0.118, 0.180, 0.460
^{133}Ba	132.905 879	7.2 yr			γ, 0.080(36), 0.276(7), 0.302(14),

TABLE 3-1 Table of nuclides (*continued*)

Nuclide		Half-life	Natural abundance, %	Thermal neutron absorption cross section, barns	Major radiations
Symbol	Mass				
					0.356(69), 0.382(8); e^-, 0.266, 0.319
^{137m}Ba		2.554 min			γ, 0.662(89); e^-, 0.624, 0.656
^{140}Ba	139.910 565	12.80 d		<20	β^-, 1.02; γ, 0.030(11), 0.163(6), 0.305(6), 0.438(5), 0.537(34)
^{141}Ce	140.908 219	32.5 d		30	β^-, 0.581; e^-, 0.104, 0.139; γ, 0.145(48)
^{144}Ce	143.913 591	284 d		1.0	β^-, 0.31; γ, 0.080(2), 0.134(11)
^{197}Au	196.966 541		100	98.8	
^{198}Au	197.968 231	2.697 d		26,000	β^-, 0.962; e^-, 0.329, 0.398; γ, 0.412(95), 0.676(1), 1.088
^{199}Au	198.968 773	3.15 d		~30	β^-, 0.46, 0.30; γ, 0.158(37), 0.208(8); e^-, 0.125, 0.145
^{197}Hg	196.967 360	65 h			γ, 0.77(18), 0.191(2), 0.268
^{203}Hg	202.972 880	46.9 d			β^-, 0.214; e^-, 0.194, 0.264, 0.275; γ, 0.279(77)
^{203}Tl	202.972 353		29.50	11	
^{205}Tl	203.973 865	3.81 yr			β^-, 0.766
^{210}Pb	209.984 187	20.4 yr			β^-, 0.061; γ, 0.047(4); α, 3.72
^{207}Bi	206.978 438	30.2 yr			γ, 0.570(98), 1.063(77), 1.771(9); e, 0.482, 0.975, 1.048
^{210}Po	209.982 876	138.40 d		<0.03	α, 5.305; γ, 0.803
^{226}Ra	226.025 360	1602 yr		20	α, 4.78, 4.60; γ, 0.186(4), 0.26, 0.42, 0.61; e^-, 0.170

TABLE 3-1 Table of nuclides (*continued*)

Nuclide				Thermal neutron absorption cross section, barns	
Symbol	Mass	Half-life	Natural abundance, %		Major radiations
^{241}Am	241.056 714	433 yr		700	α, 5.49, 5.44; γ, 0.060(36), 0.101, 0.208, 0.335, 0.37, 0.663, 0.722

ELECTRONEGATIVITY

According to Pauling, electronegativity χ is the relative attraction of an atom for the valence electrons in a covalent bond. It is proportional to the effective nuclear charge and inversely proportional to the covalent radius.

$$\chi = \frac{0.31(n+1\pm c)}{r} + 0.50$$

where n is the number of valence electrons, c is any formal valence charge on the atom and the sign before its corresponds to the sign of this charge, and r is the covalent radius. Because electronegativity is concerned with atoms in molecules rather than atoms in isolation, it is not possible to define precise electronegativity values. Pauling determined his set of values from bond energy data based on experimentally measured heats of dissociation and formation. Originally the element fluorine, whose atoms have the greatest attraction for electrons, was given an arbitrary electronegativity of 4.0. A revision of Pauling's values based on newer heat data assigns 3.9 to fluorine. A unit positive charge changes the χ value for an atom by about two-thirds of the electronegativity difference between it and the atom next on its right in the Periodic Table, and a unit negative charge similarly decreases the χ value.

The greater the difference in electronegativity, the greater is the ionic character of the bond. The amount of ionic character I is given by the expression

$$\mathrm{I} = 1 - e^{-0.25(\chi_A - \chi_B)^2}$$

The bond is fully covalent when $(\chi_A - \chi_B) < 0.5$ (and $\mathrm{I} < 6\%$). A different expression was proposed by Hannay-Smyth.*

$$\mathrm{I} = 0.46|\chi_A - \chi_B| + 0.035(\chi_A - \chi_B)^2$$

Other sets of electronegativities of the elements have been proposed. The rather direct, but somewhat limited, method of Mulliken makes use of the ionization potential IP and electron-affinity data (Table 3-3). Numerical values are obtained that coincide with values from other methods if electronegativities are calculated from

$$\chi = \frac{\mathrm{IP} + A}{5.6}$$

* Hannay-Smyth, *J. Am. Chem. Soc.*, **68**:171 (1946).

Electronegativities on the Allred-Rochow scale* are given by

$$\chi = 0.359\frac{Z_{\mathrm{eff}}}{r^2} + 0.744$$

where Z_{eff} is the effective nuclear charge and r is the atomic radius.

Using Pauling's values, electronegativities of the elements are arranged in periodic order in Table 3-2.

TABLE 3-2 Electronegativities of the elements

H													
2.2													
Li	Be							B	C	N	O	F	
1.0	1.5							2.0	2.5	3.0	3.5	4.0	
Na	Mg							Al	Si	P	S	Cl	
0.9	1.2							1.5	1.8	2.1	2.4	2.8	
K	Ca	Sc	Ti—V	Cr—Mn	Fe—Ni	Cu	Zn	Ga	Ge	As	Se	Br	
0.9	1.0	1.3	1.6	1.6	1.8	1.9	1.7	1.6	1.8	2.0	2.4	2.7	
Rb	Sr	Y	Zr—Nb	Mo—Tc	Ru—Pd	Ag	Cd	In	Sn	Sb	Te	I	
0.8	1.0	1.2	1.6	1.8	2.2	1.9	1.5	1.7	1.8	1.9	2.1	2.2	
Cs	Ba	La—Lu	Hf—Ta	W—Re	Os—Pt	Au	Hg	Tl	Pb	Bi			
0.7	0.9	1.1	1.3	1.8	2.2	2.4	1.4	1.8	1.8	1.9			

Electronegativities have important uses in chemistry in addition to predicting the amount of ionic character in a bond. The bond stretching force constant k (in units of 10^5 dynes $\cdot$ cm^{-1}) can be estimated for stable molecules exhibiting their normal covalences by the expression:

$$k = 1.67N\left(\frac{\chi_A \chi_B}{d^2}\right)^{3/4} + 0.30$$

where N is the bond order (i.e., the effective number of covalent or ionic bonds acting between the two atoms A and B) and d is the internuclear distance in angstroms.

Electronegativity is proportional to the work function ϕ, which is the energy necessary to just remove an electron from the metal surface in thermoelectric or photoelectric emission.

$$\chi = 0.44\phi - 0.15$$

* *J. Inorg. Nucl. Chem.*, **5**:264, 269 (1958).

ELECTRON AFFINITY

TABLE 3-3 Electron affinities of elements, molecules, and radicals

The electron affinity of an atom A is defined as the energy released when an atom and an electron react to form a negative ion in the gas phase at 0 K.

$$A(g) + e^- = A^-(g)$$

Data are limited to those negative ions which, by virtue of their positive electron affinity, are stable. Uncertainty in the final data figures is given in parentheses.

Source: H. Hotop and W. C. Lineberger, *J. Phys. Chem. Ref. Data*, **4**:539 (1975).

A. Elements

Element	Electron affinity, eV*	Element	Electron affinity, eV*
Aluminum	0.46(3)	Molybdenum	1.0(2)
Antimony	1.05(5)	Neon	<0
Argon	<0	Nickel	1.15(10)
Arsenic	0.80(5)	Niobium	1.0(3)
Astatine	2.8(2)	Nitrogen	−0.07(8)
Barium	<0	Osmium	1.1(3)
Beryllium	<0	Oxygen	1.462(3)
Bismuth	1.1(2)	Palladium	0.6(3)
Boron	0.28(1)	Phosphorus	0.743(10)
Bromine	3.364(4)	Platinum	2.128
Cadmium	<0	Polonium	1.9(3)
Calcium	<0	Potassium	0.5012(5)
Carbon	1.268(5)	Radon	<0
Cesium	0.4715(5)	Rare earths	≤0.5 (estimate)
Chlorine	3.615(4)	Rhenium	0.15(10)
Chromium	0.66(5)	Rhodium	1.2(3)
Cobalt	0.7(2)	Rubidium	0.4860(5)
Copper	1.226(10)	Ruthenium	1.1(3)
Fluorine	3.399(3)	Scandium	<0
Francium	(0.456)	Selenium	2.0206(3)
Gallium	0.30(15)	Silicon	1.385(5)
Germanium	1.2(1)	Silver	1.303(7)
Gold	2.3086(7)	Sodium	0.546(5)
Hafnium	>0	Strontium	<0
Helium	<0	Sulfur	2.0772(5)
Hydrogen	0.754 209(3)	Tantalum	0.6(4)
Indium	0.30(15)	Technetium	0.7(3)
Iodine	3.061(4)	Tellurium	1.9708(3)
Iridium	1.6(2)	Thallium	0.3(2)
Iron	0.25(20)	Tin	1.25(10)
Krypton	<0	Titanium	0.2(2)
Lanthanum	0.5(3)	Tungsten	0.6(4)
Lead	1.1(2)	Vanadium	0.5(2)
Lithium	0.620(7)	Xenon	<0
Magnesium	<0	Yttrium	0.0(3)
Manganese	<0	Zinc	≈0
Mercury	<0	Zirconium	0.5(3)

* To convert into $kJ \cdot mol^{-1}$ multiply by 96.48. To convert into $kcal \cdot mol^{-1}$ multiply by 23.06.

TABLE 3-3 Electron affinities of radicals and molecules (*continued*)

B. Molecules

Molecule	Electron affinity, eV*	Molecule	Electron affinity, eV*
BF_3	2.65	SF_6	1.43
p-Benzoquinone	1.34	2,3,5,6-Tetrachloro-benzoquinone	2.40
NO_2	3.91		
O_2	0.45	Tetracyanoethylene	2.88

C. Radicals

Radical	Electron affinity, eV*	Radical	Electron affinity, eV*
CH_3	1.08	OH	1.83
C_2H_5	0.89	CF_3O	1.35
C_6H_5	2.20	CH_3O	0.38
CCl_3	1.22	PH_2	1.60
CF_3	1.85	SH	2.19
CN	3.17	CH_3S	1.32
NH_2	1.12	SCN	2.17
C_6H_5NH	1.55	SeCN	2.64
$(C_6H_5)_2N$	1.19	SiF_3	3.35

* To convert into $kJ \cdot mol^{-1}$, multiply by 96.48. To convert into $kcal \cdot mol^{-1}$, multiply by 23.06.

BOND LENGTHS AND STRENGTHS

TABLE 3-4A Bond lengths between carbon and other elements

The numbers in parentheses following a numerical value represent the standard deviation of that value in terms of the final listed digit.

To convert the bond length from angstroms into nanometers, multiply by 0.1; to convert angstroms into picometers, multiply by 100.

Bond type	Bond length, Å*
Carbon-carbon	
Single bond	
Paraffinic: —C—C—	1.541(3)
In presence of —C=C— or of aromatic ring	1.53(1)
In presence of —C=O bond	1.516(5)
In presence of two carbon-oxygen double bonds	1.49(1)
In presence of two carbon-carbon double bonds	1.426(5)
Aryl—C=O	1.47(2)

TABLE 3-4A Bond lengths between carbon and other elements (*continued*)

Bond type	Bond length, Å*
Carbon-carbon (*continued*)	
Single bond (*continued*)	
In presence of one carbon-carbon triple bond: —C—C≡C—	1.460(3)
In presence of one carbon-nitrogen triple bond: —C—C≡N	1.464(5)
In compounds with tendency to dipole formation, e.g., C=C—C=O	1.44(1)
In aromatic compounds	1.395(3)
In presence of carbon-carbon double and triple bonds: —C=C—C≡C—	1.426(5)
In presence of two carbon-carbon triple bonds: —C≡C—C≡C—	1.373(4)
Double bond	
Single: —C=C—	1.337(6)
Conjugated with a carbon-carbon double bond: —C=C—C=C—	1.336(5)
Conjugated with a carbon-oxygen double bond: —C=C—C=O	1.36(1)
Cumulative: —C=C=C— or —C=C=O	1.309(5)
Triple bond	
Simple: —C≡C—	1.204(2)
Conjugated: —C≡C—C=C—, —C≡C—C=O, or —C≡C—aryl	1.206(4)

Bond type	Bond length, Å			
Carbon-halogen				
	Fluorine	Chlorine	Bromine	Iodine
Paraffinic: R—X	1.379(5)	1.767(2)	1.938(5)	2.139(1)
Olefinic: —C=C—X	1.333(5)	1.719(5)	1.89(1)	2.092(5)
Aromatic: Ar—X	1.328(5)	1.70(1)	1.85(1)	2.05(1)
Acetylenic: —C≡C—X	(1.27)	1.635(5)	1.795(10)	1.99(2)

Bond type	Bond length, Å
Carbon-hydrogen	
Paraffinic	
In methane (in CD_4, 1.092)	1.094
In monosubstituted carbon: H—C—Y	1.096(5)
In disubstituted carbon: H—C(X)(Y)—	1.073(5)

TABLE 3-4A Bond lengths between carbon and other elements (*continued*)

Bond type	Bond length, Å
Carbon-hydrogen (*continued*)	
Paraffinic (*continued*)	
In trisubstituted carbon: H—C(X)(Z)—Y	1.070(7)
Olefinic	
Simple: H—C=C—	1.083(5)
Cumultative carbon-carbon double bonds: H—C=C=C—	1.07(1)
Cumulative carbon-carbon-oxygen double bonds: H—C—C=C=O	1.08(1)
Aromatic	1.084(5)
Acetylenic (in C_2H_2, 1.059)	1.055(5)
In small rings	1.081(5)
In presence of a carbon triple bond: H—C≡C—	1.115(4)
Carbon-nitrogen	
Single bond	
Paraffinic:	
3 covalent nitrogen: RNH_2, R_2NH, R_3N	1.472(5)
4 covalent nitrogen: RNH_3^+, $R_3N—BX_3$	1.479(5)
In —C—N=	1.475(10)
In aromatic compounds	1.43(1)
In conjugated heterocyclic systems (partial double bond)	1.353(5)
In —N—C=O (partial double bond)	1.322(5)
Double bond: —C=N—	1.32
Triple bond (in CN radical, 1.1774): —C≡N	1.157(5)
Carbon-oxygen	
Single bond	
Paraffinic and saturated heterocyclic: —C—O—	1.426(5)
Strained, as in epoxides: —C——C— (bridged by O)	1.435(5)
In aromatic compounds, as Ar—OH	1.36(1)
Longer bond in carboxylic acids and esters (HCOOH, 1.312)	1.358(5)
In conjugated heterocyclics, as furan	1.371(16)

TABLE 3-4A Bond lengths between carbon and other elements (*continued*)

Bond type	Bond length, Å
Carbon-oxygen (*continued*)	
Double bond	
In CO^+	1.115
In CO	1.128
In CO_2^+	1.177
In HCO	1.198(8)
In carbonyls	1.145(10)
In aldehydes and ketones	1.215(5)
In acyl halides: R—CO—X	1.171(4)
Shorter bond in carboxylic acids and esters	1.233(5)
In zwitterion forms	1.26(1)
In O=C=	1.160(1)
In isocyanates: RN=C=O	1.17(1)
In conjugated systems, as in partial triple bond: O=C—C=C	1.215(5)
In *p*-quinones	1.15(2)
In metal acetylacetonates	1.28(2)
In calcite: $CaCO_3$	1.29(1)
Carbon-selenium	
Single bond	
Paraffinic: —C—Se—	1.98(2)
In presence of fluorine, as in perfluoro compounds: —CF—Se—	1.95(2)
Double bond	
In Se=C=, as SeCS and SeCO	1.709(3)
In CSe radical	1.67
Carbon-silicon	
Alkyl substituent: H_3C—Si or H_2C—Si	1.870(5)
Aryl substituent: aryl—Si	1.843(5)
Electronegative substituent: R—Si—X	1.854(5)
Carbon-sulfur	
Single bond	
Paraffinic: —C—S—	1.817(5)
In presence of fluorine, as in perfluoro-compounds: —CF—S—	1.835(1)
In heterocyclic systems: partial double bonds	1.718(5)
Double bond	
In S=C: thiophene, $S=CR_2$	1.71(1)
In sulfoxides and sulfones	1.80(1)
In presence of second carbon-carbon double bond: S=C—C=C—	1.555(1)
In SC radical [in CS_2^+, 1.554(5)]	1.5349(2)

TABLE 3-4A Bond lengths between carbon and other elements (*continued*)

Bond type	Bond length, Å	Bond type	Bond length, Å
	Other elements and carbon		
C—Al	2.24(4)	C—In	2.16(4)
C—As (paraffinic)	1.98(1)	C—Mo	2.08(4)
C—B	1.56(1)	C—Ni	2.107(5)
C—Be	1.93	C—Pb (alkyl)	2.30(1)
C—Bi	2.30	C—Pd	2.27(4)
C—Co	1.83(2)	C—Sb (paraffinic)	2.202(16)
C—Cr	1.92(4)	C—Sn	
C—Fe	1.84(2)	alkyl	2.143(5)
C—Ge		electronegative	
Alkyl	1.98(3)	substituent	2.18(2)
Aryl	1.945(5)	C—Te	1.904
C—Hg	2.07(1)	C—Tl	2.705(5)
in $Hg(CN)_2$	1.99(2)	C—W	2.06

TABLE 3-4B Bond lengths between elements other than carbon

Elements	Bond type	Bond length, Å	Elements	Bond type	Bond length, Å
	Boron			Hydrogen (*continued*)	
B—B	B_2H_6	1.77(1)	H—Mg	MgH	1.731
B—Br	BBr_3	1.87(2)	H—Na	NaH	1.887
B—Cl	BCl_3	1.72(1)	H—Sb	H_3Sb	1.707
B—F	BF_3, R_2BF	1.29(1)	H—Se	H_2Se	1.460
B—H	Boranes	1.21(2)	H—Sn	SnH_4	1.701
	Bridge	1.39(2)	D—Br	DBr	1.4144
B—N	Borazoles	1.42(1)	D—Cl	DCl	1.2746
B—O	$B(OH)_3$, $(RO)_3B$	1.362(5)	D—I		1.6165
			T—Br		1.4144
			T—Cl		1.2740
	Hydrogen			Nitrogen	
H—Al	AlH	1.646			
H—As	AsH_3	1.519	N—Cl	NO_2Cl	1.79(2)
H—Be	BeH	1.343	N—F	NF_3	1.36(2)
H—Br	HBr	1.408	N—H	NH_4^+	1.034(3)
H-Ca	CaH	2.002		NH_3, RNH_2	1.012
H—Cl	HCl	1.274		H_2NNH_2	1.038
H—F	HF	0.917		R—CO—NH_2	0.99(3)
H—Ge	GeH_4	1.53		HN=C=S	1.013(5)
H—I	HI	1.609	N—D	ND	1.041
H—K	KH	2.244	N—N	HN_3	1.02(1)
H—Li	LiH	2.595		R_2NNH_2	1.451(5)

TABLE 3-4B Bond lengths between elements other than carbon (*continued*)

Elements	Bond type	Bond length, Å
	Nitrogen (*continued*)	
	N_2O	1.126(2)
	N_2^+	1.116
N—O	NO_2Cl	1.24(1)
	$RO—NO_2$	1.36(2)
	NO_2	1.188(5)
N=O	N_2O	1.186(2)
	RNO_2	1.22(1)
	NO^+	1.0619
N—Si	SiN	1.572
	Oxygen	
O—H	H_2O	0.958
	ROH	0.97(1)
	OH^+	1.0289
	HOOH	0.960(5)
	D_2O	0.9575
	OD	0.9699
O—O	HO—OH	1.48(1)
	O_2^+	1.227
	O_2^-	1.26(2)
	O_2^{2-}	1.49(2)
	O_3	1.278(5)
O—Al	AlO	1.618
O—As	As_4O_6 (bridges)	1.79
O—Ba	BaO	1.940
O—Cl	ClO_2	1.484
	OCl_2	1.68
O—Mg	MgO	1.749
O—Os	OsO_4	1.66
O—Pb	PbO	1.934
	Phosphorus	
P—Br	PBr_3	2.23(1)
P—Cl	PCl_3	2.00(2)
P—F	$PFCl_2$	1.55(3)
	Phosphorus (*continued*)	
P—H	PH_3, PH_4^+	1.424(5)
P—I	PI_3	2.52(1)
P—N	Single bond	1.491
P—O	Single bond	1.447
	p^3 bonding	1.67
	sp^3 bonding	1.54(4)
P—S	p^3 bonding	2.12(5)
	sp^3 bonding	2.08(2)
	In rings	2.20(2)
P—C	Single bond	1.562
	p^3 bonding	1.87(2)
	Silicon	
Si—Br	$SiBr_4$, R_3SiBr	2.16(1)
Si—Cl	$SiCl_4$, R_3SiCl	2.019(5)
Si—F	SiF_4, R_3SiF	1.561(3)
	SF_6	1.58
Si—H	SiH_4	1.480(5)
	R_3SiH	1.476(5)
Si—I	SiI_4	2.34
	R_3SiI	2.46(2)
Si—O	R_3SiOR	1.533(5)
Si—Si	H_3SiSiH_3	2.30(2)
	Sulfur	
S—Br	$SOBr_2$	2.27(2)
S—Cl	S_2Cl_2	1.585(5)
S—F	SOF_2	1.585(5)
S—H	H_2S	1.333
	RSH	1.329(5)
	D_2S	1.345
S—O	SO_2	1.4321
	$SOCl_2$	1.45(2)
S—S	RSSR	2.05(1)

TABLE 3-5 Bond strengths

The quantity D_0(A—B) corresponds to the bond dissociation energy at 0 K, all species considered to be ideal gases, for a bond A—B which is broken through the reaction: Eq. $AB \rightarrow A + B$

where
$$D_0 = \Delta Hf_0^\circ(A) + \Delta Hf_0^\circ(B) - \Delta Hf_0^\circ(AB)$$

D_0 at 298 K, or ΔHf_{298}, is greater than D_0 at 0 K by an amount which lies between RT and $3/2\ RT$, or between 0.6 and 0.9 kcal · mol^{-1}. In polyatomic molecules this difference may be somewhat greater. It is important to note that the bond dissociation energy refers to the enthalpy change ΔHf in the dissociation process.

The numbers in parentheses following a numerical value represent the standard deviation of that value in terms of the final listed digit(s).

To convert the tabulated values (in kcal · mol^{-1}) to kJ · mol^{-1}, multiply by 4.184.

Source: T. L. Cottrell, *The Strengths of Chemical Bonds*, 2d ed., Butterworth, London, 1958; B. deB. Darwent, National Standard Reference Data Series, National Bureau of Standards, no. 31, Washington, 1970; S. W. Benson, *J. Chem. Educ.*, **42**:502 (1965); and J. A. Kerr, *Chem. Rev.* **66**:465 (1966).

Bond	D_0°, kcal · mol^{-1}	ΔHf_{298}, kcal · mol^{-1}
Boron		
$H_3B—BH_3$		35
$F_2B—F$	133(20)	
Bromine		
Br—Br	45.45(1)	46.10(1)
$Br—CH_3$	67(2)	68(2)
$Br—CH_2Br$		61(3)
$Br—CHBr_2$		62(4)
$Br—CBr_3$	49(3)	50(3)
$Br—CCl_3$	51(3)	52(3)
$Br—CF_3$		68(3)
Carbon (*continued*)		
$CH_3—CH_2CN$		73(2)
$CH_3—CH(CH_3)CN$		79(2)
$CH_3—C(C_6H_5)CN(CH_3)$		60
$C_2H_5—CH_2CN$		76.9(17)
$CH_3—CF_3$		101.2(11)
$CH_2F—CH_2F$		88(2)
$CF_3—CF_3$		97(2)
$CF_2{=}CF_2$		76(3)
$CF_3—CN$		120
$CH_2—CO$	80.6	81.9
$CH_3—CHO$		75
$CH_3CO—CF_3$		73.8

$Br-CF_2CF_3$		68.7(15)
$Br-CF_2CF_2CF_3$		66.5(15)
$Br\text{-}CHF_2$		69
$Br-Cl$	51.6(1)	52.3(1)
$Br-F$	67.2	68.1
$Br\text{-}CN$		91
$Br-CO-C_6H_5$		64
$Br-N$	68(5)	
$Br-NF_2$		53
$Br-NO$	27.8(15)	28.7(15)
$Br-O$	55.3(1)	56.2(1)
Carbon		
$HC\equiv CH$		230(2)
$H_2C=CH_2$		163
CH_3-CH_3		88
$CH_3-C(CH_3)_2CH_3$		69(2)
$CH_3-C(CH_3)_3$		80
$CH_3-C_6H_5$		93
$CH_3-CH_2C_6H_5$		72
$CH_3-CH(CH_3)C_6H_5$		71
$C_2H_5-CH_2C_6H_5$		71
$C_3H_7-CH_2C_6H_5$		67(2)
$CH_3-(CH=CH_2)$		29
$CH_3-(CH_2CH=CH_2)$		72
$CH_3-(C\equiv CH)$		117
$(CH_3)_3C-C(CH_3)_3$		67.5
$(CH_3)_3C-C(C_6H_5)_3$		15
$(CH_2=CH)-(CH=CH_2)$		100
$C_6H_5-C_6H_5$		100
$(HC\equiv C)-(C\equiv CH)$		150
CH_3-CN	119(5)	121(5)

$CH_3CO-COCH_3$		67(2)
$C_6H_5CO-COC_6H_5$		66.4
$C_6H_5CH_2CO-CH_2C_6H_5$		65.4
$C_6H_5CH_2-COOH$		68.1
$(C_6H_5CH_2)_2CH-COOH$		59.4
$NC-CN$	143(5)	144(5)
CF_3-NF_2		65(3)
CH_3-NH_2		79(3)
$C_6H_5CH_2-NH_2$		72(1)
$CH_3-NHC_6H_5$		68
$CH_3-N(CH_3)C_6H_5$		65
$C_6H_5CH_2-NHCH_3$		69(1)
$C_6H_5CH_2-N(CH_3)_2$		61(1)
$CH_3-(N=NCH_3)$		52.5
$C_2H_5-(N=NC_2H_5)$		50.0
$(CH_3)_3C-[N=NC(CH_3)_3]$		43.5
$C_6H_5CH_2-(N=NCH_2C_6H_5)$		37.6
$CF_3-(N=NCF_3)$		55.2
$H_2C=NH$		154(5)
$HC\equiv N$		224
CH_3-NO		41.8(9)
C_2H_5-NO		42.0(13)
C_3H_7-NO		40.1(18)
$(CH_3)_2CH-NO$		41.0(13)
C_4H_9-NO		51.5(10)
C_6H_5-NO		51.5(10)
Cl_3C-NO		32
F_3C-NO		31
C_6F_5-NO		50.5(10)
$NC-NO$		29(3)
CH_3-NO_2		59(3)
$C_2H_5-NO_2$		62
CH_3-OCH_3		80

TABLE 3-5 Bond strengths (*continued*)

Bond	D_0°, kcal · mol^{-1}	ΔHf_{298}, kcal · mol^{-1}
Carbon (*continued*)		
$CH_3-OC_6H_5$		91
$CH_3-OCH_2C_6H_5$		67
$C_2H_5-OC_6H_5$		51
$C_6H_5CH_2-OCOCH_3$	67	
$C_6H_5CH_2-OCOC_6H_5$		69
$CH_3CO-OCH_3$		97
$CH_3-O-SOCH_3$		67
$CH_2{=}CHCH_2-OSOCH_3$		50
$C_6H_5CH_2-OSOCH_3$		53
$C{=}O$	256.2(1)	257.3(1)
$H_2C{=}O$		175
$OC{=}O$	125.7(1)	127.2(1)
$SC{=}O$	148	150
$C{\equiv}O$		257
CH_3-SH	71(3)	73(3)
$CH_3-SC_6H_5$		68(2)
$CH_3-SCH_2C_6H_5$		59(2)
$OC-S$	72.9	74.2
$\dot{C}H_2-CH_3$		96
$\dot{C}H_2CH_2-CH_3$		25.5
$(\dot{C}H_2)_2C-CH_3$		51
$\dot{C}HCH-CH_3$		32
$\dot{C}_6H_{10}-CH_3$ (cyclohexyl ring)		27.5

Bond	D_0°, kcal · mol^{-1}	ΔHf_{298}, kcal · mol^{-1}
Chlorine (*continued*)		
$Cl-COC_6H_5$		74(3)
$Cl-Cl^+$		94
$Cl-Cl$	57.3(1)	
$Cl-ClO$	33.3(10)	
$O_3Cl-ClO_4$		58
$Cl-F$	59.5(5)	
O_3Cl-F		61
$Cl-N$		62
$Cl-NCl$		67
$Cl-NCl_2$		91
$Cl-NF_2$		*ca* 32
$Cl-NH_2$		60(6)
$Cl-NO$	37.0(15)	38.0(15)
$Cl-NO_2$	33(1)	34(1)
$Cl-O$	64(1)	
$OCl-O$	58(3)	
O_2Cl-O		48(1)
$Cl-SiCl_3$		111
$Cl-CH_3{}^+$		51
$Cl-Cl^+$		94
Fluorine		
$F-CH_3$		108(5)
$F-C(CH_3)_3$		105

[cyclohexenyl radical ring]—CH_3		35
[cyclohexadienyl radical ring]—CH_3		11.5
$(CH_3)_2C(\dot{C}H_2)-CH_3$		20
$\dot{O}CH_2-CH_3$		12
$(CH_3)_2C(\dot{O})-CH_3$		7
$\dot{C}H_2CO-CH_3$		30
$\dot{O}C-CH_3$		11
$\dot{O}_2C-CH_3$		−20
$CH_3-CH_3{}^+$		46
$CH_2-CH_3{}^+$		119
$CH_2{=}CH_2{}^+$		162
$HC{\equiv}CH^+$		223
Chlorine		
Cl—C	80(10)	80.8
Cl—CH_3		81(5)
Cl—$C(CH_3)_3$		78.5
Cl—CH_2Cl		74(3)
Cl—CCl_3		70(5)
Cl—CF_3		86.1(8)
Cl—CCl_2F		73(2)
Cl—CF_2Cl		76(2)
Cl—CF_2CF_3		82.7(17)
Cl—$(CH{=}CH_2)$		84
Cl—CN		105
Cl—COCl		78.5
Cl—$COCH_3$		83.5

F—C_6H_5		116
F—CCl_3		106(5)
F—CCl_2F		110(6)
F—$CClF_2$		117(6)
F—CF_3		125(4)
F—$COCH_3$		119
F—F	37(1)	
OF—F	64(3)	
O_2F—F	18.4	
F—N	71(10)	72(10)
F—NF	75(5)	76(5)
F—NF_2	57(2)	58(2)
F—NO	55.2(10)	56.3(10)
F—NO_2	46(5)	
F—F^+	>60	
Gallium		
$CH_3-Ga(CH_3)_3$	59.5	
Hydrogen		
H—Br	86.6(1)	87.5(1)
H—C	80	81.0(5)
H—CH		108(6)
H—CH_2	112.3(1)	113(1)
H—CH_3	102(2)	103(2)
D—CD_3	104.92(5)	
H—$(C{\equiv}CH)$		125(1)
H—$(C{=}CH_2)$		102
H—CH_2CH_3		98(1)
H—$CH_2C{\equiv}CH$		93.9(12)
H—$CH_2CH{=}CH_2$		85

TABLE 3-5 Bond strengths (*continued*)

Bond	D_0°, $kcal \cdot mol^{-1}$	ΔHf_{298}, $kcal \cdot mol^{-1}$
Hydrogen (*continued*)		
H-cyclopropyl		101(3)
$H—CH_2CH_2CH_3$		98(2)
$H—CH(CH_3)_2$		94.5
H—cyclobutyl		95(3)
$H—CH_2CH(CH_3)_2$		86
$H—CH(CH_3)CH_2CH_3$		95(1)
$H—C(CH_3)_3$		91
H—cyclopentadienyl (structure)		81(1)
$H—CH(CH=CH_2)_2$		80(1)
H—cyclopentenyl (structure)		82(1)
$H—CH(CH_2)C(CH_2)_2$ (structure: H—CH, CH₂ / C / CH₂, CH₂)		99(1)
$H—C(CH_3)_2CH=CH_2$		79
H—cyclopentyl		94.5(10)
$H—CH_2C(CH_3)_3$		100(1)
$H—C_6H_5$		103
$H—CH_2C_6H_5$		85(1)

Bond	D_0°, $kcal \cdot mol^{-1}$	ΔHf_{298}, $kcal \cdot mol^{-1}$
Hydrogen (*continued*)		
H—Cl	102.3(1)	
H—CO		30(2)
H—CHO		87(1)
H—COOH		90
$H—COCH_3$		87(1)
$H—COCH_2CH_3$		87(1)
H—tetrahydrofuranyl (structure)		92(1)
$H—COC_6H_5$		87(1)
$H—COCF_3$		91(2)
H—F	135(1)	135.8
H—H	103.25	104.19
H—D	104.07(1)	105.00
D—D	105.05(1)	105.96
H—I	70.4(1)	71.3(1)
H—N	85(2)	85(2)
H—NH	89(2)	90(2)
$H—NH_2$	103(2)	104(2)
$H—NHCH_3$		103(2)
$H—N(CH_3)_2$		95(2)
$H—NHC_6H_5$		80(3)
$H—N(CH_3)C_6H_5$		74(3)
$H—NF_2$		76(3)
$H—N_3$		85

$H-C(C_6H_5)_3$		75	$H-NO$		<49
H–(cyclohexa-2,5-dienyl)		74	$H-O$	101.3(5)	102.3(5)
H—cyclohexyl		95.5(10)	$H-OH$	118.0(2)	119.2(2)
H—cycloheptyl		92.5(10)	$H-OCH_3$		104.4(10)
H—norbornyl		97(3)	$H-OCH_2CH_3$		104.2
$H-CH_2Br$	97(5)		$H-OC(CH_3)_3$		105(1)
$H-CHBr_2$		104	$H-OCH_2C(CH_3)_3$		102.3(15)
$H-CBr_3$	88(2)	90(2)	$H-OC_6H_5$		88(5)
$H-CH_2Cl$		101	$H-ONO$		78.3(5)
$H-CHCl_2$		99.0	$H-ONO_2$		101.2(5)
$H-CCl_3$	89(3)	90(3)	$H-OOH$	88.5(20)	89.5(20)
$H-CCl_2CHCl_2$		94(2)	$H-OOCCH_3$		112(4)
$H-CCl_2CCl_3$		95(2)	$H-OOCCH_2CH_3$		110(4)
$H-CH_2F$		101(2)	$H-OOCC_3H_7$		103(4)
$H-CHF_2$		101(2)	$H-SH$	90(1)	91(1)
$H-CF_3$	105(3)	106(3)	$H-SCH_3$		ca 88
$H-CF_2Cl$		104(1)	$H-SiH_3$		94(3)
$H-CH_2CF_3$		106.7(11)	$H-Si(CH_3)_3$		90(3)
$H-CF_2CH_3$		99.5(1)	$H-SiCl_3$		91.3(14)
$H-CF_2CF_3$		103.1(15)	$\dot{C}H_2-H$		106
$H-CF_2CF_2CF_3$		104(2)	$\dot{C}H-H$		106
$H-CH_2I$		103(2)	$\dot{C}-H$		81
$H-CHI_2$		103(2)	$\dot{C}H_2CH_2-H$		39
$H-CN$	127(5)	129(5)	$\dot{O}CH_2-H$		22
$H-CH_2CN$		ca 93	$\dot{C}O-H$		19
$H-CH(CH_3)CN$		90(2)	$\dot{C}HCH-H$		43
$H-C(CH_3)_2CN$		87(2)	$H-\dot{O}$		102
$H-CH_2NH_2$		95(2)	$H-O\dot{O}$		47
$H-CH_2Si(CH_3)_3$		99(1)	$H-O\dot{C}H_2$		31
$H-CH_2COCH_3$		98.3(18)	$H-OO\dot{C}$		31
			$\dot{C}H-H$		ca 125
			$\dot{C}OCH_2-H$		43.5

TABLE 3-5 Bond strengths (*continued*)

Bond	D_0°, kcal · mol^{-1}	ΔHf_{298}, kcal · mol^{-1}
Hydrogen (*continued*)		
$\dot{C}H_2CO—H$		36
[ring structure]—H		40
[ring structure]—H		47.5
[ring structure]—H		24
$C^+—H$		85
$CH_3{}^+—H$		30
$CH_3CH_2{}^+—H$		29
$CH_2CH_3{}^+—H$		79
$H—H^+$		62
Iodine		
I—Br	41.9(1)	42.5(1)
$I—CH_3$	54(3)	55.5(30)
$I—CH_2CH_3$		53.5
$I—CH(CH_3)_2$		53
$I—C(CH_3)_2$		49.5
$I—CH_2CF_3$		56(1)
$I—CF_2CH_3$		52(1)

Bond	D_0°, kcal · mol^{-1}	ΔHf_{298}, kcal · mol^{-1}
Nitrogen		
N—N	225.07(1)	225.96(1)
$F_2N—NF_2$	20(1)	21(1)
$H_2N—NH_2$		71(2)
$H_2N—NHCH_3$		65
$H_2N—N(CH_3)_2$		63
$H_2N—NHC_6H_5$		51
$HN—N_2$		9(1)
ON—N	113.5(10)	114.9(10)
$ON—NO_2$	8.4(2)	9.5(2)
$O_2N—NO_2$	12.7(5)	13.7(5)
NN—O		40
ON—O		73
HN=NH		109(10)
HN=O		115
N≡N		226
$N—N^+$		200
$N—NO^+$		155
$NN—O^+$		56
$ON^+—O$		56
Osmium		
$O_3Os—O$		72(5)
Oxygen		
$HO—CH_3$	88.5(30)	90(3)
$HO—(CH=CH_2)$		87

$I-CF_2CF_3$		51(1)
$I-C_3F_7$		50(1)
$I-(CH{=}CHCH_3)$		41
$I-CH_3^+$		62
$I-C_6H_5$		64(1)
$I-C_6F_5$		66
$I-Cl$	49.7(1)	50.5(1)
$I-COCH_3$		52.5
$I-CN$		73(1)
$I-F$	66.4(10)	67(1)
I^+-H		70
$I-I$	35.60(1)	36.15
$I-I^+$		61
$I-NO$		17(1)
$I-NO_2$		18(1)
Lead		
$CH_3-Pb(CH_3)_3$		49.4(10)
Lithium		
$Li-H$	58	
Mercury		
$Hg-Br$	16.4(10)	17.4(10)
CH_3-HgCH_3		57.5
$C_2H_5-HgC_2H_5$		43.7(10)
$C_3H_7-HgC_3H_7$		47.1
$(CH_3)_2CH-HgCH(CH_3)_2$		40.7
$C_6H_5-HgC_6H_5$		68
$HO-CH_2CH{=}CH_2$		109
$HO-C_6H_5$		103
$HO-CH_2C_6H_5$		77
$HO-CHO$		96(3)
$HO-COCH_3$		108(5)
$HO-COCH_2CH_3$		43
$HO-Cl$		60(3)
$HO-I$		56(3)
$HO-NCH_3$		50
$HO-OC(CH_3)_3$		46(2)
$O-O$	117.97(10)	119.11
$HO-OH$	49.5(5)	51.1(5)
CF_3O-OCF_3		46
CH_3O-OCH_3		37.6(2)
$C_2H_5O-OC_2H_5$		38
$C_3H_7O-OC_3H_7$		37
$O-OF$	110.7	
$O-O_2ClF$		58
$FO-OF$		62(20)
$O{=}PBr_3$		119(5)
$O{=}PCl_3$		122(5)
$O{=}PF_3$		130(5)
$O-O^+$		168
$HO-CH_3^+$		67
Phosphorus		
$P-Br$		63,7
$P-Cl$		78.5
$P-F$		117
$P-H$		79(1)
$P-O$	141.5(10)	142.3(10)
$P-P$	115(2)	116(2)
$P{=}S$	82	

TABLE 3-5 Bond strengths (*continued*)

Bond	D_0°, $kcal \cdot mol^{-1}$	ΔHf_{298}, $kcal \cdot mol^{-1}$
Ruthenium		
O—RuO_3	104	
Selenium		
Se—Cl		58
Se—F		68
Se—O	81(23)	
Se—Se	65	
Silicon		
Si—Br	69(14)	
Si—Cl	76(12)	
Si—F		135
Si—H	74(6)	
Si—I		56
Si—N	ca 104	
Si—O	185(7)	
Si—S	147(3)	
Si—Se	134(6)	
Si—Si		42
H_3Si—SiH_3		81(4)
$(CH_3)_3Si$—$Si(CH_3)_3$		81
$(C_6H_5)_3Si$—$Si(C_6H_5)_3$		88(7)
Si—Te	122(9)	
Sulfur		
S—Cl		61
O_2S—F	16	
S—N	115	
S—O	123.6(20)	
OS—O	130.8(20)	
O_2S—O	81.9(10)	83.2(10)
S—S	101.5(15)	102.5(15)
HS—SH		65(5)
S—Te	60	
HS^+—H		104
HS—H^+		161
OS—O^+		155
Tin		
BrSn—Br		78
Br_3Sn—Br		65
C_2H_5Sn—$(C_2H_5)_3$		ca 57
Sn—Cl		76
Sn—H		61.0(7)
Sn—I		65
Sn—O	130(5)	131(5)
Sn—S	111(5)	112(5)
Xenon		
Xe—F		31(1)

Sodium			Zinc		
Na—H	47		Zn—H	19.6(5)	
Na—K	14.3		$C_2H_5Zn—C_2H_5$		ca 48
Na—Na	17.3				
Na—OH		91(3)			

BOND AND GROUP DIPOLE MOMENTS

All bonds between equal atoms are given zero values. Because of their symmetry, methane and ethane molecules are nonpolar. The principle of bond moments thus requires that the CH_3 group moment equal one H—C moment. Hence the substitution of any aliphatic H by CH_3 does not alter the dipole moment, and all saturated hydrocarbons have zero moments as long as the tetrahedral angles are maintained.

The group moment always includes the C—X bond. When the group is attached to an aromatic system, the moment contains the contributions through resonance of those polar structures postulated as arising through charge shifts around the ring.

All values for bond and group dipole moments in Tables 3-6 and 3-7 were obtained in benzene solution.

TABLE 3-6 Bond dipole moments

Bond	Moment, D*	Bond	Moment, D*
H—C		Se—C	0.7
Aliphatic	0.3	Si—C	1.2
Aromatic	0.0	Si—H	1.0
C—C	0.0	Si—N	1.55
C≡C	0.0	H—Sb	−0.08
C—O		G—As	−0.10
Ether, aliphatic	0.74	H—P	0.36
Alcohol, aliphatic	0.7	H—I	0.38
C=O		H—Br	0.78
Aliphatic	2.4	H—Cl	1.08
Aromatic	2.65	H—F	1.94
O—H	1.51	C—Te	0.6
C—S	0.9	N—F	0.17
C=S	2.0	P—I	0.3
S—H	0.65	P—Br	0.36
S—O	(0.2)	P—Cl	0.81
S=O		As—I	0.78
Aliphatic	2.8	As—Br	1.27
Aromatic	3.3	As—Cl	1.64
C—N, aliphatic	0.45	As—F	2.03
C=N	1.4	Sb—I	0.8
C≡N (nitrile)	3.6	Sb—Br	1.9
NC (isonitrile)	3.0	Sb—Cl	2.6
N—H	1.31	S—Cl	0.7
N—O	0.3	Cl—O	0.7
N=O	2.0	I—Br	1.2
N:, lone pair on sp^3N	1.0	I—Cl	1
C—P, aliphatic	0.8	Br—Cl	0.57
P—O	(0.3)	Br—F	1.3
P=O	2.7	Cl—F	0.88
P—S	0.5	Li—C	1.4
P=S	2.9	K—Cl	10.6
B—C, aliphatic	0.7	K—F	7.3
B—O	0.25		

TABLE 3-6 Bond dipole moments (*continued*)

Bond	Moment, D*	Bond	Moment, D*
Cs—Cl	10.5	Dative (coordination) bonds (*cont.*)	
Cs—F	7.9	P → O	2.9
Dative (coordination) bonds		S → O	3.0
		As → O	4.2
N → B	2.6	Se → O	3.1
O → B	3.6	Te → O	2.3
S → B	3.8	P → S	3.1
P → B	4.4	P → Se	3.2
N → O	4.3	Sb → S	4.5

* To convert debye units D into coulomb-meters, multiply by 3.33564×10^{-30}.

TABLE 3-7 Group dipole moments

Group	Moment, D*	
	Aromatic C—X	Aliphatic C—X
C—CH_3	0.37	0.0
C—C_2H_5	0.37	0.0
C—$C(CH_3)_3$	0.5	0.0
C—CH=CH_2	<0.4	0.6
C—C≡CH	0.7	0.9
C—F	1.47	1.79
C—Cl	1.59	1.87
C—Br	1.57	1.82
C—I	1.40	1.65
C—CH_2F	1.77(g)	
C—CF_3	2.54	2.32
C—CH_2Cl	1.85	1.95
C—$CHCl_2$	2.04	1.94
C—CCl_3	2.11	1.57
C—CH_2Br	1.86	1.96
C—C≡N	4.05	3.4
C—NC	3.5	3.5
C—CH_2CN	1.86	2.0
C—C=O	2.65	2.4
C—CHO	2.96	2.49
C—COOH	1.64	1.63
C—CO—CH_3	2.96	2.75
C—CO—OCH_3	1.83	1.75
C—CO—OC_2H_5	1.9	1.8
C—OH	1.6	1.7
C—OCH_3	1.28	1.28
C—OCF_3	2.36	

TABLE 3-7 Group dipole moments (*continued*)

	Moment, D*	
Group	Aromatic C—X	Aliphatic C—X
C—$OCOCH_3$	1.69	
C—OC_6H_5	1.16	1.16
C—CH_2OH	1.68	1.68
C—NH_2	1.53	1.46
C—$NHCH_3$	1.71	
C—$N(CH_3)_2$	1.58	0.86
C—$NHCOCH_3$	3.69	
C—$N(C_6H_5)_2$	(0.3)	−0.3
C—NCO	2.32	2.8
C—N_3	1.44	
C—NO	3.09	
C—NO_2	4.01	2.70
C—CH_2NO_2	3.3	3.4
C—SH	1.22	1.55
C—SCH_3	1.34	1.40
C—SCF_3	2.50	
C—SCN	3.59	3.6
C—NCS	2.9	3.3
C—SC_6H_5	1.51	1.5
C—SF_5	3.4	
C—$SOCF_3$	3.88	
(C—)$_2SO_2$	5.05	4.53
(C—)$_2SO_2CH_3$	4.73	
(C—)$_2SO_2CF_3$	4.32	
C—SeH	1.08	
C—$SeCH_3$	1.31	1.32
C—$Si(CH_3)_3$	0.44	0.4

* To convert debye units D into coulomb-meters, multiply by 3.33564×10^{-30}.

SECTION 4

PHYSICAL PROPERTIES

SOLUBILITIES

TABLE 4-1 Solubility of Gases in Water

The column (or line entry) headed "α" gives the volume of gas (in milliliters) measured at standard conditions (0°C and 760 mm or 101.325 kN $\cdot$ m^{-2}) dissolved in 1 mL of water at the temperature stated (in degrees Celsius) and when the pressure of the gas without that of the water vapor is 760 mm. The line entry "A" indicates the same quantity except that the gas itself is at the uniform pressure of 760 mm when in equilibrium with water.

The column headed "l" gives the volume of the gas (in milliliters) dissolved in 1 mL of water when the pressure of the gas plus that of the water vapor is 760 mm.

The column headed "q" gives the weight of gas (in grams) dissolved in 100 g of water when the pressure of the gas plus that of the water vapor is 760 mm.

	Acetylene		Air*		Ammonia		Bromine	
Temp. °C	α	q	$\alpha(\times 10^3)$	% oxygen in air	α	q	α	q
0	1.73	0.200	29.18	34.91	1130	89.5	60.5	42.9
1	1.68	0.194	28.42	34.87	—	—	—	—
2	1.63	0.188	27.69	34.82	—	—	54.1	38.3
3	1.58	0.182	26.99	34.78	—	—	—	—
4	1.53	0.176	26.32	34.74	1047	79.6	48.3	34.2
5	1.49	0.171	25.68	34.69	—	—	—	—
6	1.45	0.167	25.06	34.65	—	—	43.3	30.6
7	1.41	0.162	24.47	34.60	—	—	—	—
8	1.37	0.157	23.90	34.56	947	72.0	38.9	27.5
9	1.34	0.154	23.36	34.52	—	—	—	—
10	1.31	0.150	22.84	34.47	870	68.4	35.1	24.8
11	1.27	0.146	22.34	34.43	—	—	—	—
12	1.24	0.142	21.87	34.38	857	65.1	31.5	22.2
13	1.21	0.138	21.41	34.34	837	63.6	—	—
14	1.18	0.135	20.97	34.30	—	—	28.4	20.0
15	1.15	0.131	20.55	34.25	770	—	—	—
16	1.13	0.129	20.14	34.21	775	58.7	25.7	18.0
17	1.10	0.125	19.75	34.17	—	—	—	—
18	1.08	0.123	19.38	34.12	—	—	23.4	16.4
19	1.05	0.119	19.02	34.08	—	—	—	—
20	1.03	0.117	18.68	34.03	680	52.9	21.3	14.9
21	1.01	0.115	18.34	33.99	—	—	—	—
22	0.99	0.112	18.01	33.95	—	—	19.4	13.5
23	0.97	0.110	17.69	33.90	—	—	—	—
24	0.95	0.107	17.38	33.86	639	48.2	17.7	12.3
25	0.93	0.105	17.08	33.82	—	—	—	—
26	0.91	0.102	16.79	33.77	—	—	16.3	11.3
27	0.89	0.100	16.50	33.73	—	—	—	—
28	0.87	0.098	16.21	33.68	586	44.0	15.0	10.3
29	0.85	0.095	15.92	33.64	—	—	—	—
30	0.84	0.094	15.64	33.60	530	41.0	13.8	9.5
35	—	—	—	—	—	—	—	—
40	—	—	14.18	—	400	31.6	9.4	6.3
45	—	—	—	—	—	—	—	—
50	—	—	12.97	—	290	23.5	6.5	4.1
60	—	—	12.16	—	200	16.8	4.9	2.9
70	—	—	—	—	—	11.1	3.8	1.9
80	—	—	11.26	—	—	6.5	3.0	1.2
90	—	—	—	—	—	3.0	—	—
100	—	—	11.05	—	—	0.0	—	—

* Free from NH_3 and CO_2; total pressure of air + water vapor is 760 mm.

TABLE 4-1 Solubility of gases in water (*continued*)

Temp. °C	Carbon dioxide		Carbon monoxide		Chlorine		Ethane		Ethylene		Hydrogen	
	α	q	α	q	l	q	α	q	α	q	α	q
0	1.713	0.334 6	0.035 37	0.004 397	—	—	0.098 74	0.013 17	0.226	0.028 1	0.021 48	0.000 192 2
1	1.646	0.321 3	0.034 55	0.004 293	—	—	0.094 76	0.012 63	0.219	0.027 2	0.021 26	0.000 190 1
2	1.584	0.309 1	0.033 75	0.004 191	—	—	0.090 93	0.012 12	0.211	0.026 2	0.021 05	0.000 188 1
3	1.527	0.297 8	0.032 97	0.004 092	—	—	0.087 25	0.011 62	0.204	0.025 3	0.020 84	0.000 186 2
4	1.473	0.287 1	0.032 22	0.003 996	—	—	0.083 72	0.011 14	0.197	0.024 4	0.020 64	0.000 184 3
5	1.424	0.277 4	0.031 49	0.003 903	—	—	0.080 33	0.010 69	0.191	0.023 7	0.020 44	0.000 182 4
6	1.377	0.268 1	0.030 78	0.003 813	—	—	0.077 09	0.010 25	0.184	0.022 8	0.020 25	0.000 180 6
7	1.331	0.258 9	0.030 09	0.003 725	—	—	0.074 00	0.009 83	0.178	0.022 0	0.020 07	0.000 178 9
8	1.282	0.249 2	0.029 42	0.003 640	—	—	0.071 06	0.009 43	0.173	0.021 4	0.019 89	0.000 177 2
9	1.237	0.240 3	0.028 78	0.003 559	—	—	0.068 26	0.009 06	0.167	0.020 7	0.019 72	0.000 175 6
10	1.194	0.231 8	0.028 16	0.003 479	3.148	0.997 2	0.065 61	0.008 70	0.162	0.020 0	0.019 55	0.000 174 0
11	1.154	0.223 9	0.027 57	0.003 405	3.047	0.965 4	0.063 28	0.008 38	0.157	0.019 4	0.019 40	0.000 172 5
12	1.117	0.216 5	0.027 01	0.003 332	2.950	0.934 6	0.061 06	0.008 08	0.152	0.018 8	0.019 25	0.000 171 0
13	1.083	0.209 8	0.026 46	0.003 261	2.856	0.905 0	0.058 94	0.007 80	0.148	0.018 3	0.019 11	0.000 169 6
14	1.050	0.203 2	0.025 93	0.003 194	2.767	0.876 8	0.056 94	0.007 53	0.143	0.017 6	0.018 97	0.000 168 2
15	1.019	0.197 0	0.025 43	0.003 130	2.680	0.849 5	0.055 04	0.007 27	0.139	0.017 1	0.018 83	0.000 166 8
16	0.985	0.190 3	0.024 94	0.003 066	2.597	0.823 2	0.053 26	0.007 03	0.136	0.016 7	0.018 69	0.000 165 4
17	0.956	0.184 5	0.024 48	0.003 007	2.517	0.797 9	0.051 59	0.006 80	0.132	0.016 2	0.018 56	0.000 164 1
18	0.928	0.178 9	0.024 02	0.002 947	2.440	0.773 8	0.050 03	0.006 59	0.129	0.015 8	0.018 44	0.000 162 8
19	0.902	0.173 7	0.023 60	0.002 891	2.368	0.751 0	0.048 58	0.006 39	0.125	0.015 3	0.018 31	0.000 161 6
20	0.878	0.168 8	0.023 19	0.002 838	2.299	0.729 3	0.047 24	0.006 20	0.122	0.014 9	0.018 19	0.000 160 3
21	0.854	0.164 0	0.022 81	0.002 789	2.238	0.710 0	0.045 89	0.006 02	0.119	0.014 6	0.018 05	0.000 158 8
22	0.829	0.159 0	0.022 44	0.002 739	2.180	0.691 8	0.044 59	0.005 84	0.116	0.014 2	0.017 92	0.000 157 5
23	0.804	0.154 0	0.022 08	0.002 691	2.123	0.673 9	0.043 35	0.005 67	0.114	0.013 9	0.017 79	0.000 156 1
24	0.781	0.149 3	0.021 74	0.002 646	2.070	0.657 2	0.042 17	0.005 51	0.111	0.013 5	0.017 66	0.000 154 8

TABLE 4-1 Solubility of gases in water (*continued*)

Temp. °C	Carbon dioxide		Carbon monoxide		Chlorine		Ethane		Ethylene		Hydrogen	
	α	q	α	q	l	q	α	q	α	q	α	q
25	0.759	0.144 9	0.021 42	0.002 603	2.019	0.641 3	0.041 04	0.005 35	0.108	0.013 1	0.017 54	0.000 153 5
26	0.738	0.140 6	0.021 10	0.002 560	1.970	0.625 9	0.039 97	0.005 20	0.106	0.012 9	0.017 42	0.000 152 2
27	0.718	0.136 6	0.020 80	0.002 519	1.923	0.611 2	0.038 95	0.005 06	0.104	0.012 6	0.017 31	0.000 150 9
28	0.699	0.132 7	0.020 51	0.002 479	1.880	0.597 5	0.037 99	0.004 93	0.102	0.012 3	0.017 20	0.000 149 6
29	0.682	0.129 2	0.020 24	0.002 442	1.839	0.584 7	0.037 09	0.004 80	0.100	0.012 1	0.017 09	0.000 148 4
30	0.665	0.125 7	0.019 98	0.002 405	1.799	0.572 3	0.036 24	0.004 68	0.098	0.011 8	0.016 99	0.000 147 4
35	0.592	0.110 5	0.018 77	0.002 231	1.602	0.510 4	0.032 30	0.004 12	—	—	0.016 66	0.000 142 5
40	0.530	0.097 3	0.017 75	0.002 075	1.438	0.459 0	0.029 15	0.003 66	—	—	0.016 44	0.000 138 4
45	0.479	0.086 0	0.016 90	0.001 933	1.322	0.422 8	0.026 60	0.003 27	—	—	0.016 24	0.000 134 1
50	0.436	0.076 1	0.016 15	0.001 797	1.225	0.392 5	0.024 59	0.002 94	—	—	0.016 08	0.000 128 7
60	0.359	0.057 6	0.014 88	0.001 522	1.023	0.329 5	0.021 77	0.002 39	—	—	0.016 00	0.000 117 8
70	—	—	0.014 40	0.001 276	0.862	0.279 3	0.019 48	0.001 85	—	—	0.016 0	0.000 102
80	—	—	0.014 30	0.000 980	0.683	0.222 7	0.018 26	0.001 34	—	—	0.016 0	0.000 079
90	—	—	0.014 2	0.000 57	0.39	0.127	0.017 6	0.000 8	—	—	0.016 0	0.000 046
100	—	—	0.014 1	0.000 00	0.00	0.000	0.017 2	0.000 0	—	—	0.016 0	0.000 000

TABLE 4-1 Solubility of gases in water (*continued*)

Temp. °C	Hydrogen sulfide		Methane		Nitric oxide		Nitrogen*		Oxygen		Sulfur dioxide	
	α	q	α	q	α	q	α	q	α	q	l	q
0	4.670	0.706 6	0.055 63	0.003 959	0.073 81	0.009 833	0.023 54	0.002 942	0.048 89	0.006 945	79.789	22.83
1	4.522	0.683 9	0.054 01	0.003 842	0.071 84	0.009 564	0.022 97	0.002 869	0.047 58	0.006 756	77.210	22.09
2	4.379	0.661 9	0.052 44	0.003 728	0.069 93	0.009 305	0.022 41	0.002 798	0.046 33	0.006 574	74.691	21.37
3	4.241	0.640 7	0.050 93	0.003 619	0.068 09	0.009 057	0.021 87	0.002 730	0.045 12	0.006 400	72.230	20.66
4	4.107	0.620 1	0.049 46	0.003 513	0.066 32	0.008 816	0.021 35	0.002 663	0.043 97	0.006 232	69.828	19.98
5	3.977	0.600 1	0.048 05	0.003 410	0.064 61	0.008 584	0.020 86	0.002 600	0.042 87	0.006 072	67.485	19.31
6	3.852	0.580 9	0.046 69	0.003 312	0.062 98	0.008 361	0.020 37	0.002 537	0.041 80	0.005 918	65.200	18.65
7	3.732	0.562 4	0.045 39	0.003 217	0.061 40	0.008 147	0.019 90	0.002 477	0.040 80	0.005 773	62.973	18.02
8	3.616	0.544 6	0.044 13	0.003 127	0.059 90	0.007 943	0.019 45	0.002 419	0.039 83	0.005 632	60.805	17.40
9	3.505	0.527 6	0.042 92	0.003 039	0.058 46	0.007 747	0.019 02	0.002 365	0.038 91	0.005 498	58.697	16.80
10	3.399	0.511 2	0.041 77	0.002 955	0.057 09	0.007 560	0.018 61	0.002 312	0.038 02	0.005 368	56.647	16.21
11	3.300	0.496 0	0.040 72	0.002 879	0.055 87	0.007 393	0.018 23	0.002 263	0.037 18	0.005 246	54.655	15.64
12	3.206	0.481 4	0.039 70	0.002 805	0.054 70	0.007 233	0.017 86	0.002 216	0.036 37	0.005 128	52.723	15.09
13	3.115	0.467 4	0.038 72	0.002 733	0.053 57	0.007 078	0.017 50	0.002 170	0.035 59	0.005 014	50.849	14.56
14	3.028	0.454 0	0.037 79	0.002 665	0.052 50	0.006 930	0.017 17	0.002 126	0.034 86	0.004 906	49.033	14.04
15	2.945	0.441 1	0.036 90	0.002 599	0.051 47	0.006 788	0.016 85	0.002 085	0.034 15	0.004 802	47.276	13.54
16	2.865	0.428 7	0.036 06	0.002 538	0.050 49	0.006 652	0.016 54	0.002 045	0.033 48	0.004 703	45.578	13.05
17	2.789	0.416 9	0.035 25	0.002 478	0.049 56	0.006 524	0.016 25	0.002 006	0.032 83	0.004 606	43.939	12.59
18	2.717	0.405 6	0.034 48	0.002 422	0.048 68	0.006 400	0.015 97	0.001 970	0.032 20	0.004 514	42.360	12.14
19	2.647	0.394 8	0.033 76	0.002 369	0.047 85	0.006 283	0.015 70	0.001 935	0.031 61	0.004 426	40.838	11.70

TABLE 4-1 Solubility of gases in water (*continued*)

Temp. °C	Hydrogen sulfide		Methane		Nitric acid		Nitrogen*		Oxygen		Sulfur dioxide	
	α	q	α	q	α	q	α	q	α	q	l	q
20	2.582	0.384 6	0.033 08	0.002 319	0.047 06	0.006 173	0.015 45	0.001 901	0.031 02	0.004 339	39.374	11.28
21	2.517	0.374 5	0.032 43	0.002 270	0.046 25	0.006 059	0.015 22	0.001 869	0.030 44	0.004 252	37.970	10.88
22	2.456	0.364 8	0.031 80	0.002 222	0.045 45	0.005 947	0.014 98	0.001 838	0.029 88	0.004 169	36.617	10.50
23	2.396	0.355 4	0.031 19	0.002 177	0.044 69	0.005 838	0.014 75	0.001 809	0.029 34	0.004 087	35.302	10.12
24	2.338	0.346 3	0.030 61	0.002 133	0.043 95	0.005 733	0.014 54	0.001 780	0.028 81	0.004 007	34.026	9.76
25	2.282	0.337 5	0.030 06	0.002 091	0.043 23	0.005 630	0.014 34	0.001 751	0.028 31	0.003 931	32.786	9.41
26	2.229	0.329 0	0.029 52	0.002 050	0.042 54	0.005 530	0.014 13	0.001 724	0.027 83	0.003 857	31.584	9.06
27	2.177	0.320 8	0.029 01	0.002 011	0.041 88	0.005 435	0.013 94	0.001 698	0.027 36	0.003 787	30.422	8.73
28	2.128	0.313 0	0.028 52	0.001 974	0.041 24	0.005 342	0.013 76	0.001 672	0.026 91	0.003 718	29.314	8.42
29	2.081	0.305 5	0.028 06	0.001 938	0.040 63	0.005 252	0.013 58	0.001 647	0.026 49	0.003 651	28.210	8.10
30	2.037	0.298 3	0.027 62	0.001 904	0.040 04	0.005 165	0.013 42	0.001 624	0.026 08	0.003 588	27.161	7.80
35	1.831	0.264 8	0.025 46	0.001 733	0.037 34	0.004 757	0.012 56	0.001 501	0.024 40	0.003 315	22.489	6.47
40	1.660	0.236 1	0.023 69	0.001 586	0.035 07	0.004 394	0.011 84	0.001 391	0.023 06	0.003 082	18.766	5.41
45	1.516	0.211 0	0.022 38	0.001 466	0.033 11	0.004 059	0.011 30	0.001 300	0.021 87	0.002 858	—	—
50	1.392	0.188 3	0.021 34	0.001 359	0.031 52	0.003 758	0.010 88	0.001 216	0.020 90	0.002 657	—	—
60	1.190	0.148 0	0.019 54	0.001 144	0.029 54	0.003 237	0.010 23	0.001 052	0.019 46	0.002 274	—	—
70	1.022	0.110 1	0.018 25	0.000 926	0.028 10	0.002 668	0.009 77	0.000 851	0.018 33	0.001 856	—	—
80	0.917	0.076 5	0.017 70	0.000 695	0.027 00	0.001 984	0.009 58	0.000 660	0.017 61	0.001 381	—	—
90	0.84	0.041	0.017 35	0.000 40	0.026 5	0.001 13	0.009 5	0.000 38	0.017 2	0.000 79	—	
100	0.81	0.000	0.017 0	0.000 00	0.026 3	0.000 00	0.009 5	0.000 00	0.017 0	0.000 00	—	—

* Atmospheric nitrogen containing 98.815% N_2 by volume + 1.185% inert gases.

TABLE 4-1 Solubility of gases in water (*continued*)

Substance		0°	10°	20°	30°	40°	60°	80°
Argon	α	0.052 8	0.041 3	0.033 7	0.028 8	0.025 1	0.020 9	0.018 4
Helium	A	0.009 8	0.009 11	0.008 6	0.008 39	0.008 41	0.009 02	$0.009\ 42^{70°}$
Hydrogen bromide	l	612	582		$533^{25°}$		$469^{50°}$	$406^{75°}$
Hydrogen chloride	α	512	475	442	412	385	339	
Krypton	α	0.110 5	0.081 0	0.062 6	0.051 1	0.043 3	0.035 7	
Neon	A		$0.011\ 7^{9°}$	0.010 6	0.010 0	$0.009\ 48^{42°}$		$0.009\ 84^{73°}$
Nitrous oxide	A		0.88	0.63				
Ozone	$g \cdot L^{-1}$	0.039 4	$0.029\ 9^{12°}$	$0.021\ 0^{19°}$	$0.0139^{27°}$	0.004 2	0	
Radon	α	0.510	0.326	0.222	0.162	0.126	0.085	
Xenon	α	0.242	0.174	0.123	0.098	0.082		

VAPOR PRESSURES

TABLE 4-2 Vapor pressure of mercury

Temp. °C	mm of Hg	Temp. °C	mm of Hg	Temp. °C	mm of Hg
0	0.000 185	78	0.078 89	158	3.873
2	0.000 228	80	0.088 80	160	4.189
4	0.000 276	82	0.100 0	162	4.528
6	0.000 335	84	0.112 4	164	4.890
8	0.000 406	86	0.126 1	166	5.277
10	0.000 490	88	0.1413	168	5.689
12	0.000 588	90	0.1582	170	6.128
14	0.000 706	92	0.1769	172	6.596
16	0.000 846	94	0.1976	174	7.095
18	0.001 009	96	0.2202	176	7.626
20	0.001 201	98	0.2453	178	8.193
22	0.001 426	100	0.2729	180	8.796
24	0.001 691	102	0.3032	182	9.436
26	0.002 000	104	0.3366	184	10.116
28	0.002 359	106	0.3731	186	10.839
30	0.002 777	108	0.4132	188	11.607
32	0.003 261	110	0.4572	190	12.423
34	0.003 823	112	0.5052	192	13.287
36	0.004 471	114	0.5576	194	14.203
38	0.005 219	116	0.6150	196	15.173
40	0.006 079	118	0.6776	198	16.200
42	0.007 067	120	0.7457	200	17.287
44	0.008 200	122	0.8198	202	18.437
46	0.009 497	124	0.9004	204	19.652
48	0.010 98	126	0.9882	206	20.936
50	0.012 67	128	1.084	208	22.292
52	0.014 59	130	1.186	210	23.723
54	0.016 77	132	1.298	212	25.233
56	0.019 25	134	1.419	214	26.826
58	0.022 06	136	1.551	216	28.504
60	0.025 24	138	1.692	218	30.271
62	0.028 83	140	1.845	220	32.133
64	0.032 87	142	2.010	222	34.092
66	0.037 40	144	2.188	224	36.153
68	0.042 51	146	2.379	226	38.318
70	0.048 25	148	2.585	228	40.595
72	0.054 69	150	2.807	230	42.989
74	0.061 89	152	3.046		
76	0.069 93	154	3.303		
		156	3.578		

TABLE 4-2 Vapor pressure of mercury (*continued*)

Temp. °C	mm of Hg	Temp. °C	mm of Hg	Temp. °C	mm of Hg
232	45.503	302	257.78	372	994.34
234	48.141	304	269.17	374	1028.9
236	50.909	306	280.98	376	1064.4
238	53.812	308	293.21	378	1100.9
240	56.855	310	305.89	380	1138.4
242	60.044	312	319.02	382	1177.0
244	63.384	314	332.62	384	1216.6
246	66.882	316	346.70	386	1257.3
248	70.543	318	361.26	388	1299.1
250	74.375	320	376.33	390	1341.9
252	78.381	322	391.92	392	1386.1
254	82.568	324	408.04	394	1431.3
256	86.944	326	424.71	396	1477.7
258	91.518	328	441.94	398	1525.2
260	96.296	330	459.74	400	1574.1
262	101.28	332	478.13	430	2464
264	106.48	334	497.12	460	3715
266	111.91	336	516.74	490	5420
268	117.57	338	53 7.00	520	7691
270	123.47	340	557.90	550	10650
272	129.62	342	579.45	600	22.87 atm
274	136.02	344	601.69	650	35.49 atm
276	142.69	346	624.64	700	52.51 atm
278	149.64	348	648.30	750	74.86 atm
280	156.87	350	672.69	800	103.31 atm
282	164.39	352	697.83	850	138.42 atm
284	172.21	354	723.73	900*	180.92 atm
286	180.34	356	750.43	950	226.58 atm
288	188.79	358	777.92	1000	290.5 atm
290	197.57	360	806.23	1050	358.1 atm
292	206.70	362	835.38	1100	437.3 atm
294	216.17	364	865.36	1150	521.3 atm
296	226.00	366	896.23	1200	616.8 atm
298	236.21	368	928.02	1250	721.4 atm
300	246.80	370	960.66	1300	835.9 atm

* Critical point.

TABLE 4-3 Vapor pressure of water

For temperatures from −10 to 120°C

The values in the table are for water in contact with its own vapor. Where the water is in contact with air at a temperature t in degrees Celsius, the following correction must be added: Correction (when $t \leq 40$ °C) = $p(0.775 - 0.000\,313\ t)/100$; correction (when $t > 50$ °C) = $p(0.0652 - 0.000\,087\,5\ t)/100$.

t, °C	p, mmHg	t, °C	p, mmHg	t, °C	p, mmHg	t, °C	p, mmHg
−10.0	2.149	11.5	10.176	22.2	20.070	30.8	33.312
−9.5	2.236	12.0	10.518	22.4	20.316	31.0	33.695
−9.0	2.326	12.5	10.870	22.6	20.565	31.2	34.082
−8.5	2.418	13.0	11.231	22.8	20.815	31.4	34.471
−8.0	2.514	13.5	11.604	23.0	21.068	31.6	34.864
−7.5	2.613	14.0	11.987	23.2	21.324	31.8	35.261
−7.0	2.715	14.5	12.382	23.4	21.583	32.0	35.663
−6.5	2.822	15.0	12.788	23.6	21.845	32.2	36.068
−6.0	2.931	15.2	12.953	23.8	22.110	32.4	36.477
−5.5	3.046	15.4	13.121	24.0	22.387	32.6	36.891
−5.0	3.163	15.6	13.290	24.2	22.648	32.8	37.308
−4.5	3.284	15.8	13.461	24.4	22.922	33.0	37.729
−4.0	3.410	16.0	13.634	24.6	23.198	33.2	38.155
−3.5	3.540	16.2	13.809	24.8	23.476	33.4	38.584
−3.0	3.673	16.4	13.987	25.0	23.756	33.6	39.018
−2.5	3.813	16.6	14.166	25.2	24.039	33.8	39.457
−2.0	3.956	16.8	13.347	25.4	24.326	34.0	39.898
−1.5	4.105	17.0	14.530	25.6	24.617	34.2	40.344
−1.0	4.258	17.2	14.715	25.8	24.912	34.4	40.796
−0.5	4.416	17.4	14.903	26.0	25.209	34.6	41.251
0.0	4.579	17.6	15.092	26.2	25.509	34.8	41.710
0.5	4.750	17.8	15.284	26.4	25.812	35.0	42.175
1.0	4.926	18.0	15.477	26.6	26.117	35.2	42.644
1.5	5.107	18.2	15.673	26.8	26.426	35.4	43.117
2.0	5.294	18.4	15.871	27.0	26.739	35.6	43.595
2.5	5.486	18.6	16.071	27.2	27.055	35.8	44.078
3.0	5.685	18.8	16.272	27.4	27.374	36.0	44.563
3.5	5.889	19.0	16.477	27.6	27.696	36.2	45.054
4.0	6.101	19.2	16.685	27.8	28.021	36.4	45.549
4.5	6.318	19.4	16.894	28.0	28.349	36.6	46.050
5.0	6.543	19.6	17.105	28.2	28.680	36.8	46.556
5.5	6.775	19.8	17.319	28.4	29.015	37.0	47.067
6.0	7.013	20.0	17.535	28.6	29.354	37.2	47.582
6.5	7.259	20.2	17.753	28.8	29.697	37.4	48.102
7.0	7.513	20.4	17.974	29.0	30.043	37.6	48.627
7.5	7.775	20.6	18.197	29.2	30.392	37.8	49.157
8.0	8.045	20.8	18.422	29.4	30.745	38.0	49.692
8.5	8.323	21.0	18.650	29.6	31.102	38.2	50.231
9.0	8.609	21.2	18.880	29.8	31.461	38.4	50.774
9.5	8.905	21.4	19.113	30.0	31.824	38.6	51.323
10.0	9.209	21.6	19.349	30.2	32.191	38.8	51.879
10.5	9.521	21.8	19.587	30.4	32.561	39.0	52.442
11.0	9.844	22.0	19.827	30.6	32.934	39.2	53.009

TABLE 4-3 Vapor pressure of water (*continued*)

t, °C	p, mmHg	t, °C	p, mmHg	t, °C	p, mmHg	t, °C	p, mmHg
39.4	54.580	58.5	139.34	78.5	334.2	96.4	667.31
39.6	54.156	59.0	142.60	79.0	341.0	96.6	672.20
39.8	54.737	59.5	145.99	79.5	348.1	96.8	677.12
40.0	55.324	60.0	149.38	80.0	355.1	97.0	682.07
40.5	56.81	60.5	152.91	80.5	362.4	97.2	687.04
41.0	58.34	61.0	156.43	81.0	369.7	97.4	692.05
41.5	59.90	61.5	160.10	81.5	377.3	97.6	697.10
42.0	61.50	62.0	163.77	82.0	384.9	97.8	702.17
42.5	63.13	62.5	167.58	82.5	392.8	98.0	707.27
43.0	64.80	63.0	171.38	83.0	400.6	98.2	712.40
43.5	66.51	63.5	175.35	83.5	408.7	98.4	717.56
44.0	68.26	64.0	179.31	84.0	416.8	98.6	722.75
44.5	70.05	64.5	183.43	84.5	425.2	98.8	727.98
45.0	71.88	65.0	187.54	85.0	433.6	99.0	733.24
45.5	73.74	65.5	191.82	85.5	442.3	99.2	738.53
46.0	75.65	66.0	196.09	86.0	450.9	99.4	743.85
46.5	77.61	66.5	200.53	86.5	459.8	99.6	749.20
47.0	79.60	67.0	204.96	87.0	468.7	99.8	754.58
47.5	81.64	67.5	209.57	87.5	477.9	100.0	760.00
48.0	83.71	68.0	214.17	88.0	487.1	101.0	787.57
48.5	85.85	68.5	218.95	88.5	496.6	102.0	815.86
49.0	88.02	69.0	223.73	89.0	506.1	103.0	845.12
49.5	90.24	69.5	228.72	89.5	515.9	104.0	875.06
50.0	92.51	70.0	233.7	90.0	525.76	105.0	906.07
50.5	94.86	70.5	238.8	90.5	535.83	106.0	937.92
51.0	97.20	71.0	243.9	91.0	546.05	107.0	970.60
51.5	99.65	71.5	249.3	91.5	556.44	108.0	1004.42
52.0	102.09	72.0	254.6	92.0	566.99	109.0	1038.92
52.5	104.65	72.5	260.2	92.5	577.71	110.0	1074.56
53.0	107.20	73.0	265.7	93.0	588.60	111.0	1111.20
53.5	109.86	73.5	271.5	93.5	599.66	112.0	1148.74
54.0	112.51	74.0	277.2	94.0	610.90	113.0	1187.42
54.5	115.28	74.5	283.2	94.5	622.31	114.0	1227.25
55.0	118.04	75.0	289.1	95.0	633.90	115.0	1267.98
55.5	120.92	75.5	295.3	95.2	638.59	116.0	1309.94
56.0	123.80	76.0	301.4	95.4	643.30	117.0	1352.95
56.5	126.81	76.5	307.7	95.6	648.05	118.0	1397.18
57.0	129.82	77.0	314.1	95.8	652.82	119.0	1442.63
57.5	132.95	77.5	320.7	96.0	657.62	120.0	1489.14
58.0	136.08	78.0	327.3	96.2	662.45		

TABLE 4-4 Vapor pressure of deuterium oxide

t, °C	p, mmHg	t, °C	p, mmHg	t, °C	p, mmHg
0	3.65	20	15.2	80	331.6
1	3.93	30	28.0	90	495.5
2	4.29	40	49.3	100	722.2
3	4.65	50	83.6	101.43	760.0
3.8	5.05	60	136.6		
10	7.79	70	216.1		

BOILING POINTS

TABLE 4-5 Organic solvents arranged by boiling points

Name	BP, °C	Name	BP, °C
Ethylene oxide	10.6	2-Methyltetrahydrofuran	80.0
Chloroethane	12.3	Benzene	80.1
Furan	31.4	Cyclohexane	80.7
Methyl formate	31.5	Propyl formate	80.9
Diethyl ether	34.6	Acetonitrile	81.6
Propylene oxide	34.5	2-Propanol	82.4
Pentane	36.1	1,1-Dimethylethanol	82.4
Bromoethane	38.4	Cyclohexene	83.0
Dichloromethane	39.8	Diisopropylamine	83.5
Dimethoxymethane	42.3	1,2-Dichloroethane	83.7
Carbon disulfide	46.3	Thiophene	84.2
1-Isopropoxy-2-propanol	47.9	Trichloroethylene	87.2
Ethyl formate	54.2	Isopropyl acetate	88.2
Acetone	56.2	1-Bromo-2-methylpropane	91.5
Methyl acetate	56.3	2,5-Dimethylfuran	93–94
1,1-Dichloroethane	57.3	Ethyl chloroformate	94
Dichloroethylene	60.6	Allyl alcohol	96.6
Chloroform	61.2	1,2-Dichloropropane	96.8
Methanol	64.7	1-Propanol	97.2
Tetrahydrofuran	66.0	Heptane	98.4
Diisopropyl ether	68.0	1-Chloro-3-methylbutane	99
Hexane	68.7	Ethyl propionate	99.1
1-Chloro-2-methylpropane	68.9	2-Butanol	99.6
1,1,1-Trichloroethane	74.0	Formic acid	100.8
1,3-Dioxolane	74–75	Methylcyclohexane	100.9
Carbon tetrachloride	76.7	1,4-Dioxane	101.2
Ethyl acetate	77.1	Nitromethane	101.2
1-Chlorobutane	77.9	Propyl acetate	101.5
Ethanol	78.3	2-Pentanone	101.7
2-Butanone	79.6	3-Pentanone	102.0

TABLE 4-5 Organic solvents arranged by boiling points (*continued*)

Name	BP, °C	Name	BP, °C
2-Methyl-2-butanol	102.0	*o*-Xylene	144.4
1,1-Diethoxyethane	102.7	2-Methoxyethyl acetate	144.5
Butyl formate	106.6	1,1,2,2-Tetrachloroethane	146.3
2-Methyl-1-propanol	107.9	3-Heptanone	147.8
Toluene	110.6	Tribromomethane	149.6
sec-Butyl acetate	112.3	Nonane	150.8
1,1,2-Trichloroethane	113.5	2-Heptanone	151
Nitroethane	114.1	Isopropylbenzene	152.4
Pyridine	115.2	*N,N*-Dimethylformamide	153.0
3-Pentanol	115.6	Methoxybenzene	153.8
4-Methyl-2-pentanone	115.7	Ethyl lactate	154.5
1-Chloro-2,3-epoxypropane	116.1	Cyclohexanone	155.7
1-Butanol	117.7	Bromobenzene	156.2
Acetic acid	117.9	1,2,3-Trichloropropane	156.9
Isobutyl acetate	118.0	1-Hexanol	157.5
2-Pentanol	119.3	Propylbenzene	159.2
1-Bromo-3-methylbutane	119.7	Cyclohexanol	161.1
1-Methoxy-2-propanol	120.1	Bis(2-methoxyethyl)ether	160
2-Nitropropane	120.3	Isopentyl propionate	160.2
Tetrachloroethylene	121.1	2-Heptanol	160.4
Ethyl butyrate	121.6	Pentachloroethane	160.5
3-Hexanone	123	2-Furaldehyde	161.8
2,4-Dimethyl-3-pentanone	124	2,6-Dimethyl-4-heptanone	168.1
2-Methoxyethanol	124.6	4-Hydroxy-4-methyl-	
Octane	125.7	2-pentanone	169.2
Butyl acetate	126.1	2-Furanmethanol	170.0
Diethyl carbonate	126.8	Ethoxybenzene	170
2-Hexanone	127.2	2-Butoxyethanol	170.2
1-Chloro-2-propanol	127.4	Diisopentyl ether	173.4
2-Chloroethanol	128.6	Decane	174.2
3-Methyl-1-penten-2-one	129.5	1,3-Dichloro-2-propanol	174.3
1-Nitropropane	131.2	Cyclohexyl acetate	174–175
Chlorobenzene	131.7	1-Heptanol	175.8
1,2-Dibromoethane	131.7	Furfuryl acetate	175–177
4-Methyl-2-pentanol	131.7	1,3,3-Trimethyl-	
3-Methyl-1-butanol	132.0	2-oxabicyclo[2.2.2]octane	177.4
Cyclohexylamine	134.8	4-Isopropyl-	
2-Ethoxyethanol	134.8	1-methylbenzene	177.1
Ethylbenzene	136.2	Isopentyl butyrate	178.6
1-Pentanol	138	Bis(2-chloroethyl) ether	178.8
p-Xylene	138.4	2-Octanol	179
m-Xylene	139.1	1,2-Dichlorobenzene	180.4
Acetic anhydride	140.0	Ethyl acetoacetate	180.8
2,4-Pentanedione	140.6	Phenol	181.8
Isopentyl acetate	142	2-Ethyl-1-hexanol	184.3
Dibutyl ether	142.4	Aniline	184.4
4-Heptanone	143.7	Benzyl ethyl ether	185.0

TABLE 4-5 Organic solvents arranged by boiling points (*continued*)

Name	BP, °C	Name	BP, °C
Diethyl oxalate	185.4	2-(2-Ethoxyethoxy)ethyl acetate	218.5
1,2-Propanediol	188	Acetamide	221.2
Bis(2-ethoxyethyl) ether	188.4	Methyl salicylate	223.0
Dimethyl sulfoxide	189.0	Diethyl maleate	225.3
1,2-Ethanediol diacetate	190.2	1,4-Butanediol	230
Benzonitrile	191.0	Propyl benzoate	231.2
2,5-Hexanedione	191.4	1-Decanol	230.2
2-(2-Methoxyethoxy)-ethanol	194.1	Phenylacetonitrile	233.5
N,N-Dimethylaniline	194.2	Quinoline	237
1-Octanol	195.2	Tributyl borate	238.5
1,2-Ethanediol	197.3	Propylene carbonate	240
Diethyl malonate	199.3	2-Phenoxyethanol	240
Methyl benzoate	199.5	Bis(2-hydroxyethyl) ether	245
o-Toluidine	200.4	Dibutyl oxalate	245.5
p-Toluidine	200.6	Butyl benzoate	250
2-(2-Ethoxyethoxy)ethanol	202	1,2,3-Propanetriol triacetate	258–259
Acetophenone	202.1	1-Chloronaphthalene	259.3
1,2-Dibutoxyethane	203.6	Isopentyl benzoate	262
1-Phenylethanol	203.9	*trans*-Ethyl cinnamate	271.0
m-Toluidine	203.4	Bis[2-(methoxyethoxy)-ethyl]ether	275.3
Benzyl alcohol	205.5	1-Methoxy-2-nitrobenzene	277
Camphor	207	Isopentyl salicylate	277–278
1,3-Butanediol	207.5	1-Bromonaphthalene	281.1
1,2,3,4-Tetrahydro-naphthalene	207.6	Dimethyl *o*-phthalate	283.7
γ-Valerolactone	207–208	2,2′-(Ethylenedioxy)-bisethanol	285
o-Chloroaniline	208.8	Glycerol	290
Nitrobenzene	210.8	Diethyl *o*-phthalate	295
Ethyl benzoate	212.4	Benzyl benzoate	323.5
3,5,5-Trimethylcyclo-hex-2-en-1-one	215.2	Dibutyl *o*-phthalate	340.0
Naphthalene	217.7	Dibutyl decanedioate	344–345

TABLE 4-6 Molecular elevation of the boiling point

Ebullioscopic constants

Molecular weights can be determined with the relation

$$M = K_b \frac{1000 w_2}{w_1 \Delta T_b}$$

where ΔT_b is the elevation of the boiling point brought about by the addition of w_2 grams of solute to w_1 grams of solvent and K_b is the ebullioscopic constant. In the column headed "Barometric correction" is given the number of degrees for each millimeter of difference between the barometric reading and 760 mmHg to be subtracted from K_b if the pressure is lower, or added if higher, than 760 mm. In general, the effect is within experimental error if the pressure is within 10 mm of 760 mm.

Compound	Barometric correction	K_b
Acetic acid	0.000 8	3.07
Acetic anhydride		3.53
Acetone	0.000 4	1.71
Acetonitrile		1.30
Acetophenone		5.65
Aniline	0.000 9	3.52
Benzene	0.000 7	2.53
Benzonitrile		3.87
Bromobenzene	0.001 6	6.26
Bromoethane		2.53
2-Butanone		2.28
cis-2-Butene-1,4-diol		2.86
D-(+)-Camphor	0.001 5	5.611
Carbon disulfide	0.000 6	2.34
Carbon tetrachloride	0.001 3	5.03
Chlorobenzene	0.001 1	4.15
Chloroethane		1.95
Chloroform	0.000 9	3.63
Cyclohexane	0.000 7	2.79
1,2-Dibromoethane	0.001 6	6.608
1,1-Dichloroethane		3.13
1,2-Dichloroethane		3.44
Dichloromethane		2.60
Diethyl ether	0.000 5	2.02
Diethyl sulfide		3.23
Dimethoxymethane		2.125
N,N-Dimethylacetamide		3.22
Dimethyl sulfide		1.85
1,4-Dioxane		3.270
Ethanol	0.000 3	1.22
Ethoxybenzene		5.0
Ethyl acetate	0.000 7	2.77
Formic acid		2.4
Glycerol		6.52
Heptane	0.000 8	3.43
Hexane		2.75

TABLE 4-6 Molecular elevation of the boiling point (*continued*)
Ebullioscopic constants

Compound	Barometric correction	K_b
2-Hydroxybenzaldehyde		4.96
Iodoethane		5.16
Iodomethane		4.19
4-Isopropyl-1-methylbenzene		5.52
Methanol	0.000 2	0.83
Methoxybenzene		4.502
Methyl acetate	0.000 5	2.15
2-Methyl-2-butanol		2.255
3-Methyl-1-butanol		2.65
3-Methylbutyl acetate		4.83
Methyl formate		1.649
2-Methyl-1-propanol		2.166
2-Methyl-2-propanol		1.745
Naphthalene	0.001 4	5.80
Nitrobenzene		5.24
Nitroethane		2.60
Nitromethane		1.86
Octane		4.02
Pentyl acetate		4.83
Phenol	0.000 9	3.60
Piperidine		2.84
1-Propanol		1.59
Propionic acid		3.51
Propionitrile		1.87
Pyridine		2.710
Quinoline		5.84
1,1,2,2-Tetrachloroethylene		5.50
1,2,3,4-Tetrahydronaphthalene		5.582
Toluene	0.000 8	3.33
p-Toluidine		4.14
Trichloroethylene		4.43
1,1,2-Trichloro-1,2,2-trifluoroethane		5.75
Triethylamine		3.45
Water	0.000 1	0.512

TABLE 4-7 Binary azeotropic (constant-boiling) mixtures

A zeotrope is a mixture that can be separated by distillation.

A. Binary azeotropes containing water

System	BP of azeotrope, °C	Composition, wt %	
		Water	Other component
Inorganic acids			
Hydrogen bromide	126	52.5	47.5
Hydrogen chloride	108.58	79.78	20.22
Hydrogen fluoride	111.35	64.4	35.6
Hydrogen iodide	127	43	57
Hydrogen peroxide	zeotrope		
Nitric acid	120.7	32.6	67.4
Perchloric acid	203	28.4	71.6
Organic acids			
Formic acid	107.2	22.6	77.4
Acetic acid	zeotrope		
Propionic acid	99.9	82.3	17.7
Isobutyric acid	99.3	79	21
Butyric acid	99.4	81.6	18.4
Pentanoic acid	99.8	89	11
Isopentanoic acid	99.5	81.6	18.4
Perfluorobutyric acid	97	71	29
Crotonic acid	99.9	97.8	2.2
Alcohols			
Ethanol	78.17	4	96
Allyl alcohol	88.9	27.7	72.3
1-Propanol	71.7	71.7	28.3
2-Propanol	80.3	12.6	87.4
1-Butanol	92.7	42.5	57.5
2-Butanol	87.0	26.8	73.2
2-Methyl-2-propanol	79.9	11.7	88.3
1-Pentanol	95.8	54.4	45.6
2-Pentanol	91.7	36.5	63.5
3-Pentanol	91.7	36.0	64.0
2,2-Dimethyl-2-propanol	87.35	27.5	72.5
1-Hexanol	97.8	67.2	32.8
1-Octanol	99.4	90	10
Cyclopentanol	96.25	58	42
1-Heptanol	98.7	83	17
Phenol	99.52	90.8	9.2
2-Methoxyphenol	99.5	87.5	12.5

TABLE 4-7 Binary azeotropic (constant-boiling) mixtures (*continued*)

System	BP of azeotrope, °C	Composition, wt %	
		Water	Other component
	Alcohols (*continued*)		
1-Phenylphenol	99.95	98.75	1.25
Benzyl alcohol	99.9	91	9
2,3-Dimethyl-2,3-butanediol	zeotrope		
Furfuryl alcohol	98.5	80	20
	Aldehydes		
Propionaldehyde	47.5	2	98
Butyraldehyde	68	6	94
Pentanal	83	19	81
Paraldehyde	90	28.5	71.5
Furaldehyde	97.5	65	35
	Amines		
N-Methylbutylamine	82.7	15	85
Furfurylamine	99	74	26
Piperidine	92.8	35	65
Pyridine	93.6	41.3	58.7
2-Methylpyridine	93.5	48	52
3-Methylpyridine	97	60	40
4-Methylpyridine	97.35	62.8	37.2
2,6-Dimethylpyridine	96.02	51.8	48.2
Dibutylamine	97	50.5	49.5
Dihexylamine	99.8	92.8	7.2
Triallylamine	95	38	62
Tributylamine	99.65	79.7	20.3
Aniline	98.6	80.8	19.2
N-Ethylaniline	99.2	83.9	16.1
1-Methyl-2-(2-pyridyl)pyrrolidine	99.85	97.5	2.5
	Halogenated hydrocarbons		
Chloroform	56.1	2.8	97.2
Carbon tetrachloride	42.6	2.8	97.2
Trichloroethylene	73.4	17	83
Tetrachloroethylene	88.5	17.2	82.8
1,2-Dichloroethane	72	8.3	91.7
1-Chloropropane	44	2.2	97.8
1,2-Dichloropropane	78	12	88
Chlorobenzene	90.2	28.4	71.6

TABLE 4-7 Binary azeotropic (constant-boiling) mixtures (*continued*)

System	BP of azeotrope, °C	Composition, wt %	
		Water	Other component
Esters			
Ethyl formate	52.6	5	95
Isopropyl formate	65.0	3	97
Propyl formate	71.6	2.3	97.7
Isobutyl formate	80.4	7.8	92.2
Butyl formate	83.8	14.5	85.5
Isopentyl formate	90.2	21	79
Pentyl formate	91.6	28.4	71.6
Benzyl formate	99.2	80	20
Ethyl acetate	70.38	8.47	91.53
Allyl acetate	83	14.7	85.3
Isopropyl acetate	76.6	10.6	89.4
Propyl acetate	82.4	14	86
Isobutyl acetate	87.4	16.5	83.5
Butyl acetate	90.2	28.7	71.3
Isopentyl acetate	93.8	36.3	63.7
Pentyl acetate	95.2	41	59
Hexyl acetate	97.4	61	39
Phenyl acetate	98.9	75.1	24.9
Benzyl acetate	99.6	87.5	12.5
Methyl propionate	71.4	3.9	96.1
Ethyl propionate	81.2	10	90
Isopropyl propionate	85.2	19.9	80.1
Propyl propionate	88.9	23	77
Isobutyl propionate	92.75	52.2	47.8
Isopentyl propionate	96.55	48.5	51.5
Methyl butyrate	82.7	11.5	88.5
Ethyl butyrate	87.9	21.5	78.5
Propyl butyrate	94.1	36.4	63.6
Isobutyl butyrate	96.3	46	54
Butyl butyrate	97.2	53	47
Isopentyl butyrate	98.05	63.5	36.5
Methyl isobutyrate	77.7	6.8	93.2
Ethyl isobutyrate	85.2	15.2	84.8
Propyl isobutyrate	92.2	30.8	69.2
Isobutyl isobutyrate	95.5	39.4	60.6
Isopentyl isobutyrate	97.4	56.0	44.0
Methyl isopentanoate	87.2	19.2	80.8
Ethyl isopentanoate	92.2	30.2	69.8
Propyl isopentanoate	96.2	45.2	54.8
Isobutyl isopentanoate	97.4	55.8	44.2
Isopentyl isopentanoate	98.8	74.1	25.9

TABLE 4-7 Binary azeotropic (constant-boiling) mixtures (*continued*)

System	BP of azeotrope, °C	Composition, wt %	
		Water	Other component
Esters (*continued*)			
Ethyl pentanoate	94.5	40	60
Ethyl hexanoate	97.2	54	46
Methyl benzoate	99.08	79.2	20.8
Ethyl benzoate	99.4	84.0	16.0
Propyl benzoate	99.7	90.9	9.1
Butyl benzoate	99.9	94	6
Isopentyl benzoate	99.9	95.6	4.4
Ethyl phenylacetate	99.7	91.3	8.7
Methyl cinnamate	99.9	95.5	4.5
Methyl phthalate	99.95	97.5	2.5
Diethyl *o*-phthalate	99.98	98.0	2.0
Ethyl chloroacetate	95.2	45.1	54.9
Butyl chloroacetate	98.12	75.5	24.5
Methyl acrylate	71	7.2	92.8
Isobutyl carbonate	98.6	74	26
Ethyl crotonate	93.5	38	62
Methyl lactate	99	80	20
1,2-Ethanediol diacetate	99.7	84.6	15.4
Ethyl nitrate	74.35	22	78
Propyl nitrate	84.8	20	80
Isobutyl nitrate	89.0	25	75
Methyl sulfate	98.6	73	27
Ethers			
Ethyl vinyl ether	34.6	1.5	98.5
Diethyl ether	34.2	1.3	98.7
Ethyl propyl ether	59.5	4	96
Diisopropyl ether	62.2	4.5	95.5
Butyl ethyl ether	76.6	11.9	88.1
Diisobutyl ether	88.6	23	77
Dibutyl ether	92.9	33	67
Diisopentyl ether	97.4	54	46
1,1-Diethoxyethane	82.6	14.5	85.5
Diphenyl ether	99.33	96.75	3.25
Methoxybenzene	95.5	40.5	59.5
Hydrocarbons			
Pentane	34.6	1.4	98.6
Hexane	61.6	5.6	94.4
Heptane	79.2	12.9	87.1
2,2,4-Trimethylpentane	78.8	11.1	88.9

TABLE 4-7 Binary azeotropic (constant-boiling) mixtures (*continued*)

System	BP of azeotrope, °C	Composition, wt %	
		Water	Other component
Hydrocarbons (*continued*)			
Nonane	94.8	82	18
Undecane	98.85	96.0	4.0
Dodecane	99.45	98	2
Acrolein	52.4	2.6	97.4
Cyclohexene	70.8	8.93	91.07
Cyclohexane	69.5	8.4	91.6
1-Octene	88.0	28.7	71.3
Benzene	69.25	8.83	91.17
Toluene	84.1	13.5	86.5
Ethylbenzene	92.0	33.0	67.0
m-Xylene	92	35.8	64.2
Isopropylbenzene	95	43.8	56.2
Naphthalene	98.8	84	16
Ketones			
Acetone	zeotrope		
2-Butanone	73.5	11	89
2-Pentanone	83.3	19.5	80.5
Cyclopentanone	94.6	42.4	57.6
4-Methyl-2-pentanone	87.9	24.3	75.7
2-Heptanone	95	48	52
3-Heptanone	94.6	42.2	57.8
4-Heptanone	94.3	40.5	59.5
4-Hydroxy-4-methyl-2-pentanone	98.8	87.3	12.7
4-Methyl-3-penten-2-one	91.8	34.8	65.2
Nitriles			
Acetonitrile	76.5	16.3	83.7
Isobutyronitrile	82.5	23	77
Butyronitrile	88.7	32.5	67.5
Acrylonitrile	70.6	14.3	85.7
Miscellaneous			
Hydrazine	120	32.3	67.7
Acetamide	zeotrope		
Nitromethane	83.59	23.6	76.4
Nitroethane	87.22	28.5	71.5
2,5-Dimethylfuran	77.0	11.7	88.3
Trioxane	91.4	30	70
Carbon disulfide	42.6	2.8	97.2

TABLE 4-7 Binary azeotropic (constant-boiling) mixtures (*continued*)

B. Binary azeotropes containing organic acids

System	BP of azeotrope, °C	Composition, wt %	
		Acid	Other component
Formic acid			
2-Methylbutane	27.2	4	96
Pentane	34.2	20	80
Hexane	60.6	28	72
Methylcyclopentane	63.3	29	71
Cyclohexane	70.7	70	30
Methylcyclohexane	80.2	46.5	53.5
Heptane	78.2	56.5	43.5
Octane	90.5	63	37
Benzene	71.05	31	69
Toluene	85.8	50	50
o-Xylene	95.5	74	26
m-Xylene	92.8	71.8	28.2
Styrene	97.8	73	27
Iodomethane	42.1	6	94
Chloroform	59.15	15	85
Carbon tetrachloride	66.65	18.5	81.5
Trichloroethylene	74.1	25	75
Tetrachloroethylene	88.2	50	50
Bromoethane	38.2	3	97
1,2-Dibromoethane	94.7	51.5	48.5
1,2-Dichloroethane	77.4	14	86
1-Bromopropane	64.7	27	73
2-Bromopropane	56.0	14	86
1-Chloropropane	45.6	8	92
2-Chloropropane	34.7	1.5	98.5
1-Chloro-2-methylpropane	63.0	19	81
Bromobenzene	98.1	68	32
Chlorobenzene	93.7	59	41
Fluorobenzene	73.0	27	73
o-Chlorotoluene	100.2	83	17
Pyridine	127.43	61.4	38.6
2-Methylpyridine	158.0	25	75
2-Pentanone	105.3	32	68
3-Pentanone	105.4	33	67
Nitromethane	97.07	45.5	54.5
Diethyl sulfide	82.2	35	65
Diisopropyl sulfide	93.5	62	38
Dipropyl sulfide	98.0	83	17
Carbon disulfide	42.55	17	83

TABLE 4-7 Binary azeotropic (constant-boiling) mixtures (*continued*)

System	BP of azeotrope, °C	Composition, w %	
		Acid	Other component
Acetic acid			
Hexane	68.3	6.0	94.0
Heptane	91.7	23	67
Octane	105.7	53.7	46.3
Nonane	112.9	69	31
Decane	116.75	79.5	20.5
Undecane	117.9	95	5
Cyclohexane	78.8	9.6	90.4
Methylcyclohexane	96.3	31	69
Benzene	80.05	2.0	98.0
Toluene	100.6	28.1	71.9
o-Xylene	116.6	78	22
m-Xylene	115.35	72.5	27.5
p-Xylene	115.25	72	28
Ethylbenzene	114.65	66	34
Styrene	116.8	85.7	14.3
Isopropylbenzene	116.0	84	16
Triethylamine	163	67	33
Nitromethane	101.2	96	4
Nitroethane	112.4	30	70
Pyridine	138.1	51.1	48.9
2-Methylpyridine	144.1	40.4	59.6
3-Methylpyridine	152.5	30.4	69.6
4-Methylpyridine	154.3	30.3	69.7
2,6-Dimethylpyridine	148.1	22.9	77.1
Carbon tetrachloride	76	98.46	1.54
Trichloroethylene	86.5	96.2	3.8
Tetrachloroethylene	107.4	61.5	38.5
1,2-Dibromoethane	114.4	55	45
2-Iodopropane	88.3	9	91
1-Bromobutane	97.6	18	82
1-Bromo-2-methylpropane	90.2	12	88
Chlorobenzene	114.7	58.5	41.5
Trichloronitromethane	107.65	80.5	19.5
1,4-Dioxane	119.5	77	23
Diisopropyl sulfide	111.5	48	52
Propionic acid			
Heptane	97.8	2	98
Octane	120.9	21.5	78.5

TABLE 4-7 Binary azeotropic (constant-boiling) mixtures (*continued*)

System	BP of azeotrope, °C	Composition, w %	
		Acid	Other component
Acetic acid (*continued*)			
Nonane	134.3	54.0	46.0
Decane	139.8	80.5	19.5
o-Xylene	135.4	43	57
p-Xylene	132.5	34	66
1,3,5-Trimethylbenzene	139.3	77	23
Isopropylbenzene	139.0	65	35
Propylbenzene	139.5	75	25
Camphene	138.0	65	35
α-Pinene	136.4	58.5	41.5
Methoxybenzene	140.8	96	4
Pyridine	148.6	67.2	32.8
2-Methylpyridine	154.5	55.0	45.0
1,2-Dibromoethane	127.8	17.5	82.5
1-Iodo-2-methylpropane	119.5	9	91
Chlorobenzene	128.9	18	82
Dipropyl sulfide	136.5	45	55
Butyric acid			
Undecane	162.4	84.4	15.5
o-Xylene	143.0	10	90
m-Xylene	138.5	6	94
p-Xylene	137.8	5.5	94.5
Ethylbenzene	135.8	4	96
Styrene	143.5	15	85
1,2,4-Trimethylbenzene	159.5	45	55
1,3,5-Trimethylbenzene	158.0	38	62
Isopropylbenzene	149.5	20	80
Propylbenzene	154.5	28	72
Butylbenzene	162.5	75	25
Naphthalene	zeotrope		
Indene	163.7	84	16
Camphene	152.3	2.8	97.2
Methoxybenzene	152.9	12	88
Pyridine	163.2	92.0	8.0
2-Furaldehyde	159.4	42.5	57.5
1,2-Dibromoethane	131.1	3.5	96.5
1-Iodobutane	129.8	2.5	97.5
Chlorobenzene	131.75	2.8	97.2
1,4-Dichlorobenzene	162.0	57	43
o-Bromotoluene	163.0	72	28
m-Bromotoluene	163.6	79.5	20.5
p-Bromotoluene	161.5	75	25
α-Chlorotoluene	160.8	65	35
Ethyl bromoacetate	157.4	84	16
Propyl chloroacetate	160.5	40	60

TABLE 4-7 Binary azeotropic (constant-boiling) mixtures (*continued*)

System	BP of azeotrope, °C	Composition, w %	
		Acid	Other component
	Isobutyric acid		
2,7-Dimethyloctane	148.6	48	52
o-Xylene	141.0	22	78
m-Xylene	139.9	15	85
p-Xylene	136.4	13	87
Styrene	142.0	27	73
1,2,4-Trimethylbenzene	152.3	63	37
Isopropylbenzene	146.8	35	65
Propylbenzene	149.3	49	51
Camphene	148.1	45	55
D-Limonene	152.5	78	22
Methoxybenzene	149.0	42	58
Ethyl bromoacetate	153.0	40	60
Ethyl 2-oxopropionate	153.0	60	40
1,2-Dibromoethane	130.5	6.5	93.5
1-Iodobutane	128.8	7	93
1-Bromohexane	148.0	35	65
Bromobenzene	148.6	35	65
Chlorobenzene	131.5	8	92
o-Bromotoluene	153.9	85	15
α-Chlorotoluene	153.5	80	20
Diisopentyl ether	154.2	93	7
Ethyl bromoacetate	153.0	40	60

C. Binary azeotropes containing alcohol

System	BP of azeotrope, °C	Composition, w %	
		Alcohol	Other component
	Methanol		
Pentane	30.9	7	93
Cyclopentane	38.8	14	86
Cyclohexane	53.9	36.4	63.6
Methylcyclohexane	59.2	54	46
Heptane	59.1	51.5	48.5
Octane	62.8	67.5	32.5
Nonane	64.1	83.4	16.6
Benzene	57.5	39.1	60.9
Fluorobenzene	59.7	32	68
Toluene	63.5	72.5	27.5
Bromomethane	3.55	99.55	0.45
Iodomethane	37.8	95.5	4.5

TABLE 4-7 Binary azeotropic (constant-boiling) mixtures (*continued*)

System	BP of azeotrope, °C	Composition, w %	
		Alcohol	Other component
Methanol (*continued*)			
Bromodichloromethane	63.8	60	40
Chloroform	53.4	87.4	12.6
Carbon tetrachloride	55.7	79.44	20.56
Bromoethane	34.9	5.3	94.7
1,2-Dichloroethane	61.0	32	68
Trichloroethylene	59.3	38	62
1-Bromopropane	54.5	21	79
2-Bromopropane	48.6	15.0	85.0
1-Chloropropane	40.5	9.5	90.5
2-Chloropropane	33.4	6	94
2-Iodopropane	61.0	38	62
1-Chlorobutane	57.0	27	73
Isobutyl formate	64.6	95	5
Methyl acetate	53.5	19	81
Methyl acrylate	62.5	54	46
Methyl nitrate	52.5	73	27
Acetone	55.5	12.1	87.9
1,4-Dioxane	zeotrope		
Dipropyl ether	63.8	72	28
Methyl *tert*-butyl ether	51.3	14.3	85.7
Diethyl sulfide	61.2	62	38
Carbon disulfide	39.8	71	29
Thiophene	59.7	16.4	83.6
Nitromethane	64.4	9.1	90.9
Ethanol			
Pentane	34.3	5	95
Cyclopentane	44.7	7.5	92.5
Hexane	58.7	21	79
Cyclohexane	64.8	29.2	70.8
Heptane	70.9	49	51
Octane	77.0	78	22
Benzene	67.9	31.7	68.3
Fluorobenzene	70.0	75	25
Toluene	76.7	68	32
Bromodichloromethane	75.5	72	28
Iodomethane	41.2	96.8	3.2
Chloroform	59.3	93	7
Trichloronitromethane	77.5	34	66
Carbon tetrachloride	65.0	84.2	15.8
1,2-Dichloroethane	70.5	37	63
3-Chloro-1-propene	44	5	95

TABLE 4-7 Binary azeotropic (constant-boiling) mixtures (*continued*)

System	BP of azeotrope, °C	Composition, w %	
		Alcohol	Other component
Ethanol (*continued*)			
1-Bromopropane	62.8	20.5	79.5
2-Bromopropane	55.6	10.5	89.5
1-Chloropropane	45.0	6	94
2-Chloropropane	35.6	2.8	97.2
1-Iodopropane	75.4	44	56
2-Iodopropane	71.5	27	73
1-Bromobutane	75.0	43	57
1-Chlorobutane	65.7	20.3	79.7
2-Butanone	74.8	40	60
1,1-Diethoxyethane	78.0	76	24
Dipropyl ether	74.5	44	56
Acetronitrile	72.5	44	56
Acrylonitrile	70.8	41	59
Nitromethane	76.1	29	71
Carbon disulfide	42.6	91	9
Diethyl sulfide	72.6	56	44
1-Propanol			
Hexane	65.7	4	96
Cyclohexane	74.7	18.5	81.5
Methylcyclohexane	87.0	34.7	65.3
Heptane	84.6	34.7	65.3
Octane	93.9	70	30
Benzene	77.1	16.9	83.1
Toluene	92.5	51.2	48.8
o-Xylene	zeotrope		
m-Xylene	97.1	94	6
p-Xylene	96.9	92.2	7.8
Styrene	97.0	8	92
Propyl formate	80.7	3	97
Butyl formate	95.5	64	36
Propyl acetate	94.7	51	49
Ethyl propionate	93.4	48	52
Methyl butyrate	94.4	49	51
Dipropyl ether	85.7	30	70
1,1-Diethoxyethane	92.4	37	63
1,4-Dioxane	95.3	55	45
Chloroform	zeotrope		
Carbon tetrachloride	73.4	92.1	7.9
Trichloronitromethane	94.1	58.5	41.5
Iodethane	70	93	7
1,2-Dichloroethane	80.7	19	81

TABLE 4-7 Binary azeotropic (constant-boiling) mixtures (*continued*)

System	BP of azeotrope, °C	Composition, w %	
		Alcohol	Other component
1-Propanol (*continued*)			
Tetrachloroethylene	94.0	52	48
1-Bromopropane	69.7	9	91
1-Chlorobutane	74.8	18	82
Chlorobenzene	96.5	80	20
Flurobenzene	80.2	18	82
Nitromethane	89.1	48.4	51.6
1-Nitropropane	97.0	8.8	91.2
Carbon disulfide	45.7	94.5	5.5
2-Propanol			
Pentane	35.5	6	94
Hexane	62.7	23	77
Cyclohexane	69.4	32	68
Heptane	76.4	50.5	49.5
Octane	81.6	84	16
Benzene	71.7	33.7	66.3
Fluorobenzene	74.5	30	70
Toluene	80.6	69	31
Chloroform	60.8	4.2	95.8
Trichloronitromethane	81.9	35	65
Carbon tetrachloride	69.0	18	82
1,2-Dichloroethane	74.7	43.5	56.5
Iodoethane	67.1	15	85
3-Bromo-1-propene	66.5	20	80
1-Chloropropane	46.4	2.8	97.2
1-Bromopropane	66.8	20.5	79.5
2-Bromopropane	57.8	12	88
1-Iodopropane	79.8	42	58
2-Iodopropane	76.0	32	68
1-Chlorobutane	70.8	23	77
Ethyl acetate	75.3	25	75
Isopropyl acetate	81.3	60	40
Methyl propionate	76.4	37	63
Acrylonitrile	71.7	56	44
Butylamine	74.7	60	40
2-Butanone	77.5	32	68
1,1-Diethoxyethane	81.3	63	37
Ethyl propyl ether	62.0	10	90
Diisopropyl ether	66.2	14.1	85.9

TABLE 4-7 Binary azeotropic (constant-boiling) mixtures (*continued*)

System	BP of azeotrope, °C	Composition, w %	
		Alcohol	Other component
	1-Butanol		
Cyclohexane	79.8	9.5	90.5
Cyclohexene	82.0	5	95
Hexane	68.2	3.2	96.8
Methylcyclohexane	95.3	20	80
Heptane	93.9	18	82
Octane	108.5	45.2	54.8
Nonane	115.9	71.5	28.5
Toluene	105.5	27.8	72.2
o-Xylene	116.8	75	25
m-Xylene	116.5	71.5	28.5
p-Xylene	115.7	68	32
Ethylbenzene	115.9	65.1	34.9
Butyl formate	105.8	23.6	76.4
Isopentyl formate	115.9	69	31
Butyl acetate	117.2	47	53
Isobutyl acetate	114.5	50	50
Ethyl butyrate	115.7	64	36
Ethyl isobutyrate	109.2	17	83
Methyl isopentanoate	113.5	40	60
Ethyl borate	113.0	52	48
Ethyl carbonate	116.5	63	37
Isobutyl nitrate	112.8	45	55
Dibutyl ether	117.8	82.5	17.5
Diisobutyl ether	113.5	48	52
1,1-Diethoxyethane	101.0	13	87
Carbon tetrachloride	76.6	97.6	2.4
Tetrachloroethylene	110.0	68	32
2-Bromo-2-methylpropane	90.2	7	93
2-Iodo-2-methylpropane	110.5	30	70
Chlorobenzene	115.3	56	44
Paraldehyde	115.8	52	48
Hexaldehyde	116.8	77.1	22.9
Ethylenediamine	124.7	35.7	64.3
Pyridine	118.6	69	31
1-Nitropropane	115.3	32.2	67.8
Butyronitrile	113.0	50	50
Diisopropyl sulfide	112.0	45	55
	2-Methyl-2-propanol		
Cyclohexene	80.5	14.2	85.8
Cyclohexane	78.3	14	86

TABLE 4-7 Binary azeotropic (constant-boiling) mixtures (*continued*)

System	BP of azeotrope, °C	Composition, w %	
		Alcohol	Other component
2-Methyl-2-propanol (*continued*)			
Methylcyclopentane	71.0	5	95
Hexane	68.3	2.5	97.5
Methylcyclohexane	92.6	32	68
Heptane	90.8	27	73
2,5-Dimethylhexane	98.7	42	58
1,3-Dimethylcyclohexane	102.2	56	44
2,2,4-Trimethylpentane	92.0	27	73
Benzene	79.3	7.4	92.6
Chlorobenzene	107.1	63	37
Fluorobenzene	84.0	9	91
Toluene	101.2	45	55
Ethylbenzene	107.2	80	20
p-Xylene	107.1	88.6	11.4
Butyl formate	103.0	40	60
Isobutyl formate	97.4	12	88
Propyl acetate	101.0	17	83
Isobutyl acetate	107.6	92	8
Methyl butyrate	101.3	25	75
Ethyl isobutyrate	105.5	52	48
Methyl chloroacetate	107.6	12	88
Dipropyl ether	89.5	10	90
Isobutyl vinyl ether	82.7	6.2	93.8
1,1-Diethoxyethane	98.2	20	80
2-Pentanone	101.8	19	81
3-Pentanone	101.7	20	80
1,2-Dichloroethane	83.5	6.5	93.5
1-Bromobutane	95.0	21	79
1-Chlorobutane	77.7	4	96
2-Bromo-2-methylpropane	88.8	12	88
2-Iodo-2-methylpropane	104.0	36	64
1-Nitropropane	105.3	15.2	84.8
Isobutyl nitrate	105.6	36	64
Diisopropyl sulfide	105.8	73	27
3-Methyl-1-butanol			
Heptane	97.7	7	93
Octane	117.0	30	70
Tolueme	109.7	10	90
Ethylbenzene	125.7	49	51
Isopropylbenzene	131.6	94	6
Camphene	130.9	24	76

TABLE 4-7 Binary azeotropic (constant-boiling) mixtures (*continued*)

System	BP of azeotrope, °C	Composition, w %	
		Alcohol	Other component
3-Methyl-1-butanol (*continued*)			
Bromobenzene	131.7	85	15
o-Fluorotoluene	112.1	14.0	86.0
Butyl acetate	125.9	16.5	83.5
Paraldehyde	123.5	22.0	78.0
Dibutyl ether	129.8	65	35
Cyclohexanol			
o-Xylene	143.0	14	86
m-Xylene	138.9	5	95
Propylbenzene	153.8	40	60
Indene	160.0	75	25
Camphene	151.9	41	59
Cineole	160.6	92	8
Allyl alcohol			
Methylcyclohexane	85.0	42	58
Hexane	65.5	4.5	95.5
Cyclohexane	74.0	58	42
2,5-Dimethylhexane	89.3	50	50
Octane	93.4	68	32
Benzene	76.75	17.36	82.64
Toluene	92.4	50	50
Propyl acetate	94.2	53	47
Methyl butyrate	93.8	55	45
1,2-Dichloroethane	79.9	18	82
3-Iodo-1-propene	89.4	28	72
Chlorobenzene	96.2	85	15
Diethyl sulfide	85.1	45	55
Phenol			
2,7-Dimethyloctane	159.5	6	94
Decane	168.0	35	65
Tridecane	180.6	83.1	16.9
Butylbenzene	175.0	46	54
1,2,4-Trimethylbenzene	166.0	25	75
1,3,5-Trimethylbenzene	163.5	21	79
Indene	177.8	47	53
Camphene	156.1	22	78
Benzaldehyde	175.6	51.0	49.0

TABLE 4-7 Binary azeotropic (constant-boiling) mixtures (*continued*)

System	BP of azeotrope, °C	Composition, w %	
		Alcohol	Other component
Phenol (*continued*)			
1-Octanol	195.4	13	87
2-Octanol	184.5	50	50
Dipentyl ether	180.2	78	22
Diisopentyl ether	172.2	15	85
2-Methylpyridine	185.5	75.4	24.6
3-Methylpyridine	188.9	71.2	29.8
4-Methylpyridine	190.0	67.5	32.5
2,4-Dimethylpyridine	193.4	57.0	43.0
2,6-Dimethylpyridine	185.5	72.5	27.5
2,4,6-Trimethylpyridine	195.2	52.3	47.7
Aniline	185.8	41.9	58.1
Ethylene diacetate	195.5	39.2	60.8
Iodobenzene	177.7	53	47
Benzyl alcohol			
Naphthalene	204.1	60	40
D-Limonene	176.4	11	89
1,3,5-Triethylbenzene	203.2	57	43
o-Cresol	zeotrope		
m-Cresol	207.1	61	39
p-Cresol	206.8	62	38
N-Methylaniline	195.8	30	70
N,N-Dimethylaniline	193.9	6.5	93.5
N-Ethylaniline	202.8	50	50
N,N-Diethylaniline	204.2	72	28
Iodobenzene	187.8	12	88
Nitrobenzene	204.0	58	42
o-Bromotoluene	181.3	7	93
Borneol	205.1	85.8	14.2
2-Ethoxyethanol			
Methylcyclohexane	98.6	15	85
Heptane	96.5	14	86
Octane	116.0	38	62
Toluene	110.2	10.8	89.2
Ethylbenzene	127.8	48	52
p-Xylene	128.6	50	50
Styrene	130.0	55	45
Propylbenzene	134.6	80	20
Isopropylbenzene	133.2	67	33
Camphene	131.0	65	35
Propyl butyrate	133.5	72	28

TABLE 4-7 Binary azeotropic (constant-boiling) mixtures (*continued*)

System	BP of azeotrope, °C	Composition, w %	
		Alcohol	Other component
2-Butoxyethanol			
Dipentene	164.0	53	47
1,3,5-Trimethylbenzene	162.0	32	68
Butylbenzene	169.6	73.4	26.6
Camphene	154.5	30	70
o-Cresol	191.6	15	85
Phenetole	167.1	52	48
Cineole	168.9	58.5	41.5
Benzaldehyde	171.0	91	9
Diisobutyl sulfide	163.8	42	58
1,2-Ethanediol			
Heptane	97.9	3	97
Decane	161.0	23	77
Tridecane	188.0	55	45
Toluene	110.1	2.3	97.7
Styrene	139.5	16.5	83.5
Stilbene	196.8	87	13
m-Xylene	135.1	6.55	93.45
p-Xylene	134.5	6.4	93.6
1,3,5-Trimethylbenzene	156	13	87
Propylbenzene	152	19	81
Isopropylbenzene	147.0	18	82
Naphthalene	183.9	51	49
1-Methylnaphthalene	190.3	60.0	40.0
2-Methylnaphthalene	189.1	57.2	42.8
Anthracene	197	98.3	1.7
Indene	168.4	26	74
Acenaphthene	194.65	74.2	25.8
Fluorene	196.0	82	18
Camphene	152.5	20	80
Camphor	186.2	40	60
Biphenyl	192.3	66.5	33.5
Diphenylmethane	193.3	68.5	31.5
Benzyl alcohol	193.1	56	44
2-Phenylethanol	194.4	69	31
o-Cresol	189.6	27	73
m-Cresol	195.2	60	40
3,4-Dimethylphenol	197.2	89	11
Menthol	188.6	51.5	48.5
Ethyl benzoate	186.1	46.5	53.5
o-Bromotoluene	166.8	25	75
Dibutyl ether	139.5	6.4	93.6

TABLE 4-7 Binary azeotropic (constant-boiling) mixtures (*continued*)

System	BP of azeotrope, °C	Composition, w %	
		Alcohol	Other component
	1,2-Ethanediol (*continued*)		
Methoxybenzene	150.5	10.5	89.5
Diphenyl ether	193.1	60	40
Benzyl phenyl ether	195.5	87	13
Acetophenone	185.7	52	48
2,4-Dimethylaniline	188.6	47	53
N,N-Dimethylaniline	175.9	33.5	66.5
m-Toluidine	188.6	42	58
2,4,6-Trimethylpyridine	170.5	9.7	90.3
Quinoline	196.4	79.5	20.5
Tetrachloroethylene	119.1	94	6
1,2-Dibromoethane	129.8	4	96
Chlorobenzene	130.1	94.4	5.6
α-Chlorotoluene	167.0	30	70
Nitrobenzene	185.9	59	41
o-Nitrotoluene	188.5	48.5	51.5
	1,2-Ethanediol monoacetate		
Indene	180.0	20	80
1-Octanol	189.5	71	29
Phenol	197.5	65	35
o-Cresol	199.5	51	49
m-Cresol	206.5	31	69
p-Cresol	206.0	33	67
Dipentyl ether	180.8	42	58
Diisopentyl ether	170.2	28	72
m-Bromotoluene	182.0	32	68

D. Binary azeotropes containing ketones

System	BP of azeotrope, °C	Composition, w %	
		Ketone	Other component
	Acetone		
Cyclopentane	41.0	36	64
Pentane	32.5	20	80
Cyclohexane	53.0	67.5	32.5
Hexane	49.8	59	41

TABLE 4-7 Binary azeotropic (constant-boiling) mixtures (*continued*)

System	BP of azeotrope, °C	Composition, w %	
		Ketone	Other component
Acetone (*continued*)			
Heptane	55.9	89.5	10.5
Diethylamine	51.4	38.2	61.8
Methyl acetate	55.8	48.3	51.7
Diisopropyl ether	54.2	61	39
Chloroform	64.4	78.1	21.9
Carbon tetrachloride	56.1	11.5	88.5
Carbon disulfide	39.3	67	33
Ethylene sulfide	51.5	57	43
2-Butanone			
Cyclohexane	71.8	40	60
Hexane	64.2	28.6	71.4
Heptane	77.0	70	30
2,5-Dimethylhexane	79.0	95	5
Benzene	78.33	44	56
2-Methyl-2-propanol	78.7	69	31
Butylamine	74.0	35	65
Ethyl acetate	77.1	11.8	88.2
Methyl propionate	79.0	60	40
Butyl nitrite	76.7	30	70
1-Chlorobutane	77.0	38	62
Fluorobenzene	79.3	75	25

E. Miscellaneous binary azeotropes

System	BP of azeotrope, °C	Composition, w %	
		Solvent	Other component
Solvent: acetamide			
Dipentene	169.2	18	82
Biphenyl	213.0	50.5	49.5
Diphenylmethane	215.2	56.5	43.5
1,2-Diphenylethane	218.2	68	32
o-Xylene	142.6	11	89
m-Xylene	138.4	10	90
p-Xylene	137.8	8	92
Styrene	144	12	88
4-Isopropyl-1-methylbenzene	170.5	19	81

TABLE 4-7 Binary azeotropic (constant-boiling) mixtures (*continued*)

System	BP of azeotrope, °C	Composition, w %	
		Solvent	Other component
Solvent: acetamide (*continued*)			
Naphthalene	199.6	27	73
1-Methylnaphthalene	209.8	43.8	56.2
2-Methylnaphthalene	208.3	40	60
Indene	177.2	17.5	82.5
Acenaphthene	217.1	64.2	35.8
Camphene	155.5	12	88
Camphor	199.8	23	77
Benzaldehyde	178.6	6.5	93.5
3,4-Dimethylphenol	221.1	96	4
2-Methoxy-4-(2-propenyl)phenol	220.8	88	12
N-Methylaniline	193.8	14	86
N-Ethylaniline	199.0	18	82
N,N-Diethylaniline	198.1	24	76
Diphenyl ether	214.6	52	48
Safrole	208.8	32	68
Tetrachloroethylene	120.5	97.4	2.6
Solvent: aniline			
Nonane	149.2	13.5	86.5
Decane	167.3	36	64
Undecane	175.3	57.5	42.5
Dodecane	180.4	71.5	28.5
Tridecane	182.9	86.2	13.8
Tetradecane	183.9	95.2	4.8
Butylbenzene	177.8	46	54
1,2,4-Trimethylbenzene	168.6	13.5	86.5
1,3,5-Trimethylbenzene	164.3	12.0	88.0
Indene	179.8	41.5	58.5
1-Octanol	183.9	83	17
o-Cresol	191.3	8	92
Dipentyl ether	177.5	55	45
Diisopentyl ether	169.3	28	72
Hexachloroethane	176.8	66	34
Solvent: pyridine			
Heptane	95.6	25.3	74.7
Octane	109.5	56.1	43.9
Nonane	115.1	89.9	10.1
Toluene	110.1	22.2	77.8
Phenol	183.1	13.1	86.9
Piperidine	106.1	8	92

TABLE 4-7 Binary azeotropic (constant-boiling) mixtures (*continued*)

System	BP of azeotrope, °C	Composition, w %	
		Solvent	Other component
Solvent: thiophene			
Methylcyclopentane	71.5	14	86
Cyclohexane	77.9	41.2	58.8
Hexane	68.5	11.2	88.8
Heptane	83.1	83.2	16.8
2,3-Dimethylpentane	80.9	64	36
2,4-Dimethylpentane	76.6	42.7	57.3
Solvent: benzene			
Methylcyclopentane	71.7	16	84
Cyclohexene	78.9	64.7	35.3
Cyclohexane	77.6	51.9	48.1
Hexane	68.5	4.7	95.3
Heptane	80.1	99.3	0.7
2,2-Dimethylpentane	75.9	46.3	53.7
2,3-Dimethylpentane	79.4	78.8	21.2
2,4-Dimethylpentane	75.2	48.3	51.7
2,2,4-Trimethylpentane	80.1	97.7	2.3
Solvent: bis(2-hydroxyethyl) ether			
Biphenyl	232.7	48	52
Diphenylmethane	236.0	52	48
1,3,5-Trimethylbenzene	210.0	22	78
Naphthalene	212.6	22	78
1-Methylnaphthalene	277.0	45	55
2-Methylnaphthalene	225.5	39	61
Acenaphthene	239.6	62	38
Fluorene	243.0	80	20
Benzyl acetate	214.9	7	93
Bornyl acetate	223.0	18	82
Ethyl fumarate	217.1	10	90
Dimethyl *o*-phthalate	245.4	96.3	3.7
Methyl salicylate	220.6	15	85
2-Hydroxy-1-isopropyl-4-methylbenzene	232.3	13	87
1,2-Dihydroxybenzene	259.5	46	54
Safrole	225.5	33	67
Isosafrole	233.5	46	54
Benzyl phenyl ether	241.5	80	20
Nitrobenzene	210.0	10	90
m-Nitrotoluene	224.2	25	75
o-Nitrophenol	216.0	10.5	89.5
Quinoline	233.6	29	71
p-Dibromobenzene	212.9	13	87

TABLE 4-8 Ternary azeotropic mixtures

A. Ternary azeotropes containing water and alcohols

System	BP of azeotrope, °C	Composition, wt %		
		Water	Alcohol	Other component
Methanol				
Chloroform	52.3	1.3	8.2	90.5
2-Methyl-1,3-butadiene	30.2	0.6	5.4	94.0
Methyl chloroacetate	67.9	6.3	81.2	13.5
Ethanol				
Acetonitrile	72.9	1	55	44
Acrylonitrile	69.5	8.7	20.3	71.0
Benzene	64.9	7.4	18.5	74.1
Butylamine	81.8	7.5	42.5	50.0
Butyl methyl ether	62	6.3	8.6	85.1
Carbon disulfide	41.3	1.6	5.0	93.4
Carbon tetrachloride	62	4.5	10.0	85.5
Chloroform	55.3	2.3	3.5	94.2
Crotonaldehyde	78.0	4.8	87.9	7.3
Cyclohexane	62.6	4.8	19.7	75.5
1,2-Dichloroethane	66.7	5	17	78
1,1-Diethoxyethane	77.8	11.4	27.6	61.0
Diethoxymethane	73.2	12.1	18.4	69.5
Ethyl acetate	70.2	9.0	8.4	82.6
Heptane	68.8	6.1	33.0	60.9
Hexane	56.0	3	12	85
Toluene	74.4	12	37	51
Trichloroethylene	67.0	5.5	16.1	78.4
Triethylamine	74.7	9	13	78
1-Propanol				
Benzene	67	7.6	10.1	82.3
Carbon tetrachloride	65.4	5	11	84
Cyclohexane	66.6	8.5	10.0	81.5
1,1-Dipropoxyethane	87.6	27.4	51.6	21.0
Dipropoxymethane	86.4	8.0	44.8	47.2
Dipropyl ether	74.8	11.7	20.2	68.1
3-Pentanone	81.2	20	20	60
Propyl acetate	82.5	17.0	10.0	73.0
Propyl formate	70.8	13	5	82
Tetrachloroethylene	81.2	12.5	20.7	66.8
2-Propanol				
Benzene	66.5	7.5	18.7	73.8
Butylamine	83	12.5	40.5	47.0

TABLE 4-8 Ternary azeotropes containing water and alcohols (*continued*)

System	BP of azetrope, °C	Composition, wt %		
		Water	Alcohol	Other component
2-Propanol (*continued*)				
Cyclohexane	64.3	7.5	18.5	74.0
Toluene	76.3	13.1	38.2	48.7
Trichloroethylene	69.4	7	20	73
1-Butanol				
Butyl acetate	89.4	37.3	27.4	35.3
Butyl formate	83.6	21.3	10.0	68.7
Dibutyl ether	90.6	29.9	34.6	35.5
Heptane	78.1	41.4	7.6	51.0
Hexane	61.5	19.2	2.9	77.9
Nonane	90.0	69.9	18.3	11.8
Octane	86.1	60.0	14.6	25.4
2-Butanol				
Carbon tetrachloride	65	4.05	4.95	91.00
Cyclohexane	69.7	8.9	10.8	80.3
Isooctane	76.3	9	19	72
2-Methyl-1-propanol				
Isobutyl acetate	86.8	30.4	23.1	46.5
Isobutyl formate	80.2	17.3	6.7	76.0
Toluene	81.3	17.9	16.4	65.7
2-Methyl-2-propanol				
Benzene	67.3	8.1	21.4	70.5
Carbon tetrachloride	64.7	3.1	11.9	85.0
Cyclohexane	65.0	8	21	71
3-Methyl-1-butanol				
Isopentyl acetate	93.6	44.8	31.2	24.0
Isopentyl formate	89.8	32.4	19.6	48.0
Allyl alcohol				
Benzene	68.2	8.6	9.2	82.2
Carbon tetrachloride	65.2	5	11	84
Cyclohexane	66.2	8	11	81
Hexane	59.7	8.5	5.1	86.4

TABLE 4-8 Ternary azeotropes containing water and alcohols (*continued*)

B. Other ternary azeotropes

System	BP of azeotrope, °C	Composition, wt %	System	BP of azeotrope, °C	Composition, wt %
Water	32.5	0.4	Water	71.4	7.9
Acetone		7.6	Nitromethane		29.7
2-Methyl-1,3-butadiene		92.0	Heptane		62.4
Water	66	8.2	Water	80.7	17.4
Acetonitrile		23.3	Nitromethane		58.3
Benzene		68.5	Nonane		24.3
Water	67	6.4	Water	77.4	12.4
Acetronitrile		20.5	Nitromethane		44.3
Trichloroethylene		73.1	Octane		43.3
Water	68.6	3.5	Water	33.1	2.1
Acetonitrile		9.6	Nitromethane		6.5
Triethylamine		86.9	Pentane		91.4
Water	63.6	5	Water	82.8	20.6
2-Butanone		35	Nitromethane		73.3
Cyclohexane		60	Undecane		6.1
Water	55.0	4	Water	93.5	40.5
Butyraldehyde		21	Pyridine		54.5
Hexane		75	Dodecane		5.0
Water	107.6	21.3	Water	93.1	38.5
Formic acid		76.3	Pyridine		51.0
Isopentanoic acid		2.4	Undecane		10.5
Water	107.0	15.5	Water	92.3	35.5
Formic acid		66.8	Pyridine		45.5
Isobutyric acid		17.7	Decane		19.0

Water	107.6	19.5	Water	90.5	30.5
Formic acid		75.9	Pyridine		37.0
Butyric acid		4.6	Nonane		32.5
Water	107.2	18.6	Water	86.7	22.4
Formic acid		71.9	Pyridine		25.5
Propionic acid		9.5	Octane		52.0
Water	105	11.0	Water	78.6	14.0
Hydrogen bromide		10.4	Pyridine		15.5
Chlorobenzene		78.6	Heptane		70.5
Water	96.9	20.2	Acetic acid	134.4	23
Hydrogen chloride		5.3	Pyridine		55
Chlorobenzene		74.5	Acetic anhydride		22
Water	107.3	64.8	Acetic acid	134.1	31.4
Hydrogen chloride		15.8	Pyridine		38.2
Phenol		19.4	Decane		30.4
Water	116.1	54	Acetic acid	129.1	13.5
Hydrogen fluoride		10	Pyridine		25.2
Fluorosilic acid		36	Ethylbenzene		61.3
Water	75.1	11.5	Acetic acid	98.5	3.4
Nitroethane		75.1	Pyridine		10.6
Heptane		64.0	Heptane		86.0
Water	59.5	8.4	Acetic acid	128.0	20.7
Nitroethane		9.3	Pyridine		29.4
Hexane		82.3	Nonane		49.9
Water	82.4	19.1	Acetic acid	115.7	10.4
Nitromethane		68.1	Pyridine		20.1
Decane		12.8	Octane		69.5
Water	83.1	21.5	Acetic acid	132.2	17.7
Nitromethane		75.3	Pyridine		30.5
Dodecane		3.2	*o*-Xylene		51.8

TABLE 4-8 Ternary azeotropes containing water and alcohols (*continued*)

System	BP of azeotrope, °C	Composition, wt %	System	BP of azeotrope, °C	Composition, wt %
Acetic acid	129.2	10.2	Methanol	47.4	14.6
Pyridine		22.5	Methyl acetate		36.8
p-Xylene		67.3	Hexane		48.6
Acetic acid	163.0	75.0	Ethanol	63.2	10.4
2,6-Dimethylpyridine		13.8	Acetone		24.3
Undecane		11.2	Chloroform		65.3
Acetic acid	147.0	12.6	Ethanol	70.1	8
2,6-Dimethylpyridine		74.3	Acetonitrile		34
Decane		13.1	Triethylamine		58
Acetic acid	141.3	19.9	Ethanol	64.7	29.6
2-Methylpyridine		46.8	Benzene		12.8
Decane		33.3	Cyclohexane		57.6
Acetic acid	135.0	12.8	Ethanol	57.3	9.5
2-Methylpyridine		38.4	Chloroform		56.1
Nonane		48.8	Hexane		34.4
Acetic acid	121.3	3.6	1-Propanol	73.8	15.5
2-Methylpyridine		24.8	Benzene		30.4
Octane		71.6	Cyclohexane		54.2
Acetic acid	77.2	7.6	2-Propanol	69.1	31.1
Benzene		34.4	Benzene		15.0
Cyclohexane		58.0	Cyclohexane		53.9
Acetic acid	132	15	1-Butanol	77.4	4
2-Methyl-1-butanol		54	Benzene		48
Isopentyl acetate		31	Cyclohexane		48
Propionic acid	149.3	29.5	1-Butanol	108.7	11.9
2-Methylpyridine		32.0	Pyridine		20.7
Decane		38.5	Toluene		76.4

Propionic acid	140.1	16.5	1,2-Ethanediol	185.0	8.7
2-Methylpyridine		21.5	Phenol		74.6
Nonane		42.0	2,6-Dimethylpyridine		16.7
Propionic acid	123.7	4.5	1,2-Ethanediol	185.1	5.9
2-Methylpyridine		10.5	Phenol		79.1
Octane		85.0	2-Methylpyridine		15.0
Propionic acid	153.4	43.0	1,2-Ethanediol	186.4	15.9
2-Methylpyridine		40.0	Phenol		67.7
Undecane		17.0	3-Methylpyridine		16.4
Propionic acid	147.1	55.5	1,2-Ethanediol	188.6	29.5
Pyridine		26.4	Phenol		54.8
Undecane		18.1	2,4,6-Trimethylpyridine		15.7
Methanol	57.5	23	Acetone	60.8	3.6
Acetone		30	Chloroform		68.8
Chloroform		47	Hexane		27.6
Methanol	47	14.6	Acetone	49.7	51.1
Acetone		30.8	Methyl acetate		5.6
Hexane		59.6	Hexane		43.3
Methanol	53.7	17.4	Chloroform	62.0	79.7
Acetone		5.8	Ethyl formate		5.3
Methyl acetate		76.8	2-Bromopropane		15.7
Methanol	50.8	17.8	1,4-Dioxane	101.8	44.3
Methyl acetate		48.6	2-Methyl-1-propanol		26.7
Cyclohexane		33.6	Toluene		29.0

FREEZING POINTS

TABLE 4-9 Molecular lowering of the melting or freezing point

Cryoscopic constants

The cryoscopic constant K_f gives the depression of the melting point ΔT (in degrees Celsius) produced when 1 mol of solute is dissolved in 1000 g of a solvent. It is applicable only to dilute solutions for which the number of moles of solute is negligible in comparison with the number of moles of solvent. It is often used for molecular weight determinations,

$$M_2 = \frac{1000 w_2 K_f}{w_1 \Delta T}$$

where w_1 is the weight of the solvent and w_2 is the weight of the solute whose molecular weight is M_2.

Compound	K_f	Compound	K_f
Acetamide	4.04	Diphenylamine	8.60
Acetic acid	3.90	Diphenyl ether	7.88
Acetone	2.40	1,2-Ethanediamine	2.43
Ammonia	0.957	Ethoxybenzene	7.15
Aniline	5.87	Formamide	3.85
Antimony(III) chloride	17.95	Formic acid	2.77
Benzene	5.12	Glycerol	3.3 to 3.7
Benzonitrile	5.34	Hexamethylphosphoramide	6.93
Benzophenone	9.8		
Bicyclohexane	14.52	*N*-Methylacetamide	6.65
Biphenyl	8.0	2-Methyl-2-butanol	10.4
Borneol	35.8	Methylcyclohexane	14.13
Bornylamine	40.6	Methyl *cis*-9-octadecenoate	3.4
Butanedinitrile	18.26	2-Methyl-2-propanol	8.37
Camphene	31.08	Naphthalene	6.94
Camphoquinone	45.7	Nitrobenzene	6.852
D-(+)-Camphor	39.7	Octadecanoic acid	4.50
Carbon tetrachloride	29.8	2-Oxohexamethyleneimine	7.30
o-Cresol	5.60	Phenol	7.40
p-Cresol	6.96	Pyridine	4.75
Cyclohexane	20.0	Quinoline	1.95
Cyclohexanol	39.3	Succinonitrile	18.26
Cyclohexylcyclohexane	14.52	Sulfuric acid	1.86
Cyclopentadecanone	21.3	1,1,2,2-Tetrabromoethane	21.7
cis-Decahydronaphthalene	19.47	1,1,2,2-Tetrachloro-1,2-difluoroethane	
trans-Decahydronaphthalene	20.81		37.7
Dibenz[*de,kl*]anthracene	25.7	Tetramethylene sulfone	64.1
Dibenzyl ether	6.27	*p*-Toluidine	5.372
1,2-Dibromoethane	12.5	Tribromomethane	14.4
Diethyl ether	1.79	1,3,3-Trimethyl-2-oxabicyclo-[2.2.2]octane	
1,2-Dimethoxybenzene	6.38		6.7
N,N-Dimethylacetamide	4.46	Triphenylmethane	12.45
2,2-Dimethyl-1-propanol	11.0	Water	1.86
Dimethyl sulfoxide	4.07	*p*-Xylene	4.3
1,4-Dioxane	4.63		

VISCOSITY, DIELECTRIC CONSTANT, DIPOLE MOMENT, SURFACE TENSION, AND REFRACTIVE INDEX

TABLE 4-10 Viscosity, dielectric constant, dipole moment, and surface tension of selected organic substances

The temperature in degrees Celsius at which the viscosity, dielectric constant, dipole moment, and surface tension of a substance were measured is shown in this table in parentheses after the value. In some cases, the dipole moment was determined with the substance dissolved in a solvent, and the solvent used is also shown in parentheses after the temperature.

For the majority of compounds the dependence of the surface tension γ on the temperature can be given as

$$\gamma = a - bt$$

where a and b are constants and t is the temperature in degrees Celsius.
Alternate names for entries are listed in Table 1-14 at the bottom of each double page.

List of Abbreviations

B, benzene
C, CCl_4
cHx, cyclohexane
D, 1,4-dioxane
g, gas
Hx, hexane
lq, liquid

Substance	Viscosity, $mN \cdot s \cdot m^{-2}$	Dielectric constant ε	Dipole moment, D	Surface tension, $dyn \cdot cm^{-1}$	
				a	b
Acetaldehyde	0.280 (0), 0.256 (10), 0.22 (20)	21.8 (10), 21.1 (21)	2.71 (g)	23.90	0.136 0
Acetaldoxime	1.415 (20)	3 (23)	0.830 (20, lq), 0.90 (25, B)	30.1 (35)	
Acetamide	1.32 (105), 1.06 (120)	59.2 (83), 60.6 (94)	3.90 (25, B), 2.44 (30, B)	47.66	0.102 1
Acetanilide	2.22 (120), 1.90 (130)		3.65 (25, B)	46.21	0.091 2
Acetic acid	1.232 (20), 0.796 (50)	6.15 (20), 6.29 (40)	1.76 (g), 1.92 (20, B)	29.58	0.009 4

TABLE 4-10 Viscosity, dielectric constant, dipole moment, and surface tension of selected organic substances (*continued*)

Substance	Viscosity, $mN \cdot s \cdot m^{-2}$	Dielectric constant ε	Dipole moment, D	Surface tension $dyn \cdot cm^{-1}$	
				a	b
Acetic anhydride	0.907 (20), 0.699 (40)	23.3 (0), 21.2 (20)	2.8 (g), 3.15 (20, B)	35.52	0.143 6
Acetone					
(lq)	0.391 (0), 0.318 (20)	20.7 (25), 17.6 (56)	2.77 (22, B)	26.26	0.112
(g)	0.009 33 (100), 0.012 8 (225)	1.015 9 (100)	2.87		
Acetonitrile	0.397 (10), 0.329 (30)	37.5 (20), 26.6 (82)	3.97 (g), 3.47 (20, B)	29.58	0.117 8
Acetophenone	2.015 (15), 1.511 (30)	17.39 (25), 8.64 (202)	2.96 (30, B)	41.92	0.115 4
Acetyl bromide		16.2 (20)	2.45 (20, B)		
Acetyl chloride					
(lq)		16.9 (2), 15.8 (22)	2.47 (20, B)	26.7 (15)	
(g)		1.0217 (20)	2.71		
Acetylene (g)	0.010 2 (30), 0.012 6 (101)	1.001 34 (0)	0	3.42	0.193 5 (lq)
Acrylic acid	1.3 (20), 1.16 (25)			28.1 (30)	
Acrylonitrile	0.35 (20), 0.34 (25)	33.0 (20)	3.91 (g), 3.54 (25, B)	29.58	0.117 8
Allyl acetate	0.207 (30)			28.73	0.118 6
Allylamine	0.375 (25)		1.3 (25, B)	27.49	0.128 7
Allyl isothiocyanate		17.2 (18)	3.2 (20, B)	36.76	0.107 4
2-Aminoethanol	30.85 (15), 19.35 (25)	37.72 (25)	2.59 (25, D)	51.11	0.111 7
Aniline	5.30 (15), 4.40 (20), 3.18 (30)	6.89 (20), 5.93 (70)	1.53 (20, B)	44.83	0.108 5
Benzaldehyde	1.321 (25)	19.7 (0), 17.8 (20)	2.77 (20, lq)	40.72	0.109 0
Benzaldehyde oxime (mp 30)		3.8 (20)	1.2 (25, B)		
(mp 128)			1.5 (25, B)		
Benzene	0.649 (20), 0.566 (30), 0.395 (60)	2.292 (15), 2.274 (25), 1.002 8 (g)	0	28.88 (20)	27.56 (30)
Benzamide			3.42 (25, B)	47.26	0.070 5
Benzenesulfonyl chloride			4.50 (20, B)	45.48	0.111 7

Benzenethiol	1.239 (20), 1.144 (25)	4.38 (25)	1.13 (25, lq), 1.19 (20, B)	41.41	0.120 2
Benzonitrile	1.447 (15), 1.111 (30)	26.5 (20), 24.0 (40)	4.40 (g), 3.9 (20, B)	41.69	0.115 9
Benzophenone	4.79 (55), 1.38 (120)	14.60 (18), 11.4 (50)	3.09 (50, lq), 2.98 (25, B)	46.31	0.112 8
Benzoyl bromide	1.956 (20), 1.798 (25)	21.33 (20), 20.74 (25)	3.40 (20, B)	45.85	0.139 7
Benzoyl chloride		29 (0), 23 (20)	3.16 (25, B)	41.34	0.108 4
Benzyl acetate	1.399 (45)	5.1 (21)	1.80 (25, B)		
Benzyl alcohol	5.58 (20), 4.65 (30), 3.01 (45)	13.0 (20), 9.5 (70)	1.67 (25, B)	38.25	0.138 1
Benzylamine	1.59 (25)	5.5 (1), 4.6 (21)	1.15 (20, lq), 1.38 (25, B)	42.33	0.121 3
Benzyl benzoate	8.51 (25)	4.9 (20)	2.06 (30, B)	48.07	0.106 5
Benzyl butyl *o*-phthalate	65 (20)				
Benzyl chloride	1.400 (20), 1.290 (25)	7.0 (13)	1.83 (20, B)	39.92	0.122 7
Benzylethylamine		4.3 (20)			
Benzyl ethyl ether		3.9 (20)		32.83 (20)	29.97 (40)
Biphenyl		2.53 (75)	0	41.52	0.093 1
Bis(2-ethoxyethyl)ether			1.92 (25, B)	29.74	0.117 6
Bis(2-hydroxyethyl)ether	38.0 (20), 30.0 (25)	31.69 (20)	2.31 (20, B)	46.97	0.088 0
1,2-Bis(methoxyethoxy)-ethane	3.76 (20)				
Bis(2-methoxyethyl) ether	1.99 (20), 0.981 (25)		1.97 (25, B)	32.47	0.116 4
DL-Bornyl acetate		4.6 (21)	1.89 (22)		
3-Bromoaniline	6.81 (20), 3.70 (40)	13.0 (19)	2.67 (20, B)		
4-Bromoaniline	1.81 (80)	7.06 (30)	2.88 (25, B)		
Bromobenzene	1.196 (15), 0.985 (30)	5.40 (25)	1.70 (g), 1.50 (20, lq)	38.14	0.116 0
1-Bromobutane	0.633 (20), 0.597 (25)	7.88 (−10), 7.07 (20)	2.17 (g), 2.04 (20, lq)	28.71	0.112 6
DL-2-Bromobutane	1.434 (20)	8.64 (25)	2.22 (g), 2.14 (25, lq)	27.48	0.110 7
1-Bromo-2-chlorobenzene		6.80 (20)	2.15 (20, B)		
1-Bromo-3-chlorobenzene		4.58 (20)	1.52 (22, B)		
1-Bromo-4-chlorobenzene			0.1 (25, B)	40.03	0.100 2
Bromochloromethane	0.670 (20)	7.79	1.66 (25, B)	33.32 (20)	
Bromocyclohexane	2.0 (25)	11 (−65), 7.9 (25)	1.08 (25, lq), 2.3 (25, B)	36.13	0.111 7
1-Bromodecane		4.75 (1), 4.44 (25)	2.08 (20, lq), 1.90 (25, lq)	31.26	0.085 6

TABLE 4-10 Viscosity, dielectric constant, dipole moment, and surface tension of selected organic substances (*continued*)

Substance	Viscosity, $mN \cdot s \cdot m^{-2}$	Dielectric constant ε	Dipole moment, D	Surface tension $dyn \cdot cm^{-1}$ a	b
Bromodichloromethane			1.31 (25, B)	35.11	0.129 4
1-Bromododecane		4.07 (25)	2.01 (25, lq), 1.89 (25, B)	32.58	0.088 2
Bromoethane	0.397 (20), 0.348 (30)	13.6 (−60), 9.39 (20)	2.03 (g), 2.04 (20, lq)	26.52	0.115 9
1-Bromo-2-ethoxyethane				31.98	0.112 9
1-Bromo-2-ethoxypentane		6.45 (25)	2.32 (25, B)		
2-Bromo-2-ethoxypentane		6.40 (25)	2.07 (25, B)		
3-Bromo-3-ethoxypentane		8.24 (25)	2.15 (25, B)		
Bromoethylene		4.78 (25)	1.42 (g)		
Bromoform	2.152 (15), 1.741 (30)	4.39 (20)	1.00 (g), 0.92 (25, lq)	48.14	0.130 8
1-Bromoheptane		5.33 (25), 4.48 (90)	2.17 (g), 2.02 (20, lq)	30.74	0.098 2
2-Bromoheptane		6.46 (22)	2.08 (20, B)		
3-Bromoheptane		6.93 (22)	2.06 (20, B)		
4-Bromoheptane		6.81 (22)	2.06 (20, B)		
1-Bromohexadecane		3.71 (25)	1.98 (20, lq), 1.96 (25, C)	33.37	0.086 1
1-Bromohexane		6.30 (1), 5.82 (25)	2.06 (20, lq)	29.81	0.096 7
Bromomethane		9.82 (0), 1.006 8 (100, g)	1.79 (g)	26.52	0.115 9
1-Bromo-3-methylbutane		8.04 (−56), 6.05 (20)	1.95 (20, B)	28.10	0.099 6
2-Bromo-3-methylbutyric acid		6.5 (20)			
1-Bromo-2-methylpropane	0.643 (20), 0.518 (40), 3.26 (90)	7.70 (0), 7.2 (25)	1.92 (25, lq), 1.99 (20, B)	26.96	0.105 9
1-Bromonaphthalene	5.99 (15), 3.20 (40)	5.83 (25), 5.12 (20)	1.29 (25, lq)	46.44	0.101 8
1-Bromononane		5.42 (−20), 4.74 (25)	1.95 (25, lq)	31.36	0.089 4
1-Bromooctane		6.35 (−50)	1.99 (20, lq), 1.88 (25, lq)	31.00	0.092 8
1-Bromopentadecane		3.9 (20)			

1-Bromopentane		9.9 (−90), 6.32 (25)	2.21 (g), 2.09 (20, lq)	29.51	0.104 9
p-Bromophenol				48.88	0.107 0
1-Bromopropane	0.539 (15), 0.459 (30)	8.09 (25)	2.17 (g), 3.16 (20, lq)	28.30	0.121 8
2-Bromopropane	0.536 (15), 0.437 (30)	9.46 (25)	2.21 (g), 2.10 (25, lq)	26.21	0.118 3
1-Bromotetradecane		3.84 (25)	1.92 (20, lq), 1.83 (25, lq)	32.93	0.087 8
o-Bromotoluene	1.3 (25)	4.28 (58)	1.45 (20, B)	36.62	0.099 8
m-Bromotoluene		5.36 (58)	1.77 (20, B)		
p-Bromotoluene		5.49 (58)	1.95 (20, B)	36.40	0.099 7
Bromotrifluoromethane	0.15 (25)		0.65 (g)	4 (25)	
1-Bromoundecane		4.73 (−9)		31.94	0.086 1
Butane	0.007 39 (20, g), 0.008 39 (60, g)		0	14.87	0.120 6
1,3-Butanediol	130.3 (20), 89 (25)	28.8 (25)		37.8 (25)	
1,4-Butanediol	65–70 (25)	33 (15), 30 (30)	3.93 (20, lq), 2.4 (15, D)		
2,3-Butanediol	121 (25)			36 (25)	
Butanesulfonyl chloride			3.94 (25, D)	37.33	0.097 7
1-Butanethiol	0.501 (20), 0.450 (30)	5.07 (25), 4.59 (50)	1.54 (25, lq or B)	28.07	0.114 2
1,2,4-Butanetriol	1227 (25)				
1-Butanol	2.948 (20), 1.782 (40)	17.8 (20), 8.2 (118)	1.66 (g; 20, B)	27.18	0.089 8
DL-2-Butanol	3.907 (20), 0.527 (100)	16.6 (25)	1.66 (30, B)	23.47 (20)	22.62 (30)
2-Butanone	0.428 (20), 0.349 (40)	18.5 (20), 15.3 (60)	3.2 (30, lq), 2.76 (25, B)	26.77	0.112 2
2-Butanone oxime		3.4 (20)		31.89	0.102 2
1-Butene (g)	0.007 6 (20), 0.010 0 (120)	1.003 2 (20)	0.30	15.19	0.132 3 (lq)
2-Butene			0.33 (g, *cis*), 0 (g, *trans*)	16.11	0.128 9
3-Butenenitrile		28.1 (20)	4.53 (g)	31.40	0.108 5
2-Butoxyethanol	3.15 (25), 1.51 (60)	9.30 (25)	2.08 (25, B)	28.18	0.081 6
Butoxyethyne		6.62 (25)	2.05 (25, lq)		
2-(2-Butoxyethoxy)ethanol	4.76 (25)			30.0 (25)	
1-Butoxy-2-propanol	2.55 (25)			26.5 (25)	
Butyl acetate	0.734 (20), 0.688 (25)	6.85 (−73), 5.01 (20)	1.86 (22, B)	27.55	0.106 8
DL-*sec*-Butyl acetate				23.33 (22)	21.24 (42)
tert-Butyl acetate			1.91 (25, B)	24.69	0.110 2

TABLE 4-10 Viscosity, dielectric constant, dipole moment, and surface tension of selected organic substances (*continued*)

Substance	Viscosity, $mN \cdot s \cdot m^{-2}$	Dielectric constant ε	Dipole moment, D	Surface tension $dyn \cdot cm^{-1}$ a	b
Butylamine	0.681 (20)	4.88 (20)	1.00 (g), 1.22 (20, lq)	26.24	0.112 2
sec-Butylamine		4.4 (21)	1.28 (25, B)	23.75	0.105 7
tert-Butylamine			1.29 (25, B)	19.44	0.102 8
Butylbenzene	1.035 (20), 0.960 (25)	2.36 (20)	0.36 (20, lq)	31.28	0.102 5
sec-Butylbenzene		2.36 (20)	0.37 (20, lq)	30.48	0.097 9
tert-Butylbenzene		2.37 (20)	0.36 (20, lq)	30.10	0.098 5
Butyl butyrate	0.84 (25)			27.65	0.096 5
Butyl decyl *o*-phthalate	55 (20)				
N-Butyldiethanolamine	55 (25)				
4-*tert*-Butyl-2,5-dimethylphenol	8.30 (80)				
4-*tert*-Butyl-2,6-dimethylphenol	2.72 (80)				
6-*tert*-Butyl-2,4-dimethylphenol	2.10 (80)				
6-*tert*-butyl-3,4-dimethylphenol	3.50 (80)				
N-Butylethanolamine	17.4 (25)				
Butyl ethyl ether	0.421 (20), 0.397 (25)		1.24	22.75	0.104 9
Butyl formate	0.691 (20), 0.940 (0)	2.43 (80)	2.08 (26, lq), 2.03 (25, B)	27.08	0.102 6
Butyl methyl ether			1.25 (25, B)	22.17	0.105 7
2-*tert*-Butyl-4-methyphenol	2.55 (80)		1.31 (20, B)		
Butyl nitrate		13 (20)	2.99 (20, B)	30.35	0.112 6
2-(2-*sec*-Butylphenoxy)ethanol	65.1 (25)				
2-(4-*tert*-Butylphenoxy)ethanol	122.5 (25)				
Butyl propionate			1.79 (22, B)	27.37	0.099 3
4-*tert*-Butylpyridine	1.495 (20)		2.87 (25, C)	35.48	0.095 1
Butyl stearate	8.26 (25), 4.9 (50)	3.11 (30)	1.88 (24, B)	33.0 (25)	32.7 (30)
Butyl vinyl ether	0.5 (20)		1.25 (25, Hx)	21.99 (20)	

Butyraldehyde	0.455 (20), 0.367 (39)	13.4 (26)	2.45 (40, lq)	26.67	0.092 5
Butyric acid	1.540 (20), 0.980 (40)	2.97 (20)	1.65 (30, B)	28.35	0.092 0
Butyric anhydride	1.615 (20), 1.486 (25)	13 (20)		28.93 (20)	28.44 (25)
4-Butyrolactone	1.75 (25)	39.1 (20)	4.12 (25, B)		
Butyronitrile	0.624 (15), 0.515 (30)	20.3 (21)	4.07 (g), 3.6 (20, B)	29.51	0.103 7
Camphor		11.35 (20)	2.91 (20, B), 3.10 (25, B)		
Carbon disulfide	0.363 (20)	3.0 (−112), 2.64 (20)	0 (g), 0.12 (20, lq)	35.29	0.148 4
Carbon tetrachloride	0.965 (20), 0.793 (25)	2.24 (20), 2.23 (25)	0	29.49	0.122 4
Carbon tetrafluoride	0.020 (25)	1.000 6 (25, g)	0	14 (−73)	
Carvone		11 (22)	2.8 (15, B)	36.54	0.092 0
Chloroacetic acid	3.15 (50), 1.92 (75)	20 (20), 12.3 (60)	2.31 (30, B)	43.27	0.111 7
o-Chloroaniline	0.925 (25)	13.4 (25)	1.78 (20, B)	43.41	0.090 4
m-Chloroaniline		13.4 (19)	2.68 (20, B)		
p-Chloroaniline			2.99 (25, B)	48.69	0.109 9
Chlorobenzene	0.799 (20), 0.631 (40)	5.71 (20), 4.2 (120)	1.72 (g), 1.56 (20, lq)	35.97	0.119 1
1-Chlorobutane	0.469 (15)	9.07 (−30), 7.39 (20)	2.13 (g), 2.0 (20, B)	25.97	0.111 7
2-Chlorobutane	0.439 (15)	7.09 (30)	2.14 (g), 2.1 (20, B)	24.40	0.111 8
Chlorocyclohexane		10.9 (−47), 7.6 (25)	2.2 (25, B)	33.90	0.110 1
Chlorodifluoromethane	0.23 (25), 0.013 (25, g)	6.11 (24)	1.4 (g)	8 (25)	
1-Chlorododecane		4.2 (20)	2.11 (25, lq), 1.94 (20, B)	31.56	0.090 4
1-Chloro-2-2,3-epoxypropane	1.03 (25)	25.6 (1), 22.6 (22)	1.8 (25, C)	39.76	0.136 0
Chloroethane	0.279 (10)	1.013 (19, g)	2.0 (g), 1.96 (20, lq)	21.18 (5)	20.58 (10)
2-Chloroethanol	3.913 (15)	25.8 (25), 13 (132)	1.77 (g), 1.90 (25, B)	38.9 (20)	
Chloroform	0.596 (15), 0.514 (30)	4.81 (20), 4.31 (50)	1.1 (g), 1.1 (25, lq)	29.91	0.129 5
1-Chloroheptane		4.48 (20)	1.86 (22, B)	28.94	0.096 1
2-Chloroheptane		6.52 (22)	2.05 (22, B)		
3-Chloroheptane		6.70 (22)	2.06 (22, B)		
4-Chloroheptane		6.54 (22)	2.06 (22, B)		
1-Chlorohexane			1.94 (20, B)	28.32	0.103 8
Chloromethane					
(g)	0.0106 (20), 0.012 9 (80)	1.006 9 (100)	1.87		
(lq)		12.6 (−20)	1.86 (20)	19.5	0.165 0

TABLE 4-10 Viscosity, dielectric constant, dipole moment, and surface tension of selected organic substances (*continued*)

Substance	Viscosity, $mN \cdot s \cdot m^{-2}$	Dielectric constant ε	Dipole moment, D	Surface tension $dyn \cdot cm^{-1}$	
				a	b
1-Chloro-3-methylbutane		7.63 (−70), 6.05 (20)	1.94 (20, B)	25.51	0.107 6
Chloromethyl methyl ether			1.88 (C)		
1-Chloro-2-methylpropane	0.462 (20), 0.373 (40)	7.87 (−38), 6.49 (14)	2.06 (g), 2.0 (25, B)	24.40	0.109 9
2-Chloro-2-methylpropane	0.543 (15)	10.95 (0), 9.96 (20)	2.11 (g), 2.13 (25, B)	20.06 (15)	18.35 (30)
1-Chloronaphthalene	2.940 (25)	5.04 (25)	1.33 (25, lq), 1.52 (25, B)	44.12	0.103 5
o-Chloronitrobenzene		38 (50), 32 (80)	4.62 (g), 6.22 (50, lq)	48.10	0.117 1
m-Chloronitrobenzene		21 (50), 18 (80)	3.72 (g), 3.30 (50, lq)	49.71	0.141 7
p-Chloronitrobenzene		8 (120)	2.81 (g), 2.83 (90, lq)	45.84	0.104 6
1-Chlorooctane		5.05 (25)	2.14 (25, lq)	29.64	0.096 1
Chloropentafluoroethane	0.26 (25), 0.013 (25, g)		0.5 (g)	5 (25)	
1-Chloropentane	0.580 (20)	6.6 (11)	2.14 (g), 1.94 (20, B)	27.09	0.107 6
o-Chlorophenol	2.250 (45), 4.11 (25)	6.31 (25)	2.19 (g), 1.46 (20, lq)	42.5	0.112 2
m-Chlorophenol	4.722 (45), 11.55 (25)		2.19 (25, B)	43.7	0.100 9
p-Chlorophenol	4.99 (50)		2.09 (20, B)	19.51	0.087 5
1-Chloropropane	0.372 (15), 0.318 (30)	7.7 (20)	2.05 (g), 1.96 (20, B)	24.41	0.124 6
2-Chloropropane	0.335 (15), 0.299 (30)	9.82 (20)	2.17 (g), 2.1 (20, B)	21.37	0.088 3
1-Chloro-2-propanone		30 (19)	2.22 (g), 2.37 (20, Hx)		
3-Chloro-1-propene	0.347 (15)	8.2 (20)	2.0 (g), 1.8 (20, B)	25.50	0.094 6
o-Chlorotoluene		4.45 (20), 4.2 (55)	1.57 (g), 1.41 (20, lq)		
m-Chlorotoluene		5.5 (20), 5.0 (60)	1.77 (20, lq), 1.8 (22, B)		
p-Chlorotoluene		6.08 (20), 5.6 (55)	2.21 (g), 1.90 (20, lq)	34.93	0.108 2
Chlorotrifluoromethane	0.016 (25)	1.001 3 (29, g)	0.50 (g)	14 (−73)	
Chlorotrimethylsilane			2.09 (20, B)	19.51	0.087 5
Cinnamaldehyde		17 (20), 16.9 (24)	3.74 (g), 3.30 (30, lq)		
o-Cresol	3.506 (46)	11.5 (25)	2.32 (25, lq), 1.45 (25, B)	39.43	0.101 1

m-Cresol	18.42 (20), 5.057 (45)	11.8 (25)	2.39 (20, lq), 1.61 (25, B)	38.00	0.092 4
p-Cresol	5.607 (45)	9.91 (58)	2.35 (20, lq), 1.54 (20, B)	38.58	0.096 2
Crotonic acid			2.13 (30 B)		
Cyanoacetic acid		33.4 (19)			
Cycloheptanol				35.02	0.092 3
1,3-Cyclohexadiene		2.6 (−89)	0.38 (20, B)		
Cyclohexane	0.980 (20), 0.534 (60)	2.05 (15), 2.02 (25)	0	27.62	0.118 8
Cyclohexanecarboxylic acid		2.6 (31)			
1,4-Cyclohexanedione		15.0 (25)	1.41 (g), 1.3 (30, B)		
Cyclohexanol	41.07 (30), 17.19 (45)	15.0 (25), 7.24 (100)	1.86 (25, C)	35.33	0.096 6
Cyclohexanone	2.453 (15), 1.803 (30)	20 (−40), 18.2 (20)	3.11 (20, B), 3.01 (25, B)	37.67	0.124 2
Cyclohexanone oxime		3.0 (89)	0.83 (25, B)		
Cyclohexene	0.650 (20)	2.6 (−105), 2.22 (25)	0.61 (g), 0.28 (20, lq)	29.23	0.122 3
Cyclohexylamine	1.662 (20), 1.16 (49)	4.73 (20)	1.22 (20, lq), 1.26 (20, B)	34.19	0.118 8
Cyclohexylbenzene	3.681 (0)		0.62 (20, B)		
Cycloehexylmethanol		9.7 (60), 8.1 (80)	1.68 (20, B)		
o-Cyclohexylphenol		3.97 (55)			
p-Cyclohexylphenol		4.42 (131)			
Cyclooctane			0	32.02	0.109 0
Cyclopentane	0.439 (20)	1.965 (20)	0	25.53	0.146 2
Cyclopentanol		25 (−20), 18 (20)	1.72 (25, C)	35.04	0.101 1
Cyclopentanone		16 (−51)	3.30 (g), 2.93 (25, B)	35.55	0.110 0
Cyclopentene			0.98 (25, Hx)	25.94	0.149 5
p-Cymene	3.402 (20)	2.243 (20)	0 (lq)	28.83	0.087 7
cis-Decahydronaphthalene	3.381 (20)	2.18 (20)	0	32.18 (20)	31.01 (30)
trans-Decahydronaphthalene	2.128 (20)	2.17 (20)	0	29.89 (20)	28.87 (30)
Decamethylcyclopentasiloxane		2.5 (20)		19.56	0.056 5
Decamethyltetrasiloxane	1.28 (20)	2.4 (20)	0.79 (25, lq)	86.20 (25)	
Decane	0.928 (20), 0.775 (22)	1.991 (20), 1.844 (130)	0	25.67	0.092 0
1-Decanol		8.1 (20)	1.71 (20, B), 1.62 (25, B)	30.34	0.073 2
1-Decene	0.805 (20)		0.42 (20, B)	25.84	0.091 9
Diallyl sulfide		4.9 (20)	1.33 (25, B)		

TABLE 4-10 Viscosity, dielectric constant, dipole moment, and surface tension of selected organic substances (*continued*)

Substance	Viscosity, $mN \cdot s \cdot m^{-2}$	Dielectric constant ε	Dipole moment, D	Surface tension $dyn \cdot cm^{-1}$	
				a	*b*
Dibenzofuran		3.0 (100)	0.88 (25, B)		
Dibenzylamine		3.6 (20)	0.97 (20, lq), 1.02 (20, B)	43.27	0.108 6
Dibenzyl decanedioate		4.6 (25)			
Dibenzyl ether	3.711 (25)		1.39 (21, B)	38.2 (35)	
o-Dibromobenzene		7.35 (20)	2.13 (20, B)		
m-Dibromobenzene		3.80 (20)	1.5 (20, B)		
p-Dibromobenzene		2.57 (95)	0	41.84	0.100 7
1,4-Dibromobutane			2.16 (20, lq), 2.06 (20, B)	48.24	0.119 0
2,3-Dibromobutane		5.75 (25)	2.20 (g), 1.7 (25, lq)		
1,2-Dibromoethane	1.721 (20), 1.286 (40)	4.78 (25), 4.09 (131)	1.11 (g), 1.14 (20, lq)	35.43	0.142 8
cis-1,2-Dibromoethylene		7.7 (0), 7.08 (25)	1.35 (B)		
trans-1,2-Dibromoethylene		2.9 (0), 2.88 (25)	0		
1,2-Dibromoheptane		3.8 (25)	1.78 (25, D)		
2,3-Dibromoheptane		5.1 (25)	2.15 (25, B)		
3,4-Dibromoheptane		4.7 (25)	2.15 (25, B)		
Dibromomethane		7.77 (10), 6.7 (40)	1.43 (g), 1.85 (20, lq)	42.77	0.148 8
1,2-Dibromopropane	1.5 (25)	4.3 (20)	1.43 (25, B)	36.81	0.115 5
Dibromotetrafluoroethane	0.72 (25)	2.34 (25)		18.9 (20)	18.1 (25)
Dibutylamine	0.95 (20)	2.978 (20)	1.06 (20, lq), 1.05 (20, B)	26.50	0.095 2
Dibutyl decanedioate	9.03 (25)	4.54 (30)	2.64 (25, B)		
Dibutyl ether	0.602 (30)	3.06 (25)	1.18 (g), 1.19 (20, lq)	24.78	0.093 4
Dibutyl maleate	5.62 (20), 4.76 (25)		2.70 (25, B)	32.46	0.086 5
2,6-Di-*tert*-butyl-4-methylphenol	3.47 (80)		1.68 (20, B)		
Dibutyl *o*-phthalate	19.91 (20), 7.85 (45)	6.436 (30), 5.99 (45)	2.97 (20, lq), 2.85 (30, B)	33.40 (20)	
Dichloroacetic acid	3.23 (50), 1.92 (75)	8.2 (22), 7.8 (61)		37.8	0.0927

o-Dichlorobenzene	1.324 (25)	9.93 (25), 7.10 (90)	2.51 (g), 2.26 (24, B)	26.84 (20)	35.55 (30)
m-Dichlorobenzene	1.045 (23), 0.955 (33)	5.04 (25), 4.22 (90)	1.68 (g), 1.38 (24, B)	38.30	0.114 7
p-Dichlorobenzene	0.839 (55), 0.668 (79)	2.41 (50)	0	34.66	0.087 9
1,4-Dichlorobutane		8.9 (25)	2.22 (g), 2.13 (25, lq)	37.79	0.117 4
Dichlorodifluoromethane	0.26 (25), 0.013 (25, g)	2.13 (29)	0.51 (g)	9 (25)	
1,1-Dichloroethane	0.505 (25), 0.430 (30)	10.1 (18), 10.86 (16)	2.06 (g), 2.00 (25, B)	27.03	0.118 6
1, 2-Dichloroethane	0.887 (15), 0.730 (30)	12.7 (−10), 10.65 (20)	1.48 (g), 1.7 (20, B)	35.43	0.142 8
1,1-Dichloroethylene	0.442 (0), 0.358 (20)	4.67 (16)	1.30 (25, B)		
cis-1,2-Dichloroethylene	0.467 (20), 0.444 (25)	9.20 (25)	2.95 (g), 1.90 (25, B)	28 (20)	
trans-1,2-Dichloroethylene	0.423 (15), 0.404 (20)	2.14 (25)	0.70 (25, B)	25 (20)	
2,2′-Dichloroethyl ether	2.41 (20), 2.065 (25)	21.2 (20)	2.61 (20, B)	40.57	0.130 6
Dichlorofluoromethane	0.34 (25), 0.011 (25, g)	5.34 (28)	1.3 (g)	18 (25)	
Dichloromethane	0.449 (15), 0.393 (30)	9.14 (20), 1.006 5 (100, g)	1.60 (g), 1.90 (20, B)	30.41	0.128 4
2,4-Dichlorophenol			1.60 (25, B)	46.59	0.122 1
1,2-Dichloropropane	0.865 (20), 0.700 (25)	8.925 (26), 7.90 (35)	1.87 (25, B)	31.42	0.124 0
1,3-Dichloropropane			2.08 (g), 2.2 (25, B)	36.40	0.123 3
2,2-Dichloropropane	0.769 (15), 0.619 (30)	11.37 (20)	2.62 (g), 2.20 (25, B)	23.60 (20)	22.53 (30)
1,1-Dichloro-2-propanone		14 (20)			
1,2-Dichlorotetrafluoroethane	0.38 (25), 0.011 (25, g)	2.26 (25)	0.53 (g)	12 (25)	
α,α-Dichlorotoluene		6.9 (20)	2.07 (20, B), 2.05 (25, B)	41.26	0.103 5
Diethanolamine	368 (30), 196 (40)	2.81 (25)	2.84 (25, B)		
1,1-Diethoxyethane		3.80 (25)	1.08 (g)	23.46	0.103 0
1,2-Diethoxyethane	0.65 (20)		1.99 (20, B), 1.65 (25, B)		
Diethoxymethane				23.87	0.129 1
Diethylamine	0.388 (10), 0.273 (38)	3.6 (22)	0.92 (g), 1.11 (25, lq)	22.71	0.114 3
N,N-Diethylaniline	1.15 (30), 0.750 (75)	5.5 (19)	1.40 (20, lq), 1.80 (20, B)	36.59	0.104 0
Diethyl carbonate	0.868 (15), 0.748 (25)	2.82 (20)	1.07 (g), 0.91 (25, B)	28.62	0.110 0
Diethyl decanedioate		5.0 (30)	2.38 (20, lq), 2.52 (20, B)	34. 68	0.095 9
Diethyl ether	0.247 (15), 0.245 (20)	4.335 (20), 3.97 (40)	1.15 (g), 1.22 (16, lq)	18.92	0.090 8
Diethyl ethyl phosphonate	1.627 (15), 0.969 (45)	11.00 (15), 9.86 (45)	2.95 (32, lq), 2.91 (20, C)	30.63	0.097 5
Diethyl fumarate		6.5 (23)	2.40 (20, B)		
Diethyl glutarate		6.7 (30)	2.46 (30, lq)	34.34	0.101 0

TABLE 4-10 Viscosity, dielectric constant, dipole moment, and surface tension of selected organic substances (*continued*)

Substance	Viscosity, $mN \cdot s \cdot m^{-2}$	Dielectric constant ε	Dipole moment, D	Surface tension $dyn \cdot cm^{-1}$ *a*	*b*
Di(2-ethylhexyl)-2-ethylhexyl phosphonate	6.00 (45), 3.61 (65)	4.09 (45), 3.94 (65)			
Di(2-ethylhexyl) *o*-phthalate	33.67 (35), 21.40 (45)	4.91 (35), 4.77 (45)			
Diethyl maleate	3.57 (20), 3.14 (25)	8.58 (23)	2.56 (25, B)	34.67	0.103 9
Diethyl malonate	2.15 (20), 1.94 (25)	8.03 (25)	2.49 (20, lq), 2.54 (25, B)	33.91	0.104 2
Diethyl nonanedioate		5.13 (30)			
Diethyl oxalate	2.311 (15), 1.618 (30)	8.1 (21)	2.49 (20, D)	34.32	0.111 9
Diethyl *o*-phthalate	9.18 (35), 6.41 (45)	7.34 (35), 7.13 (45)	2.8 (25, B)	38.47	0.096 3
Diethyl succinate		6.64 (30)	2.3 (g), 2.37 (30, lq)	33.97	0.104 1
Diethyl sulfate		29 (20)	4.46 (25, D)	35.47	0.097 6
Diethyl sulfide	0.446 (20), 0.422 (25)	5.72 (25), 5.24 (50)	1.52 (g), 1.58 (20, B)	27.33	0.110 6
Diethyl sulfite		16 (20), 14 (50)			
Diethylzinc		2.5 (20)	0.62 (25, B)		
1,1-Difluoroethane	0.243 (21)		2.30 (g)		
1,2-Dihydroxybenzene		2.6 (−89)	2.60 (25, B)	47.6	0.084 9
1,3-Dihydroxybenzene		3.2 (18)	2.09 (44, B)	54.8	0.071 7
1,4-Dihydroxybenzene			1.4 (44, B)		
1,2-Diiodobenzene		5.7 (20)	1.70 (20, B)		
1,3-Diiodobenzene		4.3 (25)	1.22 (20, B)		
1,4-Diiodobenzene		2.9 (120)	0.19 (20, B)		
cis-1,2-Diiodoethylene		4.46 (83)	0.71 (B)		
trans-1,2-Diiodoethylene		2.19 (83)	0		
Diiodomethane	3.043 (15), 2.392 (30)	5.316 (25)	1.08 (25, B)	70.21	0.161 3
Diisobutylamine		2.7 (22)	1.10 (25, B)	24.00	0.091 2
Diisobutyl *o*-phthalate	30 (20)				

Diisopentylamine		2.5 (18)	1.48 (30, B)	26.04	0.085 8
Diisopentyl ether	1.40 (11), 1.012 (20)	2.82 (20)	0.98 (20, lq), 1.23 (25, B)	24.76	0.087 1
Diisopropylamine	0.40 (25)		1.26 (25, B)	21.83	0.107 7
Diisopropyl ether	0.379 (25)	3.88 (25)	1.13 (g), 1.26 (25, B)	19.89	0.104 8
1,2-Dimethoxybenzene	3.281 (25), 2.184 (40)	4.09 (25)	1.32 (25, B)	34.4	0.064 2
1,1-Dimethoxyethane				23.90	0.115 9
1,2-Dimethoxyethane	0.530 (10), 0.455 (25)	7.60 (10), 7.20 (25)	1.71 (25, B)	48.0 (25)	
Dimethoxymethane	0.340 (15), 0.325 (20)	2.65 (20)	0.74 (g)	23.59	0.119 9
N,N-Dimethylacetamide	2.141 (20), 0.838 (30)	37.78 (25)	3.80 (g), 4.60 (20, lq)	32.43 (30)	29.50 (50)
Dimethylamine	0.207 (15), 0.186 (25)	6.32 (0), 5.26 (25)	1.03 (g), 1.14 (25, lq)	29.50	0.126 5
N,N-Dimethylaniline	1.285 (25), 0.91 (50)	4.9 (20), 4.4 (70)	1.61 (g), 1.55 (25, B)	38.14	0.104 9
2,4-Dimethylaniline			1.40 (25, B)	39.34	0.099 6
2,2-Dimethylbutane	0.351 (25), 0.330 (30)	1.873 (25)	0	18.29	0.099 0
2,3-Dimethylbutane	0.361 (25), 0.342 (30)	1.890 (25)	0	19.38	0.100 0
2,3-Dimethyl-1-butanol				26.22	0.099 2
N,N-Dimethylbutyramide	1.271	2.00			
Dimethyl carbonate			0.90 (g), 0.96 (25, B)	31.94	0.134 3
1,1-Dimethylcyclopentane			0	23.78	0.101 6
2,2-Dimethyl-1,3-dioxolane-4-methanol	11 (20)				
Dimethyl ether	0.010 4 (60)	5.02 (25), 2.97 (110)	1.30 (g), 1.25 (25, B)	14.97	0.147 8
N,N-Dimethylformamide	0.845 (20), 0.598 (50)	38.3 (20), 36.71 (25)	3.86 (25, B)	36.76 (20)	34.40 (40)
2,4-Dimethylheptane		1.9 (20)	0	23.21	0.092 9
2,5-Dimethylheptane		1.9 (20)	0	23.21	0.092 9
2,6-Dimethylheptane		2 (20)	0	22.77	0.088 7
2,6-Dimethyl-4-heptanone	1.03 (20)		2.66 (25, C)		
Dimethyl hexanedioate	14 (20)		2.28 (20, B)	38.26	0.113 8
Dimethyl hydrogen phosphonate	1.08 (25)				
Dimethyl maleate	3.54 (20), 3.21 (25)		2.48 (25, C)	40.73	0.122 0
Dimethyl malonate		10 (20)	2.41 (20, B)	39.72	0.120 8
2,2-Dimethylpentane		1.91 (20)	0	19.94	0.095 7
2,3-Dimethylpentane	0.406 (20)	1.939 (20)	0	21.96	0.099 5

TABLE 4-10 Viscosity, dielectric constant, dipole moment, and surface tension of selected organic substances (*continued*)

Substance	Viscosity, $mN \cdot s \cdot m^{-2}$	Dielectric constant ε	Dipole moment, D	Surface tension $dyn \cdot cm^{-1}$	
				a	b
2,4-Dimethylpentane	0.361 (20)	1.914 (20)	0	20.09	0.097 2
3,3-Dimethylpentane		1.94 (20)	0	21.59	0.099 6
2,4-Dimethylphenol			1.48 (20, B), 1.98 (60, B)	34.57	0.086 9
2,5-Dimethylphenol	1.55 (80)		1.43 (20, B), 1.52 (60, B)	36.72	0.085 0
3,4-Dimethylphenol	3.00 (80)	4.8 (17)	1.77 (20, B)	35.75	0.091 0
3,5-Dimethylphenol	2.42 (80)		1.76 (20, B)	34.09	0.080 7
Dimethyl *o*-phthalate	17.2 (25), 6.41 (45)	8.25 (25), 8.11 (45)	2.8 (25, B)		
2,2-Dimethylpropane	0.328 (0), 0.303 (5)	1.80 (20), 1.678 (98)	0	12.05 (20)	10.98 (30)
N,N-Dimethylpropionamide	0.935	33.1			
2,5-Dimethylpyrazine		2.43 (20)	0		
2.3-Dimethylquinoxaline		2.3 (25)	0		
Dimethyl succinate		5.1 (20)	2.09 (20, B)	39.00	0.119 1
Dimethyl sulfate		48.3 (20), 46.4 (20)	4.31 (25, D)	41.26	0.116 3
Dimethyl sulfide	0.289 (20), 0.265 (36)	6.2 (20)	1.45 (25, B)	26.07	0.080 5
Dimethyl sulfite	0.715 (30), 0.436 (80)	22.5 (23)	2.93 (20, B)	36.48	0.125 3
Dimethyl sulfoxide	2.47 (20), 1.192 (55)	48.9 (20), 41.9 (55)	3.9 (25, B)	43.54 (20)	42.41 (30)
2,4-Dimethyltetrahydrothiophene-1,1-dioxide	9.04	29.5			
N,N-Dimethyl-*o*-toluidine		3.4 (20)	0.88 (25, B)		
N,N-Dimethyl-*p*-toluidine		3.9 (20)	1.29 (25, B)		
Dinonyl hexanedioate	37 (20)		2.53 (25, B)		
Dinonyl *o*-phthalate		4.65 (35), 4.52 (45)			
Dioctyl decanedioate		4.0 (27)			
Dioctyl *o*-phthalate		5.1 (25)	3.06 (25, C)		
1,4-Dioxane	1.439 (15), 1.087 (30)	2.24 (20), 2.21 (25)	0	36.23	0.139 1

Dipentyl ether	1.188 (15), 0.922 (30)	2.77 (25)	0.98 (20, lq), 1.24 (25, B)	26.66	0.092 5
Dipentyl *o*-phthalate	17.03 (35), 11.51 (45)	5.79 (35), 5.62 (45)	2.71 (20, lq)	32.56	0.073 9
Dipentyl sulfide		3.83 (25)	1.59 (25, B)	29.55	0.087 6
Diphenylamine	4.66 (55), 1.04 (130)	3.3 (52)	1.31 (20, C), 1.01 (25, B)	45.36	0.101 7
1,2-Diphenylethane		2.4 (110)	0 (110, lq), 0.45 (25, B)		
Diphenyl ether	2.61 (40), 2.09 (50)	3.65 (30)	1.16	28.70	0.078 0
Diphenylmethane		2.7 (18), 2.57 (26)	0.26 (30, lq), 0.3 (25, B)		
1,1-Dipropoxyethane				25.03	0.097 2
Dipropoxymethane				25.17	0.095 3
Dipropylamine	0.534 (20), 0.427 (37)	3.07 (20)	1.01 (20, lq), 1.03 (20, B)	24.86	0.102 2
Dipropyl carbonate				28.94	0.101 5
Dipropylene glycol butyl ether	4.23 (25)			28.2 (25)	
Dipropylene glycol ethyl ether	3.11 (25)			27.7 (25)	
Dipropylene glycol isopropyl ether	386 (25)			25.9 (25)	
Dipropylene glycol methyl ether	3.1 (25)			28.8 (25)	
Dipropyl ether	0.448 (15), 0.376 (30)	3.39 (26)	1.21 (g), 1.17 (30, Hx)	22.60	0.104 7
Divinyl ether		3.9 (20)	1.07 (20, lq)		
Dodecamethylcyclohexasiloxane		2.6 (20)			
Dodecamethylpentasiloxane		2.5 (20)		17.08 (25)	
Dodecane	1.508 (20), 1.378 (25)	2.05 (−10), 2.01 (20)	0	27.12	0.088 4
1-Dodecanol		5.15 (20), 6.5 (25)	1.52 (20, B)	31.25	0.0748
6-Dodecyne		2.17 (25)			
1,2-Epoxybutane	0.41 (20), 0.40 (25)		2.01 (20, B)	23.9 (20)	
Erythritol		28 (128)			
Ethane (g)	0.009 0 (20), 0.011 4 (100)	1.001 5 (0)	0	1.24	0.166 0 (lq)
1,2-Ethanediamine	1.54 (20), 1.226 (30)	16.8 (18), 14.2 (20)	1.96 (g), 1.92 (25, B)	44.77	0.139 8
1,2-Ethanediol	26.09 (15), 13.55 (30)	38.66 (20), 37.7 (25)	2.28 (g), 2.3 (25, D)	50.21	0.089 0
1,2-Ethanediol diacetate	3.13 (20)	13 (30)	2.34 (30, B)		

TABLE 4-10 Viscosity, dielectric constant, dipole moment, and surface tension of selected organic substances (*continued*)

Substance	Viscosity, $mN \cdot s \cdot m^{-2}$	Dielectric constant ε	Dipole moment, D	Surface tension $dyn \cdot cm^{-1}$ *a*	*b*
Ethanesulfonic acid				45.74	0.082 4
Ethanesulfonyl chloride			3.89 (25, B)	43.43	0.117 7
Ethanethiol	0.003 16 (g)	6.9 (15)	1.57 (g), 1.40 (20, B)	25.06	0.079 3
Ethanol	1.209 (19), 0.991 (30)	25.00 (20), 20.21 (55)	1.69 (g), 1.71 (25, B)	24.05	0.083 2
Ethoxybenzene	1.364 (15), 1.040 (30)	4.22 (20)	1.41 (g), 1.36 (25, CS_2)	35.17	0.110 4
2-Ethoxyethanol	2.04 (20), 1.85 (25)	29.6 (24)	2.24 (30, B)	30.59	0.089 7
2-(2-Ethoxyethoxy)ethanol	3.71 (25)			31.8 (25)	27.2 (75)
2-Ethoxyethyl acetate	1.025 (25)	7.567 (30)	2.25 (30, B)	31.8 (25)	
1-Ethoxy-2-methylbutane		3.96 (20)			
1-Ethoxynaphthalene		3.3 (19)			
1-Ethoxypentane		3.6 (23)			
1-Ethoxy-2-propanol	1.68 (25)			25.9 (25)	
α-Ethoxytoluene		3.9 (20)			
Ethyl acetate	0.473 (15), 0.426 (25)	6.11 (20), 5.30 (77)	1.78 (g), 1.84 (25, lq)	26.29	0.116 1
Ethyl acetoacetate	1.419 (20), 1.508 (25)	15.7 (22)	3.22 (18, B, keto form), 2.04 (−80, CS_2, enol form)	34.42	0.101 5
Ethylamine		6.94 (10)	1.40 (25, B)	22.63	0.137 2
2-(Ethylamino)ethanol	12.40 (25)				
N-Ethylaniline	2.04 (25), 1.08 (55)	5.76 (20)		39.00	0.107 0
Ethylbenzene	0.669 (20), 0.531 (40)	2.41 (20)	0.37 (25, lq)	31.48	0.109 4
Ethyl benzoate	2.407 (15), 1.751 (30)	6.02 (20)	1.95 (g), 1.93 (25, B)	37.16	0.105 9
Ethyl α-bromobutyrate		8 (20)	2.40 (25, B)		
2-Ethyl-1-butanol	8.021 (15), 5.892 (25)	6.19 (90)		25.06 (15)	24.32 (25)
Ethyl butyrate	0.771 (15), 0.613 (25)	5.10 (18)	1.74 (22, B)	26.55	0.104 5

2-Ethylbutyric acid	3.3 (20)			26.3 (20)	
Ethyl carbamate	0.916 (105), 0.715 (120)	14.2 (50)	2.59 (30, D)		
Ethyl chloroacetate		11.4 (21)	2.65 (25, B)	34.18	0.117 7
Ethyl chloroformate		11 (20)	2.56 (35, B)	28.90	0.108 4
Ethyl cinnamate	8.7 (20)	6.1 (18)	1.86 (20, B)	39.99	0.104 5
Ethyl crotonate		5.4 (20)	1.95 (24, B)	29.31	0.106 6
Ethyl cyanoacetate	3.256 (15), 2.148 (30)	26.9 (20)	4.04 (30, B)	38.80	0.109 2
Ethylcyclohexane	0.843 (20), 0.787 (25)	2.054 (20)	0 (g)	27.78	0.105 4
Ethyl dichloroacetate		12 (2), 10 (22)	2.63 (25, B)	34.89	0.115 8
N-Ethyldiethanolamine	53 (25)				
Ethyl dodecanoate		3.4 (20), 2.7 (143)	1.3 (20, lq)	30.05	0.086 3
Ethylene		1.001 44 (0)	0 (g)	−2.7	0.185 4
Ethylene carbonate	1.85 (40)	89.6 (40), 69.4 (91)	4.87 (25, B)		
Ethylenediamine	1.540 (18)	16.0 (18), 14.2 (20)	1.98 (g), 1.92 (25, B)	44.77	0.139 8
Ethylene dinitrate		28.3 (20)	3.58 (25, B)	49.1 (0)	46.7 (45)
2,2′-(Ethylenedioxy)diethanol	38 (20)	23.69 (20)	5.58 (lq)	47.33	0.088 0
Ethylene glycol	26.09 (15), 13.35 (30)	41.2 (20), 37.7 (25)	2.27 (g), 2.20 (15, lq)	50.21	0.089 0
Ethylene oxide	0.3 (0)	14 (−1)	1.88 (g), 1.92 (20, lq)	27.66	0.166 4
Ethyleneimine	0.418 (25)	18.3 (25)	1.89 (g), 1.77 (25, B)	7.9 (20)	
Ethyl formate	0.419 (15), 0.358 (30)	7.16 (25)	1.94 (g), 1.96 (25, lq)	26.47	0.131 5
Ethyl fumarate		6.5 (23)		33.90	0.105 6
Ethyl hexadecanoate		3.2 (20), 2.71 (104)	1.2 (lq)	32.86	0.085 9
2-Ethyl-1,3-hexanediol	323 (20)				
Ethyl hexanoate			1.80 (20, B)	27.73	0.096 0
2-Ethylhexanoic acid	7.7 (20)				
2-Ethyl-1-hexanol	9.8 (20)	4.41 (90)	1.74 (25, B)	30.0 (22)	
2-Ethylhexyl acetate	1.5 (20)				
Ethyl isobutyrate				25.33	0.104 6
Ethyl isopentyl ether		3.96 (20)			
Ethyl isothiocyanate		19.5 (21)	3.67 (20, B)	38.69	0.132 6
Ethyl lactate	2.44 (25)	13.1 (25)	2.4 (20, B)	30.72	0.098 3
Ethyl maleate		8.6 (23)			

TABLE 4-10 Viscosity, dielectric constant, dipole moment, and surface tension of selected organic substances (*continued*)

Substance	Viscosity, $mN \cdot s \cdot m^{-2}$	Dielectric constant ε	Dipole moment, D	Surface tension $dyn \cdot cm^{-1}$	
				a	b
Ethyl 3-methylbutyrate		4.71 (18)		25.79	0.100 6
Ethyl methyl ether			1.22 (g)	18.56	0.131 7
Ethyl methyl sulfide	0.373 (20), 0.354 (25)			27.63	0.128 6
Ethyl nitrate		19.4 (20)	2.93 (20, B)	30.81	0.134 5
Ethyl 9-octadecenoate		3.2 (25)	1.83 (20, lq)		
Ethyl 4-oxopentanoate		12 (21)			
3-Ethylpentane		1.94 (20)	0	22.52	0.103 2
Ethyl pentanoate	0.847 (20)	4.7 (18)	1.76 (28, B)	27.15	0.099 9
Ethyl pentyl ether		3.6 (23)	1.2 (20, B)	24.19	0.099 2
Ethyl phenylacetate		5.3 (21)	1.82 (30)		
Ethyl phenyl sulfide			4.08 (25, B)	39.30	0.113 1
Ethyl propionate	0.564 (15), 0.473 (30)	5.65 (19)	1.75 (22, B)	26.72	0.116 8
Ethyl propyl ether	0.323 (20), 0.225 (60)		1.16 (25, B)	21.92	0.105 4
Ethyl salicylate	1.772 (45)	7.99 (30)	2.85 (25, B)	31.00	0.109 1
Ethyl stearate		2.98 (40), 2.69 (100)	1.65 (40, lq)		
Ethyl thiocyanate		29.3 (21)	3.33 (20, B)	37.28	0.122 6
o-Ethyltoluene				32.33	0.106 0
p-Ethyltoluene		2.24 (25)	0	30.98	0.107 5
Ethyl trichloroacetate		7.8 (20)	2.56 (25, B)	32.97	0.107 3
Ethyl vinyl ether	0.2		1.26 (20, B)	19.00 (20)	
Ethynyl acetate				32.81 (20)	30.20 (40)
Fluorobenzene	0.620 (15), 0.517 (30)	5.42 (25), 4.7 (60)	1.61 (g)	29.67	0.120 4
1-Fluorohexane				23.41	0.100 1
2-Fluoro-2-methylbutane		5.89 (20)	1.92 (25, B)		

1-Fluoropentane		4.24 (20)	1.85 (25, B)	22.81	0.131 5
o-Fluorotoluene	0.680 (20), 0.601 (30)	4.22 (30), 3.9 (60)	1.35 (g), 1.26 (30, lq)		
m-Fluorotoluene	0.608 (20), 0.534 (30)	5.42 (30), 4.9 (60)	1.86 (g), 1.66 (30, lq)	32.31	0.125 7
p-Fluorotoluene	0.622 (20), 0.522 (30)	5.86 (30), 5.3 (60)	2.00 (g), 1.76 (30, lq)	30.44	0.110 9
Formamide	4.320 (15), 2.296 (30)	111.0 (20), 103.5 (40)	3.73 (g)	59.13	0.084 2
Formanilide	1.65 (120)		3.37 (25, C)	44.30	0.087 5
Formic acid	1.966 (15), 1.219 (40)	58.5 (15), 57.0 (21)	1.35 (g), 1.20 (25, B)	39.87	0.109 8
2-Furaldehyde	2.475 (0), 1.494 (25)	41.9 (20), 34.9 (50)	2.13 (25, lq), 3.63 (25, B)	46.41	0.132 7
Furan	0.380 (20), 0.361 (25)	2.95 (25)	0.66 (g), 0.67 (20, B)	24.10 (20)	23.38 (25)
Furfuryl alcohol	4.62 (25)		1.92 (25, lq)	ca 38 (20)	
Glycerol	945 (25), 134 (50)	42.5 (25)	2.68 (25, D)	63.14 (17)	62.5 (25)
Glycerol triacetate		7.2 (20)	2.73 (25, B)	37.88	0.081
Glycerol trinitrate	36.0 (20), 13.6 (40)	19 (20)	3.38 (25, B)	55.74	0.250 4
Glycerol trioleate		3.2 (26)	3.11 (23, B)	36.03	0.069 9
Glycerol tripalmitate		2.9 (65)	2.80 (23, B)	32.26	0.067 2
Glycerol tristearate		2.8 (70)	2.86 (23, B)	32.73	0.068 5
Heptanaldehyde	0.977 (15)	9.1 (20)	2.26 (40, lq), 2.58 (22, B)	28.64	0.092 0
Heptane	0.416 (20), 0.341 (40)	1.924 (20), 1.85 (70)	0	22.10	0.098 0
Heptanoic acid	3.40 (30)	2.6 (71)		29.88	0.084 8
1-Heptanol	7.014 (20), 8.53 (15)	12.1 (22)	1.73 (20, B)		
DL-2-Heptanol	5.06 (25)	9.21 (22)	1.73 (20, B)		
DL-3-Heptanol		6.9 (22)	1.73 (20, B)		
4-Heptanol		6.2 (22)	1.72 (20, B)		
2-Heptanone	0.854 (15), 0.686 (30)	11.95 (20), 8.27 (100)	2.61 (22, B)	28.76	0.105 6
3-Heptanone		12.9 (22)	2.81 (22, B)	28.24	0.101 5
4-Heptanone	0.736 (20)	12.60 (20), 9.46 (80)	2.74 (20, B)	28.11	0.106 0
1-Heptene	0.35 (20), 0.34 (25)	2.07 (20)	0.34 (20, lq)	22.28	0.099 1
Hexadecamethylcyclooctasiloxane		2.7 (20)			
Hexadecane	3.591 (22)		0	29.18	0.085 4
1-Hexadecanol		3.8 (50)	1.67 (25, B)		
1,5-Hexadiene	0.275 (20), 0.244 (36)				
2,4-Hexadiene		2.2 (25)	0.31 (25, B)		

TABLE 4-10 Viscosity, dielectric constant, dipole moment, and surface tension of selected organic substances (*continued*)

Substance	Viscosity, $mN \cdot s \cdot m^{-2}$	Dielectric constant ε	Dipole moment, D	Surface tension $dyn \cdot cm^{-1}$	
				a	*b*
Hexafluorobenzene			0	22.6 (20)	
Hexamethyldisiloxane		2.2 (20)	0.37 (25, lq)	17.01	0.076 3
Hexamethylphosphoramide	3.47 (20)	30 (20)	4.31 (25, lq)	33.8 (20)	
Hexane	0.313 (20), 0.271 (40)	1.904 (15), 1.890 (20)	0	20.44	0.102 2
Hexanedinitrile	5.99	32.45	3.8 (25, B)	47.88	0.097 3
2,4-Hexanedione				32.22	0.100 2
Hexanenitrile	1.041 (15), 0.830 (30)	17.26 (25)		29.64	0.090 7
Hexanoic acid	3.525 (15), 2.511 (30)	2.63 (71)	1.13 (25, lq)	28.05 (20)	27.55 (25)
1-Hexanol	6.203 (15), 3.872 (30)	13.3 (25), 8.5 (75)	1.55 (20, B)	27.81	0.080 1
2-Hexanone	0.584 (25)	14.6 (15)	2.68 (22, B)	28.18	0.109 2
1-Hexene	0.26 (20), 0.25 (25)	2.051 (20)	0.34 (20, lq)	20.47	0.102 7
Hexyl acetate				28.44	0.097 0
4-Hydroxy-4-methyl-2-pentanone	2.9 (25)	18.2 (25)	3.24 (20, B)	31.0 (20)	
Iodobenzene	1.774 (17), 0.488 (149)	4.62 (20)	1.71 (g), 1.3 (20, B)	41.52	0.112 3
1-Iodobutane		6.22 (20), 4.52 (130)	2.10 (g), 1.90 (20, B)	30.82	0.013 1
2-Iodobutane			2.06 (20, B)	30.32	0.105 6
1-Iodododecane		3.9 (20)	1.87 (20, C)		
Iodoethane	0.617 (15), 0.540 (30)	10.2 (−50), 7.82 (20)	1.91 (g), 1.69 (20, lq)	31.67	0.128 6
1-Iodoheptane		4.9 (22)	1.86 (22, B)	32.18	0.088 7
3-Iodoheptane		6.4 (22)	1.95 (22, B)		
1-Iodohexadecane		3.5 (20)		34.49	0.088 0
1-Iodohexane		5.37 (20)	1.94 (20, C)	31.63	0.084 5
Iodomethane	0.500 (20), 0.424 (40)	7.00 (20)	1.64 (g), 1.42 (20, B)	33.42	0.123 4
1-Iodo-3-methylbutane		5.6 (19)	1.85 (20, B)	30.37	0.091 5
2-Iodo-2-methylbutane		8.19 (20)	2.20 (20, B)		

1-Iodo-2-methylpropane	0.875 (20), 0.697 (40)	6.5 (20)	1.89 (20, B)	30.26	0.017 2
1-Iodooctane		4.6 (25)	1.80 (25, lq), 1.90 (20, C)	32.51	0.091 5
2-Iodooctane		5.8 (20)	2.07 (20, C)		
1-Iodopentane		5.81 (20)	1.90 (20, B)	31.41	0.101 4
1-Iodopropane	0.837 (15), 0.670 (30)	7.00 (20)	2.03 (g), 1.86 (20, B)	31.64	0.113 6
2-Iodopropane	0.732 (15), 0.620 (30)	7.87 (20)	2.01 (20, B)	29.35	0.110 7
p-Iodotoluene		4.4 (35)	1.72 (22, B)	39.23	0.096 5
α-Ionone		11 (18)		34.10	0.094 9
β-Ionone		12 (20)		35.36	0.095 0
Iron pentacarbonyl		2.6 (20)			
Isobutyl acetate		5.29 (20)	1.87 (22, B)	25.59	0.101 3
Isobutylamine	0.553 (25)	4.43 (21)	1.27 (25, B)	24.48	0.109 2
Isobutylbenzene		2.319 (20), 2.298 (30)	0.31 (20, lq)	29.39	0.096 1
Isobutyl butyrate		4.1 (20)		24.47	0.084 3
Isobutyl formate	0.680 (20)	6.41 (19)	1.89 (20, B)	26.14	0.112 2
Isobutyl isobutyrate				30.92	0.127 0
Isobutyl nitrate		2.7 (20)			
Isobutyl pentanoate		3.8 (19)			
Isobutyl propionate				28.97	0.116 6
Isobutyric acid	1.44 (15)	2.7 (20)	1.09 (25, lq)	26.88	0.092 0
Isobutyric anhydride		14 (20)			
Isobutyronitrile	0.551 (15), 0.456 (30)	20.4 (24)	3.61 (25, B)	24.93 (20)	23.84 (30)
Isopentyl acetate	0.872 (20), 0.790 (25)	4.81 (20), 4.63 (30)	1.84 (22, B), 1.76 (30, lq)	26.75	0.098 9
Isopentyl butyrate		4.0 (20)		27.32	0.091 8
Isopentyl pentanoate		3.6 (19)	1.8 (28, B)		
Isopentyl propionate		4.2 (20)			
Isopropyl acetate	0.559 (20)		1.86 (22, B)	24.44	0.107 2
Isopropylamine	0.36 (25)	5.45 (20)	1.45 (25, B)	19.91	0.097 2
Isopropylbenzene	0.791 (20), 0.739 (25)	2.39 (20)	0.65 (g), 0.39 (20, lq)	30.32	0.105 4
Isopropyl formate	0.512 (20)			24.56	0.114 7
1-Isopropyl-4-methylbenzene	3.402 (20), 1.600 (30)	2.24 (20)	0	29.44 (20)	
Isoquinoline	3.253 (30)	10.7 (20)	2.75 (g), 2.55 (25, B)		

TABLE 4-10 Viscosity, dielectric constant, dipole moment, and surface tension of selected organic substances (*continued*)

Substance	Viscosity, $mN \cdot s \cdot m^{-2}$	Dielectric constant ε	Dipole moment, D	Surface tension $dyn \cdot cm^{-1}$ a	b
Lactamide					
Lactic acid	40.33 (25)	22 (17)			
Lactonitrile	2.01 (30)	38 (20)		38.31	0.096 0
D-Limonene		2.4 (20)	1.57 (25, B)	29.50	0.092 9
DL-Limonene		2.3 (20)	0.63 (25, B)	29.11	0.091 3
DL-Mandelonitrile		17.8 (23)		45.90	0.098 8
Menthol	6.89 (35)		1.55 (20, B)		
2-Mercaptoethanol	3.4 (20)				
Methacrylic acid	1.32 (20)		1.65	26.5 (25)	
Methacrylonitrile	0.392 (20)		3.69 (g)	24.4 (20)	
Methane (g)	0.010 9 (20), 0.013 3 (100)	1.000 94 (0)	0	*	
Methanesulfonic acid				52.28	0.089 3
Methanethiol			1.26 (g)	28.09	0.169 6
Methanol	0.676 (10), 0.544 (25)	41.8 (−20), 33.62 (20)	1.69 (g), 1.68 (22, B)	24.00	0.077 3
o-Methoxybenzaldehyde			4.34 (20, B)	45.34	0.110 5
p-Methoxybenzaldehyde		22.3 (22), 10.4 (248)	3.26 (35, B)	44.69	0.104 7
Methoxybenzene	1.152 (15), 0.789 (30)	4.33 (25), 3.9 (70)	1.36 (g), 1.24 (20, B)	38.11	0.120 4
2-Methoxyethanol	1.72 (20), 1.60 (25)	16.93 (25), 16.0 (30)	2.04 (25, B)	33.30	0.098 4
2-(2-Methoxyethoxy)ethanol	3.48 (25), 1.61 (60)			34.8 (25)	29.9 (75)
2-Methoxyethyl acetate		8.25 (20)	2.13 (30, B)		
1-Methoxy-2-nitrobenzene			4.83 (g)	48.62	0.118 5
o-Methoxyphenol		12 (25)		41.2	0.094 3

* $38.618 - 0.1873T - 0.000356T^2$.

2-Methoxy-4-(2-propenyl)phenol	6.931 (25)		2.46 (25, B)		
o-Methoxytoluene		3.5 (20)			
m-Methoxytoluene		3.5 (20)			
p-Methoxytoluene		4.0 (20)		36.20	0.107 1
N-Methylacetamide	3.88 (30), 2.54 (45)	178.9 (30), 138.6 (60)	4.39 (20, D)	33.67 (30)	30.62 (50)
Methyl acetate	0.388 (20), 0.320 (40)	7.03 (20), 6.68 (25)	1.70 (g), 1.75 (25, B)	27.95	0.128 9
Methyl acetoacetate	1.704 (20)			34.98	0.094 4
Methyl acrylate	1.398 (20)		1.77 (25, B)		
Methylamine	0.285 (15), 0.236 (0)	11.4 (−10), 10.0 (18)	1.29 (g)	22.87	0.148 8
N-Methylaniline	2.02 (25), 1.084 (55)		1.67 (25, B)	39.32	0.097 0
Methyl benzoate	2.298 (15), 1.673 (30)	6.59 (20)	1.86 (25, B)	40.10	0.117 1
2-Methyl-1,2-butadiene	0.266 (0.3), 0.223 (20)	2.1 (25)	0.15 (g)		
2-Methylbutane	0.237 (15), 0.215 (25)	1.871 (0), 1.845 (20)	0.13 (g)	17.20	0.110 3
2-Methyl-1-butanol	5.50 (20), 1.44 (60)	14.7 (25)		21.5 (25)	
2-Methyl-2-butanol	5.48 (15), 2.81 (30)	5.82 (25)	1.72 (20, B)	24.18	0.074 8
3-Methyl-1-butanol	4.81 (15), 2.96 (30)	14.7 (25), 5.82 (130)	1.82 (25, B)	25.76	0.082 0
3-Methyl-2-butanol	3.51 (25)			23.0 (25)	
2-Methyl-1-butene		2.20 (20)	0.52 (20, lq)	18.81	0.114 8
2-Methyl-2-butene			0.11 (25, lq), 0.34 (25, B)	19.70	0.127 1
3-Methyl-1-butene		1.002 8 (100, g)	0.25 (g)	16.42	0.103 1
2-Methylbutyl acetate	0.872 (20)	4.63 (30)	1.82 (22)	26.75	0.098 9
Methyl butyrate	0.580 (20), 0.459 (40)	5.6 (20)	1.72 (22, B)	27.48	0.114 5
3-Methylbutyric acid	2.731 (15), 2.411 (20)	2.64 (20)	0.63 (25)	27.28	0.088 6
3-Methylbutyronitrile		18 (220)	3.62 (25, C)	27.58	0.082 7
Methyl chloroacetate		12.9 (21)		37.90	0.130 4
Methyl cyanoacetate	3.82 (50), 2.69 (65)	19.23 (50), 17.57 (65)		41.32	0.107 4
Methylcyclohexane	0.734 (20), 0.685 (25)	2.02 (20), 2.07 (25)	0	26.11	0.113 0
cis-2-Methylcyclohexanol	18.08 (25), 13.60 (30)	13.3*	2.58 (30, lq),* 1.95 (25, B)*	32.45	0.077 0*
trans-2-Methylcyclohexanol	37.13 (25), 25.14 (30)				

* Mixed isomers.

TABLE 4-10 Viscosity, dielectric constant, dipole moment, and surface tension of selected organic substances (*continued*)

Substance	Viscosity, $mN \cdot s \cdot m^{-2}$	Dielectric constant ε	Dipole moment, D	Surface tension $dyn \cdot cm^{-1}$	
				a	*b*
cis-3-Methylcyclohexanol	19.7 (25), 17.23 (30)	16.47 (20)	1.91	29.08	0.062 9*
trans-3-Methylcyclohexanol	25.52 (16), 15.60 (30)	8.05	1.75	28.80 (30)	
cis-4-Methylcyclohexanol	0.247 (25)	13.3*	2.70 (30, lq),* 1.9 (25, B)*	29.07	0.069 0*
trans-4-Methylcyclohexanol	0.385 (25)				
2-Methylcyclohexanone		16 (−15), 14 (20)	2.98 (25, B)	34.06	0.102 7
3-Methylcyclohexanone		18 (−80), 12 (20)	3.06 (25, B)	33.06	0.092 5
4-Methylcyclohexanone		15 (−41), 12 (20)	3.07 (25, B)	32.83	0.093 5
Methylcyclopentane	0.507 (20), 0.478 (25)	1.985 (20)	0	24.63	0.116 3
Methyl decanoate			1.65 (20, Hx)	30.33	0.091 2
Methyl dichloroacetate				37.00	0.121 9
Methyl dodecanoate			1.70 (20, Hx)	31.37	0.089 3
N-Methylformamide	1.99 (15), 1.65 (25)	200.1 (15), 182.4 (25)	3.86 (25, B)	37.96 (30)	35.02 (50)
Methyl formate	0.360 (15), 0.319 (29)	8.5 (20)	1.77 (g)	28.29	0.157 2
Methyl heptanoate				28.95	0.098 7
2-Methyl-2-heptanol		3.38 (−7), 2.46 (25)			
2-Methyl-3-heptanol		3.37 (20), 3.75 (60)	1.63 (20, B)		
2-Methyl-4-heptanol		3.30 (20), 3.65 (60)			
3-Methyl-3-heptanol		3.74 (20), 2.89 (60)			
3-Methyl-4-heptanol		9.1 (−20), 7.4 (20)			
4-Methyl-3-heptanol		5.25 (20), 4.62 (55)			
4-Methyl-4-heptanol		2.87 (20), 3.27 (60)			
Methyl hexadecanoate				31.50	0.077 5
2-Methylhexane	0.378 (20)	1.92 (20)	0	21.22	0.096 64

* Mixed isomers.

3-Methylhexane	0.372 (20)	1.93 (20)	0	21.73	0.097 0
Methyl hexanoate			1.70 (20, Hx)	28.47	0.104 5
Methyl isobutyrate	0.523 (20), 0.419 (40)		1.98 (20, B)	25.99	0.113 1
Methyl methacrylate	0.632 (20)	2.9 (20)	1.68 (25, B)	28–29 (30)	
Methyl *o*-methoxybenzoate		7.7 (21)			
Methyl *p*-methoxybenzoate		4.3 (33)			
1-Methylnaphthalene		2.7 (20)	0.23 (20, B)	39.96	0.0934
Methyl *o*-nitrobenzoate		28 (25)	3.67 (30, B)		
Methyl octadecanoate				32.20	0.77 5
2-Methyloctane		1.97 (20)	0	23.76	0.094 0
4-Methyloctane		1.97 (20)	0	24.22	0.094 0
Methyl octanoate				29.93	0.100 2
Methyl oleate	4.88 (20)	3.211 (20)		31.3 (25)	25.4 (100)
2-Methylpentane	0.310 (20), 0.295 (25)	1.88 (20)	0	19.37	0.099 7
3-Methylpentane	0.307 (25), 0.292 (30)	1.895 (20)	0	20.26	0.106 0
2-Methyl-2,4-pentanediol	34.4 (20)		2.9 (0)	33.1 (20)	
4-Methylpentanenitrile	0.980 (20), 9.843 (30)	15.5 (22)	3.53 (25, B)	28.89	0.091 7
Methyl pentanoate	0.713 (20)	4.3 (19)	1.62 (22, B)	27.85	0.104 4
2-Methyl-1-pentanol				26.98	0.081 9
3-Methyl-1-pentanol				26.92	0.078 9
4-Methyl-1-pentanol				25.93	0.074 3
2-Methyl-2-pentanol				25.07	0.086 1
3-Methyl-2-pentanol				27.14	0.091 9
4-Methyl-2-pentanol	4.074 (25)			24.67	0.082 1
2-Methyl-3-pentanol				26.43	0.091 4
3-Methyl-3-pentanol				25.48	0.088 8
4-Methyl-2-pentanone	0.585 (20), 0.522 (30)	13.11 (20), 11.78 (40)		23.64 (20)	19.62 (60)
4-Methyl-3-penten-2-one	0.879 (25)	15.6 (0), 15.1 (20)	3.20 (25, B)		
1-Methyl-1-phenylhydrazine		7.3 (19)	1.84 (15, B)		
Methyl phenyl sulfide			1.38 (20, B)	42.81	0.123 8
2-Methylpropane	0.007 44 (20, g)		0	12.83	0.123 6
2-Methylpropanenitrile	0.551 (15), 0.456 (30)	20.2	4.07 (g), 3.60 (20, B)		

TABLE 4-10 Viscosity, dielectric constant, dipole moment, and surface tension of selected organic substances (*continued*)

Substance	Viscosity, $mN \cdot s \cdot m^{-2}$	Dielectric constant ε	Dipole moment, D	Surface tension $dyn \cdot cm^{-1}$	
				a	b
2-Methyl-1-propanol	4.70 (15), 2.876 (30)	26 (−34), 17.93 (25)	2.96 (30, lq), 1.78 (20, B)	24.53	0.079 5
2-Methyl-2-propanol	3.316 (20), 2.039 (40)	10.9 (30), 8.49 (50)	1.67 (22, B)	20.02 (15)	19.10 (30)
2-Methylpropene			0.50 (g)	14.84	0.131 9
N-Methylpropionamide	6.06 (20), 3.56 (40)	185 (20), 151 (40)	3.59 (g)	31.20 (20)	29.12 (50)
Methyl propionate	0.477 (15)	6.21	1.70 (22, B)	27.58	0.125 8
2-Methylpropionic acid	1.213 (25), 1.126 (30)	2.73 (40)	1.08 (25, lq)	25.55 (20)	25.13 (25)
1-Methylpropyl acetate				25.72	0.105 4
2-Methylpropyl acetate	0.702 (20), 0.366 (78)	5.29 (20)	1.87 (22, B)	25.59	0.101 3
2-Methylpropylamine	21.7 (25)	4.43 (21)	1.27 (27)	24.48	0.109 2
2-Methylpropyl formate	0.680 (20)	6.41 (19)	1.88 (22)	26.14	0.112 2
Methyl propyl ketoxime		3.3 (20)			
2-Methylpyridine	0.805 (20), 0.710 (30)	9.8 (20)	1.96 (25, B)	36.11	0.124 3
3-Methylpyridine			2.41 (25, B)	37.35	0.115 3
4-Methylpyridine			2.60 (25, B)	37.71	0.114 1
N-Methyl-2-pyrrolidinone	1.666 (25)	32.0 (25)	4.09 (30, B)		
Methyl salicylate		9.41 (30)	2.47 (25, B)	42.15	0.117 4
Methyl tetradecanoate			1.62 (25, B)	31.00	0.080 0
2-Methyltetrahydrofuran	0.601 (0), 0.536 (10)	6.92 (0), 6.63 (10)			
Methyl thiocyanate	64.3 (0)	4.3 (19)	3.34 (20, B)	40.66	0.130 5
Morpholine	2.53 (15), 1.79 (30)	7.33 (25)	1.75 (25, lq), 1.52 (25, B)	37.63 (20)	36.24 (30)
Napththalene	0.780 (100), 0.967 (80)	2.54 (85)	0	42.84	0.110 7
1-Naphthonitrile		16 (70)			
2-Naphthonitrile		17 (70)			
o-Nitroaniline		34.5 (90)	4.28 (20, B)		

p-Nitroaniline		56.3 (160)	6.3 (25, B)	60.62	0.092 3
o-Nitroanisole			4.83 (g)	48.62	0.118 5
Nitrobenzene	2.165 (15), 1.55 (35)	34.82 (25), 24.9 (90)	4.22 (g), 3.96 (25, B)	46.34	0.115 7
m-Nitrobenzyl alcohol		22 (20)			
2-Nitrobiphenyl	12 (45)		3.82 (20, B)		
Nitroethane	0.677 (20), 0.63 (35)	28.06 (30), 27.4 (35)	3.61 (g)	35.27	0.125 5
Nitromethane	0.692 (15), 0.596 (30)	35.87 (30), 35.1 (35)	3.46 (g)	40.72	0.167 8
1-Nitro-2-methoxybenzene			4.83 (g)	48.62	0.118 5
o-Nitrophenol	2.343 (45)	17 (50)	3.14 (25, B)	47.35	0.117 4
1-Nitropropane	0.798 (25), 0.70 (35)	23.24 (30), 22.7 (35)	3.60 (g)	32.62	0.100 9
2-Nitropropane	0.750 (25)	25.52 (30)	3.76 (g)	32.18	0.115 8
N-Nitrosodimethylamine		53 (20)	4.01 (20, B)		
o-Nitrotoluene	2.37 (20), 1.63 (40)	27.4 (20), 22.0 (58)	3.72 (20, B)	44.10	0.117 4
m-Nitrotoluene	2.33 (20), 1.60 (40)	24 (20), 22 (58)	4.20 (20, B)	43.54	0.111 8
p-Nitrotoluene	1.20 (60)	22 (52)	4.47 (25, B)	42.26	0.097 4
Nonane	0.713 (20), 0.666 (25)	1.972 (20), 1.85 (110)	0	24.72	0.093 5
1-Nonanol	14.3 (20)		1.72 (20, B)	29.79	0.078 9
1-Nonene	0.620 (20), 0.586 (25)		0.59 (20, B)	24.90	0.093 8
(*Z,Z*)-9,12-Octadecadienoic acid		2.70 (70), 2.60 (120)	1.40 (18, Hx)		
Octamethylcyclotetrasiloxane	2.20 (20)	2.4 (20)	0.42 (25, lq), 0.67 (25, B)	20.19	0.081 1
Octamethyltrisiloxane	0.82 (20)	2.3 (20)	0.64 (25, lq)	67.56 (25)	
Octane	0.546 (20), 0.433 (40)	1.95 (20), 1.83 (110)	0	23.52	0.095 1
Octanenitrile	1.811 (15), 1.356 (30)	13.90 (25)		29.61	0.080 2
Octanoic acid	5.828 (20), 4.690 (25)	2.45 (20)	1.15 (25, lq)	29.2 (20)	28.7 (25)
1-Octanol	10.64 (15), 6.125 (30)	11.3 (10), 10.34 (20)	1.72 (20, B)	29.09	0.079 5
2-Octanol		8.20 (20), 6.52 (40)	1.65 (20, B)	27.96	0.082 0
2-Octanone		10.39 (20), 7.42 (100)	2.72 (15, B)		
1-Octene	0.470 (20), 0.447 (25)	2.084 (20)	0.34 (20, lq)	23.68	0.095 8
Oleic acid	38.80 (20), 27.64 (25)	2.46 (20), 2.45 (60)	1.44 (25, lq)	32.80 (20)	27.94 (90)
Oxalyl chloride		3.5 (21)	0.93 (20, B)		
2-Oxohexamethyleneimine	9 (78)		3.88 (25, B)		
4-Oxopentanoic acid				41.69	0.076 3

TABLE 4-10 Viscosity, dielectric constant, dipole moment, and surface tension of selected organic substances (*continued*)

Substance	Viscosity, $mN \cdot s \cdot m^{-2}$	Dielectric constant ε	Dipole moment, D	Surface tension $dyn \cdot cm^{-1}$	
				a	*b*
Palmitic acid		2.3 (70)			
Paraldehyde		13.9 (25)	1.91 (25, lq)	28.28	0.106 2
Parathion	15.30 (25)		4.98 (25, B)	39.2 (25)	
Pentachloroethane	2.741 (15), 2.070 (30)	3.73 (20)	0.92 (g), 0.98 (25, lq)	37.09	0.117 8
Pentadecane	2.814 (22)		0	28.78	0.085 7
cis-1,3-Pentadiene		2.32 (25)	0.50 (25, B)		
Pentanaldehyde		10.1 (17)	2.59 (20, B)	27.96	0.101 0
Pentane	0.237 (15), 0.215 (25)	2.011 (−90), 1.84 (20)	0	18.25	0.112 1
1,5-Pentanediol	128 (20)		2.45 (20, D)	43.2 (20)	
2,4-Pentanedione	0.6 (20)	25.7 (20), 17.39 (25)	3.03 (g), 2.5 (20, B)	33.28	0.114 4
Pentanenitrile	0.779 (15), 0.637 (30)	17.4 (21)	3.57 (25, B)	27.44 (20)	26.33 (30)
1-Pentanethiol		4.55 (25), 4.23 (50)	1.54 (25, lq)		
Pentanoic acid	2.359 (15), 1.774 (30)	2.66 (20)	1.61 (20, D)	28.90	0.088 7
1-Pentanol	4.650 (15), 2.987 (20)	16.9 (20), 13.9 (25)	1.71 (20, B)	27.54	0.087 4
2-Pentanol	5.130 (15), 2.780 (30)	13.82 (22)	1.66 (22, B)	25.96	0.100 4
3-Pentanol	7.337 (15), 3.306 (30)	13.02 (22)	1.64 (22, B)	24.60 (20)	23.76 (30)
2-Pentanone	0.473 (25)	15.45 (20), 11.73 (80)	2.72 (22, B)	24.89	0.065 5
3-Pentanone	0.493 (15), 0.423 (30)	19.4 (−20), 17.00 (20)	2.72 (20, B)	27.36	0.104 7
1-Pentene	0.24 (0)	2.10 (20)	0.34 (20, lq)	18.20	0.109 9
cis-2-Pentene				19.73	0.117 2
trans-2-Pentene				18.90	0.099 7
Pentyl acetate	0.924 (20), 0.862 (25)	4.75 (20)	1.72 (g), 1.91 (25, B)	27.66	0.099 4
Pentylamine	1.018 (20)	4.5 (22)	1.55 (30, B)	24.4 (13)	
Pentyl formate		6.5 (20)		28.09	0.102 3
Pentyl nitrate		9 (18)			

Phenanthrene		2.8 (20)	0		
Phenol	6.024 (35), 3.421 (50)	9.78 (60)	1.53 (20, B)	43.54	0.106 8
Phenoxyacetaldehyde		4.8 (20)			
Phenoxyacetylene		4.8 (20)	1.42 (25, lq)		
2-Phenylacetamide				46.26	0.078 8
Phenyl acetate	1.799 (45)	5.23 (20)	1.54 (22, B)		
Phenylacetonitrile	1.93 (25)	19.0 (25), 8.5 (234)	3.47 (27, B)	44.57	0.115 5
Phenylacetylene		3.0 (20)	0.72 (20, B)		
1-Phenylethanol		13 (20), 7.6 (90)	1.51 (20, B)	42.88	0.103 8
Phenylhydrazine		7.2 (21)	1.67 (25, B)	48.14	0.129 2
Phenyl isocyanate		8.8 (20)			
Phenyl isothiocyanate		10 (20)		42.73	0.108 6
1-Phenylpropene		2.7 (20)			
2-Phenylpropene		2.3 (20)			
3-Phenylpropene		2.6 (20)			
Phenyl propyl ether				34.27	0.105 6
Phenyl salicylate		6.3 (50)		45.20	0.097 6
Phosgene		4.7 (0), 4.3 (22)			
Phthalide		36 (75)			
DL-α-Pinene	1.61 (25)	2.64 (25)	0.60 (25, B)	28.35	0.094 4
L-β-Pinene	1.70 (20), 1.41 (25)	2.76 (20)		28.26	0.093 4
Piperidine	1.679 (15), 1.224 (30)	5.8 (20)	1.19 (25, B)	31.79	0.115 3
Propane (g)	0.008 1 (20), 0.010 7 (125)	1.6 (0)	0	9.22	0.087 4 (lq)
1,2-Propanediamine	1.46	10.2			
1,3-Propanediamine	17.85	9.55	1.96 (25, B)		
1,2-Propanediol	56.0 (20), 18.0 (40)	32.0 (20)	2.27 (25, D)	72.0 (25)	
1,3-Propanediol	56.0 (20), 18.0 (40)	35.0 (20)	2.52 (25, D)	47.43	0.090 3
1-Propanethiol			1.55 (25, lq)	27.38	0.127 2
2-Propanethiol			1.64 (25, lq)	24.26	0.117 4
1-Propanol	2.522 (15), 1.722 (30)	22.2 (20), 20.33 (25)	1.67 (g), 1.75 (25, B)	25.26	0.077 7
2-Propanol	2.859 (15), 1.765 (30)	18.3 (25), 16.24 (40)	1.69 (g), 1.66 (30, B)	22.90	0.078 9
2-Propenaldehyde			3.04 (g), 2.90 (25, B)		

TABLE 4-10 Viscosity, dielectric constant, dipole moment, and surface tension of selected organic substances (*continued*)

Substance	Viscosity, $mN \cdot s \cdot m^{-2}$	Dielectric constant ε	Dipole moment, D	Surface tension $dyn \cdot cm^{-1}$	
				a	*b*
Propene (g)	0.008 43 (20), 0.009 33 (50)	1.88 (20), 1.44 (90)	0.35 (g)	9.99	0.142 7 (lq)
2-Propen-1-ol	1.363 (20), 0.914 (40)	21.6 (15)	1.63 (g)	27.53	0.090 2
Propionaldehyde	0.357 (15), 0.317 (27)	18.5 (17)	2.75 (g), 2.57 (20, B)		
Propionamide			3.4 (30, B)	39.05	0.090 9
Propionic acid	1.175 (15), 0.956 (30)	3.30 (10), 3.44 (40)	1.76 (g), 1.77 (25, D)	28.68	0.099 3
Propionic anhydride	1.144 (20), 1.061 (25)	18.3 (16)		30.30 (20)	29.70 (25)
Propionitrile	0.454 (15), 0.389 (30)	22.2 (20), 24.2 (50)	4.06 (g), 3.60 (20, B)	29.63	0.115 3
Propyl acetate	0.585 (20), 0.460 (40)	5.69 (19)	1.86 (25, B)	26.60	0.112 0
Propylamine	0.343 (25)	5.31 (20)	1.17 (g), 1.36 (20, B)	24.86	0.124 3
Propylbenzene		2.37 (20), 2.351 (30)	0.35 (25, lq)	31.13	0.107 5
Propyl benzoate				36.55	0.106 9
Propyl butyrate	0.831 (20)	4.3 (20)		27.06	0.100 0
Propyl chloroacetate				32.91	0.108 3
Propylene carbonate	2.53	64.4			
Propylene oxide	0.327 (20), 0.28 (25)		2.00 (g)		
Propyleneimine	0.491 (25)		1.77 (g, *cis*), 1.60 (g, *trans*)		
Propyl formate	0.574 (20), 0.417 (40)	7.72 (19)	1.91 (22, B)	26.77	0.111 9
Propyl isobutyrate	0.831 (20)			25.83	0.101 5
Propyl nitrate		14 (18)	3.01 (20, B)	29.67	0.123 7
Propyl pentoate	1.053 (20)	4 (19)		27.72	0.098 4
Propyl propionate	0.673 (20)	4.7 (20)	1.79 (22, B)	26.85	0.105 9
Propyne			0.75 (g)	14.51	0.148 2
2-Propyn-1-ol	1.68 (20)	24.5 (20)	1.78 (25, B)	38.59	0.127 0
Pulegone		9.5 (20)	2.00 (25, B)		

Pyradazine			3.97 (35, D)	50.55	0.103 6
Pyrazine		2.8 (54)	0		
Pyridine	1.130 (10), 0.829 (30)	12.3 (25), 9.4 (116)	2.20 (20, B), 2.25 (g)	39.82	0.130 6
Pyrimidine			2.44 (35, D)	32.85	0.101 0
Pyrrole	1.352 (20), 1.233 (25)	7.48 (18), 8.13 (25)	1.80 (25, B)	39.81	0.110 0
Pyrrolidine			1.58 (20, B)	31.48	0.090 0
2-Pyrrolidone	13.3 (25)		3.55 (25, B)		
Quinoline	4.354 (15), 3.37 (25)	9.00 (25)	2.18 (25, B)	45.25	0.106 3
Safrole	2.294 (25)	3.1 (21)			
Salicylaldehyde	2.90 (20), 1.67 (45)	13.9 (20)	3.1 (30, lq), 2.86 (20, B)	45.38	0.124 2
Squalane	6.08 (20)		0		
Squalene	12 (25)		0.68 (25, B)		
D-Sorbitol		33 (80)			
Stearic acid	11.6 (70)	2.29 (70), 2.26 (100)	1.76 (25, D)		
Styrene	0.751 (20), 0.696 (25)	2.43 (25), 2.32 (75)	0.13 (25, lq)	32.0 (20)	30.98 (30)
Succinonitrile	2.591 (60), 2.008 (75)	56.5 (57), 54 (68)	3.68 (30, toluene)	53.26	0.107 9
1,1,2,2-Tetrabromoethane	13.950 (11), 9.797 (20)	8.6 (3), 7.0 (22)	1.29 (20, Hx)	52.37	0.146 3
1,1,2,2-Tetrachlorodifluoroethane	1.21 (25), 1.208 (30)	2.52 (25)		26.13	0.113 3
1,1,2,2-Tetrachloroethane	1.844 (15), 1.456 (30)	8.20 (20)	1.29 (g), 1.45 (25, Hx)	38.75	0.126 8
Tetrachloroethylene	1.932 (15), 0.798 (30)	2.30 (25)	0	32.86 (15)	31.27 (30)
Tetradecamethylcyclohepta-siloxane		2.7 (20)			
Tetradecamethylhexasiloxane		2.5 (20)	1.58 (20, lq)	17.42 (25)	
Tetradecane	2.131 (22)		0	28.30	0.086 9
Tetradecanoic acid			0.76 (25, B)	33.90	0.093 2
1-Tetradecanol		4.72 (38), 4.40 (48)	1.69 (25, C)	32.72	0.070 3
Tetraethylene glycol	44.9 (25)		5.84 (20, lq)	45 (25)	
Tetraethyllead			0.3 (20, B)	30.50	0.096 9
Tetraethylsilane			0	25.22	0.107 9
Tetraethyl silicate		4.1 (20)	1.72 (32, B)	23.63	0.097 9
Tetrahydrofuran	0.55 (20), 0.460 (25)	11.6 (−70), 7.58 (25)	1.75 (25, B)	26.5 (25)	

TABLE 4-10 Viscosity, dielectric constant, dipole moment, and surface tension of selected organic substances (*continued*)

Substance	Viscosity, $mN \cdot s \cdot m^{-2}$	Dielectric constant ε	Dipole moment, D	Surface tension $dyn \cdot cm^{-1}$ *a*	*b*
2,5-Tetrahydrofurandimethanol	225 (25)				
Tetrahydro-2-furanmethanol	6.24 (20)	13.61 (23)	2.12 (35, lq)	39.96	0.100 8
1,2,3,4-Tetrahydronaphthalene	2.202 (20), 2.003 (25)	2.76 (20)	0.60 (25, lq)	35.55	0.095 4
1,2,3,4-Tetrahydro-2-naphthol		11.7 (20), 6.7 (90)			
Tetrahydropyran	0.826 (20), 0.764 (25)	5.61 (25)	1.55 (25, B)		
Tetrahydropyran-2-methanol	11.0 (20)			34.1 (25)	
Tetrahydrothiophene-1,1-dioxide	9.87 (30)	43.3 (30)	4.81 (25, B)	35.5 (30)	
Tetrahydrothiophene oxide	52 (30), 19 (80)	42.5 (30)			
1,1,2,2-Tetramethylurea		23.06	3.47 (25, B)		
Tetranitromethane	1.76 (20)	2.32 (20)	0		
Tetrathiomethylmethane		2.82 (70)			
Thiacyclohexane				36.06 (20)	33.74 (40)
Thiacyclopentane	1.042 (20), 0.971 (25)		1.90 (25, B)	38.44	0.134 2
Thioacetic acid		12.8 (20)			
2,2′-Thiodiethanol	65.2 (20)			53.8 (20)	
Thiophene	0.662 (20), 0.353 (82)	2.76 (16), 2.57 (25)	0.55 (g), 0.52 (25, B)	34.00	0.132 8
Thymol			1.55 (25, B)	33.95	0.082 1
Toluene	0.623 (15), 0.523 (30)	2.385 (20), 2.364 (30)	0.45 (20, lq)	30.90	0.118 9
p-Toluenesulfonyl chloride				42.41	0.090 3
o-Toluidine	5.195 (15), 4.39 (20)	6.34 (18), 5.71 (58)	1.60 (25, B)	42.87	0.109 4
m-Toluidine	4.418 (15), 2.741 (30)	5.95 (18), 5.45 (58)	1.45 (25, B)	40.33	0.097 9
p-Toluidine	1.945 (45), 1.557 (60)	4.98 (54)	1.52 (25, B)	39.58	0.095 7
m-Tolunitrile			4.21 (22, B)	38.85	0.101 3
p-Tolunitrile			4.47 (20, B)	39.79	0.110 0
Tribenzylamine			0.65 (20, B)	42.41	0.095 3

Tributyl phosphite	1.9 (25)		1.92 (20, C)	27.57	0.086 5
2,2,2-Tribromoacetaldehyde		7.6 (20)	1.70 (20, B)		
Tribromoethane	2.152 (15), 1.741 (30)	4.39 (20)	0.99 (g)	48.14	0.130 8
1,2,3-Tribromopropane		6.45 (20)	1.59 (25, B)	47.99	0.126 7
Tributylamine	1.35 (25)		0.78 (25, B)	26.47	0.083 1
Tributyl borate	1.776 (20), 1.601 (25)		0.78 (25, C)	26.2 (20)	25.8 (25)
Tributyl phosphate	11.1 (15) 3.39 (25)	7.96 (30)	3.07 (25, B)	28.71	0.066 6
Trichloroacetaldehyde		7.6 (−40), 4.9 (20)	1.96 (25, B)	27.66	9.119 7
Trichloroacetic acid		4.6 (60)	1.1 (25, B, dimer)	35.4	0.089 5
Trichloroacetonitrile		7.85 (19)	1.93 (19, lq)		
1,1,1-Trichloroethane	0.903 (15), 0.725 (30)	7.1 (7), 7.52 (20)	1.79 (g), 1.6 (25, B)	28.28	0.124 2
1,1,2-Trichloroethane	0.119 (20), 0.110 (25)	8.78 (23)	1.45 (g)	37.40	0.135 1
Trichloroethylene	0.566 (20), 0.532 (25)	3.42 (16)	0.77 (30, lq), 0.95 (30, B)	29.5 (20)	28.8 (25)
Trichlorofluoromethane	0.42 (25), 0.011 (25, g)	2.28 (29)	0.45 (g), 0.49 (lq)	18 (25)	
Trichloromethylsilane	0.47 (20)		1.87 (25, B)	20.3 (20)	
2,4,6-Trichlorophenol			1.88 (25, D)	43.13	0.095 5
1,2,3-Trichloropropane		7.5 (20)	1.61 (g)	37.8 (20)	37.05 (25)
Trichlorosilane	0.332 (20), 0.316 (25)		0.86 (g), 0.98 (25, B)	20.43	0.107 6
α,α,α-Trichlorotoluene	3.07 (10), 2.55 (17)	6.9 (21)	2.17 (20, B)		
1,1,2-Trichloro-1,2,2-trifluoro-ethane	0.711 (20), 0.627 (30)	2.41 (25)		17.75 (20)	16.56 (30)
Tridecane	1.883 (20), 1.55 (23)		0	27.73	0.087 2
1-Tridecene				28.01	0.088 4
Triethanolamine	613.6 (25), 208.1 (40)	29.36 (25)	3.57 (25, B)		
Triethylaluminum		2.9 (20)			
Triethylamine	0.394 (15), 0.363 (30)	2.42 (25)	0.66 (g), 0.9 (25, B)	22.70	0.099 2
Triethylene glycol	49.0 (20), 8.5 (60)	23.7 (20)	5.58 (20, lq)	47.33	0.088 0
Triethyl phosphate	1.684 (40), 1.376 (55)	13.43 (15), 10.93 (65)	3.08 (25, B)	31.81	0.092 8
Triethyl phosphite	0.72 (25)	5.0	1.82 (25, D)	25.73	0.087 8
Trifluoroacetic acid	0.926 (20), 0.653 (40)	8.55 (20), 5.76 (50)	2.28 (g)	15.64	0.184 4
2,2,2-Trifluoroethanol	1.996 (20)		2.03 (25, cHx)	20.6 (33)	
α,α,α-Trifluorotoluene		9.2 (30), 8.1 (60)			

TABLE 4-10 Viscosity, dielectric constant, dipole moment, and surface tension of selected organic substances (*continued*)

Substance	Viscosity, $mN \cdot s \cdot m^{-2}$	Dielectric constant ε	Dipole moment, D	Surface tension $dyn \cdot cm^{-1}$	
				a	*b*
Trimethylamine	0.321 (−33)	2.4 (25)		16.24	0.113 3
1,2,3-Trimethylbenzene		2.636 (20), 2.609 (30)	0.56 (20, lq)	30.91	0.104 0
1,2,4-Trimethylbenzene	0.894 (15), 0.730 (30)	2.38 (20), 2.36 (30)	0.30 (20, lq)	31.76	0.102 5
1,3,5-Trimethylbenzene	1.154 (20)	2.28	0	29.79	0.089 7
Trimethyl borate		8 (20)	0.82 (25, C)		
2,2,3-Trimethylbutane	0.579 (20)	1.93 (20)	0	20.70	0.097 3
cis,cis-1,3,5-Trimethyl-cyclohexane	0.632 (20), 0.558 (30)				
trans-1,3,5-Trimethyl-cyclohexane	0.714 (20), 0.624 (30)				
Trimethylene sulfide	0.638 (20), 0.607 (25)		1.78 (25, B)	36.3 (20)	35.0 (30)
3,5,5-Trimethyl-1-hexanol	11.06 (25)				
2,6,8-Trimethyl-4-nonanone	1.9 (20)				
1,3,5-Trimethyl-2-oxa bicyclo[2.2.2]octane		4.57 (24)	1.54 (25, C)	32.1 (20)	31.1 (25)
2,2,3-Trimethylpentane	0.598 (20)	1.962 (20)	0	22.46	0.089 5
2,2,4-Trimethylpentane	0.502 (20)	1.940 (20)	0	20.55	0.088 8
Trimethyl phosphite	0.61 (20)		1.83 (20, C)	27.18 (20)	24.88 (40)
2,4,6-Trimethylpyridine	1.498 (20), 1.496 (25)	6.6	1.95 (25, B)		
Triphenylamine				46.2	0.095 5
Triphenyl phosphite	25.18 (15), 6.95 (45)	3.67 (45), 3.57 (65)	2.04 (25, B)		
Tripropylamine			0.58 (20, lq), 0.76 (20, B)	24.58	0.087 8
Tripropylene glycol	56.1 (25)			34 (25)	
Tripropylene glycol butyl ether	6.58 (25)			28.8 (25)	
Tripropylene·glycol ethyl ether	5.17 (25)			28.2 (25)	

Tripropylene glycol isopropyl ether	7.7 (25)			27.4 (25)	
Tripropylene glycol methyl ether	5.96 (25)			30.0 (25)	
Tris(dimethylamino) phosphine oxide	3.34 (30)	30 (20)			
Tris(4-ethylphenyl) phosphite	30.22 (15), 9.047 (45)	3.74 (15), 3.61 (45)	2.08 (25, B)		
Tris(*m*-tolyl) phosphite	37.55 (15), 9.132 (45)	3.67 (15), 3.53 (45)	1.62 (25, B)		
Tris(*p*-tolyl) phosphite	35.52 (15), 8.794 (45)	3.88 (15), 3.74 (45)	1.77 (25, B)		
Tritolyl phosphate	38.8 (35), 16.8 (55)	6.92 (40)	2.84 (40, C)	40.9 (20)	
Undecane	1.186 (20), 0.761 (50)	2.00 (20), 1.84 (150)	0	26.26	0.090 1
2-Undecanone	1.61 (30)		2.71 (15, B)		
Urea			4.59 (25, D)		
Vinyl acetate	0.421 (20)		1.79 (25, B)	23.95 (20)	22.54 (30)
o-Xylene	0.809 (20), 0.627 (40)	2.57 (20), 2.54 (30)	0.62 (g), 0.52 (25, lq)	32.51	0.110 1
m-Xylene	0.617 (20), 0.497 (40)	2.37 (20), 2.35 (30)	0.33 (20, lq), 0.37 (20, B)	31.23	0.110 4
p-Xylene	0.644 (20), 0.513 (40)	2.26 (20), 2.22 (50)	0	30.69	0.107 4
Xylitol		40 (20)			

TABLE 4-11 Viscosity, dielectric constant, dipole moment, and surface tension of selected inorganic substances

For the majority of compounds the dependence of the surface tension γ on the temperature can be given as

$$\gamma = a - bt$$

where a and b are constants and t is the temperature in degrees Celsius.

Substance	Viscosity, $mN \cdot s \cdot m^{-2}$	Dielectric constant ε	Dipole moment, D	Surface tension, $dyn \cdot cm^{-1}$	
				a	b
Air (20°C)	0.018 2	1.000 536 4			
$AlBr_3$		3.38^{100}	5.2		
Ar					
(g, 20°C)	0.022 3	1.000 517 2			
(lq)		1.538^{-191}	0	34.28	0.249 3
$AsBr_3$		8.83^{35}	1.61	54.51	0.1043
$AsCl_3$		12.6^{20}	1.59	41.67	0.097 81
AsH_3 (arsine)		2.05^{20}	0.20		
BBr_3		2.58^{0}	0	31.90	0.128 0
BCl_3			0		
BF_3			0	−2.92	0.203 0
B_2H_6 (diborane)		$1.872^{-92.5}$	0	−3.13	0.178 5
B_5H_9			2.13		
$B_3H_6N_3$ (triborotriazine)			0		
Br_2					
(g, 20°C)		1.012 8			
(lq)	1.03^{16}	3.09^{20}	0	45.5	0.182 0
BrF_3	2.22^{20}		1.1	38.30	0.099 9
BrF_5	0.62^{24}	$7.91^{24.5}$	1.51	25.24	0.109 8
Cl_2					
(g, 20°C)	0.013 2		0		
(lq)		1.91^{14}			
ClF_3	0.48^{12}	4.29^{25}	0.554	26.9	0.166 0
ClO_3F (perchloryl fluoride)			0.023	12.24	0.157 6
Co					
(g)	$0.017\ 5^{20}$	$1.000\ 70^{0}$	0.112		
(lq)				−30.20	0.207 3
CO_2					
(g, 20°C)	0.014 7	1.000 922	0		
(lq)	0.071^{20}	$1.60^{0°C,\ 50\ atm}$			
$COCl_2$		4.34^{22}	1.17	22.59	0.145 6
COF_2			0.95		
COS			0.712	12.12	0.177 9
COSe		3.47^{10}	0.73		
CS			1.98		
CS_2					
(g)		$1.002\ 9^{0}$	0		
(lq)	0.375^{20}	2.6^{20}			

TABLE 4-11 Viscosity, dielectric constant, dipole moment, and surface tension of selected inorganic substances (*continued*)

Substance	Viscosity, $mN \cdot s \cdot m^{-2}$	Dielectric constant ε	Dipole moment, D	Surface tension, $dyn \cdot cm^{-1}$	
				a	*b*
CrO_2Cl_2 [chromyl(VI) chloride]		2.6^{20}	0.47		
D_2 (deuterium)		1.277^{-253}			
DH				6.537	0.188 3
D_2O	1.098^{25}	78.25^{25}	1.87	(71.72^{20})*	(68.38^{40})*
F_2		1.54^{-202}		−16.10	0.164 6
$GaCl_3$			0.85	35.0	0.100 0
$GeCl_4$		2.430^{25}	0	(22.44^{30})*	
H_2					
(g, 20°C)	0.008 8	1.000 253 8	0		
(lq)		$1.228^{20.4\,K}$			
HBr					
(g)		$1.003\ 13^{0}$	0.82		
(lq)	0.83^{-67}	3.82^{25}		13.10	0.207 9
HCl					
(g)		$1.004\ 6^{0}$	1.08		
(lq)	0.51^{-95}	4.60^{28}			
HCN	0.206^{18}	116^{20}	2.98	(19.45^{10})*	(18.33^{20})*
HCNO (isocyanate)			1.6		
HCNS (isothiocyanate)			1.7		
HF	0.256^{0}	83.6^{0}	1.82	10.41	0.078 67
HI					
(g)		$1.002\ 34^{0}$	0.44		
(lq)		2.90^{22}			
NH_3 (azide)			0.8		
H_2O (see Table 4-12)					
H_2O_2	1.25^{20}	84.2^{0}	2.2	78.97	0.154 9
HNO_3			2.17		
H_2S					
(g)		$1.00\ 4\ 0^{0}$	0.97		
(lq)	0.412^{0}	5.93^{10}		48.95	0.175 8
H_2Se			0.24	22.32	0.148 2
H_2SO_4	24.54^{25}	100^{25}			
HSO_3Cl (chlorosulfonic acid)	2.43^{20}	60^{20}			
HSO_3F (fluorosulfonic acid	1.56^{25}	$\sim 120^{25}$			
H_2Te			<0.2	29.03	0.261 9
He					
(g, 20°C)	0.019 6	1.000 065 0	0		
Hg	1.552^{20}		0	490.6	0.204 9

* Actual values of surface tension.

TABLE 4-11 Viscosity, dielectric constant, dipole moment, and surface tension of selected inorganic substances (*continued*)

Substance	Viscosity, $mN \cdot s \cdot m^{-2}$	Dielectric constant ε	Dipole moment, D	Surface tension, $dyn \cdot 2m^{-1}$	
				a	*b*
I_2	1.98^{116}	11.1^{118}	0		
IF_5			2.18	33.16	0.131 8
Kr					
(g, 20°C)	0.025 0		<0.05		
(lq)				40.576	0.289 0
Ne (g, 20°C)	0.031 3	1.000 063 9	0		
N_2					
(g, 20°C)	0.017 6	1.000 548 0	0		
(lq)		1.454^{-203}		26.42	0.226 5
NH_3					
(g)		$1.007\ 2^{0}$	1.47		
(lq)	$0.254^{-33.5}$	$22.4^{-33.4}$		(37.91^{-50})*	(35.38^{-40})*
N_2H_4 (hydrazine)	0.97^{20}	52.9^{20}	1.75		
NO			0.153	−67.48	0.585 3
N_2O					
(g)	$0.014\ 6^{20}$	$1.001\ 13^{0}$	0.167		
(lq)		1.52^{15}		5.09	0.203 2
NO_2			0.316		
N_2O_4		2.56^{15}	0.5		
NOBr (nitrosyl bromide)		13.4^{15}	1.8		
NOCl		18.2^{12}	1.9	29.49	0.149 3
NOF			1.81	14.00	0.116 5
NO_2F (nitryl fluoride)			0.47	8.26	0.185 4
O_2					
(g, 20°C)	0.020 4	1.000 494 7	0		
(lq)		1.507^{-193}		−33.72	0.256 1
O_3			0.53	(38.1^{-183})*	
OF_2 (oxygen difluoride)			0.297		
OsO_4			0		
PBr_3		3.9^{20}	0.5	45.34	0.128 3
PCl_3		3.43^{25}	0.78	31.14	0.126 6
PCl_5		2.7^{165}	0.9		
PF_5			0		
PH_3		2.9^{15}	0.58		
PI_3		4.12^{65}	0	61.66	0.067 71
$POCl_3$	1.065^{25}	13.7^{25}	2.41	35.22	0.127 5
POF_3			1.76		
$PSCl_3$		5.8^{22}	1.42	37.00	0.127 2
$PbCl_4$		2.78^{20}			
S_2Cl_2 dimer		4.79^{15}	1.0	46.23	0.146 4

* Actual values of surface tension.

TABLE 4-11 Viscosity, dielectric constant, dipole moment, and surface tension of selected inorganic substances (*continued*)

Substance	Viscosity, $mN \cdot s \cdot m^{-2}$	Dielectric constant ε	Dipole moment, D	Surface tension, $dyn \cdot cm^{-1}$	
				a	*b*
S_2F_2					
FSSF isomer			1.45		
$S{=}SF_2$ isomer			1.03		
SF_4			0.632	12.87	0.173 4
SF_6			0	5.66	0.119 0
S_2F_{10}		2.020^{20}	0		
SO_2					
(g)	$0.012\,6^{29}$	$1.009\,3^{0}$	1.63		
(lq)		15.0^{0}		26.58	0.194 8
SO_3		3.11^{18}	0		
$SOBr_2$ (thionyl bromide)		9.06^{20}	9.11		
$SOCl_2$		9.25^{20}	1.45	36.10	0.141 6
SO_2Cl_2 (sulfuryl chloride)		9.15^{20}	1.81	32.10	0.132 8
$SbCl_3$		33.2^{75}	3.93	47.87	0.123 8
$SbCl_5$		3.22^{20}	0		
SbF_5				49.07	0.193 7
SbH_3			0.12		
SeF_4				38.61	0.127 4
SeF_6			0		
$SeOCl_2$		55^{25}	2.64		
$SiCl_4$		2.40^{16}	0	20.78	0.099 62
SiF_4			0		
SiH_4			0		
$SiHCl_3$			0.86	20.43	0.107 6
$SnBr_4$			0		
$SnCl_4$		2.89^{20}	0	29.92	0.113 4
TeF_6			0		
$TiCl_4$		2.80^{20}	0	(33.54^{20})*	(31.06^{40})*
UF_6					
(g)		$1.002\,92^{67}$	0		
(lq)		2.18^{65}		25.5	0.124 0
VCl_4		3.05^{25}	0		
$VOBr_3$		3.6^{25}			
$VOCl_3$		3.4^{25}	0.3	(36.36^{20})*	(33.60^{40})*
Xe (g, 20°C)	0.022 8	1.001 23	0		

* Actual values of surface tension.

TABLE 4-12 Refractive index, viscosity, dielectric constant, and surface tension of water at various temperatures

Temp., °C	Refractive index n_D	Viscosity, $mn \cdot s \cdot m^{-2}$	Dielectric constant ε	Surface tension, $dyn \cdot cm^{-1}$
0	1.333 95	1.770 2	87.74	75.83
5	1.333 88	1.510 8	85.76	75.09
10	1.333 69	1.303 9	83.83	74.36
15	1.333 39	1.137 4	81.95	73.62
20	1.333 00	1.001 9	80.10	72.88
21	1.332 90	0.976 4	79.73	72.73
22	1.332 80	0.953 2	79.38	72.58
23	1.332 71	0.931 0	79.02	72.43
24	1.332 61	0.910 0	78.65	72.29
25	1.332 50	0.890 3	78.30	72.14
26	1.332 40	0.870 3	77.94	71.99
27	1.332 29	0.851 2	77.60	71.84
28	1.332 17	0.832 8	77.24	71.69
29	1.332 06	0.814 5	76.90	71.55
30	1.331 94	0.797 3	76.55	71.40
35	1.331 31	0.719 0	74.83	70.66
40	1.330 61	0.652 6	73.15	69.92
45	1.329 85	0.597 2	71.51	69.18
50	1.329 04	0.546 8	69.91	68.45
55	1.328 17	0.504 2	68.35	67.71
60	1.327 25	0.466 9	66.82	66.97
65	1.326 16	0.434 1	65.32	66.23
70	1.325 11	0.405 0	63.86	65.49
75	1.323 99	0.379 2	62.43	64.75
80		0.356 0	61.03	64.01
85		0.335 2	59.66	63.28
90		0.316 5	58.32	62.54
95		0.299 5	57.01	61.80
100		0.284 0	55.72	61.80

COMBUSTIBLE MIXTURES

TABLE 4-13 Properties of combustible mixtures in air

Additional compounds can be found in National Fire Protection Association, *Fire Protection Handbook*, 14th ed., 1976.

Substance	Autoignition temperature, °C	Flammable limits, percent by volume of fuel (25°C, 760 mm)	
		Lower	Upper
Acetaldehyde	175	4.0	6.0
Acetic acid, glacial	465	5.4	16.0
Acetic anhydride	390	2.9	10.3
Acetone	465	2.6	12.8
Acetonitrile	524	4.4	16.0
Acetylene	305	2.5	100
Acrolein	235*	2.8	31.0
Acrylonitrile	481	3.0	17
Allyl alcohol	378	2.5	18.0
Allylamine	374	2.2	22
Ammonia, anhydrous	651	16	25
Aniline	615	1.3	
Benzene	560	1.3	7.1
Biscyclohexyl	245	0.7 (100°C)	5.1 (150°C)
1-Bromobutane	265	2.6 (100°C)	6.6 (100°C)
3-Bromopropene	295	4.4	7.3
Butane	405	1.9	8.5
Butanol	365	1.4	11.2
2-Butanone	516	1.8	10
1-Butene	385	1.6	10.0
3-Buten-1-ol		4.7	34
Butyl acetate	425	1.7	7.6
Butylamine	312	1.7	9.8
Butylbenzene	410	0.8	5.8
Butylene oxide		1.5	18.3
Butyl formate	322	1.7	8.2
Butyraldehyde	230	2.5	12.5
Butyric acid	450	2.0	10.0
Carbon disulfide	90	1.3	50.0
Carbon monoxide	609	12.5	74
Carbonyl sulfide		12	29
Chlorobenzene	640	1.3	7.1
2-Chloro-1,3-butadiene		4.0	20.0
1-Chlorobutane		1.8	10.1
2-Chloro-2-butene		2.3	9.3
1-Chloro-1,1-difluoroethane		6.2	17.9
2-Chloroethanol	425	4.9	15.9

* Unstable.

TABLE 4-13 Properties of combustible mixtures in air (*continued*)

Substance	Autoignition temperature, °C	Flammable limits, percent by volume of fuel (25°C, 760 mm)	
		Lower	Upper
Chloromethane	632	10.7	17.4
1-Chloropentane	260	1.6	8.6
2-Chloropropane	593	2.8	10.7
1-Chloro-1-propene		4.5	16
3-Chloro-1-propene	485	2.9	11.1
Chlorotrifluoroethylene		8.4	38.7
Crotonaldehyde	232	2.1	15.5
Cumene	425	0.9	6.5
Cyanogen		6.6	42.6
Cyclohexane	245	1.3	8
Cyclopropane	500	2.4	10.4
Decahydronaphthalene	250	0.7	4.9
Decane	210	0.8	5.4
Diborane	38–52†	0.8	88
Dibutyl ether	194	1.5	7.6
o-Dichlorobenzene	648	2.2	9.2
1,2-Dichloroethylene		9.7	12.8
Dichloropropane	557	3.4	14.5
Diisopropyl ether	443	1.4	7.9
Diethylamine	312	1.8	10.1
Diethyl ether	160	1.9	36.0
2,2-Dimethylbutane	425	1.2	7.0
Dimethyl ether		3.4	27.0
N,N-Dimethylformamide	445	1.2	7.0
1,1-Dimethylhydrazine	249	2	95
2,3-Dimethylpentane	335	1.1	6.7
2,2-Dimethylpropane	450	1.4	7.5
Dimethyl sulfide	206	2.2	19.7
Dimethyl sulfoxide	215	2.6	28.5
1,4-Dioxane	180	2.0	22.0
Divinyl ether	360	1.7	27
Ethane	515	3.0	12.5
Ethanol	365	3.3	19
2-Ethoxyethanol	235	1.8	14.0
1-Ethoxypropane		1.7	9.0
Ethyl acetate	427	2.2	11.0
Ethylamine	385	3.5	14.0
Ethylbenzene	432	1.0	6.7
Ethylcyclobutane	210	1.2	7.7
Ethylene	490	2.7	36.0
Ethyleneimine	320	3.6	46
Ethylene oxide	429	3.6	100

† Ignites in moist air.

TABLE 4-13 Properties of combustible mixtures in air (*continued*)

Substance	Autoignition temperature, °C	Flammable limits, percent by volume of fuel (25°C, 760 mm)	
		Lower	Upper
Ethyl formate	455	2.8	16.0
1,3-Ethylidene dichloride	440	6.2	16
Ethyl nitrite	90	3.0	50
Ethyl propionate	440	1.9	11
Ethyl vinyl ether	202	1.7	28
Formaldehyde	429	7.0	73
2-Furaldehyde	316	2.1	19.3
Furan		2.3	14.3
Furfuryl alcohol	491	1.8	16.3
Gasoline, 92 octane	~280	1.4	7.6
Heptane	215	1.0	6.7
Hexane	225	1.1	7.5
2-Hexanone	533	1.2	8
Hydrocyanic acid, 96%	538	5.6	40.0
Hydrogen	400	4.0	75
4-Hydroxy-4-methyl-2-pentanone	603	1.8	6.9
Isobutyl acetate	421	2.4	10.5
Isobutylbenzene	430	0.8	6.0
Isopentane	420	1.4	7.6
Isopentyl acetate	360	1.0	7.5
Isoprene	220	2	9
Isopropyl acetate	460	1.8	8
Isopropyl alcohol	399	2.0	12
Methane	540	5.4	15.0
Methanethiol		3.9	21.8
Methanol	385	6.7	36.0
2-Methoxyethyl acetate		1.7	8.2
Methyl acetate	502	3.1	16
Methyl acrylate		2.8	25
Methylamine	430	4.9	20.6
2-Methyl-2-butanol	437	1.2	9.0
3-Methyl-1-butene	365	1.5	9.1
Methylcyclohexane	250	1.2	6.7
Methyl formate	465	5.0	23
2-Methylpropene	465	1.8	9.6
4-Methyl-2-pentanone	460	1.4	7.5
2-Methylpropene	465	1.8	9.6
α-Methylstyrene	574	1.9	6.1
Methyl propionate	469	2.5	13
Nicotine	244	0.7	4.0
Nitrobenzene	482	1.8 (93°C)	
Nonane	205	0.8	2.9

TABLE 4-13 Properties of combustible mixtures in air (*continued*)

Substance	Autoignition temperature, °C	Flammable limits, percent by volume of fuel (25°C, 760 mm)	
		Lower	Upper
Octane	220	1.0	6.5
Pentanamine		2.2	22
Pentane	260	1.5	7.8
2-Pentanone	505	1.5	8.2
Pentyl acetate	360	1.1	7.5
Petroleum ether	550	1.1	5.9
Propane	450	2.2	9.5
1,3-Propanediol	371	2.6	12.5
Propanol	440	2.1	13.5
Propene	460	2.0	11.1
Propanamine	318	2.0	10.4
Propionaldehyde	207	2.9	17.0
Propyl acetate	450	2.0	8
Propylene oxide		2.8	37.0
Propyl nitrate	175	2	100
Pyridine	482	1.8	12.4
Styrene	490	1.1	6.1
Tetrahydrofuran	321	2	11.8
Tetrahydrofurfuryl alcohol	282	1.5	9.7
Tetrahydronaphthalene	385	0.8	5.0
Toluene	480	1.2	7.1
Trichlorothylene	420	12.5	90
Triethylamine		1.2	8.0
Triethylene glycol	371	0.9	9.2
Trimethylamine	190	2.0	11.6
Trioxane	414	3.6	29
Vinyl acetate	427	2.6	13.4
Vinyl butyrate		1.4	8.8
Vinyl chloride	461	3.6	33.0
Vinyl fluoride		2.6	21.7
Xylene, *m*- and *p*-	530	1.1	7.0
Xylene, *o*-	465	1.0	6.0

SECTION 5
THERMODYNAMIC PROPERTIES

ENTHALPIES AND GIBBS (FREE) ENERGIES OF FORMATION, ENTROPIES, AND HEAT CAPACITIES

The tables in this section contain values of the enthalpy and Gibbs (formerly free) energy of formation, entropy, and heat capacity at 298.15 K (25°C). No values are given in these tables for metal alloys or other solid solutions, for fused salts, or for substances of undefined chemical composition.

For a more complete listing of compounds see the tables in "Selected Values of Chemical Thermodynamical Properties," by D. D. Wagman et al., National Bureau of Standards Technical Notes 270-3, 270-4, 270-5, 270-6, 270-7, and 270-8, Washington; "JANAF Thermochemical Tables," by D. R. Stull and H. Prophet, National Bureau of Standards Publication 37, Washington; supplements to JANAF appearing in *J. Phys. Chem. Ref. Data*; D. R. Stull, E. F. Westrum, Jr., and G. C. Sinke, *The Chemical Thermodynamics of Organic Compounds*, Wiley-Interscience, New York, 1969; and I. Barin and O. Knacke, *Thermochemical Properties of Inorganic Substances*, Springer-Verlag, Berlin, 1973.

The physical state of each substance is indicated in the column headed "State" as crystalline solid (c), liquid (liq), gaseous (g), or amorphous (amorp). Solutions in water are listed as aqueous (aq).

The values of the thermodynamic properties of the pure substances given in these tables are, for the substances in their standard states, defined as follows: For a pure solid or liquid, the standard state is the substance in the condensed phase under a pressure of 1 atm. For a gas, the standard state is the hypothetical ideal gas at unit fugacity, in which state the enthalpy is that of the real gas at the same tamperature and at zero pressure.

The values of ΔHf° and ΔGf° given in the tables represent the change in the appropriate thermodynamic quantity when one gram formula weight of the substance in its standard state is formed, isothermally at the indicated temperature, from the elements, each in its appropriate standard reference state. The standard reference state at 25°C for each element has been chosen to be the standard state that is thermodynamically stable at 25°C and 1 atm pressure. The standard reference states are indicated in the tables by the fact that the values of ΔHf° and ΔGf° are exactly zero.

The values of S° represent the virtual or "thermal" entropy of the substance in the standard state at 298.15 K, omitting contributions from nuclear spins. Isotope mixing effects are also excluded except in the case of the ^1H-^2H system.

Solutions in water are designated as aqueous, and the concentration of the solution is expressed in terms of the number of moles of solvent associated with 1 mol of the solute. If no concentration is indicated, the solution is assumed to be dilute. The standard state for a solute in aqueous solution is taken as the hypothetical ideal solution of unit molality (indicated as std state, $m = 1$). In this state the partial molal enthalpy and the heat capacity of the solute are the same as in the infinitely dilute real solution (aq. ∞).

TABLE 5-1 Enthalpies and Gibbs (free) energies of formation, entropies, and heat capacities of organic compounds

Substance	State	ΔHf°, kcal · mol^{-1}	ΔGf°, kcal · mol^{-1}	S°, cal · deg^{-1} · mol^{-1}	C_p°, cal · deg^{-1} · mol^{-1}
Acenaphthene	c	16.8			
Acenaphthylene	c	44.7			
Acetaldehyde	liq	−45.96	−30.64	38.3	65.6
	g	−39.76	−31.86	63.15	13.06

TABLE 5-1 Enthalpies and Gibbs (free) energies of formation, entropies, and heat capacities of organic compounds (*continued*)

Substance	State	$\Delta Hf°$, kcal · mol^{-1}	$\Delta Gf°$, kcal · mol^{-1}	$S°$, cal · deg^{-1} · mol^{-1}	$C_p°$, cal · deg^{-1} · mol^{-1}
Acetaldoxime	c	−18.6			
	liq	−19.5			
Acetamide	c	−76.0			
Acetamidoguanidine nitrate	c	−118.1			
1-Acetamido-2-nitroguanidine	c	−46.3			
5-Acetamidotetrazole	c	−1.2			
Acetanilide	c	−50.3			
Acetic acid	liq	−115.71	−93.2	38.2	29.7
	g	−103.93	−90.03	67.52	15.90
Ionized; std state, $m = 1$	aq	−116.16	−88.29	20.7	−1.5
Nonionized; std state, $m = 1$	aq	−116.70	−94.78	42.7	
Acetic anhydride	liq	−149.14	−116.82	64.2	
	g	−137.60	−113.93	93.20	23.78
Acetone	liq	−59.18	−37.22	47.9	30.22
	g	−51.78	−36.58	70.49	17.90
Acetone glyceraldehyde	liq	−180			
Acetonitrile	liq	12.8	23.7	35.76	21.86
	g	21.00	25.24	58.19	12.48
Acetophenone	liq	−34.07	−4.06	59.62	
	g	−20.76	0.44	89.12	
Acetyl radical	g	−4.0			
N-Acetylbenzidine	c	−38.0			
Acetyl bromide	liq	−53.5			
Acetyl chloride	liq	−65.44	−49.73	48.0	28
	g	−58.30	−46.29	70.47	16.21
Acetylene	g	54.19	50.00	48.00	10.50
Std state, $m = 1$	aq	50.54	51.88	29.5	
Acetylenedicarbonitrile	liq	119.6			
	g	127.50	122.10	69.31	20.53
Acetylene dicarboxylic acid	c	−138.1			
Acetyl fluoride	liq	−112.4			
N-Acetylhydrazobenzene	c	−2.0			
o-Acetylhydroxybenzoic acid	c	−194.93			
N-Acetylimidazole	c	−28.6			
Acetyl iodide	liq	−39.3			
4-Acetylresorcinol	c	−137.1			
N-Acetyltetrazole	c	19.49			
Acridine	c	44.8			
Acrolein	liq	−29.97	−16.17		
	g	−20.50	−15.45		

TABLE 5-1 Enthalpies and Gibbs (free) energies of formation, entropies, and heat capacities of organic compounds (*continued*)

Substance	State	$\Delta Hf°$, kcal · mol^{-1}	$\Delta Gf°$, kcal · mol^{-1}	$S°$, cal · deg^{-1} · mol^{-1}	$C_p°$, cal · deg^{-1} · mol^{-1}
Acrylic acid	liq	−91.8			
	g	−80.36	−68.37	75.29	18.59
Acrylonitrile	liq	36.1			
	g	44.20	46.68	65.47	15.24
Adenine	c	23.21	71.58	36.1	
Adipic acid	c	−237.60			
	liq	−235.51	−177.17		
Aetioporphyrin I	c	−6.0			
Aetioporphyrin II	c	0.4			
α-Alanine					
D	c	−134.03	−88.23	31.6	
L	c	−133.96	−88.49	30.88	
DL	c	−134.55	−88.92	31.6	
Alanine anhydride	c	−128.0			
α-Alanylglycine					
DL	c	−185.64	−117.00	51.0	
L	c	−197.52	−127.30	46.62	
DL-Alanylphenylalanine	c	−170.2			
Alanylphenylalanyl anhydride	c	−89.3			
Allantoin (5-ureidohydantoin)	c	−171.50	−106.65	46.6	
Allomucic acid	c	−142			
Alloxan monohydrate	c	−239.08	−182.08	44.6	
Alloxantin dihydrate	c	−510.3			
Allyl radical	g	38			
1-Allyl-5-allylamino-tetrazole	c	83.7			
1-Allyl-5-aminotetrazole	c	63.4			
2-Allyl-5-aminotetrazole	c	67.6			
Allylcyclopentane	liq	−15.74			
Allyl ethyl sulfoxide	liq	−41.83			
Allyl trichloroacetate	liq	−94.5			
Amalic acid	c	−367.0			
Amarine	c	63			
p-Aminoacetophenone	c	70.2			
3-Aminoacridine	c	39.8			
5-Aminoacridine	c	38.1			
2-Aminobenzoic acid	c	−95.8			
3-Aminobenzoic acid	c	−98.2			
4-Aminobenzoic acid	c	−98.8			
2-Aminobiphenyl	c	26.8			
4-Aminobiphenyl	c	19.4			

TABLE 5-1 Enthalpies and Gibbs (free) energies of formation, entropies, and heat capacities of organic compounds (*continued*)

Substance	State	$\Delta Hf°$, kcal · mol^{-1}	$\Delta Gf°$, kcal · mol^{-1}	$S°$, cal · deg^{-1} · mol^{-1}	$C_p°$, cal · deg^{-1} · mol^{-1}
1-Aminobutane butylamine)	liq	−30.52			
	g	−22.00	11.76	86.76	28.33
2-Aminobutane (*sec*-butylamine)	g	−24.90	9.71	83.90	27.99
4-Aminobutanoic acid	c	−138.1			
2-Aminoethanesulfonic acid	c	−187.7	−134.3	36.8	33.6
Ionized; std state, $m = 1$	aq	−171.92	−121.76	47.8	
Nonionized; std state, $m = 1$	aq	−181.92	−134.12	55.7	
2-Aminohexanoic acid (norleucine)	c	−152.7			
4-Aminohexanoic acid	c	−154.5			
5-Aminohexanoic acid	c	−153.7			
6-Aminohexanoic acid	c	−152.7			
3-Amino-2-methylpropane (2-butylamine)	liq	−31.68			
5-Aminopentanoic acid	c	−144.5			
5-Aminotetrazole	c	49.7			
5-Aminotetrazole nitrate	c	−6.6			
3-Amino-1,2,4-triazole	c	18.4			
Amygdalin	c	−455			
1,2-Anyhydroglucose-3,5,6-triacetate	c	−411.7			
Aniline	liq	7.55	35.63	45.72	45.90
	g	20.76	39.84	76.28	25.91
Anisine	c	−51			
Anisoyl glycine	c	−180.9			
Anthracene	c	29.0	68.30	49.58	49.7
9,10-Anthracenedione	c	−49.6			
β-D-Arabinose	c	−252.84			
β-L-Arabinose	c	−252.84			
D-Arabonic acid-γ-lactone	c	−238.2			
L-Arginine	c	−148.66			
D-Arginine	c	−149.05	−57.43	59.9	
L-Ascorbic acid (vitamin C)	c	−278.34			
L-Asparagine	c	−188.50	−126.73	41.7	
L-Aspartic acid	c	−232.47	−174.53	40.66	
Azobenzene					
cis	c	86.7			
trans	c	76.6			

TABLE 5-1 Enthalpies and Gibbs (free) energies of formation, entropies, and heat capacities of organic compounds (*continued*)

Substance	State	$\Delta Hf°$, kcal · mol^{-1}	$\Delta Gf°$, kcal · mol^{-1}	$S°$, cal · deg^{-1} · mol^{-1}	$C_p°$, cal · deg^{-1} · mol^{-1}
Azodicarbamide	c	−69.90			
Azulene	g	66.90	84.10	80.75	30.69
Barbituric acid	c	−152.2			
Benzaldehyde	liq	−21.23	2.24		
	g	−9.57	5.85		
Benzamide	c	−48.42			
Benzanilide	c	−22.3			
1,2-Benzanthracene	c	41			
2,3-Benzanthracene	c	38.3	85.79	51.48	
1,2-Benzanthra-9,10-quinone	c	−55.4			
Benzene	liq	11.71	29.72	41.41	19.52
	g	19.82	30.99	64.34	
Benzenethiol (thiophenol)	liq	15.32	32.02	53.25	41.40
	g	26.66	35.28	80.51	25.07
Benzidine	c	16.9			
Benzil	c	−36.8			
Benzoic acid	c	−92.03	−58.62	40.05	34.97
Benzoic anhydride	c	−103.0			
Benzonitrile	g	52.30	62.33	76.73	26.07
Benzophenone	c	−8.0	33.5	58.6	
p-Benzoquinone	c	−44.33			
Benzotriazole	c	59.74			
DL-Benzoylalanine	c	−147.9			
Benzoyl bromide	liq	−25.58			
Benzoyl chloride	liq	−39.17			
Benzoyl iodide	liq	−12.31			
Benzoylphenylalanine	c	−129.6			
Benzoyl sarcosine	c	−135.7			
3,4-Benzphenanthrene	c	44.2			
Benzyl radical	g	45			
Benzyl alcohol	liq	−38.49	−6.57	51.8	
Benzyl bromide (2-bromotoluene)	liq	5.6			
Benzyl chloride	liq	−7.8			
N-Benzyldiphenylamine	c	44.2			
Benzyl ethyl sulfide	liq	−1.17			
Benzyl iodide	liq	13.8			
Benzyl mercaptan	liq	10.4			
Benzyl methyl ketone	liq	−36.30			
Benzyl methyl sulfide	liq	6.27			
Bicyclo[4.1.0]heptane	g	0.33			
Bicyclo[3.1.0]hexane	g	9.09			
Bicyclo[4.2.0]octane	g	−6.39			

TABLE 5-1 Enthalpies and Gibbs (free) energies of formation, entropies, and heat capacities of organic compounds (*continued*)

Substance	State	$\Delta Hf°$, kcal · mol^{-1}	$\Delta Gf°$, kcal · mol^{-1}	$S°$, cal · deg^{-1} · mol^{-1}	$C_p°$, cal · deg^{-1} · mol^{-1}
Bicyclo[5.1.0]octane	g	−3.85			
Bicyclopropyl	g	30.9			
Biphenyl	c	24.02	60.75	49.2	38.80
	liq	28.5	62.07	59.8	
Biphenylene	liq	84.4			
N,N′-Bisuccinimide	c	−169.5			
Brassidic acid	c	−214			
Bromal	liq	−31.13			
Bromal hydrate	c	−112			
Bromobenzene	liq	14.5	30.12	52.0	37.17
4-Bromobenzoic acid	c	−90.4			
1-Bromobutane	g	−25.65	−3.08	88.39	26.13
2-Bromobutane	liq	−37.2	−4.60		
	g	−28.70	−6.16	88.50	26.48
Bromochlorodifluoromethane	g	−112.7	−107.18	76.14	
Bromochlorofluoromethane	g	−70.5	−66.58	72.88	
Bromochloromethane	g	−12.0	−9.39	68.67	
Bromodichlorofluoromethane	g	−64.4	−58.98	78.87	
Bromodichloromethane	g	−14.0	−10.16	75.56	
Bromodifluoromethane	g	−110.8	−106.90	70.51	
Bromoethane	liq	−21.99	−6.64	47.5	24.1
	g	−15.30	−6.29	68.71	15.45
Bromoethene (vinyl bromide)	g	18.73	19.30	65.83	13.26
Bromofluoromethane	g	−60.4	−57.71	65.97	
1-Bromoheptane	liq	−52.21			
1-Bromohexane	liq	−46.42			
Bromoiodomethane	g	12.0	9.36	73.49	
Bromomethane	g	−9.02	−6.75	58.76	10.15
2-Bromo-2-methylpropane	liq	−39.3			
	g	−32.00	−6.73	79.34	27.85
1-Bromooctane	liq	−58.57			
1-Bromopentane	liq	−40.68			
	g	−30.87	−1.37	97.70	31.60
1-Bromopropane	g	−21.00	−5.37	79.08	20.66
2-Bromopropane	g	−23.20	−6.51	75.53	21.37
N-Bromosuccinimide	c	−80.35			
Bromotrichloromethane	g	−8.9	−2.96	79.55	
Bromotrifluoromethane	g	−155.1	−148.8	71.16	16.57
Brucine	c	−188.6			
1,2-Butadiene	g	38.77	47.43	70.03	19.15

TABLE 5-1 Enthalpies and Gibbs (free) energies of formation, entropies, and heat capacities of organic compounds (*continued*)

Substance	State	$\Delta Hf°$, kcal · mol^{-1}	$\Delta Gf°$, kcal · mol^{-1}	$S°$, cal · deg^{-1} · mol^{-1}	$C_p°$, cal · deg^{-1} · mol^{-1}
1,3-Butadiene	g	26.33	36.01	66.62	19.01
Butadiyne (biacetylene)	g	113.00	106.11	59.76	17.60
Butane	liq	−35.29	−3.60	55.2	
	g	−30.15	−4.10	74.12	23.29
1,2-Butanediamine	liq	−28.74			
2,3-Butanedione (diacetyl)	liq	−87.44			
1,4-Butanedithiol	liq	−25.11			
1-Butanethiol (butyl mercaptan)	liq	−29.79	0.97	65.96	
	g	−21.05	2.64	89.68	28.24
2-Butanethiol	liq	−31.13	−0.04	64.87	
	g	−23.00	1.29	87.65	28.51
1-Butanol	liq	−78.18	−38.84	54.1	42.31
	g	−65.65	−36.04	86.7	26.29
2-Butanol	liq	−81.88	−42.31	53.8	47.5
	g	−69.94	−40.06	85.6	27.08
2-Butanone (methyl ethyl ketone)	liq	−65.29	−36.18	57.08	37.98
	g	−56.26	−34.91	80.81	24.59
1-Butene	g	−0.03	17.04	73.04	20.47
2-Butene					
cis	g	−1.67	15.74	71.90	18.86
trans	g	−2.67	15.05	70.86	20.99
1-Buten-3-yne	g	72.80	73.13	66.77	17.49
tert-Butoxy radical	g	−24.7			
tert-Butyl radical	g	6.7			
N-Butylacetamide	liq	−91.02			
Butyl acetate	liq	−126.52			
tert-Butylamine	liq	−35.97			
	g	−28.65	6.90	80.76	28.67
Butylbenzene	liq	−18.67$^{18°C}$	27.50		
	g	−3.30	34.58	105.04	41.85
sec-Butylbenzene	liq	−15.87			
tert-Butylbenzene	liq	−16.90			
sec-Butyl butyrate	liq	−141.6			
Butyl chloroacetate	liq	−128.7			
Butyl 2-chlorobutyrate	liq	−156.6			
Butyl 3-chlorobutyrate	liq	−146.0			
Butyl 4-chlorobutyrate	liq	−147.7			
Butyl 2-chloropropionate	liq	−136.7			
Butyl 3-chloropropionate	liq	−133.4			
Butyl crotonate	liq	−111.8			
Butylcyclohexane	g	−50.95	13.49	109.58	49.50
Butylcyclopentane	g	−40.22	14.67	109.04	42.42

TABLE 5-1 Enthalpies and Gibbs (free) energies of formation, entropies, and heat capacities of organic compounds (*continued*)

Substance	State	$\Delta Hf°$, kcal · mol^{-1}	$\Delta Gf°$, kcal · mol^{-1}	$S°$, cal · deg^{-1} · mol^{-1}	$C_p°$, cal · deg^{-1} · mol^{-1}
Butyl dichloroacetate	liq	−131.5			
Butyl ether	liq	−156.1			
	g	−87.2	114.96	48.82	
tert-Butyl hydroperoxide	liq	−70.2			
Butyllithium	liq	−31.6			
Butyl trichloroacetate	liq	−130.6			
1-Butyne (ethyl acetylene)	g	39.48	48.30	69.51	19.46
2-Butyne (dimethyl acetylene)	g	34.97	44.32	67.71	18.63
Butyraldehyde	g	−49.00	−27.43	82.44	24.52
Butyramide	c	−87.5			
Butyric acid	liq	−127.59	−90.27	54.1	42.1
Butyronitrile	g	8.14	25.97	77.98	23.19
Caffeine (methyl theobromine)	c	−76.2			
Capric acid (decanoic acid)	c	−170.59			
Caproic acid (hexanoic acid)	liq	−139.71			
ε-Caprolactam	c	−78.54	−22.72	40.3	
Caprylic acid (octanoic acid)	liq	−151.93			
Carbazole	c	30.3			
Carboxyl radical	g	−54			
CCH radical	g	114	105	49.6	8.87
Cellobiose	c	−532.5			
Chloroacetamide	c	−80.9			
Chloroacetic acid	c, l	−122.3			
Ionized	aq	−119.81			
Nonionized; std state, $m=1$	aq	−118.92			
Chloroacetyl chloride	liq	−68.0			
2-Chlorobenzaldehyde	liq	−28.4			
3-Chlorobenzaldehyde	liq	−30.2			
4-Chlorobenzaldehyde	c	−35.1			
Chlorobenzene	liq	2.58	21.32	50.0	35.9
2-Chlorobenzoic acid	c	−95.3			
3-Chlorobenzoic acid	c	−101.2			
4-Chlorobenzoic acid	c	−102.19			
Chlorobenzoquinone	c	−52.7			
1-Chlorobutane	g	−35.20	−9.27	85.58	25.71
2-Chlorobutane	g	−38.60	−12.78	85.94	25.93
2-Chlorobutyric acid	liq	−137.6			
3-Chlorobutyric acid	liq	−133.0			

TABLE 5-1 Enthalpies and Gibbs (free) energies of formation, entropies, and heat capacities of organic compounds (*continued*)

Substance	State	$\Delta Hf°$, kcal · mol^{-1}	$\Delta Gf°$, kcal · mol^{-1}	$S°$, cal · deg^{-1} · mol^{-1}	$C_p°$, cal · deg^{-1} · mol^{-1}
4-Chlorobutyric acid	liq	−135.4			
Chlorocyclohexane	liq	−49.54			
2-Chloro-1,1-difluoroethylene	g	−79.2	−72.90	72.28	
Chlorodifluoromethane	g	−115.6	−108.1	67.12	13.35
Chloroethane (ethyl chloride)	g	−26.83	−14.46	65.91	14.97
Chloroethylene (vinyl chloride)	g	8.40	12.31	63.08	12.84
Chloroethyne	g	51	47	57.81	12.98
Chlorofluoromethane	g	−63.2	−57.11	63.16	11.24
Chloroform	liq	−31.6	−17.17	48.5	
	g	−24.60	−16.76	70.63	15.63
Chloroiodomethane	g	3.0	3.69	70.78	
Chloromethane (methyl chloride)	g	−19.59	−13.97	55.97	9.74
Chloromethyloxirane	liq	−35.48			
1-Chloro-2-methylpropane	g	−38.10	−11.87	84.56	25.93
2-Chloro-2-methylpropane	g	−43.80	−15.32	77.00	27.30
1-Chloronaphthalene	liq	13.0			
2-Chloronaphthalene	c	13.2			
1-Chloropentane	g	−41.80	−8.94	94.89	31.18
3-Chorophenol	c	−49.4			
4-Chlorophenol	c	−47.3			
1-Chloropropan-2,3-diol	liq	−125.58			
2-Chloropropan-1,3-diol	liq	−123.71			
1-Chloropropane	g	−31.10	−12.11	76.27	20.23
2-Chloropropane	g	−35.00	−14.94	72.70	20.87
3-Chloro-1-propene (allyl chloride)	g	−0.15	10.42	73.29	18.01
2-Chloropropionic acid	liq	−125.0			
3-Chloropropionic acid	c	−131.4			
N-Chlorosuccinimide	c	−85.58			
Chlorotrifluoromethane	g	−169.20	−159.38	68.16	15.98
Chlorotrinitromethane	liq	−6.54			
Chrysene	c	34.7			
Cinchonamine	c	−10.4			
Cinchonidine	c	7.1			
Cinchonine	c	7.4			
Cinnamic acid					
cis	c	−72.0			
trans	c	−80.53			
Cinnamic anhydride	c	−83.1			
Citraconic acid	c	−197.04			

TABLE 5-1 Enthalpies and Gibbs (free) energies of formation, entropies, and heat capacities of organic compounds (*continued*)

Substance	State	$\Delta Hf°$, kcal · mol^{-1}	$\Delta Gf°$, kcal · mol^{-1}	$S°$, cal · deg^{-1} · mol^{-1}	$C_p°$, cal · deg^{-1} · mol^{-1}
Citric acid	c	−369.0	−295.5	39.73	
Citric acid monohydrate	c	−439.4	−352.0	67.74	1.276
Codeine monohydrate	c	−151.2			
Coniine	liq	−57.6			
Creatine	c	−128.16	−63.32	45.3	
Creatine hydrate	c	−199.1			
Creatinine	c	−56.77	−6.97	40.10	
o-Cresol	g	−30.74	−8.86	85.47	31.15
m-Cresol	g	−31.63	−9.69	85.27	29.27
p-Cresol	g	−29.97	−7.38	83.09	29.75
m-Cresol acetate	liq	−89.41			
Crotonic acid					
cis	liq	−83			
trans	c	−102.9			
trans-Crotononitrile	g	35.77	46.22	71.31	19.62
Cyanamide	c	14.05			
1-Cyanoguanidine	c	5.4	42.9	30.90	28.40
3-Cyanopyridine	c	46.23			
5-Cyanotetrazole	c	96.1			
4-Cyanothiazole	c	52.63			
Cyclobutane	g	6.37	26.30	63.43	17.26
Cyclobutene	g	31.00	41.76	62.98	16.03
Cyclododecane	c	−73.29			
Cycloheptane	liq	−37.47	12.92	57.97	29.42
Cycloheptanone	liq	−71.5			
1,3,5-Cycloheptatriene	liq	34.22	58.09	51.30	38.90
1,3-Cyclohexadien-5-yl radical	g	49.4			
Cyclohexane	liq	−37.34	6.37	48.84	37.4
	g	−29.43	7.59	71.28	25.40
Cyclohexane-1,2-dicarboxylic acid					
cis	c	−229.7			
trans	c	−232.0			
Cyclohexanethiol	g	−22.80			
Cyclohexanol	liq	−83.22	−31.87	47.7	
Cyclohexanone	g	−55.00	−21.69	77.00	26.21
Cyclohexene	liq	−9.28	24.28	51.67	34.9
	g	−1.28	25.54	74.27	25.10
Cyclohexen-3-yl radical	g	29			
1-Cyclohexenylmethanol	liq	−91.4			
Cyclohexyl radical	g	13			
Cyclooctane	liq	−40.58	18.60	62.62	
Cyclooctanone	liq	−77.9			

TABLE 5-1 Enthalpies and Gibbs (free) energies of formation, entropies, and heat capacities of organic compounds (*continued*)

Substance	State	$\Delta Hf°$, kcal · mol^{-1}	$\Delta Gf°$, kcal · mol^{-1}	$S°$, cal · deg^{-1} · mol^{-1}	$C_p°$, cal · deg^{-1} · mol^{-1}
1,3,5,7-Cyclooctatetraene	liq	60.93	85.70	52.65	
Cyclopentadiene	g	32.00	42.86	64.00	
Cyclopentane	liq	−25.28	8.70	48.82	30.80
	g	−18.46	9.23	70.00	19.84
Cyclopentanediol-1,2					
cis	c	−115.9			
trans	c	−117.1			
Cyclopentanethiol	g	−11.45	13.63	86.38	25.79
Cyclopentanol	liq	−71.74	−30.55	49.2	
Cyclopentanone	liq	−56.24			
	g	−46.03			
Cyclopentene	liq	1.02	25.93	48.10	29.24
	g	7.87	26.48	69.23	17.95
1-Cyclopentenylmethanol	liq	8.2			
Cyclopentyl-1-thiaethane	g	−15.41			
Cyclopropane	g	12.74	24.95	56.75	13.37
Cyclopropene	g	66.0	68.42	58.38	
Cyclopropyl radical	g	55			
L-Cysteine	c	−124.5			
L-Cystine	c	−245.7			
Decahydronapthalene					
cis	liq	−52.45	16.47	63.34	55.45
trans	liq	−55.14	13.79	63.32	54.61
Decanal	g	−79.09	−15.90	138.28	57.29
Decane	liq	−71.95	−4.19	101.70	75.16
1,10-Decanediol	c	−165.74			
1-Decanethiol	liq	−66.07			
	g	−50.54	14.68	145.82	61.08
1-Decanoic acid	c	−170.59			
1-Decanol	liq	−114.6	−31.6	10.2.9	
	g	−96.0	−24.9	142.8	59.1
1-Decene	liq	−41.73	25.10	101.58	
1-Decyne	g	9.85	60.28	125.36	52.51
Deoxybenzoin	c	−16.96			
Desoxyamalic acid	c	−285.7			
Diacetamide	c	−117			
Diacetyl peroxide	liq	−127.9			
o-Diallyl phthalate	liq	−131.6			
Dialuric acid	c	−314.4			
2,6-Diaminopyridine	c	−1.56			
Diamylose	c	−850			
Diazomethane	g	46.0	52.06	58.02	12.55
Dibenzoylethane	c	−61.1			
Dibenzoylethylene	c	−27.4			

TABLE 5-1 Enthalpies and Gibbs (free) energies of formation, entropies, and heat capacities of organic compounds (*continued*)

Substance	State	$\Delta Hf°$, kcal · mol^{-1}	$\Delta Gf°$, kcal · mol^{-1}	$S°$, cal · deg^{-1} · mol^{-1}	$C_p°$, cal · deg^{-1} · mol^{-1}
Dibenzoylmethane	c	−53.6			
Dibenzoyl peroxide	c	−100			
Dibenzyl	c	10.53	62.15	64.4	61.0
Dibenzyl ketone	c	−20.1			
Dibenzyl sulfide	c	23.74			
Dibenzyl sulfone	c	−42.1			
1,2-Dibromobutane	g	−23.70	−3.14	97.70	30.38
Dibromochlorofluoro-methane	g	−55.4	−53.40	81.99	
Dibromochloromethane	g	−5.0	−4.50	78.31	
1,2-Dibromocycloheptane	liq	−37.67			
1,2-Dibromocyclohexane	liq	−38.8			
1,2-Dibromocyclooctane	liq	−41.41			
Dibromodichloromethane	g	−7.0	−4.67	83.23	
Dibromodifluoromethane	g	−102.7	−100.16	77.66	
1,2-Dibromoethane	liq	−19.4	−5.0	53.37	32.51
Dibromofluoromethane	g	−53.4	−52.84	75.70	
Dibromomethane	g	−3.53	−3.87	70.10	13.04
1,2-Dibromopropane	g	−17.40	−4.22	89.90	24.57
Dibutylborinic acid	liq	−146.3			
Dibutyl ether	g	−79.80	−21.16	119.60	48.76
Dibutylmercury	liq	−23.4			
Di-*tert*-butyl peroxide	liq	−91.0			
Dibutyl-*o*-phthalate	c	−201			
Dibutyl sulfate	liq	−216.1			
Dibutyl sulfite	liq	−165.6			
Dibutyl sulfone	c	−145.76			
Dichloroacetic acid	liq	−119.0			
Ionized	aq	−122.4			
Nonionized	aq	−120.4			
Dichloroacetylene	g	50	47	65	15.67
1,2-Dichlorobenzene	g	7.16	19.76	81.61	27.12
1,3-Dichlorobenzene	g	6.32	18.78	82.09	27.20
1,4-Dichlorobenzene	g	5.50	18.44	80.47	27.22
Dichlorodifluoromethane	g	−117.90	−108.51	71.91	17.31
1,1-Dichloroethane	liq	−38.3	−18.1	50.61	30.18
	g	−31.10	−17.52	72.91	18.25
1,2-Dichloroethane	liq	−39.49	−19.03	49.84	30.9
	g	−31.00	−17.65	73.66	18.80
1,1-Dichloroethylene	liq	−5.8	5.85	48.17	26.60
	g	0.30	5.78	68.85	16.02
cis-1,2-Dichloroethylene	liq	−6.6	5.27	47.42	27
	g	0.45	5.82	69.20	15.55
trans-1,2-Dichloroethylene	g	1.00	6.35	69.29	15.93

TABLE 5-1 Enthalpies and Gibbs (free) energies of formation, entropies, and heat capacities of organic compounds (*continued*)

Substance	State	$\Delta Hf°$, $kcal \cdot mol^{-1}$	$\Delta Gf°$, $kcal \cdot mol^{-1}$	$S°$, $cal \cdot deg^{-1} \cdot mol^{-1}$	$C_p°$, $cal \cdot deg^{-1} \cdot mol^{-1}$
Dichlorofluoromethane	g	−68.10	−60.77	70.04	14.58
Dichloromethane	liq	−29.7	−16.83	42.7	
	g	−22.80	−16.46	64.61	12.16
1,2-Dichloropropane	g	−39.60	−19.86	84.80	23.47
1,3-Dichloropropane	g	−38.60	−19.74	87.76	23.81
2,2-Dichloropropane	g	−42.00	−20.21	77.92	25.30
Dicyanoacetylene	liq	119.6			
1,4-Dicyanobutyne-2	c	87.6			
Dicyclohexadiene	liq	6.3			
Dicyclopentadiene	c	27.9			
Dicyclopentyl	liq	−41.8			
2,2-Diethoxypropane	liq	−128.83			
Diethylamine	g	−17.30	17.23	84.18	27.66
Diethylbarbituric acid (veronal)	c	−178.7			
1,2-Diethylbenzene	g	−4.53	33.72	103.81	43.63
1,3-Diethylbenzene	g	−5.22	32.67	104.99	42.27
1,4-Diethylbenzene	g	−5.32	32.95	103.73	42.10
Diethylenediamine	c	−3.2	57.4	20.5	
Diethylene glycol	liq	−150.2			
	g	−136.5		105.4	32.3
Diethyl ether (ethyl ether)	liq	−65.30	−27.88	60.5	40.8
	g	−60.26	−29.24	81.90	26.89
Diethylmercury	liq	7.1			
Diethylmethyl phosphonate	liq	−245.3			
Diethylnitramine	liq	−25.4			
Diethyl oxalate	liq	−192.51			
Diethyl peroxide	liq	−53.4			
Diethyl *o*-phthalate	liq	−186			
Diethyl selenide	liq	−23.0			
Diethyl sulfate	liq	−194.28			
Diethyl sulfite	liq	−143.50			
Diethyl sulfone	c	−123.13			
Diethyl sulfoxide	liq	−63.97			
Diethylzinc	liq	4.0			
1,2-Difluorobenzene	liq	−79.04	−59.41	53.20	38.01
1,3-Difluorobenzene	g	−74.09	−61.43	76.57	25.40
1,4-Difluorobenzene	g	−73.43	−60.43	75.43	25.55
2,2′-Difluorobiphenyl	c	−70.73			
4,4-Difluorobiphenyl	c	−70.91			
2,2-Difluorochloroethylene	g	−75.4	−69.1	72.39	17.23
1,1-Difluoroethane	g	−119.70	−105.87	67.50	16.24

TABLE 5-1 Enthalpies and Gibbs (free) energies of formation, entropies, and heat capacities of organic compounds (*continued*)

Substance	State	$\Delta Hf°$, kcal · mol^{-1}	$\Delta Gf°$, kcal · mol^{-1}	$S°$, cal · deg^{-1} · mol^{-1}	$C_p°$, cal · deg^{-1} · mol^{-1}
1,1-Difluoroethylene	g	−82.50	−76.84	63.38	14.14
Difluoromethane	g	−108.24	−101.66	58.94	10.25
Diglycylglycine	c	−230.8			
9,10-Dihydroanthracene	c	15.87			
1,2-Dihydronaphthalene	liq	18.0			
1,4-Dihydronaphthalene	liq	21.0			
4*H*-Dihydropyran	liq	−37.5			
5,12-Dihydrotetracene	c	25.44			
2,3-Dihydrothiophene	liq	12.73			
2,5-Dihydrothiophene	liq	11.31			
1,2-Dihydroxybenzene	c	−86.3	−50.20	35.9	31.6
1,3-Dihydroxybenzene	c	−87.95	−50.00	35.3	31.3
1,2-Diiodobenzene	c	41.2			
1,3-Diiodobenzene	c	44.7			
1,4-Diiodobenzene	c	38.4			
1,2-Diiodoethane	g	15.90	18.76	83.30	19.67
Diiodomethane	g	28.30	24.24	73.95	13.80
Diisopropyl ether	liq	−83.94	−21.1	70.4	
	g	−76.20	−29.13	93.27	37.83
Diisopropyl ketone	g	−74.40			
Diisopropylmercury	liq	−3.1			
1,2-Dimethoxybenzene	liq	−69.4			
Dimethoxyborane	liq	−144.5			
1,2-Dimethoxyethane	liq	−90.02			
2,2-Dimethoxypropane	liq	−108.92			
cis-α,β-Dimethylacrylic acid	c	−117.3			
Dimethyl adipate	liq	−211.9			
Dimethylamine	g	−4.50	16.25	65.24	16.50
Std state, $m = 1$	aq	−16.88	13.85	31.8	
$(CH_3)_2NH_2^+$; std state, $m = 1$	aq	−28.74	−0.80	41.2	
Dimethylaminotrimethylsilane	liq	−66.8			
N,N-Dimethylaniline	liq	8.2			
2,2-Dimethylbutane	g	−44.35	−2.20	85.62	33.91
2,3-Dimethylbutane	g	−42.49	−0.98	87.42	33.59
2,3-Dimethyl-1-butene	g	−13.32	18.89	87.39	34.29
2,3-Dimethyl-2-butene	g	−14.15	18.18	87.15	29.54
3,3-Dimethyl-1-butene	g	−10.31	23.46	82.16	30.23
2,3-Dimethyl-2-butenoic acid	c	−108.9			
Dimethylcadmium	g	9.528		72.40	31.5
Dimethylchlorosilane	liq	−79.8			

TABLE 5-1 Enthalpies and Gibbs (free) energies of formation, entropies, and heat capacities of organic compounds (*continued*)

Substance	State	$\Delta Hf°$, kcal · mol^{-1}	$\Delta Gf°$, kcal · mol^{-1}	$S°$, cal · deg^{-1} · mol^{-1}	$C_p°$, cal · deg^{-1} · mol^{-1}
1,1-Dimethylcyclohexane	liq	−52.31	6.34	63.87	
	g	−43.26	8.42	87.24	36.90
1,2-Dimethylcyclohexane					
cis	g	−41.15	9.85	89.51	37.40
trans	g	−43.02	8.24	88.65	38.00
1,3-Dimethylcyclohexane					
cis	g	−44.16	7.13	88.54	37.60
trans	g	−42.20	8.68	89.92	37.60
1,4-Dimethylcyclohexane					
cis	g	−42.22	9.07	88.54	37.60
trans	g	−44.12	7.58	87.19	37.70
1,1-Dimethylcyclopentane	g	−33.05	9.33	85.87	31.86
1,2-Dimethylcyclopentane					
cis	g	−30.96	10.93	87.51	32.06
trans	g	−32.67	9.17	87.67	32.14
1,3-Dimethylcyclopentane					
cis	g	−32.47	9.37	87.67	32.14
trans	g	−31.93	9.91	87.67	32.14
Dimethyldichlorosilane	g	−110.2		80.16	24.17
cis-2,4-Dimethyl-1,3-dioxane	liq	−111.79			
4,5-Dimethyl-1,3-dioxane	liq	−108.32			
5.5-Dimethyl-1,3-dioxane	liq	−110.53			
4,4′-Dimethyldiphenylamine	c	−2.8			
Dimethyl ether	g	−43.99	−26.99	63.83	15.73
N,N-Dimethylformamide	liq	−57.2		28.5	37.45
Dimethylfulvene	liq	21.5			
Dimethyl fumarate	liq	−174.3			
Dimethyl glutarate	liq	−205.9			
Dimethylglyoxime	c	−42.51			
2,2-Dimethylhexane	liq	−62.63	0.71	79.33	
	g	−53.71	2.56	103.06	
2,3-Dimethylhexane	liq	−60.40	2.17	81.91	
2,3-Dimethylhexane	g	−51.13	4.23	106.11	
2,4-Dimethylhexane	liq	−61.47	0.89	82.62	
	g	−52.44	2.80	106.51	
2,5-Dimethylhexane	liq	−62.26	0.60	80.96	
	g	−53.21	2.50	104.93	
3,3-Dimethylhexane	liq	−61.58	1.23	81.12	
	g	−52.61	3.17	104.70	
3,4-Dimethylhexane	liq	−60.23	2.03	82.97	
	g	−50.91	4.14	107.15	

TABLE 5-1 Enthalpies and Gibbs (free) energies of formation, entropies, and heat capacities of organic compounds (*continued*)

Substance	State	$\Delta Hf°$, kcal · mol^{-1}	$\Delta Gf°$, kcal · mol^{-1}	$S°$, cal · deg^{-1} · mol^{-1}	$C_p°$, cal · deg^{-1} · mol^{-1}
2,2-Dimethyl-3-hexene					
cis	liq	−30.22			
trans	liq	−34.64			
5,5-Dimethylhydantoin	c	−126.4			
1,1-Dimethylhydrazine	liq	11.8	49.4	47.32	39.21
1,2-Dimethylhydrazine	liq	13.3	50.8	47.60	40.88
Dimethyl ketone radical	g	−11			
Dimethyl maleate	liq	−168.2			
Dimethylmaleic anhydride	c	−139.0			
Dimethyl malonate	liq	−190.2			
Dimethylmercury	liq	14.0			
Dimethylnitramine	c	−16.9			
Dimethyl oxalate	liq	−181.0			
2,2-Dimethylpentane	g	−49.27	0.02	93.90	39.67
2,3-Dimethylpentane	g	−47.62	0.16	98.96	39.67
2,4-Dimethylpentane	g	−48.28	0.74	94.80	39.67
3,3-Dimethylpentane	g	−48.17	0.63	95.53	39.67
2,7-Dimethylphenanthrene	c	8.70			
4,5-Dimethylphenanthrene	c	21.26			
9,10-Dimethyl-phenanthrene	c	11.4			
Dimethyl *m*-phthalate	c	−171			
Dimethyl *o*-phthalate	liq	−162			
Dimethyl *p*-phthalate	c	−170			
2,2-Dimethylpropane	g	−39.67	−0.364	73.23	29.07
2,3-Dimethylpyridine	liq	4.62			
2,4-Dimethylpyridine	liq	3.85			
2,5-Dimethylpyridine	liq	4.45			
2,6-Dimethylpyridine	liq	3.02			
3,4-Dimethylpyridine	liq	4.36			
3,5-Dimethylpyridine	liq	5.36			
Dimethyl succinate	liq	−199.6			
1,1-Dimethylsuccinic acid	c	−236.08			
1,2-Dimethylsuccinic acid					
cis	c	−233.6			
trans	c	−235.1			
Dimethyl sulfate	liq	−175.23			
Dimethyl sulfite	liq	−125.07			
Dimethyl sulfone	c	−107.8	−72.3	34.77	
Dimethyl sulfoxide	liq	−48.6	−23.7	45.0	35.2
3,3-Dimethyl-2-thiabutane	liq	−37.49			
2,2-Dimethylthia-cyclopropane	liq	−5.78			

TABLE 5-1 Enthalpies and Gibbs (free) energies of formation, entropies, and heat capacities of organic compounds (*continued*)

Substance	State	$\Delta Hf°$, kcal · mol^{-1}	$\Delta Gf°$, kcal · mol^{-1}	$S°$, cal · deg^{-1} · mol^{-1}	$C_p°$, cal · deg^{-1} · mol^{-1}
2,2-Dimethyl-3-thiapentane	liq	−44.7			
2,4-Dimethyl-3-thiapentane	g	−33.76	6.48	99.30	40.45
2,3-Dinitroaniline	c	−2.8			
2,4-Dinitroaniline	c	−16.3			
2,5-Dinitroaniline	c	−10.6			
2,6-Dinitroaniline	c	−12.1			
3,4-Dinitroaniline	c	−7.8			
3,5-Dinitroaniline	c	−9.3			
2,4-Dinitroanisole	c	−44.6			
2,6-Dinitroanisole	c	−45.2			
1,2-Dinitrobenzene	c	2.06	50.56	51.7	
1,3-Dinitrobenzene	c	−4.04	44.13	52.8	
2,4-Dinitrophenol	c	−55.6			
2,6-Dinitrophenol	c	−50.2			
2,4-Dinitroresorcinol	c	−99.3			
4,6-Dinitroresorcinol	c	−105.1			
2,4-Dinitrotoluene	c	−17.1			
2,6-Dinitrotoluene	c	−12.2			
1,4-Dioxane	liq	−84.47	−44.96	46.67	
	g	−75.30	−43.21	71.65	22.48
1,3-Dioxane	liq	−89.99			
1,4-Dioxatetralin	liq	−60.9			
Dioxindole	c	−76.9			
1,3-Dioxolane	g	−71.1			
Dipentene	liq	−12.1			
N,N-Diphenylacetamide	c	−10.3			
Diphenylamine	c	31.07			
1,4-Diphenylbutadiene					
cis,cis	c	47.51			
trans,trans	c	42.73			
Diphenylbutadiyne	c	123.91			
1,4-Diphenylbutane	c	−2.36			
1,4-Diphenyl-1,4-butanedione	c	−61.24	1.87	77.6	
1,4-Diphenyl-2-butene-1,4-dione	c	−27.55	26.64	76.3	
Diphenylcarbinol	c	−25.04			
Diphenyl carbonate	c	−95.93	−42.05	66.54	
Diphenyldichlorosilane	liq	−66.5			
Diphenyl disulfide	c	35.8			
Diphenyl disulfone	c	−153.59			
1,1-Diphenylethane	liq	11.7	58.58	80.28	
1,2- Diphenylethane	liq	12.31	63.87	64.6	
1,1-Diphenylethene	liq	41.21			

TABLE 5-1 Enthalpies and Gibbs (free) energies of formation, entropies, and heat capacities of organic compounds (*continued*)

Substance	State	$\Delta Hf°$, kcal · mol^{-1}	$\Delta Gf°$, kcal · mol^{-1}	$S°$, cal · deg^{-1} · mol^{-1}	$C_p°$, cal · deg^{-1} · mol^{-1}
Diphenyl ether	liq	−3.48	34.47	69.62	
Diphenylethyne	c	74.66			
Diphenylfulvene	c	7.1			
Diphenylmercury	c	66.8			
Diphenylmethane	liq	21.25	66.19	57.2	55.7
Diphenyl sulfide	liq	39.1			
Diphenyl sulfone	c	−53.71			
Diphenyl sulfoxide	c	2.40			
Dipropyl ether	g	−70.00	−25.23	100.98	37.83
Dipropylmercury	liq	−5.0			
Dipropyl sulfate	liq	−205.22			
Dipropyl sulfite	liq	−154.52			
Dipropyl sulfone	liq	−130.94			
Dipropyl sulfoxide	liq	−78.65			
2,3-Dithiabutane	liq	−14.82	1.67	56.26	34.92
5,6-Dithiadecane	g	−37.86	12.87	136.91	55.23
3,4-Dithiahexane	liq	−28.69	2.28	72.90	
1,3-Dithian-2-thione	c	−3.1			
4,5-Dithiaoctane	liq	−40.95	4.56	89.28	
N,N-Dithiodiethylamine	liq	−29.1			
1,3-Dithiolan-2-thione	c	3.1			
Di-*p*-tolyl sulfone	c	−74.32			
Divinyl ether	g	−9.53			
Divinyl sulfone	liq	−49.5			
Dodecane	liq	−84.16	6.71	117.26	89.86
1-Dodecene	g	−39.52	32.96	147.78	64.43
1-Dodecyne	g	−0.01	64.22	143.98	63.44
Dulcitol	c	−321.9			
Eicosane	g	−108.93	28.04	223.26	110.73
Eiconsanoic acid (arachidic acid)	c	−241.9			
1-Eicosene	g	−78.93	49.03	222.26	108.15
Ergosterol	c	−188.8			
meso-Erythritol	c	−127.56	−152.12	39.9	
Ethane	g	−20.24	−7.84	54.76	12.54
1,2-Ethanedithiol	liq	−12.83			
Ethanethiol	g	−11.02	−1.12	70.77	17.37
Ethanol	liq	−66.20	−41.63	38.49	26.76
	g	−56.03	−40.13	67.54	15.64
Ethoxy radical	g	−6			
Ethyl radical	g	26.0	31	59.2	
Ethyl acetate	liq	−114.49	−79.52	62.0	
	g	−105.86	−78.25	86.70	27.16
Ethyl allyl sulfone	liq	−96.95			

TABLE 5-1 Enthalpies and Gibbs (free) energies of formation, entropies, and heat capacities of organic compounds (*continued*)

Substance	State	ΔHf°, kcal · mol^{-1}	ΔGf°, kcal · mol^{-1}	S°, cal · deg^{-1} · mol^{-1}	C_p°, cal · deg^{-1} · mol^{-1}
Ethylamine	g	−11.00	8.91	68.08	17.36
N-Ethylaniline	liq	0.9	45.10	57.2	
Ethylbenzene	liq	−2.98	28.61	60.99	
	g	7.12	31.21	86.15	30.69
2-Ethyl-1-butene	g	−12.32	19.11	90.01	31.92
Ethyl carbamate (urethane)	c	−124.4			
Ethyl crotonate	liq	−100.4			
Ethylcyclohexane	liq	−50.72	6.95	67.14	
1-Ethylcyclohexene	liq	−25.50			
Ethylcyclopentane	liq	−39.08	8.92	67.00	
	g	−30.37	10.65	90.42	31.49
Ethyldiethylcarbamate	liq	−141.6			
Ethylene	g	12.50	16.31	52.39	10.24
Ethylene carbonate	c	−138.9			
Ethylene chlorohydrin	liq	−70.6			
1,2-Ethylenediamine	liq	−15.06		50	
	aq, 200	−13.32			
Ethylenediaminetetraacetic acid	c	−420.5			
Ethylenediammonium chloride	c	−122.7			
	aq, 5000	−115.92			
Ethylene glycol (2,1-ethanediol)	liq	−108.70	−77.25	39.9	35.8
	g	−93.05	−72.77	77.33	23.20
	aq, 1	−109.01			
Ethyleneimine (azirane)	g	29.50	42.54	59.90	12.55
Ethylene oxide	g	−12.58	−3.13	57.94	11.54
2-Ethyl-1-hexanal	liq	−83.30			
2-Ethyl-2-hexanal	liq	−62.44			
3-Ethylhexane	liq	−59.88	1.79	84.95	
Ethylidenecyclohexane	liq	−21.19			
Ethyl isovalerate	liq	−136.5			
Ethyllithium	c	−14.0			
Ethylmercury bromide	c	−25.7			
Ethylmercury chloride	c	−33.7			
Ethylmercury iodide	c	−15.7			
Ethyl methyl ether	g	−51.73	−28.12	74.24	21.45
Ethyl nitrate	g	−36.80	−8.81	83.25	23.27
Ethyl nitrite	g	−24.9		24.74	23.71
3-Ethylpentane	g	−45.33	2.63	98.35	39.67
Ethyl pentanoate	liq	−132.2			
Ethyl peroxyl radical	g	(−2)			

TABLE 5-1 Enthalpies and Gibbs (free) energies of formation, entropies, and heat capacities of organic compounds (*continued*)

Substance	State	$\Delta Hf°$, kcal · mol^{-1}	$\Delta Gf°$, kcal · mol^{-1}	$S°$, cal · deg^{-1} · mol^{-1}	$C_p°$, cal · deg^{-1} · mol^{-1}
2-Ethylphenol	c	−49.91			
3-Ethylphenol	c	−51.21			
4-Ethylphenol	c	−53.63			
Ethylphosphonic acid	c	−251.3			
Ethyl propanoate	liq	−122.16	−79.16		
2-Ethylpyridine	liq	−1.2			
Ethylsuccinic acid	c	−236.4			
Ethyl thioacetate	liq	−64.01			
Ethyl β-vinylacrylate	liq	−80.8			
Ethyl vinyl ether	g	−33.63			
Ethynylbenzene (phenylacetylene)	g	78.22	86.46	76.88	27.46
Fluoranthene	c	45.75	82.60	55.09	
Fluoroacetamide	c	−118.7			
Fluoroacetic acid	c	−164.5			
Fluorobenzene	g	−27.86	−16.50	72.33	22.57
2-Fluorobenzoic acid	c	−135.67			
3-Fluorobenzoic acid	c	−139.13			
4-Fluorobenzoic acid	c	−140.00			
Fluoroethane	g	−62.90	−50.44	63.34	14.21
2-Fluoroethanol	liq	−111.3			
Fluoromethane	g	−56.80	−51.09	53.25	8.96
1-Fluoropropane	g	−67.20	−47.87	72.71	19.75
2-Fluoropropane	g	−69.00	−48.81	69.82	19.60
4-Fluorotoluene	liq	−44.80	−19.06	56.67	
Fluorotrinitromethane	liq	−52.8			
Formaldehyde	g	−27.70	−26.27	52.29	8.46
unhydrolyzed	aq	−35.9	−31.02		
Formamide	liq	−60.7			
	g	−44.5	−33.71	59.41	10.84
Formanilide	c	−36.2			
Formic acid	liq	−101.51	−86.38	30.82	23.67
	g	−90.49	−83.89	59.45	10.81
Ionized; std state, $m = 1$	aq	−101.71	−83.9	22	−21.0
Nonionized; std state, $m = 1$	aq	−101.68	−89.0	39	
Dimer	g	−195.08			
Formyl					
HCO	g	10.4	6.76	53.66	8.27
HCO^+	g	204	201	48.3	8.62
Formyl fluoride	g	−90	−88	59.0	9.66
N-Formyl-DL-leucine	c	−222.1			
Formyl urea	c	−118			
β-D-Fructose	c	−302.2			

TABLE 5-1 Enthalpies and Gibbs (free) energies of formation, entropies, and heat capacities of organic compounds (*continued*)

Substance	State	$\Delta Hf°$, kcal · mol^{-1}	$\Delta Gf°$, kcal · mol^{-1}	$S°$, cal · deg^{-1} · mol^{-1}	$C_p°$, cal · deg^{-1} · mol^{-1}
D-Fucose	c	−262.7			
Fumaric acid	c	−193.84	−156.70	39.7	
Fumaronitrile	c	64.11			
Furan	g	−8.23	0.21	63.86	15.64
Furfural	liq	−47.8			
Furfuryl alcohol	liq	−66.05	−36.85	51.50	
2-Furoic acid (pyromucic acid)	c	−119.12			
Furylacrylic acid	c	−109.7			
Furylethylene	liq	−2.5			
D-Galactonic acid	c	−384.8			
D-Galactose	c	−304.1	−219.60	49.1	
D-Glucaric acid-1,4-lactone	c	−343.2			
D-Glucaric acid-3,6-lactone	c	−343.6			
D-Gluconic acid	c	−379.3			
D-Gluconic acid-δ-lactone	c	−300.3			
D-Glucose					
α	c	−304.26	−217.6	50.7	
β	c	−302.76			
D-Glutamic acid	c	−240.19	−173.87	45.7	
L-Glutamic acid	c	−241.32	−174.78	44.98	
L-Glutamine	c	−197.3			
Glutaric acid	c	−229.44			
Glyceraldehyde	liq	−143			
Glycerol	liq	−159.76	−114.01	48.87	35.9
Glyceryl-1-acetate	liq	−217.5			
Glyceryl-1-benzoate	c	−185.80			
Glyceryl-2-benzoate	c	−184.71			
Glyceryl-1-caprate	c	−265.05			
Glyceryl-2-caprate	c	−261.90			
Glyceryl-1,3-diacetate	liq	−268.2			
Glyceryl-1-laurate	c	−277.46			
Glyceryl-2-laurate	c	−275.48			
Glyceryl-2-myristate	c	−292.31			
Glyceryl-1-palmitate	c	−306.28			
Glyceryl-1-stearate	c	−319.64			
Glyceryl triacetate	liq	−318.3			
Glyceryl trilaurate	c	−489			
Glyceryl trimyristate	c	−520			
Glyceryl trinitrate	liq	−88.6			

TABLE 5-1 Enthalpies and Gibbs (free) energies of formation, entropies, and heat capacities of organic compounds (*continued*)

Substance	State	$\Delta Hf°$, kcal · mol^{-1}	$\Delta Gf°$, kcal · mol^{-1}	$S°$, cal · deg^{-1} · mol^{-1}	$C_p°$, cal · deg^{-1} · mol^{-1}
Glycine	c	−126.22	−88.09	24.74	23.71
Ionized; std state, $m = 1$	aq	−112.28	−75.28	26.54	
Nonionized; std state, $m = 1$	aq	−122.85	−88.62	37.84	
$NH_3^+CH_2COOH$; std state, $m = 1$	aq	−123.78	−91.82	45.46	
Glycol acetal	liq	−91.1			
Glycolic acid	c	−158.6			
Glycylglycine	c	−178.51	−117.25	45.4	
Glycylphenylalanine	c	−163.9			
Glycylvaline	c	−200.0			
Glyoxal	g	−50.66			
Glyoxime	c	−21.63			
Glyoxylic acid	c	−199.7			
Guanidine	c	−13.39			
Guanidine carbonate	c	−232.10	−133.23	70.6	61.87
Guanidine nitrate	c	−92.5			
Guanidine sulfate	c	−288.0			
Guanine	c	−43.72	11.33	38.3	
Guanylurea nitrate	c	−102.1			
Heptadecane	g	−94.15	22.01	195.33	94.33
Heptadecanoic acid	c	−220.9			
1-Heptadecene	g	−64.15	43.00	194.33	91.76
1-Heptanal	g	−63.10	−20.71	110.34	40.89
Heptane	liq	−53.63	0.42	77.92	53.76
	g	−44.88	1.91	102.27	39.67
1-Heptanethiol	g	−35.76	8.65	117.89	44.68
Heptanoic acid (enanthic acid)	liq	−145.75			
1-Heptanol	liq	−95.8	−34.0	76.5	66.5
	g	−79.3	−28.9	114.8	42.7
1-Heptene	liq	−23.41	21.22	78.31	50.62
	g	−14.89	22.90	101.24	37.10
1-Heptyne	g	24.62	54.18	97.44	36.11
Hexachlorobenzene	c	−31.30	0.25	62.20	48.11
	g	−8.10	10.56	105.45	41.40
Hexachloroethane	g	−33.20	−13.13	95.30	32.68
Hexadecafluoroethylcyclohexane	liq	−799.1			
Hexadecafluoroheptane	liq	−817.6	−739.24	134.28	
	g	−808.9	−737.87	158.88	
Hexadecane	g	−89.23	20.00	186.02	88.86
Hexadecanoic acid (palmitic acid)	c	−213.3	−75.54	108.12	

TABLE 5-1 Enthalpies and Gibbs (free) energies of formation, entropies, and heat capacities of organic compounds (*continued*)

Substance	State	$\Delta Hf°$, kcal · mol^{-1}	$\Delta Gf°$, kcal · mol^{-1}	$S°$, cal · deg^{-1} · mol^{-1}	$C_p°$, cal · deg^{-1} · mol^{-1}
1-Hexadecanol (cetyl alcohol)	c, II	−163.4	−23.6	108.0	104.8
	liq	−151.86	−23.08	145.0	
1-Hexadecene	g	−59.23	40.99	185.02	86.29
Hexafluorobenzene	liq	−237.27	−211.43	66.90	52.96
	g	−228.64	−210.18	91.59	37.43
Hexafluoroethane	g	−320.90	−300.15	79.30	25.43
Hexahydroindane					
cis	g	−30.4			
trans	g	−31.4			
Hexamethylbenzene	c	−39.19	28.06	71.66	61.5
Hexamethyldisiloxane	liq	−194.7	−129.5	103.69	74.42
Hexamethylenetetramine	c	30.0	103.92	39.05	
Hexanal	g	−59.37	−23.93	101.07	35.43
Hexanamide	c	−101.48			
Hexane	liq	−47.52	−0.91	70.76	45.2
	g	−39.96	−0.06	92.83	34.20
1-Hexanethiol	g	−30.83	6.65	108.58	39.21
1-Hexanol	liq	−90.7	−36.4	69.2	56.6
	g	−75.9	−32.4	105.5	37.2
1-Hexene	liq	−17.30	19.93	70.55	43.81
	g	−9.96	20.90	91.93	31.63
2-Hexene					
cis	g	−12.51	18.22	92.37	30.04
trans	g	−12.27	18.27	90.97	31.64
3-Hexene					
cis	g	−11.38	19.84	90.73	29.55
trans	g	−13.01	18.55	89.59	31.75
1-Hexyne	g	29.55	52.24	88.13	30.65
Hippuric acid (benzoylglycine)	c	−145.63	−88.33	57.2	
Hydantoic acid	c	−179			
Hydantoin	c	−107.2			
Hydrazobenzene	c	52.9			
Hydroquinone	c	−87.08	−49.48	33.5	33.9
Hydrosorbic acid	liq	−110.2			
o-Hydroxybenzoic acid	c	−140.64	−100.7	42.6	38.03
m-Hydroxybenzoic acid	c	−139.8	−99.74	42.3	37.59
p-Hydroxybenzoic acid	c	−139.7	−99.55	42.0	37.08
β-Hydroxybutyric acid	liq	−162.3			
Hydroxyisobutyric acid	c	−177.9			
L-Hydroxyproline	c	−158.1			
8-Hydroxyquinoline	c	−19.9			
Hypoxanthene (6-oxypurine)	c	−26.47	18.39	34.8	

TABLE 5-1 Enthalpies and Gibbs (free) energies of formation, entropies, and heat capacities of organic compounds (*continued*)

Substance	State	$\Delta Hf°$, kcal · mol^{-1}	$\Delta Gf°$, kcal · mol^{-1}	$S°$, cal · deg^{-1} · mol^{-1}	$C_p°$, cal · deg^{-1} · mol^{-1}
Imidazole	c	14.5			
Indane	liq	2.56	36.04	56.01	45.47
Indene	liq	26.39	52.00	51.19	44.68
Indole	c	29.8			
Iodobenzene	g	38.85	44.88	79.84	24.08
2-Iodobenzoic acid	c	−72.2			
3-Iodobenzoic acid	c	−75.7			
4-Iodobenzoic acid	c	−75.5			
Iodocyclohexane	liq	−23.5			
Iodoethane	liq	−9.6	3.5	50.6	27.5
	g	−2.00	5.10	70.82	15.76
Iodomethane	liq	−3.29	3.61	38.9	
	g	3.29	3.72	60.64	10.54
2-Iodo-2-methylpropane	g	−17.60	5.65	81.79	28.27
1-Iodonaphthalene	liq	38.6			
2-Iodonaphthalene	c	34.5			
2-Iodophenol	c	−22.9			
3-Iodophenol	c	−22.6			
4-Iodophenol	c	−22.8			
1-Iodopropane	g	−7.30	6.68	80.32	21.48
2-Iodopropane	g	−10.00	4.80	77.55	21.53
3-Iodopropene (allyl iodide)	liq	13.7			
3-Iodopropionic acid	c	−109.9			
2-Iodotoluene	liq	18.7			
3-Iodotoluene	liq	18.9			
4-Iodotoluene	liq	16.1			
Isatin	c	−62.7			
Isobutylbenzene	liq	−16.68			
Isobutyl dichloracetate	liq	−132.4			
Isobutyl phenyl ketone	liq	−52.63			
Isobutyl trichloroacetate	liq	−132.4			
Isobutyronitrile	g	6.07	24.76	74.88	23.04
L-Isoleucine	c	−151.8	−82.97	49.71	45.00
Isopropenyl acetate	liq	−92.31			
Isopropyl radical	g	17.6			
Isopropyl acetate	liq	−124.01			
Isopropylbenzene	liq	−9.85	29.70	66.87	
(cumene)	g	0.94	32.74	92.87	36.26
Isopropyl nitrate	g	−45.65	−9.72	89.20	28.84
Isopropyl thiolacetate	liq	−71.26			
Isopropyl trichloroacetate	liq	−128.2			
Isoquinoline	c	37.9			
L-Isoserine	c	−177.8			
Isothiocyanic acid	g	30.50	26.98	59.28	11.09

TABLE 5-1 Enthalpies and Gibbs (free) energies of formation, entropies, and heat capacities of organic compounds (*continued*)

Substance	State	$\Delta Hf°$, kcal · mol^{-1}	$\Delta Gf°$, kcal · mol^{-1}	$S°$, cal · deg^{-1} · mol^{-1}	$C_p°$, cal · deg^{-1} · mol^{-1}
Itaconic acid	c	−201.06			
Ketene	g	−14.60	−14.41	57.79	12.37
α-Ketoglutaric acid	c	−245.35			
D-Lactic acid	c	−165.88		34.3	
L-Lactic acid	c	−165.89	−124.98	34.00	
	liq	−161.2	−123.84	45.9	
β-Lactose	c	−534.1	−374.52	92.3	
Lauric acid (dodecanoic acid)	c	−185.14			
D-Leucine	c	−152.36	−82.97	49.71	
L-Leucine	c	−154.6	−82.76	50.62	48.03
DL-Leucine	c	−153.14	−83.54	49.5	
DL-Leucylglycine	c	−205.7	−112.14	67.2	
Leucylglycylglycine	c	−259.6			
Levulinic acid	c	−166.6			
Levulinic lactone	liq	−76.2			
(+)-Limonene	liq	−13.0			
DL-Lysine	c	−162.2			
Maleic acid	c	−188.94	−149.40	38.1	32.36
Maleic anhydride	c	−112.08			
L-Malic acid	c	−263.78	−211.45		
DL-Malic acid	c	−264.27			
Malonamide	c	−130.5			
Malonic acid	c	−212.96			
Malonic diamide	c	−130.52			
Malononitrile	c	44.6			
Maltose	c	−530.8	−412.60		
L-Mandelic acid	c	−138.8			
D-Mannitol	c	−139.61	−225.20	57.0	
D-Mannose	c	−301.9			
Melamine (triaminotriazine)	c	−17.3	44.10	35.63	
Melezitose	c	−815			
2-Mercaptopropionic acid	liq	−111.9	−82.19	54.70	
Mesaconic acid	c	−197			
Mesoxalic acid	c	−290.7			
2,2-Metacyclophane	g	40.8			
Methane	g	−17.89	−12.15	44.52	8.54
Methanethiol (methyl mercaptan)	g	−5.49	−2.37	60.96	12.01
Methanol	liq	−57.13	−39.87	30.41	19.40
	g	−48.06	−38.82	57.29	10.49
Std state, $m = 1$	aq	−58.78			

TABLE 5-1 Enthalpies and Gibbs (free) energies of formation, entropies, and heat capacities of organic compounds (*continued*)

Substance	State	$\Delta Hf°$, kcal · mol^{-1}	$\Delta Gf°$, kcal · mol^{-1}	$S°$, cal · deg^{-1} · mol^{-1}	$C_p°$, cal · deg^{-1} · mol^{-1}
L-Methionine	c	−180.4	−120.88	55.32	
Methoxyl radical	g	(2)			
2-Methoxybenzaldehyde	c	−63.7			
3-Methoxybenzaldehyde	liq	−66.0			
4-Methoxybenzaldehyde	liq	−63.9			
Methoxybenzene (anisole)	g	−17.3			
Methoxymethyl radical	g	(−4)			
2-Methoxytetrahydropyran	liq	−105.7			
5-Methoxytetrazole	c	16.6			
Methyl (CH_3)	g	34.82	35.35	46.38	9.25
Methyl acetate	liq	−106.4			
Methyl acrylate	g	−70.10	−56.78		
Methyl allantoin (pyvurile)	c	−177.0			
Methyl allyl sulfone	liq	−91.95			
Methylamido radical (CH_3NH)	g	35			
Methylamine	g	−5.50	7.71	57.98	11.97
Std state, $m = 1$	aq	−16.77	4.94	29.5	
Methylaminolithium	c	−22.92			
N-Methylaniline	liq	7.7			
Methyl benzoate	liq	−79.8			
Methyl benzyl sulfone	c	−88.65			
2-Methylbiphenyl	liq	25.8			
3-Methylbiphenyl	liq	20.4			
4-Methylbiphenyl	c	13.2			
2-Methyl-1,3-butadiene (isoprene)	g	18.10	34.86	75.44	25.00
3-Methyl-1,2-butadiene	g	31.00	47.47	76.40	25.20
2-Methylbutane	g	−36.92	−3.54	82.12	28.39
3-Methyl-1-butanethiol	g	−27.44			
2-Methyl-2-butanethiol	liq	−38.90	0.56	69.34	
	g	−30.36	2.20	92.48	34.30
2-Methyl-1-butanol	liq	−85.2			52.6
3-Methyl-1-butanol	liq	−85.2			50.3
2-Methyl-2-butanol	liq	−90.7	−41.9	54.8	59.2
	g	−78.8	−39.5	86.7	
3-Methyl-2-butanol	liq	−87.5			55.5
2-Methyl-1-butene	g	−8.68	15.68	81.15	26.28
3-Methyl-1-butene	g	−6.92	17.87	79.70	28.35
2-Methyl-2-butene	g	−10.17	14.26	80.92	25.10
Methyl butyl sulfone	liq	−128.00			
Methyl *tert*-butyl sulfone	c	−132.8			
3-Methyl-1-butyne	g	32.60	49.12	76.23	25.02
Methyl caprate	liq	−153.07			

TABLE 5-1 Enthalpies and Gibbs (free) energies of formation, entropies, and heat capacities of organic compounds (*continued*)

Substance	State	$\Delta Hf°$, kcal · mol^{-1}	$\Delta Gf°$, kcal · mol^{-1}	$S°$, cal · deg^{-1} · mol^{-1}	$C_p°$, cal · deg^{-1} · mol^{-1}
Methyl caproate	liq	−129.10			
N-Methylcaprolactam	liq	−73.3			
5-Methylcaprolactam	c	−86.9			
7-Methylcaprolactam	c	−86.5			
Methyl caprylate	liq	−141.07			
Methyl crotonate	liq	−91.5			
Methylcyclohexane	liq	−45.45	4.86	59.26	
	g	−36.99	6.52	82.06	32.27
2-Methylcyclohexanol					
cis	liq	−93.3			
trans	liq	−99.4			
3-Methylcyclohexanol					
cis	liq	−99.5			
trans	liq	−94.3			
4-Methylcyclohexanol					
cis	liq	−98.8			
trans	liq	−103.6			
2-Methylcyclohexanone	liq	−68.8			
Methylcyclopentane	g	−25.50	8.55	81.24	26.24
1-Methylcyclopentanol	liq	−82.3			
2-Methylcyclopentanone	liq	−63.4			
1-Methylcyclopentene	g	−1.30	24.41	78.00	24.10
3-Methylcyclopentene	g	2.07	27.48	79.00	23.90
4-Methylcyclopentene	g	3.53	29.06	78.60	23.90
Methyldichlorosilane	liq	−105.9			
2-Methyl-1,3-dioxane	liq	−104.60			
4-Methyl-1,3-dioxane	liq	−99.80			
N-Methyldiphenylamine	liq	28.8			
4-Methyldiphenylamine	c	11.7			
Methylene	g	92.35	88.25	46.32	8.27
2-Methylenecyclohexanol	liq	−66.3			
2-Methylenecyclopentanol	liq	11.2			
β-Methylene-β-propiolactone (diketene)	liq	−55.72			
Methylene sulfate	c	−164.6			
1-Methyl-2-ethylbenzene	g	0.29	31.33	95.42	37.74
1-Methyl-3-ethylbenzene	g	−0.46	30.22	96.60	36.38
1-Methyl-4-ethylbenzene	g	−0.78	30.28	95.34	36.22
2-Methyl-3-ethylpentane	liq	−59.69	3.03	81.41	
	g	−50.48	5.08	105.43	
3-Methyl-3-ethylpentane	liq	−60.46	2.69	79.97	
	g	−51.28	4.76	103.48	
2-Methyl-3-ethyl-1-pentene	g	−23.97			

TABLE 5-1 Enthalpies and Gibbs (free) energies of formation, entropies, and heat capacities of organic compounds (*continued*)

Substance	State	$\Delta Hf°$, kcal · mol^{-1}	$\Delta Gf°$, kcal · mol^{-1}	$S°$, cal · deg^{-1} · mol^{-1}	$C_p°$, cal · deg^{-1} · mol^{-1}
Methyl ethyl sulfite	liq	−135.55			
Methyl ethyl sulfone	c	−116.17			
Methyl formate	liq	−90.60	−71.53	29	
	g	−83.70	−71.03	72.00	15.90
Methylglyoxal	g	−64.8			
Methylglyoxime	c	−30.3			
2-Methylheptane	liq	−60.98	0.92	84.16	
	g	−51.50	3.05	108.81	
3-Methylheptane	liq	−60.34	1.12	85.66	
	g	−50.82	3.28	110.32	
4-Methylheptane	liq	−60.17	1.87	83.72	
	g	−50.69	4.00	108.35	
Methyl heptanoate	liq	−135.54			
2-Methylhexane	liq	−54.93	−0.69	77.28	53.28
	g	−46.59	0.77	100.38	39.67
3-Methylhexane	liq	−54.35	−0.39	78.23	
	g	−45.96	1.10	101.37	39.67
Methyl hexanoate	liq	−129.11			
5-Methylhydantoin	c	−116.3			
Methylhydrazine	liq	12.9	43.0	39.66	32.25
	g	22.55	44.66	66.61	17.0
Methylidyne					
CH	g	142.00	134.02	43.72	6.97
CH^+	g	388.8	380.1	41.00	6.97
α-Methylindole	c	14.5			
Methyl isocyanide	g	35.6	39.6	58.99	12.65
1-Methyl-2-isopropylbenzene (*o*-cymene)	liq	−18.19			
1-Methyl-3-isopropylbenzene	liq	−18.69			
Hexamethylenetetramine	liq	−18.7	28.65	73.28	
Methyl isopropyl ether	g	−60.24	−28.89	80.86	26.55
Methyl isopropyl ketone	g	−62.76			
Methyl isopropyl sulfone	liq	−120.44			
Methyl isothiocyanate (CH_3NCS)	g	31.3	34.5	69.29	15.65
3-Methylisoxazole	liq	−5.0			
5-Methylisoxazole	liq	−6.4			
Methyl laurate	liq	−165.66			
Methylmercury bromide	c	−20.6			
Methylmercury chloride	c	−27.8			
Methylmercury iodide	c	−10.4			
Methyl myristate	liq	−177.80			

TABLE 5-1 Enthalpies and Gibbs (free) energies of formation, entropies, and heat capacities of organic compounds (*continued*)

Substance	State	$\Delta Hf°$, kcal · mol^{-1}	$\Delta Gf°$, kcal · mol^{-1}	$S°$, cal · deg^{-1} · mol^{-1}	$C_p°$, cal · deg^{-1} · mol^{-1}
1-Methylnaphthalene	liq	13.43	46.26	60.90	53.63
2-Methylnaphthalene	c	10.72	46.03	52.58	46.84
Methyl nitrate	liq	−38.0	−10.4	51.9	37.6
	g	−29.8	−9.4	76.1	
Methyl nitrite	g	−15.30	0.24	67.95	15.11
Methyl oleate	liq	−174.2			
Methyl pelargonate	liq	−147.29			
2-Methylpentane	g	−41.66	−1.20	90.95	34.46
3-Methylpentane	g	−41.02	−0.51	90.77	34.20
Methyl pentanoate	liq	−122.90			
2-Methyl-1-pentene	g	−12.49	18.55	91.34	32.41
3-Methyl-1-pentene	g	−10.76	20.66	90.06	34.04
4-Methyl-1-pentene	g	−10.54	21.52	87.89	30.23
2-Methyl-2-pentene	g	−14.28	17.02	90.45	30.26
3-Methyl-2-pentene					
cis	g	−13.80	17.50	90.45	30.26
trans	g	−14.02	17.04	91.26	30.26
4-Methyl-2-pentene					
cis	g	−12.03	19.63	89.23	31.92
trans	g	−12.99	19.03	88.02	33.80
Methyl phenyl sulfone	c	−82.49			
Methylphosphonic acid	c	−252			
2-Methylpropanal	g	−52.25			
2-Methylpropane	g	−32.15	−4.99	70.42	23.14
2-Methyl-1,2-propanediamine	liq	−32.00			
2-Methyl-1-propanethiol	g	−23.24	1.33	86.73	28.28
2-Methyl-2-propanethiol	g	−26.17	0.17	80.79	28.91
2-Methyl-1-propanol	g	−67.69	−39.99	85.81	26.6
2-Methyl-2-propanol	liq	−85.86	−44.14	46.10	52.61
	g	−74.67	−42.46	77.98	27.10
2-Methylpropene	g	−4.04	13.88	70.17	21.30
Methyl propyl ether	g	−56.82	−26.27	83.52	26.89
7-Methylpurine	c	51.3			
2-Methylpyridine (2-picoline)	liq	13.83	39.80	52.07	37.86
	g	24.05	42.32	77.68	23.90
3-Methylpyridine	liq	15.57	41.16	51.70	37.93
4-Methylpyridine	liq	13.58			
N-Methylpyrrolidone	liq	−62.64			
Methyl salicylate	liq	−127.1			
α-Methylstyrene	liq	16.8			
	g	27.00	49.84	91.70	34.70

TABLE 5-1 Enthalpies and Gibbs (free) energies of formation, entropies, and heat capacities of organic compounds (*continued*)

Substance	State	$\Delta Hf°$, kcal · mol^{-1}	$\Delta Gf°$, kcal · mol^{-1}	$S°$, cal · deg^{-1} · mol^{-1}	$C_p°$, cal · deg^{-1} · mol^{-1}
β-Methylstyrene					
cis	g	29.00	51.84	91.70	34.70
trans	g	28.00	51.08	90.90	34.90
Methylsuccinic acid	c	−229.02			
3-Methyl-2-thiabutane	g	−21.61	3.21	85.87	28.00
2-Methylthiacyclopentane	g	−15.12			
2-Methyl-3-thiapentane	liq	−37.3			
4-Methylthiazole	liq	16.31			
2-Methylthiophene	liq	10.75	27.35	52.22	29.43
3-Methylthiophene	liq	10.38	27.00	52.19	29.38
4-Methyluracil	c	−109.2			
Methyl valerate	liq	−122.89			
Morphine monohydrate	c	−170.1			
Mucic acid	c	−423			
Murexide	c	−289.7			
Myrcene	liq	3.5			
Myristic acid (tetradecanoic acid)	c	−199.21			
Naphthalene	c	18.0	48.05	39.89	
	g	35.6	53.44	80.22	31.68
1-Naphthol	g	−5.1			
2-Naphthol	g	−10.1			
1,4-Napthoquinone	c	−43.83			
1-Naphthyl acetate	c	−68.89			
2-Naphthyl acetate	c	−72.72			
1-Naphthylamine	c	16.2			
2-Naphthylamine	c	14.4			
Narceine dihydrate	c	−421.2			
Narcotine	c	−210.9			
Nicotine	liq	9.4			
Nitrilotriacetic acid	c		−312.5		
2-Nitroaniline	c	−3.45	42.60	42.1	39.3
3-Nitroaniline	c	−4.46	41.60	42.1	40.2
4-Nitroaniline	c	−9.91	36.10	42.1	40.4
Nitrobenzene	liq	3.80	34.95	53.6	44.4
2-Nitrobenzoic acid	c	−94.25	−46.95	49.8	
3-Nitrobenzoic acid	c	−100.25	−52.71	49.0	
4-Nitrobenzoic acid	c	−101.25	−53.07	50.2	43.3
3-Nitrobiphenyl	c	15.6			
4-Nitrobiphenyl	c	9.7			
1-Nitrobutane	g	−34.40	2.42	94.28	29.85
2-Nitrobutane	g	−39.10	−1.49	91.62	29.51
3-Nitro-2-butanol	liq	−93.2			

TABLE 5-1 Enthalpies and Gibbs (free) energies of formation, entropies, and heat capacities of organic compounds (*continued*)

Substance	State	$\Delta Hf°$, kcal · mol^{-1}	$\Delta Gf°$, kcal · mol^{-1}	$S°$, cal · deg^{-1} · mol^{-1}	$C_p°$, cal · deg^{-1} · mol^{-1}
2-Nitrodiphenylamine	c	15.4			
Nitroethane	g	−24.4	−1.17	75.39	18.69
aci form	aq	−30.7			
nitro form	aq	−32			
2-Nitroethanol	liq	−83.8			
Nitroguanidine	c	−22.1			
Nitromethane	liq	−27.03	−3.47	41.05	25.33
	g	−17.86	−1.66	65.73	13.70
1-Nitronaphthalene	c	10.2			
1-Nitropropane	g	−30.00	0.08	85.00	24.41
2-Nitropropane	g	−33.21	−3.06	83.10	24.26
4-Nitrosodiphenylamine	c	50.9			
Nonadecane	g	−104.00	26.03	213.95	105.26
1-Nonadecene	g	−74.00	47.02	212.95	102.69
1-Nonanal	g	−74.16	−17.91	128.97	51.82
Nonane	liq	−65.84	2.81	94.09	
	g	−54.74	5.93	120.86	50.60
1-Nonanethiol	g	−45.61	12.67	136.51	55.61
Nonanoic acid	liq	−157.68			
1-Nonanol	liq	−109.2	−32.4	91.3	67.50
1-Nonene	g	−24.74	26.93	119.86	48.03
Octadecane	g	−99.08	24.02	204.64	99.80
Octadecanoic acid	c	−226.5			
1-Octadecene	g	−69.08	45.01	203.64	97.22
Octafluorocyclobutane	g	−365.20	−334.33	95.69	37.32
1-Octanal	g	−69.23	−19.91	119.66	46.36
Octanamide	c	−113.1			
Octane	liq	−59.74	1.77	85.50	45.14
	g	−49.82	3.92	111.55	45.14
1-Octanethiol	g	−40.68	10.67	127.20	50.14
Octanoic acid	liq	−151.93			
1-Octanol	liq	−101.6	−34.2	90.2	77.7
2-Octanone	liq	−91.9	−33.54	89.35	65.31
1-Octene	liq	−29.52	22.49	86.15	57.65
	g	−19.82	24.91	110.55	42.56
1-Octyne	g	19.70	56.26	106.75	41.58
Oleic acid	c	−187.2			
DL-Ornithine	c	−156.0			
Oxacyclobutane (trimethylene oxide)	g	−19.25	−2.33	65.46	
Oxalic acid	c	−197.7	−166.8	28.7	
Std state, $m = 1$	aq	−197.2	−161.1	10.9	
Oxalic acid dihydrate	c	−341.0			
Oxalyl chloride	liq	−85.6			

TABLE 5-1 Enthalpies and Gibbs (free) energies of formation, entropies, and heat capacities of organic compounds (*continued*)

Substance	State	$\Delta Hf°$, kcal · mol^{-1}	$\Delta Gf°$, kcal · mol^{-1}	$S°$, cal · deg^{-1} · mol^{-1}	$C_p°$, cal · deg^{-1} · mol^{-1}
Oxamic acid	c	−160.4			
Oxamide	c	−123.0	−81.9	28.2	
Oxindole	c	−41.2			
8-Oxypurine	c	−15.4			
Palmitic acid	c	−213.10			
Papaverine	c	−120.2			
Parabanic acid	c	−138.0			
[1,8]-Paracyclophane	c	−19.6			
[2,2]-Paracyclophane	g	59.9			
[6,6]-Paracyclophane	c	−46.1			
Paraldehyde	liq	−164.2			
Pentachloroethane	g	−34.8	−16.79	91.17	28.22
Pentachlorofluoroethane	g	−75.8	−55.93	93.54	
Pentachlorophenol	c	−70.6	−34.44	60.21	48.27
Pentadecane	g	−84.31	17.98	176.71	83.40
1-Pentadecene	g	−54.31	38.97	175.71	80.82
1-Pentadecyne	g	−14.78	70.25	171.91	79.84
1,2-Pentadiene	g	34.80	50.29	79.70	25.20
1,3-Pentadiene					
cis	g	18.70	34.84	77.50	22.60
trans	g	18.60	35.07	76.40	24.70
1,4-Pentadiene	g	25.20	40.69	79.70	25.10
2,3-Pentadiene	g	33.10	49.21	77.60	24.20
Pentaerythritol	c	−220.0	−146.73	47.34	45.51
Pentaerythritol tetranitrate	c	−128.8			
Pentafluorobenzoic acid	c	−296.34			
Pentafluoroethane	g	−264.00	−246.00	79.76	22.88
Pentafluorophenol	c	−244.86			
Pentamethylbenzene	liq	−32.33	25.64	70.22	51.74
Pentamethylbenzoic acid	c	−128.13			
1-Pentanal	g	−54.45	−25.88	91.53	29.96
Pentanamide	c	−90.70			
Pentan-2,4-dione (acetylacetone)	liq	−101.33			
	g	−90.47		95.1	28.7
Pentan-1,5-dithiol	liq	−30.99			
Pentane	g	−35.00	−2.00	83.40	28.73
1-Pentanethiol	liq	−35.72	2.28	74.18	
Pentanoic acid	liq	−133.71	−89.10	62.10	50.48
1-Pentanol	liq	−85.0	−38.3	62.0	49.8
2-Pentanol	liq	−87.7			
3-Pentanol	liq	−88.5	−40.4	57.4	60.0
2-Pentanone	g	−61.82	−32.76	89.91	28.91
3-Pentanone	liq	−70.87			

TABLE 5-1 Enthalpies and Gibbs (free) energies of formation, entropies, and heat capacities of organic compounds (*continued*)

Substance	State	$\Delta Hf°$, kcal · mol^{-1}	$\Delta Gf°$, kcal · mol^{-1}	$S°$, cal · deg^{-1} · mol^{-1}	$C_p°$, cal · deg^{-1} · mol^{-1}
1-Pentene	g	−5.00	18.91	82.65	26.19
2-Pentene					
cis	g	−6.71	17.17	82.76	24.32
trans	g	−7.59	16.71	81.36	25.92
2-Pentenoic acid	liq	−106.7			
3-Pentenoic acid	liq	−103.9			
4-Pentenoic acid	liq	−102.9			
1-Pentyne	g	34.50	50.25	78.82	25.50
2-Pentyne	g	30.80	46.41	79.30	23.59
Perfluoropiperidine	liq	−482.9	−422.67	94.02	70.93
Perylene	c	43.69			
α-Phellandrene	liq	−14.3			
Phenacetin	c	−101.1			
9,10-Phenanthraquinone	c	−55.18			
Phenanthrene	c	27.3	64.12	50.6	
Phenazine	c	56.4			
Phenol	c	−39.44	−12.05	34.42	32.2
	liq	−37.80	−11.02		30.46
	g	−23.03	−7.86	75.43	24.75
Phenoxy radical	g	10			
Phenoxyacetic acid	c	−122.8			
Phenyl radical	g	71			
Phenyl acetate	liq	−80.02			
Phenylacetic acid	c	−95.3			
β-Phenyl-1-alanine, DL- and L-	c	−111.9	−50.6	51.06	48.52
Phenyl benzoate	c	−57.7			
2-Phenylbenzoic acid	c	−83.4			
Phenylboronic acid	c	−172.0			
1-Phenylcyclohexene	liq	−4.0			
Phenylcyclopropane	liq	24.7			
N-Phenyldiacetimide	c	−86.63			
p-Phenylenediamine	c	0.73			
Phenyl ethyl sulfide	liq	5.29			
DL-Phenylglyceric acid	c	−178.5			
N-Phenylglycine	c	−96.2			
a-Phenylglycine	c	−103.2			
Phenylglyoxime					
α	c	−4.9			
β	c	10.1			
Phenylglyoxylic acid	c	−115.3			
Phenylhydrazine	liq	34.03			
Phenyl methyl sulfide	liq	11.5			
N-Phenyl-2-naphthylamine	c	38.2			

TABLE 5-1 Enthalpies and Gibbs (free) energies of formation, entropies, and heat capacities of organic compounds (*continued*)

Substance	State	$\Delta Hf°$, kcal · mol^{-1}	$\Delta Gf°$, kcal · mol^{-1}	$S°$, cal · deg^{-1} · mol^{-1}	$C_p°$, cal · deg^{-1} · mol^{-1}
N-Phenylpyrrole	c	38.1			
2-Phenylpyrrole	c	34.5			
Phenyl salicylate	c	−104.3			
Phenyl thiolacetate	liq	−29.16			
Phosgene	g	−52.80	−49.42	67.82	13.79
Phthalamide	c	−104.4			
m-Phthalic acid	c	−191.91			
o-Phthalic acid	c	−186.91	−141.39	49.7	45.0
p-Phthalic acid	c	−195.05			
Phthalic anhydride	c	−110.1	−79.12	42.9	38.5
Phthalonitrile	c	65.82			
Pimelic acid	c	−241.25			
Pinene					
α	liq	−3.9			
β	liq	−1.8			
Piperazine	c	−10.90			
Piperidine	liq	−21.05			
α-Piperidone	c	−73.3	−26.79	39.4	
DL-Proline	c	−125.7			
Propadiene	g	45.92	48.37	58.30	14.10
Propane	g	−24.82	−5.63	64.58	17.59
1,2-Propanediamine	liq	−23.38			
1,2-Propanediol	liq	−119.6			
1,3-Propanediol	liq	−124.4			
1,3-Propanedithiol	liq	−18.83			
2,3-Propanedithiol	liq	−18.82			
1-Propanethiol	g	−16.22	0.52	80.40	22.65
2-Propanethiol	g	−18.22	−0.61	77.51	22.94
1-Propanol	liq	−72.66	−40.78	46.5	33.7
1-Propanol	g	−61.28	−38.67	77.61	20.82
2-Propanol	liq	−75.97	−43.09	43.16	36.06
	g	−65.11	−41.44	74.07	21.21
1,2,3-Propenetricarboxylic acid					
cis	c	−292.7			
trans	c	−294.7			
2-Propen-1-ol (allyl alcohol)	g	−31.55	−17.03	73.51	18.17
Propionaldehyde	g	−45.90	−31.18	72.83	18.80
Propionamide	c	−81.7			
Propionic acid	liq	−122.07	−91.65		
Propionic anhydride	liq	−161.53	−113.66		
Propionitrile	liq	3.5	21.31	45.25	
	g	12.10	22.98	68.50	17.46

TABLE 5-1 Enthalpies and Gibbs (free) energies of formation, entropies, and heat capacities of organic compounds (*continued*)

Substance	State	$\Delta Hf°$, kcal · mol^{-1}	$\Delta Gf°$, kcal · mol^{-1}	$S°$, cal · deg^{-1} · mol^{-1}	$C_p°$, cal · deg^{-1} · mol^{-1}
1-Propylamine	g	−17.30	9.51	77.48	22.89
2-Propylamine	liq	−26.83			
Propylbenzene	g	1.87	32.80	95.76	36.41
Propylcarbamate	c	−132.07			
Propyl chloroacetate	liq	−123.3			
Propylcyclohexane	g	−46.20	11.31	100.27	44.03
Propylcyclopentane	g	−35.39	12.57	99.73	36.96
Propylene (propene)	g	4.88	15.02	63.72	15.37
Propylene oxide	g	−22.17	−6.16	68.53	17.29
Propyl nitrate	g	−41.60	−6.53	92.10	28.99
Propyl phenyl ketone	liq	−45.14			
Propyl thiolacetate	liq	−70.29			
Propyl trichloroacetate	liq	−122.7			
Propyne (methyl acetylene)	g	44.32	46.47	59.30	14.50
Pyrazine	c	33.41			
Pyrazole	c	28.3			
Pyrene	c	27.44	64.40	53.75	56.4
Pyridazine	liq	53.74			
Pyridine	liq	23.96	43.34	42.52	31.72
	g	33.61	45.46	67.59	18.67
Pyrimidine	liq	35.04			
Pyrrole	liq	15.08			
Pyrrole-2-aldehyde	c	−24.8			
Pyrrole-2-aldoxime	c	2.9			
Pyrrolidine	liq	−9.84	25.94	48.76	
	g	−0.86	27.41	73.97	19.39
2-Pyrrolidone	c	−68.3			
Pyruvic acid	liq	−139.7	−110.75	42.9	
Quinaldine	c	39.3			
Quinhydrone	c	−19.79	−77.19	77.9	66.2
Quinidine	c	−38.3			
Quinine	c	−37.1			
Quinoline	liq	37.33	65.90	51.9	
p-Quinone	c	−44.10	−20.0	38.9	
Raffinose	c	−761			
L-Rhamnose	c	−256.5			
Rhamnose triacetate	c	−455.4			
D-Ribose	c	−251.16			
Saccharinic acid lactone	c	−249.6			
Salicylaldehyde	liq	−66.9			
Salicylaldoxime	c	−43.91			
Salicyclic acid	c	−140.9	−99.93	42.6	

TABLE 5-1 Enthalpies and Gibbs (free) energies of formation, entropies, and heat capacities of organic compounds (*continued*)

Substance	State	$\Delta Hf°$, kcal · mol^{-1}	$\Delta Gf°$, kcal · mol^{-1}	$S°$, cal · deg^{-1} · mol^{-1}	$C°_p$, cal · deg^{-1} · mol^{-1}
Sarcosine	c	−121.2			
Sebacic acid	c	−258.8			
Semicarbazide, std state, $m = 1$	aq	−39.9	−9.7	71.2	
L-Serine	c	−173.6			
Serylserine	c	−281.8			
Sorbic acid	c	−93.4			
L-Sorbose	c	−303.68	−217.10	52.8	
5,5′-Spirobis(1,3-dioxane)	c	−167.8			
Spiropentane	g	44.27	63.41	67.45	21.06
Stearic acid	c	−226.5			
Stilbene					
cis	liq	43.81			
trans	c	32.27	75.90	60.0	
Strychnine	c	−41.0			
Styrene	liq	24.83	48.37	56.78	43.64
	g	35.22	51.10	82.48	29.18
Suberic acid	c	−248.1			
Succinamide	c	−138.9			
Succinic acid	c	−224.79	−178.64	42.0	35.8
Sucrose	c	−531.9	−369.18	86.1	
L-Tartaric acid	c	−306.5			
DL-Tartaric acid	c	−308.5			
meso-Tartaric acid	c	−305.9			
Tetrabromomethane	g	19.00	15.61	85.53	21.78
Tetracene	c	37.95			
Tetrachlorobenzoquinone	c	−69.0			
1,1,1,2-Tetrachlorodifluoroethane	g	−117.1	−97.3	91.5	29.5
1,1,1,2-Tetrachloroethane	g	−35.7	−19.2	85.05	24.67
1,1,2,2-Tetrachloroethane	liq	−47.0	−22.7	59.0	39.6
	g	−36.50	−20.45	86.69	24.09
Tetrachloroethylene	g	−3.40	4.90	81.46	22.69
Tetrachloromethane	liq	−31.75	−14.97	51.67	
	g	−22.90	−12.80	74.07	19.94
1,1,2,2-Tetracyanocyclopropane	c	141			
Tetracyanoethylene	c	149.1			
Tetradecane	g	−79.38	15.97	167.40	77.93
Tetradecanoic acid	c	−199.2			
1-Tetradecene	g	−49.36	36.99	166.40	75.36
Tetraethylene glycol	liq	−234.6			
Tetraethyllead	liq	12.7	80.4	112.92	
	g	26.3			

TABLE 5-1 Enthalpies and Gibbs (free) energies of formation, entropies, and heat capacities of organic compounds (*continued*)

Substance	State	$\Delta Hf°$, kcal · mol^{-1}	$\Delta Gf°$, kcal · mol^{-1}	$S°$, cal · deg^{-1} · mol^{-1}	$C_p°$, cal · deg^{-1} · mol^{-1}
1,1,1,2-Tetrafluoroethane	g	−214.10	−197.46	75.58	20.62
Tetrafluoroethylene	g	−157.40	−149.07	71.69	19.24
Tetrafluoromethane	g	−223.0	−212.3	62.45	14.59
Tetrahydrofuran	liq	−51.67			
Tetrahydrofurfuryl alcohol	liq	−104.1			
1,2,3,4-Tetrahydro-naphthalene	liq	−6.1			
Tetrahydropyran	liq	−61.1			
1,2,5,6-Tetrahydropyridine	liq	8.0			
Tetraiodomethane	g	62.84	51.89	93.60	22.91
1,2,3,4-Tetramethylbenzene	liq	−23.0	25.49	69.45	
1,2,3,5-Tetramethylbenzene	liq	−23.54	23.58	99.55	57.5
1,2,4,5-Tetramethylbenzene	liq	−29.48	24.20	71.83	51.6
2,2,3,3-Tetramethylbutane	g	−53.99	5.26	93.06	
Tetramethyllead	liq	23.5	62.8	76.5	
	g	32.6	64.7	100.5	34.42
Tetramethylsilane	g	−68.50	−23.92	86.30	31.12
Tetramethylsuccinic acid	c	−242.0			
Tetramethylthia-cyclopropane	c	−19.84			
Tetranitromethane	liq	8.9			
1,1,1,2-Tetraphenylethane	c	53.31			
1,1,2,2-Tetraphenylethane	c	51.63			
Tetraphenylethene	c	74.46			
Tetraphenylhydrazine	c	109.4			
Tetraphenylmethane	c	59.1	137.20		
Tetrazole	c	56.7			
Thebaine	c	−63.0			
Theobromine	c	−86.4			
Thiaadamantane	c	−34.22			
2-Thiabutane	liq	−21.89	1.79	57.14	
	g	−14.25	2.73	79.62	22.73
Thiacyclobutane	g	14.61	25.69	68.17	16.57
Thiacycloheptane	g	−14.66	20.09	86.50	29.78
Thiacyclohexane	liq	−25.32	9.96	52.16	
	g	−15.12	12.68	77.26	25.86
Thiacyclopentane	liq	−17.39	8.97	49.67	
	g	−8.08	11.00	73.94	21.72
Thiacyclopropane	liq	12.41	22.52	38.84	
	g	19.65	23.16	61.01	12.83
4-Thia-5,5-dimethylhex-1-ene	liq	−21.68			
2-Thiaheptane	g	−29.34	8.39	107.73	39.10
3-Thiaheptane	g	−29.92	7.65	108.27	38.71

TABLE 5-1 Enthalpies and Gibbs (free) energies of formation, entropies, and heat capacities of organic compounds (*continued*)

Substance	State	$\Delta Hf°$, kcal · mol^{-1}	$\Delta Gf°$, kcal · mol^{-1}	$S°$, cal · deg^{-1} · mol^{-1}	$C_p°$, cal · deg^{-1} · mol^{-1}
4-Thiaheptane	liq	−40.62	5.12	80.85	
	g	−29.96	7.94	107.16	38.53
2-Thiahexane	liq	−34.15	4.08	73.49	
	g	−24.42	6.37	98.43	33.64
3-Thiahexane	liq	−34.58	3.50	73.98	
	g	−25.00	5.63	98.97	33.25
5-Thianonane	liq	−52.74	7.66	96.82	
	g	−39.99	11.76	125.76	49.46
2-Thiapentane	liq	−28.21	2.79	65.14	
	g	−19.54	4.40	88.84	28.05
3-Thiapentane	liq	−28.43	2.81	64.36	40.97
	g	−19.95	4.25	87.96	27.97
2-Thiapropane	g	−8.97	1.66	68.32	17.71
6-Thiaundecane	liq	−63.61			
Thioacetic acid	g	−43.49	−36.81	74.86	19.33
Thiohydantoic acid	c	−132.6			
Thiohydantoin	c	−59.5			
Thiolacetic acid	liq	−52.39			
β-Thiolactic acid	liq	−111.6			
Thiophene	liq	19.24	28.97	43.30	
	g	27.66	30.30	66.65	17.42
Thiosemicarbazide	c	6.0			
Thiourea	c	−21.13	5.2	27.7	
	aq, 100	−15.6			
Threonine, L- and DL-	c	−181.4			
Thymine	c	−111.9			
Thymol	c	−74.0			
Tiglic acid	c	−117.3			
Toluene	liq	2.87	27.19	52.81	37.58
	g	11.95	29.16	76.64	24.77
2-Toluenethiol	liq	10.57			
m-Toluic acid	c	−101.85			
o-Toluic acid	c	−99.55			
p-Toluic acid	c	−102.59			
o-Toluic anhydride	c	−127.5			
p-Toluic anhyride	c	−124.5			
Trehalose	c	−531.3			
2,4,6-Triamino-1,3,5-triazine	g	−17.13	42.33	74.10	20.93
2-Triazoethanol	liq	22.6			
Tribenzylamine	c	33.6			
Tribromochloromethane	g	3.0	2.17	85.36	
Tribromofluoromethane	g	−45.4	−46.14	82.65	
Tribromomethane	g	4.00	1.78	79.01	16.96

TABLE 5-1 Enthalpies and Gibbs (free) energies of formation, entropies, and heat capacities of organic compounds (*continued*)

Substance	State	$\Delta Hf°$, kcal · mol^{-1}	$\Delta Gf°$, kcal · mol^{-1}	$S°$, cal · deg^{-1} · mol^{-1}	$C_p°$, cal · deg^{-1} · mol^{-1}
Tributylamine	liq	−67.32			
Tributyl borate	liq	−286.7			
Tributylboron	liq	−83.4			
Tributyl phosphate	liq	−348			
Tributylphosphine oxide	c	−110			
Trichloroacetaldehyde	liq	−56.1			
Trichloroacetamide	c	−85.6			
Trichloroacetic acid	c	−120.7			
Ionized	aq	−123.4			
Trichloroacetyl chloride	liq	−66.4			
Trichlorobenzoquinone	c	−64.5			
1,1,1-Trichloroethane	g	−34.01	−18.21	76.49	22.07
1,1,2-Trichloroethane	g	−33.10	−18.52	80.57	21.47
Trichloroethylene	g	−1.40	4.75	77.63	19.17
Trichlorofluoromethane	g	−68.10	−58.68	74.06	18.66
Trichloromethyl	g	19	22	70.9	15.21
1,2,3-Trichloropropane	g	−44.40	−23.37	91.52	26.82
1,1,1-Tricyanoethane	c	83.9			
Tricyanoethylene	c	105.0			
Tridecane	g	−74.45	13.97	158.09	72.47
Tridecanoic acid	c	−192.8			
1-Tridecene	g	−44.45	34.96	157.09	69.89
Triethylaluminum	liq	−56.6			
Triethylamine	g	−23.80	26.36	96.90	38.46
Triethylaminoborane	liq	−47.47			
Triethyl arsenite	liq	−168.9			
Triethylarsine	liq	3.1			
Triethyl borate	liq	−250.4			
Triethylenediamine	c	−3.4	57.28	37.67	
Triethylene glycol	liq	−192.2			
Triethyl phosphate	liq	−297			
Triethylphosphine	liq	−21.3			
Triethyl phosphite	liq	−205.9			
Triethylstibine	liq	1.2			
Triethylsuccinic acid	c	−254.9			
Triethyl thionophosphate	liq	−232.5			
Trifluoroacetic acid	liq	−255.4			
Trifluoroacetonitrile	g	−118.4	−110.4	71.3	18.70
1,1,1-Trifluoroethane	g	−178.20	−162.11	68.67	18.76
2,2,2-Trifluoroethanol	liq	−207.4			
Trifluoroethylene	g	−118.50	−112.22	69.94	16.54
Trifluoroiodomethane	g	−141.0	−136.70	73.50	
Trifluoromethane	g	−165.71	−157.48	62.04	12.22

TABLE 5-1 Enthalpies and Gibbs (free) energies of formation, entropies, and heat capacities of organic compounds (*continued*)

Substance	State	$\Delta Hf°$, kcal · mol^{-1}	$\Delta Gf°$, kcal · mol^{-1}	$S°$, cal · deg^{-1} · mol^{-1}	$C_p°$, cal · deg^{-1} · mol^{-1}
Trifluoromethyl					
CF_3	g	−112.4	−109.2	63.3	11.90
CF_3^+	g	100.6	103.1	60.8	11.87
Trifluoromethylbenzene	liq	−152.40	−123.98	64.89	
	g	−143.42	−122.20	89.05	31.17
Trifluoromethylhypofluorite (CF_3OF)	g	−183	−169	77.06	18.97
DL-Trihydroxyglutaric acid	c	−356			
Triiodomethane	g	50.40	42.54	84.97	17.94
Trimethylacetic acid	liq	−134.9			
Trimethylacetic anhydride	liq	−186.4			
2,4,5-Trimethylacetophenone	liq	−60.3			
2,4,6-Trimethylacetophenone	liq	−63.9			
Trimethylaluminum	liq	−36.1		50.05	37.19
Trimethylamine	g	−5.70	23.64	69.02	21.93
Std state, $m = 1$	aq	−18.17	22.22	31.9	
Trimethylamine aluminum chloride adduct	c	−210.1			
Trimethylammonium ion Std state, $m = 1$	aq	−26.99	8.90	47.0	
Trimethyl arsenite	liq	−141.2			
Trimethylarsine	liq	−3.9			
1,2,3-Trimethylbenzene	liq	−14.01	25.68	66.40	
1,2,4-Trimethylbenzene	liq	−14.79	24.46	67.93	
1,3,5-Trimethylbenzene	liq	−15.18	24.83	65.38	
Trimethyl borate	liq	−222.9			
Trimethylboron	liq	−34.1			
2,2,3-Trimethylbutane	g	−48.95	1.02	91.61	39.33
Trimethylchlorosilane	liq	−91.8			
cis,cis-1,3,5-Trimethylcyclohexane	g	−51.48	8.10	93.30	42.93
2,2,3-Trimethylpentane	liq	−61.44	2.21	78.30	
	g	−52.61	4.09	101.62	
2,2,4-Trimethylpentane	liq	−61.97	1.65	78.40	
	g	−53.57	3.27	101.15	
2,3,3-Trimethylpentane	liq	−60.63	2.54	79.93	
	g	−51.73	4.52	103.14	
2,3,4-Trimethylpentane	liq	−60.98	2.55	78.71	
	g	−51.97	4.52	102.31	
2,4,4-Trimethyl-1-pentene	liq	−35.21	20.66	73.2	

TABLE 5-1 Enthalpies and Gibbs (free) energies of formation, entropies, and heat capacities of organic compounds (*continued*)

Substance	State	$\Delta Hf°$, kcal · mol^{-1}	$\Delta Gf°$, kcal · mol^{-1}	$S°$, cal · deg^{-1} · mol^{-1}	$C_p°$, cal · deg^{-1} · mol^{-1}
2,4,4-Trimethyl-2-pentene	liq	−34.44	21.04	74.5	
Trimethylphosphine	liq	−29.2			
Trimethylphosphine-*N*-ethylimine	liq	−35.8			
Trimethylphosphine oxide	c	−114.2			
Trimethyl phosphite	liq	−177.1			
Trimethylsilanol	liq	−130.3			
Trimethylstibine	liq	0.2			
Trimethylsuccinic acid	c	−239.2			
Trimethylsuccinic anhydride	c	−164.5			
Trimethylthiacyclopropane	liq	−14.47			
Trimethylurea	c	−79.0			
2,4,6-Trinitroanisole	c	−37.6			
1,3,5-Trinitrobenzene	c	−10.40			
Trinitromethane	c	−11.50			
1,4,5-Trinitronaphthalene	c	8.7			
1,3,8-Trinotronaphthalene	c	5.8			
2,4,6-Trinitrophenetole	c	−48.9			
2,4,6-Trinitrophenol	c	−51.23			
2,4,6-Trinitrophenylhydrazine	c	8.8			
2,4,6-Trinitrotoluene	c	−16.0			
2,4,6-Trinitro-*m*-xylene	c	−24.5			
Triphenylamine	c	$58.70^{18°C}$	120.50		
Triphenylarsine	c	74.1			
Triphenylcarbinol	c	−0.80	65.2	78.7	
Triphenylene	c	33.72	78.68	60.87	
1,1,1-Triphenylethane	c	37.56			
1,1,2-Triphenylethane	c	31.11			
Triphenylethylene	c	55.8	123.00		
Triphenylmethane	c	38.71	98.60	74.6	70.5
Triphenyl phosphate	c	−181			
Triphenylphosphine	c	55.5			
Triphenylphosphine oxide	c	−14.4			
Tripropylamine	liq	−49.51			
Tris(acetylacetonato)-chromium	c	−366.4			
1,1,1-Tris(hydroxymethyl)-ethane	c	−177.96			
Tropolone	c	−57.18			
L-Tryptophan	c	−99.8	−28.54	60.00	56.92
L-Tyrosine	c	−163.4	−92.18	51.15	51.73

TABLE 5-1 Enthalpies and Gibbs (free) energies of formation, entropies, and heat capacities of organic compounds (*continued*)

Substance	State	$\Delta Hf°$, kcal · mol^{-1}	$\Delta Gf°$, kcal · mol^{-1}	$S°$, cal · deg^{-1} · mol^{-1}	$C_p°$, cal · deg^{-1} · mol^{-1}
Undecane	liq	−78.05	5.44	109.49	
	g	−64.60	9.94	139.48	61.53
1-Undecene	g	−34.60	30.94	138.48	58.96
Urea	c	−79.71	−47.19	25.00	22.26
Std state, $m = 1$	aq	−75.95			
Urea nitrate	c	−134.8			
Urea oxalate	c	−365.3			
Uric acid	c	−147.73	−85.75	41.4	
Valine, L and DL-	c	−148.2	−85.80	42.75	40.35
Valylphenylalanine	c	−183.5			
Vinyl radical	g	63			
Vinyl bromide	g	18.7	19.3	65.90	13.27
Vinyl chloride	g	8.5	12.4	63.07	12.84
Vinylcyclohexane	liq	−21.19			
Vinylcyclopropane	liq	29.3			
2-Vinylpyridine	liq	37.2			
Xanthine	c	−90.49	−39.64	38.5	
o-Xylene	liq	−5.84	26.37	58.91	44.9
	g	4.54	29.18	84.31	31.85
m-Xylene	liq	−6.08	25.73	60.27	43.8
	g	4.12	28.41	85.49	30.49
p-Xylene	liq	−5.84	26.31	59.12	
	g	4.29	28.95	84.23	30.32
2,3-Xylenol	g	−37.57			
2,4-Xylenol	g	−38.93			
2,5-Xylenol	g	−38.63			
2,6-Xylenol	g	−38.66			
3,4-Xylenol	g	−37.42			
3.5-Xylenol	g	−38.61			
Xylitol	c	−267.32			
D-Xylose	c	−252.8			

TABLE 5-2 Heats of melting and vaporization (or sublimation) and specific heat at various temperatures of organic compounds

Abbreviations Used in the Table

ΔHm, enthalpy of melting (at the melting point) in kcal · mol^{-1}
ΔHv, enthalpy of vaporization (at the boiling point) in kcal · mol^{-1}
ΔHs, enthalpy of sublimation (at 298 K) in kcal · mol^{-1}
C_p, specific heat (at temperature specified, measured on the Kelvin scale) for physical state in existence at that temperature, expressed in cal · K^{-1} · mol^{-1}
ΔHt, enthalpy of transition (at temperature specified, measured in degrees Celsius) in kcal · mol^{-1}

Substance	ΔHm	ΔHv	ΔHs	C_p 400 K	C_p 600 K	C_p 800 K	C_p 1000 K
Acenaphthene			20.6				
Acenaphthylene			17.0				
Acetaldehyde	0.770	6.24		15.73	20.52	24.20	29.96
Acetanilide			19.3				
Acetic acid	2.80	5.663		19.52	25.15	29.08	31.99
Acetic anhydride	2.51	9.85	11.54	30.86	41.62	48.91	54.11
Acetone	1.366	6.952		22.00	29.34	34.93	39.15
Acetonitrile	1.952	7.3	7.94	14.62	18.35	21.26	23.50
ΔHt, 0.215^{-56}							
Acetophenone		9.275	13.4				
Acetyl bromide			7.9				
Acetyl chloride			7.2	18.86	23.18	26.30	28.60
Acetylene	0.900	4.05	5.1	11.97	13.73	14.93	15.92
Acetylenedicarbonitrile			6.88	22.66	25.37	27.26	28.62
Acetyl fluoride			6.0				
Acetyl iodide			7.9				
Acrylic acid		11.21	12.98	22.94	29.50	33.93	37.12

Acrylonitrile		7.8		18.36	23.11	26.43	28.88
Adenine			25.8				
Adipic acid			30.8				
α-Alanine			33.0				
Allyl ethyl sulfoxide			17.1				
Allyl trichloroacetate			12.5				
1-Aminobutane			8.50	35.44	47.30	56.01	62.54
2-Aminobutane			7.5	35.40	47.55	56.42	62.54
Aniline	2.519	10.643	13.325	34.17	46.09	53.79	69.18
Anthracene		13.5	24.7				
9,10-Anthracenedione			26.8				
Azoisopropane			8.5				
Azulene	2.89	13.26	22.8	42.15	59.32	70.59	78.24
Benzaldehyde			12				
1,2-Benzanthra-9,10-quinone			19.8				
Benzene	2.358	7.352	8.090	26.74	37.73	45.06	50.16
Benzenethiol	2.736	9.53	11.64	32.76	44.13	51.59	56.79
Benzil			23.5				
Benzilidene anil			20.5				
Benzoic acid	4.32	12.10	22.70				
Benzoic anhydride			23				
Benzonitrile	2.60	11.0	13.26	33.65	44.80	52.08	57.08
Benzophenone			22.5				
1,4-Benzoquinone			15.00				
Benzoyl bromide			14.0				
Benzoyl chloride			13.1				
Benzoyl iodide			14.8				
3,4-Benzophenanthrene			25.4				
Benzyl bromide			11.3				
Benzyl chloride			12.3				
Benzyl ethyl sulfide			13.6				
Benzyl iodide			11.3				

TABLE 5-2 Heats of melting and vaporization (or sublimation and specific heat at various temperatures of organic compounds (*continued*)

Substance	ΔHm	ΔHv	ΔHs	C_p			
				400 K	600 K	800 K	1000 K
Benzyl methyl ketone			12.78				
Benzyl methyl sulfide			12.8				
Bicyclo[4.1.0]heptane			9.14				
Bicyclo[3.1.0]hexane			7.85				
Bicyclo[4.2.0]octane			9.85				
Bicylo[5.1.0]octane			10.42				
Bicyclopropyl			8.0				
Biphenyl	4.44	10.9		52.83	73.54	86.92	96.00
Biphenylene			30.8				
Bromobenzene	2.54	9.05	10.62	30.44	40.99	47.78	52.40
4-Bromobenzoic acid			21.0				
1-Bromobutane	1.6	7.78		32.64	43.00	50.48	56.03
2-Bromobutane			8.45	33.09	43.76	51.31	56.93
Bromoethane	1.4	6.41	6.57	18.93	24.56	28.58	31.59
Bromoethene				15.91	19.83	22.50	24.46
1-Bromoheptane			12.05				
1-Bromohexane			10.91				
Bromomethane ΔHt, $0.113^{-99.4}$	1.429	5.715		11.94	14.98	17.26	19.01
2-Bromo-2-methylpropane ΔHt, $1.35^{-64.5}$, $0.25^{-41.6}$	0.47		7.4	34.93	45.58	52.65	57.74
1-Bromooctane			13.14				
1-Bromopentane	2.74	8.24		39.58	52.34	61.55	68.36
1-Bromopropane	1.56	7.14		25.70	33.66	39.41	43.70
2-Bromopropane		6.79		26.34	34.42	40.09	44.26
1,2-Butadiene	1.665	5.82	5.71	23.54	30.72	36.01	40.02
1,3-Butadiene	1.908	5.42	5.03	24.29	31.84	36.84	40.52

n-Butadiene sulfone			14.7				
Butadiyne				20.17	23.14	25.11	26.61
Butane ΔHt, $0.494^{-165.60}$	1.114	5.352	5.035	29.60	40.30	48.23	54.22
2,3-Butanedione			9.25				
1,4-Butanedithiol			13.22				
1-Butanethiol	2.500	7.702	8.73	34.95	46.54	55.68	62.95
2-Butanethiol	1.548	7.312	8.14	35.38	46.42	54.29	60.02
1-Butanol	2.24	10.31	12.52	32.80	43.90	52.11	58.26
2-Butanol		9.75	11.87	33.70	44.72	52.68	58.62
2-Butanone	2.017	7.475	8.34	29.81	39.09	46.08	51.33
1-Butene	0.920	5.238	4.81	26.04	35.14	41.80	46.82
2-Butene							
cis	1.747		5.29	24.33	33.80	40.87	46.15
trans	2.332	5.580	5.10	26.02	34.80	44.20	46.58
1-Buten-3-yne		5.439		21.26	26.67	30.40	33.16
N-Butylacetamide			18.2				
Butyl acetate			10.42				
tert-Butylamine		8.58	7.10	36.46	48.87	57.49	63.79
Butylbenzene							
stable(I)	2.682(I)		11.98	54.75	75.20	89.37	99.49
metastable(II)	2.691(II)	9.38					
sec-Butylbenzene			11.72				
tert-Butylbenzene			11.50				
sec-Butyl butyrate			11.3				
Butyl chloroacetate			12.2				
Butyl 2-chlorobutyrate			12.6				
Butyl 3-chlorobutyrate			12.7				
Butyl 4-chlorobutyrate			13.0				
Butyl 2-chloropropionate			13.0				
Butyl 3-chloropropionate			13.3				
Butyl crotonate			12.4				
sec-Butyl crotonate			11.8				

TABLE 5-2 Heats of melting and vaporization (or sublimation and specific heat at various temperatures of organic compounds (*continued*)

Substance	ΔHm	ΔHv	ΔHs	C_p			
				400 K	600 K	800 K	1000 K
Butylcyclohexane	3.384		11.96	66.00	93.10	112.30	125.70
Butylcyclopentane	2.704	9.20	11.00	57.77	80.38	97.35	114.80
N-Butyldiacetimide		8.69	15.4				
Butyl dichloroacetate			12.5				
tert-Butyl hydroperoxide			11.41				
Butylisobutylamine			10.73				
Butyl lithium			25.6				
Butyl trichloroacetate			12.8				
1-Butyne	1.441		5.67	23.87	30.83	35.95	39.84
2-Butyne	2.207	5.861	6.38	22.62	29.68	35.14	39.29
Butyraldehyde	2.654	6.340	8.05	30.20	39.60	46.60	51.70
Butyric acid	2.50		15.2				
Butyronitrile	1.2	10.04	9.53	28.39	37.07	43.48	48.22
D-Camphor	1.635	8.13					
ε-Caprolactam		14.22	19.9				
Carbazole			20.2				
Carbon disulfide	1.049	6.401					
Chloroacetic acid			18				
Chloroacetyl chloride			9.3				
2-Chlorobenzaldehyde			13.3				
Chlorobenzene	2.28	8.73	9.81	30.62	41.16	47.89	52.48
2-Chlorobenzoic acid			19.0				
3-Chlorobenzoic acid			19.6				
4-Chlorobenzoic acid			21.0				
Chlorobenzoquinone			16.5				

1-Chlorobutane		7.38	8.0	32.30	42.77	50.31	55.92
2-Chlorobutane		6.98	7.60	32.52	43.18	50.84	56.60
Chlorocyclohexane			10.4				
Chlorodifluoromethane	0.985	4.833		15.63	18.87	20.84	22.10
Chloroethane	1.064	5.892		18.54	24.28	28.39	31.48
1-Chloro-2-ethylbenzene			11.3				
1-Chloro-4-ethylbenzene			11.5				
Chloroethylene				15.56	19.61	22.35	24.35
Chloroethyne				14.39	15.97	16.98	17.75
Chlorofluoromethane				13.29	16.57	18.81	20.39
Chloroform	2.28	7.08	7.48	17.75	20.38	21.87	22.83
Chloromethane	1.537	5.147		11.52	14.66	17.04	18.86
Chloromethyloxirane			9.7				
1-Chloro-2-methylpropane			7.57	32.52	43.18	50.84	56.60
2-Chloro-2-methylpropane	0.48	6.6		34.00	44.20	51.50	57.00
ΔHt, $0.41^{-90.1}$, $1.39^{-53.6}$							
1-Chloronaphthalene			15.6				
2-Chloronaphthalene			19.6				
1-Chloropentane		7.93	9.1	39.24	52.11	61.38	68.25
3-Chlorophenol			12.7				
4-Chlorophenol			12.4				
1-Chloropropane		6.62	6.9	25.36	33.43	39.24	43.59
2-Chloropropane		6.34	6.47	25.99	34.20	39.94	44.16
3-Chloro-1-propene				22.12	28.43	32.93	36.30
Chlorotrifluoromethane				18.53	21.60	23.17	24.03
Chlorotrinitromethane			10.86				
Chrysene			28.1				
o-Cresol		10.20	18.17	39.74	52.77	61.55	68.82
m-Cresol		10.32	14.75	38.74	52.26	61.27	68.50
p-Cresol		10.32	17.67	38.65	52.10	61.11	68.48
m-Cresyl acetate			14.51				

TABLE 5-2 Heats of melting and vaporization (or sublimation and specific heat at various temperatures of organic compounds (*continued*)

Substance	ΔHm	ΔHv	ΔHs	C_p			
				400 K	600 K	800 K	1000 K
Cubane			19.2				
4-Cyanothiazole			17.67				
Cyclobutane ΔHt, $1.38^{-126.79}$	0.260	5.781	5.65	23.89	34.76	42.42	47.96
Cyclobutene				21.59	30.30	36.26	40.53
Cyclododecane			18.26				
Cycloheptane ΔHt, $1.187^{-138.4}$, $0.069^{-75.0}$, $0.108^{-60.8}$	0.450	7.93	9.21	41.82	62.42	77.03	87.40
Cycloheptanone			12.4				
1,3,5-Cycloheptatriene ΔHt, $0.561^{-119.19}$	0.277	9.250		37.13	50.07	58.58	64.58
Cyclohexane ΔHt, 1.611^{-87}	0.640	7.160	7.896	35.82	53.83	66.76	75.80
Cyclohexanol ΔHt, $1.96^{-9.7}$	0.406	10.875	12.820	41.14	59.29	72.18	81.13
Cyclohexanone		9.00	10.77	36.00	52.90	65.00	73.00
Cyclohexene ΔHt, $1.016^{-134.4}$	0.787	7.285	8.00	34.64	49.45	59.49	66.62
Cyclooctane ΔHt, $1.507^{-106.7}$, $0.114^{-89.35}$	0.576	8.58	10.36	47.82	71.00	87.30	99.01
Cyclooctanone			13.0				
1,3,5,7-cyclooctatetraene	2.695	8.700	10.30	38.45	52.77	62.23	68.88
Cyclopentadiene			6.78				
Cyclopentane ΔHt, $1.167^{-150.76}$, $0.823^{-135.08}$	0.1455	6.524	6.818	28.38	42.57	52.60	59.84
Cyclopentanethiol	1.872	8.443	9.93	34.53	48.65	58.61	65.84
Cyclopentanol			13.74				
Cyclopentanone			10.21				
Cyclopentene ΔHt, $0.115^{-186.08}$	0.804		6.71	25.08	37.19	45.78	51.94
Cyclopropane	1.301	4.793		18.31	26.15	33.57	35.39

Decahydronaphthalene							
cis ΔHt, $0.511^{-57.1}$	2.268	9.940	12.0	56.64	84.14	103.36	116.91
trans	3.455	9.260	11.6	56.78	84.20	103.40	116.93
Decanal				71.80	95.70	113.00	125.70
Decane	6.863	9.388	12.277	71.24	96.36	114.92	128.20
1-Decanethiol	7.4	11.1	15.5	76.63	102.63	122.10	136.98
Decanoic acid	7.0		28.4				
1-Decanol	9.0	11.9	18.6	74.44	99.94	118.53	132.24
1-Decene ΔHt, $1.90^{-74.8}$	3.300	9.24	12.06	67.79	91.27	108.28	120.90
1-Decyne				65.64	86.96	102.42	113.90
Deoxybenzoin			22.3				
Dibenzilidene azine			22.3				
Dibenzyl ketone			21.3				
Dibenzyl sulfide			22.3				
Dibenzyl sulfone			27.8				
1,2-Dibromobutane			10.8	36.77	46.70	53.60	58.50
1,2-Dibromocycloheptane			12.43				
1,2-Dibromocyclohexane			12.07				
1,2-Dibromocyclooctane			13.04				
1,2-Dibromoethane	2.62	8.69	9.86	23.83	29.24	32.94	35.80
1,2-Dibromoheptane			13.01				
1,2-Dibromopropane				29.74	37.63	42.91	46.74
Dibutylborinic acid			15				
Dibutyl ether		8.83	10.5	60.78	81.29	96.52	107.86
Dibutyl mercury			15.6				
Di-*tert*-butyl peroxide			7.6				
Dibutyl *o*-phthalate			21.9				
Dibutyl sulfate			18.1				
Dibutyl sulfite			16.2				
Dibutyl sulfone			24.0				
Dichloroacetyl chloride			9.4				
1,2-Dichlorobenzene	3.19	9.7	11.56	34.12	44.07	50.28	54.42

TABLE 5-2 Heats of melting and vaporization (or sublimation and specific heat at various temperatures of organic compounds (*continued*)

Substance	ΔHm	ΔHv	ΔHs	C_p			
				400 K	600 K	800 K	1000 K
1,3-Dichlorobenzene			11.44	34.18	44.09	50.29	54.42
1,4-Dichlorobenzene	4.34	9.5	15.5	34.24	44.16	50.35	54.46
2,6-Dichlorobenzoquinone			16.7				
2,2′-Dichlorobiphenyl			23.0				
4,4′-Dichlorobiphenyl			24.8				
Dichlorodifluoromethane				19.69	22.37	23.69	24.39
1,1-Dichloroethane	1.881	6.97	7.36	21.85	27.18	30.79	33.40
1,2-Dichloroethane	2.112	7.65	8.47	22.00	26.90	30.40	33.00
1,1-Dichloroethylene	1.557	6.26	6.328	18.80	22.44	24.71	26.29
1,2-Dichloroethylene							
cis	1.72	7.08	7.43	18.41	22.23	24.60	26.23
trans	1.72	6.65	6.92	18.58	22.28	24.62	26.24
Dichlorofluoromethane				16.78	19.70	21.41	22.51
Dichloromethane	1.1	6.74	6.94	14.24	17.30	19.32	20.76
1,2-Dichloropropane		7.59	8.68	28.60	36.47	41.97	46.08
1,3-Dichloropropane		8.10	9.66	28.69	36.22	41.56	45.50
2,2-Dichloropropane		7.0	7.8	30.56	38.06	43.00	46.56
Dicyanoacetylene			6.88				
2,2-Diethoxypropane			7.61				
Diethylamine			7.6	34.88	47.14	56.16	62.91
1,2-Diethylbenzene	4.01	9.42	12.61	56.01	75.66	89.54	99.49
1,3-Diethylbenzene	2.62	9.41	12.55	55.01	75.19	89.31	99.37
1,4-Diethylbenzene	2.53	9.41	12.54	54.68	74.84	89.04	99.16
Diethylene glycol		12.50	13.7				
Diethyl ether	1.745	6.38	6.516	33.01	43.92	52.26	58.51
Diethylmercury			10.7				

Diethylmethyl phosphonate			13.5				
Diethylnitramine			12.7				
Diethyl oxalate		10.04	15.2				
Diethyl peroxide			7.3				
Diethyl *o*-phthalate			21.1				
Diethyl selenide			9.3				
Diethyl sulfate			13.6				
Diethyl sulfite			11.6				
Diethyl sulfone			20.6				
Diethyl sulfoxide			14.9				
1,2-Difluorobenzene	2.640	7.699	8.65	32.76	43.33	50.12	54.72
1,3-Difluorobenzene			8.29	32.72	43.13	49.67	53.93
1,4-Difluorobenzene			8.51	32.84	43.20	49.68	53.99
2,2′-Difluorobiphenyl			22.7				
4,4′-Difluorobiphenyl			21.8				
1,1-Difluoroethane		5.1		19.93	25.70	29.70	32.57
1,1-Difluoroethylene				17.16	21.32	23.95	25.74
Difluoromethane				12.22	15.72	18.22	19.98
9,10-Dihydroanthracene			22.3				
4*H*-Dihydropyran			7.7				
5,12-Dihydrotetracene			27.7				
2,3-Dihydrothiophene			9.02				
2,5-Dihydrothiophene			9.55				
1,2-Diiodobenzene			15.5				
1,2-Diiodoethane			15.7	22.94	27.92	31.37	33.84
Diiodomethane	3.02(I) 2.88(II)		12.2	15.74	18.37	20.06	21.29
Diisopropyl ether	2.635	6.95	7.75	46.90	62.61	74.39	83.17
Diisopropyl ketone			9.93				
Diisopropylmercury			12.8				
1,2-Dimethoxybenzene			16.0				
Dimethoxyborane			6.14				

TABLE 5-2 Heats of melting and vaporization (or sublimation and specific heat at various temperatures of organic compounds (*continued*)

Substance	ΔHm	ΔHv	ΔHs	C_p 400 K	C_p 600 K	C_p 800 K	C_p 1000 K
2,2-Dimethoxypropane			7.03				
Dimethylamine	1.420	6.330	6.07	20.89	28.41	33.94	38.19
Dimethylaminotrimethylsilane			7.6				
2,2-Dimethylbutane ΔHt, $1.289^{-147.34}$ $0.068^{-132.28}$	0.138	6.287	6.618	43.70	60.00	71.40	79.70
2,3-Dimethylbutane ΔHt, $1.552^{-137.08}$	0.194	6.519	6.96	43.30	59.20	75.20	79.10
2,3-Dimethyl-1-butene		6.55	6.97	42.60	55.40	65.00	72.20
2,3-Dimethyl-2-butene ΔHt, $0.844^{-76.34}$	1.542	7.083	7.776	37.48	51.78	62.78	71.14
3,3-Dimethyl-1-butene ΔHt, $1.037^{-148.3}$	0.261	6.13	6.36	38.90	53.40	63.60	71.00
Dimethylcadmium			9.07				
1,1-Dimethylcyclohexane ΔHt, $1.430^{-120.01}$	0.495	7.79	9.043	50.70	74.10	90.70	102.20
1,2-Dimethylcyclohexane							
cis ΔHt, $1.974^{-100.6}$	0.393	8.04	9.492	51.10	74.00	90.10	101.40
trans	2.491(I) 2.508(II)	7.86	9.168	51.90	74.60	90.50	101.70
1,3-Dimethylcyclohexane							
cis	2.586	7.84	9.137	51.20	74.20	90.50	102.00
trans	2.358	8.09	9.369	51.10	73.80	89.80	101.10
1,4-Dimethylcyclohexane							
cis	2.225	8.07	9.329	51.10	73.80	89.80	101.10
trans	2.947	7.79	9.053	51.60	74.60	90.60	101.90
1,1-Dimethylcyclopentane ΔHt, $1.551^{-126.36}$	0.258	7.239	8.079	43.55	62.78	76.18	85.83
1,2-Dimethylcyclopentane							
cis ΔHt, $1.594^{-131.66}$	0.396	7.576	8.549	43.67	62.72	75.98	85.57
trans	1.713	7.375	8.259	43.71	62.66	75.84	85.43

1,3-Dimethylcyclopentane							
cis	1.761	7.265	8.200	43.71	62.66	75.84	85.43
trans	1.738	7.361	8.248	43.71	62.66	75.84	85.43
Dimethyldichlorosilane			8.2				
cis-2,4-Dimethyl-1,3-dioxane			9.53				
4,5-Dimethyl-1,3-dioxane			10.16				
5,5-Dimethyl-1,3-dioxane			9.86				
Dimethyl ether	1.180	5.141		19.02	25.16	30.04	33.79
N,N-Dimethylformamide			11.4				
Dimethylfulvene			10.6				
Dimethylglyoxime			23.2				
2,2-Dimethylhexane	1.62	7.71	8.91				
2,3-Dimethylhexane		7.94	9.27				
2,4-Dimethylhexane		7.79	9.03				
2,5-Dimethylhexane	3.096	7.80	9.05				
3,3-Dimethylhexane	1.7	7.76	8.97				
3,4-Dimethylhexane		7.95	9.32				
2,2-Dimethyl-3-hexene							
cis			8.88				
trans			8.91				
1,1-Dimethylhydrazine			8.37				
1,2-Dimethylhydrazine			9.40				
Dimethylmercury			8.26				
Dimethylnitramine			16.7				
2,2-Dimethylpentane	1.392	6.97	7.75	50.42	68.33	81.43	91.20
2,3-Dimethylpentane		7.26	8.19	50.42	68.33	81.43	91.20
2,4-Dimethylpentane	1.636	7.05	7.86	50.42	68.33	81.43	91.20
3,3-Dimethylpentane	1.689	7.09	7.89	50.42	68.33	81.43	91.20
2,7-Dimethylphenanthrene			25.5				
4,5-Dimethylphenanthrene			25.0				
9,10-Dimethylphenanthrene			28.6				
2,2-Dimethylpropane Δt, $0.616^{-133.14}$	0.752	5.438	5.205	37.55	51.21	60.78	67.80

TABLE 5-2 Heats of melting and vaporization (or sublimation and specific heat at various temperatures of organic compounds (*continued*)

Substance	ΔHm	ΔHv	ΔHs	C_p			
				400 K	600 K	800 K	1000 K
2,3-Dimethylpyridine			11.70				
2,4-Dimethylpyridine			11.42				
2,5-Dimethylpyridine			11.43				
2,6-Dimethylpyridine			11.01				
3,4-Dimethylpyridine			12.38				
3,5-Dimethylpyridine			12.04				
Dimethyl sulfate			11.6				
Dimethyl sulfite			9.6				
Dimethyl sulfone			18.4				
Dimethyl sulfoxide	1.56	12.66	12.64				
3,3-Dimethyl-2-thiabutane	2.011(I) 1.83(II)	7.523	8.57				
2,2-Dimethylthiacyclopropane			8.55				
2,2-Dimethyl-3-thiapentane	1.69	8.00	9.4				
2,4-Dimethyl-3-thiapentane	2.49	8.04	9.44	50.64	66.22	77.12	85.24
1,3-Dinitrobenzene			14.3				
2,4-Dinitrophenol			25				
2,6-Dinitrophenol			26.8				
1,1-Dinitropropane			14.93				
1,4-Dioxane ΔHt, $0.562^{-0.3}$	3.07		9.20	30.23	43.44	52.15	58.05
1,3-Dioxolan			8.5				
Dipentene			11.5				
Diphenylamine			23.1				
Diphenylchlorosilane			16.6				
Diphenyl disulfide			22.7				
Diphenyl disulfone			38.7				

1,2-Diphenylethane		12.3	20.1				
1,1-Diphenylethene			17.5				
Diphenyl ether	4.115	15.5^{25}	19.6				
Diphenylfulvene			25				
Diphenylmercury			26.95				
Diphenylmethane			19.7				
Diphenyl sulfide			16.2				
Diphenyl sulfone			25.4				
Diphenyl sulfoxide			23.2				
Dipropyl ether			8.6	46.90	62.61	74.39	83.17
Dipropylmercury			13.2				
Dipropyl sulfate			16.0				
Dipropyl sulfite			14.0				
Dipropyl sulfone			19.1				
Dipropyl sulfoxide			17.8				
2,3-Dithiabutane	2.197	8.05	9.17	26.36	32.83	37.66	41.31
5,6-Dithiadecane		11.2	15.2	68.38	89.98	105.83	117.86
3,4-Dithiahexane	2.248	9.01	10.89	40.90	52.24	60.19	65.97
1,3-Dithian-2-thione			21.85				
4,5-Dithiaoctane	3.30	10.02	12.55	44.50	71.30	83.70	93.20
N,*N*-Dithiodiethylamine			12.6				
1,3-Diothiolan-2-thione			19.56				
Di-*p*-tolyl sulfone			26.2				
Divinyl ether			6.26				
Divinyl sulfone			13.5				
Dodecane	8.57	10.43	14.65	85.13	115.04	136.76	152.90
Dodecanedioic acid			36.6				
1-Dodecene ΔHt, $1.088^{-60.2}$	4.76	10.27	14.42	8.68	109.95	130.41	145.50
Eicosane	16.70	13.74	24.1	140.65	189.78	225.28	251.60
Eicosanoic acid	17.2		48				
1-Eicosene	8.2	13.35	23.86	137.20	184.69	218.93	244.20
meso-Erythritol			32.3				

TABLE 5-2 Heats of melting and vaporization (or sublimation and specific heat at various temperatures of organic compounds (*continued*)

Substance	ΔHm	ΔHv	ΔHs	C_p			
				400 K	600 K	800 K	1000 K
Ethane	0.683	3.517	1.200	15.65	21.35	25.81	29.30
1,2-Ethanedithiol			10.68				
Ethanethiol	1.189	6.401	6.526	21.08	27.21	31.83	35.38
Ethanol	1.198	9.255	10.11	19.36	25.69	30.33	33.83
Ethyl acetate	2.505	7.720	8.63	32.84	43.65	51.01	56.05
Ethyl allyl sulfone			20.0				
Ethylamine		6.7	6.7	21.65	28.68	33.89	37.88
N-Ethylaniline			12.5				
Ethylbenzene	2.195	8.50	10.10	40.76	56.44	67.15	74.77
3-Ethyl-1-butene		6.88	7.41	40.70	54.50	64.40	71.90
Ethyl crotonate			10.6				
Ethylcyclohexane	1.992	8.20	9.67	51.60	74.10	90.10	101.30
1-Ethylcyclohexene			10.34				
Ethylcyclopentane	1.642(I) 1.889(II)	7.715	8.72	43.89	61.70	75.22	85.16
Ethylene	0.801	3.237		12.67	17.87	20.03	22.43
Ethylene carbonate	2.41		17.5				
Ethylene glycol	2.78	11.86	15.68	27.06	32.72	36.90	39.88
Ethyleneimine		7.24	7.55	16.83	23.56	28.14	31.45
Ethylene oxide	1.236	6.101	5.96	14.95	20.62	24.60	27.47
Ethyl formate	2.20	7.201					
2-Ethyl-1-hexanal			11.70				
3-Ethylhexane		8.03	9.48				
Ethylisovalerate			10.5				
Ethyllithium			27.9				
Ethylmercury bromide			18.3				

Ethylmercury chloride			18.2				
Ethylmercury iodide			19.0				
Ethyl methyl ether				26.08	34.58	41.19	46.18
Ethyl nitrate	2.04	7.92	8.67	28.73	37.07	42.72	46.69
3-Ethylpentane	2.282	7.40	8.42	50.42	68.33	81.43	91.20
Ethyl pentanoate			11.0				
2-Ethylphenol			15.20				
3-Ethylphenol			16.30				
4-Ethylphenol			19.20				
Ethylphosphonic acid			12.1				
Ethyl propanoate		8.178	9.0				
Ethyl β-vinylacrylate			11.6				
Ethyl vinyl ether			6.35				
Ethynylbenzene				35.95	48.01	55.79	61.17
Fluoranthrene			24.65				
Fluorobenzene	2.702	7.457	8.27	29.99	40.86	47.83	52.58
4-Fluorobenzoic acid			21.8				
Fluoroethane				17.71	23.56	27.82	31.00
Fluoromethane				10.56	13.83	16.45	18.44
1-Fluoropropane				24.55	32.82	38.88	43.37
2-Fluoropropane				24.72	33.14	39.14	43.55
4-Fluorotoluene	2.235	8.144	9.42	36.43	49.70	58.60	64.84
Fluorotrinitromethane			8.3				
Formaldehyde		5.85		9.38	11.52	13.37	14.81
Formic acid	3.035	5.24	11.03	12.85	16.02	18.35	19.95
Formyl							
HCO				8.73	9.79	10.75	11.49
HCO^+				9.39	10.39	11.14	11.78
Fumaric acid			32.5				
Fumaronitrile			17.2				
Furan ΔHt, $0.489^{-123.2}$	0.909	6.474	6.61	21.20	29.31	34.41	37.89
Furfuryl alcohol	3.12		15.4				

TABLE 5-2 Heats of melting and vaporization (or sublimation and specific heat at various temperatures of organic compounds (*continued*)

Substance	ΔHm	ΔHv	ΔHs	C_p			
				400 K	600 K	800 K	1000 K
2-Furoic acid			25.92				
Furylethylene			9.1				
Glycerol	4.416		20.5				
Glyceryl triacetate			19.6				
Glyceryl trinitrate			23.9				
Heptadecane ΔHt, $2.62^{11.1}$	9.67	12.64	20.6	119.83	161.75	192.08	214.60
Heptadecanoic acid	12.3						
1-Heptadecene	7.5	12.39	20.32	116.38	156.66	185.74	207.20
1-Heptanal	5.637		11.40	51.00	67.70	79.80	88.70
Heptane	3.359	7.575	8.74	50.42	68.33	81.43	91.20
1-Heptanethiol	6.067	9.5	12.06	55.81	74.60	88.91	99.98
Heptanoic acid			18.0				
1-Heptanol	3.16	11.5	16.5	53.62	71.92	85.32	95.25
1-Heptene ΔHt, $0.07^{1.36}$	2.964(I) 3.021(II)	7.43	8.52	46.97	63.24	75.09	83.90
Hexachlorobenzene	6.1		23.2	48.08	55.78	59.96	62.34
Hexachloroethane ΔHt, $1.9^{71.3}$	2.33	12.2	16.5	36.21	39.82	41.48	42.38
Hexadecafluoroethylcyclohexane			9.20				
Hexadecafluoroheptane			8.7				
Hexadecane	12.39	12.24	19.38	112.89	152.41	181.02	202.20
Hexadecanoic acid	12.8		36.9				
1-Hexadecanol ΔHt, $4.8^{44.0}$, $5.7^{49.1}$	7.8		40.5	116.09	156.00	184.90	206.30
1-Hexadecene	7.216	12.05	19.14	109.44	147.32	174.67	194.80
Hexafluorobenzene	2.770	7.571	8.61	43.88	52.55	57.62	60.63
Hexafluoroethane ΔHt, $0.893^{-169.17}$	0.642	3.860		30.01	35.60	38.40	39.87

Hexahydroindane							
cis			11.0				
trans			10.7				
Hexamethylbenzene ΔHt, $0.269^{-156.67}$, $0.422^{110.7}$	4.93		17.9	74.18	97.13	113.51	125.55
Hexmethyldisiloxane			8.9				
Hexanal				44.00	58.30	68.70	76.40
Hexanamide			22.72				
Hexane	3.126	6.896	7.54	43.47	58.99	70.36	78.89
1-Hexanethiol	4.305	8.9	11.14	48.87	65.26	77.84	87.65
Hexanoic acid	6.98	15.45	17.3				
1-Hexanol	3.68	11.6	14.8	46.68	62.58	74.25	82.92
1-Hexene	2.234	6.76	7.32	40.03	53.90	64.02	71.54
2-Hexene							
cis		6.96	7.52	38.60	53.00	63.40	71.20
trans		6.91	7.54	39.70	53.40	63.60	71.20
3-Hexene							
cis		6.86	7.47	38.50	53.20	63.50	71.20
trans		6.92	7.54	40.20	53.90	63.90	71.40
1-Hexyne				37.87	49.59	58.16	64.56
Hydroquinone			23.7				
8-Hydroxyquinoline			26.0				
Indane			11.8				
Indene			12.64				
Indole			16.7				
Iodobenzene	2.33	9.44	11.85	31.10	41.43	48.07	52.60
4-Iodobenzoic acid			21.0				
Iodocyclohexane			11.3				
Iodoethane		7.115	7.7	19.18	24.64	28.65	31.65
Iodomethane		6.52	6.63	12.33	15.28	17.47	19.17
2-Iodo-2-methylpropane	3.47		8.46	35.27	45.82	52.85	57.91
1-Iodonaphthalene			17.3				

TABLE 5-2 Heats of melting and vaporization (or sublimation and specific heat at various temperatures of organic compounds (*continued*)

				C_p			
Substance	ΔHm	ΔHv	ΔHs	400 K	600 K	800 K	1000 K
2-Iodonaphthalene			21.7				
1-Iodopropane			8.6	26.27	34.11	39.80	44.03
2-Iodopropane			8.14	26.59	34.58	40.21	44.34
3-Iodopropene			9.1				
Iodotoluene, 3- and 4-			13.0				
Isobutylbenzene			11.54				
Isobutyl dichloroacetate			12.5				
Isobutyl phenyl ketone			14.22				
Isobutyl trichloroacetate			12.7				
Isobutyronitrile	7.754	8.99	28.56	37.39	43.74	48.40	
Isopropyl acetate			8.89				
Isopropylbenezene	1.86	8.97	10.79	48.00	66.20	78.60	87.30
Isopropyl nitrate		8.35	9.27	35.96	46.81	54.13	59.26
Isopropyl trichloroacetate			12.4				
Isothiocyanic acid				12.71	14.57	15.74	16.57
Ketene			4.18	14.22	16.89	18.80	20.25
Lauric acid	8.8		31.7				
Leucine			36.0				
(+)-Limonene			11.5				
Maleic acid			26.3				
Maleic anhydride			17.1				
Malononitrile			18.9				
D-Mannitol	5.39						
Melamine			29.7				

2,2-Metacyclophane			22.0				
Methane ΔHt, $0.0187^{-248 \text{ to } -252.7}$	0.225	1.953		9.71	12.55	15.18	17.40
Methanethiol ΔHt, $0.0525^{-135.6}$	1.411	5.872	5.7	14.04	17.57	20.32	22.48
Methanol ΔHt, $0.152^{-115.8}$	0.768	8.24	8.94	12.29	16.02	19.04	21.38
4-Methoxybenzaldehyde			15.42				
Methoxybenzene			11.18				
2-Methoxytetrahydropyran			10.2				
Methyl (CH_3)				10.05	11.54	12.89	14.09
Methyl allyl sulfone			19.0				
Methylamine	1.466	6.169	5.80	14.38	18.86	22.44	25.26
Methyl benzyl sulfone			23.7				
2-Methyl-1,3-butadiene	1.155	6.191	6.32	31.80	41.40	48.00	52.90
3-Methyl-1,2-butadiene		6.51	6.68	31.00	40.30	47.20	52.40
2-Methylbutane	1.231	5.901	5.94	36.49	49.89	59.71	67.12
2-Methyl-1-butanethiol		8.0					
3-Methyl-1-butanethiol	1.78						
2-Methyl-2-butanethiol ΔHt, $1.907^{-114.0}$	0.1454	7.50	8.51	42.79	56.58	66.28	73.30
3-Methylbutanoic acid	1.750	10.32					
2-Methyl-1-butanol		10.5	12.9				
3-Methyl-1-butanol		10.54	13.0				
2-Methyl-2-butanol ΔHt, $0.47^{-127.2}$	1.06	9.6	11.9				
3-Methyl-2-butanol		9.9	12.4				
2-Methyl-1-butene	1.891	6.094	6.181	33.20	44.72	53.15	59.43
3-Methyl-1-butene	1.281	5.750	5.70	35.26	45.90	53.85	59.83
2-Methyl-2-butene	1.816	6.287	6.468	31.93	43.42	52.05	58.55
Methyl butyl sulfone			18.2				
Methyl *tert*-butyl sulfone			19.7				
3-Methyl-1-butyne		6.25	6.16	31.10	40.60	47.40	52.40
Methyl crotonate			9.8				
Methylcyclohexane	1.614	7.44	8.45	44.35	64.46	78.74	88.79
2-Methylcyclohexanol, *cis*- and *trans*-			15.1				

TABLE 5-2 Heats of melting and vaporization (or sublimation and specific heat at various temperatures of organic compounds (*continued*)

Substance	ΔHm	ΔHv	ΔHs	C_p			
				400 K	600 K	800 K	1000 K
3-Methylcyclohexanol							
cis			15.6				
trans			15.7				
4-Methylcyclohexanol							
cis			15.7				
trans			15.8				
Methylcyclopentane	1.656	6.95	7.55	36.11	52.43	64.00	72.44
1-Methylcyclopentene			7.55	32.50	46.80	57.00	64.30
3-Methylcyclopentene			7.7	32.60	47.10	57.20	64.50
4-Methylcyclopentene			7.7	32.50	47.00	57.10	64.40
Methyldichlorosilane			6.7				
2-Methyl-1,3-dioxane			9.23				
4-Methyl-1,3-dioxane			9.36				
Methylene (CH_2)				8.64	9.37	10.14	10.89
1-Methyl-2-ethylbenzene	2.38(I) 2.28(II)	9.29	11.40	48.50	65.80	78.10	86.90
1-Methyl-3-ethylbenzene	1.82(I) 1.79(II)	9.21	11.21	47.50	65.40	77.80	86.80
1-Methyl-4-ethylbenzene	3.19	9.18	11.14	47.20	65.00	77.60	86.60
2-Methyl-3-ethylpentane	2.71	7.88	9.20				
3-Methyl-3-ethylpentane	2.59	7.84	9.08				
2-Methyl-3-ethyl-1-pentene			8.98				
Methyl ethyl sulfite			10.4				
Methyl ethyl sulfone			18.6				
Methyl formate	1.800	6.75		19.50	25.20	29.10	32.00
Methylglyoxal			9.1				
2-Methylheptane	2.839	8.08	9.48				

3-Methylheptane	2.779	8.10	9.52				
4-Methylheptane	2.59	8.10	9.48				
Methyl heptanoate			12.0				
2-Methylhexane	2.195	7.33	8.32	50.42	68.33	81.43	91.20
3-Methylhexane		7.36	8.39	50.42	68.33	81.43	91.20
Methyl hexanoate			11.1				
Methylhydrazine			9.65				
Methylidyne							
CH				6.98	7.11	7.40	7.78
CH^+				6.98	7.10	7.36	7.65
1-Methyl-2-isopropylbenzene	2.39	9.17	12.10				
1-Methyl-3-isopropylbenzene	3.27	9.11	11.94				
1-Methyl-4-isopropylbenzene	2.31	9.12	12.02				
Methyl isopropyl ether			6.27	32.97	44.17	52.67	59.08
Methyl isopropyl ketone			8.82				
Methyl isopropyl sulfone			16.8				
3-Methylisoxazole			9.8				
5-Methylisoxazole			10.0				
Methylmercury bromide			16.2				
Methylmercury chloride			15.5				
Methylmercury iodide			15.6				
1-Methylnaphthalene ΔHt, $1.190^{-32.37}$	1.160	11.0		50.74	69.79	82.48	91.21
2-Methylnaphthalene ΔHt, $1.34^{15.4}$	2.808	11.0		50.50	69.31	82.03	90.86
Methyl nitrate	1.97	7.54	8.1	21.87	27.54	31.47	34.19
Methyl nitrite		5.0	5.4	18.24	23.35	26.97	29.52
2-Methylpentane	1.498	6.643	7.138	44.00	59.60	70.80	79.20
3-Methylpentane		6.711	7.236	43.47	59.00	70.40	78.90
Methyl pentanoate			10.2				
2-Methyl-1-pentene		6.71	7.29	40.80	54.40	64.40	71.80
3-Methyl-1-pentene		6.43	6.83	42.50	55.60	65.20	72.30
4-Methyl-1-pentene		6.47	6.86	38.90	52.90	63.10	70.70
2-Methyl-2-pentene		6.93	7.55	39.00	53.20	58.60	71.10

TABLE 5-2 Heats of melting and vaporization (or sublimation and specific heat at various temperatures of organic compounds (*continued*)

Substance	ΔHm	ΔHv	ΔHs	C_p			
				400 K	600 K	800 K	1000 K
3-Methyl-2-pentene							
cis		6.89	7.49	39.00	53.20	63.40	71.10
trans		7.00	7.67	39.00	53.20	63.40	71.10
4-Methyl-2-pentene							
cis		6.59	7.04	40.05	54.10	64.00	71.50
trans		6.68	7.16	41.90	54.80	64.50	71.80
Methyl phenyl sulfone			22.0				
Methylphosphonic acid			11.5				
2-Methylpropanal			7.5				
2-Methylpropane	1.085	5.089	4.57	29.77	40.62	48.49	54.40
2-Methyl-1-propanethiol	1.191	7.412	8.28	35.31	46.26	53.77	59.17
2-Methyl-2-propanethiol							
$\Delta Ht, 0.972^{-121.6}, 0.155^{-116.2}, 0.232^{-73.8}$	0.593	6.80	7.36	36.13	47.60	55.53	61.24
2-Methyl-1-propanol		9.80	12.04				
2-Methyl-2-propanol ΔHt, $0.20^{12.99}$	1.602	9.33	12.73	34.16	45.37	53.28	59.16
2-Methylpropene	1.418	5.286	4.92	26.57	35.30	41.86	46.85
Methyl propyl ether			6.6	33.01	43.92	52.26	58.51
2-Methylpyridine	2.324	8.654	10.15	31.92	44.55	53.21	59.34
3-Methylpyridine	3.389	8.932	10.62	31.82	44.47	53.12	59.23
α-Methylstyrene				44.80	60.70	71.80	79.80
β-Methylstyrene							
cis				44.80	60.70	71.80	79.80
trans				45.20	61.20	72.20	80.00
3-Methyl-2-thiabutane	2.236	7.338	8.15	34.69	46.01	54.95	62.29
2-Methylthiacyclopentane		8.7	10.1				
2-Methyl-3-thiapentane	2.08		9.2				

4-Methylthiazole			10.48				
2-Methylthiophene	2.263	8.103	9.26	29.43	39.57	46.43	51.30
3-Methylthiophene	2.518	8.186	9.44	29.38	39.34	45.95	50.59
Naphthalene	4.536	10.34	17.6	42.83	59.67	70.77	78.38
1-Naphthol			21.9				
2-Naphthol			19.8				
1,4-Naphthoquinone			17.3				
1-Naphthylamine			21.5				
2-Naphthylamine			21.1				
p-Nitroaniline	5.04		26				
Nitrobenzene	2.78	9.744					
1-Nitrobutane		9.3	11.6	37.65	50.21	59.03	65.39
2-Nitrobutane		8.8	10.48	37.61	50.46	59.44	65.96
Nitroethane		8.4	9.9	23.66	31.45	36.81	40.67
Nitromethane	2.319	8.12	9.17	16.80	21.92	25.56	28.17
1-Nitronaphthalene			25.6				
1-Nitropropane		8.8	10.37	30.72	40.87	47.96	53.06
2-Nitropropane		8.4	9.88	30.89	41.19	48.22	53.24
Nonadecane ΔHt, $3.30^{22.8}$	10.95	13.39	22.9	133.71	180.43	214.21	239.20
1-Nonadecene	8.0	13.06	22.68	130.26	175.35	207.86	231.80
1-Nonanal			17.28	64.80	86.40	101.90	113.40
Nonane ΔHt, $1.50^{-55.97}$	3.72	8.82	11.10	64.30	87.01	103.56	115.90
1-Nonanethiol	8.0	10.6		69.69	93.28	111.04	124.65
Nonanoic acid			19.7				
1-Nonanol		13.0	18.6	67.50	90.60	107.46	119.91
1-Nonene	4.3	8.68	10.88	60.85	81.93	97.22	108.50
Octadecane	14.81	13.02	21.7	126.77	171.09	203.15	226.90
Octadecanoic acid	15.1		39.8				
1-Octadecene	7.8	12.74	21.50	123.32	166.00	196.80	219.50
Octafluorocyclobutane	0.662	5.58		44.50	53.85	58.65	61.50
1-Octanal			16.28	57.90	77.00	90.90	101.00
Octanamide			26.4				

TABLE 5-2 Heats of melting and vaporization (or sublimation and specific heat at various temperatures of organic compounds (*continued*)

Substance	ΔHm	ΔHv	ΔHs	C_p			
				400 K	600 K	800 K	1000 K
Octane	4.957	8.225	9.916	57.35	77.67	92.50	103.60
1-Octanethiol	5.8	10.1		62.75	83.94	99.97	112.31
Octanoic acid	3.30	16.73	19.2				
1-Octanol	10.1	11.2	15.6	60.56	81.26	96.39	107.58
1-Octene	3.660	8.07	9.70	53.91	72.58	86.15	96.20
1-Octyne				51.75	68.28	80.30	89.20
Oxalic acid ΔHt, 0.3($\alpha \rightarrow \beta$)			23.4				
Oxalyl chloride			7.6				
Oxamide			26.8				
Palmitic acid	10.30		37				
[1.8]-Paracyclophane			26.5				
[2.2]-Paracyclophane			23.0				
[6.6]-Paracyclophane			27.5				
Paraldehyde			9.9				
Pentachloroethane	2.7	8.9	10.9	31.96	36.35	38.71	40.17
Pentachlorofluoroethane	0.449						
Pentachlorophenol			16.1				
Pentadecane ΔHt, $2.19^{-2.25}$	8.31	11.82	18.20	105.95	143.07	169.95	189.90
1-Pentadecene	6.9	11.63	17.96	102.50	137.98	163.60	182.50
1,2-Pentadiene		6.59	6.85	31.40	40.80	47.70	52.80
1,3-Pentadiene							
cis		6.60	6.77	29.50	39.90	47.00	52.20
trans		6.46	6.64	31.20	40.90	47.70	52.60
1,4-Pentadiene	1.468	6.01	6.01	31.30	40.80	47.60	52.70
2,3-Pentadiene		6.75	7.05	29.90	39.40	46.60	52.00
Pentaerythritol			34.4				

Pentaerythritol tetranitrate			36.3				
Pentafluorobenzoic acid			21.9				
Pentafluoroethane				27.20	32.94	36.12	37.98
Pentafluorophenol			16.1				
Pentamethylbenzene ΔH_t, 0.473[23.7]	2.95			65.00	86.08	101.29	112.33
1-Pentanal				37.10	49.00	57.70	64.00
Pentanamide			21.34				
Pentan-2,4-dione			10.82				
Pentane	2.008	6.16	6.32	36.53	49.64	59.30	66.55
Pentan-1,5-dithiol			14.17				
Pentanenitrile	1.130	7.98					
1-Pentanethiol	4.19	8.34	9.83	41.93	55.92	66.78	75.32
Pentanoic acid	3.850	10.53	16.6				
1-Pentanol	2.34	10.6	13.61	39.74	53.24	63.18	70.59
2-Pentanol		10.3	12.7				
3-Pentanol		10.1	12.8				
2-Pentanone		7.98	9.89	36.42	48.32	57.13	63.61
1-Pentene	1.388	6.02	6.09	33.10	44.56	52.95	59.21
2-Pentene							
cis	1.700	6.24	6.41	31.57	43.62	52.29	58.78
trans	1.996	6.23	6.38	32.67	44.02	52.45	58.81
1-Pentyne		6.63	6.79	31.10	40.40	47.10	52.20
2-Pentyne		6.99	7.35	29.20	38.70	45.90	51.40
Perylene			30.0				
α-Phellandrene			12.1				
9,10-Phenanthraquinone			21.9				
Phenanthrene		13.3	21.1				
Phenol	2.752	9.73	16.41	32.45	43.54	50.62	55.49
Phenyl acetate			13.0				
β-Phenyl-1-alanine, DL- and L-			36.8				
Phenyl benzoate			23.0				
N-Phenyldiacetimide			21.5				

TABLE 5-2 Heats of melting and vaporization (or sublimation and specific heat at various temperatures of organic compounds (*continued*)

Substance	ΔHm	ΔHv	ΔHs	C_p			
				400 K	600 K	800 K	1000 K
Phenyl ethyl sulfide			13.2				
Phenylhydrazine			14.69				
1-Phenyl-2-methylpropane	2.99	9.04	11.82				
Phenyl methyl sulfide			12.1				
Phenyl salicylate			22.0				
Phosgene							
I	1.372	5.832		15.28	16.98	17.92	18.49
II	1.335						
III	1.131						
m-Phthalic acid			25.5				
p-Phthalic acid			23.5				
Phthalic anhydride			21.19				
α-Pinene			10.7				
β-Pinene			11.1				
Propadiene		4.45		17.21	22.00	25.42	28.00
1-Propanal			7.09	23.09	30.22	35.45	39.27
Propane	0.842	4.487	3.605	22.47	30.76	36.99	41.73
Propane-2,3-dithiol			11.87				
1-Propanethiol ΔHt, $0.949^{-131.06}$	1.309	7.059	7.62	27.86	36.72	43.60	49.01
2-Propanethiol ΔHt, $0.013^{-160.6}$	1.371	6.670	7.039	28.35	37.02	43.26	47.92
1-Propanol	1.242	9.982	11.36	25.86	34.56	41.04	45.93
2-Propanol	1.293	9.510	10.85	26.78	35.76	42.13	46.82
2-Propen-1-ol			11.3	22.81	30.11	35.28	39.06
Propionic acid	1.800	7.716	13.7				
Propionic anhydride			12.6				
Propionitrile ΔHt, $0.408^{-96.19}$	1.202	7.353	8.632	21.18	27.42	32.14	35.70
1-Propylamine			7.46	28.51	37.99	44.94	50.21

Propylbenzene							
I	2.215	9.14	11.05	47.82	65.86	78.30	87.16
II	2.03						
Propyl carbamate			19.4				
Propyl chloroacetate			11.6				
Propylcyclohexane	2.479	8.62	10.78	59.10	83.80	101.20	113.40
Propylcyclopentane	2.398	8.15	9.82	50.83	71.04	86.28	97.50
Propylene	0.718	4.40		19.23	25.81	30.77	34.52
Propylene oxide	1.561	6.87	6.67	22.16	30.07	35.68	39.79
Propyl nitrate		8.58	9.70	35.79	46.49	53.87	59.08
Propyl phenyl ketone			14.51				
Propyl trichloroacetate			12.7				
Propyne		5.29		17.33	21.80	25.14	27.71
Pyrazine			13.45				
Pyrene			22.5				
Pyridazine			12.78				
Pyridine	1.979	8.39	9.61	25.42	35.72	42.49	47.17
Pyrimidine			11.95				
Pyrrole			10.80				
Pyrrolidine ΔHt, $0.129^{-66.01}$	2.050	7.89	8.98	27.33	40.31	49.35	55.84
Salicyclic acid			22.74				
Sebacic acid			38.4				
5,5′-Spirobis(1,3-dioxane)			17.4				
Spiropentane	1.538	6.39	6.58	28.55	40.10	47.91	53.51
cis-Stilbene			16.5				
Styrene	2.617	8.85	10.50	38.32	52.14	61.40	67.92
Suberic acid			34.2				
Succinic acid			28.1				
Tetrabromomethane				23.20	24.51	25.51	25.32
Tetracene			30				
Tetrachlorobenzoquinone			23.6				
1,1,1,2-Tetrachloroethane				28.36	33.28	36.24	38.17

TABLE 5-2 Heats of melting and vaporization (or sublimation and specific heat at various temperatures of organic compounds (*continued*)

Substance	ΔHm	ΔHv	ΔHs	C_p 400 K	600 K	800 K	1000 K
1,1,2,2-Tetrachloroethane		9.24	10.7	27.90	32.91	35.85	37.76
Tetrachloroethylene	2.5	8.3	9.4	25.10	27.86	29.29	30.07
Tetrachloromethane ΔHt, $1.095^{-47.9}$	0.601	7.16	7.79	21.92	23.82	24.64	25.05
Tetracyanoethylene			19.4				
Tetradecane	10.90	11.38	17.01	99.01	133.72	158.89	177.60
Tetradecanoic acid			33.4				
1-Tetradecene	6.6	11.21	16.78	95.56	128.64	152.54	170.20
Tetraethylene glycol			24				
Tetraethyllead			13.6				
1,1,1,2-Tetrafluoroethane				24.90	30.76	34.20	36.36
Tetrafluoroethylene	1.844	4.02		21.97	25.53	27.61	28.86
Tetrafluoromethane ΔHt, $0.353^{-196.92}$	0.167			17.30	20.74	22.58	23.61
Tetrahydrofuran			7.65				
Tetrahydrofurfuryl alcohol			15.9				
1,2,3,4-Tetrahydronaphthalene			13.4				
Tetrahydropyran			8.35				
Tetraiodomethane				24.00	24.94	25.31	25.49
1,2,3,4-Tetramethylbenzene	2.684	10.76	13.66	56.81	75.68	89.42	99.47
1,2,3,5-Tetramethylbenzene	2.561	10.47	13.34	55.76	74.81	88.79	99.01
1,2,4,5-Tetramethylbenzene	5.02	10.88	18	55.50	74.38	88.41	98.71
2,2,3,3-Tetramethylbutane ΔHt, $0.478^{-120.66}$	1.802	7.51	10.24				
Tetramethyllead			9.1				
Tetranitromethane			10.3				
Tetrazole			23				
2-Thiabutane	2.333	7.06	7.61	27.81	36.41	42.93	47.94
Thiacyclobutane ΔHt, $0.160^{-96.45}$	1.971	7.7	8.56	21.89	30.45	36.40	40.67

Thiacycloheptane			11.30	42.0	65.0	79.0	88.0
Thiacyclohexane ΔHt, $0.262^{-71.75}$, $1.858^{-33.14}$	0.585	8.60	10.22	35.71	52.37	64.00	72.34
Thiacyclopentane	1.757	8.28	9.28	28.95	40.04	47.66	53.14
Thiacyclopropane		6.98	7.24	16.53	21.99	25.61	28.21
4-Thia-5,5′-dimethyl-1-hexene			10.6				
2-Thiaheptane			10.88	48.67	65.02	77.59	87.41
3-Thiaheptane	2.96	8.78	10.74	48.37	64.96	77.74	87.75
4-Thiaheptane	2.90	8.76	10.64	48.21	65.13	78.45	89.05
2-Thiahexane	2.976	8.2	9.8	41.73	55.68	66.53	75.08
3-Thiahexane	2.529	8.3	9.58	41.43	55.62	66.68	75.42
5-Thianonane	4.64		12.75	62.09	83.81	100.58	113.71
2-Thiapentane	2.369	7.67	8.65	34.64	45.86	54.45	61.14
3-Thiapentane	2.845	7.59	8.55	34.65	46.11	54.91	61.79
2-Thiapropane	1.908	6.45	6.61	21.12	27.01	31.58	35.17
6-Thiaundecane			14.7				
Thioacetic acid				22.25	26.72	30.41	32.62
Thiophene ΔHt, $0.152^{-101.6}$	1.216	7.52	8.27	23.02	30.95	36.01	39.54
Thymol			21.8				
Toluene	1.586	7.93	9.08	33.48	47.20	56.61	63.32
2-Toluenethiol			12.3				
2,4,6-Triamino-1,3,5-triazine			29.5				
Tribromomethane				18.80	21.03	22.29	23.12
Tributyl phosphate			17.2				
Trichloroacetyl chloride			9.8				
Trichlorobenzoquinone			21.2				
1,1,1-Trichloroethane ΔHt, $1.79^{-48.95}$	0.45	7.96	7.76	25.72	30.68	33.73	35.81
1,1,2-Trichloroethane	2.7	8.3	9.4	25.03	30.13	33.28	35.42
Trichloroethylene		7.52	8.2	21.80	25.06	26.94	28.15
Trichlorofluoromethane				20.84	23.13	24.19	24.74
Trichloromethyl (CCl_3)				16.66	18.16	18.83	19.18
1,2,3-Trichloropropane		8.87	11.22	31.71	38.87	43.79	47.34

TABLE 5-2 Heats of melting and vaporization (or sublimation and specific heat at various temperatures of organic compounds (*continued*)

Substance	ΔHm	ΔHv	ΔHs	C_p			
				400 K	600 K	800 K	1000 K
Tricyanoethylene			19.4				
Tridecane ΔHt, $1.831^{-18.2}$	6.81	10.91	15.83	92.07	124.38	147.82	165.20
Tridecanoic acid	8.2		35.0				
1-Tridecene	6.2	10.75	15.60	88.62	119.29	141.48	157.80
Trimethylaluminum			17.5				
Triethylamine			8.29	48.70	66.10	78.56	87.80
Triethylaminoborane			14.5				
Triethyl arsenite			12.1				
Triethylarsine			10.3				
Triethyl borate			10.5				
Triethylenediamine ΔHt, $2.30^{79.8}$	1.45		14.8				
Triethylene glycol		17.07	18.9				
Triethyl phosphate			13.7				
Triethylphosphine			9.5				
Triethyl phosphite			10.0				
Triethylstibine			10.4				
1,1,1-Trifluoroethane	1.480	4.58		22.75	28.38	31.98	34.44
Trifluoroethylene				19.39	23.30	25.69	27.23
Trifluoromethane	0.970	3.99		14.61	18.16	20.35	21.76
Trifluoromethyl							
CF_2				13.74	16.17	17.50	18.25
CF_3^+				13.62	16.00	17.35	18.13
Trifluoromethylbenzene	3.29	7.80	8.98	40.59	54.20	62.75	68.45
Triodomethane	3.9		16.7	19.60	21.52	22.64	23.38
2,4,5-Trimethylacetophenone			15.1				
2,4,6-Trimethylacetophenone			14.9				
Trimethylaluminum			15.1				

Trimethylamine	1.564	5.48	5.26	28.08	38.34	45.62	50.98
Trimethyl arsenite			10.1				
Trimethylarsine			6.9				
1,2,3-Trimethylbenzene	1.955	9.57	11.73	46.90	64.00	76.70	85.90
ΔHt, $0.157^{-54.46}$, $0.319^{-42.89}$							
1,2,4-Trimethylbenzene, I	3.153	9.38	11.46	46.96	64.29	76.93	86.10
1,3,5-Trimethylbenzene							
I	2.274	9.33	11.35	46.41	64.08	76.84	86.07
II	1.932						
III	1.892						
Trimethyl borate			8.3				
Trimethylboron			4.83				
2,2,3-Trimethylbutane ΔHt, $0.586^{-157.8}$	0.540	6.92	7.65	50.83	69.61	82.73	92.32
Trimethylchlorosilane			7.2				
cis,cis-1,3,5-Trimethylcyclohexane				58.05	83.94	102.20	115.21
2,2,3-Trimethylpentane	2.06	7.65	8.82				
2,2,4-Trimethylpentane	2.20	7.41	8.40				
2,3,3-Trimethylpentane	0.205	7.73	8.90				
ΔHt, $1.850^{-109.01}$							
2,3,4-Trimethylpentane	2.215	7.82	9.01				
2,4,4-Trimethyl-1-pentene		7.5	8.5				
2,4,4-Trimethyl-2-pentene		7.8	8.9				
Trimethylphosphine			6.7				
Trimethylphosphine oxide			12.0				
Trimethyl phosphite			8.8				
Trimethylsilanol			10.9				
Trimethylstibine			7.5				
Trimethylsuccinic anhydride			17.7				
Trimethylthiacyclopropane			9.40				
2,4,6-Trinitroanisole			31.8				
1,3,5-Trinitrobenzene			23.8				
Trinitromethane			11.15				

TABLE 5-2 Heats of melting and vaporization (or sublimation and specific heat at various temperatures of organic compounds (*continued*)

Substance	ΔHm	ΔHv	ΔHs	C_p			
				400 K	600 K	800 K	1000 K
2,4,6-Trinitrophenetole			28.8				
2,4,6-Trinitrotoluene			28.3				
Triphenylarsine			23.5				
Triphenylene			28.2				
Triphenylmethane			23.9				
Triphenylphosphine			23				
Tropolone			20.0				
Undecane ΔHt, $1.64^{-36.55}$	5.28	9.92	13.47	78.18	105.80	125.69	140.60
Undecanoic acid	6.2		29.0				
1-Undecene ΔHt, $2.202^{-55.8}$	4.06	9.77	13.24	74.74	100.61	119.34	133.20
Urea			21.0				
o-Xylene	3.25	8.80	10.38	41.03	55.98	66.64	74.35
m-Xylene	2.765	8.69	10.20	40.03	55.51	66.41	74.23
p-Xylene	4.09	8.60	10.13	39.70	55.16	66.14	74.02
2,3-Xylenol			20.1				
2,4-Xylenol			15.74				
2,5-Xylenol			20.31				
2,6-Xylenol			18.07				
3,4-Xylenol			20.49				
3,5-Xylenol			19.80				

CRITICAL PHENOMENA

The *critical temperature* T_c of a gas is the temperature above which the gas cannot be liquefied no matter how high the pressure.

The *critical pressure* P_c is the lowest pressure which will liquefy the gas at its critical temperature.

The *critical molar volume* V_c is the volume of 1 mol at the critical temperature and the critical pressure. It can be computed from the critical density ρ_c as follows:

$$\frac{\text{Molecular weight in g} \cdot \text{mol}^{-1}}{\rho_c \text{ in g} \cdot \text{cm}^{-3}} = V_c \text{ in cm}^3 \cdot \text{mol}^{-1}$$

The critical pressure, critical molar volume, and critical temperature are the values of the pressure, molar volume, and thermodynamic temperature at which the densities of coexisting liquid and gaseous phases just become identical. At this critical point the *critical compressibility factor* Z_c is

$$Z_c = \frac{P_c V_c}{RT_c}$$

Since pressure, volume, and temperature are related to the corresponding critical properties, the function connecting the reduced properties becomes the same for each substance. The reduced property is expressed as a fraction of the critical property.

$$P_r = \frac{P}{P_c} \qquad V_r = \frac{V}{V_c} \qquad T_r = \frac{T}{T_c}$$

TABLE 5.3 Critical properties

Substance	T_c, K	P_c, atm	V_c, cm$^3 \cdot$ mol^{-1}
Acetaldehyde	461	55	154
Acetic acid	594.4	57.1	171.3
Acetic anydride	569	46.2	290
Acetone	508.1	46.4	209
Acetonitrile	548	47.7	173
Acetophenone	701	38	376
Acetyl chloride	508	58	204
Acetylene	308.3	60.6	113
Acrylic acid	615	56	210
Acrylonitrile	536	45	210
Air	132.5	37.2	92.7
Allene	393		
Allyl alcohol	545	56.4	203
Allyl sulfide	653		
Aluminum trichloride	629	26	261
Aminoethanol	614	44	196
Ammonia	405.6	111.3	72.5
Aniline	699	52.4	270
Anisole	368	41.2	
Anthracene	883		
Antimony tribromide	904.5	56	
Antimony trichloride	794		270

TABLE 5-3 Critical properties (*continued*)

Substance	T_c, K	P_c, atm	V_c, $cm^3 \cdot mol^{-1}$
Argon	150.8	48.1	74.9
Arsine	373.0		
Benzaldehyde	695	46	
Benzene	562.1	48.3	259
Benzoic acid	752	45	341
Benzonitrile	699.4	41.6	
Benzyl alcohol	677	46	334
Biphenyl	789	38	502
Bismuth tribromide	1219		301
Bismuth trichloride	1179	118	261
Boron pentafluoride	470		
Boron tribomide	573		280
Boron trichloride	451.9	38.2	
Boron trifluoride	260.8	49.2	
Bromine	584	102	127
Bromobenzene	670	44.6	324
Bromoethane	503.8	61.5	215
Bromomethane	464	85	
Bromopentafluorobenzene	670	44.6	
Bromotrifluoromethane	340.2	39.2	200
1,2-Butadiene	443.7	44.4	219
1,3-Butadiene	425	42.7	221
Butane	425.2	37.5	255
1-Butanol	562.9	43.6	274
2-Butanol	536.0	41.4	268
2-Butanone	535.5	41.0	267
1-Butene	419.6	39.7	240
cis-2-Butene	435.6	41.5	234
trans-2-Butene	428.6	40.5	238
3-Butenenitrile	585	39	265
1-Buten-3-yne	455	49	202
Butyl acetate	579	31	400
1-Butylamine	524	41	288
N-Butylaniline	72	28	518
Butylbenzene	660.5	28.5	497
sec-Butylbenzene	664	29.1	
tert-Butylbenzene	660	29.3	
Butyl benzoate	723	26	561
Butylcyclohexane	667	31.1	
sec-Butylcyclohexane	669	26.4	
tert-Butylcyclohexane	659	26.3	
Butyl ethyl ether	531	30	390
1-Butyne	463.7	46.5	220
2-Butyne	488.6	502	221
Butyraldehyde	524	40	278
Butyric acid	628	52.0	292
Butyronitrile	582.2	37.4	285

TABLE 5-3 Critical properties (*continued*)

Substance	T_c, K	P_c, atm	V_c, cm$^3 \cdot$ mol^{-1}
Carbon dioxide	304.2	72.8	94.0
Carbon disulfide	552	78.0	170
Carbon monoxide	132.9	34.5	93.1
Carbon tetrachloride	556.4	45.0	276
Carbon tetrafluoride	227.6	36.9	140
Carbonyl chloride (phosgene)	455	56	190
Carbonyl sulfide	375	58	140
Chlorine	417	76.1	124
Chlorine pentafluoride	415.7	51.9	230.9
Chlorine trifluoride	426.6		
Chlorobenzene	632.4	44.6	308
1-Chlorobutane	542	36.4	312
2-Chlorobutane	520.6	39	305
1-Chloro-1,1-difluoroethane	410.2	40.7	231
2-Chloro-1,1-difluoroethylene	400.5	44.0	197
Chlorodifluoromethane	369.2	49.1	165
Chloroethane	460.4	52.0	199
Chloroform	536.4	54.0	239
Chloromethane	416.3	65.9	139
2-Chloro-2-methylpropane	507	39	295
Chloropentafluoroacetone	410.7	28.4	
Chloropentafluoroethane	353.2	31.2	252
1-Chloropropane	503	45.2	254
2-Chloropropane	485	46.6	230
3-Chloropropene	514	47	234
Chlorotrifluoromethane	302.0	38.7	180
Chlorotrifluorosilane	308.5	34.2	
o-Cresol	697.6	49.4	282
m-Cresol	705.8	45.0	310
p-Cresol	704.6	50.8	277
Cyanogen	400	59	
Cyclobutane	459.9	49.2	210
Cycloheptane	589	36.7	390
Cyclohexane	553.4	40.2	308
Cyclohexanol	625	37	327
Cyclohexanone	629	38	312
Cyclohexene	560.4	42.9	292
Cyclopentane	511.6	44.5	260
Cyclopentanone	626	53	268
Cyclopentene	506.0		
Cyclopropane	397.8	54.2	170
Cymene	658		
cis-Decalin	702.2	31	
trans-Decalin	690.0	31	
Decane	617.6	20.8	603
Decanenitrile	621.9	32.1	
1-Decanol	700	22	600

TABLE 5-3 Critical properties (*continued*)

Substance	T_c, K	P_c, atm	V_c, $cm^3 \cdot mol^{-1}$
1-Decene	615	21.8	650
Decylcyclohexane	750	13.4	
Decylcyclopentane	723.8	15.0	
Deuterium			
(equilibrium)	38.3	16.28	60.4
(normal)	38.4	16.43	60.3
Deuterium bromide	361.9		
Deuterium chloride	328.4		
Deuterium hydride	35.8	14.64	62.8
Deuterium iodide	421.7		
Deuterium oxide	644.0	213.8	55.6
Diborane	289.0	39.5	
1,2-Dibromoethane	582.9	70.6	
Dibromomethane	583	71	
Dibromotetrafluoroethane	487.6	34	329
Dibutylamine	596	25	517
Dibutyl ether	580	25	500
1,2-Dichlorobenzene	697.3	40.5	360
1,3-Dichlorobenzene	684	38	359
1,4-Dichlorobenzene	685	39	372
Dichlorodifluoromethane	385.0	40.7	217
1,1-Dichloroethane	523	50	240
1,2-Dichloroethane	561	53	220
1,1-Dichloroethylene	544		
1,2-Dichloroethylene	516.5	54.4	
Dichlorofluoromethane	451.6	51.0	197
Dichloromethane	510	60.0	193
1,2-Dichloropropane	577	44	226
Dichlorosilane	449	46.1	
1,1-Dichloro-1,2,2,2-tetrafluoroethane	418.6	32.6	294
1,2-Dichloro-1,1,2,2-tetrafluoroethane	418.9	32.6	293
Diethylamine	496.6	36.6	301
1,4-Diethylbenzene	657.9	27.7	480
Diethyl disulfide	642		
Diethylene glycol	681	46	316
Diethyl ether	466.7	35.9	280
3,3-Diethylpentane	610	26.4	
Diethyl sulfide	557	39.1	318
Difluoroamine (HNF_2)	403	93	
cis-Difluorodiazine (N_2F_2)	272	70	
trans-Difluorodiazine	260	55	
1,1-Difluoroethane	386.6	44.4	181
1,1-Difluoroethylene	302.8	44.0	154
Dihexyl ether	657	18	720
Dihydrogen disulfide	572	58.3	
Dihydrogen heptasulfide	1015	33	

TABLE 5-3 Critical properties (*continued*)

Substance	T_c, K	P_c, atm	V_c, $cm^3 \cdot mol^{-1}$
Dihydrogen hexasulfide	980	36	
Dihydrogen octasulfide	1040	32	
Dihydrogen pentasulfide	930	38.4	
Dihydrogen tetrasulfide	855	43.1	
Dihydrogen trisulfide	738	50.6	
Diisopropyl ether	500	28.4	385
1,2-Dimethoxyethane	536	38.2	271
Dimethoxymethane	497		
Dimethylamine	437.6	52.4	187
N,N-Dimethylaniline	687	35.8	
2,2-Dimethylbutane	488.7	30.4	359
2,3-Dimethylbutane	499.9	30.9	358
2,3-Dimethyl-1-butene	501	32.0	343
2,3-Dimethyl-2-butene	524	33.2	351
3,3-Dimethyl-1-butene	490	32.1	340
1,1-Dimethylcyclohexane	591	29.3	416
cis-1,2-Dimethylcyclohexane	606	29.3	
trans-1,2-Dimethylcyclohexane	596	29.3	
cis-1,3-Dimethylcyclohexane	591	29.3	
trans-1,3-Dimethylcyclohexane	598	29.3	
1,1-Dimethylcyclopentane	547	34.0	360
cis-1,2-Dimethylcyclopentane	564.8	34.0	368
trans-1,2-Dimethylcyclopentane	553.2	34.0	362
Dimethyl ether	400.0	53.0	178
2,2-Dimethylhexane	549.8	25.0	478
2,3-Dimethylhexane	563.4	25.9	468
2,4-Dimethylhexane	553.5	25.2	472
2,5-Dimethylhexane	550.0	24.5	482
3,3-Dimethylhexane	562.0	26.2	443
3,4-Dimethylhexane	568.8	26.6	466
Dimethyl oxalate	628	39.2	
2,2-Dimethylpentane	520.4	27.4	416
2,3-Dimethylpentane	537.3	28.7	393
2,4-Dimethylpentane	519.7	27.0	418
3,3-Dimethylpentane	536.3	29.1	414
2,2-Dimethylpropane	433.8	31.6	303
2,2-Dimethyl-1-propanol	549	39	319
2,3-Dimethylpyridine	655.4		
2,4-Dimethylpyridine	644.2		
2,5-Dimethylpyridine	644		
2,6-Dimethylpyridine	623.7		
3,4-Dimethylpyridine	683.8		
3,5-Dimethylpyridine	667.2		
N,N-Dimethyl-*o*-toluidine	668	30.8	
1,4-Dioxane	587	51.4	238
Diphenyl ether	766	31	
Diphenylmethane	767	29.4	

TABLE 5-3 Critical properties (*continued*)

Substance	T_c, K	P_c, atm	V_c, $cm^3 \cdot mol^{-1}$
Dipropylamine	550	31	407
Dodecane	658.3	18.0	713
1-Dodecanol	679	19	718
1-Dodecene	657	18.3	
Dodecylcyclopentane	750	12.8	
Ethane	305.4	48.2	148
Ethanethiol	498.6	54.2	207
Ethanol	516.2	63.0	167
Ethoxybenzene	647.1	33.8	
Ethyl acetate	523.2	37.8	286
Ethyl acetoacetate	673		
Ethyl acrylate	552	37.0	320
Ethylamine	456	55.5	178
Ethylbenzene	617.1	35.6	374
Ethyl benzoate	697	32	451
2-Ethyl-1-butanol	418.8		
Ethyl butyrate	565.9	30.2	395
Ethyl crotonate	599		
Ethylcyclohexane	609	29.9	450
Ethylcyclopentane	369.5	33.5	375
Ethylene	282.4	49.7	129
Ethylenediamine	592.9	62.1	206
Ethylene glycol	645	76	186
Ethylene oxide	469	71.0	140
Ethyl formate	508.4	46.8	229
3-Ethylhexane	565.4	25.7	455
2-Ethylhexanol	613	27.2	494
2-Ethyl-1-methylbenzene	651	30.0	460
3-Ethyl-1-methylbenzene	637	28.0	490
4-Ethyl-1-methylbenzene	640	29.0	470
Ethyl 3-methylbutyrate	588.0		
1-Ethyl-1-methylcyclopentane	592	29.5	
Ethyl methyl ether	437.8	43.4	221
Ethyl methyl ketone	535.6	41.0	267
3-Ethyl-2-methylpentane	567.0	26.7	443
3-Ethyl-2-methylpentane	576.5	27.7	455
3-Ethyl-3-methylpentane	576.4	27.7	455
Ethyl-2-methylpropanoate	553	30	410
Ethyl methyl sulfide	533	42	
3-Ethylpentane	540.6	28.5	416
o-Ethylphenol	703.0		
m-Ethylphenol	716.4		
p-Ethylphenol	716.4		
Ethylpropanoate	546.0	33.2	345
Ethyl propyl ether	500.6	32.1	244
o-Ethyltoluene	653	31	461
m-Ethyltoluene	636	31	461

TABLE 5-3 Critical properties (*continued*)

Substance	T_c, K	P_c, atm	V_c, $cm^3 \cdot mol^{-1}$
p-Ethyltoluene	636	31	461
Ethyl vinyl ether	475	40.2	260
Fluorine	144.3	51.5	66.2
Fluorobenzene	560.1	44.9	271
Fluoroethane	375.3	49.6	169
Fluoromethane	317.8	58.0	124
Fluorotrichloromethane	471.1	43.2	248
Formaldehyde	408	65	
Formic acid	580		
Furan	490.2	54.3	218
Germanium tetrachloride	550.0	38	330
Glycerol	726	66	255
Hafnium tetrabromide	746		415
Hafnium tetrachloride	723	57.0	304
Hafnium tetraiodide	916		528
Helium-3	3.30	1.167	73.2
Helium-4	5.19	2.24	57.3
Heptadecane	733	13	1000
1-Heptadecanol	736	14	
Heptane	540.2	27.0	432
1-Heptanol	633	30	435
1-Heptene	537.2	28	440
Heptylcyclopentane	679	19.2	
Hexadecane	717	14	
1-Hexadecene	717	13.2	
Hexadecylcyclopentane	791	9.6	
1,5-Hexadiene	507	34	328
Hexafluoroethane	292.8	29.4	223.7
Hexamethylbenzene	767		
Hexane	507.4	29.3	370
1-Hexanol	610	40	381
1-Hexene	504.3	31.3	350
cis-2-Hexene	518	32.4	351
trans-2-Hexene	516	32.3	351
cis-3-Hexene	517	32.4	350
trans-3-Hexene	519.9	32.1	350
Hexylcyclopentane	660.1	21.1	
Hydrazine	653	145	96.1
Hydrogen			
(equilibrium)	32.9	12.77	65.4
(normal)	33.2	12.8	65.0
Hydrogen bromide	363.2	84.4	100.0
Hydroben chloride	324.6	82.0	81.0
Hydrogen cyanide	456.8	53.2	139
Hydrogen deuteride, *see* Deuterium hydride			

TABLE 5-3 Critical properties (*continued*)

Substance	T_c, K	P_c, atm	V_c, $cm^3 \cdot mol^{-1}$
Hydrogen fluoride	461	64	69
Hydrogen iodide	424.0	82.0	131
Hydrogen selenide	411	88	
Hydrogen sulfide	373.2	88.2	98.5
Icosane	767	11.0	
1-Icosanol	770	12.0	
Iodine	819	115	155
Iodobenzene	721	44.6	351
Iodomethane	528	65	190
Isobutyl acetate	561	30	414
Isobutylamine	516	42	284
Isobutylbenzene	650	31	480
Isobutyl butyrate	611		
Isobutylcyclohexane	659	30.8	
Isobutyl formate	551	38.3	350
Isobutyl 3-methylbutyrate	621		
Isobutyl propanoate	592		
Isobutyric acid	609	40	292
Isopropylamine	476	50	229
Isopropylbenzene	631.0	31.7	428
Isopropylcyclohexane	640	28	
Isopropylcyclopentane	601	29.6	
2-Isopropyl-1-methylbenzene	670	28.6	
3-Isopropyl-1-methylbenzene	666	29.0	
4-Isopropyl-1-methylbenzene	653	27.9	
Isoquinoline	803		
Isoxazole	552.0		
Ketene	380	64	145
Krypton	209.4	54.3	91.2
Mercury	1173	180	
Methane	190.6	45.4	99.0
Methanethiol	470.0	71.4	145
Methanol	512.6	79.9	118
Methoxybenzene (anisole)	641	41.2	
Methyl acetate	506.8	46.3	228
Methyl acrylate	536	42	265
Methylamine	430	73.6	140
N-Methylaniline	701	51.3	
Methyl benzoate	692	36	396
2-Methyl-1,3-butadiene	484	38.0	276
3-Methyl-1,2-butadiene	496	40.6	267
2-Methylbutane	460.4	33.3	306
2-Methyl-1-butanol	571	38	322
3-Methyl-1-butanol	579.5	38	329
2-Methyl-2-butanol	545	39	319
3-Methyl-2-butanone	553.4	38.0	310

TABLE 5-3 Critical properties (*continued*)

Substance	T_c, K	P_c, atm	V_c, $cm^3 \cdot mol^{-1}$
2-Methyl-1-butene	465	34.0	294
2-Methyl-2-butene	470	34.0	318
3-Methyl-1-butene	450	34.7	300
Methyl butyrate	554.4	34.3	340
3-Methylbutyric acid	634		
Methylcyclohexane	572.1	34.3	368
Methylcyclopentane	532.7	37.4	319
N-Methylethylamine	496.6	36.6	243
Methyl formate	487.2	59.2	172
2-Methylheptane	559.6	24.5	488
3-Methylheptane	563.6	25.1	464
4-Methylheptane	561.7	25.1	476
2-Methylhexane	530.3	27.0	421
3-Methylhexane	535.2	27.8	404
Methylhydrazine	567	79.3	271
Methyl isobutyrate	540.8	33.9	339
Methyl isocyanate	491	55	
1-Methylnaphthalene	772	35.2	445
2-Methylnaphthalene	761	34.6	462
2-Methylpentane	497.5	29.7	367
3-Methylpentane	504.4	30.8	367
2-Methyl-2,4-pentanediol	678	33.9	
4-Methyl-2-pentanone	571	32.3	371
2-Methyl-2-pentene	518	32.4	351
cis-3-Methyl-2-pentene	518	32.4	351
trans-3-Methyl-2-pentene	521	32.3	350
cis-4-Methyl-2-pentene	490	30	360
trans-4-Methyl-2-pentene	493	30	360
Methyl phenyl ether	641	41.2	
2-Methylpropanal	513	41	274
2-Methylpropane	408.1	36.0	263
Methyl propanoate	530.6	39.5	282
2-Methyl-1-propanol (isobutyl alcohol)	547.7	42.4	273
2-Methyl-2-propanol	506.2	39.2	275
2-Methylpropene	417.9	39.5	239
2-Methylpyridine	621		
3-Methylpyridine	645		
4-Methylpyridine	646	44	311
α-Methylstyrene	654	33.6	397
Methyl vinyl ether	436	47	205
Morpholine	618	54	253
Naphthalene	748.4	40.0	410
Neon	44.4	27.2	41.7
Niobium pentabromide	1010		469
Niobium pentachloride	807		400
Niobium pentafluoride	737	62	155
Nitric oxide	180	64	58

TABLE 5-3 Critical properties (*continued*)

Substance	T_c, K	P_c, atm	V_c, $cm^3 \cdot mol^{-1}$
Nitrobenzene	732		
Nitrogen-14	126.2	33.5	89.5
Nitrogen-15	126.3	33.5	90.4
Nitrogen dioxide (equilibrium)	431.4	100	170
Nitrogen trifluoride	234.0	44.7	
Nitromethane	588	62.3	173
Nitrosyl chloride	440	90	139
Nitrous oxide	309.6	71.5	97.4
Nitryl fluoride	349.4		
Nonadecane	756	11.0	
Nonane	594.6	22.8	548
1-Nonanol	677		546
1-Nonene	592	23.1	580
Nonylcyclopentane	710.5	16.3	
Octadecane	745	11.9	
1-Octadecanol	747	14	
1-Octadecene	739	11.2	
Octane	568.8	24.5	492
1-Octanol	658	34	490
2-Octanol	637	27	494
1-Octene	566.6	25.9	464
trans-2-Octene	580	27.3	
Octylcyclopentane	694	17.7	
Oxygen	154.6	49.8	73.4
Oxygen difluoride	215.2	48.9	97.7
Ozone	161.3	55.0	88.9
Paraldehyde	563		
Pentachloroethane	646.1		
Pentadecane	707	15	880
1-Pentadecene	704	14.4	
Pentadecylcyclopentane	780	10.1	
1,2-Pentadiene	503	40.2	276
trans-1,3-Pentadiene	496	39.4	275
1,4-Pentadiene	478	37.4	276
Pentafluorobenzene	532.0	34.7	
1,1,2*H*-Pentafluoropropane	380.11	31.0	273
Pentanal	554	35	333
Pentane	469.6	33.3	304
Pentanoic acid	651	38	340
1-Pentanol	586	38	326
2-Pentanone	564.0	38.4	301
3-Pentanone	561.0	36.9	336
1-Pentene	464.7	40.0	300
cis-2-Pentene	476	36.0	300
trans-2-Pentene	475	36.1	300
Pentyl formate	576		
1-Pentyne	493.4	40	278

TABLE 5-3 Critical properties (*continued*)

Substance	T_c, K	P_c, atm	V_c, $cm^3 \cdot mol^{-1}$
Perchloryl fluoride	368.4	53.0	161
Perfluoroacetone	357.3	28.0	
Perfluorobenzene	516.7	32.6	
Perfluorobutane	386.4	22.9	378
Perfluoro-(2-butyltetrahydrofuran)	500.3	15.9	588
Perfluorocyclobutane	388.4	27.41	260
Perfluorocyclohexane	457.2	24	
Perfluorocyclohexene	461.8		
Perfluorodecene	542.3	14.3	
Perfluoroethane	292.8	29.4	223.7
Perfluoroheptane	474.8	16.0	664
Perfluoroheptene	478.1		
Perfluorohexane	451.7	18.8	442
Perfluorohexene	454.3		
Perfluoromethylcyclohexane	486.8	23	
Perfluoronaphthalene	673.1		
Perfluorononane	524.0	15.4	
Perfluorooctane	502	16.4	
Perfluoropentane	422	20.1	
Perfluoropropane	345.1	26.5	299
Phenanthrene	878		
Phenetole	647	33.8	
Phenol	694.2	60.5	229
Phosgene	455	56	190
Phosphine	324.4	64.5	
Phosphonium chloride	322.2	72.7	
Phosphorus bromide difluoride	386		
Phosphorus chloride difluoride	362.32	44.6	
Phosphorus dibromide fluoride	527		
Phosphorus dichloride fluoride	463.0	49.3	
Phosphorus pentachloride	645		
Phosphorus trichloride	563		260
Phosphorus trifluoride	271.2	42.7	
Phosphoryl chloride difluoride	423.8	43.4	
Phosphoryl trichloride	602		
Phosphoryl trifluoride	346.5	41.8	
Phthalic anhydride	810	47	368
Piperidine	594.0	47	289
Propadiene	393	54.0	162
Propane	369.8	41.9	203
1,2-Propanediol	625	60	237
1,3-Propanediol	658	59	241
Propanoic acid	612	53.0	230
1-Propanol	536.7	51.0	218.5
2-Propanol	508.3	47.0	220
2-Propenal	506	51	
Propionaldehyde	496	47	223

TABLE 5-3 Critical properties (*continued*)

Substance	T_c, K	P_c, atm	V_c, $cm^3 \cdot mol^{-1}$
Propionitrile	564.4	41.3	230
Propyl acetate	549.4	23.9	345
Propylamine	497.0	46.8	233
Propylbenzene	638.3	31.6	440
Propylcyclopentane	603	29.6	425
Propylcyclohexane	639	27.7	
Propylene	365.0	45.6	181
Propylene oxide	482.2	48.6	186
Propyl formate	538.0	40.1	285
Propyl propanoate	578		
1-Propyne	402.4	55.5	164
Pyridine	620.0	55.6	254
Pyrrole	639.6	56	
Pyrrolidine	568.6	55.4	249
Quinoline	794.4		
Radon	376.9	62	139
Rhenium(VII) oxide	942		334
Selenium	1766		
Silane	269.6	47.8	
Silicon chloride trifluoride	307.6	34.2	
Silicon tetrachloride	507	37	326
Silicon tetrafluoride	259.1	36.7	
Silicon trichlorofluoride	438.5	35.3	
Styrene	647	39.4	
Sulfur	1314		
Sulfur dioxide	430.8	77.8	122
Sulfur hexafluoride	318.7	37.1	198
Sulfur tetrafluoride	364.0		
Sulfur trioxide	491.0	81	130
Tantalum pentabromide	974		461
Tantalum pentachloride	767		400
o-Terphenyl	891.0	38.5	769
m-Terphenyl	924.8	34.6	784
p-Terphenyl	926.0	32.8	779
1,1,2,2-Tetrachloro-1,2-difluoroethane	551	34	370
1,1,2,2-Tetrachloroethane	661.1		
Tetrachloroethylene	620	44	290
Tetradecane	694	16	830
1-Tetradecene	689	15.4	
Tetradecylcyclopentane	772	11.1	
Tetrafluoroethylene	306.4	38.9	175
Tetrafluorohydrazine	309.4	37	
Tetrahydrofuran	540.2	51.2	224
1,2,3,4-Tetrahydronaphthalene	719	34.7	
Tetrahydrothiophene	631.9		
1,2,4,5-Tetramethylbenzene	675	29	480

TABLE 5-3 Critical properties (*continued*)

Substance	T_c, K	P_c, atm	V_c, $cm^3 \cdot mol^{-1}$
2,2,3,3-Tetramethylbutane	567.8	28.3	461
2,2,3,3-Tetramethylhexane	623.1	24.8	
2,2,5,5-Tetramethylhexane	581.5	21.6	
2,2,3,3-Tetramethylpentane	607.6	27.0	
2,2,3,4-Tetramethylpentane	592.7	25.7	
2,2,4,4-Tetramethylpentane	574.7	24.5	
2,3,3,4-Tetramethylpentane	607.6	26.8	
2-Thiapropane	503.1	54.6	201
Thiophene	579.4	56.2	219
Thymol	698		
Tin(IV) chloride	591.8	37.0	351
Titanium tetrachloride	638	46	340
Toluene	591.7	40.6	316
o-Toluidine	694	37	343
m-Toludine	709	41	343
p-Toluidine	667		
Toluonitrile	723		
Tributylamine	643	18	
1,1,2-Trichloroethane	602	41	294
Trichloroethylene	571	48.5	256
Trichlorofluoromethane	471.2	43.5	248
1,2,3-Trichloropropane	651	39	348
1,2,2-Trichloro-1,1,2-trifluoroethane	487.2	33.7	304
Tridecane	675.8	17.0	780
1-Tridecene	674	16.8	
Tridecylcyclopentane	761	11.9	
Triethanolamine	787.4	24.2	
Triethylamine	535	30	390
Trifluoroacetic acid	491.3	32.2	204
1,1-Trifluoroethane	346.2	37.1	221
Trifluoromethane	298.89	47.7	133.3
Trimethylamine	433.2	40.2	254
1,2,3-Trimethylbenzene	664.5	34.1	430
1,2,4-Trimethylbenzene	649.1	31.9	430
1,3,5-Trimethylbenzene	637.3	30.9	433
2,2,3-Trimethylbutane	531.1	29.2	398
2,2,3-Trimethyl-1-butene	533	28.6	400
Trimethylchlorosilane	497.7	31.6	
1,1,2-Trimethylcyclopentane	579.5	29.0	
1,1,3-Trimethylcyclopentane	569.5	27.9	
cis,cis,trans-1,2,4-Trimethylcyclopentane	579	28.4	
cis,trans,cis-1,2,4-Trimethylcyclopentane	571	27.7	
3,3,5-Trimethylheptane	609.6	22.9	
2,2,3-Trimethylhexane	588	24.6	
2,2,4-Trimethylhexane	573.7	23.4	
2,2,5-Trimethylhexane	567.9	23.0	519
2,2,3-Trimethylpentane	563.4	26.9	436

TABLE 5-3 Critical properties (*continued*)

Substance	T_c, K	P_c, atm	V_c, $cm^3 \cdot mol^{-1}$
2,2,4-Trimethylpentane	543.9	25.3	468
2,3,3-Trimethylpentane	573.5	27.8	455
2,3,4-Trimethylpentane	566.3	26.9	461
2,2,4-Trimethyl-1,3-pentanediol	671	25.6	364.6
1*H*-Undecafluoropentane	443.9		
Undecane	638.8	19.4	660
1-Undecene	637	19.7	
Uranium hexafluoride	505.8	45.5	250
Vinyl acetate	525	43	265
Vinyl chloride	429.7	55.3	169
Vinyl fluoride	327.8	51.7	114
Vinyl formate	475	57	210
Water	647.3	217.6	56.0
Xenon	289.7	57.6	118
o-Xylene	630.2	36.8	369
m-Xylene	617.0	35.0	376
p-Xylene	616.2	34.7	379
2,3-Xylenol	722.6	48	470
2,4-Xylenol	707.6	43	509
2,5-Xylenol	723.0	48	470
2,6-Xylenol	700.9	42	509
3,4-Xylenol	729.8	49	552
3,5-Xylenol	715.6	36	611
Zirconium tetrabromide	805		415
Zirconium tetrachloride	778	56.9	319
Zirconium tetraiodide	960		528

Estimation of Critical Properties

When the critical properties are unavailable, they may be estimated employing structural contributions to estimate T_c, P_c, and V_c. Lydersen's critical-property increments* provide good estimates for T_c and P_c; Vetere's group contributions† yield reasonable estimates for V_c. The units employed are kelvins, atmospheres, and cubic centimeters per mole. Typical errors in estimated values are less than 2% for T_c but may rise up to 5% for higher-molecular-weight (greater than 100) nonpolar materials; errors are uncertain for molecules with multifunctional polar groups. Errors for estimated values of P_c and V_c are about double those for T_c.

The relations are

$$T_c = T_b[0.567 + \sum \Delta_T - (\sum \Delta_T)^2]^{-1}$$

$$P_c = M(0.34 + \sum \Delta_P)^{-2}$$

* A. L. Lydersen, Univ. Wisconsin Coll. Eng., Eng. Exp. Stn, Rep. 3, Madison, April 1955.

† A. Vetere, cited in R. C. Reid, J. M. Prausnitz, and T. K. Sherwood, *The Properties of Gases and Liquids*, 3d ed., McGraw-Hill, New York, 1977, p. 17.

$$V_c = 33 + \left[\sum_i (M_i \Delta_V)\right]^{1.029}$$

where T_b is the normal boiling point and M is the molecular weight. Group contributions are listed in Table 5-4.

TABLE 5-4 Group contributions for the estimation of critical properties

There are no increments for hydrogen. All bonds shown as free are connected with atoms other than hydrogen. Values in parentheses are based upon very few experimental values.

Group	Δ_T, K	Δ_P, atm	Δ_V, $cm^3 \cdot mol^{-1}$
Nonring increments			
$-CH_3$, $-CH_2-$	0.020	0.227	3.360 (linear chain) 2.888 (side chain)
$-\overset{\vert}{\underset{\vert}{C}}H$	0.012	0.210	3.360 (linear chain) 2.888 (side chain)
$-\overset{\vert}{\underset{\vert}{C}}-$	0.0	0.210	3.360 (linear chain) 2.888 (side chain)
$=CH_2$, $=\overset{\vert}{C}H$	0.018	0.198	2.940
$=\overset{\vert}{C}-$	0.0	0.198	2.940
$=C=$	0.0	0.198	2.908
$\equiv CH$, $\equiv C-$	0.005	0.153	2.648
$-O-$	0.021	0.16	1.075
$>C=O$	0.040	0.29	1.765
$>NH$	0.031	0.135	2.333
$>N-$	0.014	0.17	1.793
$-S-$	0.015	0.27	0.591
Ring increments			
$-CH_2-$	0.013	0.184	2.813
$-\overset{\vert}{\underset{\vert}{C}}H$	0.012	0.192	2.813
$-\overset{\vert}{\underset{\vert}{C}}-$	(−0.007)	(0.154)	2.813
$=\overset{\vert}{C}H$, $=\overset{\vert}{C}-$, $=C=$	0.011	0.154	2.538
$-O-$	(0.014)	(0.12)	0.790
$>C=O$	(0.033)	(0.2)	1.500
$>NH$	(0.024)	(0.09)	1.736
$>N-$	(0.007)	(0.13)	1.883
$-S-$	(0.008)	(0.24)	0.911

TABLE 5-4 Group contributions for the estimation of critical properties (*continued*)

Group	Δ_T, K	Δ_P, atm	Δ_V, $cm^3 \cdot mol^{-1}$
General substituents			
—F	0.018	0.224	0.770
—Cl	0.017	0.320	1.237
—Br	0.010	(0.50)	0.899
—I	0.012	(0.83)	0.702
—OH			
Alcohols	0.082	0.06	0.704
Phenols	0.031	(−0.02)	1.553
HC=O (aldehyde) (with bond up from C)	0.048	0.33	2.333
—COOH	0.085	(0.4)	1.652
—COO— (ester)	0.047	0.47	1.607
$—NH_2$	0.031	0.095	2.184
—CN	(0.060)	(0.36)	2.784
$—NO_2$	(0.055)	(0.42)	1.559
—SH	0.015	0.27	1.537
—Si— (with bonds up and down)	0.03	(0.54)	

SECTION 6
SPECTROSCOPY

ULTRAVIOLET-VISIBLE SPECTROSCOPY

Molecules with two or more isolated chromophores (absorbing groups) absorb light of nearly the same wavelength as does a molecule containing only a single chromophore of a particular type. The intensity of the absorption is proportional to the number of that type of chromophore present in the molecule. Representative chromophores are given in Table 6-1.

The solvent chosen must dissolve the sample, yet be relatively transparent in the spectral region of interest. In order to avoid poor resolution and difficulties in spectrum interpretation, a solvent should not be employed for measurements that are near the wavelength of or are shorter than the wavelength of its ultraviolet cutoff, that is, the wavelength at which absorbance for the solvent alone approaches one absorbance unit. Ultraviolet cutoffs for solvents commonly used are given in Table 6-2.

Appreciable interaction between choromophores does not occur unless they are linked directly to each other, or forced into close proximity as a result of molecular stereochemical configuration. Interposition of a single methylene group, or *meta* orientation about an aromatic ring, is sufficient to insulate chromophores almost completely from each other. Certain combinations of functional groups afford chromophoric systems which give rise to characteristic absorption bands.

Sets of empirical rules, often referred to as Woodward's rules or the Woodward–Fieser rules, enable the absorption maxima of dienes (Table 6-3) and enones and dienones (Table 6-4) to be predicted. To the respective base values (absorption wavelength of parent compound) are added the increments for the structural features or substituent groups present. When necessary, a solvent correction is also applied (Table 6-5).

Ring substitution on the benzene ring affords shifts to longer wavelengths (Table 6-6) and intensification of the spectrum. With electron-withdrawing substituents, practically no change in the maximum position is observed. The spectra of heteroaromatics are related to their isocyclic analogs, but only in the crudest way. As with benzene, the magnitude of substituent shifts can be estimated, but tautomeric possibilities may invalidate the empirical method.

When electronically complementary groups are situated *para* to each other in disubstituted benzenes, there is a more pronounced shift to a longer wavelength than would be expected from the additive effect due to the extension of the choromophore from the electron-donating group through the ring to the electron-withdrawing group. When the *para* groups are not complementary, or when the groups are situated *ortho* or *meta* to each other, disubstituted benzenes show a more or less additive effect of the two substituents on the wavelength maximum. Calculation of the principal band of selected substituted benzenes is illustrated in Table 6-7.

TABLE 6-1 Electronic absorption bands for representative chromophores

Chromophore	System	λ_{max}	ε_{max}
Acetylide	$-C\equiv C-$	175–180	6 000
Aldehyde	$-CHO$	210 280–300	strong 11–18
Amine	$-NH_2$	195	2 800
Azido	$>C=N-$	190	5 000
Azo	$-N=N-$	285–400	3–25
Bromide	$-Br$	208	300
Carbonyl	$>C=O$	195 270–285	1 000 18–30
Carboxyl	$-COOH$	200–210	50–70
Disulfide	$-S-S-$	194 255	5 500 400
Ester	$-COOR$	205	50
Ether	$-O-$	185	1 000
Ethylene	$-C=C-$	190	8 000
Iodide	$-I$	260	400
Nitrate	$-ONO_2$	270 (shoulder)	12
Nitrile	$-C\equiv N$	160	—
Nitrite	$-ONO$	220–230 300–400	1 000–2 000 10
Nitro	$-NO_2$	210	strong
Nitroso	$-NO$	302	100
Oxime	$-NOH$	190	5 000
Sulfone	$-SO_2-$	180	—
Sulfoxide	$>S=O$	210	1 500
Thiocarbonyl	$>C=S$	205	strong
Thioether	$-S-$	194 215	4 600 1 600
Thiol	$-SH$	195	1 400
	$-(C=C)_2-$ (acyclic)	210–230	21 000
	$-(C=C)_3-$	260	35 000
	$-(C=C)_4-$	300	52 000
	$-(C=C)_5-$	330	118 000

TABLE 6-1 Electronic absorption bands for representative chlromophores (*continued*)

Chromophore	System	λ_{max}	ε_{max}
	$-(C=C)_2-$ (alicyclic)	230–260	3 000–8 000
	$C=C-C\equiv C$	219	6 500
	$C=C-C=N$	220	23 000
	$C=C-C=O$	210–250 300–350	10 000–20 000 weak
	$C=C-NO_2$	229	9 500
Benzene		184 204 255	46 700 6 900 170
Diphenyl		246	20 000
Naphthalene		222 275 312	112 000 5 600 175
Anthracene		252 375	199 000 7 900
Phenanthrene		251 292	66 000 14 000
Naphthacene		272 473	180 000 12 500
Pentacene		310 585	300 000 12 000
Pyridine		174 195 257	80 000 6 000 1 700
Quinoline		227 270 314	37 000 3 600 2 750
Isoquinoline		218 266 317	80 000 4 000 3 500

TABLE 6-2 Ultraviolet cutoffs of spectrograde solvents

Absorbance of 1.00 in a 10.0 mm cell vs. distilled water

Solvent	Wavelength, nm	Solvent	Wavelength, nm
Acetic acid	260	Butyl acetate	254
Acetone	330	Carbon disulfide	380
Acetonitrile	190	Carbon tetrachloride	265
Benzene	280	1-Chlorobutane	220
1-Butanol	210	Chloroform (stabilized with ethanol)	245
2-Butanol	260		

TABLE 6-2 Ultraviolet cutoffs of spectrograde solvents (*continued*)

Solvent	Wavelength, nm	Solvent	Wavelength, nm
Cyclohexane	210	Methylene chloride	235
1,2-Dichloroethane	226	Methyl ethyl ketone	330
Diethyl ether	218	Methyl isobutyl ketone	335
1,2-Dimethoxyethane	240	2-Methyl-1-propanol	230
N,N-Dimethylacetamide	268	*N*-Methylpyrrolidone	285
N,N-Dimethylformamide	270	Nitromethane	380
Dimethylsulfoxide	265	Pentane	210
1,4-Dioxane	215	Pentyl acetate	212
Ethanol	210	1-Propanol	210
2-Ethoxyethanol	210	2-Propanol	210
Ethyl acetate	255	Pyridine	330
Ethylene chloride	228	Tetrachloroethylene	
Glycerol	207	(stabilized with thymol)	290
Heptane	197	Tetrahydrofuran	220
Hexadecane	200	Toluene	286
Hexane	210	1,1,2-Trichloro-1,2,2-	
Isobutyl alcohol	230	trifluoroethane	231
Methanol	210	2,2,4-Trimethylpentane	215
2-Methoxyethanol	210	*o*-Xylene	290
Methylcyclohexane	210	Water	191

TABLE 6-3 Absorption wavelength of dienes

Heteroannular and acyclic dienes usually display molar absorptivities in the 8000 to 20,000 range, whereas homoannular dienes are in the 5000 to 8000 range.

Poor correlations are obtained for cross-conjugated polyene systems such as

```
          C=C—
         /
    —C=C
         \
          C=C—
```

The correlations presented here are sometimes referred to as Woodward's rules or the Woodward–Fieser rules.

Base value for heteroannular or open chain diene, nm	214
Base value for homoannular diene, nm	253
Increment (in nm) for	
double bond extending conjugation	30
Alkyl substituent or ring residue	5
Exocyclic double bond	5
Polar groupings:	
-*O*-acyl	0
-*O*-alkyl	6
-*S*-alkyl	30
-Cl, -Br	5
-*N*(alkyl)$_2$	60
Solvent correction (see Table 6-5)	
Calculated wavelength =	total

TABLE 6-4 Absorption wavelength of enones and dienones

$$O{=}C{-}\overset{\alpha}{C}{=}C\begin{matrix}\beta\\ \beta\end{matrix} \qquad O{=}C{-}\overset{\alpha}{C}{=}\overset{\beta}{C}{-}\overset{\gamma}{C}{=}C\begin{matrix}\delta\\ \delta\end{matrix}$$

Base values, nm	
Acyclic α,β-unsaturated ketones	215
Acyclic α,β-unsaturated aldehyde	210
Six-membered cyclic α,β-unsaturated ketones	215
Five-membered cyclic α,β-unsaturated ketones	214
α,β-Unsaturated carboxylic acids and esters	195
Increments (in nm) for	
Double bond extending conjugation:	
Heteroannular	30
Homoannular	69
Alkyl group or ring residue:	
α	10
β	12
γ, δ	18
Polar groups:	
—OH	
α	35
β	30
γ	50
$-O-CO-CH_3$ and $-O-CO-C_6H_5$: α, β, γ, δ	6
$-OCH_3$	
α	35
β	30
γ	17
δ	31
—S—alkyl, β	85
—Cl	
α	15
β	12
—Br	
α	25
β	30
$-N(alkyl)_2$, β	95
Exocyclic double bond	5
Solvent correction (see Table 6-5)	____
Calculated wavelength =	total

TABLE 6-5 Solvent correction for ultraviolet-visible spectroscopy

Solvent	Correction, nm
Chloroform	+1
Cyclohexane	
Diethyl ether	+11
1,4-Dioxane	+5
Ethanol	0
Hexane	+11
Methanol	0
Water	−8

TABLE 6–6 Primary band of substituted benzene and heteroaromatics

In methanol

Base value: 203.5 nm

Substituent	Wavelength shift, nm	Substituent	Wavelength shift, nm
$-CH_3$	3.0	$-COOH$	25.5
$-CH{=}CH_2$	44.5	$-COO^-$	20.5
$-C{\equiv}CH$	44	$-CN$	20.5
$-C_6H_5$	48	$-NH_2$	26.5
$-F$	0	$-NH_3^+$	−0.5
$-Cl$	6.0	$-N(CH_3)_2$	47.0
$-Br$	6.5	$-NH-CO-CH_3$	38.5
$-I$	3.5	$-NO_2$	57
$-OH$	7.0	$-SH$	32
$-O^-$	31.5	$-SO-C_6H_5$	28
$-OCH_3$	13.5	$-SO_2CH_3$	13
$-OC_6H_5$	51.5	$-SO_2NH_2$	14.0
$-CHO$	46.0	$-CH{=}CH-C_6H_5$	
$-CO-CH_3$	42.0	*cis*	79
$-CO-C_6H_5$	48	*trans*	92.0
		$-CH{=}CH-COOH$, *trans*	69.5

Heteroaromatic	Base value, nm	Heteroaromatic	Base value, nm
Furan	200	Pyridine	257
Pyrazine	257	Pyrimidine	ca 235
Pyrazole	214	Pyrrole	209
Pyridazine	ca 240	Thiophene	231

TABLE 6-7 Wavelength calculation of the principal band of substituted benzene derivatives

In ethanol

Base value of parent chromophore, nm	
C_6H_5COOH or C_6H_5COO—alkyl	230
C_6H_5—CO—alkyl (or aryl)	246
C_6H_5CHO	250
Increment (in nm) for each substituent on phenyl ring	
—Alkyl or ring residue	
o-, *m*-	3
p-	10
—OH and —O— alkyl	
o-, *m*-	7
p-	25
$—O^-$	
o-	11
m-	20
p-	78*
—Cl	
o-, *m*-	0
p-	10
—Br	
o-, *m*-	2
p-	15
$—NH_2$	
o-, *m*-	13
p-	58
$—NHCO—CH_3$	
o-, *m*-	20
p-	45
$—NHCH_3$	
p-	73
$—N(CH_3)_2$	
o-, *m*-	20
p-	85

* Value may be decreased markedly by steric hindrance to coplanarity.

PHOTOLUMINESCENCE

TABLE 6-8 Fluorescence spectroscopy of some organic compounds

Compound	Solvent	pH	Excitation wavelength, nm	Emission wavelength, nm
Acenaphthene	Pentane		291	341
Acridine	CF_3COOH		358	475
Adenine	Water	1	280	375
Adenosine	Water	1	285	395
Adenosine triphosphate	Water	1	285	395
Adrenalin			295	335
p-Aminobenzoic acid	Water	8	295	345
Aminopterin	Water	7	280, 370	460
1-Aminopyrene	CF_3COOH		330, 342	415
p-Aminosalicyclic acid	Water	11	300	405
Amobarbital	Water	14	265	410
Anilines	Water	7	280, 291	344, 361
Anthracene	Pentane		420	430
Anthranilic acid	Water	7	300	405
Azaindoles	Water	10	290, 299	310, 347
Benz[*c*]acridine	CF_3COOH		295, 380	480
Benz[*a*]anthracene	Pentane		284	382
1,2-Benzanthracene			280, 340	390, 410
Benzanthrone	CF_3COOH		370, 420	550
Benzo[*b*]chrysene	Pentane		283	398
11-*H*-Benzo[*a*]fluorene	Pentane		317	340
Benzoic acid	70% H_2SO_4		285	385
3,4-Benzopyrene	Benzene		365	390, 480
Benzo[*e*]pyrene	Pentane		329	389
Benzoquinoline	CF_3COOH		280	425
Benzoxanthane	Pentane		363	418
Bromolysergic acid diethyl amide	Water	1	315	460
Brucine	Water	7	305	500
Carbazole	*N,N*-Dimethyl formamide		291	359
Chlortetracycline			355	445
Chrysene	Pentane		250, 300, 310	260, 380
Cinchonine	Water	1	320	420
Coumarin	Ethanol		280	352
Dibenzo[*a,c*]anthracene	Pentane		280	381
Dibenzo[*b,k*]chrysene	Pentane		308	428
Dibenzo[*a,e*]pyrene	Pentane		370	401
3,4,8,9-Dibenzopyrene			370, 335, 390, 410	480, 510
5,12-Dihydronaphthacene	Pentane		282	340
1,4-Diphenylbutadiene	Pentane		328	370

TABLE 6-8 Fluorescence spectroscopy of some organic compounds (*continued*)

Compound	Solvent	pH	Excitation wavelength, nm	Emission wavelength, nm
Epinephrine	Water	7	295	335
Ethacridine	Water	2	370, 425	515
Fluoranthrene	Pentane		354	464
Fluorene	Pentane		300	321
Fluorescein	Water	7–11	490	515
Folic acid	Water	7	365	450
Gentisic acid	Water	7	315	440
Griseofulvin	Water	7	295, 335	450
Guanine	Water	1	285	365
Harmine	Water	1	300, 365	400
Hippuric acid	70% H_2SO_4		270	370
Homovanillic acid	Water	7	270	315
m-Hydroxybenzoic acid	Water	12	314	430
p-Hydroxycinnamic acid	Water	7	350	440
7-Hydroxycoumarin	Ethanol		325	441
5-Hydroxyindole	Water	1	290	355
5-Hydroxyindoleacetic acid	Water	7	300	355
3-Hydroxykynurenine	Water	11	365	460
p-Hydroxymandelic acid	Water	7	300	380
p-Hydroxyphenylacetic acid	Water	7	280	310
p-Hydroxyphenylpyruvic acid	Water	7	290	345
p-Hydroxyphenylserine	Water	1	290	320
5-Hydroxytryptophan	Water	7	295	340
Imipramine	Water	14	295	415
Indoleacetic acid	Water	8	285	360
Indoles	Water	7	269, 315	355
Indomethacin	Water	13	300	410
Kynurenic acid	Water	7	325	405
		11	325	440
Lysergic acid diethylamide	Water	1	325	445
Menadione	Ethanol		335	480
9-Methylanthracene	Pentane		382	410
3-Methylcholanthrene	Pentane		297	392
7-Methyldibenzopyrene	Pentane		460	467
2-Methylphenanthrene	Pentane		257	357
3-Methylphenanthrene	Pentane		292	368
1-Methylpyrene	Pentane		336	394
4-Methylpyrene	Pentane		338	386
Naphthacene			290, 310	480, 515
1-Naphthol	0.1*M* NaOH 20% ethanol		365	480
2-Naphthol	0.1*M* NaOH 20% ethanol		365	426
Oxytetracycline			390	520
Phenanthrene	Pentane		252	362

TABLE 6-8 Fluorescence spectroscopy of some organic compounds (*continued*)

Compound	Solvent	pH	Excitation wavelength, nm	Emission wavelength, nm
Phenylalanine	Water		215, 260	282
o-Phenylenepyrene	Pentane		360	506
Phenylephrine			270	305
Picene	Pentane		281	398
Procaine	Water	11	275	345
Pyrene	Pentane		330	382
Pyridoxal	Water	12	310	365
Quinacrine	Water	11	285	420
Quinidine	Water	1	350	450
Quinine	Water	1	250, 350	450
Reserpine	Water	1	300	375
Resorcinol	Water		265	315
Riboflavin	Water	7	270, 370, 445	520
Rutin	Water	1	430	520
Salicyclic acid	Water	11	310	435
Scoparone	Water	10	350, 365	430
Scopoletin	Water	10	365, 390	460
Serotonin	3*M* HCl		295	550
Skatole	Water		290	370
Streptomycin	Water	13	366	445
p-Terphenyl	Pentane		284	338
Thiopental			315	530
Thymol	Water	7	265	300
Tocopherol	Hexane-ethanol		295	340
Tribenzo[*a,e,i*]pyrene	Pentane		384	448
Triphenylene	Pentane		288	357
Tryptamine	Water	7	290	360
Tryptophan	Water	11	285	365
Tyramine	Water	1	275	310
Tyrosine	Water	7	275	310
Uric acid	Water	1	325	370
Vitamin A	1-Butanol		340	490
Vitamin B_{12}	Water	7	275	305
Warfarin	Methanol		290, 342	385
Xanthine	Water	1	315	435
2,6-Xylenol			275	305
3,4-Xylenol			280	310
Yohimbine	Water	1	270	360
Zoxazolamine	Water	11	280	320

TABLE 6-9 Fluorescence quantum yield values

Compound	Solvent	Q_F value vs. Q_F standard	
	Q_F standard		
9-Aminoacridine	Water	0.99	
Anthracene	Ethanol	0.30	
POPOP*	Toluene	0.85	
Quinine sulfate dihydrate	1 N H_2SO_4	0.55	
	Secondary standards		
Acridine orange hydrochloride	Ethanol	0.54	Quinine sulfate
		0.58	Anthracene
1,8-ANS† (free acid)	Ethanol	0.38	Anthracene
		0.39	POPOP
1,8-ANS (magnesium salt)	Ethanol	0.29	Anthracene
		0.31	POPOP
Fluorescein	0.1 N NaOH	0.91	Quinine sulfate
		0.94	POPOP
Fluorescein, ethyl ester	0.1 N NaOH	0.99	Quinine sulfate
		0.99	POPOP
Rhodamine B	Ethanol	0.69	Quinine sulfate
		0.70	Anthracene
2,6-TNS‡ (potassium salt)	Ethanol	0.48	Anthracene
		0.51	POPOP

* POPOP *p*-bis[2-(5-phenyloxazoyl)]benzene.
† ANS, anilino-8-naphthalene sulfonic acid.
‡ TNS, 2-*p*-toluidinylnaphthalene-6-sulfonate.

TABLE 6-10 Phosphorescence spectroscopy of some organic compounds

Abbreviations Used in the Table

EPA, diethyl ether, isopentane, and ethanol (5:5:2) volume ratio

Compound	Solvent	Lifetime, s	Excitation wavelength, nm	Emission wavelength, nm
Acenaphthene	Ethanol		300	515
3-Acetylpyridine	Ethanol	0.5	395	525
Adenine	Water–methanol (9:1)	2.9	278	406
Adenosine	Ethanol	0.8	280	422
p-Aminobenzoic acid	Ethanol		305	425
2-Aminofluorene	Ethanol	4.6	380	590
6-Amino-6-methylmercapto-purine	Water–methanol (9:1)	0.66	321	456

TABLE 6-10 Phosphorescence spectroscopy of some organic compounds (*continued*)

Compound	Solvent	Lifetime, s	Excitation wavelength, nm	Emission wavelength, nm
2-Amino-4-methylpyrimidine	Ethanol	2.1	302	438
2-Amino-5-nitrobenzothiazole	EPA	0.41	375	515
2-Amino-5-nitrobiphenyl	EPA	0.56	380	520
3-L-Aminotyrosine·2HCl	Ethanol	0.8	286	398
Anthracene	Ethanol		300	462
Aspirin	EPA	2.1	240	380
Atropine	Ethanol	1.4		410
8-Azaguanine	Ethanol	1.8	282	442
Benzaldehyde	Ethanol	3.4	254	433
1,2-Benzanthracene	Ethanol	2.2	310	510
Benzimidazole	Ethanol	2.3	280	406
Benzocaine	Ethanol	3.4	310	430
1,2-Benzofluorene	Ethanol		315	502
Benzoic acid	EPA	2.4	240	400
3,4-Benzopyrene	Ethanol		325	508
Benzyl alcohol	Ethanol		219	393
6-Benzylaminopurine	Water-methanol (9:1)	2.8	286	413
Biphenyl	Ethanol	1.0	270	385
6-Bromopurine	Water-methanol (9:1)	0.5	273	420
Brucine	Ethanol	0.9	305	435
Caffeine	Ethanol	2.0	285	440
Carbazole	Ethanol	7.8	341	436
2-Chloro-4-aminobenzoic acid	Ethanol	1.0	312	337
p-Chlorophenol	Ethanol	<0.2	290	505
o-Chlorophenoxyacetic acid	Ethanol	0.7	280	518
p-Chlorophenoxyacetic acid	Ethanol	<0.5	283	396
6-Chloropurine	Water-methanol (9:1)	0.64	273	419
Chlorpromazine·HCl	Ethanol	0.3	320	490
Chlorotetracycline	Ethanol	2.7	280	410
Cocaine·HCl	Ethanol	2.7	240	400
Codeine	Ethanol	0.3	270	505
Cytidine	Water-methanol (9:1)		290	420
Desoxypyridoxine·HCl	Ethanol	1.4	290	442
Diacetylsulfanilamide	Ethanol	1.3	280	405
2,6-Diaminopurine	Water-methanol (9:1)	2.7	288	410
2,6-Diaminopurine sulfate	Ethanol	1.7	294	424
1,2,5,6-Dibenzanthracene	Ethanol	1.3	340	550
2,6-Dichloro-4-nitroaniline	EPA	0.5	368	525
2,4-Dichlorophenoxyacetic acid	Ethanol	<0.5	289	490
2,6-Diethyl-4-nitroaniline	EPA	0.66	388	525
3,4-Dihydroxymandelic acid	Ethanol	1.1	294	412

TABLE 6-10 Phosphorescence spectroscopy of some organic compounds (*continued*)

Compound	Solvent	Lifetime, s	Excitation wavelength, nm	Emission wavelength, nm
3,4-Dihydroxyphenylacetic acid	Ethanol	0.9	295	430
2,5-Dimethoxy-4-methyl-amphetamine	Water–methanol (9:1)	3.9	289	411
5,7-Dimethyl-1,2-benzacridine	Ethanol	0.6	310	555
N,*N*-Dimethyl-4-nitroaniline	EPA	0.54	398	525
N,*N*-Dimethyltryptamine	Water–methanol (9:1)	6.9	286	434
Dopamine	Ethanol	0.9	285	430
Ephedrine	Ethanol	3.6	225	390
Epinephrine	Ethanol	1.0	283	425
N-Ethylcarbazole	Ethanol	7.8	340	437
Ethyl 3-indoleacetate	Ethanol	3.3	290	440
Folic acid	Ethanol		367	425
Hippuric acid	EPA	4.9	311	450
Homovanillic acid	Ethanol	0.8	289	435
DL-5-Hydroxytryptophan	Ethanol	6.3	315	435
Indole-3-acetic acid	Ethanol	<0.5	290	438
3-Indoleacetonitrile	Ethanol	7.1	285	438
Indole-3-butanoic acid	Ethanol	0.6	284	510
Indolecarboxylic acid	Ethanol	5.5	290	429
Indole-2-propanoic acid	Ethanol	0.6	290	440
D-Lysergic acid	Water–methanol (9:1)	0.1	310	518
2-Methylcarbazole	Ethanol	8.1	333	442
N-Methylcarbazole	Ethanol	8.4	336	437
6-Methylmercaptopurine	Water–methanol (9:1)	0.6	291	420
N-Methyl-4-nitroaniline	EPA	0.5	390	522
6-Methylpurine	Water–Methanol (9:1)	3.2	272	405
Morphine	Ethanol	0.3	285	500
Naphthacene	Ethanol		300	518
Naphthalene	EPA	1.8	310	475
1-Naphthaleneacetic acid	Ethanol	2.8	295	510
1-Naphthol	Ethanol	1.1	320	475
2-Naphthoxyacetic acid	Ethanol	2.6	328	497
2-Naphthylamine	Ethanol	2.3	270	303
Niacinamide	Ethanol		270	410
Nicotine	Ethanol	5.2	270	390
5-Nitroacenaphthene	EPA		380	540
4-Nitroaniline	EPA	0.6	380	510
9-Nitroanthracene	EPA		248	488
1-Nitroanthraquinone	EPA	0.3	250	490
4-Nitrobiphenyl	EPA		330	480
3-Nitro-*N*-ethylcarbazole	EPA	0.4	315	475
2-Nitrofluorene	EPA	0.4	340	517

TABLE 6-10 Phosphorescence spectroscopy of some organic compounds (*continued*)

Compound	Solvent	Lifetime, s	Excitation wavelength, nm	Emission wavelength, nm
6-Nitroindole	EPA	0.4	372	520
1-Nitronaphthalene	EPA		340	520
2-Nitronaphthalene	EPA	0.4	260	500
4-Nitro-1-naphthylamine	EPA		400	578
4-Nitrophenol	Ethanol	<0.2	355	520
4-Nitrophenylhydrazine	EPA	0.5	390	520
4-Nitro-2-toluidine	EPA	0.5	375	520
Papaverine·HCl	Ethanal	1.5	260	480
Phenacetin	EPA			410
Phenanthrene	EPA	2.6	340	465
Phenobarbital	Ethanol	1.8	240	380
Phenylalanine	Ethanol		270	385
DL-2-Phenyllactic acid	Ethanol	5.4	262	383
Phthalylsulfathiazole	Ethanol	0.9	305	405
Procaine·HCl	Ethanol	3.5	310	430
Purine	Water–Methanol (9:1)	2.2	272	405
Pyrene	Ethanol		330	515
Pyridine	Ethanol	1.4	310	440
Pyridine-3-sulfonic acid	Ethanol	1.2	272	408
Pyridoxine·HCl	Ethanol		290	425
Quercetin	Ethanol	2.1	345	480
Quinidine sulfate	Ethanol	1.3	340	500
Quinine·HCl	Ehtnaol	1.3	340	500
Salicyclic acid	Ethanol	6.2	315	430
Strychnine phosphate	Ethanol	1.2	290	440
Sulfabenzamide	Ethanol	0.7	305	405
Sulfadiazine	Ethanol	0.7	275	410
Sulfanilamide	Ethanol	2.9	300	410
Sulfapyridine	Ethanol	1.4	310	440
Sulfathiazole	Ethanol	0.9	310	420
1,2,4,5-Tetramethylbenzene	EPA	4.5	275	390
2-Thiouracil	Ethanol	<0.5	310	430
2,4,5-Trichlorophenol	Ethanol	<0.2	305	485
2,4,5-Trichlorophenoxyacetic acid	Ethanol	1.1	295	475
Triphenylene	Ethanol	15	290	460
Tryptophan	Ethanol	1.5	295	440
Tyrosine	Ethanol	2.8	290	390
Vitamin K_1	Hexane	0.4	345	570
Vitamin K_3	Hexane	0.5	335	510
Vitamin K_5	Water–Methanol (9:1)	1.3	310	535
Warfarin	Ethanol	0.8	305	460
Yohimbine·HCl	Ethanol	7.4	290	410

INFRARED SPECTROSCOPY

TABLE 6-11 Absorption frequencies of single bonds to hydrogen

Abbreviations Used in the Table

m, moderately strong — var, of variable strength
m-s, moderate to strong — w, weak
s, strong — w-m, weak to moderately strong

Group	Band, cm^{-1}	Remarks
	Saturated C—H	
$-\overset{H}{\underset{H}{C}}-H$	2975–2950 (s) 2885–2865 (w) 1450–1260 (m)	Two or three bands usually; asymmetrical and symmetrical CH stretching, respectively. In presence of double bond adjacent to CH_3 group symmetrical band splits into two. Sensitive to adjacent negative substituents
$-\overset{H}{\underset{H}{C}}-$ acyclic	ca 2930 (s) 2870–2840 (w) 1480–1440 (m) ca 720 (w)	Frequency increased in strained systems. Symmetrical band splits into two bands when double bond adjacent. Scissoring mode Rocking mode
	Alkane residues attached to carbon	
Cyclopropane	ca 3050 (w) 540–500 470–460 (s)	CH stretching Aliphatic cyclopropanes
Cyclobutanes Cyclopentanes	580–490 (s) 595–490 (s)	Alkyl derivatives: 550–530 cm^{-1} Alkyl derivatives: 585–530 cm^{-1}
$>C(CH_3)_2$	ca 1380 (m) 1175–1165 (m) 1150–1130 (m)	A roughly symmetrical doublet If no H on central carbon, then one band at ca 1190 cm^{-1}
$-C(CH_3)_3$	1395–1385 (m) 1365 (s)	Split into two bands
Aryl-CH_3 Aryl-C_2H_5 Aryl-C_3H_7 (or C_4H_9)	390–260 (m) 565–540 (m-s) 585–565 (m)	 Two bands

TABLE 6-11 Absorption frequencies of single bonds to hydrogen (*continued*)

Group	Band, cm^{-1}	Remarks
Alkane residues attached to carbon (*continued*)		
$-(CH_2)_n-$		
$n=1$	785–770 (w–m)	Rocking vibrations
$n=2$	745–735 (w–m)	
$n=3$	735–725 (w–m)	
$n\geq 4$	725–720 (w–m)	
Alkane residues attached to miscellaneous atoms		
Epoxide C—H	ca 3050 (m–s)	
$>C\overset{NH}{\triangle}CH_2$ (aziridine)	ca 3050 (m–s)	
$-CH_2-$halogen	ca 3050 (m–s) 1435–1385 (m) 1300–1240 (s)	Halogens except fluorine
$-CHO$	2900–2800 (w) 2775–2700 (w) 1420–1370 (m)	
$-CO-CH_3$	3100–2900 (w) 1450–1400 (s) 1360–1355 (s)	
$-O-CH_3$ ethers	2835–2810 (s) 1470–1430 (m–s) ca 1030 (w–m)	 Two bands
$-O-C(CH_3)_3$	1200–1155 (s)	
$-O-CH_2-O-$	2790–2770 (m)	
$-O-CH_2-$ esters	1475–1460 (m–s) 1470–1435 (m–s)	 Acyclic esters. Frequency increased ca 30 cm^{-1} for cyclic and small ring systems.
$-O-CO-CH_3$	1450–1400 (s) 1385–1365 (s) 1360–1355 (s)	Acetate esters The high intensity of these bands often dominates this region of the spectrum.
$-CH_2-\overset{\vert}{C}=C<$	1445–1430 (m)	
$-CH_2-SO_2-$	ca 1250 (m)	

TABLE 6-11 Absorption frequencies of single bonds to hydrogen (*continued*)

Group	Band, cm^{-1}	Remarks
Alkane residues attached to miscellaneous atoms (*continued*)		
P—CH_3	1320–1280 (s)	
Se—CH_3	ca 1280 (m)	
B—CH_3	1460–1405 (m)	
	1320–1280 (m)	
Si—CH_3	1265–1250 (m-s)	
Sn—CH_3	1200–1180 (m)	
Pb—CH_3	1170–1155 (m)	
As—CH_3	1265–1240 (m)	
Ge—CH_3	1240–1230 (m)	
Sb—CH_3	1215–1195 (m)	
Bi—CH_3	1165–1145 (m)	
—CH_2—(Cd, Hg, Zn, Sn)	1430–1415 (m)	
N—CH_3 and N—CH_2—	2820–2780 (s)	
	1440–1390 (m)	Ethylenediamine complexes
N—CH_2—CH_2—N	1480–1450 (s)	Ethylenediamine complexes
N—CH_3		
Amine · HCl	1475–1395 (m)	
Amino acid · HCl	1490–1480 (m)	
Amides	1420–1405 (s)	
N—CH_2— amides	ca 1440 (m)	
S—CH_3	2990–2955 (m-s)	
	2900–2865 (m-s)	
	1440–1415 (m)	
	1325–1290 (m)	
	1030–960 (m)	
	710–685 (w-m)	
S—CH_2—	2950–2930 (m)	
	2880–2845 (m)	
	1440–1415 (m)	
	1270–1220 (s)	
—C≡CH	ca 3300 (s)	Sharp
	700–600	Bending
$>C=CH-$ (trisubstituted, one H)	3040–3010 (m)	
$>C=CH_2$	3095–3075 (m)	CH stretching sometimes obscured by much stronger bands of saturated CH groups
	2985–2970 (m)	

TABLE 6-11 Absorption frequencies of single bonds to hydrogen (*continued*)

Group	Band, cm^{-1}	Remarks
Alkane residues attached to miscellaneous atoms (*continued*)		
R H C=C H H	995–980 (s) 940–900 (s) ca 635 (s) 485–445 (m–s)	
R H C=C R H	895–885 (s) 560–530 (s) 470–435 (m)	
R H C=C H R	980–955 (s) 455–370 (m–s)	
H H C=C R R	730–655 (m) 670–455 (s)	
R H C=C R R	850–790 (m) 570–515 (s) 525–470 (s)	
$-O-CH=CH_2$	965–960 (s) 945–940 (m) 820–810 (s)	
$-S-CH=CH_2$	ca 965 (s) ca 860 (s)	
$-CO-CH=CH_2$	995–980 (s) 965–955 (m)	
$-CO-OCH=CH_2$	950–935 (s) 870–850 (s)	
$-CO-C=CH_2$	ca 930 (s)	
$-CO-OC=CH_2$	880–865	
$-O-CH=CH-$ *trans*	940–920 (s)	
$-CO-CH=CH-$ *trans*	ca 990 (s)	
Hydroxyl group O—H compounds		
Primary aliphatic alcohols	3640–3630 (s)	Only in very dilute solutions in nonpolar solvents
	1350–1260 (s)	OH bending
	1085–1030 (s)	Also broad band at 700–600 cm^{-1}

TABLE 6-11 Absorption frequencies of single bonds to hydrogen (*continued*)

Group	Band, cm^{-1}	Remarks
Hydroxyl group O—H compounds (*continued*)		
Secondary aliphatic alcohols	3625–3620 (s)	See comments under primary aliphatic alcohols
	1350–1260 (s)	
	1125–1085 (s)	Also for α-unsaturated and cyclic tertiary aliphatic alcohols
Tertiary aliphatic alcohols	3620–3610 (s)	See comments under primary aliphatic alcohols
	1410–1310 (s)	
	1205–1125 (s)	
Aryl—OH	ca 3610 (s)	See comments under primary aliphatic alcohols
	1410–1310 (s)	
	1260–1180 (s)	
	1085–1030 (s)	Also for unsaturated secondary aliphatic alcohols
Carboxylic acids	3300–2500(w–m)	Broad
	995–915 (s)	Broad diffuse band
Enol form of β-diketones	2700–2500 (var)	Broad
Free oximes	3600–3570(w–m)	Shoulder
Free hydroperoxides	3560–3530 (m)	
Peroxy acids	ca 3280 (m)	
Phosphorus acids	2700–2560 (m)	Broad
Water in solution	3710	When solution is damp
Intermolecular H bond		
Dimeric	3600–3500	Rather sharp. Absorptions arising from H bond with polar solvents also appear in this region.
Polymeric	3400–3200 (s)	Broad
Intramolecular H bond		
Polyvalent alcohols	3600–3500 (s)	Sharper than dimeric band above
Chelation	3200–2500	Broad and occasionally weak; the lower the frequency, the stronger the intramolecular bond

TABLE 6-11 Absorption frequencies of single bonds to hydrogen (*continued*)

Group	Band, cm^{-1}	Remarks
	Hydroxyl group O—H compounds (*continued*)	
Water of crystallation (solid state spectra)	3600–3100 (w)	Usually a weak band at 1640–1615 cm^{-1} also. Water in trace amounts in KBr disks shows a broad band at 3450 cm^{-1}.
	Amine, imine, ammonium, and amide N—H	
Primary amines	3550–3300 (m)	Two bands in this range
Aliphatic	1650–1560 (m)	
	1090–1020 (w–m)	
	850–810 (w–m)	With α-carbon branching at 795 cm^{-1} and strong
	495–445 (m–s)	Broad
	ca 290 (s)	Broad
Aromatic	1350–1260 (s)	Also for secondary aryl amines
	445–345	
Amino acids	3100–3030 (m)	Values for solid states; broad bands also (but not always) near 2500 and 200 cm^{-1}
	2800–2400 (m)	Number of sharp bands; dilute solution
	1625–1560 (m)	
	1550–1550 (m)	
Amino salts	3550–3100 (m)	Values for solid state
	ca 3380	Dilute solutions
	ca 3280	
Secondary amines	3550–3400 (w)	Only one band, whereas primary amines show two bands
	1580–1490 (w)	Often too weak to be noticed
	1190–1170 (m)	
	1145–1130 (m)	
	455–405 (w–m)	
Salts	ca 2500	Sharp; broad values for solid state
	ca 2400	Sharp; broad values for solid state
	1620–1560 (m–s)	
Tertiary amines		
$R_1R_2R_3NH^+$	2700–2250	Group of relatively sharp bands; broad bands in solid state
Ammonium ion	3300–3030 (s)	Group of bands
	1430–1390 (s)	

TABLE 6-11 Absorption frequencies of single bonds to hydrogen (*continued*)

Group	Band, cm^{-1}	Remarks
Amine, imine, ammonium, and amide N—H (*continued*)		
Imines =N=H	3350–3310 (w) 3490 (s) 3490 (s)	Aliphatic Aryl Pyrroles, indoles; band sharp
Imine salts	2700–2330 (m–s) 2200–1800 (m)	Dilute solutions One or more bands; useful to distinguish from protonated tertiary amines
Primary amide —$CONH_2$	ca 3500 (m) ca 3400 (m)	Lowered ca 150 cm^{-1} in the solid state and on H bonding; often several bands 3200–3050 cm^{-1}
Secondary amide —CONH—	3460–3400 (m) 3100–3070 (w)	Two bands; lowered on H bonding and in solid state. Only one band with lactams Extra band with bonded and solid-state samples
Miscellaneous R—H		
—S—H	2600–2550 (w)	Weaker than OH and less affected by H bonding
P—H	2440–2350 (m)	Sharp
P(=O)—OH	2700–2560 (m)	Associated OH
R—D	100/137 times the corresponding RH frequency	Useful when assigning RH bands; deuteration leads to a known shift to lower frequency

TABLE 6-12 Absorption frequencies of triple bonds

Abbreviations Used in the Table

m, moderately strong — var, of variable strength
m–s, moderate to strong — w–m, weak to moderately strong
s, strong

Group	Band, cm^{-1}	Remarks
Alkynes		
Terminal	3300 (s)	CH stretching
	2140–2100 (w–m)*	C≡C stretching
	1375–1225 (w–m)	
	695–575 (m–s)	Two bands if molecule has axial symmetry
	ca 630 (s)	Alkyl monosubstituted
Nonterminal	2260–2150 (var)*	Symmetrical or nearly symmetrical substitution makes the C≡C stretching frequency inactive. When more than one C≡C linkage is present, and sometimes when there is only one, there are frequently more absorption bands in this region than there are triple bonds to account for them.
R_1—C≡C—R_2	540–465 (m)	The longer the chain, the lower the frequency
Aryl—C≡C—	ca 550 (m) ca 350 (var)	
—C≡C—halogen (Cl, Br, I)	185–160 (var)	
Nitriles —C≡N	2260–2200 (var)	Stronger and toward the lower end of the range when conjugated; occasionally very weak or absent
Aliphatic	580–555 (m–s) 560–525 (m–s) 390–350 (s)	
Aromatic	580–540 (s) 430–380 (m)	
Isonitriles R—$\overset{+}{N}$≡$\overset{-}{C}$ or R—N=C:	2175–2150 (s) 2150–2115 (s) 1595	Very sensitive to changes in substituents Not found for nitriles
Cyanamides >N—C≡N ⇌ >$\overset{+}{N}$—C=$\overset{-}{N}$	2225–2210 (s)	

* Conjugation with olefinic or acetylenic groups lowers the frequency and raises the intensity. Conjugation with carbonyl groups usually has little effect on the position of absorption.

TABLE 6-12 Absorption frequencies of triple bonds (*continued*)

Group	Band, cm^{-1}	Remarks
Thiocyanates R—S—C≡N	2175–2140 (s) 404–400 (s) ca 600 (m–s)	Aryl thiocyanates at the upper end of the range, alkyl at the lower end Aliphatic derivatives
Nitrile *N*-oxides —C≡N→O	2305–2285 (s) 1395–1365 (s)	Aryl derivatives
Diazonium salts $R-\overset{+}{N}\equiv N$	2300–2230 (m–s)	
Selenocyanates R—Se—C≡N	ca 2160 (m–s) 545–520 ca 390 ca 350	

TABLE 6-13 Absorption frequencies of cumulated double bonds

Abbreviations Used in the Table

m–s, moderate to strong vs, very strong
s, strong w, weak

Group	Band, cm^{-1}	Remarks
Carbon dioxide O=C=O	2349 (s)	Appears in many spectra as a result of inequalities in path length
Isocyanates —N=C=O	2275–2250 (vs)	Position unaffected by conjugation
Isoselenocyanates —N=C=Se	2200–2000 (s) 675–605	Broad; usually two bands
Azides $-N_3$ or $-N=\overset{+}{N}=\overset{-}{N}$	2140–2030 (s) 1340–1180 (w)	 Not observed for ionic azides
—N=C=N—	2155–2130 (s)	Split into unsymmetrical doublet by conjugation with aryl groups: 2145–2125 (vs) and 2115–2105 (vs)

TABLE 6-13 Absorption frequencies of cumulated double bonds (*continued*)

Group	Band, cm^{-1}	Remarks
Isothiocyanates —N=C=S	2140–1990 (vs) 649–600 (m–s) 565–510 (m–s) 470–440 (m–s)	Broad; usually a doublet
Ketenes >C=C=O	ca 2150 (s)	
Ketenimines C=C=N—	2050–2000 (s)	
Allenes >C=C=C<	2000–1915 (m–s)	Two bands when terminal allene or when bonded to electron-attracting groups
Thionylamines —N=S=O	1300–1230 (s) 1180–1110 (s)	
Diazoalkanes $R_2C=\overset{+}{N}=\overset{-}{N}$ $-CH=\overset{+}{N}=\overset{-}{N}$	2030–2000 (s) 2050–2035 (s)	
Diazoketones $-CO-CH=\overset{+}{N}=\overset{-}{N}$	2100–2080 2075–2050	Monosubstituted Disubstituted

TABLE 6-14 Absorption frequencies of carbonyl bands

All bands quoted are strong.

Groups	Band, cm^{-1}	Remarks
Acid anhydrides **—CO—O—CO—**		
Saturated	1850–1800 1790–1740	Two bands usually separated by about 60 cm^{-1}. The higher-frequency band is more intense in acyclic anhydrides, and the lower-frequency band is more intense in cyclic anhydrides.
Aryl and α,β-unsaturated	1830–1780 1700–1710	

TABLE 6-14 Absorption frequencies of carbonyl bands (*continued*)

Groups	Band, cm^{-1}	Remarks
Acid anhydrides (*continued*)		
—CO—O—CO—		
Saturated five-ring	1870–1820 1800–1750	
All classes	1300–1050	One or two strong bands due to CO stretching
Acid chlorides —COCl		
Saturated	1815–1790	Acid fluorides higher, bromides and iodides lower
Aryl and α,β-unsaturated	1790–1750	
Acid peroxide		
CO—O—O—CO—		
Saturated	1820–1810 1800–1780	
Aryl and α,β-unsaturated	1805–1780 1785–1755	
Esters and lactones		
—CO—O—		
Saturated	1750–1735	
Aryl and α,β-unsaturated	1730–1715	
Aryl and vinyl esters C=C—O—CO—alkyl	1800–1750	The C=C stretching band also shifts to higher frequency.
Esters with electronegative α substituents; e.g., >CCl—CO—O—	1770–1745	
α-Keto esters	1755–1740	
Six-ring and larger lactones	Similar values to the corresponding open-chain esters	
Five-ring lactone	1780–1760	
α,β-Unsaturated five-ring lactone	1770–1740	When α-CH is present, there are two bands, the relative intensity depending on the solvent.
β,γ-Unsaturated five-ring lactone, vinyl ester type	ca 1800	
Four-ring lactone	ca 1820	
β-Keto ester in H bonding enol form	ca 1650	Keto from normal; chelate-type H bond causes shift to lower frequency than the normal ester. The C=C band is strong and is usually near 1630 cm^{-1}.
All classes	1300–1050	Usually two strong bands due to CO stretching

TABLE 6-14 Absorption frequencies of carbonyl bands (*continued*)

Groups	Band, cm^{-1}	Remarks
Aldehydes —CHO (See also Table 6-39 for C—H.) All values given below are lowered in liquid-film or solid-state spectra by about 10–20 cm^{-1}. Vapor-phase spectra have values raised about 20 cm^{-1}.		
Saturated	1740–1720	
Aryl	1715–1695	*o*-Hydroxy or amino groups shift this value to 1655–1625 cm^{-1} because of intramolecular H bonding.
α,β-Unsaturated	1705–1680	
$\alpha,\beta,\gamma,\delta$-Unsaturated	1680–1660	
β-Ketoaldehyde in enol form	1670–1645	Lowering caused by chelate-type H bonding
Ketones >C=O All values given below are lowered in liquid-film or solid-state spectra by about 10–20 cm^{-1}. Vapor-phase spectra have values raised about 20 cm^{-1}.		
Saturated	1725–1705	
Aryl	1700–1680	
α,β-Unsaturated	1685–1665	
$\alpha,\beta,\alpha',\beta'$-Unsaturated and diaryl	1670–1660	
Cyclopropyl	1705–1685	
Six-ring ketones and larger	Similar values to the corresponding open-chain ketones	
Five-ring ketones	1750–1740	α,β Unsaturation, $\alpha,\beta,\alpha',\beta'$ unsaturation, etc., have a similar effect on these values as on those of open-chain ketones.
Four-ring ketones	ca 1780	
α-Halo ketones	1745–1725	Affected by conformation; highest values are obtained when both halogens are in the same plane as the C=O.
α,α'-Dihalo ketones	1765–1745	
1,2-Diketones, *syn-trans*-open chains	1730–1710	Antisymmetrical stretching frequency of both C=O's. The symmetrical stretching is inactive in the infrared but active in the Raman.

TABLE 6-14 Absorption frequencies of carbonyl bands (*continued*)

Groups	Band, cm^{-1}	Remarks
Ketones $>C=O$ (*continued*)		
syn-cis-1,2-Diketones, six-ring	1760 and 1730	
syn-cis-1,2-Diketones, five ring	1775 and 1760	
o-Amino-aryl or *o*-hydroxy-aryl ketones	1655–1635	Low because of intramolecular H bonding. Other substituents and steric hindrance affect the position of the band.
Quinones	1690–1660	C=C band is strong and is usually near 1600 cm^{-1}.
Extended quinones	1655–1635	
Tropone	1650	Near 1600 cm^{-1} when lowered by H bonding as in tropolones
Carboxylic acids $—CO_2H$		
All types	3000–2500	OH stretching; a characteristic group of small bands due to combination bands
Saturated	1725–1700	The monomer is near 1760 cm^{-1}, but is rarely observed. Occasionally both bands, the free monomer, and the H-bonded dimer can be seen in solution spectra. Ether solvents give one band near 1730 cm^{-1}.
α,β-Unsaturated	1715–1690	
Aryl	1700–1680	
α-Halo-	1740–1720	
Carboxylate ions $—CO_2^-$		
Most types	1610–1550 1420–1300	Antisymmetrical and symmetrical stretching, respectively
Amides $—CO—N<$ (See also Table 6-39 for NH stretching and bending.)		
Primary $—CONH_2$		
In solution	ca 1690	Amide I; C=O stretching
Solid state	ca 1650	
In solution	ca 1600	Amide II: mostly NH bending
Solid state	ca 1640	
		Amide I is generally more intense than amide II. (In the solid state, amides I and II may overlap.)
Secondary —CONH—		
In solution	1700–1670	Amide I
Solid state	1680–1630	
In solution	1550–1510	Amide II; found in open-chain amides only

TABLE 6-14 Absorption frequencies of carbonyl bands (*continued*)

Groups	Band, cm^{-1}	Remarks
Amides —CO—N< (*continued*)		
Solid state	1570–1515	Amide I is generally more intense than amide II.
Tertiary	1670–1630	Since H bonding is absent, solid and solution spectra are much the same.
Lactams		
Six-ring and larger rings	ca 1670	
Five-ring	ca 1700	Shifted to higher frequency when
Four-ring	ca 1745	the N atom is in a bridged system
R—CO—N—C=C		Shifted +15 cm^{-1} by the additional double bond
C=C—CO—N		Shifted by up to +15 cm^{-1} by the additional double bond. This is an unusual effect by α,β unsaturation. It is said to be due to the inductive effect of the C=C on the well-conjugated CO—N system, the usual conjugation effect being less important in such a system.
Imides —CO—N—CO—		
Cyclic six-ring	ca 1710 and ca 1700	Shift of +15 cm^{-1} with α,β unsaturation
Cyclic five-ring	ca 1770 and ca 1700	
Ureas N—CO—N		
RNHCONHR	ca 1660	
Six-ring	ca 1640	
Five-ring	ca 1720	
Urethanes R—O—CO—N	1740–1690	Also shows amide II band when nonsubstituted on N
Thioesters and Acids RCO—S—R′		
RCOSH	ca 1720	α,β-Unsaturated or aryl acid or ester shifted about −25 cm^{-1}
RCOS—alkyl	ca 1690	
RCOS—aryl	ca 1710	

Intensities of Carbonyl Bands

Acids generally absorb more strongly than esters, and esters more strongly than ketones or aldehydes. Amide absorption is usually similar in intensity to that of ketones but is subject to much greater variations.

Position of Carbonyl Absorption

The general trends of structural variation on the position of C=O stretching frequencies may be summarized as follows:

1. The more electronegative the group X in the system R—CO—X—, the higher is the frequency.
2. α,β Unsaturation causes a lowering of frequency of 15 to 40 cm^{-1}, except in amides, where little shift is observed and that usually to higher frequency.
3. Further conjugation has relatively little effect.
4. Ring strain in cyclic compounds causes a relatively large shift to higher frequency. This phenomenon provides a remarkably reliable test of ring size, distinguishing clearly between four-, five-, and larger-membered-ring ketones, lactones, and lactams. Six-ring and larger ketones, lactones, and lactams show the normal frequency found for the open-chain compounds.
5. Hydrogen bonding to a carbonyl group causes a shift to lower frequency of 40 to 60 cm^{-1}. Acids, amides, enolized β-keto carbonyl systems, and *o*-hydroxyphenol and *o*-aminophenyl carbonyl compounds show this effect. All carbonyl compounds tend to give slightly lower values for the carbonyl stretching frequency in the solid state compared with the value for dilute solutions.
6. Where more than one of the structural influences on a particular carbonyl group is operating, the net effect is usually close to addititive.

TABLE 6-15 Absorption frequencies of other double bonds

Abbreviations Used in the Table

m, moderately strong; vs, very strong
m–s, moderate to strong; w, weak
var, of variable strength

Group	Band, cm^{-1}	Remarks
	Alkenes >C=C<	
Nonconjugated	1680–1620 (w–m)	May be very weak if symmetrically substituted
Conjugated with aromatic ring	1640–1610 (m)	More intense than with unconjugated double bonds
Internal (ring)	3060–2995 (m)	Highest frequencies for smallest ring
Carbons: $n=3$	ca 1665 (w–m)	
$n=4$	ca 1565 (w–m)	
$n=5$	ca 1610 (w–m)	
	1370–1340 (s)	Characteristic
$n\geq 6$	1650–1645 (w–m)	

TABLE 6-15 Absorption frequencies of other double bonds (*continued*)

Group	Band, cm^{-1}	Remarks
Alkenes $>C=C<$ (*continued*)		
Exocyclic $C=C(CH_2)_n$ $n=2$	1780–1730 (m)	
$n=3$	ca 1680 (m)	
$n \geq 4$	1655–1650 (m)	
Fulvene	1645–1630 (m) 1370–1340 (s) 790–765 (s)	
Dienes, trienes, etc.	1650 (s) and 1600 (s)	Lower-frequency band usually more intense and may hide or overlap the higher-frequency band
α, β-Unsaturated carbonyl compounds	1640–1590 (m)	Usually much weaker than the C=O band
Enol esters, enol ethers, and enamines	1700–1650 (s)	
Imines, oximes, and amidines $>C=N-$		
Imines and oximes		
Aliphatic	1690–1640 (w)	
α,β-Unsaturated and aromatic	1650–1620 (m)	
Conjugated cyclic systems	1660–1480 (var)	
	960–930 (s)	NO stretching of oximes
Imino ethers $-O-C=N-$	1690–1640 (var)	Usually a strong doublet
Imino thioethers $-S-C=N=$	1640–1605 (var)	
Imine oxides $>C=\overset{+}{N}-\overset{-}{O}$	1620–1550 (s)	
Amidines $>N-C=N-$	1685–1580 (var)	
Benzamidines Aryl$-C=N-N$	1630–1590	
Guanidine $>N-C(-N)=N-$	1725–1625 (s)	
Azines $>C=N-N=C<$	1670–1600	
Hydrazoketones $-CO-C=N-N$	1600–1530 (vs)	

TABLE 6-15 Absorption frequencies of other double bonds (*continued*)

Group	Band, cm^{-1}	Remarks
	Azo compounds —N=N—	
Azo —N=N—		
Aliphatic	ca 1575 (var)	Very weak or inactive
Aromatic		
cis	ca 1510 (w)	
trans	1440–1410 (w)	
Azoxy $-\overset{+}{N}(O^-)=N-$		
Aliphatic	1590–1495 (m–s)	
	1345–1285 (m–s)	
Aromatic	1480–1450 (m–s)	
	1340–1315 (m–s)	
Azothio $-N=\overset{+}{N}-\overset{-}{S}-$	1465–1445 (w)	
	1070–1055 (w)	
	Nitro compounds N=O	
Nitro C—NO_2		
Aliphatic	ca 1560 (s)	The two bands are due to asymmetrical and symmetrical stretching of the N=O bond. Electron-withdrawing substituents adjacent to nitro group increase the frequency of the asymmetrical band and decrease that of the symmetrical frequency.
	1385–1350 (s)	
Aromatic	1570–1485 (s)	See above remark; also bulky orthosubstituents shift band to higher frequencies. Strong H bonding shifts frequency to lower end of range.
	1380–1320 (s)	
	865–835 (s)	Strong and sometimes at ca 750 cm^{-1}
	580–520 (var)	
α,β-Unsaturated	1530–1510 (s)	
Nitroalkenes	1360–1335 (s)	
Nitrates —O—NO_2	1650–1625 (vs)	
	1285–1275 (vs)	
	870–855 (vs)	
	760–755 (w–m)	
	710–695 (w–m)	
Nitramines >N—NO_2	1630–1550 (s)	
	1300–1250 (s)	

TABLE 6-15 Absorption frequencies of other double bonds (*continued*)

Group	Band, cm^{-1}	Remarks
	Nitro compounds N=O (*continued*)	
Nitrates —O—N=O	1680–1610 (vs)	Two bands
	815–750 (s)	*Trans* form
	850–810 (s)	*Cis* form
	690–615 (s)	
Thionitrites —S—N=O	730–685 (m–s)	
Nitroso ⩾C—N=O	1600–1500 (s)	
$N-\overset{+}{N}=\overset{-}{O}$		
Aliphatic	1530–1495 (m–s)	
Aromatic	1480–1450 (m–s)	
	1335–1315 (m–s)	
Nitrogen oxides N→O		
Pyridine	1320–1230 (m–s)	
	1190–1150 (m–s)	
Pyrazine	1380–1280 (m–s)	Affected by ring substituents
	1040–990 (m–s)	
	ca 850 (m)	

Table 6-16 Absorption frequencies of aromatic bands

Abbreviations Used in the Table

m, moderately strong — var, of variable strength
m–s, moderate to strong — w–m, weak to moderately strong
s, strong

Group	Band, cm^{-1}	Remarks
Aromatic rings	ca 1600 (m)	
	ca 1580 (m)	Stronger when ring is further conjugated
	ca 1470 (m)	When substituent on ring is electron acceptor
	ca 1510 (m)	When substituent on ring is electron donor
Five adjacent H	900–860 (w–m)	
	770–730 (s)	
	720–680 (s)	
	625–605 (w–m)	
	ca 550 (w–m)	Substituents: C=C, C≡C, C≡N

Table 6-16 Absorption frequencies of aromatic bands (*continued*)

Group	Band, cm^{-1}	Remarks
1,2-Substitution	770–735 (s) 555–495 (w–m) 470–415 (m–s)	
1,3-Substitution	810–750 (s) 560–505 (m) 460–415 (m–s)	490–460 cm^{-1} when substituents are electron-accepting groups
1,4-Substitution	860–800 (s) 650–615 (w–m) 520–440 (m–s)	520–490 cm^{-1} when substituents are electron-donating groups
1,2,3-Trisubstitution	800–760 (s) 720–685 (s) 570–535 (s) ca 485	
1,2,4-Trisubstitution	900–885 (m) 780–760 (s) 475–425 (m–a)	
1,3,5-Trisubstitution	950–925 (var) 865–810 (s) 730–680 (m–s) 535–495 (s) 470–450 (w–m)	
Pentasubstitution	900–860 (m–s) 580–535 (s)	
Hexasubstitution	415–385 (m–s)	

TABLE 6-17 Absorption frequencies of miscellaneous bands

Abbreviations Used in the Table

m, moderately strong — vs, very strong
m–s, moderate to strong — w, weak
s, strong — w–m, weak to moderately strong
var, of variable strength

Group	Band, cm^{-1}	Remarks
	Ethers	
Saturated aliphatic $\geqq C{-}O{-}C\leqq$	1150–1060 (vs) 1140–900 (s)	Two peaks may be observed for branched chain, usually 1140–1110 cm^{-1}. Usually 930–900 cm^{-1}; may be absent for symmetric ethers
Alkyl-aryl $=\underset{\vert}{C}{-}O{-}C\leqq$	1270–1230 (vs) 1120–1020 (s)	=CO stretching CO stretching
Vinyl	1225–1200 (s)	Usually about 1205 cm^{-1}
Diaryl $=\underset{\vert}{C}{-}O{-}\underset{\vert}{C}=$	1200–1120 (s) 1100–1050 (s)	
Cyclic	1270–1030 (s)	
Epoxides $>C{-}C<$ (bridged by O)	1260–1240 (m–s) 880–805 (m) 950–860 (var) 865–785 (m) 770–750 (m)	 Monosubstituted *Trans* form *Cis* form Trisubstituted
Ketals and acetals	1190–1140 (s) 1195–1125 (s) 1100–1000 (s) 1060–1035 (s)	 Strongest band Sometimes obscured
Phthalanes	915–895 (s)	
Aromatic methylenedioxy	1265–1235 (s)	
	Peroxides	
–O–O–	900–830 (w) 1150–1030 (m–s) ca 1000 (m)	 Alkyl Aryl

TABLE 6-17 Absorption frequencies of miscellaneous bands (*continued*)

Group	Band, cm^{-1}	Remarks
	Sulfur compounds	
Thiols		
—S—H	2600–2450 (w)	
—CO—SH	840–830 (m)	
—CS—SH	ca 860 (s)	Broad
Thiocarbonyl		
>C=S	1200–1050 (s)	Behaves generally in manner similar to carbonyl band
>N—C(—)=S	1570–1395 1420–1260 1140–940	
—S—C(—)=S	ca 580 (s)	
Sulfoxides		
>S=O	1075–1040 (vs)	Halogen or oxygen atom bonded to sulfur increases the frequency.
	730–690 (var) 395–360 (var)	
Sulfones		
$>SO_2$	1360–1290 (vs)	Halogen or oxygen atom bonded to sulfur increases the frequency.
	1170–1120 (vs) 610–545 (m–s) 525–495 (m–s)	
Sulfonamides		
$—SO_2—N<$	1380–1330 (vs) 1170–1140 (vs) 950–860 (m) 715–700 (w–m)	
Sulfonates		
$—SO_2—O—$	1420–1330 (s) 1200–1145 (s)	May appear as doublet
Thiosulfonates		
$—SO_2—S—$	ca 1340 (vs)	
Sulfates $—O—SO_2—O—$	1415–1380 (s) 1200–1185 (s)	Electronegative substituents increase frequencies.
Primary alkyl salts	1315–1220 (s) 1140–1075 (m)	Strongly influenced by metal ion

TABLE 6-17 Absorption frequencies of miscellaneous bands (*continued*)

Group	Band, cm^{-1}	Remarks
	Sulfur compounds (*continued*)	
Sulfates $—O—SO_2—O—$ (*continued*)		
Secondary alkyl salts	1270–1210 (vs) 1075–1050 (s)	Doublet; both bands strongly influenced by metal ion
Stretching frequencies of C—S and S—S bonds		
$—S—CH_3$	710–685 (w–m)	
$—S—CH_2—$	660–630 (w–m)	
$—S—CH<$	630–600 (w–m)	
$—S—C\equiv$	600–570 (w–m)	
—S—aryl	1110–1070 (m) 710–685 (w–m)	
R—S—S—R	705–570 (w) 520–500 (w)	
Aryl—S—S—aryl	500–430 (w–m)	
Polysulfides	500–470 (w–m)	
$CH_2—S—CH_2—$	695–655 (w–m)	CSC stretching
$(R—S)_2C=O$	880–825 (s) 570–560 (var)	
—CO—S—	1035–935 (s)	
—CS—S	ca 580 (s)	
=C(S—)(S—)	1050–900 (m–s) 980–850 (m–s) 900–800 (m–s)	Monoionic Ionic 1,1-dithiolates
	Phosphorus compounds	
P—H	2455–2265 (m) 1150–965 (w–m)	Sharp. Phosphines lie in the region 2285–2265 cm^{-1}.
$—PH_2$	1100–1085 (m) 1065–1040 (w–m) 940–910 (m)	
P—alkyl	795–650 (m–s)	
P—aryl	1130–1090 (s) 750–680 (s)	
P—O—alkyl	1050–970 (s)	Broad
P—O—aryl	1240–1190 (s)	
P—O—P	970–910	Broad

TABLE 6-17 Absorption frequencies of miscellaneous bands (*continued*)

Group	Band, cm^{-1}	Remarks
	Phosphorus compounds (*continued*)	
P=O	1350–1150 (s)	May appear as doublet
P(=O)—OH	2725–2520 (w–m) 2350–2080 (w–m) 1740–1600 (w–m) 1335 (s) 1090–910 (s) 540–450 (w–m)	H-bonded; broad Broad; may be doublet for aryl acids P=O stretching
P=S	865–655 (m–s) 595–530 (var)	
P(=S)—OH	3100–3000 (w) 2360–2200 (w) 935–910 (s) 810–750 (m–s) 655585 (var)	 PO stretching P=S stretching P=S stretching
	Silicon compounds	
Si—H	2250–2100 (s) 985–800	 SiH_3 has two bands.
Si—C≡	860–760	Accompanied by CH_2 rocking
Si—CH_3	1280–1250 (s)	Sharp
Si—C_2H_5	1250–1220 (m) 1020–1000 (m) 970–945 (m)	
Si—Aryl	1125–1090 (vs)	Splits into two bands when two aryl groups are attached to one silicon atom, but has only one band when three aryl groups attached
≡Si—OH	870–820	OH deformation band
≡Si—O—Si≡	1100–1000	
≡Si—N—Si≡	940–870 (s)	
≡Si—Cl	550–470 (s) 250–150	

TABLE 6-17 Absorption frequencies of miscellaneous bands (*continued*)

Group	Band, cm^{-1}	Remarks
	Silicon compounds (*continued*)	
$>SiCl_2$	595–535 (s) 540–460 (m)	
$-SiCl_3$	625–570 (s) 535–450 (m)	
	Boron compounds	
Boranes $>BH$ or $-BH_2$	2640–2450 (m–s) 2640–2570 (m–s) 2535–2485 (m–s) 2380–2315 (s) 2285–2265 (s) 2140–2080 (w–m) 2580–2450 (m)	Free H in BH Free H in BH_2 plus second band In complexes; second band for BH_2 Bridged H Borazoles and borazines
BH_4^-	2310–2195 (s)	Two bands
B—N	1550–1330 750–635	Borazines and borazoles
B—O	1390–1310 (s) 1280–1200	BO stretching Metal orthoborates
B—Cl B—Br	1090–890 (s)	Plus other bands at lower frequencies for BX_2 and BX_3
B—F	1500–840 (var)	Isotope splitting present
XBF_2	1500–1410 (s) 1300–1200 (s)	
X_2BF	1360–1300 (s)	
BF_3 complexes	1260–1125 (s) 1030–800 (s)	Band splitting may be added to isotopic splittings.
BF_4^-	ca 1030 (vs)	

TABLE 6-17 Absorption frequencies of miscellaneous bands (*continued*)

Group	Band, cm^{-1}	Remarks
	Halogen compounds	
C—F		
Aliphatic, mono-F	1110–1000 (vs)	
	780–680 (s)	
Aliphatic, di-F	1250–1050 (vs)	Two bands
Aliphatic, poly-F	1360–1090 (vs)	Number of bands
Aromatic	1270–1100 (m)	
	680–520 (m–s)	
	420–375 (var)	
	340–240 (s)	
$—CF_3$		
Aliphatic	1350–1120 (vs)	
	780–680 (s)	
	680–590 (s)	
	600–540 (s)	
	555–505 (s)	
Aromatic	1330–1310 (m–s)	
	600–580 (s)	
C—Cl		
Primary alkanes	730–720 (s)	
	685–680 (s)	
	660–650 (s)	
Secondary alkanes	ca 760 (m)	
	675–655 (m–s)	
	615–605 (s)	
Tertiary alkanes	635–610 (m–s)	
	580–560 (m–s)	
Poly-Cl	800–700 (vs)	
Aryl:		
1,2-	1060–1035 (m)	
1,3-	1080–1075 (m)	
1,4-	1100–1090 (m)	
Chloroformates	ca 690 (s)	
	485–470 (s)	
Axial Cl	730–580 (s)	
Equatorial Cl	780–740 (s)	
C—Br		
Primary alkanes	645–635 (s)	
	565–555 (s)	
	440–430 (var)	
Secondary alkanes	620–605 (s)	
	590–575 (m–w)	
	540–530 (s)	

TABLE 6-17 Absorption frequencies of miscellaneous bands (*continued*)

Group	Band, cm^{-1}	Remarks
	Halogen compounds (*continued*)	
C—Br (*continued*)		
Tertiary alkanes	600–595 (m–s)	
	525–505 (s)	
Axial	690–550 (s)	
Equatorial	750–685 (s)	
Aryl:		
1,2-	1045–1025 (m)	
1,3-; 1,4-	1075–1065 (m)	
Other bands	400–260 (s)	
	325–175 (m–s)	
	290–225 (m–s)	
C—I		
Primary alkanes	600–585 (s)	
	515–500 (s)	
Secondary alkanes	ca 575 (s)	
	550–520 (s)	
	490–480 (s)	
Tertiary alkanes	580–560 (s)	
	510–485 (m)	
	485–465 (s)	
Aromatic	1060–1055 (m–s)	
	310–160 (s)	
	265–185	
Axial	ca 640 (s)	
Equatorial	ca 655 (s)	
	Inorganic ions	
Ammonium	3300–3030	Several bands, all strong
Cyanate	2220–2130 (s)	
Cyanide	2200–2000	
Carbonate	1450–1410	
Hydrogen sulfate	1190–1160 (s)	
	1180–1000 (s)	
	880–840 (m)	
Nitrate	1410–1350 (vs)	
	860–800 (m)	
Nitrite	1275–1230 (s)	
	835–800 (m)	Shoulder

TABLE 6-17 Absorption frequencies of miscellaneous bands (*continued*)

Group	Band, cm^{-1}	Remarks
	Inorganic ions (*continued*)	
Phosphate	1100–1000	
Sulfate	1130–1080 (s)	
Thiocyanate	ca 2050 (s)	

TABLE 6-18 Absorption frequencies in the near infrared

Values in parentheses are molar absorptivity

Class	Band, cm^{-1}	Remarks
Acetylenes	9800–9430 6580–6400 (1.0)	 Overtone of ≡CH stretching
Alcohols (nonhydrogen-bonded)	7140–7010 (2.0)	Overtone of OH stretching
Aldehydes Aliphatic	 4640–4520 (0.5)	 Combination of C=O and CH stretchings
Aromatic	ca 8000 ca 4525 ca 4445	
Formate	4775–4630 (1.0)	
Alkanes —CH_3	 9000–8350 (0.02) 5850–5660 (0.1) 4510–4280 (0.3)	
—CH_2—	9170–8475 (0.02) 5830–6640 (0.1) 4420–4070 (0.25)	
⋛CH	8550–8130 7000–6800 5650–5560	All bands very weak
Cyclopropane	6160–6060 4500–4400	
Alkenes $\diagdown C{=}C \diagup$ (with H on one carbon)	 6850–6370 (1.0)	
>C=CH_2 and —CH=CH_2	7580–7300 (0.02) 6140–5980 (0.2) 4760–4700 (1.2)	

TABLE 6-18 Absorption frequencies in the near infrared (*continued*)

Class	Band, cm^{-1}	Remarks
Alkenes (*continued*)		
H H \ / C=C / \	4760–4660 (0.15)	*Trans* isomers have no unique bands.
$-O-CH=CH_2$	6250–6040 (0.3)	
$-CO-CH=CH_2$	7580–7410 (0.02)	
	6190–5990 (0.3)	
	4820–4750 (0.2–0.5)	
Amides		
Primary	7400–6540 (0.7)	Two bands; overtone of NH stretch
	5160–5060 (3.0) 5040–4990 (0.5) 4960–4880 (0.5)	Second overtone of C=O stretch; second overtone of NH deformation; combination of C=O and NH
Secondary	7330–7140 (0.5)	Overtone of NH stretch
	5050–4960 (0.4)	Combination of NH stretch and NH bending
Amines, aliphatic		
Primary	9710–9350	Second overtone of NH stretch
	6670–6450 (0.5)	Two bands; overtone of NH stretch
	5075–4900 (0.7)	Two bands; combination of NH stretch and NH bending
Secondary	9800–9350	Second overtone of NH stretch
	6580–6410 (0.5)	Overtone of NH stretch
Amines, aromatic		
Primary	9950–9520 (0.4)	
	7040–6850 (0.2)	
	6760–6580 (1.4)	
	5140–5040 (1.5)	
Secondary	10 000–9710	
	6800–6580 (0.5)	
Aryl-H	7660–7330 (0.1)	
	6170–5880 (0.1)	Overtone of CH stretch
Carbonyl	5200–5100	
Carboxylic acids	7000–6800	
Epoxide (terminal)	6135–5960 (0.2)	
	4665–4520 (1.2)	Cyclopropane bands in same region

TABLE 6-18 Absorption frequencies in the near infrared (*continued*)

Class	Band, cm^{-1}	Remarks
Glycols	7140–7040	
Hydroperoxides		
Aliphatic	6940–6750 (2.0) 4960–4880 (0.8)	
Aromatic	7040–6760 (1.0) 4950–4850 (1.3)	Two bands
Imides	9900–9620 6540–6370	
Nitriles	5350–5200 (0.1)	
Oximes	7140–7050	
Phosphines	5350–5260 (0.2)	
Phenols		
Nonbonded	7140–6800 (3.0) 5000–4950	
Intramolecularly bonded	7000–6700	
Thiols	5100–4950 (0.05)	

TABLE 6-19 Infrared transmitting materials

Material	Wavelength range, μm	Wavenumber range, cm^{-1}	Refractive index at 2 μm
NaCl, rock salt	0.25–17	40 000–590	1.52
KBr, potassium bromide	0.25–25	40 000–400	1.53
KCl, potassium chloride	0.30–20	33 000–500	1.5
AgCl, silver chloride*	0.40–23	25 000–435	2.0
AgBr, silver bromide*	0.50–35	20 000–286	2.2
CaF_2, calcium fluoride (Irtran-3)	0.15–9	66 700–1 110	1.40
BaF_2, barium fluoride	0.20–11.5	50 000–870	1.46
MgO, magnesium oxide (Irtran-5)	0.39–9.4	25 600–1 060	1.71
CsBr, cesium bromide	1–37	10 000–270	1.67
CsI, cesium iodide	1–50	10 000–200	1.74
TlBr–TlI, thallium bromide-iodide (KRS-5)*	0.50–35	20 000–286	2.37
ZnS, zinc sulfide (Irtran-2)	0.57–14.7	17 500–680	2.26

* Useful for internal reflection work.

TABLE 6-19 Infrared transmitting materials (*continued*)

Material	Wavelength range, μm	Wavenumber range, cm^{-1}	Refractive index at 2 μm
ZnSe, zinc selenide* (vacuum deposited) (Irtran-4)	1–18	10 000–556	2.45
CdTe, cadmium telluride (Irtran-6)	2–28	5 000–360	2.67
Al_2O_3, sapphire*	0.20–6.5	50 000–1 538	1.76
SiO_2, fused quartz	0.16–3.7	62 500–2 700	
Ge, germanium*	0.50–16.7	20 000–600	4.0
Si, silicon*	0.20–6.2	50 000–1 613	3.5
Polyethylene	16–300	625–33	1.54

* Useful for internal reflection work.

TABLE 6-20 Infrared transmission characteristics of selected solvents

Transmission below 80%, obtained with a 0.10-mm cell path, is shown as shaded area

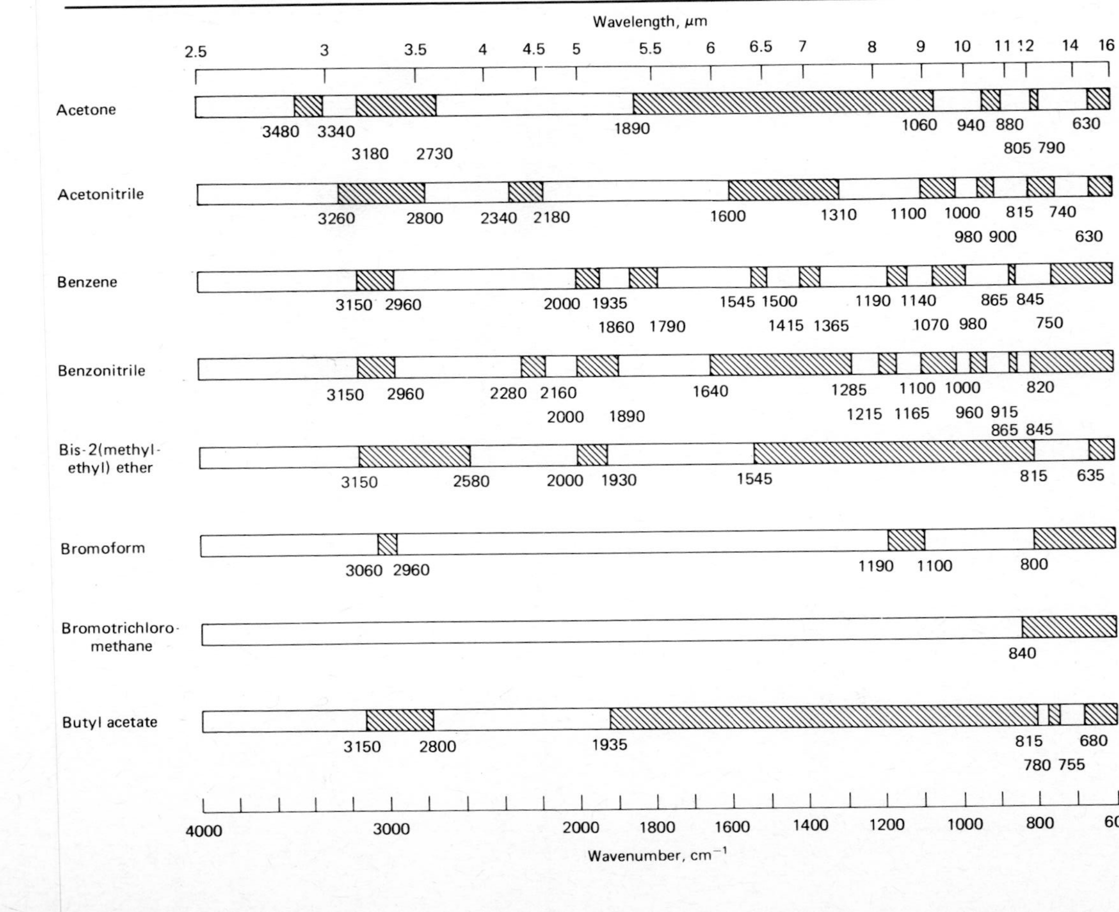

TABLE 6-20 Infrared transmission characteristics of selected solvents (*continued*)

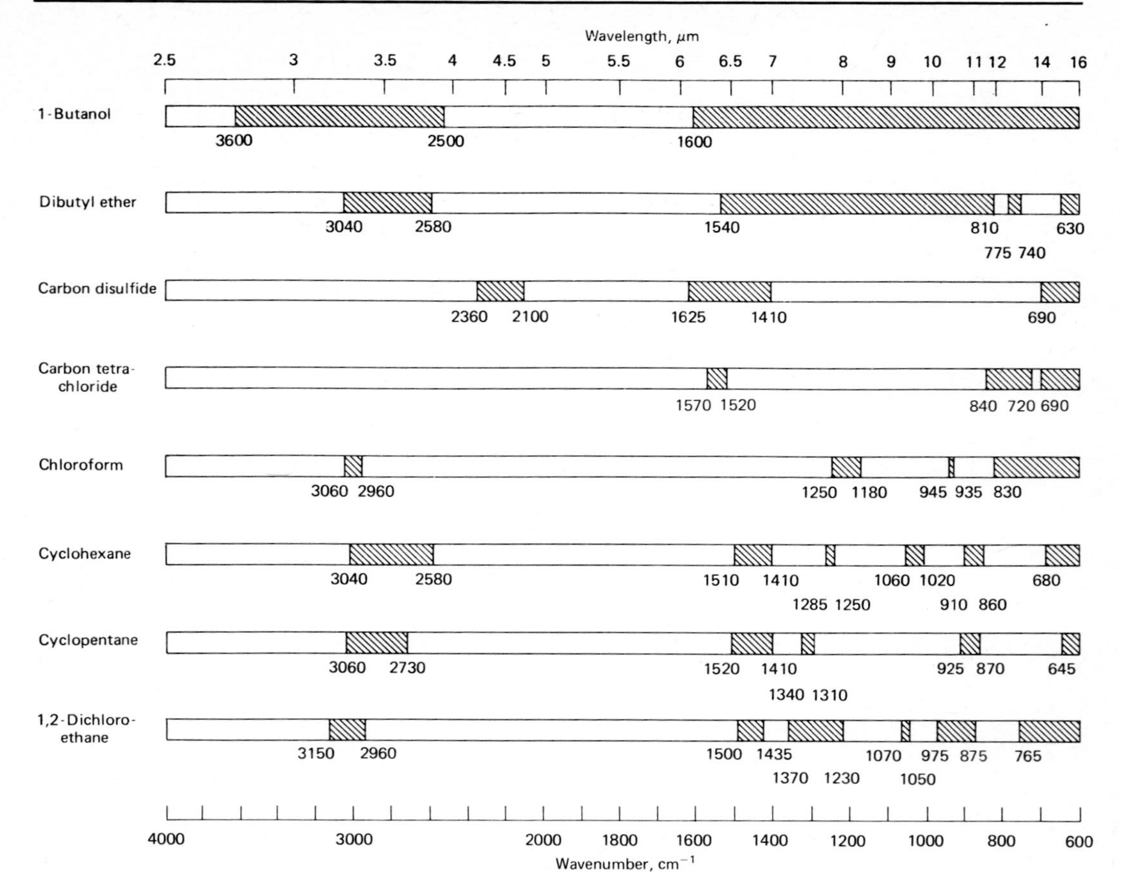

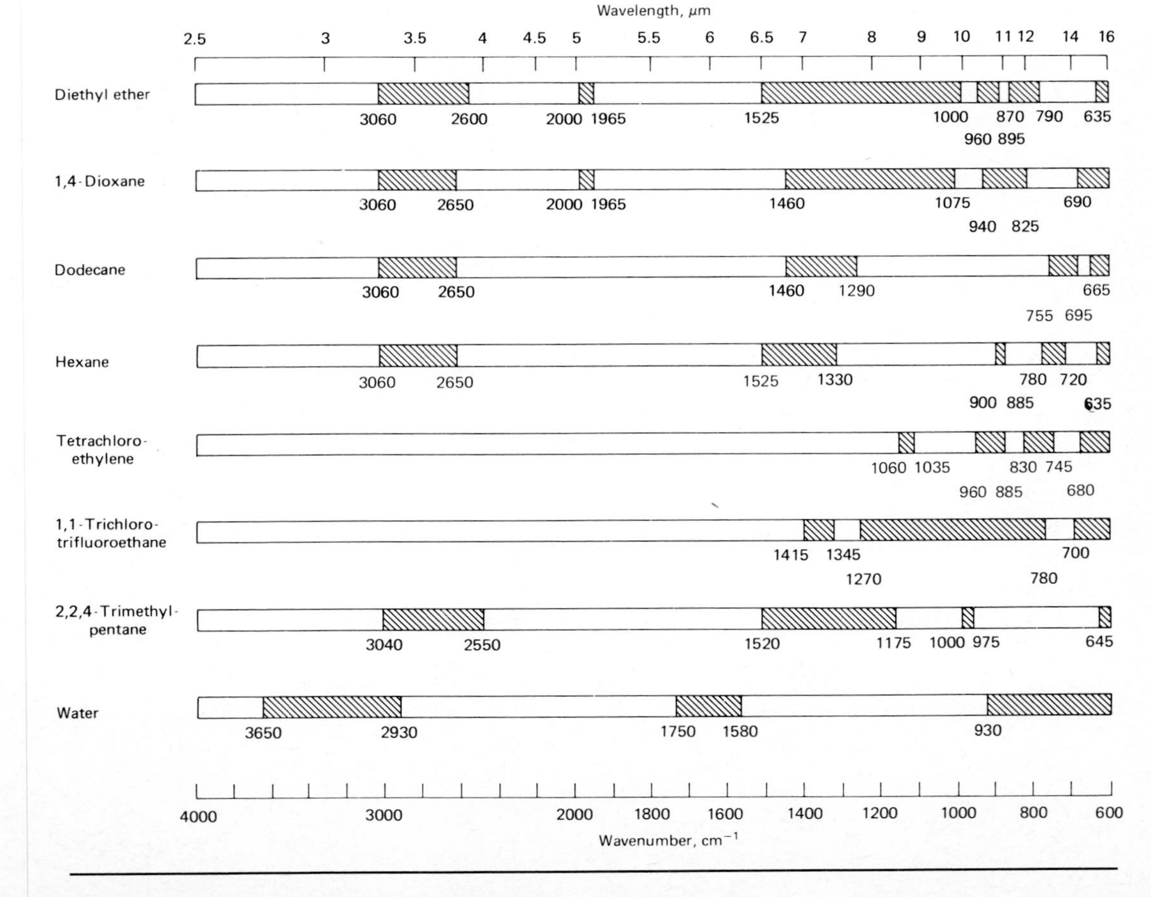
Wavelength, μm
2.5
3
3.5
4
4.5
5
5.5
6
6.5
7
8
9
10
11 12
14
16
Diethyl ether
3060 2600 2000 1965 1525 1000 870 790 635
960 895
1,4-Dioxane
3060 2650 2000 1965 1460 1075 690
940 825
Dodecane
3060 2650 1460 1290 665
755 695
Hexane
3060 2650 1525 1330 780 720
900 885 635
Tetrachloro-ethylene
1060 1035 830 745
960 885 680
1,1-Trichloro-trifluoroethane
1415 1345 700
1270 780
2,2,4-Trimethyl-pentane
3040 2550 1520 1175 1000 975 645
Water
3650 2930 1750 1580 930
4000
3000
2000
1800
1600
1400
1200
1000
800
600
Wavenumber, cm⁻¹

RAMAN SPECTROSCOPY

TABLE 6-21 Raman frequencies of single bonds to hydrogen and carbon

Abbreviations Used in the Table

m, moderately strong	vw, very weak
m–s, moderate to strong	w, weak
m–vs, moderate to very strong	w–m, weak to moderately strong
s, strong	w–m, weak to moderately strong
vs, very strong	w–vs, weak to very strong

Group	Band, cm^{-1}	Remarks
	Saturated C—H and C—C	
$-CH_3$	2969–2967 (s)	
	2884–2883 (s)	
	ca 1205 (s)	In aryl compounds
	1150–1135	In unbranched alkyls
	1060–1056	In unbranched alkyls
	975–835 (s)	Terminal rocking of methyl group
	280–220	CH_2—CH_3 torsion
$-CH_2-$	2949–2912 (s)	
	2861–2849 (s)	
	1473–1443 (m–vs)	Intensity proportional to number of CH_2 groups
	1305–1295 (s)	
	1140–1070 (m)	Often two bands; see above
	888–837 (w)	
	425–150	
	500–490	Substituent on aromatic ring
$-CH(CH_3)_2$	1350–1330 (m)	
	835–750 (s)	If attached to C=C bond, 870–800 cm^{-1}. If attached to aryl ring, 740 cm^{-1}
$-C(CH_3)_3$	1265–1240 (m)	Not seen in *tert*-butyl bromide
	1220–1200 (m)	Not seen in *tert*-butyl bromide
	760–685 (vs)	If attached to C=C or aromatic ring, 760–720 cm^{-1}
Internal tertiary carbon atom	855–805 (w)	
	455–410	
Internal quaternary carbon atom	710–680 (vs)	
	490–470	

TABLE 6-21 Raman frequencies of single bonds to hydrogen and carbon (*continued*)

Group	Band, cm^{-1}	Remarks
Saturated C—H and C—C (*continued*)		
Two adjacent tertiary carbon atoms	730–920 770–725	Often a band at 530–524 cm^{-1} indicates presence of adjacent tertiary and quaternary carbon atoms.
Dialkyl substitution at α-carbon atom	800–700 (m–s) 680–650 (vs) 605–550	
Cyclopropane	3101–3090 3038–3019 1210–1180 (s)	Shifts to 1200 cm^{-1} for monoalkyl or 1,2-dialkyl substitution and to 1320 cm^{-1} for *gem*-1,1-dialkyl substitution
Cyclobutane	1001–960 (vs)	Shifts to 933 cm^{-1} for monoalkyl, to 887 cm^{-1} for *cis*-1,3-dialkyl, and to 891 cm^{-1} plus 855 cm^{-1} (doublet) for *trans*-1,3-dialkyl substitition
Cyclopentane	900–800 (s)	
Cyclohexane	825–815 (vs) 810–795 (vs)	Boat configuration Chair configuration
Cycloheptane	ca 733	
Cyclooctane	ca 703	
$=C(CH_3)_2$	1392–1377 450–400 (vw) 270–250 (m)	
$CH_3(H)C=C(H)CH_3$ (trans)	1380–1379 492–455 (vw) 220–200 (m)	

TABLE 6-21 Raman frequencies of single bonds to hydrogen and carbon (*continued*)

Group	Band, cm^{-1}	Remarks
	Saturated C—H and C—C (*continued*)	
$(CH_3)HC=CH(CH_3)$ (trans)	1372–1368 970–952 (m) 592–545 (vw) 420–400 (m) 310–290 (m)	
$(CH_3)_2C=CH(CH_3)$	1385–1375 522–488 (w)	
$(CH_3)_2C=C(CH_3)_2$	1392–1386 690–678 (m–s) 510–485 (m) 424–388 (w)	
$\equiv C-C(=O)-C\equiv$	1170–1100 (w–m) 600–580 (m–s)	
$\equiv C-C(=O)-$	1120–1090 (m–vs) 600–510 (w–m)	Tertiary or quaternary carbon adjacent to carbonyl group lowers the frequency 300 cm^{-1}.
$-CH_2-CO-$	1420–1410 (s)	
$-CHO$	2850–2810 (m) 2720–2695 (vs)	Often appears as a shoulder
	Unsaturated C—H	
$-C\equiv C-H$	3340–3270 (w–m)	Alkyl substituents at higher frequencies; unsaturated or aryl substituents at lower frequencies
$>C=C<$ (one H)	3040–2995 (m)	
$>C=CH_2$	3095–3050 (m) 2990–2983 (s)	Asymmetric $=CH_2$ stretch Symmetric $=CH_2$ stretch

TABLE 6-21 Raman frequencies of single bonds to hydrogen and carbon (*continued*)

Group	Band, cm^{-1}	Remarks
	Unsaturated C—H (*continued*)	
H R C=C H H	1419–1415 (m) 1309–12888 (m)	Plus =CH and =CH stretching bands
H R_1 C=C H R_2	1413–1399 (m) 909–885 (m) 711–684 (w)	Plus $=CH_2$ stretching bands
R_1 R_2 C=C H H	1270–1251 (m)	Plus =CH stretching band
R_1 H C=C H R_2	1314–1290 (m)	Plus =CH stretching band
R_1 R_3 C=C R_2 H	1360–1322 (w) 830–800 (vw)	Plus =CH stretching band
	Hydroxy O—H	
Free —OH Intermolecularly bonded Aromatic —OH	3650–3250 (w) 3400–3300 (w) ca 3160 (s)	
—OH	1460–1320 (w) 1276–1205 (w–m) 1260 (w–m)	Common to all OH substituents Primary Secondary
C—C—OH primary	1070–1050 (m–s) 1030–960 (m–s) 480–430 (w–m)	CCO stretching CCO deformation
C—C—OH Secondary Tertiary	 1135–1120 (m–s) 825–815 (vs) 500–490 (w–m) 1210–1200 (m–s) 755–730 (vs) 360–350 (w–m)	
—CO—O—H	1305–1270	CO stretching

TABLE 6-21 Raman frequencies of single bonds to hydrogen and carbon (*continued*)

Group	Band, cm^{-1}	Remarks
	N—H and C—N bonds	
Amine ⪰N—H		
Associated	3400–3250 (s)	Primary amines show two bands.
Nonbonded	3550–3250 (s)	
Salts	2986–2974	Often obscured by intense CH stretching bands
$—NH_2$	1650–1590 (w–vs)	Bending
Amides		
Primary	3540–3500 (w) 3400–3380 (w)	Both bands lowered ca 150 cm^{-1} in solid state and H bonding
	1310–1250 (s)	Interaction of NH bending and CN stretching; lowered 50 cm^{-1} in nonbonded state
	1150–1095 (m)	Rocking of NH_2
Secondary	3491–3404 (m–s)	Two bands; lowered in frequency on H bonding and in solid sate
	1190–1130 (m)	
	931–865 (m–s)	
	430–395 (w–m)	
—CO—N	607–555 (m)	O=CN bending
C—N(—C)—C	1070–1045 (m)	Stretching
⪰C—N⪯		
Primary carbon	1090–1060 (m)	CN stretching
Secondary α carbon	1140–1035 (m)	Two bands but often obscured. Strong band at 800 cm^{-1}
Tertiary α carbon	1240–1020 (m)	Two bands. Strong band also at 745 cm^{-1}

TABLE 6-22 Raman frequencies of triple bonds

Abbreviations Used in the Table

m, moderately strong
m–s, moderate to strong
s, strong
s–vs, strong to very strong
vs, very strong

Group	Band, cm^{-1}	Remarks
R—C≡CH	2160–2100 (vs)	Monoalkyl substituted; C≡C stretch
	650–600 (m)	C≡CH deformation
	356–335 (s)	C≡C—C bending of monoalkyls
R_1—C≡C—R_2	2300–2190 (vs)	C≡C stretching of disubstituted alkyls; sometimes two bands
—C≡C—C≡C—	2264–2251 (vs)	
—C≡N	2260–2240 (vs)	Unsaturated nonaryl substituents lower the frequency and enhance the intensity.
	2234–2200 (vs)	Lowered ca 30 cm^{-1} with aryl and conjugated aliphatics
	840–800 (s–vs)	CCCN symmetrical stretching
	385–350 (m–s)	
	200–160 (vs)	Aliphatic nitriles
H—C≡N	2094 (vs)	
Azides $-\bar{N}-\overset{+}{N}\equiv N$	2170–2080 (s)	Asymmetric NNN stretching
	1258–1206 (s)	Symmetric NNN stretching; HN_3 at 1300 cm^{-1}
Diazonium salts $R-\overset{+}{N}\equiv N$	2300–2240 (s)	
Isonitriles $-\overset{+}{N}\equiv\bar{C}$	2146–2134	Stretching of aliphatics
	2124–2109	Stretching of aromatics
Thiocyanates —S—C≡N	2260–2240 (vs)	Stretching of C≡N
	650–600 (s)	Stretching of SC

TABLE 6-23 Raman frequencies of cumulated double bonds

Abbreviations Used in the Table

s, strong — vw, very weak
vs, very strong — w, weak

Group	Band, cm^{-1}	Remarks
Allenes		
C=C=C	2000–1960 (s)	Pseudo-asymmetric stretching
	1080–1060 (vs)	Symmetric stretching
	356	C=C=C bending
Carbodiimides (cyanamides)		
—N=C=N—	2140–2125 (s)	Asymmetric stretching of aliphatics
	2150–2100 (vs)	Asymmetric stretching of aromatics; two bands
	1460	Symmetrical stretching of aliphatics
	1150–1140 (vs)	Symmetric stretching of aryls
Cumulenes (trienes)		
C=C=C=C	2080–2030 (vs)	
	878	
Isocyanates		
—N=C=O	2300–2250 (vw)	Asymmetric stretching
	1450–1400 (s)	Symmetric stretching
Isothiocyanates		
—N=C=S	2220–2100	Two bands
	690–650	Alkyl derivatives
Ketenes		
C=C=O	2060–2040 (vs)	Pseudo-asymmetric stretching
	1130 (s)	Pseudo-symmetric stretching
	1374 (s)	Alkyl derivatives
	1120 (s)	Aryl derivatives
Sulfinylamines		
R—N=S=O	1306–1214 (w)	Asymmetric stretching
	1155–989 (s)	Symmetric stretching

TABLE 6-24 Raman frequencies of carbonyl bands

Abbreviations Used in the Table

m, moderately strong — s–vs, strong to very strong
m–s, moderate to strong — vs, very strong
s, strong — w, weak

Group	Band, cm^{-1}	Remarks
Acid anhydrides		
—CO—O—CO—		
Saturated	1850–1780 (m)	
	1771–1770 (m)	
Conjugated, noncyclic	1775	
	1720	
Acid fluorides —CO—F		
Alkyl	1840–1835	
Aryl	1812–1800	
Acid chlorides —CO—Cl		
Alkyl	1810–1770 (s)	
Aryl	1774	
	1731	
Acid bromides —CO—Br		
Alkyl	1812–1788	
Aryl	1775–1754	
Acid iodides —CO—I		
Alkyl	ca 1806	
Aryl	ca 1752	
Lactones	1850–1730 (s)	
Esters		
Saturated	1741–1725	Alkyl branching on carbon adjacent to C=O lowers frequency by 5-15 cm^{-1}.
Aryl and α,β-unsaturated	1727–1714	
Diesters		
Oxalates	1763–1761	
Phthalates	1738–1728	
C≡C—CO—O—	1716–1708	
Carbamates	1694–1688	
Aldehydes	1740–1720 (s–vs)	
Ketones		
Saturated	1725–1700 (vs)	
Aryl	1700–1650 (m)	

TABLE 6-24 Raman frequencies of carbonyl bands (*continued*)

Group	Band, cm^{-1}	Remarks
Ketones (*continued*)		
Alicyclic		
$n = 4$	1782 (m)	
$n = 5$	1744 (m)	
$n \geq 6$	1725–1699 (m)	
Carboxylic acids		
Mono-	1686–1625 (s)	These α-substituents increase the frequency: F, Cl, Br, OH.
Poly-	1782–1645	Solid state; often two bands
	1750–1710	In solution; very broad band
Amino acids	1743–1729	
Carboxylate ions	1690–1550 (w)	
	1440–1340 (vs)	
Amino acid anion	1743–1729	
	1600–1570 (w)	Often masked by water deformation band near 1630 cm^{-1}
Amides (see also Table 6-21)		
Primary		
Associated	1686–1576 (m–s)	
	1650–1620 (m)	
Nonbonded	1715–1675 (m)	
	1620–1585 (m)	
Secondary		
Associated	1680–1630 (w)	Both *cis* and *trans* forms
	1570–1510 (w)	*Trans* form
	1490–1440	*Cis* form
Nonbonded	1700–1650	Both *cis* and *trans* forms
	1550–1500	*Trans* form (no *cis* band)
Tertiary	1670–1630 (m)	
Lactams	1750–1700 (m)	

TABLE 6-25 Raman frequencies of other double bonds

Abbreviations Used in the Table

m, moderately strong	vs, very strong
m–s, moderate to strong	w, weak
s, strong	s–vs, strong to very strong
w–m, weak to moderately strong	

Group	Band, cm^{-1}	Remarks
	Alkenes $>C=C<$	
$>C=C<$	1680–1576 (m–s)	General range
H R_1 C=C H H	1648–1638 (vs)	C=C stretching
H R_1 C=C H R_2	ca 1650 (vs) 270–252 (w)	C=C stretching C=C—C skeletal deformation
R_1 R_2 C=C H H	ca 1660 (vs) 970–952 (w)	C=C stretching Asymmetric CC stretching
R_1 H C=C H R_2	1676–1665 (s)	C—C stretching
R_1 R_3 C=C R_2 H	1678–1664 (vs) 522–488 (w)	C=C stretching C=C—C skeletal deformation
R_1 R_3 C=C R_2 R_4	1680–1665 (s) 690–678 (m–s) 510–485 (m) 424–388 (w)	C=C stretching Symmetrical CC stretching Skeletal deformation Skeletal deformation

Haloalkene	X = fluorine	X = chlorine	X = bromine	X-iodine
		$>C=C<$ stretch of haloalkanes		
$H_2C=CHX$	1654	1603–1601	1596–1593	1581
$HXC=CHX$				
cis	1712	1590–1587	1587–1583	1543
trans	1694	1578–1576	1582–1581	1537
$H_2C=CX_2$	1728	1616–1611	1593	
$X_2C=CHX$	1792	1589–1582	1552	
$X_2C=CX_2$	1872	1577–1571	1547	1465 (solid)

TABLE 6-25 Raman frequencies of other double bonds (*continued*)

Group	Band, cm^{-1}	Remarks
	$>C{=}N-$ bonds	
Aldimines (azomethines) $\ce{H(R_1)C=N-R_2}$	1673–1639 1405–1400 (s)	Dialkyl substituents at higher frequency; diaryl substituents at lower end of range
Aldoximines and Ketoximes $>C{=}N{-}OH$	1680–1617 (vs) 1335–1330 (w)	
Azines $>C{=}N{-}N{=}C<$	1625–1608 (s)	
Hydrazones $H(R_1)C{=}N{-}N(H)R_2$	1660–1610 (s–vs)	
Imido ethers $(O)(-)C{=}NH$	1658–1648	NH stretching at 3360–3327 cm^{-1}
Semicarbazones and thiosemicarbazones $>C{=}N{-}N(H){-}C({=}O \text{ (or S)}){-}NH_2$	1665–1642 (vs) 1620–1610 (vs)	Aliphatic. Thiosemicarbazones fall in lower end of range. Aromatic derivatives
	Azo compounds $-N{=}N-$	
$-N{=}N-$	1580–1570 (vs) 1442–1380 (vs) 1060–1030 (vs)	Nonconjugated Conjugated to aromatic ring CN stretching in aryl compounds
	Nitro compounds $N{=}O$	
Alkyl nitrites	1660–1620 (s)	$N{=}O$ stretching
Alkyl nitrates	1635–1622 (w–m) 1285–1260 (vs) 610–562 (m)	Asymmetric NO_2 stretching Symmetric NO_2 stretching NO_2 deformation

TABLE 6-25 Raman frequencies of other double bonds (*continued*)

Group	Band, cm^{-1}	Remarks
	Nitro compounds N=O (*continued*)	
Nitroalkanes		
Primary	1560–1548 (m–s)	
	1395–1370 (s)	Sensitive to substituents attached to CNO_2 group
	915–898 (m–s)	
	894–873 (m–s)	
	618–609 (w)	
	640–615 (w)	Shoulder
	494–472 (w–m)	Broad; useful to distinguish from secondary nitroalkanes
Secondary	1553–1547 (m)	
	1375–1360 (s)	
	908–868 (m)	
	863–847 (s)	
	625–613 (m)	
	560–516 (s)	Sharp band
Tertiary	1543–1533 (m)	
	1355–1345 (s)	
Nitrogen oxides		
$\geqq\overset{+}{N}\rightarrow\bar{O}$	1612–1602 (s)	
	1252 (m)	
	1049–1017 (s)	
	835 (s)	
	541 (w)	
	469 (w)	

TABLE 6-26 Raman frequencies of aromatic compounds

Abbreviations Used in the Table

m, moderately strong
m–s, moderate to strong
m–vs, moderate to very strong
s, strong
s–vs, strong to very strong
var, of variable strength
vs, very strong
w, weak
w–m, weak to moderately strong

Group	Band, cm^{-1}	Remarks
	Common features	
Aromatic compounds	3070–3020 (s)	CH stretching
	1630–1570 (m–s)	C—C stretching
	Substitution patterns of the benzene ring	
Monosubstituted	1180–1170 (w–m)	
	1035–1015 (s)	
	1010–990 (vs)	Characteristic feature; found also with 1,3- and 1,3,5-substitutions
	630–605 (w)	
1,2-Disubstituted	1230–1215 (m)	
	1060–1020 (s)	Characteristic feature
	740–715 (m)	Lowered 60 cm^{-1} for halogen substituents
1,3-Disubstituted	1010–990 (vs)	Characteristic feature
	750–640 (s)	
1,4-Disubstituted	1230–1200 (s–vs)	
	1180–1150 (m)	
	830–750 (vs)	Lower frequency with Cl substituents
	650–630 (m–w)	
Isolated hydrogen	1379 (s–vs)	
	1290–1200 (s)	
	745–670 (m–vs)	
	580–480 (s)	Characteristic feature
1,2,3-Trisusbstituted	1100–1050 (m)	
	670–500 (vs)	The lighter the mass of the substituent, the higher the frequency.
	490–430 (w)	
1,2,4-Trisubstituted	750–650 (vs)	Lighter mass at higher frequencies
	580–540 (var)	
	500–450 (var)	

TABLE 6-26 Raman frequencies of aromatic compounds (*continued*)

Group	Band, cm^{-1}	Remarks
	Substitution patterns of the benzene ring (*continued*)	
1,3,5-Trisubstituted	1010–990 (vs)	
Completeely substituted	1296 (s) 550 (vs) 450 (m) 361 (m)	
	Other aromatic compounds	
Naphthalenes	1390–1370 1026–1012 767–762 535–512 519–512	Ring breathing α or β substituents β substituents α substituents β substituents
Disubstituted napthalenes	773–737 (s) 726–705 (s) 690–634 (s) 608 575–569 544–537	1,2-; 1,3-; 2,3-; 2,6-; 2,7- 1,3-; 1,4-(two bands); 1,6-; 1,7- (two bands) 1,2-; 1,4-(two bands); 1,5-; 1,8- (two bands) 1,3- 1,2-; 1,3-; 1,6- 1,2-; 1,7-; 1,8-
Anthracenes	1415–1385	Ring breathing

TABLE 6-27 Raman frequencies of sulfur compounds

Abbreviations Used in the Table

m, moderately strong	s–vs, strong to very strong
m–s, moderate to strong	vs, very strong
s, strong	w–m, weak to moderately strong

Group	Band, cm^{-1}	Remarks
—S—H	2590–2560 (s)	SH stretching for both aliphatic and aromatic
>C=S	1065–1050 (m) 735–690 (vs)	 Solid state
>S=O In $(RO_2)_2SO$ In $(R_2N)_2SO$	 1209–1198 1108	 One or two bands

TABLE 6-27 Raman frequencies of sulfur compounds (*continued*)

Group	Band, cm^{-1}	Remarks
⪢S=O (*continued*)		
In R_2SO	1070–1010 (w–m)	Broad
SOF_2	1308	
$SOCl_2$	1233	
$SOBr_2$	1121	
$-SO_2-$	1330–1260 (m–s)	Asymmetric SO_2 stretching
	1155–1110 (s)	Symmetric SO_2 stretching
	610–540 (m)	Scissoring mode of aryls
	512–485 (m)	Scissoring mode of alkyls
$-SO_2-N<$	ca 1322 (m)	Asymmetric SO_2 stretching
	1163–1138 (s)	Symmetric SO_2 stretching
	524–510 (s)	Scissoring mode
$-SO_2-O$	1363–1338 (w–m)	SO_2 stretching. Aryl substituents occur at higher range.
	1192–1165 (vs)	
	589–517 (w–m)	Scissoring (two bands). Aryl substituents occur at higher range of frequencies.
$-SO_2-S-$	1334–1305 (m–s)	
	1128–1126 (s)	
	559–553 (m–s)	
$X-SO_2-X$	1412–1361 (w–m)	
	(F) (Cl)	
	1263–1168 (s)	
	(F) (Cl)	
	596–531 (s)	
$-O-SO_2-O-$	1388–1372 (s)	
	1196–1188 (vs)	
$-O-C(=S)-S-$	670–620 (vs)	C=S stretching
	480–450 (vs)	CS stretching
⋜C–SH	920 (m)	C–SH deformation of aryls
	850–820 (m)	
⋜C–S–	752 (vs), 731 (vs)	With vinyl group attached
	742–722 (m–s)	With CH_3 attached
	698 (w), 678 (s)	With allyl group attached
	693–639 (s)	Ethyl or longer alkyl chain
	651–610 (s–vs)	Isopropyl group attached
	589–585 (vs)	*tert*-Butyl group attached

TABLE 6-27 Raman frequencies of sulfur compounds (*continued*)

Group	Band, cm^{-1}	Remarks
⋛C—S— (*continued*)		
$(CH_2)_n$ S (ring)		
$n = 2$	1112	
$n = 4$	688	
$n = 5$	659	
⋛C—$(S—S)_n$—C⋚	715–620 (vs)	Two bands; CS stretching
	525–510 (vs)	Two bands; SS stretching
Didi-*n*-alkyl disulfides	576 (s)	CS stretching
Di-*tert*-butyl disulfide	543 (m)	SS stretching
Trisulfides	510–480 (s)	SS stretching

TABLE 6-28 Raman frequencies of ethers

Abbreviations Used in the Table

m, moderately strong — var, of variable strength
s, strong — vs, very strong

Group	Band, cm^{-1}	Remarks
⋛C—O—C⋚		
Aliphatic	1200–1070 (m)	Asymmetrical COC stretching. Symmetrical substitution gives higher frequencies
	930–830 (s)	Symmetrical COC stretching
	800–700 (s)	Braching at α carbon gives higher frequencies.
	550–400	
Aromatic	1310–1210 (m)	
	1050–1010 (m)	
⋛C—O—C(—)(—)—O—C⋚	1145–1129 (m)	
	900–800 (vs)	
	537–370 (s)	
	396–295	
>C — C< with bridging O (epoxide)	1280–1240 (s)	Ring breathing
—O—O—	800–770 (var)	
$(CH_2)_n$ O (ring) $n = 3$	1040–1010 (s)	
$n = 4$	920–900 (s)	
$n = 5$	820–800 (s)	

TABLE 6-29 Raman frequencies of halogen compounds

Abbreviations Used in the Table

m–s, moderate to strong — var, of variable strength
s, strong — vs, very strong

Group	Band, cm^{-1}	Remarks
C—F	1400–870	Correlations of limited applicability because of vibrational coupling with stretching
C—Cl	350–290 (s)	CCCl bending; general
Primary	660–650 (vs)	
Secondary	760–605 (s)	May be one to four bands
Tertiary	620–540 (var)	May be one to three bands
=C—Cl	844–564 438–396 381–170	
$=CCl_2$	601–441 300–235	
C—Br	690–490 (s) 305–258 (m–s)	Often several bands; primary at higher range of frequencies. Tertiary has very strong band at ca 520 cm^{-1}.
=C—Br	745–565 356–318 240–115	
$=CBr_2$	467–265 185–145	
C—I	663–595 309 154–85	
=C—I	ca 180	Solid state
$=CI_2$	ca 265 ca 105	Solid state Solid state

TABLE 6-30 Raman frequencies of miscellaneous compounds

Abbreviations Used in the Table

m, moderately strong — vs, very strong
s, strong — vvs, very very strong

Group	Band, cm^{-1}	Remarks
C—As	570–550 (vs)	CAs stretching
	240–220 (vs)	CAsC deformation
C—Pb	480–420 (s)	CPb stretching
C—Hg	570–510 (vvs)	CHg stretching
C—Si	1300–1200 (s)	CSi stretching
C—Sn	600–450 (s)	CSn stretching
P—H	2350–2240 (m)	PH stretching
	Heterocyclic rings	
Trimethylene oxide	1029	
Trimethylene imine	1026	
Tetrahydrofuran	914	
Pyrrolidine	899	
1,3-Dioxolane	939	
1,4-Dioxane	834	
Piperidine	815	
Tetrahydropyran	818	
Morpholine	832	
Piperazine	836	
Furan	1515–1460	2-Substituted
	1140	
Pyrazole	1040–990	
Pyrrole	1420–1360 (vs)	
	1144	
Thiophene	1410 (s)	
	1365 (s)	
	1085 (vs)	
	1035 (s)	
	832 (vs)	
	610 (s)	
Pyridine	1030 (vs)	
	990 (vs)	

NUCLEAR MAGNETIC RESONANCE

TABLE 6-31 Nuclear properties of the elements

In the following table the magnetic moment μ is in multiples of the nuclear magneton $\mu_N(eh/4\pi Mc)$ with diamagnet correction, the spin I is in multiples of $h/2\pi$, and the electric quadrupole moment Q is in multiples of 10^{-28} squa meters. Nuclei with spin 1/2 have no quadrupole moment. The sign of μ and Q is uncertain for those nuclides f which no sign is given. Sensitivity is for equal number of nuclei at constant field. NMR frequency at any magnetic fie is the entry for column 5 multiplied by the value of the magnetic field in kilogauss. For example, in a magnetic fie of 14.0924 kG, protons (^{1}H) will precess at a frequency of 4.25760 × 14.0924 kG = 60.000 MHz. In a magnetic field 23.4924 kG, protons will precess at 4.25760 × 23.4924 kG = 100.00 MHz.

Nuclide	Natural abundance %	Spin I	Sensitivity at constant field relative to ^{1}H	NMR frequency for a 1000-G field, MHz	Magnetic moment μ/μ_N, J · T^{-1}	Electric quadrupole moment Q 10^{-28} m^2
^{1}n		−1/2	0.322	2.916 70	−1.913 12	
^{1}H	99.985	1/2	1.000	4.257 60	+2.792 78	
^{2}H	0.015	1	0.009 64	0.653 57	+0.857 42	+0.002 8
^{3}H		1/2	1.21	4.541 31	+2.978 9	
^{3}He	0.000 13	−1/2	0.443	3.243 38	−2.127 6	
^{6}Li	7.42	1	0.008 51	0.626 55	+0.822 03	−0.000 8
^{7}Li	92.58	3/2	0.294	1.654 65	+3.256 36	−0.04
^{9}Be	100	−3/2	0.013 9	0.598 27	−1.177 45	0.05
^{10}B	19.7	3	0.019 9	0.457 4	+1.800 6	+0.111
^{11}B	80.3	3/2	0.165	1.365 95	+2.688 5	+0.041
^{13}C	1.108	1/2	0.015 9	1.070 54	+0.702 4	
^{14}N	99.635	1	0.001 01	0.307 6	+0.403 75	+0.01
^{15}N	0.365	−1/2	0.001 04	0.431 5	−0.283 1	
^{17}O	0.037	−5/2	0.029 1	0.577 39	−1.893 7	−0.004
^{19}F	100	1/2	0.834	4.005 43	+2.628 8	
^{21}Ne	0.257	−3/2	0.027 2	0.336 11	−0.661 76	+0.09
^{22}Na		3	0.018 1	0.443 4	1.746	
^{23}Na	100	3/2	100	1.126 21	+2.217 40	+0.10
^{24}Na		4	0.001 15	0.322	1.690	
^{25}Mg	10.11	−5/2	0.026 8	0.260 6	−0.855 4	+0.22
^{27}Al	100	5/2	0.207	1.109 40	+3.641 3	+0.15
^{29}Si	4.71	−1/2	0.078 5	0.845 8	−0.555 26	
^{31}P	100	1/2	0.066 4	1.723 8	+1.131 7	
^{33}S	0.76	3/2	0.002 26	0.326 6	+0.643 5	−0.055
^{35}S		3/2	0.008 50	0.508		+0.038
^{35}Cl	75.53	3/2	0.004 71	0.417 1	+0.821 81	−0.080
^{36}Cl		2	0.012 1	0.489 3	+1.285 3	−0.10
^{37}Cl	24.47	3/2	0.002 72	0.347 2	+0.684 07	−0.006 2
^{39}K	93.22	3/2	0.000 508	0.198 64	+0.391 43	+0.049
^{40}K	0.011 8	4	0.005 21	0.247 0	−1.298 1	−0.061
^{41}K	6.77	3/2	0.000 083 9	0.109 03	+0.214 9	+0.060
^{43}Ca	0.145	7/2	0.063 9	0.286 54	−1.317 2	
^{45}Sc	100	7/2	0.301	1.034 34	+4.755 9	−0.22
^{47}Ti	7.32	−5/2	0.002 10	0.239 97	−0.788 46	+0.29
^{49}Ti	5.46	−7/2	0.003 76	0.240 04	−1.104 14	+0.24

TABLE 6-31 Nuclear properties of the elements (*continued*)

Nuclide	Natural abundance, %	Spin I	Sensitivity at constant field relative to ^1H	NMR frequency for a 1000-G field, MHz	Magnetic moment μ/μ_N, $J \cdot T^{-1}$	Electric quadrupole moment Q, 10^{-28} m^2
^{50}V	0.25	6	0.055 3	0.424 3	+3.347 0	0.06
^{51}V	99.75	7/2	0.383	1.119 22	+5.148 5	−0.05
^{53}Cr	9.55	3/2	0.000 10	0.240 63	−0.473 5	+0.03
^{55}Mn	100	5/2	0.178	1.055 42	+3.449	+0.4
^{57}Fe	2.17	1/2	0.000 033 3	0.138	+0.090 42	
^{59}Co	100	7/2	0.281	1.007 2	+4.616	+0.38
^{61}Ni	1.25	3/2	0.003 50	0.380 48	−0.749 8	+0.16
^{63}Cu	69.1	3/2	0.093 8	1.128 5	+2.222 8	−0.211
^{65}Cu	30.9	3/2	0.116	1.209 0	+2.381 2	−0.195
^{67}Zn	4.11	5/2	0.002 86	0.266 3	+0.875 24	+0.16
^{69}Ga	60.2	3/2	0.069 3	1.021 88	+2.014 5	+0.19
^{71}Ga	39.8	3/2	0.142	1.298 40	+2.559 7	+0.12
^{75}As	100	3/2	0.025 1	0.729 2	+1.439	+0.29
^{77}Se	7.58	1/2	0.006 97	0.811 8	+0.534	
^{79}Br	50.52	3/2	0.078 6	1.066 9	+2.105 5	+0.37
^{81}Br	49.48	3/2	0.098 4	1.149 8	+2.269 6	+0.31
^{87}Rb	27.85	3/2	0.177	1.292 3	+2.750 0	+0.13
^{93}Nb	100	9/2	0.482	1.040 48	+6.167	−0.22
^{113}In	4.23	−1/2	0.345	0.931 2	−0.622 5	
^{119}Sn	8.58	−1/2	0.051 8	1.586 8	−1.046 1	
^{121}Sb	57.25	5/2	0.160	1.019 2	+3.359 2	−0.28
^{123}Sb	42.75	7/2	0.045 7	0.551 9	+2.546 6	−0.36
^{125}Te	6.99	−1/2	0.031 6	1.345 3	−0.887 2	
^{127}I	100	5/2	0.093 5	0.851 7	+2.809 1	−0.79
^{129}Xe	26.44	−1/2	0.021 2	1.177 79	−0.776 8	
^{195}Pt	33.8	1/2	0.009 94	0.915 23	+0.602 2	
^{199}Hg	16.84	1/2	0.005 72	0.761 2	+0.504 15	
^{203}Tl	29.50	1/2	0.187	2.433 2	+1.611 5	
^{207}Pb	21.7	1/2	0.009 13	0.889 8	+0.578 3	

TABLE 6-32 Proton chemical shifts

Values are given on the officially approved δ scale; $\tau = 10.00 - \delta$.

Abbreviations Used in the Table

R, alkyl group Ar, aryl group

Substituent group	Methyl protons	Methylene protons	Methine proton
HC—C—CH_2	0.95	1.20	1.55
HC—C—NR_2	1.05	1.45	1.70
HC—C—C=C	1.00	1.35	1.70
HC—C—C=O	1.05	1.55	1.95
HC—C—NRAr	1.10	1.50	1.80

TABLE 6-32 Proton chemical shifts (*continued*)

Substituent group	Methyl protons	Methylene protons	Methine proton
HC—C—H(C=O)R	1.10	1.50	1.90
HC—C—(C=O)NR_2	1.10	1.50	1.80
HC—C—(C=O)Ar	1.15	1.55	1.90
HC—C—(C=O)OR	1.15	1.70	1.90
HC—C—Ar	1.15	1.55	1.80
HC—C—OH	1.20	1.50	1.75
HC—C—OR	1.20	1.50	1.75
HC—C—C≡CR	1.20	1.50	1.80
HC—C—C≡N	1.25	1.65	2.00
HC—C—SR	1.25	1.60	1.90
HC—C—OAr	1.30	1.55	2.00
HC—C—O(C=O)R	1.30	1.60	1.80
HC—C—SH	1.30	1.60	1.65
HC—C—(S=O)R and HC—C—SO_2R	1.35	1.70	
HC—C—NR_3^+	1.40	1.75	2.05
HC—C—O—N=O	1.40		
HC—C—O(C=O)CF_3	1.40	1.65	
HC—C—Cl	1.55	1.80	1.95
HC—C—F	1.55	1.85	2.15
HC—C—NO_2	1.60	2.05	2.50
HC—C—O(C=O)Ar	1.65	1.75	1.85
HC—C—I	1.75	1.80	2.10
HC—C—Br	1.80	1.85	1.90
HC—CH_2	0.90	1.30	1.50
HC—C=C	1.60	2.05	
HC—C≡C	1.70	2.20	2.80
HC—(C=O)OR	2.00	2.25	2.50
HC—(C=O)NR_2	2.00	2.25	2.40
HC—SR	2.05	2.55	3.00
HC—O—O	2.10	2.30	2.55
HC—(C=O)R	2.10	2.35	2.65
HC—C≡N	2.15	2.45	2.90
HC—I	2.15	3.15	4.25
HC—CHO	2.20	2.40	
HC—Ar	2.25	2.45	2.85
HC—NR_2	2.25	2.40	2.80
HC—SSR	2.35	2.70	
HC—(C=O)Ar	2.40	2.70	3.40
HC—SAr	2.40		
HC—NRAr	2.60	3.10	3.60
HC—SO_2R and HC—(SO)R	2.60	3.05	
HC—Br	2.70	3.40	4.10
HC—NR_3^+	2.95	3.10	3.60
HC—NH(C=O)R	2.95	3.35	3.85
HC—SO_3R	2.95		
HC—Cl	3.05	3.45	4.05
HC—OH and HC—OR	3.20	3.40	3.60

TABLE 6-32 Proton chemical shifts (*continued*)

Substituent group	Methyl protons	Methylene protons	Methine proton
HC—PAr_3	3.20	3.40	
HC—NH_2	3.50	3.75	4.05
HC—O(C=O)R	3.65	4.10	4.95
HC—OAr	3.80	4.00	4.60
HC—O(C=O)Ar	3.80	4.20	5.05
HC—O(C=O)CF_1	3.95	4.30	
HC—F	4.25	4.50	4.80
HC—NO_2	4.30	4.35	4.60
Cyclopropane		0.20	0.40
Cyclobutane		2.45	
Cyclopentane		1.65	
Cyclohexane		1.50	1.80
Cycloheptane		1.25	

Substituent group	Proton shift	Substituent group	Proton shift
HC≡CH	2.35	HO—C=O	10–12
HC≡CAr	2.90	HO—SO_2	11–12
HC≡C—C=C	2.75	HO—Ar	4.5–6.5
HAr	7.20	HO—R	0.5–4.5
HCO—O	8.1	HS—Ar	2.8–3.6
HCO—R	9.4–10.0	HS—R	1–2
HCO—Ar	9.7–10.5	HN—Ar	3–6
HO—N=C (oxime)	9–12	HN—R	0.5–5

Saturated heterocyclic ring systems

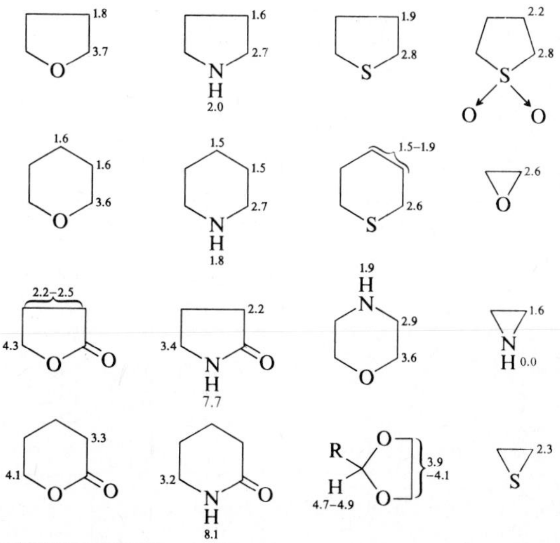

TABLE 6-32 Proton chemical shifts (*continued*)

Substituent group	Methyl protons	Methylene protons	Methine proton
	Unsaturated cyclic systems		

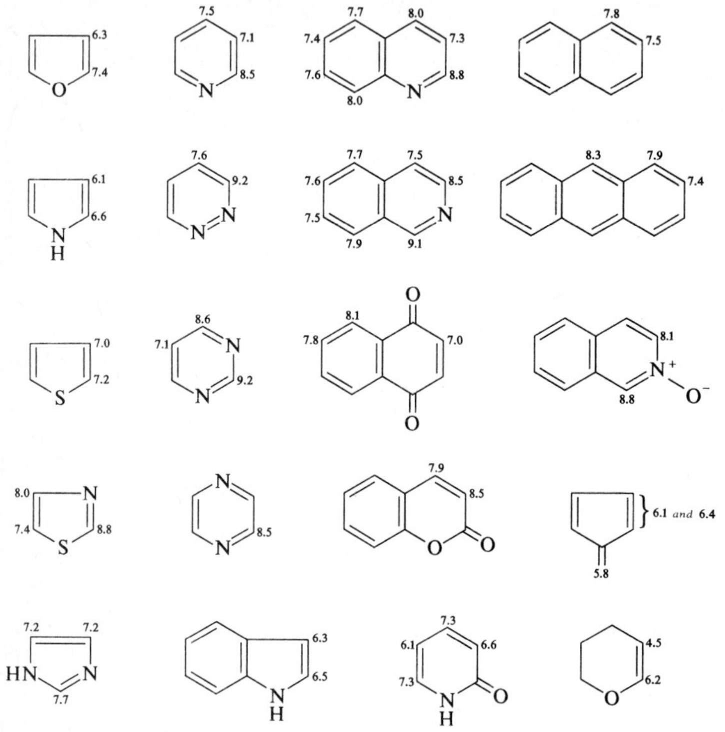

TABLE 6-33 Estimation of chemical shift for protons of $-CH_2-$ and $>CH-$ groups

$\delta_{CH_2}=0.23+C_1+C_2$ $\delta_{CH}=0.23+C_1+C_2+C_3$

X*	C	X*	C	X*	C
$-CH_3$	0.5	$-SR$	1.6	$-OR$	2.4
$-CF_3$	1.1	$-C\equiv C-Ar$	1.7	$-Cl$	2.5
$>C=C<$	1.3	$-CN$	1.7	$-OH$	2.6
$-C\equiv C-R$	1.4	$-CO-R$	1.7	$-N=C=S$	2.9
$-COOR$	1.5	$-I$	1.8	$-OCOR$	3.1
$-NR_2$	1.6	$-Ph$	1.8	$-OPh$	3.2
$-CONR_2$	1.6	$-Br$	2.3		

* R, alkyl group; Ar, aryl group; Ph, phenyl group.

TABLE 6-34 Estimation of chemical shift of proton attached to a double bond

Positive Z values indicate a downfield shift, and an arrow indicates the point of attachment of the substituent group to the double bond.

$$\delta_{C=C\diagdown H} = 5.25 + Z_{gem} + Z_{cis} + Z_{trans}$$

(R_{cis} and H on one carbon, R_{trans} and T_{gem} on the other: $R_{cis}(R_{trans})C=C(H)(T_{gem})$)

R	Z_{gem}, ppm	Z_{cis}, ppm	Z_{trans}, ppm
→H	0	0	0
→alkyl	0.45	−0.22	−0.28
→alkyl—ring (5- or 6-member)	0.69	−0.25	−0.28
→CH_2O—	0.64	−0.01	−0.02
→CH_2S—	0.71	−0.13	−0.22
→CH_2X (X: F, Cl, Br)	0.70	0.11	−0.04
→$CH_2N\lessdot$	0.58	−0.10	−0.08
↗C=C (isolated)	1.00	−0.09	−0.23
↗C=C (conjugated)	1.24	0.02	−0.05
→C≡N	0.27	0.75	0.55
→C≡C—	0.47	0.38	0.12
↗C=O (isolated)	1.10	1.12	0.87
↗C=O (conjugated)	1.06	0.91	0.74
→COOH (isolated)	0.97	1.41	0.71
→COOH (conjugated)	0.80	0.98	0.32
→COOR (isolated)	0.80	1.18	0.55
→COOR (conjugated)	0.78	1.01	0.46
→C(H)=O	1.02	0.95	1.17
→C(N<)=O	1.37	0.98	0.46
→C(Cl)=O	1.11	1.46	1.01
→OR (R: aliphatic)	1.22	−1.07	−1.21
→OR (R: conjugated)	1.21	−0.60	−1.00
→OCOR	2.11	−0.35	−0.64
→CH_2—C(—)=O; →CH_2—C≡N	0.69	−0.08	−0.06
→CH_2—aromatic ring	1.05	−0.29	−0.32
→F	1.54	−0.40	−1.02
→Cl	1.08	0.18	0.13
→Br	1.07	0.45	0.55
→I	1.14	0.81	0.88
→N—R (R: aliphatic)	0.80	−1.26	−1.21

TABLE 6-34 Estimation of chemical shift of proton attached to a double bond (*continued*)

R	Z_{gem}, ppm	Z_{cis}, ppm	Z_{trans}, ppm
→N—R (R: conjugated)	1.17	−0.53	−0.99
→N—C=O	2.08	−0.57	−0.72
→aromatic	1.38	0.36	−0.07
→CF_3	0.66	0.61	0.32
→aromatic (*o*-substituted)	1.65	0.19	0.09
→SR	1.11	−0.29	−0.13
→SO_2	1.55	1.16	0.93

TABLE 6-35 Chemical shifts in monosubstituted benzene

$\delta = 7.27 + \Delta_i$

Substituent	Δ_{ortho}	Δ_{meta}	Δ_{para}
NO_2	0.94	0.18	0.39
CHO	0.58	0.20	0.26
COOH	0.80	0.16	0.25
$COOCH_3$	0.71	0.08	0.20
COCl	0.82	0.21	0.35
CCl_3	0.8	0.2	0.2
$COCH_3$	0.62	0.10	0.25
CN	0.26	0.18	0.30
$CONH_2$	0.65	0.20	0.22
$\overset{+}{N}H_3$	0.4	0.2	0.2
CH_2X*	0.0–0.1	0.0–0.1	0.0–0.1
CH_3	−0.16	−0.09	−0.17
CH_2CH_3	−0.15	−0.06	−0.18
$CH(CH_3)_2$	−0.14	−0.09	−0.18
$C(CH_3)_2$	−0.09	0.05	−0.23
F	−0.30	−0.02	−0.23
Cl	0.01	−0.06	−0.08
Br	0.19	−0.12	−0.05
I	0.39	−0.25	−0.02
NH_2	−0.76	−0.25	−0.63
OCH_3	−0.46	−0.10	−0.41
OH	−0.49	−0.13	−0.2
OCOR	−0.2	0.1	−0.2
$NHCH_3$	−0.8	−0.3	−0.6
$N(CH_3)_2$	−0.60	−0.10	−0.62

* X = Cl, alkyl, OH, or NH_2.

TABLE 6-36 Proton spin coupling constants

Structure	J, Hz	Structure	J, Hz
$>C(H)(H)$ (geminal)	12–15	Aziridine (NH) cis	2
		trans	6
		gem	4
$>CH-CH<$ (free rotation)	6–8	Furan 2-3	1.8
		3-4	3.5
		2-4	0–1
$>CH-OH$ (no exchange)	5	2-5	1–2
$>CH-NH$	4–8	Thiophene 2-3	5–6
$>CH-SH$	6–8	3-4	3.5–5.0
		2-4	1.5
		2-5	3.4
$>CH-C(H)=O$	1–3	Fluorobenzene (H–F) *o*	6–12
		m	4–8
		p	1.5–2.5
$-N=C(H)(H)$	8–16	Fluorotoluene (CH_3–F) *o*	2.5
		m	1.5
$H_t(H_c)C=C(H_g)H$ gem	0–3	*p*	0
cis	6–14	Cyclohexane *a–a*	8–10
trans	11–18	*a–e*	2–3
$H_c(H_t)C=C(H_g)CH<$ cis	0.5–3	*e–e*	2–3
trans	0.5–3		
gem	4–10	Cyclopentane *cis*	4–6
		trans	4–6
$>C=CH-CH=C<$	10–13	Cyclobutane *cis*	8
$=CH-C(H)=O$	6	*trans*	8
		Cyclopropane *cis*	9–11
		trans	6–8
		gem	4–6
$-CH_2-C\equiv C-CH<$	0–3	Benzene (H–H) *o*	6–10
$>CH-C\equiv CH$	0–3	*m*	1–3
H–C=C–H (ring) 3-member	0–2	*p*	0–1
4-member	2–4		
5-member	5–7		
6-member	6–9	Naphthalene 1-2	8–9
7-member	10–13	2-3	6
Oxirane *cis*	4–5	Pyridine 2-3	5–6
trans	3	3-4	7–9
gem	5–6	2-4	1–2
Thiirane *cis*	0	3-5	1–2
trans	7	2-5	0–1
gem	6	2-6	0–1

TABLE 6-36 Proton spin coupling constants (*cotinued*)

Structure		J, Hz	Structure		J, Hz
Pyrrole (N–H; positions 1–5)	1-2	2-3	$H_tH_gC{=}CH_cF$	gem	72-90
	1-3	2-3		cis	−3 to 20
	2-3	2-3		trans	12-40
	3-4	3-4			
	2-4	1-2	$FC{=}C-CH_3$		2-4
	2-5	1-3			
$H-C-F$		45-52	$HC{=}C-CF$		0-6
			$HC{\equiv}CF$		21
			Cyclohexane F_a, F_e, H_e, H_a	a-a	34
				a-e	12
$CH-CF$	gauche	0-12		e-e, e-e	<5-8
	trans	10-45			

TABLE 6-37 Proton chemical shifts of reference compounds

Relative to tetramethylsilane

Compound	δ, ppm	Solvent(s)
Sodium acetate	1.90	D_2O
1,2-Dibromoethane	3.63	$CDCl_3$
1,1,2,2-Tetrachloroethane	5.95	$CDCl_3$; CCl_4
1,4-Benzoquinone	6.78	$CDCl_3$; CCl_4
1,4-Dichlorobenzene	7.23	CCl_4
1,3,5-Trinitrobenzene	9.21	DMSO-d_6*
	9.55	$CHCl_3$

* DMSO, dimethyl sulfoxide.

TABLE 6-38 Solvent positions of residual protons in incompletely deuterated solvents

Relative to tetramethylsilane

Solvent	Group	δ, ppm
Acetic-d_3 acid-d_1	Methyl	2.05
	Hydroxyl	11.5*
Acetone-d_6	Methyl	2.057
Acetonitrile-d_3	Methyl	1.95

TABLE 6-38 Solvent positions of residual protons in incompletely deuterated solvents (*continued*)

Solvent	Group	δ, ppm
Benzene-d_6	Methine	6.78
tert-Butanol-d_1 $(CH_3)_3COD$	Methyl	1.28
Chloroform-d_1	Methine	7.25
Cyclohexane-d_{12}	Methylene	1.40
Deuterium oxide	Hydroxyl	4.7*
Dimethyl-d_6-formamide-d_1	Methyl	2.75; 2.95
	Formyl	8.05
Dimethyl-d_6 sulfoxide	Methyl	2.51
	Absorbed water	3.3*
1,4-Dioxane-d_8	Methylene	3.55
Hexamethyl-d_{18}-phosphoramide	Methyl	2.60
Methanol-d_4	Methyl	3.35
	Hydroxyl	4.8*
Dichloromethane-d_2	Methylene	5.35
Pyridine-d_5	C-2 Methine	8.5
	C-3 Methine	7.0
	C-4 Methine	7.35
Toluene-d_8	Methyl	2.3
	Methine	7.2
Trifluoroacetic acid-d_1	Hydroxyl	11.3*

* These values may vary greatly, depending upon the solute and its concentration.

TABLE 6-39 Carbon-13 chemical shifts

Values given in ppm on the δ scale, relative to tetramethylsilane

Substituent group	Primary carbon	Secondary carbon	Tertiary carbon	Quaternary carbon
Alkanes				
C—C	5–30	25–45	23–58	28–50
C—O	45–60	42–71	62–78	73–86
C—N	13–45	44–58	50–70	60–75
C—S	10–30	22–42	55–67	53–62
C—halide (I to Cl)	3–25	3–40	34–58	35–75

Substituent group	δ, ppm	Substituent group	δ, ppm
Cyclopropane	−5–5	Alcohols R—OH	45–87
Cycloalkane C_4-C_{10}	5–25	Ethers R—O—R	57–87
Mercaptanes	5–70	Nitro $R—NO_2$	60–78
Amines:		Alkynes:	
$R_2N—C$	20–70	HC≡CR	63–73
Aryl—N	128–138	RC≡CR	72–95
Sulfoxides, sulfones	35–55	Acetals, ketals	88–112

TABLE 6-39 Carbon-13 chemical shifts (*continued*)

Substituent group	δ, ppm	Substituent group	δ, ppm
Thiocyanates R—SCN	96–118	Esters:	
Alkenes:		Saturated	158–165
$H_2C=$	100–122	α,β-Unsaturated	165–176
$R_2C=$	110–150	Isocyanides R—NC	162–175
Heteroaromatics:		Carboxylic acids:	
C=N	100–152	Nonconjugated	162–165
C_α	142–160	Conjugated	165–184
Cyanates R—OCN	105–120	Salts (anion)	175–195
Isocyanates R—NCO	115–135	Ketones:	
Isothiocyanates R—NCS	115–142	α-Halo	160–200
Nitriles, cyanides	117–124	Nonconjugated	192–202
Aromatics:		α,β-Unsaturated	202–220
Aryl-C	125–145	Imides	165–180
Aryl-P	119–128	Acyl chlorides R—CO—Cl	165–183
Aryl-N	128–138	Thioureas	165–185
Aryl-O	133–152	Aldehydes:	
Azomethines	145–162	α-Halo	170–190
Carbonates	159–162	Nonconjugated	182–192
Ureas	150–170	Conjugated	192–208
Anhydrides	150–175	Thioketones R—CS—R	190–202
Amides	154–178	Carbonyl $M(CO)_n$	190–218
Oximes	155–165	Allenes =C=	197–205

TABLE 6-39 Carbon-13 chemical shifts (*continued*)

Saturated heterocyclic ring systems

O 39.5	S 18.7	NH 18.2	O 22.9 72.6
S 29.7 27.5	O 26.0 68.0	S 31.2 31.7	NH 25.7 47.1
O 24.9 27.7 69.5	S 26.6 27.8 29.1	NH 25.9 27.8 47.9	O O 66.5
O O 92.8 65.9 26.6	S (O, O) 22.8 51.5	N–CH_3 24.4 56.7 48.0	O 47.3 CH_3 18.1 47.6

TABLE 6-39 Carbon-13 chemical shifts (*continued*)

Unsaturated cyclic systems

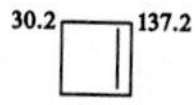

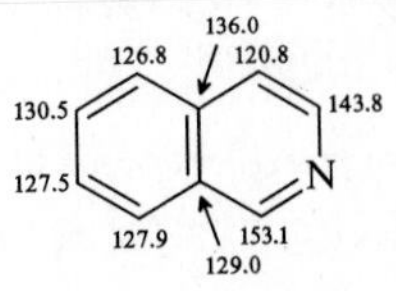

30.2 137.2

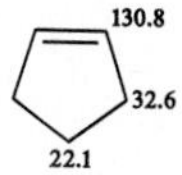

130.8 32.6 22.1

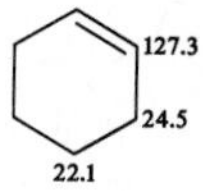

127.3 24.5 22.1

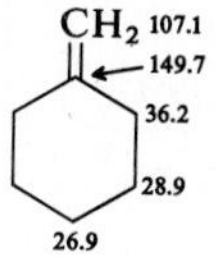

CH_2 107.1 149.7 36.2 28.9 26.9

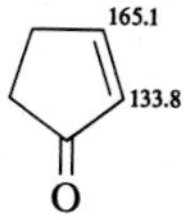

165.1 133.8 O

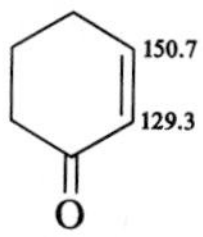

150.7 129.3 O

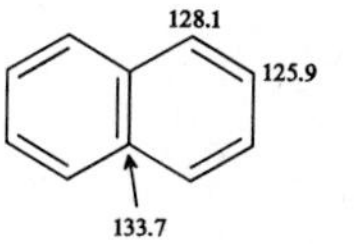

128.1 125.9 133.7

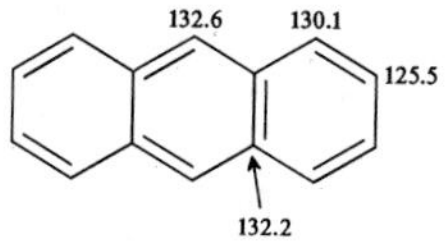

132.6 130.1 125.5 132.2

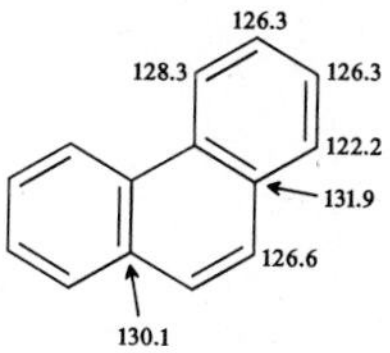

126.3 128.3 126.3 122.2 131.9 126.6 130.1

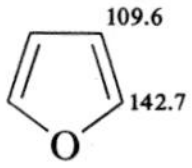

109.6 142.7 O

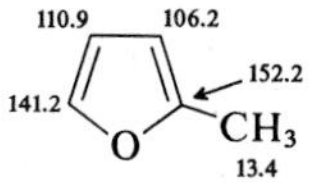

110.9 106.2 152.2 141.2 O CH_3 13.4

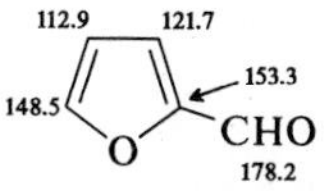

112.9 121.7 153.3 148.5 O CHO 178.2

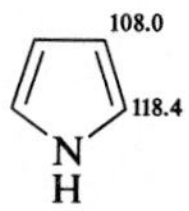

108.0 118.4 N H

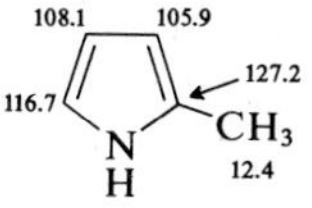

108.1 105.9 127.2 116.7 N H CH_3 12.4

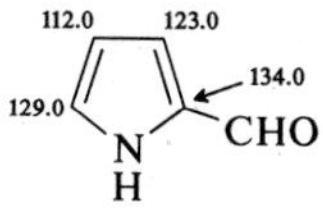

112.0 123.0 134.0 129.0 N H CHO

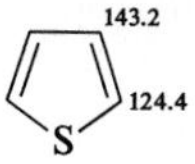

143.2 124.4 S

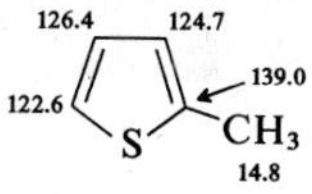

126.4 124.7 139.0 122.6 S CH_3 14.8

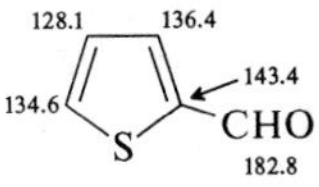

128.1 136.4 143.4 134.6 S CHO 182.8

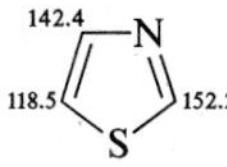

142.4 N 118.5 152.2 S

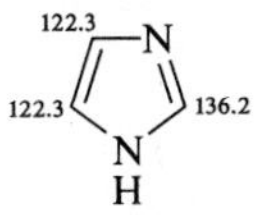

122.3 N 122.3 136.2 N H

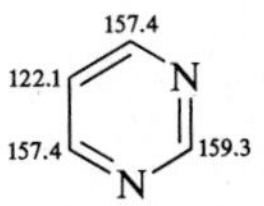

157.4 122.1 N 157.4 159.3 N

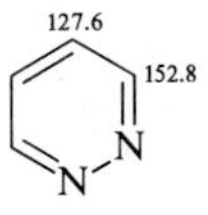

127.6 152.8 N N

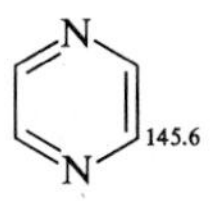

N 145.6 N

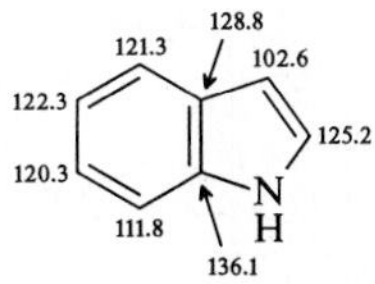

128.8 121.3 102.6 122.3 125.2 120.3 N H 111.8 136.1

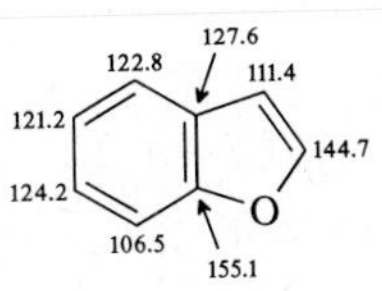

127.6 122.8 111.4 121.2 144.7 124.2 O 106.5 155.1

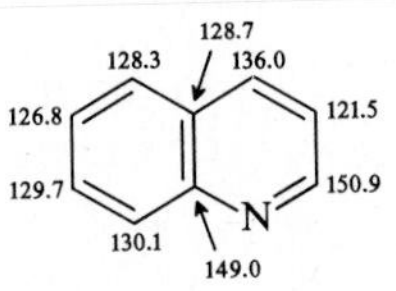

128.7 128.3 136.0 126.8 121.5 129.7 N 150.9 130.1 149.0

136.0 126.8 120.8 130.5 143.8 127.5 N 127.9 153.1 129.0

TABLE 6-39 Carbon-13 chemical shifts (*continued*)

Saturated alicyclic ring systems

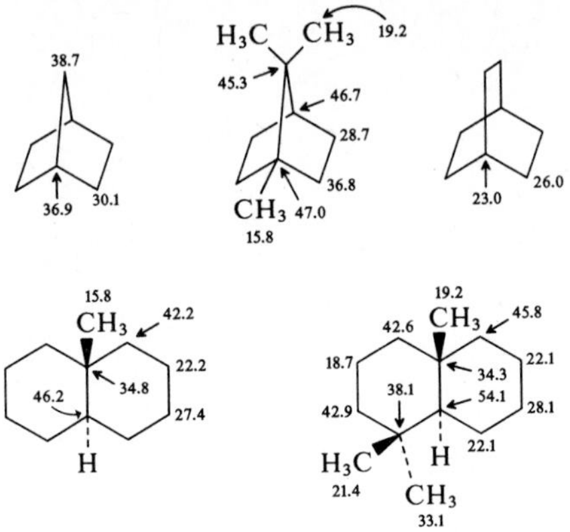

38.7
36.9 30.1

H_3C CH_3 19.2
45.3
46.7
28.7
36.8
CH_3 47.0
15.8

23.0 26.0

15.8
CH_3 42.2
22.2
34.8
46.2
27.4
H

19.2
42.6 CH_3 45.8
18.7 22.1
34.3
38.1 54.1
42.9 28.1
22.1
H_3C H
21.4 CH_3
33.1

TABLE 6-40 Estimation of chemical shifts of alkane carbons

Relative to tetramethylsilane

Positive terms indicate a downfield shift.

$$\delta_C = -2.6 + 9.1n_\alpha + 9.4n_\beta - 2.5n_\gamma + 0.3n_\delta + 0.1n_\varepsilon \text{ (plus any correction factors)}$$

where n_α is the number of carbons bonded directly to the ith carbon atom and n_β, n_γ, n_δ, and n_ε are the number of carbon atoms two, three, four, and five bonds removed. The constant is the chemical shift for methane.

Chain branching*	Correction factor	Chain branching*	Correction factor
1°(3°)	−1.1	4°(1°)	−1.5
1°(4°)	−3.4	2°(4°)	−7.2
2°(3°)	−2.5	3°(3°)	−9.5
3°(2°)	−3.7	4°(2°)	−8.4

* 1° signifies a CH_3— group; 2°, a —CH_2— group; 3°, a >CH— group; and 4°, a >C< group. 1°(3°) signifies a methyl group bound to a >CH— group, and so on.

Examples: For 3-methylpentane, $CH_3—CH_2—CH(CH_3)—CH_2—CH_3$,

$$\delta_{C=2} = -2.6 + 9.1(2) + 9.4(2) - 2.5 - 1(1)[2°(3°)] = 29.4$$

$$\delta_{C=3} = -2.6 + 9.1(3) + 9.4(2) + (2)[3°(2°)] = 36.2$$

TABLE 6-41 Effect of substituent groups on alkyl chemical shifts

These increments are added to the shift value of the appropriate carbon atom as calculated from Table 6-40.

Straight: $\underset{}{Y}-\underset{\alpha}{CH_2}-\underset{\beta}{CH_2}-CH_3$ Branched: $-\underset{\gamma}{CH_2}-\underset{\beta}{CH_2}-\underset{\alpha}{\overset{Y}{\overset{|}{C}H}}-\underset{\beta}{CH_2}-\underset{\gamma}{CH_2}-$

	α carbon		β carbon		
Substituent group Y*	Straight	Branched	Straight	Branched	γ carbon
—CO—OH	20.9	16	2.5	2	−2.2
$—COO^-$ (anion)	24.4	20	4.1	3	−1.6
—CO—OR	20.5	17	2.5	2	−2
—CO—Cl	33	28		2	
$—CO—NH_2$	22	2.5			−0.5
—CHO	31		0		−2
—CO—R	30	24	1	1	−2
—OH	48.3	40.8	10.2	7.7	−5.8
—OR	58	51	8	5	−4
$—O—CO—NH_2$	51		8		
—O—CO—R	51	45	6	5	−3
—C—CO—Ar	53				
—F	68	63	9	6	−4
—Cl	31.2	32	10.5	10	−4.6
—Br	20.0	25	10.6	10	−3.1
—I	−8	4	11.3	12	−1.0
$—NH_2$	29.3	24	11.3	10	−4.6
$—NH_3^+$	26	24	8	6	−5
—NHR	36.9	31	8.3	6	−3.5
$—NR_2$	42		6		−3
$—NR_3^+$	31		5		−7
$—NO_2$	63	57	4	4	
—CN	4	1	3	3	−3
—SH	11	11	12	11	−6
—SR	20		7		−3
$—CH{=}CH_2$	20		6		−0.5
$—C_6H_5$	23	17	9	7	−2
$—C{\equiv}CH$	4.5		5.5		−3.5

* R, alkyl group; Ar, aryl group.

TABLE 6-42 Estimation of chemical shift of carbon attached to a double bond

The olefinic carbon chemical shift is calculated from the equation

$$\delta_C = 123.3 + 10.6n_\alpha + 7.2n_\beta - 7.9n_{\alpha'} - 1.8n_{\beta'} \text{ (plus any steric correction terms)}$$

where n is the number of carbon atoms at the particular position, namely,

$$\begin{matrix} \beta & \alpha & \alpha' & \beta' \\ \text{C}- & \text{C}= & \text{C}- & \text{C} \end{matrix}$$

Substituents on both sides of the double bond are considered separately. Additional vinyl carbons are treated as if they were alkyl carbons. The method is applicable to alicyclic alkenes; in small rings carbons are counted twice, i.e., from both sides of the double bond where applicable. The constant in the equation is the chemical shift for ethylene. The effect of other substituent groups is tabulated below.

Substituent group	β	α	α'	β'
—OR	2	29	−39	−1
—OH	6			−1
—O—CO—CH_3	−3	18	−27	4
—CO—CH_3		15	6	
—CHO		13.6	13.2	
—CO—OH		5.2	9.1	
—CO—OR		6	7	
—CN		−15.4	14.3	
—F		24.9	−34.3	
—Cl	−1	3.3	−5.4	2
—Br	0	−7.2	−0.7	2
—I		−37.4	7.7	
—C_6H_5		12	−11	

Substituent pair		Steric correction term
α,α'	*trans*	0
α,α'	*cis*	−1.1
α,α	*gem*	−4.8
α',α'		+2.5
β,β		+2.3

TABLE 6-43 Carbon-13 chemical shifts in substituted benzenes

$\delta_C = 128.5 + \Delta$

Substituent group	Δ_{C-1}	Δ_{ortho}	Δ_{meta}	Δ_{para}
$-CH_3$	9.3	0.8	−0.1	−2.9
$-CH_2CH_3$	15.6	−0.4	0	−2.6
$-CH(CH_3)_2$	20.2	−2.5	0.1	−2.4
$-C(CH_3)_3$	22.4	−3.1	−0.1	−2.9
$-CH_2O-CO-CH_3$	7.7	0	0	0
$-C_6H_5$	13.1	−1.1	0.4	−1.2
$-CH=CH_2$	9.5	−2.0	0.2	−0.5
$-C\equiv CH$	−6.1	3.8	0.4	−0.2
$-CH_2OH$	12.3	−1.4	−1.4	−1.4
$-CO-OH$	2.1	1.5	0	5.1
$-COO^-$ (anion)	8	1	0	3
$-CO-OCH_3$	2.1	1.1	0.1	4.5
$-CO-CH_3$	9.1	0.1	0	4.2
$-CHO$	8.6	1.3	0.6	5.5
$-CO-Cl$	4.6	2.4	1	6.2
$-CO-CF_3$	−5.6	1.8	0.7	6.7
$-CO-C_6H_5$	9.4	1.7	−0.2	3.6
$-CN$	−15.4	3.6	0.6	3.9
$-OH$	26.9	−12.7	1.4	−7.3
$-OCH_3$	31.4	−14.0	1.0	−7.7
$-OC_6H_5$	29.2	−9.4	1.6	−5.1
$-O-CO-CH_3$	23.0	−6.4	1.3	−2.3
$-NH_2$	18.0	−13.3	0.9	−9.8
$-N(CH_3)_2$	22.4	−15.7	0.8	−11.5
$-N(C_6H_5)_2$	19	−4	1	−6
$-NHC_6H_5$	14.6	−10.7	0.7	−7.7
$-NH-CO-CH_3$	11.1	−9.9	0.2	−5.6
$-NO_2$	20.0	−4.8	0.9	5.8
$-F$	34.8	−12.9	1.4	−4.5
$-Cl$	6.2	0.4	1.3	−1.9
$-Br$	−5.5	3.4	1.7	−1.6
$-I$	−32.2	9.9	2.6	−1.4
$-CF_3$	−9.0	−2.2	0.3	3.2
$-NCO$	5.7	−3.6	1.2	−2.8
$-SH$	2.3	1.1	1.1	−3.1
$-SCH_3$	10.2	−1.8	0.4	−3.6
$-SO_2-NH_2$	15.3	−2.9	0.4	3.3
$-Si(CH_3)_3$	13.4	4.4	−1.1	−1.1

TABLE 6-44 Carbon-13 chemical shifts in substituted pyridines*

$\delta_C(k) = C_k + \Delta_i$

Substituent group	$C_2 = C_6 = 149.6$ Δ_{C-2} or Δ_{C-6}	Δ_{23}	Δ_{24}	Δ_{25}	Δ_{26}
$-CH_3$	9.1	−1.0	−0.1	−3.4	−0.1
$-CH_2CH_3$	14.0	−2.1	0.1	−3.1	0.2
$-CO-CH_3$	4.3	−2.8	0.7	3.0	−0.2
$-CHO$	3.5	−2.6	1.3	4.1	0.7
$-OH$	14.9	−17.2	0.4	−3.1	−6.8
$-OCH_3$	15.3	−13.1	2.1	−7.5	−2.2
$-NH_2$	11.3	−14.7	2.3	10.6	−0.9
$-NO_2$	8.0	−5.1	5.5	6.6	0.4
$-CN$	−15.8	5.0	−1.7	3.6	1.9
$-F$	14.4	−14.7	5.1	−2.7	−1.7
$-Cl$	2.3	0.7	3.3	−1.2	0.6
$-Br$	−6.7	4.8	3.3	−0.5	1.4

Substituent group	Δ_{32}	$C_3 = C_5 = 124.2$ Δ_{C-3} or Δ_{C-5}	Δ_{34}	Δ_{35}	
$-CH_3$	1.3	9.0	0.2	−0.8	−2.3
$-CH_2CH_3$	0.3	15.0	−1.5	−0.3	−1.8
$-CO-CH_3$	0.5	−0.3	−3.7	−2.7	4.2
$-CH0$	2.4	7.9	0	0.6	5.4
$-OH$	−10.7	31.4	−12.2	1.3	−8.6
$-NH_2$	−11.9	21.5	−14.2	0.9	−10.8
$-CN$	3.6	−13.7	4.4	0.6	4.2
$-Cl$	−0.3	8.2	−0.2	0.7	−1.4
$-Br$	2.1	−2.6	2.9	1.2	−0.9
$-I$	7.1	−28.4	9.1	2.4	0.3

Substituent group	$\Delta_{42} = \Delta_{46}$	$\Delta_{43} = \Delta_{45}$	$C_4 = 136.2$ Δ_{C-4}
$-CH_3$	0.5	0.8	10.8
$-CH_2CH_3$	0	−0.3	15.9
$-CH{=}CH_2$	0.3	−2.9	8.6
$-CO-CH_3$	1.6	−2.6	6.8
$-CHO$	1.7	−0.6	5.5
$-NH_2$	0.9	−13.8	19.6
$-CN$	2.1	2.2	−15.7
$-Br$	3.0	3.4	−3.0

* May be used for disubstituted, polyheterocyclic, and polynuclear systems if deviations due to steric and mesomeric effects are allowed for.

TABLE 6-45 Carbon-13 chemical shifts of carbonyl group

$X-C(=O)-Y$

X	Y	δ_C	X	Y	δ_C
H—	$-CH_3$	199.7	CH_3-	$-CH{=}CH_2$	196.9
H—	$-CCl_3$	175.3	CH_3-	$-C_6H_5$	197.6
H—	$-NH_2$	165.5	CH_3-	$-CH_2-CO-CH_3$	201.9 (keto)
H—	$-N(CH_3)_2$	162.4			191.4 (enol)
H—	2-Furyl	153.3	CH_3-	$-CH_2CHO$	167.7
H—	2-Pyrrolyl	134.0	CH_3-	$-C_6H_5-CH_3$	196 (*m*, *p*)
H—	2-Thienyl	143.3			199 (*o*)
$(CH_3)_2CH-$	—OH	184.8	CH_3-	$-2,6\text{-}(CH_3)_2C_6H_5$	206
C_6H_5-	—OH	172.6	CH_3-	—OH	178
CF_3-	—OH	163.0	CH_3-	$-O^-$ (anion)	181.5
CCl_3-	—OH	168.0	CH_3-	$-OCH_3$	170.7
$CH_3CH(NH_2)-$	—OH	176.5	CH_3-	$-O-CH{=}CH_2$	167.7
CF_3-	$-OCH_2CH_3$	158.1	CH_3-	$-O-CH(CH_3)_2$	170.3
H_2N-	$-OCH_2CH_3$	157.8	CH_3-	$-O-CO-CH_3$	167.3
2-Furyl	$-OCH_3$	159.1	CH_3-	$-NH_2$	172.7
$(CH_3)_2N-$	$-C_6H_5$	170.8	CH_3-	$-NHCH_3$	172
$CH_2{=}CHCH_2O-CO-$	$-OCH_2CH{=}CH_2$	157.6	CH_3-	$-N(CH_3)_2$	169.5
CH_3CH_2-	$-CH_2CH_3$	211.4	CH_3-	—Cl	169.6
CH_3-CH_2-	$-O-CO-CH_2CH_3$	170.3	CH_3-	—Br	165.6
CH_3-	$-CH_3$	205.8	CH_3-	—I	158.9
CH_3-	$-CH_2CH_3$	207			

$(CH_2)_n\ C{=}O$ (cyclic ketones)

n	δ_C
3	207.9
4	218.2
5	211.3
6	211.4
7	216.0

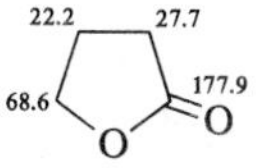

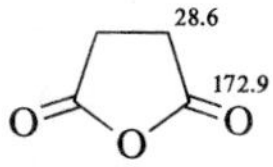

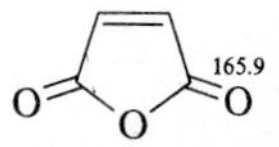

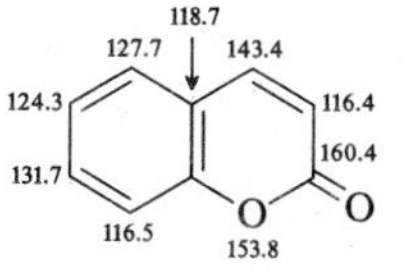

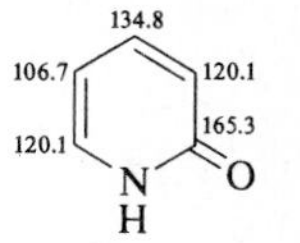

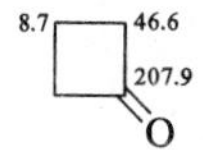

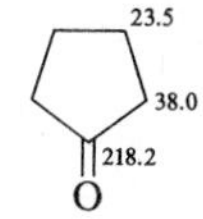

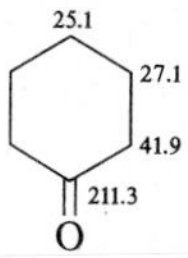

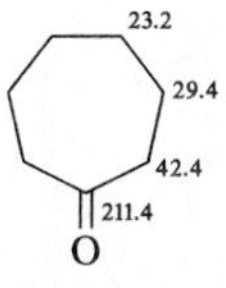

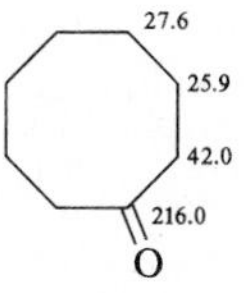

TABLE 6-46 One-bond carbon-hydrogen spin coupling constants

Structure	J_{CH}, Hz
$H{-}CH_3$	125.0
$H{-}CH_2CH_3$	124.9
$CH_3{-}\underline{CH_2}{-}CH_3$	119.2
$H{-}C(CH_3)_2$	114.2
$H{-}CH_2CH_2OH$	126.9
$H{-}CH_2CH{=}CH_2$	122.4
$H{-}CH_2C_6H_5$	129.4
$H{-}CH_2C{\equiv}CH$	132.0
$H{-}CH_2CN$	136.1
$H{-}CH(CN)_2$	145.2
$H{-}CH_2{-}$halogen	149–152
$H{-}CHF_2$	184.5
$H{-}CHCl_2$	178.0
$H{-}CH_2NH_2$	133.0
$H{-}CH_2NH_3^+$	145.0
$H{-}CH_2OH$ (or $H{-}CH_2OR$)	140–141
$H{-}CH(OR)_2$	161–162
$H{-}C(OR)_3$	186
$H{-}C(OH)R_2$	143
$H{-}CH_2NO_2$	146.0
$H{-}CH(NO_2)_2$	169.4
$H{-}CH_2COOH$	130.0
$H{-}CH(COOH)_2$	132.0
$H{-}CH{=}CH_2$	156.2
$H{-}C(CH_3){=}C(CH_3)_2$	148.4
$H{-}CH{=}C($*tert*-$C_4H_9)_2$	152
$H{-}C($*tert*-$C_4H_9){=}C($*tert*-$C_4H_9)_2$	143
Methylenecycloalkane C_4-C_7	153–155
$H{-}CH{=}C{=}CH_2$	168
$H{-}C(C_6H_5){=}CH(C_6H_5)$ *cis*	155
trans	151
Cyclopropene	220
H_t, H_g $C{=}C$ H_c, F: *gem*	200
cis	159
trans	162
H_t, H_g $C{=}C$ H_c, Cl: *gem*	195
cis	163
trans	161
H_t, H_g $C{=}C$ H_c, CHO: *gem*	162
cis	157
trans	162
H_t, H_g $C{=}C$ H_c, CN: *gem*	177
cis	163
trans	165
H, OH $C{=}N$ CH_3: *cis*	163
trans	177
$H{-}CH{=}O$; $CH_3{-}\underline{CH}{=}O$	172
$H_2N{-}CH{=}O$	188.3
$(CH_3)_2N{-}\underline{CH}{=}O$	191
$H{-}COOH$	222
$H{-}COO^-$ (anion)	195
$H{-}CO{-}OCH_3$	226
$H{-}CO{-}F$	267
$CH_3CH_2{-}O{-}\underline{CHO}$	225.6
$Cl_3{-}CHO$	207
$H{-}C{\equiv}CH$	249
$H{-}C{\equiv}CCH_3$	248
$H{-}C{\equiv}CC_6H_5$	251
$H{-}C{\equiv}CCH_2OH$	241
$H{-}CN$	269
Cyclopropane	161
Cyclobutane	136
Cyclopentane	131
Cyclohexane	123
Tetrahydrofuran 2,5	149
3,4	133
1,4-Dioxane	145
Benzene	159
Fluorobenzene 2,6	155
3,5	163
4	161
Bromobenzene 2,6	171
3,5	164
4	161
Benzonitrile 2,6	173
3,6	166
4	163
Nitrobenzene 2,6	171
3,5	167
4	163
Mesitylene	154
Pyridine ring (structure) 2,6	170
3,5	163
4	152

TABLE 6-46 One-bond carbon-hydrogen spin coupling constants (*continued*)

Structure		J_{CH}, Hz	Structure		J_{CH}, Hz
2,4,6-Trimethylpyridine		158	N, N H (imidazole ring)	2	208
N H (pyrrole ring)	2,5	183		4	199
	3,4	170	N, N, N H (triazole ring)		205
O (furan ring)	2,5	201	N–N, N, N H (tetrazole ring)		216
	3,4	175			
S (thiophene ring)	2,5	185			
	3,4	167			
N, N H (pyrazole ring)	3,5	190			
	4	178			

TABLE 6-47 Two-bond carbon-hydrogen spin coupling constants

Structure	$^2J_{CH}$, Hz	Structure		$^2J_{CH}$, Hz
$\underline{C}H_3-CH_2-H$	−4.5	$(CH_2)_n\ C=CH_2$ (ring)	$n=4$	4.2
$\underline{C}Cl_3-CH_2-H$	5.9		$n=5$	5.2
$ClCH_2-CH_2Cl$	−3.4		$n=6$	5.5
$Cl_2CH-CHCl_2$	1.2	$(\underline{H})(Cl)C=\underline{C}(H)(Cl)$	*cis*	16.0
$\underline{C}H_3-C\underline{H}O$	26.7		*trans*	0.8
$CH_2=CH_2$	−2.4	$H\underline{C}\equiv C\underline{H}$		49.3
$(C\underline{H}_3)_2\underline{C}=O$	5.5	$C_6H_5O-\underline{C}\equiv C\underline{H}$		61.0
$CH_2=\underline{C}H-C\underline{H}=O$	26.9	$HC\equiv\underline{C}-C\underline{H}O$		33.2
$(C_2H_5)_2\underline{C}H-C\underline{H}O$	26.9	$Cl\underline{C}H_2-C\underline{H}O$		32.5
$H_2NCH=C\underline{H}-\underline{C}HO$	6.0	$Cl_2\underline{C}H-C\underline{H}O$		35.3
$H_2NCH-\underline{C}H-C\underline{H}O$	20.0	$Cl_3\underline{C}-C\underline{H}O$		46.3
C_6H_6	1.0	$C_6H_5-C\equiv\underline{C}-C\underline{H}_3$		10.8

TABLE 6-48 Carbon-carbon spin coupling constants

Structure*	J_{CC}, Hz	Structure	J_{CC}, Hz
H_3C-CH_3	35	$H_3C-CH_2NH_2$	37
H_3C-CHR_2	37	$C-C=O$	38–40
H_3C-CH_2Ar	34	$\underline{C}-\underline{C}-C=O$	36
H_3C-CH_2CN	33	$\underline{C}-\underline{C}-Ar$	43
$H_3C-CH_2-CH_2OH$		$C-CO-O^-$ (anion)	52
C-1, C-2	38	$C-CO-N$	52
C-2, C-3	34	$C-CO-OH$	57

TABLE 6-48 Carbon-carbon spin coupling constants (*continued*)

Structure*	J_{CC}, Hz	Structure	J_{CC}, Hz
C—CO—OR	59	$C_6H_5NH_2$	
C—CN	52-57	1-2	61
C—C≡C $^2J_{CC} = 11.8$	67	2-3	58
$H_2C{=}CH_2$	68	3-4	57
$>\underline{C}=\underline{C}-CO-OH$	70-71	$^3J_{2\text{-}5}$	7.9
$>\underline{C}=\underline{C}-CN$	71	$C_6H_5CH_3$	44
$>\underline{C}=\underline{C}-Ar$	67-70	Pyridine	
C_6H_6	57	2-3	54
$C_6H_5NO_2$		3-4	56
1-2	55	$^3J_{2\text{-}5}$	14
2-3, 3-4	56	Furan	69
$^3J_{2\text{-}5}$	7.6	Pyrrole	69
C_6H_5I		Thiophene	64
1-2	60	$H_2\underline{C}=\underline{C}=C(CH_3)_2$	100
2-3	53	—C≡C—	170-176
3-4	58	**Structure**	**$^2J_{CC}$, Hz**
$^3J_{2\text{-}5}$	8.6		
$C_6H_5-OCH_3$			
2-3	58	$\underline{C}H_3-CO-\underline{C}H_3$	16
3-4	56	$\underline{C}H_3-C\equiv\underline{C}H$	11.8
		$\underline{C}H_3CH_2-\underline{C}N$	33

* R, alkyl group; Ar, aryl group.

TABLE 6-49 Carbon-fluorine spin coupling constants

Structure*	J_{CF}, Hz	Structure*	J_{CF}, Hz
F H C H H	−158	F F C F CH_3	−271
F H C F H	−235	F H C H Ar	−165
F F C F H	−274	$F-CH_2CH_2-$ or $F-CR_3$	−167
		p-$F-C_6H_4-OR$	−237
		p-$F-C_6H_4-R$	−241
		p-$F-C_6H_4-CF_3$	−252
		p-$F-C_6H_4-CO-CH_3$	−253
F F C F F	−259	p-$F-C_6H_4-NO_2$	−257
		$F-C_6H_5$	
		$^2J_{CF} = 21.0$	−244
		$^3J_{CF} = 7.7$	
		$^4J_{CF} = 3.4$	

TABLE 6-49 Carbon-fluorine spin coupling constants (*continued*)

Structure*	J_{CF}, Hz	Structure*	J_{CF}, Hz
$F_2C{=}CH_2$	−287	$F(H)C(F)CH_2OH$	−241
$F_2C{=}O$	−308	$F_2C(F)CH_2OH$	−278
$F(R)C{=}O$	−353	$F_2C(F)OCF_3$	−265
$F(H)C{=}O$	−369	$F_2C(F)CO{-}CH_3$	−289

* Ar, aryl group; R, alkyl group.

TABLE 6-50 Carbon-13 chemical shifts of deuterated solvents

Relative to tetramethylsilane

Solvent	Group	δ, ppm
Acetic-d_3 acid-d_1	Methyl	20.0
	Carbonyl	205.8
Acetone-d_6	Methyl	28.1
	Carbonyl	178.4
Acetonitrile-d_3	Methyl	1.3
	Carbonyl	117.7
Benzene-d_6		128.5
Carbon disulfide		193
Carbon tetrachloride		97
Chloroform-d_1		77
Cyclohexane-d_{12}		25.2
Dimethyl sulfoxide-d_6		39.5
1,4-Dioxane-d_6		67
Formic-d_1 acid-d_1	Carbonyl	165.5
Methanol-d_4		47–49
Methylene chloride-d_2		53.8
Nitromethane-d_3		57.3
Pyridine-d_5	C_3, C_5	123.5
	C_4	135.5
	C_2, C_6	149.9

TABLE 6-51 Carbon-13 spin coupling constants with various nuclei

Nuclei	Structure	^1J, Hz	^2J, Hz	^3J, Hz	^4J, Hz
^2H	$CDCl_3$	32			
	$CD_3—CO—CD_3$	20			
	$(CD_3)_2SO$	22			
	C_6D_6	26			
^7Li	CH_3Li	15			
^{11}B	$(C_6H_5)_4B^-$	49		3	
^{14}N	$(CH_3)_4N^+$	10			
	CH_3NC	8			
^{29}Si	$(CH_3)_4Si$	52			
^{31}P	$(CH_3)_3P$	14			
	$(C_4H_9)_3P$	11	12	5	
	$(C_6H_5)_3P$	12	20	7	0
	$(CH_3)_4P^+$	56			
	$(C_4H_9)_4P^+$	48	4	15	
	$(C_6H_5)_4P^+$	88	11	13	3
	$R(RO)_2P=O$	142	5-7		
	$(C_4H_9O)_3P=O$		6	7	
^{77}Se	$(CH_3)_2Se$	62			
	$(CH_3)_3Se^+$	50			
^{113}Cd	$(CH_3)_2Cd$	513, 537			
^{119}Sn	$(CH_3)_4Sn$	340			
	$(CH_3)_3SnC_6H_5$	474	37	47	11
^{125}Te	$(CH_3)_2Te$	162			
^{199}Hg	$(CH_3)_2Hg$	687			
	$(C_6H_5)_2Hg$	1186	88	102	18
^{207}Pb	$(CH_3)_2Pb$	250			
	$(C_6H_5)_4Pb$	481	68	81	20

TABLE 6-52 Boron-11 chemical shifts

Values given in ppm on the δ scale, relative to $B(OCH_3)_3$

Structure	δ, ppm	Structure	δ, ppm
R_3B	−67 to −68	$C_6H_5BCl_2$	−36
Ar_3B	−43	$C_6H_5B(OH)_2$	−14
BF_3	24	$C_6H_5B(OR)_2$	−10
BCl_3	−12	$M(BH_4)$	55-61
BBr_3	−6	$B(BF_4)$	19-20
BI_3	41		
$B(OH)_3$	36	NH—BH HB NH NH—BH	−12
$B(OR)_3$	0-1		
$B(NR_2)_3$	−13		

TABLE 6-52 Boron-11 chemical shifts (*continued*)

Structure	δ, ppm	Structure	δ, ppm	
H H H B B H N H R_2	37	R_2O(or ROH)$\cdot BCl_3$	−7 to −8	
		R_2O(or ROH)$\cdot BBr_3$	23–24	
		R_2O(or ROH)$\cdot BI_3$	74–82	
H NR_2 H B B H NR_2 H	15	(pyridine) $N\cdot BBr_3$	24	
$(CH_3)_2N—B(CH_3)_2$	62	**Boranes**		
Addition complexes		B_2H_6	1	
$R_2O\cdot BH_3$	18–19	B_4H_{10}		
$R_3N\cdot BH_3$	25	(BH_2)	25	
$R_2NH\cdot BH_3$	33	(BH)	60	
(pyridine) $N\cdot BH_3$	31		Base	Apex
		B_5H_9	31	70
		B_5H_{11}	−16	50
R_2O(or ROH)$\cdot BF_3$	17–19	$B_{10}H_{14}$	7	54

TABLE 6-53 Nitrogen-15 (or nitrogen-14) chemical shifts

Values given in ppm on the δ scale, relative to NH_3 liquid

Substituent group	δ, ppm	Substituent group	δ, ppm
Aliphatic amines		Ureas	
Primary	1–59	Aliphatic	63–84
Secondary	7–81	Aryl	105–108
Tertiary	14–44	Sulfonamides	79–164
Cyclo, primary	29–44	Amides	
Aryl amines	40–100	HCO—NHR	
Aryl hydrazines	40–100	R = primary	100–115
Piperidines,		R = secondary	104–148
decahydroquinolines	30–82	R = tertiary	96–133
Amine cations		HCO—NH—Aryl	138–141
Primary	19–59	RCO—NHR or	103–130
Secondary	40–74	RCO—NR_2	
Tertiary	30–67	RCO—NH—Aryl	131–136
Quaternary	43–70	Aryl—CO—H—Aryl	ca 126
Enamines, tertiary type		Guanidines	
Alkyl	29–82	Amino	30–60
Cycloalkyl	55–104	Imino	166–207
Aminophosphines	59–100	Thioureas	85–111
Amine *N*-oxides	95–122	Thioamides	135–154

TABLE 6-53 Nitrogen-15 (or nitrogen-14) chemical shifts (*continued*)

Substituent group	δ, ppm	Substituent group	δ, ppm
Cyanamides		Diazo	
R_2N—	−12 to −38	Internal	226-303
—CN	175-200	Terminal	315-440
Carbodiimides	95-120	Nitrilium ions	123-150
Isocyanates		Azinium ions	185-220
Alkyl, primary	14-32	Azine *N*-oxides	230-300
Alkyl, secondary and tertiary	54-57	Nitrones	270-285
Aryl	ca 46	Imides	170-178
Isothiocyanates	90-107	Imimes	310-359
Azides	52-80	Oximes	340-380
	108-122	Nitramines	
	240-260	Amine	252-280
Lactams	113-122	—NO_2	328-355
Hydrazones		Nitrates	310-353
Amino	141-167	*gem*-Polynitroalkanes	310-353
Imino	319-327	Nitro	
Cyanates	155-182	Aryl	350-382
Nitrile *N*-oxides, fulminates	195-225	Alkyl	372-410
Isonitriles		Hetero, unsaturated	354-367
Alkyl, primary	162-178	Azoxy	330-356
Alkyl, secondary	191-199	Azo	504-570
Aryl	ca 180	Nitrosamines	222-250
Nitriles			525-550
Alkyl	235-241	Nitrites	555-582
Aryl	258-268	Thionitrites	720-790
Thiocyanates	265-280	Nitroso	
Diazonium		Aliphatic amines, NO	535-560
Internal	222-230	Aryl	804-913
Terminal	315-322		

TABLE 6-53 Nitrogen-15 (or nitrogen-14) chemical shifts (*continued*)

Substituent group	δ, ppm	Substituent group	δ, ppm
Saturated cyclic systems			
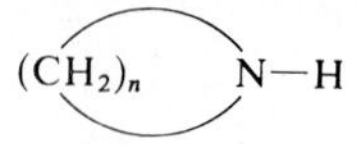 $(CH_2)_n$ N—H		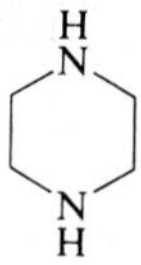 H N N H	35.5
$n = 2$	−8.5		
$n = 3$	25.3		
$n = 4$	36.7		
$n = 5$	37.7	N	7.5 (in C_6H_6) 18.0 (in H_2O)
O N H	32.1	N H	
		cis	42.4
		trans	52.9

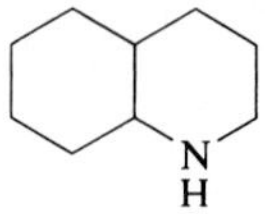

Unsaturated cyclic systems

N H 145	N N H 245	N N H 205	N N N H 240
N 316, N 316, 244 N H	N 251 O	N 383 O	N 323 S
N 298 S	270 N—N 351, N 207, 351 N H	N 317	N N 396
N N 295	N N 331	N N N 383	N N N N 381
N H 133	N N H 191	N 301, N H 179	

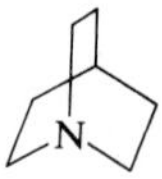

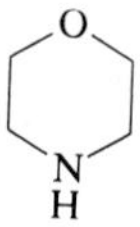

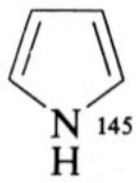

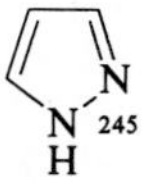

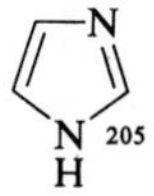

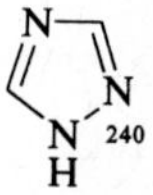

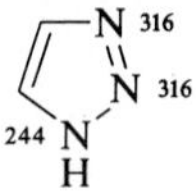

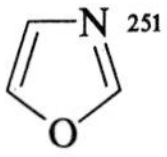

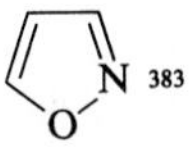

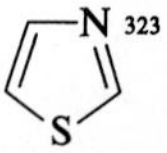

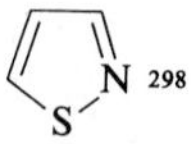

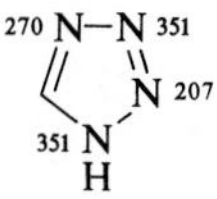

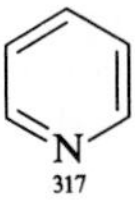

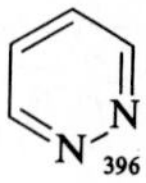

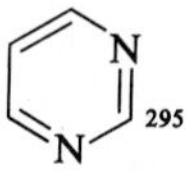

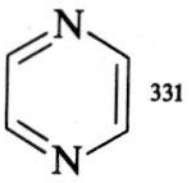

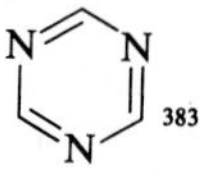

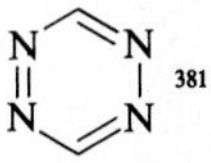

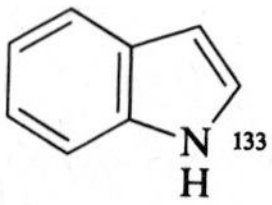

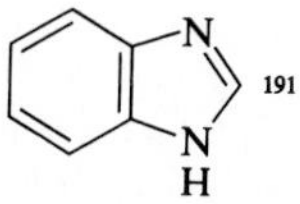

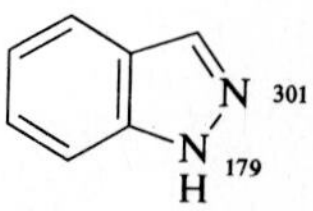

TABLE 6-53 Nitrogen-15 (or nitrogen-16) chemical shifts (*continued*)

Unsaturated cyclic systems (*continued*)

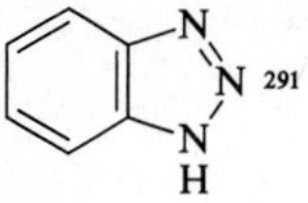

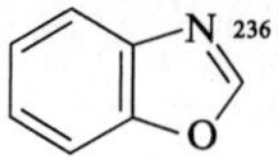

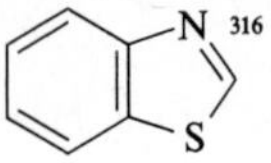

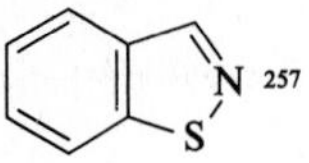

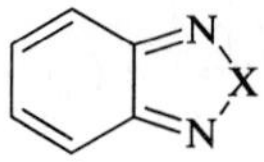

N 291
N
N
H

N 236
O

N 316
S

N 257
S

N
X
N

X	δ, ppm
O	517
S	331
Se	373

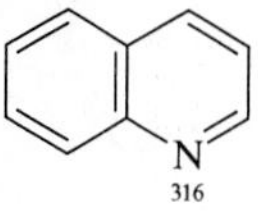

N
316

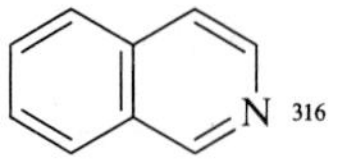

N 316

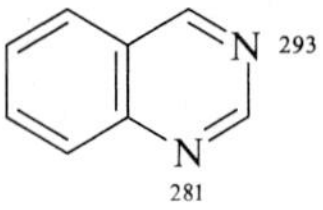

N 293
N
281

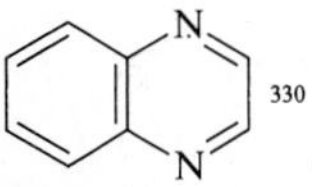

N
330
N

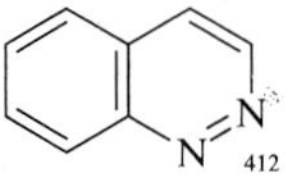

N
N
412

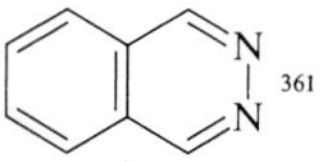

N
361
N

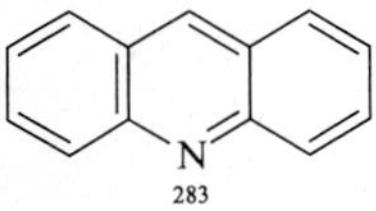

N
283

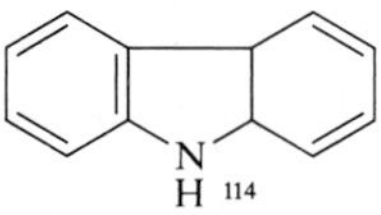

N
H 114

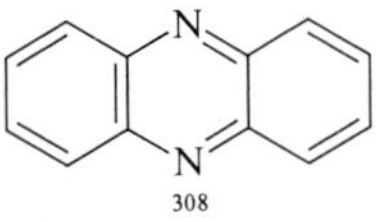

N
N
308

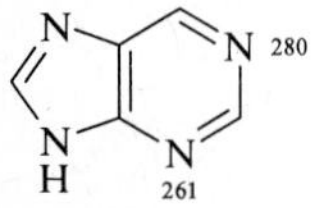

N
N 280
N
H
N
261

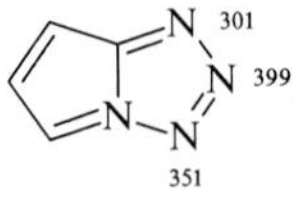

N 301
N 399
N
N
351

TABLE 6-54 Nitrogen-15 chemical shifts in monosubstituted pyridine

$\delta = 317.3 + \Delta_i$

Substituent	$\Delta_{C\text{-}2}$	$\Delta_{C\text{-}3}$	$\Delta_{C\text{-}4}$
$-CH_3$	−0.4	0.3	−8.0
$-CH_2CH_3$	−1.8		−6.6
$-CH(CH_3)_2$	−5.1		−5.9
$-C(CH_3)_3$	−2.5		−5.8
$-CN$	−0.9	−0.8	10.6
$-CHO$	10	11	29
$-CO-CH_3$	−9	15	11
$-CO-OCH_2CH_3$	11.8		−5
$-OCH_3$	−49	0	−23
$-OH$	−126	−2	−118
$-NO_2$	−23	1	22
$-NH_2$	−45	10	−46
$-F$	−42	−18	
$-Cl$	−4	4	−6
$-Br$	2	8	7

TABLE 6-55 Nitrogen-15 chemical shifts for standards

Values given in ppm, relative to NH_3 liquid at 23°C

Substance	δ, ppm	Conditions
Nitromethane (neat)	380.2	For organic solvents and acidic aqueous solutions
Potassium (or sodium) nitrate (saturated aqueous solution)	376.5	For neutral and basic aqueous solutions
$C(NO_2)_4$	331	For nitro compounds
$(CH_3)_2$—CHO (neat)	103.8	For organic solvents and aqueous solutions
$(C_2H_5)_4N^+Cl^-$	64.4	Saturated aqueous solution
$(CH_3)_4N^+Cl^-$	43.5	Saturated aqueous solution
NH_4Cl	27.3	Saturated aqueous solution
NH_4NO_3	20.7	Saturated aqueous solution
NH_3	0.0	Liquid, 25°C
	−15.9	Vapor, 5 atm

TABLE 6-56 Nitrogen-15 to hydrogen-1 spin coupling constants

Structure	*J*, Hz	Structure	*J*, Hz
R—NH_2 and R_2NH	61-67	O=C(R)—N(H)(R)	88-92
Aryl—NH_2	78	Pyrrole	97
p-CH_3O—aryl—NH_2	79	HC≡NH^+	133-136
p-O_2N—aryl—NH_2	90-93	>P—NH_2	82-90
Amine salts (alkyl and aryl)	73-76	$(R_3Si)_2NH$	67
Aryl—NHOH	79	CF_3—S—NH_2	81
Aryl—$NHCH_3$	87	$(CF_3—S)_2NH$	99
Aryl—$NHCH_2F$	90	Pyridinium ion	90
Aryl—$NHNH_2$	90	Quinolinium ion	96
p-O_2N—aryl—$NHNH_2$	99		
Aryl—SO_2—NH_2	81		
Aryl—SO_2—NHR	86		
O=C(H)—N: H_{syn} (to —CO—)	88		
H_{anti}	92-93		

TABLE 6-57 Nitrogen-15 to carbon-13 spin coupling constants

Structure	J, Hz	Structure	J, Hz
Alkyl amines	4-4.5	Alkyl—NO_2	11
Cyclic alkyl amines	2-2.5	R—CN	18
Alkyl amines protonated	4-5	$CH_3—\overset{+}{N}≡\overset{-}{C}$	
Aryl amines	10-14	$H_3C—N$	10
Aryl amines protonated	9	—N≡C	9
$CH_3CO—NH_2$	14-15	Diaryl azoxy	
$H_2N—CO—NH_2$	20	*anti*	18
Aryl—NO_2	15	*syn*	13

TABLE 6-58 Nitrogen-15 to fluorine-19 spin coupling constants

Structure	J, Hz	Structure	J, Hz
NF_3	155	Pyridine	
F_4N_2	164	2-F	52
FNO_2	158	3-F	4
F_3NO	190	2,6-di-F	37
$F_3C—O—NF_2$	164-176	Pyridinium ion	
$FCO—NF_2$	221	2-F	23
$(NF_4)^+SbF_6^-$	323	3-F	3
$(NF_4)^+AsF_6^-$	328	Quinoline, 8-F	3
$(N_2F)^+AsF_6^-$	459	Aniline	
$F_3C—NO_2$	215	2-F	0
F(trans)N=NF $(^2J = 10)$	190	3-F	0
		4-F	1.5
		Anilinium ion	
		2-F	1.4
F(cis)N=NF $(^2J = 52)$	203	3-F	0.2
		4-F	0

TABLE 6-59 Fluorine-19 chemical shifts

Values given in ppm on the δ scale, relative to CCl_3F

Substituent group	δ, ppm	Substituent group	δ, ppm
$—SO_2—F$	−67 to −42	$R—CF_2Cl$	61-71
	(aryl)(alkyl)	$\geq C—CF_3$ and aryl—CF_3	56-73
—CO—F	−29 to −20	$—CS—CF_3$	70
≥N—CO—F	−5	$\geq CF—CF_3$	71-73
Aryl—CF_2Cl	49	$—S—CF_3$	41
$—CF_2I$	56	$—S—CF_2—S—$	39
$—CF_2Br$	63	$\geq P—CF_3$	46-66

TABLE 6-59 Fluorine-19 chemical shifts (*continued*)

Substituent group	δ, ppm	Substituent group	δ, ppm
$>N-CF_3$	40–58	Perfluorocycloalkane	131–138
$>N-CF_2-C$	85–127	$>C\underline{F}-CF_3$	163–198
$-O-CF_2-R$	70–91	$>C\underline{F}(CF_3)_2$	180–191
$-O-CF_2-C\underline{F}_3$	70–91	$-CFH-$	198–231
$-CH_2-CF_3$	76–77	$-CFH_2$	235–244
$HO-CO-CF_3$	77	$F_2C=CF_2$	133
$-CHF-C\underline{F}_3$	81	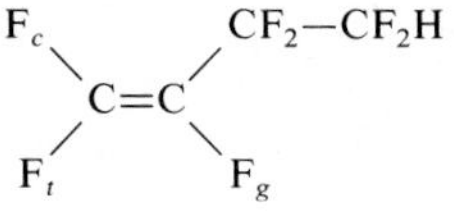 $(F_c)(F_t)C=C(CF_2-CF_2H)(F_g)$	
$-CF_2-C\underline{F}_3$	78–88	*cis*	108
$-CS-F$	81	*trans*	92
$CF_3-C-N<$	84–96	*gem*	192
$-CO-CF_2-C\underline{F}_3$	83	$H_2C=C(F_1)-C(F_2)=C(F_3)H$	
$-CF_2-$	86–126	F-1	126
$-CF_2Br$	91	F-2	155
$-C-CF_2-S-$	91–98	F-3	162
$-CF=$	180–192	$Cl\underline{F}C=CH-CF_3$	61
$-C\underline{F}_2-CF_3$	111	Cycloalkenes	
$-CO-CF_2-$	116–131	$=CF-C\underline{F}_2-C(CF_3 \text{ or } H)-$	101–113
$-C(\text{halide})-CF_2-$	119–128	$-CF_2-C\underline{F}_2-C(CF_3 \text{ or } CH_3)=$	110–114
$-C\underline{F}_2-CF_3$	121–125	$-CF_2-C\underline{F}_2-CH=$	113–116
$-CF_2-CF_2-$	121–129	$-CF_2-C\underline{F}_2-CF=$	119–122
$-CF_2-CH_2-$	122–133	$Aryl-F$	113
$-C\underline{F}_2-CHF_2$	128–132	$C_{10}H_7-F$	
$-CF_2H$	136–143	F-1	127
[cyclopropane ring] F_2	151–156	F-2	114
[cyclobutane ring] F_2	147	$C_6H_5-C_6H_4-F$	
[cyclopentene ring] F_2	96–133	F-2	117
[cyclopentene ring] F	159	F-3	113
Cyclohexane-F	210 (axial) to 240 (equatorial)	F-4	109
		C_6F_6	163

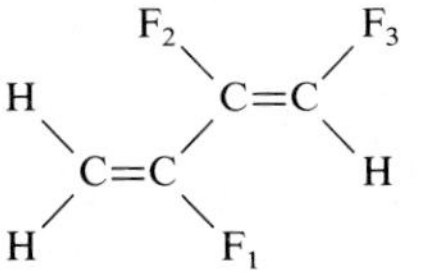

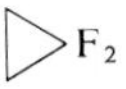

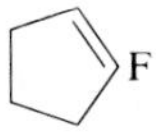

TABLE 6-60 Fluorine-19 chemical shifts for standards

Substance	Formula	δ, ppm
Trichlorofluoromethane	$CFCl_3$	0.0
α,α,α-Trifluorotoluene	$C_6H_5CF_3$	63.8
Trifluoroacetic acid	CF_3COOH	76.5
Carbon tetrafluoride	CF_4	76.7
Fluorobenzene	C_6H_5F	113.1
Perfluorocyclobutane	C_4F_8	138.0

TABLE 6-61 Fluorine-19 to fluorine-19 spin coupling constants

Structure	J_{FF}, Hz
F_2C cycloalkane	
gem	212–260
Unsaturated compounds $>C=C<$	
gem	30–90
trans	115–130
cis	9–58
Aromatic compounds, monocyclic	
ortho	18–22
meta	0–7
para	12–15
Alkanes	
$C\underline{F}Cl_2{-}C\underline{F}_2{-}CFCl_2$	6
$C\underline{F}Cl_2{-}C\underline{F}_2{-}CCl_3$	5
$C\underline{F}_2Cl{-}C\underline{F}_2{-}CF_2Cl$	1
$C\underline{F}_3{-}C\underline{F}_2{-}CF_2Cl$ (or $-CF_3$)	<1
$CF_3{-}C\underline{F}_2{-}C\underline{F}_2Cl$	2
$C\underline{F}_3{-}CF_2{-}C\underline{F}_2Cl$	9
$C\underline{F}_3{-}CF_2{-}C\underline{F}_3$	7

TABLE 6-62 Silicon-29 chemical shifts

Values given in ppm on the δ scale relative to tetramethylsilane

Substituent group X in $(CH_3)_{4-n}SiX_n$	*n*			
	1	2	3	4
$-F$	35	9	−52	−109
$-Cl$	30	32	13	−19
$-Br$	26	20	−18	−94
$-I$	9	−34	−18	−346
$-H$	−19	−42	−65	−93
$-C_2H_5$	2	5	7	8

TABLE 6-62 Silicon-29 chemical shifts (*continued*)

Substituent group X in $(CH_3)_{4-n}SiX_n$	*n* = 1	2	3	4
$-C_6H_5$	−5	−9	−12	
$-CH{=}CH_2$	−7	−14	−21	−23
−Oalkyl	14-17	−3 to −6	−41 to −45	−79 to −83
−Oaryl	17	−6	−54	−101
−O−CO−alkyl	22	4	−43	−75
$-N(CH_3)_2$	6	−2	−18	−28

Structure	δ, ppm
Hydrides	
H_3Si-	−39 to −60
$-H_2Si-$	−5 to −37
$HSi\equiv$	−2 to −39
Silicates	
Orthosilicate anions	−69 to −72
Silicon in end position	−77 to −81
Silicon in middle	−85 to −89
Branching silicons	−93 to −97
Cross-linked silicons	−107 to −120
Methyl siloxanes	
$(CH_3)_3Si-O-$ (end position)	6-8
$(CH_3)_2Si(O-)_2$ (middle)	−18 to −23
$CH_3Si(H)(O-)_2$ (middle)	−35 to −36
$CH_3Si(O-)_3$ (branching)	−65 to −66
$Si(O-)_4$ (cross-linked)	−105 to −110
Polysilanes	
$F_3Si-SiF_3$	−74
$Cl_3Si-SiCl_3$	−8
$(CH_3O)_3Si-Si(OCH_3)_3$	−53
$(CH_3)_3Si-Si(CH_3)_3$	−20
$(CH_3)_2\underline{Si}[Si(CH_3)_3]_2$	−48
$H\underline{Si}[Si(CH_3)_3]_3$	−117
$\underline{Si}[Si(CH_3)_3]_4$	−135

TABLE 6-63 Phosphorus-31 chemical shifts

Values given in ppm on the δ scale, relative to 85% H_3PO_4

Structure	Identical atoms attached directly to phosphorus	Non-identically substituted phosphorus		
		$R = CH_3$	$R = C_2H_5$	$R = C_6H_5$
P_4	461			
PR_3		62	20	6
PHR_2		99	56	41
PH_2R		164	128	122
PH_3	241			
PF_3	−97			
PRF_2			−168	−207

TABLE 6-63 Phosphorus-31 chemical shifts (*continued*)

Structure	Identical atoms attached directly to phosphorus	Non-identically substituted phosphorus		
		$R=CH_3$	$R=C_2H_5$	$R=C_6H_5$
PCl_3	−220			
$PRCl_2$		−192	−196	−162
PR_2Cl		−94	−119	−81
PBr_3	−227			
$PRBr_2$		−184	−194	−152
PR_2Br		−91	−116	−71
PI_3	−178			
$P(CN)_3$	136			
$P(SiR_3)_3$		251		
$P(OR)_3$		−141	−139	−127
$P(OR)_2Cl$		−169	−165	−157
$P(OR)Cl_2$		−114	−177	−173
$P(SR)_3$		−125	−115	−132
$P(SR)_2Cl$		−188	−186	−183
$P(SR)Cl_2$		−206	−211	−204
$P(SR)_2Br$				−184
$P(SR)Br_2$		−204		
$P(NR_2)_3$		−123	−118	
$P(NR_2)Cl_2$		−166	−162	−151
$PR(NR_2)_2$		−86	−100	−100
$PR_2(NR_2)$		−39	−62	
$F_2P{-}PF_2$	−226			
$Cl_2P{-}PCl_2$	−155			
$I_2P{-}PI_2$	−170			
$PH_2^- K^+$	255			
$P(CF_3)_3$	3			
P_4O_6	−113			

Structure	Identical atoms attached directly to phosphorus	Non-identically substituted phosphorus		
		$X = F$	$X = Cl$	$X = Br$
$P(NCO)_3$	−97			
$P(NCO)_2X$		−128	−128	−127
$P(NCO)X_2$		−131	−166	
$P(NCS)_3$	−86			
$P(NCS)_2X$			−114	−112
$P(NCS)X_2$			−155	−153

Structure	Identical atoms attached directly to phosphorus	Non-identically substituted phosphorus		
		$R = CH_3$	$R = C_2H_5$	$R = C_6H_5$
$O{=}PR_3$		−36	−48	−25
$O{=}PHR_2$		−63		−23
$O{=}PF_3$	36			

TABLE 6-63 Phosphorus-31 chemical shifts (*continued*)

Structure	Identical atoms attached directly to phosphorus	Non-identically substituted phosphorus		
		$R=CH_3$	$R=C_2H_5$	$R=C_6H_5$
$O{=}PRF_2$		−27	−29	−11
$O{=}PCl_3$	−2			
$O{=}PRCl_2$		−45	−53	−34
$O{=}PR_2Cl$		−65	−77	−43
$O{=}P(OR)_3$		−1	1	18
$O{=}P(OR)_2Cl$		−6	−3	6
$O{=}P(OR)Cl_2$		−6	−6	−2
$O{=}PH(OR)_2$		−19	−15	
$O{=}PR_2(OC_2H_5)$		−50	−52	−31
$O{=}PR(OC_2H_5)_2$		−30	−33	−17
$O{=}P(NR_2)_3$		−23	−24	−2
$O{=}PR_2(NR_2)$		−44		−26
$O{=}P(OR)_2NH_2$		−15	−12	−3
$O{=}P(OR)_2(NCS)$			19	29
$O{=}P(SR)_3$		−66	−61	−55
$O{=}PBr_3$	103			
$O{=}P(NCO)_3$	41			
$O{=}P(NCS)_3$	62			
$O{=}P(NH_2)_3$	−22			

Structure	Identical atoms attached directly to phosphorus	Structure	Identical atoms attached directly to phosphorus
PF_5	35		
$PF_6^-\ H^+$	144		
PBr_5	101		
$P(OC_2H_5)_5$	71		
PO_4^{3-}	−6	—O—P(=O)(OR)—O— (middle group)	ca 18
$O{=}P[OSi(CH_3)_3]_3$	33		
$H_4P_2O_7$	11		
Phosphonates	−24 to −2		
Phosphonium cations			
Alkyl	−43 to −32		
Aryl	−35 to −18	—O—P(=O)(O—P (etc.))—O— (branch group)	ca 30
$(O_3P{-}PO_3)^{4-}$	−9		
Polyphosphates			
O=P(OR)$_2$—O— (end group)	ca 6		

TABLE 6-63 Phosphorus-31 chemical shifts (*continued*)

Structure	Identical atoms attached directly to phosphorus	Non-identically substituted phosphorus		
		$R=CH_3$	$R=C_2H_5$	$R=C_6H_5$
$S=PR_3$		−59	−55	−43
$S=PCl_3$	−29			
$S=PRCl_2$		−80	−94	−75
$S=PR_2Cl$		−87	−109	−80
$S=PBr_3$	112			
$S=PRBr_2$		−21	−42	−20
$S=PR_2Br$		−64	−98	
$S=P(OR)_3$		−73	−68	−53
$S=P(OR)Cl_2$		−59	−56	−54
$S=P(OR)_2Cl$		−73	−68	−59
$S=PH(OR)_2$		−74	−69	−59
$S=P(SR)_3$		−98	−92	−92
$S=P(NH_2)_3$	−60			
$S=P(NR_2)_3$		−82	−78	
$Se=P(OR)_3$		−78	−71	−58
$Se=P(SR)_3$		−82	−76	
$P(OR)_5$			71	86
PRF_4		30	30	42
PR_2F_3		−9	−6	

TABLE 6-64 Phosphorus-31 spin coupling constants

Substituent group	J_{PH}, Hz	Substituent group	J_{PH}, Hz
>PH	180–225	>P—N—CH	8–25
$—PH_2^-$	134	>P—C—CH	0–4
RPH_2	160–210	>P—(phenyl)	
$>P—CH_3$	1–6		
$>P—CH_2—$	14		
>P—C(Hα)=C(Hβ)(Hγ)		*ortho*	7–10
		meta	2–4
		$O=PHR_2$	210–500
α	12–22	O—PH(S)R	490–540
β	30–40	O_2PHR	500–575
γ	14–20	$O_2PH(N)$	560–630
$(Halogen)_2P—CH$	16–20	$O_2PH(S or Se)$	630–655
>P—NH	10–28	O_3PH	630–760
$>P—O—CH_3$	11–15		
$>P—O—CH_2—R$	6–10	S(or Se)=P(<)—H	490–650
$>P—O—CHR_2$	3–7		
>P—SCH	5–20	$S(or Se)=PHR_2$	420–454

TABLE 6-64 Phosphorus-31 spin coupling constants (*continued*)

Substituent group	J_{PH}, Hz
O=P—CH_3	7–15
O=P—CH=C	15–30
O=P—CH—Aryl(or C=O)	15–30
$(Halogen)_2$P—N—CH	9–18
S=P—CH	11–15
⩸P—CH_3^+	12–17
⩸P—H^+	490–600

Substituent group	J_{PP}, Hz
>P—F	1320–1420
	(1F) (3F)
RPF_2	1140–1290
R_2PF	1020–1110
RP(N)F	920–985
	(alkyl) (aryl)
(—O)(—O)PF	1225–1305
(OCN)PF	1310
>N—P(F)<	1100–1200
>P—CF	60–90
>P—C_6H_4—F	
ortho	0–60
meta	1–7
para	0–3

Substituent group		J_{PF}, Hz
PF (5-coordinate)	axial	600–860
	equatorial	800–1000
O=P—CF		110–113
O=P—F		980–1190
P̲—O—P—F̲		2

Substituent group	J_{PB}, Hz
H_3B—P—N<	80

Substituent group	J_{PP}, Hz
>P—P<	220–400
O=P—P=O	330–500
S=P—P=S	15–500
>P—C—P<	ca 70
>P—O—P<	20–40
>P—S—P<	86–90
O=P—O—P=O	15–25
O=P—N(H)—P=O	8–30
(P—N)$_3$ ring: P—N, N, P, P—N	5–66
>P=N—P=N—	5–65

ELECTRON SPIN RESONANCE

TABLE 6-65 Spin-spin coupling (hyperfine splitting constants)

Values of coupling constant a_i given in gauss

Involves protons unless otherwise indicated.

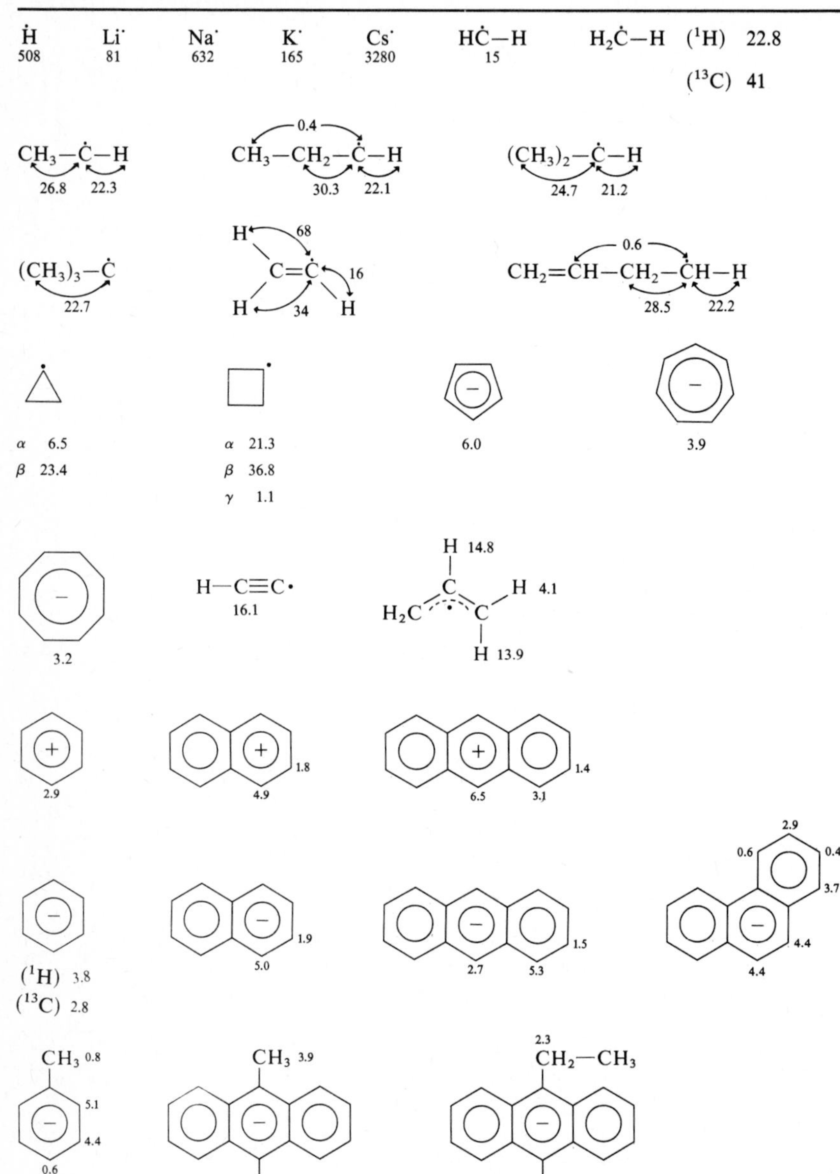

TABLE 6-65 Spin-spin coupling (hyperfine splitting constants) (*continued*)

H—Ċ—H 16.4 (benzyl radical; ring 5.2, 1.8, 6.2)

CH_3 8.0 … CH_3 (9,10-dimethylanthracene cation)

3.8 CH_2—CH_3 … CH_2—CH_3 (9,10-diethylanthracene cation)

15.2 H—Ċ—OH 0.5 (ring 5.2, 4.6, 1.6, 1.6, 5.9)

O=C—H 8.7 (benzaldehyde anion; ring 3.4, 4.7, 0.7, 1.3, 6.5)

O=C—CH_3 6.7 (acetophenone anion; ring 3.7, 4.3, 0.9, 1.1, 6.6)

O=C—H 5.6 (4-cyanobenzaldehyde anion; ring 2.7, 3.1, 0.7, 0.2; CN)

O=C—CH_3 3.9 (4-cyanoacetophenone anion; ring 2.2, 2.6, 0.7, 0.4; CN)

O=C—CH_3 (3-cyanoacetophenone anion; ring 1.2, 5.7, 1.9, 8.0; NC)

O=C—CH_3 (3-cyanoacetophenone anion; ring 4.7, 2.8, 0.9, 4.7; CN)

O=C—CH_3 0.8 (4-nitroacetophenone anion; ring 0.5, 0.5, 2.9, 2.7; NO_2(^{14}N) 5.9)

O=C—CH_3 2.8 (1,4-diacetylbenzene anion; ring 1.3, 1.7; O=C—CH_3)

O=C—CH_3 2.9 (ring 1.7, 1.0; H_3C—C=O)

1.4 H—C=O (4-nitrobenzaldehyde anion; ring 0.4, 0.2, 2.3, 3.0; NO_2 5.1)

CH_3 0.8, C=O (ring 0.5, 0.5, 2.7, 2.9; NO_2 5.9)

Ȯ (phenoxyl; ring 1–2, 1–2; CH_3 12.0)

Ȯ (phenoxyl; CH_2—CH_3, 10.2)

Ȯ (phenoxyl; CH_2—CH_2—CH_3, 8.7)

Ȯ (phenoxyl; CH 6.0, H_3C, CH_3)

Ȯ (H_3C, CH_3; 6.5)

Ȯ (H_2C—H_2C, CH_3—CH_2; 5.7)

Ȯ (CH_3—HC—CH_3, CH_3—CH 3.7—CH_3)

H_3C—CH_2—C(=Ȯ)—C(=O)—Ph (3.4)

(H_3C)$_2$CH—C(=Ȯ)—C(=O)—Ph (1.5)

^{14}N—O· (^{14}N) 14·3

·$C^{14}N$ (^{14}N) 174

TABLE 6-65 Spin-spin coupling (hyperfine splitting constants) (*continued*)

Nitrobenzene radical anion: NO_2 10.0; 3.4; 1.1; 4.0

o-Dinitrobenzene radical anion: NO_2 3.2; 0.4; 1.6; 1.6; 0.1

m-Dinitrobenzene radical anion: NO_2 4.7; 4.2; 3.1; 1.0; 4.2; NO_2 4.7

p-Dinitrobenzene radical anion: NO_2 1.7; 1.1

NO_2 14.9; *tert*-butyl; 2.5; 1.1; 1.1; 2.5

NO_2 18.8; *tert*-butyl; *tert*-butyl; 0.9; 0.9; *tert*-butyl

NO_2 17.8; H_3C; CH_3 0.9; 0.9; 0.9; 1.4

Pyrazine radical anion: (^1H) 2.6; (^{14}N) 7.1

NO_2 12.2; 3.4; 1.1; N 1.1; H; H 1.1

NO_2 11.7; 3.3; 1.1; N 1.1; H_3C; CH_3 0.7

Pyrazine radical cation: (^1H) 3.3; (^{14}N) 7.6

Stilbene radical anion: H 4.4; C=C; H; 1.9; 3.0; 0.3; 0.8; 3.9

Azobenzene radical anion: N=N 4.8; 2.8; 2.0; 0.8; 0.8; 2.8

$O{=}\dot{C}{-}H$ 137

$\dot{C}{-}C({=}O){-}OH$ 1–2.5

$(HO)(HOOC)\dot{C}{-}CH_3$ 17.1

$(HO)(HOOC)\dot{C}{-}CH_2OH$ 10.1

$(HO)(HOOC)\dot{C}{-}CH(OH)(COOH)$ 3.9

$CH_3{-}\dot{N}O_2$ 11.4 25.8

$CH_3{-}CH_2{-}\dot{N}O_2$ 8.3 9.8 25.5

$CH_3{-}C(H)(\dot{N}O_2){-}CH_3$: H 4.6; CH_3 1.8; $\dot{N}O_2$ 25.2

R(H)C=N–O· : 26.2 (*anti*) 29.2; 6.2 (*syn*) 31.6

Quinone oxime radical: 1.2; 3.7; N 33.0; O

Cyclohexanone radical anion: 10.6 (all positions)

4-Methylcyclohexanone radical anion: 13.5 axial; 7.5 equatorial; CH_3

TABLE 6-65 Spin-spin coupling (hyperfine splitting constants) (*continued*)

Radical		
Cyclopentanone ring with O (carbonyl) and − charge; 6.0 (all positions)		
1,4-Dimethoxybenzene cation (CH_3–O– ring (+) –O–CH_3)	3.5 (*trans*)	3.3 (*cis*)
	3.0	2.7
	1.5	1.8
	Ring H	

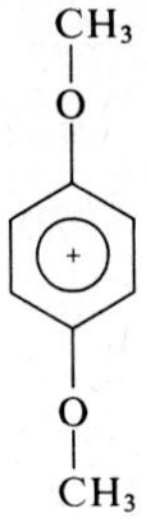

IONIZATION POTENTIALS

TABLE 6-66 Ionization potentials of molecular species

1 eV = 23.061 kcal · mol^{-1}

Values in parentheses are uncertainties in the final figure(s).

Species	Ionization potential, eV	Species	Ionization potential, eV
Diborane(6)	12.0	2-Methyl-1-propene	9.23(2)
Pentaborane(9)	10.5	Cyclobutane	10.58
Hexaborane(10)	9.3(1)	Butane	10.63(3)
Trimethylborane	8.8(2)	Isobutane	10.57
Triethylborane	9.0(2)	Cyclopentadiene	8.97
Methane	12.6	1,2-Pentadiene	9.42
CD_4	12.888	1,3-Pentadiene	8.68
Acetylene	11.4	1,4-Pentadiene	9.58
C_2D_2	11.416(6)	2,3-Pentadiene	8.68
Ethylene	10.5	2-Methyl-1,4-butadiene	8.845(5)
Ethane	11.5	Cyclopentene	9.01(1)
Propyne	10.36	1-Pentene	9.50(2)
Allene	10.16(2)	*cis*-2-Pentene	9.11
Cyclopropene	9.95	*trans*-2-Pentene	9.06
Cyclopropane	10.09(2)	2-Methyl-1-butene	9.12(2)
Propane	11.1	3-Methyl-1-butane	9.51(3)
1,2-Butadiene	9.57(2)	3-Methyl-2-butene	8.69(2)
1,3-Butadiene	9.07	Cyclopentane	10.53(5)
1-Butyne	10.18(1)	Pentane	10.35
2-Butyne	9.9(1)	Isopentane	10.32
1-Butene	9.6	Neopentane	10.35
cis-2-Butene	9.13	Benzene	9.24
trans-2-Butene	9.13	Hexa-1,3-diene-5-yne	9.50

TABLE 6-66 Ionization potentials of molecular species (*continued*)

Species	Ionization potential, eV	Species	Ionization potential, eV
1,3-Hexadiyne	9.25	2,2,4-Trimethylpentane	9.86
1,4-Hexadiyne	9.75	2,2,3,3-Tetramethylbutane	9.79
1,5-Hexadiyne	10.35	Indene	8.81
2,4-Hexadiyne	9.75	β-Methylstyrene	8.35(1)
1-Methylcyclopentadiene	8.43(5)	Propylbenzene	8.72(1)
2-Methylcyclopentadiene	8.46(5)	Isopropylbenzene	8.69(1)
Cyclohexene	8.72	1,2,3-Trimethylbenzene	8.48
1-Hexene	9.45(2)	1,2,4-Trimethylbenzene	8.27
2,3-Dimethyl-2-butene	8.30	1,3,5-Trimethylbenzene	8.4
Cyclohexane	9.8	Naphthalene	8.12
Hexane	10.18	Azulene	7.42
2-Methylpentane	10.12	Butylbenzene	8.69(1)
3-Ethylbutane	10.08	*sec*-Butylbenzene	8.68(1)
2,2-Dimethylbutane	10.06	*tert*-Butylbenzene	8.68(1)
2,3-Dimethylbutane	10.02	1,2,3,5-Tetramethylbenzene	8.47(5)
Toluene	8.82(1)	1,2,4,5-Tetramethylbenzene	8.03
Cycloheptatriene	8.5	*cis*-Decalin	9.61(2)
Bicyclo[2.2.1]heptane	8.67	*trans*-Decalin	9.61(2)
Bicyclo[3.2.0]heptane	9.37	1-Methylnaphthalene	7.96(1)
1,2-Dimethylcyclopentadiene	8.1(1)	2-Methylnaphthalene	7.955(10)
5,5-Dimethylcyclopentadiene	8.22(5)	Pentamethylbenzene	7.92(2)
1,3-Cycloheptadiene	8.55	Hexamethylcyclopentadiene	7.74(5)
Norbornene	8.95(15)	Biphenyl	8.27(1)
4-Methylcyclohexene	8.91(1)	Hexamethylbenzene	7.85(2)
Methylcyclohexane	9.85(3)	Fluorene	8.63
Heptane	9.90(5)	Diphenylacetylene	8.85(5)
Phenylacetylene	8.815(5)	Anthracene	7.55
Styrene	8.47(2)	Phenanthrene	8.1
Cyclooctatetraene	8.0	1,2-Benzanthracene	8.01
Cubane	8.74(15)	1-Phenyldodecane	9.05(10)
Ethylbenzene	8.76(1)	3-Phenyldodecane	8.95(10)
o-Xylene	8.56	7-Phenyltridecane	8.91(10)
m-Xylene	8.58	1-Phenylicosane	9.34(10)
p-Xylene	8.44	2-Phenylicosane	9.22(10)
7-Methylcycloheptatriene	8.39(10)	3-Phenylicosane	8.95(10)
1-Methylspiroheptadiene	8.02(10)	4-Phenylicosane	9.01(10)
6-Methylspiroheptadiene	8.4(1)	5-Phenylicosane	9.04(10)
1,2,3-Trimethylcyclopentadiene	7.96(5)	7-Phenylicosane	8.97(10)
		9-Phenylicosane	9.06(10)
1,5,5-Trimethylcyclopentadiene	8.0(1)	N_2	15.576
		NH_3	10.2
4-Vinylcyclohexene	8.93(2)	N_2H_2	9.85(10)
cis-1,2-Dimethylcyclohexane	10.08(2)	N_2H_4	8.74(6)
trans-1,2-Dimethylcyclohexane	10.08(3)	HCN	13.8
		C_2N_2	13.6

TABLE 6-66 Ionization potentials of molecular species (*continued*)

Species	Ionization potential, eV	Species	Ionization potential, eV
Methylamine	8.97	*N,N*-Dimethyl-*p*-toluidine	7.33
Acetonitrile	12.2	Tripropylamine	7.23
Ethyleneimine	9.94(15)	*N*-Butylaniline	7.53
Ethylamine	8.86(2)	*N,N*-Diethylaniline	6.99
Dimethylamine	8.24(2)	*N,N*-Dimethyl-4-ethylaniline	7.38
Acrylonitrile	10.91(1)	*N,N*,2,4-Tetramethylaniline	7.17
Propionitrile	11.84(2)	*N,N*,2,6-Tetramethylaniline	7.22
Propylamine	8.78(2)	*N,N*-3,5-Tetramethylaniline	7.25
Isopropylamine	8.72(3)	*N,N*-Diethyl-4-toluidine	6.93
Trimethylamine	7.82(2)	*N,N*-Dimethyl-4-isopropylaniline	7.41
3-Butenonitrile	10.39(1)		
Pyrrole	8.20(1)	Diphenylamine	7.25(3)
Butyronitrile	11.67(5)	*N,N*-Dipropylaniline	6.96
Pyrrolidine	8.41	*N,N*-Dimethyl-4-*tert*-butylaniline	7.43
Butylamine	8.71(3)		
sec-Butylamine	8.70	*N,N*-Dibutylaniline	6.95
Isobutylamine	8.70	Triphenylamine	6.86(3)
tert-Butylamine	8.64	Diazirine	10.18(5)
Diethylamine	8.01(1)	Diazomethane	8.999(1)
Pyridine	9.3	Methylhydrazine	8.00(6)
Aniline	7.7	1,1-Dimethylhydrazine	7.67(5)
2-Methylpyridine	9.02(3)	1,2-Dimethylhydrazine	7.75(10)
3-Methylpyridine	9.04(3)	*o*-Diazine	9.9
4-Methylpyridine	9.04(3)	*m*-Diazine	9.9
Cyclohexylamine	8.86	*p*-Diazine	9.8
Dipropylamine	7.84(2)	1,1-Diethylhydrazine	7.59(5)
Diisopropylamine	7.73(3)	1-Butyl-1-methylhydrazine	7.62(5)
Triethylamine	7.50(2)	*p*-Bis(dimethylamino)benzene	6.9
Benzonitrile	9.705(10)	Methyl azide	9.5(1)
N-Methylaniline	7.32	O_2	12.063(1)
m-Toluidine	7.50(2)	O_3	12.3(1)
2,3-Dimethylpyridine	8.85(2)	Water (and D_2O)	12.6
2,4-Dimethylpyridine	8.85(3)	H_2O_2	11.0
2,6-Dimethylpyridine	8.85(2)	CO	14.013(4)
Phenylacetonitrile	9.4(5)	CO_2	13.769(30)
3-Methylbenzonitrile	9.66(5)	NO	9.25
4-Methylbenzonitrile	9.76	N_2O	12.894
N-Ethylcyclohexylamine	7.56	NO_2	9.79
N,N-		Formaldehyde	10.88
Dimethylcyclohexylamine	7.12	Methanol	10.84
Dibutylamine	7.69(3)	Acetaldehyde	10.2
N-Propylaniline	7.54	Ethylene oxide	10.6
N-Ethyl-*N*-methylaniline	7.37	Ethanol	10.49
N,N-Dimethyl-*o*-toluidine	7.37	Dimethyl ether	9.98
N,N-Dimethyl-*m*-toluidine	7.35	Propenal	10.10(1)

TABLE 6-66 Ionization potentials of molecular species (*continued*)

Species	Ionization potential, eV	Species	Ionization potential, eV
Propionaldehyde	9.98	Diphenyl ether	8.82(5)
Acetone	9.69	Benzophenone	9.4
Allyl alcohol	9.67(5)	4-Methylbenzophenone	9.13(5)
Methyl vinyl ether	8.93(2)	Formic acid	11.05(1)
Propylene oxide	10.22(2)	Acetic acid	10.69(3)
Trimethylene oxide	9.667(5)	Methyl formate	10.815(5)
1-Propanol	10.1	Propionic acid	10.24(1)
2-Propanol	10.15	Ethyl formate	10.61(1)
Furan	8.89	Methyl acetate	10.27(2)
2-Butenal	9.73(1)	Dimethoxymethane	10.00(5)
Butyraldehyde	9.86(2)	Vinyl acetate	9.19(5)
2-Methylpropionaldehyde	9.74(3)	2,3-Butanedione	9.24(3)
2-Butanone	9.5	Butanoic acid	10.16(5)
Tetrahydrofuran	9.42	Isobutyric acid	10.02(5)
1-Butanol	10.04	Propyl formate	10.54(1)
Diethyl ether	9.6	Ethyl acetate	10.11(2)
Cyclopentanone	9.26(1)	Methyl propionate	10.15(3)
Dihydropyran	8.34(1)	1,4-Dioxane	9.13(3)
Pentanal	9.82(5)	1,1-Dimethoxyethane	9.65(3)
3-Methylbutyraldehyde	9.71(5)	2-Furaldehyde	9.21(1)
2-Pentanone	9.37(2)	2,4-Pentanedione	8.87(3)
3-Methyl-2-butanone	9.30(2)	Butyl formate	10.50(2)
3-Pentanone	9.32(1)	Isobutyl formate	10.46(2)
Cyclopentanone	9.25(1)	Propyl acetate	10.04(3)
Phenol	8.51	Isopropyl acetate	9.99(1)
4-Methyl-3-penten-2-one	9.08(3)	Ethyl propionate	10.00(2)
Cyclohexanone	9.14(1)	Methyl butyrate	10.07(3)
2-Hexanone	9.35	Methyl isobutyrate	9.98(2)
4-Methyl-2-pentanone	9.30	Diethoxymethane	9.70(5)
3,3-Dimethyl-2-butanone	9.17(3)	1,4-Quinone	9.67(2)
Dipropyl ether	9.27(5)	Butyl acetate	9.56(3)
Diisopropyl ether	9.20(5)	Isobutyl acetate	9.97
Benzaldehyde	9.52	*sec*-Butyl acetate	9.91(3)
Tropone	9.68(2)	Benzoic acid	9.73(9)
Benzyl alcohol	9.14(5)	*p*-Hydroxybenzaldehyde	9.32(2)
Methoxybenzene	8.21(2)	α-Hydroxyacetophenone	9.33(5)
m-Cresol	8.52(5)	Methyl benzoate	9.35(6)
2-Heptanone	9.33(3)	*p*-Methoxybenzaldehyde	8.60(3)
Acetophenone	9.27(3)	*m*-Hydroxyacetophenone	8.67(5)
4-Methylbenzaldehyde	9.33(5)	*p*-Hydroxyacetophenone	8.70(3)
Benzyl methyl ether	8.85(3)	α-Methoxyacetophenone	8.60(5)
Ethyl phenyl ether	8.13(2)	*m*-Methoxyacetophenone	8.53(5)
3-Methylanisole	8.31(5)	*p*-Methoxyacetophenone	8.62(5)
Propiophenone	9.27(5)	Methyl *p*-methylbenzoate	8.94(4)
3-Methylacetophenone	9.15(5)	*p*-Hydroxybenzophenone	8.59(5)

TABLE 6-66 Ionization potentials of molecular species (*continued*)

Species	Ionization potential, eV	Species	Ionization potential, eV
Phenyl benzoate	8.98(5)	N_2F_4	12.04(10)
Benzil	8.78(5)	OF_2	13.6
Methyl methoxyacetate	9.56(5)	XeF_2	11.5(2)
Methyl *p*-methoxybenzoate	8.43(4)	Fluoromethane	12.85(1)
Diphenyl carbonate	9.01(5)	Fluoroethylene	10.37
Acetamide	9.77(2)	Fluorobenzene	9.2
N,N-Dimethylformamide	9.12(2)	1,2-Difluorobenzene	9.31
N-Methylacetamide	8.90(2)	1,4-Difluorobenzene	9.15
NN-Dimethylacetamide	8.81(3)	Trifluoroethylene	10.14
N,N-Diethylformamide	8.89(2)	3,3,3-Trifluoro-1-propene	10.9
2-Pyridinecarboxaldehyde	9.75(5)	*o*-Fluorophenol	8.66(1)
4-Pyridinecarboxaldehyde	10.12(5)	PH_3	9.98
N,N-Diethylacetamide	8.60(2)	PF_3	9.71
Phenyl isocyanate	8.77(2)	Methylphosphine	9.72(15)
Benzamide	9.4(2)	Ethylphosphine	9.47(50)
p-Aminobenzaldehyde	8.25(2)	Trimethylphosphine	8.6(2)
p-Methoxyaniline	7.82	Triphenylphosphine	7.36(5)
Acetanilide	8.39(10)	S_6	9.7
m-Aminoacetophenone	8.09(5)	S_7	9.2(3)
p-Aminoacetophenone	8.17(2)	Hydrogen sulfide	10.4
α-Cyanoacetophenone	9.56(5)	Carbon disulfide	10.080
Nitromethane	11.1	Sulfur dioxide	12.34(2)
Nitroethane	10.88(5)	Methanethiol	9.440(5)
1-Nitropropane	10.81(3)	Ethylene sulfide	8.87(15)
2-Nitropropane	10.71(5)	Ethanethiol	9.285(5)
Nitrobenzene	9.92	Dimethyl sulfide	8.685(5)
m-Nitrotoluene	9.65(5)	Propylene sulfide	8.6(2)
p-Nitrotoluene	9.87	1-Propanethiol	9.195
o-Nitroaniline	8.66	Ethyl methyl sulfide	8.55(1)
m-Nitroaniline	8.7	Thiophene	8.860(5)
p-Nitroaniline	8.85	Methyl 1-propenyl sulfide	8.7(2)
Ethyl nitrate	11.22	1-Butanethiol	9.14(2)
Propyl nitrate	11.07(2)	Diethyl sulfide	8.430(5)
p-Nitrophenol	9.52	Methyl propyl sulfide	8.80(15)
p-Nitrobenzaldehyde	10.27(1)	Isopropyl methyl sulfide	8.7(2)
m-Nitroacetophenone	9.89(5)	Thiophenol	8.32(1)
p-Nitroacetophenone	10.07(2)	2-Ethylthiophene	8.8(2)
Methyl *p*-nitrobenzoate	10.20(3)	Dipropyl sulfide	8.5
F_2	15.7	Methyl phenyl sulfide	8.9
HF	15.77(2)	2-Propylthiophene	8.6(2)
BF_3	15.5	2-Butylthiophene	8.5(2)
C_2F_4	10.12	Dimethyl disulfide	8.46(3)
Hexafluorobenzene	9.97	Diethyl disulfide	8.27(3)
trans-N_2F_2	13.1(1)	COS	11.17(1)
NF_3	13.2(2)	SO_2F_2	13.3(1)

TABLE 6-66 Ionization potentials of molecular species (*continued*)

Species	Ionization potential, eV	Species	Ionization potential, eV
Methyl isothiocyanate	9.25(3)	*p*-Dichlorobenzene	8.95
Methyl thiocyanate	10.065(10)	Chloroform	11.42(3)
Ethyl isothiocyanate	9.14(3)	Trichloroethylene	9.45
Ethyl thiocyanate	9.89(1)	1,1,2,2-Tetrachloroethane	11.10(5)
Phenyl isothiocyanate	8.520(5)	CNCl	12.49(4)
Tolyl thiocyanate	9.06(5)	CF_3Cl	12.91(3)
Thiourea	8.50(5)	Chlorotrifluoroethylene	10.4(2)
1-Methylthiourea	8.29(5)	Chloropentafluorobenzene	10.4(1)
1-Vinylthiourea	8.29(5)	Dichlorodifluoromethane	12.31(5)
1,1-Dimethylthiourea	8.34(5)	$CF_3CCl{=}CClCF_3$	10.36(1)
1,3-Dimethylthiourea	8.17(5)	Trichlorofluoromethane	11.77(2)
1,1,3-Trimethylthiourea	7.93(5)	CF_3CCl_3	11.78(3)
Tetramethylthiourea	7.95(5)	$CFCl_2CF_2Cl$	11.99(2)
CH_3COSH	10.00(2)	ClO_3F	13.6(2)
Cl_2	11.48(1)	1-Bromo-1-propene	9.30(5)
HCl	12.74	1-Bromopropane	10.18(1)
CCl_4	11.47(1)	2-Bromopropane	10.075(10)
Tetrachloroethylene	9.32(1)	1-Bromobutane	10.125(10)
PCl_3	9.91	2-Bromobutane	9.98(1)
Chloromethane	11.3	1-Bromo-2-methylpropane	10.09(2)
Chloroethane	10.97	2-Bromo-2-methylpropane	9.89(3)
Chloroethylene	9.996	1-Bromopentane	10.10(2)
1-Chloro-1-propyne	9.9(1)	Bromobenzene	8.98(2)
1-Chloropropane	10.82(3)	*o*-Bromotoluene	8.78(1)
2-Chloropropane	10.78(2)	*m*-Bromotoluene	8.81(2)
1-Chlorobutane	10.67(3)	*p*-Bromotoluene	8.67(2)
2-Chlorobutane	10.65(3)	Dibromomethane	10.49(2)
1-Chloro-2-methylpropane	10.66(3)	*cis*-1,2-Dibromoethylene	9.45
2-Chloro-2-methylpropane	10.61(3)	*trans*-1,2-Dibromoethylene	9.46
Chlorobenzene	9.07	1,1-Dibromoethane	10.19(3)
α-Chlorotoluene	9.19(5)	1,3-Dibromopropane	10.07(2)
o-Chlorotoluene	8.83(2)	Bromoform	10.51(2)
m-Chlorotoluene	8.83(2)	Tribromoethylene	9.27
p-Chlorotoluene	8.69(2)	Cyanogen bromide	11.95(8)
endo-5-Chloro-2-norbornene	9.10(15)	Bromotrifluoromethane	11.89
exo-5-Chloro-2-norbornene	9.15(15)	2-Bromopyridine	9.65(5)
Dichloromethane	11.35(2)	4-Bromopyridine	9.94(5)
cis-1,2-Dichloroethylene	9.65	Acetyl bromide	10.55(5)
trans-1,2-Dichloroethylene	9.64	Methyl bromoacetate	10.37(5)
1,2-Dichloroethane	11.12(5)	CF_2BrCH_2Br	10.83(1)
2,3-Dichloro-1-propene	9.82(3)	Bromochloromethane	10.77(1)
1,2-Dichloropropane	10.87(5)	1-Bromo-2-chloroethane	10.63(3)
1,3-Dichloropropane	10.85(5)	Bromodichloromethane	10.88(5)
o-Dichlorobenzene	9.06	Bromotrimethylsilane	10.24(2)
m-Dichlorobenzene	9.12(1)	I_2	9.28(2)

TABLE 6-66 Ionization potentials of molecular species (*continued*)

Species	Ionization potential, eV	Species	Ionization potential, eV
HI	10.39	$CF_3CF_2CF_2CH_2Cl$	11.84(2)
ICI	10.31(2)	Dichlorofluoromethane	12.39(20)
IBr	9.98(3)	Chlorotrimethylsilane	10.58(4)
Iodomethane	9.54	Trichloromethylsilane	11.36(3)
Iodoethane	9.33	Trichlorovinylsilane	10.79(2)
1-Iodopropane	9.26(1)	Trichloroethylsilane	10.74(4)
2-Iodopropane	9.17(2)	Trichloroisopropylsilane	10.28(10)
1-Iodobutane	9.21(1)	$C_2H_5V(CO)_4$	8.2(3)
2-Iodobutane	9.09(2)	$Cr(CO)_6$	8.03(3)
1-Iodo-2-methylpropane	9.18(2)	$C_2H_5Mn(CO)_3$	8.3(4)
2-Iodo-2-methylpropane	9.02(2)	$Fe(CO)_5$	7.95(3)
1-Iodopentane	9.19(1)	$Ni(CO)_4$	8.28(3)
Iodobenzene	8.73	$Mo(CO)_6$	8.12(3)
o-Iodotoluene	8.62(1)	$W(CO)_6$	8.18(3)
m-Iodotoluene	8.61(3)	As_4	9.07(7)
p-Iodotoluene	8.50(1)	Arsine	10.03
RuO_4	12.33(23)	$AsCl_3$	11.7(1)
2-Chloropyridine	9.91(5)	Trimethylarsine	8.3(1)
4-Chloropyridine	10.15(5)	Triphenylarsine	7.34(7)
Acetyl chloride	11.02(5)	Br_2	10.54(3)
1-Chloro-2-propanone	9.99	HBr	11.62(3)
2-Chlorophenol	9.28	BrCl	11.1(2)
4-Chlorophenol	9.07	Bromomethane	10.53
Benzoyl chloride	9.70(1)	Bromoethylene	9.80
4-Chlorobenzaldehyde	9.61(1)	Bromoethane	10.29
α-Chloroacetophenone	9.5	1-Bromo-1-propyene	10.1(1)
p-Chloroacetophenone	9.47(5)	OsO_4	12.97(12)
Methyl chloroacetate	10.53(5)	Dimethylmercury	9.0
4-Methoxybenzoyl chloride	8.87(5)	Diethylmercury	8.5(1)
4-Chlorobenzoyl chloride	9.58(3)	Diisopropylmercury	7.6(1)
cis-Chlorofluoroethylene	9.86	CH_3HgCl	11.5(2)
trans-Chlorofluoroethylene	9.87	Triphenylbismuth	7.3(1)
o-Chlorofluorobenzene	9.155(10)	Stibine	9.58
m-Chlorofluorobenzene	9.21(1)	Triphenylstibine	7.3(1)
p-Chlorofluorobenzene	9.43(2)	Tetramethylstannane	8.25(15)
Chlorodifluoromethane	12.45(5)	Tetramethylplumbane	8.0(4)
1-Chloro-1,1-difluoroethane	11.98(1)	Tetramethylgermane	9.2(2)

TABLE 6-67 Ionization potentials of radical species

1 eV = 23.061 kcal · mol^{-1}

Values in parentheses are uncertainties in the final figure(s).

Species	Ionization potential, eV	Species	Ionization potential, eV
BH	9.77(5)	*tert*-Pentyl	7.1(1)
BH_2	11.4(2)	Neopentyl	8.3(1)
BF	11.3	Benzyne	9.6
C_2	12.0(6)	Cyclohexyl	7.7
C_3	12.6	Benzyl	7.76(8)
CH	11.1(2)	Cycloheptatrienyl	6.24(1)
CH_2	10.396(3)	1-Methylnaphthyl	7.35
CH_3	9.83	2-Methylnaphthyl	7.56(5)
CD_3	9.832(2)	$(CH_3)_2CCN$	9.15(10)
C_2H_3	9.4	*m*-Nitrobenzyl	8.56(10)
C_2H_5	8.4	OH	13.17(10)
$HC{\equiv}CCH_2$	8.25	HO_2	11.53(2)
Allyl	8.15	CHO	9.8
Cyclopropyl	8.05	CH_3CO	10.3
C_3H_6	9.73	C_6H_5O	8.84
Propyl	8.1	CF_2	11.8
Isopropyl	7.5	NF_2	11.9
C_4H_2	10.2(1)	CH_2F	9.35
C_4H_4	9.87	CHF_2	9.45
Cyclobutyl	7.88(5)	HS	10.5(1)
$CH_3CH{=}CHCH_2$	7.71(5)	CH_3S	8.06(10)
$CH_2{=}C(CH_3)CH_2$	8.03(5)	C_6H_5S	8.63(10)
Butyl	8.64(5)	CCl_3	8.78(5)
sec-Butyl	7.93(5)	CH_2Cl	9.32
Isobutyl	8.35(5)	$CHCl_2$	9.30
tert-Butyl	7.42(7)	NH_2	11.3
Cyclopentyl	7.79(2)		

SECTION 7
PHYSICOCHEMICAL RELATIONSHIPS

LINEAR FREE ENERGY RELATIONSHIPS

Many equilibrium and rate processes can be systematized when the influence of each substituent on the reactivity of substrates is assigned a characteristic constant σ and the reaction parameter ρ is known or can be calculated. The Hammett equation

$$\log \frac{K}{K^\circ} = \sigma\rho$$

describes the behavior of many *meta*- and *para*-substituted aromatic species. In this equation K° is the acid dissociation constant of the reference in aqueous solution at 25°C and K is the corresponding constant for the substituted acid. Separate sigma values are defined by this reaction for *meta* and *para* substituents and provide a measure of the total electronic influence (polar, inductive, and resonance effects) in the absence of conjugation effects. Sigma constants are not valid for substituents *ortho* to the reaction center because of anomalous (mainly steric) effects. The inductive effect is transmitted about equally to the *meta* and *para* positions. Consequently, σ_m is an approximate measure of the size of the inductive effect of a given substituent and $\sigma_p - \sigma_m$ is an approximate measure of a substituent's resonance effect. Values of Hammett sigma constants are listed in Table 7-1.

Taft sigma values σ^* perform a similar function with respect to aliphatic and alicyclic systems. Values of σ^* are listed in Table 7-1.

The reaction parameter ρ depends upon the reaction series but not upon the substituents employed. Values of the reaction parameter for some aromatic and aliphatic systems are given in Tables 7-2 and 7-3.

Since substituent effects in aliphatic systems and in *meta* positions in aromatic systems are essentially inductive in character, σ^* and σ_m values are related by the expression $\sigma_m = 0.217\sigma^* - 0.106$. Substituent effects fall off with increasing distance from the reaction center; generally a factor of 0.36 corresponds to the interposition of a $-CH_2-$ group, which enables σ^* values to be estimated for $R-CH_2-$ groups not otherwise available.

Two modified sigma constants have been formulated for situations in which the substituent enters into resonance with the reaction center in an electron-demanding transition state (σ^+) or for an electron-rich transition state (σ^-). σ^- constants give better correlations in reactions involving phenols, anilines, and pyridines and in nucleophilic substitutions. Values of some modified sigma constants are given in Table 7-4.

TABLE 7-1 Hammett and Taft substituent constants

Substituent	Hammett constants		Taft constant
	σ_m	σ_p	σ^*
$-AsO_3H^-$	−0.09	−0.02	0.06
$-B(OH)_2$	0.01	0.45	
$-Br$	0.39	0.23	2.84
$-CH_2Br$			1.00
m-BrC_6H_4-		0.09	
p-BrC_6H_4-		0.08	
$-CH_3$	−0.07	−0.17	0.0
$-CH_2CH_3$	−0.07	−0.15	−0.10
$-CH_2CH_2CH_3$	−0.05	−0.15	−0.12
$-CH(CH_3)_2$	−0.07	−0.15	−0.19

TABLE 7-1 Hammett and Taft substituent constants (*continued*)

Substituent	Hammett constants σ_m	Hammett constants σ_p	Taft constant σ^*
$-CH_2CH_2CH_2CH_3$	−0.07	−0.16	−0.13
$-CH_2CH(CH_3)_2$	−0.07	−0.12	−0.13
$-CH(CH_3)CH_2CH_3$		−0.12	−0.19
$-C(CH_3)_3$	−0.10	−0.20	−0.30
$-CH_2CH_2CH_2CH_2CH_3$			−0.25
$-CH_2CH_2CH(CH_3)_2$			−0.17
$-CH_2C(CH_3)_3$		−0.23	−0.12
$-CH_2CH_2CH_2CH_2CH_2CH_2CH_3$			−0.37
Cyclopropyl—	−0.07	−0.21	
Cyclohexyl—			−0.15
—3,4-$(CH_2)_2$ (fused)		−0.26	
—3,4-$(CH_2)_3$— (fused ring)		−0.48	
—3,4-$(CH)_4$— (fused ring)	0.06	0.04	
$-CH{=}CH_2$	0.02		0.56
$-CH{=}C(CH_3)_2$			0.19
$-CH{=}CHCH_3$, *trans*			0.36
$-CH_2-CH{=}CH_2$			0.0
$-CH{=}CHC_6H_5$	0.14	−0.05	0.41
$-C{\equiv}CH$	0.21	0.23	2.18
$-C{\equiv}CC_6H_5$	0.14	0.16	1.35
$-CH_2-C{\equiv}CH$			0.81
$-C_6H_5$	0.06	−0.01	0.60
p-$CH_3C_6H_4$—		−0.5	
Naphthyl— (both 1- and 2-)			0.75
$-CH_2C_6H_5$		0.46	0.22
$-CH_2CH_2-C_6H_5$			−0.06
$-CH(CH_3)C_6H_5$			0.37
$-CH(C_6H_5)_2$			0.41
$-CH_2-C_{10}H_7$			0.44
2-Furoyl—			0.25
3-Indolyl—			−0.06
2-Thienyl—			1.31
2-Thienylemethylene—			0.31
$-CHO$	0.36	0.22	
$-COCH_3$	0.38	0.50	1.65
$-COCH_2CH_3$		0.48	
$-COCH(CH_3)_2$		0.47	
$-COC(CH_3)_3$		0.32	
$-COCF_3$	0.65		3.7
$-COC_6H_5$	0.34	0.46	2.2
$-CONH_2$	0.28	0.36	1.68
$-CONHC_6H_5$			1.56
$-CH_2COCH_3$			0.60
$-CH_2CONH_2$			0.31
$-CH_2CH_2CONH_2$			0.19

TABLE 7-1 Hammett and Taft substituent constants (*continued*)

Substituent	Hammett constants σ_m	Hammett constants σ_p	Taft constant σ^*
$-CH_2CH_2CH_2CONH_2$			0.12
$-CH_2CONHC_6H_5$			0.0
$-COO^-$	−0.1	0.0	−1.06
$-COOH$	0.36	0.43	2.08
$-CO-OCH_3$	0.32	0.39	2.00
$-CO-OCH_2CH_3$	0.37	0.45	2.12
$-CH_2CO-OCH_3$			1.06
$-CH_2CO-OCH_2CH_3$			0.82
$-CH_2COO$			−0.06
$-CH_2CH_2COOH$	−0.03	−0.07	
$-Cl$	0.37	0.23	2.96
$-CCl_3$	0.47		2.65
$-CHCl_2$			1.94
$-CH_2Cl$	0.12	0.18	1.05
$-CH_2CH_2Cl$			0.38
$-CH_2CCl_3$			0.75
$-CH_2CH_2CCl_3$			0.25
$-CH{=}CCl_2$			1.00
$-CH_2CH{=}CCl_2$			0.19
$p\text{-}ClC_6H_4-$		0.08	
$-F$	0.34	0.06	3.21
$-CF_3$	0.43	0.54	2.61
$-CHF_2$			2.05
$-CH_2F$			1.10
$-CH_2CF_3$			0.90
$-CH_2CF_2CF_2CF_3$			0.87
$-C_6F_5$	−0.12	−0.03	
$-Ge(CH_3)_3$		0.0	
$-Ge(CH_2CH_3)_3$		0.0	
$-H$	0.00	0.00	0.49
$-I$	0.35	0.28	2.46
$-CH_2I$			0.85
$-IO_2$	0.70	0.76	
$-N_2^+$	1.76	1.91	
$-N_3$ (azide)	0.33	0.08	2.62
$-NH_2$	−0.16	−0.66	0.62
$-NH_3^+$	1.13	1.70	3.76
$-CH_2-NH_2$			0.50
$-CH_2-NH_3^+$			2.24
$-NH-CH_3$	−0.30	−0.84	
$-NH-C_2H_5$	−0.24	−0.61	
$-NH-C_4H_9$	−0.34	−0.51	
$-NH(CH_3)_2^+$			4.36
$-NH_2-CH_3^+$	0.96		3.74
$-NH_2-C_2H_5^+$	0.96		3.74

TABLE 7-1 Hammett and Taft substituent constants (*continued*)

Substituent	Hammett constants		Taft constant
	σ_m	σ_p	σ^*
$-N(CH_3)_3^+$	0.88	0.82	4.55
$-N(CH_3)_2$	−0.2	−0.83	0.32
$-CH_2-N(CH_3)_3^+$			1.90
$-N(CF_3)_2$	0.45	0.53	
$p-H_2N-C_6H_6-$		−0.30	
$-NH-CO-CH_3$	0.21	0.00	1.40
$-NH-CO-C_2H_5$			1.56
$-NH-CO-C_6H_5$	0.22	0.08	1.68
$-NH-CHO$	0.25		1.62
$-NH-CO-NH_2$	0.18		1.31
$-NH-OH$	−0.04	−0.34	
$-NH-CO-OC_2H_5$	0.33		1.99
$-CH_2-NH-CO-CH_3$			0.43
$-NH-SO_2-C_6H_5$			1.99
$-NH-NH_2$	−0.02	−0.55	
$-CN$	0.56	0.66	3.30
$-CH_2-CN$	0.17	0.01	1.30
$-NO$		0.12	
$-NO_2$	0.71	0.78	4.0
$-CH_2-NO_2$			1.40
$-CH_2-CH_2-NO_2$			0.50
$-CH{=}CHNO_2$	0.33	0.26	
m-$O_2N-C_6H_4$		0.18	
p-$O_2N-C_6H_4$		0.24	
$(NO_2)_3C_6H_2-$ (picryl)	0.43	0.41	
$-N(CO-CH_3)(CO-C_6H_5)$			1.37
$-N(CO-CH_3)$(naphthyl)			1.65
$-O^-$	−0.71	−0.52	
$-OH$	0.12	−0.37	1.34
$-O-CH_3$	0.12	−0.27	1.81
$-O-C_2H_5$	0.10	−0.24	1.68
$-O-C_3H_7$	0.00	−0.25	1.68
$-O-CH(CH_3)_2$	0.05	−0.45	1.62
$-O-C_4H_9$	−0.05	−0.32	1.68
$-O-$cyclopentyl			1.62
$-O-$cyclohexyl	0.29		1.81
$-O-CH_2-$cyclohexyl	0.18		1.31
$-O-C_6H_5$	0.25	−0.32	2.43
$-O-CH_2-C_6H_5$		−0.42	
$-OCF_3$	0.40	0.35	
3,4-$O-CH_2-O-$		−0.27	
3,4-$O-(CH_2-)_2O-$		−0.12	
$-O-CO-CH_3$	0.39	0.31	
$-ONO_2$			3.86
$-O-N{=}C(CH_3)_2$			1.81

TABLE 7-1 Hammett and Taft substituent constants (*continued*)

Substituent	Hammett constants σ_m	Hammett constants σ_p	Taft constant σ^*
$-ONH_3^+$			2.92
$-CH_2-O^-$			0.27
$-CH_2-OH$	0.08	0.08	0.31
$-CH_2-O-CH_3$			0.52
$-CH(OH)-CH_3$			0.12
$-CH(OH)-C_6H_5$			0.50
p-$HO-C_6H_4-$		−0.24	
p-$CH_3O-C_6H_4-$		−0.10	
$-CH_2-CH(OH)-CH_3$			−0.06
$-CH_2-C(OH)(CH_3)_2$			−0.25
$-P(CH_3)_2$	0.1	0.05	
$-P(CH_3)_3^+$	0.8	0.9	
$-P(CF_3)_2$	0.6	0.7	
$-PO_3H^-$	0.2	0.26	
$-PO(OC_2H_5)_2$	0.55	0.60	
$-SH$	0.25	0.15	1.68
$-SCH_3$	0.15	0.00	1.56
$-S(CH_3)_2^+$	1.0	0.9	
$-SCH_2CH_3$	0.23	0.03	1.56
$-SCH_2CH_2CH_3$			1.49
$-SCH_2CH_2CH_2CH_3$			1.44
$-S-$cyclohexyl			1.93
$-SC_6H_5$	0.30		1.87
$-SC(C_6H_5)_3$			0.69
$-SCH_2C_6H_5$			1.56
$-SCH_2CH_2C_6H_5$			1.44
$-CH_2SH$	0.03		0.62
$-CH_2SCH_2C_6H_5$			0.37
$-SCF_3$	0.40	0.50	
$-SCN$	0.63	0.52	3.43
$-S-CO-CH_3$	0.39	0.44	
$-S-CONH_2$	0.34		2.07
$-SO-CH_3$	0.52	0.49	
$-SO-C_6H_5$			3.24
$-CH_2-SO-CH_3$			1.33
$-SO_2-CH_3$	0.60	0.68	3.68
$-SO_2-CH_2CH_3$			3.74
$-SO_2-CH_2CH_2CH_3$			3.68
$-SO_2-C_6H_5$	0.67		3.55
$-SO_2-CF_3$	0.79	0.93	
$-SO_2-NH_2$	0.46	0.57	
$-CH_2-SO_2-CH_3$			1.38
$-SO_3^-$	0.05	0.09	0.81
$-SO_3H$		0.50	
$-SeCH_3$	0.1	0.0	

TABLE 7-1 Hammett and Taft substituent constants (*continued*)

Substituent	Hammett constants		Taft constant
	σ_m	σ_p	σ^*
—Se—cyclohexyl			2.37
—SeCN	0.67	0.66	3.61
$—Si(CH_3)_3$	−0.04	−0.07	−0.81
$—Si(CH_2CH_3)_3$		0.0	
$—Si(CH_3)_2C_6H_5$			−0.87
$—Si(CH_3)_2—O—Si(CH_3)_3$			−0.81
$—CH_2Si(CH_3)_3$	−0.16	−0.22	−0.25
$—CH_2CH_2Si(CH_3)_3$			−0.25
$—Sn(CH_3)_3$		0.0	
$—Sn(CH_2CH_3)_3$		0.0	

TABLE 7-2 pK_a° and rho values for Hammett equation

Acid	pK_a°	ρ
Arenearsonic acids		
pK_1	3.54	1.05
pK_2	8.49	0.87
Areneboronic acids (in aqueous 25% ethanol)	9.70	2.15
Arenephosphonic acids		
pK_1	1.84	0.76
pK_2	6.97	0.95
α-Arylaldoximes	10.70	0.86
Benzeneseleninic acids	4.78	1.03
Benzenesulfonamides (20°C)	10.00	1.06
Benzenesulfonanilides (20°C)		
$X—C_6H_4—SO_2—NH—C_6H_5$	8.31	1.16
$C_6H_5—SO_2—NH—C_6H_4—X$	8.31	1.74
Benzoic acids	4.21	1.00
Cinnamic acids	4.45	0.47
Phenols	9.92	2.23
Phenylacetic acids	4.30	0.49
Phenylpropiolic acids (in aqueous 35% dioxane)	3.24	0.81
Phenylpropionic acids	4.45	0.21
Phenyltrifluoromethylcarbinols	11.90	1.01
Pyridine-1-oxides	0.94	2.09
2-Pyridones	11.65	4.28
4-Pyridones	11.12	4.28
Pyrroles	17.00	4.28
5-Substituted pyrrole-2-carboxylic acids	2.82	1.40
Thiobenzoic acids	2.61	1.0
Thiophenols	6.50	2.2
Trifluoroacetophenone hydrates	10.00	1.11
5-Substituted topolones	6.42	3.10

TABLE 7-2 pK_a° and rho values for Hammett equation (*continued*)

Acid	pK_a°	ρ
Protonated cations of		
Acetophenones	−6.0	2.6
Anilines	4.60	2.90
C-Aryl-*N*-dibutylamidines (in aqueous 50% ethanol)	11.14	1.41
N,N-Dimethylanilines	5.07	3.46
Isoquinolines	5.32	5.90
1-Naphthylamines	3.85	2.81
2-Naphthylamines	4.29	2.81
Pyridines	5.18	5.90
Quinolines	4.88	5.90

TABLE 7-3 pK_a° and rho values for Taft equation

Acid	pK_a°	ρ
RCOOH	4.66	1.62
RCH_2COOH	4.76	0.67
RC≡C—COOH	2.39	1.89
$H_2C{=}C(R){-}COOH$	4.39	0.64
$(CH_3)_2C{=}C(R){-}COOH$	4.65	0.47
cis-$C_6H_5{-}CH{=}C(R){-}COOH$	3.77	0.63
trans-$C_6H_5{-}CH{=}C(R){-}COOH$	4.61	0.47
$R{-}CO{-}CH_2{-}COOH$	4.12	0.43
HON=C(R)—COOH	4.84	0.34
RCH_2OH	15.9	1.42
$RCH(OH)_2$	14.4	1.42
$R_1CO{-}NHR_2$	22.0	3.1*
$CH_3CO{-}C(R){=}C(OH)CH_3$	9.25	1.78
$CH_3CO{-}CH(R){-}CO{-}OC_2H_5$	12.59	3.44
R—CO—NHOH	9.48	0.98
$R_1R_2C{=}NOH$ (R_1, R_2 not acyl groups)	12.35	1.18
$(R)(CH_3CO)C{=}NOH$	9.00	0.94
$RC(NO_2)_2H$	5.24	3.60
RSH	10.22	3.50
RCH_2SH	10.54	1.47
R—CO—SH	3.52	1.62
Protonated cations of		
RNH_2	10.15	3.14
R_1R_2NH	10.59	3.23
$R_1R_2R_3N$	9.61	3.30
R_1R_2PH	3.59	2.61
$R_1R_2R_3P$	7.85	2.67

* σ^* for R_1CO and R_2.

TABLE 7-4 Special Hammett sigma constants

Substituent	σ_m^+	σ_p^+	σ_p^-
$-CH_3$	−0.07	−0.31	−0.17
$-C(CH_3)_3$	−0.06	−0.26	
$-C_6H_5$	0.11	−0.18	
$-CF_3$	0.52	0.61	0.74
$-F$	0.35	−0.07	0.02
$-Cl$	0.40	0.11	0.23
$-Br$	0.41	0.15	0.26
$-I$	0.36	0.14	
$-CN$	0.56	0.66	0.88
$-CHO$			1.13
$-CONH_2$			0.63
$-COCH_3$			0.85
$-COOH$	0.32	0.42	0.73
$-CO-OCH_3$	0.37	0.49	0.66
$-CO-OCH_2CH_3$	0.37	0.48	0.68
$-N_2^+$			3.2
$-NH_2$	0.16	−1.3	−0.66
$-N(CH_3)_2$		−1.7	
$-N(CH_3)_3^+$	0.36	0.41	
$-NH-CO-CH_3$		−0.60	
$-NO_2$	0.67	0.79	1.25
$-OH$		−0.92	
$-O^-$			−0.81
$-OCH_3$	0.05	−0.78	−0.27
$-SF_5$			0.70
$-SCF_3$			0.57
$-SO_2CH_3$			1.05
$-SO_2CF_3$			1.36

SECTION 8

ELECTROLYTES, ELECTROMOTIVE FORCE, AND CHEMICAL EQUILIBRIUM

EQUILIBRIUM CONSTANTS

TABLE 8-1 pK_a values of organic materials in water at 25°C

Ionic strength μ is zero unless otherwise indicated. Protonated cations are designated by (+1), (+2), etc., after the pK_a value; neutral species by (0), if not obvious; and negatively charged acids by (−1), (−2), etc.

Substance	pK_1	pK_2	pK_3	pK_4
Abietic acid	7.62			
Acetamide	−0.37(+1)			
Acetamidine	1.60(+1)			
N-(2-Acetamido)-2-aminoethanesulfonic acid (20°C)	6.88			
2-Acetamidobenzoic acid	3.63			
3-Acetamidobenzoic acid	4.07			
4-Acetamidobenzoic acid	4.28			
2-(Acetamido)butanoic acid	3.716			
N-(2-Acetamido)iminodiacetic acid (20°C)	6.62			
3-Acetamidopyridine	4.37(+1)			
Acetanilide	0.4(+1)	13.39(0)$^{40°C}$		
Acetic acid	4.756			
Acetic acid-*d* (in D_2O)	5.32			
Acetoacetic acid (18°C)	3.58			
Acetohydrazine	3.24(+1)			
Acetone oxime	12.2			
2-Acetoxybenzoic acid (acetylsalicyclic acid)	3.48			
3-Acetoxybenzoic acid	4.00			
4-Acetoxybenzoic acid	4.38			
Acetylacetic acid (18°C)	3.58			
N-Acetyl-α-alanine	3.715			
N-Acetyl-β-alanine	4.455			
2-Acetylaminobutanoic acid	3.72			
3-Acetylaminopropionic acid	4.445			
2-Acetylbenzoic acid	4.13			
3-Acetylbenzoic acid	3.83			
4-Acetylbenzoic acid	3.70			
2-Acetylcyclohexanone	14.1			
N-Acetylcysteine (30°C)	9.52			
Acetylenedicarboxylic acid	1.75	4.40		
N-Acetylglycine	3.670			
N-Acetylguanidine	8.23(+1)			
N-α-Acetyl-L-histidine	7.08			
Acetylhydroxamic acid (20°C)	9.40			
N-Acetyl-2-mercaptoethylamine	9.92(SH)			
4-Acetyl-β-mercaptoisoleucine (30°C)	10.30			
2-Acetyl-1-naphthol (30°C)	13.40			
N-Acetylpenicillamine (30°C)	9.90			

Table 8-1 pK_a values of organic materials in water at 25°C (*continued*)

Substance	pK_1	pK_2	pK_3	pK_4
2-Acetylphenol	9.19			
4-Acetylphenol	8.05			
2-Acetylpyridine	2.643(+1)			
3-Acetypyridine	3.256(+1)			
4-Acetylpyridine	3.505(+1)			
Aconitine	8.11(+1)			
Acridine	5.60(+1)			
Acrylic acid	4.26			
Adenine	4.17(+1)	9.75(0)		
Adeninedeoxyriboside-5′-phosphoric acid	——	4.4	6.4	
Adenine-*N*-oxide	2.69(+1)	8.49(0)		
Adenosine	3.5(+1)	12.34(0)		
Adenosine-5′-diphosphoric acid	——	4.2(−1)	7.20(−2)	
Adenosine-2′-phosphoric acid	3.81(+1)	6.17(0)		
Adenosine-3′-phosphoric acid	3.65(0)	5.88(−1)		
Adenosine-5′-phosphoric acid	3.74(0)	6.05(−1)	13.06(−2)	
Adenosine-5′-triphosphoric acid	——	4.00(−1)	6.48(−2)	
Adipamic acid (adipic acid monoamide)	4.629			
Adipic acid	4.418	5.412		
α-Alanine	2.34(+1)	9.87(0)		
β-Alanine	3.55(+1)	10.238(0)		
α-Alanine, methyl ester (μ=0.10)	7.743(+1)			
β-Alanine, methyl ester (μ=0.10)	9.170(+1)			
N-D-Alanyl-α-D-alanine (μ=0.1)	3.32(+1)	8.13(0)		
N-L-Alanyl-α-L-alanine (μ=0.1)	3.32(+1)	8.13(0)		
N-L-Alanyl-α-D-alanine	3.12(+1)	8.30(0)		
N-α-Alanylglycine	3.11(+1)	8.11(0)		
Alanylglycylglycine	3.190(+1)	8.15(0)		
β-Alanylhistidine	2.64	6.86	9.40	
Albumin (bovine serum (μ=0.15)	10–10.3			
2-Aldoxime pyridine	3.42(+1)	10.22(0)		
Alizarin Black SN	5.79	12.8		
Alizarin-3-sulfonic acid	5.54	11.01		
Allantoin	8.96			
Allothreonine	2.108(+1)	9.096(0)		
Alloxanic acid	6.64			
Allylacetic acid	4.68			
Allylamine	9.69(+1)			
5-Allylbarbituric acid	4.78(+1)			
5-Allyl-5-(-methylbutyl)barbituric acid	8.08			
2-Allylphenol	10.28			
1-Allylpiperidine	9.65(+1)			
2-Allylpropionic acid	4.72			
3-Amidotetrazoline	3.95(+1)			

Table 8-1 pK_a values of organic materials in water at 25°C (*continued*)

Substance	pK_1	pK_2	pK_3	pK_4
2-Aminoacetamide	7.95(+1)			
Aminoacetonitrile	5.34(+1)			
9-Aminoacridine (20°C)	9.95(+1)			
4-Aminoantipyrine	4.94(+1)			
2-Aminobenzenesulfonic acid	2.459(0)			
3-Aminobenzenesulfonic acid	3.738(0)			
4-Aminobenzenesulfonic acid	3.227(0)			
2-Aminobenzoic acid	2.09(+1)	4.79(0)		
3-Aminobenzoic acid	3.07(+1)	4.79(0)		
4-Aminobenzoic acid	2.41(+1)	4.85(0)		
2-Aminobenzoic acid, methyl ester	2.36(+1)			
3-Aminobenzoic acid, methyl ester	3.58(+1)			
4-Aminobenzoic acid, methyl ester	2.45(+1)			
3-Aminobenzonitrile	2.75(+1)			
4-Aminobenzonitrile	1.74(+1)			
4-Aminobenzophenone	2.15(+1)			
2-Aminobenzothiazole (20°C)	4.48(+1)			
2-Aminobenzoylhydrazide	1.85	3.47	12.80	
2-Aminobiphenyl	3.78(+1)			
3-Aminobiphenyl	4.18(+1)			
4-Aminobiphenyl	4.27(+1)			
4-Amino-3-bromomethylpyridine	7.47(+1)			
4-Amino-3-bromopyridine (20°C)	7.04(+1)			
2-Aminobutanoic acid	2.286(+1)	9.830(0)		
3-Aminobutanoic acid	——	10.14(0)		
4-Aminobutanoic acid	4.031(+1)	10.556(0)		
2-Aminobutanoic acid, methyl ester (μ=0.1)	7.640(+1)			
4-Aminobutanoic acid, methyl ester (μ=0.1)	9.838(+1)			
D-(+)-2-Amino-1-butanol	9.52(+1)			
3-Amino-*N*-butyl-3-methyl-2-butanone oxime	9.09(+1)			
4-Aminobutylphosphonic acid	2.55	7.55	10.9	
2-Amino-*N*-carbamoylbutanoic acid	3.886(+1)			
4-Amino-*N*-carbamoylbutanoic acid	4.683(+1)			
2-Amino-*N*-carbamoyl-2-methylpropanoic acid	4.463			
1-Amino-1-cycloheptanecarboxylic acid	2.59(+1)	10.46(0)		
1-Amino-1-cyclohexanecarboxylic acid	2.65(+1)	10.03(0)		
2-Amino-1-cyclohexanecarboxylic acid	3.56(+1)	10.21(0)		
1-Aminocyclopentane	10.65(+1)			
1-Aminocyclopropane	9.10(+1)			
10-Aminodecylphosphonic acid	——	8.0	11.25	

Table 8-1 pK_a values of organic materials in water at 25°C (*continued*)

Substance	pK_1	pK_2	pK_3	pK_4
10-Aminodecylsulfonic acid	2.65(+1)			
1-Amino-2-di(aminomethyl)butane	3.58(+3)	8.59(+2)	9.66(+1)	
2-Amino-*N*,*N*-dihydroxyethyl-2-hydroxyl-1,3-propanediol	6.484(+1)			
2-Amino-*N*,*N*-dimethylbenzoic acid	1.63(+1)	8.42(0)		
4-Amino-2,5-dimethylphenol	5.28(+1)	10.40(0)		
4-Amino-3,5-dimethylpyridine (20°C)	9.54(+1)			
12-Aminododecanoic acid	4.648(+1)			
2-Aminoethane-1-phosphoric acid	5.838	10.64		
1-Aminoethanesulfonic acid	−0.33	9.06		
2-Aminoethanesulfonic acid	1.5	9.061		
2-Aminoethanethiol (cysteamine) (μ=0.01)	8.23(+1)			
2-Aminoethanol (ethanolamine)	9.50(+1)			
2-[2-(2-Aminoethyl)aminoethyl]pyridine	3.50	6.59	9.51	
2-Amino-2-ethyl-1-butanol	9.82(+1)			
3-(2-Aminoethyl)indole	——	10.2		
3-Amino-*N*-ethyl-3-methyl-2-butanone oxime	9.23(+1)			
N-(2-Aminoethyl)morpholine	4.06(+2)	9.15(+1)		
p-(2-Aminoethyl)phenol	9.3	10.9		
2-Aminoethylphosphonic acid	2.45(+1)	7.0(0)	10.8(−1)	
N-(2-Aminoethyl)piperidine (30°C)	6.38	9.89		
2-(2-Aminoethyl)pyridine (μ=0.5)	4.24(+2)	9.78(+1)		
4-Amino-3-ethylpyridine (20°C)	9.51(+1)			
N-(2-Aminoethyl)pyrrolidine (30°C)	6.56(+2)	9.74(+1)		
2-Aminofluorine	10.34(+1)			
2-Amino-D-β-glucose (μ=0.05)	2.20(+1)	9.08(0)		
2-Amino-*N*-glycylbutanoic acid	3.155(+1)	8.331(0)		
7-Aminoheptanoic acid	4.502			
2-Aminohexanoic acid	2.335(+1)	9.834(0)		
6-Aminohexanoic acid	4.373(+1)	10.804(0)		
C-Amino-*C*-hydrazinocarbonylmethane	2.38(+2)	7.69(+1)		
2-Amino-3-hydroxybenzoic acid	2.5(+1)	5.192(0)	10.118(OH)	
L-2-Amino-3-hydroxybutanoic acid (threonine)	2.088(+1)	9.100(0)		
DL-2-Amino-4-hydroxybutanoic acid (μ=0.1)	2.265(+1)	9.257(0)		
DL-4-Amino-3-hydroxybutanoic acid (μ=0.1)	3.834(+1)	9.487(0)		
2-Amino-2′-hydroxydiethyl sulfide	9.27(+1)			
4-Amino-2-hydroxypyrimidine (cytosine)	4.58(+1)	12.15(0)		
3-Amino-*N*-isopropyl-3-methyl-2-butanone oxime	9.09(+1)			

Table 8-1 pK_a values of organic materials in water at 25°C (*continued*)

Substance	pK_1	pK_2	pK_3	pK_4
4-Amino-3-isopropylpyridine (20°C)	9.54(+1)			
1-Aminoisoquinoline (20°C, μ=0.01)	7.62(+1)			
3-Aminoisoquinoline (20°C, μ=0.005)	5.05(+1)			
4-Aminoisoxazolidine-3-one	7.4(+1)			
Aminomalonic acid	3.32(+1)	9.83(0)		
DL-2-Amino-4-mercaptobutanoic acid	2.22(+1)	8.87(0)	10.86(SH)	
2-Amino-3-mercapto-3-Methylbutanoic acid	1.8(+1)	7.9(0)	10.5(SH)	
2-Amino-6-methoxybenzothiazole	4.50(+1)			
3-Amino-4-methylbenzenesulfonic acid	3.633			
4-Amino-3-methylbenzenesulfonic acid	3.125			
2-Amino-4-methylbenzothiazole	4.7(+1)			
1-Amino-3-methylbutane	10.64(+1)			
3-Amino-3-methyl-2-butanone oxime	9.09(+1)			
3-Amino-*N*-methyl-3-methyl-2-butanone oxime	9.23(+1)			
2-Amino-3-methylpentanoic acid	2.320(+1)	9.758(0)		
3-Aminomethyl-6-methylpyridine (30°C)	8.70(+1)			
Aminomethylphosphonic acid	2.35	5.9	10.8	
2-Amino-2-methyl-1,3-propanediol	8.801			
2-Amino-2-methyl-1-propanol	9.694(+1)			
2-Amino-2-methylpropanoic acid	2.357(+1)	10.205(0)		
(2-Aminomethyl)pyridine (μ=0.5)	2.31(+2)	8.79(+1)		
2-Amino-3-methylpyridine	7.24(+1)			
4-Amino-3-methylpyridine	9.43(+1)			
2-Amino-4-methylpyridine	7.48(+1)			
2-Amino-5-methylpyridine	7.22(+1)			
2-Amino-6-methylpyridine	7.41(+1)			
2-Amino-4-methylpyrimidine (20°C)	4.11(+1)			
Aminomethylsulfonic acid	5.75(+1)			
N-Aminomorpholine	4.19(+1)			
4-Amino-1-naphthalenesulfonic acid	2.81			
1-Amino-2-naphthalenesulfonic acid	1.71			
1-Amino-3-naphthalenesulfonic acid	3.20			
1-Amino-5-naphthalenesulfonic acid	3.69			
1-Amino-6-naphthalenesulfonic acid	3.80			
1-Amino-7-naphthalenesulfonic acid	3.66			
1-Amino-8-naphthalenesulfonic acid	5.03			
2-Amino-1-naphthalenesulfonic acid	2.35			
2-Amino-4-naphthalenesulfonic acid	3.79			
2-Amino-6-naphthalenesulfonic acid	3.79	8.94		
2-Amino-8-naphthalenesulfonic acid	3.89			

Table 8-1 pK_a values of organic materials in water at 25°C (*continued*)

Substance	pK_1	pK_2	pK_3	pK_4
3-Amino-1-naphthoic acid	2.61	4.39		
4-Amino-2-naphthoic acid	2.89	4.46		
8-Amino-2-naphthol	4.20(+1)			
DL-2-Aminopentanoic acid (DL-norvaline)	2.318(+1)	9.808		
3-Aminopentanoic acid	4.02(+1)	10.399(0)		
4-Aminopentanoic acid	3.97(+1)	10.46(0)		
5-Aminopentanoic acid	4.20(+1)	9.758(0)		
5-Aminopentanoic acid, ethyl ester	10.151			
2-Aminophenol	9.28	9.72		
3-Aminophenol	9.83	9.87		
4-Aminophenol	8.50	10.30		
4-Aminophenylacetic acid (20°C)	3.60	5.26		
2-Aminophenylarsonic acid	ca 2	3.77	8.66	
3-Aminophenylarsonic acid	ca 2	4.02	8.92	
4-Aminophenylarsonic acid	ca 2	4.02	8.62	
3-Aminophenylboric acid	4.46	8.81		
4-Aminophenylboric acid	3.71	9.17		
4-Aminophenyl (4-chlorophenyl) sulfone	1.38			
2-Aminophenylphosphonic acid	—	4.10	7.29	
3-Aminophenylphosphonic acid	—	—	7.16	
4-Aminophenylphosphonic acid	—	—	7.53	
1-Amino-1,2,3-propanetricarboxylic acid (μ=2.2)	2.10(+1)	3.60(0)	4.60(−1)	9.82(−2)
3-Aminopropanoic acid	3.551(+1)	10.235(0)		
1-Amino-1-propanol	9.96(+1)			
DL-2-Amino-1-propanol	9.469(+1)			
3-Amino-1-propanol	9.96(+1)			
3-Aminopropene	9.691(+1)			
3-Amino-*N*-propyl-3-methyl-2-butanone oxime	9.09(+1)			
2-Aminopropylsulfonic acid	—	9.15		
2-Aminopyridine	6.71(+1)			
3-Aminopyridine	6.03(+1)			
4-Aminopyridine	9.114(+1)			
2-Aminopyridine-1-oxide	2.58(+1)			
3-Aminopyridine-1-oxide	1.47(+1)			
4-Aminopyridine-1-oxide	3.54(+1)			
8-Aminoquinaldine	4.86(+1)			
2-Aminoquinoline (20°C, μ=0.01)	7.34(+1)			
3-Aminoquinoline (20°C, μ=0.01)	4.95(+1)			
4-Aminoquinoline (20°C, μ=0.01)	9.17(+1)			
5-Aminoquinoline (20°C, μ=0.01)	5.46(+1)			
6-Aminoquinoline (20°C, μ=0.01)	5.63(+1)			
8-Aminoquinoline (20°C, μ=0.01)	3.99(+1)			
4-Aminosalicyclic acid	1.991(+1)	3.917(0)	13.74	

Table 8-1 pK_a values of organic materials in water at 25°C (*continued*)

Substance	pK_1	pK_2	pK_3	pK_4
5-Aminosalicyclic acid	2.74(+1)	5.84(0)		
2-Amino-3-sulfopropanoic acid	1.89(+1)	8.70(0)		
4-Amino-2,3,5,6-tetramethylpyridine (20°C)	10.58(+1)			
5-Amino-1,2,3,4-tetrazole (20°C)	1.76	6.07		
2-Aminothiazole (20°C)	5.36(+1)			
1-Amino-3-thiobutane (30°C)	9.18(+1)			
5-Amino-3-thio-1-pentanol (30°C)	9.12(+1)			
2-Aminothiophenol	<2(+1)	7.90(0)		
2-Amino-4,4,4-trifluorobutanoic acid		8.171(0)		
3-Amino-4,4,4-trifluorobutanoic acid		5.831(0)		
3-Amino-2,4,6-trinitroluene		9.5(+1)		
Angiotensin II	10.37			
Anhydroplatynecine	9.40			
Aniline	4.60(+1)			
2-Anilinoethylsulfonic acid	3.80(+1)			
3-Anilinoethylsulfonic acid	4.85(+1)			
Anthracene-1-carboxylic acid	3.68			
Anthracene-2-carboxylic acid	4.18			
Anthracene-9-carboxylic acid	3.65			
Anthraquinone-1-carboxylic acid (20°C)	3.37			
Anthraquinone-2-carboxylic acid (20°C)	3.42			
9,10-Anthraquinone monoxime	9.78			
9,10-Anthraquinone-1-sulfonic acid	0.27			
9,10-Anthraquinone-2-sulfonic acid	0.38			
Antipyrine	1.45(+1)			
Apomorphine (15°C)		8.92		
D-(−)-Arabinose	12.34			
L-(+)-Arginine		8.994(+1)	12.47(−1)	
Arsenazo III [pK_5 10.5(−4); pK_6 12.0(−5)]		1.2	2.7	7.9(−3)
Arsenoacetic acid		4.67	7.68	
Arsenoacrylic acid		4.23	8.60	
Arsenobutanoic acid		4.92	7.64	
2-Arsenocrotonic acid		4.61	8.75	
3-Arsenocrotonic acid		4.03	8.81	
Arsenopentanoic acid		4.89	7.75	
L-(+)-Ascorbic acid (vitamin C)	4.17	11.57		
L-(+)-Asparagine		8.80(0)		
L-Asparaginylglycine		4.53	9.07	
D-Aspartic acid		3.87(0)	10.00(−)	
Aspartic diamide (μ=0.2)	7.00			
Aspartylaspartic acid		3.40	4.70	8.26
α-Aspartylhistidine (38°C, μ=0.1)		3.02	6.82	7.98
β-Aspartylhistidine (38°C, μ=0.1)		2.95	6.93	8.72

Table 8-1 pK_a values of organic materials in water at 25°C (*continued*)

Substance	pK_1	pK_2	pK_3	pK_4
N-Aspartyl-*p*-tyrosine (μ=0.01)		3.57	8.92	10.23(OH)
Aspidospermine	7.65			
Atropine (17°C)	4.35(+1)			
1-Azacycloheptane	11.11(+1)			
1-Azacyclooctane	11.1(+1)			
Azetidine	11.29(+1)			
Aziridine	8.04(+1)			
Barbituric acid		8.372(0)		
m-Benzbetaine	3.217(+1)			
p-Benzbetaine	3.245(+1)			
Benzenearsonic acid (22°C)		8.48(−1)		
Benzene-1-arsonic acid-4-carboxylic acid		4.22 (COOH)	5.59	
Benzeneboronic acid	13.7			
Benzene-1-carboxylic acid-2-phosphoric acid		3.78	9.17	
Benzene-1-carboxylic acid-3-phosphoric acid		4.03	7.03	
Benzene-1-carboxylic acid-4-phosphoric aic	1.50	3.95	6.89	
Benzenediazine	11.08(+1)			
1,3-Benzenedicarboxylic acid (isophthalic acid)	3.62(0)	4.60(−1)		
1,4-Benzenedicarboxylic acid (terephthalic acid)	3.54(0)	4.46(−1)		
1,3-Benzenedicarboxylic acid mononitrile	3.60(0)			
1,4-Benzenedicarboxylic acid mononitrile	3.55(0)			
Benzenehexarboxylic acid (pK_5 6.32; pK_6 7.49)	0.68	2.21	3.52	5.09
Benzenepentacarboxylic acid (pK_5 6.46)	1.80	2.73	3.96	5.25
Benzenesulfinic acid	1.50			
Benzenesulfonic acid	2.554			
1,2,3,4-Benzenetetracarboxylic acid	2.05	3.25	4.73	6.21
1,2,3,5-Benzenetetracarboxylic acid	2.38	3.51	4.44	5.81
1,2,4,5-Benzenetetracarboxylic acid	1.92	2.87	4.49	5.63
1,2,3-Benzenetricarboxylic acid	2.88	4.75	7.13	
1,2,4-Benzenetricarboxylic acid	2.52	3.84	5.20	
1,3,5-Benzenetricarboxylic acid	2.12	4.10	5.18	
Benzil-α-dioxime	12.0			
Benzilic acid	3.09			
Benzimidazole	5.53(+1)	12.3(0)		
Benzohydroxamic acid (20°C)	8.89(0)			

Table 8-1 pK_a values of organic materials in water at 25°C (*continued*)

Substance	pK_1	pK_2	pK_3	pK_4
Benzoic acid	4.204			
5,6-Benzoquinoline (20°C)	5.00(+1)			
7,8-Benzoquinoline (20°C)	4.15(+1)			
1,4-Benzoquinone monoxime	6.20			
Benzosulfonic acid	0.70			
1,2,3-Benzotriazole	8.38(+1)			
1-Benzoylacetone	8.23			
Benzoylamine	9.34(+1)			
2-Benzoylbenzoic acid	3.54			
Benzoylglutamic acid	3.49	4.99		
N-Benzoyglycine (hippuric acid	3.65			
Benzoylhydrazine	3.03(+2)	12.45(+1)		
Benzoylpyruvic acid	6.40	12.10		
3-Benzoyl-1,1,1-trifluoroacetone	6.35			
Benzylamine	9.35(+1)			
Benzylamine-4-carboxylic acid	3.59	9.64		
2-Benzyl-2-phenylsuccinic acid (20°C)	3.69	6.47		
2-Benzylpyridine	5.13(+1)			
4-Benzylpyridine-1-oxide	−1.018(+)			
1-Benzylpyrrolidine	9.51(+1)			
2-Benzylpyrrolidine	10.31(+1)			
Benzylsuccinic acid (20°C)	4.11	5.65		
3-(Benzylthio)propanoic acid	4.463			
Berberine (18°C)	11.73(+1)			
Betaine	1.832(+1)			
Biguanide	2.96(+2)	11.51(+1)		
2,2′-Biimidazolyl (μ=0.3)	5.01(+1)			
2-Biphenylcarboxylic acid	3.46			
(1,1′-Biphenyl)-4,4′-diamine	3.63(+2)	4.70(+1)		
Bis(2-aminoethyl) ether (30°C)	8.62(+2)	9.59(+1)		
N,N′-Bis(2-aminoethyl)-ethylenediamine (20°C)	3.32(+4)	6.67(+3)	9.20(+2)	9.92(+1)
N,N-Bis(2-hydroxyethyl)-2-aminoethane sulfonic acid (BES) (20°C)	7.15			
N,N-Bis(2-hydroxyethyl)glycine (bicine) (20°C)	8.35			
Bis(2-hydroxyethyl)iminotris (hydroxymethyl)-methane (bis-tris)	6.46(+1)			
1,3-Bis[tris(hydroxymethyl) methylamino]propane (20°C)	6.80(+1)			
Bromoactic acid	2.902			
2-Bromoaniline	2.53(+1)			

Table 8-1 pK_a values of organic materials in water at 25°C (*continued*)

Substance	pK_1	pK_2	pK_3	pK_4
3-Bromoaniline	3.53(+1)			
4-Bromoaniline	3.88(+1)			
2-Bromobenzoic acid	2.85			
3-Bromobenzoic acid	3.810			
4-Bromobenzoic acid	3.99			
2-Bromobutanoic acid (35°C)	2.939			
erythro-2-Bromo-3-chlorosuccinic acid (19°C, μ=0.1)	1.4	2.6		
threo-2-Bromo-chlorosuccinic acid (19°C, μ=0.1)	1.5	2.8		
trans-2-Bromocinnamic acid	4.41			
3-Bromo-4-(dimethylamino)pyridine (20°C)	6.52(+1)			
2-Bromo-4,6-dinitroaniline	−6.94(+1)			
3-Bromo-2-hydroxymethylbenzoic acid (20°C)	3.28			
6-Bromo-2-hydroxymethylbenzoic acid (20°C)	2.25			
7-Bromo-8-hydroxyquinoline-5-sulfonic acid	2.51	6.70		
3-Bromomandelic acid	3.13			
3-Bromo-4-methylaminopyridine (20°C)	7.49(+1)			
(2-Bromomethyl)butanoic acid	3.92			
Bromomethylphosphonic acid	1.14	6.52		
2-Bromo-6-nitrobenzoic acid	1.37			
2-Bromophenol	8.452			
3-Bromophenol	9.031			
4-Bromophenol	9.34			
2-(2′-Bromophenoxy)acetic acid	3.12			
2-(3′-Bromophenoxy)acetic acid	3.09			
2-(4′-Bromophenoxy) acetic acid	3.13			
2-Bromo-2-phenylacetic acid	2.21			
2-(Bromophenyl)acetic acid	4.054			
4-(Bromophenyl)acetic acid	4.188			
4-Bromophenylarsonic acid	3.25	8.19		
4-Bromophenylphosphinic acid (17°C)	2.1			
2-Bromophenylphosphonic acid	1.64	7.00		
3-Bromophenylphosphonic acid	1.45	6.69		
4-Bromophenylphosphonic acid	1.60	6.83		
3-Bromophenylselenic acid	4.43			
4-Bromophenylselenic acid	4.50			
2-Bromopropanoic acid	2.971			
3-Bromopropanoic acid	3.992			
Bromopropynoic acid	1.855			

Table 8-1 pK_a values of organic materials in water at 25°C (*continued*)

Substance	pK_1	pK_2	pK_3	pK_4
2-Bromopyridine	0.71(+1)			
3-Bromopyridine	2.85(+1)			
4-Bromopyridine	3.71(+1)			
3-Bromoquinoline	2.69(+1)			
Bromosuccinic acid	2.55	4.41		
2-Bromo-*p*-tolylphosphonic acid	1.81	7.15		
Brucine (15°C)	2.50(+2)	8.16(+1)		
2-Butanamine (*sec*-butylamine)	10.56(+1)			
1,2-Butanediamine	6.399(+2)	9.388(+1)		
1,4-Butanediamine	9.35(+2)	10.82(+1)		
2,3-Butanediamine	6.91(+2)	10.00(+1)		
1,2,3,4-Butanetetracarboxylic acid	3.43	4.58	5.85	7.16
cis-2-Butenoic acid (isocrotonic acid)	4.44			
trans-2-Butenoic acid (*trans*-crotonic acid) (35°C)	4.676			
3-Butenoic acid (vinylacetic acid)	4.68			
3-Butoxybenzoic acid (20°C)	4.25			
Butylamine	10.64(+1)			
tert-Butylamine	10.685(+1)			
4-*tert*-Butylaniline	3.78(+1)			
N-*tert*-Butylaniline	7.10(+1)			
Butylarsonic acid (18°C)	4.23	8.91		
2-*tert*-Butylbenzoic acid	3.57			
3-*tert*-Butylbenzoic acid	4.199			
4-*tert*-Butylbenzoic acid	4.389			
N-Butylethylenediamine	7.53(+2)	10.30(+1)		
N-Butylglycine	2.35(+1)	10.25(0)		
tert-Butylhydroperoxide	12.80			
1-(*tert*-Butyl)-2-hydroxybenzene	10.62			
1-(*tert*-Butyl)-3-hydroxybenzene	10.119			
1-(*tert*-Butyl)-4-hydroxybenzene	10.23			
Butylmethylamine	10.90(+1)			
2-Butyl-1-methyl-2-pyrroline	11.84(+1)			
4-*tert*-Butylphenylactic acid	4.417			
Butylphosphinic acid	3.41			
tert-Butylphosphinic acid	4.24			
tert-Butylphosphonic acid	2.79	8.88		
1-Butylpiperidine (μ=0.02)	10.43(+1)			
2-*tert*-Butylpyridine	5.76(+1)			
3-*tert*-Butylpyridine	5.82(+1)			
4-*tert*-Butylpyridine	5.99(+1)			
2-*tert*-Butylthiazole (μ=0.1)	3.00(+1)			
4-*tert*-Butylthiazole (μ=0.1)	3.04(+1)			
2-Butyn-1,4-dioic acid	1.75	4.40		
2-Butynoic acid (tetrolic acid)	2.620			
Butyric acid	4.817			
4-Butyrobetaine (20°C)	3.94(+1)			

Table 8-1 pK_a values of organic materials in water at 25°C (*continued*)

Substance	pK_1	pK_2	pK_3	pK_4
Caffeine (40°C)	10.4			
Calcein (pK_5>12)	<4	5.4	9.0	10.5
Calmagite	8.14	12.35		
D-Camphoric acid	4.57	5.10		
Canaline	2.40	3.70	9.20	
Canavanine	2.50(+2)	6.60(+1)	9.25(0)	
N-Carbamoylacetic acid	3.64			
N-Carbamoyl-α-D-alanine	3.89(+1)			
N-Carbamoyl-β-alanine	4.99(+1)			
DL-*N*-Carbamoylalanine	3.892(+1)			
N-Carbamoylglycine	3.876			
2-Carbamoylpyridine (20°C)	2.10(+1)			
3-Carbamoylpyridine	3.328(+1)			
4-Carbamoylpyridine (20°C)	3.61(+1)			
β-Carboxymethylaminopropanoic acid	3.61(+1)	9.46(0)		
Chloroacetic acid	2.867			
N-(2′-Chloroacetyl)glycine	3.38(0)			
cis-3-Chloroacrylic acid (18°C, μ=0.1)	3.32			
trans-3-chloroacrylic acid (18°C, μ=0.1)	3.65			
2-Chloraniline	2.64(+1)			
3-Chloroaniline	3.52(+1)			
4-Chloroaniline	3.99(+1)			
2-Chlorobenzoic acid	2.877			
3-Chlorobenzoic acid	3.83			
4-Chlorobenzoic acid	3.986			
2-Chlorobutanoic acid	2.86			
3-Chlorobutanoic acid	4.05			
4-Chlorobutanoic acid	4.50			
2-Chloro-3-butenoic acid	2.54			
3-Chlorobutylarsonic acid (18°C)	3.95	8.85		
trans-2′-Chlorocinnamic acid	4.234			
trans-3′-Chlorocinnamic acid	4.294			
trans-4′-Chlorocinnamic acid	4.413			
2-Chlorocrotonic acid	3.14			
3-Chlorocrotonic acid	3.84			
Chlorodifluoroacetic acid	0.46			
1-Chloro-1,2-dihydroxybenzene	8.522			
1-Chloro-2,6-dimethyl-4-hydroxybenzene	9.549			
4-Chloro-2,6-dinitrophenol	2.97			
2-Chloroethylarsonic acid	3.68	8.37		
3-Chlorohexyl-1-arsonic acid (18°C)	3.51	8.31		
2-Chloro-3-hydroxybutanoic acid	2.59			
3-Chloro-2-(hydroxymethyl)benzoic acid (20°C)	3.27			

Table 8-1 pK_a values of organic materials in water at 25°C (*continued*)

Substance	pK_1	pK_2	pK_3	pK_4
6-Chloro-2-(hydroxymethyl)benzoic acid (20°C)	2.26			
7-Chloro-8-hydroxyquinoline-5-sulfonic acid	2.92	6.80		
2-Chloroisocrotonic acid	2.80			
3-Chloroisocrotonic acid	4.02			
3-Chlorolactic acid	3.12			
3-Chloromandelic acid	3.237			
3-Chloro-4-methoxyphenyl-phosphonic acid	2.25	6.7		
3-Chloro-4-methylaniline	4.05(+1)			
4-Chloro-*N*-methylaniline	3.9(+1)			
4-Chloro-3-methylphenol	9.549			
Chloromethylphosphonic acid	1.40	6.30		
2-Chloro-2-methylpropanoic acid	2.975			
2-Chloro-6-nitroaniline	−2.41(+1)			
4-Chloro-2-nitroaniline	−1.10(+1)			
2-Chloro-3-nitrobenzoic acid	2.02			
2-Chloro-4-nitrobenzoic acid	1.96			
2-Chloro-5-nitrobenzoic acid	2.17			
2-Chloro-6-nitrobenzoic acid	1.342			
4-Chloro-2-nitrophenol	6.48			
2-Chlorophenol	8.55			
3-Chlorophenol	9.10			
4-Chlorophenol	9.43			
(4-Chloro-3-nitrophenoxy)acetic acid	2.959			
2-Chloro-4-nitrophenylphosphonic acid	1.12	6.14		
3-Chloropentyl-1-arsonic acid (18°C)	3.71	8.77		
2-Chlorophenoxyacetic acid	3.05			
3-Chlorophenoxyacetic acid	3.07			
4-Chlorophenoxyacetic acid	3.10			
4-Chlorophenoxy-2-methylacetic acid	3.26			
2-Chlorophenylacetic acid	4.066			
3-Chlorophenylacetic acid	4.140			
4-Chlorophenylacetic acid	4.190			
2-Chlorophenylalanine	2.23(+1)	8.94(0)		
3-Chlorophenylalanine	2.17(+1)	8.91(0)		
DL-4-Chlorophenylalanine	2.08(+1)	8.96(0)		
4-Chlorophenylarsonic acid	3.33	8.25		
2-Chlorophenylphosphonic acid	1.63	6.98		
3-Chlorophenylphosphonic acid	1.55	6.65		
4-Chlorophenylphosphonic acid	1.66	6.75		
3-(2′-Chlorophenyl)propanoic acid	4.577			
3-(3′-Chlorophenyl)propanoic acid	4.585			
3-(4′-Chlorophenyl)propanoic acid	4.607			

Table 8-1 pK_a values of organic materials in water at 25°C (*continued*)

Substance	pK_1	pK_2	pK_3	pK_4
3-Chlorophenylselenic acid	4.47			
4-Chlorophenylselenic acid	4.48			
4-Chloro-1,2-phthalic acid	1.60			
2-Chloropropanoic acid	2.84			
3-Chloropropanoic acid	3.992			
2-Chloropropylarsonic acid (18°C)	3.76	8.39		
3-Chloropropylarsonic acid (18°C)	3.63	8.53		
Chloropropynoic acid	1.845			
2-Chloropyridine	0.49(+1)			
3-Chloropyridine	2.84(+1)			
4-Chloropyridine	3.83(+1)			
7-Chlorotetracycline	3.30(+1)	7.44	9.27	
4-Chloro-2-(2′-thiazolylazo)phenol	7.09			
4-Chlorothiophenol	5.9			
N-Chloro-*p*-toluenesulfonamide	4.54(+1)			
3-Chloro-*o*-toluidine	2.49(+1)			
4-Chloro-*o*-toluidine	3.385(+1)			
5-Chloro-*o*-toluidine	3.85(+1)			
6-Chloro-*o*-toludine	3.62(+1)			
Chrome Azurol S	2.45	4.86	11.47	
Chrome Dark Blue	7.56	9.3	12.4	
Cinchonine	5.85(+2)	9.92(+1)		
cis-Cinnamic acid	3.879			
trans-Cinnamic acid	4.438			
Citraconic acid	2.29(0)	6.15(−1)		
Citric acid	3.128	4.761	6.396	
L-(+)-Citrulline	2.43(+1)	9.41(0)		
Cocaine	8.41(+1)			
Codeine	7.95(+1)			
Colchicine	1.65(+1)			
Coniine (μ=0.5)	11.24(+1)			
Creatine (40°C)	3.28(+1)			
Creatinine	3.57(+1)			
o-Cresol	10.26			
m-Cresol	10.00			
p-Cresol	10.26			
Cumene hydroperoxide	12.60			
Cupreine	7.63(+1)			
Cyanamide	10.27			
Cyanoacetic acid	2.460			
Cyanoacetohydrazide	2.34(+2)	11.17(+1)		
2-Cyanobenzoic acid	3.14			
3-Cyanobenzoic acid	3.60			
4-Cyanobenzoic acid	3.55			
4-Cyanobutanoic acid	4.44			
trans-1-Cyanocyclohexane-2-carboxylic acid	3.865			
4-Cyano-2,6-dimethylphenol	8.27			

Table 8-1 pK_a values of organic materials in water at 25°C (*continued*)

Substance	pK_1	pK_2	pK_3	pK_4
4-Cyano-3,5-dimethylphenol	8.21			
2-Cyanoethylamine	7.7(+1)			
N-(2-Cyano)ethylnorcodeine	5.68(+1)			
Cyanomethylamine	5.34(+1)			
2-Cyano-2-methyl-2-phenylacetic acid	2.290			
1-Cyanomethylpiperidine	4.55(+1)			
2-Cyano-2-methylpropanoic acid	2.422			
3-Cyanophenol	8.61			
o-Cyanophenoxyacetic acid	2.98			
m-Cyanophenoxyacetic acid	3.03			
p-Cyanophenoxyacetic acid	2.93			
2-Cyanopropanoic acid	2.37			
3-Cyanopropanoic acid	3.99			
2-Cyanopyridine	−0.26(+1)			
3-Cyanopyridine	1.45(+1)			
4-Cyanopyridine	1.90(+1)			
Cyanuric acid	6.78			
Cyclobutanecarboxylic acid	4.785			
1,1-Cyclobutanedicarboxylic acid	3.13	5.88		
cis-1,2-Cyclobutanedicarboxylic acid	3.90	5.89		
trans-1,2-Cyclobutanedicarboxylic acid	3.79	5.61		
cis-1,3-Cyclobutanedicarboxylic acid	4.04	5.31		
trans-1,3-Cyclobutanedicarboxylic acid	3.81	5.28		
Cyclohexanecarboxylic acid	4.90			
1,1-Cyclohexanediacetic acid	3.49	6.96		
cis-1,2-Cyclohexanediacetic acid (20°C)	4.42	5.45		
trans-1,2-Cyclohexanediacetic acid (20°C)	4.38	5.42		
cis-1,2-Cyclohexanediamine	6.43(+2)	9.93(+1)		
trans-1,2-Cyclohexanediamine	6.34(+2)	9.74(+1)		
1,1-Cyclohexanedicarboxylic acid	3.45	4.11		
cis-1,2-Cyclohexanedicarboxylic acid (20°C)	4.34	6.76		
trans-1,2-Cyclohexanedicarboxylic acid (20°C)	4.18	5.93		
cis-1,3-Cyclohexanedicarboxylic acid (16°C)	4.10	5.46		
trans-1,3-Cyclohexanedicarboxylic acid (19°C)	4.31	5.73		
trans-1,4-Cyclohexanedicarboxylic acid (16°C)	4.18	5.42		
1,3-Cyclohexanedione	5.26			
cis,cis-1,3,5-Cyclohexanetriamine	6.9(+3)	8.7(+2)	10.4(+1)	

Table 8-1 pK_a values of organic materials in water at 25°C (*continued*)

Substance	pK_1	pK_2	pK_3	pK_4
Cyclohexanonimine	9.15			
cis-4-Cyclohexene-1,2-dicarboxylic acid (20°C)	3.89	6.79		
trans-4-Cyclohexene-1,2-dicarboxylic acid (20°C)	3.95	5.81		
Cyclohexylacetic acid	4.51			
Cyclohexylamine	10.64(+1)			
2-(Cyclohexylamino)ethanesulfonic acid (CHES) (20°C)	9.55			
3-Cyclohexylamino-1-propanesulfonic acid (CAPS) (20°C)	10.40			
4-Cyclohexylbutanoic acid	4.95			
Cyclohexylcyanoacetic acid	2.367			
1,2-Cyclohexylenedinitriloacetic acid (μ=0.1)	2.4	3.5	6.16	12.35
3-Cyclohexylpropanoic acid	4.91			
2-Cyclohexylpyrrolidine	10.76(+1)			
2-Cyclohexyl-2-pyrroline	7.91(+1)			
Cyclohexylthioacetic acid	3.488			
Cyclopentanecarboxylic acid	4.905			
cis-Cyclopentane-1-carboxylic acid-2-acetic acid	4.40	5.79		
trans-Cyclopentane-1-carboxylic acid-2-acetic acid	4.39	5.67		
Cyclopentane-1,2-diamine-*N,N',N'*-tetraacetic acid (μ=0.1)	—	—	—	10.20
Cyclopentane-1,1-dicarboxylic acid	3.23	4.08		
cis-Cyclopentane-1,2-dicarboxylic acid	4.43	6.67		
trans-Cyclopentane-1,2-dicarboxylic acid	3.96	5.85		
cis-Cyclopentane-1,3-dicarboxylic acid	4.26	5.51		
trans-Cyclopentane-1,3-dicarboxylic acid	4.32	5.42		
Cyclopentylamine	10.65(+1)			
1,1-Cyclopentyldiacetic acid	3.80	6.77		
cis-Cyclopentyl-1,2-diacetic acid	4.42	5.42		
trans-Cyclopentyl-1,2-diacetic acid	4.43	5.43		
Cyclopropanecarboxylic acid	4.827			
Cyclopropane-1,1-dicarboxylic acid	1.82	5.43		
cis-Cyclopropane-1,2-dicarboxylic acid	3.33	6.47		
trans-Cyclopropane-1,2-dicarboxylic acid	3.65	5.13		
Cyclopropylamine	9.10(+1)			

Table 8-1 pK_a values of organic materials in water at 25°C (*continued*)

Substance	pK_1	pK_2	pK_3	pK_4
5-Cyclopropyl-1,2,3,4-tetrazole	4.90(+1)			
L-Cysteic acid (3-sulfo-L-alanine)	1.89(+1)	8.7(0)		
L-(+)-Cysteine	1.71(+1)	8.39(0)	10.70(SH)	
L-(+)-Cysteine, ethyl ester	6.69 (NH_3^+)	9.17(SH)		
L-(+)-Cysteine, methyl ester	6.56 (NH_3^+)	8.99(SH)		
L-Cysteinyl-L-asparagine	2.97	7.09	8.47	
L-Cystine (35°C)	1.6(+2)	2.1(+1)	8.02(0)	8.71(−1)
Cystinylglycylglycine (35°C)	3.12	3.21	6.01	6.87
Cytidine	4.08(+1)	12.24(0)		
Cytidine-2′-phosphoric acid	0.8(+1)	4.36(0)	6.17(−1)	
Cytidine-3′-phosphoric acid	0.80(+1)	4.31(0)	6.04(−1)	13.2(sugar)
Cytidine-5′-phosphoric acid	—	4.39(0)	6.62(−1)	
Cytosine	4.58(+1)	12.15(0)		
Decanedioic acid (sebacic acid)	4.59	5.59		
Dehydroascorbic acid (20°C)	3.21	7.92	10.3	
2′-Deoxyadenosine (μ=0.1)	3.8(+1)			
Deoxycholic acid	6.58			
2-Deoxyglucose	12.52			
2-Deoxyguanosine (μ=0.1)	2.5(+1)			
5-Desoxypyridoxal (μ=0)	4.17(+1)	8.14(OH)		
1,1-Diacetic acid semicarbazide (30°C, μ=0.1)	2.96	4.04		
Diacetylacetone	7.42			
Diallylamine (μ=0.02)	9.29(+1)			
5,5-Diallylbarbituric acid	7.78(0)			
1,3-Diamino-2-aminomethylpropane	6.44(+3)	8.56(+2)	10.38(+1)	
3,5-Diaminobenzoic acid	5.30			
1,3-Diamino-*N*,*N*′-bis-(2-aminoethyl)propane (μ=0.5)	6.01(+4)	7.26(+3)	9.49(+2)	10.23(+1)
2,4-Diaminobutanoic acid (20°C)	1.85(+2)	8.24(+1)	10.40(0)	
2,2′-Diaminodiethyl sulfide (30°C)	8.84(+2)	9.64(+1)		
1,8-Diamino-3,6-dithiooctane (30°C)	8.43(+2)	9.31(+1)		
2,7-Diaminooctanedioic acid (20°C, μ=0.1)	1.84(+2)	2.64(+1)	9.23(0)	9.89(−1)
1,8-Diamino-3,6-octanedione (30°C)	8.60(+2)	9.57(+1)		
1,8-Diamino-3-oxa-6-thiooctane	8.54(+2)	9.46(+1)		
2,3-Diaminopropanoic acid (μ=0.1)	1.33(+2)	6.674(+1)	9.623(0)	
2,3-Diaminopropanoic acid, methyl ester (μ=0.1)	4.412(+1)	8.250(0)		
1,3-Diamino-2-propanol (20°C)	7.93(+2)	9.69(+1)		
2,5-Diaminopyridine (20°C)	2.13(+2)	6.48(+1)		
1,4-Diazabicyclo[2.2.2]octane	2.90(+2)	8.60(+1)		
Dibenzylamine	8.52(+1)			
Dibenzylsuccinic acid (20°C)	3.96	6.66		

Table 8-1 pK_a values of organic materials in water at 25°C (*continued*)

Substance	pK_1	pK_2	pK_3	pK_4
Dibromoacetic acid	1.39			
3,5-Dibromoaniline	2.35(+1)			
3,5-Dibromophenol	8.056			
2,2-Dibromopropanoic acid	1.48			
2,3-Dibromopropanoic acid	2.33			
rac-2,3-Dibromosuccinic acid (20°C)	1.43	2.24		
meso-2,3-Dibromosuccinic acid (20°C)	1.51	2.71		
3,5-Dibromo-*p*-L-tyrosine	2.17(+1)	6.45(0)	7.60(−1)	
Dibutylamine	11.25(+1)			
Di-*sec*-butylamine	10.91(+1)			
2,6-Di-*tert*-butylpyridine	3.58(+1)			
rac-2,3-Di-*tert*-butylsuccinic acid (μ=0.1)	3.58	10.2		
1,12-Dicarboxydodecaborane	9.07	10.23		
Dichloroacetic acid	1.26			
Dichloroacetylacetic acid	2.11			
3,5-Dichloroaniline	2.37(+1)			
1,3-Dichloro-2,5-dihydroxybenzene (μ=0.65)	7.30	9.99		
2,5-Dichloro-3,6-dihydroxy-*p*-benzoquinone	1.09	2.42		
Dichloromethylphosphonic acid	1.14	5.61		
2,4-Dichloro-6-nitroaniline	−3.00(+1)			
2,5-Dichloro-4-nitroaniline	−1.74(+1)			
2,6-Dichloro-4-nitroaniline	−3.31(+1)			
2,3-Dichlorophenol	7.44			
2,4-Dichlorophenol	7.85			
2,6-Dichlorophenol	6.78			
3,4-Dichlorophenol	8.630			
3,5-Dichlorophenol	8.179			
2,4-Dichlorophenoxyacetic acid (2,4-D)	2.64			
4,6-Dichlorophenoxy-2-methylacetic acid	3.13			
3,6-Dichlorophthalic acid	1.46			
2,2-Dichloropropanoic acid	2.06			
2,3-Dichloropropanoic acid	2.85			
rac-2,3-Dichlorosuccinic acid (20°C)	1.43	2.81		
meso-2,3-Dichlorosuccinic acid	1.49	2.97		
3,5-Dichloro-*p*-tyrosine	2.12	6.47	7.62	
2-Dicyanoethylamine	5.14(+1)			
2,2-Dicyanopropanoic acid	−2.8			
Dicyclohexylamine	11.25(+1)			
Dicyclopentylamine	10.93(+1)			
Didodecylamine	10.99(+)			

Table 8-1 pK_a values of organic materials in water at 25°C (*continued*)

Substance	pK_1	pK_2	pK_3	pK_4
Diethanolamine	8.88(+1)			
Di(ethoxyethyl)amine	8.47(+1)			
3,5-Diethoxyphenol	9.370			
3-(Diethoxyphosphinyl)benzoic acid	3.65			
4-(Diethoxyphosphinyl)benzoic acid	3.60			
3-(Diethoxyphosphinyl)phenol	8.66			
4-(Diethoxyphosphinyl)phenol	8.28			
Diethylamine	10.8(+1)			
2-(Diethylamino)ethyl-4-aminobenzoate	8.85(+1)			
α-(Diethylamino)toluene	9.44(+1)			
N,N-Diethylaniline	6.56(+1)			
5,5-Diethylbarbituric acid (veronal)	8.020(0)			
N,N-Diethylbenzylamine	9.48(+1)			
Diethylbiguanide (30°C)	2.53(+1)	11.68(0)		
Diethylenetriamine	4.42(+3)	9.21(+2)	10.02(+1)	
Diethylenetriaminepentaacetic acid (pK_5, 10.58)	1.80(0)	2.55(−1)	4.33(−2)	8.60(−3)
N,N-Diethylethylenediamine	7.70(+2)	10.46(+1)		
2,2-Diethylglutaric acid	3.62	7.12		
N,N-Diethylglycine	2.04(+1)	10.47(0)		
Diethylglycolic acid (18°C)	3.804			
Diethylmalonic acid	2.151	7.417		
Diethylmethylamine	10.43(+1)			
rac-2,3-Diethylsuccinic acid	3.63	6.46		
meso-2,3-Diethylsuccinic acid	3.54	6.59		
N,N-Diethyl-*o*-toluidine	7.18(+1)			
Difluoroacetic acid	1.33			
3,3-Difluoroacrylic acid	3.17			
Diglycolic acid	2.96			
Diguanidine	12.8			
Dihexylamine	11.0(+1)			
Dihydroarecaidine	9.70			
Dihydroarecaidine, methyl ester	8.39			
Dihydrocodeine	8.75(+1)			
Dihydroergonovine	7.38(+1)			
α-Dihydrolysergic acid	3.57	8.45		
γ-Dihydrolysergic acid	3.60	8.71		
α-Dihydrolysergol	8.30			
β-Dihydrolysergol	8.23			
Dihydromorphine	9.35			
3,4-Dihydroxyalanine	2.32(+1)	8.68(0)	9.87(−1)	
1,2-Dihydroxyanthraquinone-3-sulfonic acid (alizarin-3-sulfonic acid)	—	5.54(−1)	11.01(−2)	
3,4-Dihydroxybenzaldehyde	7.55			

Table 8-1 pK_a values of organic materials in water at 25°C (*continued*)

Substance	pK_1	pK_2	pK_3	pK_4
1,2-Dihydroxybenzene (pyrocatechol) (μ=0.1)	9.356(0)	12.98(−1)		
1,3-Dihydroxybenzene (resorcinol)	9.44(0)	12.32(−1)		
1,4-Dihydroxybenzene (hydroquinone)	9.91(0)	12.04(−1)		
4,5-Dihydroxybenzene-1,3-disulfonic acid	—	—	7.66(−2)	12.6(−3)
2,3-Dihydroxybenzoic acid (30°C)	2.98	10.14		
2,4-Dihydroxybenzoic acid (β-resorcyclic acid)	3.29	8.98		
2,5-Dihydroxybenzoic acid	2.97	10.50		
2,6-Dihydroxybenzoic acid	1.30			
3,4-Dihydroxybenzoic acid	4.48	8.67	11.74	
3,5-Dihydroxybenzoic acid	4.04			
2,5-Dihydroxy-*p*-benzoquinone	2.71	5.18		
3,4-Dihydroxy-3-cyclobutene-1,2-dione	0.541	3.480		
2,3-Dihydroxy-2-cyclopenten-1-one (20°C)	4.72			
1,4-Dihydroxy-2,6-dinitrobenzene	4.42	9.14		
Di(2,2′-hydroxyethyl)amine	8.8(+1)			
N,N-Di(2-hydroxyethyl)glycine	8.333			
Dihydroxymaleic acid	1.10			
Dihydroxymalic acid	1.92			
1,3-Dihydroxy-2-methylbenzene (μ=0.65)	10.05	11.64		
2,2-Di(hydroxymethyl)-3-hydroxypropanoic acid	4.460			
2,4-Dihydroxy-5-methylpyrimidine	9.90			
2,4-Dihydroxy-6-methylpyrimidine	9.52			
1,4-Dihydroxynaphthalene (26°C, μ=0.65)	9.37	10.93		
1,2-Dihydroxy-3-nitrobenzene	6.68			
1,2-Dihydroxy-4-nitrobenzene (μ=0.1)	6.701			
2,4-Dihydroxy-1-phenylazobenzene (μ=0.1)	11.98			
2,4-Dihydroxyoxazolidine	6.11(+1)			
2,4-Dihydroxypteridine	<1.3	7.92		
2,6-Dihydroxypurine	7.53(0)	11.84(−1)		
2,4-Dihydroxypyridine (20°C)	1.37(+1)	6.45(0)	13(−1)	
Dihydroxytartaric acid	1.95	4.00		
1,4-Dihydroxy-2,3,5,6-tetramethylbenzene (μ=0.65)	11.25	12.70		
3,5-Diiodoaniline	2.37(+1)			
2,5-Diiodohistamine	2.31(+2)	8.20(+1)	10.11(0)	
2,5-Diiodohistidine (μ=0.1)	2.72	8.18	9.76	

Table 8-1 pK$_a$ values of organic materials in water at 25°C (*continued*)

Substance	pK$_1$	pK$_2$	pK$_3$	pK$_4$
3,5-Diiodophenol	8.103			
3,5-Diiodotyrosine	2.117(+1)	6.479(0)	7.821(−1)	
Diisopropylmalonic acid	2.124	8.848		
Dilactic acid	2.955			
threo-1,4-Dimercapto-2,3-butanediol	8.9			
meso-2,3-Dimercaptosuccinic acid	2.71	3.48	8.89(SH)	10.79(SH)
3,5-Dimethoxyaniline	3.86(+1)			
2,6-Dimethoxybenzoic acid	3.44			
1,10-Dimethoxy-3,8-dimethyl-4,7-phenanthroline	7.21			
Di(2-methoxyethyl)amine	9.51(+1)			
3,5-Dimethoxyphenol	9.345			
(3,4-Dimethoxy)phenylacetic acid	4.333			
Dimethylamine	10.77(+1)			
4-Dimethylaminobenzaldehyde	1.647(+1)			
N,N-Dimethylaminocyclohexane	10.72(+1)			
4-Dimethylamino-2,3-dimethyl-1-phenyl-3-pyrazolin-5-one	4.18(+1)			
4-Dimethylamino-3,5-dimethylpyridine (20°C)	8.15(+1)			
2-(Dimethylamino)ethanol	9.26(+1)			
2-[2-(Dimethylamino)ethyl]pyridine	3.46(+2)	8.75(+1)		
3-(Dimethylaminoethyl)pyridine	4.30(+2)	8.86(+1)		
4-(Dimethylaminoethyl)pyridine	4.66(+2)	8.70(+1)		
4-(Dimethylamino)-3-ethylpyridine (20°C)	8.66(+1)			
4-(Dimethylamino)-3-isopropylpyridine (20°C)	8.27(+1)			
2-(Dimethylaminomethyl)pyridine	2.58(+2)	8.12(+1)		
3-(Dimethylaminomethyl)pyridine	3.17(+2)	8.00(+1)		
4-(Dimethylaminomethyl)pyridine	3.39(+2)	7.66(+1)		
4-(Dimethylamino)-3-methylpyridine (20°C)	8.68(+1)			
4-(Dimethylaminophenyl)phosphonic acid	2.0(+1)	4.2	7.35	
3-(Dimethylaminopropanoic acid	9.85(+1)			
4-(Dimethylamino)pyridine (20°C)	6.09(+1)			
N,N-Dimethylaniline	5.15(+1)			
2,3-Dimethylaniline	4.70(+1)			
2,4-Dimethylaniline	4.89(+1)			
2,5-Dimethylaniline	4.53(+1)			
2,6-Dimethylaniline	3.95(+1)			
3,4-Dimethylaniline	5.17(+1)			
3,5-Dimethylaniline	4.765(+1)			
N,N-Dimethylaniline-4-phosphonic acid (17°C)	2.0(+1)	4.2	7.39	
Dimethylarsinic acid (cacodylic acid)	6.273			

Table 8-1 pK_a values of organic materials in water at 25°C (*continued*)

Substance	pK_1	pK_2	pK_3	pK_4
1,3-Dimethylbarbituric acid	4.68(+1)			
2,3-Dimethylbenzoic acid	3.771			
2,4-Dimethylbenzoic acid	4.217			
2,5-Dimethylbenzoic acid	3.990			
2,6-Dimethylbenzoic acid	3.362			
3,4-Dimethylbenzoic	4.41			
3,5-Dimethylbenzoic acid	4.302			
N,N-Dimethylbenzylamine	9.02(+1)			
Dimethylbiguanide	2.77(+1)	11.52		
2,2-Dimethylbutanoic acid (18°C)	5.03			
Dimethylchlorotetracycline (μ=0.01)	3.30(+1)			
2,6-Dimethyl-4-cyanophenol	8.27			
3,5-Dimethyl-4-cyanophenol	8.21			
5,5-Dimethyl-1,3-cyclohexanedione	5.15			
cis-3,3-Dimethyl-1,2-cyclopropanedicarboxylic acid	2.34	8.31		
trans-3,3-Dimethyl-1,2-cyclopropanedicarboxylic acid	3.92	5.32		
3,5-Dimethyl-4-(dimethylamino)-pyridine (20°C)	8.12(+1)			
2,2-Dimethyl-1,3-dioxane-4,6-dione	5.1			
1,1-Dimethylethanethiol (μ=0.1)	11.22			
N,N-Dimethylethylenediamine-*N,N*-diacetic acid	6.63	9.53		
N,N'-Dimethylethylenediamine-*N,N'*-diacetic acid	7.40	10.16		
N,N-Dimethylethylenediamine-*N,N'*-diacetic acid	5.99	9.97		
N,N-Dimethylglycine	2.146(+1)	9.940(0)		
Dimethylglycolic acid (18°C)	4.04			
N,N-Dimethylglycylglycine	3.11(+1)	8.09(0)		
Dimethylglyoxime	10.60			
5,5-Dimethyl-2,4-hexanedione	10.01			
5,5-Dimethylhydantoin	9.19			
2,4-Dimethyl-8-hydroxyquinoline	6.20(+1)	10.60(0)		
3,4-Dimethyl-8-hydroxyquinoline	5.80(+1)	10.05(0)		
2,4-Dimethyl-8-hydroxyquinoline-7-sulfonic acid	3.20 (NH^+)	10.14(OH)		
Dimethylhydroxytetracycline	7.5	9.4		
2,4-Dimethylimidazole	8.38(+1)			
Dimethylmalic acid	3.17	6.06		
2,2-Dimethylmalonic acid	3.17	6.06		
3,5-Dimethyl-4-(methylamino) pyridine (20°C)	9.96(+1)			
2,3-Dimethylnaphthalene-1-carboxylic acid	3.33			

Table 8-1 pK_a values of organic materials in water at 25°C (*continued*)

Substance	pK_1	pK_2	pK_3	pK_4
2,6-Dimethyl-4-nitrophenol	7.190			
3,5-Dimethyl-4-nitrophenol	8.245			
α,α-Dimethyloxaloacetic acid	1.77	4.62		
3,3-Dimethylpentanedioic acid	3.70	6.34		
2,2-Dimethylpentanoic acid	4.969			
4,4-Dimethylpentanoic acid (18°C)	4.79			
2,3-Dimethylphenol	10.50			
2,4-Dimethylphenol	10.58			
2,5-Dimethylphenol	10.22			
2,6-Dimethylphenol	10.59			
3,4-Dimethylphenol	10.32			
3,5-Dimethylphenol	10.15			
2,6-Dimethylphenoxyacetic acid	3.356			
Dimethylphenylsilylacetic acid	5.27			
N,N'-Dimethylpiperazine	4.630(+2)	8.539(+1)		
1,2-Dimethylpiperidine	10.22			
cis-2,6-Dimethylpiperidine	11.07(+1)			
2,2-Dimethylpropanoic acid (pivalic acid)	5.031			
2,2'-Dimethylpropylphosphonic acid	2.84	8.65		
2,4-Dimethylpyridine (2,4-lutidine)	6.74(+1)			
2,5-Dimethylpyridine (2,5-lutidine)	6.43(+1)			
2,6-Dimethylpyridine (2,6-lutidine)	6.71(+1)			
3,4-Dimethylpyridine (3,4-lutidine)	6.47(+1)			
3,5-Dimethylpyridine (3,5-lutidine)	6.09(+1)			
2,4-Dimethylpyridine-1-oxide	1.627(+1)			
2,5-Dimethylpyridine-1-oxide	1.208(+1)			
2,6-Dimethylpyridine-1-oxide	1.366(+1)			
3,4-Dimethylpyridine-1-oxide	1.493(+1)			
3,5-Dimethylpyridine-1-oxide	1.181(+1)			
2,3-Dimethylquinoline	4.94(+1)			
2,6-Dimethylquinoline	5.46(+1)			
meso-2,2-Dimethylsuccinic acid	3.77	5.936		
rac-2,2-Dimethylsuccinic acid	3.93	6.20		
D-2,3-Dimethylsuccinic acid	3.82	5.93		
meso-2,3-Dimethylsuccinic acid	3.67	5.30		
rac-2,3-Dimethylsuccinic acid	3.94	6.20		
2,4-Dimethylthiazole (μ=0.1)	3.98			
2,5-Dimethylthiazole (μ=0.1)	3.91			
4,5-Dimethylthiazole (μ=0.1)	3.73			
N,N-Dimethyl-*o*-toluidine	5.86(+1)			
N,N-Dimethyl-*p*-toluidine	7.24(+1)			
2,4-Dinitroaniline	−4.25(+1)			
2,6-Dinitroaniline	−5.23(+1)			
3,5-Dinitroaniline	0.229(+1)			
2,3-Dinitrobenzoic acid	1.85			
2,4-Dinitrobenzoic acid	1.43			

Table 8-1 pK_a values of organic materials in water at 25°C (*continued*)

Substance	pK_1	pK_2	pK_3	pK_4
2,5-Dinitrobenzoic acid	1.62			
2,6-Dinitrobenzoic acid	1.14			
3,4-Dinitrobenzoic acid	2.82			
3,5-Dinitrobenzoic acid	2.85			
1,1-Dinitrobutane (20°C)	5.90			
1,1-Dinitrodecane	3.60			
1,1-Dinitroethane (20°C)	5.21			
Dinitromethane (20°C)	3.60			
1,1-Dinitropentane	5.337			
2,4-Dinitrophenol	4.08			
2,5-Dinitrophenol	5.216			
2,6-Dinitrophenol	3.713			
3,4-Dinitrophenol	5.424			
3,5-Dinitrophenol	6.732			
2,4-Dinitrophenylacetic acid	3.50			
1,1-Dinitropropane (20°C)	5.5			
2,6-Dioxo-1,2,3,6-tetrahydro-4-pyrimidinecarboxylic acid (orotic acid)	1.8(+1)	9.55(0)		
Diphenylacetic acid	3.939			
Diphenylamine	0.9(+1)			
2,2-Diphenylglutaric acid (20°C)	3.91	5.38		
1,3-Diphenylguanidine	10.12			
2,2-Diphenylheptanedioic acid (20°C)	4.28	5.39		
2,2-Diphenylhexanedioic acid (20°C)	4.17	5.40		
3,3-Diphenylhexanedioic acid	4.22	5.19		
Diphenylhydroxyacetic acid (35°C)	3.05			
Diphenylketimine	6.82			
2,2-Diphenylnonanedioic acid (20°C)	4.33	5.38		
meso-2,2-Diphenylsuccinic acid	3.48			
rac-2,2-Diphenylsuccinic acid	3.58			
2,2-Diphenylsuccinic acid, 1-methyl ester (20°C)	4.47			
2,2-Diphenylsuccinic acid, 4-methyl ester (20°C)	3.900			
Diphenylthiocarbazone	4.50	15		
Dipropylamine	10.91(+1)			
Dipropylenetriamine	7.72(+3)	9.56(+2)	10.65(+1)	
2,2-Dipropylglutaric acid	3.688	7.31		
Dipropylmalonic acid	2.04	7.51		
2,2′-Dipyridyl	−0.52(+2)	4.352(+1)		
2,3′-Dipyridyl (20°C)	1.52(+2)	4.42(+1)		
2,4′-Dipyridyl (20°C)	1.19(+2)	4.77(+1)		
3,3′-Dipyridyl (20°C, μ=0.2)	3.0(+2)	4.60(+1)		
3,4′-Dipyridyl (20°C, μ=0.2)	3.0(+2)	4.85(+1)		
4,4′-Dipyridyl	3.17(+2)	4.82(+1)		

Table 8-1 pK_a values of organic materials in water at 25°C (*continued*)

Substance	pK_1	pK_2	pK_3	pK_4
Dithiodiacetic acid (18°C)	3.075	4.201		
1,4-Dithioerythritol	9.5			
Dithiooxamide (rubeanic acid)	10.89			
Dulcitol	13.46			
Ecgonine	10.91			
Emetine	7.36(+1)	8.23(0)		
Epinephrine enantiomorph	9.39(+1)			
Epinephrine, pseudo	9.53(+1)			
Ergometrinine	7.32(+1)			
Ergonovine	6.73(+1)			
Eriochrome Black T	6.3	11.55		
1,2-Ethanediamine	6.85(+2)	9.92(+1)		
Ethane-1,2-diamino-*N*,*N*′-dimethyl-*N*,*N*′-diacetic acid (20°C)	6.047(0)	10.068(−1)		
1,2-Ethanedithiol	8.96	10.54		
Ethanethiol (μ=0.015)	10.61			
Ethoxyacetic acid (18°C)	3.65			
2-Ethoxyaniline (*o*-phenetidine)	4.47(+1)			
3-Ethoxyaniline	4.17(+1)			
4-Ethoxyaniline	5.25(+1)			
2-Ethoxybenzoic acid (20°C)	4.21			
3-Ethoxybenzoic acid (20°C)	4.17			
4-Ethoxybenzoic acid (20°C)	4.80			
Ethoxycarbonylethylamine	9.13(+1)			
2-Ethoxyethanethiol	9.38			
2-Ethoxyethylamine	6.26(+1)			
2-Ethoxyphenol	10.109			
3-Ethoxyphenol	9.655			
(4-Ethoxyphenyl)phosphonic acid	2.06	7.28		
4-Ethoxypyridine	6.67(+1)			
Ethyl acetoacetate	10.68			
3-Ethylacrylic acid	4.695			
N-Ethylalanine	2.22(+1)	10.22(0)		
Ethylamine	10.63(+1)			
(3-Ethylamino)phenylphosphonic acid	1.1(+1)	4.90(0)	7.24(−1)	
N-Ethylaniline	5.11(+1)			
2-Ethylaniline	4.42(+1)			
3-Ethylaniline	4.70(+1)			
4-Ethylaniline	5.00(+1)			
Ethylarsonic acid (18°C)	3.89	8.35		
Ethylbarbituric acid	3.69(+1)			
2-Ethylbenzimidazole (μ=0.16)	6.27(+1)			
2-Ethylbenzoic acid	3.79			
4-Ethylbenzoic acid	4.35			
Ethylbiguanide	2.09(+1)	11.47(0)		

Table 8-1 pK_a values of organic materials in water at 25°C (*continued*)

Substance	pK_1	pK_2	pK_3	pK_4
2-Ethylbutanoic acid (20°C)	4.710			
S-Ethyl-L-cysteine (μ=0.1)	2.03(+1)	8.60(0)		
Ethylenebiguanide (30°C)	1.74	2.88	11.34	11.76
Ethylenebis(thioacetic acid) (18°C)	3.382(0)	4.352(−1)		
Ethylenediamine-*N,N'*-diacetic acid	6.42	9.46		
Ethylenediamine-*N,N*-dimethyl-*N',N'*-diacetic acid	6.047	10.068		
Ethylenediamine-*N',N*-dipropanoic acid (30°C)	6.87	9.60		
Ethylenediamine-*N,N,N',N'*-tetraacetic acid (μ=0.1)	1.99	2.67	6.16	10.26
Ethylenediamine-*N,N,N',N'*-tetrapropanoic acid (30°C)	3.00	3.43	6.77	9.60
Ethylene glycol	14.22			
Ethyleneimine	8.04(+1)			
cis-Ethylene oxide dicarboxylic acid	1.93	3.92		
trans-Ethylene oxide dicarboxylic acid	1.93	3.25		
N-Ethylethylenediamine	7.63(+2)	10.56(+1)		
N-Ethylglycine (μ=0.1)	2.34(+1)	10.23(0)		
3-Ethylglutaric acid	4.28	5.33		
Ethyl hydroperoxide	11.80			
Ethyl hydrogen malonate	3.55			
3-Ethyl-2-hydroxypyridine	5.00(+1)			
Ethylmalonic acid	2.90(0)	5.55(−1)		
N-Ethyl mercaptoacetamide	8.14(SH)			
Ethyl 2-mercaptoacetate	7.95(SH)			
Ethyl 3-mercaptopropanoate	9.48(SH)			
3-Ethyl-4-(methylamino)pyridine (20°C)	9.90(+1)			
5-Ethyl-5-(1-methylbutyl)barbituric acid	8.11(0)			
Ethyl methyl ketoxime	12.45			
Ethylmethylmalonic acid	2.86(0)	6.41(−1)		
1-Ethyl-2-methylpiperidine	10.66(+1)			
3-Ethyl-6-methylpyridine (20°C)	6.51(+1)			
3-Ethyl-4-methylpyridine-1-oxide	−1.534(+1)			
5-Ethyl-2-methylpyridine-1-oxide	−1.288(+1)			
1-Ethyl-2-methyl-2-pyrroline	11.84(+1)			
Ethylmorphine (15°C)	8.08			
Ethyl nitroacetate	5.85			
3-Ethylpentane-2,4-dione	11.34			
2-Ethylpentanoic acid (18°C)	4.71			
5-Ethyl-5-pentylbarbituric acid	7.960			
2-Ethylphenol	10.2			
3-Ethylphenol	10.07			
4-Ethylphenol	10.0			

Table 8-1 pK_a values of organic materials in water at 25°C (*continued*)

Substance	pK_1	pK_2	pK_3	pK_4
4-Ethylphenylacetic acid	4.373			
5-Ethyl-5-phenylbarbituric acid	7.445			
Ethylphosphinic acid	3.29			
Ethylphosphonic acid	2.43	8.05		
1-Ethylpiperidine (μ=0.01)	10.45(+1)			
2,2-Ethylpropylglutaric acid	3.511			
Ethylpropylmalonic acid	3.14	7.43		
2-Ethylpyridine	5.89(+1)			
3-Ethylpyridine (20°C)	5.80(+1)			
4-Ethylpyridine	5.87(+1)			
Ethyl 3-pyridinecarboxylate	3.35(+1)			
Ethyl 4-pyridinecarboxylate	3.45(+1)			
2-Ethylpyridine-1-oxide	−1.19(+1)			
3-Ethylpyridine-1-oxide	−0.965(+1)			
Ethylpyrrolidine	10.43(+1)			
2-Ethyl-2-pyrroline	7.87(+1)			
Ethylsuccinic acid	4.08(0)			
S-Ethylthioacetic acid	5.06			
N-Ethyl-*o*-toluidine	4.92(+1)			
N-Ethylveratramine	7.40(+1)			
β-Eucaine	9.35(+1)			
Fluoroacetic acid	2.586			
2-Fluoroacrylic acid	2.55			
2-Fluoroaniline	3.20(+1)			
3-Fluoroaniline	3.58(+1)			
4-Fluoroaniline	4.65(+1)			
2-Fluorobenzoic acid	3.27			
3-Fluorobenzoic acid	3.865			
4-Fluorobenzoic acid	4.14			
Fluoromandelic acid	4.244			
2-Fluorophenol	8.73			
3-Fluorophenol	9.29			
4-Fluorophenol	9.89			
2-Fluorophenoxyacetic acid	3.08			
3-Fluorophenoxyacetic acid	3.08			
4-Fluorophenoxyacetic acid	3.13			
4-Fluorophenylacetic acid	4.25			
2′-Fluorophenylalanine	2.14(+1)	9.01(0)		
3′-Fluorophenylalanine	2.10(+1)	8.98(0)		
4-Fluorophenylalanine	2.13(+1)	9.05(0)		
2-Fluorophenylphosphonic acid	1.64	6.80		
3-Fluorophenylselenic acid	4.34			
4-Fluorophenylselenic acid	4.50			
2-Fluoropyridine	−0.44(+1)			
3-Fluoropyridine	2.97(+1)			
5-Fluorouracil	8.00(0)	ca 13(−1)		

Table 8-1 pK_a values of organic materials in water at 25°C (*continued*)

Substance	pK_1	pK_2	pK_3	pK_4
Folic acid (pteroylglutamic acid)	8.26			
Formic acid	3.751			
N-Formylglycine	3.43			
2-Formyl-3-hydroxypyridine (20°C)	3.40(+1)	6.95(OH)		
4-Formyl-3-hydroxypyridine	4.05(+1)	6.77(OH)		
2-Formyl-3-methoxypyridine (20°C)	3.89(+1)	12.95		
Formyl-3-methoxypyridine (20°C)	4.45(+1)	11.7		
D-(−)-Fructose	12.03			
Fumaric acid	3.10	4.60		
2-Furancarboxylic acid (2-furoic acid)	3.164			
D-(+)-Galactose	12.35			
Galactose-1-phosphoric acid	1.00	6.17		
Glucoascorbic acid	4.26	11.58		
D-Gluconic acid	3.86			
α-D-(+)-Glucose	12.28			
α-D-Glucose-1-phosphate	1.11(0)	6.504(−1)		
trans-Glutaconic acid	3.77	5.08		
D-(−)-Glutamic acid	2.162(+1)	4.272(0)	9.358(−1)	
L-Glutamic acid	2.13(+1)	4.31(0)	9.76(−1)	
Glutamic acid, 1-ethyl ester	3.85(+1)	7.84(0)		
Glutamic acid, 5-ethyl ester	2.15(+1)	9.19(0)		
L-Glutamine (μ=0.2)	2.15(+1)	9.00(0)		
Glutaric acid	3.77	6.08		
Glutaric acid monoamide	4.600(0)			
Glutarimide	11.43			
Glutathione	2.12(+1)	3.53(0)	8.66	9.12
DL-Glyceric acid	3.64			
Glycerol	14.15			
Glyceryl-1-phosphoric acid	——	6.656(−1)		
Glyceryl-2-phosphoric acid	1.335(0)	6.650(−1)		
Glycine	2.351(+1)	9.70(0)		
Glycine amide	8.03(+1)			
Glycine, ethyl ester	7.66(+1)			
Glycine hydroxamic acid	7.10	9.10		
Glycine, methyl ester	7.59(+1)			
Glycine-*O*-phenylphosphorylserine	2.96	8.07		
Glycolic acid	3.831			
N-Glycyl-α-alanine	3.15 (+1)	8.33(0)		
Glycylalanylalanine	3.38(+1)	8.10(0)		
N-Glycylasparagine	2.942			
Glycyclaspartic acid	2.81(+1)	4.45(0)	8.60(−1)	
Glycyl-DL-glutamine (18°C)	2.88(+1)	8.33(0)		
N-Glycylglycine	3.126(+1)	8.252(0)		
Glycylglycylcysteine (35°C)	2.71	2.71	7.94	7.94
Glycylglycylglycine	3.225(+1)	8.090(0)		

Table 8-1 pK_a values of organic materials in water at 25°C (*continued*)

Substance	pK_1	pK_2	pK_3	pK_4
Glycyl-L-histidine (μ=0.16)	6.79	8.20		
Glycylisoleucine	8.00			
N-Glycyl-L-leucine	3.180(+1)	8.327(0)		
Glycyl-*O*-phosphorylserine	2.90	6.02	8.43	
L-Glycylproline (μ=0.1)	2.81(+1)	8.65(0)		
N-Glycylsarcosine (μ=0.1)	2.98(+1)	8.55(0)		
N-Glycylserine	2.98(+1)	8.38(0)		
Glycylserylglycine	3.32	7.99		
Glycyltyrosine	2.93	8.45	10.49	
Glycylvaline	3.15	8.18		
Glyoxaline	7.03(+1)			
Glyoxylic acid	3.30(0)			
Guanidineacetic acid	2.82(+1)			
Guanine	3.3(+1)	9.2	12.3	
Guanine deoxyriboside-3′-phosphoric acid	—	2.9	6.4	9.7
Guanosine	1.9(+1)	9.25(0)	12.33(OH)	
Guanosine-5′-diphosphoric acid (μ=0.1; p$K_5$5 9.6)	—	—	2.9	6.3
Guanosine-3′-phosphoric acid	0.7	2.3	5.92	9.38
Guanosine-5′-phosphoric acid (μ=0.1)	—	2.4	6.1	9.4
Guanosine-5′-triphosphoric acid [μ=0.1; pK_5 7.10(−3); pK_6 9.3(−4)]	—	—	—	3.0(−2)
Guanylurea	1.80	8.20		
Harmine (20°C)	7.61(+1)			
Heptafluorobutanoic acid	0.17			
4,4,5,5,6,6,6-Heptafluorohexanoic acid	4.18			
4,4,5,5,6,6,6-Heptafluoro-2-hexenoic acid	3.23			
Heptanedioic acid (pimelic acid)	4.484	5.424		
2,4-Heptanedione	8.43(keto); 9.15(enol)			
Heptanoic acid	4.893			
Heroin	7.6(+1)			
2,4-Hexadienoic acid (sorbic acid)	4.77			
1,1,1,3,3,3-Hexafluoro-2,2-propanediol	8.801			
1,1,1,3,3,3-Hexafluoro-2-propanol	9.42			
Hexahydroazepine	11.07			
Hexamethyldisilazine	7.55			
1,2,3,8,9,10-Hexamethyl-4,7-phenanthroline (20°C)	7.26			
1,6-Hexanediamine	9.830(+2)	10.930(+1)		

Table 8-1 pK_a values of organic materials in water at 25°C (*continued*)

Substance	pK_1	pK_2	pK_3	pK_4
1,6-Hexanedioic acid	4.418	5.412		
2,4-Hexanedione	8.49 (enol); 9.32 (keto)			
2,2′,4,4′,6,6′-Hexanitrodiphenylamine	5.42 (+1)			
Hexanoic acid (20°C)	4.849			
trans-2-Hexenoic acid	4.74			
trans-3-Hexenoic acid	4.72			
3-Hexen-4-oic acid	4.58			
4-Hexen-5-oic acid	4.74			
Hexylamine	10.64(+1)			
Hexylarsonic acid	4.16	9.19		
Hexylphosphonic acid	2.6	7.9		
DL-Histidine	1.82(+2)	6.00(+1)	9.16(0)	
Histidine amide (μ=0.2)	5.78(+2)	7.64(+1)		
Histidine, methyl ester (μ=0.1)	5.01(+2)	7.23(+1)		
Histidylglycine	2.40(+2)	5.80(+1)	7.82(0)	
Histidylhistidine (μ=0.16)	5.40(+2)	6.80(+1)	7.95(0)	
DI-Homatropine	9.7(+1)			
DI-Homocysteine	2.222(+1)	8.87	10.86	
Homocysteine (μ=0.1)	1.593(+2)	2.523(+1)	8.676(0)	9.413(−1)
Hydantoin	9.12			
Hydrastine	6.23(+1)			
Hydrazine-*N,N*-diacetic acid	<0.1	2.8	3.8	
Hydrazine-*N′,N′*-diacetic acid	2.40	3.12	7.32	
4-Hydrazinocarbonylpyridine (20°C)	1.82	3.52	10.79	
N-Hydroxyacetamide	9.40			
2′-Hydroxyacetophenone	9.90			
3′-Hydroxyacetophenone	9.19			
4′-Hydroxyacetophenone	8.05			
1-Hydroxyacridine (15°C)	5.72			
2-Hydroxyacridine (15°C)	5.62			
3-Hydroxyacridine (15°C)	5.30			
α-Hydroxyasparagine	2.28(+1)	7.20(0)		
β-Hydroxyasparagine	2.09(+1)	8.29(0)		
Hydroxyaspartic acid	1.91(+1)	3.51(0)	9.11(−1)	
2-Hydroxybenzaldehyde (salicylaldehyde)	8.34			
3-Hydroxybenzaldehyde	9.00			
4-Hydroxybenzaldehyde	7.620			
2-Hydroxybenzaldehyde oxime	1.37(+1)	9.18	12.11	
2-Hydroxybenzamide	8.36			
2-Hydroxybenzenemethanol (2-hydroxybenzyl alcohol)	9.92			
3-Hydroxybenzenemethanol	9.83			
4-Hydroxybenzenemethanol	9.82			

Table 8-1 pK$_a$ values of organic materials in water at 25°C (*continued*)

Substance	pK$_1$	pK$_2$	pK$_3$	pK$_4$
4-Hydroxybenzenesulfonic acid	——	9.055(−1)		
2-Hydroxybenzohydroxamic acid	5.19			
2-Hydroxybenzoic acid (salicyclic acid)	2.98	12.38		
3-Hydroxybenzoic acid	4.076	9.85		
4-Hydroxybenzoic acid	4.582	9.23		
4-Hydroxybenzonitrile	7.95			
2-Hydroxy-5-bromobenzoic acid	2.61			
2-Hydroxybutanoic acid (30°C)	3.65			
L-3-Hydroxybutanoic acid (30°C)	4.41			
4-Hydroxybutanoic acid (30°C)	4.71			
2-Hydroxy-5-chlorobenzoic acid	2.63			
trans-2′-Hydroxycinnamic acid	4.614			
trans-3′-Hydroxycinnamic acid	4.40			
10-Hydroxycodeine	7.12			
cis-2-Hydroxycyclohexane-1-carboxylic acid	4.796			
trans-2-Hydroxycyclohexane-1-carboxylic acid	4.682			
cis-3-Hydroxycyclohexane-1-carboxylic acid	4.602			
trans-3-Hydroxycyclohexane-1-carboxylic acid	4.815			
cis-4-Hydroxycyclohexane-1-carboxylic acid	4.836			
trans-4-Hydroxycyclohexane-1-carboxylic acid	4.687			
1-Hydroxy-2,4-dihydroxymethylbenzene	9.79			
N-(Hydroxyethyl)biguanide	2.8(+2)	11.53(+1)		
N-(2-Hydroxyethyl)ethylenediamine	7.21(+2)	10.12(+1)		
N′-(2-Hydroxyethyl)ethylene-diamine-*N*,*N*,*N*′-triacetic acid	2.39	5.37	9.93	
N-(2-Hydroxyethyl)iminodiacetic acid (μ=0.1)	2.2	8.65		
N-(2-Hydroxyethyl)piperazine-*N*′-ethansulfonic acid (20°C)	7.55			
4′-(2-Hydroxyethyl)-1′-piperazine-propanesulfonic acid (20°C)	8.00			
2-Hydroxyethyltrimethylamine	8.94(+1)			
L-β-Hydroxyglutamic acid	2.09	4.18	9.20	
1-Hydroxy-4-hydroxymethylbenzene	9.84			
5-Hydroxy-2-(hydroxymethyl)-4*H*-pyran-4-one	7.90	8.03		
3-Hydroxy-2-hydroxymethylpyridine (20°C, μ=0.2)	5.00(+1)	9.07(OH)		
3-Hydroxy-4-hydroxymethylpyridine (20°C, μ=0.2)	5.00(+1)	8.95(OH)		

Table 8-1 pK_a values of organic materials in water at 25°C (*continued*)

Substance	pK_1	pK_2	pK_3	pK_4
8-Hydroxy-7-iodoquinoline-5-sulfonic acid	2.51(0)	7.417(−1)		
Hydroxylysine (38°C, μ=0.1)	2.13(+2)	8.62(+1)	9.67(0)	
2-Hydroxy-3-methoxybenzaldehyde	7.912			
3-Hydroxy-4-methoxybenzaldehyde (isovanillin)	8.889			
4-Hydroxy-3-methoxybenzaldehyde (vanillin)	7.396			
4-Hydroxy-3-methoxybenzoic acid	4.355			
1-Hydroxy-2-methoxybenzylamine	8.70(+1)	10.52(0)		
2-Hydroxy-1-methoxybenzylamine	8.89(+1)	10.52(0)		
3-Hydroxy-2-methoxybenzylamine	8.94(+1)	10.42(0)		
2-Hydroxymethyl-2-benzeneacetic acid	4.12			
(2-Hydroxy-5-methylbenzene)-methanol	10.15			
2-Hydroxy-3-methylbenzoic acid	2.99			
2-Hydroxy-4-methylbenzoic acid	3.17			
2-Hydroxy-5-methylbenzoic acid	4.08			
2-Hydroxy-6-methylbenzoic acid	3.32			
2-Hydroxy-2-methylbutanoic acid (18°C)	3.991			
3-Hydroxy-2-methylbutanoic acid (18°C)	4.648			
4-Hydroxy-4-methylpentanoic acid (18°C)	4.873			
1-Hydroxymethylphenol	9.95			
Hydroxymethylphosphoric acid	1.91	7.15		
2-Hydroxy-2-methylpropanoic acid (μ=0.1)	3.717			
2-Hydroxy-4-methylpyridine	4.529(+1)			
8-Hydroxy-2-methylquinoline	5.55(+1)	10.31(0)		
8-Hydroxy-4-methylquinoline	5.56(+1)	10.00(0)		
8-Hydroxy-2-methylquinoline-5-sulfonic acid	4.80(0)	9.30(−1)		
8-Hydroxy-4-methylquinoline-7-sulfonic acid	4.78(0)	10.01(−1)		
8-Hydroxy-6-methylquinoline-5-sulfonic acid	4.20(0)	8.7(−1)		
2-Hydroxy-1-naphthoic acid (20°C)	3.29	9.68		
2-Hydroxy-2-nitrobenzoic acid	2.23			
2-Hydroxy-3-nitrobenzoic acid	1.87			
2-Hydroxy-5-nitrobenzoic acid	2.12			
2-Hydroxy-6-nitrobenzoic acid	2.24			
2-Hydroxy-4-nitrophenylphosphonic acid	1.22	5.39		
8-Hydroxy-7-nitroquinoline-5-sulfonic acid	1.94(0)	5.750(−1)		

Table 8-1 pK$_a$ values of organic materials in water at 25°C (*continued*)

Substance	pK$_1$	pK$_2$	pK$_3$	pK$_4$
3-Hydroxy-4-nitrotoluene (μ=0.1)	7.41			
4-Hydroxypentanoic acid (18°C)	4.686			
4-Hydroxy-3-pentenoic acid	4.30			
3-Hydroxyphenazine (15°C)	2.67			
4-Hydroxyphenylarsonic acid	3.89	8.37 (phenol)	10.05	
3-Hydroxyphenylboric acid	8.55	10.84		
2-Hydroxy-2-phenylpropanoic acid	3.532			
2-(2-Hydroxyphenyl)pyridine (20°C)	4.19(+1)	10.64		
trans-4-Hydroxyproline	1.818(+1)	9.662(0)		
Hydroxypropanedioic acid (tartronic acid)	2.37	4.74		
2-Hydroxypropanoic acid	3.858			
1-Hydroxy-2-propylbenzene	10.50			
4-Hydroxypteridine	1.3(+1)	7.89(0)		
2-Hydroxypyridine	1.25(+1)	11.62(0)		
3-Hydroxypyridine	4.80(+1)	8.72(0)		
4-Hydroxypyridine	3.23(+1)	11.09(0)		
2-Hydroxypyridine-*N*-oxide	−0.62(+1)	5.97(0)		
2-Hydroxypyrimidine	2.24(+1)	9.17(0)		
4-Hydroxypyrimidine	1.85(+1)	8.59(0)		
8-Hydroxyquinazoline	3.41(+1)	8.65(0)		
2-Hydroxyquinoline (20°C)	−0.31(+1)	11.74		
3-Hydroxyquinoline (20°C)	4.30(+1)	8.06(0)		
4-Hydroxyquinoline (20°C)	2.27(+1)	11.25(0)		
5-Hydroxyquinoline (20°C)	5.20(+1)	8.54(0)		
6-Hydroxyquinoline (20°C)	5.17(+1)	8.88(0)		
7-Hydroxyquinoline (20°C)	5.48(+1)	8.85(0)		
8-Hydroxyquinoline(20°C)	4.91(+1)	9.81(0)		
8-Hydroxyquinoline-5-sulfonic acid	4.092(+1)	8.776(0)		
DL-Hydroxysuccinic acid (malic acid)	3.458	5.097		
L-Hydroxysuccinic acid	3.40	5.05		
Hydroxytetracycline	3.27(+1)	7.32(0)	9.11(−1)	
5-Hydroxy-1,2,3,4-tetrazole	3.32			
4-Hydroxy-3-(2′-thiazolyazo)toluene	8.36			
2-Hydroxytoluene	10.33			
3-Hydroxytoluene	10.10			
4-Hydroxytoluene	10.276			
4-Hydroxy-α,α,α-trifluorotoluene	8.675			
1-Hydroxy-2,4,6-trihydroxymethylbenzene	9.56			
Hydroxyuracil	8.64			
Hydroxyvaline	2.55(+1)	9.77(0)		
Hyoscyamine	9.68(+1)			
Hypoxanthene	1.79(+1)	8.91(0)	12.07(−1)	
Hypoxanthine	5.3			

Table 8-1 pK_a values of organic materials in water at 25°C (*continued*)

Substance	pK_1	pK_2	pK_3	pK_4
Imidazole	6.993(+1)	10.58(0)		
Imidazolidinetrione (parabanic acid)	6.10			
4-(4-Imidazolyl)butanoic acid (μ=0.1)	4.26(+1)	7.62(0)		
2-(4-Imidazolyl)ethylamine	5.784(+2)	9.756(+1)		
3-(4-Imidazolyl)propanoic acid (μ=0.16)	3.96(+1)	7.57(0)		
3,3′-Iminobispropanoic acid	4.11(0)	9.61(−1)		
3,3′-Iminobispropylamine (30°C)	8.02(+2)	9.70(+1)	10.70(0)	
2,2′-Iminodiacetic acid (diglycine) (30°C, μ=0.1)	2.54(0)	9.12(−1)		
4-Indanol	10.32			
Indole-3-acetic acid	4.75			
Inosine	ca 1.5(+1)	8.96(0)	12.36	
Inosine-5′-phosphoric acid	1.54(0)	6.66(−1)		
Inosine-5′-triphosphoric acid [pK_5 7.68(−4)]	—	—	2.2(−2)	6.92(−3)
Iodoacetic acid	3.175			
2-Iodoaniline	2.54(+1)			
3-Iodoaniline	3.58(+1)			
4-Iodoaniline	3.82(+1)			
2-Iodobenzoic acid	2.86			
3-Iodobenzoic acid	3.86			
4-Iodobenzoic acid	4.00			
5-Iodohistamine	4.06(+2) (imidazole)	9.20(+1) (NH_3^+)	11.88(0) (imino)	
7-Iodo-8-hydroxyquinoline-5-sulfonic acid	2.514	7.417		
Iodomandelic acid	3.264			
Iodomethylphosphoric acid	1.30	6.72		
2-Iodophenol	8.464			
3-Iodophenol	8.879			
4-Iodophenol	9.200			
2-Iodophenoxyacetic acid	3.17			
3-Iodophenoxyacetic acid	3.13			
4-Iodophenoxyacetic acid	3.16			
2-Iodophenylacetic acid	4.038			
3-Iodophenylacetic acid	4.159			
4-Iodophenylacetic acid	4.178			
2-Iodophenylphosphoric acid	1.74	7.06		
2-Iodopropanoic acid	3.11			
3-Iodopropanoic acid	4.08			
2-Iodopyridine	1.82(+1)			
3-Iodopyridine	3.25(+1)			
4-Iodopyridine (20°C)	4.02(+1)			
Isoasparagine	2.97(+1)	8.02(0)		
Isobutylacetic acid (18°C)	4.79			

Table 8-1 pK_a values of organic materials in water at 25°C (*continued*)

Substance	pK_1	pK_2	pK_3	pK_4
Isobutylamine	10.41(+1)			
Isochlorotetracycline	3.1(+1)	6.7(0)	8.3(−1)	
Isocreatine	2.84(+1)			
Isogluatamine	3.81(+1)	7.88(0)		
Isohistamine ($\mu = 0.1$)	6.036(+2)	9.274(+1)		
L-Isoleucine	2.318(+1)	9.758(0)		
Isolysergic acid	3.33(0)	8.46(NH)		
Isopilocarpine (15°C)	7.18(+1)			
2-(Isopropoxy)benzoic acid (20°C)	4.24			
3-(Isopropoxy)benzoic acid (20°C)	4.15			
4-(Isopropoxy)benzoic acid (20°C)	4.68			
Isopropylamine	10.64(+1)			
N-Isopropylaniline	5.50(+1)			
5-Isopropylbarbituric acid	4.907(+1)			
2-Isopropylbenzene acid	3.64			
4-Isopropylbenzoic acid	4.36			
N-Isopropylglycine ($\mu = 0.1$)	2.36(+1)	10.06(0)		
Isopropylmalonic acid	2.94	5.88		
Isopropylmalonic acid mononitrile	2.401			
3-Isopropyl-4-(methylamino)pyridine (20°C)	9.96(+1)			
3-Isopropylpentanedioic acid	4.30	5.51		
4-Isopropylphenylacetic acid	4.391			
Isopropylphosphinic acid	3.56			
Isopropylphosphonic acid	2.66	8.44		
2-Isopropylpyridine	5.83(+1)			
3-Isopropylpyridine (20°C)	5.72(+1)			
4-Isopropylpyridine	6.02(+1)			
DL-Isoproterenol	8.64(+1)			
Isoquinoline	5.40(+1)			
Isoretronecanol	10.83			
L-Isoserine ($\mu = 0.16$)	2.72(+1)	9.25(0)		
Isothiocyanatoacetic acid	6.62			
L-(+)-Lactic acid	3.858			
L-Leucine	2.328(+1)	9.744(0)		
Leucine amide	7.80(+1)			
Leucine, ethyl ester ($\mu = 0.1$)	7.57(+1)			
L-Leucyl-L-asparagine	3.00(+1)	8.12(0)		
L-Leucyl-L-glutamine	2.99(+1)	8.11(0)		
DL-Leucylglycine	3.25(+1)	8.28(0)		
Leucylisoserine (20°C)	3.188(+1)	8.207(0)		
D-Leucyl-L-tyrosine	3.12(+1)	8.38(0)	10.35(−1)	
L-Leucyl-L-tyrosine	3.46(+1)	7.84(0)	10.09(−1)	
Lysergic acid	3.44(+1)	7.68(0)		
L-(+)-Lysine	2.18(+2)	8.95(+1)	10.53(0)	
Lysine, methyl ester ($\mu = 0.1$)	6.965(+1)	10.251(0)		

Table 8-1 pK_a values of organic materials in water at 25°C (*continued*)

Substance	pK_1	pK_2	pK_3	pK_4
L-Lysyl-L-alanine	3.22(+1)	7.62(0)	10.70(−1)	
L-Lysyl-D-alanine	3.00(+1)	7.74(0)	10.63(−1)	
Lysylglutamic acid	2.93(+2)	4.47(+1)	7.75(0)	10.50(+1)
L-Lysyl-L-lysine ($\mu = 0.1$)	3.01(+2)	7.53(+1)	10.05(0)	10.01(−1)
L-Lysyl-D-lysine ($\mu = 0.1$)	2.85(+2)	7.53(+1)	9.92(0)	10.89(−1)
L-Lysyl-L-lysyl-L-lysine ($\mu = 0.1$)	3.08(+2)	7.34(+1)	9.80(0)	10.54(−1)
L-Lysyl-D-lysyl-L-lysine ($\mu = 0.1$)	2.91(+2)	7.29(+1)	9.79(0)	10.54(−1)
L-Lysyl-D-lysyl-lysine ($\mu = 0.1$)	2.94(+2)	7.15(+1)	9.60(0)	10.38(−1)
α-D-Lyxose	12.11			
Maleic acid	1.910	6.33		
Malonamic acid	3.641(0)			
Malonic acid	2.826	5.696		
Malonitrile (cyanoacetic acid)	2.460			
Mandelic acid	3.411			
D-(+)-Mannose	12.08			
Mercaptoacetic acid (thioglycolic acid)	3.60(0)	10.56(SH)		
2-Mercaptobenzoic acid (20°C)	4.05(0)			
2-Mercaptobutanoic acid	3.53(0)			
Mercaptodiacetic acid	3.32	4.29		
2-Mercaptoethanesulfonic acid (20°C)		9.5(−1)		
2-Mercaptoethanol	9.88			
2-Mercaptoethylamine	8.27(+1)	10.53(0)		
2-Mercaptohistidine	1.84(+1)	8.47(0)	11.4(SH)	
Mercapto-*S*-phenylacetic acid ($\mu = 0.1$)	3.9			
2-Mercaptopropane ($\mu = 0.1$)	10.86			
3-Mercapto-1,2-propanediol ($\mu = 0.5$)	9.43			
2-Mercaptopropanoic acid	4.32(0)	10.20(SH)		
3-Mercaptopropanoic acid	——	10.84(SH)		
2-Mercaptopyridine (20°C)	−1.07(+1)	10.00(0)		
3-Mercaptopyridine (20°C)	2.26(+1)	7.03(0)		
4-Mercaptopyridine (20°C)	1.43(+1)	8.86(0)		
2-Mercaptoquinoline (20°C)	−1.44(+1)	10.21(0)		
3-Mercaptoquinoline (20°C)	2.33(+1)	6.13(0)		
4-Mercaptoquinoline (20°C)	0.77(+1)	8.83(0)		
Mercaptosuccinic acid	3.30(0)	4.94(−1)	10.94(SH)	
Mesitylenic acid	4.32			
Mesoxaldialdehyde	3.60			
Methacrylic acid	4.66			
Methanethiol	10.70			
DL-Methionine	2.13(+1)	9.28(0)		
2-(*N*-Methoxyacetamido)pyridine	2.01(+1)			
3-(*N*-Methoxyacetamido)pyridine	3.52(+1)			
4-(*N*-Methoxyacetamido)pyridine	4.62(+1)			
Methoxyacetic acid	3.570			

Table 8-1 pK$_a$ values of organic materials in water at 25°C (*continued*)

Substance	pK$_1$	pK$_2$	pK$_3$	pK$_4$
3-Methoxy-D-α-alanine	2.037(+1)	9.176(0)		
2-Methoxyaniline	4.53(+1)			
3-Methoxyaniline	4.20(+1)			
4-Methoxyaniline	5.36(+1)			
2-Methoxybenzoic acid	4.09			
3-Methoxybenzoic acid	4.08			
4-Methoxybenzoic acid	4.49			
N,N-Methoxybenzylamine	9.68(+1)			
2-Methoxycarbonylaniline	2.23(+1)			
3-Methoxycarbonylaniline	3.64(+1)			
4-Methoxycarbonylaniline	2.38(+1)			
Methoxycarbonylmethylamine	7.66(+1)			
2-Methoxycarbonylpyridine	2.21(+1)			
3-Methoxycarbonylpyridine	3.13(+1)			
4-Methoxycarbonylpyridine	3.26(+1)			
trans-2-Methoxycinnamic acid	4.462			
trans-3-Methoxycinnamic acid	4.376			
trans-4-Methoxycinnamic acid	4.539			
2-Methoxyethylamine	9.45(+1)			
2-Methoxy-4-nitrophenylphosphonic acid	1.53	6.96		
2-Methoxyphenol	9.99			
3-Methoxyphenol	9.652			
4-Methoxyphenol	10.20			
(2′-Methoxy)phenoxyacetic acid	3.231			
(3′-Methoxy)phenoxyacetic acid	3.141			
(4′-Methoxy)phenoxyacetic acid	3.213			
4′-Methoxyphenylacetic acid	4.358			
(4-Methoxyphenyl)phosphinic acid (17°C)	2.35			
(2-Methoxyphenyl)phosphonic acid	2.16	7.77		
(4-Methoxyphenyl)phosphonic acid (17°C)	2.4	7.15		
3-(2′-Methoxyphenyl)propanoic acid	4.804			
3-(3′-Methoxyphenyl)propanoic acid	4.654			
3-(4′-Methoxyphenyl)propanoic acid	4.689			
3-Methoxyphenylselenic acid	4.65			
4-Methoxyphenylselenic acid	5.05			
2-Methoxy-4-(2-propenyl)phenol	10.0			
2-Methoxypyridine	3.06(+1)			
3-Methoxypyridine	4.91(+1)			
4-Methoxypyridine	6.47(+1)			
4-Methoxy-2-(2′-thiazoylazo)phenol	7.83			
2-Methylacrylic acid (18°C)	4.66			
N-Methylalanine	2.22(+1)	10.19(0)		
O-Methylallothreonine ($\mu = 0.1$)	1.92(+1)	8.90(0)		
Methylamine	10.62(+1)			

Table 8-1 pK$_a$ values of organic materials in water at 25°C (*continued*)

Substance	pK$_1$	pK$_2$	pK$_3$	pK$_4$
2-(*N*-Methylamino)benzoic acid	1.93(+1)	5.34(0)		
3-(*N*-Methylamino)benzoic acid	—	5.10(0)		
4-(*N*-Methylamino)benzoic acid	—	5.05		
Methylaminodiacetic acid (20°C)	2.146	10.088		
2-(Methylamino)ethanol	9.88(+1)			
2-(2-Methylaminoethyl)pyridine (30°C)	3.58(+2)	9.65(+1)		
2-(Methylaminomethyl)-6-methylpyridine ($\mu = 0.5$)	3.03(+2)	9.15(+1)		
2-(Methylaminomethyl)pyridine (30°C)	2.92(+2)	8.82(+1)		
4-Methylamino-3-methylpyridine (20°C)	9.83(+1)			
(3-Methylamino)phenylphosphonic acid	1.1(+1)	4.72(+1)	7.30(−1)	
(4-Methylamino)phenylphosphonic acid	—	—	7.85(−1)	
3-(Methylamino)pyridine (30°C)	8.70(+1)			
4-(Methylamino)pyridine (20°C)	9.65(+1)			
4-(Methylamino)-2,3,5,6-tetramethylpyridine (20°C)	10.06(+1)			
N-Methylaniline	4.85(+1)			
Methylarsonic acid (18°C)	3.41	8.18		
1-Methylbarbituric acid	4.35(+1)			
5-Methylbarbituric acid	3.386(+1)			
2-(*N*-Methylbenzamido)pyridine	1.44(+1)			
3-(*N*-Methylbenzamido)pyridine	3.66(+1)			
4-(*N*-Methylbenzamido)pyridine	4.68(+1)			
2-Methylbenzimidazole ($\mu = 0.16$)	6.29(+1)			
2-Methylbenzoic acid (*o*-toluic acid)	3.90			
3-Methylbenzoic acid	4.269			
4-Methylbenzoic acid	4.362			
N-Methyl-1-benzoylecgonine	8.65			
Methylbiguanidine	3.00(+2)	11.44(+1)		
2-Methyl-2-butanethiol	11.35			
2-Methylbutanoic acid	4.761			
3-Methylbutanoic acid (20°C)	4.767			
(*E*)-2-Methyl-2-butendioic acid (mesaconic acid)	3.09	4.75		
3-Methyl-2-butenoic acid	5.12			
(*E*)-2-Methyl-2-butenoic acid (tiglic acid)	4.96			
(*Z*)-2-Methyl-2-butenoic acid (angelic acid)	4.30			
4-Methylcarboxylphenol	8.47			
(*E*)-2-Methylcinnamic acid	4.500			

Table 8-1 pK_a values of organic materials in water at 25°C (*continued*)

Substance	pK_1	pK_2	pK_3	pK_4
(*E*)-3-Methylcinnamic acid	4.442			
(*E*)-4-Methylcinnamic acid	4.564			
1-Methylcyclohexane-1-carboxylic acid	5.13			
cis-2-Methylcyclohexane-1-carboxylic acid	5.03			
trans-2-methylcyclohexane-1-carboxylic acid	5.73			
cis-3-methylcyclohexane-1-carboxylic acid	4.88			
trans-3-Methylcyclohexane-1-carboxylic acid	5.02			
cis-4-Methylcyclohexane-1-carboxylic acid	5.04			
trans-4-Methylcyclohexane-1-carboxylic acid	4.89			
2-Methylcyclohexyl-1,1-diacetic acid	3.53	6.89		
3-Methylcyclohexyl-1,1-diacetic acid	3.49	6.08		
4-Methylcyclohexyl-1,1,1-diacetic acid	3.49	6.10		
3-Methylcyclopentyl-1,1-diacetic acid	3.79	6.74		
S-Methyl-L-cysteine	8.97			
N-Methylcytidine	3.88			
5-Methylcytidine	4.21			
N-Methyl-2′-deoxycytidine	3.97			
5-Methyl-2′-deoxycytidine	4.33			
2-Methyl-3,5-dinitrobenzoic acid	2.97			
5-Methyldipropylenetriamine (30°C)	6.32(+3)	9.19(+2)	10.33(+1)	
2,2′-Methylenebis(4-chlorophenol)	7.6	11.5		
2,2′-Methylenebis(4,6-dichlorophenol)	5.6	10.56		
Methylenebis(thioacetic acid (18°C)	3.310	4.345		
3,3′-(Methylenedithio)dialanine	2.200(+1)	8.16(0)		
Methylenesuccinic acid	3.85	5.45		
N-Methylethylamine	4.23(+1)			
N-Methylethylenediamine	6.86(+1)	10.15(+1)		
α-Methylglucoside	13.71			
3-Methylglutaric acid	4.24	5.41		
N-Methylglycine (sarcosine)	2.12(+1)	10.20(0)		
5-Methyl-2,4-heptanedione	8.52(enol); 9.10(keto)			
5-Methyl-2,4-hexanedione	8.66(enol); 9.31(keto)			
5-Methyl-4-hexenoic acid	4.80			
3-Methylhistamine	5.80(+1)	9.90(0)		
1-Methylhistidine	1.69	6.48	8.85	
2-Methylhistidine (18°C)	1.7	7.2	9.5	

Table 8-1 pK_a values of organic materials in water at 25°C (*continued*)

Substance	pK_1	pK_2	pK_3	pK_4
2-Methyl-8-hydroxyquinoline ($\mu = 0.005$)	4.58(+1)	11.71(0)		
4-Methyl-8-hydroxyquinoline	4.67(+1)	11.62(0)		
1-Methylimidazole	7.06(+1)			
4-Methylimidazole	7.55(+1)			
N-Methyliminodiacetic acid	2.15	10.09		
S-Methylisothiourea	9.83(+1)			
O-Methylisourea	9.72(+1)			
Methylmalonic acid	3.07	5.87		
2-(*N*-Methylmethane-sulfonamido)pyridine	1.73(+1)			
3-(*N*-Methylmethane-sulfonamido)pyridine	3.94(+1)			
4-(*N*-Methylmethane-sulfonamido)pyridine	5.14(+1)			
2-Methyl-6-methyl-aminopyridine (20°C)	3.17(+1)	8.84(0)		
3-Methyl-4-methyl-aminopyridine (20°C)	—	9.84(0)		
4-Methyl-2,2′-(4-methylpyridyl)pyridine	5.32(+1)			
N-Methylmorpholine	7.13(+1)			
2-Methyl-1-naphthoic acid	3.11			
N-Methyl-1-naphthylamine	3.70(+1)			
2-Methyl-4-nitrobenzoic acid	1.86			
2-Methyl-6-nitrobenzoic acid	1.87			
1-Methyl-2-nitroterephthalic acid	3.11			
4-Methyl-2-nitroterephthalic acid	1.82			
3-Methylpentanedioic acid	4.25	5.41		
3-Methylpentane-2,4-dione	10.87			
2-Methylpentanoic acid	4.782			
3-Methylpentanoic acid	4.766			
4-Methylpentanoic acid	4.845			
cis-3-Methyl-2-pentenoic acid	5.15			
trans-3-Methyl-2-pentenoic acid	5.13			
4-Methyl-2-pentenoic acid	4.70			
4-Methyl-3-pentenoic acid	4.60			
6-Methyl-1,10-phenanthroline	5.11(+1)			
(2-Methylphenoxy)acetic acid	3.227			
(3-Methylphenoxy)acetic acid	3.203			
(4-Methylphenoxy)acetic acid	3.215			
(2-Methylphenyl)acetic acid (18°C)	4.35			
(4-Methylphenyl)acetic acid	4.370			
5-Methyl-5-phenylbarbituric acid	8.011(0)			
3-(2-Methylphenyl)propanoic acid	4.66			
3-(3-Methylphenyl)propanoic acid	4.677			
3-(4-Methylphenyl)propanoic acid	4.684			

Table 8-1 pK$_a$ values of organic materials in water at 25°C (*continued*)

Substance	pK$_1$	pK$_2$	pK$_3$	pK$_4$
1-Methyl-2-phenylpyrrolidine	8.80			
5-Methyl-1-phenyl-1,2,3-triazole-4-carboxylic acid	3.73			
Methylphosphinic acid	3.08			
Methylphosphonic acid	2.38	7.74		
3-Methyl-*o*-phthalic acid	3.18			
4-Methyl-*o*-phthalic acid	3.89			
N-Methylpiperazine (μ=0.1)	4.94(+2)	9.09(+1)		
2-Methylpiperazine	5.62(+2)	9.60(+1)		
N-Methylpiperidine	10.19(+1)			
2-Methylpiperidine	10.95(+1)			
3-Methylpiperidine	11.07(+1)			
4-Methylpiperidine (μ=0.5)	11.23(+1)			
2-Methyl-1,2-propanediamine	6.178(+2)	9.420(+1)		
2-Methyl-2-propanethiol	11.2			
2-Methylpropanoic acid	4.853			
2-Methyl-2-propylamine	10.682(+1)			
2-Methyl-2-propylglutaric acid	3.626			
2-Methylpyridine	5.96(+1)			
3-Methylpyridine	5.68(+1)			
4-Methylpyridine	6.00(+1)			
Methyl 4-pyridinecarboxylate	3.26(+1)			
6-Methylpyridine-2-carboxylic acid	5.83			
2-Methylpyridine-1-oxide	1.029(+1)			
3-Methylpyridine-1-oxide	10.921(+1)			
4-Methylpyridine-1-oxide	1.258(+1)			
O-Methylpyridoxal (μ=0.16)	4.74			
Methyl-2-pyridyl ketoxime	9.97			
1-Methyl-2-(3-pyridyl)pyrrolidine	3.41	7.94		
1-Methylpyrrolidine	10.46(+1)			
1-Methyl-3-pyrroline	9.88(+1)			
5-Methylquinoline	4.62(+1)			
Methylsuccinic acid	4.13	5.64		
Methylsulfonylacetic acid	2.36			
3-Methylsulfonylaniline	2.68(+1)			
4-Methylsulfonylaniline	1.48(+1)			
3-Methylsulfonylbenzoic acid	3.52			
4-Methylsulfonylbenzoic acid	3.64			
4-Methylsulfonyl-3,5-dimethylphenol	8.13			
3-Methylsulfonylphenol	9.33			
4-Methylsulfonylphenol	7.83			
1-Methyl-1,2,3,4-tetrahydro-3-pyridinecarboxylic acid (arecaidine; isoguvacine)	9.07			
5-Methyl-1,2,3,4-tetrazole	3.32			
2-Methylthiazole (μ = 0.1)	3.40(+1)			
4-Methylthiazole (μ = 0.1)	3.16(+1)			

Table 8-1 pK_a values of organic materials in water at 25°C (*continued*)

Substance	pK_1	pK_2	pK_3	pK_4
5-Methylthiazole ($\mu = 0.1$)	3.03(+1)			
Methylthioacetic acid	3.72			
4-Methylthioaniline	4.40(+1)			
2-Methylthioethylamine (30°C)	9.18(+1)			
Methylthioglycolic acid	7.68			
3-(*S*-Methylthio)phenol	9.53			
4-(*S*-Methylthio)phenol	9.53			
2-Methylthiopyridine (20°C)	3.59(+1)			
3-Methylthiopyridine (20°C)	4.42(+1)			
4-Methylthiopyridine (20°C)	5.94(+1)			
5-Methylthio-1,2,3,4-tetrazole	4.00(+1)			
O-Methylthreonine	2.02(+1)	9.00(0)		
O-Methyltyrosine	2.21(+1)	9.35(0)		
1-Methylxanthine	7.70	12.0		
3-Methylxanthine	8.10	11.3		
7-Methylxanthine	8.33	ca 13		
9-Methylxanthine	6.25			
Morphine (20°C)	7.87(+1)	9.85(0)		
Morpholine	8.492(+1)			
2-(*N*-Morpholino)ethanesulfonic acid (MES) (20°C)	6.15			
3-(*N*-Morpholino)-2-hydroxy-propanesulfonic acid (37°C)	6.75			
3-(*N*-Morpholino)propanesulfonic acid (20°C)	7.20			
Murexide	0.0	9.20	10.50	
Myosmine	5.26			
1-Naphthalenecarboxylic acid (1-naphthoic acid)	3.695			
2-Naphthalenecarboxylic acid	4.161			
1-Naphthol (20°C)	9.30			
2-Naphthol (20°C)	9.57			
Naphthoquinone monoxime	8.01			
1-Naphthylacetic acid	4.236			
2-Naphthylacetic acid	4.256			
1-Naphthylamine	3.92(+1)			
2-Naphthylamine	4.11(+1)			
1-Naphthylarsonic acid	3.66	8.66		
1-Naphthylsulfonic acid	0.57			
Narceine (15°C)	3.5(+1)	9.3		
Narcotine	6.18(+1)			
Nicotine	3.15(+1)	7.87(0)		
Nicotyrine	4.76(+1)			
Nitrilotriacetic acid (NTA) (20°C)	1.65	2.94	10.33	
Nitroacetic acid	1.68			
2-Nitroaniline	−0.28(+1)			

Table 8-1 pK_a values of organic materials in water at 25°C (*continued*)

Substance	pK_1	pK_2	pK_3	pK_4
3-Nitroaniline	2.46(+1)			
4-Nitroaniline	1.01(+1)			
2-Nitrobenzene-1,4-dicarboxylic acid	1.73			
3-Nitrobenzene-1,2-dicarboxylic acid	1.88			
4-Nitrobenzene-1,2-dicarboxylic acid	2.11			
2-Nitrobenzoic acid	2.18			
3-Nitrobenzoic acid	3.46			
4-Nitrobenzoic acid	3.441			
trans-2-Nitrocinnamic acid	4.15			
trans-3-Nitrocinnamic acid	4.12			
trans-4-Nitrocinnamic acid	4.05			
Nitroethane	8.57			
2-Nitrohydroquinone	7.63	10.06		
N-Nitroiminodiacetic acid	2.21	3.33		
3-Nitromesitol	8.984			
Nitromethane	10.21			
1-Nitro-6,7-phenanthroline ($\mu = 0.2$)	3.23(+1)			
5-Nitro-1,10-phenanthroline	3.232(+1)			
6-Nitro-1,10-phenanthroline	3.23(+1)			
2-Nitrophenol	7.222			
3-Nitrophenol	8.360			
4-Nitrophenol	7.150			
(2-Nitrophenoxy)acetic acid	2.896			
(3-Nitrophenoxy)acetic acid	2.951			
(4-Nitrophenoxy)acetic acid	2.893			
2-Nitrophenylacetic acid	4.00			
3-Nitrophenylacetic acid	3.97			
4-Nitrophenylacetic acid	3.85			
2-Nitrophenylarsonic acid	3.37	8.54		
3-Nitrophenylarsonic acid	3.41	7.80		
4-Nitrophenylarsonic acid	2.90	7.80		
7-(4-Nitrophenylazo)-8-hydroxy-5-quinolinesulfonic acid	3.14(0)	7.495(−1)		
3-Nitrophenylphosphonic acid	1.30	6.27		
4-Nitrophenylphosphonic acid	1.24	6.23		
3-(2′-Nitrophenyl)propanoic acid	4.504			
3-(4′-Nitrophenyl)propanoic acid	4.473			
3-Nitrophenylselenic acid	4.07			
4-Nitrophenylselenic acid	4.00			
1-Nitropropane	8.98			
2-Nitropropane	7.675			
2-Nitropropanoic acid	3.79			
2-Nitropyridine ($\mu = 0.02$)	−2.06(+1)			
3-Nitropyridine ($\mu = 0.02$)	0.79(+1)			
4-Nitropyridine ($\mu = 0.02$)	1.23(+1)			
N-Nitrosoiminodiacetic acid	2.28	3.38		
4-Nitrosophenol	6.48			

Table 8-1 pK_a values of organic materials in water at 25°C (*continued*)

Substance	pK_1	pK_2	pK_3	pK_4
Nitrourea	4.15(+1)			
1,9-Nonanedioic acid (azelaic acid)	4.53	5.40		
Nonanoic acid (pelargonic acid)	4.95			
DL-Norleucine	2.335(+1)	9.834(0)		
Novocaine	8.85(+1)			
2,2,3,3,4,4,5,5-Octafluoropentanoic acid	2.65			
1,8-Octanedioic acid (suberic acid)	4.512	5.404		
Octanoic acid (caprylic acid)	4.895			
Octopine-DD	1.35	2.30	8.68	11.25
Octopine-LD	1.40	2.30	8.72	11.34
Octylamine	10.65(+1)			
L-(+)-Ornithine	1.94(+2)	8.65(+1)	10.76(0)	
Oxalic acid	1.271	4.272		
3,6-Oxaoctanedioic acid ($\mu = 1.0$)	3.055	3.676		
Oxoacetic acid	3.46			
2-Oxabutanedioic acid (oxaloacetic acid)	2.56	4.37		
2-Oxobutanoic acid	2.50			
5-Oxohexanoic acid (5-ketohexanoic acid) (18°C)	4.662			
3-Oxo-1,5-pentanedioic acid	3.10			
4-Oxopentanoic acid (levulinic acid)	4.59			
2-Oxopropanoic acid (pyruvic acid)	2.49			
Oxytetracycline	3.10(+1)	7.26	9.11	
Papaverine	5.90(+1)			
Pentamethylenebis(thioacetic acid) (18°C)	3.485	4.413		
3,3-Pentamethylenepentanedioic acid	3.49	6.96		
1,5-Pentanediamine	10.05(+2)	10.916(+1)		
2,4-Pentanedione	8.24(enol); 8.95(keto)			
1-Pentanoic acid (valeric acid)	4.842			
2-Pentenoic acid	4.70			
3-Pentenoic acid	4.52			
4-Pentenoic acid	4.677			
Pentylarsonic acid	4.14	9.07		
N-Pentylveratramine	7.28(+1)			
Perhydrodiphenic acid (20°C)	4.96	6.68		
Perlolidine (18°C)	4.01	11.39		
Peroxyacetic acid	8.20			
1,7-Phenanthroline	4.30(+1)			
1,10-Phenanthroline	4.857(+1)			
6,7-Phenanthroline	4.857(+1)			
Phenazine	1.2(+1)			

Table 8-1 pK_a values of organic materials in water at 25°C (*continued*)

Substance	pK_1	pK_2	pK_3	pK_4
Phenethylthioacetic acid	3.795			
Phenol	9.99			
Phenol-3-phosphoric acid	1.78	7.03	10.2	
Phenol-4-phosphoric acid	1.99	7.25	9.9	
Phenolphthalein	9.4			
3-Phenolsulfonic acid	——	9.05(−1)		
Phenosulsulfonephthalein	7.9			
Phenoxyactic acid	3.171			
2-Phenoxybenzoic acid	3.53			
3-Phenoxybenzoic acid	3.95			
4-Phenoxybenzoic acid	4.52			
5-Phenoxy-1,2,3,4-tetrazole	3.49(+1)			
Phenylacetic acid	4.312			
L-3-Phenyl-α-alanine	2.16(+1)	9.31(0)		
3-Phenyl-α-alanine, methyl ester	7.05(+1)			
Phenylalanylarginine ($\mu = 0.01$)	2.66(+1)	7.57(0)	12.40(−1)	
Phenylalanylglycine ($\mu = 0.01$)	3.10(+1)	7.71(0)		
7-Phenylazo-8-hydroxy-5-quinolinesulfonic acid	3.41(0)	7.850(−1)		
5-Phenylbarbituric acid	2.544(+1)			
2-Phenyl-2-benzylsuccinic acid	3.69	6.47		
1-Phenylbiguanide	2.13(+2)	10.76(+1)		
4-Phenylbutanoic acid	4.757			
Phenylbutazone	4.5(+1)			
2-Phenylenediamine	<2(+2)	4.47(+1)		
3-Phenylenediamine	2.65(+2)	4.88(+1)		
4-Phenylenediamine	3.29(+2)	6.08(+1)		
2-Phenylethylamine	9.83(+1)			
β-Phenylethylboronic acid	10.0			
DL-α-Phenylglycine	1.83(+1)	4.39(0)		
Phenylguanidine	10.77(+1)			
Phenylhydrazine	5.20(+1)			
2-Phenyl-3-hydroxypropanoic acid	3.53			
3-Phenyl-3-hydroxypropanoic acid	4.40			
Phenyliminodiacetic acid (20°C)	2.40	4.98		
Phenylmalonic acid	2.58	5.03		
Phenylmethanethiol	10.70			
2-Phenyl-2-phenethylsuccinic acid (20°C)	3.74	6.52		
2-Phenylphenol	9.55			
3-Phenylphenol	9.63			
4-Phenylphenol	9.55			
Phenylphosphinic acid (17°C)	2.1			
Phenylphosphonic acid	1.83	7.07		
O-Phenylphosphorylserine	2.13(+1)	8.79		
O-Phenylphosphorylserylglycine	3.18(+1)	6.95(0)		
O-Phenylphosphoryl-L-seryl-L-leucine	3.16(+1)	7.12(0)		

Table 8-1 pK_a values of organic materials in water at 25°C (*continued*)

Substance	pK_1	pK_2	pK_3	pK_4
N-Phenylpiperazine ($\mu = 0.1$)	8.71(+1)			
2-Phenylpropanoic acid	4.38			
3-Phenylpropanoic acid (35°C)	4.664			
3-Phenyl-1-propylamine	10.39(+1)			
Phenylpropynoic acid (35°C)	2.269			
Phenylselenic acid	4.79			
Phenylselenoacetic acid ($\mu = 0.1$)	3.75			
β-Phenylserine ($\mu = 0.16$)	8.79(0)			
Phenylsuccinic acid (20°C)	3.78	5.55		
Phenylsulfenylacetic acid	2.66			
Phenylsulfonylacetic acid	2.44			
5-Phenyl-1,2,3,4-tetrazole	4.38(+1)			
1-Phenyl-1,2,3-triazole-4-carboxylic acid	2.88			
1-Phenyl-1,2,3-triazole-4,5-dicarboxylic acid	2.13	4.93		
Phosphoramidic acid	3.08	8.63		
O-Phosphorylethanolamine	5.838(+1)	10.638(0)		
O-Phosphorylserylglycine	3.13	5.41	8.01	
O-Phosphoryl-L-seryl-L-leucine	3.11	5.47	8.26	
Phosphoserine	2.08	5.65	9.74	
Phthalamide	3.79(0)			
Phthalazine	3.47(+1)			
o-Phthalic acid	2.950	5.408		
Phthalimide	9.90(0)			
Physostigmine	1.76(+1)	7.88(0)		
Picric acid (2,4,6-trinitrophenol) (18°C)	0.419			
Pilocarpine	1.3(+1)	6.85(0)		
Piperazine	5.333(+2)	9.781(+1)		
1,4-Piperazinebis(ethanesulfonic acid) (20°C)	6.80			
Piperazine-2-carboxylic acid	1.5	5.41	9.53	
Piperdine	11.123(+1)			
2-Piperidinecarboxylic acid	2.12(+1)	10.75(0)		
3-Piperidinecarboxylic acid	3.35(+1)	10.64(0)		
4-Piperidinecarboxylic acid	3.73(+1)	10.72(0)		
1-(2-Piperidinyl)-2-propanone (15°C)	9.45			
Piperine (15°C)	1.98(+1)			
Proline	1.952(+1)	10.640(0)		
1,2-Propanediamine	6.607(+2)	9.720(+1)		
1,3-Propanediamine	8.49(+2)	10.47(+1)		
1-Propanethiol	10.86			
1,2,3-Propanetriamine	3.72(+3)	7.95(+2)	9.59(+1)	
1,2,3-Propanetricarboxylic acid	3.67	4.87	6.38	
Propanoic acid	4.874			
Propenoic acid	4.247			
N-Propionyglycine	3.718(0)			

Table 8-1 pK$_a$ values of organic materials in water at 25°C (*continued*)

Substance	pK$_1$	pK$_2$	pK$_3$	pK$_4$
2-Propoxybenzoic acid (20°C)	4.24			
3-Propoxybenzoic acid (20°C)	4.20			
4-Propoxybenzoic acid (20°C)	4.78			
N-Propylalanine	2.21(+1)	10.19(0)		
Propylamine	10.568(+1)			
Propylarsonic acid (18°C)	4.21	9.09		
Propylenimine	8.18(+1)			
N-Propylglycine ($\mu = 0.1$)	2.38(+1)	10.03(0)		
L-Propylglycine	3.19(+1)	8.97(0)		
Propylmalonic acid	2.97	5.84		
Propylphosphinic acid	3.46			
Propylphosphonic acid	2.49	8.18		
2-Propylpyridine	6.30(+1)			
N-Propylveratramine	7.20(+1)			
2-Propynoic acid	1.887			
Pseudoecgonine	9.70			
Pseudoisocyanine ($\mu = 0.2$)	4.59(+2)			
Pseudotropine	9.86(+1)			
Pteroylglutamic acid	8.26			
Purine	2.52(+1)	8.92(0)		
Pyrazine	0.6(+1)			
Pyrazinecarboxamide	0.5(+1)			
Pyrazole	2.61(+1)			
Pyridazine	2.33(+1)			
Pyridine	5.17(+1)			
Pyridine-d_5	5.83(+1)			
2-Pyridinealdoxime	3.56(+1)	10.17(0)		
3-Pyridinealdoxime	4.07(+1)	10.39(0)		
4-Pyridinealdoxime	4.73(+1)	10.03(0)		
2-Pyridinecarbaldehyde	3.84(+1)			
3-Pyridinecarbaldehyde	3.80(+1)			
4-Pyridinecarbaldehyde	4.74(+1)			
3-Pyridinecarbamide (nicotinamide)	3.33(+1)			
3-Pyridinecarbonitrile	1.35(+1)			
Pyridine-2-carboxylic acid (picolinic acid)	1.01(+1)	5.29(0)		
Pyridine-3-carboxylic acid (nicotinic acid)	2.07(+1)	4.75(0)		
Pyridine-4-carboxylic acid (isonicotinic acid)	1.84(+1)	4.86(0)		
Pyridine-2,3-dicarboxylic acid	2.36(+1)	7.08(0)		
Pyridine-2,4-dicarboxylic acid	2.23(+1)	7.02(0)		
Pyridine-2,6-dicarboxylic acid	2.16(+1)	6.92(0)		
Pyridine-1-oxide	0.688(+1)			
Pyridoxal	4.20(+1)	8.66(ring OH)		
Pyridoxal-5-phosphate ($\mu = 0.15$)	<2.5	4.14	6.20	8.69

Table 8-1 pK_a values of organic materials in water at 25°C (*continued*)

Substance	pK_1	pK_2	pK_3	pK_4
Pyridoxamine ($\mu = 0.1$)	3.37(+2)	8.01(+1)	10.13(ring OH)	
Pyridoxamine-5-phosphate ($\mu = 0.15$; pK_5 10.92)	2.5	3.69	5.76	8.61
Pyridoxine (vitamin B_6) (18°C)	5.00(+1)	8.96(ring OH)		
3-(2′-Pyridyl)alanine	1.37(+2)	4.02(+1)	9.22(0)	
3-(3′-Pyridyl)alanine	1.77(+2)	4.64(+1)	9.10(0)	
2-(2′-Pyridyl)benzimidazole ($\mu = 0.16$)	5.58(+1)			
2-(2′-Pyridyl)imidazole ($\mu = 0.005$)	8.98(+1)			
4-(2′-Pyridyl)imidazole ($\mu = 0.1$)	5.49(+1)			
Pyrimidine	1.30(+1)			
2,4(1*H*,3*H*)-Pyrimidinedione (uracil)	0.6(+1)	9.46(0)		
2,4,5,6(1*H*,3*H*)-Pyrimidinetetrone-5-oxime	4.57(0)			
Pyrocatecholsulfonephthaleine	7.82	9.76	11.73	
Pyroxilidine	11.11(+1)			
Pyrrole-1-carboxylic acid	4.45			
Pyrrole-2-carboxylic acid	4.45			
Pyrrole-3-carboxylic acid	4.453			
Pyrrolidine	11.305(+1)			
Pyrrolidine-2-carboxylic acid (proline)	1.952(+1)	10.640(0)		
2-[2-(*N*-Pyrrolidinyl)ethyl]pyridine	3.60(+2)	9.39(+1)		
3-[2-(*N*-Pyrrolidinyl)ethyl]pyridine	4.28(+2)	9.28(+1)		
4-[2(*N*-Pyrrolidinyl)ethyl]pyridine	4.65(+2)	9.27(+1)		
2-(1-Pyrrolidinylmethyl)pyridine	2.54(+1)	8.56(+1)		
3-(1-Pyrrolidinylmethyl)pyridine	3.14(+2)	8.36(+1)		
4-(1-Pyrrolidinylmethyl)pyridine	3.38(+2)	8.16(+1)		
3-Pyrroline	−0.27(+1)			
Quinidine	4.0(+1)	8.54(0)		
Quinine	4.11(+1)	8.52(0)		
Quinoline	4.80(+1)			
Quinoxaline	0.72(+1)			
D-Raffinose	12.74			
Riboflavin (vitamin B_2) ($\mu = 0.01$)	ca −0.2	9.69		
α-D-Ribofuranose	12.11			
D-Ribose-5′-phosphonic acid	—	6.70(−1)	13.05(−2)	
D-Saccharic acid	5.00(0)			
Saccharin (*o*-benzoic sulfimide)	2.32			
Sarcosine	2.12(+1)	10.20(0)		

Table 8-1 pK_a values of organic materials in water at 25°C (*continued*)

Substance	pK_1	pK_2	pK_3	pK_4
Sarcosine amide	8.35(+1)			
Sarcosine dimethylamide	8.86(+1)			
Sarcosine methylamide	8.28(+1)			
Sarcosylglycine ($\mu = 0.16$)	3.15(+1)	8.56(0)		
Sarcosylleucine	3.15(+1)	8.67(0)		
Sarcosylsarcosine	2.92(+1)	9.15(0)		
Sarcosylserine	3.17(+1)	8.63(0)		
3-Selenosemicarbazide ($\mu = 0.1$)	0.8(+1)			
Semicarbazide ($\mu = 0.1$)	3.53(+1)			
L-Serine	2.186(+1)	9.208(0)		
Serine, methyl ester ($\mu = 0.1$)	7.03(+1)			
Serylglycine ($\mu = 0.15$)	2.10(+1)	7.33(0)		
L-Seryl-L-leucine	3.08(+1)	7.45(0)		
Solanine	7.34(+1)			
D-Sorbitol (17.5°C)	13.60			
L-(−)-Sorbose (18°C)	11.55			
Sparteine	4.49(+1)	11.76(0)		
Spinaceamine ($\mu = 0.1$)	4.895(+2)	8.90(+1)		
Spinacine	1.649(+2)	4.936(+1)	8.663(0)	
L-Strychnine (15°C)	2.50	8.20		
Succinamic acid (succinic acid monoamide)	4.39(0)			
Succinic acid	4.207	5.635		
DL-Succinimide	9.623			
β-(4′-Sulfaminophenyl)alanine	1.99(+1)	8.64(0)	10.26(−1)	
3-Sulfamylbenzoic acid	3.54			
4-Sulfamylbenzoic acid	3.47			
4-Sulfamylphenylphosphoric acid	1.42	6.38	10.0	
Sulfanilamide	10.43(+1)			
Sulfoacetic acid	——	4.0		
3-Sulfobenzoic acid	——	3.78		
4-Sulfobenzoic acid	——	3.72		
3-Sulfophenol	0.39	9.07		
4-Sulfophenol	0.58	8.70		
2-Sulfopropanoic acid	1.99			
5-Sulfosalicyclic acid	2.49	12.00		
Sylvic acid	7.62			
D-Tartaric acid	3.036	4.366		
meso-Tartaric acid	3.22	4.81		
Tetracycline ($\mu = 0.005$)	3.30(+1)	7.68	9.69	
Tetradehydroyohimbine	10.59(+1)			
Tetraethylenepentamine [$\mu = 0.1$; pK_5 9.67(+1)]	2.98(+5)	4.72(+4)	8.08(+3)	9.10(+2)
1,4,5,6-Tetrahydro-1,2-dimethylpyridine	11.38(+1)			
1,4,5,6-Tetrahydro-2-methylpyridine	9.53(+1)			

Table 8-1 pK_a values of organic materials in water at 25°C (*continued*)

Substance	pK_1	pK_2	pK_3	pK_4
cis-Tetrahydronaphthalene-2,3-dicarboxylic acid (20°C)	3.98	6.47		
trans-Tetrahydronaphthalene-2,3-dicarboxylic acid (20°C)	4.00	5.70		
5,6,7,8-Tetrahydro-1-naphthol	10.28			
5,6,7,8-Tetrahydro-2-naphthol	10.48			
Tetrahydroserpentine	10.55(+1)			
2,3,5,6-Tetramethylbenzoic acid	3.415			
Tetramethylenebis(thioacetic acid) (18°C)	3.463	4.423		
Tetramethylenediamine	9.22(+2)	10.75(+1)		
N,N,N′,N′-Tetramethylethylenediamine	2.20(+2)	6.35(+1)		
2,3,5,6-Tetramethyl-4-methylaminopyridine	0.07(+1)			
2,2,6,6-Tetramethylpiperidine ($\mu = 0.5$)	1.24(+1)			
2,3,5,6-Tetramethylpyridine (20°C)	7.90(+1)			
Tetramethylsuccinic acid	3.50	7.28		
1,2,3,4-Tetrazole	4.90			
Thebaine	7.95(+1)			
2-Thenoyltrifluoroacetone	5.70(0)			
Theobromine	0.68(+1)	7.89		
Theophylline	<1(+1)	8.80		
Thiazoline	2.53(+1)			
Thioacetic acid	3.33			
o-Thiocresol	6.64			
m-Thiocresol	6.58			
p-Thiocresol	6.52			
Thiocyanatoacetic acid	2.58			
2,2′-Thiodiacetic acid	3.32	4.29		
4,4′-Thiodibutanoic acid (18°C)	4.351	5.275		
3,3′-Thiodipropanoic acid (18°C)	4.085	5.075		
3-Thio-*S*-methylcarbazide ($\mu = 0.1$)	7.563(+1)			
1-Thionylcarboxylic acid	3.53			
2-Thionylcarboxylic acid	4.10			
2-Thiophenecarboxylic acid (30°C)	3.529			
3-Thiophenecarboxylic acid (3-thenoic acid)	4.10			
Thiophenol	6.50			
3-Thiosemicarbazide ($\mu = 0.1$)	1.5(+1)			
3-Thiosemicarbazide-1,1-diacetic acid (30°C)	2.94	4.07		
Thiourea	2.03(+1)			
Thorin	3.7	8.3	11.8	
Thymidine	9.79	12.85		
p-Toluenesulfinic acid	1.7			

Table 8-1 pK$_a$ values of organic materials in water at 25°C (*continued*)

Substance	pK$_1$	pK$_2$	pK$_3$	pK$_4$
Toluhydroquinone	10.03	11.62		
o-Toluidine	4.45(+1)			
m-Toluidine	4.71(+1)			
p-Toluidine	5.08(+1)			
o-Tolylacetic acid (18°C)	4.36			
p-Tolyacetic acid (18°C)	4.36			
o-Tolylarsonic acid	3.82	8.85		
m-Tolylarsonic acid	3.82	8.60		
p-Tolylarsonic acid	3.70	8.68		
o-Tolylphosphonic acid	2.10	7.68		
m-Tolyphosphonic acid	1.88	7.44		
p-Tolyphosphonic acid	1.84	7.33		
3-Tolylselenic acid	4.80			
4-Tolylselenic acid	4.88			
Triacetylmethane	5.81			
Triallylamine	8.31(+1)			
1,3,5-Triazine-2,4,6-triol	7.20	11.10		
1*H*-1,2,3-Triazole	——	9.26		
1*H*-1,2,4-Triazole	2.386(+1)	9.972		
1,2,3-Triazole-4-carboxylic acid	3.22	8.73		
1,2,3-Triazole-4,5-dicarboxylic acid	1.86	5.90	9.30	
1,2,4-Triazolidine-3,5-dione (urazole)	5.80			
Tribromoacetic acid	−0.147			
2,4,6-Tribromobenzoic acid	1.41			
Trichloroacetic acid	0.52			
Trichloroacrylic acid	1.15			
3,3,3-Trichlorolactic acid	2.34			
Trichloromethylphosphonic acid	1.63	4.81		
2,4,5-Trichlorophenol	7.37			
3,4,5-Trichlorophenol	7.839			
Tricine (20°C)	8.15			
Triethanolamine	7.76(+1)			
Triethylamine	10.72(+1)			
Triethylenediamine	4.18(+2)	8.19(+1)		
Triethylenetetramine (20°C)	3.32(+4)	6.67(+3)	9.20(+2)	9.92(+1)
Triethylsuccinic acid	2.74			
Trifluoroacetic acid	0.50			
Trifluoroacrylic acid	1.79			
4,4,4-Trifluoro-2-aminobutanoic acid	1.600(+1)	8.169(0)		
4,4,4-Trifluoro-3-aminobutanoic acid	2.756(+1)	5.822(0)		
4,4,4-Trifluorobutanoic acid	4.16			
α,α,α-Trifluoro-*m*-cresol	8.950			
4,4,4-Trifluorocrotonic acid	3.15			
5,5,5-Trifluoroleucine	2.045(+1)	8.942(0)		
3-(Trifluoromethyl)aniline	3.5(+1)			
4-(Trifluoromethyl)aniline	2.6(+1)			
3-Trifluoromethylphenol	8.950			

Table 8-1 pK_a values of organic materials in water at 25°C (*continued*)

Substance	pK$_1$	pK$_2$	pK$_3$	pK$_4$
5-Trifluoromethyl-1,2,3,4-tetrazole	1.70			
6,6,6-Trifluoronorleucine	2.164(+1)	9.463(0)		
5,5,5-Trifluoronorvaline	2.042(+1)	8.916(0)		
5,5,5-Trifluoropentanoic acid	4.50			
3,3,3-Trifluoropropanoic acid	3.06			
4,4,4-Trifluorothreonine	1.554(+1)	7.822(0)		
4,4,4-Trifluorovaline	1.537(+1)	8.098(0)		
1,2,3-Trihydroxybenzene (pyrogallol)	9.03(0)	11.63(−1)		
1,3,5-Trihydroxybenzene (phloroglucinol)	8.45(0)	8.88(−1)		
2,4,6-Trihydroxybenzoic acid	1.68(0)			
3,4,5-Trihydroxybenzoic acid	4.19(0)	8.85(−1)		
3,4,5-Trihydroxycyclohex-1-ene-1-carboxylic acid [D-(−)-shikimic acid]	4.15			
2,4,6-Tri(hydroxymethyl)phenol	9.56			
Triisobutylamine	10.42(+1)			
Trimethylamine	9.80(+1)			
3-(Trimethylamino)phenol	8.06			
4-(Trimethylamino)phenol	8.35			
2,4,6-Trimethylaniline	4.38(+1)			
2,4,6-Trimethylbenzoic acid	3.448			
Trimethylenebis(thioacetic acid) (18°C)	3.435	5.383		
2,3,4-Trimethylphenol	10.59			
2,4,5-Trimethylphenol	10.57			
2,4,6-Trimethylphenol	10.88			
3,4,5-Trimethylphenol	10.25			
2,3,6-Trimethylpyridine ($\mu = 0.5$)	7.60(+1)			
2,4,6-Trimethylpyridine	7.43(+1)			
2,4,6-Trimethylpyridine-1-oxide	1.990(+1)			
3-(Trimethylsilyl)benzoic acid	4.089			
4-(Trimethylsilyl)benzoic acid	4.192			
2,4,5-Trimethylthiazole ($\mu = 0.1$)	4.55			
2,4,6-Trinitroaniline (picramide)	−10.23(+1)			
2,4,6-Trinitrobenzene acid	0.654			
2,2,2-Trinitroethanol	2.36			
Trinitromethane (20°C)	0.17			
Triphenylacetic acid	3.96			
Tripropylamine	10.66(+1)			
Tris(2-hydroxyethyl)amine	7.762(+1)			
Tris(hydroxymethyl)aminomethane (TRIS)	8.08(+1)			
2-[Tris(hydroxymethyl)methyl amino]-1-ethanesulfonic acid (TES)	7.50			

Table 8-1 pK_a values of organic materials in water at 25°C (*continued*)

Substance	pK_1	pK_2	pK_3	pK_4
3-[Tris(hydroxymethyl)methyl amino]-1-propanesulfonic acid (TAPS) (20°C)	8.4			
N-[Tris(hydroxymethyl)methyl]-glycine (tricine)	2.023(+1)	8.135		
Tris(trimethylsilyl)amine	4.70(+1)			
Trithiocarbonic acid (20°C)	2.64			
Tropacocaine (15°C)	9.88(+1)			
3-Tropanol (tropine)	10.33(+1)			
Trypsin ($\mu = 0.1$)	6.25			
L-Tryptophan	2.38(+1)	9.39(0)		
DL-Tyrosine	2.18(+1)	9.21(0)	10.47(OH)	
Tyrosine amide	7.48	9.89		
Tyrosine, ethyl ester	7.33	9.80		
Tyrosylarginine ($\mu = 0.01$)	2.65(+1)	7.39(0)	9.36(−1)	11.62(−2)
Tyrosyltyrosine	3.52(+1)	7.68(0)	9.80(−1)	10.26(−2)
α-Ureidobutanoic acid	3.886(0)			
γ-Ureidobutanoic acid	4.683(0)			
β-Ureidopropanoic acid	4.487(0)			
Uric acid	5.40	5.53		
Uridine	9.30			
Uridine-5′-diphosphoric acid	7.16			
Uridine-5′-phosphoric acid (5′-uridylic acid)	6.63			
Uridine-5′-triphosphoric acid	7.58			
DL-Valine	2.286(+1)	9.719(0)		
L-Valine	2.296(+1)	9.79(0)		
Valine amide ($\mu = 0.2$)	8.00			
L-Valine, methyl ester	7.49(+1)			
L-Valylglycine	3.23(+1)	8.00(0)		
Vetramine	7.49(+1)			
Veratrine	8.85(+1)			
Vinylmethylamine	9.69(+1)			
2-Vinylpyridine	4.98(+1)			
4-Vinylpyridine	5.62(+1)			
Vitamin B_{12}	7.64(+1)			
Xanthine (40°C)	0.68(+1)			
Xanthosine	<2.5(+1)	5.67(0)	12.00(−1)	
Xylenol Orange [pK_5 10.46(−4); pK_6 12.28(−5)]	—	2.58(−1)	3.23(−2)	6.37(−3)
D-(+)-Xylose	12.15(0)			
Zincon	—	4	7.85	15

TABLE 8-2 Proton-transfer reactions of inorganic materials in water at 25°C

Protonated cations are designated by (+1), (+2), etc. after the pK_a values.

Substance	Formula	pK_1	pK_2	pK_3	pK_4
Aluminic acid (alumina)	H_3AlO_3	11.2			
Amidophosphoric acid	$H_2NPO(OH)_2$	3.3	8.28		
Aminodisulfonic acid	$HN(SO_3H)_2$			8.50	
Ammonium ion	NH_4^+	9.24			
Arsenic acid	H_3AsO_4	2.25	6.77	11.53	
Arsenous acid	$HAsO_2$ or $HAs(OH)_4$	9.23			
Boric acid, ortho-	H_3BO_3	9.236	12.74		
Boric acid, etra-	$H_2B_4O_7$	4	9		
Carbonic acid	CO_2+H_2O	6.35	10.53		
	(without including dehydration constant)	3.76	10.329		
	CO_2+D_2O (solvent)	6.77	11.076		
Chloric acid	$HClO_3$	−1.58			
Chlorous acid	$HClO_2$	2.021			
Chlorosulfonic acid	$HOSO_2Cl$	−10.43			
Chromic acid	H_2CrO_4	−0.98	6.50		
Cyanic acid	HOCN	3.47			
Deuterium oxide	D_2O (solvent)	14.87			
Diamidophosphoric acid	$(H_2N)_2PO_2H$	4.83			
Dithionic acid	$H_2S_2O_6$	−3.4	−0.2		
Dithionous acid	$H_2S_2O_4$	0.35	2.45		
Ferricyanic acid	$H_3Fe(CN)_6$	<1			
Ferrocyanic acid	$H_2(Fe(CN)_6)^{2-}$			2.57	4.35
Fluorophosphoric acid	$FPO(OH)_2$		4.79		
Hexapolyphosphoric acid	$H_8P_6O_{19}$	ca 2.1	2.19	5.98	8.13
Hydrazinium(2+) ion (20°C)	$^+H_3NNH_3^+$	−0.88(+2)	7.956(+1)		
Hydrazinosulfuric acid	H_2NNHSO_3H	3.85			
Hydrazoic acid	HN_3	4.64			
Hydrocyanic acid	HCN	9.21			
Hydrogen bromide	HBr	−20.68			
Hydrogen chloride	HCl	−6.1			
Hydrogen fluoride	HF	3.17			
Hydrogen iodide	HI	−9.5			
Hydrogen peroxide	H_2O_2	11.58			
Hydrogen polysulfide (20°C)	H_2S_4	3.8	6.3		
Hydrogen selenide	H_2Se	3.89	11.0		
Hydrogen sulfide	H_2S	6.96	12.90		
Hydrogen telluride (20°C)	H_2Te	2.64	11–12		
Hydroperoxy radical	$HO_2^{\cdot} = H^+ + O_2^-$	4.45			
Hydroxide radical	$OH^{\cdot}$	11.9			
Hydroxylamine-*N*,*N*-di sulfonic acid	$HON(SO_3H)_2$			11.85	
Hydroxylamine-*N*-sulfonic acid	$HONH{-}OSO_2H$		ca 12.5		

TABLE 8-2 Proton-transfer reactions of inorganic materials in water at 25°C (*continued*)

Substance	Formula	pK_1	pK_2	pK_3	pK_4
Hydroxylammonium ion	$HONH_3^+$	5.98			
Hypobromous acid	HBrO	8.597			
Hypochlorous acid	HClO	7.54			
Hypoiodous acid	HIO	10.64			
Hyponitrous acid	HON=NOH	7.05	11.54		
Hypophosphoric acid (20°C)	$H_4P_2O_6$	2	2.19		
Hypophosphorus acid	HPH_2O_2	1.23			
Hyposulfurous acid	$H_2S_2O_4$	0.35	2.45		
Imidodiphosphoric acid	$(HO)_2PO{-}NH{-}PO(OH)_2$	ca 2	2.85	7.08	9.72
Iodic acid (30°C)	HIO_3	0.815			
Nitramide	O_2NNH_2	6.48			
Nitric acid	HNO_3	1.38			
Nitrous acid	HNO_2	3.14			
Osmic acid	H_2OsO_5 (mainly OsO_4)	12.0	14.5		
Perchloric acid	$HClO_4$ (completely dissociated up to 10 *M*)				
Periodic acid, para-	H_5IO_6	1.55	8.27		
Permanganic acid	$HMnO_4$	−2.25			
Peroxide radical	$HO_2^{\cdot}$	4.90			
Peroxoboric acid	$H_3BO_3+H_2O_2=(H_2BO_3{\cdot}H_2O_2)^-+H^+$	7.91			
Peroxochromic âcid	H_2CrO_5	4.30			
Peroxomonosulfuric acid	H_2SO_5	1.0	9.3		
Perxenic acid	H_4XeO_6	ca 2	ca 6	ca 10	
Phosphoric acid, ortho-	H_3PO_4	2.148	7.198	12.38	
Deuterated	D_3PO_3	2.420	7.201		
Phosphoric acid, di-	$H_4P_2O_7$	0.91	2.10	6.70	9.38
Phosphorous acid (20°C)	H_2PHO_3	1.20	6.70		
Selenic acid	H_2SeO_4	−3	1.74		
Selenous acid	H_2SeO_3	2.27	7.78		
Silicic acid	H_2SiO_3	9.77	11.80		
Sulfamic acid	$HOSO_2NH_2$	0.988			
Sulfuric acid	H_2SO_4	ca −3	1.987		
Sulfurous acid	SO_2+H_2O (includes dehydration constant)	1.89	7.20		
Telluric acid	H_6TeO_6	7.70	10.99		
Tellurous acid	H_2TeO_3	2.46	7.7		
Tetraperoxochromic acid (30°C)	H_3CrO_8	7.16			
Tetrapolyphosphoric acid (pK_5 6.63; pK_6 8.34)	$H_6P_4O_{13}$			1.3	2.23
Thiocyanic acid	HSCN	0.95			
Thiosulfuric acid	$H_2S_2O_3$	0.60	1.5–1.7		
Trimetaphosphoric acid	$H_3P_3O_4$			2.0	

TABLE 8-2 Proton-transfer reactions of inorganic materials in water at 25°C (*continued*)

Substance	Formula	pK_1	pK_2	pK_3	pK_4
Tripolyphosphoric acid ($\mu > 1$)* (pK_5 9.26)	$H_5P_3O_{10}$	−0.51	1.20	2.30	6.61
Trithiocarbonic acid (20°C)	H_2CS_3	2.68	8.18		
Tungstic acid (20°C)	H_2WO_4	ca 3.5	ca 4.6		
Vanadic acid	H_3VO_4	3.78	7.8	13.0	
Water	H_2O	14.003			
Xenon trioxide	XeO_3 (aqueous) = $HXeO_4^- + H^+$	10.8			

* Ionic strength.

TABLE 8-3 Selected equilibrium constants in aqueous solution at various temperatures

Abbreviations Used in the Table

(+1), protonated cation	(−1), singly ionized anion	pK_{auto}, negative logarithm (base 10) of autoprotolysis constant
(0), neutral molecule	(−2), doubly ionized anion	pK_{sp}, negative logarithm (base 10) of solubility product

Substance	Temperature, °C									
	0	5	10	15	20	25	30	35	40	50
Acetic acid (0)	4.780	4.770	4.762	4.758	4.757	4.756	4.757	4.762	4.769	4.787
DL-*N*-Acetylalanine (+1)		3.699	3.699	3.703	3.708	3.715	3.725	3.733	3.745	3.774
β-Acetylaminopropionic (+1)		4.479	4.465	4.465	4.449	4.445	4.444	4.443	4.445	4.457
N-Acetylglycine (+1)		3.682	3.676	3.673	3.667	3.670	3.673	3.678	3.685	3.706
α-Alanine										
(+1)	2.42		2.39		2.35	2.34	2.33	2.33	2.33	2.33
(0)	10.59		10.29		10.01	9.87	9.74	9.62	9.49	9.26
2-Aminobenzenesulfonic acid (0), pK_2	2.633	2.591	2.556	2.521	2.448	2.459	2.431	2.404	2.380	2.338
3-Aminobenzenesulfonic acid (0), pK_2	4.075	4.002	3.932	3.865	3.799	3.738	3.679	3.622	3.567	3.464
4-Aminobenzenesulfonic acid (0), pK_2	3.521	3.457	3.398	3.338	3.283	3.227	3.176	3.126	3.079	2.989
3-Aminobenzoic acid (0)					4.90	4.79	4.75		4.68	4.60
4-Aminobenzoic acid (0)					4.95	4.85	4.90		4.95	5.10
2-Aminobutyric acid										
(+1)			2.334			2.286		$2.289^{37.5°C}$		2.297
(0)			10.530			9.380		$9.518^{37.5°C}$		9.234
4-Aminobutyric acid										
(+1)			4.057	4.046	4.038	4.031	4.027	4.025	4.027	4.032
(0)			11.026	10.867	10.706	10.556	10.409	10.269	10.114	9.874

2-Aminoethylsulfonic acid (0)			9.452	9.316	9.186	9.061	8.940	8.824	8.712	9.499
2-Amino-3-methylpentanoic acid										
(+1)	2.365$^{1°C}$		2.338$^{12.5°C}$			2.320		2.317$^{37.5°C}$		2.332
(0)	10.460$^{1°C}$		10.100$^{12.5°C}$			9.758		9.439$^{37.5°C}$		9.157
2-Amino-2-methyl-1,3-propanediol	9.612	9.433	9.266	9.104	8.951	8.801	8.659	8.519	8.385	8.132
2-Amino-2-methylpropionic acid										
(+1)	2.419$^{1°C}$		2.380$^{12.5°C}$			2.357		2.351$^{37.5°C}$		2.356
(0)	10.960$^{1°C}$		10.580$^{12.5°C}$			10.205		9.872$^{37.5°C}$		9.561
2-Aminopentanoic acid										
(+1)	2.376$^{1°C}$		2.347			2.318			2.309	2.313
(0)	10.508$^{1°C}$			10.154$^{12.5°C}$		9.808		9.490$^{37.5°C}$		9.198
3-Aminopropionic acid										
(+1)	3.656	3.627		3.583		3.551		3.524	3.517	
(0)	11.000	10.830		10.526		10.235		9.963	9.842	
4-Aminopyridine (+1)	9.873	9.704	9.549	9.398	9.252	9.114	8.978	8.846	8.717	8.477
Ammonium ion (+1)	10.081	9.904	9.731	9.564	9.400	9.425	9.093	8.947	8.805	8.539
Arginine										
(+1)	1.914	1.885	1.870	1.849	1.837	1.823	1.814	1.801	1.800	1.787
(0)	9.718	9.563	9.407	9.270	9.123	8.994	8.859	8.739	8.614	8.385
Barbituric acid										
(+1)				3.969	3.980	4.02	4.00	4.008	4.017	4.032
(0)				8.493	8.435	8.372	8.302	8.227	8.147	7.974
Benzoic acid (0)		4.231	4.220	4.215	4.206	4.204	4.203	4.207	4.219	4.223
Boric acid (0)	9.508	9.439	9.380	9.327	9.280	9.236	9.197	9.161	9.132	9.080
Bromoacetic acid (0)				2.875	2.887	2.902	2.918	2.936		
3-Bromobenzoic acid (0)				3.818	3.813	3.810	3.808	3.810	3.813	
4-Bromobenzoic acid (0)				4.011	4.005	3.99	4.001	4.001	4.003	
Bromopropynoic acid (0)			1.786	1.814	1.839	1.855	1.879	1.900	1.919	
3-*tert*-Butylbenzoic acid (0)				4.266	4.231	4.199	4.170	4.143	4.119	
4-*tert*-Butybenzoic acid (0)				4.463	4.425	4.389	4.354	4.320	4.287	
2-Butynoic acid (0)			2.618	2.626	2.611	2.620	2.618	2.621	2.631	

TABLE 8-3 Selected equilibrium constants in aqueous solutions at various temperatures (*continued*)

Substance	Temperature, °C									
	0	5	10	15	20	25	30	35	40	50
Butyric acid (0)	4.806	4.804	4.803	4.805	4.810	4.817	4.827	4.840	4.854	4.885
DL-*N*-Carbamoylalanine (+1)		3.898	3.894	3.891	3.890	3.892	3.896	3.902	3.908	3.931
N-Carbamoylglycine (+1)		3.911	3.900	3.889	3.879	3.876	3.874	3.873	3.875	3.888
Carbon dioxide + water										
(0)	6.583	6.517	6.465	6.429	6.382	6.365	6.327	6.31	6.296	6.297
(−1)	10.627	10.558	10.499	10.431	10.377	10.33	10.290	10.25	10.220	10.172
Chloroacetic acid (0)				2.845	2.856	2.867	2.883	2.900		
3-Chlorobenzoic acid (0)				3.838	3.831	3.83	3.825	3.826	3.829	
4-Chlorobenzoic acid (0)				4.000	3.991	3.986	3.981	3.980	3.981	
Chloropropynoic acid (0)			1.766	1.796	1.820	1.845	1.864	1.879	1.893	
Citric acid										
(0)	3.220	3.200	3.176	3.160	3.142	3.128	3.116	3.109	3.099	3.095
(−1)	4.837	4.813	4.797	4.782	4.769	4.761	4.755	4.751	4.750	4.757
(−2)	6.393	6.386	6.383	6.384	6.388	6.396	6.406	6.423	6.439	6.484
Cyanoacetic acid (0)		2.445	2.447	2.452	2.460	2.460	2.482	2.496	2.511	
2-Cyano-2-methylpropionic acid										
(0)		2.342	2.360	2.379	2.400	2.422	2.446	2.471	2.498	
5,5-Diethylbarbituric acid (0)	8.40	8.30	8.22	8.169	8.094	8.020	7.948	7.877	7.808	7.673
Diethylmalonic acid										
(0)			2.129	2.136	2.144	2.151	2.160	2.172	2.187	
(−1)			7.400	7.401	7.408	7.417	7.428	7.441	7.457	
2,3-Dimethylbenzoic acid (0)				3.663	3.687	3.771	3.726	3.762	3.788	
2,4-Dimethylbenzoic acid (0)				4.154	4.187	4.217	4.244	4.268	4.290	
2,5-Dimethylbenzoic acid (0)				3.911	3.954	3.990	4.020	4.045	4.065	
2,6-Dimethylbenzoic acid (0)				3.234	3.304	3.362	3.409	3.445	3.472	

3,5-Dimethylbenzoic acid (0)				4.292	4.299	4.302	4.304	4.306	4.306	
N,N′-Dimethylethyleneamine-*N,N′*-diacetic acid										
(0)	6.294		6.169		6.047		5.926		5.803	
(−1)	10.446		10.268		10.068		9.882		9.684	
N,N-Dimethylglycine (0)		10.34		10.14		9.94		9.76		
3,5-Dinitrobenzoic acid (0)			2.60		2.73		2.85		2.96	3.07
2-Ethylbutyric acid (0)	4.623		4.664		4.710	4.751	4.758		4.812	4.869
5-Ethyl-5-phenylbarbituric acid (0)				7.592	7.517	7.445	7.377	7.311	7.248	7.130
Fluoroactic acid (0)				2.555	2.571	2.586	2.604	2.624		
Formic acid (0)	3.786	3.772	3.762	3.757	3.753	3.751	3.752	3.758	3.766	3.782
2-Furancarboxylic acid (0)						3.164	3.200	3.216	3.239	
Glucose-1-phosphate (0)		6.506	6.500	6.499	6.500	6.504	6.510	6.519	6.531	6.561
Glycerol-1-phosphoric acid (−1)		6.642	6.641	6.643	6.648	6.656	6.666	6.679	6.695	6.733
Glycerol-2-phosphoric acid										
(0)		1.223	1.245	1.271	1.301	1.335	1.372	1.413	1.457	1.554
(−1)		6.657	6.650	6.646	6.646	6.650	6.657	6.666	6.679	6.712
Glycine										
(+1)			2.397	2.380	2.36	2.351	2.34	2.33	2.327	2.32
(0)		10.34	10.193	10.044	9.91	9.780	9.65	9.53	9.412	9.19
Glycolic acid (0)	3.875		$3.844^{12.5°C}$			3.831		$3.833^{37.5°C}$		3.849
Glycylasparagine (+1)		2.968	2.958	2.952	2.943	2.942	2.942	2.944	2.947	2.959
N-Glycylglycine (+1)	3.201					3.126				3.159
			$8.594^{12.5°C}$			8.252		$7.948^{37.5°C}$		7.668
Hexanoic acid (0)	4.840		4.839		4.849		4.865		4.890	4.920
Hydrogen cyanide (0)			9.63	9.49	9.36	9.21	9.11	8.99	8.88	
Hydrogen peroxide (0)	12.23			11.86	11.75	11.65	11.55	11.45		11.21
Hydrogen sulfide										
(0)		7 33	7.24	7.13	7.05	6.97	6.90	6.82	6.79	6.69
(−1)		13.5		13.2		12.90	12.75	12.6		
4-Hydroxybenzoic acid (0)				4.596	4.586	4.582	4.577	4.576	4.578	

TABLE 8-3 Selected equilibrium constants in aqueous solutions at various temperatures (*continued*)

Substance	Temperature, °C									
	0	5	10	15	20	25	30	35	40	50
Hydroxylamine (0)				6.186	6.063	5.948		5.730		
2-Hydroxy-1-naphthoic acid										
(0)					3.29		3.24		3.19	3.26
(−1)					9.68		9.65		9.61	9.58
4-Hydroxyproline										
(+1)	$1.900^{1°C}$		$1.850^{12.5°C}$			1.818		$1.798^{37.5°C}$		1.796
(0)	$10.274^{1°C}$		$9.958^{12.5°C}$			9.662		$9.394^{37.5°C}$		9.138
2-Hydroxypropionic acid (0)	3.880	3.873	3.868	3.861	3.857	3.858	3.861	3.867	3.873	3.895
DL-2-Hydroxysuccinic acid										
(0)	3.537	3.520	3.494	3.482	3.472	3.458	3.452	3.446	3.444	3.445
(−1)	5.119	5.108	5.098	5.096	5.096	5.097	5.099	5.104	5.117	5.149
Hypobromous acid (0)				8.83		8.60		8.47	$8.37^{45°C}$	
Hypochlorous acid (0)	7.82	7.75	7.69	7.63	7.58	7.54	7.50	7.46		7.05
Imidazole (+1)	7.581	7.467	7.334	7.216	7.103	6.993	6.887	6.784	6.685	6.497
Iodoacetic acid (0)				3.143	3.158	3.175	3.193	3.213		
DL-Isoleucine										
(+1)	2.365		$2.338^{12.5°C}$			2.318		$2.317^{37.5°C}$		2.332
(0)	10.460		$10.100^{12.5°C}$			9.758		$9.439^{37.5°C}$		9.157
Isopropylmalonic acid, mononitrile (0)		2.299	2.320	2.343	2.365	2.401	2.427	2.452	2.481	
Lactic acid (0)	3.880	3.873	3.868	3.862	3.857	3.858	3.861	3.867	3.873	3.895
Lead sulfate, pK_{sp}	8.01			7.87		7.80		7.73		7.63
DL-Leucine										
(+1)	$2.383^{1°C}$		$2.348^{12.5°C}$			2.328		$2.327^{37.5°C}$		2.333
(0)	$10.458^{1°C}$		$10.095^{1.5°C}$			9.744		$9.434^{37.5°C}$		9.142

Malonic acid (−1)	5.670	5.665	5.667	5.673	5.683	5.696	5.710	5.730	5.753	5.803
Mannose (0)			12.45			12.08			11.81	
Mercury(I) chloride, pK_{sp}			18.65	18.48	18.27	17.88		16.79		
Methanol (solvent), pK_{auto}		17.12		16.84		16.71	16.65	16.53		
Methylamine (+1)	11.496		11.130		10.787	10.62	10.466		10.161	9.876
Methylaminodiacetic acid										
(0)	2.138		2.142		2.146		2.150		2.154	
(−1)	10.474		10.287		10.088		9.920		9.763	
3-Methylbenzoic acid (0)				4.303	4.285	4.269	4.256	4.244	4.235	
4-Methylbenzoic acid (0)				4.390	4.376	4.362	4.349	4.336	4.322	
3-Methylbutyric acid (0)	4.726		4.742		4.767		4.794		4.831	4.871
4-Methylpentanoic acid (0)	4.827		4.827		4.837		4.853		4.879	4.908
5-Methyl-5-phenylbarbituric acid										
(0)				8.104	8.057	8.011	7.966	7.922	7.879	7.797
2-Methylpropionic acid (0)	4.825		4.827		4.840	4.853	4.886		4.918	4.955
2-Methyl-2-propylamine (+1)		11.439	11.240	11.048	10.862	10.682	10.511	10.341		
Nitric acid (0)	−1.65					−1.38				−1.20
Nitrilotriacetic acid										
(0)	1.69		1.65		1.65		1.66		1.67	
(−1)	2.95		2.95		2.94		2.96		2.98	
(−2)	10.59		10.45		10.33		10.23			
4-Nitrobenzoic acid (0)				3.448	3.444	3.441	3.441	3.442	3.445	
Nitrous acid (0)				3.244	3.177	3.138		3.100		
DL-Norleucine										
(+1)	2.394		$2.356^{12.5°C}$			2.335		$2.324^{37.5°C}$		2.328
(0)	10.564		$10.190^{12.5°C}$			9.834		$9.513^{37.5°C}$		9.224
Oxalic acid (−1)	4.210	4.216	4.227	4.240	4.254	4.272	4.295	4.318	4.349	4.409
2,4-Pentanedione (0)	9.07					8.95			8.90	
Pentanoic acid (0)	4.823		4.763		4.835	4.842	4.851		4.861	4.906
Phenylalanine (0)			9.75			9.31			8.96	
Phosphoric acid (0)	2.056	2.073	2.088	2.107	2.127	2.148	2.171	2.196	2.224	2.277
(−1)	7.313	7.282	7.254	7.231	7.213	7.198	7.189	7.185	7.181	7.183

TABLE 8-3 Selected equilibrium constants in aqueous solutions at various temperatures (*continued*)

Substance	Temperature, °C									
	0	5	10	15	20	25	30	35	40	50
o-Phthalic acid										
(0)	2.925	2.927	2.931	2.937	2.943	2.950	2.958	2.967	2.978	3.001
(−1)	5.432	5.418	5.410	5.405	5.405	5.408	5.416	5.427	5.442	5.485
Piperidine (+1)	11.963	11.786	11.613	11.443	11.280	11.123	10.974	10.818	10.670	10.384
Proline										
(+1)	2.011		$1.964^{12.5°C}$			1.952		$1.950^{37.5°C}$		1.958
(0)	11.296		$10.972^{12.5°C}$			10.640		$10.342^{37.5°C}$		10.064
Propenoic acid (0)				4.267	4.250	4.247	4.249	4.267	4.301	
N-Propionylglycine (+1)		3.728	3.723	3.718	3.716	3.718	3.721	3.725	3.731	3.750
Propynoic acid (0)			1.791	1.829	1.867	1.887	1.940	1.932	1.963	
Pyrrolidine (+1)	12.17	11.98	11.81	11.63	11.43	11.30	11.15	10.99	10.84	11.56
Serine										
(+1)	$2.296^{1°C}$		$2.232^{12.5°C}$			2.186		$2.154^{37.5°C}$		2.132
(0)	$9.880^{1°C}$		$9.542^{12.5°C}$			9.208		$8.904^{37.5°C}$		8.628
Silver bromide, pK_{sp}		13.33		12.83	12.57	12.30	12.07	11.83	11.61	11.19
Silver chloride, pK_{sp}		10.595		10.152		9.749		9.381	9.21	8.88
Succinic acid										
(0)	4.285	4.263	4.245	4.232	4.218	4.207	4.198	4.191	4.188	4.186
(−1)	5.674	5.660	5.649	5.642	5.639	5.635	6.541	5.647	5.654	5.680
Sulfuric acid (−1)	1.778	$1.812^{4.3°C}$		1.894		1.987	2.05	2.095	2.17	2.246
Sulfurous acid (0)	1.63		1.74			1.89		1.98		2.12
D-Tartaric acid										
(0)	3.118	3.095	3.075	3.057	3.044	3.036	3.025	3.019	3.018	3.021
(−1)	4.426	4.407	4.391	4.381	4.372	4.366	4.365	4.367	4.372	4.391
2,3,5,6-Tetramethylbenzoic (0)				3.310	3.367	3.415	3.453	3.483	3.505	

Threonine										
(+1)	$2.200^{1°C}$		$2.132^{12.5°C}$			2.088		$2.070^{37.5°C}$		2.055
(0)	$9.748^{1°C}$		$9.420^{12.5°C}$			9.100		$8.812^{37.5°C}$		8.548
o-Toluidine (0)				4.58	4.495	4.45	4.345	4.28	4.20	
1,2,4-Triazole										
(+1)				2.451	2.418	2.386	2.327			
(0)				10.205	10.083	9.972	9.768			
3,4,5-Trihydroxybenzoic acid (0)					4.19		4.30		4.38	4.53
Tris(2-hydroxyethyl)amine (+1)	8.290	8.173	8.067	7.963	7.861	7.762	7.666	7.570	7.477	7.299
2,4,6-Trimethylbenzoic (0)				3.325	3.391	3.448	3.498	3.541	3.577	
3-Trimethylsilylbenzene acid (0)				4.142	4.116	4.089	4.060	4.029	3.996	
4-Trimethylsilylbenzoic acid (0)				4.270	4.230	4.192	4.155	4.119	4.084	
β-Ureidopropionic acid (0)		4.514	4.505	4.497	4.490	4.487	4.486	4.486	4.488	4.500
DL-Valine										
(+1)	2.320		$2.297^{12.5°C}$			2.296		$2.292^{37.5°C}$		2.310
(0)	10.413		$10.064^{12.5°C}$			9.719		$9.405^{37.5°C}$		9.124

TABLE 8-4 Indicators for aqueous acid-base titrations

This table lists some selected indicators. The pH range or transition interval given in the third column may vary appreciably from one observer to another, and, in addition, it is affected by ionic strength, temperature, and illumination; consequently only approximate values can be given. They should be considered to refer to solutions having low ionic strengths and a temperature of about 25°C. In the fourth column the $pK_a(-\log K_a)$ of the indicator as determined spectrophotometrically is listed. In the fifth column the wavelength of maximum absorption is given first for the acidic and then for the basic form of the indicator, and the same order is followed in giving the colors in the sixth column. The abbreviations used to describe the colors of the two forms of the indicator are as follows:

B, blue	O, orange	P, purple
V, violet	C, colorless	R, red
Y, yellow	G, green	O-Br, orange-brown

Indicator	Chemical name	pH range	pK_a	λ_{max}, *nm*	Color change
Cresol red (acid range)	*o*-Cresolsulfone-phthalein	0.2–1.8			R–Y
Cresol purple (acid range)	*m*-Cresolsulfonephthalein	1.2–2.8	1.51	533, ——	R–Y
Thymol blue (acid range)	Thymolsulfonephthalein	1.2–2.8	1.65	544, 430	R–Y
Tropeolin OO	Diphenylamino-*p*-benzene sodium sulfonate	1.3–3.2	2.0	527, ——	R–Y
2,6-Dinitrophenol	2,6-Dinitrophenol	2.4–4.0	3.69		C–Y
2,4-Dinitrophenol	2,4-Dinitrophenol	2.5–4.3	3.90		C–Y
Methyl yellow	Dimethylaminoazobenzene	2.9–4.0	3.3	508, ——	R–Y
Methyl orange	Dimethylaminoazobenzene sodium sulfonate	3.1–4.4	3.40	522, 464	R–O
Bromophenol blue	Tetrabromophenolsulfone-phthalein	3.0–4.6	3.85	436, 592	Y–BV
Bromocresol green	Tetrabromo-*m*-cresol-sulfonepthalein	4.0–5.6	4.68	444, 617	Y–B
Methyl red	*o*-Carboxybenzeneazo-dimethylaniline	4.4–6.2	4.95	530, 427	R–Y
Chlorophenol red	Dichlorophenolsulfone-phthalein	5.4–6.8	6.0	——, 573	Y–R
Bromocresol purple	Dibromo-*o*-cresolsulfone-phthalein	5.2–6.8	6.3	433, 591	Y–P
Bromophenol red	Dibromophenolsulfone-phthalein	5.2–6.8		——, 574	Y–R
p-Nitrophenol	*p*-Nitrophenol	5.3–7.6	7.15	320, 405	C–Y
Bromothymol blue	Dibromothymolsulfone-phthalein	6.2–7.6	7.1	433, 617	Y–B
Neutral red	Aminodimethylaminotolu-phenazonium chloride	6.8–8.0	7.4		R–Y
Phenol red	Phenolsulfonephthalein	6.4–8.0	7.9	433, 558	Y–R
m-Nitrophenol	*m*-Nitrophenol	6.4–8.8	8.3	——, 570	C–Y
Cresol red	*o*-Cresolsulfonephthalein	7.2–8.8	8.2	434, 572	Y–R
m-Cresol purple	*m*-Cresolsulfonephthalein	7.6–9.2	8.32	——, 580	Y–P
Thymol blue	Thymolsulfonephthalein	8.0–9.6	8.9	430, 596	Y–B
Phenolphthalein	Phenolphthalein	8.0–10.0	9.4	——, 553	C–R

TABLE 8-4 Indicators for aqueous acid-base titrations (*continued*)

Indicator	Chemical name	pH range	pK_a	λ_{max}, *nm*	Color change
α-Naphtholbenzein	α-Naphtholbenzein	9.0–11.0			Y–B
Thymolphthalein	Thymolphthalein	9.4–10.6	10.0	—, 598	C–B
Alizarin Yellow R	5-(*p*-Nitrophenylazo)-salicyclic acid, Na salt	10.0–12.0	11.16		Y–V
Tropeolin O	*p*-Sulfobenzenazo-resorcinol	11.0–13.0			Y-O-Br
Nitramine	2,4,6-Trinitrophenyl-methylnitroamine	10.8–13.0			C–O-Br

TABLE 8-5 National Bureau of Standards (U.S.) reference pH buffer solutions

Source: R. G. Bates, *J. Res. Natl. Bur. Stand.* (*U.S.*), **66A**:179 (1962) and B. R. Staples and R. G. Bates, ibid, **73A**:37 (1969).

Temperature °C	Secondary standard 0.05 *M* K tetraoxalate	KH tartrate (saturated at 25°C)	0.05 *M* KH_2 citrate	0.05 *M* KH phthalate	0.025 *M* KH_2PO_4, 0.025 *M* Na_2HPO_4	0.0087 *M* KH_2PO_4, 0.0302 *M* Na_2HPO_4	0.01 *M* $Na_2B_4O_7$	0.025 *M* $NaHCO_3$, 0.025 *M* Na_2CO_3	Secondary standard $Ca(OH)_2$ (saturated at 25°C)
0	1.666		3.860	4.003	6.984	7.534	9.464	10.317	13.423
5	1.668		3.840	3.999	6.951	7.500	9.395	10.245	13.207
10	1.670		3.820	3.998	6.923	7.472	9.332	10.179	13.003
15	1.672		3.802	3.999	6.900	7.448	9.276	10.118	12.810
20	1.675		3.788	4.002	6.881	7.429	9.225	10.062	12.627
25	1.679	3.557	3.776	4.008	6.865	7.413	9.180	10.012	12.454
30	1.683	3.552	3.766	4.015	6.853	7.400	9.139	9.966	12.289
35	1.688	3.549	3.759	4.024	6.844	7.389	9.102	9.925	12.133
38	1.691	3.548		4.030	6.840	7.384	9.081		12.043
40	1.694	3.547	3.753	4.035	6.838	7.380	9.068	9.889	11.984
45	1.700	3.547		4.047	6.834	7.373	9.038		11.841
50	1.707	3.549	3.749	4.060	6.833	7.367	9.011	9.828	11.705
55	1.715	3.554		4.075	6.834		8.985		11.574
60	1.723	3.560		4.091	6.836		8.962		11.449
70	1.743	3.580		4.126	6.845		8.921		
80	1.766	3.609		4.164	6.859		8.885		
90	1.792	3.650		4.205	6.877		8.850		
95	1.806	3.674		4.227	6.886		8.833		
Dilution value $\Delta pH_{1/2}$	+0.186	+0.049	0.024	+0.052	+0.080	+0.070	+0.01	0.079	−0.28

TABLE 8-6 Compositions of National Bureau of Standards (U.S.) standard pH buffer solutions

Air weight of material per liter of buffer solution

Standard	Weight, g
$KH_3(C_2O_4)_2 \cdot 2H_2O$, 0.05 *M*	12.61
Potassium hydrogen tartrate, about 0.034 *M*	Saturated at 25°C
Potassium hydrogen phthalate, 0.05 *M*	10.12
Phosphate (solution 1)	
KH_2PO_4, 0.025 *M*	3.39
Na_2HPO_4, 0.025 *M*	3.53
Phosphate (solution 2)	
KH_2PO_4, 0.008665 *M*	1.179
Na_2HPO_4, 0.03032 *M*	4.30
$Na_2B_4O_7 \cdot 10H_2O$, 0.01 *M*	3.80
Carbonate	
$NaHCO_3$, 0.025 *M*	2.10
Na_2CO_3, 0.025 *M*	2.65
$Ca(OH)_2$, about 0.0203 *M*	Saturated at 25°C

Standard Reference pH Buffer Solutions

The buffer value for the National Bureau of Standards (U.S.) reference pH buffer solutions is given below:

Buffer solution	KH tartrate	0.05 *M* KH_2 citrate	0.05 *M* KH phthalate	0.025 *M* KH_2PO_4, 0.025 *M* Na_2HPO_4	0.0087 *M* KH_2PO_4, 0.0302 *M* Na_2HPO_4	0.01 *M* $Na_2B_4O_7$	0.025 *M* $NaHCO_3$, 0.025 *M* Na_2CO_3
Buffer value β	0.027	0.034	0.016	0.029	0.016	0.020	0.029

For the secondary pH reference standards, the buffer value is 0.070 for potassium tetroxalate and 0.09 for calcium hydroxide.

To prepare the standard pH buffer solutions recommended by the National Bureau of Standards (U.S.), the indicated weights of the pure materials in Table 8-6 should be dissolved in water of specific conductivity not greater than 5 micromhos. The tartrate, phthalate, and phosphates can be dried for 2 h at 110°C before use. Potassium tetroxalate and calcium hydroxide need not be dried. Fresh-looking crystals of borax should be used. Before use, excess solid potassium hydrogen tartrate and calcium hydroxide must be removed. Buffer solutions pH 6 or above should be stored in plastic containers and should be protected from carbon dioxide with soda-lime traps. The solutions should be replaced within 2 to 3 weeks, or sooner if formation of mold is noticed. A crystal of thymol may be added as a preservative.

Buffer Solutions Other Than Standards

The range of the buffering effect of a single weak acid group is approximately one pH unit on either side of the pK_a. The ranges of some useful buffer systems are collected in Table 8-7. After all the components have been brought together, the pH of the resulting solution should be determined at the temperature to be employed with reference to standard reference solutions. Buffer components should be compatible with other components in the system under study; this is particularly significant for buffers employed in biological studies. Check tables of formation constants to ascertain whether metal-binding character exists.

When there are two or more acid groups per molecule, or a mixture is composed of several overlapping acids, the useful range is larger. Universal buffer solutions consist of a mixture of acid groups which overlap such that successive pK_a values differ by 2 pH units or less. The Prideaux-Ward mixture comprises phosphate, phenyl acetate, and borate plus HCl and covers the range from 2 to 12 pH units. The McIlvaine buffer is a mixture of citric acid and Na_2HPO_4 that covers the range from pH 2.2 to 8.0. The Britton-Robinson system consists of acetic acid, phosphoric acid, and boric acid plus NaOH and covers the range from pH 4.0 to 11.5. A mixture composed of Na_2CO_3, NaH_2PO_4, citric acid, and 2-amino-2-methyl-1,3-propanediol covers the range from pH 2.2 to 11.0.

TABLE 8-7 pH values of buffer solutions for control purposes

Materials*	pH range
Glycine and HCl	1.0–3.7
Citrate and HCl	1.3–4.7
p-Toluenesulfonate and *p*-toluenesulfonic acid	1.1–3.3
Formate and HCl	2.8–4.6
Succinic acid and borax	3.0–5.8
Phenyl acetate and HCl	3.5–5.0
Acetate and acetic acid	3.7–5.6
Succinate and succinic acid	4.8–6.3
2-(*N*-Morpholino)ethanesulfonic acid and NaOH	5.2–7.1
2,2-Bis(hydroxymethyl)-2,2′,2″-nitrilotriethanol and HCl	5.8–7.2
KH_2PO_4 and borax	5.8–9.2
N-Tris(hydroxymethyl)methyl-2-aminoethanesulfonic acid and NaOH	6.8–8.2
KH_2PO_4 and Na_2HPO_4	6.1–7.5
N-2-Hydroxyethylpiperazine-*N*′-2-ethanesulfonic acid and NaOH	6.9–8.3
Triethanolamine and HCl	6.9–8.5
Diethylbarbiturate (veronal) and HCl	7.0–8.5
Tris(hydroxymethyl)aminomethane and HCl	7.2–9.0
N-Tris(hydroxymethyl)methylglycine and HCl	
N,*N*-Bis(2-hydroxyethyl)glycine and HCl	
Borax and HCl	7.6–8.9
Glycine and NaOH	8.2–10.1
Ammonia (aqueous) and NH_4Cl	8.3–9.2
Ethanolamine and HCl	8.6–10.4
Borax and NaOH	9.4–11.1
Carbonate and hydrogen carbonate	9.2–11.0
Na_2HPO_4 and NaOH	11.0–12.0

General directions for the preparation of buffer solutions of varying pH but fixed ionic strength are given by Bates.* Preparation of McIlvaine buffered solutions at ionic strengths of 0.5 and 1.0 and Britton-Robinson solutions of constant ionic strength have been described by Elving et al.† and Frugoni,‡ respectively.

* Bates, *Determination of pH, Theory and Practice*, Wiley, New York, 1964, pp. 121–122.
† Elving, Markowitz, and Rosenthal, *Anal. Chem.*, **28**:1179 (1956).
‡ Frugoni, *Gazz. Chim. Ital.*, **87**:403 (1957).

REFERENCE ELECTRODES

TABLE 8-8 Potentials of reference electrodes (in volts) as a function of temperature

Liquid-junction potential included

Temp., °C	0.1 *M* KCl, calomel*	1.0 *M* KCl, calomel*	3.5 *M* KCl, calomel*	Saturated KCl, calomel*	1.0 *M* KCl, Ag/AgCl†	1.0 *M* KBr, Ag/AgBr‡	1.0 *M* KI, Ag/AgI§
0	0.3367	0.2883		0.25918	0.23655	0.08128	−0.14637
5					0.23413	0.07961	−0.14719
10	0.3362	0.2868	0.2556	0.25387	0.23142	0.07773	−0.14822
15	0.3361			0.2511	0.22857	0.07572	−0.14942
20	0.3358	0.2844	0.2520	0.24775	0.22557	0.07349	−0.15081
25	0.3356	0.2830	0.2501	0.24453	0.22234	0.07106	−0.15244
30	0.3354	0.2815	0.2481	0.24118	0.21904	0.06856	−0.15405
35	0.3351			0.2376	0.21565	0.06585	−0.15590
38	0.3350		0.2448	0.2355			
40	0.3345	0.2782	0.2439	0.23449	0.21208	0.06310	−0.15788
45					0.20835	0.06012	−0.15998
50	0.3315	0.2745		0.22737	0.20449	0.05704	−0.16219
55					0.20056		
60	0.3248	0.2702		0.2235	0.19649		
70					0.18782		
80				0.2083	0.1787		
90					0.1695	0.0251	

* Bates et al., *J. Res. Natl. Bur. Stand.*, **45**:418 (1950).
† Bates and Bower, *J. Res. Natl. Bur. Stand.*, **53**:283 (1954).
‡ Hetzer, Robinson, and Bates, *J. Phys. Chem.*, **66**:1423 (1962).
§ Hetzer, Robinson, and Bates, *J. Phys. Chem.*, **68**:1929 (1964).

Temp., °C	125	150	175	200	225	250	275
1.0 *M* KCl, Ag/AgCl*	0.1330	0.1032	0.0708	0.0348	−0.0051	−0.054	−0.090
1.0 *M* KBr, Ag/AgBr†	−0.0048	−0.0312	−0.0612	−0.0951			

* Greeley et al., *J. Phys. Chem.*, **64**:652 (1960).
† Towns et al., *J. Phys. Chem.*, **64**:1861 (1960).

The values of several additional reference electrodes at 25°C are listed:

Reference electrode	Potential, V
Ag/AgCl, saturated KCl	0.198
Ag/AgCl, 0.1 *M* KCl	0.288
Hg/HgO, 1.0 *M* NaOH	0.140
Hg/HgO, 0.1 *M* NaOH	0.165
Hg/Hg_2SO_4, saturated K_2SO_4 (22°C)	0.658
Hg/$HgSO_4$, saturated KCl	0.655

TABLE 8-9 Potentials of reference electrodes (in volts) at 25°C for water–organic solvent mixtures

Electrolyte solution of 1 M HCl

Solvent, wt %	Methanol, Ag/AgCl	Ethanol, Ag/AgCl	2-Propanol, Ag/AgCl	Acetone, Ag/AgCl	Dioxane, Ag/AgCl	Ethylene glycol, Ag/AgCl	Methanol, calomel	Dioxane, calomel
5			0.2180	0.2190		0.2190		
10	0.2153	0.2146	0.2138	0.2156		0.2160		
20	0.2090	0.2075	0.2063	0.2079	0.2031	0.2101	0.255	0.2501
30		0.2003				0.2036		
40	0.1968	0.1945		0.1859		0.1972	0.243	
45					0.1635			0.2104
50		0.1859		0.158				
60	0.1818	0.173				0.1807		
70		0.158			0.0659		0.216	0.1126
80	0.1492	0.136						
82					−0.0614			−0.0014
90	0.1135	0.196		−0.034				
94.2	0.0841							
98		0.0215						
99							0.103	
100	−0.0099	−0.0081		−0.53				

ELECTRODE POTENTIALS

TABLE 8-10 Potentials of selected half-reactions at 25°C

This table is a summary of oxidation-reduction half-reactions arranged in order of decreasing oxidation strength and is useful for selecting reagent systems.

Abbreviations Used in the Table

g, gas liq, liquid s, solid

Half-reaction		$E°$, V
$F_2(g)+2H^++2e^-$	$=2HF$	3.06
$O_3+2H^++2e^-$	$=O_2+H_2O$	2.07
$S_2O_8^{2-}+2e^-$	$=2SO_4^{2-}$	2.01
$Ag^{2+}+e^-$	$=Ag^+$	2.00
$H_2O_2+2H^++2e^-$	$=2H_2O$	1.77
$MnO_4^-+4H^++3e^-$	$=MnO_2(s)+2H_2O$	1.70
$Ce(IV)+e^-$	$=Ce(III)$ (in 1 M $HClO_4$)	1.61
$H_5IO_6+H^++2e^-$	$=IO_3^-+3H_2O$	1.6
Bi_2O_4(bismuthate)$+4H^++2e^-$	$=2BiO^++2H_2O$	1.59
$BrO_3^-+6H^++5e^-$	$=\frac{1}{2}Br_2+3H_2O$	1.52
$MnO_4^-+8H^++5e^-$	$=Mn^{2+}+4H_2O$	1.51
$PbO_2+4H^++2e^-$	$=Pb^{2+}+2H_2O$	1.455
Cl_2+2e^-	$=2Cl^-$	1.36
$Cr_2O_7^{2-}+14H^++6e^-$	$=2Cr^{3+}+7H_2O$	1.33
$MnO_2(s)+4H^++2e^-$	$=Mn^{2+}+2H_2O$	1.23
$O_2(g)+4H^++4e^-$	$=2H_2O$	1.229
$IO_3^-+6H^++5e^-$	$=\frac{1}{2}I_2+3H_2O$	1.20
$Br_2(liq)+2e^-$	$=2Br^-$	1.065
$ICl_2^-+e^-$	$=\frac{1}{2}I_2+2Cl^-$	1.06
$VO_2^++2H^++e^-$	$=VO^{2+}+H_2O$	1.00
$HNO_2+H^++e^-$	$=NO(g)+H_2O$	1.00
$NO_3^-+3H^++2e^-$	$=HNO_2+H_2O$	0.94
$2Hg^{2+}+2e^-$	$=Hg_2^{2+}$	0.92
$Cu^{2+}+I^-+e^-$	$=CuI$	0.86
Ag^++e^-	$=Ag$	0.799
$Hg_2^{2+}+2e^-$	$=2Hg$	0.79
$Fe(III)+e^-$	$=Fe^{2+}$	0.771
$O_2(g)+2H^++2e^-$	$=H_2O_2$	0.682
$2HgCl_2+2e^-$	$=Hg_2Cl_2(s)+2Cl^-$	0.63
$Hg_2SO_4(s)+2e^-$	$=2Hg+SO_4^{2-}$	0.615
$H_3AsO_4+2H^++2e^-$	$=HAsO_2+2H_2O$	0.581
$Sb_2O_5+6H^++4e^-$	$=2SbO^++3H_2O$	0.559
$I_3^-+2e^-$	$=3I^-$	0.545
Cu^++e^-	$=Cu$	0.52
$VO^{2+}+2H^++e^-$	$=V^{3+}+H_2O$	0.337
$Fe(CN)_6^{3-}+e^-$	$=Fe(CN)_6^{4-}$	0.36
$Cu^{2+}+2e^-$	$=Cu$	0.337
$UO_2^{2+}+4H^++2e^-$	$=U^{4+}+2H_2O$	0.334
$BiO^++2H^++3e^-$	$=Bi+H_2O$	0.32
$Hg_2Cl_2(s)+2e^-$	$=2Hg+2Cl^-$	0.2676

TABLE 8-10 Potentials of selected half-reactions at 25°C (*continued*)

Half-reaction		$E°$, V
$AgCl(s)+e^-$	$=Ag+Cl^-$	0.2223
$SbO^++2H^++3e^-$	$=Sb+H_2O$	0.212
$CuCl_3^{2-}+e^-$	$=Cu+3Cl^-$	0.178
$SO_4^{2-}+4H^++2e^-$	$=SO_2(aq)+2H_2O$	0.17
$Sn^{4+}+2e^+$	$=Sn^{2+}$	0.154
$S+2H^++2e^-$	$=H_2S(g)$	0.141
$TiO^{2+}+2H^++e^-$	$=Ti^{3+}+H_2O$	0.10
$S_4O_6^{2-}+2e^-$	$=2S_2O_3^{2-}$	0.08
$AgBr(s)+e^-$	$=Ag+Br^-$	0.071
$2H^++2e^-$	$=H_2$	0.0000
$Pb^{2+}+2e^-$	$=Pb$	−0.126
$Sn^{2+}+2e^-$	$=Sn$	−0.136
$Agl(s)+e^-$	$=Ag+I^-$	−0.152
$Mo^{3+}+3e^-$	$=Mo$	ca −0.2
$N_2+5H^++4e^-$	$=H_2NNH_3^+$	−0.23
$Ni^{2+}+2e^-$	$=Ni$	−0.246
$V^{3+}+e^-$	$=V^{2+}$	−0.255
$Co^{2+}+2e^-$	$=Co$	−0.277
$Ag(CN)_2^-+e$	$=Ag+2CN^-$	−0.31
$Cd^{2+}+2e^-$	$=Cd$	−0.403
$Cr^{3+}+e^-$	$=Cr^{2+}$	−0.41
$Fe^{2+}+2e^-$	$=Fe$	−0.440
$2CO_2+2H^++2e^-$	$=H_2C_2O_4$	−0.49
$H_3PO_3+2H^++2e^-$	$=H_3PO_2+H_2O$	−0.50
$U^{4+}+e^-$	$=U^{3+}$	−0.61
$Zn^{2+}+2e^-$	$=Zn$	−0.763
$Cr^{2-}+2e^-$	$=Cr$	−0.91
$Mn^{2+}+2e^-$	$=Mn$	−1.18
$Zr^{4+}+4e^-$	$=Zr$	−1.53
$Ti^{3+}+3e^-$	$=Ti$	−1.63
$Al^{3+}+3e^-$	$=Al$	−1.66
$Th^{4+}+4e^-$	$=Th$	−1.90
$Mg^{2+}+2e^-$	$=Mg$	−2.37
$La^{3+}+3e^-$	$=La$	−2.52
Na^++e^-	$=Na$	−2.714
$Ca^{2+}+2e^-$	$=Ca$	−2.870
$Sr^{2+}+2e^-$	$=Sr$	−2.89
K^++e^-	$=K$	−2.925
Li^++e^-	$=Li$	−3.045

TABLE 8-11 Half-wave potentials (vs. saturated calomel electrode) of organic compounds at 25°C

The solvent systems in this table are listed below:

A acetonitrile and a perchlorate salt such as $LiClO_4$ or a tetraalkyl ammonium salt
B acetic acid and an alkali acetate, often plus a tetraalkyl ammonium iodide
C 0.05 to 0.175 *M* tetraalkyl ammonium halide and 75% 1,4-dioxane
D buffer plus 50% ethanol (EtOH)

Abbreviations Used in the Table

Bu, butyl — *M*, molar — MeOH, methanol
Et, ethyl — Me, methyl — PrOH, propanol
EtOH, ethanol

Compound	Solvent system	$E_{1/2}$
	Unsaturated aliphatic hydrocarbons	
Acrylonitrile	C but 30% EtOH	−1.94
Allene	C	−2.29
1,3-Butadiene	A	−2.03
	C	−2.59
1,3-Butadiyne	C	−1.89
1-Buten-2-yne	C	−2.40
1,4-Cyclohexadiene	A	−1.6
Cyclohexene	A	−1.89
1,3,5,7-Cyclooctatetraene	B	−1.42
	C	−1.51
Diethyl fumarate	B, pH 4.0	−0.84
Diethyl maleate	B, pH 4.0	−0.95
2,3-Dimethyl-1,3-butadiene	A	−1.83
Dimethylfulvene	C	−1.89
Diphenylacetylene	C	−2.20
1,1-Diphenylethylene	B	−1.52
	C	−2.19
Ethyl methacrylate	0.1 *N* LiCl+25% EtOH	−1.9
2-Methyl-1,3-butadiene	A	−1.84
2-Methyl-1-butene	A	−1.97
1-Piperidino-4-cyano-4-phenyl-1,3-butadiene	$LiClO_4$ in dimethylformamide	−0.16
trans-Stilbene	B	−1.51
Tetrakis(dimethylamino)ethylene	A	−0.75
	Aromatic hydrocarbons	
Acenaphthene	A	−0.95
	B	−1.36
	C	−2.58
Anthracene	A	−0.84
	B	−1.20
	C	−1.94

TABLE 8-11 Half-wave potentials (vs. saturated calomel electrode) of organic compounds at 25°C (*continued*)

Compound	Solvent system	$E_{1/2}$
	Aromatic hydrocarbons (*continued*)	
Azulene	A	−0.71
	C	−1.66, −2.26, −2.56
1,2-Benzanthracene	C	−2.03, −2.54
2,3-Benzanthracene	A	−0.54, −1.20
Benzene	A	−2.08
1,2-Benzo[*a*]pyrene	A	−0.76
Biphenyl	A	−1.48
	B	−1.91
	C	−2.70
Chrysene	A	−1.22
1,2,5,6-Dibenzanthracene	A	−1.00, −1.26
1,2-Dihydronaphthalene	C	−2.57
9,10-Dimethylanthracene	A	−0.65
2,3-Dimethylnaphthalene	A	−1.08, −1.34
9,10-Diphenylanthracene	A	−0.92
Fluorene	A	−1.25
	B	−1.65
	C	−2.65
Hexamethylbenzene	A	−1.16
	B	−1.52
Indan	A	−1.59, −2.02
Indene	A	−1.23
	C	−2.81
1-Methylnaphthalene	A	−1.24
	B	−1.53
	C	−2.46
2-Methylnaphthalene	A	−1.22
	B	−1.55
	C	−2.46
Naphthalene	A	−1.34
	B	−1.72
Pentamethylbenzene	A	−1.28
	B	−1.62
Phenanthrene	A	−1.23
	B	−1.68
	C	−2.46, −2.71
Phenylacetylene	C	−2.37
Pyrene	A	−1.06, −1.24
trans-Stilbene	B	−1.51
	C	−2.26
Styrene	C	−2.35

TABLE 8-11 Half-wave potentials (vs. saturated calomel electrode) of organic compounds at 25°C (*continued*)

Compound	Solvent system	$E_{1/2}$
	Aromatic hydrocarbons (*continued*)	
1,2,3,5-Tetramethylbenzene	A	−1.50, −1.99
1,2,4,5-Tetramethylbenzene	A	−1.29
Tetraphenylethylene	C	−2.05
1,4,5,8-Tetraphenylnaphthalene	A	−1.39
Toluene	A	−1.98
1,2,3-Trimethylbenzene	A	−1.58
1,2,4-Trimethylbenzene	A	−1.41
1,3,5-Trimethylbenzene	A	−1.50
	B	−1.90
Triphenylene	A	−1.46, −1.55
Triphenylmethane	C	−1.01, −1.68, −1.96
o-Xylene	A	−1.58, −2.04
m-Xylene	A	−1.58
p-Xylene	A	−1.56
	Aldehydes	
Acetaldehyde	B, pH 6.8–13	−1.89
Benzaldehyde	McIlvaine buffer, pH 2.2	−0.96, −1.32
Bromoacetaldehyde	pH 8.5	−0.40
	pH 9.8	−1.58, −1.82
Chloroacetaldehyde	Ammonia buffer, pH 8.4	−1.06, −1.66
Cinnamaldehyde	Buffer + EtOH, pH 6.0	−0.9, −1.5, −1.7
Crotonaldehyde	B, pH 1.3–2.0	−0.92
	Ammonia buffer, pH 8.0	−1.30
Dichloroacetaldehyde	Ammonia buffer, pH 8.4	−1.03, −1.67
3,7-Dimethyl-2,6-octadienal	0.1 *M* Et_4NI	−1.56, −2.22
Formaldehyde	0.05 *M* KOH+0.1 *M* KCl, pH 12.7	−1.59
2-Furaldehyde	pH 1–8	−0.86−0.07 pH
	pH 10	−1.43
Glucose	Phosphate buffer, pH 7	−1.55
Glyceraldehyde	Britton-Robinson buffer, pH 5.0	−1.47
	Britton-Robinson buffer, pH 8.0	−1.55
Glycolaldehyde	0.1 *M* KOH, pH 13	−1.70
Glyoxal	B, pH 3.4	−1.41
4-Hydroxybenzaldehyde	Britton-Robinson buffer, pH 1.8	−1.16
	Britton-Robinson buffer, pH 6.8	−1.45
4-Hydroxy-2-methoxybenzaldehyde	McIlvaine buffer, pH 2.2	−1.05
	McIlvaine buffer, pH 5.0	−1.16, −1.36
	McIlvaine buffer, pH 8.0	−1.47

TABLE 8-11 Half-wave potentials (vs. saturated calomel electrode) of organic compounds at 25°C (*continued*)

Compound	Solvent system	$E_{1/2}$
	Aldehydes (*continued*)	
o-Methoxybenzaldehyde	Britton-Robinson buffer, pH 1.8	−1.02
	Britton-Robinson buffer, pH 6.8	−1.49
p-Methoxybenzaldehyde	Britton-Robinson buffer, pH 1.8	−1.17
	Britton-Robinson buffer, pH 6.8	−1.48
Methyl glyoxal	A, pH 4.5	−0.83
m-Nitrobenzaldehyde	Buffer + 10% EtOH, pH 2.0	−0.28, −1.20
Phthalaldehyde	Buffer, pH 3.1	−0.64, −1.07
	Buffer, pH 7.3	−0.89, −1.29
2-Propenal (acrolein)	pH 4.5	−1.36
	pH 9.0	−1.1
Propionaldehyde	0.1 *M* LiOH, pH 13	−1.93
Pyrrole-2-carbaldehyde	0.1 *M* HCl+50% EtOH	−1.25
Salicylaldehyde	McIlvaine buffer, pH 2.2	−0.99, −1.23
	McIlvaine buffer, pH 5.0	−1.20, −1.30
	McIlvaine buffer, pH 8.0	−1.32
Trichloroacetaldehyde	Ammonia buffer, pH 8.4	−1.35, −1.66
	0.1 *M* KCl+50% EtOH	−1.55
	Ketones	
Acetone	B, pH 9.3	−1.52
	C	−2.46
Acetophenone	D + McIlvaine buffer, pH 4.9	−1.33
	D + McIlvaine buffer, pH 7.2	−1.58
	D + McIlvaine buffer, pH 1.3	−1.08
7*H*-Benz[*de*]anthracen-7-one	0.1 *N* H_2SO_4+75% MeOH	−0.96
Benzil	D + McIlvaine buffer, pH 1.3	−0.27
	D + McIlvaine buffer, pH 4.9	−0.50
Benzoin	D + McIlvaine buffer, pH 1.3	−0.90
	D + McIlvaine buffer, pH 8.6	−1.49
Benzophenone	D + McIlvaine buffer, pH 1.3	−0.94
	D + McIlvaine buffer, pH 8.6	−1.36
Benzoylacetone	Buffer, pH 2.6	−1.60
	Buffer, pH 5.3 and pH 7.6	−1.68
	Buffer, pH 9.7	−1.72
Bromoacetone	0.1 *M* LiCl	−0.29
2,3-Butanedione	0.1 *M* HCl	−0.84
3-Buten-2-one	0.1 *M* KCl	−1.42
Butyrophenone	0.1 *M* NH_4Cl+50% EtOH	−1.55
D-Carvone	0.1 *M* Et_4NI + 80% EtOH	−1.71
Chloroacetone	0.1 *M* LiCl	−1.18
Coumarin	McIlvaine buffer, pH 2.0	−0.95
	McIlvaine buffer, pH 5.0	−1.11, −1.44

TABLE 8-11 Half-wave potentials (vs. saturated calomel electrode) of organic compounds at 25°C (*continued*)

Compound	Solvent system	$E_{1/2}$
	Ketones (*continued*)	
Cyclohexanone	C	−2.45
cis-Dibenzoylethylene	D, pH 1	−0.30
	D, pH 11	−0.62, −1.65
trans-Dibenzoylethylene	D, pH 1	−0.12
	D, pH 11	−0.57, −1.52
Dibenzoylmethane	D, pH 1.3	−0.59
	D, pH 11.3	−1.30, −1.62
9,10-Dihydro-9-oxoanthracene	D, pH 2.0	−0.93
1,5-Diphenyl-1,5-pentanedione	A	−2.10
1,5-Diphenylthiocarbazone	D, pH 7.0	−0.6
Flavanone	Acetate buffer + Me_4NOH + 50% 2-PrOH, pH 6.1	−1.30
	Acetate buffer + Me_4NOH + 50% 2-PrOH, pH 9.6	−1.51
Fluorescein	Acetate buffer, pH 2.0	−0.50
	Phthalate buffer, pH 5.0	−0.65
	Borate buffer, pH 10.1	−1.18, −1.44
Fructose	0.02 *M* LiCl	−1.76
Girard derivatives of aliphatic ketones	pH 8.2	−1.52
o-Hydroxyacetophenone	D, pH 5	−1.36
p-Hydroxyacetophenone	D, pH 5	−1.46
1,2,3-Indantrione (ninhydrin)	Britton-Robinson buffer, pH 2.5	−0.67, −0.83
	Britton-Robinson buffer, pH 4.5	−0.73, −1.01
	Britton-Robinson buffer, pH 6.8	−0.10, −0.90, −1.20
	Britton-Robinson buffer, pH 9.2	−1.35
α-Ionone	C	−1.59, −2.08
Isatin	Phosphate buffer + citrate buffer, pH 2.9	−0.3, −0.5
	Phosphate buffer + citrate buffer, pH 4.3	−0.3, −0.5, −0.8
	Phosphate buffer + citrate buffer, pH 5.4	−0.8
4-Methyl-3,5-heptadien-2-one	A	−0.64
4-Methyl-2,6-heptanedione	A	−1.28
4-Methyl-3-penten-2-one	D + McIlvaine buffer, pH 1.3	−1.01
	D + McIlvaine buffer, pH 11.3	−1.60
4-Phenyl-3-buten-2-one	D, pH 1.3	−0.72
	D, pH 8.6	−1.27
Phthalide	0.1 *M* Bu_4NI + 50% dioxane	−0.20

TABLE 8-11 Half-wave potentials (vs. saturated calomel electrode) of organic compounds at 25°C (*continued*)

Compound	Solvent system	$E_{1/2}$
	Ketones (*continued*)	
Phthalimide	pH 4.2	−1.1, −1.5
	pH 9.7	−1.2, −1.4
Pulegone	C	−1.74
Quinalizarin	Phosphate buffer + 1% EtOH, pH 8.0	−0.56
Testosterone	D + Britton-Robinson buffer, pH 2.6	−1.20
	D + Britton-Robinson buffer, pH 5.8	−1.40
	D + Britton-Robinson buffer, pH 8.8	−1.53, −1.79
	Quinones	
Anthraquinone	Acetate buffer + 40% dioxane, pH 5.6	−0.51
	Phosphate buffer + 40% dioxane, pH 7.9	−0.71
o-Benzoquinone	Britton-Robinson buffer, pH 7.0	+0.20
	Britton-Robinson buffer, pH 9.0	+0.08
2,3-Dimethylnaphthoquinone	D, pH 5.4	−0.22
1,2-Naphthoquinone	Phosphate buffer, pH 5.0	−0.03
	Phosphate buffer, pH 7.0	−0.13
1,4-Naphthoquinone	Britton-Robinson buffer, pH 7.0	−0.07
	Britton-Robinson buffer, pH 9.0	−0.19
	Acids	
Acetic acid	A	−2.3
Acrylic acid	pH 5.6	−0.85
Adenosine-5′-phosphoric acid	$HClO_4 + KClO_4$, pH 2.2	−1.13
4-Aminobenzenesulfonic acid	0.05 *M* Me_4NI	−1.58
3-Aminobenzoic acid	pH 5.6	−0.67
Anthranilic acid	pH 5.6	−0.67
Ascorbic acid	Birtton-Robinson buffer, pH 3.4	+0.17
	Britton-Robinson buffer, pH 7.0	−0.06
Barbituric acid	Borate buffer, pH 9.3	−0.04
Benzoic acid	A	−2.1
Benzoylformic acid	Britton-Robinson buffer, pH 2.2	−0.48
	Britton-Robinson buffer, pH 5.5	−0.85, −1.26
	Britton-Robinson buffer, pH 7.2	−0.98, −1.25
	Britton-Robinson buffer, pH 9.2	−1.25

TABLE 8-11 Half-wave potentials (vs. saturated calomel electrode) of organic compounds at 25°C (*continued*)

Compound	Solvent system	$E_{1/2}$
	Acids (*continued*)	
Bromoacetic acid	pH 1.1	−0.54
2-Bromopropionic acid	pH 2.0	−0.39
Crotonic acid	C	−1.94
Dibromoacetic acid	pH 1.1	−0.03, −0.59
Dichloroacetic acid	pH 8.2	−1.57
5,5-Diethylbarbituric acid	Borate buffer, pH 9.3	0.00
Flavanol	D, pH 5.6	−1.25
	D, pH 7.7	−1.40
Folic acid	Britton-Robinson buffer, pH 4.6	−0.73
Formic acid	0.1 *M* KCl	−1.66
Fumaric acid	HCl + KCl, pH 2.6	−0.83
	Acetate buffer, pH 4.0	−0.93
	Acetate buffer, pH 5.9	−1.20
2,4-Hexadienedioic acid	Acetate buffer, pH 4.5	−0.97
Iodoacetic acid	pH 1	−0.16
Maleic acid	Britton-Robinson buffer, pH 2.0	−0.70
	Britton-Robinson buffer, pH 4.0	−0.97
	Britton-Robinson buffer, pH 6.0	−1.11, −1.30
	Britton-Robinson buffer, pH 10.0	−1.51
Mercaptoacetic acid	B, pH 6.8	−0.38
Methacrylic acid	D + 0.1 *M* LiCl	−1.69
Nitrobenzoic acids	Buffer + 10% EtOH, pH 2.0	−0.2, −0.7
Oxalic acid	B, pH 5.4–6.1	−1.80
2-Oxo-1,5-pentanedioic acid	HCl + KCl, ph 1.8	−0.59
	Ammonia buffer, pH 8.2	−1.30
2-Oxopropionic acid	Britton-Robinson buffer, pH 5.6	−1.17
	Britton-Robinson buffer, pH 6.8	−1.22, −1.53
	Britton-Robinson buffer, pH 9.7	−1.51
Phenolphthalein	Phthalate buffer, pH 2.5	−0.67
	Phthalate buffer, pH 4.7	−0.80
	D, pH 9.6	−0.98, −1.35
Picric acid	pH 4.2	−0.34
	pH 11.7	−0.36, −0.56, −0.96
1,2,3-Propenetricarboxylic acid	pH 7.0	−2.1
Trichloroacetic acid	Ammonia buffer, pH 8.2	−0.84, −1.57
	Phosphate buffer, pH 10.4	−0.9, −1.6
3,4,5-Trihydroxybenzoic acid	Phosphate buffer, pH 2.9	+0.50
	Phosphate buffer, pH 8.8	+0.1
p-Aminophenol	Britton-Robinson buffer, pH 6.3	+0.14
	Britton-Robinson buffer, pH 8.6	−0.04
	Britton-Robinson buffer, pH 12.0	−0.16

TABLE 8-11 Half-wave potentials (vs. saturated calomel electrode) of organic compounds at 25°C (*continued*)

Compound	Solvent system	$E_{1/2}$
	Acids (*continued*)	
o-Chlorophenol	pH 5.6	−0.63
m-Chlorophenol	pH 5.6	−0.73
p-Chlorophenol	pH 5.6	−0.65
o-Cresol	pH 5.6	−0.56
m-Cresol	pH 5.6	−0.61
p-Cresol	pH 5.6	−0.54
1,2-Dihydroxybenzene	pH 5.6	−0.35
1,3-Dihydroxybenzene	pH 5.6	−0.61
1,4-Dihydroxybenzene	pH 5.6	−0.23
o-Methoxyphenol	pH 5.6	−0.46
m-Methoxyphenol	pH 5.6	−0.62
p-Methoxyphenol	pH 5.6	−0.41
1-Naphthol	A	−0.74
2-Naphthol	A	−0.82
1,2,3-Trihydroxybenzene	Britton-Robinson buffer, pH 3.1	+0.35
	Britton-Robinson buffer, pH 6.5	+0.10
	Britton-Robinson buffer, pH 9.5	−0.10
	Halogen compounds	
Bromobenzene	A	−1.98
	C	−2.32
1-Bromobutane	C	−2.27
Bromoethane	C	−2.08
Bromomethane	C	−1.63
1-Bromonaphthalene (also 2-bromonaphthalene)	A	−1.55, −1.60
3-Bromo-1-propene	C	−1.29
p-Bromotoluene	A	−1.72
Carbon tetrachloride	C	−0.78, −1.71
Chlorobenzene	A	−2.07
Chloroform	C	−1.63
Chloromethane	C	−2.23
3-Chloro-1-propene	C	−1.91
α-Chlorotoluene	C	−1.81
p-Chlorotoluene	A	−1.76
N-Chloro-*p*-toluenesulfonamide	0.5 *M* K_2SO_4	−0.13
9,10-Dibromoanthracene	A	−1.15, −1.47
p-Dibromobenzene	C	−2.10
1,2-Dibromobutane	D + 1% Na_2SO_3	−1.45
Dibromoethane	C	−1.48
meso-2,3-Dibromosuccinic acid	Acetate buffer, pH 4.0	−0.23, −0.89
Dichlorobenzenes	C	−2.5

TABLE 8-11 Half-wave potentials (vs. saturated calomel electrode) of organic compounds at 25°C (*continued*)

Compound	Solvent system	$E_{1/2}$
	Halogen compounds (*continued*)	
Dichloromethane	C	−1.60
Diiodomethane	C	−1.12, −1.53
Hexabromobenzene	C	−0.8, −1.5
Hexachlorobenzene	C	−1.4, −1.7
Iodobenzene	A	−1.72
Iodoethane	C	−1.67
Iodomethane	A	−2.12
	C	−1.63
Tetrabromomethane	C	−0.3, −0.75, −1.49
Tetraidomethane	C	−0.45, −1.05, −1.46
Tribromomethane	C	−0.64, −1.47
α,α,α-Trichlorotoluene	C	−0.68, −1.65, −2.00
	Nitro and nitroso compounds	
1,2-Dinitrobenzene	Phthalate buffer, pH 2.5	−0.12, −0.32, −1.26
	Borate buffer, pH 9.2	−0.38, −0.74
1,3-Dinitrobenzene	Phthalate buffer, pH 2.5	−0.17, −0.29
	Borate buffer, pH 9.2	−0.46, −0.68
1,4-Dinitrobenzene	Phthalate buffer, pH 2.5	−0.12, −0.33
	Borate buffer, pH 9.2	−0.35, −0.80
Methyl nitrobenzoates	Buffer + 10% EtOH, pH 2.0	−0.20 to −0.25 −0.68 to −0.74
p-Nitroacetophenone	Britton-Robinson buffer, pH 2.2	−0.16, −0.61, −1.09
	Britton-Robinson buffer, pH 10.0	−0.51, −1.40, −1.73
o-Nitroaniline	0.03 *M* LiCl + 0.02 *M* benzoic acid in EtOH	−0.88
m-Nitroaniline	Britton-Robinson buffer, pH 4.3	−0.3, −0.8
	Briton-Robinson buffer, pH 7.2	−0.5
	Britton-Robinson buffer, pH 9.2	−0.7
p-Nitroaniline	pH 2.0	−0.36
	Acetate buffer, pH 4.6	−0.5
o-Nitroanisole	Buffer + 10% EtOH, pH 2.0	−0.29, −0.58

TABLE 8-11 Half-wave potentials (vs. saturated calomel electrode) of organic compounds at 25°C (*continued*)

Compound	Solvent system	$E_{1/2}$
	Nitro and nitroso compounds (*continued*)	
p-Nitroanisole	Buffer + 10% EtOH, pH 2.0	−0.35, −0.64
1-Nitroanthraquinone	Britton-Robinson buffer, pH 7.0	−0.16
Nitrobenzene	HCl + KCl + 8% EtOH, pH 0.5	−0.16, −0.76
	Phthalate buffer, pH 2.5	−0.30
	Borate buffer, pH 9.2	−0.70
Nitrocresols	Britton-Robinson buffer, pH 2.2	−0.2 to −0.3
	Britton-Robinson buffer, pH 4.5	−0.4 to −0.5
	Britton-Robinson buffer, pH 8.0	−0.6
Nitroethane	Britton-Robinson buffer + 30% MeOH, pH 1.8	−0.7
	Britton-Robinson buffer + 30% MeOH, pH 4.6	−0.8
2-Nitrohydroquinone	Phosphate buffer + citrate buffer, pH 2.1	−0.2
	Phosphate buffer + citrate buffer, pH 5.2	−0.4
	Phosphate buffer + citrate buffer, pH 8.0	−0.5
Nitromethane	Britton-Robinson buffer + 30% MeOH, pH 1.8	−0.8
	Britton-Robinson buffer + 30% MeOH, pH 4.6	−0.85
o-Nitrophenol	Britton-Robinson buffer + 10% EtOH, pH 2.0	−0.23
	Britton-Robinson buffer + 10% EtOH, pH 4.0	−0.4
	Britton-Robinson buffer + 10% EtOH, pH 8.0	−0.65
	Britton-Robinson buffer + 10% EtOH, pH 10.0	−0.80
m-Nitrophenol	Britton-Robinson buffer + 10% EtOH, pH 2.0	−0.37
	Britton-Robinson buffer + 10% EtOH, pH 4.0	−0.40
	Britton-Robinson buffer + 10% EtOH, pH 8.0	−0.64
	Britton-Robinson buffer + 10% EtOH, pH 10.0	−0.76
p-Nitrophenol	Britton-Robinson buffer + 10% EtOH, pH 2.0	−0.35
	Britton-Robinson buffer + 10% EtOH, pH 4.0	−0.50
	Britton-Robinson buffer + 10% EtOH, pH 8.0	−0.82
1-Nitropropane	Britton-Robinson buffer + 30% MeOH, pH 1.8	−0.73

TABLE 8-11 Half-wave potentials (vs. saturated calomel electrode) of organic compounds at 25°C (*continued*)

Compound	Solvent system	$E_{1/2}$
	Nitro and nitroso compounds (*continued*)	
1-Nitropropane (*continued*)	Britton-Robinson buffer + 30% MeOH, pH 8.6	−0.88
	Britton-Robinson buffer + 30% MeOH, pH 8.0	−0.95
2-Nitropropane	McIlvaine buffer, pH 2.1	−0.53
	McIlvaine buffer, pH 5.1	−0.81
Nitrosobenzene	McIlvaine buffer, pH 6.0	−0.03
	McIlvaine buffer, pH 8.0	−0.14
1-Nitroso-2-naphthol	D + buffer, pH 4.0	+0.02
	D + buffer, pH 7.0	−0.20
	D + buffer, pH 9.0	−0.31
N-Nitrosophenylhydroxylamine	pH 2.0	−0.84
o-Nitrotoluene	Phthalate buffer, pH 2.5	−0.35, −0.66
	Phthalate buffer, pH 7.4	−0.60, −1.06
m-Nitrotoluene (also *p*-nitrotoluene)	Phthalate buffer, pH 2.5	−0.30, −0.53
	Phthalate buffer, pH 7.4	−0.58, −1.06
Tetranitromethane	pH 12.0	−0.41
1,3,5-Trinitrobenzene	Phthalate buffer, pH 4.1	−0.20, −0.29, −0.34
	Borate buffer, pH 9.2	−0.34. −0.48, −0.65
	Heterocyclic compounds containing nitrogen	
Acridine	D, pH 8.3	−0.80, −1.45
Cinchonine	B, pH 3	−0.90
2-Furanmethanol	Britton-Robinson buffer, pH 2.0	−0.96
	Britton-Robinson buffer, pH 5.8	−1.38, −1.70
2-Hydroxyphenazine	Britton-Robinson buffer, pH 4.0	−0.24
8-Hydroxyquinoline	B, pH 5.0	−1.12
	Phosphate buffer, pH 8.0	−1.18, −1.71
3-Methylpyridine	D + 0.1 *M* LiCl	−1.76
4-Methylpyridine	D + 0.1 *M* LiCl	−1.87
Phenazine	Phosphate buffer + citrate buffer, pH 7.0	−0.36
Pyridine	Phosphate buffer + citrate buffer, pH 7.0	−1.75
Pyridine-2-carboxylic acid	B, pH 4.1	−1.10
	B, pH 9.3	−1.48, −1.94
Pyridine-3-carboxylic acid	0.1 *M* HCl	−1.08
Pyridine-4-carboxylic acid	Britton-Robinson buffer, pH 6.1	−1.14
	pH 9.0	−1.39, −1.68

TABLE 8-11 Half-wave potentials (vs. saturated calomel electrode) of organic compounds at 25°C (*continued*)

Compound	Solvent system	$E_{1/2}$
	Heterocyclic compounds containing nitrogen (*continued*)	
Pyrimidine	Citrate buffer, pH 3.6	−0.92, −1.24
	Ammonia buffer, pH 9.2	−1.54
Quinoline-8-carboxylic acid	pH 9	−1.11
Quinoxaline	Phosphate buffer + citrate buffer, pH 7.0	−0.66, −1.52
	Azo, hydrazine, hydroxylamine, and oxime compounds	
Azobenzene	D, pH 4.0	−0.20
	D, pH 7.0	−0.50
Azoxybenzene	Buffer + 20% EtOH, pH 6.3	−0.30
Benzoin 1-oxime	Buffer, pH 2.0	−0.88
	Buffer, pH 5.6	−1.08
	Buffer, pH 8.2	−1.67
Benzoylhydrazine	0.13 *M* NaOH, pH 13.0	−0.30
Dimethylglyoxime	Ammonia buffer, pH 9.6	−1.63
Hydrazine	Britton-Robinson buffer, pH 9.3	−0.09
Hydroxylamine	Britton-Robinson buffer, pH 4.6	−1.42
	Britton-Robinson buffer, pH 9.2	−1.65
Oxamide	Acetate buffer	−1.55
Phenylhydrazine	McIlvaine buffer, pH 2	+0.19
	0.13 *M* NaOH, pH 13.0	−0.36
Phenylhydroxylamine	McIlvaine buffer + 10% EtOH, pH 2	−0.68
	McIlvaine buffer + 10 EtOH, pH 4–10	−0.33 0.061 pH
Salicylaldoxime	Phosphate buffer, pH 5.4	−1.02
Thiosemicarbazide	Borate buffer, pH 9.3	−0.26
Thiourea	0.1 *M* sulfuric acid	+0.02
	Indicators and dyestuffs	
Brilliant Green	HCl + KCl, pH 2.0	−0.2, −0.5
Indigo carmine	pH 2.5	−0.24
Indigo disulfonate	pH 7.0	−0.37
Malachite Green G	HCl + KCl, pH 2.0	−0.2, −0.5
Metanil yellow	Phosphate buffer + 1% EtOH, pH 7.0	−0.51
Methylene blue	Britton-Robinson buffer, pH 4.9	−0.15
	Britton-Robinson buffer, pH 9.2	−0.30

TABLE 8-11 Half-wave potentials (vs. saturated calomel electrode) of organic compounds at 25°C (*continued*)

Compound	Solvent system	$E_{1/2}$
	Indicators and dyestuffs (*continued*)	
Methylene green	Phosphate buffer + 1% EtOH, pH 7.0	−0.12
Methyl orange	Phosphate buffer + 1% EtOH, pH 7.0	−0.51
Morin	D, pH 7.6	−1.7
Neutral red	Britton-Robinson buffer, pH 2.0	−0.21
	Britton-Robinson buffer, pH 7.0	−0.57
	Peroxide	
Ethyl peroxide	0.02 *M* HCl	−0.2

SECTION 9

LABORATORY MANIPULATIONS

COOLING

TABLE 9-1 Cooling mixtures

The table below gives the lowest temperature that can be obtained from a mixture of the inorganic salt with finely shaved dry ice. With the organic substances, dry ice (−78°C) in small lumps can be added to the solvent until a slight excess of dry ice remains or liquid nitrogen (−196°C) can be poured into the solvent until a slush is formed that consists of the solid-liquid mixture at its melting point.

Substance	Dry ice, g/100 g	Temperature, °C
$CaCl_2 \cdot 6H_2O$	41	−9.0
	81	−21.5
	123	−40.3
	143	−55
NH_4Cl	25	−15.4
NaBr	66	−28
$MgCl_2$	85	−34

Substance	Temperature, °C	Substance	Temperature, °C
Ethylene glycol	−13	Acetone-CO_2	−77
1,2-Dichlorobenzene	−17	Ethyl acetate	−84
Carbon tetrachloride	−22.9	2-Butanone	−87
Bromobenzene	−31	Hexane	−95
Methoxybenzene	−37	Methanol	−98
Chlorobenzene	−45	Carbon disulfide	−112
Bis(2-ethoxyethyl) ether	−44	Bromoethane	−119
N-Methylaniline	−57	Pentane	−130
p-Cymene	−68	2-Methylbutane	−160

HUMIDIFICATION AND DRYING

A saturated aqueous solution in contact with an excess of a definite solid phase at a given temperature will maintain constant humidity in an enclosed space. Table 9-2 gives a number of salts suitable for this purpose. The aqueous tension (in millimeters of Hg) of a solution at a given temperature is found by multiplying the decimal fraction of the humidity by the aqueous tension at 100% humidity for the specific temperature. For example, the aqueous tension of a saturated solution of NaCl at 20°C is $0.757 \times 17.54 = 13.28$ mmHg and at 80°C is $0.764 \times 355.1 = 271.3$ mmHg.

TABLE 9-2 Humidity (%) maintained by saturated solutions of various salts at specified temperatures

Solid phase	Temperature, °C						
	10	20	25	30	40	60	80
$K_2Cr_2O_7$			98.0				
K_2SO_4	98	97	97	96	96	96	
KNO_3	95	93	92.5	91	88	82	
KCl	88	85.0	84.3	84	81.7	80.7	79.5
KBr	86	84	80.7		79.6	79.0	79.3
NaCl	76	75.7	75.3	74.9	74.7	74.9	76.4
$NaNO_3$	77	75	73.8	72.8	71.5	67.5	65.5
KI					66.8	63.1	60.8
$NaNO_2$		66	65	63.0	61.5	59.3	58.9
$Na_2CrO_4 \cdot 4H_2O$				64.6	61.8	55.6	56.2
$NaBr \cdot 2H_2O$	58	57.9	57.7		52.4	49.9	50.0
$Na_2Cr_2O_7 \cdot 2H_2O$	58	55	54		53.6	55.2	56.0
$Mg(NO_3)_2 \cdot 6H_2O$	57	55	52.9	52	49	43	
$K_2CO_3 \cdot 2H_2O$	47	44	42.8	42	40		
$NaI \cdot 2H_2O$		47		36.4	32.3	25.3	23.2
$MgCl_2 \cdot 6H_2O$	34	33	33.0	33	32	30	
$CaCl_2 \cdot 6H_2O$	38	32.6	29	26			
$KF \cdot 2H_2O$				27.4	22.8	21.0	22.8
$KC_2H_3O_2 \cdot 1.5H_2O$	24	23	22.5	22	20		
$LiCl \cdot H_2O$	13	12	11.1	12	11	11	
KOH	13	9	8	7	6	5	
Aqueous tension at 100% humidity, mmHg	9.21	17.54	23.76	31.82	55.32	149.4	355.1

Solid phase	Temperature °C	Humidity, %
KF	100	22.9
KI	100	56.2
$(NH_4)_2SO_4$	20–30	81.1
	108	75
$BaCl_2 \cdot 2H_2O$	25	90.2
NaF	100	96.6

TABLE 9-3 Drying agents

Drying agent	Most useful for	Residual water, mg H_2O per liter of dry air (25°C)	Grams water removed per gram of desiccant	Regeneration, °C
Al_2O_3	Hydrocarbons	0.002–0.005	0.2	175 (24 h)
$Ba(ClO_4)_2$ [a]	Inert gas streams	0.6–0.8	0.17	140
BaO	Basic gases: hydrocarbons, aldehydes, alcohols	0.0007–0.003	0.12	1000
CaC_2 [b]	Ethers		0.56	Impossible
$CaCl_2$ [c]	Inert organics	0.1–0.2	0.15 (1 H_2O) 0.30 (2 H_2O)	250
CaH_2 [d]	Hydrocarbons, ethers, amines, esters, higher alcohols	1×10^{-5}	0.85	Impossible
CaO	Ethers, esters, alcohols, amines	0.01–0.003	0.31	Difficult, 1000
$CaSO_4$	Most organic substances	0.005–0.07	0.07	225
Dow Desiccant 812 [e]	Most materials	(5–200 ppm)		No
K_2CO_3	Most materials except acids and phenols		0.16	158
KOH	Amines	0.01–0.9		Impossible

$LiAlH_4$ [f]	Hydrocarbons		1.9	Impossible
$Mg(ClO_4)_2$ [a]	Gas streams	0.0005–0.002	0.24	250 (high vacuum)
MgO	All but acidic compounds	0.008	0.45	800
$MgSO_4$	Most organic compounds	1–12	0.15–0.75	Not feasible
Molecular sieves				
4X	Molecules with effective diameter > 4Å	0.001	0.18	250
5X	Molecules with effective diameter > 5Å	0.001	0.18	250
9.5% Na-Pb alloy [d]	Hydrocarbons, ethers	(For solvents only)	0.08	Impossible
Na_2SO_4	Ketones, acids, alkyl and aryl halides	12	1.25	150
P_2O_5	Gas streams; not suitable for alcohols, amines, ketones, or amines	2×10^{-5}	0.5	Not feasible
Silica gel	Most organic amines	0.002–0.07	0.2	200–350
Sulfuric acid	Air and inert gas streams	0.003–0.008	Indefinite	Not feasible

[a] May form explosive mixtures on contact with organic material.
[b] Explosive C_2H_2 formed.
[c] Drying action slow.
[d] H_2 formed.
[e] Used for column drying of organic liquids.
[f] Strong reductant.

TABLE 9-4 Solvents of chromatographic interest

Solvent	Boiling point, °C	Solvent strength parameter		Viscosity, $mN \cdot s \cdot m^{-2}$ (20°C)	Refractive index (20°C)	UV cutoff, nm
		$e°$ (SiO_2)	$e°$ (Al_2O_3)			
Fluoroalkanes			−0.25		1.25	
Pentane	36	0.0	0.0	$0.24^{15°C}$	1.358	210
Hexane	69	0.0	0.0	0.31	1.375	210
2,2,4-Trimethylpentane	99		0.01	0.50	1.392	215
Decane	174		0.04	0.93	1.412	210
Cyclohexane	81	−0.05	0.04	0.98	1.426	210
Cyclopentane	49		0.05	0.44	1.407	210
Diisobutylene	101		0.06		1.411	
1-Pentene	30		0.08	$0.24^{0°C}$	1.371	
Carbon disulfide	46	0.14	0.15	0.36	1.626	380
Carbon tetrachloride	77	0.14	0.18	0.97	1.466	265
1-Chlorobutane	78		0.26	0.43	1.402	220
1-Chloropentane	98		0.26	0.58	1.412	225
o-Xylene	144		0.26	0.81	1.505	290
Diisopropyl ether	68		0.28	$0.38^{25°C}$	1.369	220
2-Chloropropane	35		0.29	0.33	1.378	225
Toluene	111		0.29	0.59	1.497	286
1-Chloropropane	47		0.30	0.35	1.389	225
Chlorobenzene	132		0.40	0.80	1.525	
Benzene	80	0.25	0.32	0.65	1.501	280
Bromoethane	38		0.37	0.40	1.424	

Diethyl ether	35	0.38	0.38	0.25	1.353	218
Diethyl sulfide	92		0.38	0.45	1.443	290
Chloroform	62	0.26	0.40	0.57	1.443	245
Dichloromethane	41		0.42	0.44	1.425	235
4-Methyl-2-pentanone	116		0.43	$0.42^{15°C}$	1.396	335
Tetrahydrofuran	66		0.45	0.55	1.407	220
1,2-Dichloroethane	84		0.49	0.80	1.445	228
2-Butanone	80		0.51	$0.42^{15°C}$	1.379	330
1-Nitropropane	131		0.53	$0.80^{25°C}$	1.402	380
Acetone	56	0.47	0.56	0.32	1.359	330
1,4-Dioxane	101	0.49	0.56	$1.44^{15°C}$	1.420	215
Ethyl acetate	77	0.38	0.58	0.45	1.372	255
Methyl acetate	56		0.60	$0.48^{15°C}$	1.362	260
1-Pentanol	138		0.61	4.1	1.410	210
Dimethyl sulfoxide	189		0.62	2.47	1.478	265
Aniline	184		0.62	4.40	1.586	
Diethylamine	56		0.63	0.33	1.386	275
Nitromethane	101		0.64	0.67	1.394	380
Acetonitrile	82	0.50	0.65	0.37	1.344	190
Pyridine	115		0.71	0.97	1.510	330
2-Butoxyethanol	170		0.74	$3.15^{25°C}$	1.420	220
1-Propanol	97		0.82	2.25	1.386	210
2-Propanol	82		0.82	2.50	1.377	210
Ethanol	78		0.88	1.20	1.361	210
Methanol	65		0.95	0.59	1.328	210
Ethylene glycol	198		1.11	21.8	1.432	210
Acetic acid	118		large	1.23	1.372	260
Water	100		large	1.00	1.333	191

TABLE 9-5 Solvents having the same refractive index and the same density at 25°C

Solvent 1	Solvent 2	Refractive index		Density, g/mL	
		1	2	1	2
Acetone	Ethanol	1.357	1.359	0.788	0.786
Ethyl formate	Methyl acetate	1.358	1.360	0.916	0.935
Ethanol	Propionitrile	1.359	1.363	0.786	0.777
2,2-Dimethylbutane	2-Methylpentane	1.366	1.369	0.644	0.649
2-Methylpentane	Hexane	1.369	1.372	0.649	0.655
Isopropyl acetate	2-Chloropropane	1.375	1.376	0.868	0.865
3-Butanone	Butyraldehyde	1.377	1.378	0.801	0.799
Butyraldehyde	Butyronitrile	1.378	1.382	0.799	0.786
Dipropyl ether	Butyl ethyl ether	1.379	1.380	0.753	0.746
Propyl acetate	Ethyl propionate	1.382	1.382	0.883	0.888
Propyl acetate	1-Chloropropane	1.382	1.386	0.883	0.890
Butyronitrile	2-Methyl-2-propanol	1.382	1.385	0.786	0.781
Ethyl propionate	1-Chloropropane	1.382	1.386	0.888	0.890
1-Propanol	2-Pentanone	1.383	1.387	0.806	0.804
Isobutyl formate	1-Chloropropane	1.383	1.386	0.881	0.890
1-Chloropropane	Butyl formate	1.386	1.387	0.890	0.888
Butyl formate	Methyl butyrate	1.387	1.391	0.888	0.875
Methyl butyrate	2-Chlorobutane	1.392	1.395	0.875	0.868
Butyl acetate	2-Chlorobutane	1.392	1.395	0.877	0.868
4-Methyl-2-pentanone	Pentanonitrile	1.394	1.395	0.797	0.795
4-Methyl-2-pentanone	1-Butanol	1.394	1.397	0.797	0.812
2-Methyl-1-propanol	Pentanonitrile	1.394	1.395	0.798	0.795
2-Methyl-1-propanol	2-Hexanone	1.394	1.395	0.798	0.810
2-Butanol	2,4-Dimethyl-3-pentanone	1.395	1.399	0.803	0.805
2-Hexanone	1-Butanol	1.395	1.397	0.810	0.812
Pentanonitrile	2,4-Dimethyl-3-pentanone	1.395	1.399	0.795	0.805
2-Chlorobutane	Isobutyl butyrate	1.395	1.399	0.868	0.860
Butyric acid	2-Methoxyethanol	1.396	1.400	0.955	0.960
1-Butanol	3-Methyl-2-pentanone	1.397	1.398	0.812	0.808
1-Chloro-2-methylpropane	Isobutyl butyrate	1.397	1.399	0.872	0.860
1-Chloro-2-methylpropane	Pentyl acetate	1.397	1.400	0.872	0.871
Methyl methacrylate	3-Methyl-2-pentanone	1.398	1.398	0.795	0.808
Triethylamine	2,2,3-Trimethylpentane	1.399	1.401	0.723	0.712
Butylamine	Dodecane	1.399	1.400	0.736	0.746
Isobutyl butyrate	1-Chlorobutane	1.399	1.401	0.860	0.875
1-Nitropropane	Propionic anhydride	1.399	1.400	0.995	1.007
Pentyl acetate	1-Chlorobutane	1.400	1.400	0.871	0.881
Pentyl acetate	Tetrahydrofuran	1.400	1.404	0.871	0.885
Dodecane	Dipropylamine	1.400	1.400	0.746	0.736
1-Chlorobutane	Tetrahydrofuran	1.401	1.404	0.871	0.885
Isopentanoic acid	2-Ethoxyethanol	1.402	1.405	0.923	0.926
Dipropylamine	Cyclopentane	1.403	1.404	0.736	0.740
2-Pentanol	4-Heptanone	1.404	1.405	0.804	0.813

TABLE 9-5 Solvents having the same refractive index and the same density at 25°C (*continued*)

Solvent 1	Solvent 2	Refractive index		Density, g/mL	
		1	2	1	2
3-Methyl-1-butanol	Hexanonitrile	1.404	1.405	0.805	0.801
3-Methyl-1-butanol	4-Heptanone	1.404	1.405	0.805	0.813
Hexanonitrile	4-Heptanone	1.405	1.405	0.801	0.813
Hexanonitrile	1-Pentanol	1.405	1.408	0.801	0.810
Hexanonitrile	2-Methyl-1-butanol	1.405	1.409	0.801	0.815
4-Heptanone	1-Pentanol	1.405	1.408	0.813	0.810
2-Ethoxyethanol	Pentanoic acid	1.405	1.406	0.926	0.936
2-Heptanone	1-Pentanol	1.406	1.408	0.811	0.810
2-Heptanone	2-Methyl-1-butanol	1.406	1.409	0.811	0.815
2-Heptanone	Dipentyl ether	1.406	1.410	0.811	0.799
2-Pentanol	3-Isopropyl-2-pentanone	1.407	1.409	0.804	0.808
1-Pentanol	Dipentyl ether	1.408	1.410	0.810	0.799
2-Methyl-1-butanol	Dipentyl ether	1.409	1.410	0.815	0.799
Isopentyl isopentanoate	Allyl alcohol	1.410	1.411	0.853	0.847
Dipentyl ether	2-Octanone	1.410	1.414	0.799	0.814
2,4-Dimethyldioxane	3-Chloropentene	1.412	1.413	0.935	0.932
2,4-Dimethyldioxane	Hexanoic acid	1.412	1.415	0.935	0.923
Diethyl malonate	Ethyl cyanoacetate	1.412	1.415	1.051	1.056
3-Chloropentene	Octanoic acid	1.413	1.415	0.932	0.923
2-Octanone	1-Hexanol	1.414	1.416	0.814	0.814
2-Octanone	Octanonitrile	1.414	1.418	0.814	0.810
3-Octanone	3-Methyl-2-heptanone	1.414	1.416	0.830	0.818
3-Methyl-2-heptanone	1-Hexanol	1.415	1.416	0.818	0.814
3-Methyl-2-heptanone	Octanonitrile	1.415	1.418	0.818	0.810
1-Hexanol	Octanonitrile	1.416	1.418	0.814	0.810
Dibutylamine	Allylamine	1.416	1.419	0.756	0.758
Allylamine	Methylcyclohexane	1.419	1.421	0.758	0.765
Butyrolactone	1,3-Propanediol	1.434	1.438	1.051	1.049
Butyrolactone	Diethylmaleate	1.434	1.438	1.051	1.064
2-Chloromethyl-2-propanol	Diethyl maleate	1.436	1.438	1.059	1.064
N-Methylmorpholine	Dibutyl decanedioate	1.436	1.440	0.924	0.932
1,3-Propanediol	Diethyl maleate	1.438	1.438	1.049	1.064
Methyl salicylate	Diethyl sulfide	1.438	1.442	0.836	0.831
Methyl salicylate	1-Butanethiol	1.438	1.442	0.836	0.837
1-Chlorodecane	Mesityl oxide	1.441	1.442	0.862	0.850
Diethylene glycol	Formamide	1.445	1.446	1.128	1.129
Diethylene glycol	Ethylene glycol diglycidyl ether	1.445	1.447	1.128	1.134
Formamide	Ethylene glycol diglycidyl ether	1.446	1.447	1.129	1.134
2-Methylmorpholine	Cyclohexanone	1.446	1.448	0.951	0.943

TABLE 9-5 Solvents having the same refractive index and the same density at 25°C (*continued*)

Solvent 1	Solvent 2	Refractive index		Density, g/mL	
		1	2	1	2
2-Methylmorpholine	1-Amino-2-propanol	1.446	1.448	0.951	0.961
Dipropylene glycol monoethyl ether	Tetrahydrofurfuryl alcohol	1.446	1.450	1.043	1.050
1-Amino-2-methyl-2-pentanol	2-Butylcyclohexanone	1.449	1.453	0.904	0.901
2-Propylcyclohexanone	4-Methylcyclohexanol	1.452	1.454	0.923	0.908
Carbon tetrachloride	4,5-Dichloro-1,3-dioxolan-2-one	1.459	1.461	1.584	1.591
N-Butyldiethanolamine	Cyclohexanol	1.461	1.465	0.965	0.968
D-α-Pinene	*trans*-Decahydronaphthalene	1.464	1.468	0.855	0.867
Propylbenzene	*p*-Xylene	1.490	1.493	0.858	0.857
Propylbenzene	Toluene	1.490	1.494	0.858	0.860
Phenyl 1-hydroxyphenyl ether	1,3-Dimorpholyl-2-propanol	1.491	1.493	1.081	1.094
Phenetole	Pyridine	1.505	1.507	0.961	0.978
2-Furanmethanol	Thiophene	1.524	1.526	1.057	1.059
m-Cresol	Benzaldehyde	1.542	1.544	1.037	1.041

TABLE 9-6 McReynolds' constants for stationary phases in gas chromatography

The McReynolds' constants listed are differences in retention index units between the reference compound run on squalane and on the other phases listed. The last entry in the table shows the absolute retention indices for the reference compounds on squalane. Reference compounds are (1) benzene, (2) 1-butanol, (3) 2-pentanone, (4) 1-nitropropane, and (5) pyridine. (Note that Rohrschneider's constants are based on these reference compounds and may differ slightly from the McReynolds' constants. The reference compounds for Rohrschneider's constants are (1) benzene, (2) ethanol, (3) 2-butanone, (4) nitromethane, and (5) pyridine.) The minimum temperature is that at which normal gas-liquid chromatography (GLC) behavior is expected. Below that temperature, the phase will be a solid or an extremely viscous gum. The maximum temperature is that above which the bleed rate will be excessive.

Liquid phase	Chemical type	Similar liquid phases	Temperature, °C		Reference compounds					
			Minimum	Maximum	1	2	3	4	5	Sum
Squalane	(2,6,10,15,19,23-Hexamethyltetracosane		20	150	0	0	0	0	0	0
Paraffin oil					9	5	2	6	11	33
Apolane-87	(24,24-Diethyl-19,29-dioctadecylheptatetracontane		30	280	21	10	3	12	25	71
Apiezon L			50	250	32	22	15	32	42	143
SE 30		SP-2100, SF 96	50	350	15	53	44	64	41	217
OV-101	Poly(dimethylsiloxane)	OV-1, DC 200, DC 410	50	350	17	57	45	67	43	229
OV-73	Poly(diphenyldimethylsiloxane), 5% : 95%	SE 52	0	325	32	72	65	98	67	334
SE 54	Poly(diphenylvinyldimethylsiloxane), 5% : 1% : 94%		50	300	33	72	66	99	67	337
OV-3	Poly(diphenyldimethylsiloxane), 10% : 90%		0	350	44	86	81	124	88	423
Dexsil 300	Poly(carboranemethylsiloxane)		50	500	47	80	103	148	96	474
Kel F Wax				150	55	67	114	143	116	495
Apiezon H				300	59	86	81	151	129	506
Dexsil 400	Carborane and methylphenylsilicone		50	500	72	108	118	166	123	587

TABLE 9-6 McReynolds' constants for stationary phases in gas chromatography (*continued*)

Liquid phase	Chemical type	Similar liquid phases	Temperature, °C		Reference compounds					
			Minimum	Maximum	1	2	3	4	5	Sum
OV-7	Poly(diphenyldimethylsiloxane), 20% : 80%	DC 550	20	350	69	113	111	171	128	592
Di(2-ethylhexyl) sebacate			0	125	72	168	108	180	125	653
Diisodecyl adipate				175	71	171	113	185	128	668
Decyl octyl adipate					79	179	119	193	134	704
Bis(2-ethylhexyl)-tetrachlorophthalate			0	150	112	150	123	168	181	734
Diisodecyl phthalate			0	175	84	173	137	218	155	767
Dinonyl phthalate			20	150	83	183	147	231	159	803
OV-11	Poly(diphenyldimethylsiloxane), 35% : 65%	DC 710	0	350	107	149	153	228	190	827
Dioctyl phthalate			20	125	92	186	150	236	167	831
Hallcomid M-18			40	150	79	268	130	222	146	845
OV-17	Poly(diphenyldimethylsiloxane), 50% : 50%		0	325	119	158	162	243	202	884
Dexsil 410	Carborane and methylcyanoethylsilicone		50	500	72	286	174	249	171	952
UCON LB-550-X			0	200	118	271	158	243	206	996
Span 80			15	150	97	266	170	216	268	1017
OV-22	Poly(diphenyldimethylsiloxane), 65% : 35%		0	350	160	188	191	283	253	1075
Polypropylene glycol			0	150	128	294	173	264	226	1085
Didecyl phthalate			10	175	136	255	213	320	235	1159
OV-25	Poly(diphenyldimethylsiloxane), 75% : 25%		0	350	178	204	208	305	280	1175
Polyphenyl ether OS-138 (6 rings)			0	225	182	233	228	313	293	1249

Neopentyl glycol sebacate		HI-EFF-3CP	50	225	172	327	225	344	326	1394
Squalene			0	100	152	341	328	329	344	1404
UCON 50-HB-280X			0	200	177	362	227	351	302	1419
Tricresyl phosphate			20	125	176	321	250	374	299	1420
Sucrose acetate isobutyrate			0	200	172	330	251	378	295	1426
QF-1			0	250	144	233	355	463	305	1500
OV-210	Poly(trifluoropropylsiloxane)	SP-2401, FS 1265	0	275	146	238	358	468	310	1520
OV-215	Poly(trifluoropropylmethyl-siloxane)	XE 60	0	275	149	240	363	478	315	1545
UCON 50-HB-2000	Emulphor ON-870		0	200	202	394	253	392	341	1582
Triton X-100			0	200	203	399	268	402	362	1634
UCON 50-HB-5100			0	200	214	418	278	421	375	1706
Siponate DS-10			0	200	99	569	320	344	388	1720
Tween 80			0	150	227	430	283	438	396	1747
XE-60	Poly(cyanoethylphenyl-methylsiloxane)		0	250	204	381	340	493	367	1785
OV-225	Poly(cyanopropylphenyl-methylsiloxane)		0	265	228	369	338	492	386	1813
Neopentyl glycol adipate		HI-EFF-3AP	50	225	232	421	311	461	424	1849
UCON 75-H-90000	Igepal CO-880		100	200	255	452	299	470	406	1882
Triton X-305			0	200	262	467	314	488	430	1961
Neopentyl glycol succinate		HI-EFF-3BP	50	230	272	469	366	539	474	2120
Igepal CO 990			100	200	298	508	345	540	475	2166
Carbowax 20M	Poly(ethylene glycol)	FFAP, SP-2300	25	275	322	536	368	572	510	2308
Epon 1001			50	225	284	489	406	539	601	2319
Carbowax 4000			60	200	325	551	375	582	520	2353
Ethylene glycol isophthalate		HI-EFF-2EP	100	225	326	508	425	607	561	2427
Ethylene glycol adipate		HI-EFF-2AP	100	225	372	576	453	655	617	2673
Butane-1,4-diol succinate		HI-EFF-4BP	50	225	369	591	457	661	629	2707
Phenyldiethanolamine succinate		HI-EFF-10BP	0	200	386	555	472	674	654	2741

TABLE 9-6 McReynolds' constants for stationary phases in gas chromatography (*continued*)

Liquid phase	Chemical type	Similar liquid phases	Temperature, °C		Reference compounds					
			Minimum	Maximum	1	2	3	4	5	Sum
Diethylene glycol adipate		HI-EFF-1AP, LAC-1-R-296, SP-2330	25	275	378	603	460	665	658	2764
Carbowax 1540			50	175	371	639	453	666	641	2770
Hyprose SP-80			0	175	336	742	492	639	727	2936
SILAR-7CP			0	250	440	638	605	844	673	3200
ECNSS-M			30	200	421	690	581	803	732	3227
EGSS-X			90	200	484	710	585	831	778	3388
Ethylene glycol phthalate		HI-EFF-2GP	100	200	453	697	602	816	872	3410
SILAR-9CP			0	250	489	725	631	910	778	3536
SILAR-10C		SP-2340	25	275	523	757	659	942	801	3682
Diethylene glycol succinate		HI-EFF-1BP, LAC-3-R-728	20	200	499	751	593	840	860	3543
Tetrahydroxyethylenediamine		THEED	0	150	463	942	626	801	893	3725
Tetracyanoethylated pentaerythritol			30	175	526	782	677	920	837	3742
Ethylene glycol succinate		HI-EFF-2BP	100	200	537	787	643	903	889	3759
1,2,3,4-Tetrakis-(2-cyanoethoxy)butane			110	200	617	860	773	1048	941	4239
1,2,3,4,5,6-Hexakis(2-cyanoethoxy)cyclohexane			125	150	567	825	713	978	901	3984
1,2,3-Tris-(2-cyanoethoxy)propane			0	175	593	857	752	1028	915	4145
N,N-Bis(2-cyanoethyl)-formamide			0	125	690	991	853	1110	1000	4644
OV-275	Dicyanoallylsilicone		25	250	781	1006	885	1177	1089	4938
Absolute retention index values on squalane for reference compounds					653	590	627	652	699	

McReynolds' Constants

The *Kovats retention indices* (R.I.) indicate where compounds will appear on a chromatogram with respect to unbranched alkanes injected with the sample. By definition, the R.I. for pentane is 500, for hexane is 600, for heptane is 700, and so on, regardless of the column used or the operating conditions, although the exact conditions and column must be specified, such as liquid loading, particular support used, and any pretreatment. For example, suppose that on a 20% squalane column at 100°C, the retention times for hexane, benzene, and octane are found to be 15, 16, and 25 min, respectively. On a graph of ln t'_R (naperian logarithm of the adjusted retention time) of the alkanes versus their retention indices, a R.I. of 653 for benzene is read off the graph. The number 653 for benzene (see the last line of Table 9-6 in the column headed "1" under "Reference compounds") means that it elutes halfway between hexane and heptane on a logarithmic time scale. If the experiment is repeated with a dinonyl phthalate column, the R.I. for benzene is found to be 736 (lying between heptane and octane), which implies that dinonyl phthalate will retard benzene slightly more than squalane will; that is, dinonyl phthalate is slightly more polar than squalane by $\Delta I = 83$ units (the entry in Table 9-6 for dinoyl phthalate in the column headed "1" under "Reference compounds"). The difference gives a measure of solute-solvent interaction due to all intermolecular forces other than London dispersion forces. The latter are the principal solute-solvent effects with squalane.

Now the overall effects due to hydrogen bonding, dipole moment, acid-base properties, and molecular configuration can be expressed as

$$\sum \Delta I = ax' + by' + cz' + du' + es'$$

where $x' = \Delta I$ for benzene (the column headed "1" in Table 9-6, intermolecular forces typical of aromatics and olefins), $y' = \Delta I$ for 1-butanol (the column headed "2" in Table 9-6, electron attraction typical of alcohols, nitriles, acids, and nitro and alkyl monochlorides, dichlorides and trichlorides), $z' = \Delta I$ for 2-pentanone (the column headed "3" in Table 9-6, electron repulsion typical of ketones, ethers, aldehydes, esters, epoxides, and dimethylamino derivatives), $u' = \Delta I$ for 1-nitropropane (the column headed "4" in Table 9-6, typical of nitro and nitrile derivatives), and $s' = \Delta I$ for pyridine (or dioxane) (the column headed "5" in Table 9-6).

SECTION 10

POLYMERS, RUBBERS, FATS, OILS AND WAXES

POLYMERS

Polymers are mixtures of macromolecules with similar structures and molecular weights that exhibit some average characteristic properties. In some polymers long segments of linear polymer chains are oriented in a regular manner with respect to one another. Such polymers have many of the physical characteristics of crystals and are said to be *crystalline.* Polymers that have polar functional groups show a considerable tendency to be crystalline. Orientation is aided by alignment of dipoles on different chains. Van der Waals' interactions between long hydrocarbon chains may provide sufficient total attractive energy to account for a high degree of regularity within the polymers.

Irregularities such as branch points, comonomer units, and cross-links lead to *amorphous* polymers. They do not have true melting points but instead have glass transition temperatures at which the rigid and glass like material becomes a viscous liquid as the temperature is raised.

Elastomers. Elastomers is a generic name for polymers that exhibit rubberlike elasticity. Elastomers are soft yet sufficiently elastic that they can be stretched several hundred percent under tension. When the stretching force is removed, they retract rapidly and recover their original dimensions.

Polymers that soften or melt and then solidify and regain their original properties on cooling are called *thermoplastic.* A thermoplastic polymer is usually a single strand of linear polymer with few if any cross-links.

Thermosetting Polymers. Polymers that soften or melt on warming and then become infusible solids are called *thermosetting.* The term implies that thermal decomposition has not taken place. Thermosetting plastics contain a cross-linked polymer network that extends through the finished article, making is stable to heat and insoluble in organic solvents. Many molded plastics are shaped while molten and are then heated further to become rigid solids of desired shapes.

Synthetic Rubbers. Synthetic rubbers are polymers with rubberlike characteristics that are prepared from dienes or olefins. Rubbers with special properties can also be prepared from other polymers, such as polyacrylates, fluorinated hydrocarbons, and polyurethanes.

Structural Differences. Polymers exhibit structural differences. A *linear* polymer consists of long segments of single strands that are oriented in a regular manner with respect to one another. *Branched* polymers have substituents attached to the repeating units that extend the polymer laterally. When these units participate in chain propagation and link together chains, a *cross-linked* polymer is formed. A *ladder* polymer results when repeating units have a tetravalent structure such that a polymer consists of two backbone chains regularly cross-linked at short intervals.

Generally polymers involve bonding of the most substituted carbon of one monomeric unit to the least substituted carbon atom of the adjacent unit in a *head-to-tail* arrangement. Substituents appear on alternate carbon atoms. *Tacticity* refers to the configuration of substituents relative to the backbone axis. In an *isotactic* arrangement, substituents are on the same plane of the backbone axis; that is, the configuration at each chiral center is identical.

```
 Y  Y  Y  Y
 |  |  |  |
—C—C—C—C—
```

In a *syndiotactic* arrangement, the substituents are in an ordered alternating sequence, appearing alternately on one side and then on the other side of the chain, thus

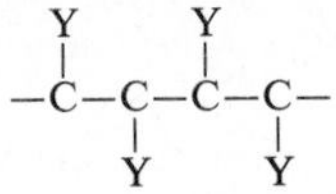

```
 Y     Y
 |     |
-C--C--C--C-
    |     |
    Y     Y
```

In an *atactic* arrangement, substituents are in an unordered sequence along the polymer chains.

Copolymerization. Copolymerization occurs when a mixture of two or more monomer types polymerizes so that each kind of monomer enters the polymer chain. The fundamental structure resulting from copolymerization depends on the nature of the monomers and the relative rates of monomer reactions with the growing polymer chain. A tendency toward alternation of monomer units is common.

—X—Y—X—Y—X—Y—

Random copolymerization is rather unusual. Sometimes a monomer which does not easily form a homopolymer will readily add to a reactive group at the end of a growing polymer chain. In turn, that monomer tends to make the other monomer much more reactive.

In *graft copolymers* the chain backbone is composed of one kind of monomer and the branches are made up of another kind of monomer.

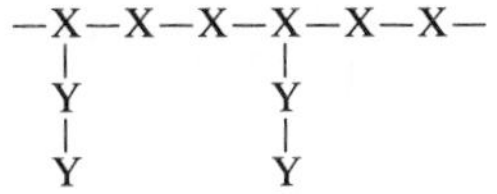

```
-X-X-X-X-X-X-
 |     |
 Y     Y
 |     |
 Y     Y
```

The structure of a *block copolymer* consists of a homopolymer attached to chains of another homopolymer.

—XXXX—YYY—XXXX—YYY—

Configurations around any double bond give rise to *cis* and *trans* stereoisomerism.

ADDITIVES TO POLYMERS

Antioxidants

Antioxidants markedly retard the rate of autoxidation throughout the useful life of the polymer. Chain-terminating antioxidants have a reactive —NH or —OH functional group and include compounds such as secondary aryl amines or hindered phenols. They function by transfer of hydrogen to free radicals, principally to peroxy radicals. Butylated hydroxytoluene is a widely used example.

Peroxide-decomposing antioxidants destroy hydroperoxides, the sources of free radicals in polymers. Phosphites and thioesters such as tris(nonylphenyl) phosphite, distearyl pentaerythritol diphosphite, and dialkyl thiodipropionates are examples of peroxide-decomposing antioxidants.

Antistatic Agents

External antistatic agents are usually quaternary ammonium salts of fatty acids and ethoxylated glycerol esters of fatty acids that are applied to the plastic surface. Internal antistatic agents are compounded into plastics during processing. Carbon blacks provide a conductive path through the bulk of the plastic. Other types of internal agents must bloom to the surface after compounding in order to be active. These latter materials are ethoxylated fatty amines and ethoxylated glycerol esters of fatty acids, which often must be individually selected to match chemically each plastic type.

Antistatic agents require ambient moisture to function. Consequently their effectiveness is dependent on the relative humidity. They provide a broad range of protection at 50% relative humidity. Much below 20% relative humidity, only materials which provide a conductive path through the bulk of the plastic to ground (such as carbon black) will reduce electrostatic charging.

Chain-Transfer Agents

Chain-transfer agents are used to regulate the molecular weight of polymers. These agents react with the developing polymer and interrupt the growth of a particular chain. The products, however, are free radicals that are capable of adding to monomers and initiating the formation of new chains. The overall effect is to reduce the average molecular weight of the polymer without reducing the rate of polymerization. Branching may occur as a result of chain transfer between a growing but rather short chain with another and longer polymer chain. Branching may also occur if the radical end of a growing chain abstracts a hydrogen from a carbon atom four or five carbons removed from the end. Thiols are commonly used as chain-transfer agents.

Coupling Agents

Coupling agents are molecular bridges between the interface of an inorganic surface (or filler) and an organic polymer matrix. Titanium-derived coupling agents interact with the free protons at the inorganic interface to form organic monomolecular layers on the inorganic surface. The titanate-coupling-agent molecule has six functions:

$$\overset{1}{(\mathrm{RO})_m}\text{—}\overset{2}{\mathrm{Ti}}\text{—}(\overset{3}{\mathrm{O}}\text{—}\overset{4}{\mathrm{Y}}\text{—}\overset{5}{\mathrm{R}^2}\text{—}\overset{6}{\mathrm{Z}})_n$$

where

Type	m	n
Monoalkoxy	1	3
Coordinate	4	2
Chelate	1	2

Function 1 is the attachment of the hydrolyzable portion of the molecule to the surface of the inorganic (or proton-bearing) species.

Function 2 is the ability of the titanate molecule to transesterify.

Function 3 affects performance as determined by the chemistry of alkylate, carboxyl, sulfonyl, phenolic, phosphate, pyrophosphate, and phosphite groups.

Function 4 provides van der Waals' entanglement via long carbon chains.

Function 5 provides thermoset reactivity via functional groups such as methacrylates and amines.

Function 6 permits the presence of two or three pendent organic groups. This allows all functionality to be controlled to the first-, second-, or third-degree levels.

Silane coupling agents are represented by the formula

$$Z{-}R{-}SiY_3$$

where Y represents a hydrolyzable group (typically alkoxy); Z is a functional organic group, such as amino, methacryloxy, epoxy; and R typically is a small aliphatic linkage that serves to attach the functional organic group to silicon in a stable fashion. Bonding to surface hydroxy groups of inorganic compounds is accomplished by the $-SiY_3$ portion, either by direct bonding of this group or more commonly via its hydrolysis product $-Si(OH)_3$. Subsequent reaction of the functional organic group with the organic matrix completes the coupling reaction and establishes a covalent chemical bond from the organic phase through the silane coupling agent to the inorganic phase.

Flame Retardants

Flame retardants are thought to function via several mechanisms, dependent upon the class of flame retardant used. Halogenated flame retardants are thought to function principally in the vapor phase either as a diluent and heat sink or as a free-radical trap that stops or slows flame propagation. Phosphorus compounds are thought to function in the solid phase by forming a glaze or coating over the substrate that prevents the heat and mass transfer necessary for sustained combustion. With some additives, as the temperature is increased, the flame retardant acts as a solvent for the polymer, causing it to melt at lower temperatures and flow away from the ignition source.

Mineral hydrates, such as alumina trihydrate and magnesium sulfate heptahydrate, are used in highly filled thermoset resins.

Foaming Agents (Chemical Blowing Agents)

Foaming agents are added to polymers during processing to form minute gas cells throughout the product. Physical foaming agents include liquids and gases. Compressed nitrogen is often used in injection molding. Common liquid foaming agents are short-chain aliphatic hydrocarbons in the C_5 to C_7 range and their chlorinated or fluorinated analogs.

The chemical foaming agent used varies with the temperature employed during processing. At relatively low temperatures (15 to 200°C), the foaming agent is often 4,4′-oxybis(benzenesulfonylhydrazide) or *p*-toluenesulfonylhydrazide. In the midrange (160 to 232°C), either sodium hydrogen carbonate or 1,1′azobisformamide is used. For the high range (200 to 285°C), there are *p*-toluenesulfonyl semicarbazide, 5-phenyltetrazole and analogs, and trihydrazinotriazine.

Inhibitors

Inhibitors slow or stop polymerization by reacting with the initiator or the growing polymer chain. The free radical formed from an inhibitor must be sufficiently unreactive that it does not function as a chain-transfer agent and begin another growing chain. Benzoquinone is a typical free-radical chain inhibitor. The resonance-stabilized free radical usually dimerizes or disproportionates to produce inert products and end the chain process.

Lubricants

Materials such as fatty acids are added to reduce the surface tension and improve the handling qualities of plastic films.

Plasticizers

Plasticizers are relatively nonvolatile liquids which are blended with polymers to alter their properties by intrusion between polymer chains. Diisooctyl phthalate is a common plasticizer. A plasticizer must be compatible with the polymer to avoid bleeding out over long periods of time. Products containing plasticizers tend to be more flexible and workable.

Ultraviolet Stabilizers

2-Hydroxybenzophenones represent the largest and most versatile class of ultraviolet stabilizers that are used to protect materials from the degradative effects of ultraviolet radiation. They function by absorbing ultraviolet radiation and by quenching electronically excited states.

Hindered amines, such as 4-(2,2,6,6-tetramethylpiperidinyl) decanedioate, serve as radical scavengers and will protect thin films under conditions in which ultraviolet absorbers are ineffective. Metal salts of nickel, such as dibutyldithiocarbamate, are used in polyolefins to quench singlet oxygen or electronically excited states of other species in the polymer. Zinc salts function as peroxide decomposers.

Vulcanization and Curing

Originally, vulcanization implied heating natural rubber with sulfur, but the term is now also employed fur curing polymers. When sulfur is employed, sulfide and disulfide cross-links form between polymer chains. This provides sufficient rigidity to prevent *plastic flow.* Plastic flow is a process in which coiled polymers slip past each other under an external deforming force; when the force is released, the polymer chains do not completely return to their original positions.

Organic peroxides are used extensively for the curing of unsaturated polyester resins and the polymerization of monomers having vinyl unsaturation. The —O—O— bond is split into free radicals which can initiate polymerization or cross-linking of various monomers or polymers.

TABLE 10-1 Plastic families

Acetals	Alloys (*continued*)
Acrylics	Acrylonitrile-butadiene-styrene-polycarbonate alloy (ABS-PC)
Poly(methyl methacrylate) (PMMA)	Allyls
Poly(acrylonitrile)	Allyl-diglycol-carbonate polymer
Alkyds	Diallyl phthalate (DAP) polymer
Alloys	Cellulosics
Acrylic-poly(vinyl chloride) alloy	Cellulose acetate resin
Acrylonitrile-butadiene-styrene-poly(vinyl chloride) alloy (ABS-PVC)	Cellulose-acetate-propionate resin

TABLE 10-1 Plastic families (*continued*)

Cellulosics (*continued*)
Cellulose-acetate-butyrate resin
Cellulose nitrate resin
Ethyl cellulose resin
Rayon
Chlorinated polyether
Epoxy
Fluorocarbons
Poly(tetrafluoroethylene) (PTFE)
Poly(chlorotrifluoroethylene) (PCTFE)
Perfluoroalkoxy (PFA) resin
Fluorinated ethylene-propylene (FEP) resin
Poly(vinylidene fluoride) (PVDF)
Ethylene-chlorotrifluoroethylene copolymer
Ethylene-tetrafluoroethylene copolymer
Poly(vinyl fluoride) (PVF)
Melamine formaldehyde
Melamine phenolic
Nitrile resins
Phenolics
Polyamides
Nylon 6
Nylon 6/6
Nylon 6/9
Nylon 6/12
Nylon 11
Nylon 12
Aromatic nylons
Poly(amide-imide)
Poly(aryl ether)
Polycarbonate (PC)
Polyesters
Poly(butylene terephthalate) (PBT) [aso called polytetramethylene terephthalate (PTMT)]
Poly(ethylene terephthalate) (PET)
Unsaturated polyesters (SMC, BMC)
Butadiene–maleic acid copolymer (BMC)
Styrene–maleic acid copolymer (SMC)
Polyimide
Poly(methylpentene)
Polyolefins (PO)
Low-density polyethylene (LDPE)
High-density polyethylene (HDPE)
Ultrahigh-molecular-weight polyethylene (UHMWPE)
Polypropylene (PP)
Polybutylene (PB)
Polyallomers
Poly(phenylene oxide)
Poly(phenylene sulfide) (PPS)
Polyurethanes
Silicones
Styrenics
Polystyrene(PS)
Acrylonitrile-butadiene-styrene (ABS) copolymer
Sytrene-acrylonitrile (SAN) copolymer
Styrene-butadiene copolymer
Sulfones
Polysulfone (PSF)
Poly(ether sulfone)
Poly(phenyl sulfone)
Thermoplastic elastomers
Polyolefin
Polyester
Block copolymers
Styrene-butadiene block copolymer
Styrene-isoprene block copolymer
Styrene-ethylene block copolymer
Styrene-butylene block copolymer
Urea formaldehyde
Vinyls
Poly(vinyl chloride) (PVC)
Poly(vinyl acetate) (PVAC)
Poly(vinylidene chloride)
Poly(vinyl butyrate) (PVB)
Poly(vinyl formal)
Poly(vinyl alcohol) (PVAL)

FORMULAS AND KEY PROPERTIES OF PLASTIC MATERIALS

Acetals

Homopolymer

Acetal homopolymers are prepared from formaldehyde and consist of high-molecular-weight linear polymers of formaldehyde.

$$\mathrm{H{-}\overset{\overset{\displaystyle H}{|}}{C}{=}O} \rightarrow \left[-\overset{\overset{\displaystyle H}{|}}{\underset{\underset{\displaystyle H}{|}}{C}}-O- \right]_n$$

The good mechanical properties of this homopolymer result from the ability of the oxymethylene chains to pack together into a highly ordered crystalline configuration as the polymers change from the molten to the solid state.

Key properties include high melt point, strength and rigidity, good frictional properties, and resistance to fatigue. Higher molecular weight increases toughness but reduces melt flow.

Copolymer

Acetal copolymers are prepared by copolymerization of 1,3,5-trioxane with small amounts of a comonomer. Carbon-carbon bonds are distributed randomly in the polymer chain. These carbon-carbon bonds help to stabilize the polymer against thermal, oxidative, and acidic attack.

Acrylics

Poly(methyl methacrylate)

The monomer used for poly(methyl methacrylate), 2-hydroxy-2-methylpropanenitrile, is prepared by the following reaction:

$$CH_3{-}\underset{\underset{\displaystyle O}{\|}}{C}{-}CH_3 + HCN \rightarrow CH_3{-}\overset{\overset{\displaystyle OH}{|}}{\underset{\underset{\displaystyle CN}{|}}{C}}{-}CH_3$$

2-Hydroxy-2-methylpropanenitrile is then reacted with methanol (or other alcohol) to yield methacrylate ester. Free-radical polymerization is initiated by peroxide or azo catalysts and produce poly(methyl methacrylate) resins having the following formula:

$$\left[-CH_2{-}\overset{\overset{\displaystyle CH_3}{|}}{\underset{\underset{\displaystyle COOCH_3}{|}}{C}}{-} \right]_n$$

Key properties are improved resistance to heat, light, and weathering. This polymer is unaffected by most detergents, cleaning agents, and solutions of inorganic acids, alkalies, and aliphatic hydrocarbons. Poly(methyl methacrylate) has light transmittance of 92% with a haze of 1 to 3% and its clarity is equal to glass.

Poly(methyl acrylate)

The monomer used for preparing poly(methyl acrylate) is produced by the oxidation of propylene. The resin is made by free-radical polymerization initiated by peroxide or azo catalysts and has the following formula:

$$\left[\begin{array}{l} -CH_2-\underset{|}{C}H- \\ \qquad\quad COOCH_3 \end{array} \right]_n$$

Resins vary from soft, elastic, film-forming materials to hard plastics.

Poly(acrylic acid) and Poly(methacrylic acid)

Glacial acrylic acid and glacial methacrylic acid can be polymerized to produce water-soluble polymers having the following structures:

$$\left[-CH_2-\underset{\displaystyle COOH}{\underset{|}{C}H}- \right]_n \quad \left[-CH_2-\overset{\displaystyle CH_3}{\overset{|}{\underset{\displaystyle COOH}{\underset{|}{C}}}}- \right]_n$$

These monomers provide a means for introducing carboxyl groups into copolymers. In copolymers these acids can improve adhesion properties, improve freeze-thaw and mechanical stability of polymer dispersions, provide stability in alkalies (including ammonia), increase resistance to attack by oils, and provide reactive centers for cross-linking by divalent metal ions, diamines, or epoxides.

Functional Group Methacrylate Monomers

Hydroxyethyl methacrylate and dimethylaminoethyl methacrylate produce polymers having the following formulas:

$$\left[-CH_2-\overset{\displaystyle CH_3}{\overset{|}{\underset{\displaystyle COOCH_2CH_2OH}{\underset{|}{C}}}}- \right]_n \quad \left[-CH_2-\overset{\displaystyle CH_3}{\overset{|}{\underset{\displaystyle COOCH_2CH_2(CH_3)_2}{\underset{|}{C}}}}- \right]_n$$

The use of hydroxyethyl (also hydroxypropyl) methacrylate as a monomer permits the introduction of reactive hydroxyl groups into the copolymers. This offers the possibility for subsequent cross-linking with an HO-reactive difunctional agent (diisocyanate, diepoxide, or melamine-formaldehyde resin). Hydroxyl groups promote adhesion to polar substrates.

Use of dimethylaminoethyl (also *tert*-butylaminoethyl) methacrylate as a monomer permits the introduction of pendent amino groups which can serve as sites for secondary cross-linking, provide a way to make the copolymer acid-soluble, and provide anchoring sites for dyes and pigments.

Poly(acrylonitrile)

Poly(acrylonitrile) polymers have the following formula:

$$\left[-CH_2-\underset{\displaystyle CN}{\underset{|}{C}H}- \right]_n$$

Alkyds

Alkyds are formulated from polyester resins, cross-linking monomers, and fillers of mineral or glass. The unsaturated polyester resins used for thermosetting alkyds are the reaction products of polyfunctional organic alcohols (glycols) and dibasic organic acids.

Key properties of alkyds are dimensional stability, colorability, and arc track resistance. Chemical resistance is generally poor.

Alloys

Polymer alloys are physical mixtures of structurally different homopolymers or copolymers. The mixture is held together by secondary intermolecular forces such as dipole interaction, hydrogen bonding, or van der Waals' forces.

Homogeneous alloys have a single glass transition temperature which is determined by the ratio of the components. The physical properties of these alloys are averages based on the composition of the alloy.

Heterogeneous alloys can be formed when graft or block copolymers are combined with a compatible polymer. Alloys of incompatible polymers can be formed if an interfacial agent can be found.

Allyls

Diallyl Phthalate (and Diallyl 1,3-Phthalate)

These allyl polymers are prepared from

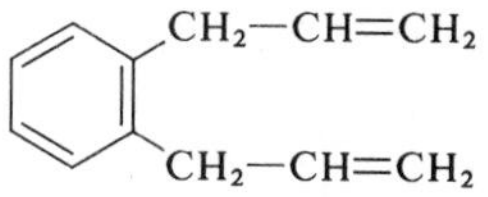

$$C_6H_4(CH_2{-}CH{=}CH_2)_2$$

These resulting polymers are solid, linear, internally cyclized, thermoplastic structures containing unreacted allylic groups spaced at regular intervals along the polymer chain.

Molding compounds with mineral, glass, or synthetic fiber filling exhibit good electrical properties under high humidity and high temperature conditions, stable low-loss factors, high surface and volume resistivity, and high arc and track resistance.

Cellulosics

Cellulose Triacetate

Cellulose triacetate is prepared according to the following reaction:

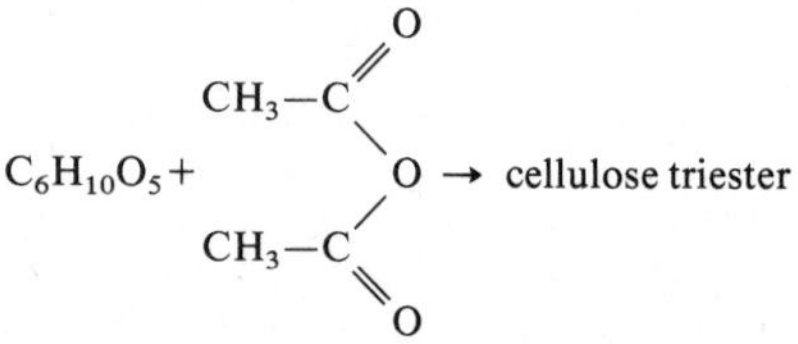

$$C_6H_{10}O_5 + (CH_3{-}C(=O))_2O \rightarrow \text{cellulose triester}$$

Because cellulose triacetate has a high softening temperature, it must be processed in solution. A mixture of dichloromethane and methanol is a common solvent.

Cellulose triacetate sheeting and film have good gauge uniformity and good optical clarity. Cellulose triacetate products have good dimensional stability and resistance to water and have good folding endurance and burst strength. It is highly resistant to solvents such as acetone. Cellulose triacetate products have good heat resistance and a high dielectric constant.

Cellulose Acetate, Propionate, and Butyrate

Cellulose acetate is prepared by hydrolyzing the triester to remove some of the acetyl groups; the plastic-grade resin contains 38 to 40% acetyl. The propionate and butyrate esters are made by substituting propionic acid and its anhydride (or butyric acid and its anhydride) for some of the acetic acid and acetic anhydride. Plastic grades of cellulose-acetate-propionate resin contain 39 to 47% propionyl and 2 to 9% acetyl; cellulose-acetate-butyrate resins contain 26 to 39% butyryl and 12 to 15% acetyl.

These cellulose esters form tough, strong, stiff, hard plastics with almost unlimited color possibilities. Articles made from these plastics have a high gloss and are suitable for use in contact with food.

Cellulose Nitrate

Cellulose nitrate is prepared according to the following reaction:

$$C_6H_{10}O_5 + HNO_3 \rightarrow [-C_6H_7O_2(OH)(ONO_2)_2-]_n$$

The nitrogen content for plastics is usually about 11%, for lacquers and cement base it is 12%, and for explosives it is 13%. The standard plasticizer added is camphor.

Key properties of cellulose nitrate are good dimensional stability, low water absorption, and toughness. Its disadvantages are its flammability and lack of stability to heat and sunlight.

Ethyl Cellulose

Ethyl cellulose is prepared by reacting cellulose with caustic to form caustic cellulose, which is then reacted with chloroethane to form ethyl cellulose. Plastic-grade material contains 44 to 48% ethoxyl.

Although not as resistant as cellulose esters to acids, it is much more resistant to bases. An outstanding feature is its toughness at low temperatures.

Rayon

Viscose rayon is obtained by reacting the hydroxy groups of cellulose with carbon disulfide in the presence of alkali to give xanthates. When this solution is poured (spun) into an acid medium, the reaction is reversed and the cellulose is regenerated (coagulated).

Epoxy

Epoxy resin is prepared by the following condensation reaction:

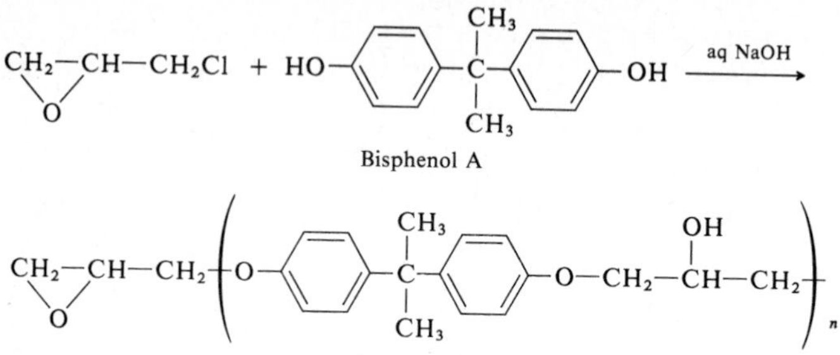

$$\underset{\diagdown\, O \,\diagup}{CH_2-CH}-CH_2Cl + HO-C_6H_4-\underset{CH_3}{\overset{CH_3}{\overset{|}{\underset{|}{C}}}}-C_6H_4-OH \xrightarrow{\text{aq NaOH}}$$

Bisphenol A

$$\underset{\diagdown\, O \,\diagup}{CH_2-CH}-CH_2 \left(O-C_6H_4-\underset{CH_3}{\overset{CH_3}{\overset{|}{\underset{|}{C}}}}-C_6H_4-O-CH_2-\overset{OH}{\overset{|}{CH}}-CH_2 \right)_n$$

The condensation leaves epoxy end groups that are then reacted in a separate step with nucleophilic compounds (alcohols, acids, or amines). For use as an adhesive, the epoxy resin and the curing resin (usually an aliphatic polyamine) are packaged separately and mixed together immediately before use.

Epoxy novolac resins are produced by glycidation of the low-molecular-weight reaction products of phenol (or cresol) with formaldehyde. Highly cross-linked systems are formed that have superior performance at elevated temperatures.

Fluorocarbon

Poly(tetrafluoroethylene)

Poly(tetrafluoroethylene) is prepared from tetrafluoroethylene and consists of repeating units in a predominantly linear chain:

$$F_2C=CF_2 \rightarrow [-CF_2-CF_2-]_n$$

Tetrafluoroethylene polymer has the lowest coefficient of friction of any solid. It has remarkable chemical resistance and a very low brittleness temperature (-100°C). Its dielectric constant and loss factor are low and stable across a broad temperature and frequency range. Its impact strength is high.

Fluorinated Ethylene-Propylene Resin

Polymer molecules of fluorinated ethylene-propylene consist of predominantly linear chains with this structure:

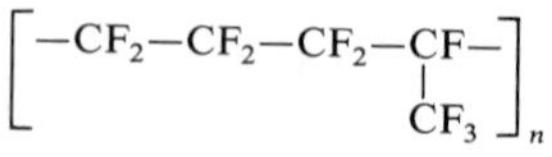

$$\left[-CF_2-CF_2-CF_2-\underset{CF_3}{\underset{|}{CF}}- \right]_n$$

Key properties are its flexibility, translucency, and resistance to all known chemicals except molten alkali metals, elemental fluorine and fluorine precursors at elevated temperatures, and concentrated perchloric acid. It withstands temperatures from −270° to 250°C and may be sterilized repeatedly by all known chemical and thermal methods.

Perfluoroalkoxy Resin

Perfluoroalkoxy resin has the following formula:

$$\left[-CF_2-CF_2-\underset{\underset{\displaystyle R}{|}}{\underset{\displaystyle O}{\underset{|}{C}}}F-CF_2-CF_2- \right]_n \qquad \text{where R is } -C_nF_{2n+1}$$

It resembles polytetrafluoroethylene and fluorinated ethylene propylene in its chemical resistance, electrical properties, and coefficient of friction. Its strength, hardness, and wear resistance are about equal to the former plastic and superior to that of the latter at temperatures above 150°C.

Poly(vinylidene fluoride)

Poly(vinylidene fluoride) consists of linear chains in which the predominant repeating unit is

$$[-CH_2-CF_2-]_n$$

It has good weathering resistance and does not support combustion. It is resistant to most chemicals and solvents and has greater strength, wear resistance, and creep resistance than the preceding three fluorocarbon resins.

Poly(1-chloro-1,2,2-trifluoroethylene)

Poly(1-chloro-1,2,2-trifluoroethylene consists of linear chains in which the predominant repeating unit is

$$\left[-CF_2-\underset{\underset{\displaystyle Cl}{|}}{C}F- \right]_n$$

It possesses outstanding barrier properties to gases, especially water vapor. It is surpassed only by the fully fluorinated polymers in chemical resistance. A few solvents dissolve it at temperatures above 100°C, and it is swollen by a number of solvents, especially chlorinated solvents. It is harder and stronger than perfluorinated polymers, and its impact strength is lower.

Ethylene-Chlorotrifluoroethylene Copolymer

Ethylene-chlorotrifluoroethylene copolymer consists of linear chains in which the predominant 1:1 alternating copolymer is

$$\left[-CH_2-CH_2-CF_2-\underset{\underset{\displaystyle Cl}{|}}{C}F- \right]_n$$

This copolymer has useful properties from cryogenic temperatures to 180°C. Its dielectric constant is low and stable over a broad temperature and frequency range.

Ethylene-Tetrafluoroethylene Copolymer

Ethylene-tetrafluoroethylene copolymer consists of linear chains in which the repeating unit is

$$[-CH_2-CH_2-CF_2-CF_2-]_n$$

Its properties resemble those of ethylene-chlorotrifluoroethylene copolymer.

Poly(vinyl fluoride)

Poly(vinyl fluoride) consists of linear chains in which the repeating unit is

$$[-CH_2-CHF-]_n$$

It is used only as a film, and it has good resistance to abrasion and resists staining. It also has outstanding weathering resistance and maintains useful properties from −100 to 150°C.

Nitrile Resins

The principal monomer of nitrile resins is acrylonitrile (see "Polyacrylonitrile"), which constitutes about 70% by weight of the polymer and provides the polymer with good gas barrier and chemical resistance properties. The remainder of the polymer is 20 to 30% methylacrylate (or styrene), with 0 to 10% butadiene to serve as an impact-modifying termonomer.

Melamine Formaldehyde

The monomer used for preparing melamine formaldehyde is formed as follows:

$$C_3N_3(NH_2)_3 + 6\ HCHO \longrightarrow C_3N_3[N(CH_2OH)_2]_3$$

Hexamethylolmelamine

Hexamethylolmelamine can further condense in the presence of an acid catalyst; ether linkages can also form (see "Urea Formaldehyde"). A wide variety of resins can be obtained by careful selection of pH, reaction temperature, reactant ratio, amino monomer, and extent of condensation. Liquid coating resins are prepared by reacting methanol or butanol with the initial methylolated products. These can be used to produce hard, solvent-resistant coatings by heating with a variety of hydroxy, carboxyl, and amide functional polymers to produce a cross-linked film.

Phenolics

Phenol-Formaldehyde Resin

Phenol-formaldehyde resin is prepared as follows:

$$C_6H_5OH + H_2C{=}O \rightarrow [-C_6H_2(OH)CH_2-]_n$$

One-Stage Resins. The ratio of formaldehyde to phenol is high enough to allow the thermosetting process to take place without the addition of other sources of cross-links.

Two-Stage Resins. The ratio of formaldehyde to phenol is low enough to prevent the thermosetting reaction from occurring during manufacture of the resin. At this point the resin is termed *novolac* resin. Subsequently, hexamethylenetetramine is incorporated into the material to act as a source of chemical cross-links during the molding operation (and conversion to the thermoset or cured state).

Polyamides

Nylon 6, 11, and 12

This class of polymers is polymerized by addition reactions of ring compounds that contain both acid and amine groups on the monomer.

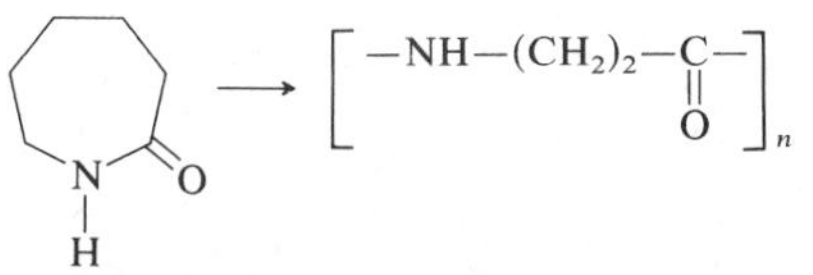

$$\text{(ring: N–H, C=O)} \longrightarrow \left[-NH-(CH_2)_2-\underset{\underset{O}{\|}}{C}- \right]_n$$

Nylon 6 is polymerized from 2-oxohexamethyleneimine (6 carbons); nylon 11 and 12 are made this way from 11- and 12-carbon rings, respectively.

Nylon 6/6, 6/9, and 6/12

As illustrated below, nylon 6/6 is polymerized from 1,6-hexanedioic acid (six carbons) and 1,6-hexanediamine (six carbons).

$$HOOC-(CH_2)_4-COOH \;+\; H_2N-CH_2-(CH_2)_4-CH_2-NH_2 \rightarrow$$

1,6-Hexanedioic acid 1,6-Hexanediamine

$$\left[-NH-(CH_2)_6-BH-\underset{\underset{O}{\|}}{C}-(CH_2)_4-\underset{\underset{O}{\|}}{C}- \right]_n$$

Poly(hexamethylene 1,6- hexanediamide)

Other nylons are made this way from different combinations of monomers to produce types 6/9, 6/10, and 6/12.

Nylon 6 and 6/6 possess the maximum stiffness, strength, and heat resistance of all the types of nylon. Type 6/6 has a higher melt temperature, whereas type 6 has a higher impact resistance and better processibility. At a sacrifice in stiffness and heat resistance, the higher analogs of nylon are useful primarily for improved chemical resistance in certain environments (acids, bases, and zinc chloride solutions) and for lower moisture absorption.

Aromatic nylons, $[-NH-C_6H_4-CO-]_n$, (also called aramids) have specialty uses because of their improved clarity.

Poly(amide-imide)

Poly(amide-imide) is the condensation polymer of 1,2,4-benzenetricarboxylic anhydride and various aromatic diamines and has the general structure:

$$\left[-R-\overset{H}{\overset{|}{N}}-\underset{O}{\underset{\|}{C}}-C_6H_3\left\langle \begin{matrix} \overset{O}{\overset{\|}{C}} \\ \underset{O}{\underset{\|}{C}} \end{matrix} \right\rangle N- \right]_n$$

It is characterized by high strength and good impact resistance, and retains its physical properties at temperatures up to to 260°C. Its radiation (gamma) resistance is good.

Polycarbonate

Polycarbonate is a polyester in which dihydric (or polyhydric) phenols are joined through carbonate linkages. The general-purpose type of polycarbonate is based on 2,2-bis(4′-hydroxybenzene)propane (bisphenol A) and has the general structure:

$$\left[-C_6H_4-\overset{CH_3}{\underset{CH_3}{\overset{|}{\underset{|}{C}}}}-C_6H_4-O-\overset{O}{\overset{\|}{C}}-O- \right]_n$$

Polycarbonates are the toughest of all thermoplastics. They are window-clear, amazingly strong and rigid, autoclavable, and nontoxic. They have a brittleness temperature of −135°C.

Polyester

Poly(butylene terephthalate)

Poly(butylene terephthalate) is prepared in a condensation reaction between dimethyl terephthalate and 1,4-butanediol and its repeating unit has the general structure

$$\left[-O-CH_2CH_2CH_2CH_2-O-C_6H_4-\overset{O}{\overset{\|}{C}}- \right]_n$$

This thermoplastic shows good tensile strength, toughness, low water absorption, and good frictional properties, plus good chemical resistance and electrical properties.

Poly(ethylene terephthalate)

Poly(ethylene terephthalate) is prepared by the reaction of either terephthalic acid or dimethyl terephthalate with ethylene glycol, and its repeating unit has the general structure

$$\left[-O-CH_2CH_2-O-C_6H_4-\overset{O}{\overset{\|}{C}}- \right]_n$$

The resin has the ability to be oriented by a drawing process and crystallized to yield a high-strength product.

Unsaturated Polyesters

Unsaturated polyesters are produced by reaction between two types of dibasic acids, one of which is unsaturated, and an alcohol to produce an ester. Double bonds in the body of the unsaturated dibasic acid are obtained by using maleic anhydride or fumaric acid.

PCTA Copolyester

Poly(1,4-cyclohexanedimethylene terephthalic acid) (PCTA) copolyester is a polymer of cyclohexanedimethanol and terephthalic acid, with another acid substituted for a portion of the trephthalic acid otherwise required. It has the following formula:

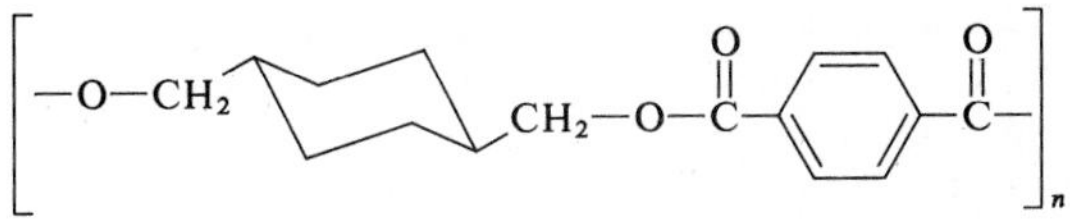

$$\left[-O-CH_2-C_6H_{10}-CH_2-O-\overset{O}{\overset{\|}{C}}-C_6H_4-\overset{O}{\overset{\|}{C}}-\right]_n$$

Polyimides

Polyimides have the following formula:

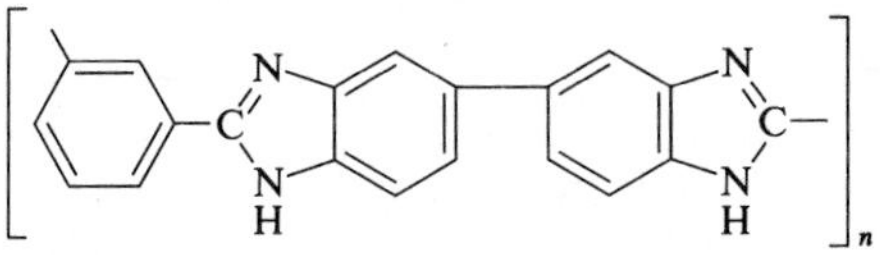

$$\left[C_6H_5-C\overset{N}{\underset{NH}{\diagdown}}C_6H_3-C_6H_3\overset{N}{\underset{NH}{\diagdown}}C-\right]_n$$

They are used as high-temperature structural adhesives since they become rubbery rather than melt at about 300°C.

Poly(methylpentene)

Poly(methylpentene) is obtained by a Ziegler-type catalytic polymerization of 4-methyl-1-pentene.

Its key properties are its excellent transparency, rigidity, and chemical resistance, plus its resistance to impact and to high temperatures. It withstands repeated autoclaving, even at 150°C.

Polyolefins

Polyethylene

Polymerization of ethylene results in an essentially straight-chain high-molecular-weight hydrocarbon.

$$CH_2{=}CH_2 \rightarrow [-CH_2-CH_2-]_n$$

Branching occurs to some extent and can be controlled. Minimum branching results in a "high-density" polyethylene because of its closely packed molecular chains. More branching gives a less compact solid known as "low-density" polyethylene.

A key property is its chemical inertness. Strong oxidizing agents eventually cause some oxidation, and some solvents cause softening or swelling, but there is no known solvent for polyethylene at room temperature. The brittleness temperature is −100°C for both types. Polyethylene has good low-temperature toughness, low water absorption, and good flexibility at subzero temperatures.

Polypropylene

The polymerization of propylene results in a polymer with the following structure:

$$CH_2{=}CH{-}CH_3 \rightarrow \left[-CH_2-\underset{\displaystyle CH_3}{\underset{|}{CH}}- \right]_n$$

The desired form in homopolymers is the isotactic arrangement (at least 93% is required to give the desired properties). Copolymers have a random arrangement. In block copolymers a secondary reactor is used where active polymer chains can further polymerize to produce segments that use ethylene monomer.

Polypropylene is translucent and autoclavable and has no known solvent at room temperature. It is slightly more susceptible to strong oxidizing agents than polyethylene.

Polybutylene

Polybutylene is composed of linear chains having an isotactic arrangement of ethyl side groups along the chain backbone.

$$CH_2{=}CH{-}CH_2{-}CH_3 \rightarrow \left[-CH_2-\underset{\displaystyle \underset{\displaystyle CH_3}{\underset{|}{CH_2}}}{\underset{|}{CH}}- \right]_n$$

It has a helical conformation in the stable crystalline form.

Polybutylene exhibits high tear, impact, and puncture resistance. It also has low creep, excellent chemical resistance, and abrasion resistance with coilability.

Ionomer

Ionomer is the generic name for polymers based on sodium or zinc salts of ethylene–methacrylic acid copolymers in which interchain ionic bonding, occurring randomly between the long-chain polymer molecules, produces solid-state properties.

The abrasion resistance of ionomers is outstanding, and ionomer films exhibit optical clarity. In composite structures ionomers serve as a heat-seal layer.

Poly(phenylene sulfide)

Poly(phenylene sulfide) has the following formula:

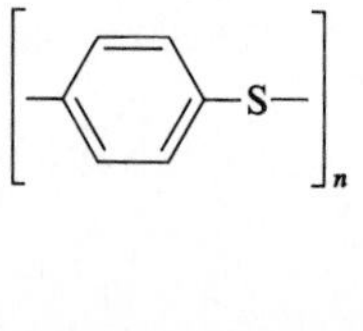

$$\left[-C_6H_4-S- \right]_n$$

The recurring *para*-substituted benzene rings and sulfur atoms form a symmetrical rigid backbone.

The high degree of crystallization and the thermal stability of the bond between the benzene ring and sulfur are the two properties responsible for the polymer's high melting point, thermal stability, inherent flame retardance, and good chemical resistance. There are no known solvents of poly(phenylene sulfide) that can function below 205°C.

Polyurethane

Foams

Polyurethane foams are prepared by the polymerization of polyols with isocyanates.

$$H{+}O{-}CH_2{-}CH_2{+}_nOH + \text{excess } CH_3C_6H_3(N{=}C{=}O)_2 \longrightarrow$$

$$O{=}C{=}N{-}C_6H_3(CH_3){-}NH{-}\overset{O}{\overset{\|}{C}}{-}\left[O{-}CH_2{-}CH_2\right]_n{-}O{-}\overset{O}{\overset{\|}{C}}{-}NH{-}C_6H_3(CH_3){-}N{=}C{=}O$$

Commonly used isocyanates are toluene diisocyanate, methylene diphenyl isocyanate, and polymeric isocyanates. Polyols used are macroglycols based on either polyester or polyether. The former [poly(ethylene phthalate) or poly(ethylene 1,6-hexanedioate)] have hydroxyl groups that are free to react with the isocyanate. Most flexible foam is made from 80/20 toluene diisocyanate (which refers to the ratio of 2,4-toluene diisocyanate to 2,6-toluene diisocyanate). High-resilience foam contains about 80% 80/20 toluene diisocyanate and 20% poly(methylene diphenyl isocyanate), while semiflexible foam is almost always 100% poly(methylene diphenyl isocyanate). Much of the latter reacts by trimerization to form isocyanurate rings.

Flexible foams are used in mattresses, cushions, and safety applications. Rigid and semi-flexible foams are used in structural applications and to encapsulate sensitive components to protect them against shock, vibration, and moisture. Foam coatings are tough, hard, flexible, and chemically resistant.

Elastrometric Fiber

Elastometric fibers are prepared by the polymerization of polymeric polyols with diisocyanates.

$$\begin{array}{l} CH_2{-}({-}OCH_2CH_2{-}O{-})_xH \\ | \\ CH{-}({-}O{-}CH_2CH_2{-}O{-})_yH \\ | \\ CH_2{-}({-}OCH_2CH_2{-}O{-}_zH \end{array} + CH_3C_6H_3(N{=}C{=}O)_2 \longrightarrow \text{essentially linear polymers}$$

Polymeric polyols Diisocyanate

The structure of elastometric fibers is similar to that illustrated for polyurethane foams.

Silicones

Silicones are formed in the following multistage reaction:

$$R_2SiCl_2 + 2H_2O \rightarrow R_2Si(OH)_2 + 2HCl$$
$$\downarrow$$
$$[-Si(R)_2-O-]_n$$

The silanols formed above are unstable and undergo dehydration. On polycondensation, they give polysiloxanes (or silicones) which are characterized by their three-dimensional branched-chain structure. Various organic groups introduced within the polysiloxane chain impart certain characteristics and properties to these resins.

Methyl groups impart water repellency, surface hardness, and noncombustibility.

Phenyl groups impart resistance to temperature variations, flexibility under heat, resistance to abrasion, and compatibility with organic products.

Vinyl groups strengthen the rigidity of the molecular stucture by creating easier cross-linkage of molecules.

Methoxy and alkoxy groups facilitate cross-linking at low temperatures.

Oils and gums are nonhighly branched or straight-chain polymers whose viscosity increases with the degree of polycondensation.

Styrenics

Polystyrene

Polystyrene has the following formula:

$$\left[-CH_2-\underset{\displaystyle C_6H_5}{CH}-\right]_n$$

Polystyrene is rigid with excellent dimensional stability, has good chemical resistance to aqueous solutions, and is an extremely clear material.

Impact polystyrene contains polybutadiene added to reduce brittleness. The polybutadiene is usually dispersed as a discrete phase in a continuous polystyrene matrix. Polystyrene can be grafted onto rubber particles, which assures good adhesion between the phases.

Acrylonitrile-Butadiene-Styrene (ABS) Copolymers

This basic three-monomer system can be tailored to yield resins with a variety of properties. Acrylonitrile contributes heat resistance, high strength, and chemical resistance. Butadiene contributes impact strength, toughness, and retention of low-temperature properties. Styrene contributes gloss, processibility, and rigidity. ABS polymers are composed of discrete polybutadiene particles grafted with the styrene-acrylonitrile copolymer; these are dispersed in the continuous matrix of the copolymer.

Styrene-Acrylonitrile (SAN) Copolymers

SAN resins are random, amorphous copolymers whose properties vary with molecular weight and copolymer composition. An increase in molecular weight or in acrylonitrile content generally enhances the physical properties of the copolymer but at some loss in ease of processing and with a slight increase in polymer color.

SAN resins are rigid, hard, transparent thermoplastics which process easily and have good dimensional stability—a combination of properties unique in transparent polymers.

Sulfones

Below are the formulas for three polysulfones.

$$\left[-C_6H_4-C(CH_3)_2-C_6H_4-O-C_6H_4-SO_2-C_6H_4-O-\right]_n$$

Polysulfone

$$\left[-O-C_6H_4-SO_2-C_6H_4-\right]_n$$

Poly(ester sulfone)

$$\left[-O-C_6H_4-SO_2-C_6H_4-O-C_6H_4-C_6H_4-\right]_n$$

Poly(phenyl sulfone)

The isopropylidene linkage imparts chemical resistance, the ether linkage imparts temperature resistance, and the sulfone linkage imparts impact strength. The brittleness temperature of polysulfones is −100°C. Polysulfones are clear, strong, nontoxic, and virtually unbreakable. They do not hydrolyze during autoclaving and are resistant to acids, bases, aqueous solutions, aliphatic hydrocarbons, and alcohols.

Thermoplastic Elastomers

Polyolefins

In these thermoplastic elastomers the hard component is a crystalline polyolefin, such as polyethylene or polypropylene, and the soft portion is composed of ethylene-propylene rubber. Attractive forces between the rubber and resin phases serve as labile cross-links. Some contain a chemically cross-linked rubber phase that imparts a higher degree of elasticity.

Styrene-Butadiene-Styrene Block Copolymers

Styrene blocks associate into domains that form hard regions. The midblock, which is normally butadiene, ethylene-butene, or isoprene blocks, forms the soft domains. Polystyrene domains serve as cross-links.

Polyurethanes

The hard portion of polyurethane consists of a chain extender and polyisocyanate. The soft component is composed of polyol segments.

Polyesters

The hard portion consists of copolyester, and the soft portion is composed of polyol segments.

Vinyl

Poly(vinyl chloride) (PVC)

Polymerization of vinyl chloride results in the formation of a polymer with the following formula:

$$CH_2{=}CHCl \rightarrow \left[-CH_2-\underset{\displaystyle Cl}{\underset{|}{CH}}- \right]_n$$

When blended with phthalate ester plasticizers, PVC becomes soft and pliable.

Its key properties are good resistance to oils and a very low permeability to most gases.

Poly(vinyl acetate)

Poly(vinyl acetate) has the following formula:

$$\left[-CH_2-\underset{\displaystyle O-CO-CH_3}{\underset{|}{CH}}- \right]_n$$

Poly(vinyl acetate) is used in latex water paints because of its weathering, quick-drying, recoatability, and self-priming properties. It is also used in hot-melt and solution adhesives.

Poly(vinyl alcohol)

Poly(vinyl alcohol) has the following formula:

$$\left[-CH_2-\underset{\displaystyle OH}{\underset{|}{CH}}- \right]_n$$

It is used in adhesives, paper coating and sizing, and textile warp size and finishing applications.

Poly(vinyl butyral)

Poly(vinyl butyral) is prepared according to the following reaction:

$$\left[-CH_2-\underset{\displaystyle OH}{\underset{|}{CH}}- \right]_n + CH_3CH_2CH_2CHO \rightarrow \left[\begin{array}{l} -CH_2-CH-CH_2-CH- \\ \qquad\quad | \qquad\qquad\quad | \\ \qquad\quad O-CH-\!\!-O \\ \qquad\qquad\quad | \\ \qquad\qquad\quad CH_2-CH_2-CH_3 \end{array} \right]_n$$

Its key characteristics are its excellent optical and adhesive properties. It is used as the interlayer film for safety glass.

Poly(vinylidene chloride)

Poly(vinylidene chloride) is prepared according to the following reaction:

$$CH_2{=}CCl_2 + CH_2{=}CHCl \rightarrow [-CH_2-CCl_2-CH_2-CHCl-]_n$$

Random copolymer

Urea Formaldehyde

The reaction of urea with formaldehyde yields the following products, which are used as monomers in the preparation of urea formaldehyde resin.

$$H_2N-CO-NH_2 + H_2CO \rightarrow H_2N-CO-NH-CH_2OH$$
$$+ HOCH_2-NH-CO-NH-CH_2OH$$

The reaction conditions can be varied so that only one of these monomers is formed. 1-Hydroxymethylurea and 1,3-bis(hydroxymethyl)urea condense in the presence of an acid catalyst to produce urea formaldehyde resins. A wide variety of resins can be obtained by careful selection of the pH, reaction temperature, reactant ratio, amino monomer, and degree of polymerization. If the reaction is carried far enough, an infusible polymer network is produced.

Liquid coating resins are prepared by reacting methanol or butanol with the initial hydroxymethylureas. Ether exchange reactions between the amino resin and the reactive sites on the polymer produce a cross-linked film.

TABLE 10-2 Properties of commercial plastics

Properties	Acetal				
	Homopolymer	Copolymer	20% glass-reinforced homopolymer	25% glass-reinforced copolymer	21% poly(tetrafluoroethylene)-filled homopolymer
Physical					
Melting temperature, °C					
Crystalline	175	175	181	175	181
Amorphous					
Specific gravity	1.42	1.41	1.56	1.61	1.54
Water absorption (24 h), %	0.25–0.40	0.22	0.25	0.29	0.20
Dielectric strength, $KV \cdot mm^{-1}$	19.7	19.7	19.3	22.8	15.7
Electrical					
Volume (dc) resistivity, ohm-cm	10^{15}	10^{15}	5×10^{14}		3×10^{16}
Dielectric constant (60 Hz)	3.7	3.7	3.9		3.1
Dielectric constant (10^6 Hz)	3.7	3.7	3.9		3.1
Dissipation (power) factor (60 Hz)					
Dissipation factor (10^6 Hz)	0.005	0.005	0.005		0.005
Mechanical					
Compressive modulus, $10^3\ lb \cdot in^{-2}$	670	450			

Compressive strength, rupture or 1% yield, 10^3 lb · in^{-2}	5.29	16 (10% yield)	18 (10% yield)	17 (10% yield)	13 (10% yield)
Elongation at break, %	25–75	40–75	7	3	15–22
Flexural modulus at 23°C, 10^3 lb · in^2	380–430	375	730	1100	340–350
Flexural strength, rupture or yield, 10^3 lb · in^{-2}	14	13	15	28	
Hardness, Rockwell (or Shore)	M94	M78	M90	M79	M78
Impact strength (Izod) at 23°C, J · m^{-1}	69–123	53–80	43	96	37–64
Tensile modulus, 10^3 lb · in^{-2}	520	410	1000	1250	
Tensile strength at break, 10^3 lb · in^{-2}	10	10	8.5	18.5	7.6
Tensile yield strength, 10^3 lb · in^{-2}	9.5–12	8.5			6.9–7.6
Thermal					
Burning rate, mm · min^{-1}	27.9				
Coefficient of linear thermal expansion, 10^{-6}°C	100	85	36–81		75
Deflection temperature under flexural load (264 lb · in^{-2}), °C	124	110	157	163	100
Maximum recommended service temperature, °C	84				
Specific heat, cal · g^{-1}	0.35				
Thermal conductivity, W · m^{-1} · K^{-1}	0.23	0.23			

TABLE 10-2 Properties of commercial plastics (*continued*)

Properties	Acrylic				Alkyd, molded	Alloy	
	Poly(methyl methacrylate)	Cast sheet	Impact-modified	Heat-resistant		Acrylic poly(vinyl chloride) alloy	Acrylonitrile-butadiene-styrene-poly(vinyl chloride) alloy
<u>Physical</u>							
Melting temperature, °C							
Crystalline	90–105	90–105	80–100	100–125		105	
Amorphous							
Specific gravity	1.17–1.20	1.18–1.20	1.11–1.18	1.16–1.19	2.22–2.24		
Water absorption (24 h), %	0.1–0.4	0.2–0.4	0.2–0.8	0.2–0.3		0.06	
Dielectric strength, $KV \cdot mm^{-1}$	15.7–19.9	17.7–21.7	15.0–19.9	15.7–19.9		>15.7	19.7
<u>Electrical</u>							
Volume (dc) resistivity, ohm-cm	$>10^{14}$	$>10^{14}$					
Dielectric constant (60 Hz)	3.3–4.5	3.5–4.5			3.8–5.0		
Dielectric constant (10^6 Hz)		3.0–3.5			3.6–4.7		
Dissipation (power) factor (60 Hz)		0.04–0.06			0.012–0.026		
Dissipation factor (10^6 Hz)		0.02–0.03			0.01–0.016		
<u>Mechanical</u>							
Compressive modulus, 10^4 lb · in^{-2}	370–460	390–475	240–370	350–460		330–400	

Compressive strength, rupture or 1% yield, 10^3 lb · in^{-2}	12–18	11–19	4–14	17	16–20	8.4	
Elongation at break, %	2–10	2–7	20–70	3–5		100	
Flexural modulus at 23°C, 10^3 lb · in^{-2}	420–460	390–475	200–380	460–500		330–400	340
Flexural strength, rupture or yield, 10^3 lb · in^{-2}	13–19	12–17	7–13	12–16		10.7	9.6
Hardness, Rockwell (or Shore)	M85–M105	M80–M100	R105–R120	M95–M105	E76	R99–R105	R100
Impact strength (Izod) at 23°C, J · m^{-1}	16–27	16–21	43–133	16–21	27–240	800	560
Tensile modulus, 10^3 lb · in^{-2}	380–450	350–450	200–400	350–460		330–335	330
Tensile strength at break, 10^3 lb · in^{-2}	7–11	8–11	5–9	10	4.5–6.5	6.5	5.8
Tensile yield strength, 10^3 lb · in^{-2}					10–13		
Thermal							
Burning rate, mm · min^{-1}		0.5–2.2			Self-extinguishing		
Coefficient of linear thermal expansion, 10^{-6}°C	50–90	50–90	50–80	50–60	40–55		46
Deflection temperature under flexural load (264 lb · in^{-2}), °C	74–99	71–102	74–95	88–104	177–204	71	
Maximum recommended service temperature, °C		60–71			220		
Specific heat, cal · g^{-1}	0.36	0.35					
Thermal conductivity, W · m^{-1} · K^{-1}	0.17–0.25	0.17–0.25	0.17–0.21	0.19			

TABLE 10-2 Properties of commercial plastics (*continued*)

Properties	Alloy	Allyl			Cellulosic		
	Polycarbonate acrylonitrile-butadiene-styrene alloy	Allyl-diglycol-carbonate polymer	Diallyl phthalate molding		Cellulose acetate		Cellulose-acetate-butyrate resin
			Glass-filled	Mineral-filled	Sheet	Molding	Sheet
Physical							
Melting temperature, °C							
Crystalline		Thermoset	Thermoset	Thermoset	230	230	140
Amorphous	150						
Specific gravity	1.12–1.20	1.3–1.4	1.7–2.0	1.65–1.85	1.27–1.34	1.29–1.34	1.15–1.22
Water absorption (24 h), %	0.21–0.24	0.2	0.12–0.35	0.2–0.5	2–7	1.7–6.5	0.9–2.2
Dielectric strength, $kV \cdot mm^{-1}$	17.7	15.0	15.7–17.7	15.7–17.7	11–24	9–24	9–18
Electrical							
Volume (dc) resistivity, ohm-cm					10^{10}–10^{13}	10^{10}–10^{13}	10^{10}–10^{12}
Dielectric constant (60 Hz)					3.4–7.4	3.5–7.5	3.7–4.3
Dielectric constant (10^6 Hz)					3.2–7.0	3.2–7.0	3.3–3.8
Dissipation (power) factor (60 Hz)					0.01–0.06	0.01–0.06	0.01–0.04
Dissipation factor (10^6 Hz)					0.01–0.06	0.01–0.10	0.01–0.04
Mechanical							
Compressive modulus, $10^3\ lb \cdot in^{-2}$		300					

Compressive strength, rupture or 1% yield, 10^3 lb · in^{-2}	11	21–23	25–35	20–32	22–33	25–36	
Elongation at break, %	10–15		3–5	3–5	17–40	6–40	50–100
Flexural modulus at 23°C, 10^3 lb · in^{-2}	300–400	250–330	1200–1500	1000–1400			740–1300
Flexural strength, rupture or yield, 10^3 lb · in^{-2}	13.0–13.7	6–13	9–20	8.5–11	6–10	2–16	4–9
Hardness, Rockwell (or Shore)	R117	M95–M100	E80–E87	E61	R85–R120	R100–R123	R50–R95
Impact strength (Izod) at 23°C, J · m^{-1}	560	11–21	21–800	16–43	107–454	53–214	133–288
Tensile modulus, 10^3 lb · in^{-2}	370–380	300	1400–2200	1200–2200			200–250
Tensile strength at break, 10^3 lb · in^{-2}	7.0–7.3	5–6	6–11	5–8	4.5–8.0	1.9–9.0	2.6–6.9
Tensile yield strength, 10^3 lb · in^{-2}	8.5				2.2–7.4	4.1–7.6	
Thermal							
Burning rate, mm · min^{-1}						1.3–3.8	1.3–3.8
Coefficient of linear thermal expansion, 10^{-6}°C	63–67	5.4–9.6	0.68–2.4	2.8	100–150	80–180	110–170
Deflection temperature under flexural load (264 lb · in^{-2}), °C	104–116	60–88	165–288+	160–288	44–91	51–98	49–58
Maximum recommended service temperature, °C							
Specific heat, cal · g^{-1}					0.3–0.4	0.3–0.42	0.3–0.4
Thermal conductivity, W · m^{-1} · K^{-1}	0.25–0.38	0.20–0.21	0.21–0.63	0.30–1.04	0.17–0.34	0.17–0.34	0.17–0.34

TABLE 10-2 Properties of commercial plastics (*continued*)

Properties	Cellulosic					Epoxy	
						Bisphenol	
	Cellulose-acetate butyrate resin, molding	Cellulose-acetate-propionate resin, molding	Ethyl cellulose	Cellulose nitrate	Chlorinated polyether	Glass-fiber-reinforced	Mineral-filled
Physical							
Melting temperature, °C							
Crystalline	140	190	135		125	Thermoset	Thermoset
Amorphous							
Specific gravity	1.15–1.22	1.17–1.24	1.09–1.17	1.35–1.40	1.4	1.6–2.0	1.6–2.1
Water absorption (24 h), %	0.9–2.2	1.2–2.8	0.8–1.8			0.04–0.20	0.03–0.20
Dielectric strength, $kV \cdot mm^{-1}$	9–13	12–17.7	13.8–19.7			9.8–15.7	9.8–15.7
Electrical							
Volume (dc) resistivity, ohm-cm	10^{10}–10^{12}			10^{10}			
Dielectric constant (60 Hz)	3.5–6.4			7.0–7.5			
Dielectric constant (10^6 Hz)	3.2–6.2		3.01	6.6			
Dissipation (power) factor (60 Hz)	0.01–0.04						
Dissipation factor (10^6 Hz)	0.01–0.04						
Mechanical							
Compressive modulus, $10^3 lb \cdot in^{-2}$						3000	

Compressive strength, rupture or 1% yield, 10^3 lb · in^{-2}	2.1–7.5	2.4–7.0		2.1–8.0		18,000–40,000	18,000–40,000
Elongation at break, %	40–88	29–100	5–40	40–45	600–800	4	
Flexural modulus at 23°C, 10^3 lb · in^{-2}	90–300	120–350				2–4.5	
Flexural strength, rupture or yield, 10^3 lb · in^{-2}	1.8–9.3	2.9–11.4	4–12	9–11	5	8–30	6–18
Hardness, Rockwell (or Shore)	R31–R116	R10–R122	R50–R115	R95–R115	R100	M100–M112	M100–M112
Impact strength (Izod) at 23°C, J · m^{-1}	53–582	27 to no break	21	267–374	21	16–533	16–22
Tensile modulus, 10^3 lb · in^{-2}	50–200	60–215		190–220		3	
Tensile strength at break, 10^3 lb · in^{-2}	2.6–6.9	2.0–7.8	2–8	7–8	1.5–1.8	5–20	4–10
Tensile yield strength, 10^3 lb · in^{-2}							
Thermal							
Burning rate, mm · min^{-1}	1.3–3.8				Self-extinguishing		
Coefficient of linear thermal expansion, 10^{-6}°C	110–170	110–170	100–200	80–120	6.6	11–50	20–60
Deflection temperature under flexural load (264 lb · in^{-2}), °C	44–94	44–109	45–88	60–71	185	107–260	107–260
Maximum recommended service temperature, °C					255		
Specific heat, cal · g^{-1}	0.3–0.4			0.31–0.41			
Thermal conductivity, W · m^{-1} · K^{-1}	0.17–0.30	0.17–0.30	0.16–0.30	0.23		0.17–0.42	0.17–1.48

TABLE 10-2 Properties of commercial plastics (*continued*)

Properties	Epoxy			Fluorocarbon			
	Casting resin		Novolac resin	Poly(tetrafluoroethylene)		Poly(chloro-trifluoro-ethylene)	Perfluoroalkoxy
	Unfilled	Flexible	Mineral-filled	Granular	Glass-fiber-reinforced		
Physical							
Melting temperature, °C							
Crystalline	Thermoset	Thermoset	Thermoset	327	327	220	310
Amorphous							
Specific gravity	1.11–1.40	1.05–1.35	1.7–2.1	2.14–2.20	2.2–2.3	2.1–2.2	2.12–2.17
Water absorption (24 h), %	0.08–0.15	0.27–0.50	0.05–0.2	0.01		0.03	
Dielectric strength, $kV \cdot mm^{-1}$	11.8–19.7	9.3–15.8	11.8–13.8	18.9	12.6	19.7–23	19.7
Electrical							
Volume (dc) resistivity, ohm-cm	10^{12}–10^{17}			10^{18}		10^{18}	
Dielectric constant (60 Hz)	3.5–5.0			2.1		2.3–2.7	
Dielectric constant (10^6 Hz)	3.5–5.0			2.1		2.3–2.5	
Dissipation (power) factor (60 Hz)				0.0002		0.001	
Dissipation factor (10^6 Hz)				0.0002		0.005	
Mechanical							
Compressive modulus, $10^3\ lb \cdot in^{-2}$				60			

Compressive strength, rupture or 1% yield, 10^3 lb · in^{-2}	15–25	1–14	30	1.7		4.6–7.4	
Elongation at break, %	3–6	20–70	2–4	200–400	200–300	80–250	300
Flexural modulus at 23°C, 10^3 lb · in^{-2}			2000	80	235	120	
Flexural strength, rupture or yield, 10^{-3} lb · in^{-2}	13–21	1–13	16–20		2	7.4–9.3	
Hardness, Rockwell (or Shore)	M80–M110			(D50–D55)	(D60–D70)	R75–R95	(D64)
Impact strength (Izod) at 23°C, J · m^{-1}	10.7–53	187–267	21	160	144	133–160	No break
Tensile modulus, 10^3 lb · in^{-2}	350	1–350		58–80		150–300	
Tensile strength at break, 10^3 lb · in^{-2}	4–13	2–10	6–12	2–5	2–2.7	4.5–6	4–4.3
Tensile yield strength, 10^3 lb · in^{-2}			30				
Thermal							
Burning rate, mm · min^{-1}				Self-extinguishing	Self-extinguishing	Self-extinguishing	
Coefficient of linear thermal expansion, 10^{-6}°C	45–65	20–100	22–30	100	77–100	70	
Deflection temperature under flexural load (264 lb · in^{-2}), °C	46–288	23–121	149–260	121 (66 lb · in^{-2})		126 (66 lb · in^{-2})	74 (66 lb · in^{-2}
Maximum recommended service temperature, °C				260		200	
Specific heat, cal · g^{-1}				0.25		0.22	
Thermal conductivity, W · m^{-1} · K^{-1}	0.17–0.21			0.25	0.34–0.40	0.19–0.22	0.25

TABLE 10-2 Properties of commercial plastics (*continued*)

Properties	Fluorocarbon					Melamine formaldehyde	
	Fluorinated ethylene-propylene resin	Poly(vinylidene fluoride)	Ethylene-tetrafluoroethylene copolymer		Ethylene-chlorotrifluoro-ethylene copolymer	Cellulose-filled	Glass-fiber-reinforced
			Unfilled	Glass-fiber-reinforced			
Physical							
Melting temperature, °C							
Crystalline	275	156	270	270	245	Thermoset	Thermoset
Amorphous							
Specific gravity	2.14–2.17	1.75–1.78	1.7	1.8	1.68	1.47–1.52	1.5–2.0
Water absorption (24 h), %	<0.01	0.04–0.06	0.03	0.02	0.01	0.1–0.8	0.09–1.3
Dielectric strength, kV · mm^{-1}	20–24	10	16	17	19	11–16	5–15
Electrical							
Volume (dc) resistivity, ohm-cm							
Dielectric constant (60 Hz)	2.1	8–9	2.6		2.6		
Dielectric constant (10^6 Hz)	2.1	8–9	2.6		2.6		
Dissipation (power) factor (60 Hz)		High					
Dissipation factor (10^6 Hz)		High					
Mechanical							
Compressive modulus, 10^3 lb · in^{-2}		120	120	1200	240		

Compressive strength, rupture or 1% yield, 10^3 lb · in^{-2}	2.2	8.7–10	7.1	10		33–45	20–35
Elongation at break, %	250–330	25–500	100–400	8	200–300	0.6–1.0	0.6
Flexural modulus at 23°C, 10^3 lb · in^{-2}	80–95	200	200	950	240	1100	
Flexural strength, rupture or yield, 10^3 lb · in^{-2}		8.6–11	5.5	10.7	7	9–16	14–23
Hardness, Rockwell (or Shore)	(D60–D65)	(D80)	R50 (D75)	R74	R95	M115–M125	M115
Impact strength (Izod) at 23°C, J · m^{-1}	No break	192–214	No break	480	No break	11–21	32–961
Tensile modulus, 10^3 lb · in^{-2}	50	120	120	1200	240	1.1–1.4	1.6–2.4
Tensile strength at break, 10^3 lb · in^{-2}	2.7–3.1	5.5–7.4	6.5	12	7	5–13	5–10.5
Tensile yield strength, 10^3 lb · in^{-2}							
Thermal							
Burning rate, mm · min^{-1}	Not combustible	Not combustible	Not combustible	Not combustible	Not combustible	Self-extinguishing	Self-extinguishing
Coefficient of linear thermal expansion, 10^{-6}°C	83–105	85	59	10–32	80	40–45	15–28
Deflection temperature under flexural load (264 lb · in^{-2}), °C	70 (66 lb · in^{-2})	80–90	71	210	77	177–199	190–204
Maximum recommended service temperature, °C	205	150				210	
Specific heat, cal · g^{-1}	0.28						
Thermal conductivity, W · m^{-1} · K^{-1}	0.25	0.19–0.24	0.24		0.16	0.27–0.41	0.41–0.49

TABLE 10-2 Properties of commercial plastics (*continued*)

Properties	Melamine phenolic, woodflour- and cellulose-filled	Nitrile	Phenolic				
			Unfilled	Woodflour-filled	Glass-fiber-reinforced	Cellulose-filled	Mineral-filled
Physical							
Melting temperature, °C							
Crystalline	Thermoset		Thermoset	Thermoset	Thermoset	Thermoset	Thermoset
Amorphous		95					
Specific gravity	1.5–1.7	1.15	1.24–1.32	1.37–1.46	1.69–2.0	1.38–1.42	1.42–1.84
Water absorption (24 h), %	0.3–0.65	0.28	0.1–0.36	0.3–1.2	0.03–1.2	0.5–0.9	0.1–0.3
Dielectric strength, $kV \cdot mm^{-1}$	8.7–12.8	8.7–9.5	9.8–15.8	10.2–15.8	5.5–15.8	11.8–15	7.9–13.8
Electrical							
Volume (dc) resistivity, ohm-cm		1.9×10^{15}	1×10^{12} to 7×10^{12}				
Dielectric constant (60 Hz)			6.5–7.5				
Dielectric constant (10^6 Hz)			4.0–5.5				
Dissipation (power) factor (60 Hz)			0.10–0.15				
Dissipation factor (10^6 Hz)			0.04–0.05				
Mechanical							
Compressive modulus, 10^3 $lb \cdot in^{-2}$							

Compressive strength, rupture or 1% yield, 10^3 lb · in^{-2}	26–30	12	18–32	25–31	26–70	22–31	22.5–34.6
Elongation at break, %	0.4–0.8	3–4	1.5–2.0	0.4–0.8	0.2	1–2	0.1–0.5
Flexural modulus at 23°C, 10^3 lb · in^{-2}	1000–1200	500–590	700–1500	1000–1200	2000–33,000	900–1300	1000–2000
Flexural strength, rupture or yield, 10^3 lb · in^{-2}	8–10	14	11–17	7–14	15–60	5.5–11	11–14
Hardness, Rockwell (or Shore)	E95–E100	M72–M76	M93–M120	M100–M115	E54–E101	M95–M115	E88
Impact strength (Izod) at 23°C, J · m^{-1}	11–21	80–256	13–21	11–32	27–960	21–59	14–19
Tensile modulus, 10^3 lb · in^{-2}	800–1700	510–580	700–1500	800–1700	1900–3300		2400
Tensile strength at break, 10^3 lb · in^{-2}	6–8	9	6–9	5–9	7–18	3.5–6.5	6–9.7
Tensile yield strength, 10^3 lb · in^{-2}			12–15				
Thermal							
Burning rate, mm · min^{-1}			Self-extinguishing				
Coefficient of linear thermal expansion, 10^{-6}°C	10–40	66	68	30–45	8–21	20–31	19–26
Deflection temperature under flexural load (264 lb · in^{-2}), °C	140–154	73	74–80	149–188	177–316	149–177	320–246
Maximum recommended service temperature, °C							
Specific heat, cal · g^{-1}							
Thermal conductivity, W · m^{-1} · K^{-1}	0.17–0.30	0.26	0.15	0.17–0.34	0.34–0.59	0.25–0.38	0.42–0.57

TABLE 10-2 Properties of commercial plastics (*continued*)

	Polyamide						
	Nylon 6				Nylon 6/6		
Properties	Molding and extrusion	30–35% glass-fiber-reinforced	High-impact copolymer	Molding	33% glass-fiber-reinforced	Molybdenum disulfide-filled	Nylon 6/6–nylon 6 copolymer
Physical							
Melting temperature, °C							
Crystalline	216	216	216	265	265	265	240
Amorphous							
Specific gravity	1.12–1.14	1.35–1.42	1.08–1.17	1.13–1.15	1.38	1.15–1.17	1.08–1.14
Water absorption (24 h), %	2.9	1.2	1.3–1.5	1.0–1.3	1.0	0.8–1.1	1.5–2.0
Dielectric strength, $kV \cdot mm^{-1}$	15.8	15.8	22	24		14	15.8
Electrical							
Volume (dc) resistivity, ohm-cm	10^{12}			10^{12}–10^{15}			10^{10}
Dielectric constant (60 Hz)	9.8			4.0			16
Dielectric constant (10^6 Hz)	3.7			3.6			4
Dissipation (power) factor (60 Hz)	0.14			0.01–0.02			0.4
Dissipation factor (10^6 Hz)	0.12			0.02–0.03			0.1
Mechanical							
Compressive modulus, $10^3\ lb \cdot in^{-2}$	250						

Property							
Compressive strength, rupture or 1% yield, 10^3 lb · in^{-2}	13–16	19		15 (yield)	24.9	12.5	
Elongation at break, %	30–100	3–6	150–270	60	3	15	40
Flexural modulus at 23°C, 10^3 lb · in^{-2}	390	1500	110–320	420	1300	450	150–410
Flexural strength, rupture or yield, 10^3 lb · in^{-2}	14	33	5–12	17	41	17	
Hardness, Rockwell (or Shore)	R119	M101	R81–R110	R120	M100	R119	R119
Impact strength (Izod) at 23°C, J · m^{-1}	32–53	160	96 to no break	43–53	117	240	37
Tensile modulus, 10^3 lb · in^{-2}	380	1450				550	150–410
Tensile strength at break, 10^3 lb · in^{-2}	11.8	25	7.5–11	12	28	13.7	7.4–12.4
Tensile yield strength, 10^3 lb · in^{-2}	8			8			
Thermal							
Burning rate, mm · min^{-1}	Self-extinguishing	Self-extinguishing	Self-extinguishing	Self-extinguishing	Self-extinguishing	Self-extinguishing	Self-extinguishing
Coefficient of linear thermal expansion, 10^{-6}°C	80–90	20–30	30–40	80	15–20	54	
Deflection temperature under flexural load (264 lb · in^{-2}), °C	68–85	210	45–54	75	249	127	77
Maximum recommended service temperature, °C	107			135			
Specific heat, cal · g^{-1}	0.4			0.4			
Thermal conductivity, W · m^{-1} · K^{-1}	0.24	0.24		0.24	0.22		

TABLE 10-2 Properties of commercial plastics (*continued*)

Properties	Polyamide						Poly(amide-imide), unfilled
	Nylon 6/9, molding and extrusion	Nylon 6/12		Nylon 11, molding and extrusion	Nylon 12, molding and extrusion	Aromatic nylon (aramid), molded and unfilled	
		Molding	30–35% glass-fiber-reinforced				
Physical							
Melting temperature, °C							
Crystalline	205	217	217	194	179	275	
Amorphous							275
Specific gravity	1.08–1.10	1.06–1.08	1.31–1.38	1.03–1.05	1.01–1.02	1.30	1.40
Water absorption (24 h), %	0.5	0.4	0.2	0.3	0.25	0.6	0.28
Dielectric strength, $kV \cdot mm^{-1}$	24	16	21	17	18	31	24
Electrical							
Volume (dc) resistivity, ohm-cm		10^{15}			10^{14}		
Dielectric constant (60 Hz)		4.0			3.8		
Dielectric constant (10^6 Hz)		3.5			3.0		
Dissipation (power) factor (60 Hz)		0.02			0.07		
Dissipation factor (10^6 Hz)		0.02			0.04		
Mechanical							
Compressive modulus, $10^3 lb \cdot in^{-2}$				180		290	413

Property							
Compressive strength, rupture or 1% yield, 10^3 lb · in^{-2}		2.4			7.5	30	40
Elongation at break, %	1125	150	4	300	300	5	12–18
Flexural modulus at 23°C, 10^3 lb · in^{-2}	290	290	1120	150	165	640	664
Flexural strength, rupture or yield, 10^3 lb · in^{-2}					1.5	25.8	30
Hardness, Rockwell (or Shore)	R111	R114	E40–E50	R108	R106–R109	E90	E78
Impact strength (Izod) at 23°C, J · m^{-1}	59	53	139	96	107–300	75	133
Tensile modulus, 10^3 lb · in^{-2}	275	290	1200	185	180		730
Tensile strength at break, 10^3 lb · in^{-2}	8.5	8.8	24	8	8–9	17.5	26.9
Tensile yield strength, 10^3 lb · in^{-2}		8.8					
Thermal							
Burning rate, mm · min^{-1}				Self-extinguishing			
Coefficient of linear thermal expansion, 10^{-6}°C		90		55–100	67–100	40	36
Deflection temperature under flexural load (264 lb · in^{-2}), °C	57–60	82	93–218	54	54	260	274
Maximum recommended service temperature, °C				100–120			260
Specific heat, cal · g^{-1}		0.4		0.58			
Thermal conductivity, W · m^{-1} · K^{-1}		0.22		0.34	0.22	0.22	0.25

TABLE 10-2 Properties of commercial plastics (*continued*)

Properties	Poly(aryl ether), unfilled	Polycarbonate		Thermoplastic polyester			
				Poly(butylene terephthalate)		Poly(ethylene terephthalate)	
		Low viscosity	30% glass-fiber-reinforced	Unfilled	30% glass-fiber-reinforced	Unfilled	30% glass-fiber-reinforced
Physical							
Melting temperature, °C							
Crystalline				232–267	232–267	245	245
Amorphous	160	140	150				
Specific gravity	1.14	1.2	1.4	1.31–1.38	1.52	1.34–1.39	1.27
Water absorption (24 h), %	0.25	0.15	0.14	0.08–0.09	0.06–0.08	0.1–0.2	0.05
Dielectric strength, $kV \cdot mm^{-1}$	17	15	19	16–22	18–22		22
Electrical							
Volume (dc) resistivity, ohm-cm		2×10^{16}	$>10^{16}$		10^{16}	10^{16}	
Dielectric constant (60 Hz)		3.17	3.35				
Dielectric constant (10^6 Hz)		2.96	3.31			3.25	
Dissipation (power) factor (60 Hz)		0.0009	0.011				
Dissipation factor (10^6 Hz)		0.010	0.007				
Mechanical							
Compressive modulus, $10^3\ lb \cdot in^{-2}$		350	1300				

Compressive strength, rupture or 1% yield, 10^3 lb · in^{-2}		12.5	18	8.6–14.5	18–23.5	11–15	25
Elongation at break, %	80	110	3–5	50–300	2–4	50–300	3
Flexural modulus at 23°C, 10^3 ib · in^{-2}	300	340	1100	330–400	1100–1200	35–450	1440
Flexural strength, rupture or yield, 10^3 lb · in^{-2}	11	13.5	23	12–16.7	26–29	14–18	33.5
Hardness, Rockwell (or Shore)	R117	M70	M92	M68–M78	M90	M94–M101	M100
Impact strength (Izod) at 23°C, J · m^{-1}	427	14	107	43–53	69–85	13–32	101
Tensile modulus, 10^3 lb · in^{-2}	320	345	1250	280	1300	400–600	1440
Tensile strength at break, 10^3 lb · in^{-2}	7.5	9.5	19	8.2	17–19	8.5–10.5	23
Tensile yield strength, 10^3 lb · in^{-2}		9.0					
Thermal							
Burning rate, mm · min^{-1}		Self-extinguishing	Self-extinguishing				
Coefficient of linear thermal expansion, 10^{-6}°C	65	68	22	60–95	25	65	29
Deflection temperature under flexural load (264 lb · in^{-2}), °C	149	138–145	146	50–85	220	38–41	224
Maximum recommended service temperature, °C		143					
Specific heat, cal · g^{-1}		0.3				0.27	
Thermal conductivity, W · m^{-1} · K^{-1}	0.30	0.20	0.22	0.18–0.30	0.30	0.15	

TABLE 10-2 Properties of commercial plastics (*continued*)

	Thermoplastic polyester		Thermosetting and alkyd polyester				
	Aromatic polyester		Unsaturated polyester		Alkyd molding compounds		
Properties	Extrusion-transparent	Injection molding	Styrene–maleic acid copolymer, low-shrink	Butadiene–maleic acid copolymer	Putty, mineral-filled	Glass-fiber-reinforced	Polyimide, unfilled
Physical							
Melting temperature, °C							
Crystalline			Thermoset	Thermoset	Thermoset	Thermoset	
Amorphous	81						310–365
Specific gravity		1.39					1.36–1.43
Water absorption (24 h), %		0.01					0.24
Dielectric strength, $kV \cdot mm^{-1}$		14					22
Electrical							
Volume (dc) resistivity, ohm-cm							$>10^{16}$
Dielectric constant (60 Hz)							
Dielectric constant (10^6 Hz)							3–4
Dissipation (power) factor (60 Hz)							
Dissipation factor (10^6 Hz)							
Mechanical							
Compressive modulus, $10^3\ lb \cdot in^{-2}$					2000–3000		

Property							
Compressive strength, rupture or 1% yield, 10^3 lb · in^{-2}		10	15–30	14–30	12–38	15–36	30–40
Elongation at break. %	225	7–10	3–5				8–10
Flexural modulus at 23°C, 10^3 lb · $^{-2}$	290	700	1000–2500		2000	2000	450–500
Flexural strength, rupture or yield, 10^3 lb · in^{-2}	10.6	12	9–35	16–24	6–17	8.5–26	19–28.8
Hardness, Rockwell (or Shore)	R105		40–70 (Barcol)	50–60 (Barcol)	E98	E95	E52–E99
Impact strength (Izod) at 23°C, J · m^{-1}	101		133–800	214–694	16–27	27–854	80
Tensile modulus, 10^3 lb · in^{-2}		300	1000–2500	1500–2500	500–3000		300
Tensile strength at break, 10^3 lb · in^{-2}	6	11	4.5–20	5–10	3–9	4–9.5	10.5–17.1
Tensile yield strength, 10^3 lb · in^{-2}	7						12.5
Thermal							
Burning rate, mm · min^{-1}							
Coefficient of linear thermal expansion, 10^{-6}°C		29	6–30		20–50	15–33	45–56
Deflection temperature under flexural load (264 lb · in^{-2}), °C	63	282	190–260	160–177	177–260	204–260	277–360
Maximum recommended service temperature, °C							
Specific heat, cal · g^{-1}							0.27
Thermal conductivity, W · m^{-1} · K^{-1}		0.29		0.76–0.93	0.51–0.89	0.6–0.89	0.10–0.11

TABLE 10-2 Properties of commercial plastics (*continued*)

Properties	Poly(methyl pentene), unfilled	Polyolefin					
		Polyethylene					
		Low-density	Medium-density	High-density	Ultra high-molecular-weight	Glass-fiber-reinforced, high-density	Ethylene-vinyl acetate copolymer
Physical							
Melting temperature, °C							
Crystalline	230–240						65–90
Amorphous		95–130	120–140	120–140	125–135	120–140	
Specific gravity	0.84	0.910–0.925	0.926–0.94	0.941–0.965	0.94	1.28	0.92–0.95
Water absorption (24 h), %	0.01	<0.01	<0.01	<0.01	<0.01	0.02	0.05–0.13
Dielectric strength, $kV \cdot mm^{-1}$		18–39	18–39	18–39	28	20	24–30
Electrical							
Volume (dc) resistivity, ohm-cm		$>10^{15}$	$>10^{15}$	$<10^{15}$			
Dielectric constant (60 Hz)		2.3	2.3	2.3			
Dielectric constant (10^6 Hz)		2.3	2.3	2.3			
Dissipation (power) factor (60 Hz)		<0.0005	<0.0005	<0.0005			
Dissipation factor (10^6 Hz)		<0.0005	<0.0005	<0.0005			
Mechanical							
Compressive modulus, $10^3\ lb \cdot in^{-2}$	114–171						

Property							
Compressive strength, rupture or 1% yield, 10^3 lb · in^{-2}	5–6.6			2.7–3.6		7	
Elongation at break, %	10–50	90–800	50–600	20–130	450–525	1.5	550–900
Flexural modulus at 23°C, 10^3 lb · in^{-2}	110–260	8–60	60–115	100–260	130–140	800	1–20
Flexural strength, rupture or yield, 10^3 lb · in^{-2}	4–6.5					11	
Hardness, Rockwell (or Shore)	L67–L74	(D40–D51)	(D50–D60)	R30–R50	R50	R75	
Impact strength (Izod) at 23°C, J · m^{-1}	16–64	No break	27–854	27–1068	No break	59	No break
Tensile modulus, 10^3 lb · in^{-2}	160–280	14–38	25–55	60–180			20–120
Tensile strength at break, 10^3 lb · in^{-2}	3.5–4	0.6–2.3	1.2–3.5	3.1–5.5	5.6	9	1.4–2.8
Tensile yield strength, 10^3 lb · in^{-2}		0.8–1.2	1.0–2.2	3–4	3.1–4.0		
Thermal							
Burning rate, mm · min^{-1}		1.0	1.0	1.0			
Coefficient of linear thermal expansion, 10^{-6}°C	117	100–220	140–160	110–130	130	48	160–200
Deflection temperature under flexural load (264 lb · in^{-2}), °C	41	32–41	41–49	43–54	43–49	121	34
Maximum recommended service temperature, °C	175	70	93	200			
Specific heat, cal · g^{-1}		0.55	0.55	0.46–0.55			
Thermal conductivity, W · m^{-1} · K^{-1}	0.17	0.34	0.34–0.42	0.46–0.51		0.46	

TABLE 10-2 Properties of commercial plastics (*continued*)

Properties	Polyolefin					Poly(phenylene sulfide)	
		Polypropylene					
	Polybutylene extrusion	Homopolymer	Copolymer	Impact copolymer	Polyallomer	Injection molding	40% glass-fiber-reinforced
Physical							
Melting temperature, °C							
Crystalline	126	168	160–168		120–135	290	290
Amorphous							
Specific gravity	0.91–0.925	0.90–0.91	0.89–0.905	0.90	0.90	1.3	1.6
Water absorption (24 h), %	0.01–0.02	0.01–0.03	0.03	<0.03	<0.01	<0.02	0.05
Dielectric strength, kV · mm^{-1}	18	24	24	24	31	15	18
Electrical							
Volume (dc) resistivity, ohm-cm		10^{17}	10^{17}	10^{17}			
Dielectric constant (60 Hz)		2.2–2.6	2.3				
Dielectric constant (10^6 Hz)		2.2–2.6	2.3	2.3			
Dissipation (power) factor (60 Hz)		<0.0005	0.0001–0.0005				
Dissipation factor (10^6 Hz)		0.0005–0.002	0.0001–0.002	0.0003			
Mechanical							
Compressive modulus, 10^3 lb · in^{-2}	31	150–300					

Property							
Compressive strength, rupture or 1% yield, 10^3 lb · in^{-2}		5.5–8.0	3.5–8.0			16	21
Elongation at break, %	300–380	100–600	200–700	8–20	400–500	1–2	1
Flexural modulus at 23°C, 10^3 lb · in^{-2}	45–50	170–250	130–200	130–190	70–110	550	1700
Flexural strength, rupture or yield, 10^3 lb · in^{-2}	2–2.3	6–8	5–7			14	29
Hardness, Rockwell (or Shore)		R80–R102	R50–R96	R40–R90	R50–R85	R123	R123
Impact strength (Izod) at 23°C, J · m^{-1}	No break	21–53	53–1068	80–900	91–203	<27	75
Tensile modulus, 10^3 lb · in^{-2}	30–40	165–225	100–170			480	1100
Tensile strength at break, 10^3 lb · in^{-2}	3.8–4.4	4.5–6	4–5.5		3–3.8	9.5	19.5
Tensile yield strength, 10^3 lb · in^{-2}	1.7–2.5	4.5–5.4	3.5–4.3	2.5–3.1	3–3.4		
<u>Thermal</u>							
Burning rate, mm · min^{-1}							
Coefficient of linear thermal expansion, 10^{-6}°C	128–150	81–100	68–95	60–90	83–100	49	22
Deflection temperature under flexural load (264 lb · in^{-2}), °C	54–60	48–57	45–57	90–105 (66 lb · in^{-2})	51–56	135	249
Maximum recommended service temperature, °C		160	240	140–160			
Specific heat, cal · g^{-1}		0.44–0.46	0.45–0.50	0.45–0.50			
Thermal conductivity, W · m^{-1} · K^{-1}	0.22	0.12	0.15–0.17	0.12–0.17	0.09–0.17	0.29	0.29

TABLE 10-2 Properties of commercial plastics (*continued*)

Properties	Polyurethane			Silicone			Styrenic
	Casting resin		Thermoplastic elastomer	Cast resin, flexible	Mineral- and/or glass-filled	Epoxy molding and encapsulating compound	Polystyrene
	Liquid	Unsaturated					Crystal
Physical							
Melting temperature, °C							
Crystalline	Thermoset	Thermoset		Thermoset	Thermoset	Thermoset	
Amorphous			120–160				85–105
Specific gravity	1.1–1.5	1.05	1.05–1.25	0.99–1.5	1.8–1.94	1.84	1.04–1.05
Water absorption (24 h), %	0.02–1.5	0.1–0.2	0.7–0.9				0.03–0.10
Dielectric strength, $kV \cdot mm^{-1}$	12–20		13–25	22	8–15	10	24
Electrical							
Volume (dc) resistivity, ohm-cm	10^{11}–10^{15}		10^{11}–10^{13}	10^{14}–10^{15}			$>10^{16}$
Dielectric constant (60 Hz)	4.0–7.5		5.4–7.6	2.7–4.2			
Dielectric constant (10^6 Hz)							2.5
Dissipation (power) factor (60 Hz)							
Dissipation factor (10^6 Hz)							
Mechanical							
Compressive modulus, 10^3 lb $\cdot$ in^{-2}	10–100		4-9				